PUMP HANDBOOK

EDITED BY

Igor J. Karassik
William C. Krutzsch
Warren H. Fraser

Worthington Pump Inc.

Joseph P. Messina

Public Service Electric and Gas Company
New Jersey Institute of Technology

SECOND EDITION

McGRAW-HILL BOOK COMPANY

*New York   St. Louis   San Francisco   Auckland   Bogotá
Hamburg   London   Madrid   Mexico
Montreal   New Delhi   Panama   Paris   São Paulo
Singapore   Sydney   Tokyo   Toronto*

Library of Congress Cataloging in Publication Data
Main entry under title:

Pump handbook.

Includes index.
1. Pumping machinery—Handbooks, manuals, etc.
I. Karassik, Igor J., 1911–
TJ900.P79   1985   621.6   84-9708
ISBN 0-07-033302-5

ISBN 0-07-033302-5

    67890   DOCDOC   93210

The editors for this book were Harold B. Crawford and Susan Thomas,
the design supervisor was Mark E. Safran, the designer was Delgado
Design Associates, and the production supervisor was Teresa F. Leaden.
It was set in Caledonia by University Graphics, Inc.

Printed and bound by R. R. Donnelley & Sons Company.

In memory of our good friend and colleague
William C. Krutzsch

CONTENTS

List of Contributors / *xi*
Preface to the Second Edition / *xvii*
Preface to the First Edition / *xix*
SI Units—A Commentary / *xxi*

Chapter 1   Introduction: Classification and Selection of Pumps   1.1

Chapter 2   Centrifugal Pumps   2.1

2.1   Centrifugal Pump Theory / 2.3
2.2   Centrifugal Pump Construction / 2.33
    2.2.1   Centrifugal Pumps: Major Components / 2.33
    2.2.2   Centrifugal Pump Packing / 2.114
    2.2.3   Centrifugal Pump Mechanical Seals / 2.127
    2.2.4   Centrifugal Pump Injection-Type Shaft Seals / 2.159
    2.2.5   Centrifugal Pump Oil Film Journal Bearings / 2.166
2.3   Centrifugal Pump Performance / 2.193
    2.3.1   Centrifugal Pumps: General Performance Characteristics / 2.194
    2.3.2   Centrifugal Pump Hydraulic Performance and Diagnostics / 2.266
    2.3.3   Centrifugal Pump Mechanical Performance, Instrumentation, and Diagnostics / 2.274
    2.3.4   Centrifugal Pump Minimum Flow Control Systems / 2.305
2.4   Centrifugal Pump Priming / 2.319

Chapter 3   Displacement Pumps   3.1

3.1   Power Pumps / 3.3
3.2   Steam Pumps / 3.25
3.3   Diaphragm Pumps / 3.49
3.4   Screw Pumps / 3.57
3.5   Rotary Pumps / 3.79
3.6   Displacement Pump Performance, Instrumentation, and Diagnostics / 3.107
3.7   Displacement Pump Flow Control / 3.119

Chapter 4   Jet Pumps   4.1

Chapter 5   Materials of Construction   5.1

5.1   Materials of Construction of Metallic Pumps / 5.3
5.2   Materials of Construction of Nonmetallic Pumps / 5.15

Chapter 6   Pump Drivers   6.1

6.1   Prime Movers / 6.3
      6.1.1   Electric Motors and Motor Controls / 6.3
      6.1.2   Steam Turbines / 6.26
      6.1.3   Engines / 6.44
      6.1.4   Hydraulic Turbines / 6.61
      6.1.5   Gas Turbines / 6.71
6.2   Speed-Varying Devices / 6.81
      6.2.1   Eddy-Current Couplings / 6.81
      6.2.2   Single-Unit Adjustable-Speed Electric Drives / 6.92
      6.2.3   Fluid Couplings / 6.110
      6.2.4   Gears / 6.125
      6.2.5   Adjustable-Speed Belt Drives / 6.145
6.3   Power Transmission Devices / 6.153
      6.3.1   Pump Couplings and Intermediate Shafting / 6.153
      6.3.2   Magnetic Drives / 6.166
      6.3.3   Hydraulic Pump and Motor Power Transmission Systems / 6.172

Chapter 7   Pump Controls and Valves   7.1

Chapter 8   Pumping Systems   8.1

8.1   General Characteristics of Pumping Systems and System-Head Curves / 8.3
8.2   Branch-Line Pumping Systems / 8.77
8.3   Waterhammer / 8.85
8.4   Pump Noise / 8.101

Chapter 9   Pump Services   9.1

9.1   Water Supply / 9.3
9.2   Sewage Treatment / 9.23
9.3   Drainage and Irrigation / 9.41
9.4   Fire Pumps / 9.51
9.5   Steam Power Plant / 9.61
9.6   Chemical Industry / 9.83
9.7   Petroleum Industry / 9.101
9.8   Pulp and Paper Mills / 9.121
9.9   Food and Beverage / 9.137
9.10   Mining / 9.143
9.11   Marine / 9.151
9.12   Hydraulic Presses / 9.167
9.13   Refrigeration, Heating, and Air Conditioning / 9.179
9.14   Pumped Storage / 9.187
9.15   Nuclear / 9.203
       9.15.1   Nuclear Electric Generation / 9.203
       9.15.2   Nuclear Pump Seismic Qualifications / 9.223
9.16   Metering / 9.235
9.17   Solids Pumping / 9.239
       9.17.1   Hydraulic Transport of Solids / 9.239
       9.17.2   Construction of Solids-Handling Centrifugal Pumps / 9.261
       9.17.3   Construction of Solids-Handling Displacement Pumps / 9.269
9.18   Oil Wells / 9.277
9.19   Cryogenic Liquefied Gas Service / 9.295
9.20   Water Pressure Booster Systems / 9.303

Chapter 10   Intakes and Suction Piping   10.1

10.1   Intakes, Suction Piping, and Strainers / 10.3
10.2   Intake Modeling / 10.35

Chapter 11   Selecting and Purchasing Pumps   11.1

Chapter 12   Installation, Operation, and Maintenance   12.1

Chapter 13   Pump Testing   13.1

Appendix   Technical Data   A.1

Index follows appendix

LIST OF CONTRIBUTORS

Arnold, Conrad L., B.S. (E.E.)   SUBSECTION 6.2.3   FLUID COUPLINGS
Director of Engineering, American Standard Industrial Division, Detroit, MI

Ashton, Robert D., B.S. (E.T.M.E.)   SUBSECTION 2.2.4   CENTRIFUGAL PUMP INJECTION-TYPE
SHAFT SEALS
*Manager, Proposal Applications, Byron Jackson Pump Division, Borg-Warner Industrial
Products, Inc., Long Beach, CA*

Beck, Wesley W., B.S. (C.E.), P.E.   CHAPTER 13   PUMP TESTING
*Hydraulic Consulting Engineer, Denver, CO. Formerly with the Chief Engineers Office of the
U.S. Bureau of Reclamation*

Benjes, H. H., B.S. (C.E.), P.E.   SECTION 9.2   SEWAGE TREATMENT
Retired Partner, Black & Veatch, Engineers-Architects, Kansas City, MO

Bergeron, Wallace L., B.S. (E.E.)   SUBSECTION 6.1.2   STEAM TURBINES
Senior Market Engineer, Elliott Company, Jeannette, PA

Birgel, W. J., B.S. (E.E.)   SUBSECTION 6.2.1   EDDY-CURRENT COUPLING
President, VS Systems, Inc., St. Paul, MN

Birk, John R., B.S. (M.E.), P.E.   SECTION 9.6   CHEMICAL INDUSTRY
Consultant, Senior Vice President (retired), The Duriron Company, Inc., Dayton, OH

Brennan, James R., B.S. (M.I.E.)   SECTION 3.4   SCREW PUMPS
Manager of Engineering, Transamerica DeLaval Inc., Pyramid Pump Division, Monroe, NC

Brennan, John R., B.S. (Ch.E.)   SECTION 9.18   OIL WELLS
Consultant, National Supply Company, Division of Armco, Inc., Los Nietos, CA

Buse, Fred, B.S. (Marine Engineering)   SECTION 3.1   POWER PUMPS
*Chief Engineer, Centrifugal Pumps, Standard Pump Division, Ingersoll-Rand Company,
Allentown, PA*

Clopton, D. E., B.S. (C.E.), P.E.   SECTION 9.1   WATER SUPPLY
*Assistant Project Manager, Water Quality Division, URS/Forrest and Cotton, Inc.,
Consulting Engineers, Dallas, TX*

Costigan, James L., B.S. (Chem.)   SECTION 9.9   FOOD AND BEVERAGE
Sales Manager, Tri-Clover Division, Ladish Company, Kenosha, WI

Czarnecki, G. J., B.Sc., M.Sc. (Tech.)   SECTION 3.4   SCREW PUMPS
Chief Engineer (retired), Transamerica DeLaval Inc., Pyramid Pump Division, Monroe, NC

DiMasi, Mario, B.S. (M.E.), M.B.A.   SECTION 9.4   FIRE PUMPS
District Manager, Peerless Pump, Union, NJ

DiVona, A. A., B.S. (M.E.)   SUBSECTION 6.1.1   ELECTRIC MOTORS AND MOTOR CONTROLS
Account Executive, Industrial Sales, Westinghouse Electric Corporation, Hillside, NJ

Dolan, A. J., B.S. (E.E.), M.S. (E.E.), P.E.   SECTION 6.1.1   ELECTRIC MOTORS AND MOTOR
CONTROLS
Fellow District Engineer, Westinghouse Electric Corporation, Hillside, NJ

Dornaus, Wilson L., B.S. (C.E.), P.E.   SECTION 10.1   IINTAKES, SUCTION PIPING, AND
STRAINERS
Pump Consultant, Lafayette, CA

Eller, David, B.S. (A.E.), P.E.   SUBSECTION 6.3.3   HYDRAULIC PUMP AND MOTOR POWER-
TRANSMISSION SYSTEMS
President and Chief Engineer, M&W Pump Corporation, Deerfield Beach, FL

*Elvitsky, A. W., B.S. (M.E.), M.S. (M.E.), P.E.   SECTION 9.7   PETROLEUM INDUSTRY
Vice President and Chief Engineer, United Centrifugal Pumps, San Jose, CA

Foster, W. E., B.S. (C.E.), P.E.   SECTION 9.2   SEWAGE TREATMENT
Partner, Black & Veatch, Engineers-Architects, Kansas City, MO

Fraser, Warren H., B.M.E.   SUBSECTION 2.3.2   CENTRIFUGAL PUMP HYDRAULIC PERFORMANCE
CHARACTERISTICS AND DIAGNOSTICS; SEECTION 5.1 MATERIALS OF CONSTRUCTION OF METALLIC PUMPS
Chief Design Engineer, Worthington Pump Group, McGraw-Edison Company, Harrison, NJ

Freeborough, Robert M., B.S. (Min.E.)   SECTION 3.2   STEAM PUMPS
Manager, Parts Marketing, Worthington Corporation, Timonium, MD

Gandhi, R. L., B.E. (C.E.), M.E. (Public Health)   SUBSECTION 9.17.1   HYDRAULIC TRANSPORT
OF SOLIDS
Chief Slurry Engineer, Bechtel Petroleum, Inc., San Francisco, CA

Giddings, J. F., Diploma, Mechanical, Electrical, and Civil Engineering   SECTION
9.8   PULP AND PAPER MILLS
Development Manager, Parsons & Whittemore, Lyddon, Ltd., Croydon, England

Glanville, Robert H., M.E.   SECTION 9.16   METERING
Vice President Engineering, BIF, A Unit of General Signal, Providence, RI

*Gunther, F. J., B.S. (M.E.), M.S. (M.E.)   SECTION 6.1.3   ENGINES
Late Sales Engineer, Waukesha Motor Company, Waukesha, WI

Haentjens, W. D., B.M.E., M.S. (M.E.), P.E.   SECTION 9.10   MINING
President and Chief Engineer, Barrett, Haentjens & Company, Hazleton, PA

Hatch, Jan A. SUBSECTION 6.3.2   MAGNETIC DRIVES   Marketing Manager, The Konto
Company, Inc., Orange, MA

*Deceased.

Heisler, S. I., B.S. (M.E.), P.E.   CHAPTER 11   SELECTING AND PURCHASING PUMPS
Manager of Supplier Performance, Bechtel Power Corporation, San Francisco, CA

Hill, R. A., B.S. (M.E.)   SUBSECTION 9.17.1   HYDRAULIC TRANSPORT OF SOLIDS
Engineering Manager, Slurry Technology, Bechtel Petroleum, Inc., San Francisco, CA

Honeycutt, F. G., Jr., B.S. (C.E.) P.E.   SECTION 9.1   WATER SUPPLY
*Assistant Vice President and Head Water Quality Division, URS/Forrest and Cotton, Inc.,
   Consulting Engineers, Dallas, TX*

Ingram, James H., B.S. (M.E.), M.S. (M.E.)   SUBSECTION 2.3.3   CENTRIFUGAL PUMP
   MECHANICAL PERFORMANCE, INSTRUMENTATION, AND DIAGNOSTICS
Senior Engineering Specialist, Monsanto Fibers and Intermediates Company, Texas City, TX

Jackson, Charles, B.S. (M.E.), A.A.S. (Electronics), P.E.   SUBSECTION 2.3.3.   CENTRIFUGAL
   PUMP MECHANICAL PERFORMANCE, INSTRUMENTATION, AND DIAGNOSTICS
Distinguished Fellow, Monsanto Corporate Engineering, Texas City, TX

Jekat, Walter K., Dipl.-Ing.   SECTION 2.1   CENTRIFUGAL PUMP THEORY
Designer of Turbomachinery, Pump Consultant to KSB, Munich, Germany

Jumpeter, Alex M., B.S. (Ch.E.)   CHAPTER 4   JET PUMPS
*Engineering Manager, Process Equipment, Schutte and Koerting Company, Cornwells
   Heights, PA*

Karassik, Igor J., B.S. (M.E.), M.S. (M.E.), P.E.   SUBSECTION 2.2.1   CENTRIFUGAL
   PUMPS: MAJOR COMPONENTS; SECTION 2.4   CENTRIFUGAL PUMP PRIMING SECTION 9.5   STEAM
   POWER PLANT; CHAPTER 12   INSTALLATION, OPERATION, AND MAINTENANCE
*Chief Consulting Engineer, Worthington Group, McGraw-Edison Company, Basking Ridge,
   NJ*

Kittredge, C. P., B.S. (C.E.), Doctor of Technical Science (M.E.)   SUBSECTION
   2.3.1   CENTRIFUGAL PUMPS: GENERAL PERFORMANCE CHARACTERISITCS
Consulting Engineer, Princeton, NJ

Kron, H. O., B.S. (M.E.), P.E.   SUBSECTION 6.2.4   GEARS
Executive Vice President, Philadelphia Gear Corporation, King of Prussia, PA

*Krutzsch, W. C., B.S. (M.E.), P.E.   SI UNITS—A COMMENTARY; CHAPTER 1
   INTRODUCTION: CLASSIFICATION, AND SELECTION OF PUMPS
*Late Director, Research and Development, Engineered Products, Worthington Pump Group,
   McGraw-Edison Company, Harrison, NJ*

Landon, Fred K., B.S. (Aero.E.), P.E.   SUBSECTION 6.3.1   PUMP COUPLINGS AND
   INTERMEDIATE SHAFTING
Manager, Engineering, Rexnord, Inc., Houston, TX

Larsen, Johannes, B.S. (C.E.), M.S. (M.E.)   SECTION 10.2   INTAKE MODELING
Lead Research Engineer, Hydraulic Structures, Alden Research Laboratory, Holden, MA

Lippincott, J. K., B.S. (M.E.)   SECTION 3.4   SCREW PUMPS
*Vice President, General Manager, Transamerica DeLaval Inc., Pyramid Pump Division,
   Monroe, NC*

Little, C. W., Jr., B.E. (E.E.), D. Eng.   SECTION 3.5   ROTARY PUMPS
*Former Vice President, General Manager, Manufactured Products Division, Waukesha
   Foundry Company, Waukesha, WI*

---

* Deceased.

Messina, Joseph P., B.S. (M.E.), M.S. (C.E.), P.E. SECTION 8.1 GENERAL CHARACTERISTICS OF PUMPING SYSTEMS AND SYSTEM-HEAD CURVES; SECTION 8.2 BRANCH-LINE PUMPING SYSTEMS; APPENDIX: TECHNICAL DATA
Principal Staff Engineer, Public Service Electric and Gas Company, Newark, NJ; Adjunct Instructor, New Jersey Institute of Technology, Newark, NJ; Pump Consultant

Netzel, James P., B.S. (M.E.) SUBSECTION 2.2.2 CENTRIFUGAL PUMP PACKING; SUBSECTION 2.2.3 CENTRIFUGAL PUMP MECHANICAL SEALS
Chief Engineer, John Crane-Houdaille, Inc., Morton Grove, IL

Norton, Robert D., B.S. (M.E.), P.E. SECTION 5.2 MATERIALS OF CONSTRUCTION OF NONMETALLIC PUMPS
Consulting Engineer, Everett Process Systems, Lansdale, PA

Nuta, D., B.S. (C.E.), M.S. (Applied Mathematics and Computer Science), P.E. SUBSECTION 9.15.2 NUCLEAR PUMP SEISMIC QUALIFICATIONS
Associate Consulting Engineer, Ebasco Services, Inc., New York, NY

O'Keefe, W., A.B., P.E. SUBSECTION 2.3.4 CENTRIFUGAL PUMP MINIMUM FLOW CONTROL SSTEMS; CHAPTER 7 PUMP CONTROLS AND VALVES
Editor, Power Magazine, McGraw-Hill Publications Company, New York, NY

*Olson, Richard G., M.E., M.S., P.E. SECTION 6.1.5 GAS TURBINES
Late Marketing Supervisor, International Turbine Systems, Turbodyne Corporation, Minneapolis, MN

Padmanabhan, Mahadevan, B.S. (C.E.), M.S. (C.E.), Ph.D., P.E. SECTION 10.2 INTAKE MODELING
Lead Research Engineer, Hydraulic Machinery, Alden Research Laboratory, Holden, MA

Parmakian, John, B.S. (M.E.), M.S. (C.E.), P.E. SECTION 8.3 WATERHAMMER
Consulting Engineer, Boulder, CO

Peacock, James H., B.S. (Met.E.) SECTION 9.6 CHEMICAL INDUSTRY
Manager, Materials Division, The Duriron Company, Inc., Dayton, OH

Potthoff, E. O., B.S. (E.E.), P.E. SUBSECTION 6.2.2 SINGLE-UNIT ADJUSTABLE-SPEED ELECTRIC DRIVES
Industrial Engineer (retired), Industrial Sales Division, General Electric Company, Schenectady, NY

Ramsey, Melvin A., M.E., P.E. SECTION 9.13 REFRIGERATION, HEATING, AND AIR CONDITIONING
Consulting Engineer, Schenectady, NY

*Rich, George R., B.S. (C.E.), C.E., D.Eng., P.E. SECTION 9.14 PUMPED STORAGE
Late Director, Senior Vice President, Chief Engineer, Chas. T. Main, Inc., Boston, MA

Robertson, John S., B.S. (C.E.), P.E. SECTION 9.3 DRAINAGE AND IRRIGATION
Chief, Electrical and Mechanical Branch, Engineering and Construction, Headquarters, U.S. Army Corps of Engineers

Rupp, Warren E. SECTION 3.3 DIAPHRAGM PUMPS
President, The Warren Rupp Company, Mansfield, OH

Shapiro, Wilbur, B.S., M.S. SUBSECTION 2.2.5 CENTRIFUGAL PUMP OIL FILM JOURNAL BEARINGS
Senior Staff Consultant, Mechanical Technology Incorporated, Latham, NY

° Deceased.

Shikasho, Satoru, B.S. (M.E.), P.E.   SECTION 9.20   WATER PRESSURE BOOSTER SYSTEMS
Chief Product Engineer, Packaged Products, ITT Bell & Gossett, Morton Grove, IL

Smith, L. R. SECTION 9.19   CRYOGENIC LIQUEFIED GAS SERVICE
Retired, formerly of J. C. Carter Company, Costa Mesa, CA

Smith, Will, B.S. (M.E.), M.S. (M.E.), P.E.   SECTION 3.7   DISPLACEMENT PUMP FLOW
   CONTROL; SUBSECTION 9.17.3   CONSTRUCTION OF SOLIDS-HANDLING DISPLACEMENT PUMPS
*Engineering Product Manager, Custom Pump Operations, Worthington Division, McGraw-
   Edison Company, Harrison, NJ*

Snoek, P. E., B.S. (C.E.), M.S. (C.E.), M.B.A.   SUBSECTION 9.17.1   HYDRAULIC TRANSPORT OF
   SOLIDS
Manager Slurry Technology, Bechtel Petroleum, Inc., San Francisco, CA

Snyder, Milton B., B.S. (B.A.)   SUBSECTION 6.2.5   ADJUSTABLE-SPEED BELT DRIVES
Sales Engineer, Master-Reeves Division, Reliance Electric Company, Columbus, IN

°Soete, George W., B.S. (M.E.), M.S. (M.E.), P.E.   SECTION 9.11   MARINE
Late Engineering Supervisor, DeLaval Turbine Inc., Turbine Division, Trenton, NJ

Sparks, Cecil R., B.S. (M.E.), M.S. (M.E.), P.E.   SECTION 8.4   PUMP NOISE
Director of Engineering Physics, Southwest Research Institute, San Antonio, TX

Szenasi, Fred R., B.S. (M.E.), M.S. (M.E.), P.E.   SECTION 3.6   DISPLACEMENT-PUMP
   PERFORMANCE, INSTRUMENTATION, AND DIAGNOSTICS; SECTION 8.4   PUMP NOISE
Senior Project Engineer, Engineering Dynamics Inc., San Antonio, TX

Tullo, C. J., P.E.   SECTION 2.4   CENTRIFUGAL PUMP PRIMING
*Chief Engineer (retired), Centrifugal Pump Engineering, Worthington Pump, Inc., Harrison,
   NJ*

Wachel, J. C., B.S. (M.E.), M.S. (M.E.)   SECTION 3.6   DISPLACEMENT PUMP PERFORMANCE,
   INSTRUMENTATION, AND DIAGNOSTICS; SECTION 8.4   PUMP NOISE
Manager of Engineering, Engineering Dynamics, Inc., San Antonio, TX

Wepfer, W. M., B.S. (M.E.), P.E.   SUBSECTION 9.15.1   NUCLEAR ELECTRICAL GENERATION
*Consulting Engineer, formerly Manager, Pump Design, Westinghouse Electric Corporation,
   Pittsburgh, PA*

Whippen, Warren G., B.S. (M.E.), P.E.   SUBSECTION 6.1.4   HYDRAULIC TURBINES
Manager Product Development, Hydro-Turbine Division, Allis-Chalmers, York, PA

Wilson, G., B.S. (M.E.), P.E.   SUBSECTION 9.17.2   CONSTRUCTION OF SOLIDS-HANDLING
   CENTRIFUGAL PUMPS
Manager, Research and Development, Goulds Pumps, Inc., Baldwinsville, NY

Zeitlin, A. B., M.S. (M.E.), Dr.-Eng. (E.E.), P.E.   SECTION 9.12   HYDRAULIC PRESSES
President, Press Technology Corporation, Mamaroneck, NY

°Deceased.

PREFACE TO THE SECOND EDITION

Once more, the dubious honor of writing a preface has been bestowed upon me by my three co-editors. And while they were perfectly willing to share the pluses and minuses of collective editorship, they refused to engage in collective "prefaceship," if I may be allowed to coin a word. At best, they reserved for themselves the right of looking over my shoulder and criticizing the spirit of levity with which I chose to approach the task for which they had unanimously volunteered me. I should add parenthetically that the preface to the first edition (which you can read on the following pages) is actually my fourth draft; the first three were judged too irreverent by my co-editors. (I have preserved these first three drafts for whoever inherits my collection of unpublished material.)

Assuming that my co-editors are more charitable this time, or alternately that our publisher is pressed for time, what follows (if not what precedes) will appear more or less as written.

First of all, we would like to assure the readers of this second edition of the *Pump Handbook* that it is not merely a slightly warmed-over version of the first edition, with such errata as we have spotted corrected and with a few insignificant changes and additions. Actually, the task of rewriting and editing the material in a form that would correspond to what was planned for this second edition proved to be a monumental, not to say awesome, undertaking.

To begin with, in concert with the publishers, it was decided that all data given here would appear in both USCS and SI units. This was not as simple a task as it may appear, for the reason that "absolute" pure SI units do not lend themselves too well to the scale of numbers generally encountered in industrial processes. To give but one example, the pascal, which is the SI unit of pressure, corresponds to 0.000145 lb/in^2, and even the kilopascal is only 0.145 lb/in^2. While this might be a reasonably satisfactory unit for scientific work, the case is hardly such for centrifugal pumps used in everyday life.

This led us to choose what might be called a modified set of SI units, all as explained in *SI Units—A Commentary* on page xxi. Even conveying this desirable concept of a practical set of SI units to the authors of the various sections proved to be somewhat difficult. As a result, we have permitted these authors some leeway in their specific choice, understanding full well that what is desirable in one industry may differ from the preferred choice in another.

We decided that a number of sections and subsections in the first edition could benefit by being significantly expanded. This, for instance, is the case with the following:

2.2.1   Centrifugal Pumps: Major Components

2.3.1   Centrifugal Pumps: General Performance Characteristics

2.4   Centrifugal Pump Priming

8.1   General Characteristics of Pumping Systems and System-Head Curves

8.4   Pump Noise

9.4   Fire Pumps

9.15.1   Nuclear Electric Generation

9.17.1   Hydraulic Transport of Solids

10.1   Intakes, Suction Piping, and Strainers

App.   Technical Data

At the same time, we felt that some material originally included in the subsection on Centrifugal Pumps: Major Components should be excised from there and treated in greater depth separately. This expanded coverage includes the following:

2.2.2   Centrifugal Pump Packing

2.2.3   Centrifugal Pump Mechanical Seals

2.2.4   Centrifugal Pump Injection-Type Shaft Seals

2.2.5   Centrifugal Pump Oil Film Journal Bearings

Finally, a large amount of subject matter has been added to the second edition:

2.3.2   Centrifugal Pump Hydraulic Performance and Diagnostics

2.3.3   Centrifugal Pump Mechanical Performance, Instrumentation, and Diagnostics

2.3.4   Centrifugal Pump Minimum Flow Control Systems

3.3   Diaphragm Pumps

3.6   Displacement Pump Performance, Instrumentation, and Diagnostics

3.7   Displacement Pump Flow Control

5.2   Materials of Construction of Nonmetallic Pumps

6.3.2   Magnetic Drives

6.3.3   Hydraulic Pump and Motor Power Transmission Systems

9.15.2   Nuclear Pump Seismic Qualifications

9.17.3   Construction of Solids-Handling Displacement Pumps

9.18   Oil Wells

9.19   Cryogenic Liquefied Gas Service

9.20   Water Pressure Booster Systems

10.2   Intake Modeling

In brief, the editors have attempted to increase the usefulness of this handbook. The extent to which we have achieved this objective, we will leave to the judgment of our readers.

IGOR J. KARASSIK

PREFACE
TO THE FIRST EDITION

Considering that I had written the prefaces of the three books published so far under my name, my colleagues thought it both polite and expedient to suggest that I prepare the preface to this handbook, coedited by the four of us. Except for the writing of the opening paragraph of an article, a preface is the most difficult assignment that I know. Certainly the preface to a handbook should do more than describe minutely and in proper order the material which is contained therein.

Yet I submit that the saying "a book should not be judged by its cover" should be expanded by adding "and not by its preface." If the reader will accept this disclaimer, I can proceed.

As will be stated in Section 1, "Introduction and Classification of Pumps," it can rightly be claimed that no machine and very few tools have had as long a history in the service of man as the pump, or have filled as broad a need in his life. Every process which underlies our modern civilization involves the transfer of liquids from one level of pressure or static energy to another. Thus pumps have played an essential role in our life ever since the dawn of civilization.

Thus it is that a constantly growing population of technical personnel is in need of information that will help it in either designing, selecting, operating, or maintaining pumping equipment. There has never been a dearth of excellent books and articles on the subject of pumps. But the editors and the publisher felt that a need existed for a handbook on pumps which would present this information in a compact and authoritative form. The format of a handbook permits a selection of the most versatile group of contributors, each an expert on his particular subject, each with a background of experience which makes him particularly knowledgeable in the area assigned to him.

This handbook deals first with the theory, construction details, and performance characteristics of all the major types of pumps—centrifugal pumps, power pumps, steam pumps, screw and rotary pumps, jet pumps, and many of their variants. It deals with prime movers, couplings, controls, valves, and the instruments used in pumping systems. It treats in detail the systems in which pumps operate and the characteristics of these systems. And because of the many services in which pumps have to be applied, a total of 21 different services—ranging from water supply, through steam power plants, construction, marine applications, and refrigeration to metering and solids pumping—are examined and described in detail, again by a specialist in each case.

Finally, the handbook provides information on the selection, purchasing, installation, operation, testing, and maintenance of pumps. An appendix provides a variety of technical data useful to anyone dealing with pumping equipment.

We are greatly indebted to the men who supplied the individual sections which make up this handbook. We hope that our common task will have produced a handbook which will help its user to make a better and more economical pump installation than he would have done without it, to install equipment that will perform more satisfactorily and for longer uninterrupted periods, and when trouble occurs, to diagnose it quickly and accurately. If this handbook does all this, the contributors, its editors, and its publisher will be pleased and satisfied.

No doubt a few readers will look for subject matter that they will not find in this handbook. Into the making of decisions on what to include and what to leave out must always enter an element of personal opinion; therefore we will feel some responsibility for their disappointment. But we submit that it was quite impossible to include even everything we had wanted to cover. As to our possible sins of commission, they are obviously unknown to us at this writing. We can only promise that we shall correct them if an opportunity is afforded us.

IGOR J. KARASSIK

SI UNITS A COMMENTARY

W.C. KRUTZSCH

Since the publication of the first edition of this handbook in 1976, the involvement of the world in general, and of the United States in particular, with the SI system of units has changed rather substantially. As an illustration of this change, one need only consider the differences in automotive manufacturing practices. In 1976 the U.S. automotive industry was well aware of the emergence of this system and was initiating steps to embrace it, but there was little evidence in its product that this was the case. Today, by contrast, you need only try to remove a nut or cap screw from your car with your faithful old wrenches to realize that a change has been made. They no longer fit! Now you need metric wrenches, even on cars made in the United States.

Closer to home, the publisher of this volume, who was content in 1976 to accept a token acknowledgment of the SI system in the book, has now insisted that such units be provided throughout as a supplement to the United States customary system of units (USCS). Accordingly, this entire volume has been produced with dual units, which should make it easier to work with, particularly for readers in metric countries, who will no longer find it necessary to make either approximate mental transpositions or exact mathematical conversions. To accomplish this, the editors and authors were not only required to expend more time and effort on the text than in the first edition but were also obligated to make some difficult decisions about the units to be employed. How these decisions were arrived at and how they have affected the remainder of the book will be reviewed in this commentary.

The designation SI is the official abbreviation, in any language, of the French title "Le Système International d'Unités," given by the 11th General Conference on Weights and Measures (sponsored by the International Bureau of Weights and Measures) in 1960 to a coherent system of units selected from metric systems. This system of units has since been adopted by the International Organization for Standardization (ISO) as an international standard.

The SI system consists of seven basic units, two supplementary units, a series of derived units, and a series of approved prefixes for multiples and submultiples of the foregoing. The names and definitions of the basic and supplementary units are contained in Tables 1a and 1b of the Appendix. Table 2 lists the units and Table 3 the prefixes. Table 10 provides conversions of USCS to SI units.

The decision making begins immediately upon recognition of the difference in magnitude between USCS and SI units in the two most fundamental quantities to be established in determining the performance of any pump: pressure and capacity. Let us begin with pressure.

The standard SI unit of pressure, the pascal, equal to one newton° per square meter†, is a minuscule value relative to the pound per square inch (1 lb/in² = 6,894.757 Pa) or to the old, established metric unit of pressure the kilogram per square centimeter (1 kgf/cm² = 98,066.50 Pa). In order to eliminate the necessity for dealing with significant multiples of these already large numbers when describing the pressure ratings of modern pumps, different sponsoring groups have settled on two competing proposals. One group supports selection of the kilopascal, a unit which does provide a numerically reasonable value (1 lb/in² = 6.894757 kPa) and is a rational multiple of a true SI unit. The other group, equally vocal, supports the bar (1 bar = 10⁵ Pa). This support is based heavily on the fact that the value of this special derived unit is close to one atmosphere. It is important, however, to be aware that it is not exactly equal to a standard atmosphere (101, 325.0 Pa) or to the so-called metric atmosphere (1 kgf/cm² = 98,066.50 Pa) but is close enough to be confused with both.

As yet there is no consensus about which of these units should be used as the standard. Accordingly, both are used, often in the same metric country. Since the world cannot agree and since we must all live with the world as it is, the editors concluded that restricting usage to one or the other would be arbitrary, grossly artificial, and not in the best interests of the reader. We therefore have permitted individual authors to use what they are most accustomed to, and both units will be encountered in the text.

Units of pressure are utilized to define both the performance and the mechanical integrity of displacement pumps. For kinetic pumps, however, which are by far the most significant industrial pumps, pressure is used only to describe rated and hydrostatic values, or mechanical integrity. Performance is generally measured in terms of total head, expressed as feet in USCS units and as meters in SI units. This sounds straightforward enough until a definition of head, including consistent units, is attempted. Then we encounter the dilemma of mass versus force, or weight.

The total head developed by a kinetic pump, or the head contained in a vertical column of liquid, is actually a measure of the internal energy added to or contained in the liquid. The units used to define it could be energy per unit volume, or energy per unit mass, or energy per unit weight. If we select the last, we arrive conveniently, in USCS units, as foot-pounds per pound, or simply feet. In SI units, the terms would be newton-meters per newton, or simply meters. In fact, however, metric countries weigh objects in kilograms, not newtons, and so the SI term for head may be defined at places in this volume in terms of kilogram-meters per kilogram, even though this does not conform strictly to SI rules.

Similar ambiguity is observed with the units of capacity, except here there may be even more variations. The standard SI unit of capacity is the cubic meter per second, which is indeed a very large value (1 m³/s = 15,850.32 U.S. gal/min) and is therefore really only suitable for very large pumps. Recently, some industry groups have suggested that a suitable alternative might be the liter per second (1 l/s = 10⁻³ m³/s = 15.85032 U.S. gal/min), while others have maintained strong support for the traditional metric unit of capacity the cubic meter per hour (1 m³/h = 4.402867 U.S. gal/min). All of these units will be encountered in the text.

The value for the unit of horsepower (hp) used throughout this book and in the United States is the equivalent of 550 foot pounds (force) per second, or 0.74569987 kilowatts (kW). The horsepower used herein is approximately 1.014 times greater than the metric horsepower, which is equivalent to 0.735499 kilowatts. Whenever the rating of an electric motor is given in this book in horsepower, it is the output rating. The equivalent output power in kilowatts is shown in parentheses.

Variations in SI units have arisen because of differing requirements in various user industry groups, and our editors and authors have had little choice but to reflect those variations. We sincerely hope that this decision to be realistic will not also invite unwanted confusion.

In any event, this commentary will not be the final word on SI units as they relate to pumps and pump systems, since the entire subject is obviously still far from settled. Practices will continue to change, and the reader will have to remain alert to further variations of national and international practices in this area.

° The newton (symbol N) is the SI unit of force, equal to that which, when applied to a body having a mass of 1 kg, gives it an acceleration of 1 m/s².

† In countries using the SI system exclusively, the correct spelling is *metre*. This book uses the spelling meter in deference to prevailing U.S. practice.

PUMP HANDBOOK

INTRODUCTION: CLASSIFICATION

AND

SELECTION

OF

PUMPS

W. C. Krutzsch

INTRODUCTION

Only the sail can contend with the pump for the title of the earliest invention for the conversion of natural energy to useful work, and it is doubtful that the sail takes precedence. Since the sail cannot, in any event, be classified as a machine, the pump stands essentially unchallenged as the earliest form of machine which substituted natural energy for human muscular effort.

The earliest pumps of which knowledge exists are variously known, depending upon which culture recorded their description, as Persian wheels, waterwheels, or norias. These devices were all undershot waterwheels containing buckets which filled with water when they were submerged in a stream and which automatically emptied into a collecting trough as they were carried to their highest point by the rotating wheel. Similar waterwheels have continued in existence in parts of the Orient even into the twentieth century.

The best known of the early pumps, the Archimedean screw, also persists into modern times. It is still being manufactured for low-head applications where the liquid is frequently laden with trash or other solids.

Perhaps most interesting, however, is the fact that with all the technological development which has occurred since ancient times, including the transformation from water power through other forms of energy all the way to nuclear fission, the pump remains probably the second most common machine in use, exceeded in numbers only by the electric motor.

Since pumps have existed for so long and are in such widespread use, it is hardly surprising that they are produced in a seemingly endless variety of sizes and types and are applied to an apparently equally endless variety of services. While this variety has contributed to an extensive body of periodical literature, it has also tended to preclude the publication of comprehensive works. With the preparation of this handbook, an effort has been made to provide just such a comprehensive treatment.

But even here it has been necessary to impose a limitation on subject matter. It has been necessary to exclude material uniquely pertinent to certain types of auxiliary pumps which lose their identity to the basic machine they serve and where the user controls neither the specification, purchase, nor operation of the pump. Examples of such pumps would be those incorporated into automobiles or domestic appliances. Nevertheless, these pumps do fall within classifications and types covered in the handbook, and basic information on them may therefore be obtained herein once the type of pump has been identified. Only specific details of these highly proprietary applications are omitted.

Such extensive coverage has required the establishment of a systematic method of classifying pumps. Although some rare types may have been overlooked in spite of all precautions, and obsolete types that are no longer of practical importance have been deliberately omitted, principal classifications and subordinate types are covered in the following section.

CLASSIFICATION OF PUMPS

Pumps may be classified on the basis of the applications they serve, the materials from which they are constructed, the liquids they handle, and even their orientation in space. All such classifications, however, are limited in scope and tend to substantially overlap each other. A more basic system of classification, the one used in this handbook, first defines the principle by which energy is added to the fluid, goes on to identify the means by which this principle is implemented, and finally delineates specific geometries commonly employed. This system is therefore related to the pump itself and is unrelated to any consideration external to the pump or even to the materials from which it may be constructed.

Under this system, all pumps may be divided into two major categories: (1) *dynamic*, in which energy is continuously added to increase the fluid velocities within the machine to values in excess of those occurring at the discharge such that subsequent velocity reduction within or beyond the pump produces a pressure increase, and (2) *displacement*, in which energy is periodically added by application of force to one or more movable boundaries of any desired number of enclosed, fluid-containing volumes, resulting in a direct increase in pressure up to the value required to move the fluid through valves or ports into the discharge line.

Dynamic pumps may be further subdivided into several varieties of centrifugal and other special-effect pumps. Figure 1 presents in outline form a summary of the significant classifications and subclassifications within this category.

Displacement pumps are essentially divided into reciprocating and rotary types, depending on the nature of movement of the pressure-producing members. Each of these major classifications may be further subdivided into several specific types of commercial importance, as indicated in Fig. 2.

Definitions of the terms employed in Figs. 1 and 2, where they are not self-evident, and illustrations and further information on classifications shown are contained in the appropriate sections of the handbook.

SELECTION OF PUMPS

Given the variety of pumps which is evident from the foregoing system of classification, it is conceivable that an inexperienced person might well become somewhat bewildered in trying to

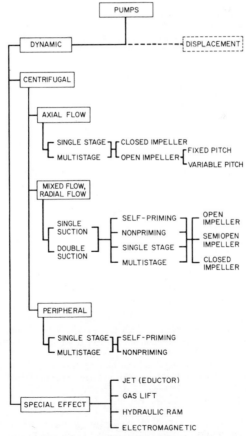

FIG. 1   Classification of dynamic pumps.

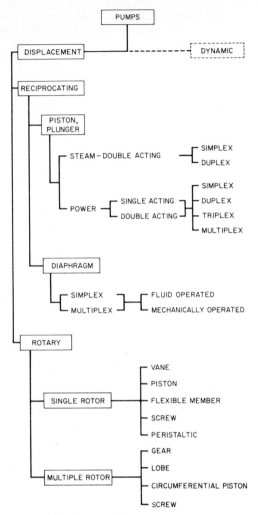

FIG. 2   Classification of displacement pumps.

determine the particular type to use in meeting most effectively the requirements for a given installation. Recognizing this, the editors have incorporated in Chap. 11, "Selecting and Purchasing Pumps," a guide which provides the reader with reasonable familiarity regarding the details which must be established by or on behalf of the user in order to assure an adequate match between system and pump.

Supplementing the information contained in Chap. 11, the sections on centrifugal, rotary, and reciprocating pumps also provide valuable insights into the capabilities and limitations of each of these classes. None of these, however, provide a concise comparison between the various types, and Fig. 3 has been included here to do just that, at least for the basic, common criteria of pressure and capacity.

The lines plotted in Fig. 3 for each of the three pump classes represent the upper limits of pressure and capacity currently available commercially throughout the free world. At or close to the limits shown, there may be only a few available sources, and pumps may well be specially

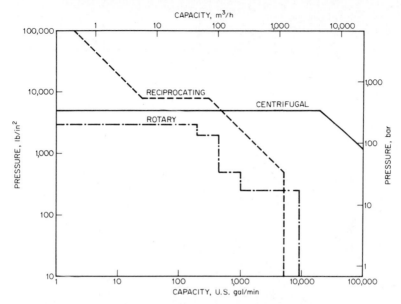

FIG. 3 Approximate upper limit of pressure and capacity by pump class.

engineered to meet performance requirements. At lower values of pressure and capacity, well within the envelopes of coverage, pumps may be available from dozens of sources as preengineered, or standard, products. Note also that reciprocating pumps run off the pressure scale while centrifugals run off the capacity scale. For the former, some highly specialized units are obtainable at least up to 150,000 lb/in² (10,350 bar)° and perhaps slightly higher. For the latter, custom-engineered pumps would probably be available up to about 3,000,000 U.S. gal/min (680,000 m³/h), at least for pressures below 10 lb/in² (0.69 bar).

Given that the liquid can be handled by any of the three basic types and given conditions within the coverage areas of all three, the most economic order of consideration for a given set of conditions would generally be centrifugal, rotary, and reciprocating, in that order. In many cases, however, either the liquid may not be suitable for all three or other considerations—such as self-priming or air-handling capabilities, abrasion resistance, control requirements, or variations in flow—may preclude the use of certain pumps and limit freedom of choice. Nevertheless, it is hoped that the information in Fig. 3 will be a useful adjunct to that contained elsewhere in this volume.

° 1 bar = 10⁵ Pa. For a discussion of bar, see *SI Units—A Commentary* in the front matter.

CENTRIFUGAL PUMPS

SECTION 2.1
CENTRIFUGAL PUMP THEORY

WALTER K. JEKAT

INTRODUCTION

The fluid motions in a centrifugal pump are complex. The velocity vectors are not parallel to the walls of the fluid passages, and appreciable secondary motions occur near the impeller discharge and in the diffusion section. These details of the true fluid motions are not well understood.

Most practical pump design is based on a one-dimensional approximation that neglects all secondary motions and treats the main flow on the basis of available flow areas and conduit wall directions. Where this leads to obvious errors, correction factors (such as the slip factor) are introduced. Although flow areas are generally calculated under the assumption of uniform velocity over the cross section, there is sometimes a flow factor applied which recognizes the blockage effect of the boundary layers. This is particularly important for the throat area of volute or diffuser.

Because of the widespread use and considerable success of the one-dimensional analysis, the following text presents it in a generalized, though abbreviated, form. The generalization has been accomplished by the extensive use of the concept of specific speed.

Much work has been done refining the basic centrifugal pump theory by two- or three-dimensional considerations. The mathematical complexities are great, and testing the detailed flow structure is difficult. Important features of the flow, such as secondary motions, are still neglected. Therefore only moderate success has been had in elucidating certain details of the flow in a centrifugal pump. No comprehensive theory which would permit the complete hydrodynamic design has evolved.

Users of centrifugal pumps will not normally wish to go deeply into the design, but they can profit from familiarity with pump theory by being able to establish the rotative speed and main dimensions of an optimally designed pump. This knowledge will help in evaluating the offers of various manufacturers. In addition, the ability to ascertain reasonably accurate efficiencies for the required pumps without the manufacturers' offers is advantageous.

NOMENCLATURE

A = sum of areas between vanes, in^2 (mm^2)

D = diameter, in (mm)

H = total head, ft (m)

L = flow friction loss, ft (m)

N = rotative speed, rpm

NPSH = net positive suction head, ft (m)

N_S = specific speed, rpm $\sqrt{\text{gpm}}/\text{ft}^{0.75}$ (rpm $\sqrt{\text{m}^3/\text{s}}/\text{m}^{0.75}$)

P = power, hp (kW)

Q = capacity, gpm (m^3/s or m^3/h when so indicated)

S = suction specific speed, rpm $\sqrt{\text{gpm}}/\text{ft}^{0.75}$ (rpm $\sqrt{\text{m}^3/\text{s}}/\text{m}^{0.75}$)

T = torque, lbf·ft (N·m)

sp. gr. = specific gravity

a = area between two vanes, in^2 (mm^2)

b = width, in (mm)

c = absolute velocity, ft/s (m/s)

e = sum of widths of shrouds at outer impeller diameter, in (mm)

g = gravity constant, 32.2 ft/s^2 (9.81 m/s^2)

h = static head, ft (m)

h = wall distance, in (mm)

m = length along streamline, starting from inlet, in (mm)

n = distance from shroud to point on orthogonal, in (mm)

p = pressure, lb/in^2 (kPa)

r = radius, in (mm)

s = vane thickness, in (mm)

t = clearance between impeller and volute tongue, in (mm)

u = peripheral velocity, ft/s (m/s)

w = relative velocity, ft/s (m/s)

y = elevation, ft (m)

z = number of vanes

α = angle between c and u, deg

β = angle between w and u, deg

γ = density, lb/ft^3 (kg/l)

ϵ = angle between streamline direction and axis of rotation, deg

η = efficiency

θ = angular distance from radial line rotating with impeller, rad

μ = slip factor

ν = kinematic viscosity, ft^2/s (m^2/s)

ρ = radius of curvature, in (mm)

φ = central angle, deg

ψ = head coefficient

ω = angular velocity, s^{-1}

Subscripts

DF = disk friction

 H = hydraulic

 H = hub

 I = impeller

 L = leakage

 M = mechanical

 S = suction

 S = shaft

 V = volumetric

 V = volute

W = water

 a = at shroud

av = average

 b = at hub

 m = meridional

 m = mean

 p = at pressure side of vane

 r = refers to arbitrary radius

 s = at suction side of vane

th = theoretical

thr = at throat of volute or diffuser

 u = denotes peripheral component

 θ = denotes circumferential component

 I = beginning of confined vane channel

 II = end of confined vane channel

BASIC HYDRAULIC RELATIONSHIPS ⸻

Bernoulli's Equation for a Stationary Conduit Assuming no flow losses, the total head H is the same for any point along a streamline:

in USCS units
$$H = \frac{144p}{\gamma} + \frac{c^2}{2g} + y = \text{constant}$$

in SI units
$$H = \frac{0.102p}{\gamma} + \frac{c^2}{2g} + y = \text{constant}$$

(1)

The individual terms in Bernoulli's equation (1) have the following physical meaning:

$144p/\gamma$ = static pressure head ($0.102p/\gamma$ = static pressure head)

$c^2/2g$ = dynamic head

y = elevation

Applying Bernoulli's equation to two stations in a conduit (Fig. 1) yields

in USCS units
$$\frac{144p}{\gamma} + \frac{c_1^2}{2g} + y_1 = \frac{144p_0}{\gamma} + \frac{c_0^2}{2g} + y_0$$

in SI units
$$\frac{0.102p_1}{\gamma} + \frac{c_1^2}{2g} + y_1 = \frac{0.102p_0}{\gamma} + \frac{c_0^2}{2g} + y_0$$

(2)

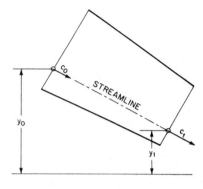

FIG. 1   Illustration of Bernoulli's equation.

When flow friction losses are included, Eq. 2 reads

in USCS units
$$\frac{144p_1}{\gamma} + \frac{c_1^2}{2g} + y_1 = \frac{144p_0}{\gamma} + \frac{c_0^2}{2g} + y_0 - L_{0-1}$$

in SI units
$$\frac{0.102p_1}{\gamma} + \frac{c_1^2}{2g} + y_1 = \frac{0.102p_0}{\gamma} + \frac{c_0^2}{2g} + y_0 - L_{0-1}$$

(3)

where L_{0-1} is the friction loss in feet (meters) incurred from station 0 to station 1.

EXAMPLE   The static pressure at the impeller inlet (station 1), when taking suction from a large reservoir (station 0 at surface), is

in USCS units
$$p_1 = p_0 - \frac{1}{144}\left(\gamma y_1 - \gamma y_0 + \frac{\gamma c_1^2}{2g} + \gamma L_{0-1}\right)$$

in SI units
$$p_1 = p_0 - \frac{1}{0.102}\left(\gamma y_1 - \gamma y_0 + \frac{\gamma c_1^2}{2g} + \gamma L_{0-1}\right)$$

The flow velocity c_0 at the surface of the reservoir is taken to be zero.

For water at 68°F (20°C), the density γ is 62.3 lb/ft^3 (1.0 kg/l) and the above equation becomes

in USCS units
$$p_1 = p_0 - \frac{y_0 - y_1}{2.31} - \frac{c_1^2}{148.8} - \frac{L_{0-1}}{2.31}$$

in SI units
$$p_1 = p_0 - \frac{y_1 - y_0}{0.102} - \frac{c_1^2}{2.00} - \frac{L_{0-1}}{0.102}$$

Euler's Equation for an Impeller   The torque required to drive an impeller is equal to the change of moment of momentum of the fluid passing through the impeller:

in USCS units
$$T = \frac{\gamma Q/449}{g} \frac{r_2 c_{u3}' - r_1 c_{u0}'}{12}$$

in SI units
$$T = \gamma Q(r_2 c_{u3}' - r_1 c_{u0}')$$

(4)

Station 0 is just in front of the vane inlets, whereas station 3 is right after the ends of the vanes (Fig. 2). The flow velocities are given a prime designation, which signifies that they are actual, rather than "theoretical," velocities (more about this on p. 2.9 under the heading "Velocity Triangles"). Equation 4 neglects impeller disk friction, leakage losses, seal friction, and all other mechanical losses.

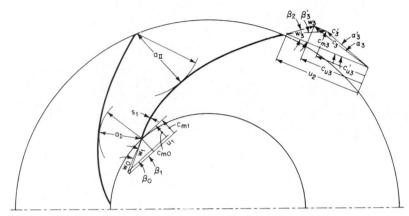

FIG. 2　Velocity triangles.

The power input is

in USCS units
$$P = \frac{\omega T}{550} = \frac{\gamma Q}{449 \times 550g}(u_2 c'_{u3} - u_1 c'_{u0})$$

in SI units
$$P = \omega T = \gamma Q(u_2 c'_{u3} - u_1 c'_{u0})$$

(5)

If there were no hydraulic losses, then the pump would produce the theoretical total head:

in USCS units
$$H_{th} = \frac{449 \times 550P}{\gamma Q} = \frac{1}{g}(u_2 c'_{u3} - u_1 c'_{u0})$$

in SI units
$$H_{th} = \frac{P}{g\gamma Q} = \frac{1}{g}(u_2 c'_{u3} - u_1 c'_{u0})$$

(6)

The actually produced total head H is smaller:

$$H = \eta_H H_{th}$$

(7)

$$H = \frac{\eta_H}{g}(u_2 c'_{u3} - u_1 c'_{u0})$$

(8)

The hydraulic efficiency η_H takes account of all flow losses inside the pump. It does not include disk friction, leakage, and mechanical losses. Often there is no prerotation ($c'_{u0} = 0$), and the total head becomes

$$H = \frac{\eta_H}{g} u_2 c'_{u3}$$

(9)

Bernoulli's Equation for an Impeller　The difference in static head between the two stations 0 and 3 (located on a common streamline) of an impeller can be found by subtracting from the theoretical total head as given by Euler's equation (6):

$$H_{th} = \frac{1}{g}(u_2 c'_{u3} - u_1 c'_{u0})$$

the difference in the dynamic head as given by Eq. 1:

$$\frac{(c_0')^2}{2g} - \frac{(c_3')^2}{2g}$$

and the flow losses incurred from 0 to 3. The static head difference is then

$$h_3 - h_0 = \frac{1}{g}(u_2 c_{u3}' - u_1 c_{u0}') - \frac{(c_0')^2}{2g} + \frac{(c_3')^2}{2g} - L_{0-3} \tag{10}$$

By observing the trigonometric relationships between the various velocities, Eq. 10 can be converted to

$$h_3 - h_0 = \frac{1}{2g}(u_2^2 - u_1^2) + \frac{(w_0')^2}{2g} - \frac{(w_3')^2}{2g} - L_{0-3} \tag{11}$$

which is sometimes called Bernoulli's equation for an impeller.

For no prerotation we have $(w_0')^2 = u_1^2 + c_{m0}^2$ and obtain

$$h_3 - h_0 = \frac{1}{2g}[u_2^2 - (w_3')^2 + c_{m0}^2] - L_{0-3} \tag{12}$$

Instead of using the impeller loss L_{0-3}, an impeller efficiency η_1 can be defined by

$$h_3 - h_0 = \frac{\eta_1}{2g}[u_2^2 - (w_3')^2 + c_{m0}^2] \tag{13}$$

Overall Efficiency   The overall efficiency (short form, efficiency) is simply the ratio of water power to shaft power:

$$\eta = \frac{P_W}{P_S} \tag{14}$$

The water power

in USCS units

$$P_W = QH\frac{\text{sp. gr.}}{3960}$$

in SI units

$$P_W = 9.80\,QH\text{ sp. gr.} \tag{15}$$

is the power which would be required if the desired head at the desired capacity could be produced without any losses whatsoever. The specific gravity for water at 39°F (4°C) is 1.0.

The shaft power can be added up such that

$$P_S = P_W + P_H + \left(\frac{1}{\eta_v} - 1\right)(P_W + P_H) + P_{DF} + P_M \tag{16}$$

where P_H stands for the power consumed by the hydraulic losses. From Eq. 7 we have

$$\eta_H = \frac{H}{H_{\text{th}}} = \frac{P_W}{P_W + P_H} \tag{17}$$

The third term,

$$\left(\frac{1}{\eta_v} - 1\right)(P_W + P_H)$$

represents the power wasted because of the internal leakage flow Q_L (from the impeller discharge through the wearing rings or vane front clearances and thrust-equalization holes to the impeller inlet). The definition of the volumetric efficiency η_v is

$$\eta_v = \frac{Q}{Q + Q_L} \tag{18}$$

The P_{DF} term is the power required to overcome the impeller-disk friction. The P_M term is the sum of all mechanical power losses, such as those caused by bearings and seals.

Combining Eqs. 14, 16, and 17, we arrive at

$$\eta = \frac{1}{\dfrac{1}{\eta_H \eta_v} + \dfrac{P_{DF}}{P_W} + \dfrac{P_M}{P_W}} \tag{19}$$

With this equation the efficiency can be calculated from estimates of the individual losses. Such estimates are given under "Design Procedures" on p. 2.14.

VELOCITY TRIANGLES

The velocity of a fluid element is represented by a vector. The length of the vector gives the magnitude of the velocity in feet per second, and the direction of the vector is tangential to the streamline.

The fluid velocity in a stationary conduit (inlet or discharge pipe, inlet guide vanes, diffuser vanes, volute) is measured in reference to an earthbound coordinate system. Therefore it is called *absolute velocity* and denoted c'. The prime signifies an *actual* flow velocity. Those *idealized* velocity vectors, which are calculated by assuming perfect guidance of the flow by vanes or walls, are given no prime. Failure to distinguish clearly between actual and idealized velocities produces an erroneous application of basic impeller theory and consequently a faulty design.

The fluid velocity in an impeller channel can be represented by either the absolute velocity c' or the *relative velocity* w'. The coordinate system for the relative velocity rotates with the impeller angular velocity $\omega = u/r$. Again, no prime is assigned to the idealized velocity. The idealized relative velocity w is easily calculated by dividing the flow per vane channel by the cross-sectional area of the channel.

The absolute velocity can be considered as the resultant of the relative velocity and the local impeller peripheral speed. Velocity triangles provide much information on the design under consideration and should be drawn for every calculated station. Figure 2 gives an example. The vectors of the absolute velocity and the relative velocity end at the same corner of the triangle. The peripheral velocity vector u starts at the beginning of the absolute velocity vector and goes to the beginning of the relative velocity vector.

The *meridional velocity* c_m is the component in the meridional plane of the absolute as well as the relative velocity. It always forms a right angle with u. For strictly radial flow the meridional velocity is also the radial component of the absolute and relative velocity. Likewise, for strictly axial flow it is the axial component.

The peripheral components of the absolute velocity c' and the relative velocity w' are denoted c_u' and w_u'. The prime is omitted if idealized velocities are used. The angle between the absolute velocity c' and the peripheral direction is α'. If taken right after the impeller vane endings, the angle α_3' gives the direction with which the fluid enters the diffusing system. The relative velocity w' forms with the peripheral direction the angle β', which is only approximately equal to the vane angle β. By definition, the idealized relative velocity w forms with the peripheral direction the vane angle β.

Slip Factor   Often, idealized velocity triangles are drawn by assuming perfect guidance of the flow by the vanes. The vane angle β is substituted for the flow angle β'. The angle between the idealized absolute velocity c and the peripheral direction is now denoted α. The meridional velocity c_m is normally assumed to remain the same (therefore the prime was not used before). We know that the relative velocity w and the absolute velocity c are not actual velocities, but they are utilized in the design of impellers because they are much easier to calculate than actual flow velocities. However, the results must be corrected to those which would be obtained if the actual flow velocities were employed.

Deviation of the fluid from the vane direction is especially important at the impeller discharge since this deviation reduces the peripheral component of the absolute impeller discharge velocity. This causes a proportionate reduction in head as well as power input. With the usual backward-leaning vanes ($\beta_2 < 90°$), the flow angle β_3' is smaller than the vane angle β_2. The phenomenon is often called *slip* and is a consequence of the nonuniform velocity distribution across the impeller channels, boundary-layer accumulation, and flow separation if it occurs. Accurate prediction of slip is very difficult since the causes of slip cannot be predetermined in a practical manner. This is the greatest drawback of the existing centrifugal impeller theory. The way in which this problem is currently treated is first to define a slip factor,

$$\mu = \frac{c'_{u3}}{c_{u3}} \tag{20}$$

and then to make the assumption that the slip factor is a function of only the number of vanes, the vane discharge angle, and sometimes the impeller-radius ratio r_2/r_1. All other details of the impeller design are not taken into account. The construction of a slip factor formula is usually based on theoretical reasoning. Often empirical coefficients bring the formula into agreement with available test data. Numerous slip factor formulas have been proposed since Stodola[1] published the first, a theoretical one:

$$\mu = 1 - \frac{\pi \sin \beta_2}{z} \tag{21}$$

This formula was originally derived for the case of zero flow through a centrifugal impeller but has since frequently been used for the design flow. It has enjoyed appreciable acceptance in American pump practice.

In Europe the slip factor formula by Pfleiderer[2]

$$\mu = \frac{1}{1 + a\left(1 + \dfrac{\beta_2}{60}\right)\left(\dfrac{r_2^2}{zS}\right)} \tag{22}$$

is in widespread use, where S is the static moment of the mean streamline

$$S = \int_{r1}^{r2} r \, dx \tag{23}$$

which calls for a simple graphical integration in the meridional plane. Figure 3 identifies the terms in Eq. 23. For a cylindrical vane

$$S = \int_{r1}^{r2} r \, dr = \tfrac{1}{2}(r_2^2 - r_1^2)$$

and the slip factor

$$\mu = \frac{1}{1 + \dfrac{a}{z}\left(1 + \dfrac{\beta_2}{60}\right)\dfrac{2}{1 - (r_1^2/r_2^2)}} \tag{24}$$

For radius ratios r_1/r_2 below ½, experience teaches that the slip factor does not increase anymore. For such small radius ratios the slip factor for $r_1/r_2 = $ ½ should be used.

Turbomachinery designers have long recognized that the slip factor is affected not only by the impeller configuration but also by the interaction of the diffusing system with the impeller. Pfleiderer,[2] on the basis of a series of tests, takes this into account by keying the coefficient a in Eqs. 22 and 24 to the casing design:

Volute: $a = 0.65$ to 0.85

Vaned diffuser: $a = 0.6$

Vaneless diffuser: $a = 0.85$ to 1.0

Once we have accumulated some test data, we can make better predictions of the slip factor for subsequent similar designs. First, we combine Eqs. 9 and 20 into

$$H = \mu \frac{\eta_H}{g} u_2 c_{u3} \tag{25}$$

Then we obtain the peripheral component c_{u3} of the idealized absolute velocity from the impeller-discharge velocity triangle. Furthermore, an estimate of the hydraulic efficiency (for instance, as described under "Hydraulic Efficiency" on p. 2.20) is made. Entering the tested total pump head into Eq. 25 yields the slip factor. A decisive advantage is gained if such empirical slip factors are applied to families of centrifugal pumps, which are similar in more detail than is taken into account by the slip factor formulas.

Head Determined from Impeller-Discharge Velocity Triangles

From the impeller-discharge velocity triangle it is apparent that

$$c_{u3} = u_2 - c_{m3} \cot \beta_2 \tag{26}$$

From Eq. 20 we have

$$c'_{u3} = \mu c_{u3}$$

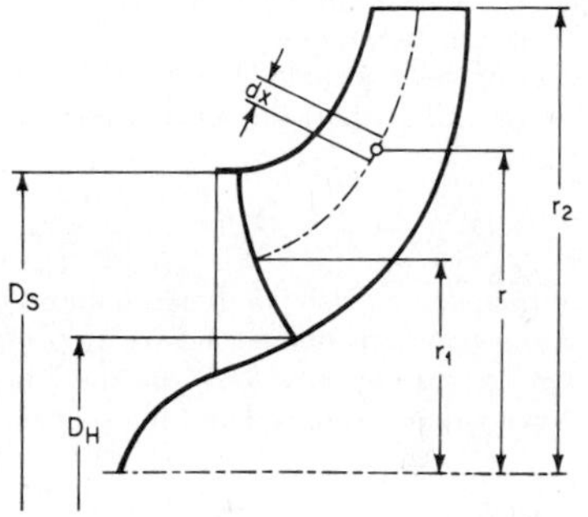

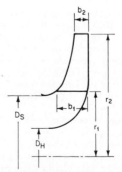

FIG. 3   Impeller with vanes extended into axial inlet.

Entering these two expressions into Euler's equation (9),

$$H = \frac{\eta_H}{g} u_2 c'_{u3}$$

we obtain

$$H = \mu \eta_H \frac{u_2^2}{g} \left(1 - \frac{c_{m3}}{u_2} \cot \beta_2 \right) \tag{27}$$

This equation enables us to ascertain the effect of the two design parameters c_{m3}/u_2 and β_2 on the total head. If c_{m3}/u_2 is reduced, the head increases. The effect of the vane discharge angle is not so straightforward. If β_2 is increased, cot β_2 is decreased and the head becomes larger. But the slip factor formulas (Eqs. 21 and 22) indicate that larger angles β_2 decrease the slip factor μ. Therefore the effects of changes in β_2 are partly canceled out in Eq. 27. Indeed, some tests showed that varying the blade angle β_2 from 20 to 30° had only a negligible effect on the pump head. If, though, vane discharge angles as large as 90° are employed, an appreciable head increase can be expected.

The pump head as defined by Eq. 27 can be made dimensionless by dividing by $u_2^2/2g$, resulting in the head coefficient

$$\psi = \frac{H}{u_2^2/2g} = 2\mu \eta_H \left(1 - \frac{c_{m3}}{u_2} \cot \beta_2 \right) \tag{28}$$

The meridional velocity c_{m3} is calculated right after the impeller discharge with

in USCS units $$c_{m3} = \frac{Q/449}{2\pi r_2 b_3/144}$$

in SI units $$c_{m3} = \frac{Q \times 10^6}{2\pi r_2 b_3} \tag{29}$$

where b_3 is taken equal to b_2, the vane width at the impeller discharge (Fig. 4).

FIG. 4   Impeller with cylindrical vanes.

SIMILITUDE

A large number of centrifugal pumps, for greatly varying capacities, heads, and rotative speeds, have been built and tested. Because their efficiencies range from, say, 15 to over 90%, it became necessary to determine whether low efficiencies were always due to poor design or whether some unfavorable conditions of service existed which precluded good performance. Conversely, it was important to find out whether certain conditions of service were conducive to high efficiencies. Furthermore, it was desirable to be able to group conditions of service in such a way that a large number of designs could be lumped together into a single expression. Dimensional analysis suggested for this purpose the specific speed

$$\frac{N\sqrt{Q}}{(gH)^{0.75}}$$

That the specific speed is a dimensionless group can be verified by entering consistent dimensions, such as revolutions per second for the rotative speed, cubic feet per second (cubic meters per second) for the capacity, and feet (meters) for the head.

Normally g is dropped and the specific speed is written

$$N_S = \frac{N\sqrt{Q}}{H^{0.75}} \tag{30}$$

In American centrifugal pump practice the following inconsistent but generally accepted dimensions are used: revolutions per minute for rotative speed, gallons per minute for capacity, and feet for the head. In SI units the specific speed is referred to as N_{Sm}, with the capacity in cubic meters per second and the head in meters. Thus, $N_S = 51.65 N_{Sm}$. It is in these forms that the specific speed is given in this section. That the resultant value for the specific speed now has a dimension does not diminish its utility. In principle it remains nondimensional.

In 1947 Wislicenus[3] published a plot of the "approximate statistical averages" of the efficiencies of a large number of commercial centrifugal pumps versus specific speed for values above $N_S = 500$ ($N_{Sm} = 10$). While it became obvious that a correlation existed, it was necessary to introduce an additional parameter having something to do with the scale of the pump. At that time a capacity between 100 and 10,000 gpm (23 and 2300 m^3/h) was selected, but with only 100 and 200 gpm (23 and 45 m^3/h) extended down to $N_S = 500$ ($N_{Sm} = 10$). This graph was quickly accepted in the United States and became the criterion by which designers judged the merits of their designs. Indeed, it was often referred to as "the chart," and efficiency read from it was called "chart efficiency."

An updated version of this chart based on the evaluation of 528 new test points is given in Fig. 5. The only test results used were those which form a cluster of points or single points of unquestionable accuracy. The chart was extended down to $N_S = 180$ ($N_{Sm} = 3.5$) and $Q = 30$ gpm (7 m^3/h) over the whole range from $N_S = 180$ to 3000 ($N_{Sm} = 3.5$ to 58). In addition, efficiencies are given for 5 and 10 gpm (1.1 and 2.3 m^3/h) at specific speeds from 100 to 500 ($N_{Sm} = 2$ to 10).

Just as in the original Wislicenus chart, efficiencies start dropping drastically at specific speeds below 1000 ($N_{Sm} = 20$). Also, smaller capacities exhibit lower efficiencies than higher capacities at all specific speeds. In comparison to the Wislicenus chart, however, the newer designs (Fig. 5) show higher efficiencies for low specific speed, especially below $N_S = 1000$ ($N_{Sm} = 20$), and low capacities, especially below 200 gpm (45 m^3/h). Apparently, these are the regions where improvements were possible. No appreciable efficiency improvements have been achieved in the intervening 25 years for pumps of high specific speed and large capacity. For instance, at $N_S = 2500$ ($N_{Sm} = 48$) and $Q = 10,000$ gpm (2273 m^3/h), we still register $\eta = 0.89$ as a representative value.

Designers used to be happy when a design exceeded chart efficiency. Figure 5 will still permit this because the curves do not take into account the few pumps of exceptionally high efficiencies.

The utility of specific speed as a significant group for the prediction of efficiency rests upon the premise that geometrically similar pumps, operated at the same specific speed, have geometrically similar velocity triangles. Consequently, the ratios of the flow velocities to the impeller peripheral speed are the same. Were it not for the effects of scale, their relative losses would also be the same. Scale affects losses because of its influence upon flow friction and leakage losses.

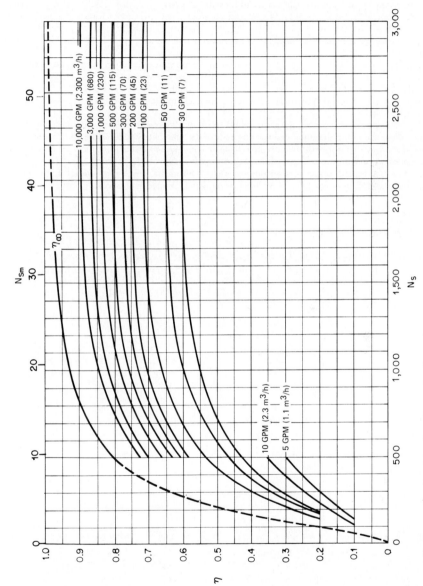

FIG. 5 Efficiency as a function of specific speed and capacity.

A chart such as Fig. 5 is intellectually disappointing; to represent scale, a dimensional parameter—the capacity Q—appears in an otherwise dimensionless plot (η versus N_S). It would seem more appropriate to use a dimensionless parameter, perhaps a Reynolds number, but the situation is complicated by the fact that surface roughness and clearances are not usually scaled properly. It is an empirical fact that, in all known attempts to use a single parameter for scale, capacity has correlated best with test results. To a large degree this is supported by another empirical finding, that the hydraulic efficiency is mainly dependent upon the capacity (see "Hydraulic Efficiency" on p. 2.20).

The specific speed chart (Fig. 5) is convenient to use because designers who want to predict efficiency quickly need to know only the head and capacity, which must be specified by the conditions of service. The only choice to be made is rotative speed. Several such choices (for instance, 1,800 and 3,600 rpm) may be entered in order to ascertain the probable effect upon efficiency.

The selection of the specific speed is found to be insufficient, however, since nothing is yet known about the pump design which would produce the efficiency shown in Fig. 5. For this, it is necessary to have an approximation of the velocity triangles. In order to reduce the amount of information to be transmitted, only the most important—the velocity triangle after the impeller discharge—is specified (assuming that the impeller inlet and diffusing system design follow common practice; see "Impeller Inlet," "Volute Casing," and "Vaned Diffuser").

DESIGN PROCEDURES

Design Parameters Specifying Impeller-Discharge Velocity Triangle
Figure 6 gives the band of commonly used values of c_{m3}/u_2. This parameter decreases with smaller specific speed, tending toward zero at zero specific speed. If values higher than those shown are attempted at low specific speeds, such small vane discharge widths would result that the usual manufacture of the impellers by casting would be difficult.

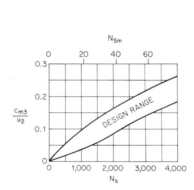

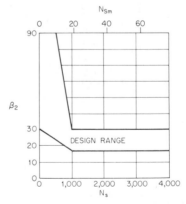

FIG. 6 c_{m3}/u_2 versus specific speed.

FIG. 7 Impeller discharge angle versus specific speed.

Specifying c_{m3}/u_2 ties the meridional velocity (taken right after the impeller discharge) to the impeller peripheral speed. If, in addition, one flow angle is known, then the whole impeller-discharge velocity triangle can be drawn. It is customary to specify, instead of a flow angle, the vane discharge angle, since the latter is one of the design choices which must be made. The range of commonly used values of β_2 is given in Fig. 7.

Since the vane discharge angle is not a flow angle, it is necessary to know the slip factor in order to complete the velocity triangle. This factor is given in an implicit manner by supplying the usual number of vanes z, which is four to eight for specific speeds up to 4000 ($N_{Sm} = 77$).

The three design parameters c_{m3}/u_2, β_2, and z are therefore sufficient to describe the impeller-discharge velocity triangle in an approximate manner. The values plotted in Figs. 6 and 7 originate from pumps designed by many engineers from several organizations. It is therefore to be expected that the values form bands rather than arrange themselves neatly along lines.

In addition to their influence upon the design point efficiency, the three design parameters have an effect upon the shape of the head capacity curve. It rises toward the shutoff if smaller values of β_2 and z and larger values of c_{m3}/u_2 are selected. When this is a consideration, one can select the extreme values within the bands in Figs. 6 and 7.

Impeller Inlet   The impeller inlet design is based on the inlet flow angle β_0 (taken at the outer diameter of the inlet). The station 0 is located right in front of the vane inlet tip. The flow angle β_0 is selected between 10 and 25°, independent of specific speed. An often used flow angle is 17°, which is a compromise between efficiency and cavitation. For best impeller efficiency, the angle β_0 should be larger; for lower NPSH, it should be smaller.

It is convenient to make $NPSH$ dimensionless by using the concept of suction specific speed[4] (see also Subsec. 2.3.1):

$$S = \frac{N\sqrt{Q}}{NPSH^{0.75}}$$

where N = rpm, Q = gpm, and $NPSH$ = ft, or

$$S_m = \frac{N\sqrt{Q}}{NPSH^{0.75}}$$

where N = rpm, Q = m^3/s, and $NPSH$ = m, such that

$$S = 51.65 S_m \tag{31}$$

A typical value for suction specific speed is $S = 9000$ ($S_m = 174$), which can be achieved with centrifugal impellers of good manufacture having a flow angle of about 17° and approximately five to seven vanes. Many commercial pumps have lower suction specific speeds, in the range from 5000 to 7000 ($S_m = 97$ to 136). On the other hand, boiler-feed and, especially, condensate pumps require suction specific speeds as high as 12,000 to 18,000 ($S_m = 232$ to 348). To reach such values, the flow angle is taken as low as 10° and the number of vanes is reduced to as few as four. Fewer vanes (as well as thinner vanes) are beneficial because they reduce the blockage effect. A disadvantage of low flow angles (and consequently large inlet diameters) is that the pump is more likely to run rough at part load, especially below 50% capacity.

Should the available NPSH be so low that the required suction specific speed is above about 18,000 ($S_m = 348$), then a separate axial flow impeller of special design—an inducer—is used ahead of the centrifugal impeller.[5,6] Its flow angle is typically between 5 and 10°. The vane angle is about 3 to 5° larger.[7] The number of vanes is often only two, and not more than four. The vane thickness is made as small as possible.

The suction diameter D_S (Figs. 3 and 4) is

in USCS units
$$D_S = 4.54 \left(\frac{Q}{kN \tan \beta_0}\right)^{1/3}$$

in SI units
$$D_S = 2897 \left(\frac{Q}{kN \tan \beta_0}\right)^{1/3} \tag{32}$$

whereby the hub ratio k is

$$k = 1 - \left(\frac{D_H}{D_S}\right)^2 \tag{33}$$

For vanes which extend into the axial portion of the inlet (Fig. 3), the flow angle β_0 varies along the leading edge. Assuming uniform meridional velocity over the radius, the flow angle is determined from

$$\tan \beta_{0(r)} = \tan \beta_0 \frac{r_1}{r} \tag{34}$$

The flow angle, and consequently the vane angle, are assumed to be constant for vanes with leading edges parallel to the axis (Fig. 4).

The vane angle β_1 is made larger than the flow angle β_0 in order to compensate for the blockage of the vanes.[2] From this approach follows (Fig. 2):

$$\tan \beta_{1(r)} = \frac{\tan \beta_{0(r)}}{1 - (zs_1/2\pi r \sin \beta_{1(r)})} \tag{35}$$

The equation is used for vanes according to Fig. 3 as well as for vanes according to Fig. 4 ($r = r_1$).

For vanes with leading edges parallel to the axis, it remains to determine the inlet width b_1 (Fig. 4). This can be done by assuming a constant and equal meridional velocity for the axial and radial portion of the inlet passage, resulting in

$$b_1 = \frac{k}{4} D_S \frac{D_S}{D_1} \tag{36}$$

whereby D_S/D_1 is between 0.8 and 1.0.

Impeller Vane Layout   The selection of the impeller discharge and that of the inlet design have so far been treated independently. Actually they should be taken together, for they greatly influence the whole vane shape and velocity distribution in the impeller.

In order to judge whether a particular vane shape is likely to lead to high impeller efficiency, it is most helpful to consider the distribution of the relative velocity along the flow path. Once the flow has entered the impeller vane system, it is not confined by a channel until it reaches the section a_I (Fig. 2). In order to avoid the danger of flow separation, the vane angle is increased only slightly up to section a_I. From there on to section a_{II} (Fig. 2), the flow is confined. The mean relative velocities in the two sections are

in USCS units $w_I = \dfrac{Q/449}{za_I/144}$ and $w_{II} = \dfrac{Q/449}{za_{II}/144}$

in SI units $w_I = \dfrac{Q}{za_I} \times 10^6$ and $w_{II} = \dfrac{Q}{za_{II}} \times 10^6$

The ratio of the relative velocities is

$$\frac{w_{II}}{w_I} = \frac{a_I}{a_{II}}$$

Normally the cross-sectional area of the channel increases gradually from a_I to a_{II}. The reduction of the relative velocity causes an increase of static head, as indicated by Eq. 11. There are limitations to the area increase if flow separation is to be avoided. For high hydraulic efficiency, the area ratio is selected as $a_{II}/a_I = 1.0$ to 1.3.

The determination of a_I and a_{II} is simple provided cylindrical vanes (Fig. 4) are used. Only the meridional section (Fig. 4) and an axial view (Fig. 2) are required. The vane layout depends mainly upon the selected angles β_1 and β_2. The recommended ranges for these two angles overlap. Therefore the outlet vane angle is occasionally equal to the inlet vane angle. In such a case, the vane takes the shape of a logarithmic spiral. Its coordinates (Fig. 8) can be calculated from

$$r = r_1 e^{(\pi\varphi/180)\tan\beta} \tag{37}$$

If β_2 is different from β_1, Eq. 37 of the logarithmic spiral can still be utilized for the vane layout by a stepwise variation of the vane angle β.

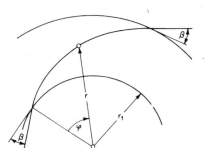

FIG. 8   Logarithmic spiral.

For twisted vanes (Fig. 3), areas a_I and a_{II} can be obtained only after considerable labor has been expended on the complicated vane layout. In order to reduce the number of trials, it is desirable to be able to calculate the areas between vanes, at least approximately, before starting the first vane layout.

For Francis-type vanes (Fig. 18), with mixed flow in the vane inlet area and a small hub, a useful approximation is

$$A_I = za_I \approx \pi r_1^2 \sin \beta_{1m} \tag{37a}$$

The mean inlet vane angle β_{1m} is calculated from Eq. 35 by entering the mean radius

$$r_{1m} = \left(\frac{r_1^2 + r_H^2}{2}\right)^{1/2} \tag{38}$$

If the front shroud in the discharge region is inclined, as is usual for Francis-type impellers, the discharge area between vanes can be approximated by

$$A_{II} = za_{II} \approx b_2(2\pi r_2 \sin \beta_2 - zs_2) \tag{39}$$

If the front shroud in the discharge region is not inclined, about 25% should be subtracted from Eq. 39.

To make sure that the flow encounters no sudden changes of flow area and vane angle, it is recommended that their values, as determined from the vane layout, be plotted against the radius or length of flow path.

Volute Casing   A large part of the dynamic head $(c_3')^2/2g$ of the fluid issuing from the impeller is converted to static head in the following diffusing system. The most popular one for centrifugal pumps is the volute.

The design of the volute starts preferably with the calculation of the area of the throat (see Fig. 9):

in USCS units
$$A_{thr} = \frac{Q/449}{c_{thr}/144}$$
$$\tag{40}$$

in SI units
$$A_{thr} = \frac{Q}{c_{thr}} \times 10^6$$

The average throat velocity c_{thr} is determined from

$$\frac{c_{thr}}{c_{u3}'} = \frac{r_2}{r_4} C \tag{41}$$

for frictionless flow $C = 1$, according to the law of constant angular momentum. Experience has shown that $C = 1$ is indeed a good design value for volutes of large pumps or very smooth volutes of medium-size pumps. For commercially cast volutes of medium size and small pumps, the value $C = 0.9$ gives a reasonable approximation.

The distance r_4 of the center of the throat section from the axis is (Fig. 9):

$$r_4 \approx r_2 + t + r_{thr} \tag{42}$$

whereby the throat section is assumed to be circular.

The clearance t between impeller and tongue of volute has a bearing on efficiency and pressure pulsations. If the highest possible efficiency is desired, the clearance is made as small as $\frac{1}{32}$ in (1 mm), but such a small clearance produces larger pressure pulsations and has led to impeller failure in some cases. As a remedy, the clearance is then increased to about $\frac{1}{8}$ in (3 mm). A reasonably safe value for the tongue clearance, still allowing good efficiency, is 5 to 10% of the impeller radius.

The calculation of the throat area can be aided by initially assuming a throat velocity c_{thr}, for example, from the ratio c_{thr}/u_2 as given by Fig. 10, which shows a band of empirical values plotted versus the specific speed. The sources are Refs. 8 and 9, in the range of N_S from 200 to 2000 (N_{Sm} = 4 to 39) augmented by data collected by the author. With the initially assumed throat velocity

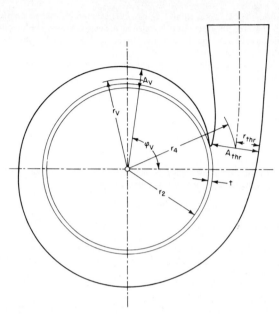

FIG. 9 Volute casing.

c_{thr} and tongue clearance t, the throat area A_{thr} and the throat distance r_4 can be calculated. Then Eq. 41 is applied as a check, and if necessary a correction is made for the throat area.

An alternate method for the calculation of the throat area was originally developed by Anderson.[9] It makes use of the area ratio $y = A_{II}/A_{thr}$, where $A_{II} = za_{II}$ is the total outlet area between impeller vanes (Fig. 2). Worster[10] has verified that, for the condition of constant angular momentum, a relation exists between y and N_S. Figure 11 is a plot of empirical data for $1/y = A_{thr}/A_{II}$ versus N_S. The band results from the author's data and shows the considerable range of area ratios which are used in actual pump designs. The line in the middle of the band is from Anderson.[9] The condition of constant angular momentum is approximately fulfilled for the lower limit of A_{thr}/A_{II} in Fig. 11. High efficiency can be expected between this limit and Anderson's line.

Commercial pump lines make use of the fact that large variations in A_{thr}/A_{II} are tolerable if impellers of different widths are used in the same volute casing or if the same impeller is used in volutes of different throat areas. The intent is to arrive at new combinations of head and capacity without manufacturing additional hardware.

It is not recommended that the curves in Fig. 11 be extrapolated to lower specific speeds. The resultant impeller discharge widths would, in many cases, be too small for manufacturing. Actual successful low-specific-speed pumps have much wider impeller discharges.

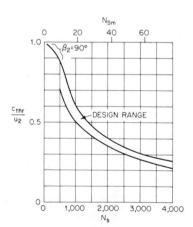

FIG. 10 c_{thr}/u_2 versus specific speed.

The intermediate volute areas

in USCS units

$$A_V = \frac{Q/449}{c_V/144} \frac{\varphi_V}{360}$$

(43)

in SI units
$$A_V = \frac{Q}{c_V} \frac{\varphi_V}{360} \times 10^6$$

can be calculated on the basis of constant angular momentum:

$$\frac{c_V}{c'_{u3}} = \frac{r_2}{r_V} C \tag{44}$$

Stepanoff[8] prefers to use a constant mean velocity for all volute sections, which results in volute areas being proportional to the central angle φ_V:

$$A_V = A_{\text{thr}} \frac{\varphi_V}{360} \tag{45}$$

The resultant intermediate volute areas are larger than those calculated using the constant angular momentum method. Still, there is very little, if any, detrimental effect on the efficiency. Even a constant area collector, instead of a volute, reduces the efficiency at the design point by no more than 1 to 2 points.

The volute is followed by a conical diffuser, making the transition to the discharge pipe. The total included angle of the diffuser can be taken from 6 to 10°. The volute sections may have shapes different from circles. For instance, rectangles or trapezoids are often used. These shapes make it possible for the entrance width of the volute to be appreciably greater than the impeller discharge width b_2 (the volute entrance width may be as large as two or three times the impeller discharge width). Such loose-fitting volutes are common design practice in the United States, but much less so in Europe. According to tests[11] made in Germany, they produce slightly better efficiency, and their great advantage is that they are easier to cast (because of the better core support) and easier to clean after casting.

Volutes cause radial thrust, small at the design point but increasing as we approach larger and, especially, smaller capacities. The reason for the radial thrust is that the pressure at the periphery of the volute entrance is no longer constant, as it should be at the design point. Twin volutes, vaned diffusers, and, to a lesser degree, concentric casings reduce the radial thrust. Agostinelli and his coauthors[12] give numerical values based on extensive tests and also describe a semiconcentric casing design in which the collector is circular for a portion of the angular distance from the tongue before the radial extent of the casing wall is increased toward the throat section. This simple design can have radial thrust as low as that of the more complicated twin volute.

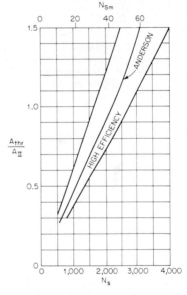

FIG. 11 $A_{\text{thr}}/A_{\text{II}}$ versus specific speed.

Vaned Diffuser

Instead of a volute, a vaned diffuser may be used. It consists of a number of vanes set around the impeller. The flow from the vaned diffuser is collected in a volute or circular casing and discharged through the outlet pipe.

The design of the vaned diffuser is analogous to that of the volute. Instead of one throat section, there are now several (Fig. 12). Again, Eq. 41,

$$\frac{c_{\text{thr}}}{c'_{u3}} = \frac{r_2}{r_4} C$$

can be used to calculate the throat velocity c_{thr}. The distance r_4 of the center of the throat from the axis is now smaller than for the volute pump of equal specific speed. Therefore one would expect the throat velocity of a diffuser to be larger. This is actually not the case, however, because the factor C is smaller for vaned diffusers. A typical value is $C = 0.8$. The ratios c_{thr}/u_2 in Fig. 10 can also be used for vaned diffusers.

From the throat on, the area of the vane channel increases progressively, so that further, though slight, pressure increase takes place. The centerline of the vane channel after the throat may be straight or curved. The straight diffusing channel is slightly more efficient but results in a larger casing. The vane surface from the vane inlet tip to the throat can be shaped like a volute, but an approximation by a circular arc is entirely permissible. The vane entrance angle is made 3 to 5° larger than α_3'. Moderate deviations do not matter; what matters most is the proper choice of the throat area. The side walls of the vaned diffuser are preferably parallel, but in some designs they spread apart with increasing radius. A good choice for the number of diffuser vanes is one more than the number of impeller vanes. In this way, circulation around the diffuser vanes, resulting from uneven impeller channel discharge, is minimized.

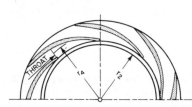

FIG. 12 Vaned diffuser.

Hydraulic Efficiency   The hydraulic efficiency depends, naturally, very much upon the design and execution of the flow passages. Nevertheless, if one considers only pumps of above-average performance and plots their hydraulic efficiency (at the design point) versus capacity, the points arrange themselves in a narrow band. The value of the hydraulic efficiency can be approximated by

in USCS units
$$\eta_H \approx 1 - \frac{0.8}{Q^{0.25}}$$

in SI units
$$\eta_H \approx 1 - \frac{0.071}{Q^{0.25}}$$

(46)

This equation is plotted in Fig. 13. The specific speed has only a small influence on the hydraulic efficiency. At $N_S = 500$, the hydraulic efficiency seems to be about 2 points lower than at $N_S =$

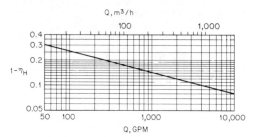

FIG. 13   Hydraulic efficiency versus capacity.

2000 ($N_{Sm} = 39$). This possible variation has been neglected in Eq. 46 since it is not greater than the scatter of the test data.

There are many pumps in service which have, because of mediocre design or manufacture, hydraulic efficiencies 5 points lower than given by Eq. 46. Even hydraulic efficiencies 10 points lower have been tolerated in some cases, especially for small capacities.

Volumetric Efficiency   The definition of the volumetric efficiency is given by Eq. 18:

$$\eta_V = \frac{Q}{Q + Q_L}$$

To calculate the leakage flow, the details of the individual pump design must be known. Standard methods can be found in Refs. 2 and 8. For an approximate prediction of η_V at the design point, Fig. 14 can be used. It shows the volumetric efficiency as a function of specific speed and flow for pumps of normal construction. For instance, for a pump handling 1000 gpm (227 m³/h) at a specific speed of 2500 ($N_{Sm} = 48$), Fig. 14 estimates $\eta_V \approx 0.98$. For 100 gpm (23 m³/h) at $N_S = 500$ ($N_{Sm} = 10$), the estimate is $\eta_V \approx 0.9$.

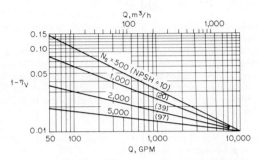

FIG. 14 Volumetric efficiency as a function of specific speed and capacity.

Mechanical Losses The mechanical losses can be calculated only when the design details of bearings and seals are available. For an approximate prediction of the ratio of mechanical power loss to water power, Fig. 15 can be used. It shows the general trends, in that the mechanical power loss relative to the water power increases with decreasing specific speed and capacity.

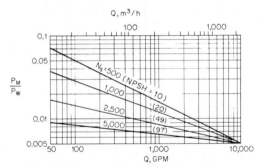

FIG. 15 Ratio of mechanical power loss to water power as a function of specific speed and capacity.

Disk Friction The outer surfaces of a rotating impeller are subject to friction with the surrounding fluid. There have been a number of attempts to predict this power loss on the basis of tests with rotating disks,[13-15] therefore the expression *disk friction*. Pfleiderer[2] writes on the basis of tests by Schultz-Grunow[13]:

in USCS units $\qquad P_{DF} = 163(10^{-17})N^3D^4(D + 5e)$ $\qquad\qquad$ (47)

in SI units $\qquad P_{DF} = 1149(10^{-25})N^3D^4(D + 5e)$

valid for $Re = \omega(D/2)^2/\nu = 7.0(10^5)$. The disk friction power increases approximately with $Re^{-0.2}$. Formula 47 aims to account not only for the friction on the two sides of the impeller but also for the friction on the outer cylindrical surfaces of the shrouds (e is the sum of the width of two shrouds). References 8 and 14 state that Eq. 47 gives values on the high side.

The ratio of disk friction power to water power, as a function of the specific speed, can be derived by utilizing Eqs. 15, 28, 30, and 47:

in USCS units

$$\frac{P_{DF}}{P_W} = \frac{1.38(10^5)}{N_S^2 \psi^{2.5}} \frac{D + 5e}{D}$$

(48)

in SI units

$$\frac{P_{DF}}{P_W} = \frac{51.8}{N_{Sm}^2 \psi^{2.5}} \frac{D + 5e}{D}$$

In view of the smaller value obtained below in "Impeller-Disk Friction Deduced from Maximum Efficiency," it is recommended that only 70 to 80% of Eq. 48 be used.

The conditions in a centrifugal pump are more complicated than in a simple disk-friction experiment. Very probably an appreciable portion (according to Ref. 16, almost 50%) of the disk friction loss is recovered as contribution to the pump head, if the rotating flow induced by disk friction is allowed to enter the casing freely. The whirl velocity c_{u3}' of the flow issuing from the impeller reduces the friction on the outer cylindrical impeller surfaces.[16] The wearing-ring leakage causes a radial flow which tends to reduce disk friction.

Maximum Efficiency   It is of some interest to estimate the maximum efficiency at a given specific speed. Such an estimate can be made on the basis of Fig. 5 by extrapolating the capacity to infinity. A great deal of the efficiency increase with capacity results from the rise in hydraulic efficiency. The approximate Formula 46 for the hydraulic efficiency

in USCS units

$$\eta_H \approx 1 - \frac{0.8}{Q^{0.25}}$$

in SI units

$$\eta_H \approx 1 - \frac{0.071}{Q^{0.25}}$$

suggests, therefore, plotting the value $1 - \eta$ versus $1/Q^{0.25}$ (Fig. 16). The pump efficiency η is taken from the specific speed chart, Fig. 5. Extrapolating to $1/Q^{0.25} = 0$ in Fig. 16 provides an

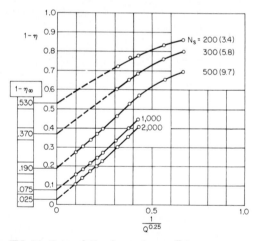

FIG. 16   Extrapolation for maximum efficiency.

estimate of the efficiency at infinite capacity—the maximum efficiency η_∞. Between $N_S = 500$ and 2000 ($N_{Sm} = 9.7$ and 39), the extrapolation is reliable and can be represented by

in USCS units

$$\eta_\infty = \frac{1}{1 + (7,800/N_S^{5/3})}$$

(49)

in SI units
$$\eta_\infty = \frac{1}{1 + (10.89/N_{Sm}^{5/3})}$$

This is the upper line denoted η_∞ in Fig. 5.

Impeller Disk Friction Deduced from Maximum Efficiency   At infinite capacity, the hydraulic efficiency is unity according to Eq. 46. The volumetric efficiency reaches the same value and P_M/P_W vanishes. Equation 19 for the pump efficiency becomes simply

$$\eta_\infty = \frac{1}{1 + (P_{DF}/P_W)} \tag{50}$$

When this is equated with the previous estimate of the maximum efficiency (Eq. 49), the following expression for the impeller disk friction results:

in USCS units
$$\frac{P_{DF}}{P_W} = \frac{7800}{N_S^{5/3}}$$
$$\tag{51}$$

in SI units
$$\frac{P_{DF}}{P_W} = \frac{10.89}{N_{Sm}^{5/3}}$$

Since Eq. 49 is reliable only from $N_S = 500$ to 2000 ($N_{Sm} = 9.7$ to 39), the same limitation applies to Eq. 51. The numerical values are approximately 70% of those determined via Eq. 48, which is based on Pfleiderer's disk friction formula (Eq. 47).

Above specific speeds of 2000 ($N_{Sm} = 39$), the disk friction power is relatively small and can be approximated by

$$\frac{P_{DF}}{P_W} \approx 0.02$$

The impeller disk friction (Eq. 51), as deduced from the maximum efficiency and therefore indirectly from the performance of actual pumps, does include those effects of pump design and operation which are not represented in the usual rotating disk experiments: partial recovery of impeller disk friction energy in the casing, reduction of outer shroud friction due to impeller swirl, and wearing-ring flow. Provided typical pump design and manufacturing procedures are followed, Eq. 51 will probably provide a more realistic estimate of the effective disk friction loss of centrifugal pumps than will the formulas based solely on rotating disk experiments. As written, Eq. 51 is limited to water at room temperature and those fluids which have similar kinematic viscosity. For other viscosities, the rotating disk experiments suggest a correction be made by multiplying Eq. 51 by $(\nu/\nu_{water})^{0.2}$.

COMPREHENSIVE EXAMPLE

CONDITIONS OF SERVICE

H = 200 ft (60.96 m)
Q = 1000 gpm (0.063 m³/s)
$NPSH$ = 40 ft available (12.19 m)

SELECTION OF SPEED   We try the four-pole motor speed of 1780 rpm and calculate, from Eq. 30, the specific speed:

in USCS units
$$N_S = \frac{N\sqrt{Q}}{H^{0.75}} = \frac{1780\sqrt{1{,}000}}{200^{0.75}} = 1057$$

in SI units
$$N_{Sm} = \frac{1780\sqrt{0.063}}{(60.96)^{0.75}} = 20.48$$

According to the specific speed chart of Fig. 5, the pump efficiency to be expected is 0.785. The required suction specific speed (Eq. 31) is:

in USCS units
$$S = \frac{N\sqrt{Q}}{\text{NPSH}^{0.75}} = \frac{1780\sqrt{1000}}{40^{0.75}} = 3537$$

in SI units
$$S_m = \frac{1780\sqrt{0.063}}{(12.19)^{0.75}} = 68.5$$

a low value for normal inlet design. Therefore we try the two-pole motor speed of 3560 rpm and calculate:

in USCS units
$$N_S = \frac{3560\sqrt{1000}}{200^{0.75}} = 2114$$

in SI units
$$N_{Sm} = \frac{3560\sqrt{0.063}}{(60.96)^{0.75}} = 40.96$$

in USCS units
$$S = \frac{3560\sqrt{1000}}{40^{0.75}} = 7075$$

in SI units
$$S_m = \frac{3560\sqrt{0.063}}{(12.19)^{0.75}} = 137$$

The expected pump efficiency is 0.83, which would reduce the input power by 5.4%. The suction specific speed is still below 9000, the value mentioned under the heading "Impeller Inlet" on p. 2.15. Because of the appreciable efficiency gain and the reduction of pump size, we select the speed

$$N = 3560 \text{ rpm}$$

IMPELLER-DISCHARGE VELOCITY TRIANGLE   Having fixed the specific speed at 2114 ($N_{Sm} = 41$), we obtain $c_{m3}/u_2 = 0.12$ from Fig. 6, $\beta_2 = 25°$ from Fig. 7, and $z = 7$, to be evaluated by slip factor formulas. The slip factor according to Stodola's formula (Eq. 21) is

$$\mu = 1 - \frac{\pi \sin \beta_2}{z} = 1 - \frac{\pi \sin 25°}{7} = 0.81$$

Pfleiderer's formula (Eq. 24) gives

$$\mu = \frac{1}{1 + \dfrac{a}{z}\left(1 + \dfrac{\beta_2}{60}\right)\dfrac{2}{1 - (r_1^2/r_2^2)}} = \frac{1}{1 + (0.65/7) \times 1.42(2/0.75)} = 0.74$$

whereby we took the lower value for the volute $a = 0.65$. The radius ratio was tentatively estimated to be equal to or below $r_1/r_2 = 0.5$. We accept $\mu = 0.74$ since Stodola's $\mu = 0.81$ appears high from experience with similar pumps. The hydraulic efficiency is estimated from Eq. 46 or from Fig. 13:

in USCS units
$$\eta_H \approx 1 - \frac{0.8}{Q^{0.25}} = 1 - \frac{0.8}{1000^{0.25}} = 0.86$$

in SI units
$$\eta_H \approx 1 - \frac{0.071}{Q^{0.25}} = 1 - \frac{0.071}{(0.063)^{0.25}} = 0.86$$

The head coefficient can now be calculated from Eq. 28:

$$\psi = 2\mu\eta_H\left(1 - \frac{c_{m3}}{u_2}\cot \beta_2\right) = 2 \times 0.74 \times 0.86(1 - 0.12 \cot 25°) = 0.945$$

and the necessary impeller peripheral speed, also from Eq. 28, is

in USCS units $\qquad u_2 = \sqrt{\dfrac{2gH}{\psi}} = \sqrt{\dfrac{64.4 \times 200}{0.945}} = 116.8$ ft/s

in SI units $\qquad u_2 = \sqrt{\dfrac{2gH}{\psi}} = \sqrt{\dfrac{19.62 \times 60.96}{0.945}} = 35.6$ m/s

With this, we calculate the meridional velocity right after the impeller discharge:

in USCS units $\qquad c_{m3} = \dfrac{c_{m3}}{u_2} u_2 = 0.12 \times 116.8 = 14.0$ ft/s

in SI units $\qquad c_{m3} = \dfrac{c_{m3}}{u_2} u_2 = 0.12 \times 35.6 = 4.27$ m/s

The peripheral component of the absolute impeller-discharge velocity without slip is calculated with the geometric relationships of Fig. 2:

in USCS units $\qquad c_{u3} = u_2 - c_{m3} \cot \beta_2 = 116.8 - 14.0 \cot 25° = 86.8$ ft/s

in SI units $\qquad c_{u3} = u_2 - c_{m3} \cot \beta_2 = 35.6 - 4.27 \cot 25° = 26.46$ m/s

With slip, we have, according to Eq. 20,

in USCS units $\qquad c'_{u3} = \mu c_{u3} = 0.74 \times 86.8 = 64.2$ ft/s

in SI units $\qquad c'_{u3} = \mu c_{u3} = 0.74 \times 26.46 = 19.57$ m/s

Now all the ingredients for drawing the impeller-discharge velocity triangle in Fig. 17 are available. The triangle reveals that the absolute flow angle α'_3, under which the flow leaves the impeller, is 12.3°. This is also roughly the angle for which the tongue of the volute is to be designed.

VELOCITIES IN FT/S (m/s)

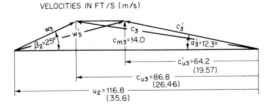

FIG. 17 Comprehensive example; impeller-discharge velocity triangles.

IMPELLER-DISCHARGE DIMENSIONS The impeller diameter is

in USCS units $\qquad D_2 = \dfrac{60 \times u_2 \times 12}{\pi N} = \dfrac{60 \times 116.8 \times 12}{\pi \times 3560} = 7.52$ in

in SI units $\qquad D_2 = \dfrac{60 \times u_2 \times 1000}{\pi N} = \dfrac{60 \times 35.6 \times 1000}{\pi \times 3560} = 191$ mm

and the impeller radius $r_2 = D_2/2 = 3.76$ in (95.5 mm). The impeller discharge width is calculated from Eq. 29:

in USCS units $\qquad b_3 = \dfrac{Q/449}{2\pi r_2 c_{m3}/144} = \dfrac{1000/449}{2\pi \times 3.76 \times 14.0/144} = 0.97$ in in SI units $\qquad b_3 = \dfrac{Q \times 10^6}{2\pi r_2 c_{m3}} = \dfrac{0.063 \times 10^6}{2\pi \times 95.5 \times 4.27} = 24.6 \text{ mm}$

IMPELLER INLET DIAMETER   Since NPSH requirements are not critical, we select a normal flow angle of

$$\beta_0 = 17°$$

An axial inlet shall be assumed with a small hub which does not protrude far enough to block the inlet flow. The hub ratio according to Eq. 33 is then

$$k = 1 - \left(\frac{D_H}{D_S}\right)^2 = 1 - \left(\frac{0}{D_S}\right)^2 = 1.0$$

Equation 32 gives the suction diameter:

in USCS units $D_S = 4.54 \left(\dfrac{Q}{kN \tan \beta_0}\right)^{1/3} = 4.54 \left(\dfrac{1000}{1.0 \times 3560 \tan 17°}\right)^{1/3} = 4.42 \text{ in}$

in SI units $\qquad D_S = 2897 \left(\dfrac{Q}{kN \tan \beta_0}\right)^{1/3} = 2897 \left(\dfrac{0.63}{1.0 \times 3560 \tan 17°}\right)^{1/3} = 112 \text{ mm}$

and the inlet radius $r_1 = r_S = D_S/2 = 2.21$ in (56 mm).

With the inlet radius and the previously calculated impeller-discharge dimensions, the impeller profile can now be drawn as in Fig. 18. The mean inlet radius, as defined by Eq. 38, is

in USCS units $\qquad r_{1m} = \left(\dfrac{r_1^2 + r_H^2}{2}\right)^{1/2} = \left(\dfrac{2.21^2 + 0^2}{2}\right)^{1/2} = 1.56 \text{ in}$

in SI units $\qquad r_{1m} = \left(\dfrac{r_1^2 + r_H^2}{2}\right)^{1/2} = \left(\dfrac{56^2 + 0^2}{2}\right)^{1/2} = 39.6 \text{ mm}$

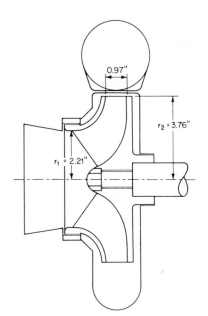

FIG. 18   Comprehensive example; impeller profile with volute.

While Pfleiderer[2] does not specifically say so, it is proper to take r_{1m} for r_1 in his equation (Eq. 24) for the slip factor if the vanes are extended into the axial inlet. The radius ratio r_1/r_2, under which no further increase of the slip factor occurs, would then become r_{1m}/r_2. In our case

in USCS units
$$\frac{r_{1m}}{r_2} = \frac{1.56}{3.76} = 0.41$$

in SI units
$$\frac{r_{1m}}{r_2} = \frac{39.6}{95.5} = 0.41$$

Therefore the previously made assumption, that it is under 0.5, is fulfilled.

INLET VANE ANGLES   The inlet vane thickness s_1 shall be $\frac{3}{16}$ = 0.1875 in (4.8 mm). The inlet vane angle at outer radius is calculated from Eq. 35:

in USCS units $$\tan \beta_1 = \frac{\tan \beta_{0(r)}}{1 - \dfrac{zs_1}{2\pi r_1 \sin \beta_{1(r)}}} = \frac{\tan 17°}{1 - \dfrac{7 \times 0.1875}{2\pi \times 2.21 \sin \beta_1}} = 0.409$$

$$\tan \beta_1 = \frac{\tan \beta_{0(r)}}{1 - \dfrac{zs_1}{2\pi r_1 \sin \beta_{1(r)}}} = \frac{\tan 17°}{1 - \dfrac{7 \times 4.8}{2\pi \times 56 \sin \beta_1}} = 0.409$$

in SI units $\beta_1 = 22.2°$

Some iteration was necessary because β_1 appears on both sides of Eq. 35.

By the same procedure, one can calculate that, at the mean radius $r_{1m} = 1.56$ in, the mean inlet vane angle is

$$\beta_{1m} = 30.5°$$

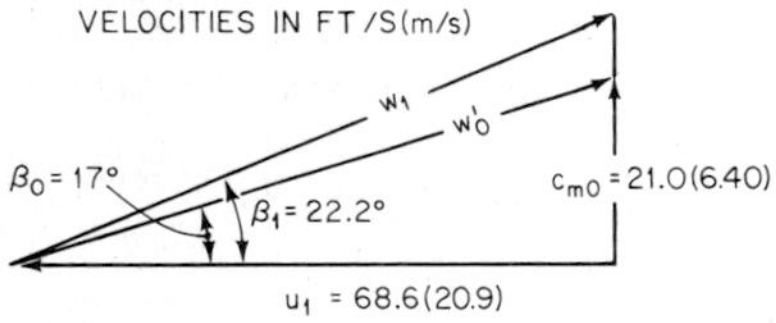

VELOCITIES IN FT /S(m/s)

$\beta_0 = 17°$ w_1 w_0' $c_{m0} = 21.0(6.40)$

$\beta_1 = 22.2°$

$u_1 = 68.6(20.9)$

FIG. 19 Comprehensive example; impeller inlet velocity triangles.

IMPELLER INLET VELOCITY TRIANGLE   In order to draw the inlet velocity triangle (Fig. 19), the following calculations are performed:

The meridional velocity ahead of the vanes:

in USCS units $$c_{m0} = \frac{Q/449}{\pi r_1^2/144} = \frac{1000/449}{\pi \times 2.21^2/144} = 21.0 \text{ ft/s}$$

in SI units $$c_{m0} = \frac{Q \times 10^6}{\pi r_1^2} = \frac{0.063 \times 10^6}{\pi \times 56^2} = 6.39 \text{ m/s}$$

The peripheral vane velocity at the eye:

in USCS units $$u_1 = \frac{\pi r_1 N}{30 \times 12} = \frac{\pi \times 2.21 \times 3560}{360} = 68.6 \text{ ft/s}$$

in SI units $$u_1 = \frac{\pi r_1 N}{30 \times 1000} = \frac{\pi \times 56 \times 3560}{30,000} = 20.88 \text{ m/s}$$

FLOW AREAS BETWEEN VANES   The inlet area between vanes is approximately, from Eq. 37a,

in USCS units $A_I = za_1 \approx \pi r_1^2 \sin \beta_{1m} = \pi \times 2.21 \sin 30.5° = 7.77 \text{ in}^2$

in SI units $A_I = za_1 \approx \pi r_1^2 \sin \beta_{1m} = \pi \times 56^2 \times \sin 30.5° = 5000 \text{ mm}^2$

The discharge area between vanes, assuming a vane thickness $s_2 = \%_6$ in $= 0.1875$ in (4.8 mm), is approximately, from Eq. 39,

in USCS units $\qquad A_{II} = z a_{II} \approx b_2 (2\pi r_2 \sin \beta_2 - z s_2)$

$$= 0.97(2\pi \times 3.76 \sin 25° - 7 \times 0.1875) = 8.41 \text{ in}^2$$

in SI units $\qquad A_{II} = z a_{II} \approx b_2 (2\pi r_2 \sin \beta_2 - z s_2)$

$$= 24.6(2\pi \times 95.5 \sin 25° - 7 \times 4.8) = 5411 \text{ mm}^2$$

With the above, the area ratio becomes

$$\frac{A_{II}}{A_I} = \frac{8.41}{7.77} = 1.08$$

which is well below 1.3 and therefore acceptable. The areas between vanes must be checked by the actual vane layout.

VOLUTE CASING   Because of its simplicity, a single volute shall be used. The throat velocity c_{thr} is tentatively selected from Fig. 10 for $N_S = 2114$:

$$\frac{c_{thr}}{u_2} \approx 0.35$$

in USCS units $\qquad c_{thr} = \dfrac{c_{thr}}{u_2} u_2 \approx 0.35 \times 116.8 = 40.9 \text{ ft/s}$

in SI units $\qquad c_{thr} = \dfrac{c_{thr}}{u_2} u_2 \approx 0.35 \times 35.6 = 12.46 \text{ m/s}$

The tentative throat area from Eq. 40 is

in USCS units $\qquad A_{thr} = \dfrac{Q/449}{c_{thr}/144} = \dfrac{1000/449}{40.9/144} = 7.84 \text{ in}^2$

in SI units $\qquad A_{thr} = \dfrac{Q \times 10^6}{c_{thr}} = \dfrac{0.063 \times 10^6}{12.46} = 5056 \text{ mm}^2$

Assuming a circular throat section, its radius is

in USCS units $\qquad\qquad\qquad\qquad r_{thr} = 1.58 \text{ in}$

in SI units $\qquad\qquad\qquad\qquad r_{thr} = 40.1 \text{ mm}$

The tongue distance is selected to be 7% of the impeller radius:

in USCS units $\qquad\qquad\qquad\qquad t = 0.07 \times 3.76 = 0.26 \text{ in}$

in SI units $\qquad\qquad\qquad\qquad t = 0.07 \times 95.5 = 6.7 \text{ mm}$

The distance of the throat center from the axis is, from Eq. 42,

in USCS units $\qquad r_4 \approx r_2 + t + r_{thr} = 3.76 + 0.26 + 1.58 = 5.60 \text{ in}$

in SI units $\qquad r_4 \approx r_2 + t + r_{thr} = 95.5 + 6.7 + 40.1 = 142.3 \text{ mm}$

Now the tentative choice of the throat velocity is checked by applying Eq. 41 in the form

in USCS units $\qquad C = \dfrac{c_{thr}}{c'_{u3}} \dfrac{r_4}{r_2} = \dfrac{40.9}{64.2} \dfrac{5.60}{3.76} = 0.95$

in SI units $\qquad C = \dfrac{c_{thr} r_4}{c'_{u3} r_2} = \dfrac{12.46}{19.57} \dfrac{142.3}{95.5} = 0.95$

According to the comments under "Volute Casing," this appears to be a very reasonable factor C.

A further verification is obtained by calculating

in USCS units
$$\frac{A_{thr}}{A_{II}} = \frac{7.84}{8.41} = 0.93$$

in SI units
$$\frac{A_{thr}}{A_{II}} = \frac{5056}{5411} = 0.93$$

and entering this value into Fig. 11 at $N_S = 2114$ ($N_{Sm} = 41$). This point falls into the region marked High Efficiency.

There is now enough assurance to make the tentative choice of the throat velocity permanent and to proceed with the volute design. If the flow factor C had been larger than 1.0 or smaller than 0.9, then the calculation would have to be repeated with an improved choice for c_{thr}. The intermediate volute areas are first approximated by Stepanoff's equation (Eq. 45):

$$A_V = A_{thr} \frac{\varphi_V}{360}$$

The intermediate volute areas are assumed to be circular, and their radii and distances from the axis are calculated. The results are listed in Table 1. Some designers would stop here and dimension the volute according to this table. We will go further and apply the law of constant angular

TABLE 1   Volute Dimensions

φ_V	A_V, in^2 (mm^2)	r, in (mm)	r_V, in (mm)
0	0 (0)	0 (0)	4.02 (102.2)
45	0.98 (632)	0.56 (14.2)	4.58 (116.4)
90	1.96 (1264)	0.79 (20.1)	4.81 (122.3)
135	2.94 (1896)	0.97 (24.6)	4.99 (126.8)
180	3.92 (2528)	1.11 (28.2)	5.13 (130.4)
225	4.90 (3160)	1.25 (31.8)	5.27 (134.0)
270	5.88 (3792)	1.36 (34.5)	5.38 (136.7)
315	6.86 (4424)	1.48 (37.6)	5.50 (139.8)
360	7.84 (5056)	1.58 (40.1)	5.60 (142.3)

momentum to the intermediate volute areas also (Eq. 44). To get the calculation started, we will first use as an approximation the distances r_V from Table 1. The flow factor shall remain constant at $C = 0.95$. Equation 44 states

in USCS units
$$c_V = \frac{r_2}{r_V} Cc'_{u3} = \frac{3.76 \times 0.95 \times 64.2}{r_V} = \frac{229.3}{r_V}$$

in SI units
$$c_V = \frac{r_2}{r_V} Cc'_{u3} = \frac{95.5 \times 0.95 \times 19.57}{r_V} = \frac{1775}{r_V}$$

and from 43

in USCS units
$$A_V = \frac{Q/449}{c_V/144} \frac{\varphi_V}{360} = \frac{1000/449}{c_V/144} \frac{\varphi_V}{360} = 0.891 \frac{\varphi_V}{c_V}$$

in SI units
$$A_V = \frac{Q \times 10^6}{c_V} \frac{\varphi_V}{360} = \frac{0.063 \times 10^6}{c_V} \frac{\varphi_V}{360} = 175 \frac{\varphi_V}{c_V}$$

The newly calculated intermediate areas result in new center distances r_V, which are then compared with the assumed values. The results are listed in Table 2. The calculated center distances

TABLE 2 Volute Dimensions

φ_V	Assumed r_V, in (mm)	c_V, ft/s (m/s)	A_V, in² (mm²)	r, in (mm)	Calculated r_V, in (mm)
0	4.02 (102.2)	57.04 (17.39)	0 (0)	0 (0)	4.02 (102.2)
45	4.58 (116.4)	50.06 (15.26)	0.80 (516)	0.5 (12.7)	4.52 (114.8)
90	4.81 (122.3)	47.67 (14.53)	1.68 (1084)	0.73 (18.5)	4.75 (120.7)
135	4.99 (126.8)	45.95 (14.00)	2.62 (1690)	0.91 (23.1)	4.93 (125.2)
180	5.13 (130.4)	44.70 (13.62)	3.59 (2316)	1.07 (27.2)	5.09 (129.3)
225	5.27 (134.0)	43.51 (13.26)	4.61 (2963)	1.21 (30.7)	5.23 (132.8)
270	5.38 (136.7)	42.62 (12.99)	5.64 (3638)	1.34 (34.0)	5.36 (136.1)
315	5.50 (139.8)	41.69 (12.71)	6.73 (4341)	1.46 (37.1)	5.49 (139.4)
360	5.60 (142.3)	40.90 (12.47)	7.84 (5056)	1.58 (40.1)	5.60 (142.3)

r_V are close to the assumed values. A further iteration step could be performed by entering the newly calculated center distances as assumed values into the above procedure. For the present example, the resultant changes of the volute dimension would be negligible.

The volute can now be drawn using the above listed section radii r and distances r_V of the volute centers from the axis (Figs. 18 and 20).

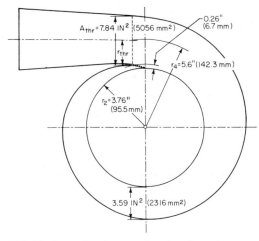

FIG. 20 Comprehensive example; volute dimensions.

LOSS ESTIMATES, RESULTANT EFFICIENCY, AND SHAFT POWER   The following estimates can be made for the losses:

Hydraulic efficiency from Eq. 46 (or from Fig. 13):

in USCS units
$$\eta_H \approx 1 - \frac{0.8}{Q^{0.25}} = 1 - \frac{0.8}{1000^{0.25}} = 0.86$$

in SI units
$$\eta_H \approx 1 - \frac{0.071}{Q^{0.25}} = 1 - \frac{0.071}{(0.063)^{0.25}} = 0.86$$

Volumetric efficiency from Fig. 14 for $N_S = 2114$ ($N_{Sm} = 41$) and $Q = 1000$ gpm (227 m³/h):

$$\eta_V \approx 0.98$$

Ratio of mechanical losses to water power from Fig. 15 for $N_S = 2114$ ($N_{Sm} = 41$) and $Q = 1000$ gpm (227 m^3/h):

$$\frac{P_M}{P_W} \approx 0.012$$

Ratio of impeller disk friction to water power from Eq. 51:

in USCS units $\qquad \dfrac{P_{DF}}{P_W} = \dfrac{7800}{N_S^{5/3}} = \dfrac{7800}{2114^{5/3}} = 0.023$

in SI units $\qquad \dfrac{P_{DF}}{P_W} = \dfrac{10.89}{N_{Sm}^{5/3}} = \dfrac{10.89}{41^{5/3}} = 0.023$

Entering the above estimates into Eq. 19 gives the pump efficiency:

$$\eta = \frac{1}{\dfrac{1}{\eta_H \eta_V} + \dfrac{P_{DF}}{P_W} + \dfrac{P_M}{P_W}} \approx \frac{1}{\dfrac{1}{0.86 \times 0.98} + 0.023 + 0.012} = 0.82$$

The pump efficiency calculated from estimated losses is close to the value of 0.83 read from Fig. 5.

With $\eta = 0.82$, the power necessary to operate the pump is (Eqs. 14 and 15):

in USCS units $P_S = \dfrac{1}{\eta} QH \dfrac{\text{sp. gr.}}{3960} = \dfrac{1000 \times 200 \times 1}{0.82 \times 3960} = 61.6$ hp

in SI units $\qquad P_S = \dfrac{1}{\eta} 9.80 QH \text{ sp. gr.} = \dfrac{9.80 \times 0.063 \times 60.96 \times 1}{0.82} = 45.9$ kW

REFERENCES

1. Stodola, A.: *Steam and Gas Turbines*, Peter Smith, Gloucester, Mass., 1945.

2. Pfleiderer, C.: *Die Kreiselpumpen*, Springer-Verlag, Berlin, 1961.

3. Wislicenus, G. F.: *Fluid Mechanics of Turbomachinery*, Dover, New York, 1965.

4. Wislicenus, G. F., R. M. Watson, and I. J. Karassik: "Cavitation Characteristics of Centrifugal Pumps Described by Similarity Considerations," *Trans. ASME* **61**:17 (1939).

5. Ross, C. C., and G. Banerian: "Some Aspects of High Suction Specific Speed Inducers," *Trans. ASME* **78**:1715 (1956).

6. Jekat, W. K.: "A New Approach to the Reduction of Pump Cavitation: The Hubless Inducer," *J. Basic Eng., Trans ASME*, ser. D **89**:125 (1967).

7. Jekat, W. K.: "Reynolds Number and Incidence-Angle Effects on Inducer Cavitation," *ASME Paper* 66-WA/FE-31, 1966.

8. Stepanoff, A. J.: *Centrifugal and Axial Flow Pumps*, Wiley, New York, 1957.

9. Anderson, H. H.: "Modern Developments in the Use of Large Single-Entry Centrifugal Pumps," *Proc. Inst. Mech. Eng. (London)* **169**(6) (1955).

10. Worster, R. C.: "The Interaction of Impeller and Volute in Determining the Performance of a Centrifugal Pump," British Hydromechanics Research Association, RR 679, 1960.

11. Bergen, J.: "Untersuchungen an einer Kreiselpumpe mit verstellbarem Spiralgehäuse," *Konstrukt.* **22**(12) (1970).

12. Agostinelli, A., D. Nobles, and C. R. Mockridge: "An Experimental Investigation of Radial Thrust in Centrifugal Pumps," *J. Eng. Power, Trans. ASME*, ser. A **82**:120 (1960).

13. Schultz-Grunow, F.: "Der Reibungswiderstand rotierender Scheiben," *Z. Angew. Math. Mech.* **15**(4) (1935).

14. Pantell, K.: "Versuche über Scheibenreibung," *Forsch. Gebiet Ingenieurw.* **16**(4) (1949/50).

15. Daily, J. W., and R. E. Nece: "Chamber-Dimension Effects on Induced Flow and Frictional Resistance of Enclosed Rotating Disks," *J. Basic Eng., Trans. ASME*, ser. D **82**:217 (1960).

16. Bennett, T. P., and R. C. Worster: "The Friction on Rotating Disks and the Effect on Net Radial Flow and Externally Applied Whirl," British Hydromechanics Research Association, RR 691, 1961.

17. Flügel, G.: "Ein neues Verfahren der graphischen Integration, angewandt auf Strömungen," doctoral dissertation, Oldenbourg, Munich, 1914.

18. Stanitz, J. D., and V. D. Prian: "A Rapid Approximate Method for Determining Velocity Distribution on Impeller Blades of Centrifugal Compressors," *NACA TN* 2421, 1951.

19. Wu, C.-H.: "A General Theory of Three-Dimensional Flow in Subsonic and Supersonic Turbomachines of Axial-, Radial-, and Mixed-Flow Types," *NACA TN* 2604, 1952.

20. Smith, K. J., and J. T. Hamrick: "A Rapid Approximate Method for the Design of Hub Shroud Profiles of Centrifugal Impellers of Given Blade Shape," *NACA TN* 3399, 1955.

21. Stockman, N. O., and J. L. Kramer: "Method for Design of Pump Impellers Using a High-Speed Computer," *NASA TN* D-1562, 1963.

22. Katsanis, T.: "Use of Arbitrary Quasi-Orthogonals for Calculating Flow Distribution in a Turbomachine," *J. Eng. Power, Trans. ASME*, ser. A (1966).

FURTHER READING

Horlock, J. H., and H. Marsh, Fluid Mechanics of Turbomachines: A Review, *Int. J. Heat Fluid Flow (1982)*.

Neal, A. N., "Through-Flow Analysis of Pumps and Fans," NEL report no. 669, National Engineering Laboratory, Glasgow, 1980.

SECTION 2.2
CENTRIFUGAL PUMP CONSTRUCTION

2.2.1
CENTRIFUGAL PUMPS: MAJOR COMPONENTS

IGOR J. KARASSIK

CLASSIFICATION AND NOMENCLATURE _____

A centrifugal pump consists of a set of rotating vanes, enclosed within a housing or casing and used to impart energy to a fluid through centrifugal force. Thus, stripped of all refinements, a centrifugal pump has two main parts: (1) a rotating element, including an impeller and a shaft, and (2) a stationary element made up of a casing, stuffing box, and bearings.

In a centrifugal pump the liquid is forced, by atmospheric or other pressure, into a set of rotating vanes. These vanes constitute an impeller which discharges the liquid at its periphery at a higher velocity. This velocity is converted to pressure energy by means of a volute (Fig. 1) or by a set of stationary diffusion vanes (Fig. 2) surrounding the impeller periphery. Pumps with volute casings are generally called *volute pumps,* while those with diffusion vanes are called *diffuser pumps.* Diffuser pumps were once quite commonly called *turbine pumps,* but this term has recently been more selectively applied to the vertical deep-well centrifugal diffuser pumps usually referred to as *vertical turbine pumps.* Figure 1 shows the path of the liquid passing through an end-suction volute pump operating at rated capacity (capacity at which best efficiency is obtained).

Impellers are classified according to the major direction of flow in reference to axis of rotation. Thus, centrifugal pumps may have

1. Radial-flow impellers (Figs. 25, 34, 35, 36, and 37)

2. Axial-flow impellers (Fig. 29)

3. Mixed-flow impellers, which combine radial- and axial-flow principles (Figs. 27 and 28)

Impellers are further classified as

1. Single-suction, with a single inlet on one side (Figs. 25, 33, and 37)

2. Double-suction, with water flowing to the impeller symmetrically from both sides (Figs. 26, 27, and 38)

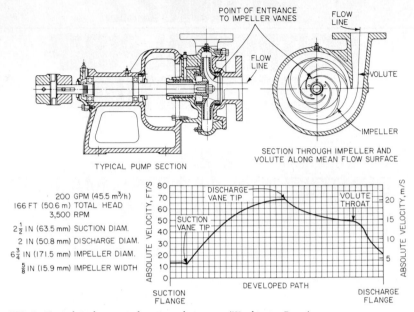

FIG. 1   Typical single-stage end-suction volute pump. (Worthington Pump)

The mechanical construction of the impellers gives a still further subdivision into

1. Enclosed, with shrouds or side walls enclosing the waterways (Figs. 25, 26, 27, etc.)

2. Open, with no shrouds (Figs. 29, 33, 34, etc.)

3. Semiopen or semienclosed (Fig. 36)

A pump in which the head is developed by a single impeller is called a *single-stage pump.* Often the total head to be developed requires the use of two or more impellers operating in series, each taking its suction from the discharge of the preceding impeller. For this purpose, two or more single-stage pumps can be connected in series or all the impellers can be incorporated in a single casing. The unit is then called a *multistage pump.*

The mechanical design of the casing provides the added pump classification of *axially split* or *radially split,* and the axis of rotation determines whether the pump is a horizontal or vertical unit. Horizontal-shaft centrifugal pumps are classified still further according to the location of the suction nozzle:

1. End-suction (Figs. 1, 8, 11, 13, etc.)

2. Side-suction (Figs. 7, 10, etc.)

3. Bottom-suction (Fig. 15)

4. Top-suction (Fig. 22)

Some pumps operate in air, with the liquid coming to and being conducted away from the pumps by piping. Other pumps, most often ver-

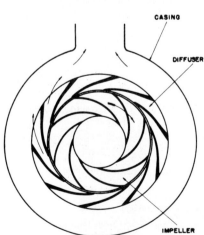

FIG. 2   Typical diffuser-type pump.

tical types, are submerged in their suction supply. Vertical-shaft pumps are therefore called either *dry-pit* or *wet-pit* types. If the wet-pit pumps are axial-flow, mixed-flow, or vertical-turbine types, the liquid is discharged up through the supporting drop or column pipe to a discharge point above or below the supporting floor. These pumps are consequently designated as *aboveground discharge* or *belowground discharge units*.

Figures 3, 4, and 8 show typical constructions of a horizontal double-suction volute pump, the bowl section of a single-stage axial-flow propeller pump, and a vertical dry-pit single-suction volute pump, respectively. Names recommended by the Hydraulic Institute for various parts are given in Table 1.

CASINGS AND DIFFUSERS

The Volute Casing Pump This pump (Fig. 1) derives its name from the spiral-shaped casing surrounding the impeller. This casing section collects the liquid discharged by the impeller and converts velocity energy to pressure energy.

A centrifugal pump volute increases in area from its initial point until it encompasses the full 360° around the impeller and then flares out to the final discharge opening. The wall dividing the initial section and the discharge nozzle portion of the casing is called the tongue of the volute or the "cutwater." The diffusion vanes and concentric casing of a diffuser pump fulfill the same function as the volute casing in energy conversion.

In propeller and other pumps in which axial-flow impellers are used, it is not practical to use a volute casing; instead, the impeller is enclosed in a pipelike casing. Generally diffusion vanes are used following the impeller proper, but in certain extremely low-head units these vanes may be omitted.

TABLE 1 Recommended Names of Centrifugal Pump Parts

Item no.	Name of part[a]	Item no.	Name of part[a]
1	Casing	33	Bearing housing (outboard)
1A	Casing (lower half)	35	Bearing cover (inboard)
1B	Casing (upper half)	36	Propeller key
2	Impeller	37	Bearing cover (outboard)
4	Propeller	39	Bearing bushing
6	Pump shaft	40	Deflector
7	Casing ring	42	Coupling (driver half)
8	Impeller ring	44	Coupling (pump half)
9	Suction cover	46	Coupling key
11	Stuffing box cover	48	Coupling bushing
13	Packing	50	Coupling lock nut
14	Shaft sleeve	52	Coupling pin
15	Discharge bowl	59	Handhole cover
16	Bearing (inboard)	68	Shaft collar
17	Gland	72	Thrust collar
18	Bearing (outboard)	78	Bearing spacer
19	Frame	85	Shaft enclosing tube
20	Shaft sleeve nut	89	Seal
22	Bearing lock nut	91	Suction bowl
24	Impeller nut	101	Column pipe
25	Suction head ring	103	Connector bearing
27	Stuffing box cover ring	123	Bearing end cover
29	Lantern ring (seal cage)	125	Grease (oil) cup
31	Bearing housing (inboard)	127	Seal pipe (tubing)
32	Impeller key		

[a]These parts are called out in Figs. 3, 4, and 8.

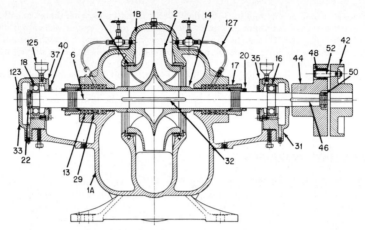

FIG. 3 Horizontal single-stage double-suction volute pump. (Numbers refer to parts listed in Table 1.) (Worthington Pump)

The diffuser is seldom applied to a single-stage radial-flow pump. Except for certain high-pressure multistage pump designs, the major application of diffusion vane pumps is in vertical turbine pumps and in single-stage low-head propeller pumps (Fig. 4).

Radial Thrust   In a *single-volute pump casing* design (Fig. 5), uniform or near uniform pressures act on the impeller when the pump is operated at design capacity (which coincides with the best efficiency). At other capacities, the pressures around the impeller are not uniform (Fig. 6) and there is a resultant radial reaction (F). A detailed discussion of the radial thrust and of its magnitude is presented in Subsec. 2.3.1. Note that the unbalanced radial thrust increases as capacity decreases from that at the design flow.

For any percentage of capacity, this radial reaction is a function of total head and of the width and diameter of the impeller. Thus a high-head pump with a large impeller diameter will have a much greater radial reaction force at partial capacities than a low-head pump with a small impeller diameter. A zero radial reaction is not often realized; the minimum reaction occurs at design capacity.

Although the same tendency for unbalance exists in the diffuser-type pump, the reaction is limited to a small arc repeated all around the impeller. As a result, the individual reactions cancel each other.

In a centrifugal pump design, shaft diameter as well as bearing size can be affected by the allowable deflection as determined by the shaft span, impeller weight, radial reaction forces, and torque to be transmitted.

Formerly, standard designs compensated for radial reaction forces of a magnitude encountered at capacities in excess of 50% of design capacity for maximum-diameter impeller for the pump. For sustained operation

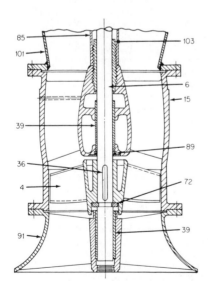

FIG. 4   Vertical wet-pit diffuser pump bowl. (Numbers refer to parts listed in Table 1.) (Worthington Pump)

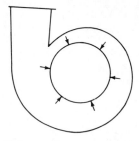

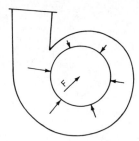

FIG. 5 Uniform casing pressures exist at design capacity, resulting in zero radial reaction.

FIG. 6 At reduced capacities uniform pressures do not exist in a single-volute casing, resulting in a radial reaction F.

at lower capacities, the pump manufacturer, if properly advised, would supply a heavier shaft, usually at a much higher cost. Sustained operation at extremely low flows without the manufacturer being informed at the time of purchase is a much more common practice today. The result: broken shafts, especially on high-head units.

Because of the increasing application of pumps which must operate at reduced capacities, it has become desirable to design standard units to accommodate such conditions. One solution is to use heavier shafts and bearings. Except for low-head pumps in which only a small additional load is involved, this solution is not economical. The only practical answer is a casing design that develops a much smaller radial reaction force at partial capacities. One of these is the *double-volute* casing design, also called *twin-volute* or *dual-volute* design.

The application of the double-volute design principle to neutralize radial reaction forces at reduced capacity is illustrated in Fig. 7. Basically, this design consists of two 180° volutes; a passage external to the second joins the two into a common discharge. Although a pressure unbalance exists at partial capacity through each 180° arc, the two forces are approximately equal and opposite. Thus little if any radial force acts on the shaft and bearings. See also Subsec. 2.3.1.

The double-volute design has many hidden advantages. For example, in large-capacity medium- and high-head single-stage vertical pumps, the rib forming the second volute and separating it from the discharge waterway of the first volute strengthens the casing (Fig. 8).

Individual stages of a multistage pump can be made double volute as illustrated in Fig. 9. The kinetic energy of the water discharged from the impeller must be transformed into pressure energy and must then be turned back 180° to enter the impeller of the next stage. The double volute therefore also acts as a return channel. The back view in Fig. 9 shows this, as well as the guide vanes used to straighten the flow into the next stage.

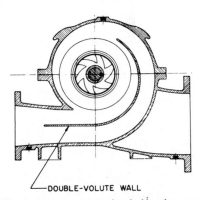

DOUBLE-VOLUTE WALL

FIG. 7 Transverse view of a double-volute casing pump.

Solid and Split Casings   *Solid casing* implies a design in which the discharge waterways leading to the discharge nozzle are all contained in one casting, or fabricated piece. The casing must have one side open so that the impeller may be introduced into it. As the sidewalls surrounding the impeller are in reality part of the casing, a solid casing, strictly speaking, cannot be used, and designs normally called solid casing are really radially split (Figs. 1, 11, 12, and 13).

A *split casing* is made of two or more parts fastened together. The term *horizontally split* had regularly been used to describe pumps with a casing divided by a horizontal plane through the shaft centerline, or axis (Fig. 10). The term *axially split* is now preferred. Since both the suction

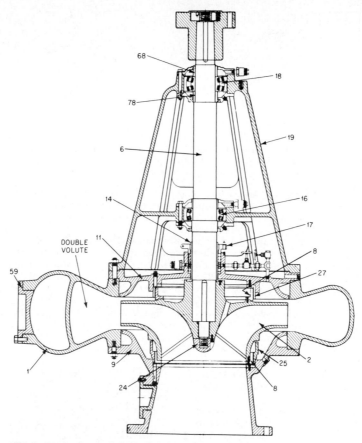

FIG. 8 Sectional view of a vertical-shaft end-suction pump with a double-volute casing. (Numbers refer to parts listed in Table 1.) (Worthington Pump)

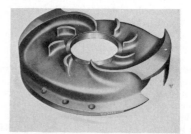

FIG. 9 Double volute of a multistage pump: front view (left) and back view (right). (Worthington Pump)

FIG. 10 Axially split casing horizontal-shaft double-suction volute pump. (Worthington Pump)

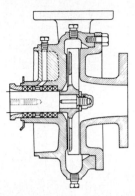

FIG. 11 End-suction pump with open impeller and removable suction cover. (Worthington Pump)

and discharge nozzles are usually in the same half of the casing, the other half may be removed for inspection of the interior without disturbing the bearings or the piping. Like its counterpart *horizontally split*, the term *vertically split* is poor terminology. It refers to a casing split in a plane perpendicular to the axis of rotation. The term *radially split* is now preferred.

End-Suction Pumps Most end-suction single-stage pumps are made of one-piece solid casings. At least one side of the casing must be open so the impeller can be assembled in the pump; thus a cover is required for that side. If the cover is on the suction side, it becomes the casing sidewall and contains the suction opening (Fig. 1). This is called the *suction cover* or *casing suction head*. Other designs are made with stuffing box covers (Fig. 12), and still others have both casing suction covers and stuffing box covers (Figs. 8 and 13).

For general service, the end-suction single-stage pump design is extensively used for small pumps up to 4- and 6-in (102- and 152-mm) discharge size for both motor-mounted and coupled types. In all these the small size makes it feasible to cast the volute and one side integrally. Whether the stuffing box side or the suction side is made integrally with the casing is usually determined by the most economical pump design. For larger pumps, especially those for special service, such as sewage handling, there is a demand for pumps of both rotations. A design with separate suction and stuffing box heads permits use of the same casing for either rotation if the flanges on the two sides are made identical. There is also a demand for vertical pumps that can be disassembled by removal of the rotor and bearing assembly from the top of the casing. Many horizontal applications of the pumps of the same line, however, require partial dismantling from the suction side. Such lines are most adaptable when they have separate suction and stuffing box covers.

Casing Construction for Open- and Semiopen-Impeller Pumps In the inexpensive or open- or semiopen-impeller pump, the impeller rotates within close clearance of the pump casing or suction cover (Fig. 11). If the intended service is severe, a side plate is mounted within the casing to provide a renewable close-clearance guide to the liquid flowing through the impeller (Fig. 12). One of the advantages of using side plates is that abra-

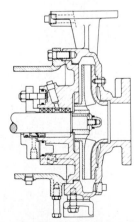

FIG. 12 End-suction pump with semiopen impeller and renewable side plate. (Worthington Pump)

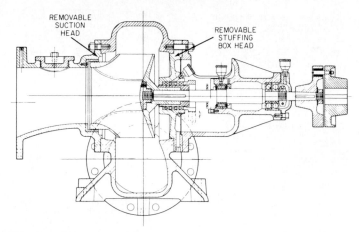

FIG. 13   End-suction pump with removable suction and stuffing box heads. (Worthington Pump)

sion-resistant material, such as stainless steel, may be used for the impeller and side plate while the casing itself may be of a less costly material. Although double-suction semiopen-impeller pumps are seldom used today, they were common in the past and were generally made with side plates.

In order to maintain pump efficiency, a close running clearance is required between the front unshrouded face of the open or semiopen impeller and the casing, suction cover, or side plate. Pump designs provide either jackscrews or shims to adjust the position of the thrust bearing housing (and, as a result, the axial position of the shaft and impeller) relative to the bearing frame.

Prerotation and Stop Pieces
Improper entrance conditions and inadequate suction approach shapes may cause the liquid column in the suction pipe to spiral for some distance ahead of the impeller entrance. This phenomenon is called *prerotation*, and it is attributed to various operational and design factors in both vertical and horizontal pumps.

Prerotation is usually harmful to pump operation because the liquid enters between the impeller vanes at an angle other than that allowed for in the design. This frequently lowers the net effective suction head and the pump efficiency. Various means are used to avoid prerotation both in the construction of the pump and in the design of the suction approaches.

Practically all horizontal single-stage double-suction pumps and most multistage pumps have a suction volute that guides the liquid in a streamline flow to the impeller eye. The flow comes to the eye at right angles to the shaft and separates unequally on the two sides of the shaft. Moving from the suction nozzle to the impeller eye, the suction waterways reduce in area, meeting in a projecting section of the sidewall dividing the two sections. This dividing projection is called a *stop piece*. To prevent prerotation in end-suction pumps, a radial-fin stop piece projecting toward the center may be cast into the suction nozzle wall.

Nozzle Locations
The discharge nozzle of end-suction single-stage horizontal pumps is usually in a top-vertical position (Figs. 1, 11, and 12). However, other nozzle positions may be obtained, such as top-horizontal, bottom-horizontal, or bottom-vertical. Figure 14 illustrates the

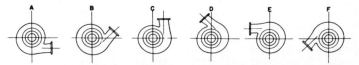

FIG. 14   Possible positions of discharge nozzles for a specific design of end-suction solid-casing horizontal-shaft pump. Rotation illustrated is counterclockwise from suction end.

flexibility available in discharge nozzle location. Sometimes the pump frame, bearing bracket, or baseplate may interfere with the discharge flange, prohibiting a bottom-horizontal or bottom-vertical discharge nozzle position. In other instances, solid casings cannot be rotated for various nozzle positions because the stuffing box seal connection would become inaccessible.

Practically all double-suction axially split casing pumps have a side discharge nozzle and either a side- or a bottom-suction nozzle. If the suction nozzle is placed on the side of the pump casing with its axial centerline (Fig. 10), the pump is classified as a *side-suction pump*. If its suction nozzle points vertically downward (Fig. 15), the pump is called a *bottom-suction pump*. Single-stage bottom-suction pumps are rarely made in sizes below 10-in (254-mm) discharge nozzle diameter.

Special nozzle positions can sometimes be provided for double-suction axially split casing pumps to meet special piping arrangements, for example, a vertically split casing with bottom suction and top discharge in the same half of the casing. Such special designs are generally costly and should be avoided.

FIG. 15   Bottom-suction axially split casing single-stage pump. (Worthington Pump)

Centrifugal Pump Rotation

Because suction and discharge nozzle locations are affected by pump rotation, it is very important to understand how the direction of rotation is defined. According to Hydraulic Institute standards, rotation is defined as clockwise or counterclockwise by looking at the driven end of a horizontal pump or looking down on a vertical pump, although some manufacturers still designate rotation of a horizontal pump from its outboard end. To avoid misunderstanding, clockwise or counterclockwise rotation should always be qualified by including the direction from which one looks at the pump.

The terms *inboard end*—the end closest to the driver—and *outboard end*—the end farthest away—are used only with horizontal pumps. The terms lose their significance with dual-driven pumps. Any centrifugal pump casing pattern may be arranged for either clockwise or counterclockwise rotation, except for end-suction pumps, which have integral heads on one side. These require separate directional patterns.

Casing Handholes

Casing handholes are furnished primarily on pumps handling sewage and stringy materials that may become lodged on the impeller suction vane edges or on the tongue of the volute. The holes permit removal of this material without completely dismantling the pump. End-suction pumps used primarily for liquids of this type are provided with handholes or access to the suction side of the impellers. These access points are located on the suction head or in the suction elbow. Handholes are also provided in drainage, irrigation, circulating, and supply pumps if foreign matter may become lodged in the waterways. On very large pumps, manholes provide access to the interior for both cleaning and inspection.

Mechanical Features of Casings

Most single-stage centrifugal pumps are intended for service at moderate pressures and temperatures. As a result, pump manufacturers usually design a special line of pumps for high operating pressures and temperatures rather than make their standard line unduly expensive by making it suitable for too wide a range of operating conditions.

If axially split casings are subject to high pressure, they tend to "breathe" at the split joint, leading to misalignment of the rotor and, even worse, leakage. For such conditions, internal and external ribbing is applied to casings at the points subject to greatest stress. In addition, whereas most pumps are supported by feet at the bottom of the casing, high temperatures require centerline support so that, as the pump becomes heated, expansion will not cause misalignment.

Series Units

For large-capacity medium-high-head conditions that require such an arrangement, two single-stage double-suction pumps may be connected in series on one baseplate with a single driver. Such an arrangement was very common in waterworks applications for heads of 250 to 400 ft (76 to 122m). One series arrangement uses a double-extended shaft motor in the middle, driving two pumps connected in series by external piping. In a second type, a standard motor is used with one pump having a double-extended shaft. This latter arrangement may have limited applications because the shaft of the pump next to the motor must be strong enough to transmit the total pumping horsepower. If the total pressure generated by such a series unit is relatively high, the casing of the second pump could require ribbing.

   Higher heads per stage are more common today, and series units are generally used only for much higher ranges of total head.

MULTISTAGE PUMP CASINGS

While the majority of single-stage pumps are of the volute-casing type, both volute and diffuser casings are used in multistage pump construction. Because a volute casing gives rise to radial thrust, axially split multistage casings generally have staggered volutes so that the resultant of the individual radial thrusts is balanced out (Figs. 16 and 17). Both axially and radially split casings are used for multistage pumps. The choice between the two designs is dictated by the design pressure, with 2000 lb/in^2 (138 bar)° being the average upper limit for 3600-rpm axially split casing pumps.

Axially Split Casings   Regardless of the arrangement of the stages in the casing, it is necessary to connect the successive stages of a multistage pump. In the low and medium pressure and capacity range, these interstage passages are cast integrally with the casing proper (Figs. 18 and 19). As the pressures and capacities increase, the desire to maintain as small a casing diameter as possible, coupled to the necessity of avoiding sudden changes in the velocity or the direction of flow, leads to the use of external interstage passages cast separately from the pump casing. They are formed in the shape of loops, bolted or welded to the casing proper (Fig. 20).

Interstage Construction for Axially Split Casing Pumps   A multistage pump inherently has adjoining chambers subjected to different pressures, and means must be made available to isolate these chambers from one another so that the leakage from high to low pressure will take place only at the clearance joints formed between the stationary and rotating elements of the pump and so that this leakage will be kept to a minimum. The isolating wall used to sep-

°1 bar = 10^5 Pa. For a discussion of bar, see *SI Units—A Commentary* in the front matter.

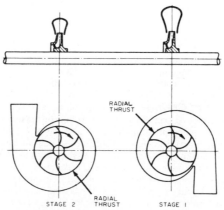

FIG. 16   Arrangement of multistage volute pump for radial-thrust balance.

FIG. 17   Horizontal flange of an axially split casing six-stage pump. (Worthington Pump)

FIG. 18   Two-stage axially split casing volute pump for small capacities and pressures up to 200 to 250 lb/in² (14 to 17.5 bar). (Worthington Pump)

FIG. 19   Two-stage axially split casing volute pump for pressures up to 400 lb/in² (28 bar). (Worthington Pump)

FIG. 20 Six-stage axially split casing volute pump for pressures up to 1300 lb/in² (90 bar). (Worthington Pump)

arate two adjacent chambers of a multistage pump is called a *stage piece, diaphragm,* or *interstage diaphragm.* The stage piece may be formed of a single piece, or it may be fitted with a renewable stage piece bushing at the clearance joint between the stationary stage piece and the part of the rotor immediately inside the former. The stage pieces, which are usually solid, are assembled on the rotor along with impellers, sleeves, bearings, and similar components. To prevent the stage pieces from rotating, a locked tongue-and-groove joint is provided in the lower half of the casing. Clamping the upper casing half to the lower half securely holds the stage piece and prevents rotation.

The problem of seating a solid stage piece against an axially split casing is one which has given designers much trouble. First, there is a three-way joint; second, this seating must make the joint tight and leakproof under a pressure differential without resorting to bolting the stage piece directly to the casing.

To overcome this problem, it is wise to make a pump which has a small-diameter casing so that when the casing bolting is pulled tight, there is a seal fitting of the two casing halves adjacent to the stage piece. The small diameter likewise helps to eliminate the possibility of a stage piece cocking and thereby leaving a clearance on the upper half casing when it is pulled down. No matter how rigidly the stage piece may be located in the lower half casing, there must be a sliding fit between the seat face of the stage piece and that in the upper half casing in order that the upper half casing may be pulled down. Each stage piece, furthermore, must be arranged so that the pressure differential developed by the pump will tend to seat the piece tightly against the casing rather than open up the joint.

We have said that axially split casing pumps are used for working pressure up to 2000 lb/in² (138 bar). High-pressure piping systems, of which these pumps form a part, are inevitably made of steel because this material has the valuable property of yielding without breaking. Considerable piping strains are unavoidable, and these strains, or at least a part thereof, are transmitted to the pump casing. The latter consists essentially of a barrel that is split axially, flanged at the split, and fitted with two necks which serve as inlet and discharge openings. When piping stresses exist, these necks, being the weakest part of the casing, are in danger of breaking off if they cannot yield. Steel is therefore the safest material for pump casings whenver the working pressures in the pump are in excess of 1000 lb/in² (70 bar).

This brings up a very important feature in the design of the suction and discharge flanges. While raised-face flanges are perfectly satisfactory for steel-casing pumps, their use is extremely dangerous with cast iron pumps. This danger arises from the lack of elasticity in cast iron, which leads to flange breakage when the bolts are being tightened up, the fulcrum of the bending moment being located inward of the bolt circle. As a result, it is essential to avoid raised-face flanges with cast iron casings as well as the use of a raised-face flange pipe directly against a flat-face cast iron pump flange. Suction flanges should obviously be suitable for whatever hydrostatic test pressure is applied to the pump casing. Therefore, if this pressure is over 500 lb/in² (35 bar), a 300-lb/in² (21-bar) series suction flange would be used.

The location of the pump casing foot support is not critical in smaller pumps operating at pressures under 250 lb/in² (17 bar) and at moderate temperatures (Fig. 18). The unit is extremely small, and very little distortion is bound to occur. However, for larger units operating at higher pressures and perhaps at higher temperatures, it is important that the casing be supported as close as possible to the horizontal centerline and immediately below the bearings (Figs. 19 and 20).

Radially Split Double-Casing Pumps   The oldest form of radially split casing multistage pump is that commonly called the *ring-casing* or the "doughnut" type. When it was found necessary to use more than one stage for the generation of higher pressures, two or more single-stage units of the prevalent radially split casing type were assembled and bolted together.

In later designs, the individual stage sections and separate suction and discharge heads were held together with large throughbolts. These pumps, still an assembly of bolted-up sections, presented serious dismantling and reassembly problems because suction and discharge connections had to be broken each time the pump was opened. The double-casing pump retained the advantages of the radially split casing design and solved the dismantling problem.

The basic principle consists in enclosing the working parts of a multistage centrifugal pump in an inner casing and building a second casing around this inner casing. The space between the two casings is maintained at the discharge pressure of the last pump stage. The construction of the inner casing follows one of two basic principles: (1) axial splitting (Fig. 21) or (2) radial splitting (Fig. 22).

The double-casing pump with radially split inner casing is an evolution of the ring-casing pump, with added provisions for ease of dismantling. The inner unit is generally constructed exactly as a ring-casing pump. After assembly, it is inserted and bolted inside a cylindrical casing which supports it and leaves it free to expand under temperature changes. In Fig. 23 the inner assembly of such a pump is being inserted into the outer casing. Figure 24 shows the external appearance of this type of pump. The suction and discharge nozzles form an integral part of the outer casing, and the internal assembly of the pump can be withdrawn without disturbing the piping connections.

Hydrostatic Pressure Tests   It is standard practice for the manufacturer to conduct hydrostatic tests for the parts of a pump which contain fluid under pressure. This means that the pump casing and, where applicable, such other parts as suction or stuffing box covers are assembled with the internal parts, such as the pump rotor, removed and are then subjected to a hydrostatic test, generally for about 5 min. Such a test demonstrates that the casing containment is sound and that there is no leakage of fluid to the exterior.

Additional tests may be conducted by the pump manufacturer to determine the soundness of internal partitions separating areas of the pump operating under different pressures.

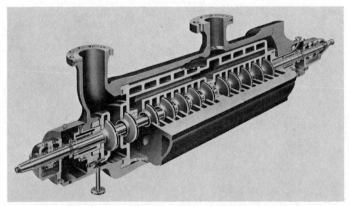

FIG. 21   Double-casing multistage pump with axially split inner casing. (Allis-Chalmers)

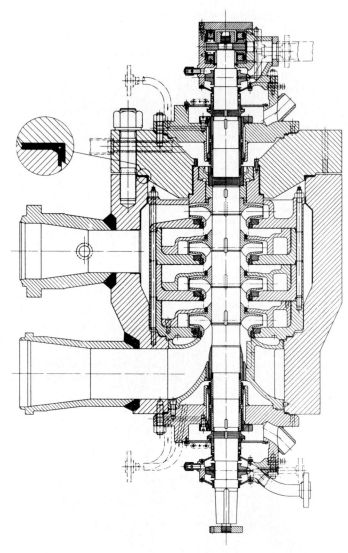

FIG. 22 Double-casing multistage pump with radially split inner casing. (Worthington Pump)

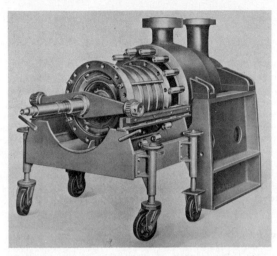

FIG. 23 Rotor assembly of radially split double-casing pump being inserted into its outer casing. (Worthington Pump)

The definition of the applicable hydrostatic pressure for these tests varies. The most generally accepted definition is that given by the Hydraulic Institute Standards (13th edition):

Each part of the pump which contains fluid under pressure shall be capable of withstanding a hydrostatic test at not less than the greater of the following:

- *150% of the pressure which will occur in that part when the pump is operated at rated conditions for the given application of the pump*
- *125% of the pressure which would occur in that part when operating at rated pump rpm for the given application, but with the pump discharge valve closed*

FIG. 24 Multistage radially split double-casing pump. (Worthington Pump)

IMPELLERS

In a single-suction impeller, the liquid enters the suction eye on one side only. As a double-suction impeller is, in effect, two single-suction impellers arranged back to back in a single casing, the liquid enters the impeller simultaneously from both sides, while the two casing suction passageways are connected to a common suction passage and a single suction nozzle.

For the general service single-stage axially split casing design, a double-suction impeller is favored because it is theoretically in axial hydraulic balance and because the greater suction area of a double-suction impeller permits the pump to operate with less net absolute suction head. For small units, the single-suction impeller is more practical for manufacturing reasons, as the waterways are not divided into two very narrow passages. It is also sometimes preferred for structural reasons. End-suction pumps with single-suction overhung impellers have both first-cost and maintenance advantages not obtainable with double-suction impellers. Most radially split casing pumps therefore use single-suction impellers. Because an overhung impeller does not require the extension of a shaft into the impeller suction eye, single-suction impellers are preferred for pumps handling suspended matter, such as sewage. In multistage pumps, single-suction impellers are almost universally used because of the design and first-cost complexity that double-suction staging introduces.

Impellers are called radial vane or radial flow when the liquid pumped is made to discharge radially to the periphery. Impellers of this type usually have a specific speed below 4200 (2600) if single suction and below 6000 (3700) if double suction. Specific speed is discussed in detail in Subsec. 2.3.1. Units for specific speed used here are (in USCS) revolutions per minute, gallons per minute, and feet; in SI, they are revolutions per minute, liters per second, and meters. Impellers can also be classified by the shape and form of their vanes:

1. The straight-vane impeller (Figs. 25, 34, 35, 36, and 37)
2. The Francis-vane or screw-vane impeller (Figs. 26 and 27)
3. The mixed-flow impeller (Fig. 28)
4. The propeller or axial-flow impeller (Fig. 29)

In a *straight-vane radial impeller*, the vane surfaces are generated by straight lines parallel to the axis of rotation. These are also called single-curvature vanes. The vane surfaces of a *Francis-vane radial impeller* have double curvature. An impeller design that has both a radial-flow and an axial-flow component is called a *mixed-flow impeller*. It is generally restricted to single-suction designs with a specific speed above 4200 (2600). Types with lower specific speeds are called Francis-vane impellers. Mixed-flow impellers with a very small radial-flow component are usually referred to as *propellers*. In a true propeller, or *axial-flow impeller*, the flow strictly parallels the axis of rotation. In other words, it moves only axially.

An *inducer* is a low-head axial-flow impeller with few blades which is placed in front of a conventional impeller. The hydraulic characteristics of an inducer are such that it requires con-

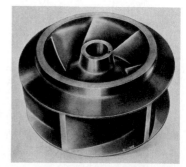

FIG. 25 Straight-vane radial single-suction closed impeller. (Worthington Pump)

FIG. 26 Francis-vane radial double-suction closed impeller. (Worthington Pump)

FIG. 27 High-specific-speed Francis-vane radial double-suction closed impeller. (Worthington Pump)

FIG. 28 Open mixed-flow impeller. (Worthington Pump)

siderably less NPSH than a conventional impeller. Both inducer and impeller are mounted on the same shaft and rotate at the same speed (Fig. 30). The main purpose of the inducer is not to generate an appreciable portion of the total pump head but to increase the suction pressure of a conventional impeller. Inducers are therefore used to reduce the NPSH requirements of a given pump or to permit the pump to operate at higher speeds with a given available NPSH. (For a further discussion of inducers, see Sec. 2.1 and Subsec. 2.3.1.)

The relation of single-suction impeller profiles to specific speed is shown in Fig. 31. Classification of impellers according to their vane shape is naturally arbitrary inasmuch as there is much overlapping in the types of impellers used in the different types of pumps. For example, impellers in single- and double-suction pumps of low specific speed have vanes extending across the suction eye. This provides a mixed flow at the impeller entrance for low pickup losses at high rotative speeds but allows the discharge portion of the impeller to use the straight-vane principle. In pumps of higher specific speed operating against low heads,

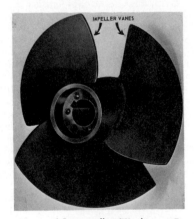

FIG. 29 Axial-flow impeller. (Worthington Pump)

impellers have double-curvature vanes extending over the full vane surface. They are, therefore, full Francis-type impellers. The mixed-flow impeller, usually a single-suction type, is essentially one-half of a double-suction high-specific-speed Francis-vane impeller.

In addition, many impellers are designed for specific applications. For instance, the conventional impeller design with sharp vane edges and restricted areas is not suitable for handling liquids containing rags, stringy materials, and solids like sewage because it will become clogged. Special nonclogging impellers with blunt edges and large waterways have been developed for such services (Fig. 32). For pumps up to the 12- to 16-in (305- to 406-mm) size, these impellers have only two vanes. Larger pumps normally use three or four vanes.

Another impeller design used for paper pulp pumps (Fig. 33) is fully open and nonclogging and has screw and radial streamlined vanes. The screw-conveyor end projects far into the suction nozzle, permitting the pump to handle high-consistency paper pulp stock.

Impeller Mechanical Types Mechanical design also determines impeller classification. Accordingly, impellers may be (1) completely open, (2) semiopen, or (3) closed.

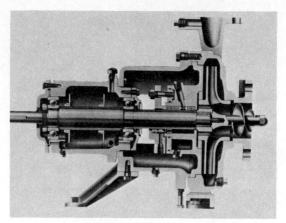

FIG. 30 Centrifugal pump with conventional impeller preceded by an inducer. (Worthington Pump)

Strictly speaking, an *open impeller* (Figs. 34 and 35) consists of nothing but vanes attached to a central hub for mounting on the shaft without any form of sidewall or shroud. The disadvantage of this impeller is structural weakness. If the vanes are long, they must be strengthened by ribs or a partial shroud. Generally, open impellers are used in small, inexpensive pumps. One advantage of open impellers is that they are better suited for handling liquids containing stringy materials. It is also sometimes claimed that they are better suited for handling liquids containing suspended matter because the solids in such matter are more likely to clog in the space between the rotating shrouds of a closed impeller and the stationary casing walls. It has been demonstrated, however, that closed impellers do not clog easily, thus disproving the claim for the superiority of the open-impeller design. In addition, the open impeller is much more sensitive to wear than the closed impeller and therefore its efficiency deteriorates rather rapidly.

The open impeller rotates between two side plates, between the casing walls of the volute or between the stuffing box head and the suction head. The clearance between the impeller vanes and the sidewalls allows a certain amount of water slippage. This slippage increases as wear increases. To restore the original efficiency, both the impeller and the side plate must be replaced. This, incidentally, involves a much larger expense than would be entailed in closed-impeller pumps, where simple rings form the leakage joint.

The *semiopen impeller* (Fig. 36) incorporates a single shroud, usually at the back of the impeller. This shroud may or may not have pump-out vanes, which are vanes located at the back of the impeller shroud (Fig. 37). This function is to reduce the pressure at the back hub of the impeller and to prevent foreign matter from lodging in back of the impeller and interfering with the proper operation of the pump and of the stuffing box.

The *closed impeller* (Figs. 25 to 27), which is almost universally used in centrifugal pumps handling clear liquids, incorporates shrouds or enclosing sidewalls that totally enclose the impeller waterways from the suction eye to the periphery. Although this design prevents the liquid slippage that occurs between an open or semiopen impeller and its side plates, a running joint must be provided between the impeller and the casing to separate the discharge and suction chambers of the pump. This running joint is usually formed by a relatively short cylindrical surface on the impeller shroud that rotates within a slightly larger stationary cylindrical surface. If one or both surfaces are made renewable, the leakage joint can be repaired when wear causes excessive leakage.

If the pump shaft terminates at the impeller so that the latter is supported by bearings on one side, the impeller is called an *overhung impeller*. This type of construction is the best for end-suction pumps with single-suction impellers.

Impeller Nomenclature   The inlet of an impeller just before the section at which the vanes start is called the suction eye (Fig. 38). In a closed-impeller pump, the suction eye diameter is

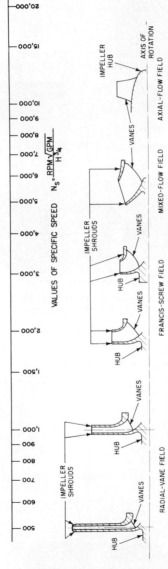

FIG. 31 Variations in impeller profiles with specific speed and approximate range of specific speed for the various types. (To convert specific speed in USCS units to SI, multiply by 0.61123).

VALUES OF SPECIFIC SPEED

$$N_s = \frac{RPM \sqrt{GPM}}{H^{3/4}}$$

RADIAL-VANE FIELD

FRANCIS-SCREW FIELD

MIXED-FLOW FIELD

AXIAL-FLOW FIELD

IMPELLER SHROUDS

IMPELLER SHROUDS

IMPELLER HUB

AXIS OF ROTATION

VANES

HUB

FIG. 32 Phantom view of radial-vane nonclogging impeller. (Worthington Pump)

FIG. 33 Paper plup impeller. (Worthington Pump)

FIG. 34 Open impellers. Notice that the impellers at left and right are strengthened by a partial shroud. (Worthington Pump)

FIG. 35 Open impeller with partial shroud. (Worthington Pump)

FIG. 36 Semiopen impeller. (Worthington Pump)

FIG. 37 Front and back views of an open impeller with partial shroud and pump-out vanes on back side. (Worthington Pump)

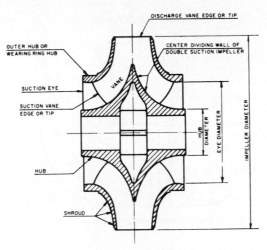

FIG. 38   Parts of a double-suction impeller.

taken as the smallest inside diameter of the shroud. In determining the area of the suction eye, the area occupied by the impeller shaft hub is deducted.

The *hub* is the central part of the impeller which is bored out to receive the pump shaft. The expression, however, is also frequently used for the part of the impeller which rotates in the casing fit or in the casing wearing ring. It is then referred to as the *outer impeller hub* or the *wearing-ring hub of the shroud.*

WEARING RINGS

Wearing rings provide an easily and economically renewable leakage joint between the impeller and the casing. A leakage joint without renewable parts is illustrated in Fig. 39. To restore the original clearances of such a joint after wear occurs, the user must either (1) build up the worn surfaces by welding, metal spraying, or other means and then true up the part or (2) buy new parts.

The new parts are not very costly in small pumps, especially if the stationary casing element

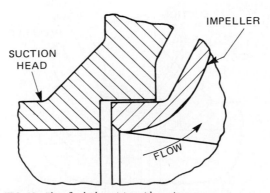

FIG. 39   Plain flat leakage joint with no rings.

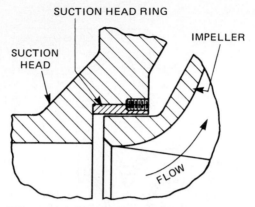

FIG. 40   Single flat-casing-ring construction.

is a simple suction cover. This is not true for larger pumps or where the stationary element of the leakage joint is part of a complicated casting. If the first cost of a pump is of prime importance, it is more economical to provide for remachining both the stationary parts and the impeller. Renewable casing and impeller rings can then be installed (Figs. 40, 41, 42, etc.). Nomenclature for the casing or stationary part forming the leakage joint surface is as follows: (1) *casing ring* (if mounted in the casing), (2) *suction cover ring* or *suction head ring* (if mounted in a suction cover or head), and (3) *stuffing box cover ring* or *head ring* (if mounted in the stuffing box cover or head). Some engineers like to identify the part further by adding the word *wearing*, as, for example, *casing wearing ring*. A renewable part for the impeller wearing surface is called the *impeller ring*. Pumps with both stationary and rotating rings are said to have *double-ring* construction.

Wearing-Ring Types   There are various types of wearing-ring designs, and the selection of the most desirable type depends on the liquid being handled, the pressure differential across the leakage joint, the rubbing speed, and the particular pump design. In general, centrifugal pump designers use that ring construction which they have found to be most suitable for each particular pump service.

The most common ring constructions are the flat type (Figs. 40 and 41) and the L type. The leakage joint in the former is a straight annular clearance. In the L-type ring (Fig. 43), the axial clearance between the impeller and the casing ring is large and so the velocity of the liquid flowing into the stream entering the suction eye of the impeller is low. The L-type casing rings shown

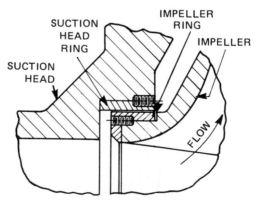

FIG. 41   Double flat-ring construction.

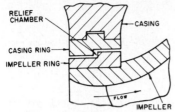

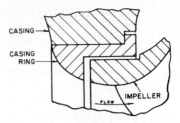

FIG. 42 Step-type leakage joint with double rings.

FIG. 43 An L-type nozzle casing ring.

in Figs. 43 and 44 have the additional function of guiding the liquid into the impeller eye; they are called *nozzle rings*. Impeller rings of the L type shown in Fig. 44 also furnish protection for the face of the impeller wearing-ring hub.

Some designers favor labyrinth-type rings (Figs. 45 and 46), which have two or more annular leakage joints connected by relief chambers. In leakage joints involving a single unbroken path, the flow is a function of the area and the length of the joint and of the pressure differential across the joint. If the path is broken by relief chambers (Figs. 42, 45, and 46), the velocity energy in the jet is dissipated in each relief chamber, increasing the resistance. As a result, with several relief chambers and several leakage joints for the same actual flow through the joint, the area and hence the clearance between the rings can be greater than for an unbroken, shorter leakage joint.

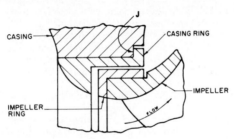

FIG. 44 Double rings, both of L type.

The single labyrinth ring with only one relief chamber (Fig. 45) is often called an *intermeshing ring*. The *step-ring type* (Fig. 42) utilizes two flat-ring elements of slightly different diameters over the total leakage joint width with a relief chamber between the two elements. Other ring designs also use some form of relief chamber. For example, one commonly used in small pumps has a flat joint similar to that in Fig. 40, but with one surface broken by a number of grooves.

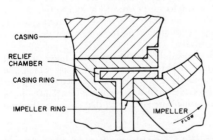

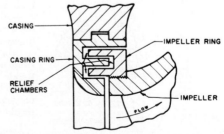

FIG. 45 Single labyrinth, intermeshing type. Double-ring construction with nozzle-type casing ring.

FIG. 46 Labyrinth-type rings in double-ring construction.

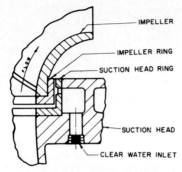

FIG. 47 Water-flushed wearing ring.

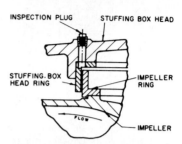

FIG. 48 Wearing ring with inspection port for checking clearance.

These act as relief chambers to dissipate the jet velocity head, thereby increasing the resistance through the joint and decreasing the leakage.

For raw water pumps in waterworks service and for larger pumps on sewage service, in which the liquid contains sand and grit, *water-flushed rings* have been used (Fig. 47). Clear water under a pressure greater than that on the discharge side of the rings is piped to the inlet and distributed by the cored passage, the holes through the stationary ring, and the groove to the leakage joint. Ideally the clear water should fill the leakage joint, with some flow to the suction and discharge sides to prevent any sand or grit from getting into the clearance space.

In large pumps [roughly 36 in (900 mm) or larger discharge size], particularly vertical end-suction single-stage volute pumps, size alone permits some refinements not found in smaller pumps. One example is the inclusion of inspection ports for measuring ring clearance (Fig. 48). These ports can be used to check the impeller centering after the original installation as well as to observe ring wear without dismantling the pump.

The lower rings of large vertical pumps handling liquids containing sand and grit on intermittent service are highly subject to wear. During shutdown periods the grit and sand settle out and naturally accumulate in the region in which these rings are installed, as it is the lowest point on the discharge side of the pump. When the pump is started again, this foreign matter is washed into the joint all at once and causes wear. To prevent this action in medium and large pumps a dam-type ring is often used (Fig. 49). Periodically the pocket on the discharge side of the dam can be flushed out.

One problem with the simple water-flushed ring is the failure to get uniform pressure in the stationary ring groove. If pump size and design permit, two sets of wearing rings arranged in tandem and separated by a large water space (Fig. 50) provide the best solution. The large water space allows uniform distribution of the flushing water to the full 360° degrees of each leakage joint. Because ring 2 is shorter and because a greater clearance is used there than at ring 1, equal flow can be made to take place to the discharge pressure side and to the suction pressure side. This design also makes it easier to harden the surfaces with stellite or to flame-plate them with tungsten carbide.

For pumps handling gritty or sandy water, the ring construction should provide an apron on which the stream leaving the leakage joint can impinge, as sand or grit in the jet will erode any surface it hits. Thus a form of L-type casing ring similar to that shown in Fig. 49 should be used.

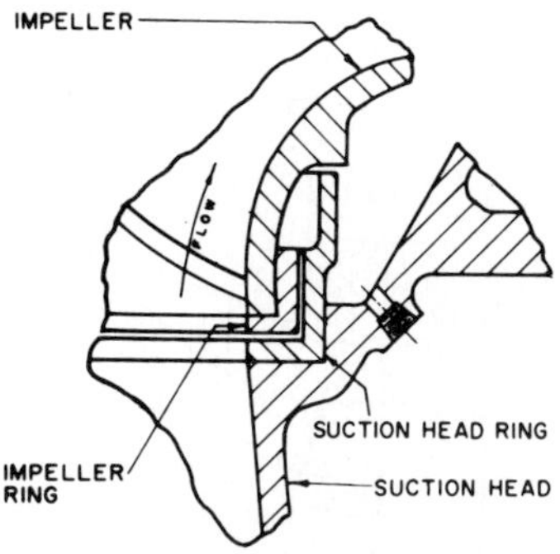

FIG. 49 Dam-type ring construction.

Wearing-Ring Location In a few designs used by one or two sewage pump manufacturers, leakage is

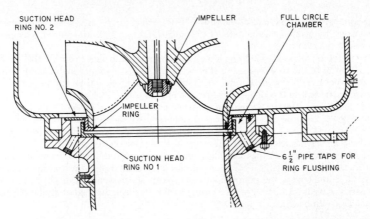

FIG. 50   Two sets of rings with space between for flushing water. (Worthington Pump)

controlled by an axial clearance (Fig. 51). Generally, this design requires a means of adjusting the shaft position for proper clearance. Then, if uniform wear occurs over the two surfaces, the original clearance can be restored by adjusting the position of the impeller. There is a limit to the amount of wear that can be compensated for since the impeller must be nearly central in the casing waterways.

Leakage joints with axial clearance are not overly popular for double-suction pumps because a very close tolerance is required in machining the fit of the rings in reference to the centerline of the volute waterways. Joints with radial clearances, however, allow some shifting of the impeller for centering. The only adverse effect is a slight inequality in the lengths of the leakage paths on the two sides.

So far, this discussion has treated only those leakage joints located adjacent to the impeller eye or at the smallest outside shroud diameter. There have been designs where the leakage joint has been at the periphery of the impeller. In a vertical pump this design is advantageous because the space between the joint and the suction waterways is open and so sand or grit cannot collect. Because of rubbing speed and because the impeller diameters used in the same casing vary over a wide range, the design is impractical in regular pump lines.

Mounting of Stationary Wearing Rings   In small single-suction pumps with suction heads, a stationary wearing ring is usually pressed into a bore in the head and may or may not be further locked by several set screws located half in the head and half in the ring (Fig. 41). Larger pumps often use an L-type ring with the flange held against a face on the head. In axially split casing pumps, the cylindrical casing bore (in which the casing ring will be mounted) should be slightly larger than the outside diameter of the ring. Unless some clearance is provided, distortion of the ring may occur when the two casing halves are assembled. However, the joint between the casing ring and the casing must be tight enough to prevent leakage. This is usually provided by a radial metal-to-metal joint (Fig. 43) so arranged that the discharge pressure presses the ring against the casing surface.

As it is not desirable for the casing ring of an axially split casing pump to be pinched by the casing, the ring will not be held tightly enough to prevent its rotation unless special provisions are made to keep it in place. One means of accomplishing this is to place a pin in the casing that will project into a hole bored in the ring, or, conversely, to provide a pin in the ring that will fit into a hole bored in the casing or into a recess at the casing split joint.

Another favorite method is to have a tongue on the casing ring that extends around 180° and engages a

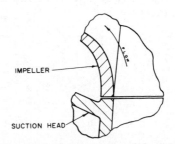

FIG. 51   Leakage joint with axial clearance.

corresponding groove in one-half of the casing. This method can be used with casing rings having a central flange by making the diameter of the flange larger for 180° and cutting a deeper groove in that half of the casing.

Many methods are used for holding impeller rings on the impeller. Probably the simplest is to rely on a press fit of the ring on the impeller or, if the ring is of proper material, on a shrink fit. Designers do not usually feel that a press fit is sufficient and often add several machine screws or set screws located half in the ring and half in the impeller, as in Fig. 41.

Some designers prefer to thread the impeller and ring and then screw the ring onto the impeller (Fig. 46). Some even favor the use of right- and left-hand threads for the different sides so that rotation will tighten the ring on the impeller. Generally some additional locking device is used rather than relying solely on the friction grip of the ring on the impeller.

In the design of impeller rings, consideration has to be given to the stretch of the ring caused by centrifugal force, especially if the pump is of a high-speed design for the capacity involved. For example, some boiler-feed pumps operate at speeds which would cause the rings to become loose if only a press fit was used. For such pumps shrink fits should be used or, preferably, impeller rings should be eliminated.

Wearing-Ring Clearances   Typical clearance and tolerance standards for nongalling wearing-joint metals in general service pumps are shown in Fig. 52. They apply to the following combinations: (1) bronze with a dissimilar bronze, (2) cast iron with bronze, (3) steel with bronze, (4) Monel metal with bronze, and (5) cast iron with cast iron. If the metals gall easily (like the chrome steels), the values given should be increased by about 0.002 to 0.004 in.

In multistage pumps, the basic diameter clearance should be increased by 0.003 in for larger rings. The tolerance indicated is positive for the casing ring and negative for the impeller hub or impeller ring.

In a single-stage pump with a joint of nongalling components, the correct machining dimension for a casing-ring diameter of 9.000 in would be 9.000 plus 0.003 and minus 0.000 in; for the

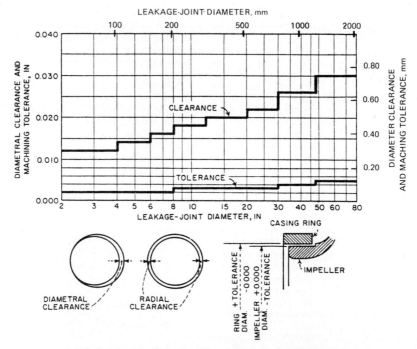

FIG. 52   Wearing-ring clearances for single-stage pumps using nongalling materials.

impeller hub or ring, the values would be 9.000 minus 0.018 (or 8.982) plus 0.000 and minus 0.003 in. Diametral clearances would be between 0.018 and 0.024 in. Obviously, clearances and tolerances are not translated into SI units be merely using conversion multipliers; they are rounded off as they are in the USCS system. The SI values for the USCS examples given above are

Increase for galling metals	0.05 to 0.10 mm
Increase for multistage pumps	0.08 mm

For a single-stage pump with a casing ring diameter of 230 mm, the machining dimensions would be

Casing ring	230 plus 0.08 and minus 0.00 mm
Impeller hub	230 minus 0.50 (229.5) plus 0.00 and minus 0.08 mm
Diametral clearances	between 0.50 and 0.66 mm

Naturally the manufacturer's recommendation for ring clearances and tolerances should be followed.

AXIAL THRUST

Axial Thrust in Single-Stage Pumps with Closed Impellers
The pressures generated by a centrifugal pump exert forces on both stationary and rotating parts. The design of these parts balances some of these forces, but separate means may be required to counterbalance others.

Axial hydraulic thrust on an impeller is the sum of the unbalanced forces acting in the axial direction. As reliable large-capacity thrust bearings are now readily available, axial thrust in single-stage pumps remains a problem only in larger units. Theoretically, a double-suction impeller is in hydraulic axial balance, with the pressures on one side equal to and counterbalancing the pressures on the other (Fig. 53). In practice, this balance may not be achieved for the following reasons:

1. The suction passages to the two suction eyes may not provide equal or uniform flows to the two sides.

2. External conditions, such as an elbow located too close to the pump suction nozzle, may cause unequal flow to the two suction eyes.

3. The two sides of the discharge casing waterways may not be symmetrical, or the impeller may be located off-center. These conditions will alter the flow characteristics between the impeller shrouds and the casing, causing unequal pressures on the shrouds.

4. Unequal leakage through the two leakage joints can upset the balance.

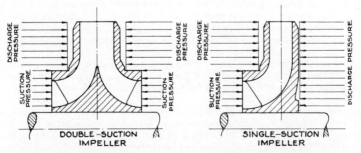

FIG. 53 Origin of pressures acting on impeller shrouds to produce axial thrust.

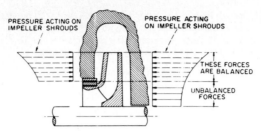

FIG. 54 Pressure distribution on front and back shrouds of single-suction impeller with shaft through impeller eye.

Combined, these factors can create axial unbalance. To compensate for this, all centrifugal pumps, even those with double-suction impellers, incorporate thrust bearings.

The ordinary single-suction closed radial-flow impeller with the shaft passing through the impeller eye (Fig. 53) is subject to axial thrust because a portion of the front wall is exposed to suction pressure and thus relatively more backwall surface is exposed to discharge pressure. If the discharge chamber pressure was uniform over the entire impeller surface, the axial force acting toward the suction would be equal to the product of the net pressure generated by the impeller and the unbalanced annular area.

Actually, pressure on the two single-suction closed impeller walls is not uniform. The liquid trapped between the impeller shrouds and casing walls is in rotation, and the pressure at the impeller periphery is appreciably higher than at the impeller hub. Although we need not be concerned with the theoretical calculations for this pressure variation, Fig. 54 describes it qualitatively. Generally speaking, axial thrust toward the impeller suction is about 20 to 30% less than the product of the net pressure and the unbalanced area.

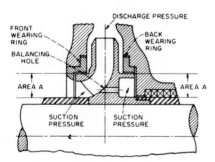

FIG. 55 Balancing axial thrust of single-suction impeller by means of wearing ring on back side and balancing holes.

To eliminate the axial thrust of a single-suction impeller, a pump can be provided with both front and back wearing rings; to equalize thrust areas, the inner diameter of both rings is made the same (Fig. 55). Pressure approximately equal to the suction pressure is maintained in a chamber located on the impeller side of the back wearing ring by drilling balancing holes through the impeller. Leakage past the back wearing ring is returned to the suction area through these holes. However, with large single-stage single-suction pumps, balancing holes are considered undesirable because leakage back to the impeller suction opposes the main flow, creating disturbances. In such pumps, a piped connection to the pump suction replaces the balancing holes. Another way to eliminate or reduce axial thrust in single-suction impellers is to use pump-out vanes on the back shroud. The effect of these vanes is to reduce the pressure acting on the back shroud of the impeller (Fig. 56). This design, however, is generally used only in pumps handling gritty liquids, where it keeps the clearance space between the impeller back shroud and the casing free of foreign matter.

So far, our discussion of the axial thrust has been limited to single-suction closed impellers with a shaft passing through the impeller eye and located in pumps with two stuffing boxes, one on either side of the impeller. In these pumps, suction-pressure magnitude does not affect the resulting axial thrust.

On the other hand, axial forces acting on an overhung impeller with a single stuffing box (Fig. 57) are definitely affected by suction pressure. In addition to the unbalanced force found in a single-suction two-box design (Fig. 54), there is an axial force equivalent to the product of the shaft area through the stuffing box and the difference between suction and atmospheric pressure. This force acts toward the impeller suction when the suction pressure is less than atmospheric or in the opposite direction when it is higher than atmospheric.

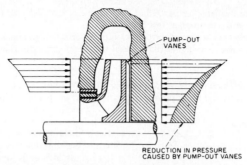

FIG. 56 Pump-out vanes used in a single-suction impeller to reduce axial thrust.

When an overhung impeller pump handles a suction lift, the additional axial force is very low. For example, if the shaft diameter through the stuffing box is 2 in (50.8 mm) (area = 3.14 in² or 20.26 cm²) and if the suction lift is 20 ft (6.1 m) of water equivalent to an absolute pressure of 6.06 lb/in² (0.42 bar abs), the axial force caused by the overhung impeller and acting toward the suction will be only 27 lb (121 N). On the other hand, if the suction pressure is 100 lb/in² (6.89 bar), the force will be 314 lb (1405 N) and will act in the opposite direction. Therefore, as the same pump may be used under many conditions over a wide range of suction pressures, the thrust bearing of pumps with single-suction overhung impellers must be arranged to take thrust in either direction. They must also be selected with sufficient thrust capacity to counteract forces set up under the maximum suction pressure established for that particular pump.

This extra thrust capacity may become quite significant in certain special cases, as, for instance, with boiler circulating pumps. These are usually of the single-suction single-stage overhung impeller type and may be exposed to suction pressures as high as 2800 lb/in² gage (193 bar). If such a pump has a shaft diameter of 6 in (15.24 cm), the unbalanced thrust would be as much as 77,500 lb (346,770 N), and the thrust bearing has to be capable of counteracting this.

Axial Thrust of Single-Suction Semiopen Radial-Flow Impellers The axial thrust generated in semiopen impellers is higher than that in closed impellers. This is illustrated in Fig. 58, which shows that the pressure on the open side of the impeller varies from essentially the discharge pressure at the periphery (at diameter D_2) to the suction pressure at the impeller eye (at diameter D_1). The pressure distribution on the back shroud is essentially the same as that illustrated in Fig. 54, varying from discharge pressure at the periphery to some portion of this pressure at the impeller hub. This latter pressure is, of course, substantially higher than the suction pressure. The unbalanced portion of the axial thrust on the impeller is represented by the cross-hatched area in Fig. 58.

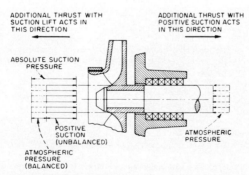

FIG. 57 Axial thrust problem with single-suction overhung impeller and single stuffing box.

One of the means available for partially balancing this increased axial thrust is to provide the back shroud with pump-out vanes, as in Figs. 37 and 56.

Fully open impellers or semiopen impellers with a portion of the back shroud removed produce an axial thrust somewhat higher than closed impellers and somewhat lower than semiopen impellers.

Axial Thrust of Mixed-Flow and Axial-Flow Impellers

The axial thrust in radial impellers is produced by the static pressures on the impeller shrouds. Axial-flow impellers have

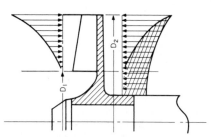

no shrouds, and the axial thrust is created strictly by the difference in pressure on the two faces of the impeller vanes. In addition, there may be a difference in pressure acting on the two shaft hub ends, one generally subject to discharge pressure and the other to suction pressure.

With mixed-flow impellers, axial thrust is a combination of forces caused by the action of the vanes on the liquid and those arising from the difference in the pressures acting on the various surfaces. Wearing rings are often provided on the back of mixed-flow impellers, with either balancing holes through the impeller hub or an external balancing pipe leading back to the suction.

FIG. 58   Axial thrust in a semiopen single-suction impeller.

Except for very large units and in certain special applications, the maximum axial thrust developed by mixed-flow and axial-flow impellers is not significant because the operating heads of these pumps are relatively low. This thrust is carried by thrust bearings with the necessary load capacity.

Axial Thrust in Multistage Pumps

Most multistage pumps are built with single-suction impellers in order to simplify the design of the interstage connections. Two obvious arrangements are possible for the single-suction impellers:

1. Several single-suction impellers may be mounted on one shaft, each having its suction inlet facing in the same direction and its stages following one another in ascending order of pressure (Fig. 59). The axial thrust is then balanced by a hydraulic balancing device.

2. An even number of single-suction impellers may be used, one-half facing in one direction and the other half facing in the opposite direction. With this arrangement, axial thrust on the first half is compensated by the thrust in the opposite direction on the other half (Fig. 60). This mounting of single-suction impellers back to back is frequently called *opposed impellers*.

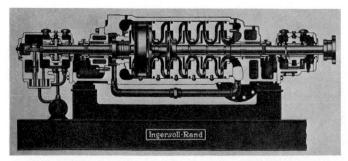

FIG. 59   Multistage pump with single-suction impellers facing in one direction and hydraulic balancing device. (Ingersoll-Rand)

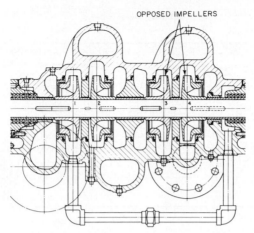

FIG. 60 Four-stage pump with opposed impellers. (Worthington Pump)

An uneven number of single-suction impellers may be used with this arrangement, provided the correct shaft and interstage bushing diameters are used to give the effect of a hydraulic balancing device that will compensate for the hydraulic thrust on one of the stages.

It is important to note that the opposed impeller arrangement completely balances axial thrust only under the following conditions:

1. The pump must be provided with two stuffing boxes.
2. The shaft must have a constant diameter.
3. The impeller hubs must not extend through the interstage portion of the casing separating adjacent stages.

Except for some special pumps that have an internal and enclosed bearing at one end, and therefore only one stuffing box, most multistage pumps fulfill the first condition. Because of structural requirements, however, the last two conditions are not practical. A slight residual thrust is usually present in multistage opposed impeller pumps and is carried on the thrust bearing.

HYDRAULIC BALANCING DEVICES _____

If all the single-suction impellers of a multistage pump face in the same direction, the total theoretical hydraulic axial thrust acting toward the suction end of the pump will be the sum of the individual impeller thrusts. The thrust magnitude (in pounds) will be approximately equal to the product of the net pump pressure (in pounds per square inch) and the annular unbalanced area (in square inches). Actually the axial thrust turns out to be about 70 to 80% of this theoretical value.

Some form of hydraulic balancing device must be used to balance this axial thrust and to reduce the pressure on the stuffing box adjacent to the last-stage impeller. This hydraulic balancing device may be a balancing drum, a balancing disk, or a combination of the two.

Balancing Drums The balancing drum is illustrated in Fig. 61. The balancing chamber at the back of the last-stage impeller is separated from the pump interior by a drum that is either keyed or screwed to the shaft and rotates with it. The drum is separated by a small radial clearance from the stationary portion of the balancing device, called the *balancing-drum head,* which is fixed to the pump casing.

The balancing chamber is connected either to the pump suction or to the vessel from which

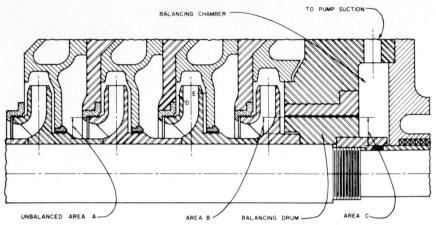

FIG. 61   Balancing drum.

the pump takes its suction. Thus the back pressure in the balancing chamber is only slightly higher than the suction pressure, the difference between the two being equal to the friction losses between this chamber and the point of return. The leakage between the drum and the drum head is, of course, a function of the differential pressure across the drum and of the clearance area.

The forces acting on the balancing drum in Fig. 61 are the following:

1. Toward the discharge end: the discharge pressure multiplied by the front balancing area (area *B*) of the drum.

2. Toward the suction end: the back pressure in the balancing chamber multiplied by the back balancing area (area *C*) of the drum.

The first force is greater than the second, thereby counterbalancing the axial thrust exerted upon the single-suction impellers. The drum diameter can be selected to balance axial thrust completely or within 90 to 95%, depending on the desirability of carrying any thrust-bearing loads.

It has been assumed in the preceding simplified description that the pressure acting on the impeller walls is constant over their entire surface and that the axial thrust is equal to the product of the total net pressure generated and the unbalanced area. Actually this pressure varies somewhat in the radial direction because of the centrifugal force exerted upon the water by the outer impeller shroud (Fig. 54). Furthermore, the pressures at two corresponding points on the opposite impeller faces (*D* and *E*, Fig. 61) may not be equal because of variation in clearance between the impeller wall and the casing section separating successive stages. Finally, pressure distribution over the impeller wall surface may vary with head and capacity operating conditions.

This pressure distribution and design data can be determined by test quite accurately for any one fixed operating condition, and an effective balancing drum could be designed on the basis of the forces resulting from this pressure distribution. Unfortunately, varying head and capacity conditions change the pressure distribution, and as the area of the balancing drum is necessarily fixed, the equilibrium of the axial forces can be destroyed.

The objection to this is not primarily the amount of the thrust, but rather that the direction of the thrust cannot be predetermined because of the uncertainty about internal pressures. Still it is advisable to predetermine normal thrust direction, as this can influence external mechanical thrust-bearing design. Because 100% balance is unattainable in practice and because the slight but predictable unbalance can be carried on a thrust bearing, the balancing drum is often designed to balance only 90 to 95% of total impeller thrust.

The balancing drum satisfactorily balances the axial thrust of single-suction impellers and reduces pressure on the discharge-side stuffing box. It lacks, however, the virtue of automatic compensation for any changes in axial thrust caused by varying impeller reaction characteristics. In effect, if the axial thrust and balancing drum forces become unequal, the rotating element will tend to move in the direction of the greater force. The thrust bearing must then prevent excessive movement of the rotating element. The balancing drum performs no restoring function until such time as the drum force again equals the axial thrust. This automatic compensation is the major feature that differentiates the balancing disk from the balancing drum.

Balancing Disks   The operation of the simple balancing disk is illustrated in Fig. 62. The disk is fixed to and rotates with the shaft. It is separated by a small axial clearance from the balancing disk head, which is fixed to the casing. The leakage through this clearance flows into the balancing chamber and from there either to the pump suction or to the vessel from which the pump takes its suction. The back of the balancing disk is subject to the balancing chamber back pressure, whereas the disk face experiences a range of pressures. These vary from discharge pressure at its smallest diameter to back pressure at its periphery. The inner and outer disk diameters are chosen so that the difference between the total force acting on the disk face and that acting on its back will balance the impeller axial thrust.

If the axial thrust of the impellers should exceed the thrust acting on the disk during operation, the latter is moved toward the disk head, reducing the axial clearance between the disk and the disk head. The amount of leakage through the clearance is reduced so that the friction losses in the leakage return line are also reduced, lowering the back pressure in the balancing chamber. This lowering of pressure automatically increases the pressure difference acting on the disk and moves it away from the disk head, increasing the clearance. Now the pressure builds up in the balancing chamber, and the disk is again moved toward the disk head until an equilibrium is reached.

To assure proper balancing disk operation, the change in back pressure in the balancing chamber must be of an appreciable magnitude. Thus, with the balancing disk wide open with respect to the disk head, the back pressure must be substantially higher than the suction pressure to give a resultant force that restores the normal disk position. This can be accomplished by introducing a restricting orifice in the leakage return line that increases back pressure when leakage past the disk increases beyond normal. The disadvantage of this arrangement is that the pressure on the stuffing box packing is variable—a condition that is injurious to the life of the packing and therefore to be avoided. The higher pressure that can occur at the packing is also undesirable.

Combination Balancing Disk and Drum   For the reasons just described, the simple balancing disk is seldom used. The combination balancing disk and drum (Fig. 63) was developed to obviate the shortcomings of the disk while retaining the advantage of automatic compensation for axial thrust changes.

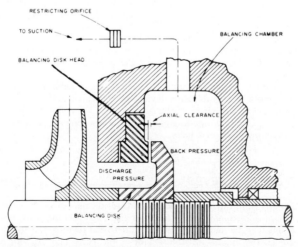

FIG. 62   Simple balancing disk.

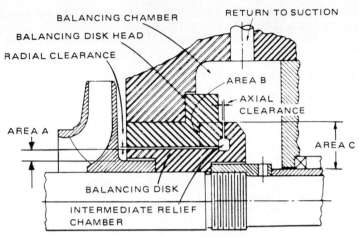

FIG. 63 Combination balancing disk and drum.

The rotating portion of this balancing device consists of a long cylindrical body that turns within a drum portion of the disk head. This rotating part incorporates a disk similar to the one previously described. In this design, radial clearance remains constant regardless of disk position, whereas the axial clearance varies with the pump rotor position. The following forces act on this device:

1. Toward the discharge end: the sum of the discharge pressure multiplied by area A, plus the average intermediate pressure multiplied by area B.

2. Toward the suction end: the back pressure multiplied by area C.

Whereas the position-restoring feature of the simple balancing disk required an undesirably wide variation of the back pressure, it is now possible to depend upon a variation of the intermediate pressure to achieve the same effect. Here is how it works: When the pump rotor moves toward the suction end (to the left in Fig. 63) because of increased axial thrust, the axial clearance is reduced and pressure builds up in the intermediate relief chamber, increasing the average value of the intermediate pressure acting on area B. In other words, with reduced leakage, the pressure drop across the radial clearance decreases, increasing the pressure drop across the axial clearance. The increase in intermediate pressure forces the balancing disk toward the discharge end until equilibrium is reached. Movement of the pump rotor toward the discharge end would have the opposite effect, increasing the axial clearance and the leakage and decreasing the intermediate pressure acting on area B.

There are now in use numerous hydraulic balancing device modifications. One typical design separates the drum portion of a combination device into two halves, one preceding and the second following the disk (Fig. 64). The virtue of this arrangement is a definite cushioning effect at the intermediate relief chamber, thus avoiding too positive a restoring action, which might result in the contacting and scoring of the disk faces.

SHAFTS AND SHAFT SLEEVES

The basic function of a centrifugal pump shaft is to transmit the torques encountered in starting and during operation while supporting the impeller and other rotating parts. It must do this job with a deflection less than the minimum clearance between rotating and stationary parts. The loads involved are (1) the torques, (2) the weight of the parts, and (3) both radial and axial hydraulic forces. In designing a shaft, the maximum allowable deflection, the span or overhang,

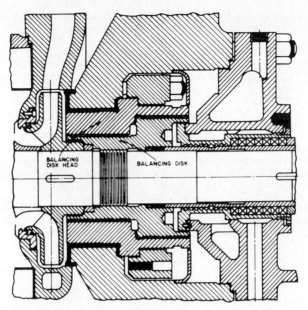

FIG. 64 Combination balancing disk and drum with disk located in center portion of drum. (Worthington Pump)

and the location of the loads all have to be considered, as does the critical speed of the resulting design.

Shafts are usually proportioned to withstand the stress set up when a pump is started quickly, for example, when the driving motor is thrown directly across the line. If the pump handles hot liquids, the shaft is designed to withstand the stress set up when the unit is started cold without any preliminary warmup.

Critical Speeds　Any object made of an elastic material has a natural period of vibration. When a pump rotor or shaft rotates at any speed corresponding to its natural frequency, minor unbalances will be magnified. These speeds are called the *critical speeds*.

In conventional pump designs, the rotating assembly is theoretically uniform around the shaft axis and the center of mass should coincide with the axis of rotation. This theory does not hold for two reasons. First, there are always minor machining or casting irregularities; second, there are variations in metal density of each part. Thus, even in vertical-shaft machines having no radial deflection caused by the weight of the parts, this eccentricity of the center of mass produces centrifugal force and therefore a deflection when the assembly rotates. At the speed at which the centrifugal force exceeds the elastic restoring force, the rotor will vibrate as though it were seriously unbalanced. If it is run at that speed without restraining forces, the deflection will increase until the shaft fails.

Rigid- and Flexible-Shaft Designs　The lowest critical speed is called the first critical speed; the next highest is called the second, and so forth. In centrifugal pump nomenclature, a *rigid shaft* means one with an operating speed lower than its first critical speed; a *flexible shaft* is one with an operating speed higher than its first critical speed. Once an operating speed has been selected, relative shaft dimensions must still be determined. In other words, it must be decided whether the pump will operate above or below the first critical speed.

Actually, the shaft critical speed can be reached and passed without danger because frictional forces tend to restrain the deflection. These forces are exerted by the surrounding liquid, the stuffing box packing, and the various internal leakage joints acting as internal liquid-lubricated bearings. Once the critical speed is passed, the pump will run smoothly again up to the second speed corresponding to the natural rotor frequency, and so on to the third, fourth, and all higher critical speeds.

Designs rated for 1750 rpm (or lower) are usually of the rigid-shaft type. On the other hand, high-head 3600 rpm (or higher) multistage pumps, such as those in boiler-feed service, are frequently of the flexible-shaft type. It is possible to operate centrifugal pumps above their critical speeds for the following two reasons: (1) very little time is required to attain full speed from rest (the time required to pass through the critical speed must therefore be extremely short) and (2) the pumped liquid in the stuffing box packing and the internal leakage joints acts as a restraining force on the vibration.

Experience has proved that, although it was usually assumed necessary to use shafts of such rigidity that the first critical speed is at least 20% above the operating speed, equally satisfactorily results can be obtained with lighter shafts with a first critical speed of about 60 to 75% of the operating speed. This, it is felt, is a sufficient margin to avoid any danger caused by operation close to the critical speed.

Influence of Shaft Deflection To understand the effect of critical speed upon the selection of shaft size, consider the fact that the first critical speed of a shaft is linked to its static deflection. Shaft deflection depends upon the weight of the rotating element (w), the shaft span (l), and the shaft diameter (d). The basic formula is

$$f = \frac{wl^3}{CEI}$$

where f = deflection, in (m)
  w = weight of the rotating element, lb (N)
  l = shaft span, in (m)
  C = coefficient depending on shaft-support method and load distribution
  E = modulus of elasticity of shaft materials, lb/in^2 (N/m^2)
  I = moment of inertia ($\pi \, d^4/64$) in^4 (m^4)

This formula is given in its most simplified form, that is, for a shaft of constant diameter. If the shaft is of varying diameter (the usual situation), deflection calculations are much more complex. A graphical deflection analysis is then the most practical answer.

This formula solves only for static deflection, the only variable that affects critical speed calculations. The actual shaft deflection—which must be determined to establish minimum permissible internal clearances—must take into account all transverse hydraulic reactions on the rotor, the weights of the rotating element, and other external loads, such as belt pull.

It is not necessary to calculate exact deflection to make a relative shaft comparison. Instead, a factor can be developed that will be representative of relative shaft deflections. As a significant portion of rotor weight is in the shaft, and as methods of bearing support and modulus of elasticity are common to similar designs, deflection f can be shown as follows:

$$f = \text{function of } \frac{(ld^2)(l^3)}{d^4}$$

or
$$f = \text{function of } \frac{l^4}{d^2}$$

In other words, pump deflection varies approximately as the fourth power of shaft span and inversely as the square of shaft diameter. Therefore the lower the l^4/d^2 factor for a given pump, the lower the unsupported shaft deflection, essentially in proportion to this factor.

For practical purposes, the first critical speed N_c can be calculated as

in USCS units $\qquad N_c = \dfrac{187.7}{\sqrt{f} \text{ (in)}}$ rpm

in SI units $\qquad N_c = \dfrac{946}{\sqrt{f} \text{ (mm)}}$ rpm

To maintain internal clearances at the wearing rings, it is usually desirable to limit shaft deflection under most adverse conditions to between 0.005 and 0.006 in (0.127 and 0.152 mm). It follows that a shaft design with deflection of 0.005 to 0.006 in (0.127 to 0.152 mm) will have a first critical speed of 2400 to 2650 rpm. This is the reason for using rigid shafts for pumps that operate at 1750 rpm or lower. Multistage pumps operating at 3600 rpm or higher use shafts of equal stiffness (for the same purpose of avoiding wearing ring contact). However, their corresponding critical speed is about 25 to 40% less than their operating speed.

Lomakin Effect   All the above refers to the behavior of a rotor and its shaft operating in air. In reality, the rotor operates immersed in the liquid being pumped, and this liquid flows through one or more of the small annular areas created by clearances separating regions in the pump under different pressures, such as at the wearing rings, interstage bushings, or balancing devices. This flow of liquid creates what is called a hydrodynamic bearing effect and essentially transforms the rotor from one supported at two bearings external to the pump to one with several additional internal bearings lubricated by the liquid pumped. This phenomenon is generally called the Lomakin effect.

The result of the Lomakin effect is that the deflection of the shaft when a pump is running is reduced somewhat from the value calculated for the shaft operating in air and the critical speed is increased. Advantage is sometimes taken of this effect—particularly in the design of some multistage pumps—to permit the use of longer and more slender shafts. Whether this is sound practice remains a very controversial subject. The supportive effect of the hydrodynamic bearings depends on (a) the pressure differential, and therefore disappears completely when the pump is at rest, and (b) the clearance, and therefore decreases substantially as the internal clearances increase with erosive or contact wear. Thus, contact between rotating and stationary parts will take place every time a pump is started if the internal clearances are initially less than the shaft deflection in air. This contact will again take place as the pump coasts down after being stopped. Furthermore, as wear takes place at the running joints, the shaft assumes a deflection closer and closer to its deflection in air, unsupported by the Lomakin effect.

In view of all these facts, it is recommended that pump users acquaint themselves not only with the calculated shaft deflections with a pump running and in new condition but also with the shaft deflections in air. This way, they can compare these with the internal clearances.

Shaft Sizing   Shaft diameters are usually larger than what is actually needed to transmit the torque. A factor that assures this conservative design is the requirement for ease of rotor assembly.

The shaft diameter must be stepped up several times from the end of the coupling to its center to facilitate impeller mounting (Fig. 65). Starting with the maximum diameter at the impeller mounting, there is a stepdown for the shaft sleeve and another for the external shaft nut, followed by several more for the bearings and the coupling. Therefore the shaft diameter at the impellers exceeds that required for torsional strength at the coupling by at least an amount sufficient to provide all intervening stepdowns.

One frequent exception to shaft oversizing at the impeller occurs in units consisting of two double-suction single-stage pumps operating in series, one of which is fitted with a double-extended shaft. As this pump must transmit the total horsepower for the entire series unit, the shaft diameter at its inboard bearing may have to be larger than normal.

Shaft design of end-suction overhung-impeller pumps presents a somewhat different problem. One method for reducing shaft deflection at the impeller and stuffing box—where concentricity of running fits is extremely important—is to considerably increase shaft diameter between the bearings.

Except in certain smaller sizes, centrifugal pump shafts are protected against wear, erosion, and corrosion by renewable shaft sleeves. In very small pumps, however, shaft sleeves present a certain disadvantage. As the sleeve cannot appreciably contribute to shaft strength, the shaft itself must be designed for the full maximum stress. Shaft diameter is then materially increased by the addition of the sleeve, as the sleeve thickness cannot be decreased beyond a certain safe minimum. The impeller suction area may therefore become dangerously reduced, and if the eye diameter is increased to maintain a constant eye area, the liquid pickup speed must be increased unfavorably. Other disadvantages accrue from greater hydraulic and stuffing box losses caused by increasing the effective shaft diameter out of proportion to the pump size.

To eliminate these shortcomings, very small pumps frequently use shafts of stainless steel or

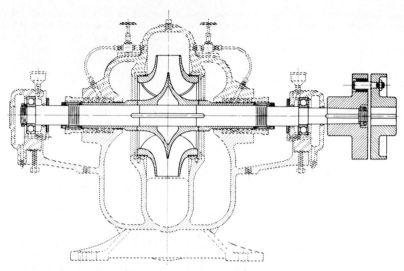

FIG. 65 Rotor assembly of a single-stage double-suction pump. (Worthington Pump)

some other material that is sufficiently resistant to corrosion and wear not to need shaft sleeves. One such pump is illustrated in Fig. 66. Manufacturing costs, of course, are much less for this type of design, and the cost of replacing the shaft is about the same as the cost of new sleeves (including installation).

Shaft Sleeves Pump shafts are usually protected from erosion, corrosion, and wear at stuffing boxes, leakage joints, internal bearings, and in the waterways by renewable sleeves.

The most common shaft sleeve function is that of protecting the shaft from wear at a stuffing box. Shaft sleeves serving other functions are given specific names to indicate their purpose. For example, a shaft sleeve used between two multistage pump impellers in conjunction with the interstage bushing to form an interstage leakage joint is called an *interstage* or *distance sleeve*.

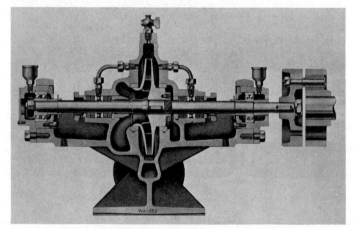

FIG. 66 Section of a small centrifugal pump with no shaft sleeves. (Worthington Pump)

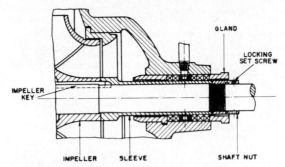

FIG. 67 Sleeve with external locknut and impeller key extending into sleeve to prevent slip.

In medium-size centrifugal pumps with two external bearings on opposite sides of the casing (the common double-suction and multistage varieties), the favored shaft sleeve construction uses an external shaft nut to hold the sleeve in axial position against the impeller hub. Sleeve rotation is prevented by a key, usually an extension of the impeller key (Fig. 67). If the axial thrust exceeds the frictional grip of the impeller on the shaft, it is transmitted through the sleeve to the external shaft nut.

In larger high-head pumps, a high axial load on the sleeve is possible and a design similar to that shown in Fig. 68 may be favored. This design has the commercial advantages of simplicity and low replacement cost. Some manufacturers favor the sleeve shown in Fig. 69, in which the impeller end of the sleeve is threaded and screwed to a matching thread on the shaft. A key cannot be used with this type of sleeve, and right- and left-hand threads are substituted so that the frictional grip of the packing on the sleeve will tighten it against the impeller hub. In the sleeve designs shown in Figs. 67 and 68, right-hand threads are usually used for all shaft nuts because keys prevent the sleeve from rotating. As a safety precaution, the external shaft nuts and the sleeve itself use set screws for a locking device.

In pumps with overhung impellers, various types of sleeves are used. Often stuffing boxes are placed close to the impeller, and the sleeve protects the impeller hub from wear (Fig. 70). As a portion of the sleeve in this design fits directly onto the shaft, the impeller key can be used to prevent sleeve rotation. Part of the sleeve is clamped between the impeller and a shaft shoulder to maintain the axial position of the sleeve.

In designs with a metal-to-metal joint between the sleeve and the impeller hub, operation under a positive suction head often starts liquid leakage into the clearance between shaft and sleeve. For a pump operating under negative suction head, the various clearances may cause slight

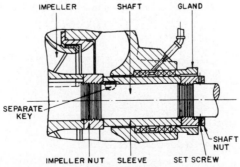

FIG. 68 Sleeve with internal impeller nut, external shaft-sleeve nut, and separate key for sleeve.

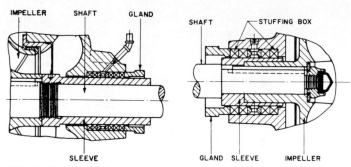

FIG. 69 Sleeve threaded onto shaft, with no external locknut.

air leakage into the pump. Usually this leakage is not important; however, it occasionally causes trouble, and a sleeve design with a leakage seal may then become desirable. One possible arrangement is shown in Fig. 71. The design shown in Fig. 72 is used for high-temperature process pumps. The contact surface of the sleeve and shaft is ground at a 45° angle. That end of the sleeve is locked, but the other is free to expand with temperature changes.

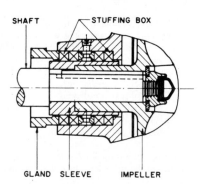

FIG. 70 Sleeve for pumps with overhung impeller hubs extending into stuffing box.

Material for Stuffing Box Sleeves

Stuffing box shaft sleeves are surrounded in the stuffing box by packing; the sleeve must be smooth so that it can turn without generating too much friction and heat. Thus the sleeve materials must be capable of taking a very fine finish, preferably a polish. Cast iron is therefore not suitable. A hard bronze is generally used for pumps handling clear water, but chrome or other stainless steels are sometimes preferred. For pumps subject to grit, hardened chrome or other stainless steels give good results. For more severe conditions, stellited sleeves are often used and occasionally sleeves that are chromium-plated at the packing area. Ceramic-coated sleeves with the coating applied by flame plating are also used for some severe services. Sleeves made entirely of a hardened chrome steel are usually the most economical and satisfactory.

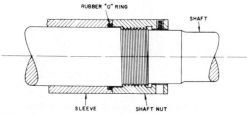

FIG. 71 Seal arrangement for shaft sleeve to prevent leakage along the shaft.

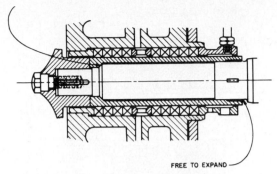

FREE TO EXPAND

FIG. 72   Sleeve with 45° bevel contacting surface.

STUFFING BOXES

Stuffing boxes have the primary function of protecting the pump against leakage at the point where the shaft passes out through the pump casing. If the pump handles a suction lift and the pressure at the interior stuffing box end is below atmospheric, the stuffing box function is to prevent air leakage into the pump. If this pressure is above atmospheric, the function is to prevent liquid leakage out of the pump.

For general service pumps, a stuffing box usually takes the form of a cylindrical recess that accommodates a number of rings of packing around the shaft or shaft sleeve (Figs. 73 and 74). If sealing the box is desired, a lantern ring or seal cage (Fig. 75) is used to separate the rings of packing into approximately equal sections. The packing is compressed to give the desired fit on the shaft or sleeve by a gland that can be adjusted in an axial direction. The bottom or inside end of the box may be formed by the pump casing (Fig. 70), a throat bushing (Fig. 73), or a bottoming ring (Fig. 74).

For manufacturing reasons, throat bushings are widely used on smaller pumps with axially split casings. Throat bushings are always solid rather than split. The bushing is usually held from rotation by a tongue-and-groove joint locked in the lower half of the casing.

Lantern Rings (Seal Cages)   When a pump operates with negative suction head, the inner end of the stuffing box is under vacuum and air tends to leak into the pump. For this type of service, packing is usually separated into two sections by a lantern ring, or seal cage (Fig. 73).

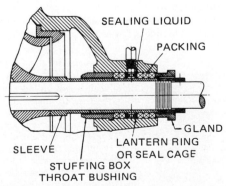

SEALING LIQUID

PACKING

GLAND

SLEEVE

LANTERN RING
OR SEAL CAGE

STUFFING BOX
THROAT BUSHING

FIG. 73   Conventional stuffing box with throat bushing.

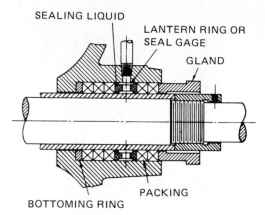

SEALING LIQUID

LANTERN RING OR
SEAL GAGE

GLAND

PACKING

BOTTOMING RING

FIG. 74 Conventional stuffing box with bottoming ring.

Water or some other sealing fluid is introduced under pressure into the space, causing flow of sealing fluid in both axial directions. This construction is useful for pumps handling flammable or chemically active and dangerous liquids since it prevents outflow of the pumped liquid. Lantern rings are usually axially split for ease of assembly.

Some installations involve variable suction conditions, the pump operating part of the time with head on suction and part of the time with suction lift. When the operating pressure inside the pump exceeds atmospheric pressure, the liquid lantern ring becomes inoperative (except for lubrication). However, it is maintained in service so that when the pump is primed at starting, all air can be excluded.

Sealing Liquid Arrangements When a pump handles clean, cool water, stuffing box seals are usually connected to the pump discharge or, in multistage pumps, to an intermediate stage. An independent supply of sealing water should be provided if any of the following conditions exist:

1. A suction lift in excess of 15 ft (4.5 m)
2. A discharge pressure under 10 lb/in^2 (0.7 bar)
3. Hot water (over 250°F or 120°C) being handled without adequate cooling (except for boiler-feed pumps, in which lantern rings are not used)
4. Muddy, sandy, or gritty water being handled
5. The pump is a hot-well pump
6. The liquid being handled is other than water—such as acid, juice, molasses, or sticky liquids—without special provision in the stuffing box design for the nature of the liquid

If the suction lift exceeds 15 ft (4.5 m), excessive air infiltration through the stuffing boxes may make priming difficult unless an independent seal is provided. A discharge pressure under 10 lb/in^2 (0.7 bar) may not provide sufficient sealing pressure. Hot-well (or condensate) pumps operate with as much as 28-inHg (710 mmHg) vacuum, and air infiltration would take place when the pumps are standing idle on standby service.

When sealing water is taken from the pump discharge, an external connection may be made through small-diameter piping (Fig. 76) or internal passages. In

FIG. 75 Lantern ring (also called seal cage). some pumps these connections are arranged so that a

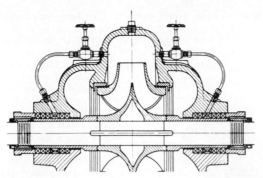

FIG. 76 Piping connections from pump discharge to seal cages.

sealing liquid can be introduced into the packing space through an internal drilled passage either from the pump casing or from an external source (Fig. 77). When the liquid pumped is used for sealing, the external connection is plugged. If an external sealing liquid source is required, it is connected to the external pipe tap with a socket-head pipe plug inserted at the internal pipe tap.

It is sometimes desirable to locate the lantern ring with more packing on one side than on the other. For example, on gritty-water service, a lantern ring location closer to the inner portion of the pump would divert a greater proportion of sealing liquid into the pump, thereby keeping grit from working into the box. An arrangement with most of the packing rings between the lantern ring and the inner end of the stuffing box would be applied to reduce dilution of the pumped liquid.

Some pumps handle water in which there are small, even microscopic, solids. Using water of this kind as a sealing liquid introduces the solids into the leakage path, shortening the life of the packing and sleeves. It is sometimes possible to remove these solids by installing small pressure filters in the sealing water piping from the casing to the stuffing box.

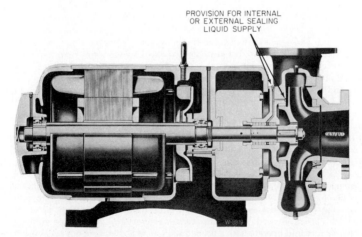

PROVISION FOR INTERNAL
OR EXTERNAL SEALING
LIQUID SUPPLY

FIG. 77 End-suction pump with provision for internal or external sealing-liquid supply. (Worthington Pump)

Filters ultimately get clogged, though, unless they are frequently backwashed or ótherwise cleaned out. This disadvantage can be obviated by using a cyclone (or centrifugal) separator. The operating principle of the cyclone separator is based on the fact that if liquid under pressure is introduced tangentially into a vortexing chamber, centrifugal force will make it rotate in the chamber, creating a vortex. Particles heavier than the liquid in which they are carried will hug the outside wall of the vortexing chamber and the liquid in the center of the chamber will be relatively free of foreign matter. The action of such a separator is illustrated on Fig. 78. Liquid piped from the pump discharge or from an intermediate stage of a multistage pump is piped to inlet tap A, which is drilled tangentially to the cyclone bore. The liquid containing solids is directed downward to the apex of the cone at outlet tap B and is piped to the suction or to a low-pressure point in the system. The cleaned liquid is taken off at the center of the cyclone at outlet tap C and is piped to the stuffing box.

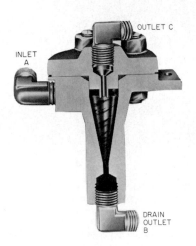

Sand that will pass through a No. 40 sieve will be 100% eliminated in a cyclone separator, with supply pressures as low as 20 lb/in^2 (1.4 bar). With 100-lb/in^2 (7-bar) supply pressure, 95% of the particles of 5-micron size will be eliminated.

FIG. 78 Illustration of the principle of cyclone separators. (Borg-Warner)

Most city ordinances require that some form of backflow preventer be interposed between city water supply lines and connections to equipment where backflow or siphoning could contaminate drinking water supply. This is the case, for instance, with an independent sealing supply to stuffing boxes of sewage pumps. Quite a variety of backflow preventers are available. In most cases, the device consists of two spring-loaded check valves in series and a spring-loaded, diaphragm-actuated, differential-pressure relief valve located in the zone between the check valves (Fig. 79).

In normal operation, both check valves remain open as long as there is a demand for sealing water. The differential-pressure relief valve remains closed because of the pressure drop past the first check valve. If the pressure downstream of the device increases, tending to reverse the direction of flow, both check valves close and prevent backflow. If the second check valve is prevented from closing tightly, the leakage past it increases the pressure between the two check valves, the relief valve opens, and water is discharged to atmosphere. Thus the relief valve operates automatically to maintain the pressure between the two check valves lower than the supply pressure.

Some local ordinances prohibit any connection between the city water line and a sewage or process liquid line. In such cases an open tank under atmospheric pressure is installed into which city water can be admitted and from which a small pump can deliver the required quantity of sealing water. Such a water-sealing supply unit (Fig. 80) can be installed in a location from which it can serve a number of pumps.

The tank is equipped with a float valve to feed and regulate the water level so that contamination of the city water supply is prevented. A small close-coupled pump is mounted directly on the tank and maintains a constant pressure of clear water at the stuffing box seals of the battery of pumps it serves. A small recirculation line is provided from the close-coupled pump discharge back to the tank to prevent operation at shutoff. The discharge pressure of the small supply pump is set by the maximum sealing pressure required at any of the pumps served. Supply at the individual stuffing boxes is then regulated by setting small control valves in each individual line.

If clean, cool water is not available (as with some drainage, irrigation, or sewage pumps), grease or oil seals are often used. Most pumps for sewage service have a single stuffing box subject to discharge pressure and operate with a flooded suction. It is therefore not necessary to seal these pumps against air leakage, but forcing grease or oil into the sealing space at the packing helps to exclude grit. Figure 81 shows a typical weighted grease sealer.

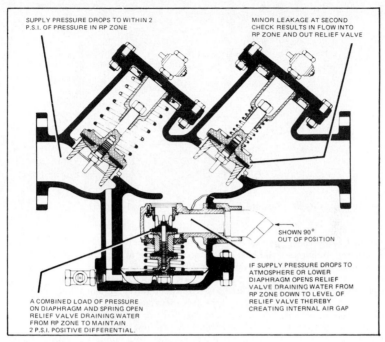

SUPPLY PRESSURE DROPS TO WITHIN 2 P.S.I. OF PRESSURE IN RP ZONE

MINOR LEAKAGE AT SECOND CHECK RESULTS IN FLOW INTO RP ZONE AND OUT RELIEF VALVE

SHOWN 90° OUT OF POSITION

IF SUPPLY PRESSURE DROPS TO ATMOSPHERE OR LOWER DIAPHRAGM OPENS RELIEF VALVE DRAINING WATER FROM RP ZONE DOWN TO LEVEL OF RELIEF VALVE THEREBY CREATING INTERNAL AIR GAP

A COMBINED LOAD OF PRESSURE ON DIAPHRAGM AND SPRING OPEN RELIEF VALVE DRAINING WATER FROM RP ZONE TO MAINTAIN 2 P.S.I. POSITIVE DIFFERENTIAL.

FIG. 79 Backflow preventer. (Hersey Products)

Automatic grease or oil sealers that exert pump discharge pressure in a cylinder on one side of a plunger, with light grease or oil on the other side, are available for sewage service. The oil or grease line is connected to the stuffing box seal, which is at about 80% of the discharge pressure. As a result, there is a slow flow of grease or oil into the pump when the unit is in operation. No flow takes place when the pump is out of service. Figure 82 shows an automatic grease sealer mounted on a vertical sewage pump.

Water-Cooled Stuffing Boxes High temperatures or pressures complicate the problem of maintaining stuffing box packing. Pumps in these more difficult services are usually provided

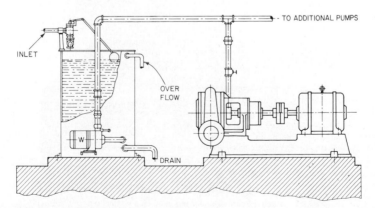

TO ADDITIONAL PUMPS

INLET

OVER FLOW

W

DRAIN

FIG. 80 Water seal unit. (Worthington Pump)

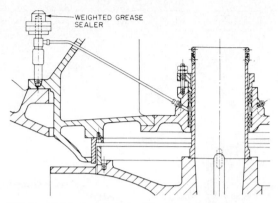

FIG. 81   Weighted grease sealer. (Worthington Pump)

with jacketed, water-cooled stuffing boxes. The cooling water removes heat from the liquid leaking through the stuffing box and heat generated by friction in the box, thus improving packing service conditions. In some special cases, oil or gasoline may be used in the cooling jackets instead of water. Two water-cooled stuffing box designs are commonly used. The first (Fig. 83) provides cored passes in the casing casting. These passages, which surround the stuffing box, are arranged with in-and-out connections. The second type uses a separate cooling chamber combined with the stuffing box proper, with the whole assembly inserted into and bolted to the pump casing (Fig. 84). The choice between the two is based on manufacturing preferences.

Stuffing box pressure and temperature limitations vary with pump type because it is generally not economical to use expensive stuffing box construction for infrequent high-temperature or high-pressure applications. Therefore, whenever the manufacturer's stuffing box limitations for a given pump are exceeded, the only solution is the application of pressure-reducing devices ahead of the stuffing box.

FIG. 82   Automatic grease sealer mounted on a vertical pump. (Zimmer & Francescon)

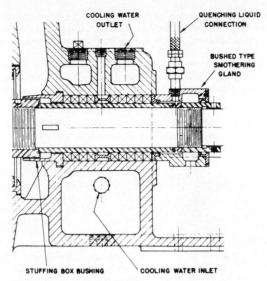

FIG. 83 Water-cooled stuffing box with cored water passage cast in casing.

Pressure-Reducing Devices   Essentially, pressure-reducing devices consist of a bushing or meshing labyrinth ending in a relief chamber located between the pump interior and the stuffing box. The relief chamber is connected to some suitable low-pressure point in the installation, and the leakage past the pressure-reducing device is returned to this point. If the pumped liquid must be salvaged, as with treated feedwater, it is returned to the pumping cycle. If the liquid is expendable, the relief chamber can be connected to a drain.

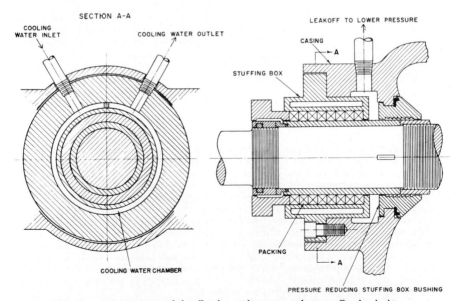

FIG. 84   Separate water-cooled stuffing box with pressure-reducing stuffing box bushing.

There are many different pressure-reducing device designs. Figure 84 illustrates a design for limited pressures. A short serrated stuffing box bushing is inserted at the bottom of the stuffing box, followed by a relief chamber. The leakage past the serrated bushing is bled off to a low-pressure point.

With relatively high-pressure units, intermeshing labyrinths may be located following the balancing device and ahead of the stuffing box. Piping from the chamber following pressure-reducing devices should be amply sized so that as wear increases leakage, piping friction will not increase stuffing box pressure.

Stuffing Box Glands   Stuffing box glands may assume several forms, but basically they can be classified into two groups: solid glands and split glands (Fig. 85). Split glands are made in halves so that they may be removed from the shaft without dismantling the pump, thus providing more working space when the stuffing boxes are being repacked. Split glands are desirable for pumps that have to be repacked frequently, especially if the space between the box and the bearing is restricted. The two halves are generally held together by bolts, although other methods are also used. Split glands are generally a construction refinement rather than a necessity, and they are rarely used in smaller pumps. They are commonly furnished for large single-stage pumps, for some multistage pumps, and for refinery pumps.

FIG. 85   Split stuffing box gland.

Another common refinement is the use of swing bolts in stuffing box glands. Such bolts may be swung to the side, out of the way, when the stuffing box is being repacked.

Stuffing box leakage into the atmosphere might, in some services, seriously inconvenience or even endanger the operating personnel—for example, when such liquids as hydrocarbons are being pumped at vaporizing temperatures or temperatures above their flash point. As this leakage cannot always be cooled sufficiently by a water-cooled stuffing box, smothering glands are used (Fig. 83). Provision is made in the gland to introduce a liquid—either water or another hydrocarbon at low temperature—that mixes intimately with the leakage, lowering its temperature or, if the liquid is volatile, absorbing it.

Stuffing box glands are usually made of bronze, although cast iron or steel may be used for all iron-fitted pumps. Iron or steel glands are generally bushed with a nonsparking material like bronze in refinery service to prevent the ignition of flammable vapors by the glands sparking against a ferrous metal shaft or sleeve.

Stuffing Box Packing   See Sec. 2.2.2.

MECHANICAL SEALS

The conventional stuffing box design and composition packing are impractical to use for sealing a rotating shaft for many conditions of service. In the ordinary stuffing box, the sealing between the moving shaft or shaft sleeve and the stationary portion of the box is accomplished by means of rings of packing forced between the two surfaces and held tightly in place by a stuffing box gland. The leakage around the shaft is controlled merely by tightening up or loosening the gland studs. The actual sealing surfaces consist of the axial rotating surfaces of the shaft or shaft sleeve and the stationary packing. Attempts to reduce or eliminate all leakage from a conventional stuffing box increase the gland pressure. The packing, being semiplastic, forms more closely to the shaft and tends to cut down the leakage. After a certain point, however, the leakage continues no matter how tightly the gland studs are brought up. The frictional horsepower increases rapidly at this point, the heat generated cannot be properly dissipated, and the stuffing box fails to function. Even before this condition is reached, the shaft sleeves may be severely worn and scored, so that it becomes impossible to pack the stuffing box satisfactorily.

These undesirable characteristics prohibit the use of packing as the sealing medium between rotating surfaces if the leakage is to be held to an absolute minimum under severe pressure. The condition, in turn, automatically eliminates use of the axial surfaces as the sealing surfaces, for a semiplastic packing is the only material that can always be made to form about the shaft and compensate for the wear. Another factor that makes stuffing boxes unsatisfactory for certain applications is the relatively small lubricating value of many liquids frequently handled by centrifugal pumps, such as propane or butane. These liquids actually dissolve the lubricants normally used to impregnate the packing. Seal oil must therefore be introduced into the lantern gland or a packed box to lubricate the packing and give it reasonable life. With these facts in mind, designers have attempted to produce an entirely different type of seal with wearing surfaces other than the axial surfaces of the shaft and packing.

This form of seal, called the mechanical seal, is a later development than regular stuffing boxes but has found general acceptance in those pumping applications in which the shortcomings of packed stuffing boxes have proved excessive. Fields in which the packed boxes gave good service, however, have shown little tendency to replace them with mechanical seals.

Principles and Construction of Mechanical Seals See Sec. 2.2.3.

Limitations on Seal Applications The substitution of mechanical seals for packed stuffing boxes is not always an unmixed blessing, and for some services—for example, where conditions cause the pumped liquid to form crystals after temperature changes or on settling—the mechanical seal is not as desirable as packing. If mechanical seals are used for such services, it is very important to provide adequate flushing. Another condition unfavorable to mechanical seals is a pump service with long idle periods, when the pump may even be drained. The flexible materials used in the seal may harden, or slight rusting may occur, thereby possibly causing the seal to stick and become damaged on restarting. Finally, mechanical seals are still subject to failure on occasion, and their failure may be more rapid than that of conventional packing. If packing fails, the pump can usually be kept running by temporary adjustments until it is convenient to shut it down. If a mechanical seal fails, the pump must be shut down at once in nearly every case.

As both packed boxes and mechanical seals are subject to wear, neither are perfect. Whether one or the other should be used depends on the specific application. In some cases, both give good service and the choice between them becomes a matter of personal preference or first cost. Table 2 adds other comments and summarizes the advantages and disadvantages of packing and mechanical seals.

Injection-Type Shaft Seals See Sec. 2.2.4.

BEARINGS

The function of bearings in centrifugal pumps is to keep the shaft or rotor in correct alignment with the stationary parts under the action of radial and transverse loads. Bearings that give radial positioning to the rotor are known as *line bearings*, and those that locate the rotor axially are called *thrust bearings*. In most applications the thrust bearings actually serve both as thrust and radial bearings.

Types of Bearings Used All types of bearings have been used in centrifugal pumps. Even the same basic design of pump is often made with two or more different bearings, required either by varying service conditions or by the preference of the purchaser. In most pumps, however, either antifriction or oil film (sleeve-type) bearings are used today.

In horizontal pumps with bearings on each end, the bearings are usually designated by their location as *inboard* and *outboard*. Inboard bearings are located between the casing and the coupling. Pumps with overhung impellers have both bearings on the same side of the casing so that the bearing nearest the impeller is called inboard and the one farthest away outboard. In a pump provided with bearings at both ends, the thrust bearing is usually placed at the outboard end and the line bearing at the inboard end.

The bearings are mounted in a housing that is usually supported by brackets attached or integral to the pump casing. The housing also serves the function of containing the lubricant necessary for proper operation of the bearing. Occasionally the bearings of very large pumps are supported in housings that form the top of pedestals mounted on soleplates or on the pump bedplate. These are called *pedestal bearings*.

TABLE 2 Comparison of Packing and Mechanical Seals

Advantages	Disadvantages
Packing	
1. Low initial cost 2. Easily installed as rings and glands are split 3. Good reliability at medium pressures and shaft speeds 4. Can handle large axial movement (e.g., thermal expansion of stuffing box versus shaft) 5. Can be used in rotating or reciprocating applications 6. Leakage increases gradually, giving adequate warning of impending breakdown	1. Relatively high leakage 2. Requires regular maintenance 3. Wear of shaft sleeve can be relatively high 4. Power losses may be high
Mechanical seals	
1. Very low leakage 2. Require no maintenance 3. Eliminate sleeve wear 4. Very good reliability 5. Can handle higher pressures and speeds 6. Easily applied to carcinogenic, toxic, or radioactive liquids	1. High initial cost 2. Easily installed but require some disassembly of pump (couplings, etc.)

SOURCE: Crane Packing Co.

Because of the heat generated by the bearing or the heat in the liquid being pumped, some means other than radiation to the surrounding air must occasionally be used to keep the bearing temperature within proper limits. If the bearings have a forced-feed lubrication system, cooling is usually accomplished by circulating the oil through a separate water-to-oil cooler. Otherwise a jacket through which a cooling liquid is circulated is usually incorporated as part of the housing.

Pump bearings may be rigid or self-aligning. A self-aligning bearing will automatically adjust itself to a change in the angular position of the shaft. In babbitted or sleeve bearings, the name *self-aligning* is applied to bearings that have a spherical fit of the sleeve in the housing. In antifriction bearings, the name is applied to bearings the outer race of which is spherically ground or the housing of which provides a spherical fit.

Although double-suction pumps are theoretically in hydraulic balance, this balance is rarely realized in practice, and so even these pumps are provided with thrust bearings. A centrifugal pump, being a product of the foundry, is subject to minor irregularities that may cause differences in the eddy currents set up on the two sides of the impeller. As this disturbance can create an axial hydraulic thrust, some form of thrust bearing that is capable of taking thrust in either direction is necessary to maintain the rotor in its proper position.

The thrust capacity of the bearing of a double-suction pump is usually far in excess of the probable imbalance caused by irregularities. This provision is made because (1) unequal wear of the rings and other parts may cause imbalance and (2) the flow of the liquid into the two suction eyes may be unequal and cause imbalance because of an improper suction-piping arrangement.

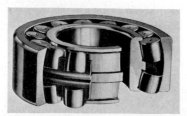

FIG. 86 Self-aligning spherical roller bearing. (SKF)

Antifriction Bearings   The most common antifriction bearings used on centrifugal pumps are the various types of ball bearings. Roller bearings are used less often, although the spherical roller bearing (Fig. 86) is used frequently for large shaft sizes, for which there is a limited choice of ball bearings. As most roller bearings are suitable only for radial loads, their use on centrifugal pumps tends to be limited to applications in which they are not required to carry a combined radial and thrust load.

Ball Bearings   As the coefficient of rolling friction is less than that of sliding friction, one must not consider a ball bearing in the same light as a sleeve bearing. In the former, the load is carried on a point contact of the ball with the race, but the point of contact does not rub or slide over the race and no appreciable heat is generated. Furthermore, the point of contact is constantly changing as the ball rolls in the race, and the operation is practically frictionless. In the sleeve bearing, there is a constant rubbing of one surface over another, and the friction must be reduced by the use of a lubricant.

Ball bearings operated at an absolutely constant speed theoretically require no lubricant. However, no speed can be called absolutely constant, for the conditions affecting the speed always vary slightly. For instance, a motor with a full-load speed rated at 3510 rpm might vary in speed in the course of a minute from 3505 to 3515 rpm. Each variation in speed causes the balls in a ball bearing to lag or lead the race because of their inertia. Consequently, a very slight, almost immeasurable sliding action takes place. Another limiting condition is that the hardest of metals suffers minute deformation on carrying load, thus upsetting perfect point contact and adding another slight sliding action. For these reasons ball bearings must be given some lubrication.

Ball thrust bearings are built to carry heavy loads by pure rolling motion on an angular contact. As thrust load is axial, it is equally distributed to all the balls around the race, and the individual load on each ball is only a very small fraction of the total thrust load. In such bearings it is essential that the balls be equally spaced, and for this purpose a retaining cage is used between the balls and between the inner and outer races. This cage carries no load, but the contact between it and the ball produces sliding friction that generates a small amount of heat. It is for this reason that ball thrust bearings are generally water-jacketed.

Types and Applications   Pump designers have a wide variety of antifriction bearings to choose from. Each type has characteristics that could make it a good or bad selection for a specific application. Although several types might sometimes be acceptable, it is best for purchasers to leave the choice to the manufacturer. For example, some purchasers specify double-row bearings whatever the size or type of pump, even though single-row bearings are often equally suitable or better.

The most common ball bearings used on centrifugal pumps are (1) single-row deep-groove, (2) double-row deep-groove, (3) double-row self-aligning, and (4) angular-contact, either single- or double-row. All except the double-row self-aligning bearings are capable of carrying thrust loads as well as radial loads.

Sealed ball bearings, adapter ball bearings, and other modifications have also found special applications. Sealed prelubricated bearings require special attention if the unit in which they are installed is not operated for a long period of time (for instance, one kept in stock or storage). The shaft should be turned over occasionally, say once every three months, to agitate the lubricant and maintain a film coating of the balls of such units.

The self-aligning ball bearing (Fig. 87) is the most serviceable bearing for heavy loads, high speeds, long bearing spans, and no end thrust. For this reason it is ideally adapted for service as a line bearing on a centrifugal pump. Its double row of balls run in fixed grooves in the inner, or shaft, race; its outer race is ground to a spherical seat. Any slight vibration or shaft deflection is therefore taken care of by this bearing, which operates as a pivot. In lightly constructed pumps, a self-aligning bearing will also compensate for the slight misalignment caused by the "breathing" that takes place in the casing when pressure is developed.

The self-aligning ball bearing has proved very satisfactory for high speeds and has long life, even with long bearing spans. It has very

FIG. 87 Self-aligning double-row ball bearing. (SKF)

FIG. 88   Single-row deep-groove ball bearing. (SKF)   **FIG. 89**   Double-row deep-groove ball bearing. (SKF)

little thrust capacity, however, and is not used for combined radial and thrust loads in centrifugal pumps. For large shafts, the self-aligning spherical roller bearing (Fig. 86) is used instead, for it can carry such loads with a considerable thrust component.

The single-row deep-groove ball bearing (Fig. 88) is the most commonly used bearing on centrifugal pumps, except for the larger sizes. It is good for both radial thrust and combined loads but requires careful alignment between the shaft and the housing in which the bearing is mounted. It is sometimes used with seals built into the bearing in order to exclude dirt, retain lubricant, or both.

The double-row deep-groove ball bearing—in effect, two single-row bearings placed side by side—has greater capacity for both radial and thrust loads (Fig. 89). It is used quite commonly if the loading is more than that permitted by the single-row bearing.

The angular-contact ball bearing operates on a principle that makes it good for heavy thrust loads. The single-row type (Fig. 90) is good for thrust in only one direction, whereas the double-row type (Fig. 91), which is basically two single-row bearings placed back to back, can carry thrust in either direction. Two single-row angular-contact bearings are frequently matched and the faces of the races ground by the manufacturer so they can be used in tandem for large one-directional thrust loads or back to back (Fig. 92) for two-directional thrust loads. The two bearings are sometimes locked together by recessing the inner races and pressing them onto a short ring (Fig. 93). If two separate angular-contact bearings are used, care must be taken to mount them correctly on the shaft.

The single-row angular-contact ball bearing can be used singly on centrifugal pumps only if the thrust is always in one direction. The field of application of this type of bearing is thus limited primarily to vertical pumps. Another very interesting application is the use of two such bearings in an end-suction pump to take care of axial thrust in either direction. This arrangement permits a certain amount of axial adjustment of the impeller in its volute, accomplished by loosening up one bearing nut and tightening the other. Unfortunately, such adjustment is extremely delicate and requires a first-class mechanic; its commercial practicability is therefore somewhat limited.

The double-row angular-contact bearing—or its equivalent: a matched pair mounted back to back—has been found very satisfactory for pumps capable of a high thrust load in either direction. Some pump manufacturers standardize on this bearing for many applications.

FIG. 90   Single-row angular-contact ball bearing. (New Departure)   **FIG. 91**   Double-row angular-contact ball bearing. (New Departure)

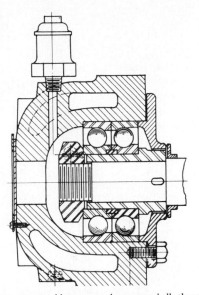

FIG. 92 Two single-row angular-contact bearings mounted back to back to act as a double-row bearing. (New Departure)

FIG. 93 Double-row angular-contact ball thrust bearing that is grease-lubricated and water-cooled.

Lubrication of Antifriction Bearings In the layout of a line of centrifugal pumps, the choice of the lubricant for the pump bearings is dictated by application requirements, by cost considerations, and sometimes by the preferences of a group of purchasers committed to the major portion of the output of that line. For example, in vertical wet-pit condenser circulating pumps, water is the lubricant of choice, in preference to grease or oil. If oil or grease was used in such pumps and the lubricant leaked into the pumping system, the condenser operation might be seriously affected because the tubes would become coated with the lubricant.

Most centrifugal pumps for refinery service are supplied with oil-lubricated bearings because of the insistence of refinery engineers on this feature. In the marine field, on the other hand, the preference lies with grease-lubricated bearings. For very high pump operating speeds (5000 rpm and above), oil lubrication is found to be the most satisfactory.

For highly competitive lines of small pumps, the main consideration is cost and so the most economical lubricant is chosen, depending upon the type of bearing used.

Ball bearings used in centrifugal pumps are usually grease-lubricated, although some services use oil lubrication. In grease-lubricated bearings, the grease packed into the bearing is thrown out by the rotation of the balls, creating a slight suction at the inner race. (Even if the grade of grease is relatively light, it is still a semisolid and flows slowly. As heat is generated in the bearing, however, the flow of the grease is accelerated until the grease is thrown out at the outer race by the rotation.) As the expelled grease is cooled by contact with the housing and thus attracted to the inner race, there is a continuous circulation of grease to lubricate and cool the bearing. This method of lubrication requires a minimum amount of attention and has proved itself very satisfactory.

As housings of bearings in vertical pumps require seals to prevent the escape of the lubricant, grease is usually preferred, for it lessens the chance of leakage (Fig. 94).

A bearing fully packed with grease prevents proper grease circulation in itself and its housing. As a rough rule, therefore, it is recommended that only one-third of the void spaces in the housing be filled. An excess amount of grease will cause the bearing to heat up, and grease will flow out of the seals to relieve the situation. Unless the excess grease can escape through the seal or through the relief cock that is used on many large units, the bearing will probably fail early.

In oil-lubricated ball bearings, a suitable oil level must be maintained in the housing. This

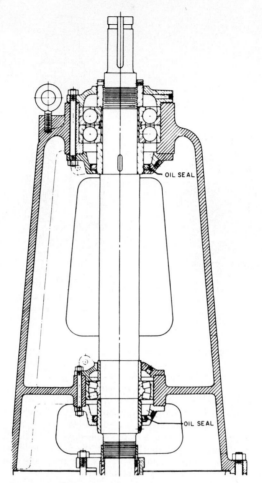

FIG. 94   Ball bearing with seal in vertical pump. The seal guards against escape of grease.

level should be at about the center of the lowermost ball of a stationary bearing. It may be achieved by a dam and an oil slinger to maintain the level behind the dam and thereby increase the leeway in the amount of oil the operator must keep in the housing. Oil rings are sometimes used to supply oil to the bearings from the bearing housing reservoir (Fig. 95). In other designs, a constant-level oiler is used (Fig. 96).

Because of the advantages of interchangeability, some pump lines are built with bearing housings that can be adapted to either oil or grease lubrication with a minimum of modification (Fig. 97).

Oil Film or Sleeve Bearings   See Subsec. 2.2.5.

COUPLINGS

Centrifugal pumps are connected to their drivers through couplings of one sort or another, except for close-coupled units, in which the impeller is mounted on an extension of the shaft of the driver.

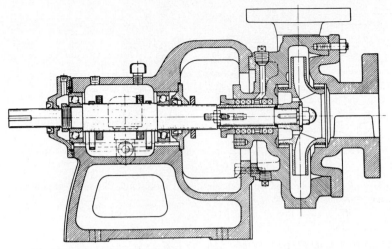

FIG. 95 Ball bearing pump with oil rings.

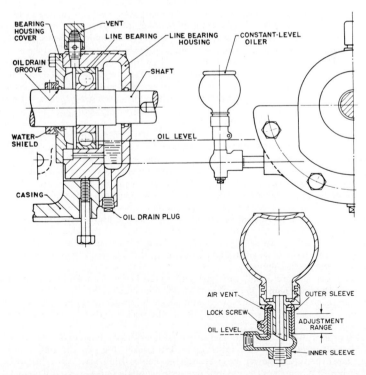

FIG. 96 Constant-level oiler.

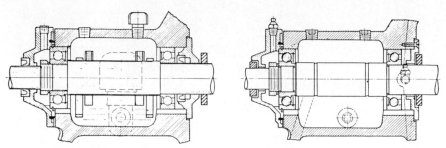

FIG. 97   Ball bearings arranged (left) with oil rings in the housing and (right) for grease lubrication.

Because couplings may be used with both centrifugal and positive displacement pumps, they are discussed separately in Sec. 6.3.

BEDPLATES AND OTHER PUMP SUPPORTS

For very obvious reasons, it is desirable that pumps and their drivers be removable from their mountings. Consequently, they are usually bolted and doweled to machined surfaces that in turn are firmly connected to a foundation. To simplify the installation of horizontal-shaft units, these machined surfaces are usually part of a common bedplate on which either the pump or the pump and its driver have been prealigned.

Bedplates   The primary function of a pump bedplate is to furnish mounting surfaces for the pump feet that can be rigidly attached to the foundation. Mounting surfaces are also necessary for the feet of the pump driver or drivers and for the feet of any independently mounted power transmission device. Although such surfaces could be provided by separate bedplates or by individually planned surfaces, it would be necessary to align these separate surfaces and fasten them to the foundation with the utmost care. Usually this method requires in-place mounting in the field as well as drilling and tapping for the holding-down bolts after all parts have been aligned. To minimize such field work, coupled horizontal-shaft pumps are usually purchased with a continuous base extending under the pump and its driver; ordinarily, both these units are mounted and aligned at the place of manufacture.

Although such bases are designed to be quite rigid, they deflect if improperly supported. It is therefore necessary to support them on a foundation that can supply the required rigidity. Furthermore, as the base can be sprung out of shape by improper handling during transit, it is imperative that the alignment be carefully rechecked during erection and prior to starting the unit.

As the unit size increases, so do the size, weight, and cost of the base required. The cost of a prealigned base for most large units exceeds the cost of the field work necessary to align individual bedplates or soleplates and to mount the component parts. Such bases are therefore used only if appearances require them or if their function as a drip collector justifies the additional cost. Even in fairly small units, the height at which the feet of the pump and the other elements are located may differ considerably. A more rigid and nice-looking installation can frequently be obtained by using individual bases or soleplates and building up the foundation to various heights under the separate portions of equipment.

Cast iron baseplates are usually provided with a raised edge or raised lip around the base to prevent dripping or draining onto the floor (Fig. 98). The base itself is sloped toward one end to collect the drainage for further disposal. A drain pocket is provided near the bottom of the slope, usually with a mesh screen. A tapped connection in the pocket permits piping the drainage to a convenient point.

Bedplates fabricated of steel plate and structural steel shapes, now used very extensively, do not easily permit incorporation of a raised lip or drip pocket (Figs. 99 and 100). From a utility viewpoint, however, the customary use of bearing brackets as drip pockets (to collect leakage from the stuffing boxes) now makes the use of a raised lip bedplate unnecessary in horizontal pumps

FIG. 98 Horizontal-shaft centrifugal pump and driver on cast iron bedplate. (Worthington Pump)

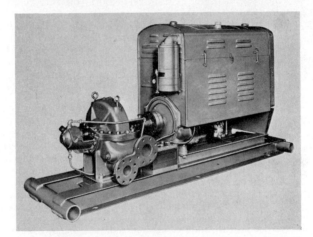

FIG. 99 Pump and internal combustion engine mounted on portable steel skid base. (Worthington Pump)

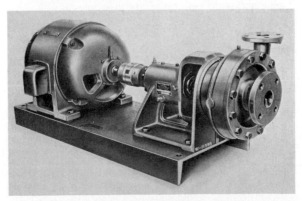

FIG. 100 Small centrifugal pump on structural steel bedplate made of a simple channel. (Worthington Pump)

except in applications where the pump handles a cold liquid in a moist atmosphere and thus must contend with a considerable condensation on its surface.

Soleplates Soleplates are cast iron or steel pads located under the feet of the pump or its driver and embedded in the foundation. The pump or its driver is doweled and bolted to them. Soleplates are customarily used for vertical dry-pit pumps and also for some of the larger horizontal units to save the cost of the large bedplates otherwise required.

Centerline Support For operation at high temperatures, the pump casing must be supported as near to its horizontal centerline as possible in order to prevent excessive strains caused by temperature differences. Such strain might seriously disturb the alignment of the unit and eventually damage it. Centerline construction is usually employed in boiler-feed, refinery, or hot-water circulating pumps (Fig. 101). The exact temperature at which centerline support construction becomes mandatory varies from 250 to 350°F (121 to 177°C).

FIG. 101 Single-stage pump with centerline support. (Worthington Pump)

Horizontal Units Using Flexible Pipe Connections The foregoing discussion of bedplates and supports for horizontal-shaft units assumed their application to pumps with piping setups that do not impose hydraulic thrusts on the pumps. If flexible pipe connections or expansion joints are desirable in the suction or discharge piping of a pump (or in both), the pump manufacturer should be so advised for several reasons. First, the pump casing will be required to withstand various stresses caused by the resultant hydraulic thrust load. Although this is rarely a limiting or dangerous factor, it is best that the manufacturer have the opportunity to check the strength of the pump casing. Second, the resulting hydraulic thrust has to be transmitted from the pump casing through the casing feet to the bedplate or soleplate and then to the foundation. Usually horizontal-shaft pumps are merely bolted to their bases or soleplates, and so any tendency to displacement is resisted only by the frictional grip of the casing feet on the base and by relatively small dowels. If flexible pipe joints are used, this attachment may not be sufficent to withstand the hydraulic thrust. If high hydraulic thrust loads are to be encountered, therefore, the pump feet must be keyed to the base or supports. Similarly, the bedplate or supporting soleplates must be of a design that will permit transmission of the load to the foundation.

VERTICAL PUMPS

Vertical-shaft pumps fall into two classifications, dry-pit and wet-pit. Dry-pit pumps are surrounded by air, and the wet-pit types are either fully or partially submerged in the liquid handled.

Vertical Dry-Pit Pumps Dry-pit pumps with external bearings include most small, medium, and large vertical sewage pumps; most medium and large drainage and irrigation pumps for medium and high head; many large condenser circulating and water supply pumps; and many marine pumps. Sometimes the vertical design is preferred (especially for marine pumps) because it saves floor space. At other times it is desirable to mount a pump at a low elevation because of suction conditions, and it is then also preferable or necessary to have the pump driver at a high elevation. The vertical pump is normally used for very large capacity applications because it is more economical than the horizontal type, all factors considered.

Many vertical dry-pit pumps are basically horizontal designs with minor modifications (usually in the bearings) to adapt them for vertical-shaft drive. This is not true of small- and medium-sized sewage pumps, however; in these units a purely vertical design is the most popular. Most of these sewage pumps have elbow suction nozzles (Figs. 102–104) because their suction supply is usually taken from a wet well adjacent to the pit in which the pump is installed. The suction elbow usually contains a handhole with a removable cover to provide easy access to the impeller.

To dismantle one of these pumps, the stuffing box head must be unbolted from the casing after the intermediate shaft or the motor and motor stand have been removed. The rotor assembly is drawn out upward, complete with the stuffing box head, the bearing housing, and the like. This rotor assembly can then be completely dismantled at a convenient location.

Vertical-shaft installations of single-suction pumps with a suction elbow are commonly furnished with either a pedestal or a base elbow (Fig. 102), both of which can be bolted to soleplates or even grouted in. The grouting arrangement is not desirable unless there is full assurance that the pedestal or elbow will never be disturbed or that the grouted space is reasonably regular and the grout will separate from the pump without excessive difficulty.

Vertical single-suction pumps with bottom suction are commonly used for larger sewage, water supply, or condenser circulating applications. Such pumps are provided with wing feet that are bolted to soleplates grouted in concrete pedestals or piers (Fig. 105). Sometimes the wing feet may be grouted right in the pedestals. These must be suitably arranged to provide proper access to any handholes in the pump and to allow clearance for the elbow suction nozzles if these are used.

FIG. 102 Small vertical sewage pump with intermediate shafting. (Worthington Pump)

If a vertical pump is applied to condensate service or some other service for which the eye of the impeller must be vented to prevent vapor binding, a pump with a bottom single-inlet impeller is not desirable because it does not permit effective venting. Neither does a vertical pump employing a double-suction impeller (Fig. 106). The most suitable design for such applications incorporates a top single-inlet impeller (Fig. 107).

If the driver of a vertical dry-pit pump can be located immediately above the pump, it is often supported on the pump itself (Fig. 104). The shafts of the pump and driver may be connected by a flexible coupling, which requires that each have its own thrust bearing. If the pump shaft is rigidly coupled to the driver shaft or is an extension of the driver shaft, a common thrust bearing is used, normally in the driver.

Although the driving motors are frequently mounted right on top of the pump casing, one important reason for the use of the vertical-shaft design is the possibility of locating the motors at an elevation sufficiently above the pumps to prevent accidental flooding of the motors. The pump and its driver may be separated by an appreciable length of shafting, which may require steady bearings between the two units. Subsection 6.3.1 discusses the construction and arrangement of the shafting used to connect vertical pumps to drivers located some distance above the pump elevation.

Bearings for vertical dry-pit pumps and for intermediate guide bearings are usually antifriction grease-lubricated types to simplify the problem of retaining a lubricant in a housing with a shaft projecting vertically through it. Larger units, for which antifriction bearings are not available or desirable, use self-oiling babbitt steady bearings with spiral oil grooves (Figs. 108 and 109). Figure 109 illustrates a vertical dry-pit pump design with a single-sleeve line bearing. The pump is connected by a rigid coupling to its motor (not shown), which is provided with a line and a thrust bearing.

FIG. 104 Vertical sewage pump with direct-mounted motor. (Worthington Pump)

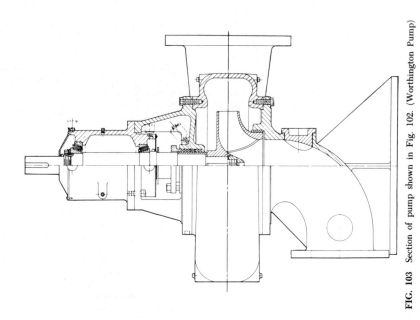

FIG. 103 Section of pump shown in Fig. 102. (Worthington Pump)

2.92

FIG. 105   Vertical bottom-suction volute pumps with piloted couplings. (Worthington Pump)

Vertical dry-pit centrifugal pumps are structurally similar to horizontal-shaft pumps. It is to be noted, however, that many of the very large vertical single-stage single-suction (usually bottom) volute pumps that are preferred for large storm water pumpage, drainage, irrigation, sewage, and water supply projects have no comparable counterpart among horizontal-shaft units. The basic U-section casing of these pumps, which is struc-turally weak, often requires the use of heavy ribbing to provide sufficient rigidity. In com-parable water turbine practice, a set of vanes (called a *speed ring*) is employed between the casing and runner to act as a strut. Although the speed ring does not adversely affect the operation of a water turbine, it would function basically as a diffuser in a pump because of the inherent hydraulic limitations of that construc-tion. Some high-head pumps of this type have been made in the twin-volute design. The wall separating the two volutes acts as a stengthen-ing rib for the casing, thus making it easier to design a casting strong enough for the pressure involved.

FIG. 106   Vertical double-suction volute pump with direct-mounted motor. (Worthington Pump)

BASES AND SUPPORTS FOR VERTICAL PUMPING EQUIP-MENT   Vertical-shaft pumps, like horizontal-shaft units, must be firmly supported. Depend-ing upon the installation, the unit may be sup-ported at one or several elevations. Vertical units are seldom supported from walls, but even that type of support is sometimes encountered.

Occasionally a nominal horizontal-shaft pump design is arranged with a vertical shaft and a wall used as the supporting foundation. Regular horizontal-shaft units can be used for this purpose without modification, except that the bedplate is attached to a wall. Careful

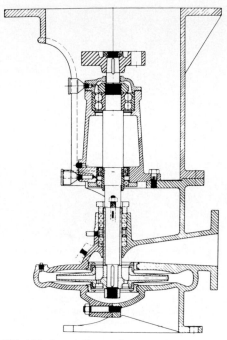

FIG. 107 Section of vertical pump with top-suction inlet impeller. (Worthington Pump)

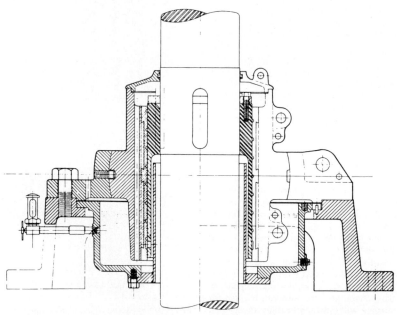

FIG. 108 Self-oiling babbitt steady bearing for large vertical shafting. (Worthington Pump)

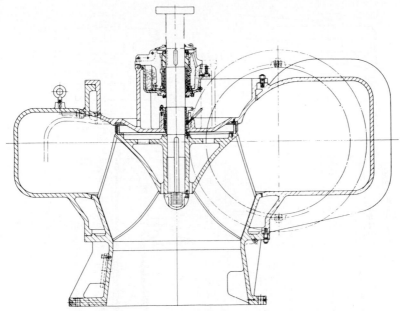

FIG. 109 Section of large vertical bottom-suction volute pump with single-sleeve bearing. (Worthington Pump)

attention must be given to the arrangement of the pump bearings to prevent escape of lubricant.

Installations of double-suction single-stage pumps with the shaft in the vertical position are relatively rare, except in some marine or navy applications. Hence manufacturers have very few standard pumps of this kind arranged so that a portion of the casing forms the support (to be mounted on soleplates). Figure 106 shows such a pump, which also has a casing extension to support the driving motor.

Vertical Wet-Pit Pumps Vertical pumps intended for submerged operation are manufactured in a great number of designs, depending mainly upon the service for which they are intended. Small pumps of this type are often referred to as sump pumps. Wet-pit centrifugal pumps can be classified in the following manner:

1. Vertical turbine pumps
2. Propeller or modified propeller pumps
3. Volute pumps

VERTICAL TURBINE PUMPS Vertical turbine pumps were originally developed for pumping water from wells and have been called *deep-well pumps, turbine well pumps,* and *borehole pumps.* As their application to other fields has increased, the name *vertical turbine pumps* has been generally adopted by manufacturers. (This is not too specific a designation because the term *turbine pump* has been applied in the past to any pump employing a diffuser. There is now a tendency to designate pumps using diffusion vanes as *diffuser pumps* to distinguish them from *volute pumps.* As that designation becomes more universal, applying the term *vertical turbine pumps* to the construction formerly called *turbine well pumps* will become more specific.)

The largest fields of application for the vertical turbine pump are pumping from wells for irrigation and other agricultural purposes, for municipal water supply, and for industrial water supplies, processing, circulating, refrigerating, and air conditioning. This type of pump has also been used for brine pumping, mine dewatering, oil field repressuring, and other purposes.

These pumps have been made for capacities as low as 10 or 15 gpm (2 or 3 m³/h) and as high

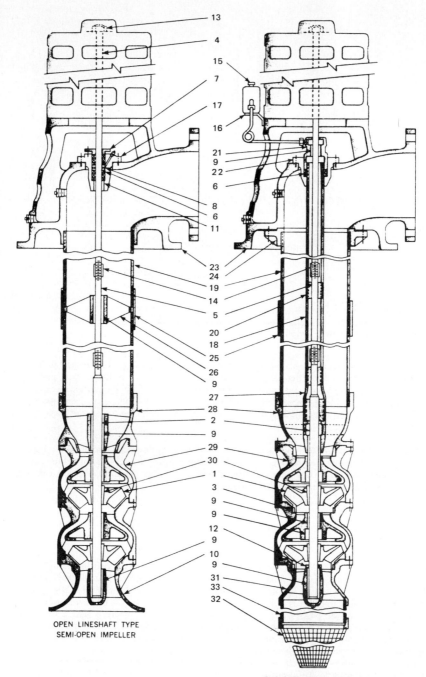

13
4
15
7
17
16
21
9
22
6
8
6
11
23
24
19
14
5
20
18
25
26
9
27
28
2
9
29
30
1
3
9
9
12
9
10
9
31
33
32

OPEN LINESHAFT TYPE
SEMI-OPEN IMPELLER

ENCLOSED LINESHAFT TYPE
ENCLOSED IMPELLER

2.96

as 25,000 gpm (5700 m³/h) or more and for heads up to 1000 ft (300 m). Most applications naturally involve the smaller capacities. The capacity of the pumps used for bored wells is naturally limited by the size of the well as well as by the rate at which water can be drawn without lowering its level to a point of insufficient pump submergence.

Vertical turbine pumps should be designed with a shaft that can be readily raised or lowered from the top to permit proper positioning of the impeller in the bowl. An adequate thrust bearing is also necessary to support the vertical shafting, the impeller, and the hydraulic thrust developed when the pump is in service. As the driving mechanism must also have a thrust bearing to support its vertical shaft, it is usually provided with one large enough to carry the pump parts as well. For these two reasons, the hollow-shaft motor or gear is most commonly used for vertical turbine pump drive. In addition, these pumps are sometimes made with their own thrust bearings to allow for belt drive or for drive through a flexible coupling by a solid-shaft motor, gear, or turbine. Dual-driven pumps usually employ an angle gear with a vertical motor mounted on its top.

The design of vertical pumps illustrates how a centrifugal pump can be specialized to meet a specific application. Figure 110 illustrates a turbine design with closed impellers and enclosed-line shafting and another turbine design with closed impellers and open-line shafting.

The bowl assembly, or section, consists of the suction case (also called suction head or inlet vane), the impeller or impellers, the discharge bowl, the intermediate bowl or bowls (if more than one stage is involved), the discharge case, the various bearings, the shaft, and miscellaneous parts, such as keys, impeller-locking devices, and the like. The column pipe assembly consists of the column pipe, the shafting above the bowl assembly, the shaft bearings, and the cover pipe or bearing retainers. The pump is suspended from the driving head, which consists of the discharge elbow (for aboveground discharge), the motor or driver and support, and either the stuffing box (in open-shaft construction) or the assembly for providing tension on the cover pipe and introducing lubricant into it. Belowground discharge is taken from a tee in the column pipe, and the driving head functions principally as a stand for the driver and support for the column pipe.

Liquid in a vertical turbine pump is guided into the impeller by the suction case or head. This may be a tapered section for attachment of a conical strainer or suction pipe, or it may be a bell mouth (Fig. 110).

Semiopen and enclosed impellers are both commonly used. For proper clearances in the various stages, the semiopen impeller requires more care in assembly on the impeller shaft and more accurate field adjustment of the vertical shaft position in order to obtain the best efficiency. Enclosed impellers are favored over semiopen ones, moreover, because wear on the latter reduces capacity, which cannot be restored unless new impellers are installed. Normal wear on enclosed impellers does not affect impeller vanes, and worn clearances may be restored by replacing wearing rings. The thrust produced by semiopen impellers may be as much as 150% of that by enclosed impellers.

Occasionally in power plants the maximum water level that can be carried in the condenser hot well will not give adequate NPSH for a conventional horizontal condensate pump mounted on the basement floor, especially if the unit has been installed in a space originally allotted for a smaller pump. Building a pit for a conventional horizontal condensate pump or a vertical dry-pit pump that will provide sufficient submergence involves considerable expense. Pumps of the design shown in Fig. 111 have become quite popular in such applications. This is basically a vertical turbine pump mounted in a tank (often called a can) that is sunk into the floor. The length of the pump has to be such that sufficient NPSH will be available for the first-stage impeller design, and the diameter and length of the tank have to allow for proper flow through the space between the pump and tank and then for a turn and flow into the bell mouth. Installing this design in an existing plant is naturally much less expensive than making a pit because the size of the hole necessary to install the tank is much smaller. The same basic design has also been applied to pumps

FIG. 110 (at left) Sections of vertical turbine pump: (1) impeller, (2) pump shaft, (3) impeller ring, (4) head shaft, (5) line shaft, (6) packing, (7) gland, (8) lantern ring, (9) bearing bushing, (10) suction bell, (11) stuffing box bushing, (12) protecting collar, (13) shaft adjusting nut, (14) shaft coupling, (15) lubricator, (16) lubricator bracket, (17) stuffing box, (18) shaft enclosing tube, (19) column pipe, (20) enclosed line shaft bearing, (21) tubing nut, (22) tubing tension plate, (23) surface discharge head, (24) top column flange, (25) column pipe coupling, (26) open line shaft retainer bearing, (27) tubing adapter, (28) discharge case, (29) immediate bowl, (30) impeller lock collet, (31) suction case, (32) strainer, (33) suction pipe. (Adapted from Hydraulic Institute Standards, 13th ed., copyright Hydraulic Institute)

handling volatile liquids that are mounted on the operating floor and not provided with sufficient NPSH.

PROPELLER PUMPS   Originally the term *vertical propeller pump* was applied to vertical wet-pit diffuser or turbine pumps with a propeller or axial-flow impellers, usually for installation in an open sump with a relatively short setting (Figs. 112 and 113). Operating heads exceeding the capacity of a single-stage axial-flow impeller might call for a pump of two or more stages or a single-stage pump with a lower specific speed and a mixed-flow impeller. High enough operating heads might demand a pump with mixed-flow impellers and two or more stages. For lack of a more suitable name, such high-head designs have usually been classified as propeller pumps also.

Although vertical turbine pumps and vertical modified propeller pumps are basically the same

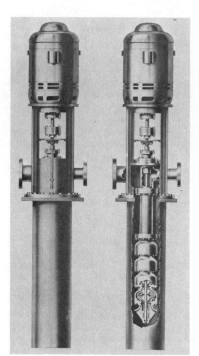

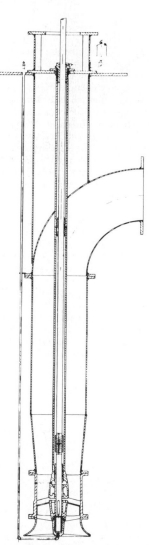

FIG. 111   Vertical turbine can pump for condensate service. (Worthington Pump)

FIG. 112   Section of vertical propeller pump with belowground discharge. (Worthington Pump)

mechanically and could even be of the same specific speed hydraulically, a basic turbine pump design is suitable for a large number of stages, whereas a modified propeller pump design is basically intended for a maximum of two or three stages.

Most wet-pit drainage, low-head irrigation, and storm water installations employ conventional propeller or modified propeller pumps. These pumps have also been used for condenser circulating service, but a specialized design dominates this field. As large power plants are usually located in heavily populated areas, they frequently have to use badly contaminated water (both fresh and salt) as a cooling medium. Such water quickly shortens the life of fabricated steel. Cast iron, bronze, or an even more corrosion-resistant cast metal must therefore be used for the column pipe assembly. This requirement means a very heavy pump if large capacities are involved. To avoid the necessity of lifting this large mass for maintenance of the rotating parts, some designs (one of which is illustrated in Fig. 114) are built so that the impeller, diffuser, and shaft assembly can be removed from the top without disturbing the column pipe assembly. These designs are commonly designated as *pullout* designs.

Like vertical turbine pumps, propeller and modified propeller pumps have been made with both open- and enclosed-line shafting. Except for condenser circulating service, enclosed shafting—using oil as a lubricant but with a grease-lubricated tail bearing below the impeller—seems to be favored. Some pumps handling condenser circulating water use enclosed shafting but with water (often from another source) as the lubricant, thus eliminating any possibility of oil getting into the circulating water and coating the condenser tubes.

Propeller pumps have open propellers. Modified propeller pumps with mixed-flow impellers are made with both open and closed impellers.

VOLUTE PUMPS   A variety of wet-pit pumps are available. The liquid pumped—clean water, sewage, abrasive liquids or slurries—dictates whether a semiopen or an enclosed impeller is used, whether the shafting is open or closed to the liquid pumped, and whether the bearings are submerged or located above the liquid.

Figure 115 illustrates a single-volute pump with a single-suction enclosed nonclog impeller, no pump-out vanes or wearing-ring joints on the back side of the impeller, and enclosed shafting. The pump is designed to be suspended from an upper floor by means of a drop pipe and for pumping sewage or other solids-laden liquids. To seal against leakage along the shaft at the point where it passes through the casing, a stuffing box is provided. The design of the stuffing box may be either that shown in Fig. 116, which uses rings of packing and a spring-loaded gland, or that shown in Fig. 117, which uses U-cup packing requiring no gland. The pump shown in Fig. 115 uses two sleeve bearings above the impeller. The bottom bearing is grease-lubricated, and a seal is provided to prevent grease leakage as well as to keep out any grit. The upper bearing connects the shaft cover pipe to the pump bearing bracket and the upper end of the cover pipe to the floorplate. This bearing is gravity-feed oil-lubricated. If intermediate bearings are required, they are also supported by the cover pipe and are oil-lubricated. The pump thrust is carried by the motor, which may be either hollow-shaft or solid-shaft construction. The latter type requires the use of a rigid coupling between the pump and motor shafts.

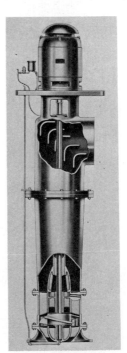

FIG. 113 Vertical propeller pump with belowground discharge. (Peerless Pump)

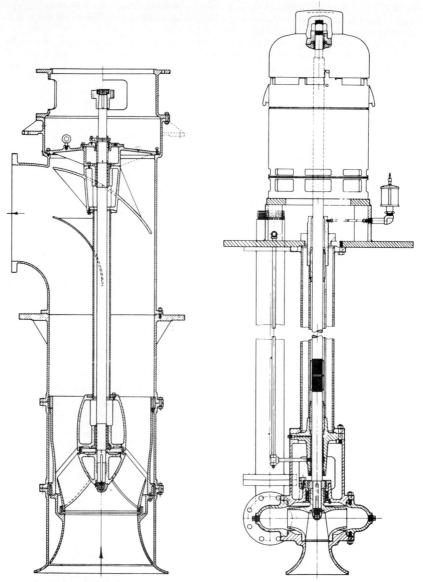

FIG. 114 Section of vertical modified propeller pump with removable bowl and shafting assembly. (Worthington Pump)

FIG. 115 Section of vertical wet-pit, nonclogging, pump. (Worthington Pump)

A design which uses open shafting and no stuffing box at the pump casing and incorporates its own thrust bearing is shown in Fig. 118. The impeller shown in Fig. 118 is of the vortex type (sometimes called *recessed impeller*, Fig. 119), which is suitable for pumping heavy concentrations of solid material (such as sludges or slurries) or in certain food processing applications, but other types of impeller can be substituted. Pumped liquid leakage from the casing is relieved back to the suction through holes in the support pipe. The stuffing box at the driver floor elevation is used only when gastight construction is desired. The lower and any intermediate sleeve bearings are grease-lubricated as shown, but gravity-feed oil lubrication is also available in other designs.

The upper antifriction thrust bearing is grease-lubricated. A solid shaft motor and a flexible shaft are used.

Figure 120 illustrates what is called a cantilever-shaft pump, which has the unique feature of having no bearings below the liquid surface. The shaft is exposed to the liquid pumped. External antifriction grease-lubricated bearings are provided above the floorplate and are properly spaced to support the rigid shaft. They carry both the thrust and the radial load. A flexible coupling is used between the pump and the solid-shaft motor. The stuffing box at the floorplate may be eliminated if holes are provided in the drop pipe to maintain the liquid level in the pipe even with the sump liquid level. Either semiopen or enclosed impellers may be used.

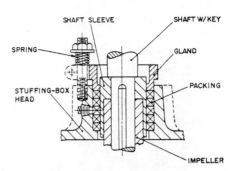

FIG. 116   Typical stuffing box arrangement.

A very interesting design of wet-pit pump is shown in Fig. 121. It uses a single-stage double-suction impeller in a twin-volute casing. Because the axial thrust is balanced, the thrust bearing need carry only the weight of the rotating element. The pump requires no stuffing box or mechanical seal. The shaft is entirely enclosed, and the bearings are externally lubricated, either with oil or with water. The lower bearing receives its lubrication from an external pipe connection.

The term *sump pump* ordinarily conveys the idea of a vertical wet-pit pump that is suspended from a floorplate or sump cover or supported by a foot on the bottom of a well, that is motor-driven and automatically controlled by a float switch, and that is used to remove drainage collected in a sump. The term does not indicate a specific construction, for both diffuser and volute designs are used; these may be single-stage or multistage and have open or closed impellers of a wide range of specific speeds.

For very small capacities driven by fractional-horsepower motors, *cellar drainers* can be used. These are small and usually single-stage volute pumps with single-suction impellers (either top or bottom suction) supported by a foot on the casing; the motor is supported well above the impeller

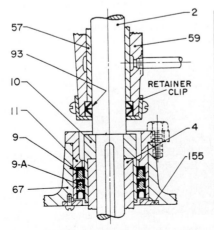

STATIONARY PARTS		
REF. NO.	NO. OF PCS.	NAME OF PARTS
9	3	U-CUP SEAL RING
9-A	2	SPACERS
11	1	SEAL RING ADAPTER
57	1	LOWER PUMP BEARING
59	1	BEARING BRACKET
67	1	STUFFING-BOX HEAD
93	2	OIL SEAL
155	1	SEAL-RING RETAINER

ROTATING PARTS		
REF. NO.	NO. OF PCS.	NAME OF PARTS
2	1	SHAFT W/KEY
4	1	IMPELLER
10	1	SHAFT SLEEVE

FIG. 117   Stuffing box arrangement with U-cup packing. (Worthington Pump)

by some form of column enclosing the shaft. These drainers are made as complete units, including float, float switch, motor, and strainers (Fig. 122).

The larger sump pumps are usually standardized but obtainable in any length, with covers of various sizes (on which a float switch may be mounted), and the like. Duplex units, that is, two pumps on a common sump cover (sometimes with a hole for access to the sump), are often used (Fig. 123). Such units may operate their pumps in a fixed order, or a mechanical or electric alternator may be used to equalize their operation.

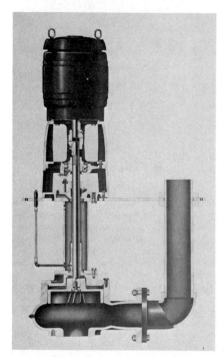

FIG. 118 Vertical wet-pit volute pump with open shafting, no stuffing box, and its own thrust bearing. (Aurora Pump)

APPLICATION OF VERTICAL WET-PIT PUMPS Like all pumps, the vertical wet-pit pump has advantages and disadvantages. One advantage is that installation does not require a separate dry pit to collect the pumped liquid. If the impeller (first-stage impeller in multistage pumps) is submerged, there is no priming problem and the pump can be automatically controlled without fear of its ever running dry. Moreover, the available NPSH is greater (except in closed tanks) and often permits a higher rotative speed for the same service conditions. A second advantage is that the motor or driver can be located at any desired height above any flood level.

The mechanical disadvantages are the following: (1) possibility of freezing when idle, (2) possibility of damage by floating objects if the unit is installed in an open ditch or similar installation, (3) inconvenience of lifting out and dismantling for inspection and repairs, no matter how small, and (4) relative short life of the pump bearings unless the water and bearing design are ideal. In summary, the vertical wet-pit pump is the best pump available for some applications, not ideal but the most economical for other installations, a poor choice for some, and the least desirable for still others.

Axial Thrust in Vertical Pumps with Single-Suction Impellers The subject of axial thrust in horizontal pumps has already been discussed in this section. When pumps are

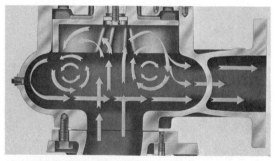

FIG. 119 Principle of operation of vortex pump shown in Fig. 118. (Aurora Pump)

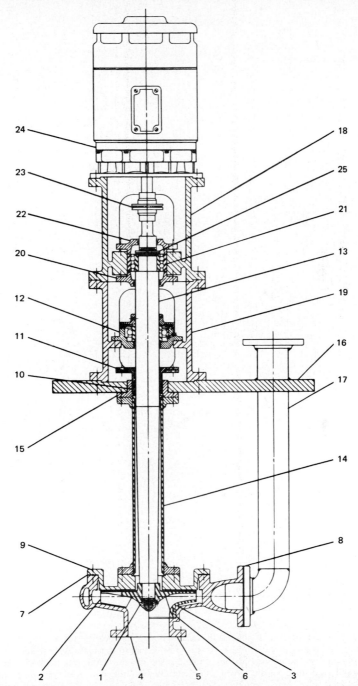

FIG. 120 Vertical wet-pit cantilever-type volute pump: (1) impeller nut, (2) open impeller, (3) closed impeller, (4) open casing, (5) closed casing, (6) casing wearing ring, (7) casing gasket, (8) discharge flange gasket, (9) back cover, (10) packing, (11) packing gland, (12) radial bearing, (13) shaft, (14) supporting pipe, (15) separate stuffing box, (16) sump cover, (17) discharge pipe, (18) upper motor pedestal, (19) lower motor pedestal, (20) lower thrust bearing cover, (21) thrust bearing, (22) upper thrust bearing cover, (23) coupling, (24) motor, (25) bearing locknut and washer. (Laurence Pump & Engine)

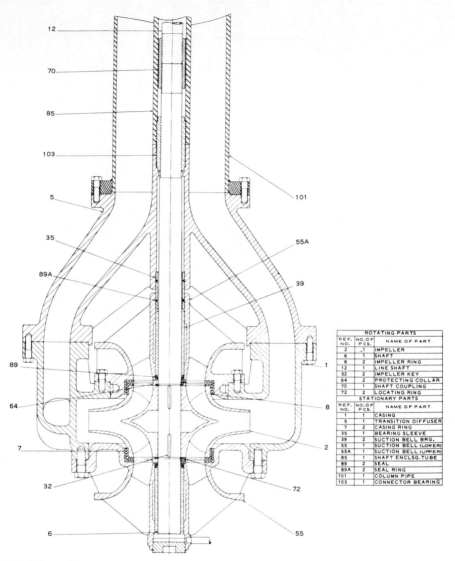

ROTATING PARTS		
REF. NO.	NO. OF PCS.	NAME OF PART
2	1	IMPELLER
6	1	SHAFT
8	2	IMPELLER RING
12	1	LINE SHAFT
32	2	IMPELLER KEY
64	2	PROTECTING COLLAR
70	1	SHAFT COUPLING
72	2	LOCATING RING
STATIONARY PARTS		
REF. NO.	NO. OF PCS.	NAME OF PART
1	1	CASING
5	1	TRANSITION DIFFUSER
7	2	CASING RING
35	1	BEARING SLEEVE
39	2	SUCTION BELL BRG.
55	1	SUCTION BELL (LOWER)
55A	1	SUCTION BELL (UPPER)
85	1	SHAFT ENCLSG.TUBE
89	2	SEAL
89A	2	SEAL RING
101	1	COLUMN PIPE
103	1	CONNECTOR BEARING

FIG. 121   Double-suction wet-pit pump. (Worthington Pump)

installed in a vertical position, there are additional factors which need be taken into account when determining the amount and direction of thrust to be absorbed in the thrust bearings.

The first and most significant of these factors is the weight of the rotating parts, which is a constant downward force for any given pump, completely independent of the pump operating capacity or total head. Since in most cases the single-suction impellers of vertical pumps are mounted with the suction eye facing downward, the normal hydraulic axial thrust is exerted downward and the weight of the rotating parts is additive to this axial thrust.

The second factor involves the dynamic force (or change in momentum) caused by the change in the direction of flow, from vertical to either horizontal or partly horizontal, as the pumped liquid flows through the impeller. This force acts upward and balances a small portion of the hydraulic downthrust and of the rotor weight. The magnitude of this force for water having a specific weight (force) of 62.34 lb/ft^3 (9.79 kN/m^3) is

in USCS units $\qquad F_u = \dfrac{KV_e^2 A_e}{2g(2.31)}$

in SI units $\qquad F_u = \dfrac{KV_e^2 A_e}{2g(1.02)}$

where F_u = upthrust, lb (N)
$\quad\ \ K$ = constant related to impeller type
$\quad\ \ V_e$ = velocity in impeller eye,
$\qquad\ $ ft/s (m/s)
$\quad\ \ A_e$ = net eye area, in^2 (cm^2)
$\quad\ \ g$ = 32.2 ft/s^2 (9.81 m/s^2)

The constant K is 1.0 for a fully radial-flow impeller, less than 1.0 for a mixed-flow impeller, and essentially zero for a fully axial-flow impeller.

Under normal operating conditions, the upward thrust caused by the change of momentum is hardly significant in comparison with the downward thrust caused by the unbalanced pressures acting on the single-suction impeller. Consider the example of an impeller with the following characteristics:

Capacity	2500 gpm (568 m^3/h)
Total head	231 ft (70.4 m)
Net pressure	100 lb/in^2 (6.89 bar)
Unbalanced eye area	40 in^2 (258 cm^2)

FIG. 122 Cellar-drainer sump pump. (Sta-Rite Products)

If we neglect the effect of the pressure distribution on the shrouds of the impeller, the downward thrust is

in USCS units $\qquad\qquad\qquad 100 \times 40 = 4000$ lb

in SI units $\qquad\qquad 6.89 \times 10^5 \times \dfrac{258}{10,000} = 17,800$ N

The upward force caused by the change of momentum is

in USCS units $\qquad \dfrac{1.0 \times \left(\dfrac{2500 \times 0.32}{40}\right)^2 \times 40}{64.4 \times 2.31} = 107.5$ lb

in SI units $\qquad \dfrac{1 \times \left(\dfrac{568 \times 10,000}{258 \times 3600}\right)^2 \times 258}{2 \times 9.81 \times 1.02} = 481$ N

This is certainly negligible relative to the downward hydraulic axial thrust.

The situation is quite different, however, during the start-up of a vertical pump. Although the motor may get up to full speed in just a few seconds, it takes a certain amount of time for the total head to increase from zero to that corresponding to normal operating capacity. Consequently the pump will be operating in a very high capacity range. Since the upward force caused by the change of momentum varies as the square of the capacity while the downward axial thrust caused

FIG. 123   Duplex sump pump. (Economy Pump)

by pressure differences is very low, there can be a momentary *net* upward force, or upthrust. This means that thrust bearings intended to accommodate the axial thrust of vertical pumps must be capable of accommodating some thrust in the upward direction in addition to the normal downward thrust. This is particularly true for pumps with relatively low heads per stage and with short settings because in these units the rotor weight does not at all times compensate for the upthrust from the change in momentum.

Particular care must be exercised in defining the range of vertical movement allowed for the thrust bearings because any such movement must remain within the displacement limits of any mechanical seal used in the pump.

In rather rare cases, a close-coupled vertical pump is operated at such a high capacity that a continuous net upthrust is generated. Such operation can damage the pump because the line shaft is operated in compression and may buckle, causing vibration and bearing wear. The manufacturer's comments should be invited if such operation is contemplated.

Shaft Elongation in Vertical Pumps   The elongation of a vertical pump shaft is caused by three separate phenomena: (1) the tensile stress caused by the weight of the rotor, (2) the tensile stress caused by the axial thrust, and (3) the thermal expansion of the shaft.

In most cases, the tensile stress created by the axial thrust is several times greater than that created by the weight. In a typical example of a 16,000-gpm (3636-m³/h) pump designed for a 175-ft (53.3 m) head and 50 ft (15.25 m) long, the elongation caused by the weight of a 1600-lb (726 kg) impeller will be of the order of 0.0033 in (0.084 mm). The elongation caused by the axial thrust will be approximately 0.0315 in (0.8 mm).

The elongation caused by thermal expansion has to be considered from two angles. First, if the shaft and the stationary parts are built of materials which have essentially the same coefficient of expansion, both will expand equally and no significant relative elongation will take place. Second, if the pumps are operated in essentially the same range of temperatures in which they were assembled, no significant relative expansion will take place, even if dissimilar coefficients of expansion are involved. Whatever the case, the pump manufacturers take these factors into consideration by providing the necessary vertical end-play between the stationary and rotating pump components.

Load on Foundations of Vertical Pumps

If the motor support is integral with the pump discharge column, as in Figs. 110, 112, and 113, and if the hydraulic thrust is carried by the motor thrust bearing, this thrust is not additive to the deadweight of the pump and of its motor plus the weight of the water contained in the pump, insofar as the load on the foundations is concerned. This is because the pump and motor mounted in this fashion form a self-contained entity and all internal forces and stresses are balanced within this entity. If the pump and motor are supported separately, however, as in Fig. 124, and are joined by a rigid coupling which transmits the pump hydraulic thrust to the motor thrust bearing, the foundations will carry the following loads when the pump is running:

1. Foundation supporting the motor: the weight of the motor plus the weight of the pump rotor plus the axial hydraulic thrust developed by the pump

2. Foundation supporting the pump: the weight of the pump stationary parts plus the weight of the water in the pump, including the water in the discharge pipe supported by the foundation, less the axial hydraulic thrust developed by the pump

Since when the pump is idle the foundation supporting the pump is not benefited by the reduction in load equivalent to the axial hydraulic thrust, both running and idle operating conditions must be considered in this case.

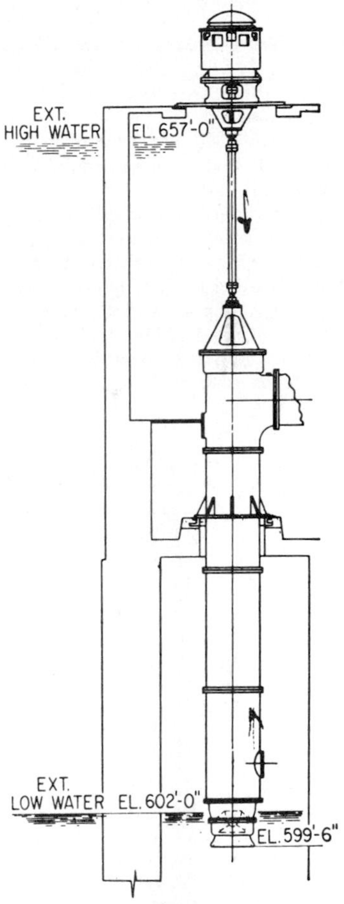

FIG. 124 Vertical pump and driving motor supported separately. (Worthington Pump)

Typical Arrangements of Vertical Pumps

A pump is only part of a pumping system. The hydraulic design of the system external to the pump will affect the overall economy of the installation and can easily have an adverse effect upon the performance of the pump itself. Vertical pumps are particularly susceptible because the small floor space occupied by each unit offers the temptation to reduce the size of the station by placing the units closer together. If the size is reduced, the suction arrangement may not permit the proper flow of water to the pump suction intake. Recommended arrangements for vertical pumps are discussed in Sec. 10.1.

SPECIAL PURPOSE PUMPS

In addition to the more or less general purpose pumps described on the preceding pages, there are literally hundreds of designs of centrifugal pumps intended for very specific applications.

While it is impossible to describe every one of these special purpose designs, many of them are discussed and illustrated in Secs 9.1 through 9.23, which cover a variety of pump services. There are, however, several specific designs which are seeing rapidly increasing usage and so are discussed in greater detail here.

Submersible Motor-Driven Wet-Pit Pumps   The installation of conventional vertical wet-pit pumps with the motor located above the liquid level may require a considerable length of drive shafting, particularly in the case of deep settings. The addition of this shafting, of the many line bearings, and possibly of an external lubrication system may represent a major portion of the total installed cost of the pumping unit. Furthermore, shaft alignment becomes more critical and shaft elongation and power losses increase rapidly as the setting is increased, especially for deep-well pumps.

A great variety of submersible motors have been developed to obviate these shortcomings. They are described in Subsec. 6.1.1, and a classification of the various types of such motors is presented in Fig. 20 of that subsection. Submersible wet-pit pumps eliminate the need for extended shafting, shaft couplings, a stuffing box, a subsurface motor stand, and, in some cases, an expensive pump house. Both vertical turbine and volute-type wet-pit pumps may be so driven.

Figures 125 and 126 illustrate, respectively, the external appearance and a cross section of a vertical turbine pump driven by a submersible motor located at the bottom of the pump. The pump suction is through a perforated strainer located between the motor and the first-stage

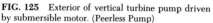

FIG. 125   Exterior of vertical turbine pump driven by submersible motor. (Peerless Pump)

FIG. 126   Cross-sectional view of submersible pump shown in Fig. 125. (Peerless Pump)

impeller bowl. There is, of course, no shafting above the pump and the pump-and-motor unit is supported by the discharge pipe only. No external lubrication is required. The motor is completely enclosed and oil-filled and is provided with a thrust bearing to carry the pump downthrust. A mechanical seal is provided at the motor shaft extension, which is connected to the pump shaft with a rigid coupling. Only a discharge elbow and the electric cable connection are seen above the surface support plate. On occasion this type of pump is used horizontally as a booster pump in a pipeline, and in such cases the elbow at the discharge is eliminated.

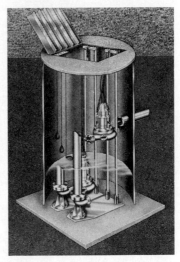

Vertical wet-pit volute-type sump pumps can be obtained with close-coupled submersible motors for drainage, sewage, process, and slurry services. Figure 127 illustrates dual submersible sewage pumps in a belowground collecting tank. The pumps (Fig. 128) are supported by guide rails which make it possible to lower and raise the pumps by means of a chain hoist. During this operation, the discharge pipe is connected and disconnected without dewatering the tank. Other arrangements use foot-supported pumps with rigid discharge piping.

Motors used for this type of construction are usually hermetically sealed, employing a double mechanically sealed oil chamber with a moisture-sensing probe to detect any influx of conductive liquid past the outer seal. Controls

FIG. 127 Dual submersible sewage pumps in belowground collecting tank. (Worthington Pump)

to start and stop the pump motors can be either an air compressor bubbler system or level-sensing mercury switches which tilt when floated (Fig. 127).

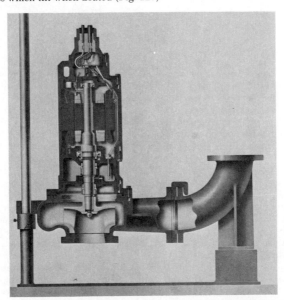

FIG. 128 Section of submersible sewage pump shown in Fig. 127. (Worthington Pump)

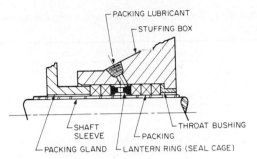

FIG. 2   Common packing arrangement.

sleeve has a coated material for a hard-wear surface, the sleeve must also have good thermal shock resistance.

Lantern Rings (Seal Cages)   When an application requires that a lubricant be introduced to the packing, a lantern ring is used to distribute the flow (Fig. 2). This ring is used at or near the center of the packing installation. For ease of assembly, most lantern rings are axially split. Materials of construction range from metal to TFE (tetrafluoroethylene). TFE lantern rings are usually filled with glass or with glass and molybdenum disulfide and are inherently self-lubricating and will not score the shaft. See Subsec. 2.2.1 also.

Stuffing Box Gland Plates   All mechanical packings are mechanically loaded in the axial direction by the stuffing box gland (Fig. 2). In cases where leakage of the process liquid is dangerous or can vaporize and create a hazard to operating personnel, a smothering gland is used to introduce a neutral liquid at lower temperatures (Fig. 3). A sufficient quantity of quenching liquid should be used to eliminate the danger from the liquid being pumped. The neutral liquid circulated in the gland mixes with the leakage and carries it to a safe place for disposal. Close clearances in the gland control leakage of the combined liquids to atmosphere. This quench can also be used to protect the packing from any wear through abrasion, because the leakage cannot vaporize and leave behind abrasive crystals.

Glands are usually made of bronze. However, cast iron or steel may be used for all-iron pumps. When iron or steel glands are used, they are normally bushed with a nonsparking material like bronze. Also see Subsec. 2.2.1.

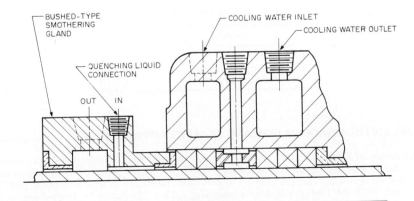

FIG. 3   Smothering gland and water-cooled stuffing box.

TABLE 4 Leakage to Prevent Packing Burning and Sleeve
Scoring

Pressure, lb/in² (kPa)	Leakage, drops/min	(cc/min)
0–60 (0–400)	60	4
61–100 (401–700)		190
101–250 (701–1700)		470

Leakage and Power Consumption The basic operating parameters for compression
packing are the *PV* (pressure × velocity) factor and the projected bearing area of the assembly.
Together they determine the rate of heat generation for the system. Some leakage of product
being sealed or of packing lubricant is necessary to keep the packing from burning up or scoring
the shaft sleeve. Minimum values for leakage for different packing pressures are given in Table
4. Grafoil and carbon filament compression packings can be used with reduced leakage rates, or
in dry gastight pump stuffing boxes, as the developed heat is dissipated through the packing and
pump housing.

The exact pressure *P* between the shaft sleeve and packing is a function of the pressure distribution over the length of the packing and the axial loading from the gland. For ease of calculation in determining the *PV* value, the gage pressure of the liquid at the packing is multiplied
by π × packing ID × rpm. The heat generation at the packing can then be estimated as

$$Q = \frac{f\pi^2 PND^2 L}{CJ}$$

where *Q* = heat generated, Btu/min (W)
    f = coefficient of friction
    P = liquid pressure at packing, lb/in² gage (Pa)
    N = shaft speed, rpm
    D = sleeve OD or packing ID, in (m)
    L = sleeve length covered by packing, in (m)
    C = 12 (60 for SI units)
    J = mechanical equivalent of heat = 778 ft·lb/Btu (1 N·m/s·W)

The coefficient of friction for various packings at a pressure of 100 lb/in² (689 kPa) are given
in Table 5. The heat generated by the packing must be removed by the leakage through the
packing.

TABLE 5 Typical Coefficients of Friction

Material	*f*
Plain cotton	0.22
TFE-impregnated asbestos	0.17
Grease-lube asbestos	0.10

APPLICATION INFORMATION AND SEALING ARRANGEMENTS _____

Materials of Construction Basically, stuffing box packing is a pressure breakdown device.
In order for a packed stuffing box to operate properly, the correct packing must be applied and
the appropriate design features included in the system. There are naumerous types of packing
materials available. Each is suited for a particular service. These may be grouped into three
categories:

FIG. 4   Graphited asbestos packing in continuous coil form. (John Crane-Houdaille, Inc.)

FIG. 5   Nonasbestos packing. (John Crane-Houdaille, Inc.)

1. *Asbestos Packing.* Asbestos prelubricated with graphite, grease, or inert oil is an excellent general service material. This soft material is also suited for cold and hot water applications to temperatures of 450°F (232°C). Pressures above 200 lb/in² (1380 kPa) limit the material to moderate surface speeds of 1000 ft/min (5 m/s). An example of graphited asbestos packing in continuous coil form is shown in Fig. 4.

2. *Nonasbestos Packing.* Since asbestos is a proven carcinogen, there has been increased interest in other types of packing materials. These include cotton, TFE filament, aramid fiber, graphite yarn, and Grafoil,° each of which may be impregnated with a lubricant. With the exception of Grafoil, all materials are of interlaced construction for greater flexibility (Fig. 5). Packing made with this type of construction remains intact even if individual yarns wear away at the inside diameter of the packing.

   Grafoil rings are available cut from sheet stock and laminated or made from ribbon tape. Laminated rings must have an interference fit with the shaft as they have poor radial compression. Ribbon tape rings are easily compressed to the shaft and effect a better seal. Grafoil packing is available in spilt or endless rings (Fig. 6) or ribbon tape for ease of maintenance and installation. For operating limits of these types of packing, see Table 6.

3. *Metallic Packing.* The basic materials of construction are lead or babbitt, aluminum, and copper. Used in wire or foil form, they have a flexible core of either asbestos or plastic. The packing is impregnated with graphite or oil lubricant (Figs. 7 and 8). Babbitt is used in water and oil services at temperatures up to 450°F (229°C) and pressures to 250 lb/in² (1551 kPa). Copper foil is used with water and low-sulphur oils. Aluminum is used for oil service and heat transfer fluids. Both copper and aluminum have a temperature range to 1000°F (537°C) and pressures

° A registered trademark of Union Carbide.

FIG. 6   Grafoil packing rings. (John Crane-Houdaille, Inc.)

FIG. 7   Metallic packing in spiral form. (John Crane-Houdaille, Inc.)

TABLE 6 Service Limitations of Common Packing Materials[a]

Packing material	Pressure (max)[b] lb/in² gage (kPa gage)	PV rating (max)[c], lb/in² gage·fpm (bar·m/s)	Temp. (max)[d] °F (°C)	pH range	Comments
Cotton	100 (689)	188,000 (65.8)	150 (65.6)	5–7	Nonabrasive material; for cold water and dilute salt solutions
Flax/ramie	100 (689)	188,000 (65.8)	150 (65.6)	5–7	High wet strength and excellent resistance to fungi and rotting; for cold water and dilute salt solutions
Plastic	100 (689)	188,000 (65.8)	600 (315.5)	4–8	Excellent sealing qualities; reacts well to gland adjustments; can extrude at higher pressures if not backed up by braided or asbestos packing
	250 (1723)	471,000 (165)	150 (65.6)		
Asbestos, grease- or oil-impregnated	100 (689)	188,000 (65.8)	750 (398.8)	4–8	For hot or cold water, brine, oil, mild caustics, solvents, and acids
	250 (1723)	471,000 (165)	500 (260)		
Asbestos, TFE-impregnated	250 (1723)	471,000 (165)	500 (260)	2–10	For mild chemicals and solvents
Lead	250 (1723)	471,000 (165)	450 (232.2)	2–10	Shaft sleeve must have Brinell hardness of 500 or more; for hot oils and boiler-feed water

Aluminum or copper	250 (1723)	471,000 (165)	750 (398.8)	3–10	Shaft sleeve must have Brinell hardness of 500 or more; for hot oils and boiler-feed water
TFE filament	250 (1723)	471,000 (165)	500 (260)	0–14	For corrosive liquids and food service; usually requires slightly higher break in leakage
Aramid fiber	250 (1723)	471,000 (165)	500 (260)	3–10	Strong resilient packing; maximum speed 1900 fpm (9.6 m/s); good in abrasives and chemicals
Graphite/carbon filament	250 (1723)	471,000 (165)	750 (398.8)	0–14	For corrosive liquids and high-temperature applications
Grafoil	250 (1723)	471,000 (165)	750 (398.8)	0–14	Excellent conductor of heat from the sealing surfaces; operates with minimum leakage; excellent radiation resistance

[a]Continuous lubrication introducted at the lantern ring. Table is only a guide. Consult packing manufacturer with complete operating conditions for exact recommendation.
[b]Pressure relates to the operating pressure at the stuffing box.
[c]PV data based on a 2-in (5.08-cm) shaft at 1750 and 3600 rpm.
[d]Temperature is the product temperature.

FIG. 8   Metallic packing in ring form. (John Crane-Houdaille, Inc.)

FIG. 9   Combination of hard and soft packing. (John Crane-Houdaille, Inc.)

to 250 lb/in² (1551 kPa). Babbitt is not suitable for running against brass or bronze shaft sleeves, and when copper or aluminum is used, the sleeve should be hardened to 500 Brinell or more.

Additional service limitations for common packing materials can be found in Table 6.

ENVIRONMENTAL LIMITATIONS

Pressure   Every pumping application will result in either a positive or negative pressure at the throat of the pump stuffing box. A positive pressure will force the liquid pumped through the packing to the atmosphere side of the pump. Higher pressures will result in greater leakage from the pump. This results in excessive tightening of the gland, which causes accelerated wear of the shaft or shaft sleeve and packing. For pressures at the stuffing box greater than 75 lb/in² (517 kPa), some means of throttling the pressure should be considered: A combination of hard and soft rings die-formed to exact stuffing box bore and sleeve dimensions can be used (Fig. 9). Harder rings at the inboard end of the box and at the lantern ring and gland break down the pressure and prevent the extrusion of the packing.

If the packing itself cannot be used to break down the pressure, then a throttle bushing must be used (Fig. 10). This is a typical arrangement for a vertical turbine pump, where the stuffing box is subject to discharge pressure of the pump. Here the throttle bushing is used to bring the liquid to almost suction pressure, and most of the leakage through the bushing is bled back to suction. When the suction pressure is less than atmospheric, as in condensate pump service, an orifice is used in the piping back to suction to maintain pressure above atmospheric in the stuffing box. This prevents air leakage into the pump and excessive flow and wear at the bushing.

When the pressure at the stuffing box is atmospheric or just below during normal operation, a bypass from pump casing discharge through an orifice can be used to inject liquid into the lantern ring. The sealing liquid will flow partly into the pump and partly out to atmosphere, thereby preventing air from entering at the stuffing box. This arrangement is commonly used to handle clean, cool water. In some pumps these connections are arranged so that liquid can be introduced into the lantern ring through internally drilled passages.

When the negative pressure is very low, for example, when the suction lift is in excess of 15 ft (4.6 m), an independent injection of 10 to 25 lb/in² (70 to 172 kPa) higher than atmospheric must be used on the pump. Otherwise, priming may be difficult. Hot-well pumps or condensate pumps operate with as much as 28 in (0.7 m) vacuum, and air leakage into the pump would occur even on standby service. Here, a continuous injection of clean, cool water is required, or, alternatively, a cross connection of sealing water to another operating pump will provide sealing pressure as long as one pump is operating. A lantern ring is provided for this, as shown in Fig. 10.

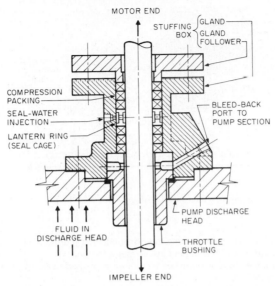

FIG. 10 Compression packing with throttle bushing for pressure breakdown.

Temperature   The control of temperature at the stuffing box is an important factor in promoting the life of the packing. Even though packings are rated for high product temperatures, cooling in most cases is desirable. The heat developed at the packing must also be removed. Rules of thumb are:

* For light service conditions with pressures to 15 lb/in^2 (103 kPa) and temperatures to 200°F (90°C), no cooling is required as leakage through the packing is adequate.

* For medium service conditions with pressures to 50 lb/in^2 (345 kPa) and temperatures to 250°F (118°C), lantern ring cooling is desirable; 1 gpm (3.78 l/min) at 5 lb/in^2 (34 kPa) below process pressure.

* For high service conditions with pressures to 100 lb/in^2 (690 kPa) and temperatures to 300°F (131°C), lantern ring cooling is desirable; 1 gpm (3.78 l/min) at 5 lb/in^2 (34 kPa) below process pressure plus water-cooled stuffing box (Fig. 3).

   When cooling water is required, it is circulated through the stuffing box at the lantern ring. For horizontal-shaft pumps, inlet should be at the bottom, with the outlet at the top of the stuffing box. Some of the coolant may flow to the process liquid and some to atmosphere. The outboard packing rings seal only the cool liquid.

   If the product to be sealed will solidify, then the packing box will have to be heated before the pump is started. This can be accomplished by steam or electric tracing of the pump stuffing box.

Abrasives   Liquids that contain abrasives in the form of *suspended solids*, such as sand and dirt, will shorten the life of the packing. Particles will imbed themselves in the packing and will begin to wear the shaft or shaft sleeve. Abrasives can be eliminated at the sealing surfaces by injecting a clean liquid into the lantern ring. The injection may take the form of a bypass line from the pump discharge through a filter or centrifugal separator (see Subsec. 2.2.1). Where necessary, the clean injection may also be from an external source.

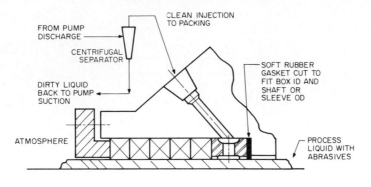

FIG. 11 Clean injection through a centrifugal separator to keep abrasives out of stuffing box.

To keep the abrasives from the packing in horizontal-shaft pumps, the injection may be made directly to the inboard end of the stuffing box through the lantern ring (Fig. 11). A soft rubber gasket between the lantern ring and the box shoulder can be used to limit the flow of clean liquid and the ingress of abrasives.

When separators and filters cannot be used, an injection from an external source must be considered. Two rings of packing are located between the lantern ring and the inboard end of the stuffing box to keep product dilution to a minimum. External flushing should be injected into the stuffing box at a pressure 10 to 25 lb/in² (70 to 172 kPa) greater than the pressure at the inboard end of the box from the liquid being pumped.

Dissolved solids in the liquid being pumped may also create a wear problem. Here, an increase or decrease in stuffing box temperature may be necessary to keep solids in solution.

INSTALLING CONTINUOUS COIL PACKING

1. Loosen and remove the gland from the stuffing box.

2. Using a packing puller, begin to remove the old packing rings.

3. Remove the split lantern ring and then continue removing packing with the puller.

4. After the packing has been removed, check the sleeve for scoring and nicks. If the shaft sleeve or shaft cannot be cleaned up, it must be replaced. Check the size of the stuffing box bore and the shaft sleeve or shaft diameter to determine what size packing should be used.

5. After the size of the packing has been determined, wrap the packing tightly around a mandrel, which should be the same size as the pump shaft or sleeve. The number of coils should be sufficient to fill the stuffing box. Cut the packing along one side to form individual rings.

6. Before beginning the assembly of any packing material, be sure to read all instructions from the manufacturer. Assemble the split packing rings on the pump. Each ring should be seated individually, with the split ends staggered 90° and the gland tightened to seat and fully compress the ring. Be sure the lantern ring is reinstalled correctly at the flush connection. Then back off the gland and retighten only finger-tight. The exception to this procedure is that TFE packing should be installed one ring at a time but not seated because TFE packings have high thermal expansion.

7. Allow excess leakage during break-in to avoid the possibility of rapid expansion of the packing, which could score the shaft sleeve or shaft so that leakage could not be controlled.

8. Leakage should be generous upon start-up. If the packing begins to overheat at start-up, stop the pump and loosen the packing until leakage is obtained. Restart only if the packing is leaking.

TABLE 7 Packing Troubles, Causes, and Cures

Trouble	Cause	Cure
No liquid delivered by pump	Lack of prime (packing loose or defective, allowing air to leak into suction)	Tighten or replace packing and prime pump.
Not enough liquid delivered by pump	Air leaking into stuffing box	Check for leakage through stuffing box while operating—if no leakage after reasonable gland adjustment, new packing may be needed or Lantern ring may be clogged or displaced and may need centering in line with sealing liquid connection. or Sealing liquid line may be clogged. or Shaft or shaft sleeve below packing may be badly scored, allowing air to be sucked into pump.
	Defective packing	Replace packing and check smoothness of shaft or shaft sleeve.
Not enough pump pressure	Defective packing	As for preceding.
Pump works for a while and quits	Air leaks into stuffing box	As for preceding.
Pump takes too much power	Packing too tight	Release gland pressure and retighten reasonably. Keep leakage flowing—if none, check packing, sleeve, or shaft.
Pump leaks excessively at stuffing	Defective packing	Replace worn packing. Replace packing damaged by lack of lubrication.
	Wrong type of packing	Replace packing not properly installed or run in. Replace improper packing with correct grade for liquid being handled.
	Shaft or shaft sleeve scored	Put in lathe and machine-true and smooth or replace.
Stuffing box overheats	Packing too tight	Release gland pressure.
	Packing not lubricated	Release gland pressure and replace all packing if any burnt or damaged.
	Wrong grade of packing	Check with pump or packing manufacturer for correct grade.

TABLE 7 Packing Troubles, Causes, and Cures (*Continued*)

Trouble	Cause	Cure
	Insufficient cooling water to jackets	Check for open supply line valves or clogged line.
	Stuffing box improperly packed	Repack.
Packing wears too fast	Shaft or shaft sleeve worn or scored	Remachine or replace.
	Insufficient or no lubrication	Repack, making sure packing loose enough to allow some leakage.
	Packing packed improperly	Repack properly, making sure all old packing removed and box clean.
	Wrong grade of packing	Check with pump or packing manufacturer.
	Pulsating pressure on external seal liquid line	Remove cause of pulsation.

CAUSES FOR SHORT PACKING LIFE

In order for packing to operate properly, the equipment must be in good condition. Shafts should be checked for runout and eccentricity to be sure they are within the manufacturer's recommended tolerances. Surfaces in contact with the packing should be finished to the correct smoothness and tolerance. Table 7 lists common troubles that affect packing life. Causes and possible cures are also given.

FURTHER READING

Berzinsa, A.: "Finer Points of Compression Packing Help with Centrifugal Pumps," *Power*, June 1974, pp. 58–59.

Crane Packing Company: *Engineered Fluids Sealing, Materials, Design, and Applications: An Information Compendium of Technical Papers*, Morton Grove, Ill. 1979.

Elonka, S.: "Packing," Power Practical Manual, *Power*, March 1955, pp. 107–130.

Kristle, O. M.: "How to Get Longer Life out of Pump Packing and Shaft Sleeve," *Water and Sewage Works*, June 1967, pp. 199–201.

Neale, M. J.: "Packing Glands," in *Tribology Handbook* (Sec. A36), Wiley, New York, 1973.

2.2.3
CENTRIFUGAL PUMP MECHANICAL SEALS

JAMES P. NETZEL

A mechanical seal may be used in a conventional pump stuffing box to seal any number of liquids at various speeds, pressures, and temperatures. The mechanical seal has found acceptance in pump applications where packed stuffing boxes have proved inadequate in service life or leakage. Pumps designed for severe service conditions or shorter stuffing boxes require mechanical seals.

DESIGN FUNDAMENTALS _____

Basic Components The sealing surfaces are perpendicular to the shaft, with contact between the primary and mating rings to achieve a dynamic seal. The primary ring is flexibly mounted in the seal head assembly, which usually rotates with the pump shaft, and the mating ring is usually fixed to the pump gland plate. Each of the sealing planes is lapped flat to eliminate any visible leakage. Wear occurs at the seal faces from sliding contact between the primary and mating rings. The amount of wear is small, as a film of the liquid sealed is maintained between the sealing faces. Normally the mating surfaces of the seal are of dissimilar materials and held in contact with a spring. The preload from the spring is required to produce the initial seal. The spring pressure holds the primary and mating rings together during shutdown or when there is a lack of liquid pressure.

Any seal installation is made up of two assemblies (Fig. 1). The seal head assembly includes the primary ring and its associated component parts. The mating ring assembly includes those parts required for the mating ring to function. During operation, one of the components is in motion. Normally, the seal head is rotating and the other component is stationary. There are three points of sealing common to all mechanical seal installations:

1. At the mating surfaces of the primary and mating rings
2. Between the rotating component and the shaft or sleeve
3. Between the stationary component and the gland plate

When a seal is installed on a sleeve, there is an additional point of sealing between the shaft

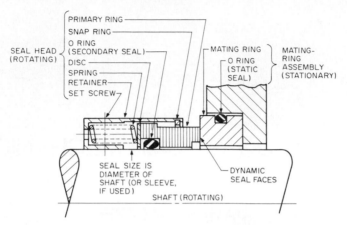

FIG. 1   Mechanical seal.

and sleeve. Certain mating ring designs may also require an additional seal between the gland plate and the stuffing box.

The secondary seal, between the rotating seal component and the shaft or sleeve, must be partially dynamic. As the seal faces wear, the primary ring must move slightly forward. Because of vibration from the machinery, shaft runout, and thermal expansion of the shaft against the pump casing, the secondary seal must move along the shaft. This is not a static seal in the assembly. Flexibility in sealing is achieved from such secondary seal forms as bellows, O ring, wedge, or V ring. Most seal designs fix the seal head to the shaft and provide for a positive drive to the primary ring.

The mating ring is usually a separate replaceable part. A static seal is used to prevent leakage between the mating ring and the gland plate. The static seal and the mating ring form the mating ring assembly. Although mechanical seals may differ in various physical aspects, they are fundamentally the same in principle. The wide variation in design is a result of the many methods used to provide flexibility, ease of installation, and economy.

Seal Balance   The greatest concern for the seal user is the dynamic contact between the mating seal surfaces. The performance of this contact determines the effectiveness of the seal. If the seal load at the faces is too high, the liquid film between the seal rings could be squeezed out or vaporized. An unstable condition would result, with a high wear rate of the sealing planes. The power would increase as the friction increased with solid contact. Seal face materials also have a bearing limit, which should not be exceeded. Seal balancing can avoid these conditions and lead to a more efficient installation.

The pressure in any stuffing box acts equally in all directions and forces the primary ring against the mating ring. Pressure acts only on the annular area a_c (Fig. 2a) so that the closing force in pounds (newtons) on the seal face is

$$F_c = pa_c$$

where p = stuffing box pressure, lb/in^2 (N/m^2)
a_c = hydraulic closing area, in^2 (m^2)

The pressure in pounds per square inch (newtons per square meter) between the primary and mating rings is

$$P'_f = \frac{F_c}{a_o} = \frac{pa_c}{a_o}$$

where a_o = hydraulic opening area (seal face area), in^2 (m^2).

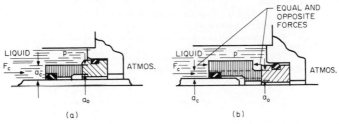

FIG. 2 Hydraulic pressure acting on the primary ring: (*a*) unbalanced, (*b*) balanced.

To relieve the pressure at the seal face, the relationship between opening and closing force can be controlled. If a_o is held constant and a_c decreased by a shoulder on a sleeve or seal hardware, the seal face pressure can be lowered (Fig. 2*b*). This is called seal balancing. A seal without a shoulder in the design is an unbalanced seal. A balanced seal is designed to operate with the shoulder.

The ratio of hydraulic closing area to seal face area is defined as seal balance *b*:

$$b = \frac{a_c}{a_o}$$

Seals can be balanced for pressures at the outside diameter of the seal faces, as shown in Fig. 2*b*. This is typical for a seal mounted inside the stuffing box. Seals installed outside the stuffing box can be balanced for pressure at the inside diameter of the seal faces. In special cases, seals can be double-balanced for pressure at both the outside and the inside diameter of the seal. Seal balance can range from 0.65 to 1.35, depending on operating conditions.

Face Pressure   As relative motion takes place between the seal planes, a liquid film develops. The generation of this film is believed to be the result of surface waviness in the individual sealing planes. Pressure and thermal distortion—as well as antirotation devices, such as drive pins, keys, or dents, used in the seal design—have an influence on surface waviness and on how the film develops between the sliding surfaces. Hydraulic pressure develops in the seal face, which tends to separate the sealing planes. The pressure distribution, referred to as a pressure wedge (Fig. 3), may be considered as linear, concave, or convex. The actual face pressure P_f in pounds per square inch (newtons per square meter) is the sum of the hydraulic pressure P_h and the spring pressure P_{sp} designed into the mechanical seal. The face pressure P_f is a further refinement of P_f', which does not take into account the liquid film pressure or the mechanical load of the seal:

$$P_f = P_h + P_{sp}$$

where $P_h = \Delta p(b - k)$, lb/in^2 (N/m^2)
$\quad \Delta p$ = pressure differential across seal face, lb/in^2 (N/m^2)
$\quad\;\; b$ = seal balance
$\quad\;\; k$ = pressure gradient factor (Table 1)

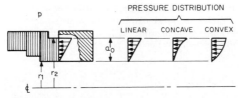

FIG. 3 Pressure wedge.

TABLE 1 Pressure Gradient Factors for Various Services

Liquid sealed	k
Light-specific-gravity liquids	0.3
Water-base solutions	0.5
Oil-base solutions	0.7

The mechanical pressure for a seal design is

$$P_{sp} = \frac{F_{sp}}{a_o} \text{ lb/in}^2 \text{ (N/m}^2)$$

where F_{sp} = seal spring load, lb (N)
  a_o = seal face area, in^2 (m^2)

Then the actual face pressure can be expressed as

$$P_f = \Delta p(b - k) + P_{sp}$$

The actual face pressure is used in the estimate of the operating pressure and velocity for a given seal installation.

Pressure-Velocity   As the sealing planes move relative to each other, they are affected by the actual face pressure and rotational speed. The product of the two, pressure times velocity, is referred to as PV and is defined as the power N_f per unit area with a coefficient of friction of unity:

$$PV = \frac{N_f}{a_o}$$

For seals, the equation for PV can be written as follows:

$$PV = P_f V_m = [\Delta p(b - k) + P_{sp}]V_m$$

where V_m = velocity at the mean face diameter d_m, ft/min (m/s).
  The PV for a given seal installation may be compared with values developed by seal manufacturers as a measure of adhesive wear.

Power Consumption   The PV value also enables the seal user to estimate the power loss at the seal with the following equation:

$$N_f = (PV)fa_o \text{ ft·lb/min (N·m/s)}$$

where f is the coefficient of friction.
  As a rule of thumb, the power to start a seal is generally five times the running value. The coefficients of friction for various common seal face materials are given in Table 2. These coefficients were developed with water as a lubricant at an operating PV value of 100,000 lb/in^2·ft/min (35.03 bar·m/s).° Values in oil would be slightly higher as a result of the viscous shear of the fluid film at the seal faces.

EXAMPLE   A pump having a 2-in (50.8-mm) diameter sleeve at the stuffing box is fitted with a balanced seal of this size and mean diameter. The seal operates in water at 300 lb/in^2 (20.68 bar), 3600 rpm, and ambient temperatures. The materials of construction are carbon and tungsten carbide. Determine the PV value and power loss of the seal.

Given
  $\Delta p = 300$ lb/in^2 (20.68 bar)

°1 bar = 10^5 Pa. For a discussion of bar, see *SI Units—A Commentary* in the front matter.

TABLE 2 Coefficient of Friction for Various Seal Face Materials[a]

Sliding materials		Coefficient of friction
Rotating	Stationary	
Carbon-graphite (resin filled)	Cast iron	0.07
	Ceramic	0.07
	Tungsten carbide	0.07
	Silicon carbide	0.02
	Silicon carbide converted carbon	0.015
Silicon carbide	Tungsten carbide	0.02
Silicon carbide converted carbon		0.05
Silicon carbide		0.02
Tungsten carbide		0.08

[a]Water lubrication, $PV = 100,000 \dfrac{\text{lb}}{\text{in}^2} \cdot \dfrac{\text{ft}}{\text{min}}$ (35.03 bar·m/s).

SOURCE: John Crane-Houdaille, Inc.

$$b = 0.75$$
$$k = 0.5 \text{ (Table 1)}$$
$$d_m = 2 \text{ in (50.8 mm)}$$
$$P_{sp} = 25 \text{ lb/in}^2 \text{ (1.72 bar)}$$
$$V_m = \frac{\pi}{12} \times 2 \times 3600 = 1885 \text{ ft/min} \left(\frac{\pi \times 50.8 \times 3600}{1000 \times 60} = 9.57 \text{ m/s} \right)$$
$$a_o = 0.4 \text{ in}^2 \text{ (0.000258 m}^2\text{)}$$
$$f = 0.07 \text{ (Table 2)}$$

in USCS units $PV = [300(0.75 - 0.5) + 25](1885) = 188,400 \dfrac{\text{lb}}{\text{in}^2} \cdot \dfrac{\text{ft}}{\text{min}}$

$$N_f = (188,400)(0.07)(0.4) = 5275 \frac{\text{ft·lb}}{\text{min}} = 0.16 \text{ hp}$$

in SI units $PV = [20.68(0.75 - 0.5) + 1.72](9.57) = 66 \text{ bar·m/s} = 66 \times 10^5 \text{ N/m}^2 \cdot \text{m/s}$

$$N_f = (66 \times 10^5)(0.07)(2.58 \times 10^{-4}) = 119 \text{ N·m/s} = 119 \text{ W}$$

Temperature Control   Control of temperature at the seal faces is desirable because wear is a direct function of temperature. Heat at the seal faces also causes thermal distortion, which will contribute to increased seal leakage. Many applications require some type of cooling.

The temperature of the sealing surfaces is a function of the heat generated by the seal plus the heat gained or lost to the pumpage. The heat generated at the faces from sliding contact is the mechanical power consumption of the seal being transferred into heat. Therefore,

$$Q_s = C_1 N_f = C_1(PVfa_o)$$

where Q_s = heat input from the seal, Btu/h (W)
 C_1 = 0.077 for USCS units and 1 for SI units

If the heat is removed at the same rate it is produced, the temperature will not increase. If the amount of heat removed is less than that generated, the seal face temperature will increase to a point where seal face damage will occur. Estimated values for heat input are given in Figs. 4 and 5.

Heat removal from a single seal is accomplished by a seal flush. The seal flush is usually a bypass from the discharge line on the pump or an injection from an external source. The flow rate for cooling can be found by:

$$\text{gpm (m}^3/\text{h)} = \frac{Q_s}{C_2(\text{sp. ht.})(\text{sp. gr.})\Delta T}$$

where Q_s = seal heat, Btu/h (W)
          C_2 = 500 in USCS units and 1000 in SI units
   sp. ht. = specific heat of coolant, Btu/lb·°F (J/kg·K)
   sp. gr. = specific gravity of coolant
      ΔT = temperature rise, °F (K)

When handling liquids at elevated temperatures, the heat input from the process must be considered in the calculation of coolant flow. Then

$$Q_{\text{net}} = Q_p + Q_s$$

The heat load Q_p from the process can be determined from Fig. 6.

EXAMPLE   Determine the net heat input for a 4-in (102-mm) diameter balanced seal in water at 1800 rpm. Pressure and temperature are 400 lb/in^2 (2758 kPa) and 170°F (76.7°C). From Fig. 5,

in USCS units Q_s = (3500 Btu/h/1000 rpm) × 1800 = 6300 Btu/h

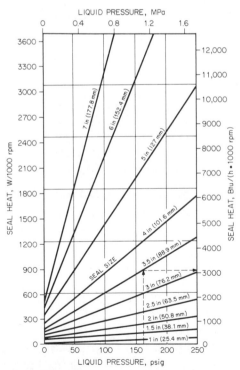

FIG. 4   Unbalanced seal heat generation. (John Crane-Houdaille, Inc.)

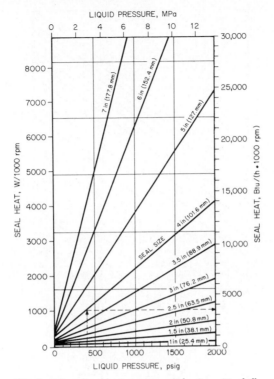

FIG. 5   Balanced seal heat generation. (John Crane-Houdaille, Inc.)

$$Q_s \; = \; (1025 \; \text{W/1000 rpm}) \times 1800 \; = \; 1845 \; \text{W}$$

From Fig. 6, assuming that the stuffing box will be cooled to 70°F (21°C) and that the temperature difference between the stuffing box and pumpage is 100 °F (37.8°C)

in USCS units $Q_p \; = \; 255 \; \text{Btu/h}$

in SI units $Q_p \; = \; 75 \; \text{W}$

$$Q_{\text{net}} \; = \; 6555 \; \text{Btu/h} \; (1920 \text{W})$$

The total heat input may be used to estimate the required flow to the seal. When multiple seals are used in a pump stuffing box, the heat load from each seal must be considered, as well as any heat soak from the process.

Different methods are used to supply cool liquid to the pump stuffing box (Fig. 7). When the liquid is clean, an internal flush connection at (A) can be used to cool the seal. When the liquid is dirty, an external flush at (B) can be used. This will allow the flush, a bypass from discharge line, to pass through a filter or centrifugal separator. The seal faces will be flushed with clean, cool liquid. Increased pressure from the flush provides positive circulation and prevents flashing at the seal faces caused by the heat generation.

When a pump handles liquids near their boiling point, additional cooling of the stuffing box is required. A typical arrangement to accomplish this is shown in Fig. 8. This seal is equipped with a pumping ring and heat exchanger. The pumping ring acts as a miniature pump, causing the liquid to flow through the outlet piping at the top of the seal chamber. The liquid passes through the heat exchanger and returns directly to the faces at the bottom inlet in the end plate. As the liquid is circulated, heat is removed from the seal and stuffing box. A closed loop system

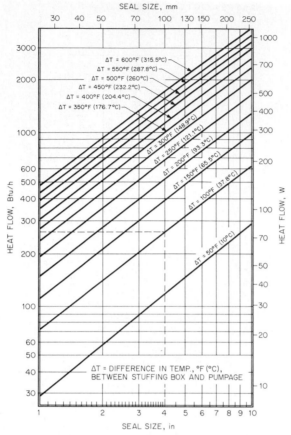

FIG. 6 Heat soak from process when water is used for lubrication. (John Crane-Houdaille, Inc.)

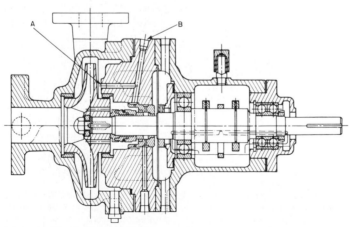

FIG. 7 Cooling circulation to mechanical seal: (*A*) internal circulation plug port, (*B*) external circulation plug port.

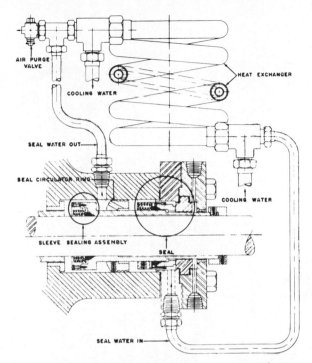

FIG. 8 Mechanical seal with external cooling arrangement. (John Crane-Houdaille, Inc.)

is commonly used on hot-water pumps. This method is extremely efficient since the coolant is circulated only in the stuffing box and does not reduce the temperature of the liquid in the pump.

On process pumps with stuffing boxes of small cross-sectional area, a pumping ring may be mounted on the outside diameter of the seal head (Fig. 9). Here this screw-type ring, referred to as an axial-flow pumping ring, can be used to circulate coolant when multiple seals are used.

It should be noted that during periods of shutdown different wear problems may exist because the seal faces may be too cold. The product being pumped may be a solution that can crystallize or solidify at ambient temperatures. For these applications the seal faces may have to be preheated before starting, to avoid damage to the seal.

Seal Lubrication/Leakage The effective operation of a seal depends on the lubricating film between the sliding surfaces of the primary and mating rings. The escape of liquid from the film to the low-pressure or atmosphere side of the seal is the leakage. Small distortions or motions transmitted to the seal faces will influence seal lubrication as well as leakage.

Despite the fact that the surfaces of a seal have been lapped to precision flatness, each develops a natural surface waviness during fabrication of the seal, as shown in Fig. 10 a. When relative motion begins, the surface waviness will grow and a hydrodynamic film will develop between the sealing planes. The surface waviness during stable operation (Fig. 10b) will be larger than the initial waviness of a new surface, as a result of the heat being developed. The film allows operation of the seal with minimum wear and power loss. Leakage will also be minimal.

If the load on the seal faces increases, the surface will grow and patches of the sealing plane will break through the hydrodynamic film. When this occurs, the seal is operating in a region of thermoelastic instability (Fig. 10c). Heavy leakage will occur as portions of the surface which have broken through the film increase in temperature, turning bright red. The liquid film will begin to carbonize, vaporize, or flash, producing more leakage and wear of the seal parts. Flashing from the pump stuffing box may be heard above the equipment noise as a spitting or sputtering sound.

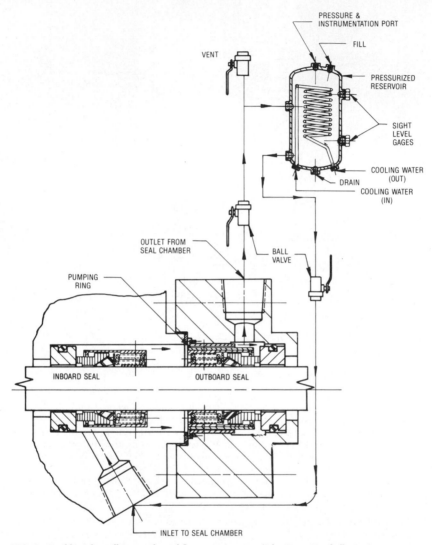

FIG. 9   Double seal installation with axial-flow pumping ring. (John Crane-Houdaille, Inc.)

After the seal has cooled to room temperature, the spots will appear as depressions in the sealing planes (Fig. 10*d*) and may not be in any relation to the waviness of the cooled surface. Sealing difficulty from thermoelastic instability can be overcome by incorporating hydropads into the sealing planes. Some seal manufacturers use this method to provide interface cooling directly to the seal surfaces (Fig. 11).

Leakage is affected by the parallelism of the sealing planes, angular misalignment, coning (negative face rotation), thermal distortion (positive face rotation), shaft runout and whirl, axial vibration, and fluctuating pressure. For parallel faces which take into account seal geometry only, the theoretical leakage in cubic centimeters per hour can be estimated from

$$Q = -C_3 \frac{h^3(P_2 - P_1)}{\mu \ln(R_2/R_1)}$$

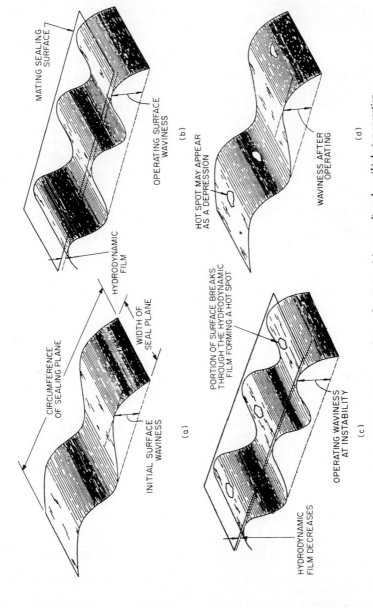

FIG. 10 Mechanical seal face waviness for different conditions of operation. (*a*) new sealing plane, (*b*) during operation without surface instability, (*c*) during operation with instability, (*d*) after instability when sliding system comes to rest.

MATING SEALING SURFACE

OPERATING SURFACE WAVINESS

(b)

HYDRODYNAMIC FILM

CIRCUMFERENCE OF SEALING PLANE

WIDTH OF SEAL PLANE

INITIAL SURFACE WAVINESS

(a)

HOT SPOT MAY APPEAR AS A DEPRESSION

WAVINESS AFTER OPERATING

(d)

PORTION OF SURFACE BREAKS THROUGH THE HYDRODYNAMIC FILM FORMING A HOT SPOT

OPERATING WAVINESS AT INSTABILITY

HYDRODYNAMIC FILM DECREASES

(c)

where $C_3 = 2.13 \times 10^{10}$ in USCS units and 1.88×10^9 in SI units

$$h = \text{face gap, in (m)}$$
$$P_2 = \text{pressure at face ID, lb/in}^2 \text{ (N/m}^2)$$
$$P_1 = \text{pressure at face OD, lb/in}^2 \text{ (N/m}^2)$$
$$\mu = \text{dynamic viscosity, cP (N·s/m}^2)$$
$$R_2 = \text{outer face radius, in (m)}$$
$$R_1 = \text{inner face radius, in (m)}$$

FIG. 11 Primary ring with hydropads for face lubrication. (John Crane-Houdaille, Inc.)

Negative leakage indicates flow from the face outer diameter to the inner diameter. The effect of centrifugal force from one of the rotating sealing planes is very small and can be neglected in normal pump applications. The gap between the seal faces is a function of the materials of construction, flatness, and the liquid being sealed. The face gap can range from 20×10^{-6} to 50×10^{-6} in $(0.508 \times 10^{-6}$ to 1.27×10^{-6} m).

CLASSIFICATION OF SEALS BY ARRANGEMENT

Sealing arrangements may be classified into two groups:

1. Single seal installations
   a. Internally mounted
   b. externally mounted
2. Multiple seal installations
   a. Double seals
   b. Tandem seals

Single seals are used in most applications. This is the simplest seal arrangement with the least number of parts. An installation may be referred to as inside-mounted or outside-mounted, depending on whether the seal is mounted inside or outside the pump stuffing box (Fig. 12). The most common installation is an inside-mounted seal. Here the liquid under pressure acts with the spring load to keep the seal faces in contact.

Outside-mounted seals are considered for low-pressure applications. An external seal installation is used to minimize corrosion that might occur if the metal parts of the seal were directly exposed to the liquid being sealed.

Multiple seals are used in applications requiring

1. A neutral liquid for lubrication
2. Improved corrosion resistance
3. A buffered area for plant safety

Double seals consist of two single seals back to back, with the primary rings facing in opposite directions in the stuffing box chamber. The neutral liquid, at pressure higher than that of the liquid being pumped, lubricates the seal faces (Fig. 13). The inboard seal keeps the liquid being pumped from entering the pump stuffing box. Both inboard and outboard seals prevent the loss of neutral lubricating liquid.

Double seals may be used in an opposed arrangement. Two seals are mounted face to face, with the primary sealing rings rotating on a common mating ring (Fig. 14). The neutral liquid is circulated between the seals at a pressure lower than that of the process liquid. The inboard seal is similar to a single inside-mounted seal and carries the full differential pressure of the pump stuffing box to the neutral liquid. The outboard seal carries only the pressure of the neutral liquid

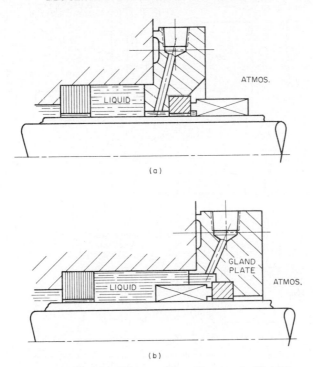

FIG. 12 Single seal installations: (*a*) outside-mounted, (*b*) inside-mounted.

to atmosphere. The purpose of this arrangement is to fit a seal installation having a shorter axial length than is possible with back-to-back double seals and still form a buffered area for plant safety.

Tandem seals are arranged with two single seals mounted in the same direction (Fig. 15). The outboard seal and neutral liquid create a buffer zone between the liquid being pumped and the atmosphere. Normally, the pressure differential from the liquid being sealed and atmosphere is taken across the inboard seal, while the neutral lubricating liquid is at atmospheric pressure. The liquid in the outboard seal chamber may be circulated to remove seal heat. Tandem seals are used on toxic or flammable liquids, which require a buffered or safety zone.

Package seals are an extension of other seal arrangements. A package seal requires no special measurements prior to seal installation. For a single seal, the seal package consists of the gland

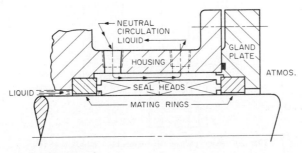

FIG. 13 Double seals.

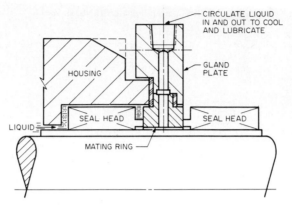

FIG. 14 Opposed double seals.

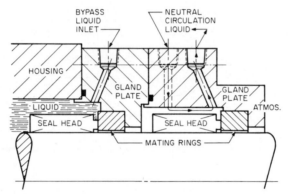

FIG. 15 Tandem seals.

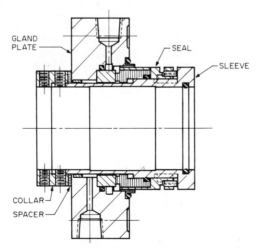

FIG. 16 Single-package seal assembly. (John Crane-Houdaille, Inc.)

plate, sleeve, and drive collar (Fig. 16). A spacer is provided on most package seals to properly set the seal faces. The spacer is removed after the drive collar has been locked to the shaft and the gland plate bolted to the pump.

CLASSIFICATION OF SEALS BY DESIGN

There are four seal classification groups:

1. Unbalanced or balanced
2. Rotating or stationary seal head
3. Single-spring or multiple-spring construction
4. Pusher or nonpusher secondary seal design

The selection of an unbalanced or balanced seal is determined by the pressure in the pump stuffing box. Balance is a way of controlling the contact pressure between the seal faces and power generated by the seal. When the percentage of balance b (ratio of hydraulic closing area to seal face area) is 100 or greater, the seal is referred to as unbalanced. When the percentage of balance for a seal is less than 100, the seal is balanced. Figure 17 illustrates common unbalanced and balanced seals.

The selection of a rotating or stationary seal is determined by the speed of the pump shaft. A seal that rotates with the shaft is a rotating seal assembly (Figs. 1, 16, and 17). When the mating ring rotates with the shaft, the seal is stationary (Fig. 18). Rotating seal heads are common in industry for normal pump shaft speeds. As a rule of thumb, when the shaft speed exceeds 5000 ft/min (25.4 m/s), stationary seals are required. Higher-speed applications require a rotating mating ring to keep unbalanced forces, which may result in seal vibration, to a minimum.

The selection of a single-spring or multiple-spring seal head construction is determined by the space limits and the liquid sealed. Single-spring seals are most often used with bellows seals to load the seal faces (Fig. 19a). The advantage of this type of construction is that the openness of design makes the spring a nonclogging component of the seal assembly. The coils are made of a large-diameter spring wire and, as a result, can withstand a great degree of corrosion.

Multiple-spring seals require a shorter axial space. Face loading is accomplished by a combination of springs placed about the circumference of the shaft (Fig. 19b). Most multiple-spring designs are used with assemblies having O rings or wedges as secondary seals.

Pusher-type seals are defined as seal assemblies in which the secondary seal is moved along the shaft by the mechanical load of the seal and the hydraulic pressure in the stuffing box. The designation applies to seals which use an O ring, wedge, or V ring. A typical construction is illustrated in Fig. 20. The gland plate with its mating ring is fitted to the pump stuffing box and is the stationary member of the seal assembly. There are three points of sealing:

1. Between the gland plate and face of the stuffing box by means of a gasket
2. Between the mating ring and the primary ring through surface contact
3. Between the primary ring and shaft or sleeve by an O ring

The primary ring, with a hardened metal surface, rotates with the shaft and is held against the stationary ring by the compression ring through loading of the O ring. The compression ring supports a nest of springs that are connected at the opposite end by a collar, which is fixed to the shaft. The primary ring is flexibly mounted to take up any shaft deflection or equipment vibration. The collar is fixed to the shaft by set screws.

Another pusher-type seal is illustrated in Fig. 21. When elastomers cannot be used in the product, then a wedge made of Teflon or Grafoil must be considered. A metal retainer locked to the shaft by (A) provides positive drive through the shaft and to the primary ring (F) through drive dents (D) which fit corresponding grooves. The seal between the primary ring and shaft or sleeve is made by a wedge (E), which is preloaded by multiple springs (B). The spring load is

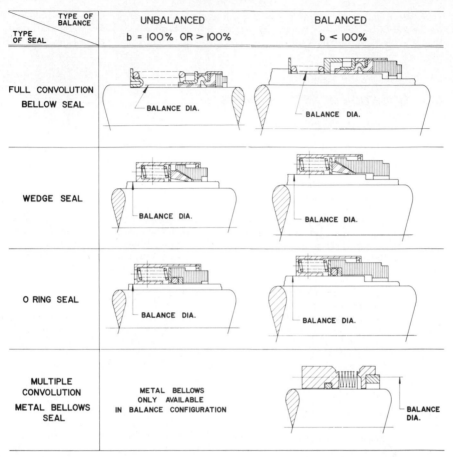

TYPE OF SEAL \ TYPE OF BALANCE	UNBALANCED b = 100% OR > 100%	BALANCED b < 100%
FULL CONVOLUTION BELLOW SEAL	BALANCE DIA.	BALANCE DIA.
WEDGE SEAL	BALANCE DIA.	BALANCE DIA.
O RING SEAL	BALANCE DIA.	BALANCE DIA.
MULTIPLE CONVOLUTION METAL BELLOWS SEAL	METAL BELLOWS ONLY AVAILABLE IN BALANCE CONFIGURATION	BALANCE DIA.

FIG. 17   Common unbalanced and balanced seals.

distributed uniformly by a metal disc (C). The primary ring (G) contacts the mating ring (H) to form the dynamic seal.

Another form of pusher seal is illustrated in Fig. 22. This external split seal is popular for applications involving sewage or slurries. The basic seal parts are located outside the stuffing box so that the split members can be replaced without dismantling the pump or installing new shaft sleeves. The gland sleeve (1) is fitted to the stuffing box (2) and is sealed against the stuffing box bore by an O ring (3). The split end seal (4) is fitted into the gland sleeve and held in place by a nut (5). The antirotation stud is fitted to the stuffing box and passes through the guide lugs (7) on the gland sleeve. This assembly moves axially in the stuffing box according to the pressure exerted by the adjustable springs (8) acting on the gland nut and abutment plate through the yoke plate (9). These springs force the end seal against the cone (10), which rotates with the shaft and is sealed by the cone O ring (11) to provide the running seal.

Nonpusher seals are defined as seal assemblies in which the secondary seal is not forced along the shaft by the mechanical load or hydraulic pressure in the stuffing box. Instead, all movement is taken up by the bellows convolution. This definition applies to those seals which use half-, full-, and multiple-convolution bellows as a secondary seal.

The half-convolution bellows seals are always made of an elastomer (Fig. 23). The tail of the bellows is held to the shaft by a drive band. This squeeze fit seals the shaft and allows the unit to

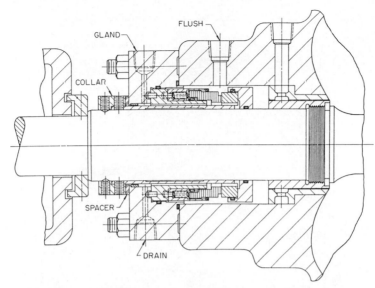

FIG. 18   Stationary seal with rotating mating ring. (John Crane-Houdaille, Inc.)

rotate with the shaft. Positive drive is accomplished through the drive band, retainer, and primary ring by a series of slots and dents. A static seal is created at the back of the primary ring and front of the bellows. This type of seal is used for light-duty service conditions.

The full-convolution bellows seal is illustrated in Fig. 24. The tail of the bellows is held to the shaft by a drive band. The squeeze fit seals the shaft and allows the unit to rotate with the shaft. The drive for the seal assembly is similar to that of the half-convolution seal. Static sealing is accomplished at the front of the bellows and the back of the primary ring. The heavier full-convolution bellows design can tolerate greater shaft motion and runout to pressures of 1200 lb/in^2 (8.28 MPa) gage.

Multiple-convolution bellows seals are necessary to add flexibility to those secondary seal materials that cannot be used in any other shape. The mechanical characteristics of Teflon and metals require multiple-convolution designs.

A Teflon bellows assembly is illustrated in Fig. 25. Because of the large cross-sectional area, these types of seals are mounted outside the stuffing box. Pressure at the inside diameter of the seal helps keep the faces closed. Small springs on the atmosphere side of the seal supply the mechanical load to keep the seal faces closed initially.

Metal bellows seals come in various designs. The seal head is of all-metal construction, with either carbon or tungsten carbide insert as a seal face. The insert is a shrink fit to the assembly. Figure 26 illustrates a stationary seal head with a rotating mating ring. The seal head is mounted in the gland plate and held in place with cap screws. High-temperature elastomers and Grafoil extend the temperature limits of this type of seal design to 750°F (400°C). In refinery service, seals with rotating mating rings are preferred. To promote good seal life, the liquid to be sealed must be at least 50°F (10°C) below its atmospheric boiling point. When temperatures are above 450°F (232°C), a nitrogen gas or steam quench at the atmosphere side of the seal should be used to prevent carbonization of the product being sealed.

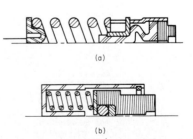

(a)

(b)

FIG. 19   Comparison of (*a*) single-spring and (*b*) multiple-spring seals.

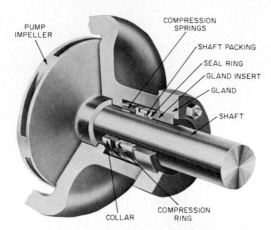

FIG. 20 O-ring-type mechanical seal. (Durametallic)

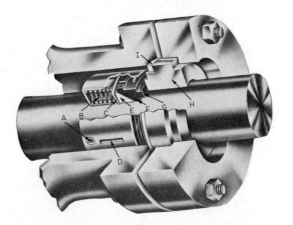

FIG. 21 Wedge type mechanical seal. (John Crane-Houdaille, Inc.)

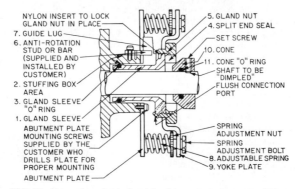

FIG. 22 External mechanical seal for slurry or sewage service. (Johns-Manville)

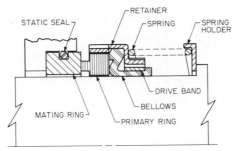

FIG. 23 Half-convolution bellows seal. (John Crane-Houdaille, Inc.)

FIG. 24 Full-convolution bellows seal. (John Crane-Houdaille, Inc.)

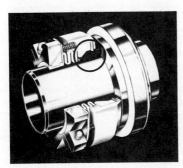

FIG. 25 Teflon bellows seal. (John Crane-Houdaille, Inc.)

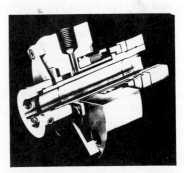

FIG. 26 Stationary metal bellows seal. (John Crane-Houdaille, Inc.)

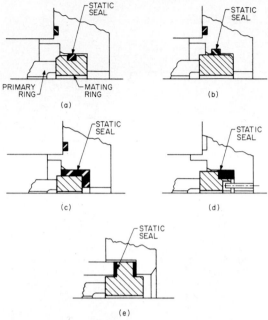

FIG. 27   Common mating ring assemblies: (*a*) grooved O ring, (*b*) square section, (*c*) cup-mounted, (*d*) floating, (*e*) clamped-in.

COMMON TYPES OF MATING RINGS

The purpose of the mating ring is to provide a sealing surface with which the primary ring can make contact, creating a dynamic seal. Since wear will occur at the point of contact, the mating ring must be a replaceable member of the seal assembly. This is accomplished by using such static seal forms as O rings and gaskets (Fig. 27). The material of construction for the static seal determines the temperature limit for the assembly. The mating ring is normally made of hard metal or ceramic.

MATERIALS OF CONSTRUCTION

All component parts of a seal are selected on the basis of their corrosion resistance to the liquid being sealed. The NACE (National Association of Corrosion Engineers) Corrosion Handbook provides corrosion rates for many materials of construction for mechanical seals used with a wide variety of liquids and gases. When the corrosion rate is greater than 2 mils (0.05 mm) per year, double seals that keep the hardware items of the seal in a neutral liquid should be selected to reduce corrosion. In this design only the inside diameter of the mating ring, primary ring, and secondary seal are exposed to the corrosive liquid and should be constructed of corrosion-resistant materials, such as ceramic, carbon, and Teflon. Common materials of construction are given in Table 3. Table 4 lists the properties of common seal face materials.

The operating temperature is a primary consideration in the design of the secondary and static seals in the assembly. These parts must retain their flexibility throughout the life of the seal, as flexibility is necessary to retain the liquid at the secondary seal as well as to allow a degree of freedom for the primary ring to follow the mating ring. The usable temperature limits for common secondary and static seal materials are given in Table 5.

TABLE 3 Common Materials of Construction for Mechanical Seals

Component	Material of construction
Secondary seals	
O ring	Nitrile, ethylene, propylene, chloroprene
Bellows	Nitrile, ethylene, propylene, chloroprene
Wedge	Fluorocarbon resin, Grafoil[a]
Metal bellows	Stainless steel, nickel-base alloy
Primary ring	Carbon, metal-filled carbon, tungsten carbide, silicon carbide, siliconized carbon, bronze
Hardware (retainers, disk, snap ring, set screws, springs)	18-8 stainless steel, 316 stainless steel, nickel base alloys, titanium
Mating ring	Ceramic, cast iron, tungsten carbide, silicon carbide

[a]A registered trademark of Union Carbide.

An additional consideration in the selection of the primary and mating ring materials in sliding contact is their PV limitation. This value is an indication of how well the material combination will resist adhesive wear, which is the dominant wear in mechanical seals. Limiting PV values for various face combinations are given in Table 6. Each limiting value has been developed for a wear rate which provides an equivalent seal life of two years. A PV value for an individual application may be compared with the limiting PV value for the materials used to determine satisfactory service. These values apply to aqueous solutions at 120°F (49°C). For lubricating liquids such as oil, values 60% higher can be used.

GLAND PLATES AND PIPING ARRANGEMENTS

The success or failure of a seal installation can often be traced to the proper selection of the gland plate and associated piping arrangement. The purpose of the gland plate, or end plate, is to hold the mating ring assembly to the pump. The plate is also a pressure-containing component of the assembly. The alignment of one of the sealing surfaces, and possibly a bushing, is dependent on the fit of the gland plate to the stuffing box. To ensure proper installation, the API (American Petroleum Institute) specification requires a register fit with the inside or outside diameter of the stuffing box. This is an industry standard. In addition, the gland plate at the stuffing box must completely confine the static seal.

There are three basic gland plate constructions, as shown in Fig. 28.

A *plain* gland plate is used where seal cooling is provided internally through the pump stuffing box and where the liquid to be sealed is not considered hazardous to the plant environment and will not crystallize or carbonize at the atmospheric side of the seal.

A *flush* gland plate is used where internal cooling is not available. Here, coolant (liquid sealed or liquid from an external source) is directed to the seal faces where the seal heat is generated.

A *flush-and-quench* gland plate is required on those applications which need direct cooling as well as a quench fluid at the atmospheric side of the seal. The purpose of the quench fluid, which may be a liquid, gas, or steam, is to prevent the buildup of any carbonized or crystallized material along the shaft. When properly applied, a seal quench can increase the life of a seal installation by eliminating the loss of seal flexibility due to hangup.

A *flush, vent, and drain* gland plate is used where seal leakage needs to be controlled. Flammable vapors leaking from the seal can be vented to a flair and burned off, while nonflammable liquid leakage can be directed to a safe sump.

Figure 29 illustrates some *restrictive devices* used in the gland when quench, or vent-and-drain, connections are used. These bushings may be pressed in place, as shown in Fig. 29a, or allowed to float, as in Figs. 29b and 29c. Floating bushings allow for closer running fits with the shaft because such bushings are not restricted at their outside diameter. Small packing rings can also be used for a seal quench, as shown in 29d.

TABLE 4 Properties of Common Seal Face Materials

Property			Ceramic		Carbides		Carbon			
	Cast iron	Ni-resist	85% (Al_2O_3)	99% (Al_2O_3)	Tungsten (6% Co)	Silicon Si-C	Resin	Babbitt	Bronze	Si-C Conv.
Modulus of elasticity, $\times 10^6$ lb/in² ($\times 10^3$ MPa)	13–15.95 (90–110)	10.5–16.9 (72–117)	32 (221)	50 (245)	90 (621)	48–57 (331–393)	2.5–4.0 (17.2–27.6)	1.04–4.1 (7.2–28.3)	2.9–4.4 (20–30)	2–2.3 (13.8–15.9)
Tensile strength, $\times 10^3$ lb/in² (MPa)	65–120 (448–827)	20–45 (138–310)	20 (138)	39 (269)	123.25 (850)	20.65 (142)	4.5–9 (31–62)	8–8.6 (55–59)	7.5–9 (52–62)	2 (14)
Coefficient of thermal expansion $\times 10^{-6}$ in/in·°F (cm/cm·K)	6.6 (11.88)	6.5–6.8 (11.7–12.24)	3.9 (7.02)	4.3 (7.74)	2.53 (4.55)	1.88 (3.38)	2.3–3.4 (4.14–6.12)	2.1–2.7 (3.78–4.86)	2.4–3.1 (4.32–5.58)	2.4–3.2 (4.32–5.76)
Thermal conductivity, Btu·ft/h·ft²·°F (W/m·K)	23–29 (39.79–50.17)	25–28 (43.25–48.44)	8.5 (14.70)	14.5 (25.08)	41–48 (70.93–83.04)	41–60 (70.93–103.8)	3.8–12 (6.57–20.76)	6–9 (10.38–15.57)	8–8.5 (13.84–14.70)	30 (51.9)
Density, lb/in³ (kg/m³)	0.259–0.268 (7169–7418)	0.264–0.268 (7307–7418)	0.123 (3405)	0.137 (3792)	0.59 (16,331)	0.104 (2879)	0.064–0.069 (1771–1910)	0.083–0.112 (2297–3100)	0.083–0.097 (2297–2685)	0.067–0.070 (1854–1938)
Hardness	Brinell		Rockwell A			Rockwell 45N	Shore			Rockwell 15T
	217–269	131–183	87	87	92	86–88	80–105	60–95	70–92	90

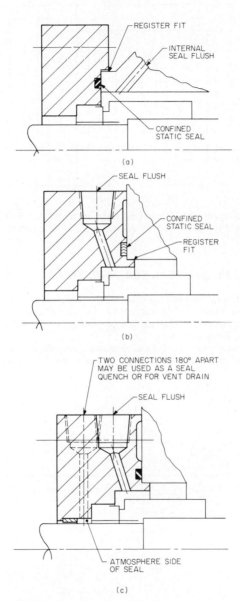

FIG. 28 Basic gland plate designs: (*a*) plain gland plate, (*b*) flush gland plate, (*c*) quench, or vent-and-drain, gland plate.

TABLE 5 Temperature Limits of Secondary Seal Materials

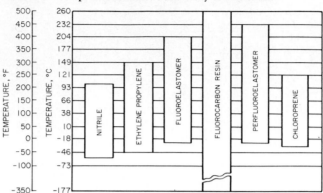

A piping arrangement, or plan, defines how a seal installation will be cooled or, in some cases, heated. Commonly used systems have been defined by API and are shown in Figs. 30 and 31.

INSTALLING THE MECHANICAL SEAL AND TROUBLE SHOOTING _____

A successful seal installation requires operation of the pump within the manufacturer's specification. Relative movement between the seal parts or shaft sleeve usually indicates that mechanical motion has been transmitted to the seal parts from misalignment (angular or parallel), end play, or radial runout of the pump (Fig. 32).

Angular misalignment results when the mating ring is not square with the shaft and will cause excessive movement of internal seal parts as the primary ring follows the out-of-square mating

DESCRIPTION		COMMENTS
(a) THROTTLE BUSHING	ATMOS. SEAL ID 0.0125 in (3.17 X 10⁻⁴m) RADIAL CLEARANCE	USED ON VENT AND DRAIN DESIGNS OF NONSPARKING MATERIALS. MEETS API SPECIFICATION.
(b) FLOATING THROTTLE BUSHING	0.0025 in (6.35 X 10⁻⁵m) RADIAL CLEARANCE	MAY BE USED ON QUENCH AND VENT AND DRAIN GLANDS. SPRING LOADED TO FLOAT WITH SHAFT. MADE OF NONSPARKING MATERIALS. MEETS API SPECIFICATION. REQUIRES MORE SPACE THAN FIXED BUSHING.
(c) FLOATING BUSHING		SOFT PACKING SIZED TO SHAFT DIAMETER USED ON QUENCH AND VENT AND DRAIN GLANDS. SPRING LOADED. NO ADJUSTMENTS REQUIRED. EXCELLENT DRY RUN.
(d) PACKING RINGS		CREATES POSITIVE SEAL ON QUENCH DESIGNS. REQUIRES SOME ADJUSTMENT DURING OPERATION.

FIG. 29 Common restrictive devices used with quench, or vent-and-drain, gland plates.

TABLE 6 Frequently Used Seal Face Materials and Their *PV* Limitations

Sliding materials		PV limit, $\dfrac{\text{lb}}{\text{in}^2} \cdot \dfrac{\text{ft}}{\text{min}}$ (bar·m/s)	Comments
Rotating	Stationary		
Carbon-graphite	Ni-resist	100,000 (35.03)	Better thermal shock resistance than ceramic
	Ceramic (85% Al_2O_3)	100,000 (35.03)	Poor thermal shock resistance and much better corrosion resistance than Ni-resist
	Ceramic (99% Al_2O_3)	100,000 (35.03)	Better corrosion resistance than 85% Al_2O_3 ceramic
	Tungsten carbide (6% Co)	500,000 (175.15)	With bronze-filled carbon-graphite, PV is up to 100,00 $\dfrac{\text{lb}}{\text{in}^2} \cdot \dfrac{\text{ft}}{\text{min}}$ (35.03 bar·m/s)
	Tungsten carbide (6% Ni)	500,000 (175.15)	Ni binder for better corrosion resistance
	Silicon carbide converted carbon	500,000 (175.15)	Good wear resistance; thin layer of Si-C makes relapping questionable
	Silicon carbide (solid)	500,000 (175.15)	Better corrosion resistance than tungsten carbide but poorer thermal shock resistance
Carbon-graphite		50,000 (17.51)	Low PV, but very good against face blistering
Ceramic		10,000 (3.50)	Good service on sealing paint pigments
Tungsten carbide		120,000 (42.04)	PV is up to 185,000 $\dfrac{\text{lb}}{\text{in}^2} \cdot \dfrac{\text{ft}}{\text{min}}$ (64.8 bar·m/s) with two grades that have different % of binder
Silicon carbide converted carbon		500,000 (175.15)	Excellent abrasion resistance; more economical than solid silicon carbide
Silicon carbide (solid)		500,000 (175.15)	Excellent abrasion resistance, good corrosion resistance, and moderate thermal shock resistance

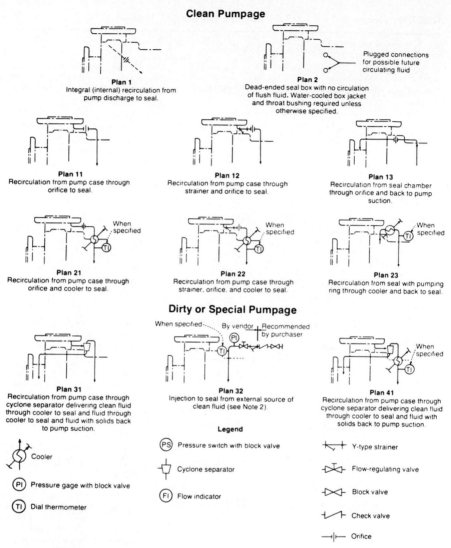

Clean Pumpage

Plan 1
Integral (internal) recirculation from
pump discharge to seal.

Plan 2
Dead-ended seal box with no circulation
of flush fluid. Water-cooled box jacket
and throat bushing required unless
otherwise specified.

Plugged connections
for possible future
circulating fluid

Plan 11
Recirculation from pump case through
orifice to seal.

Plan 12
Recirculation from pump case through
strainer and orifice to seal.

Plan 13
Recirculation from seal chamber
through orifice and back to pump
suction.

Plan 21
Recirculation from pump case through
orifice and cooler to seal.

When specified

Plan 22
Recirculation from pump case through
strainer, orifice, and cooler to seal.

When specified

Plan 23
Recirculation from seal with pumping
ring through cooler and back to seal.

When specified

Dirty or Special Pumpage

When specified · · By vendor | Recommended
by purchaser

Plan 31
Recirculation from pump case through
cyclone separator delivering clean fluid
through cooler to seal and fluid through
cooler to seal and fluid with solids back
to pump suction.

Plan 32
Injection to seal from external source of
clean fluid (see Note 2).

Plan 41
Recirculation from pump case through
cyclone separator delivering clean fluid
through cooler to seal and fluid with
solids back to pump suction.

When specified

Legend

Cooler

(PI) Pressure gage with block valve

(TI) Dial thermometer

(PS) Pressure switch with block valve

Cyclone separator

(FI) Flow indicator

Y-type strainer

Flow-regulating valve

Block valve

Check valve

Orifice

NOTES:
1. These plans represent commonly used systems. Other variations and systems are available and should be
specified in detail by the purchaser or mutually agreed upon by the purchaser and the vendor.
2. For Plan 32, the purchaser shall specify the fluid characteristics, and the vendor shall specify the volume
(gallons per minute) and pressure (pounds per square inch gage) required.

FIG. 30   Piping for mechanical seals from API Standard 610, 6th ed. (Reproduced by courtesy of the American
Petroleum Institute)

ring. This movement will fret the sleeve or seal hardware on pusher types of seal designs. Angular
misalignment may also occur from a stuffing box which has been distorted by piping strain developed at operating temperatures. Here, damage to the wearing rings may also be found if the
pump stuffing box has been distorted.

Parallel misalignment results when the stuffing box is not properly aligned with the rest of the

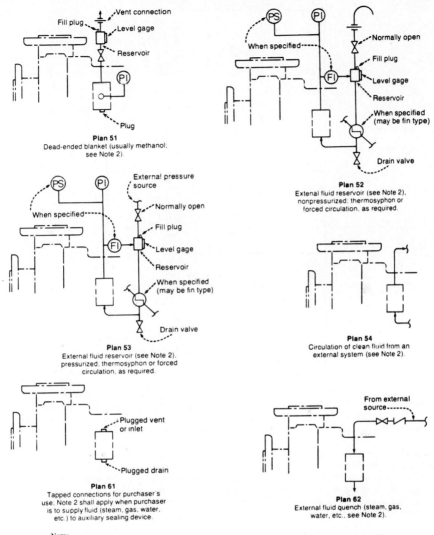

Plan 51
Dead-ended blanket (usually methanol;
see Note 2).

Plan 52
External fluid reservoir (see Note 2),
nonpressurized; thermosyphon or
forced circulation, as required.

Plan 53
External fluid reservoir (see Note 2),
pressurized; thermosyphon or forced
circulation, as required.

Plan 54
Circulation of clean fluid from an
external system (see Note 2).

Plan 61
Tapped connections for purchaser's
use. Note 2 shall apply when purchaser
is to supply fluid (steam, gas, water,
etc.) to auxiliary sealing device.

Plan 62
External fluid quench (steam, gas,
water, etc., see Note 2).

NOTES
1. These plans represent commonly used systems. Other variations and systems are available and should be specified in detail by the purchaser or mutually agreed upon by the purchaser and the vendor.
2. The purchaser shall specify the fluid characteristics when supplemental seal fluid is provided. The vendor shall specify the volume (gallons per minute) and pressure (pounds per square inch gage) required, where these are factors.
3. See Figure D-2 for explanation of symbols not specified here.

FIG. 31 Piping for throttle bushing, auxiliary seal device, tandem seals, or double seals from API Standard 610, 6th ed. (Reproduced by courtesy of the American Petroleum Institute)

pump. No seal problems will occur unless the shaft strikes the inside diameter of the mating ring. If damage has occurred, there will also be damage to the bushing at the bottom of the box at the same location as the mating ring.

Excessive axial endplay can damage the seal surfaces and also cause fretting. If the seal is continually being loaded and unloaded, abrasives can penetrate the seal faces and cause prema-

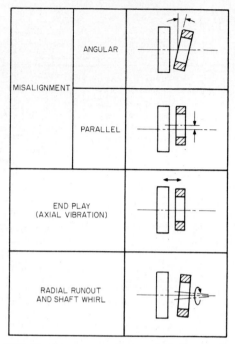

FIG. 32 Common types of motion that influence seal performance.

ture wear of the primary and mating rings. Thermal damage in the form of heat checking in the seal faces as a result of excessive endplay can occur if the seal is operated below working height.

Radial runout in excess of limits established by the pump manufacturer will cause excessive vibration at the seal. This vibration, coupled with small amounts of the other types of motion which have been defined, will shorten seal life.

Instructions and seal drawings should be reviewed to determine the installation dimension, or spacing, required to ensure that the seal is at its proper working height (Fig. 33). The installation reference can be determined by locating the face of the box on the surface of the sleeve and then measuring along the sleeve after it has been removed from the unit. It is not necessary to use this procedure if a step in the sleeve or collar has been designed into the assembly to provide for proper seal setting. Assembly of other parts of the seal will bring the unit to its correct working height.

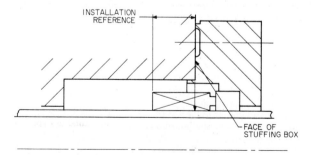

FIG. 33 Typical installation reference dimensions.

TABLE 7 Checklist for Identifying Causes of Seal Leakage

Symptom	Possible causes	Corrective procedures
Seal spits and sputters ("face popping") in operation	Seal fluid vaporizing at seal interfaces	Increase cooling of seal faces Check for proper seal balance with seal manufacturer Add bypass flush line if not in use Enlarge bypass flush line and/or orifices in gland plate Check for seal interface cooling with seal manufacturer
Seal drips steadily	Faces not flat Carbon graphite seal faces blistered Seal faces thermally distorted	Check for incorrect installation dimensions Check for improper materials or seals for the application Improve cooling flush lines Check for gland plate distortion due to overtorquing of gland bolts Check gland gasket for proper compression Clean out foreign particles between seal faces; relap faces if necessary Check for cracks and chips at seal faces; replace primary and mating rings
	Secondary seals nicked or scratched during installation O rings overaged Secondary seals hard and brittle from compression set Secondary seals soft and sticky from chemical attack	Replace secondary seals Check for proper lead in chamfers, burrs, etc. Check for proper seals with seal manufacturer Check with seal manufacturer for other material
	Spring failure Hardware damaged by erosion Drive mechanisms corroded	Replace parts Check with seal manufacturer for other material
Seal squeals during operation	Amount of liquid inadequate to lubricate seal faces	Add bypass flush line if not in use Enlarge bypass flush line and/or orifices in gland plate
Carbon dust accumulates on outside of gland ring	Amount of liquid inadequate to lubricate seal faces Liquid film evaporating between seal faces	Add bypass flush line if not in use Enlarge bypass flush line and/or orifices in gland plate Check for proper seal design with seal manufacturer if pressure in stuffing box is excessively high

TABLE 7   Checklist for Identifying Causes of Seal Leakage (*Continued*)

Symptom	Possible causes	Corrective procedures
Seal leaks	Nothing appears to be wrong	Refer to list under "Seal drips steadily" Check for squareness of stuffing box to shaft Align shaft, impeller, bearing, etc., to prevent shaft vibration and/or distortion of gland plate and/or mating ring
Seal life is short	Abrasive fluid	Prevent abrasives from accumulating at seal faces Add bypass flush line if not in use Use abrasive separator or filter
	Seal running too hot	Increase cooling of seal faces Increase bypass flush line flow Check for obstructed flow in cooling lines
	Equipment mechanically out of line	Align Check for rubbing of seal on shaft

Package shaft seals (Fig. 16) can be assembled with relative ease, for just the bolts in the gland plate and set screws need to be fastened to the stuffing box and shaft. After the seal spacer is removed, the unit is ready to operate.

To assemble a mechanical seal to a pump, a spacer coupling is required. If the pump is packed but may later be converted to mechanical seals, a spacer coupling should be included in the design.

Since a seal has precision-lapped faces and secondary seal surfaces are critical in the assembly, installation to the equipment should be kept as clean as possible. All lead edges on sleeves and glands should have sufficient chamfers to facilitate installation.

When mechanical seals are properly applied, there should be no static leakage and, under normal conditions, the amount of dynamic leakage should range from none to just a few drops per minute. Under full vacuum, a mechanical seal is used to prevent air from leaking into the pump. If excessive leakage occurs, the cause must be identified and corrected. Causes for seal leakage with possible corrections are listed in Table 7. Also, Fig. 34 illustrates the most common failures found in mechanical seals. Further information on seal leakage and related condition of seal parts may be found in the works listed in "Further Reading."

FURTHER READING

Abar, J. W.: "Failures of Mechanical Face Seals," in *Metals Handbook,* Vol. 10, 8th ed., American Society for Metals, Metals Park, Ohio, 1975.

American Petroleum Institute: Centrifugal Pumps for General Refinery Services, API Standard 610, 6th ed., Washington, D.C., 1980.

Crane Packing Company: *Engineered Fluid Sealing: Materials, Design, and Applications*, Morton Grove, Ill., 1979.

Crane Packing Company: *Identifying Causes of Seal Leakage*, S-2031 and Bulletin, Morton Grove, Ill., 1979.

1. FULL CONTACT PATTERN

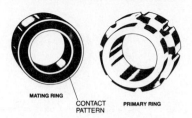

MATING RING

CONTACT PATTERN

PRIMARY RING

OBSERVATION:
Typical contact pattern for a **non-leaking** seal. Full contact on the mating ring surface through 360°. Little or no measurable wear on either seal ring. If leakage is present with this type face pattern, the secondary seals must be examined.

2. CONING (NEGATIVE ROTATION)

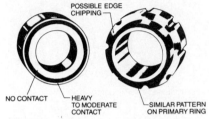

POSSIBLE EDGE CHIPPING

NO CONTACT

HEAVY TO MODERATE CONTACT

SIMILAR PATTERN ON PRIMARY RING

OBSERVATION:
Heavy contact on the mating ring pattern at the outside diameter of the seal. Fades away to no visible contact at the inside diameter of contact pattern. Possible edge chipping on the outside diameter of primary ring.

3. THERMAL DISTORTION (POSITIVE ROTATION)

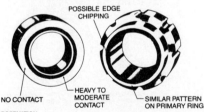

POSSIBLE EDGE CHIPPING

NO CONTACT

HEAVY TO MODERATE CONTACT

SIMILAR PATTERN ON PRIMARY RING

OBSERVATION:
Heavy contact on the mating ring pattern at the inside diameter of the seal. Fades away to no visible contact at the outside diameter of contact pattern. Possible edge chipping on the inside diameter of the primary ring.

4. MECHANICAL DISTORTION

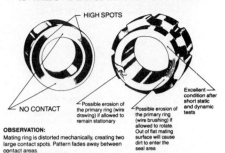

HIGH SPOTS

NO CONTACT

Possible erosion of the primary ring (wire drawing) if allowed to remain stationary

Possible erosion of the primary ring (wire brushing) if allowed to rotate.
Out of flat mating surface will cause dirt to enter the seal area

Excellent condition after short static and dynamic tests

OBSERVATION:
Mating ring is distorted mechanically, creating two large contact spots. Pattern fades away between contact areas.

5. MECHANICAL DISTORTION

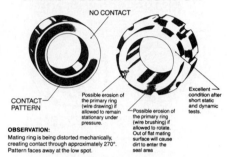

NO CONTACT

CONTACT PATTERN

Possible erosion of the primary ring (wire drawing) if allowed to remain stationary under pressure.

Possible erosion of the primary ring (wire brushing) if allowed to rotate.
Out of flat mating surface will cause dirt to enter the seal area

Excellent condition after short static and dynamic tests.

OBSERVATION:
Mating ring is being distorted mechanically, creating contact through approximately 270°. Pattern faces away at the low spot.

6. MECHANICAL DISTORTION

NO CONTACT

CONTACT ONLY AT HIGH SPOTS

OBSERVATION:
Mating ring is being distorted mechanically, creating contact at bolts. High spots are at each bolt location.

FIG. 34 Identifying causes of seal leakage. (John Crane-Houdaille, Inc.)

Hamaker, J. B.: "Mechanical Seal Lubrication Systems," 1977 ASLE Education Program—Fluid Sealing Course, American Society of Lubrication Engineers and Crane Packing Company, Morton Grove, Ill., May 1977.

Hamner, N. E. (compiler), *Corrosion Data Survey*, 5th ed., National Association of Corrosion Engineers, Houston, 1975.

Netzel, J. P.: "Surface Disturbances in Mechanical Face Seals from Thermoelastic Instability," American Society of Lubrication Engineers, 35th Annual Meeting, Anaheim, Calif., May 5, 1980.

Schoenherr, K.: "Design Terminology for Mechanical Face Seals," *SAE Transactions* **74**(650301) (1966).

7. HIGH WEAR OR THERMALLY DISTRESSED SURFACE

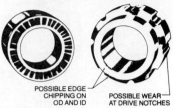

POSSIBLE EDGE CHIPPING ON OD AND ID

POSSIBLE WEAR AT DRIVE NOTCHES

OBSERVATION:
High wear of mating or thermally distressed surface (heat checking) through 360°. High primary ring wear with carbon deposits on atmosphere side of seal. Possible edge chipping of primary ring due to opening and closing of seal faces.

8. SECTION OF THERMALLY DISTRESSED SURFACE

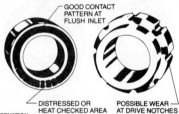

GOOD CONTACT PATTERN AT FLUSH INLET

DISTRESSED OR HEAT CHECKED AREA

POSSIBLE WEAR AT DRIVE NOTCHES

OBSERVATION:
Thermally distressed area approximately 1/3 of the contact pattern. Distressed area 180° from inlet of seal flush. High primary ring wear with possible carbon deposits on atmosphere side of seal.

9. PATCHES OF THERMALLY DISTRESSED SURFACE

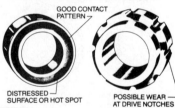

GOOD CONTACT PATTERN

DISTRESSED SURFACE OR HOT SPOT

POSSIBLE WEAR AT DRIVE NOTCHES

OBSERVATION:
Patches of thermally distressed surface (heat checking). 2, 3, 4, 5 or 6 hot spots are possible. High primary ring wear with possible carbon deposits on atmosphere side of seal. Failure due to hot spots (thermal asperities) is likely to occur on light specific gravity liquids at high speeds and pressures.

10. HIGH WEAR AND GROOVING

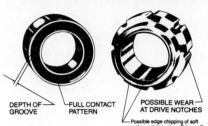

DEPTH OF GROOVE

FULL CONTACT PATTERN

POSSIBLE WEAR AT DRIVE NOTCHES

Possible edge chipping of soft carbon primary rings. Edges will be rounded for hard material of construction like tungsten carbide.

OBSERVATION:
High wear of the mating ring. Primary ring has grooved the mating ring evenly through 360°.

11. OUT-OF-SQUARE MATING RING

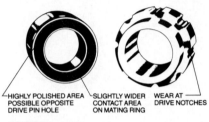

HIGHLY POLISHED AREA POSSIBLE OPPOSITE DRIVE PIN HOLE

SLIGHTLY WIDER CONTACT AREA ON MATING RING

WEAR AT DRIVE NOTCHES

OBSERVATION:
Contact pattern through 360° slightly larger than primary ring face width. High spot may be present on the mating ring opposite a drive pin hole. Mating ring without static seals will rock or move in gland plate or holder.

12. WIDE CONTACT PATTERN

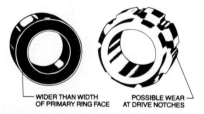

WIDER THAN WIDTH OF PRIMARY RING FACE

POSSIBLE WEAR AT DRIVE NOTCHES

OBSERVATION:
Contact pattern considerably wider on the mating ring than the face width of the primary ring.

FIG. 34　Identifying causes of seal leakage (*Continued*)

Schoenherr, K., and R. L. Johnson, "Seal Wear," in *Wear Control Handbook* (M. Peterson and W. Winer, eds.), American Society of Mechanical Engineers, New York, 1980.

Snapp, R. B.: "Theoretical Analysis of Face Type Seals with Varying Radial Face Profiles," 64-WA/LUB 6, American Society of Mechanical Engineers, New York, 1964.

2.2.4
CENTRIFUGAL PUMP INJECTION-TYPE SHAFT SEALS

ROBERT D. ASHTON

Injection-type shaft seals (sometimes called *packless stuffing boxes*) are designed to control leakage from hot-water pumps. Cool water is injected into each seal to either suppress or regulate the hot leakage, which would otherwise flash upon reaching the outside of the pump.

Injection-type shaft seals provide high reliability and yet require little maintenance. They are used primarily in power plant boiler-feed and reactor-feed pump applications where shaft peripheral speeds are high (3600 rpm and up) and pumping temperatures are greater than 250°F (120°C). Under these conditions, conventional packing or mechanical-seal-type stuffing boxes may not be suitable or desirable.

Injection shaft seals are either *serrated throttle bushings* or *floating ring seal* designs that regulate the flow, temperature, and pressure of the controlled leakage. The flow of the cool injection and of any hot water in the seals depends on operating pressures but is restricted by close seal clearances and regulated by injection control valves.

The operating temperature of the seals is controlled by allowing cool injection water to surround the outside of the seal. Ports in the seal allow the cool injection water access to the shaft to either overcome or mix with hot water in the seal so that the resulting seal leakage is cool.

The seal designs must provide sufficient pressure breakdown between pump suction or balance device chamber pressures inside the pump and atmospheric conditions outside the pump. Upon reaching the outside of the seal, the cool leakage is piped away by gravity drain for eventual return to the power plant feedwater system.

SERRATED THROTTLE BUSHINGS

The construction of a serrated (sometimes called *labyrinth* or *grooved*) bushing (Fig. 1) varies from one pump manufacturer to another. However, the serrated designs basically involve a rotating shaft running with reasonably small clearance, 0.002 to 0.003 in per inch (0.002 to 0.003 mm per millimeter) of shaft diameter, within a solid stainless steel (typically 12% chrome) stationary bushing installed in the pump casing end cover. Grooved serrations are applied to the hardened stainless steel rotating surface (usually a shaft sleeve) or to the stationary bushing or to both rotating and stationary surfaces to effect a high throttling action or stuffing box leakage reduction.

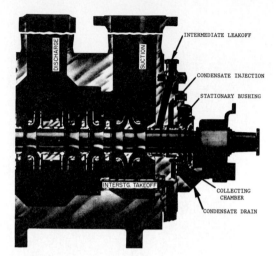

FIG. 1 Modern double-case boiler-feed pump with serrated throttle bushing shaft seals (Byron Jackson Pump Division, Borg-Warner)

The serrations create a more effective pressure breakdown than a smooth axial surface and allow increased running clearances to let foreign particles, such as grit, pass through or flush out in the grooves. The increased clearances allow for more tolerance for any displacement between rotating and stationary components that might result from assembly misalignment or from distortions induced by transient temperatures. The grooved running surface area at clearances is greatly reduced relative to that for a smooth axial surface, and therefore possible metal surface contact between rotating and stationary parts is reduced during transient operation.

Seal leakage could be reduced by decreasing the running clearance in the injection seal, but the smaller clearance would limit the ability of the serrated seal to conform to radial shaft movement. Reduced clearances in any injection seal design may lead to problems with thermal distortions caused by loss of available cool injection water and may also increase galling by particles trapped in the seal running fits.

FLOATING RING DESIGN

A segmented throttle bushing made of many floating rings will allow conformity with radial shaft movement and smaller running clearances to reduce seal leakage. The construction of a floating ring stuffing box (Fig. 2) varies from one pump manufacturer to another, but the unit basically consists of a stack of stainless steel (typically 17% chrome) rings (instead of a stationary bushing) running against a smooth hardened stainless steel (typically 12 to 17% chrome) rotating shaft sleeve surface.

The separate rings are all contained within a housing installed in the pump casing end cover. Each ring is loaded against an adjacent stainless steel ring spacer so that a stationary seal is produced in the axial direction but the ring is still able to move radially with the shaft. Axial loading of the rings is provided by hydraulic pressure during operation and by springs during idle pump periods. The rings are locked against rotation usually by pegs and/or a slot arrangement.

A small radial clearance, 0.001 to 0.0015 in per inch (0.001 to 0.0015 mm per millimeter) of shaft diameter, is provided between the rings and shaft sleeve to provide adequate throttling of seal leakage across the limited number of seal rings. The length of each ring varies with the diameter of the injection-type seal but is generally about ½ in (13 mm). The radial clearance allows for

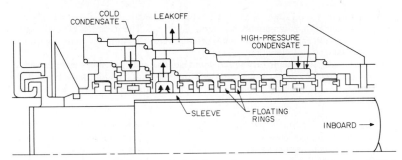

FIG. 2   Floating ring seal design with shaft sleeve. Pegs prevent rotation of spring-loaded rings but let them float radially. (From *Power*, August 1980, McGraw-Hill, New York, copyright 1980)

a reduced injection flow but increases the sensitivity of the seal to foreign particles in the seal leakage.

The individual seal rings float to a certain degree and find their own position relative to the shaft. Their multiplicity reduces the effect of any angular displacement between the rotating and stationary components that might arise from errors in original assembly or from distortions caused by temperature changes during pump load variations.

Some injection-type shaft seal designs use a combination of serrations and floating rings in the same stuffing box. For either type of seal, careful maintenance during assembly or disassembly requires accurate alignment of rotating and stationary components as well as careful handling to avoid scratching of components. Cleanliness is of utmost importance with shaft seals.

CONDENSATE INJECTION REQUIREMENTS

Because boiler-feed and reactor-feed pumps normally have high feedwater temperatures, 250 to 500°F (120 to 260°C), with consequent vapor pressures higher than external atmospheric conditions, the seal leakage must be cooled to avoid flashing in the stuffing box or outside the pump. This cooling is usually accomplished by injecting cold condensate from a power plant condensate pump directly into the seal to cool the stuffing box components as well as the seal leakage.

The amount of condensate injection water required will depend on several factors, including (1) whether the design is serrated or floating ring, (2) type of seal control system, (3) diameter, clearance, and rotating speed of running fit, and (4) internal pump pressure and temperature. A typical example of stuffing box flows is as follows: for a 5500-rpm boiler-feed pump with a temperature-controlled seal system (described later) with a serrated 5-in (127-mm) diameter running seal with about 0.015 in (0.38 mm) of diametrical clearance, the approximate flows are

1. Total injection per seal: 5 to 15 gpm (0.3 to 1.0 l/s)
2. Leakage from pump internal to seal: 0 to 5 gpm (0 to 0.3 l/s)
3. Drainage from each seal: 8 to 20 gpm (0.5 to 1.3 l/s)

Given a seal with a floating ring design (having typically one-half the clearance of seals with a serrated bushing design), the above conditions would require only 50 to 60% of the injection water needed by the serrated bushing design with the larger clearance.

During pump standby periods, higher injection and leakage flows are required because of reduced seal throttling caused by low or zero speed, by high internal pump pressures, or by a pump balance device that induces higher internal leakage during standby. The injection flows to both stuffing boxes on a multistage pump may not be the same as a result of pump balance action at one end of the pump.

Cold condensate, 85 to 110°F (30 to 45°C), is usually available from the power plant condensate pump or secondary condensate booster pump discharge. The variation in condensate supply
pressure relative to internal boiler-feed pump pressure makes it necessary to use injection control
systems for satisfactory seal operation.

Condensate injection systems vary to accommodate the wide range of pressures and temperatures found in different power plant feedwater system layouts. The four basic types of injection
flow systems are (1) manual, (2) pressure-controlled, (3) temperature-controlled, and (4) intermediate-leakoff.

Manual System   A manual flow control requires setting and readjusting a valve by hand for
each stuffing box each time there is a change in pump operating conditions due to varying plant
loads. While it is possible to use such a control system, it is not usually recommended because of
the inability to automatically compensate for rapid changing conditions.

Pressure-Controlled System   The simplest automatic condensate injection system is a
pressure-controlled shaft seal where cold condensate maintained at a pressure greater than boiler-
feed pump suction pressure is injected into the central portion of the seal. A small portion of this
injection water flows inward into the pump proper; most of the injected water flows out of the
seal into a collection chamber adjacent to the pump bearing bracket. This leakage in the collection
chamber, which is vented to atmosphere, is drained by gravity for eventual return to the main
feedwater system.

The pressure-controlled system has a pneumatic injection control valve. This valve, in the
condensate injection line, is governed by an air signal received from a differential pressure control
monitor which maintains the preset pressure differential, 10 to 25 lb/in^2 (70 to 170 kN/m^2),
between injection pressure and internal pump pressure. This system is not always favored, however, because of a feedback instability tendency resulting from operating pressure changes affecting valve position, which changes pressure, etc. Another unfavorable factor for hot-water service
is the introduction of cold condensate into a hot pump at all times, including pump idle periods,
affecting pump prewarming conditions.

Temperature-Controlled System   The temperature-controlled seal system (Fig. 3)
throttles the injection flow to maintain a preset seal drain temperature. This system has a tem-
perature-sensing probe in each seal drain line. Each probe is connected to an indicating temperature controller, which provides an air signal to a pneumatic control valve in the condensate injection line for control of the seal injection flow rate.

The cold condensate is injected into the stuffing box central portion and allowed to mix with

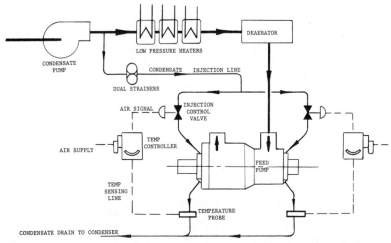

FIG. 3   Temperature-controlled system regulating condensate injection to feed pump
shaft seals.

hot water entering the seal from inside the pump. The drain temperature is maintained at a preset 140 to 150°F (60 to 66°C) to preclude flashing in the stuffing box or in the drains. Note that with this system some hot water enters the seal; therefore cold condensate does *not* enter the hot pump and does not adversely affect pump warming conditions, especially during extended idle periods. The required condensate injection pressure is at least equal to the internal stuffing box pressure plus interim frictional loss between the condensate supply source and the point of hot-water mixture. Note that this system *may* allow satisfactory operation even when condensate supply pressure is nearly equal to boiler-feed pump suction pressure. In addition to providing rapid response to variations in operating pump conditions, this type of control will always supply just enough injection water to maintain the recommended drainage temperature.

Intermediate Leakoff System   The intermediate-leakoff type of shaft seal system (Fig. 4) has many variations but basically uses a bleedoff from a central portion of the stuffing box. This system may be used to reduce internal stuffing box pressure if high boiler-feed pump suction pressure exists. To create a positive leakoff flow, the intermediate bleedoff flow is piped back to a plant feedwater system low-pressure point, such as a plant condenser, heater, or booster pump suction, where the pressure is less than boiler-feed pump suction pressure. However, the back pressure of the leakoff destination must be above the bleedoff vapor pressure to suppress flashing in the leakoff lines. This back pressure may be the leakoff destination pressure or may be created by an orifice or a valve.

Cold condensate injection into the stuffing box is controlled by a pressure differential monitor maintaining a preset pressure above the bleedoff pressure or by a stuffing box drain temperature control. Note that the cold condensate injection pressure need not equal or overcome a high feed pump suction pressure. The condensate injection temperature must still be 85 to 110°F (30 to 45°C) to keep the drain temperature below flashing condition. Condensate injection shaft seals without intermediate bleedoff but subject to suction pressures in excess of about 250 lb/in² (1725 kN/m²) must be extremely long for proper pressure breakdown. Longer shaft seals require thicker pump case end covers, affecting pump cost, and a longer rotating element, which could adversely affect rotor dynamics. The intermediate-leakoff shaft seal is very effective where there is high feed pump suction pressure imposed by boiler-feed booster pumps or in a closed feedwater system, with no deaerating open heater, wherein condensate pump discharge can be fully imposed on the feed pump at low plant loads.

In plant systems with feedwater heaters between booster pump and feed pump, high pressure condensate from the cooler booster pump is injected into the seal to allow a cooler intermediate leakoff to help prevent flashing (see Fig. 2).

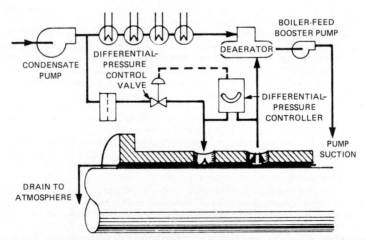

FIG. 4   Typical intermediate-leakoff shaft seal system. (From *Power*, September 1980, McGraw-Hill, New York, copyright 1980)

INJECTION SOURCES

Many power plant feedwater systems with deaerating direct-contact heaters (open cycle) usually have the boiler-feed pump drawing water directly from the deaerator. These open cycles with constant-speed condensate pumps always have cold condensate supply pressures in excess of boiler-feed pump suction pressure because of the increased available condensate pump head at low system flows and the interim system frictional loss at high system flows.

If a boiler-feed booster pump or a closed feedwater system with no open heaters and resultant higher boiler-feed pump suction pressures is used, the pump manufacturer may elect to require condensate injection seal water booster pumps or may use an intermediate-leakoff packless shaft seal with condensate injection overcoming only the intermediate leakoff pressure. If the condensate pumps are variable-speed units which allow the condensate injection pressure to drop to equal feed pump suction pressure at low loads, with little or no feedwater system frictional drop, then condensate injection seal water pumps are required. For this situation, at least one pump manufacturer offers an optional pumping ring configuration in their packless seal that can increase the seal water pressure in the stuffing box to overcome pump suction pressure.

AUXILIARY EQUIPMENT

The condensate injection piping should be conservatively sized based on maximum injection flow requirements to obtain a low pressure drop between injection source and seal injection control valve. These control valves may be equipped with limit stops to prevent full closure and allow continuous cool injection to the seals under almost all operating conditions. In some installations isolating lines are furnished around the valves to allow continued injection flow even during control valve maintenance. The valves can also be designed to remain open during a failure, such as loss of station air to the pneumatic controls, and to close only with air supply. The air supply filter regulators for each control must be furnished with relatively dry clean air at station supply pressure.

The condensate injection supply to the seals must be clear and free of foreign matter to prevent damage to stuffing box components. It is therefore necessary to install filters in the injection line prior to the control valve. To keep damaging fine mill scale, oxide particles, abrasives, and other materials from entering the small seal clearances, several pump manufacturers recommend 100-mesh (150-micron) dual strainers. If dual strainers with isolating valves are used, each filter can be cleaned without interrupting injection flow during pump operation. Pressure gages should be installed before and after each filter to permit the operator to monitor filter pressure drop. A differential pressure switch and alarm for each filter are preferable to alert the operator to clean the strainer when pressure drop becomes excessive.

The condensate injection shaft seals should always be filled with cool water before and during pump operation, even during reverse pump rotation. Some pump manufacturers stipulate that condensate injection *must* be continuous without any interruption during all operation modes.

The clearances in the condensate injection shaft seal may double over the service life of internal wearing parts. With double clearances, the leakage will approximately double. This factor should be considered when sizing the return drain piping back to the plant condenser if frictional losses are to be kept to a minimum. The drain line should be pitched at least ¼ in per foot (20 mm per meter). The collecting chamber at the pump stuffing box is vented to atmosphere, and the only head available to evacuate the chamber is the static head between the pump and the point of return. This head must always be well in excess of the frictional losses (even after the leakage is doubled); otherwise the drains may back up, the collection chambers overflow, and the adjacent bearing brackets flood, with subsequent possible intrusion of water into the pump bearings and lubricating oil.

The seal collection chambers have especially large connections to assure proper drainage, provided no back pressure exists. Two types of condensate drain systems may be used to dispose of the drain coming from the collecting seal chambers. One system uses traps that are piped directly to the plant condenser if sufficient static head exists for positive drain flow. The second system collects the drain in a condensate storage tank into which various other drains (from other pumps shaft seals, etc.) are also directed. As this vented storage tank is under atmospheric pressure, it must be set at a reasonable elevation below the pump centerline so that the static elevation difference will overcome frictional losses in the drain piping. A separate condensate transfer pump, under control of the storage tank liquid level control system, may then pump the condensate drains from the storage tank into the plant condenser. The storage tank should have its own overflow protection system that allows outside drainage if for some reason proper drainage cannot be achieved. For example, the top of the tank vent pipe should be below the pump centerline to help preclude the possibility of drainage backing up to the level of the pump seal collection chambers. Note that this storage tank should also be large enough for adequate drainage collection to help prevent backups.

PACKLESS SHAFT SEALS WITHOUT INJECTION

The packless shaft seals which have been so successfully applied to boiler-feed, reactor-feed, and booster pump service are applicable to a number of other services. For instance, they are very suitable for cold-water condensate booster pumps and for high-pressure pumps applied to hydraulic descaling or hydraulic-press work. In such services there is no need to bring in injection supply water to the breakdown seals (unless the pumped water is not clear and free of gritty material), for the water handled by the pump is already cold with no danger of flashing as it leaves the pump stuffing boxes.

2.2.5
CENTRIFUGAL PUMP OIL FILM JOURNAL BEARINGS

WILBUR SHAPIRO

PRINCIPLES OF OPERATION

A journal bearing is essentially a viscous pump, and it derives load capacity by pumping the lubricant through a small clearance region. Referring to Fig. 1, the fluid is dragged along by the rotating journal. To generate pressure, the resistance to pumping must increase in the direction of flow. This is accomplished by a movement of the journal such that the clearance distribution takes on the form of a tapered wedge in the direction of rotation, as shown in Fig. 1.

The eccentricity e is the total displacement of the journal from its concentric position. The attitude angle γ in Fig. 1 is the angle between the load direction and the line of centers. Note that, because of the necessity to form a converging wedge, the displacement of the journal is not along a line that is coincident with the load vector. A positive pressure is produced in the converging region of the clearance. Downstream of the minimum film thickness, which occurs along the line of centers, the film becomes divergent. The resistance decreases in the direction of pumping, and either negative pressures occur or the air in the lubricant gasifies or cavitates and a region of atmospheric pressure occurs in the bearing area. This phenomenon is known as fluid film bearing cavitation. It should be clearly distinguished from other forms of cavitation that take place in pumps, such as in the impeller, for example. Here the fluid is traveling at high velocity and the inertia forces on each fluid element dominate. Implosions occur in the impeller and can cause damage.

In a bearing, the viscous forces dominate and each fluid particle moves at constant velocity in proportion to the net shearing forces on it. Thus, cavitation in a bearing is more of a change of phase of the lubricant which occurs in a region of lower pressure that permits the release of entrained gases. Generally, bearing cavitation does not cause damage.

Regimes of Lubrication Whether or not a fluid film can be formed by journal rotation is dependent on several factors, including surface speed, viscosity, and load capacity. There is a parameter[1] that is often used to determine a particular regime of lubrication: $ZN/\overline{P}$

where Z = viscosity of lubricant, cP (Pa·s)
  N = rotating speed, rpm
  $\overline{P}$ = average pressure of the bearing, lb/in^2 (bar)°

° 1 bar = 10^5 Pa. For a discussion of bar, see *SI Units—A Commentary* in front matter.

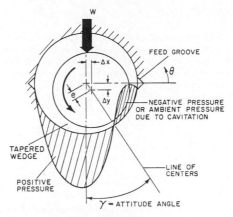

FIG. 1 Two-groove cylindrical bearing.

A plot of coefficient of friction versus $ZN/\overline{P}$ generally has the form of Fig. 2. At low values of $ZN/\overline{P}$, there is a combination of viscosity, speed, and load that places a bearing in a boundary lubricated regime, where typical coefficients of friction are 0.08 to 0.14. Boundary lubrication implies intimate contact between the opposed surfaces. As the value of the parameter increases as a result of increased speed, increased viscosity or lowered load, there is a dramatic reduction in coefficient of friction. In this region, there is a mixed film lubrication and the coefficient of friction varies between 0.02 and 0.08. By mixed film lubrication is meant a situation where the journal is partly surrounded by a fluid film and partly supported by rubbing contact between the opposed members. As $ZN/\overline{P}$ increases further, a situation of full fluid film lubrication prevails. A general rule of thumb is that $ZN/\overline{P}$ should be 30 (0.44) or greater for a fluid film to be generated. Note that as $ZN/\overline{P}$ continues to increase beyond the full film demarcation, the coefficient of friction rises but at a relatively low rate and generally remains in regions of low coefficients of friction.

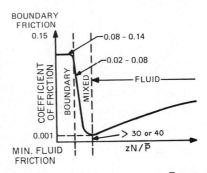

FIG. 2 Coefficient of friction versus $ZN/\overline{P}$. (Ref. 1)

EXAMPLE

$$Z = 30 \text{ cP} \ (0.03 \text{ Pa} \cdot \text{s})$$

$$N = 150 \text{ rpm}$$

$$\overline{P} = 200 \text{ lb/in}^2 \ (13.6 \text{ bar})$$

$$\frac{ZN}{\overline{P}} = \frac{30 \times 150}{200} = 22.5$$

The bearing is not fluidborne and is operating in the mixed film regime.
At what speed will the bearing become hydrodynamic?
For hydrodynamic operation $ZN/\overline{P} = 30$. Therefore

$$N = \frac{30\overline{P}}{Z} = \frac{30 \times 200}{30} = 200 \text{ rpm}$$

THEORETICAL FOUNDATIONS

The foundation of fluid film bearing analysis emanates from the boundary layer theory of fluid mechanics. The governing differential equation was first formulated by Osborne Reynolds in 1886 and is known as Reynolds' equation in his honor. It has been only in the last 20 years or so that general solutions have been obtained, and this has been primarily due to the use of numerical methods applied to the digital computer. References 2 and 3 go into the details of contemporary numerical solutions and are recommended for those interested in the analytical aspects of lubrication.

Principal Assumptions   Reynolds' equation can be derived from the Navier-Stokes equation of fluid mechanics, and there are a number of textbooks available that comprehensively describe the derivation.[4] The primary assumptions are

1. Laminar flow conditions prevail, and the fluids obey a Newtonian shear stress distribution, where the shear stress is proportional to the velocity gradient.

2. Inertial forces, resulting from acceleration of the liquid, are small relative to the viscous shear forces and may be neglected.

3. The pressure across the film is constant since the fluid films are so thin.

4. The height of the fluid film is very small relative to other geometric dimensions, and so the curvature of the fluid film can be ignored.

5. The viscosity of the liquid remains constant. In most cases, this is a reasonable assumption since it has been repeatedly demonstrated that, if the average viscosity is used, little error is introduced and the complexity of the analysis is considerably reduced.

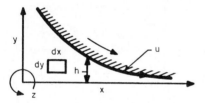

FIG. 3   Fluid control volume.

Derivation of Reynolds' Equation of Lubrication   Assume that the rotating journal has a peripheral velocity U. Consider an elemental volume in the clearance space of the bearing and establish equilibrium (Fig. 3). Note that since inertial forces are neglected, the volume is in equilibrium by the pressure and shear forces acting upon it and there is no acceleration. As shown in Fig. 4, p is the pressure and τ is the shear stress acting upon the volume. Summing forces in the x direction

$$p \, dy \, dz - \left(p + \frac{\partial p}{\partial x} dx \right) dy \, dz - \tau \, dx \, dz + \left(\tau + \frac{\partial \tau}{\partial y} dy \right) dx \, dz = 0$$

$$-\frac{\partial p}{\partial x} dx \, dy \, dz + \frac{\partial \tau}{\partial y} dx \, dy \, dz = 0 \tag{1}$$

or

$$\frac{\partial p}{\partial x} = \frac{\partial \tau}{\partial y} \tag{2}$$

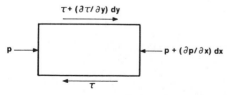

FIG. 4   Force equilibrium on fluid element.

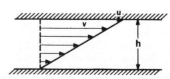

FIG. 5   Laminar velocity distribution across film.

For a Newtonian fluid in laminar flow, the shear stress is directly related to the velocity gradient with the proportionality constant being the absolute viscosity μ (Fig. 5):

$$\tau = \mu \frac{dv}{dy}$$

$$\frac{\partial \tau}{\partial y} = \mu \frac{\partial^2 v}{\partial y^2}$$

(3)

where v is the fluid velocity.

Substituting Eq. 3 into Eq. 2, we obtain

$$\frac{\partial p}{\partial x} = \mu \frac{\partial^2 v}{\partial y^2}$$

(4)

Integrating with respect to y twice produces the following equation:

$$v = \frac{1}{\mu} \frac{\partial p}{\partial x} \frac{y^2}{2} + C_1 y + C_2$$

(5)

where C_1, and C_2 are constaints of integration.

The boundary conditions are

$$v = 0, \ y = 0 \text{ and}$$
$$v = u, \ y = h$$

(6)

Substituting the boundary conditions of Eq. 6 into Eq. 5 results in the following expression for v:

$$v = \frac{1}{2\mu} \frac{\partial p}{\partial x} (y^2 - h^2 y) + \frac{uy}{2}$$

(7)

The velocity in the z direction would be similar except that the surface velocity term would be omitted becauses there is no surface velocity in the z direction. In addition, the pressure gradient would be with respect to z.

Now let us consider the flow across the film due to this velocity. Note that Eq. 7 is the velocity computed in the x direction, which is in the direction of rotation of the journal:

$$q_x = \int_0^h v_x \, dy = \int_0^h \left[\frac{1}{2\mu} \frac{\partial p}{\partial x} (y^2 - h^2 y) + \frac{uy}{2} \right] dy$$

(8)

After integrating,
$$q_x = -\frac{1}{12\mu} \frac{\partial p}{\partial x} h^3 + \frac{1}{2} uh$$

(9)

Note that q_x is the flow per unit width across the film. The flow in the axial direction is

$$q_z = -\frac{1}{12\mu} \frac{\partial p}{\partial z} h^3$$

(10)

Now let us consider a flow balance through an elemental volume across the film (Fig. 6). The net outflow through the volume equals the net reduction in volume per unit time:

$$\left(q_x + \frac{\partial q_x}{\partial x} dx \right) dz - q_x \, dz + \left(q_z + \frac{\partial q_z}{\partial z} dz \right) dx - q_z \, dx = -\frac{\partial h}{\partial t} dx \, dz$$

(11)

Thus

$$\frac{\partial q_x}{\partial x} + \frac{\partial q_z}{\partial z} = -\frac{\partial h}{\partial t}$$

(12)

Substituting Eqs. 9 and 10 into Eq. 12, we obtain

$$-\frac{\partial}{\partial x} \left(\frac{1}{12\mu} h^3 \frac{\partial p}{\partial x} \right) - \frac{\partial}{\partial z} \left(\frac{1}{12\mu} h^3 \frac{\partial p}{\partial z} \right) = -\frac{\partial h}{\partial t} - \frac{u}{2} \frac{\partial h}{\partial x}$$

(13)

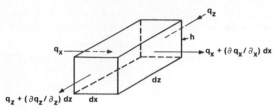

FIG. 6 Flow balance across control volume.

and the final equation becomes

$$\frac{\partial}{\partial x}\left(\frac{h^3}{\mu}\frac{\partial p}{\partial x}\right) + \frac{\partial}{\partial z}\left(\frac{h^3}{\mu}\frac{\partial p}{\partial z}\right) = 6\mu\frac{\partial h}{\partial x} + 12\frac{\partial h}{\partial t} \tag{14}$$

Equation 14 is the general form of Reynolds' equation used for laminar two-dimensional lubrication problems.

Reynolds' equation is a flow balance equation. The left-hand side represents pressure-induced flows in the x and z directions through the differential element. The first term on the right-hand side represents shear flow of the fluid induced by the surface velocity of the journal u. Note that this term contains the derivative of clearance with respect to distance. If this term is zero, then there is zero pressure produced by hydrodynamic action; the term $\partial h/\partial x$ is the mathematical representation of the tapered wedge. The second term on the right-hand side refers to a time rate of change of the film thickness, which can be translated to a normal velocity of the center of the journal. It produces pressure by a fluid velocity normal to the bearing surfaces that attempts to squeeze fluid out of a restricted clearance space. This phenomenon is called the squeeze film effect in bearing terminology. Since it is proportional to the velocity of the center of the journal, it is the phenomenon that produces viscous damping in a bearing.

The solution to Reynolds' equation (Eq. 14) provides the pressure at all points in the bearing. The application of the digital computer has enabled rapid solution of Reynolds' equation over a grid network representing the bearing area.[2,3]

Once the pressures have been obtained, numerical integration is applied to determine the performance parameters, i.e., load capacity:

$$w = \int\int prd\theta \, dr \tag{15}$$

flow across any circumferential line:

$$q_\theta = \oint\left(-\frac{1}{12\mu}\frac{\partial p}{r\partial\theta}h^3 + \frac{1}{2}uh\right)dz \tag{16}$$

flow across any axial line:

$$qz = \oint\left(-\frac{1}{12\mu}\frac{\partial p}{\partial z}h^3\right)rd\theta \tag{17}$$

The viscous frictional moment is obtained by integrating the shear stress over the area and can be shown to be

$$M_f = \int\int\left[\frac{1}{r}\frac{\partial p}{\partial\theta}\frac{h}{2} + \frac{\mu r\omega}{h}\right]r^2d\theta \, dz \tag{18}$$

where ω = journal surface speed, rad/s.

Typical computer program output includes

- Pressure distribution throughout grid network
- Load capacity
- Side leakage and carryover flows

- Viscous power losses
- Righting moments due to misalignments
- Attitude angles
- Cross-coupled spring and damping coefficients due to displacements and velocity perturbations of the journal center
- Clearance distribution

Turbulence   Equation 14 is for laminar conditions. For very high speed bearings operation beyond the turbulent regime may occur and Reynolds' equation must be modified. Turbulent theory has been developed, and there is information in the literature that permits performance predictions of turbulent bearings.[5,6]

The onset of turbulence is determined by examining the bearing Reynolds number, which is the ratio of inertia to viscous forces and is defined as

$$Re = \frac{\rho u h}{\mu} \tag{19}$$

where Re = Reynolds number
ρ = fluid density, lb·s^2/in^4 (kg·s^2/mm^4)
u = surface velocity, in/s (mm/s)
h = local film thickness, in (mm)
μ = viscosity, lb·s/in^2 (Pa·s)

A reasonable approximation is to use the concentric clearance c for h. In terms of N rpm and journal diameter D, the Reynolds number is

$$Re = \rho \frac{\pi D N}{60} \frac{c}{\mu} \tag{20}$$

The criterion for turbulence in journal bearings is that $Re \geq 1000$.

EXAMPLE   As an example, consider the following:

Journal diameter D = 5 in (127 mm)
Bearing length L = 5 in (127 mm)
Radial clearance c = 0.0025 in (0.064 mm)
Operating speed N = 5000 rpm
Lubricant viscosity μ = 2 × 10^{-6} lb·s/in^2 (14 × 10^{-3} Pa·s)
Lubricant density ρ = 7.95 × 10^{-5} lb·s^2/in^4 (8.66 × 10^{-11} kg·s^2/mm^4)

$$Re = \rho \frac{\pi D N}{60} \frac{c}{\mu}$$

$$= 7.95 \times 10^{-5} \times \frac{\pi \times 5 \times 5000}{60} \times \frac{0.0025}{2 \times 10^{-6}} = 130$$

Thus, the bearing is operating in the laminar regime. To become turbulent (assuming constant viscosity), the operating speed would have to increase to approximately 38,500 rpm.

The example cited is for a relatively high-speed, oil-lubricated bearing for pump applications. In general, most pump bearings operate in the laminar regime. Exceptions might occur when water is used as the lubricant because it its much less viscous than oil.

Evaluation of Frictional Losses   It is often desirable to obtain a quick estimate of viscous drag losses that the journal bearings produce. If we consider shear forces again, we return to the laminar flow equation:

$$F = \mu A \frac{u}{h} \tag{21}$$

where F = viscous shear force, lb (N)
  μ = viscosity, lb·s/in^2 (Pa·s)
  A = surface area, in^2 (mm^2)
  u = journal velocity, in/s (mm/s)
  h = film thickness, in (mm)

To obtain friction, we multiply both sides of Eq. 21 by the journal radius R; then the viscous frictional moment is

$$M = \mu A R \frac{u}{h}$$

The frictional horsepower loss is

$$FHP = \frac{NM}{63,000}$$

where N = rotating speed, rpm
  M = moment, lb·in (N·m)

Also

$$A = \text{surface area of bearing} = \pi DL, \text{ in}^2 \text{ (mm}^2\text{)}$$

$$u = \text{surface speed} = \frac{\pi DN}{60} \text{ in/s (mm/s)}$$

Substituting, we obtain

$$FHP = \frac{\mu \, L \, D^3 N^2}{766,000h} \tag{22}$$

In obtaining approximate losses for estimation purposes, the concentric clearance c is substituted for the local film thickness h.

If we consider the previous example, where D = 5 in (127 mm), L = 5 in (127 mm), c = 0.0025 in (0.064 mm), N = 5000 rpm, and μ = 2 × 10^{-6} lb·s/in^2 (14 × 10^{-3} Pa·s), the horsepower loss is

$$FHP = \frac{(2 \times 10^{-6})(5)(5)^3(5000)^2}{(766,000)(0.0025)} = 16.32 \text{ hp}$$

Note that for thrust bearings the frictional horsepower loss is

$$FHP = \frac{\mu N^2}{h} \frac{OD^4 - ID^4}{6.127 \times 10^6} \tag{23}$$

where OD = outside diameter
  ID = inside diameter

A general rule of thumb is that the frictional horsepower in a thrust bearing is approximately twice that in a journal bearing.

BEARING TYPES

Cylindrical Bearing The most common type of journal bearing is the plain cylindrical bushing shown schematically in Fig. 1. It may be split and have lubricating feed grooves at the parting line. A ramification is to incorporate axial grooves to enable better cooling and to improve whirl stability (described in more detail below in the discussion of cylindrical bearing with axial grooves). The principal advantages of cylindrical bearings are (1) simple construction and (2) high load capacity relative to other bearing configurations.

This type of bearing also has several disadvantages:

- **Whirl Instability.** It is prone to subsynchronous whirling at high speeds and also at low loads. Whirling is an orbiting of the journal (shaft) center in the bearing, a motion that is superimposed upon the normal journal rotation. The orbital frequency is approximately half the rotating speed of the shaft. The expression *half-frequency whirl* is commonly used. The reason for the occurrence of this whirl and more details concerning bearing dynamics are presented in the discussion of bearing dynamics.
- **Viscous Heat Generation.** Because of the generally large and uninterrupted surface area of this bearing, it generates more viscous power loss than some other types.
- **Contamination.** The cylindrical bearing is more susceptible to contamination problems than other types because contaminants that are dragged in at the leading edge of the bearing cannot easily dislodge because of the absence of grooves or other escape paths.

The advantages of simplicity and load capacity make the plain journal a leading candidate for most applications, but performance should be carefully investigated for whirl instability and potential thermal problems.

Cylindrical bearings are generally used for medium speed (500 in/s [200 mm/sec] surface speed) and medium to heavy load applications (250 to 400 lb/in^2 [17 to 28 bar] on projected area).

Cylindrical Bearing with Axial Grooves   A typical configuration of this type of bearing is a plain cylindrical bearing with four equally spaced longitudinal grooves extending most of the way through the bearing. There is usually a slight land area at either end of the groove to force the inlet flow to each groove into the bearing clearance region (Fig. 7) rather than out the groove

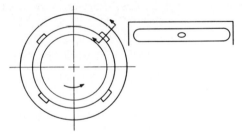

FIG. 7   Cylindrical bearing with axial grooving.

ends. This configuration is a little less simple than the plain cylindrical bearing, and because the grooves consume some land area, this configuration has less load capacity than the plain bushing. Since oil is fed into each of the axial grooves, this bearing requires more inlet flow but also will run cooler than the plain bushing. The grooves act as convenient outlets for any contaminants in the lubricant, and thus the grooved bearing can tolerate more contamination than the plain cylindrical bearing.

In general, this bearing can be considered as an alternate to a plain bearing if the former can correct a whirl or overheating problem.

Elliptical and Lobe Bearings   Elliptical and lobe bearings have noncircular geometries. Figure 8 shows two types of three-lobe bearings, with the clearance distribution exaggerated so that the lobe geometry is easily discernible. An elliptical bearing is simply a two-lobe bearing with the major axis along the horizontal axis.

The lobe bearing shown in Fig. 8a is a symmetric lobe bearing, where the minimum concentric clearance occurs in the center of each lobed region. Thus, at the leading edge region a converging clearance produces positive pressure, but downstream of the minimum film thickness there occurs a divergent film thickness distribution with resultant negative, or cavitation, pressures.

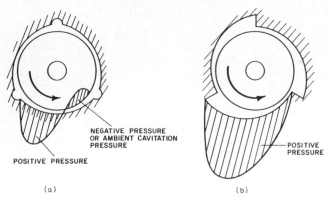

FIG. 8   (*a*) Symmetric lobe bearing; (*b*) canted lobe bearing.

The canted lobe shown in Fig. 8*b*, on the other hand, generally develops positive pressure throughout the lobe because the bearing is constructed with a completely converging film thickness in each lobed region. This design has excellent whirl resistance (superior to that of the symmetric lobe bearing) and reasonably good load capability. A 2:1 ratio between leading and trailing edge concentric clearance is generally a reasonable compromise with respect to performance.

Elliptical and lobe bearings are often used because they provide better resistance to whirl than do cylindrical configurations. They do so because they have multiple load-producing pads that assist in preventing large-attitude angles and cross coupling (see discussion of bearing dynamics, p. 2.184). They are generally used for high-speed, low-load applications where whirl might be a problem.

Elliptical, or two-lobe, bearings generally have poor horizontal stiffness because of the large clearances along the major diameter of the ellipse. The split elliptical configuration, however, is easier to manufacture than the other types because it is two cylindrical bearing halves with material removed along the parting line. The other types of lobe bearings are complicated to manufacture. Lobe bearings are usually clearance- and tolerance-sensitive.

Tilting-Pad Bearing   Tilting-pad bearings (Fig. 9) are used extensively, especially in high-speed applications, because of their whirl-free characteristics. They are the most whirl-free of all bearing configurations.

An important geometric variable for tilting-pad bearings is the preload ratio, defined as follows:

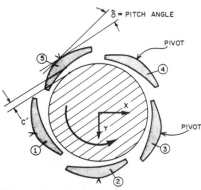

FIG. 9   Five-pad tilting-pad bearing.

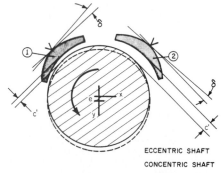

PAD 1 c'/c < 1.0 CONVERGING CLEARANCE
PAD 2 c'/c > 1.0 DIVERGING CLEARANCE

FIG. 10   Tilting-pad bearing preload.

$$\text{Preload ratio} = PR = \frac{c - c'}{c} = 1 - \frac{c'}{c} \qquad (24)$$

where c = machined clearance
c' = concentric pivot film thickness

The variable c' is an installed clearance and is dependent upon the radial position of the pivot. In Fig. 10 are shown two pads. Pad 1 has been installed such that the preload ratio is less than 1. For pad 2 the preload ratio is 1. The solid line represents the position of the journal in the concentric position. The dashed portion of the journal represents its position when load is applied to the bottom pads (not shown). Pad 1 is operating with a good converging wedge even though the journal is moving away from it. Pad 2, on the other hand, is operating with a completely diverging film, which means that it is totally unloaded. Thus, bearings with installed pad preload ratios of 1 or greater will operate with unloaded pads, which reduces overall stiffness of the bearing and results in a deterioration of stability because the unloaded pads do not aid in resisting cross-coupling influences. In the unloaded position they are also subject to flutter instability and to a phenomenon known as *leading edge lockup*, where the leading edge is forced against the shaft and is maintained in that position by the frictional interaction of the shaft and the pad. This is especially prevalent in bearings that operate with low-viscosity lubricants, such as gas or water bearings. Thus, it is important to design bearings with preload, although for manufacturing reasons it is common practice to produce bearings without preload.

Tilting-pad bearings have some other characteristics that are both positive and negative:

- They are not as clearance-sensitive as most other bearings.
- Because the pads can move, they can operate safely at lower minimum film thickness than other bearings.
- They do not provide as much squeeze film damping as rigid configurations.
- Generally they are more expensive than other bearings.
- For high-speed applications, their pivot contacts can be subjected to fretting corrosion.

Hybrid Bearings A hybrid bearing, schematically shown in Fig. 11, derives load capacity from two sources: (1) the normal hydrodynamic pressure generation and (2) an external high-pressure supply that introduces oil into recesses machined into the bearing surface via restrictors

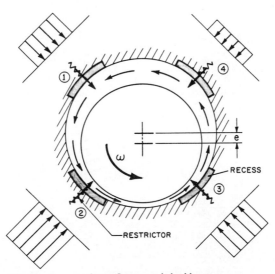

FIG. 11 Cross-coupling influences in hybrid bearings.

(orifices or capillaries upstream of the recesses). External pressure significantly enhances load capacity; also, these bearings have excellent low- or zero-speed load capability. They are sometimes used as start-up devices to lift off the rotor. When self-sustaining hydrodynamic speeds are attained, external pressure is shut off. The characteristics of externally pressurized, or hybrid, bearings are summarized as follows:

- High load and stiffness capability.
- External flow assists in cooling.
- Clearances and tolerances generally more liberal than in hydrodynamic bearings.
- Requires external fluid-supply systems.
- Applied when there is not sufficient generating speed or where high load capacity and stiffness are required.
- Sometimes applied to prevent whirl. However, rotational speed can unbalance recess pressures, introduce cross coupling, and promote whirl.

STEADY STATE PERFORMANCE

Computer-generated performance was obtained for most of the bearing types previously discussed. Information was plotted in nondimensional format so that there are no restrictions on operating conditions, lubricant properties, etc. Use of the charts will be subsequently demonstrated by numerical example.

Viscosity   One of the key parameters in determining performance of a bearing is the lubricant viscosity. Viscosity characteristics of commonly used SAE (Society of Automotive Engineers) grades of oil are shown in Fig. 12. The units of viscosity are microreyns, where the reyn has the units of $lb \cdot s/in^2$ and comes from the ratio of shear stress to the velocity gradient across the film, as indicated by Eq. 21. Other units of viscosity are centipoises, Saybolt seconds universal (SSU), and centistokes. Conversion factors are as follows:

$$\mu \text{ (reyns)} = Z \times 1.45 \times 10^{-7} \tag{25}$$

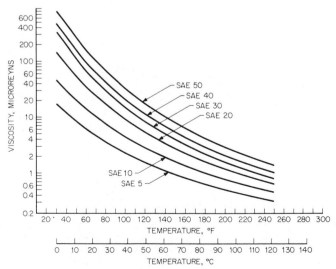

FIG. 12   Viscosity characteristics for SAE oil grades.

$$\nu \text{ (centistokes)} = 0.22 \text{ (SSU)} - \frac{180}{\text{SSU}} \tag{26}$$

$$Z \text{ (centipoises)} = \nu \text{ (centistokes)} \times SG \text{ (sp. gr.)} \tag{27}$$

Performance Curves   Performance plots were generated for the following types of bearings:

- *Two-groove cylindrical bearing.*
- *Symmetric three-lobe bearing.* Each lobe was offset such that in the concentric position the minimum film thickness in the center of each lobed region was one-half of the machined clearance c (see definition below). The pads are each 110° in angular extent.
- *Canted three-lobe bearing.* The lobing was canted such that in the concentric position the leading edge clearance was twice the trailing edge and the trailing edge film thickness (minimum) in the concentric position was $0.5c$, where c equals the machined clearance. The pads are each 110° in angular extent.
- *Tilting-pad bearing.* The tilting-pad bearing for which information was obtained was a five-pad bearing with a 60° pad and a preload ratio of 30%.

Two length/diameter ratios were examined for each type of bearing: $L/D = 0.5$ and 1.0. Definition of the nondimensional parameters are as follows:

$$W = \text{nondimensional load parameter} = wc^2/6\mu\omega RL^3 \tag{28}$$

$$P = \text{nondimensional viscous power loss parameter} = 1100cp/\mu(\omega RL)^2 \tag{29}$$

$$Q = \text{nondimensional flow parameter} = 2q/0.26\omega RLc \tag{30}$$

$$HM = \text{nondimensional minimum film thickness} = h_M/c \tag{31}$$

where w = bearing load capacity, lb (N)
c = reference clearance (machined clearance = radius of bearing − radius of shaft), in (mm)
μ = absolute viscosity, reyns (lb·s/in²) (cP)
ω = shaft or journal rotational speed, rad/s
R = shaft radius, in (mm)
L = bearing length, in (mm)
p = viscous power loss, hp (kW)
q = flow, gpm (m³/h)
q_i = inlet flow to leading edge of bearing (for multipad bearings, equals the sum of inlet flow to each pad), gpm (m³/h)
q_s = side leakage flow, or flow out of the bearing ends (for multipad bearings, equals the sum of side leakage flow of each pad), gpm (m³/h)
h_M = minimum film thickness in bearing, in (mm)

Performance curves are shown in Figs. 13 through 17.

At times, the nondimensional data can be confusing and lead to erroneous judgments. For example, the nondimensional power loss P is greater for an $L/D = 0.5$ than for an $L/D = 1.0$. However, when the dimensional value of power loss is being computed, the nondimensional value is multiplied by L^2; therefore the power loss for $L/D = 1.0$ will be, as expected, greater than for $L/D = 0.5$. If the reader uses the data as presented, the dimensional information will prove consistent.

To make comparisons among the bearings using the nondimensional data is not strictly proper because there may be slight inconsistencies in preloads and the bearings will not be operating at the same average viscosity, etc. Subsequently, dimensional data derived from the performance curves will be compared. However, comparisons of the nondimensional information will provide an indication of performance parameters among the bearings. Comparisons were made at equal values of the load parameter W and results are indicated in Table 1.

Comparisons of the different bearing types should be made only at the same L/D ratio because of the anomalies (discussed above) of nondimensional parameters that occur at different L/D

TABLE 1 Comparative Results of Bearing Types at $W = 0.2$

Bearing type	$L/D = 0.5$				$L/D = 1$			
	P	HM	QS	QI	P	HM	QS	QI
Two-groove cylindrical	1.13	0.36	0.78	1.81	0.72	0.25	0.50	1.50
Three-lobe	1.80	0.32	0.79	1.85	1.05	0.22	0.22	1.65
Canted three-lobe	1.52	0.31	1.42	3.30	0.91	0.22	0.90	2.58
Tilting-pad	1.50	0.30	0.44	2.82	0.90	0.20	0.16	1.50

QI = nondimensional inlet flow = $2q_t/0.26\ wRLc$
QS = nondimensional side leakage flow = $2q_s/0.26\ wRLc$

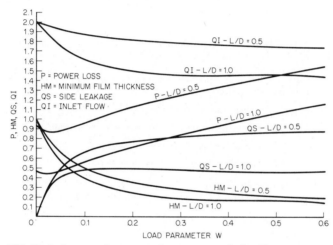

FIG. 13 Performance characteristics for two-groove cylindrical bearings.

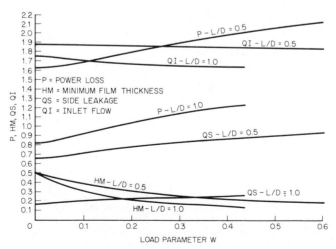

FIG. 14 Performance characteristics for three-lobe bearings.

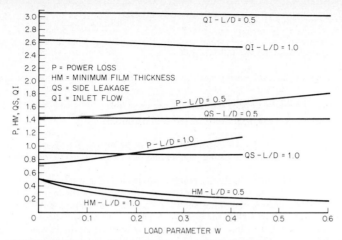

FIG. 15 Performance characteristics for canted three-lobe bearings.

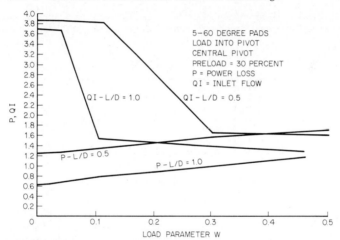

FIG. 16 Performance characteristics for tilting-pad bearings.

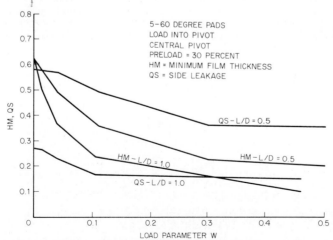

FIG. 17 Performance characteristics for tilting-pad bearings.

2.179

ratios. If we assume that all the reference variables that go into the nondimensional parameters are identical, we can establish the following conclusions:

- The two-groove cylindrical bearing has the highest film thickness and thus the highest load capacity.
- The symmetric three-lobe bearing has the highest power loss.
- The canted three-lobe bearing has the greatest flow requirements.

Note that these comparisons were made on the basis of steady-state performance only; the major reason for applying lobe and tilting-pad bearings is to avoid dynamic instabilities.

Heat Balance   Performance is based upon the assumption of a uniform viscosity in the fluid film. Since the viscosity is a strong function of temperature and since the temperature rise of the lubricant due to viscous heat generation is not known a priori, an iterative procedure is required to determine the average viscosity in the film. To determine an average viscosity, there must be a simplified heat balance in the film. The assumption is made that all the viscous heat generated in the film is absorbed by the lubricant as it flows through the film and produces a temperature rise.

Figure 18 shows a developed view of the bearing surface and the parameters involved with conducting the heat balance. The lubricant enters a pad or bearing at the leading edge with an

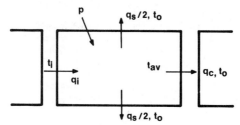

FIG. 18   Heat balance in a fluid film.

inlet flow q_i and an inlet temperature t_i. As the lubricant enters the bearing surface and flows through, it is exposed to viscous shear, which adds heat to the fluid. In Fig. 18, the viscous shear is indicated as a power input p. Some of the fluid flows out of the sides of the bearing, represented by $q_s/2$, and some of the fluid, q_c, is carried either over to the grooving of the next pad or back to the inlet groove for a single pad bearing. Although the fluid temperature, and thus the viscosity, change along the length of the pad, it is assumed that the temperature in the film increases to some value t_o, for both the side leakage and carry-over fluids. Then the heat balance is conducted as follows:

$$\text{Heat added to fluid by viscous shear} = \text{heat absorbed by fluid} \qquad (32)$$
$$pJ = q_s\rho C_p(t_o - t_i) = q_c\rho c_p(t_o - t_i) = (q_s + q_c)\rho C_p(t_o - t_i)$$

Since by continuity of flow,

$$q_i = q_s + q_c \qquad (33)$$

$$pJ = q_i\,\rho C_p\,\Delta t \qquad (34)$$

$$\Delta t = \frac{pJ}{q_i\,\rho C_p} \qquad (35)$$

where p = viscous heat generation, hp
J = heat equivalent constant = 0.7069 Btu/hp·s (J/kg·s)
ρ = specific weight of flow, lb/in³ (kg/m³)

C_p = specific heat of fluid, Btu/lb·F° (J/kg·C°)
q_i = bearing inlet flow, in³/s (mm³/s)
Δt = temperature rise, F° (C°)

If q_i is in gallons per minute (1 in³/s = 0.26 gpm) = 16.39 cm³/s)

$$\Delta t = \frac{0.7069(0.26)p}{q_i \rho C_p} = \frac{0.1838p}{q_i \rho C_p} \tag{36}$$

Before proceeding to a sample problem, a word about the flows q_i and q_s and the general heat balance philosophy. The flow required by the bearing to prevent starvation is q_i. The minimum make-up flow, or the flow that is lost from the ends of the bearing, is the side leakage q_s. Theoretically then, the only flow that need be supplied to the bearing is q_s. However, if this were true, then the heat balance formulation would require another balance between the carry-over flow q_c at some temperature t_o and the make-up flow q_s coming into the pad at temperature t_i. It would be found that, if we supplied only q_s to the bearing, excessive temperatures would result. In most instances the amount of flow supplied to a bearing exceeds both the inlet flow and the side leakage flow by a significant margin. Designers often determine the flow to be supplied to a bearing system by a bulk temperature rise of the total flow entering the inlet pipe and exiting the exhaust pipe. Thus, the entire bearing is treated as a black box, and the total flow q_T to the bearing system for a given bulk temperature rise Δt_t is

$$q_T = \frac{pJ}{\rho C_p \, \Delta t_t} \tag{37}$$

Normally, Δt_t is selected between 20 and 40°F. Since q_T will exceed q_i, not all the flow will enter the film. Some will surround the bearing or flow through the feed groove and act as a cooling medium for heat transfer through the bearing walls or for cooling the fluid that does enter the bearing surface, q_i.

EXAMPLE   This example demonstrates the use of the design curves and how to perform a simplified heat balance in bearing analysis.

Given

    Bearing type: two-groove cylindrical

    Bearing length L = 3 in (76 mm)

    Bearing diameter D = 6 in (152 mm)

    L/D = 0.5

    N = 1800 rpm; ω = 1800 × $\pi/30$ = 188.5 rad/s

    Bearing load w = 6000 lb (2721 kg)

    Inlet temperature t_i = 120°F (49°C)

    Lubricant = SAE 20

    Bearing radial clearance c = 0.003 in (0.076 mm)

(NOTE: A general rule of thumb is that c/R = 0.001.)
    Assume an average lubricant temperature of 130°F − μ (SAE 20 at 130°F) = 4.5 × 10⁻⁶ reyn (lb·s/in²) (31 × 10⁻³ Pa·s)

$$\rho = 0.0307 \text{ lb/in}^3 \ (849.7 \text{ kg/m}^3) \text{ (average value for most oils)}$$

$$C_p = 0.5 \text{ Btu/lb·F}° \ (2093 \text{ J/kg·C}°) \text{ (average value for most oils)}$$

Compute

$$W = \frac{wc^2}{6\mu\omega RL^3} = \frac{(6000)(0.003^2)}{(6)(4.5 \times 10^{-6})(188.5)(3)(3^3)} = 0.1309$$

From Fig. 13 and Eq. 29,

$$P = 1.0 = \frac{1100cp}{\mu(\omega RL)^2}$$

$$p = \frac{(1)(4.5 \times 10^{-6})(188.5 \times 3 \times 3)^2}{(1100)(0.003)} = 3.92 \text{ hp } (2.95 \text{ kW})$$

From Fig. 13 and Eq. 30,

$$Q_I = 1.85 = \frac{2q_i}{0.26\omega RLC}$$

$$q_i = \frac{(1.85)(0.26)(188.5)(3)(3)(0.003)}{2} = 1.224 \text{ gpm } (4.63 \text{ l/min})$$

From Eq. 36,

$$\Delta t = \frac{0.1838p}{q_i \rho C_p} = \frac{(0.1838)(3.92)}{(1.224)(0.0307)(0.5)} = 38.35 \text{ F}° \ (21.3 \text{ C}°)$$

$$t_{av} = \frac{t_i + (t_i + \Delta t)}{2} = \frac{2t_i + \Delta t}{2} = \frac{(2)(120) + 38.85}{2} = 139.2°\text{F} \ (59.6°\text{C})$$

The assumed t_{av} of 130°F was apparently not high enough, and therefore we must repeat the calculations with a higher value of assumed t_{av}.

Assume $t_{av} = 138°$F, $\mu = 3.8 \times 10^{-6}$ lb·s/in² (26.2 × 10⁻³ Pa·s) and repeat the calculations:

$$W = \frac{5.8945 \times 10^{-7}}{\mu} = 0.1552$$

$$P = 1.08 = (1.1466 \times 10^{-6}) \left(\frac{p}{\mu} \right)$$

$$p = \frac{(1.08)(3.8 \times 10^{-6})}{1.1466 \times 10^{-6}} = 3.579 \text{ hp } (2.67 \text{ kW})$$

$$QI = 1.83 = \frac{q_i}{0.6616}$$

$$q_i = 1.211 \text{ gpm } (4.58 \text{ l/min})$$

$$\Delta t = 11.974 \frac{p}{q_i} = \frac{(11.974)(3.579)}{1.211} = 35.38 \text{ F}° \ (19.65 \text{ C}°)$$

$$t_{av} = \frac{(2)(120) + 35.38}{2} = 137.7°\text{F} \ (58.7°\text{C})$$

which for all practical purposes equals the assumed value of $t_{av} = 138°$F (58.9°C).

In conducting the iterative procedure for determining the average fluid temperature, it is sometimes helpful to plot points on a viscosity chart. Referring to Fig. 19, suppose point 1 is the resultant average viscosity and temperature of the initial guess and point 2 represents the result of the second guess. Then, by drawing a straight line between points 1 and 2 and establishing where it intersects the lubricant viscosity curve, we can determine convergence to the proper result of average viscosity and temperature.

Now that we have convergence, we can determine the remaining variables, which are minimum film thickness H_M and side leakage Q_S from Fig. 13:

$$HM = 0.41 = \frac{h_M}{c}$$

$$h_M = (0.41)(0.003) = 0.00123 \text{ in } (0.031 \text{ mm})$$

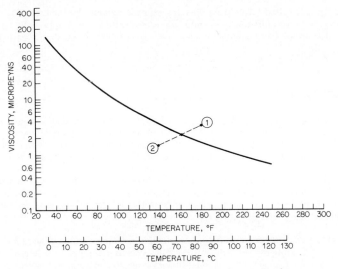

FIG. 19 Graphical determination of average viscosity for SAE 20 oil.

$$QS = 0.74 = \frac{2q_s}{0.26\omega RLC} = \frac{q_s}{0.6616}$$

$$q_s = (0.6616)(0.74) = 0.490 \text{ gpm } (1.85 \text{ l/min})$$

A summary of total bearing performance is as follows:

Load w = 6000 lb (2721 kg)

Miminum film thickness h_M = 0.00123 in (0.031 mm)

Viscous power loss p = 3.579 (2.669 kW)

Inlet flow q_i = 1.211 gpm (4.58 l/min)

Side leakage q_s = 0.490 gpm (1.85 l/min)

Fluid temperature rise Δt = 35.38 F° (19.65 C°)

We can repeat the same calculation for the other types of bearings, using Figs. 14 through 17. The results are shown in Table 2.

TABLE 2 Comparative Bearing Performance

Bearing type	Δt, °F (C°)	t_{av}, °F (°C)	p, hp	q_i, gpm (cm³/s)	q_s, gpm (cm³/s)	h_M, mils (mm)
Two-groove cylindrical	35 (19.44)	138 (58.9)	3.58	1.21 (76.7)	0.490 (31.05)	1.23 (0.0312)
Symmetric three-lobe	49 (27.22)	144 (62.2)	4.97	1.22 (77.3)	0.516 (32.7)	0.96 (0.0244)
Canted three-lobe	31 (17.22)	135 (57.2)	5.20	2.01 (127.4)	0.940 (59.6)	1.05 (0.0267)
Tilting-pad	26 (14.44)	134 (56.7)	4.88	2.26 (143.2)	0.311 (19.71)	1.02 (0.0259)

The following values were used: w = 6000 lb (2721 kg), SAE 20 oil, L = 3 in (76 mm), D = 3 in (76 mm), N = 1800 rpm, t_i = 120°F (48.9 °C), c = 0.003 in (0.076 mm)

The two-groove cylindrical bearing operates with the highest film thickness. The symmetric three-lobe has the lowest film thickness and highest temperature rise. It appears that, on the basis of steady-state performance, manufacture, and cost, the two-groove cylindrical bearing is the best choice.

BEARING DYNAMICS[7]

Fluid film bearings can significantly influence the dynamics of rotating shafts. They are a primary source of damping and thus can inhibit vibrations. Alternatively, they provide a mechanism for self-excited rotor whirl. Whirl is manifest as an orbiting of the journal at a subsynchronous frequency, usually close to one-half the rotating speed. Whirl is usually destructive and must be avoided.

It is the purpose of this section to provide some insight into bearing dynamics, present some background on analytical methods and representations, and discuss some particular bearings and factors that can influence dynamic characteristics. Dynamic performance data and sample problems are presented for several bearing types.

The Concept of Cross Coupling
As mentioned in the opening paragraphs of this subsection, a journal bearing derives load capacity from viscous pumping of the lubricant through a small clearance region. To generate pressure, the resistance to pumping must increase in the direction of fluid flow. This is accomplished by a movement of the journal such that the clearance distribution takes on the form of a tapered wedge in the direction of rotation, as shown in Fig. 1.

The attitude angle γ in Fig. 1 is the angle between the load direction and the line of centers. Thus, the displacement of the journal is not along a line that is coincident with the load vector, and a load in one direction causes not only displacements in that direction but orthogonal displacements as well.

Similarly, a displacement of the journal in the bearing will cause a load opposing the displacement and a load orthogonal to it. Thus, there are strong cross-coupling influences introduced by the mechanism by which a bearing operates. The concept of cross coupling is significant in dynamic characteristics.

It is the cross-coupling characteristics of a journal bearing that can promote self-excited instabilities in the form of bearing whirl. Motion in one direction produces orthogonal forces which in turn cause orthogonal motion. The process continues, and an orbital motion of the journal results. This orbital motion is generally in the same direction as shaft rotation and subsynchronous in frequency. Half-frequency whirl is a self-excited phenomenon and does not require external forces to promote it.

Cross-Coupled Spring and Damping Coefficients
For dynamic considerations, a convenient representation of bearing characteristics is by cross-coupled spring and damping coefficients. These are obtained as follows (refer to Fig. 1):

1. The equilibrium position to support the given load is established by computer solution of Reynolds' equation.

2. A small displacement to the journal is applied in the y direction. A new solution of Reynolds' equation is obtained, and the resulting forces in the x and y directions are produced. The spring coefficients are as follows:

$$K_{xy} = \frac{\Delta F_x}{\Delta y} \tag{38}$$

$$K_{yy} = \frac{\Delta F_y}{\Delta y} \tag{39}$$

where K_{xy} = stiffness in x direction due to y displacement
ΔF_x = difference in x forces between displaced and equilibrium positions
Δy = displacement from equilibrium position in y direction

K_{yy} = stiffness in y direction due to y displacement

ΔF_y = difference in y forces between displaced and equilibrium positions

3. The journal is returned to its equilibrium position and an x displacement applied. Similar reasoning produces K_{xx} and K_{yx}.

The cross-coupled damping coefficients are produced in a like manner, except instead of displacements in the x and y direction, velocities in these directions are consecutively applied with the journal in the equilibrium position. The mechanism for increasing load capacity is squeeze film in which the last term on the right-hand side of Eq. 14 is actuated.

Thus for most fixed bearing configurations, there are eight coefficients: four spring and four damping.

The total force on the journal is

$$F_i = K_{ij}x_j + D_{ij}\dot{x}_j$$

where (40)

$$F_i = \text{force in the } i\text{th direction.}$$

Repeated subscripts imply summation:

$$K_{ij}x_j = K_{ix} + K_{iy}y + \cdots$$

It should be realized that the cross-coupled spring and damping coefficients represent a linearization of bearing characteristics. When they are used, the equilibrium position should be accurately determined, as the coefficients are valid only about a small displacement region encompassing the equilibrium position of the journal. This is true because the spring and damping coefficients remain constant only about a small region of the equilibrium position.

Consider the two-groove cylindrical bearing shown in Fig. 1, with the geometric and operating conditions indicated in Table 3.

The computer solution (also performance curves, Fig. 13) produces the following results:

Bearing load w	= 20,780 lb (9424 kg)
Power loss	= 15.51 hp (11.56 kW)
Minimum film thickness h_M	= 0.00125 in (0.032 mm)
Side leakage q_s	= 0.941 gpm (3.56 l/min)

The spring and damping coefficients are:

Spring coefficients, lb/in (kg/mm):
$$\begin{bmatrix} K_{xx} & K_{xy} \\ K_{yx} & K_{yy} \end{bmatrix} = \begin{bmatrix} -12.14 \times 10^6 & -4.64 \times 10^6 \\ 28.3 \times 10^6 & -20.41 \times 10^6 \end{bmatrix}$$

Damping coefficients, lb·s/in kg·s/in:
$$\begin{bmatrix} D_{xx} & D_{xy} \\ D_{yx} & D_{yy} \end{bmatrix} = \begin{bmatrix} -2.85 \times 10^4 & 2.66 \times 10^4 \\ -2.69 \times 10^4 & -1.11 \times 10^5 \end{bmatrix}$$

TABLE 3 Two-Groove Cylindrical Bearing Geometry And Operating Conditions

Journal diameter D	= 5 in (127 mm)
Bearing length L	= 5 in (127 mm)
Active pad angle θ_p	= 160° (10° grooves on either side)
Radial clearance c	= 0.0025 in (0.064 mm)
Operating speed N	= 5000 rpm
Lubricant viscosity μ	= 2×10^{-6} lb·s/in^2
	(13.79×10^{-3} Pa·s)
Eccentricity ratio ϵ	= 0.5
Load direction is vertical downward	

The negative signs imply a positive stiffness because the restoring load is opposite the applied load. Note that for this bearing configuration there is very strong cross coupling, evidenced by the magnitude of the off-diagonal terms.

Critical Mass   The cross-coupled spring and damping coefficients provide a convenient way of representing a bearing in a stability analysis. They reduce the fluid film bearing to a spring-mass system (Fig. 20), and consequently stability and dynamics problems are simplified considerably.

Consider a journal of mass M operating in a bearing. The journal can be considered to have two degrees of freedom, x and y. The governing equations are

$$M\ddot{x} + D_{xx}\dot{x} + D_{xy}\dot{y} + K_{xx}x + K_{xy}y = 0 \tag{41}$$

$$M\ddot{y} + D_{yy}\dot{y} + D_{yx}\dot{x} + K_{yy}y + K_{yx}x = 0 \tag{42}$$

Assume a sinusoidal response to the form

$$x = x_0 e^{\beta t} \tag{43}$$

$$x = y_0 e^{\beta t} \tag{44}$$

where β is a complex variable:

$$\beta = \alpha + i\omega \tag{45}$$

By Euler expansion of $e^{\beta t}$, another way to write Eqs. 43 and 44 is

$$x = x_0 e^{\alpha t} (\cos \omega t + i \sin \omega t) \tag{46}$$

$$y = y_0 e^{\alpha t} (\cos \omega t + i \sin \omega t) \tag{47}$$

An interpretation of α and ω is shown in Fig. 21. The real part of $\beta = \alpha$ is called the growth, or attenuation, factor. The imaginary part is the frequency of vibration. A positive real part means that response to a disturbance grows in time. The growth factor is similar to the logarithmic mean decrement, which is common in vibration theory:

$$\alpha\tau = \ln \frac{x_{n+1}}{x_n} \tag{48}$$

where τ = period of vibration
x_{n+1} = amplitude at time $n + 1$
x_n = amplitude at time n

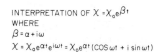

INTERPRETATION OF $X = X_0 e^{\beta t}$
WHERE
$\beta = \alpha + i\omega$
$X = X_0 e^{\alpha t} e^{i\omega t} = X_0 e^{\alpha t} (\cos \omega t + i \sin \omega t)$

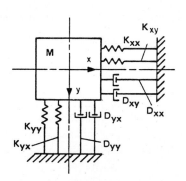

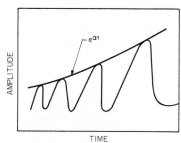

REAL PART α = GROWTH FACTOR
IMAGINARY PART ω = FREQUENCY

FIG. 20   Point mass representation of a bearing supported on cross-coupled springs and dampers.

FIG. 21   Interpretation of growth factor α and orbital frequency ω.

Thus, the growth factor α is a measure of the growth or decay of the journal to a small disturbance. A positive growth factor implies an instability.

The solutions to Eqs. 41 and 42 are obtained by substituting Eqs. 43 and 44, which produces the following:

$$\begin{bmatrix} (M\beta^2 + D_{xx}\beta + K_{xx}) & (D_{xy}\beta + K_{xy}) \\ (\beta D_{yx} + K_{yx}) & (M\beta^2 + \beta D_{yy} + K_{yy}) \end{bmatrix} \begin{Bmatrix} x_0 \\ y_0 \end{Bmatrix} = \{0\} \tag{49}$$

To obtain a solution, the determinant of the coefficient matrix must vanish. Expansion produces a polynomial in β that can be solved for the roots of β, which in turn provide the growth factors and frequencies of vibration.

It is possible to obtain a closed-form solution of Eqs. 41 and 42 for the critical mass and resulting orbital frequency. The critical mass M is defined as that mass above which an instability will occur. At the threshold of instability the real part of $\beta = \alpha$ goes to zero and $\beta = i\omega$.

Substituting into Eq. 49, expanding the determinant, and separating real and imaginary components produces the following equations:

$$\underset{A}{M^2\omega^4} - \underset{}{M(K_{yy} + K_{xx})\omega^2} + \underset{B}{(D_{yx}D_{xy} - D_{xx}D_{yy})\omega^2} + \underset{C}{K_{xx}K_{yy} - K_{xy}K_{yx}} = 0 \tag{50}$$

$$\underset{D}{M(D_{yy} + D_{xx})\omega^2} + \underset{E}{(D_{xy}K_{yx} + D_{yx}K_{xy} - D_{xx}K_{yy} - D_{yy}K_{xx})} = 0 \tag{51}$$

The two equations can be solved for M and ω:

$$M = \frac{BED}{E^2 - AED + CD^2} \tag{52}$$

$$\omega = \sqrt{\frac{-(AED + E^2 + CD^2)}{BD^2}} \tag{53}$$

Thus, if the cross-coupled coefficients are known, it is possible to determine the critical mass and the orbital frequency. If the mass acting on the bearing exceeds or equals the critical value, then an instability will occur.

Dynamic Stability of Various Bearing Types   There are several parameters that can be used to establish the stability characteristics of a particular type of bearing. The most significant is the critical mass, derived above. It is also possible to get some feel for stability from purely steady-state performance by examining the bearing attitude angle. The larger the attitude angle, the more cross-coupling influence and the worse the stability characteristics. Interpreting attitude angles, however, can prove misleading. In Fig. 22 are shown plots of the attitude angle γ versus the load parameter W for some of the different types of bearings previously described. The contradiction in these results is that the symmetric three-lobe bearing has higher attitude angles than the two-groove cylindrical bearing even though, as will be subsequently demonstrated, it has superior stability characteristics. The reason for the higher attitude angle is that the bearing is cavitating in the diverging region of the loaded pad, which results in a large shift in the journal. Whirl motion, however, is prevented by the accompanying lobes.

A more direct and accurate approach to establishing bearing stability is to determine the critical mass acting on the bearing. Figure 23 shows results obtained for the fixed bearing geometries previously discussed. For any particular bearing geometry, if the critical mass attributable to the bearing exceeds that of the data plotted, the bearing will be unstable. Thus, if the plotted point falls to the right of the bearing line, the bearing is stable; if it falls to the left, the bearing is unstable.

The critical mass is defined as follows:

$$M = m\omega Rc^3 / 24\mu L^5 \tag{54}$$

where M = nondimensional critical mass
$\quad m$ = dimensional critical mass, lb·s²/in (kg·s²/mm)
$\quad \omega$ = shaft speed, rad/s
$\quad R$ = shaft radius, in (mm)

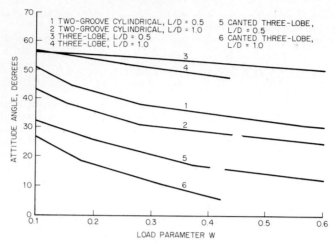

FIG. 22 Attitude angle versus load parameter for various bearing types.

c = bearing machined radial clearance, in (mm)
μ = lubricant viscosity, lb·sec/in² (Pa·s)
L = bearing length, in (mm)

Examination of the curves in Fig. 23 clearly indicates the superiority of the canted three-lobe bearing and the inferiority of the cylindrical configuration.

EXAMPLE   Consider a high-speed bearing for which

N = 60,000 rpm
$\omega = \dfrac{60{,}000\pi}{30}$ = 6283 rad/s

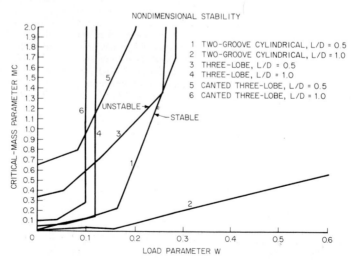

FIG. 23 Critical mass versus load parameter for various bearing types.

TABLE 4 Stability Comparison Chart

Bearing type	M	m, lb·s²/in (kg·s²/mm)	w', lb (kg)[a]
Two-groove cylindrical	0.02	0.038 (6.79 × 10⁻⁴)	14.7 (6.7)
Symmetric three-lobe	0.35	0.668 (11.9 × 10⁻³)	257.9 (117.1)
Canted three-lobe	0.68	1.299 (23.2 × 10⁻³)	501.4 (227.6)

$$m = \frac{24M\mu h^5}{\omega Rc^3} = \frac{M(24)(10^{-6})(0.5^5)}{(6283)(0.5)(0.0005^3)} = 1.91M$$

[a]The w' represents the maximum mass in weight units that could act on the bearing.

$D = 1$ in (25.4 mm)
$L = 0.5$ in (12.7 mm)
$\mu = 1 \times 10^{-6}$ lb·s/in² (6.9 × 10⁻³ Pa·s)
$C = 0.0005$ in (0.013 mm)
$w = 100$ lb (45.35 kg)

$$W = \frac{wC^2}{6\mu\omega RL^3} = \frac{(100)(0.0005^2)}{(6)(1 \times 10^{-6})(6283)(0.5)(0.5^3)} = 0.0106$$

The value of M is obtained from Fig. 22 and indicated in Table 4 for the various bearing types considered. Dimensional units are also given.

Table 4 clearly indicates that the lobe bearings can permit significantly more attributable mass than can the cylindrical bearing. If a symmetric rotor was being supported by two bearings, then the cylindrical bearings would go unstable if half the weight of the rotor exceeded 14.7 lb (6.67 kg). The half weights go to 258 lb (117 kg) and 501 lb (227.2 kg) for the symmetric three-lobe and canted three-lobe bearings, respectively.

No mention has been made of the tilting-pad bearing because, for all practical purposes, these bearings are always stable.

Operating Conditions That Affect Bearing Stability

CAVITATION   Figure 1 shows a pressure distribution in a journal bearing. Positive pressure is generated in the converging wedge because the journal is pumping fluid through a restriction. On the downstream side of the minimum film thickness, the journal is pumping fluid out of a diverging region. In this region the pressure decreases. Either the pressure becomes negative, which is defined as pressure below ambient pressure, or the film cavitates and decreases to atmospheric pressure as the lubricant releases entrapped air.

With respect to stability, cavitation is a more desirable condition than the development of negative pressure. From examination of Fig. 1, it is seen that the negative pressure pulls the journal in an orthogonal direction and increases the cross coupling. The more eccentric the bearing, the larger the negative pressure or cavitated region.

In lightly loaded bearings that are pressure-fed, negative pressures can occur because pressures in the divergent region have not approached atmospheric pressure. Cavitation does not occur, and the bearing is prone to instability. Thus, the feed pressure and load on a bearing are two additional parameters that affect stability.

Lobe bearings are often used because of their excellent antiwhirl characteristics. Figure 9 shows two types of lobe bearings: symmetric and canted. The symmetric lobe bearing is designed so that, in the concentric position, the minimum film thickness occurs at the center of each lobe. Note that this permits a region of converging film followed by a region of diverging film. Thus, depending upon the ambient pressure, it is possible to have negative pressures in a symmetric lobe bearing. Under very high ambient conditions, a symmetric lobe bearing can go unstable.

The canted lobe bearing is designed to have a completely converging wedge and positive pressure throughout its arc length. Its stability characteristics are superior to those of the symmetric lobe bearing. Its steady-state characteristics are also superior.

HYBRID BEARINGS   At times, externally pressurized bearings have been resorted to for stability improvement. The philosophy is that externally pressurized bearings are not subject to high attitude angles, as can be the case with hydrodynamic journal bearings. Although this is generally true, hybrid bearings can still be subject to considerable cross coupling. Figure 11 shows a schematic arrangement of a hybrid bearing. Oil is fed through restrictors from an external source into pocket recesses. From there it exits into the clearance region between recesses. Lubricant is also pumped into and out of recesses by the rotating shaft by viscous drag in the same manner as in a purely hydrodynamic bearing.

Consider recess 2 in Fig. 11. Oil is pumped from the shaft via a converging wedge, and it augments the pressure in the recess provided by the external system. The net result is a higher pressure in the bearing domain covered by recess 2 than would occur without rotation. Now consider recess 3. Here the journal is pumping fluid out of the recess into a diverging film, so that the hydrodynamic action tends to reduce the pressure in this recess domain. By similar reasoning, it can be shown that recess 4 operates at a lower pressure than recess 1. The net result of these variations in pressure due to rotation is that cross-coupling forces are introduced and the hybrid bearing may not prevent instability.

BEARING MATERIALS AND FAILURE MODES

Materials   The most common material used for oil-lubricated fluid film bearings is babbitt. Tin- and lead-based babbitts are relatively soft materials and offer the best insurance against shaft damage. They also allow embedded dirt and contaminants without significant damage.

There are two types of babbitt in common use. One has a tin base (86 to 88%), with about 3 to 8% copper and 4 to 14% antimony; the other has a lead base with a maximum of 20% tin and about 10 to 15% antimony. The remainder is principally lead. The physical properties of babbitt are shown in Table 5. The primary limitations of babbitt are operating temperature (300°F [140°C] max) and fatigue strength. The chemical composition of various babbitt alloys are indicated in Table 6.

Tin-based babbitts have better characteristics than lead-based babbitts; they have better corrosion resistance, are less likely to wipe under poor conditions of lubrication, and can be bonded more easily than lead-based materials. Because of cost considerations, however, lead-based babbitts are widely used. The more widely used is the SAE 15 alloy containing 1% arsenic (Table 6).

Tin-based babbitts do not experience as many corrosion problems as lead-based babbitts. SAE 11 babbitts (containing 8% antimony and 8% copper) are used extensively for industrial applications.

Babbitts are bonded to a backing shell of another material, such as steel or bronze, because they are not a good structural material. The thinner the babbitt layer, the greater the fatigue resistance. In automotive applications, where resistance to fatigue is important, babbitt thickness is from 0.001 to 0.005 in (0.02 to 0.12 mm). For pump applications, the thickness varies from $\frac{1}{32}$ to $\frac{1}{8}$ in (0.8 to 3 mm). The thicker layers provide good conformity and embedability.

TABLE 5   Properties of Bearing Alloys

Bearing material	Brinell hardness at room temperature	Brinell hardness at 30°F (17°C)	Minimum Brinell hardness of shaft	Load carrying capacity, lb/in² (kg/m²)	Max operating temperature, °F (°C)
Tin-based babbitt	20–30	6–12	150 or less	800–1500 ($5.62 \times 10^5 - 10.55 \times 10^5$)	300 (149)
Lead-based babbitt	15–20	6–12	150 or less	800–1200 ($5.62 \times 10^5 - 8.44 \times 10^5$)	300 (149)

SOURCE: Ref. 8

TABLE 6 Percent Composition[a] of Babbitts with SAE Classifications of 11 to 15

	SAE no. (similar ASTM spec.)				
Element	11 (none)	12 (B23, alloy 2)	13 (none)	14 (B23, alloy 7)	15 (B23, alloy 15)
Tin (min)	86.0	88.2	5.0–7.0	9.2–10.8	0.9–1.2
Antimony	6.0–7.5	7.0–8.0	9.0–11.0	14.0–16.0	14.0–15.5
Lead	0.5	0.5	Remainder	Remainder	Remainder
Copper	5.0–6.5	3.0–4.0	0.5	0.5	0.5
Iron	0.08	0.08	—	—	—
Arsenic	0.1	0.1	0.25	0.6	0.8–1.2
Bismuth	0.08	0.08	—	—	—
Zinc	0.005	0.005	0.005	0.005	0.005
Aluminum	0.005	0.005	0.005	0.005	0.005
Cadmium	—	—	0.05	0.05	0.05
Others	0.2	0.2	0.2	0.2	0.2

[a]Percentage of minor constituents represent limiting values except as noted.

SOURCE: *SAE Handbook*, Society of Automotive Engineers, New York, 1960, p. 201.

Failure Modes   The failure modes most commonly found are fatigue, wiping, overheating, corrosion, and wear.

Fatigue occurs because of cyclic loads normal to the bearing surface. Figure 24 shows babbit fatigue in a 7-in (178-mm) diameter journal bearing from a steam turbine.

Wiping results from surface-to-surface contact and smears the babbitt, as shown in Fig. 25. Usual causes of wiping are bearing overload, insufficient rotational speed to form a film, and loss of lubricant.

Overheating is manifest by discoloration of the surface and cracking of the babbitt material.

Corrosion is failure by chemical action. It is more common with lead-based babbitts, which react with acids in the lubricant. Figure 26 shows corrosion damage of a 5-in. (127-mm) diameter lead-based babbitt bearing.

Wear results from contaminants in the film and is evidenced by scoring marks that may be localized or persist around a large circumferential region of the bearing.

FIG. 24 Babbitt fatigue in a 7-in (178-mm) diameter turbine bearing. (Westinghouse Electric Corp. *Photo originally reproduced in* Ref. 8, p. 18–19.)

FIG. 25 Bearing wipe on a 3-in (76-mm) diameter bearing due to temporary loss of lubricant. (Westinghouse Electric Corp. *Photo originally reproduced in* Ref. 8, p. 18–25)

FIG. 26 Corrosion on a 5-in (127-mm) diameter lead-based babbitt bearing. (Westinghouse Electric Corp. *Photo originally reproduced in* Ref. 8, p. 18–21.)

REFERENCES

1. Fuller, D. D.: *Theory and Practice of Lubrication for Engineers*, Wiley, New York, 1956.

2. Castelli, V., and W. Shapiro: "Improved Method of Numerical Solution of the General Incompressible Fluid-Film Lubrication Problem," *Trans. ASME, J. Lub. Technol.*, April, 1967, pp. 211–218.

3. Castelli, V., and J. Pirvics: "Review of Methods in Gas Bearing Film Analysis," *Trans. ASME, J. Lub. Technol.*, October 1968, pp. 777–792.

4. Pinkus, O., and B. Sternlicht: *Theory of Hydrodynamic Lubrication*, McGraw-Hill, New York, 1961.

5. Ng, C. W., and C. H. T. Pan: "A Linearized Turbulent Lubrication Theory," *Trans. ASME, J. Basic Eng.*, Series D, Vol. 87, 1965, p. 675.

6. Elrod, H. G., and C. W. Ng: "A Theory for Turbulent Films and Its Application to Bearings," *Trans. ASME, J. Lub. Technol.*, July 1967, p. 346.

7. Shapiro, W., and R. Colsher: "Dynamic Characteristics of Fluid Film Bearings," *Proc. Sixth Turbomachinery Symposium*, sponsored by the Gas Turbine Laboratories, Department of Mechanical Engineering, Texas A&M University, College Station, Texas, December 1977.

8. O'Connor, J. J., J. Boyd, and E. A. Avallone: *Standard Handbook of Lubrication Engineering*, McGraw-Hill, New York, 1968.

SECTION 2.3
CENTRIFUGAL PUMP PERFORMANCE

2.3.1
CENTRIFUGAL PUMPS: GENERAL PERFORMANCE CHARACTERISTICS

C. P. KITTREDGE

DEFINITIONS

The most important operating characteristcs are the *capacity Q*, the *head H*, the *power P*, and the *efficiency η*. Variables which influence these are the *speed n* and the impeller or wheel *diameter D*. It has been assumed that all other dimensions of the impeller and casing are fixed. The *specific speed n_s* is a parameter which classifies impellers according to geometry and operating characteristics. The value at the operating conditions corresponding to best efficiency is called the specific speed of the impeller and, usually, is the value of interest.

Units   The systems of measure used in this subsection are the U.S. Customary System, hereafter abbreviated USCS, and the International System, hereafter abbreviated SI. *Numerical values* are given in USCS followed by the SI equivalents in parentheses. A discussion of SI units will be found in the front matter of this handbook. In this subsection, the pound (lb) is exclusively the pound-force (USCS) and the kilogram (kg) is exclusively the kilogram mass (SI). Other units are specified where used.

Capacity   The *pump capacity Q* is the volume of liquid per unit time delivered by the pump. In USCS measure it is usually expressed in gallons° per minute (gpm) or, for very large pumps, in cubic feet per second (ft^3/s). In SI measure the corresponding units are liters per second (l/s) and cubic meters per second (m^3/s). If the capacity is measured at a location *m*, where the specific weight of the liquid, $γ_m$, is different from the specific weight in the inlet flange, $γ_s$, then the capacity is given by $Q = Q_m(γ_m/γ_s)$.

Head   The *pump head H* represents the net work done on a unit weight of liquid in passing from the inlet or suction flange *s* to the discharge flange *d*. It is given by

$$H = \left(\frac{p}{γ} + \frac{V^2}{2g} + Z\right)_d - \left(\frac{p}{γ} + \frac{V^2}{2g} + Z\right)_s \tag{1}$$

The term $p/γ$, called the *pressure head* or *flow work*, represents the work required to move a unit weight of liquid across an arbitrary plane perpendicular to the velocity vector *V* against the pressure *p*. The term $V^2/2g$, called the *velocity head*, represents the kinetic energy of a unit

---

°United States: 1 gal = 231 in^3; United Kingdom: 1 gal = 277.418 in^3.

weight of liquid moving with velocity V. The term Z, called the *elevation head* or *potential head*, represents the potential energy of a unit weight of liquid with respect to the chosen datum.

The first parenthetical term in Eq. 1 represents the *discharge head*, h_d, and the second, the *inlet* or *suction head* h_s. The difference is variously called the *pump head, pump total head,* or *total dynamic head*.

The unit of H is customarily feet of liquid pumped in USCS and meters of liquid pumped in SI. This requires that consistent units be used for all quantities in the equation. Typical units would be

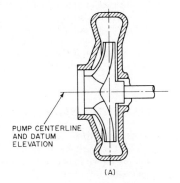

- Static pressure p, lb/ft^2 (Pa)°
- Specific weight γ, lb/ft^3 (N/m^3)
- Average liquid velocity V, ft/s (m/s), where $V = Q/A$ and A = cross-sectional area of flow passage, ft^2 (m^2)
- Acceleration of gravity, g, ft/s^2 (m/s^2), assumed to be 32.174 ft/s^2 (9.8067 m/s^2) since corrections for latitude and elevation above or below mean sea level usually are small
- Elevation Z, ft (m) above or below the datum

PUMP CENTERLINE AND DATUM ELEVATION

(A)

The standard datum for horizontal-shaft pumps is a horizontal plane through the centerline of the shaft (Fig. 1a). For vertical-shaft pumps the datum is a horizontal plane through the entrance eye of the first-stage impeller (Fig 1b) if single suction or through the centerline of the first stage impeller (Fig. 1c) if double suction. Since pump head is the difference between the discharge and suction heads, it is not necessary that the standard datum be used, and any convenient datum may be selected for computing the pump head.

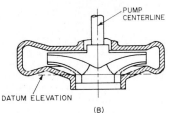

PUMP CENTERLINE

DATUM ELEVATION

(B)

Power   In USCS, the pump output is customarily given as liquid horsepower (lhp) or as water horsepower if water is the liquid pumped. It is given by

$$lhp = \frac{QH(\text{sp. gr.})}{3960} \qquad (2)$$

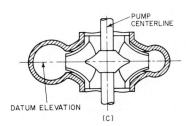

PUMP CENTERLINE

DATUM ELEVATION

(C)

FIG. 1   Elevation datum for defining pump head. (Ref. 24)

where Q is in gallons per minute, H is in feet, and sp. gr. is specific gravity. If Q is in cubic feet per second, the equation becomes

$$lhp = \frac{QH(\text{sp. gr.})}{8.82} \qquad (3)$$

In SI, the power P in watts (W) is given by

$$P = 9797QH(\text{sp. gr.}) \qquad (4)$$

°In SI, the unit of pressure is the pascal (Pa), equal to one newton per square meter.

where Q is in cubic meters per second and H is in meters.

When Q is in liters per second and H is in meters

$$P = 9.797QH \text{ (sp. gr.)} \tag{5}$$

Efficiency   The pump efficiency η is the liquid horsepower divided by the power input to the pump shaft. The latter usually is called the brake horsepower (bhp). The efficiency may be expressed as a decimal or multiplied by 100 and expressed as percent. In this subsection, the efficiency will always be the decimal value unless otherwise noted. Some pump-driver units are so constructed that the actual power input to the pump is difficult or impossible to obtain. Typical of these is the "canned" pump for volatile or dangerous liquids. In such case, only an *overall efficiency* can be obtained. If the driver is an electric motor, this is called the *wire-to-liquid efficiency* or, when water is the liquid pumped, the *wire-to-water efficiency*.

CHARACTERISTIC CURVES

Ideal Pump and Liquid   An ideal impeller contains a very large number of vanes of infinitesimal thickness. The particles of an ideal liquid move exactly parallel to such vane surfaces without friction. As shown in Sec. 2.1, an analysis of the power transmitted to an ideal liquid by such an impeller leads to

$$H_e = \frac{u_2 c_{u2}}{g} - \frac{u_1 c_{u1}}{g} \tag{6}$$

often called *Euler's pump equation*. The Euler head is the work done on a unit of weight of the liquid by the impeller. In Fig. 2, u is the peripheral velocity of any point on a vane and w is the velocity of a liquid particle relative to the vane as seen by an observer attached to and moving with the vane. The absolute velocity c of a liquid particle is the vector sum of u and w. The subscripts 1 and 2 refer to the entrance and exit cross sections of the flow passages.

The second term on the right-hand side in Euler's equation frequently is small relative to the first term and may be neglected so that Eq. 6 becomes

$$H_e \cong \frac{u_2 c_{u2}}{g} \tag{7}$$

As shown in Fig. 2*b*, the absolute velocity vector c may be resolved into the meridian, or radial, velocity c_m and the peripheral velocity c_u. From the geometry of the figure

$$c_{u2} = u_2 - \frac{c_{m2}}{\tan \beta_2} \tag{8}$$

which, substituted into Eq. 7, gives

$$H_e = \frac{u_2^2}{g} - \frac{u_2 c_{m2}}{g \tan \beta_2} \tag{9}$$

Neglecting leakage flow, the meridian velocity c_m must be proportional to the capacity Q. With the additional assumption of constant impeller speed, Eq. 9 becomes

$$H_e = k_1 - k_2 Q \tag{10}$$

in which k_1 and k_2 are constants, with the value of k_2 dependent on the value of the vane angle β_2. Figure 3 shows the Euler head-capacity characteristics for the three possible conditions on the vane angle at exit β_2. The second right-hand term in Eq. 6 may be treated in like manner to the foregoing and included in Eqs. 9 and 10. The effect on Fig. 3 would be to change the value of $H_e = u_2^2/g$ at $Q = 0$ and the slopes of the lines, but all head-capacity characteristics would remain straight lines.

Real Pump and Liquid   The vanes of real pump impellers have finite thickness and are relatively widely spaced. Investigations[1,2] have shown that the liquid does not flow parallel to the vane surfaces even at the point of best efficiency, so that the conditions required for Euler's equa-

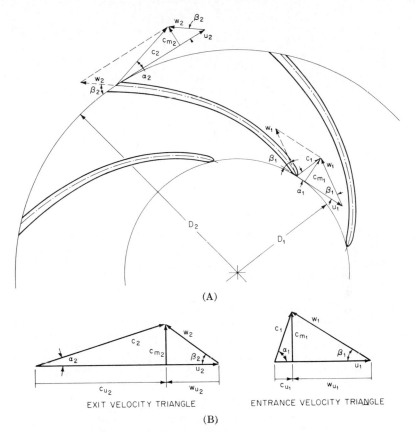

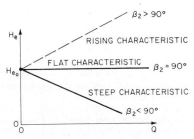

FIG. 2   Velocity diagrams for radial-flow impellers.

tion are not fulfilled. The head is always less than predicted by Euler's equation, and the head-capacity characteristic frequently is an irregular curve. Analysis has so far failed to predict the characteristics of real pumps with requisite accuracy, and so graphical representation of actual tests is in common use. It is customary to plot the head, power, and efficiency as functions of capacity for a constant speed. An example is shown in Fig. 4.

Pumps are designed to operate at the point of best efficiency. The head, power, and capacity at best efficiency, often called the *normal* values, are indicated in this subsection by H_n, P_n, and Q_n respectively. Sometimes a pump may be operated continuously at a capacity slightly above or below Q_n. In such case, the actual operating point is called the *rated* or *guarantee* point if the manufacturer specified this capacity in the guarantee. It is unusual to operate a pump continuously at a capacity at which the efficiency is much below the maximum value. Apart from the unfavorable economics, the pump may be severely damaged by continued off-design operation, as described later.

Backward-Curved Vanes, $\beta_2 < 90°$ Figure 4 shows the characteristics of a double-suction pump with backward-curved vanes, $\beta_2 = 23°$. The

FIG. 3   Euler's head-capacity characteristics.

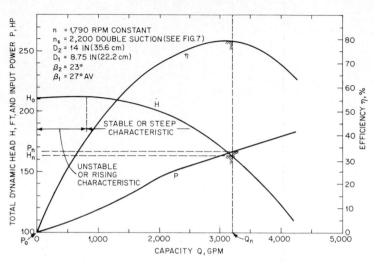

FIG. 4 Typical pump characteristics, backward-curved vanes (ft × 0.3048 = m; hp × 745.7 = W; gpm × 0.06309 = l/s).

impeller discharged into a single volute casing, and the specific speed was $n_s \approx 2200$ (1347) at best efficiency. At shutoff ($Q = 0$), Eq. 9 predicts Euler's head H_e to be 374 ft (114 m), whereas the pump actually developed about 210 ft (64 m), and this head remained nearly constant for the range $0 < Q < 1000$ gpm (63 l/s). For $Q > 1000$ gpm (63 l/s), the head decreased with increasing capacity but not in the linear fashion predicted by Euler's equation. At best efficiency, where $Q_n = 3200$ gpm (202 l/s), Eq. 9 predicts $H_e = 281$ ft (86 m), whereas the pump actually developed $H_n = 164$ ft (50 m).

Radial Vanes, $\beta_2 = 90°$

Large numbers of radial-vane pumps are used in many applications, from cellar drainers, cooling-water pumps for internal combustion engines, and other applications where low first cost is more important than high efficiency to highly engineered pumps designed for very high heads. The impellers are rarely more than 6 in (15 cm) in diameter, but the speed range may be from a few hundred to 30,000 rpm or more. The casings usually are concentric with the impellers and have one or more discharge nozzles that act as diffusers. The impellers usually are open, with from three to six flat vanes. The clearance between vanes and casing is relatively large for easy assembly. Such pumps exhibit a flat head-capacity curve from shutoff to approximately 75% of best efficiency capacity, and beyond this flow the head-capacity curve is steep. The pumps develop a higher head, up to 8000 ft (2400 m) per stage, than pumps with backward-curved vanes, but the efficiency of the former usually is lower. (See also Subsection 2.2.1.)

Figure 5 shows the characteristics of a pump as reported by Rupp.[3] The impeller was fully shrouded, $D = 5.25$ in (13.3 cm), and fitted with 30 vanes of varying length. The best efficiency, $\eta = 55\%$, was unusually high for the specific speed $n_s = 475$, as may be seen from Fig. 6. The head-capacity curve showed a rising (unstable) characteristic for $0 < Q < 25$ gpm (1.6 l/s) and a steep characteristic for $Q > 25$ gpm (1.6 l/s).

Figure 7 shows the characteristics of a pump as reported by Barske.[4] The impeller was open, $D \approx 3$ in (7.6 cm), and fitted with six radial tapered vanes. The effective β_2 may have been slightly greater than 90° due to the taper. At 30,000 rpm the best efficiency was more than 35% at $n_s = 355$ (217), which is much higher than for a conventional pump of this specific speed and capacity (Fig. 6). The head-capacity curve showed a nearly flat characteristic over most of the usable range, as predicted by Euler's equation, but the head was always lower than the Euler head H_e. The smooth concentric casing was fitted with a single diffusing discharge nozzle. When two or more nozzles were used, the head-capacity curves showed irregularities at low capacities and became steep at high capacities.

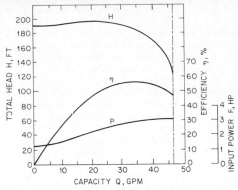

FIG. 5 Pump characteristics, radial vanes (ft × 0.3048 = m; hp × 745.7 = W; gpm × 0.06309 = l/s). (Ref. 3)

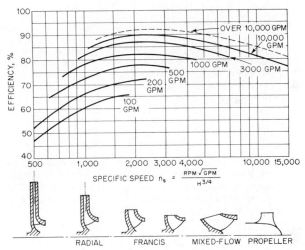

FIG. 6 Pump efficiency versus specific speed and size (gpm × 0.06309 = l/s). (Worthington Pump)

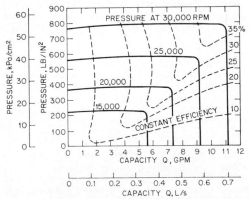

FIG. 7 Pump characteristics, radial vanes (lb/in² × 6.894 = kPa; gpm × 0.06309 = l/s). (Ref. 4)

Manson[5] has reported performance characteristics for jet engine fuel pumps having straight radial vanes in enclosed impellers. The head curves showed unstable characteristics at low capacities and steep characteristics at higher capacities. The best efficiency reported was 54.7% for an impeller diameter of 3.300 in (8.382 cm) and speed n = 28,650 rpm.

Forward-Curved Vanes, $\beta_2 > 90°$

Pumps with forward-curved vanes have been proposed,[6] but the research necessary to achieve an efficient design appears never to have been carried out. Tests have been made of conventional double-suction pumps with the impellers mounted in the reversed position but with rotation correct for the volute casing.

Table 1 shows the pertinent results for six different pumps. Both capacity and efficiency were drastically reduced, and there was only a modest increase in head for five of the six pumps. The sixth pump showed a 38% increase in head over that obtained with the impeller correctly mounted. Published estimates[7,8] of the head-capacity curves to be expected from reversed impellers predict an unstable characteristic at the low end of the capacity range and a steep characteristic at the high end of the range.

PERFORMANCE PARAMETERS

Affinity Laws Consider a series of geometrically similar pumps. It may be shown by a dimensional analysis that three important dimensionless parameters describing pump performance are $D\omega/\sqrt{gH}$, $Q/D^2\sqrt{gH}$, and $P/D^2\rho(gH)^{3/2}$, where ω is the pump speed in radians per unit time, ρ is the mass density of the liquid, and g is the acceleration of gravity, all expressed in the system of measure selected for the other variables. At a common operating point, say the best efficiency point,

$$\frac{D\omega}{\sqrt{gH}} = C_1 \qquad \frac{Q}{D^2\sqrt{gH}} = C_2 \qquad \frac{P}{D^2\rho(gH)^{3/2}} = C_3 \qquad (11)$$

for the geometrically similar series. The C's depend slightly on the efficiency, which in turn may depend on the size of the pump, but it is customary to assume them to be constants. Since gravity is very nearly constant over the surface of the earth, it may be combined with the C's. Also, it is common practice to restrict consideration to cold water and allow C_3 to absorb ρ. Equations 11 may then be written

$$\frac{Dn}{\sqrt{H}} = k_3 \qquad \frac{Q}{D^2\sqrt{H}} = k_4 \qquad \frac{P}{D^2H^{3/2}} = k_5 \qquad (12)$$

Equations 12 are known as the *affinity laws*. Since the parameters are no longer dimensionless, any convenient system of units may be used but, once selected, no change in the units may be made. Usually, in the United States H = pump head in feet, Q = capacity in gallons per minute, P = power in horsepower, n = pump speed in revolutions per minute, and D = impeller diameter in either feet or inches. Obviously the numerical values of the k's depend upon the system of

TABLE 1 Effects of Reversed Mounting of Impeller

Number of stages	Specific speed per stage n_s, USCS (SI)	Percent of normal shutoff head	Percent of normal values at best efficiency			
			Head	Capacity	Power	Efficiency
2	828 (507)	86	111	65	104	71
2	1024 (627)	82	112	88	145	68
1	1240 (759)	75	105	38.5	68.5	59
1	1430 (876)	82	106	69.7	138	53.5
1	2570 (1574)	74.5	117	62	138	52.5
1	2740 (1678)	77.5	138	61.5	180	47

SOURCE: Worthington Pump.

measure and units employed. Sometimes k_3, k_4, and k_5 are called, respectively, the *unit speed*, *unit capacity*, and *unit power*.

Specific Speed The diameter may be eliminated from the first two of Eqs. 12 to obtain the specific speed:

$$n_s = k_3 \sqrt{k_4} = \frac{n\sqrt{Q}}{H^{3/4}} \tag{13}$$

In the United States it is usual to express n in revolutions per minute, Q in gallons per minute, and H in feet. In SI, n_s is called *type number*, usually with n in revolutions per minute, Q in liters per second, and H in meters. The foregoing are the only units used for n_s in this subsection. The specific speed° n_s is a useful dimensional parameter for classifying the overall geometry and performance characteristics of impellers. In Eq. 13, Q is taken in U.S. practice as the full pump capacity for either single-suction or double-suction impellers, unlike the case of suction specific speed (Eq. 28). European practice is to use one-half capacity for both.

It has been common practice to use the term *specific speed of a pump* to mean the specific speed of the impeller. This is particularly confusing with multistage pumps, for which only the specific speed of a single stage is significant. With multistage pumps, the head of a single stage is to be used in Eq. 13 and not the total head of the pump. The head of a single stage is almost always the total pump head divided by the number of stages. Figure 6 shows efficiency as a function of specific speed and capacity for well-designed pumps. Also shown are typical impeller profiles which correspond approximately to the specific speeds indicated. For a series of *geometrically similar* impellers operated at dynamically similar conditions, that is, at identical specific speeds, the affinity laws become

$$\frac{Q_1}{Q_2} = \frac{n_1 D_1^3}{n_2 D_2^3} \tag{14a}$$

$$\frac{H_1}{H_2} = \frac{n_1^2 D_1^2}{n_2^2 D_2^2} \tag{14b}$$

$$\frac{P_1}{P_2} = \frac{n_1^3 D_1^5}{n_2^3 D_2^5} \tag{14c}$$

where subscripts 1 and 2 refer to any two noncavitating operating conditions. *See the discussion on diameter reduction which follows for the affinity laws to be used when an impeller is reduced in diameter.*

Classification of Curve Shapes A useful method for comparing characteristics of pumps of different specific speeds is to normalize on a selected operating condition, usually best efficiency. Thus,

$$q = \frac{Q}{Q_n} \qquad h = \frac{H}{H_n} \qquad p = \frac{P}{P_n} \tag{15}$$

where the subscript n designates values for the best efficiency point. Figures 8, 9, and 10 show approximate performance curves normalized on the conditions of best efficiency and for a wide range of specific speeds. These curves are applicable to pumps of any size because absolute magnitudes have been eliminated. In Fig. 8, curves 1 and 2 exhibit a *rising head* or *unstable* characteristic where the head increases with increasing capacity over the lower part of the capacity range. This may cause instability at heads greater than the shutoff value, particularly if two or more pumps are operated in parallel. Curve 3 exhibits an almost constant head at low capacities and is often called a *flat* characteristic. Curves 4 to 7 are typical of a *steep* or *stable head* characteristic, in which the head always decreases with increasing capacity. Although the shape of the head-capacity curve is primarily a function of the specific speed, the designer has some control

---

°The dimensionless specific speed $\omega_s = \omega\sqrt{Q}/(gH)^{3/4}$, obtained by eliminating D from the first two of Eqs. 11, is rarely used.

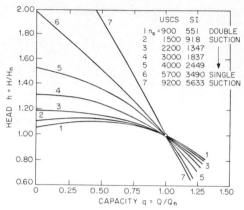

FIG. 8 Head-capacity curves for several specific speeds. (Ref. 12)

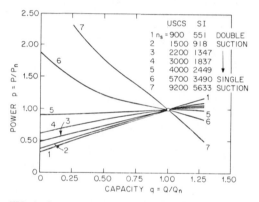

FIG. 9 Power-capacity curves for several specific speeds. (Ref. 12)

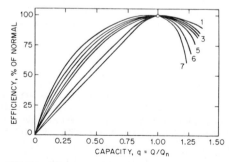

FIG. 10 Efficiency-capacity curves for several specific speeds. (Ref. 12)

through selection of the vane angle β_2, number of impeller vanes z, and capacity coefficient $\phi = c_{m2}/u_2$, as described in Sec. 2.1 (see also Fig. 2). For pumps having a single-suction specific speed approximately 5000 (3060) and higher, the power is maximum at shutoff and decreases with increasing capacity. This may require an increase in the power rating of the driving motor over that required for operation at normal capacity.

Efficiency The efficiency η is the product of three efficiencies:

$$\eta = \eta_m \eta_v \eta_h \tag{16}$$

The mechanical efficiency η_m accounts for the bearing, stuffing box, and all disk-friction losses including those in the wearing rings and balancing disks or drums if present. The volumetric efficiency η_v accounts for leakage through the wearing rings, internal labyrinths, balancing devices, and glands. The hydraulic efficiency η_h accounts for liquid friction losses in all through-flow passages, including the suction elbow or nozzle, impeller, diffusion vanes, volute casing, and the crossover passages of multistage pumps. Figure 11 shows an estimate of the losses from various sources in double-suction single-stage pumps having at least 12-in (30-cm) discharge pipe diameter. Minimum losses and hence maximum efficiencies are seen to be in the vicinity of $n_s \approx 2500$ (1530), which agrees with Fig. 6.

Effects of Pump Speed Increasing the impeller speed increases the efficiency of centrifugal pumps. Figure 7 shows a gain of about 15% for an increase in speed from 15,000 to 30,000 rpm. The increases are less dramatic at lower speeds. For example, Ippen[9] reported about 1% increase in the efficiency of a small pump, $D = 8$ in (20.3 cm) and $n_s = 1992$ (1220), at best efficiency, for an increase in speed from 1240 to 1880 rpm. Within limits, the cost of the pump and driver usually decreases with increasing speed. Abrasion and wear increase with increasing speed, particularly if the liquid contains solid particles in suspension. The danger of cavitation damage usually increases with increasing speed unless certain suction requirements can be met, as described later.

Effects of Specific Speed Figures 6 and 11 show that maximum efficiency is obtained in the range 2000 (1225) $< n_s <$ 3000 (1840), but this is not the only criterion. Pumps for high heads and small capacities occupy the range 500 (300) $< n_s <$ 1000 (600). At the other extreme, pumps for very low heads and large capacities may have $n_s = 15,000$ (9200) or higher. For given head and capacity, the pump having the highest specific speed that will meet the requirements

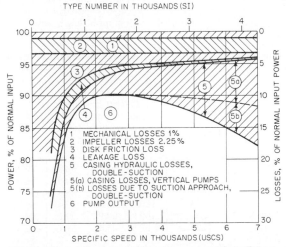

FIG. 11 Power balance for double-suction pumps at best efficiency. (Ref. 12)

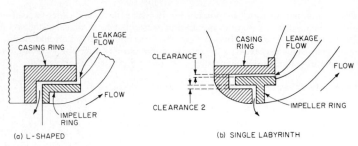

FIG. 12 Typical wearing rings.

probably will be the smallest and least expensive. However, Eq. 13 shows that it will run at the highest speed and be subject to maximum wear and cavitation damage, as previously mentioned.

Effects of Clearance

WEARING-RING CLEARANCE   Details of wearing-ring construction are given in Sec. 2.2. Schematic outlines of two designs of rings are shown in Fig. 12. The L-shaped construction shown in Fig. 12a is very widely used with the close clearance between the cylindrical portions of the rings. Leakage losses increase and pump performance falls off as the rings wear. Table 2 shows some of the effects of increasing the clearance of rings similar to Fig. 12a.

The labyrinth construction shown in Fig. 12b has been used to increase the leakage path without increasing the axial length of the rings. If the pressure differential across these rings is high enough, the pump shaft may take on lateral vibrations with relatively large amplitude and long period, which can cause serious damage. One remedy for these vibrations is to increase clearance 2 in Fig. 12b relative to clearance 1, at the expense of an increased leakage flow. High-pressure breakdown through plain rings may cause vibration,[10] but this is not usually a serious problem. It is considered good practice to replace or repair wearing rings when the nominal clearance has doubled. The presence of abrasive solids in the liquid pumped may be expected to increase wearing-ring clearances rapidly.

VANE-TIP CLEARANCE   Many impellers are made without an outer shroud and rely on close running clearances between the vane tips and the casing to hold leakage across the vane tips to a minimum. Although this construction usually is not used with pumps having specific speeds less than about

TABLE 2   Effects of Increased Wearing-Ring Clearance on Centrifugal Pump Performance

Specific speed n_s, USCS (SI)	Design head, ft (m)	Ring clearance, % of normal value	Percent of values at best efficiency with normal ring clearance				Percent of values at shutoff with normal ring clearance	
			Q	H	P	η	H_0	P_0
2100 (1286)	63 (19.2)	178	100	98.3	98.9	99.4	97.0	100
....	......	356	100	97.5	99.0	98.5	93.6	98.2
....	......	688	100	96.0	98.9	97.1	91.2	94.8
....	......	1375	100	94.3	97.4	96.8	88.8	92.5
3500 (2143)	65 (19.8)	354	100	90.0	99.1	90.8	85.0	96.2
4300 (2633)	41 (12.5)	7270	62	65.5	81.7	49.8	44.3	106
4800 (2939)	26 (7.9)	5220	96	78.8	89.2	84.8	78.2	83.3

SOURCE: Worthington Pump.

6000 (3670), Wood et al.[11] have reported good results with semiopen impellers at 1800 (1100) $\leq n_s \leq$ 4100 (2500). It appears that both head and efficiency increase with decreasing tip clearance and are quite sensitive to rather small changes in clearance. Reducing the tip clearance from about 0.060 in (1.5 mm) to about 0.010 in (0.25 mm) may increase the efficiency by as much as 10%. Abrasive solids in the liquid pumped probably will increase tip clearances rapidly.

MODIFICATIONS TO IMPELLER AND CASING

Diameter Reduction   To reduce cost, pump casings usually are designed to accommodate several different impellers. Also, a variety of operating requirements can be met by changing the outside diameter of a given radial impeller. Euler's equation (Eq. 7) shows that the head should be proportional to $(nD)^2$ provided that the exit velocity triangles (Fig. 2b) remain similar before and after cutting, with w_2 always parallel to itself as u_2 is reduced. This is the usual assumption and leads to

$$\frac{Q_1}{Q_2} = \frac{n_1 D_1}{n_2 D_2} \tag{17a}$$

$$\frac{H_1}{H_2} = \frac{n_1^2 D_1^2}{n_2^2 D_2^2} \tag{17b}$$

$$\frac{P_1}{P_2} = \frac{n_1^3 D_1^3}{n_2^3 D_2^3} \tag{17c}$$

which apply only to a given impeller with altered D and constant efficiency but *not* to a geometrically similar series of impellers. The assumptions on which Eqs. 17 were based are rarely if ever fulfilled in practice, so that exact predictions by the equations should not be expected. Diameter reductions greater than from 10 to 20% of the original full diameter of the impeller are rarely made.

RADIAL DISCHARGE IMPELLERS   Impellers of low specific speed may be cut successfully provided the following items are kept in mind:

1. The angle β_2 may change as D is reduced, but this usually can be corrected by filing the vane tips. (See the discussion on vane-tip filing which follows.)
2. Tapered vane tips will be thickened by cutting and should be filed to restore the original shape. (See the discussion on vane-tip filing, p. 0.00.)
3. Bearing and stuffing box friction remain constant, but disk friction should decrease with decreasing D.
4. The length of flow path in the pump casing is increased by decreasing D.
5. Since c_{m1} is smaller at the reduced capacity, the inlet triangles no longer remain similar before and after cutting, and local flow separation may take place near the vane entrance tips.
6. The second right-hand term in Eq. 6 was neglected in arriving at Eqs. 17, but it may represent a significant decrease in head as D is reduced.
7. Some vane overlap should be maintained after cutting. Usually the initial vane overlap decreases with increasing specific speed, so that the higher the specific speed, the less the allowable diameter reduction.

Most of the losses are approximately proportional to Q^2, and hence to D^2 by Eqs. 17. Since the power output decreases approximately as D^3, it is reasonable to expect the maximum efficiency to decrease as the wheel is cut, and this often is the case. By Eqs. 13 and 17, the product $n_s D$ should remain constant so that the specific speed at best efficiency increases as the wheel diameter is reduced (Table 3).

The characteristics of the pump shown in Fig. 13 may be used to illustrate reduction of diameter at constant speed. Starting with the best efficiency point and $D = 16\frac{5}{16}$ in (41.4 cm), let it be

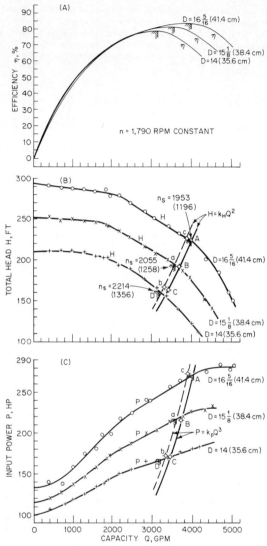

FIG. 13 Diameter reduction of radial-flow impeller (gpm × 0.06309 = l/s). D is measured in inches (centimeters).

required to reduce the head from $H = 224.4$ to $H' = 192.9$ ft (68.4 m to 58.8 m) and to determine the wheel diameter, capacity, and power for the new conditions. Since the speed is constant, Eqs. 17 may be written

$$H = k_H Q^2 \qquad \text{and} \qquad P = k_p Q^3 \qquad (18)$$

where k_H and k_p may be obtained from the known operating conditions at $D = 16\frac{5}{16}$ in (41.4 cm). Plot a few points for assumed capacities and draw the curve segments as shown by the solid lines in Figs. 13*b* and 13*c*. Then, from Eqs. 17,

$$D' = D\sqrt{H'/H} \qquad Q' = Q\sqrt{H'/H} \qquad P' = P(H'/H)^{3/2} \qquad (19a)$$

TABLE 3 Predicted Characteristics at Different Impeller Diameters on a Radial-Flow Pump

D, in (cm)	Test values					Predicted from D = 16⅜ in (41.4 cm)			Predicted from D = 15⅛ in (38.4 cm)		
	Q, gpm (l/s)	H, ft (m)	P, hp (kW)	n_s	n_sD	Q', gpm (l/s)	H', ft (m)	P', hp (kW)	Q', gpm (l/s)	H' ft (m)	P', hp (kW)
16⅜ (41.4)	4000 (252)	224.4 (68.4)	270.4 (201.6)	1953 (1196)	31,860 (49,508)	....	....	....	3883 (245) −2.93% error[a]	227.3 (69.3) 1.29% error[a]	272.2 (203.0) 0.68% error[a]
15⅛ (38.4)	3600 (227.1)	195.4 (59.6)	217.0 (161.8)	2055 (1258)	31,080 (48,318)	3709 (234) 3.02% error[a]	192.9 (58.8) −1.28% error[a]	215.5 (160.7) −0.69% error[a]	....	....	....
14 (35.6)	3200 (201.9)	163.6 (49.9)	167.2 (124.7)	2214 (1356)	31,000 (48,261)	3433 (217) 7.28% error[a]	165.3 (50.4) 1.03% error[a]	170.9 (127.4) 2.33% error[a]	3332 (210.2) 4.13% error[a]	167.4 (51.0) 2.33% error[a]	172.1 (128.3) 2.93% error[a]

[a]Error in predicting best efficiency point.

2.207

$$D' = D(Q'/Q) \qquad H' = H(Q'/Q)^2 \qquad P' = P(Q'/Q)^3 \qquad (19b)$$

from which $D = 15\frac{1}{8}$ in (38.4 cm), $Q = 3709$ gpm (234 l/s), and $P' = 215.5$ hp (160.7 kW). In Fig. 13, the initial conditions were at points A and the computed conditions after cutting at points B. The test curve for $D = 15\frac{1}{8}$ in (38.4 cm) shows the best efficiency point a at a lower capacity than predicted by Eqs. 19, but the head-capacity curve satisfies the predicted values very closely. The power-capacity prediction was not quite as good. Table 3 and Fig. 13 give actual and predicted performance for three impeller diameters. The error in predicting the best efficiency point was computed by (predicted value minus test value) (100)/(test value). As the wheel diameter was reduced, the best efficiency point moved to a lower capacity than predicted by Eqs. 19 and the specific speed increased, showing that dynamically similar operating conditions were not maintained.

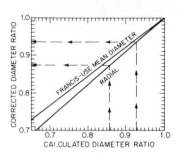

FIG. 14 Corrections for calculated impeller diameter reductions. (Ref. 12)

Wheel cutting should be done in two or more steps with a test after each cut to avoid too large a reduction in diameter. Figure 14 shows an approximate correction, given by Stepanoff,[12] which may be applied to the ratio D'/D as computed by Eqs. 17 or 19. The accuracy of the correction decreases with increasing specific speed. Figure 15 shows a correction proposed by Rütschi[13] on the basis of extensive tests on low-specific-speed pumps. The corrected diameter reduction ΔD is the diameter reduction $D - D'$ given by Eqs. 19 and multiplied by k from Fig. 15. The shaded area in Fig. 15 indicates the range of scatter of the test points operating at or near maximum efficiency. Near shutoff the values of k were smaller and at maximum capacity the values of k were larger than shown in Fig. 15. Table 4 shows the results of applying Figs. 14 and 15 to the pump of the preceding example.

There is no independent control of Q and H in impeller cutting, although Q may be increased somewhat by underfiling the vane tips as described later. The discharge and power will automatically adjust to the values at which the pump head satisfies the system head-capacity curve.

MIXED-FLOW IMPELLERS   Diameter reduction of mixed-flow impellers is usually done by cutting a maximum at the outside diameter D_o and little or nothing at the inside diameter D_i, as shown in Fig. 16. Stepanoff[14] recommends that the calculations be based on the average diameter $D_{av} = (D_o + D_i)/2$ or estimated from the vane-length ratio FK/EK or GK/EK in Fig. 16d. Figure 16 shows a portion of the characteristics of a mixed-flow impeller on which two cuts were made as shown in Fig. 16. The calculations were made by Eqs. 19 using the mean diameter $D_m = \sqrt{(D_o^2 + D_i^2)/2}$ instead of the outside diameter in each case. The predictions and test results are

TABLE 4   Impeller Diameter Corrections

D before cutting	in	16.3125	16.3125
D' predicted by Eqs. (19)	in	15.125	14.000
D' corrected by Fig. 14	in	15.25[a]	14.26[b]
D' corrected by Fig. 15	in	15.60[c]	14.93[d]

[a] $D'/D = 15.125/16.3125 = 0.927$; by Fig. 14, corrected $D'/D = 0.935$. Corrected $D' = (0.935)(16.3125) = 15.25$ in.

[b] $D'/D = 14.000/16.3125 = 0.858$; by Fig. 14, corrected $D'/D = 0.874$. Corrected $D' = (0.874)(16.3125) = 14.26$ in.

[c] $D - D' = 16.3125 - 15.125 = 1.1875$; by Fig. 15, $K \simeq 0.6$ at $n_s = 1,953$. Corrected $D - D' = (0.6)(1.1875) = 0.7125$ and corrected $D' = 16.3125 - 0.7125 = 15.60$ in.

[d] $D - D' = 16.3125 - 14.000 = 2.3125$; by Fig. 15, $K \simeq 0.6$ at $n_s = 1,953$. Corrected $D - D' = (0.6)(2.3125) = 1.3875$ and corrected $D' = 16.3125 - 1.3875 = 14.93$ in.

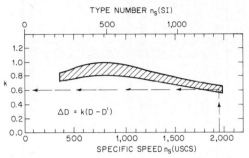

FIG. 15 Corrections for calculated impeller diameter reductions. (Ref. 13)

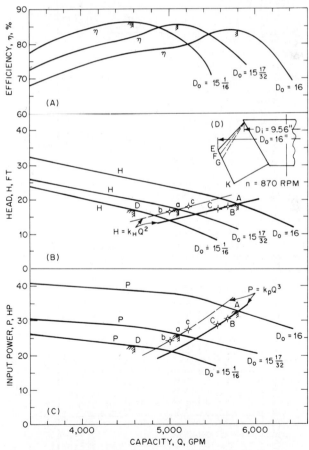

FIG. 16 Diameter reduction of mixed-flow impeller (ft × 0.3048 = m; in × 2.540 = cm; gpm × 0.06309 = l/s; hp × 0.7457 = kW).

TABLE 5 Predicted Characteristics at Different Impeller Diameters on a Mixed-Flow Pump

D, in (cm)	D_m, in (cm)	Test values					Predicted from $D = 16.00$ (40.64 cm), $D_m = 13.17$ (33.45 cm)			Predicted from $D = 15.53$ (39.45 cm), $D_m = 12.89$ (32.74 cm)		
		Q, gpm (l/s)	H, ft (m)	P, hp (kW)	n_s	$n_s D_m$	Q', gpm (l/s)	H', ft (m)	P', hp (kW)	Q', gpm (l/s)	H', ft (m)	P', hp (kW)
16 (40.64)	13.17 (33.45)	5800 (365.9)	18.6 (5.67)	32.5 (24.2)	7385 (4522)	97,300 (151,300)	· · · · · · · ·	· · · · · · · ·	· · · · · · · ·	5210 (329) −11.3 error[a]	17.9 (5.46) −3.91 error[a]	27.4 (20.4) −18.8 error[a]
15 17/32 (39.45)	12.89 (32.74)	5100 (321.8)	17.1 (5.21)	25.7 (19.2)	7385 (4522)	95,200 (148,050)	5680 (358.4) 10.2 error[a]	17.8 (5.43) 4.50 error[a]	30.4 (22.7) 15.5 error[a]			
15 1/16 (38.26)	12.62 (32.05)	4600 (290.2)	16.8 (5.12)	22.6 (16.9)	7100 (4347)	89,600 (139,300)	5560 (350.8) 17.3 error[a]	17.1 (5.21) 1.79 error[a]	28.6 (21.3) 22.4 error[a]	5000 (315.5) 8.00 error[a]	16.4 (5.00) −2.44 error[a]	24.1 (18.0) 6.22 error[a]

[a]Error in predicting best efficiency point.

shown in Fig. 16 and Table 5. It is clear that the actual change in the characteristics far exceeded the predicted values. Except for the use of the mean diameters, the procedure was essentially the same as that described for Fig. 13, and all points and curves are similarly labeled. The corrections given in Fig. 14 would have made very little difference in the computed diameter reductions, and those of Fig. 15 were not applicable to impellers having specific speeds greater than $n_s = 2000$ (1225). In this case, the product $n_s D_m$ did not remain constant and the maximum efficiency increased as D_m was reduced. Although the changes in diameter were small, the area of vane removed was rather large for each cut. The second cut eliminated most of the vane overlap.

The characteristics of mixed-flow impellers can be changed by cutting, but very small cuts may produce a significant effect. The impellers of propeller pumps are not usually subject to diameter reduction.

Shaping Vane Tips   If the discharge tips of the impeller vanes are thick, performance usually can be improved by filing over a sufficient length of vane to produce a long, gradual taper. Chamfering, or rounding, the discharge tips may increase the losses and should never be done. Reducing the impeller diameter frequently increases the tip thickness.

OVERFILING   This is shown at B in Fig. 17, and the unfiled vane is shown at A. Usually there is little or no increase in the vane spacing d before and d_F after filing, so that the discharge area is practically unchanged. Experience indicates that any change in the angle β_2 due to overfiling usually produces a negligible change in performance.

UNDERFILING   This is shown at C in Fig. 17. If properly done, underfiling will increase the vane spacing from d to d_F and hence the discharge area, which lowers the average meridian velocity c_{m2} at any given capacity Q. The angle β_2 usually is increased slightly. Figure 18A and Eq. 7 show that the head and consequently the power increase at the same capacity. The maximum efficiency usually is improved and may be moved to a higher capacity. At the same head, Fig. 18B shows that both c_{m2} and the capacity will increase. The change both in the area and in c_{m2} may increase the capacity by as much as 10%. Table 6 shows the results of tests before and after underfiling the impeller vanes of nine different pumps. In general, they confirm the foregoing predictions based on changes in area and in the velocity triangles.

INLET VANE TIPS   If the inlet vane tips are blunt, as shown at D in Fig. 17, the cavitation characteristics may be improved by sharpening them, as shown at E. In this case overfiling increases the

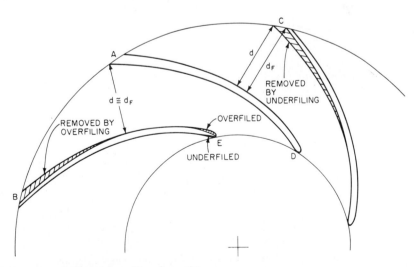

FIG. 17   Underfiling and overfiling of vane tips.

TABLE 6 Changes in Performance When Impeller Vanes are Underfiled

Specific speed n_s, USCS (SI)	No. of stages	Impeller diameter D_2, in (cm)	Change in vane spacing d_F/d	$\dfrac{H_F - H}{H}$, $\%^b$	Changes at best efficiency point after filing, %				
					H_0^c	H	Q	P	η
862 (528)	1	11½ (29.21)	1.13	4	1.5	4	0	9	−5
945 (579)	2	8⅜ (21.27)	1.05	5.5	0	3	4	−0.5	7.5^d
1000 (612)	4	9⅛ (23.18)	1.055	5.5	0	5.5	0	3	2.5
1080 (661)	2	9¹³⁄₁₆ (24.92)	1.08	10	2.5	10	0	6.8	3
1525 (934)	2	10½ (26.67)	1.035	3.2	3	0.5	4.5	3.5	1.2
1950 (1194)	1	22 (55.88)	1.02	1.5	1	1.5	0	1.5	0
3300^a (2021)	1	12 (30.48)	. . . . .	7.8	2.5	7.8	0	8.5	−0.6
3450^a (2112)	1	30½ (77.47)	. . . . .	6.5	0.5	6.5	0	7.8	−2.2
4300 (2633)	1	41¼ (104.8)	. . . . .	5		5	0	0.5	4.5^e

^{a}Double suction.
^{b}H_F is head after filing.
^{c}Shutoff head.
^{d}Due in part to changes in pump casing.
^{e}Due in part to rounding of inlet vane tips.

SOURCE: Worthington Pump.

effective flow area, which reduces c_{m1} for a given capacity. If·more area is needed, it may be advantageous to cut back part of the vane and sharpen the leading edge. Overfiling tends to increase β_1, which is incompatible with a decrease in c_{m1} (Fig. 19). The increase in β_1 increases the angle of attack of the liquid approaching the vane. In Figs. 2 and 19, w_1 is tangent to the centerline of the vane at entrance and w_0 is the velocity of the approaching liquid as seen by an observer moving with the vane. The angle δ between w_0 and w_1 is the angle of attack, which increases to δ_F after overfiling. Although the increase in δ tends to increase the opportunity for

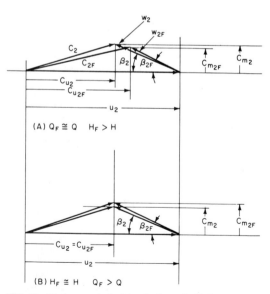

FIG. 18 Discharge velocity triangles for underfiled vanes.

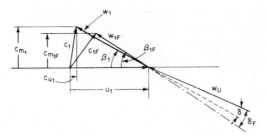

FIG. 19 Inlet velocity triangles for overfiled vanes.

the liquid to separate from the low-pressure face of the vane, this disadvantage usually is outweighed by the improvement that results when c_{m1} is reduced.

CASING TONGUE   The casing tongue forms part of the throat of the discharge nozzle of many volute casings (Fig. 68). Frequently the throat area is small enough to act as a throttle and reduce the maximum capacity otherwise obtainable from the impeller. Cutting back the tongue increases the throat area and increases the maximum capacity. The head-capacity characteristic is then said to *carry out farther*. Shortening the discharge nozzle may increase the diffusion losses a little and result in a slightly lower efficiency.

CAVITATION

The formation and subsequent collapse of vapor-filled cavities in a liquid due to dynamic action are called *cavitation*. The cavities may be bubbles, vapor-filled pockets, or a combination of both. The local pressure must be at or below the vapor pressure of the liquid for cavitation to begin, and the cavities must encounter a region of pressure higher than the vapor pressure in order to collapse. Dissolved gases often are liberated shortly before vaporization begins. This may be an indication of impending cavitation, but true cavitation requires vaporization of the liquid. Boiling accomplished by the addition of heat or the reduction of static pressure without dynamic action of the liquid is arbitrarily excluded from the definition of cavitation. With mixtures of liquids, such as gasoline, the light fractions tend to cavitate first.

When a liquid flows over a surface having convex curvature, the pressure near the surface is lowered and the flow tends to separate from the surface. *Separation* and *cavitation* are completely different phenomena. Without cavitation, a separated region contains turbulent eddying liquid at pressures higher than the vapor pressure. When the pressure is low enough, the separated region may contain a vapor pocket which fills from the downstream end,[15] collapses, and forms again many times each second. This causes noise and, if severe enough, vibration. Vapor-filled bubbles usually are present and collapse very rapidly in any region where the pressure is above the vapor pressure. Knapp[16] found the life cycle of a bubble to be on the order of 0.003 s.

Bubbles which collapse on a solid boundary may cause severe mechanical damage. Shutler and Mesler[17] photographed bubbles which distorted into toroidal-shaped rings during collapse and produced ring-shaped indentations in a soft metal boundary. The bubbles rebounded following the initial collapse and caused pitting of the boundary. Pressures on the order of 10^4 atm have been estimated[18] during collapse of a bubble. All known materials can be damaged by exposure to bubble collapse for a sufficiently long time. This is properly called *cavitation erosion*, or *pitting*. Figure 20 shows extensive damage to the suction side of pump impeller vanes after about three months' operation with cavitation. At two locations, the pitting has penetrated deeply into the ⅜-in (9.5-mm) thickness of stainless steel. The unfavorable inlet flow conditions, believed to have been the cause of the cavitation, were at least partly due to elbows in the approach piping. Modifications in the approach piping and the pump inlet passages reduced the cavitation to the point that impeller life was extended to several years.

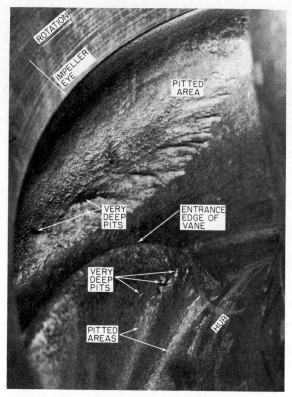

FIG. 20   Impeller damaged by cavitation. (Transamerica DeLaval, Ref. 67)

It has been postulated that high temperatures and chemical action may be present at bubble collapse, but any damaging effects due to them appear to be secondary to the mechanical action. It seems possible that erosion by foreign materials in the liquid and cavitation pitting may augment each other. Controlled experiments[19,20] with water indicated that the damage to metal depends on the liquid temperature and was a maximum at about 100 to 120°F (38 to 49°C). Cavitation pitting, as measured by weight of the boundary material removed per unit time, frequently increases with time. Cast iron and steel boundaries are particularly vulnerable. Controlled experiments have shown that cavitation pitting in metals such as aluminum, steel, and stainless steel depends strongly on the velocity of the fluid in the undisturbed flow past the boundary. On the basis of tests of short duration, Knapp[15] reported that damage to annealed aluminum increased approximately as the sixth power of the velocity of the undisturbed flow past the surface. Hammitt[21] found a more complicated relationship between velocity and damage, as shown in Fig. 21. It seems clear that once cavitation begins it will increase rapidly with increasing velocities. Fre-

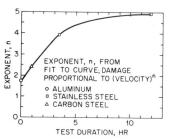

FIG. 21   Cavitation damage exponent versus test time for several materials in water. (Ref. 21)

quently the rate of pitting accelerates with elapsed time. Comprehensive discussions of cavitation and erosion together with additional references are given by Preece.[22]

Centrifugal pumps begin to cavitate when the suction head is insufficient to maintain pressures above the vapor pressure throughout the flow passages. The most sensitive areas usually are the low-pressure sides of the impeller vanes near the inlet edge and the front shroud, where the curvature is greatest. Axial-flow and high-specific-speed impellers without front shrouds are especially sensitive to cavitation on the low-pressure sides of the vane tips and in the close tip-clearance spaces. Sensitive areas in the pump casing include the low-pressure side of the tongue and the low-pressure sides of diffusion vanes near the inlet edges. As the suction head is reduced, all existing areas of cavitation tend to increase and additional areas may develop. Apart from the noise and vibration, cavitation damage may render an impeller useless in as little as a few weeks of continuous operation. In multistage pumps cavitation usually is limited to the first stage, but Kovats[23] has pointed out that second and higher stages may cavitate if the flow is reduced by lowering the suction head (submergence control). Cavitation tends to lower the axial thrust of an impeller. This could impair the balancing of multistage pumps with opposed impellers. A reduction in suction pressure may cause the flow past a balancing drum or disk to cavitate where the liquid discharges from the narrow clearance space. This may produce vibration and damage resulting from contact between fixed and running surfaces.

Net Positive Suction Head   The *net positive suction head* (NPSH), h_{sv}, is a statement of the *minimum* suction conditions required to prevent cavitation in a pump. The *required*, or *minimum*, NPSH must be determined by test and usually will be stated by the manufacturer. The *available* NPSH at installation must be at least equal to the required NPSH if cavitation is to be prevented. Increasing the available NPSH provides a margin of safety against the onset of cavitation. Figure 22 and the following symbols will be used to compute the NPSH:

p_a = absolute pressure in atmosphere surrounding gage, Fig. 22

p_s = gage pressure indicated by gage or manometer connected to pump suction at section *s-s*; may be positive or negative

p_t = absolute pressure on free surface of liquid in closed tank connected to pump suction

p_{vp} = vapor pressure of liquid being pumped corresponding to the temperature at section *s-s* (if liquid is a mixture of hydrocarbons, p_{vp} must be measured by the *bubble point* method)

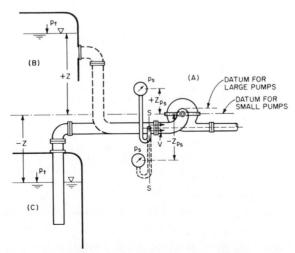

FIG. 22   Definition sketch for computing NPSH.

h_f = lost head due to friction in suction line between tank and section s-s.

V = average velocity at section s-s

Z, Z_{ps} = vertical distances defined by Fig. 22; may be positive or negative

γ = specific weight of liquid at pumping temperature

It is satisfactory to choose the datum for small pumps as shown in Figs. 1 and 22, but with large pumps the datum should be raised to the elevation where cavitation is most likely to start. For example, the datum for a large horizontal-shaft propeller pump should be taken at the highest elevation of the impeller-vane tips. The available NPSH is given by

$$h_{sv} = \frac{p_a - p_{vp}}{\gamma} + \frac{p_s}{\gamma} + Z_{ps} + \frac{V^2}{2g} \tag{20}$$

or

$$h_{sv} = \frac{p_t - p_{vp}}{\gamma} + Z - h_f \tag{21}$$

Consistent units must be chosen so that each term in Eqs. 20 and 21 represents feet (or meters) of the liquid pumped. Equation 20 is useful for evaluating the results of tests. Equation 21 is useful for estimating available NPSH during the design phase of an installation. In Eq. 20, the first term represents the height of a liquid barometer, h_b, containing the liquid being pumped and the sum of the remaining terms represents the suction head h_s of Eq. 1. Therefore

$$h_{sv} = h_b + h_s \tag{22}$$

Usually, a positive value of h_s is called a *suction head* and a negative value of h_s is called a *suction lift*.

Figures 23 to 29, taken from the Hydraulic Institute Standards,[24] are useful in determining suction conditions so that cavitation may be avoided in pumps of good commercial design. Pumps of special design may exceed the limits set by the charts. Restrictions on the use of these charts are stated in the legends and should be observed carefully. Figures 23 to 26 relate specific speed, total head of first stage, and total suction head or total suction lift at sea level for pumps handling clear cold water, that is, temperature not exceeding 85°F (29.4°C). Figures 27 and 28 relate NPSH, capacity, and speed for pumps handling clear hot water, that is, temperature higher than 85°F (29.4°C). These are especially useful with boiler-feed pumps. Figure 29 relates NPSH, capacity, and speed for condensate pumps of not more than three stages with shaft through the eye of the first-stage impeller. Note that there are separate capacity scales for single- and double-suction impellers. This chart may be applied to single-suction overhung impellers by dividing the specified capacity by 1.2 for $Q \leq 400$ gpm (25.2 l/s) and by 1.15 for $Q > 400$ gpm (25.2 l/s). The data of Figs. 23 to 26 may be applied to other temperatures and elevations, as shown in the following example.

EXAMPLE Given a double-suction pump with $H = 200$ ft (61 m) on the first stage and n_s = 2300 (1408), Fig. 23 shows a permissible suction lift of 10 ft (3 m), so that $h_{s1} = -10$ ft (-3 m). Determine h_s if this pump is installed at 1500 ft (457 m) elevation, where the atmospheric pressure is $p_{a2} = 13.92$ lb/in^2 (95.98 kPa) abs, and the temperature of the water being pumped is 180°F (82.2°C).

At sea level and $T_1 = 85$°F (29.4°C), steam tables give $p_{a1} = 14.70$ lb/in^2 (104.4 kPa) abs, $p_{vp1} = 0.60$ lb/in^2 (4.14 kPa) abs and $v_{f1} = 1/\gamma_1 = 0.01609$ ft^3/lb (0.001004 m^3/kg).

At 180°F (82.2°C), steam tables give $p_{vp2} = 7.51$ lb/in^2 (51.78 kPa) abs and $v_{f2} = 1/\gamma_2$ = 0.01651 ft^3/lb (0.001031 m^3/kg).

The respective barometric heights are then

$$h_{b1} = (p_{a1} - p_{vp1})v_{f1} = (144)(14.70 - 0.60)(0.01609) = 32.67 \text{ ft } (9.958 \text{ m})$$

and

$$h_{b2} = (p_{a2} - p_{vp2})v_{f2} = (144)(13.92 - 7.51)(0.01651) = 15.24 \text{ ft } (4.645 \text{ m})$$

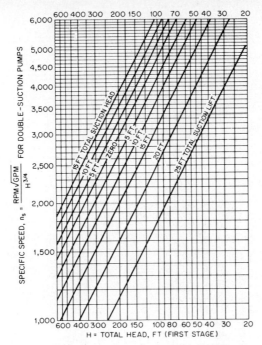

FIG. 23 Upper limits of specific speeds for double-suction pumps handling clear water at 85°F (29.4°C) at sea level (ft × 0.3048 = m). (Ref. 24)

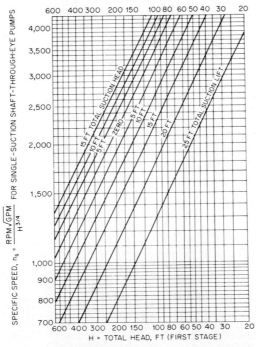

FIG. 24 Upper limits of specific speeds for single-suction shaft-through-eye pumps handling clear water at 85°F (29.4°C) at sea level (ft × 0.3048 = m). (Ref. 24)

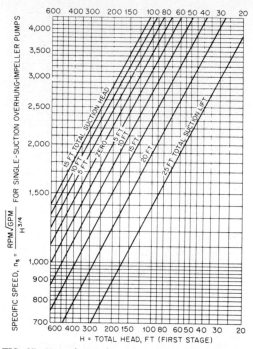

FIG. 25 Upper limits of specific speeds for single-suction overhung-impeller pumps handling clear water at 85°F (29.4°C) at sea level (ft × 0.3048 = m). (Ref. 24)

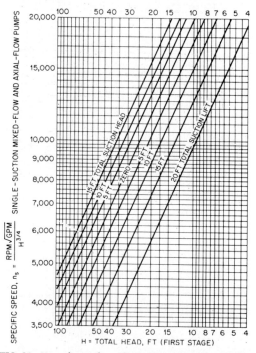

FIG. 26 Upper limits of specific speeds for single-suction and mixed-flow and axial-flow pumps handling clear water at 85°F (29.4°C) at sea level (ft × 0.3048 = m). (Ref. 24)

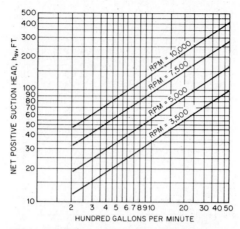

FIG. 27 NPSH for centrifugal hot-water pumps, single suction (ft × 0.3048 = m; gpm × 0.06309 = l/s). (Ref. 24)

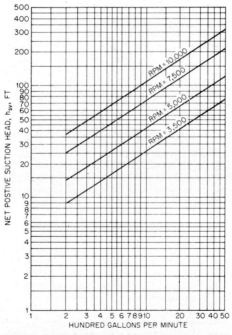

FIG. 28 NPSH for centrifugal hot-water pumps, double suction, first stage (ft × 0.3048 = m; gpm × 0.06309 = l/s). (Ref. 24)

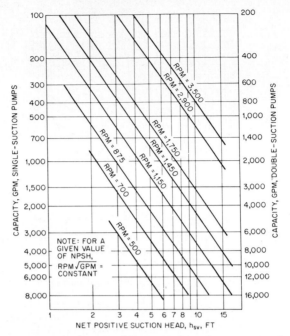

FIG. 29 Capacity and speed limitations for condensate pumps with shaft through eye of impeller (ft × 0.3048 = m; gpm × 0.06309 = l/s). (Ref. 24)

By Eq. 21,

$$h_{sv} = h_{b1} + h_{s1} = h_{b2} + h_{s2}$$

from which

$$h_{s2} = h_{b1} - h_{b2} + h_{s1} = 32.67 - 15.24 - 10 = 7.43 \text{ ft (2.26 m)}$$

the *suction head* required to prevent cavitation.

A recent revision of the Hydraulic Institute Standards (14th ed.) has replaced Figs. 23 to 29 with the recommendation that to avoid cavitation, pump speeds be limited to suction specific speeds not exceeding 8500 (5200). See discussion of suction specific speed which follows.

Cavitation Tests The very faint noise produced by incipient cavitation is almost always masked by machinery noise, and so special facilities and procedures are required for accurate testing. Figure 22c shows a suitable arrangement in which either or both p_i and $-Z$ can be varied to control h_s and h_{sv}. A head-capacity curve is obtained with ample h_{sv} to prevent cavitation. The test is repeated at a reduced constant value of h_{sv} such that cavitation will take place at some capacity beyond which values of head will fall below those without cavitation. Figure 30 shows a series of such tests on a low-specific-speed pump. The capacity at which the head curve becomes approximately vertical is called the *cutoff capacity*, an example of which is shown at NPSH = 9 ft (2.74 m) in Fig. 30. Cavitation is assumed to impend at a capacity above which the two head curves just begin to separate. Point C in Fig. 30 is typical of this condition. A useful cavitation test is to hold Q constant and vary h_{sv}. The results when H is plotted as a function of h_{sv} will be similar to the curve of Fig. 32 and will be discussed later. References 24 and 25 may be consulted for details of other test procedures.

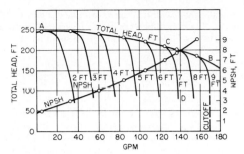

FIG. 30   Test of a 1.5-in (3.81-cm) single-stage pump at 3470 rpm, on water, 70°F (21°C) (ft × 0.3048 = m; gpm × 0.06309 = l/s). (Ref. 12)

Thoma Cavitation Parameter σ All the terms in Eqs. 20 to 22 may be made dimensionless by dividing each by the pump head H. The resulting parameter

$$\sigma = \frac{h_{sv}}{H} \tag{23}$$

has been very useful in predicting the onset of cavitation. Equation 23 is Thoma's[26,27] similarity law for cavitation in pumps and turbines. When the vapor pressure is reached at any point in the flow, cavitation impends and $\sigma = \sigma_C$, which is called the *critical sigma*. Experimentally determined values of $\sigma \approx \sigma_C$ are plotted against specific speed in Fig. 31. The scatter of the points is partly due to the difficulty in determining the onset of cavitation. For a given specific speed, the larger the value of the available σ, the safer the pump against cavitation.

Detecting the Onset of Cavitation   Geometrically similar pumps operated at the same specific speeds should have the same values of σ_C provided that disturbing influences, such as vapor pockets, do not destroy the dynamic similarity of the flow patterns. Figure 32 shows a curve, typical of a test to determine the onset of cavitation, in which Q is held constant and h_{sv} decreased

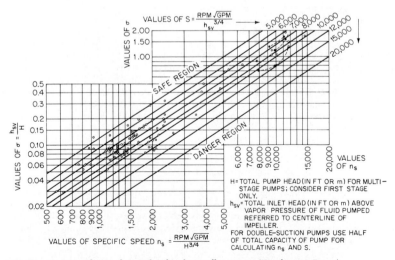

FIG. 31   Cavitation limits of centrifugal and propeller pumps. (Worthington Pump)

until a drop in head indicates that cavitation has taken place. Point C is the critical point and marks the onset of cavitation. It is very difficult to determine $(h_{sv})_C$ directly, and the usual practice is to use a point C_1, where the head is lower than at C by a small but measurable amount ΔH, as a pseudo critical point. Then, by Eq. 23,

$$\sigma_{C1} = \frac{(h_{sv})_C - \Delta(h_{sv})_{C1}}{H} = \sigma_C - \frac{\Delta(h_{sv})_{C1}}{H} \tag{24}$$

Equation 24 may be applied to the case where $(h_{sv})_{C1} > (h_{sv})_C$ by changing the sign of $\Delta(h_{sv})_{C1}$. A common problem is the determination of σ_C or $(h_{sv})_C$ at a particular speed from tests on the same pump at a different speed or from tests of a geometrically similar pump. Two of the criteria that may be used are (1) hold $\Delta H/H$ constant for both cases and (2) hold ΔH constant regardless of the value of H. These require separate and different treatment.

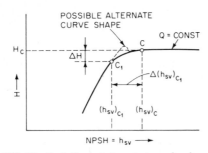

FIG. 32 Cavitation test at constant speed and constant capacity.

Stepanoff,[28] on the basis of reasonable assumptions about the volume of vapor generated by limited cavitation, has demonstrated that, if $\Delta H/H$ is constant, then $\Delta h_{sv}/H$ is also constant for small values of ΔH. If the criterion is to keep $\Delta H/H$ constant, then the Thoma similarity criterion holds and σ is the same for both cases. If the criterion is to keep ΔH constant, the Thoma criterion is violated and a different similarity relationship, one due to Tenot,[29] is required. Applied to the case of controlled cavitation, Tenot's equation is

$$\frac{\sigma_C - \sigma_1}{\sigma_C - \sigma_2} = \frac{H_2}{H_1} \tag{25}$$

where the subscripts 1 and 2 refer to two determinations of the point C_1 in Fig. 32 with different heads H_1 and H_2. Tenot's equation may be solved for σ_C and, together with Eq. 23, gives

$$\sigma_C = \frac{\sigma_2 H_2 - \sigma_1 H_1}{H_2 - H_1} = \frac{h_{sv2} - h_{sv1}}{H_2 - H_1} \tag{26}$$

For accurately determining σ_C, the quantity $h_{sv2} - h_{sv1}$ should be as large as possible.

Figure 33 shows the results of cavitation tests on a small pump reported by Krisam.[30] Portions of the head-capacity curves for a variety of constant speeds are shown by solid lines, and the values corresponding to maximum efficiency are indicated by the dashed curve. The required h_{sv} reported by Krisam is shown by curve AC. The values of σ computed for the end points of the curve differ slightly from those reported by Krisam, which were 0.093 at 1600 rpm and 0.061 at 2800 rpm. The two dot-dash curves show h_{sv} predicted on a basis of σ = constant. The equation of curve AB is $h_{sv} = 0.101H$ and predicts safe values of h_{sv} for all speeds. The equation of curve DC is $h_{sv} = 0.064H$. This lies well below the test curve. Assuming that the test curve represents the minimum safe values of h_{sv}, tests at the lowest head and speed should be used for extrapolation purposes to avoid the onset of cavitation at any higher head and speed. To determine σ_C, the following data may be read from the test curves: N = 1600 rpm, H = 46 ft (14 m), h_{sv} = 4.63 ft (1.41 m), N = 2800 rpm, H = 144 ft (43.9 m), h_{sv} = 9.16 ft (2.79 m). Then by Eq. 26,

$$\sigma_C = \frac{9.16 - 4.63}{144 - 46} = 0.046$$

The equation of curve EF in Fig. 33 is $h_{sv} = 0.046H$ and shows that curve AC has some margin of safety. Unless cavitation can be tolerated, values of $h_{sv} = \sigma_C H$ should not be used because unexpected changes in operating conditions may produce values of $\sigma < \sigma_C$ with accompanying cavitation. A further use of Eq. 26 is to compute h_{sv} for any desired speed, such as 2200 rpm, for which H = 89 ft (27.1 m). Using the previous data

$$\sigma_C = 0.046 = \frac{9.2 - h_{sv}}{144 - 89}$$

from which h_{sv} = 6.7 ft (2.04 m).

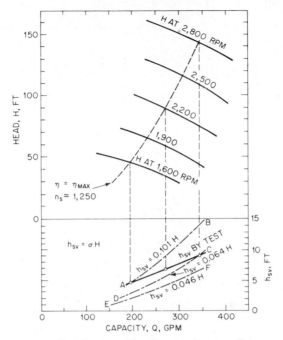

FIG. 33   Effect of pump speed on the value of σ at best efficiency (ft $\times$ 0.3048 = m; gpm $\times$ 0.06309 = l/s). (Ref. 30)

Rütschi[13] has found that σ_C depends on the hydraulic efficiency, as shown in Fig. 34. The hydraulic efficiency usually increases with increasing Reynolds number, which may be defined by

$$ Re = \frac{D\sqrt{H}}{\nu} \tag{27} $$

where D = impeller diameter in ft (m), H = pump head in ft (m), and ν = kinematic viscosity of liquid pumped in ft^2/s (m^2/s). It would be reasonable to expect that lower values of σ and h_{sv} might be used with larger pumps, for which the Reynolds numbers are larger, but this appears not to be true for large pumped storage units, as pointed out below in the discussion of suction specific speed.

Suction Specific Speed S   The *suction specific speed* S[31] may be obtained by replacing H in Eq. 13 by h_{sv}:

$$ S = \frac{N\sqrt{Q}}{(h_{sv})^{3/4}} \tag{28} $$

Note that Q = *half the discharge* of a double-suction impeller when computing S. Equations 13, 23, and 28 may be combined to yield

$$ \sigma = \left(\frac{n_s}{S}\right)^{4/3} \tag{29} $$

$$ n_s = S(\sigma)^{3/4} \tag{30} $$

Figure 31 shows lines of constant S. Values of S computed from Figs. 23 to 26 lie within the range $7500 < S < 11,000$ ($4600 < S < 6700$). The 14th edition of the Hydraulic Institute Standards acknowledges this and recommends a maximum S value of 8500 (5200) be used to limit pump speed. Higher values may apply to special designs or service conditions, such as an inducer ahead

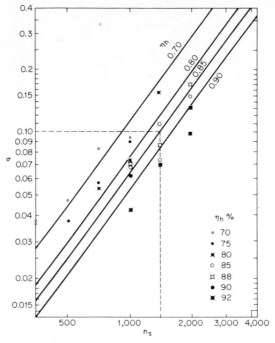

FIG. 34   Cavitation parameter σ versus specific speed for different efficiencies. (Ref. 13)

of the first-stage impeller. For a given specific speed, the lower the value of S, the safer the pump against cavitation. Experience with large European pumped storage installations has shown that cavitation began at $S \approx 6000$ (3670), and this value is recommended for large pumps. Figure 35 shows a summary of data and formulas that may be useful with commercial pumps.

German practice differs considerably from that in the United States in computing suction specific speed. Pfleiderer[32] defined a hub correction k as

$$k = 1 - \left(\frac{d_h}{D_o}\right)^2 \tag{31}$$

where d_h = the hub diameter and D_o = the diameter of the suction nozzle, in any consistent units. The suction specific speed S_G is defined as

$$S_G = \frac{(n/100)^2 Q}{k h_{sv}{}^{3/2}} \tag{32}$$

where n is measured in rpm, Q in cubic meters per second per impeller inlet, and h_{sv} in meters of liquid pumped. It follows that

$$S = 5164 \sqrt{S_G k} \tag{33}$$

NPSH for Liquids Other Than Cold Water   Field experience, together with carefully controlled laboratory experiments, has indicated that pumps handling hot water or certain liquid hydrocarbons may be operated safely with less NPSH than would normally be required for cold water. This may lower the cost of an installation appreciably, particularly in the case of refinery pumps. A theory for this has been given by Stepanoff and others.[14,33–35] Figure 36 shows the results of cavitation tests on two liquids for constant capacity and constant pump speed. No cavitation is present at point C or at $h_{sv} > h_{svc}$. With cold deaerated water, lowering h_{sv} slightly below h_{svc}

SPECIFIC SPEED LIMIT VS. TOTAL HEAD WITH ZERO SUCTION HEAD
85°F (29.4°C), 1.0 S.G. WATER, SEA LEVEL EQUIVALENT TO 32.6 FT (9.94 m) NPSH
8,000 (4,900), SUCTION SPECIFIC SPEED

SPECIFIC SPEED RANGE	1,000 – 5,000 (600 – 3,000)	1,000 – 5,000 (600 – 3,000)	5,000 – 9,000 (3,000 – 5,500)	9,000 – 13,000 (5,500 – 8,000)
CORRESPONDING MAX. TOTAL HEAD RANGE IN FT OR (m) THAT CAN BE PUMPED WITH 32.6 FT (9.94 m) NPSH	821 – 96 (250 – 29)	516 – 61 (157 –18.6)	61 – 28 (18.6 – 8.5)	28 – 17 (8.5 – 5.2)

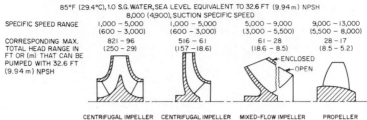

CENTRIFUGAL IMPELLER DOUBLE-SUCTION ENCLOSED DESIGN	CENTRIFUGAL IMPELLER SINGLE-SUCTION ENCLOSED DESIGN	MIXED-FLOW IMPELLER SINGLE-SUCTION ENCLOSED & OPEN DESIGN	PROPELLER SINGLE-SUCTION OPEN DESIGN

FORMULAS:

FOR SAME Q, S, h_{sv}: $n_{\text{DOUBLE SUCTION}} = \sqrt{2}\, n_{\text{SINGLE SUCTION}} = 1.414\, n_{\text{SINGLE SUCTION}}$

FOR SAME Q, H, h_{sv}: $(n_s)_{\text{DOUBLE SUCTION}} = \sqrt{2}\, (n_s)_{\text{SINGLE SUCTION}} = 1.414\, (n_s)_{\text{SINGLE SUCTION}}$

FOR SAME Q, n_s, h_{sv}: $H_{\text{DOUBLE SUCTION}} = 1.587\, H_{\text{SINGLE SUCTION}}$

FOR SAME Q, n: $(h_{sv})_{\text{DOUBLE SUCTION}} = 0.630\, (h_{sv})_{\text{SINGLE SUCTION}}$

$$n_{\text{MAX}} = S(h_{sv})^{3/4}/\sqrt{Q} \cong 8000\,(h_{sv})^{3/4}/\sqrt{Q}$$

Q = GPM (L/S); H = FT (m) OF FLUID PUMPED; n = RPM; h_{sv} = FT (m) OF FLUID PUMPED

FIG. 35 Summary for commercial pumps.

produces limited cavitation and a decrease in pump head ΔH to point C_1, but Δh_{svw} usually is negligible. With hot water ($T \geqq 100°F = 37.8°C$) or with many liquid hydrocarbons, a much larger decrease in h_{sv} will be required to produce the same drop in head ΔH that was shown by the cold water test. The NPSH *reduction*, or NPSH *adjustment*, is $\Delta H_{sv} \approx \Delta h_{svl}$. In practice, ΔH has been limited to $\Delta h \leqq 0.03H$, for which there is a negligible sacrifice in performance. The pumps usually are made of stainless steel or other cavitation-resistant materials, and the small localized production of vapor bubbles is not enough to cause damage.

Chart for NPSH Reductions

A composite chart of NPSH *reductions* for deaerated hot water and certain gas-free liquid hydrocarbons is shown in Fig. 37. The curves of vapor pressure in pounds per square inch absolute versus Fahrenheit temperature and the curves of constant NPSH reduction were based on laboratory tests with the liquids shown. Pending further experience with pumps operated at reduced NPSH, Fig. 37 should be used subject to the following limitations:

1. No NPSH reduction should exceed 50% of the NPSH required by the pump for cold water or 10 ft (3.0 m), whichever is smaller.

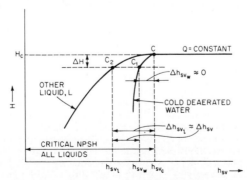

FIG. 36 Cavitation tests with different liquids at constant speed and constant capacity. (Ref. 34)

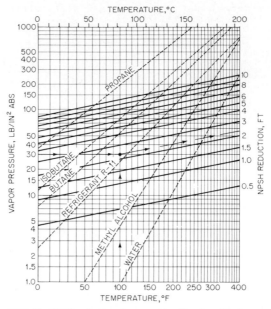

FIG. 37 NPSH reductions for pumps handling liquid hydrocarbons and hot water. This chart has been constructed from tests data obtained by using the liquids shown. For applicability to other liquids, refer to the text (ft × 0.3048 = m; lb/in² × 6.895 = kPa). (Ref. 24, Fig. 61, p. 99)

2. NPSH may have to be increased *above* the normal cold-water value to avoid unsatisfactory operation when (*a*) *entrained* air or other noncondensable gas is present in the liquid or (*b*) *dissolved* air or other noncondensable gas is present in the liquid and the absolute suction pressure is low enough to permit release of the gas from solution.

3. The vapor pressure of hydrocarbon mixtures should be determined by the bubble point method at pumping temperature (see Ref. 25). Do not use the Reid vapor pressure or the vapor pressure of the lightest fraction.

4. If the suction system may be susceptible to transient changes in absolute pressure or temperature, a suitable margin of safety in NPSH should be provided. This is particularly important with hot water and may exceed the reduction that would otherwise apply with steady-state conditions.

5. Although experience has indicated the reliability of Fig. 37 for hot water and the liquid hydrocarbons shown, its use with other liquids is not recommended unless it is clearly understood that the results must be accepted on an experimental basis.

USE OF FIGURE 37   Given a fluid having a vapor pressure of 30 lb/in² (210 kPa) abs at 100°F (37.8°C). Follow the arrows on the key shown on the chart and obtain an NPSH reduction of about 2.3 ft (0.70 m). Since this does not correspond to one of the liquids for which vapor pressure curves are shown on the chart, the use of this NPSH reduction should be considered a tentative value only. Given a pump requiring 16-ft (4.9-m) cold-water NPSH at the operating capacity; the pump is to handle propane at 55°F (12.8°C). Figure 37 shows the vapor pressure to be about 105 lb/in² (733 kPa) abs and the NPSH reduction to be about 9.5 ft (2.9 m). Since this exceeds 8 ft (2.44 m) which is half the cold-water NPSH, the recommended NPSH for the pump handling propane is half the cold-water NPSH, or 8 ft (2.44 m). If the temperature of the propane in the previous example is reduced to 14°F (−10°C), Fig. 37 shows the vapor pressure to be 50 lb/in² (349 kPa) abs and the NPSH reduction to be about 5.7 ft (1.74 m), which is less than half the cold-water NPSH. The NPSH required for pumping propane at $14°F$ $(-10°C)$ is then $16 - 5.7 = 10.3 \approx 10$ ft $(4.87 - 1.74 = 3.13 \approx 3$ m).

Reduction of Cavitation Damage

Once the pump has been built and installed,* there is little that can be done to reduce cavitation damage. As previously mentioned, sharpening the leading edges of the vanes by filing may be beneficial. Stepanoff[12] has suggested cutting back part of the vanes in the impeller eye together with sharpening the tips, for low-specific-speed pumps, as a means of reducing the inlet velocity c_1 and thus lowering σ. Although a small amount of prerotation or prewhirl in the direction of impeller rotation may be desirable,[36] excessive amounts should be avoided. This may require straightening vanes ahead of the impeller and rearranging the suction piping to avoid changes in direction or other obstructions. The cavitation damage to the impeller shown in Fig. 20 was believed to have been at least partly due to bad flow conditions produced by two 90° elbows in the suction piping. The planes of the elbows were at 90° to each other, and this arrangment should be avoided.

Straightening vanes in the impeller inlet may increase the NPSH requirement at all capacities. Three or four radial ribs equally spaced around the inlet and extending inward about one-quarter of the inlet diameter are effective against excessive prerotation and may require less NPSH than full-length vanes. This is very important with axial-flow pumps, which are apt to have unfavorable cavitation characteristics at partial capacities. Operation near the best efficiency point usually minimizes cavitation.

The admission of a small amount of air into the pump suction tends to reduce cavitation noise.[7] This rarely is done, however, because it is difficult to inject the right amount of air under varying head and capacity conditions and frequently there are objections to mixing air with the liquid pumped.

If a new impeller is required because of cavitation, the design should take into account the most recent advances described in the literature. Gongwer[37] has suggested (1) the use of ample fillets where the vanes join the shrouds, (2) sharpened leading edges of vanes, (3) reduction of β_1 in the immediate vicinity of the shrouds, and (4) raking the leading edges of the vanes forward out of the eye. Increasing the number of vanes for propeller pumps lowers σ for a given submergence. A change in the impeller material may be very beneficial, as described below.

Resistance of Materials to Cavitation Damage

Table 7 shows the relative resistance of several metals to cavitation pitting produced by magnetostriction vibration. It will be seen that cast iron, the most commonly used material for impellers, has relatively little pitting resistance relative to bronze and stainless steel, which are readily cast and finished. Elastomeric coatings have been found to be highly resistant to cavitation pitting. Table 8 shows the relative merits of several elastomers which were tested on a rotating disk at 150 ft/s (45.7 m/s). The best of the elastomers were even more resistant to cavitation damage than stellite 6B, which leads the list of metals in Table 7. The value of such coatings has been known for a long time, but only recently has it appeared possible to secure an adequate bond between the coating and the metal. Polyurethane and neoprene, which show high resistance to cavitation pitting and may be applied in liquid form, should be considered if other methods of reducing cavitation damage cannot be used.

Inducers

It is sometimes difficult or impossible to provide the required NPSH for an otherwise acceptable pump. The required NPSH can be provided by a small booster pump, called an *inducer*, placed ahead of the first-stage impeller.

Figure 38 shows a typical inducer driven directly from the pump shaft. The inducer blades are designed to operate with low NPSH and to provide enough head to meet the NPSH requirements of the first-stage impeller. Inducers can operate with varying degrees of cavitation since the collapse of the vapor bubbles is distributed over a relatively large vane area. The hubless inducer described in Ref. 38 ejects cavitation bubbles into a central vaneless core, where they collapse without damage.

Supercavitating inducers, in which only the high-pressure faces of the blades are in contact with the liquid, have been used for special applications with highly volatile liquids. The jet inducer[39] consists of one or more liquid jets directed axially toward the eye of the suction impeller.

---

*Sometimes it is possible to lower the pump, and this should be considered before other alterations are made.

TABLE 7 Cavitation Erosion Resistance of Metals

Alloy	Magnetostriction weight loss after 2 h, mg
Rolled stellite[a]	0.6
Welded aluminum bronze	3.2
Cast aluminum bronze	5.8
Welded stainless steel (2 layers, 17 Cr–7 Ni)	6.0
Hot rolled stainless steel (26 Cr–13 Ni)	8.0
Tempered rolled stainless steel (12 Cr)	9.0
Cast stainless steel (18 Cr–8 Ni)	13.0
Cast stainless steel (12 Cr)	20.0
Cast manganese bronze	80.0
Welded mild steel	97.0
Plate steel	98.0
Cast steel	105.0
Aluminum	124.0
Brass	156.0
Cast iron	224.0

[a]Despite the high resistance of this material to cavitation damage, it is not suitable for ordinary use because of its comparatively high cost and the difficulty encountered in machining and grinding.

SOURCE: Ref. 68.

The liquid for the jets usually is supplied through a bypass line from the pump discharge (see also Ref. 40).

The inducers described in Refs. 41 and 42 contribute not more than 5% of the total pump head. Although the efficiency of the inducer alone is low, the reduction in overall pump efficiency is not significant. Since this type of inducer causes pre-rotation, a careful match between inducer and suction impeller is required. In vertical multistage pumps, where a long shaft can be better supported, a vaned diffuser may be inserted between the inducer and the first-stage impeller. Such an arrangement is very beneficial for operation at reduced capacity. Refs. 41 and 42 show that a suitable inducer-impeller combination can operate at about 50% of the NPSH required for the impeller alone at capacities not exceeding the normal value. The NPSH requirement increases rapidly for capacities above normal, and so operation in this range should be avoided.

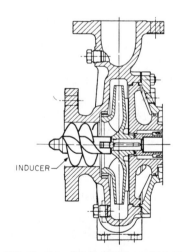

INDUCER

FIG. 38 Pump fitted with inducer. (Ref. 41)

Entrained Air Air or other gases may enter the impeller inlet from several sources. The immediate effect usually will be a drop in capacity and power. This will be followed by loss of prime if more gas is present than the impeller can handle. Air may be released from solution or enter through leaks in the suction piping. Stuffing box air leakage may be prevented by lantern rings supplied with liquid from the pump discharge. If the pump takes water from a sump with a free surface, a vortex may form from the free surface to the impeller inlet. The remedy may be the introduction of one or more baffle plates or even major changes in the sump. For information on proper sump design and the prevention of

TABLE 8 Cavitation Erosion Resistance of Elastomeric Coatings

Material	Subtype	Thickness of coating in rotating-disk cavitation tests, in (mm)	Cavitation test exposure period, h	Degree of erosion after exposure period, 150 ft/s (45.7 m/s)
Neoprene solvent base,	A	0.030 (0.76)	24	Slight
brush-applied	B	0.025 (0.64)	17	Slight
Neoprene, cured sheet, cold-bonded	. . .	0.062 (1.57)	14	None
Neoprene, *in situ* cured and bonded	. . .	0.060 (1.52)	10½	None
Polyurethane, liquid	A	0.062 (1.57)	12	Slight
	B	0.018 (0.46)	12	Severe
	C	0.062 (1.57)	12	None
Polyurethane cured	A	0.060 (1.52)	14	None
sheet, cold-bonded	B	0.062 (1.57)	12	Severe
Polysulfide, liquid	. . .	0.062 (1.57)	12	Severe
Polysiloxane, liquid	. . .	0.062 (1.57)	7	Severe
Butyl, cured sheet, cold-bonded	. . .	0.060 (1.52)	2¼	Severe
Butyl, *in situ* cured and bonded	. . .	0.060 (1.52)	12	Severe
Cis-polybutadiene (98%) cured sheet, cold bonded	. . .	0.060 (1.52)	10	None
Polybutadiene (polysulfide modified) *in situ* cured and bonded	. . .	0.060 (1.52)	13	Severe
Styrene-butadiene copolymer, *in situ* cured and bonded (SBR)	. . .	0.060 (1.52)	24	None
Natural rubber, cured sheet, cold-bonded	. . .	0.062 (1.57)	10	None
Natural rubber, *in situ* cured and bonded	. . .	0.060 (1.52)	16	Severe

SOURCE: Ref. 68.

air-entraining vortices, see Secs. 10.1 and 10.2, pp. 457 and 460 of Ref. 7, and Ref. 43. It is sometimes permissible to inject a small amount of air into the pump suction to reduce the noise and damage from cavitation caused by inadequate NPSH or recirculation in the impeller (see Subsec. 2.3.2).

PUMP SELECTION

Many manufacturers issue *pump selection charts*, which show performance data for their commercial line of pumps. If pumping requirements can be met by one of these, there is usually a considerable saving over the cost of a custom-designed unit. A typical selection chart is shown in Fig. 39. Each numbered field covers the range of head and capacity that can be had with good efficiency from the pump bearing that particular designation.

Let it be required to pump 5000 gpm (315 l/s) of cold water against a head of 180 ft (54.9 m) at sea level. Figure 39, curve A, shows that the A1015L pump running at 1775 rpm will meet the requirement. Figure 40, which is typical of performance curves accompanying pump selection charts, shows that the impeller diameter should be about 14.6 in (37.1 cm). The efficiency should be slightly over 89%, and about 260 hp (194 kW) will be required to drive the pump. Since

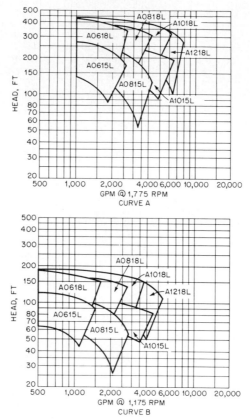

FIG. 39 Pump selection chart (ft $\times$ 0.3048 = m; gpm $\times$ 0.06309 = l/s). (Transamerica DeLaval)

this impeller will be about midway between the minimum and maximum diameter, the upper curves show that about 21 ft (6.40 m) of NPSH must be provided, which is equivalent to a 12-ft (3.7-m) maximum suction lift. A reasonable number of such charts will cover a complete line of pumps and supply all the data usually needed to select a pump for a given set of operating conditions.

STARTING CENTRIFUGAL PUMPS

Priming   Centrifugal pumps usually are completely filled with the liquid to be pumped *before starting*. When so filled with liquid, the pump is said to be *primed*. Pumps have been developed to start with air in the casing and then be primed.[44] This procedure is unusual with low-specific-speed pumps but is sometimes done with propeller pumps.[12] In many installations, the pump is at a lower elevation than the supply and remains primed at all times. This is customary for pumps of high specific speed and all pumps requiring a positive suction head to avoid cavitation.

Pumps operated with a suction lift may be primed in any of several ways. A relatively inexpensive method is to install a special type of check valve, called a *foot valve*, on the inlet end of the suction pipe and prime the pump by filling the system with liquid from any available source. Foot valves cause undesirable frictional loss and may leak enough to require priming before each

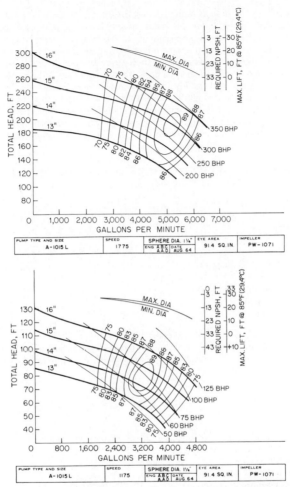

FIG. 40 Pump characteristics (in × 2.54 = cm; in² × 6.452 = cm²; ft × 0.3048 = m; gpm × 0.06309 = l/s; hp × 0.7457 = kW). (Transamerica DeLaval)

starting of the pump. A better method is to close a valve in the discharge line and prime by evacuating air from the highest point of the pump casing. Many types of vacuum pumps are available for this service. A priming chamber is a tank which holds enough liquid to keep the pump submerged until pumping action can be initiated. Self-priming pumps usually incorporate some form of priming chamber in the pump casing. Section 2.4 and Ref. 7 may be consulted for further details.

Torque Characteristics of Drivers Centrifugal pumps of all specific speeds usually have such low starting torques (turning moments) that an analysis of the starting phase of operation seldom is required. Steam and gas turbines have high starting torques so that no special starting procedures are necessary when they are used to drive pumps. If a pump is directly connected to an internal combustion engine, the starting motor of the engine should be made adequate to start both driver and pump. If the starter does not have enough torque to handle both units, a clutch must be provided to uncouple the pump until the driver is started.

Electric motors are the most commonly used drivers for centrifugal pumps. Direct current motors and alternating current induction motors usually have ample starting torque for all pump installations, provided the power supply is adequate. Many types of reduced voltage starters are available[7] to limit the inrush current to safe values for a given power supply. Synchronous motors are often used with large pumps because of their favorable power-factor properties. They are started as induction motors and run as such up to about 95% of synchronous speed. At this point, dc field excitation is applied and the maximum torque the motor can then develop is called the *pull-in torque,* which must be enough to accelerate the motor and connected inertia load to synchronous speed in about 0.2 s if synchronous operation is to be achieved. Centrifugal pumps usually require maximum torque at the normal operating point, and this should be considered in selecting a driver, particularly a synchronous motor, to be sure that the available pull-in torque will bring the unit to synchronous speed.

Torque Requirements of Pumps   The *torque,* or turning moment, for a pump may be estimated from the power curve in USCS units by

$$M = \frac{5252\ P}{n} \tag{34}$$

and in SI units by

$$M = \frac{9549\ P}{n}$$

where M = pump torque, lb·ft (N·m)
P = power, hp (W)
n = speed, rpm

Equation 34 makes no allowance for accelerating the rotating elements or the liquid in the pump. If a 10% allowance for accelerating torque is included, the constant should be correspondingly increased. The time Δt required to change the pump speed by an amount $\Delta n = n_2 - n_1$ is given by

$$\Delta t = \frac{I \Delta n}{k(M_m - M)} \tag{35}$$

where Δt = time, s
I = moment of inertia (flywheel effect) of all rotating elements of driver, pump, and liquid, lb·ft² (kg·m²)
Δn = change in speed, rpm
k = 307 in USCS (9.545 in SI)
M_m = driver torque, lb·ft (N·m)
M = pump torque, lb·ft (N·m) (Eq. 34)

The inertia I of the driver and pump usually can be obtained from the manufacturers of the equipment. The largest permissible Δn for accurate calculation will depend on how rapidly M_m and M vary with speed. The quantity $M_m - M$ should be nearly constant over the interval Δn if an accurate estimate of Δt is to be obtained. Torque-speed characteristics of electric motors may be obtained from the manufacturers.

Horizontal-shaft pumps fitted with plain bearings and packed glands require a *breakaway torque* of about 15% of M_n, the torque at the normal operating point, to overcome the static friction. This may be reduced to about 10% of M_n if the pump is fitted with antifriction bearings. The breakaway torque may be assumed to decrease linearly with speed to nearly zero when the speed reaches 15 to 20% of normal. Construction of torque-speed curves requires a knowledge of the pump characteristics at normal speed as well as details of the entire pumping system. Some typical examples taken from Refs. 7 and 69 are given below. The following forms of the affinity laws (Eqs. 17) are useful in constructing the various performance curves:

$$Q_2 = Q_1 \frac{n_2}{n_1} \tag{36}$$

$$H_2 = H_1 \left(\frac{n_2}{n_1}\right)^2 = H_1 \left(\frac{Q_2}{Q_1}\right)^2 \qquad (37)$$

$$P_2 = P_1 \left(\frac{n_2}{n_1}\right)^3 \qquad (38)$$

$$M_2 = M_1 \left(\frac{n_2}{n_1}\right)^2 \qquad (39)$$

where Q = capacity
  n = speed
  H = head
  P = power
  M = torque

in any consistent units of measure. Once speeds n_1 and n_2 are chosen, the subscripts 1 and 2 refer to corresponding points on the characteristic curves for these speeds.

LOW-SPECIFIC-SPEED PUMPS   Figure 41 shows the constant-speed characteristics of a pump having $n_s \approx 1740$ (1065) at best efficiency. This pump usually would be started with a valve in the discharge line closed. During the starting phase, the pump operates at shutoff with $P_1 = 25.8$ hp (19.2 kW) and $n_1 = 1770$ rpm. Then, by Eq. 34, $M_1 = 76.6$ lb·ft (103.9 N·m). These values may be used in Eq. 39 to evaluate starting torques M_2 at as many speeds n_2 as desired and plotted in Fig. 42 as curve BCD. Section AB of the starting torque curve is an estimate of the breakaway torque. If the discharge valve is now opened, the speed remains nearly constant but the torque increases as the capacity and power increase. If the normal operating point is $Q_n = 1400$ gpm (88.3 l/s) and $P_n = 53.2$ hp (39.7 kW), the motor torque will be $M_n = 158$ lb·ft (214 N·m) by Eq. 34. The vertical line DE in Fig. 42 shows the change in torque produced by opening the discharge valve.

Instead of starting the pump with the discharge valve closed, let the pump be started with a check valve in the discharge line held closed by a static head of 100 ft (30.5 m). The frictional head in the system may be represented by kQ^2. The value of k may be estimated from the geometry of the system or from a frictional-loss measurement at any convenient flowrate Q, preferably near the normal capacity Q_n. In this example, $Q_n = 1400$ gpm (88.33 l/s) and $k = 14.4/10^6$ (0.00109). The curve labeled system head 1 in Fig. 41 was computed from $H = 100 + (14.4/$

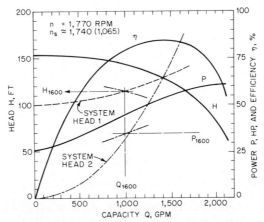

FIG. 41 Characteristics of a 6-by-8 double-suction pump at 1770 rpm (ft × 0.3048 = m; gpm × 0.06309 = l/s; hp × 0.7457 = kW). (Ref. 7)

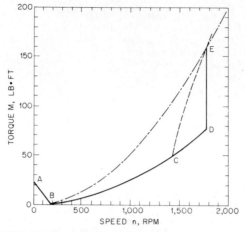

FIG. 42 Torque characteristics of 6-by-8 pump shown in
Fig. 41 (lb·ft × 1.356 = N·m). (Ref. 7)

$10^6)Q^2$ ft ($H = 30.48 + 0.00109Q^2$ m) and intersects the head-capacity curve at $H_n = 128$ ft
(39.01 m) and $Q_n = 1400$ gpm (88.33 l/s). The normal shutoff head is $H_1 = 153$ ft (46.6 m) at
$n_1 = 1770$ rpm. By Eq. 37, the pump will develop a shutoff head $H_2 = 100$ ft (30.5 m) at $n_2 =$
1430 rpm. By Eq. 39, the torque at 1430 rpm will be 50 lb·ft (68 N·m), corresponding to point
C in Fig. 42. The portion ABC of the starting torque curve has already been constructed. Trial-
and-error methods must be used to obtain the portion CE of the starting torque curve.

  The auxiliary curves in Fig. 43 are useful in constructing the CE portion of the starting torque
curve. Select a value of n_2 intermediate between 1427 and 1770 rpm, say $n_2 = 1600$ rpm. In Fig.
41, read values of Q_1, H_1, and P_1 for speed $n_1 = 1770$ rpm. By Eqs. 36 and 37, determine values
of Q_2 and H_2 and plot as shown in Fig. 41 until an intersection with the system-head 1 curve is
obtained which provides Q_{1600} corresponding to $n = 1600$ rpm. By Eq. 38, determine value of

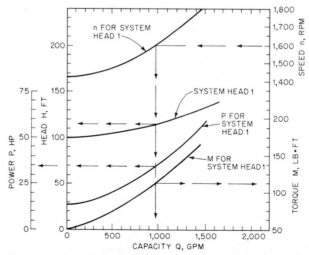

FIG. 43 Analysis of 6-by-8 pump shown in Fig. 41 (ft × 0.3048 = m; gpm
× 0.06309 = l/s; lb·ft × 1.356 = N·m; hp × 0.7457 = kW). (Ref. 7).

P_2 and plot as shown in Fig. 41 until an intersection is obtained with the Q_{1600} line which provides P_{1600} corresponding to $n = 1600$ rpm. Equation 34 is now used to obtain M_{1600}, which is one point on the desired starting torque curve. The process is repeated for various speeds n_2 until the curve CE in Fig. 42 can be drawn. The complete starting torque curve for this example is $ABCE$ in Fig. 41 with steady-state operation at point E.

Assume that the pump of the preceding examples is installed in a system having zero static head but a long pipeline with friction head given by $H = (65.4/10^6)Q^2$ ft ($H = 0.00500Q^2$ m), as shown in Fig. 41 by the curve labeled system head 2. The valve in the discharge line is assumed to open instantaneously when power is first applied to the pump. The procedure described to construct curve CE of the preceding example must now be used together with the system-head 2 curve of Fig. 41 to obtain the curve BE of Fig. 42. The complete starting torque curve for this example is ABE in Fig. 42, with steady-state operation at point E.

The inertia of the fluid in the system was neglected in solving the previous examples. Some of the power must be used to accelerate the liquid, and this may be appreciable in the case of a long pipeline. Low-specific speed pumps, which are used with long pipelines, have rising power-capacity curves with minimum power at shutoff and maximum power at normal capacity. Experience has shown that the starting torque–speed curves computed by neglecting the inertia of the liquid are conservative, so that inertia effects need not be included. The inertia effect of the liquid does slow the starting operation. If the time required to reach any event, such as a particular speed or capacity, is required, the inertia of the liquid should be considered, but including it greatly increases the difficulty of computation. References 45 through 48 give general methods for handling problems involving liquid transients.

High-specific-speed pumps have falling power-capacity curves with maximum power at shutoff and minimum power at normal capacity. Neglecting the inertia of the liquid probably will result in too low a value for the computed starting torque for such pumps. If liquid inertia is to be included, consult Refs. 44 to 47 and Sec. 8.1.

MEDIUM- AND HIGH-SPECIFIC-SPEED PUMPS   Figure 44 shows constant-speed characteristics for a medium-specific-speed pump, $n_s = 4570$ (2798) at best efficiency. The shutoff power is the same as the power at best efficiency, and the starting torque–speed curve is but little affected by the method of starting, as shown by Fig. 45.

Figure 46 shows the constant-speed characteristics of a high-specific-speed propeller pump, $n_s \approx 12{,}000$ (7350) at best efficiency. Figure 47 shows the starting torque–speed curve when the pump is started against a static head of 14 ft (4.3 m) and a friction head of 1 ft (0.3 m) at $Q_n = 12{,}500$ gpm (789 l/s). The system was assumed full of water with a closed check valve at the outlet end of the short discharge pipe. The methods of computation for Figs. 45 and 47 were the same as for Fig. 42.

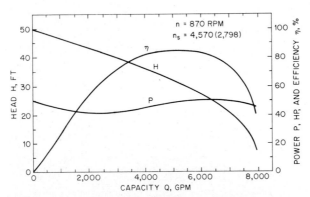

FIG. 44   Characteristics of a 16-in (40.6-cm) volute pump with mixed-flow impeller with flat power characteristic (ft $\times$ 0.3048 = m; gpm $\times$ 0.06309 = l/s; hp $\times$ 0.7457 = kW). (Ref. 7)

Sometimes propeller pumps are started with the pump submerged but with the discharge column filled with air. In such a case, the torque-time characteristic for the driver must be known and a step-by-step calculation carried out. If the discharge column is a siphon initially filled with air, the starting torque may exceed the normal running torque during some short period of the starting operation. If the pump is driven by a synchronous motor, it is particularly important to investigate the starting torque in the range of 90 to 100% of normal speed to make sure that the pull-in torque of the motor is not exceeded. For additional information regarding starting high-specific-speed pumps discharging through long and/or large-diameter systems, see Sec. 8.1.

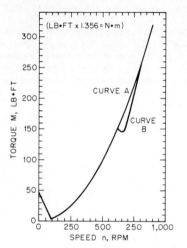

FIG. 45 Torque characteristics of pump shown in Fig. 44. (Ref. 7)

Miscellaneous Requirements Pumps handling hot liquids should be warmed up to operating temperature before being started unless they have been especially designed for quick starting. Failure to do this may cause serious damage to wearing rings, seals, and any hydraulic balancing device that may be present. A careful check of the installation should be made before starting new pumps, pumps that have had a major overhaul, or pumps that have been standing idle for a long time. It is very important to follow the manufacturer's instructions when starting boiler-feed pumps. If these are unavailable, Ref. 7 may be consulted. Ascertain that the shaft is not frozen, that the direction of rotation is correct, preferably with the coupling disengaged, and that bearing lubrication and gland cooling water meet normal requirements. Failure to do this may result in damage to the pump or driver.

A pump may run backwards at runaway speed if the discharge valve fails to close following shutdown. Any attempt to start the pump from this condition will put a prolonged overload on the motor. Figure 48 shows one example of the torque-speed transient for a pump, $n_s = 1700$ (1040), started from a runaway reversed speed while normal pump head was maintained between the suction and discharge flanges. In most practical cases, water hammer effects would make this

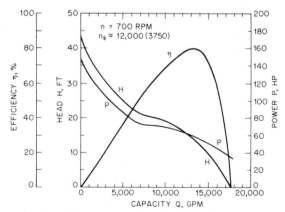

FIG. 46 Characteristics of a 30-in (76.2-cm) discharge propeller pump at 700 rpm (ft × 0.3048 = m; gpm × 0.06309 = l/s; hp × 0.7457 = kW). (Ref. 7)

transient even more unfavorable than Fig. 48 indicates. The duration of such a transient will always be much longer than the normal starting time, and so protective devices would probably disconnect the motor from the power supply before normal operation could be achieved. Consult Sec. 8.1 for additional information on this subject.

CAPACITY REGULATION

Capacity variation ordinarily is accomplished by a change in pump head, speed, or both simultaneously. The capacity and power input of pumps with specific speeds up to about 4000 (2450)

double suction increase with decreasing head, so that the drivers of such pumps may be overloaded if the head falls below a safe minimum value. Increasing the head of high-specificspeed pumps decreases the capacity but *increases* the power input. The drivers of these pumps should either be able to meet possible load increases or be equipped with suitable overload protection. Capacity regulation by the various methods given below may be manual or automatic (see also Refs. 7, 12, 32, 36, and 49).

Discharge Throttling This is the cheapest and most common method of capacity modulation for low- and medium-specificspeed pumps. Usually its use is restricted to such pumps. Partial closure of any type of valve in the discharge line will increase the system head so that the system-head curve will intersect the head-capacity curve at a smaller

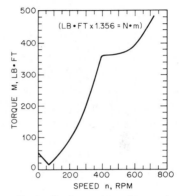

FIG. 47 Torque characteristics of pump shown in Fig. 46 (Ref. 7)

capacity, as shown in Fig. 49. Discharge throttling moves the operating point to one of lower efficiency, and power is lost at the throttle valve. This may be important in large installations, where more costly methods of modulation may be economically attractive. Throttling to the point of cutoff may cause excessive heating of the liquid in the pump. This may require a bypass to maintain the necessary minimum flow or use of different method of modulation. This is particularly important with pumps handling hot water or volatile liquids, as previously mentioned. Refer to Sec. 8.2 for information regarding the sizing of a pump bypass.

Suction Throttling If sufficient NPSH is available, some power can be saved by throttling in the suction line. Jet engine fuel pumps frequently are suction throttled[5] because discharge

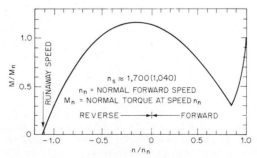

FIG. 48 Torque characteristics of a double-suction pump, n_s ≈ 1700 (1040), from reversed runaway speed to normal forward speed. (Ref. 7)

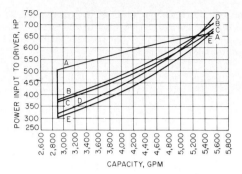

FIG. 49 Power requirements of two double-suction pumps in series operated at constant head and variable capacity. Total H_n = 382 ft (116 m) for both pumps at 1800 rpm (gpm $\times$ 0.06309 = l/s; hp $\times$ 0.7457 = kW). (Ref. 50)

Curve *AA:* constant speed with discharge throttling

Curve *BB:* synchronous motor with variable-speed hydraulic coupling on each pump

Curve *CC:* variable-speed wound-rotor induction motor

Curve *DD:* dc motor with rectifier and shunt field control

Curve *EE:* synchronous motor with variable-speed constant-efficiency mechanical speed reducer

throttling may cause overheating and vaporization of the liquid. At very low capacity, the impellers of these pumps are only partly filled with liquid, so that the power input and temperature rise are about one-third the values for impellers running full with discharge throttling. The capacity of condensate pumps frequently is submergence-controlled,[7] which is equivalent to suction throttling. Special design reduces cavitation damage of these pumps to a negligible amount.

Bypass Regulation   All or part of the pump capacity may be diverted from the discharge line to the pump suction or other suitable point through a bypass line. The bypass may contain one or more metering orifices and suitable control valves. Metered bypasses are commonly used with boiler-feed pumps for reduced-capacity operation, mainly to prevent overheating. There is a considerable power saving if excess capacity of propeller pumps is bypassed instead of using discharge throttling.

Speed Regulation   This can be used to minimize power requirements and eliminate overheating during capacity modulation. Steam turbines and internal combustion engines are readily adaptable to speed regulation at small extra cost. A wide variety of variable-speed mechanical, magnetic, and hydraulic drives are available, as well as both ac and dc variable-speed motors. Usually variable-speed motors are so expensive that they can be justified only by an economic study of a particular case. Figure 49 shows a study by Richardson[50] of power requirements with various drivers wherein substantial economies in power may be obtained from variable-speed drives.

Regulation by Adjustable Vanes   Adjustable guide vanes ahead of the impeller have been investigated and found effective with a pump of specific speed n_s = 5700 (3490). The vanes produced a positive prewhirl which reduced the head, capacity, and efficiency. Relatively little regulation was obtained from the vanes with pumps having n_s = 3920 (2400) and 1060 (649). Adjustable outlet diffusion vanes have been used with good success on several large European storage pumps for hydroelectric developments. Propeller pumps with adjustable-pitch blades have been investigated with good success. Wide capacity variation was obtained at constant head and with relatively little loss in efficiency. These methods are so complicated and expensive that they probably will have very limited application in practice. Reference 36 may be consulted for further discussion and bibliography.

Air Admission   Admitting air into the pump suction has been demonstrated as a means of capacity regulation, with some saving in power over discharge throttling. Usually air in the pumped liquid is undesirable, and there is always the danger that too much air will cause the pump to lose its prime. The method has rarely been used in practice but might be applicable to isolated cases.

PARALLEL AND SERIES OPERATION

Two or more pumps may be arranged for parallel or series operation to meet a wide range of requirements in the most economical manner. If the pumps are close together, that is, in the same station, the analysis given below should be adequate to secure satisfactory operation. If the pumps are widely separated, as in the case of two or more pumps at widely spaced intervals along a pipeline, serious pressure transients may be generated by improper starting or stopping procedures. The analysis of such cases may be quite complicated, and Refs. 46 to 48 should be consulted for methods of solution.

Parallel Operation   Parallel operation of two or more pumps is a common method of meeting variable-capacity requirements. By starting only those pumps needed to meet the demand, operation near maximum efficiency can usually be obtained. The head-capacity characteristics of the pumps need not be identical, but pumps with unstable characteristics may give trouble unless operation only on the steep portion of the characteristic can be assured. Care should be taken to see that no one pump, when combined with pumps of different characteristics, is forced to operate at flows less than the minimum required to prevent recirculation. See discussion on operation at other than normal capacity which follows. Multiple pumps in a station provide spares for emergency service and for the downtime needed for maintenance and repair.

The possibility of driving two pumps from a single motor should always be considered, as it usually is possible to drive the smaller pumps at about 40% higher speed than a single pump of twice the capacity. The saving in cost of the higher-speed motor may largely offset the increased cost of two pumps and give additional flexibility of operation.

One of the first steps in planning for multiple-pump operation is to draw the system-head curve, as shown in Fig. 50. The system head consists of the static head H_s and the sum H_f of the pipe-friction head and the head lost in the valves and fittings (see Secs. 8.1 and 8.2). The head-capacity curves of the various pumps are plotted on the same diagram, and their intersections

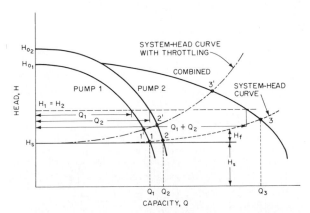

FIG. 50   Head-capacity curves of pumps operating in parallel.

with the system-head curve show possible operating points. *Combined head-capacity curves* are drawn by adding the capacities of the various combinations of pumps for as many values of the head as necessary. The intersection of any combined H-Q curve with the system-head curve is an operating point. Figure 50 shows two head-capacity curves and the combined curve. Points 1, 2, and 3 are possible operating conditions. Additional operating points may be obtained by changing the speed of the pumps or by increasing the system-head loss by throttling. Any number of pumps in parallel may be included on a single diagram, although separate diagrams for different combinations of pumps may be preferable.

The overall efficiency η of pumps in parallel is given by

$$\eta = \frac{H(\text{sp. gr.})}{k} \times \frac{\Sigma Q}{\Sigma P}$$

where H = head, ft (m)
   sp. gr. = specific gravity of the liquid
     k = 3960 USCS (0.1021 SI)
    ΣQ = sum of the pump capacities, gpm (l/s)
    ΣP = total power supplied to all pumps, hp (W)

Series Operation   Pumps are frequently operated in series to supply heads greater than those of the individual pumps. The planning procedure is similar to the case of pumps in parallel. The system-head curve and the individual head-capacity curves for the pumps are plotted as shown in Fig. 51. The pump heads are added as shown to obtain the combined head-capacity curve. In this example, pump 2 operating alone will deliver no liquid because its shutoff head is less than the system static head.

There are two possible operating points, 1 and 2, as shown by the appropriate intersections with the system-head curve. As with parallel operation, other operating points could be obtained by throttling or by changing the pump speeds. The overall efficiency of pumps in series is given by

$$\eta = \frac{Q(\text{sp. gr.})}{k} \times \frac{\Sigma H}{\Sigma P}$$

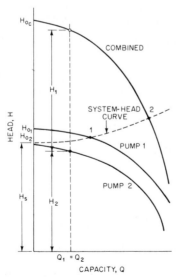

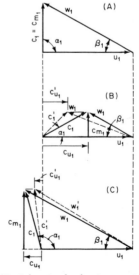

FIG. 51   Head-capacity curves of pumps operating in series.

FIG. 52   Inlet triangles showing prerotation.

wherein the symbols are the same as for parallel operation. It is important to note that the stuffing box pressure of the second pump is increased by the discharge pressure of the first pump. This may require a special packing box for the second pump with leakoff to the suction of the first pump. The higher suction pressure may increase both the first cost and the maintenance costs of the second pump.

OPERATION AT OTHER THAN THE NORMAL CAPACITY

Centrifugal pumps usually are designed to operate near the point of best efficiency, but many applications require operation over a wide range of capacities, including shutoff, for extended periods of time. Pumps for such service are available but may require special design and construction at higher cost. Noise, vibration, and cavitation may be encountered at low capacities. Large radial shaft forces at shutoff as well as lack of through flow to provide cooling may cause damage or breakage to such parts as shafts, bearings, seals, glands, and wearing rings of pumps not intended for such service. Some of the phenomena associated with operation at other than normal capacity are described below.

Prerotation Figure 52A shows an inlet velocity triangle for operation at best efficiency. The capacity is proportional to $c_{m1} = c_1$, which is perpendicular to the peripheral velocity u_1. The relative velocity w_1 is assumed parallel to the vane at the vane inlet angle β_1. There is no prerotation, and this is often called "shockless" entrance.

Fig. 52B shows an inlet velocity triangle for a capacity *less* than that at best efficiency. The solid-line vectors show the flow of an ideal liquid, which never separates from the vanes. Assuming no guide vanes ahead of the impeller, the average real-liquid velocities will be approximately as shown by the broken-line vectors. The average absolute liquid velocity c_1' has a *prerotation* component c_{u1}' parallel to and in the direction of the peripheral velocity u_1. As throttling is increased, c_{m1} decreases and c_{u1}' increases but probably never reaches u_1 in magnitude.

Fig. 52C shows an inlet velocity triangle for a capacity *greater* than that at best efficiency. As before, the average real-liquid velocities will be approximately as shown by the broken-line vectors, assuming no guide vanes ahead of the impeller. The *prerotation* velocity c_{u1}' is in the direction opposite the peripheral velocity u_1. This prerotation is limited by the maximum capacity attainable (largest c_{m1}).

In pumps equipped with long, straight suction nozzles but no suction elbow, prerotation has been detected over considerable distances upstream from the impeller eye. Suction pressures measured at wall taps where prerotation is present are always higher than the true average static pressure across the measuring section. This means that the pump head as determined from wall taps is less than it would be if true average static pressures were measured.

Recirculation and Separation There is a small flow from impeller discharge to suction through the wearing rings and any hydraulic balancing device present. This takes place at all capacities but does not usually contribute to raising the liquid temperature very much unless operation is near shutoff.

When the capacity has been reduced by throttling (or as a result of an increase in system head), a secondary flow called *recirculation* begins. Recirculation is a flow reversal at the suction and/or at the discharge tips of the impeller vanes. All impellers have a critical capacity at which recirculation occurs. The capacities at which suction and discharge recirculation begin can be controlled to some extent by design, but recirculation cannot be eliminated.

Suction recirculation is the reversal of flow at the impeller eye. A portion of the flow is directed out of the eye at the eye diameter, as shown in Fig. 53, and travels upstream with a rotational velocity approaching the peripheral velocity of the diameter. A rotating annulus of liquid is produced upstream from the impeller inlet, and through the core of this annulus passes an axial flow corresponding to the output capacity of the pump. The high shear rate between the rotating annulus and the axial flow through the core produces vortices which form and collapse, producing noise and cavitation in the suction of the pump.

Discharge recirculation is the reversal of flow at the discharge tips of the impeller vanes, as shown in Fig. 54. The high shear rate between the inward and outward relative velocities produces vortices that cavitate and usually attack the pressure side of the vanes.

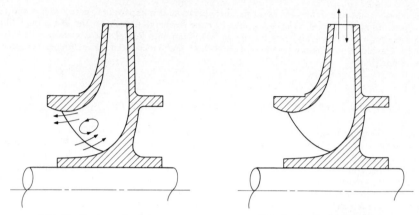

FIG. 53 Suction recirculation.   **FIG. 54** Discharge recirculation.

The capacity at which suction recirculation occurs is directly related to the design suction specific speed S of the pump. The higher the suction specific speed, the closer will be the beginning of recirculation to the capacity at best efficiency. Figure 55 shows the relation between the suction specific speed and suction recirculation for pumps up to 2500 (1530) specific speed, and Fig. 56 shows the same relation for pumps up to 10,000 (6123) specific speed.

For water pumps rated at 2500 gpm (158 l/s) and 150 ft (45.7 m) total head or less, the minimum operating flows can be as low as 50% of the suction recirculation values shown for continuous operation and as low as 25% for intermittent operation. For hydrocarbons, the minimum operating flows can be as low as 60% of the suction recirculation values shown for continuous operation and as low as 25% for intermittent operation.[51,52]

Fischer and Thoma[1] conducted a visual and photographic investigation of a radial-flow centrifugal pump, n_s = 1400 (857) at best efficiency. *Separation* was detected on the low-pressure faces of the vanes at all capacities, and the separated regions nearly filled the flow channels as shutoff was approached. Some separation was found on the high-pressure faces of the vanes at

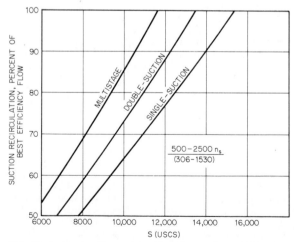

FIG. 55 Suction specific speed S at best efficiency flow, single suction or one side of double suction (to obtain S in SI units, multiply by 0.6123).

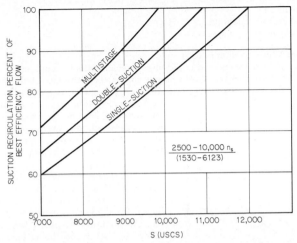

FIG. 56 Suction specific speed S at best efficiency flow, single suction or one side of double suction (to obtain S in SI units, multiply by 0.6123).

capacities above normal. In all cases the separated regions were filled with turbulent eddying fluid. Near shutoff, the fluid in the separated zones surged back and forth parallel to the vanes, with outward flow in some channels matched by inward flow in the others. At shutoff, each impeller channel was filled with small continually changing eddies generally rotating in sense opposite that of the impeller. Other investigators have reported backflow along the pressure sides of the impeller vanes near the discharge tips at $Q < Q_n$.

The high turbulence produced by recirculation and separation accounts for most of the power consumed at shutoff. This may vary from about 30% of the normal power for pumps of very low specific speed to nearly three times the normal power for propeller pumps. Separation and, possibly, cavitation may take place on the casing tongue or diffusion vanes at very low capacities. Operation near shutoff causes not only excessive heating but also vibration and cavitation, which may cause serious mechanical damage.

Temperature Rise Under steady-state conditions, friction and the work of compression increase the temperature of the liquid as it flows from suction to discharge. A further temperature increase may arise from liquid returned to the pump suction through wearing rings, a balancing device, or a minimum-flow bypass line that protects the pump when operating at or near shutoff.

Assuming that all heat generated remains in the liquid, the temperature rise ΔT in °F is

$$\Delta T = \frac{H}{778 \, C_p \eta} \tag{40}$$

where H = total head, ft
C_p = specific heat of the liquid, Btu/lb·°F
η = pump efficiency, decimal value

The computation has been facilitated by Bush,[53] who replaced Eq. 40 by

$$\Delta T = \frac{H}{778 \, (1.0) \, \eta - C} \tag{41}$$

wherein C_p has been set equal to unity and values of the correction factor C are as given in Fig. 57.

In SI units, the temperature rise ΔT in °C is

$$\Delta T = \frac{9.807 H}{C_p \eta} \tag{42}$$

where H = total head, m

C_p = specific heat of the liquid, J/kg·°C

η = pump efficiency, decimal value

Values of C_p in SI units are given in Ref. 54.

General service pumps handling cold liquids may be able to stand a temperature rise as great as 100 °F (56 °C). Most modern boiler-feed pumps may safely withstand a temperature rise of 50°F (28°C). The NPSH required to avoid cavitation or to prevent flashing of hot liquid returned to the pump suction may be the controlling factor. Minimum flow may be dictated by other factors, such as recirculation and unbalanced radial and axial forces on the impeller. Axial forces can be the controlling factor with single-stage double-suction pumps.

It is especially important to protect even small pumps handling hot liquids from operation at shutoff. This is usually done by providing a bypass line fitted with a metering orifice to maintain the minimum required flow through the pump. In the case of boiler-feed pumps, the bypass flow usually is returned to one of the feedwater heaters. Unless especially designed for cold starting, pumps handling hot liquids should be warmed up gradually before being put into operation.

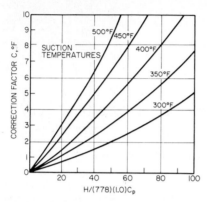

FIG. 57 Correction factor for water at elevated temperatures. (Ref. 53)

Radial Thrust   The pressure distribution at discharge from the impeller is very rarely uniform around the periphery, regardless of the casing design or operating point. This leads to a radial force on the pump shaft called the *radial thrust* or *radial reaction*. The radial thrust F_r in pounds (newtons) is

$$F_r = kK_r(\text{sp. gr.})HD_2 b_2 \tag{43}$$

where k = 0.433 USCS (9790 SI)

K_r = experimentally determined coefficient

sp. gr. = specific gravity of the liquid pumped (equal to unity for cold water)

H = pump head, ft (m)

D_2 = outside diameter of impeller, in (m)

b_2 = breadth of impeller at discharge, including shrouds, in (m)

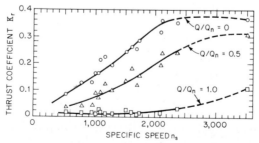

FIG. 58 K_r as a function of specific speed and capacity for single-volute pumps (to obtain n_s in SI units, multiply by 0.6123). (Ref. 55)

Values of K_r determined by Agostinelli et al.[55] for single-volute pumps are given in Fig. 58 as functions of specific speed and capacity. The magnitude and direction of F_r on the pump shaft may be estimated from Fig. 59, but Eq. 43 probably will be more accurate for determining the magnitude of the force. The radial thrust usually is minimum near $Q = Q_n$, the capacity at best efficiency, but rarely goes completely to zero. Near shutoff, F_r usually is maximum and may be a considerable force on the shaft in high-head pumps.

The radial thrust can be made much smaller throughout the entire capacity range by using a double volute (twin volute) or a concentric casing, and these designs should be considered, particularly if the pump must operate at small capacities. Figures 60 to 62 compare radial forces generated by three types of casings: a standard volute, a double volute, and a modified concentric casing. The last-named casing was concentric with the impeller for 270° from the tongue and then enlarged in the manner of a single volute to form the discharge nozzle. The magnitude and direction of F_r on the pump shaft for the modified concentric casing may be estimated from Fig. 63. The direction of F_r on the pump shaft with a double volute was somewhat random but in the general vicinity of the casing tongue. Radial forces on pumps fitted with diffusion vanes usually are rather small, although they may be significant near shutoff.

EXAMPLE   Consider a single-stage centrifugal pump, n_s = 2000 (1225) at best efficiency, handling cold water, sp. gr. = 1.0. Estimate the radial thrust on the impeller at half the normal capacity when fitted with (a) a single volute, (b) a modified concentric casing, and (c) a double volute. Impeller dimensions are D_2 = 15.125 in (38.4 cm) and b_2 = 2.5 in (6.35 cm). The shutoff head is H = 252 ft (76.8 m), and the head at half capacity is H = 244 ft (74.4 m).

Solution   (a) K_r = 0.2 from Fig. 58. By Eq. 43

$$F_r = (0.433)(0.2)(1.0)(244)(15.125)(2.5) = 799 \text{ lb } (3554 \text{ N})$$

Estimating between the curves for n_s = 2370 (1451) and 1735 (1062) in Fig. 59, the direction of F_r on the shaft should be about 65° to 70° from the casing tongue in the direction of rotation.

(b) Use Fig. 61, n_s = 2120 (1298), which is nearest to n_s = 2000 (1225), to find the radial

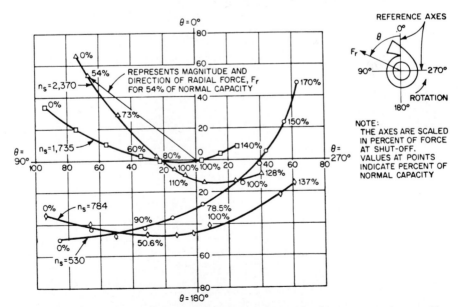

FIG. 59   Polar plot showing direction of resultant radial forces for single-volute pumps at various capacities and specific speeds. To obtain n_s in SI units, multiply by 0.6123 (Ref. 55)

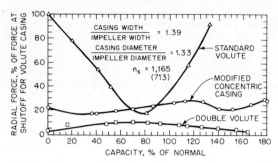

FIG. 60 Comparison of the effect of three casing designs on radial forces for $n_s = 1165$ (713). (Ref. 55)

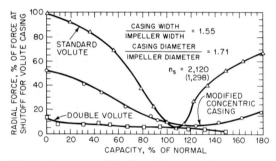

FIG. 61 Comparison of the effect of three casing designs on radial forces for $n_s = 2120$ (1298). (Ref. 55)

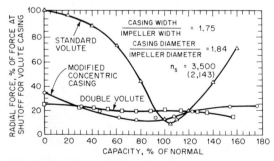

FIG. 62 Comparison of effect of three casing designs on radial forces for $n_s = 3500$ (2143). (Ref. 55)

2.246

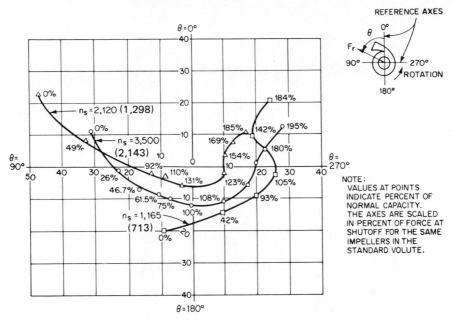

FIG. 63 Polar plot showing direction of resultant radial forces for modified concentric casings at various capacities and specific speeds. The casings were concentric for 270° from the tongue. (Ref. 55)

force for a modified concentric casing at half capacity, which is about 33% of shutoff value for a single-volute casing. From Fig. 58, for a single-volute casing at shutoff, $K_r = 0.34$ and

$$F_r = (0.433)(0.34)(1.0)(252)(15.125)(2.5) = 1403 \text{ lb (6241 N)}$$

Then, for a modified concentric casing at half capacity,

$$F_r = (0.33)(1403) = 463 \text{ lb (2059 N)}$$

From Fig. 63, the direction of F_r should be about 75 to 80° from the casing tongue in the direction of rotation.

(c) From Fig. 61, the radial force for a double-volute casing is about 8% of the shutoff value for the single-volute casing, and so

$$F_r = (0.08)(1403) = 112 \text{ lb (499 N)}$$

According to Agostinelli et al.,[55] the direction of the radial thrust in double-volute casings was found to be generally toward the casing tongue. Stepanoff[12] has found this direction to follow approximately that in single-volute casings (see also Biheller[56]).

ABNORMAL OPERATION

Complete Pump Characteristics Many types of abnormal operation involve reversed pump rotation, reversed flow direction, or both, and special tests are required to cover these modes of operation. Several methods of organizing the data have been proposed, and each has certain advantages. The Thoma diagrams shown in Figs. 64 and 65[45] are easily understood and are truly *complete characteristics diagrams* because all possible modes of operation are covered (see also Ref. 57).

Figure 66 shows schematic cross sections of the two pumps tested by Swanson[58] for which characteristics are given in Fig. 65.

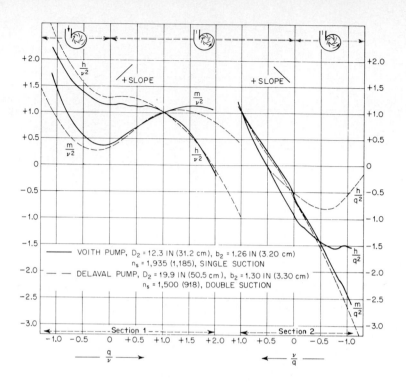

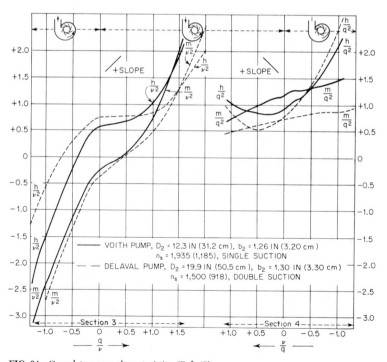

FIG. 64 Complete pump characteristics. (Ref. 45)

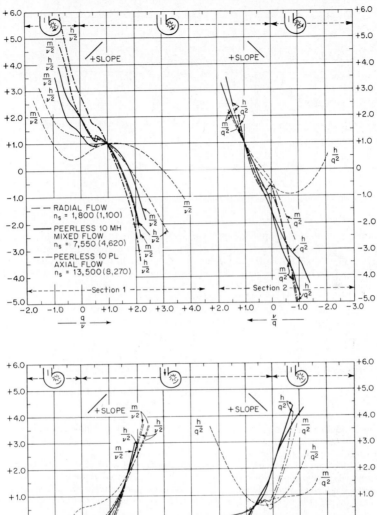

FIG. 65 Complete pump characteristics. (Ref. 45)

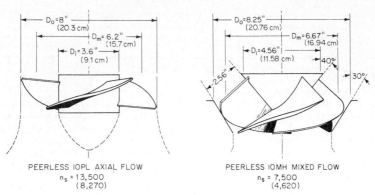

PEERLESS IOPL AXIAL FLOW
n_s = 13,500
(8,270)

PEERLESS IOMH MIXED FLOW
n_s = 7,500
(4,620)

FIG. 66   Schematic cross sections of high-specific-speed pumps. (Ref. 45)

The *Karman circle diagram* (Ref. 59) attempted to show the complete characteristics as a four-quadrant contour plot of surfaces representing head and torque with speed and capacity as base coordinates. Since the head and torque tend to infinity in two zones of operation, another diagram would be required to show the complete pump characteristics. The data presented in such a diagram are, nevertheless, adequate for almost all requirements. One example of a circle diagram is given in Fig. 67. Other examples may be found in Refs. 12, 58, and 59.

Frequently, tests with negative head and torque have been omitted so that only half of the usual circle diagram could be shown. This has been called a *three-quadrant plot*, but the information necessary to predict an event, such as possible water-column separation following a power failure, is lacking.

Power-Failure Transient   A sudden power failure which leaves a pump and driver running free may cause serious damage to the system. Except for rare cases where a flywheel is provided, the pump and driver usually have a rather small moment of inertia, and so the pump will slow down rapidly. Unless the pipeline is very short, the inertia of the liquid will maintain a strong forward flow while the decelerating pump acts as a throttle valve. The pressure in the discharge line falls rapidly and, under some circumstances, may go below atmospheric pressure, both at the pump discharge and at any points of high elevation along the pipeline. The minimum pressure head which occurs during this phase of the motion is called the *downsurge*, and it may be low enough to cause vaporization followed by complete separation of the liquid column. There have been cases where pipelines have collapsed under the external atmospheric pressure during separation. When the liquid columns rejoin, following separation, the shock pressures may be sufficient to rupture the pipe or the pump casing. Closing a valve in the discharge line will only worsen the situation, and so valves having programmed operation should be closed very little, if at all, before reverse flow begins.

Reversed flow may be controlled by valves or by arranging to have the discharge pipe empty while air is admitted at or near the outlet. If reversed flow is not checked, it will bring the pump to rest and then accelerate it with reversed rotation. Eventually the pump will run as a turbine at the runaway speed corresponding to the available static head diminished by the frictional losses in the system. However, while reversed flow is being established, the reversed speed may reach a value considerably in excess of the steady-state runaway speed. Maximum reversed speed appears to increase with increasing efficiency and increasing specific speed of the pump. Calculations indicate maximum reversed speeds more than 150% of normal speed for n_s = 1935 (1185) and η = 84.1%.[45] This should be considered in selecting a driver, particularly if it is a large electric motor.

There will be a pressure increase, called the *upsurge*, in the discharge pipe during reversed flow. The maximum upsurge usually occurs a short time before maximum reversed speed is reached and may cause a pressure as much as 60% or more above normal at the pump discharge. A further discussion of power-failure transients is given in Sec. 9.4.

ANALYSIS OF TRANSIENT OPERATION   The data of Figs. 64 and 65 have been presented in a form suitable for general application to pumps having approximately the same specific speeds as those tested. The symbols are $h = H/H_n$, $q = Q/Q_n$, $m = M/M_n$, and $\nu = n/n_n$, wherein H, Q, M, and n represent instantaneous values of head, capacity, torque, and speed respectively and the subscript n refers to the values at best efficiency for normal constant-speed pump operation. Any consistent system of units may be used. According to the affinity laws (Eqs. 17), q is proportional to ν, and h and m are proportional to ν^2. Thus the affinity laws are incorporated in the scales of the diagrams. Figures 64 and 65 are divided into sections for convenience in reading data from the curves. The curves of sections 1 and 3 extend to infinity as q/ν increases without limit in either the positive or negative direction. This difficulty is eliminated by sections 2 and 4, where the curves are plotted against ν/q, which is zero when q/ν becomes infinite.

Usually any case of transient operation would begin at or near the point $q/\nu = \nu/q = 1$, which appears in both sections 1 and 2 of Figs. 64 and 65. The detailed analysis of transient behavior is beyond the scope of this treatise. An analytical solution by the rigid column method, in which the liquid is assumed to be a rigid body, is given in Ref. 45. Friction is easily included, and the results are satisfactory for many cases. The same reference includes a semigraphical solution to allow for elastic waves in the liquid, but friction must be neglected. Graphical solutions including both elastic waves and friction are discussed in Refs. 45 to 48. Computer solutions of a variety of transient problems are discussed in Ref. 47. These offer considerable flexibility in the analysis once the necessary programs have been prepared.

Some extreme conditions of abnormal operation can be estimated at points where the curves of Figs. 64 and 65 cross the zero axes and are listed in Table 9. The data for columns 2 to 4 of Table 9 were read from sections 2 of Figs. 64 and 65, and the data for columns 6 and 7 were read from sections 3. Column 8 was computed from column 7 by assuming $h = 1$. Let the pump having $n_s = 1500$ (918) deliver cold water with normal head $H_n = 100$ ft (30.48 m), and let the center of the discharge flange be 4 ft (1.2 m) above the free surface in the supply sump. The discharge pressure head following power failure may be estimated by assuming the inertia of the rotating elements to be negligible relative to the inertia of the liquid in the pipeline. Then $q = 1$ and, from column 2 of Table 9, the downsurge pressure head is $(-0.22)(100) - 4 = -26$ ft (−7.9 m), which is not low enough to cause separation of the water column.

Actually the downsurge would be less than this because of the effects of inertia and friction, which have been neglected. If this pump were stopped suddenly by a shaft seizure or by an obstruction fouling the impeller, column 4 of Table 9 shows the downsurge to be $(-0.55)(100) - 4 = -59$ feet (−18 m), which would cause water column separation and, probably, subsequent water hammer. If, following power failure, the pump were allowed to operate as a no-load turbine under the full normal pump head, column 8 of Table 9 shows the runaway speed would be 1.14 times the normal pump speed. The steady-state runaway speed usually would be less than this because the effective head would be decreased by friction but higher speeds would be reached during the transient preceding steady-state operation. Column 8 shows that runaway speeds increase with increasing specific speed.

TABLE 9   Abnormal Operating Conditions of Several Pumps

Specific speed n_s, USCS (SI)	Downsurge				Runaway turbine		
	Free-running $m/q^2 = 0$		Locked-rotor $\nu/q = 0$		$m/\nu^2 = 0$		$h = 1$
	h/q^2	ν/q	h/q^2	m/q^2	q/ν	h/ν^2	ν
1	2	3	4	5	6	7	8
1500 (918)[a]	−0.22	0.36	−0.55	−0.53	0.46	0.77	1.14
1800 (1100)	−0.24	0.31	−0.60	−0.44	0.56	0.75	1.16
1935 (1185)	−0.36	0.32	−0.94	−0.60	0.41	0.66	1.23
7550 (4620)	−0.23	0.56	−1.57	−1.38	0.99	0.38	1.62
13,500 (8270)	−0.12	0.67	−0.96	−0.60	1.08	0.33	1.73

[a]Double-suction.

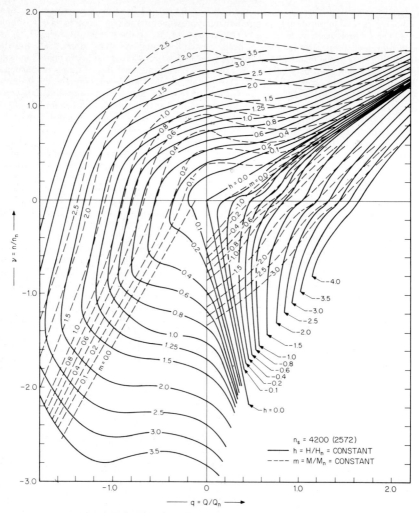

FIG. 67a Circle diagram of pump characteristics. (Courtesy Combustion Engineering)

Incorrect Rotation Correct rotation of the driver should be verified before it is coupled to the pump (Fig. 68). Sections 3 of Figs. 64 and 65 show that reversed rotation might produce some positive head and capacity with pumps of low specific speed, but at very low efficiency. It is unlikely that positive head would be produced by reversed rotation of a high-specific-speed pump.

Reversed Impeller Some double-suction impellers can be mounted reversed on the shaft. If the impeller is accidentally reversed, as at *B* in Fig. 68, the capacity and efficiency probably will be much reduced and the power consumption increased. Care should be taken to prevent this, as the error might go undetected in some cases until the driver was damaged by overload. Table 1 shows performance data for six pumps with reversed impellers. At least one of these would overload the driver excessively.

Further discussion of abnormal operating conditions may be found in Refs. 7, 12, and 36.

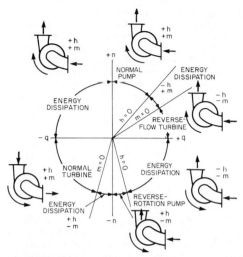

FIG. 67b   Explanatory diagram for Fig. 67a.

Vibration   Vibration caused by flow through wearing rings and by cavitation has been discussed in the foregoing and some remedies indicated. Vibration due to unbalance is not usually serious in horizontal units but may be of major importance in long vertical units, where the discharge column is supported at only one or two points. The structural vibrations may be quite complicated and involve both natural frequencies and higher harmonics. Vibration problems in vertical units should be anticipated during the design stage. If vibration is encountered in existing units, the following steps may help to reduce it: (1) dynamically balance all rotating elements of both pump and motor; (2) increase the rigidity of the main support and of the connection between the motor and the discharge column; (3) change the stiffness of the discharge column to raise or lower natural frequencies as required. A portable vibration analyzer may be helpful in this undertaking. Kovats[60] has discussed the analysis of this problem in some detail.

PREDICTION OF EFFICIENCY FROM MODEL TESTS

Many pumps used in pumped storage power plants and water supply projects are so large and expensive that extensive use is made of small models to determine the best design. It is often necessary to estimate the efficiency of a prototype pump, as a part of the guarantee, from the performance of a geometrically similar model. A model and prototype are said to operate under

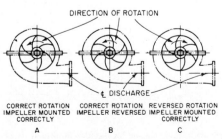

DIRECTION OF ROTATION

℄ DISCHARGE

CORRECT ROTATION IMPELLER MOUNTED CORRECTLY	CORRECT ROTATION IMPELLER REVERSED	REVERSED ROTATION IMPELLER MOUNTED CORRECTLY
A	B	C

FIG. 68   Pump assembly and rotation. (Ref. 7)

dynamically similar conditions when $Dn/\sqrt{H} = D'n'/\sqrt{H'}$, where D = impeller diameter, n = pump speed, and H = pump head in any consistent units of measure. Primed quantities refer to the model and unprimed quantities to the prototype. Dynamic similarity is a prerequisite to model-prototype testing so that losses that are proportional to the squares of fluid velocities, called *kinetic losses,* will scale directly with size and not change the efficiency. Surface frictional losses are boundary-layer phenomena which depend on Reynolds number, $Re = D\sqrt{H}/\nu$, where ν = kinematic viscosity of the liquid pumped. Reynolds numbers increase with increasing size, and, within limits, surface frictional-loss coefficients decrease with increasing Reynolds number. This leads to a gain in efficiency with increasing size. Computational difficulties have forced an empirical approach to the problem. Details of the development of a number of formulas are given in Ref. 61.

Moody-Staufer Formula   In 1925, L. F. Moody and F. Staufer independently developed a formula that was later modified by Pantell to the form

$$\left(\frac{1 - \eta_h}{1 - \eta_h'}\right)\left(\frac{\eta_h'}{\eta_h}\right) = \left(\frac{D'}{D}\right)^n \tag{44}$$

where η_h = hydraulic efficiency, discussed previously, and D = impeller diameter. Primed quantities refer to the model, and η is a constant to be determined by tests. The model must be tested with the same liquid that will be used in the prototype, that is, cold water in most practical cases. The original formula contained a correction for head, which is negligible if $H' \geq 0.8H$, and this requirement is now virtually mandatory in commercial practice. The meager information available indicates $0.2 \geq n \geq 0.1$ approximately, with the higher value currently favored. Improvements in construction and testing techniques very likely will move η toward the lower value in the future. The Moody-Staufer formula has been widely used since first publication. In practice, both η_h' and η_h usually are replaced by the overall efficiencies η' and η, respectively, because of the difficulty in determining proper values for the mechanical and volumetric efficiencies (see Eq. 16).

Rütschi Formulas   The general form of several empirical formulas due to K. Rütschi[61] and others was originally given as

$$\eta_h = \frac{\mathbf{f}}{\mathbf{f}'}\,\eta_h' \tag{45}$$

where η_h and η_h' are the hydraulic efficiencies of the prototype and model, respectively, and $\mathbf{f}$ and $\mathbf{f}'$ are values of an empirical $\mathbf{f}$ function for both the prototype and the model. The $\mathbf{f}$ function was obtained from tests of six single-stage pumps, $n_s < 2000$ (1225), and is shown in Fig. 69 based on the *eye diameters* D_o of the pumps in *millimeters.* Thus the $\mathbf{f}$ function depends on actual size in addition to scale ratio. The extrapolated portion of the curve, shown dashed in Fig. 69, checked well with values for a model and large prototype, shown by E' and E, respectively. One of several formulas which have been proposed to fit the curve in Fig. 69 is, in SI units,

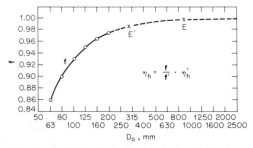

FIG. 69   The $\mathbf{f}$ function for the Rütschi formula (to obtain D_o in inches, multiply by 0.03937). (Refs. 13 and 61)

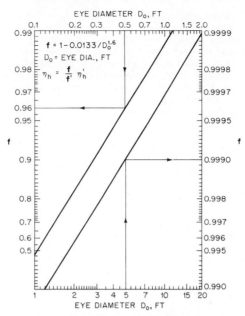

EYE DIAMETER D_o, FT

$f = 1 - 0.0133/D_o^{1.6}$
D_o = EYE DIA., FT
$\eta_h = \dfrac{f}{f'} \eta_h'$

EYE DIAMETER D_o, FT

FIG. 70 Chart for the solution of the Rütschi formula (to obtain D_o in meters, multiply by 0.3048). (Ref. 61)

$$\mathbf{f} = 1 - \frac{3.15}{D_o^{1.6}} \tag{46}$$

where the eye diameter D_o is in centimeters or, in USCS units,

$$\mathbf{f} = 1 - \frac{0.0133}{D_o^{1.6}} \tag{47}$$

where D_o is in feet. Figure 70 gives a graphical solution of Eq. 47. The Society of German Engineers (VDI) has adopted a slightly modified version of the Rütschi formula as standard. Rütschi later recommended, in discussion of Ref. 61, that the internal efficiency $\eta_i = \eta/\eta_m$ be used instead of the hydraulic efficiency η_h in Eq. 45. The mechanical efficiency η_m will probably be very high for both model and prototype for most cases of interest, and so good results should be obtained if the overall efficiency is used in Eq. 45.

Model-prototype geometric similarity should include surface finish and wearing ring or tip clearances, but this may be difficult or impossible to achieve. Anderson (Ref. 40, pp. 49–56) has proposed a method which includes a correction for dissimilarity in surface finish. Other studies in progress will include corrections for dissimilarity in both surface finish and clearances.

OPERATION OF PUMPS AS TURBINES

Centrifugal pumps may be used as hydraulic turbines in some cases where low first cost is paramount. Since the pump has no speed-regulating mechanism, considerable speed variation must be expected unless the head and load remain very nearly constant. Some speed control could be obtained by throttling the discharge automatically, but this would increase the cost, and the power lost in the throttle valve would lower the overall efficiency.

Pump Selection   Once the head, speed, and power output of the turbine have been specified, it is necessary to select a pump which, when used as a turbine, will satisfy the requirements.
Assuming that performance curves for a series of pumps are available,° a typical set of such curves
should be normalized using the head, power, and capacity of the best efficiency point as normal
values. These curves will correspond to the right-hand part of section 1 of either Fig. 64 or 65. In
normalizing the power P, let $p = P/P_n$ and the curve of p/ν^3 will be identical with the curve of
m/ν^2 in Fig. 64 or 65. The normalized curves may be compared with the curves in sections 1 of
Figs. 64 and 65 to determine which curves best represent the characteristics of the proposed
pump. Once a choice has been made, the approximate turbine performance can be obtained from
the corresponding figure of Figs. 71 to 74.

EXAMPLE   Assume that the turbine specifications are $H_T = 20$ ft, $P_T = 12.75$ hp and $n =$
580 rpm and that the characteristics of the DeLaval L10/8 pump are representative of a series
of pumps from which a selection can be made. The turbine discharge Q_T, in gallons per minute after substituting the above values is

$$Q_T = \frac{3960 P_T}{H_T \eta_T} = \frac{2520}{\eta_T} \tag{48}$$

where η_T is turbine efficiency, shown in Fig. 71. Only the normal values Q_n, H_n, P_n, etc., are
common to the curves of both Figs. 64 and 71 so that these alone can be used in selecting the
required pump. Values of ν/q, h/q^2, and η_T are read from the curves of Fig. 71 and corresponding values of Q computed by Eq. 48. Values of Q_n and H_n are then given by

$$Q_n = Q(\nu/q) \tag{49}$$

$$H_n = \frac{H(\nu/q)^2}{h/q^2} = \frac{20(\nu/q)^2}{h/q^2} \tag{50}$$

For example, in Fig. 71 at $\nu/q = 0.700$, read $h/q^2 = 0.595$ and $\eta_T = 0.805$. By Eq. 48,
$Q_T = 2520/0.805 = 3130$ gpm; by Eq. 49, $Q_n = (3130)(0.700) = 2190$ gpm; and by Eq. 50,
$H_n = (20)(0.700)^2/0.595 = 16.5$ ft. In a similar manner, the locus of the best efficiency points
for an infinite number of pumps, each having the same characteristics as shown in Figs. 64
and 71, is obtained, and each pump would satisfy the turbine requirements. This locus of best
efficiency points is plotted as curve A in Fig. 75.

The head-capacity curve for a DeLaval L 16/14 pump having a 17-in-diameter impeller
tested at 720 rpm is shown as curve B in Fig. 75. The best efficiency point was found to be at

---

°It is assumed that turbine mode characteristics of the proposed pump are not available when the initial
selection is made.

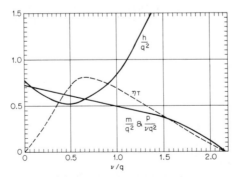

FIG. 71 Dimensionless characteristic curves for
DeLaval L 10/8 pump, constant-discharge turbine operation. (Ref. 62)

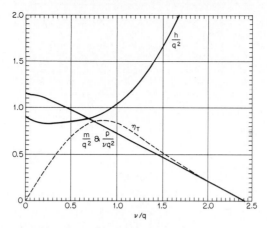

FIG. 72 Dimensionless characteristic curves for Voith pump, constant-discharge turbine operation. (Ref. 62)

Q_n = 3500 gpm and H_n = 37.2 ft. The locus of the best efficiency points for this pump for different speeds and impeller diameters is given by Eqs. 17 as

$$H_n = 37.2 \left(\frac{Q_n}{3500} \right)^2 = \frac{3.04}{10^6} Q_n^2 \tag{51}$$

and is shown by curve C in Fig. 75. Curve C intersects curve A at two points, showing that the L 16/14 pump satisfies the turbine requirements. Only the intersection at the higher turbine efficiency is of interest. At this point, Q_n = 2490 gpm and H_n = 18.8 ft. Since the turbine speed was specified to be 580 rpm, the required impeller diameter is given by Eqs. 17 as

$$D = \frac{(17)(2490/3500)}{580/720} = 15 \text{ in}$$

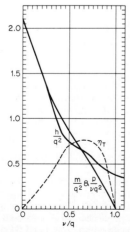

FIG. 73 Dimensionless characteristic curves for Peerless 10MH pump, constant-discharge turbine operation. (Ref. 62)

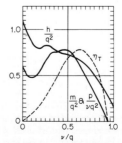

FIG. 74 Dimensionless characteristic curves for Peerless 10PL pump, constant-discharge turbine operation. (Ref. 62)

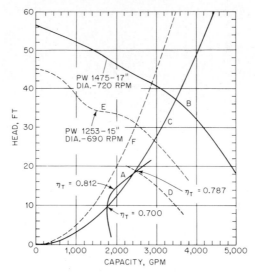

FIG. 75 Head-capacity curves for example of pump selection (ft $\times$ 0.3048 = m; gpm $\times$ 0.06309 = l/s). (Ref. 62)

or by

$$D = \frac{17\sqrt{18.8/37.2}}{580/720} = 15 \text{ in}$$

The computed head-capacity curve for the 15-in-diameter impeller at 580 rpm is shown as curve D in Fig. 75. The turbine discharge is 3200 gpm from Eq. 48 with $\eta_T = 0.787$.

The optimum solution would be to have curve C tangent to curve A at the point corresponding to maximum turbine efficiency, in this case $_{\eta T} = 0.812$. Since this would require a smaller pump, curve E in Fig. 75 shows a head-capacity curve for a K 14/12 pump, which was the next smaller pump in the series. Curve F, the locus of the best efficiency points, does not intersect curve A, showing that the smaller pump will not satisfy the turbine requirements. It is important to note that the turbine head, 20 ft, specified for this example was assumed to be the net head from inlet to outlet flange of the pump when installed and operated as a turbine.

The procedure outlined above should lead to the selection of a pump large enough to provide the required power. However, it probably will be necessary to apply the affinity laws over such wide ranges of the variables that the usual degree of accuracy should not be expected. Considerable care should be exercised if it becomes necessary to interpolate between the curves of Figs. 71 to 74. The computed performance will very likely differ from the results of subsequent tests. The curves of Fig. 71 may be converted to show the constant-head characteristics of the L 16/14 pump when installed and operated as a turbine. Details of the method of computation are given in Ref. 62, and the computed characteristics are shown in Fig. 76.

A comprehensive study by Acres American[63,64] led to a computer program to aid in selecting a pump to meet specified requirements when operating in the turbine mode. Known pump characteristics are entered in the program according to a specified format. The computer compares them with stored characteristics of pumps for which turbine mode characteristics are known and provides estimated turbine mode characteristics for the proposed pump. Vols. I and III of Ref. 63 describe the method of computation and give complete instructions for using the program.

A simplified method of pump selection has been proposed.[65,66] Equating the output power of the turbine P_T to the input power of the pump P_P led to $Q_T H_T \eta_T = Q_P H_P / \eta_P$. It was then

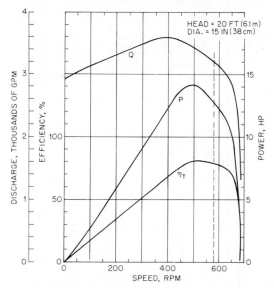

FIG. 76 Computed constant-head turbine characteristics for DeLaval L 16/14 pump (gpm × 0.06309 = l/s; hp × 0.7457 = kW). (Ref. 62)

assumed that $\eta_T \cong \eta_P = \eta$ and that the capacity and head may be separately computed from $Q_T\eta = Q_P$ and $H_T\eta = H_P$. If Q_T and H_T were known, η was assumed and the computed values of Q_P and H_P were used to enter a set of pump selection charts, similar to Fig. 40, to estimate the pump size and required impeller diameter. If P_T and H_T were specified, Eq. 48 could be used to obtain $Q_P = 3960P_T/H_T$, where H is in feet of liquid handled, Q is in gallons per minute, and P is in horsepower. A difficulty with this method is the assumption of the correct value of η. If the assumed value is too low, the pump selected may be too small to meet the requirements. If the assumed value is too high, the pump selected may be unnecessarily expensive and may operate at relatively poor efficiency as a turbine. Although the method described previously is more complicated, it avoids a direct assumption of η and gives an indication of the approximate efficiency at which the turbine will operate.

VORTEX PUMPS

A typical vortex pump is shown in Fig. 77.° The ability of this type of pump to handle relatively large amounts of suspended solids as well as entrained air or gas more than offsets the relatively low efficiency. Table 10 lists performance data for four typical vortex pumps and four radial-flow centrifugal pumps of nearly the same head and capacity. Figure 78 shows the head-capacity characteristics of a typical vortex pump with impellers of different diameter together with curves of constant efficiency and constant NPSH. Power-capacity curves for the same impellers are shown in Fig. 79.

Curves for a conventional radial-flow pump have been added for comparison in Figs. 78 and 79. Note that the head of the vortex pump does not decrease as rapidly with increasing capacity as does the head of the conventional pump. The power requirement of the vortex pump increases almost linearly with increasing capacity, whereas the power required by a conventional pump of about the same specific speed reaches a maximum and then decreases with increasing capacity.

° See also pages 2.100 to 2.101, 9.27, and 9.31.

TABLE 10 Characteristics of Typical Vortex Pumps and Comparable Radial-Flow Centrifugal Pumps at 1750 rpm

Impeller	Capacity, gpm (l/s)	Total head, ft (m)	Specific speed, USCS (SI)	Suction specific speed, USCS (SI)	Impeller diameter, in (mm)	Sphere diameter,[a] in (mm)	Efficiency[b] at $Q = Q_n$, %	Shutoff power,[b] $\%P_n$	Power[b] at $1.5Q_n$, $\%P_n$	Shutoff head,[b] $\%H_n$	Head[b] at $1.5Q_n$, $\%H_n$	NPSHR[b] at Q_n, ft (m)	NPSHR[b] at $1.5Q_n$, ft (m)
Vortex	100 (6.3)	24 (7.3)	1614 (988)	7700 (4715)	5⅞ (149)	2 (51)	39.5	37.5	131	120	71	3 (0.9)	8 (2.4)
Radial	108 (6.8)	26 (7.9)	1580 (967)	6430 (3937)	5⅞ (149)	⅝ (16)	59	50	108	144	50	4 (1.2)	7.5 (2.3)
Vortex	200 (12.6)	61 (18.6)	1134 (694)	7400 (4531)	8⅜ (213)	2 (51)	45	45	129	123	82	5 (1.5)	8 (2.4)
Radial	225 (14.2)	60 (18.3)	1218 (746)	6450 (3949)	8⅜ (213)	¾ (19)	61	50	130	125	65	6.5 (2.0)	14 (3.4)
Vortex	850 (53.6)	108 (32.9)	1523 (933)	9820 (6013)	11½ (292)	4 (102)	59	49	134	120	83	9 (2.7)	15 (4.6)
Radial	900 (56.8)	102 (31.1)	1636 (1002)	8690 (5321)	11½ (292)	⅞ (22)	76	47	113	135	44	11 (3.4)	21 (6.1)
Vortex	1050 (66.2)	150 (45.7)	1323 (810)	7088 (4340)	13 (330)	3¾ (95)	56	48	137	115	87	16 (4.9)	20 (6.1)
Radial	1250 (78.9)	154 (46.9)	1415 (866)	— —	13 (330)	1 1/32 (26)	83	52	123	116	45	11 (3.4)	- (-)
Vortex average			1399 (856)				50	45	133	120	81		
Radial average			1462 (895)				70	50	119	130	51		

[a]Diameter of largest sphere that will pass through pump.
[b]Subscript n designates values at best efficiency point.

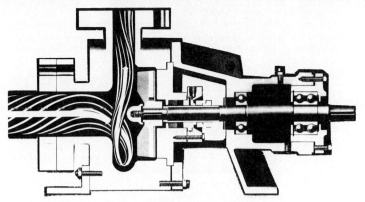

FIG. 77 Vortex pump. (Courtesy Fybroc Division, METPRO)

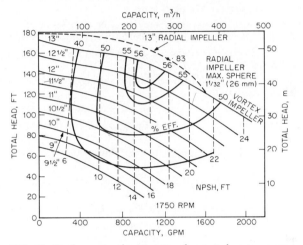

FIG. 78 Head-capacity characteristics of a typical vortex pump. Curves show approximate characteristics when pumping clear water (in × 2.54 = cm). (Courtesy Duriron)

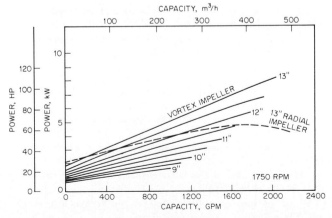

FIG. 79 Power-capacity characteristics of a typical vortex pump. Curves show approximate characteristics when pumping clear water (in × 2.54 = cm). (Courtesy Duriron)

Thus if the motor of the vortex pump has been selected to match the power required at the normal capacity for best efficiency, it will be overloaded if the pump is operated much beyond that point.

REFERENCES

1. Fischer, K., and D. Thoma: "Investigation of the Flow Conditions in a Centrifugal Pump," *Trans ASME* **54**:141 (1932).

2. Osborne, W. C., and D. A. Morelli: "Head and Flow Observations on a High-Efficiency Free Centrifugal-Pump Impeller," *Trans ASME* **72**:999 (1950).

3. Rupp, W. E.: High Efficiency Low Specific Speed Centrifugal Pump, U.S. Patent No. 3,205,828, Sept. 14, 1965.

4. Barske, U. M.: "Development of Some Unconventional Centrifugal Pumps," *Proc. Inst. Mech. Eng. (London)* **174**(11):437 (1960).

5. Manson, W. W.: "Experience with Inlet Throttled Centrifugal Pumps, Gas Turbine Pumps," *Cavitation in Fluid Machinery*, Symposium Publication, ASME, pp. 21–27, 1972.

6. Wislicenus, G. F.: "Critical Considerations on Cavitation Limits of Centrifugal and Axial-Flow Pumps," *Trans ASME* **78**:1707 (1956).

7. Karassik, I. J., and R. Carter: *Centrifugal Pumps: Selection, Operation and Maintenance,* McGraw-Hill, New York, 1960.

8. Holland, F. A., and F. S. Chapman: *Pumping of Liquids,* Reinhold, New York, 1966.

9. Ippen, A. T.: "The Influence of Viscosity on Centrifugal Pump Performance," *Trans. ASME* **68**(8):823 (1946).

10. Black, H. F., and D. N. Jensen: "Effects of High-Pressure Ring Seals on Pump Rotor Vibrations," ASME Paper No. 71-WA/FE-38, 1971.

11. Wood, G. M., H. Welna, and R. P. Lamers: "Tip-Clearance Effects in Centrifugal Pumps," *Trans. ASME, J. Basic Eng.,* ser. D, **89**:932 (1965).

12. Stepanoff, A. J.: *Centrifugal and Axial Flow Pumps,* 2d ed., Wiley, New York, 1967.

13. Rütschi, K.: Untersuchungen an Spiralgehäusepumpen verschiedener Schnelläufigkeit, *Schweiz. Arch Angew. Wiss. Tech.* **17**(2):33 (1951).

14. Stepanoff, A. J.: *Pumps and Blowers: Two Phase Flow,* Wiley, New York, 1965.

15. Knapp, R. T.: "Recent Investigations of the Mechanics of Cavitation and Cavitation Damage," *Trans. ASME* **77**:1045 (1955).

16. Knapp, R. T.: "Cavitation Mechanics and Its Relation to the Design of Hydraulic Equipment," James Clayton Lecture, *Proc. Inst. Mech. Eng. (London),* sec. A, **166**:150 (1952).

17. Shutler, N. D., and R. B. Mesler: "A Photographic Study of the Dynamics and Damage Capabilities of Bubbles Collapsing near Solid Boundaries," *Trans. ASME, J. Basic Eng.,* ser. D, **87**:511 (1965).

18. Hickling, R., and M. S. Plesset: "The Collapse of a Spherical Cavity in a Compressible Liquid," Division of Engineering and Applied Sciences, Report No. 85-24, California Institute of Technology, March 1963.

19. Hickling, R.: "Some Physical Effects of Cavity Collapse in Liquids," *Trans. ASME, J. Basic Eng.,* ser. D, **88**:229 (1966).

20. Plesset, M. S.: "Temperature Effects in Cavitation Damage," *Trans. ASME, J. Basic Eng.,* ser. D, **94**:559 (1972).

21. Hammitt, F. G.: "Observations on Cavitation Damage in a Flowing System," *Trans. ASME, J. Basic Eng.,* ser. D, **85**:347 (1963).

22. Preece, C. M. (ed.): *Treatise on Materials Science and Technology,* vol. 16, *Erosion,* Academic Press, New York, 1979.

23. Kovats, A.: *Design and Performance of Centrifugal and Axial Flow Pumps and Compressors*, Macmillan, New York, 1964.

24. *Hydraulic Institute Standards for Centrifugal, Rotary & Reciprocating Pumps*, 13th ed., Hydraulic Institute, Cleveland, Ohio, 1975.

25. "Centrifugal Pumps," PTC 8.2-1965, American Society of Mechanical Engineers, New York, 1965.

26. Bischoff, A.: "Untersuchungen über das Verhalten einer Kreiselpumpe bei Betrieb im Kavitationsbereich," *Mitt. Hydraul. Inst. Tech. Hochsch. München*, 8:48 (1936).

27. Thoma, D.: "Verhalten einer Kreiselpumpe beim Betrieb im Hohlsog—(Kavitations-) Bereich," *Z. Ver. Deut Ing.* 81(33):972 (1937).

28. Stepanoff, A. J.: "Cavitation in Centrifugal Pumps with Liquids Other Than Water," *Trans. ASME, J. Eng. Power*, ser. A, 83:79 (1961).

29. Tenot, M. A.: "Phénomènes de la Cavitation," *Mem. Soc. Ing. Civils France Bull.*, pp. 377–480, May and June, 1934.

30. Krisam, F.: "Neue Erkenntnisse im Kreiselpumpenbau," *Z. Ver. Deut. Ing.*, 95(11/12):320 (1953).

31. Wislicenus, G. F., R. M. Watson, and I. J. Karassik: "Cavitation Characteristics of Centrifugal Pumps Described by Similarity Considerations," *Trans. ASME* 61:17 (1939); 62:155 (1940).

32. Pfleiderer, C.: *Die Kreiselpumpen*, 5th ed., Springer-Verlag OHG, Berlin, 1961.

33. Stahl, H. A., and A. J. Stepanoff: "Thermodynamic Aspects of Cavitation in Centrifugal Pumps," *Trans. ASME* 78:1691 (1956).

34. Salemann, V.: "Cavitation and NPSH Requirements of Various Liquids," *Trans. ASME, J. Basic Eng.*, ser. D, 81:167 (1959).

35. Stepanoff, A. J.: "Cavitation Properties of Liquids," *Trans. ASME, J. Eng. Power*, ser. A, 86:195 (1964).

36. Lazarkiewicz, S., and A. T. Troskolański: *Impeller Pumps*, Pergamon Press, New York, 1965.

37. Gongwer, C. A.: "A Theory of Cavitation Flow in Centrifugal-Pump Impellers," *Trans. ASME*, 63:29 (1941).

38. Jekat, W. K., "A New Approach to the Reduction of Pump Cavitation: The Hubless Inducer," *Trans. ASME, J. Basic Eng.*, ser. D, 89:125 (1967).

39. Lewis, R. A., "An Experimental Analysis of a Jet Inducer with Multiple Nozzles Calibrated in Liquid Mercury and in Water," *Cavitation in Fluid Machinery*, Symposium Publication, ASME, 1965, p. 109.

40. Anderson, H. H., *Centrifugal Pumps*, Trade and Technical Press, Surrey, England, 1980, pp. 69–70.

41. Grohmann, M., "Extend Pump Application with Inducers," *Hydrocarbon Processing*, December 1979, p. 121.

42. Grohmann, M., "The Inducer: An Advanced Suction Impeller Design Helps Reduce Cost of Pump and Plant," *Pump World* 4(3):4 (1978).

43. Doolin, J. H., "Centrifugal Pumps and Entrained Air Problems," *Pump World*, 4(3): (1978).

44. *Mechanical Engineering* 93(6):39 (1971).

45. Kittredge, C. P.: "Hydraulic Transients in Centrifugal Pump Systems," *Trans. ASME*, 78(6):1307 (1956).

46. Parmakian, J.: *Waterhammer Analysis*, Prentice-Hall, Englewood Cliffs, N.J., 1955.

47. Streeter, V. L., and E. B. Wylie: *Hydraulic Transients*, McGraw-Hill, New York, 1967.

48. Bergeron, L.: *Waterhammer in Hydraulics and Wave Surges in Electricity*, Wiley, New York, 1961.

49. Addison, H.: *Centrifugal and Other Rotodynamic Pumps*, 3d ed., Chapman & Hall, London, 1966.

50. Richardson, C. A.: "Economics of Electric Power Pumping," *Allis-Chalmers Elec. Rev.* **9**:20 (1944).
51. Fraser, W. H., "Flow Recirculation in Centrifugal Pumps," *10th Turbomachinery Symposium 1981*, p. 95; Texas A&M University, College Station, Tex., 1981.
52. Fraser, W. H., "Recirculation in Centrifugal Pumps," *Materials of Construction of Fluid Machinery and Their Relationship to Design and Performance*, ASME, November 1981, pp. 65–86.
53. Bush, A. R., "Calculate Temperature Rise through Boiler-Feed Pumps," *Power* **107**(4):69 (1963).
54. Haywood, R. W., *Thermodynamic Tables in SI (Metric) Units*, Cambridge University Press, Cambridge, Mass. 1968.
55. Agostinelli, A., D. Nobles, and C. R. Mockridge: "An Experimental Investigation of Radial Thrust in Centrifugal Pumps," *Trans. ASME, J. Eng. Power*, ser. A, **82**:120 (1960).
56. Biheller, H. J.: "Radial Force on the Impeller of Centrifugal Pumps with Volute, Semivolute, and Fully Concentric Casings," *Trans. ASME, J. Eng. Power*, ser. A, **87**:319 (1965).
57. Donksy, B.: "Complete Pump Characteristics and the Effects of Specific Speeds on Hydraulic Transients," *Trans. ASME, J. Basic Eng.*, ser. D, **83**:685 (1961).
58. Swanson, W. M.: "Complete Characteristic Circle Diagrams for Turbomachinery," *Trans. ASME* **75**:819 (1953).
59. Knapp, R. T.: "Complete Characteristics of Centrifugal Pumps and Their Use in the Prediction of Transient Behavior," *Trans. ASME* **59**:683 (1937): **60**:676 (1938).
60. Kovats, A.: "Vibration of Vertical Pumps," *Trans. ASME, J. Eng. Power*, ser. A, **84**:195 (1962).
61. Kittredge, C. P.: "Estimating the Efficiency of Prototype Pumps from Model Tests," *Trans. ASME, J. Eng. Power*, ser. A, **90**:129, 301 (1968).
62. Kittredge, C. P.: "Centrifugal Pumps Used as Hydraulic Turbines," *Trans. ASME, J. Eng. Power*, ser. A, **83**:74 (1961).
63. Acres American Inc. for U.S. Department of Energy, Idaho National Engineering Laboratory, *Small Hydro Plant Development Program*, vols. I, II, and III, subcontract No. K-1574, October 1980, available from National Technical Information Service, U.S. Department of Commerce, Springfield, Vir. 22161.
64. Lawrence, J. D., and L. Pereira, "Innovative Equipment for Small-Scale Hydro Developments," Waterpower '81, an international conference on hydropower, *Proceedings*, vol. II, Washington, D.C., June 22–24, 1981, pp. 1622–1639.
65. Childs, S. M.: "Convert Pumps to Turbines and Recover HP," *Hydrocarbon Process. Petrol. Refiner* **41**:(10)173 (1962).
66. Hancock, J. W.: "Centrifugal Pump or Water Turbine," *Pipe Line News*, June 1963, p. 25.
67. Pilarczyk, K., and V. Rusak: "Application of Air Model Testing in the Study of Inlet Flow in Pumps," *Cavitation in Fluid Machinery*, Symposium Publication, ASME, p. 91, 1965.
68. Kallas, D. H., and J. Z. Lichtman: "Cavitation Erosion," in *Environmental Effects on Polymeric Materials* (D. V. Rosato and R. T. Schwartz, eds.), Wiley-Interscience, New York, 1968, pp. 223–280.

FURTHER READING

Carter, R.: "How Much Torque Is Needed to Start Centrifugal Pumps?" *Power* **94**(1):88 (1950).
Church, A. H.: *Centrifugal Pumps and Blowers*, Wiley, New York, 1944.
Gartmann, H.: DeLaval Engineering Handbook, 3d ed., McGraw-Hill, New York, 1970.

Hydraulic Institute Standards for Centrifugal, Rotary and Reciprocating Pumps, 14th ed., Hydraulic Institute, Cleveland, Ohio, 1983.

Moody, L. F., and T. Zowski: "Hydraulic Machinery," sec. 26 of *Handbook of Applied Hydraulics* (Davis and Sorenson, eds.), 3d ed., McGraw-Hill, New York, 1969.

Spannhake, W.: *Centrifugal Pumps, Turbines, and Propellers,* Technology Press, MIT, Cambridge, Mass., 1934.

Wislicenus, G. F.: *Fluid Mechanics of Turbomachinery,* McGraw-Hill, New York, 1947.

2.3.2
CENTRIFUGAL PUMP HYDRAULIC PERFORMANCE AND DIAGNOSTICS

WARREN H. FRASER

Any successful mathematical model of the mechanics of head generation in centrifugal pumps should do more than just make accurate predictions of pump performance; it should also be capable of identifying the cause of operational difficulties. Unlike mechanical malfunctions, which can be detected, analyzed, and corrected, many of the problems caused by hydraulic forces cannot be corrected in the mechanical sense—they are the unavoidable side effects of head generation in a rotating pressure field. For example, a thrust bearing may fail from lack of lubrication, from misalignment, or because an underrated bearing has been used. These are mechanical failures that can be corrected. A thrust bearing may also fail from a complex pattern of dynamic loading that reflects pressure pulsations of high intensity and a broad spectrum of frequencies during operation at reduced flow. This is an example of a failure from hydraulic causes and can be corrected only by resorting to an oversized bearing or modifying the operation of the pumping system to avoid low-flow operation.

Whenever a consistent correlation can be made between the known dynamics of head generation and operational problems, it becomes possible to devise a strategy to improve operation and reduce mechanical failures. The most significant problems caused by hydraulic dynamic forces can be listed as follows.

CAVITATION

Cause and Effect   Cavitation is the formation of vapor bubbles in any flow that is subjected to an ambient pressure equal to or less than the vapor pressure of the liquid being pumped. Cavitation damage is the loss of metal produced by the collapse of the vapor bubbles against the metal surfaces of the impeller or casing. NPSH is the net positive suction head required to prevent cavitation.

Diagnosis from Pump Operation   Pump operation in the presence of cavitation will reduce both the total head and the output capacity. A steady crackling noise in and around the pump suction indicates cavitation. A random crackling noise with high-intensity knocks indicates suction recirculation rather than cavitation from inadequate NPSH.

Diagnosis from Visual Examination of Surface Damage   Cavitation damage from inadequate NPSH occurs on the low-pressure or the visible surface of the impeller inlet vane.

Instrumentation   A suction gage or manometer in the pump suction can be used to determine whether the NPSH available is equal to or greater than the NPSH required from the manufacturer's rating curve.

Corrective Procedures   If the additional NPSH cannot be supplied, then the capacity of the pump should be reduced to the point where the required NPSH is equal to or less than the available NPSH. If this is not possible, then a small amount of air can be bled into the pump suction to reduce the damage associated with cavitation.

SUCTION AND DISCHARGE RECIRCULATION

Cause and Effect   Recirculation occurs at reduced flows and is the reversal of a portion of the flow back through the impeller. Recirculation at the inlet of the impeller is known as *suction recirculation*. Recirculation at the outlet of the impeller is *discharge recirculation*. Suction and discharge recirculation can be very damaging to pump operation and should be avoided for continuous operation.

Diagnosis from Pump Operation   Suction recirculation will produce a loud crackling noise in and around the suction of the pump. Recirculation noise is of greater intensity than the noise from low-NPSH cavitation and is a random knocking sound. Discharge recirculation will produce the same characteristic sound as suction recirculation except that the highest intensity is in the discharge volute or diffuser.

Diagnosis from Visual Examination   Suction and discharge recirculation produce cavitation damage to the pressure side of the impeller vanes. Viewed from the suction of the impeller, the pressure side would be the invisible, or underside, of the vane. Figure 1 shows how a mirror can be used to examine the pressure side of the inlet vane for cavitation damage from suction recirculation. Damage to the pressure side of the vane from discharge recirculation is shown in

FIG. 1   Examining the pressure side of the inlet vanes for suction recirculation damage.

Fig. 2. Guide vanes in the suction may show cavitation damage from impingement of the backflow from the impeller eye during suction recirculation. Similarly, the tongue or diffuser vanes may show cavitation damage on the impeller side from operation in discharge recirculation.

Instrumentation   The presence of suction or discharge recirculation can be determined by monitoring the pressure pulsations in the suction and in the discharge casing. Piezoelectric transducers installed as close to the impeller as possible in the suction and in the discharge of the pump can be used to detect pressure pulsations. The data may be analyzed with a spectrum analyzer coupled to an XY plotter to produce a record of the pressure pulsations versus frequency for selected flows. Figure 3 shows a typical plot of pressure pulsations versus capacity. As can be seen, a sudden increase in the magnitude of pressure pulsations indicates the onset of recirculation.

The onset of suction recirculation can be determined by an impact head tube installed at the impeller eye, as shown in Fig. 4. With the tube directed into the eye, the reading in the normal pumping range is the suction head minus the velocity head at the eye. At the point of suction recirculation, however, the flow reversal from the eye impinges on the head tube with a rapid rise in the gage reading.

Corrective   Procedures   Every impeller design has specific recirculation characteristics. These characteristics are inherent in the design and cannot be changed without modifying the design. An analysis of the symptoms associated with recirculation should consider the following as possible corrective procedures:

FIG. 2   Damage to the pressure side of the vane from discharge recirculation.

1. Increase the output capacity of the pump.

2. Install a bypass between the discharge and the suction of the pump.

3. Bleed air into the suction of the pump to reduce the intensity of the noise, vibration, and cavitation damage.

4. Substitute a harder material for the impeller to reduce the rate of cavitation damage.

AXIAL THRUST

Cause and Effect   Axial thrust is the thrust imposed in the direction of the shaft. It may occur in either the inboard or the outboard direction and is usually composed of a dynamic cyclic component superimposed on a steady-state load in either direction. The dynamic cyclic component increases in the recirculation zone and may impose excessive stresses in the shaft, with eventual failure from metal fatigue in tension. The static component may impose an excessive load in the thrust bearing, causing unacceptable bearing temperatures. The majority of thrust bearing failures are caused by fatigue failure of the bearing components from the dynamic cyclic axial loads.

Diagnosis from Pump Operation   High axial loads usually produce high thrust bearing temperatures and short thrust bearing life.

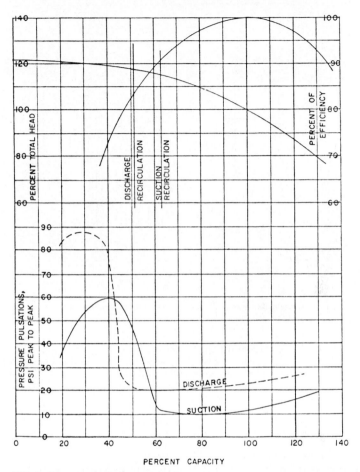

FIG. 3   Pressure pulsations versus capacity.

Diagnosis from Visual Examination of Damage

BEARING DAMAGE   Static thrust in excess of the bearing rating will cause cracking of the balls or rollers and of the race in antifriction bearings and metal scoring of the shoes in tilting-pad bearings. Bearing failure from dynamic loading in excess of the bearing rating will cause fatigue failures of the balls or rollers and race in antifriction bearings.

It is important to differentiate clearly between static load failure and fatigue failure. This can be done by examination of a cross section of the failure under the microscope. Fatigue failure from dynamic loading will show a hammering effect caused by points of impact. Fatigue failure from excessive static loading will show metal fatigue without the hammering effect of impact loading.

SHAFT FAILURE   Shaft failure at the outboard, or unloaded, end of the shaft in multistage or double-suction pumps is a fatigue failure in tension resulting from the high cyclic stresses induced in the shaft when the pump is operated in the discharge recirculation zone. Axial cyclic stresses can be reduced by increasing the pump output or, if this is not possible, by installing a recirculation line to bypass approximately 25% of the rated flow of the pump.

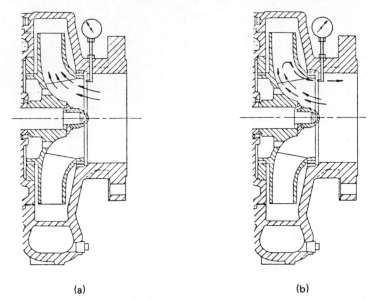

(a)   (b)

FIG. 4   Installation of impact head tube to detect suction recirculation during (a) normal flow and (b) recirculation flow.

Instrumentation   Displacement-type pickups should be used to determine the axial movement of the shaft relative to the bearing housing. The deflection of the thrust bearing housing can be determined by seismic instruments. Axial loading of the tilting-shoe type of thrust bearing can be monitored by a load cell permanently installed in the leveling plate. A typical installation is shown in Fig. 5.

Corrective Procedures   To determine the most effective procedure to correct axial thrust problems, it is necessary to determine whether the loads are static or dynamic or a combination of both. If it is a static failure, the thrust can usually be reduced by restoration of the internal clearances. Most shaft and bearing failures from axial thrust, however, are fatigue failures. If the failure is a fatigue failure, the loading can be decreased by increasing the capacity of the pump. If this is not possible, shaft failures can be reduced by substituting a shaft material of higher endurance limit. Antifriction bearing failures can best be corrected by substituting a tilting-shoe type of thrust bearing. The high cyclic axial forces are better absorbed in the oil film of the tilting-shoe bearing than in the rolling element bearing.

RADIAL THRUST

Cause and Effect   Radial thrust is the thrust imposed on the pump rotor and directed toward the center of rotation of the shaft. The forces are usually composed of a dynamic cyclic component superimposed on a steady-state load. The dynamic cyclic component increases rapidly at low-flow operation when the pump is operating in the recirculation zone. The static load also increases with low- and high-flow operation, with the minimum value at or near the maximum efficiency capacity.

Diagnosis from Pump Operation   High radial thrust is difficult to determine from pump operation. Persistent packing or mechanical seal problems may indicate excessive shaft deflection from radial loads. As in the case of high axial loads, high radial loads may produce high bearing temperatures with reduced life.

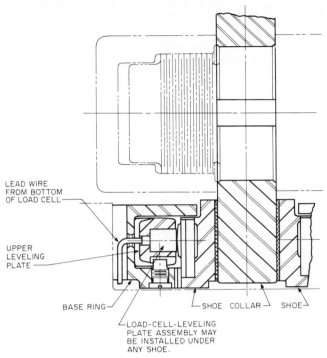

LEAD WIRE
FROM BOTTOM
OF LOAD CELL

UPPER
LEVELING
PLATE

BASE RING

SHOE COLLAR SHOE

LOAD-CELL-LEVELING
PLATE ASSEMBLY MAY
BE INSTALLED UNDER
ANY SHOE.

FIG. 5 Load cell to monitor axial load.

Diagnosis from Visual Examination of Damage

BEARING DAMAGE Static radial loads in excess of the bearing rating will cause cracking of the balls or rollers and the races in antifriction bearings. In the case of sleeve bearings, the bearing metal will be worn in one direction only and the journal will be worn uniformly. If the opposite is true (that is, the bearing is worn uniformly and the journal excessively in one direction), then the cause of the failure is a bent shaft and not excessive bearing loads.

SHAFT FAILURES Shaft failures from excessive radial loads usually occur at the midpoint of the shaft span in double-suction or multistage pumps. In the case of end-suction pumps, shaft failures usually occur at the shoulder of the shaft, where the impeller hub joins the shaft sleeve.

Instrumentation It is difficult to devise instrumentation to determine excessive radial loading of the shaft and bearings. Temperature rise of the bearings may or may not be symptomatic of excessive radial loading. High bearing temperatures may occur from misalignment, inadequate lubrication, or excessive axial loading of the thrust bearing, and these causes should be eliminated before concluding that the radial loads are excessive.

Corrective Procedures Most bearing and shaft failures are caused by excessive radial loads when the pump operates at low flow rates. Radial loads can be reduced by operating the pump at higher capacities or by installing a bypass from the discharge back to the pump suction. For pumps handling water, the life of the shaft can be extended by substituting a 13% chrome steel shaft for carbon steel. An austenitic steel of the 300 series should not be used, as the endurance limit of the 300 series steels is less than that of the chrome steels in fresh water. For liquids other than water, the endurance limit of the shaft material in the liquid being pumped is the determining factor in the shaft life.

PRESSURE PULSATIONS

Cause and Effect   Pressure pulsations are present in both the suction and the discharge of any centrifugal pump. The magnitude and frequencies of the pulsations depend upon the design of the pump, the head produced by the pump, the response of the suction and discharge piping, and the point of operation of the pump on its characteristic curve. The observed frequencies in the discharge may be the running frequency, the vane passing frequency, or multiples of each. In addition, random frequencies with pressure pulsations higher than either the rotating or the vane passing frequencies have been observed. The cause of these random frequency pulsations is not known, but they should not be dismissed as spurious or irrelevant data in any analysis of symptomatic operational problems.

The observed frequencies in the pump suction are much lower than in the discharge. Typical frequencies are in the order of 5 to 25 cycles/s, and they do not appear to bear any direct relation to the rotational speed of the pump or the vane passing frequency.

Diagnosis from Pump Operation   In most pumping installations of 1000 ft (305 m) or less of head per stage, there is little outward manifestation of pressure pulsations during normal pumping operation. Other than for specialized applications, such as white water pumps for paper machines (where the discharge pressure pulsations may affect the quality of the paper) or quiet pumps in marine service, there are few external symptoms of internal pressure pulsations. For high-head pumps, however, suction and discharge pressure pulsations may cause instability of pump controls, vibration of suction and discharge piping, and high levels of pump noise.

Diagnosis from Visual Examination of Damage   In the case of high-head pumps, any failure of internal pressure-containing members should be investigated on the presumption

TABLE 1   Corrective Procedures for Various Problems

Problem	Corrective procedure
1. Vibration of suction or discharge piping	*a.* Search for responsive resonant frequencies in the piping or supports. If any part of the system responds to the frequency of the pressure pulsations, alter the system to shift the resonant frequencies.
	b. If possible, increase the output of the pump by changing the mode of operation or by installing a bypass from the discharge to the suction of the pump.
	c. If the piping responds to the vane passing frequency of the pump, the impellers can be replaced with a unit containing either one fewer or one more vane.
2. Instability of pump controls	*a.* If possible, increase the output of the pump by changing the mode of operation or by installing a bypass from the discharge to the suction of the pump.
	b. Install acoustical filters to reduce the magnitude of the pressure pulsations.
3. Fatigue failure of internal pressure-containing components of the pump	*a.* If possible, increase the output of the pump by changing the mode of operation or by installing a bypass from the discharge to the suction of the pump.
	b. Redesign the failed components to reduce the induced cyclic stresses to below the endurance limit of the material.
	c. If the spectral analysis shows that the maximum pressure pulsations correspond to the vane passing frequency of the impeller, then the impeller can be replaced by one having either one fewer or one more vane of the same design.

that the failures are fatigue failures from internal pressure pulsations. Examination of the fracture will determine whether the failure is a fatigue failure or not. If it is a fatigue failure, the cause can usually be traced to high cyclic stress induced in the pressure-containing member from high-frequency pressure pulsations.

Instrumentation   Pressure pulsations are usually measured with piezoelectric pressure transducers and recorded as peak-to-peak pressure pulsations over a broad frequency band. Recorded on tape or strip charts, a spectral analysis may be performed for any operating condition.

Corrective Procedures   A spectral analysis of the pressure pulsations at the suction and at the discharge of the pump is necessary before a strategy for corrective procedures can be developed. Once the spectral analysis is available, problems associated with pressure pulsations can usually be reduced by implementing the procedures shown in Table 1.

2.3.3
CENTRIFUGAL PUMP MECHANICAL PERFORMANCE, INSTRUMENTATION, AND DIAGNOSTICS

CHARLES JACKSON
JAMES H. INGRAM

MECHANICAL PERFORMANCE _____

The mechanical performance of a pump would imply only the rotating mechanical masses, with no consideration given to hydraulic (process) effects. The rotating masses (impellers, sleeves, nuts, coupling, bearings, seals, etc.) can be examined as pure mechanics. A person concerned with mechanical performance should be intimately familiar with pump design, construction, and maintenance to be successful.

In discussing the mechanical performance of centrifugal pumps, two examples will be used. The first will be a horizontal, 500-hp (373-kW), single-stage (overhung impeller) American Petroleum Institute (API) process pump. The second will be a six-stage, horizontal, 1000-hp (746-kW), multistage boiler-feed pump.

Normally, the rotor dynamics will involve (a) a review of the shaft stiffness of the bearings and structure, (b) a mass model of the rotor, and (c) a critical speed analysis with mode shapes of the rotor or shaft.

SINGLE-STAGE PUMP _____

An $8 \times 6 \times 13$ pump is operating on water at 3550 rpm with a design flow of 2500 gpm (567 m^3/h) at 600 ft (183 m) total head, 1.0 sp. gr., requiring approximately 500 hp (373 kW). The pump operated extremely rough, and the bearings and bearing housing failed. The impeller weighs 61.4 lb (27.9 kg).

If the impeller is fitted on the shaft with an eccentricity of 0.002 in (0.051 mm), a calculated centrifugal force Fe of 44 lb (196 N) would cause a deflection of 0.0026 in (0.066 mm) from

$$y = wl^3/3EI$$

where w = weight (force) of impeller + Fe, lb (N)
  l = length of overhang, in (m)
  E = modulus of elasticity, lb/in^2 (N/m^2 or Pa)
  I = shaft section moment of inertia, in^4 (mm^4)

This pump has a 5212 line bearing and tandem mounted (DB) 7311DB angular contact thrust bearings (40° contact angle). An extremely loose fit of the radial bearing in the bearing housing could cause the outer race to spin, which could cause a vibration equal to twice the rotation frequency. Interference fitting could lead to radial bearings' accepting thrust (for which many are not designed) from thermal expansion of the shaft or from the thrust bearing.

Frequencies Generated The following data and definitions are needed to compute the frequencies generated by defective bearings[1]:

  rpm = revolutions per minute
  rps = revolutions per second
  FTF = fundamental train frequency, Hz
  BPF_I = ball passing frequency of inner race, Hz
  BPF_O = ball passing frequency of outer race, Hz
  BSF = ball spin frequency, Hz
  Bd = ball or roller diameter, in (mm)
  Nb = number of balls or rollers
  Pd = pitch diameter, in (mm)
  $\emptyset$ = contact angle

The formulas are

$$\text{rps} = \frac{\text{rpm}}{60}$$

$$FTF = \frac{\text{rps}}{2}\left(1 - \frac{Bd}{Pd}\cos\emptyset\right)$$

$$BPF_I = \left(\frac{Nb}{2}\text{rps}\right)\left(1 + \frac{Bd}{Pd}\cos\emptyset\right)$$

$$BPF_O = \left(\frac{Nb}{2}\text{rps}\right)\left(1 - \frac{Bd}{Pd}\cos\emptyset\right)$$

$$BSF = \left(\frac{Pd}{2Bd}\text{rps}\right)\left[1 - \left(\frac{Bd}{Pd}\right)^2\cos^2\emptyset\right]$$

The pitch diameter is the diameter measured across the bearing from ball or roller center to ball or roller center. The contact angle is measured from a line perpendicular to the shaft to the point at which the balls or rollers contact the race. The contact angle of a deep groove ball bearing is zero.

It is necessary to distinguish between the ball frequency and the impeller vane passing frequency, which is 17,750 cpm (5 vanes × 3550 rpm × 1 casing cutwater) for this example. The mode shape of the pump shaft is conical (pivotal) in the first mode of a cantilevered shaft mount. The stiffness map of the rotor looks like that shown in Fig. 1.

This pump has two design faults, as can be seen from the stiffness map. The first is that an excessively large coupling is used. This heavy overhung mass at the coupling forces the first shaft resonance to be very near the pump operating speed. Normally, the shaft in this type of pump is considered to be "rigid," i.e., operating safely below the first undamped shaft resonance. In this case, the pump is affected by two negatively additive errors. The heavy mass coupling effect is compounded by a weak baseplate that is not properly grouted, leaving a void under the pump supports. In rotor (shaft) supports, two spring supports in series reciprocally add, similar to electric resistors in parallel (Table 1).

TABLE 1   Logic of Spring Equivalent Stiffness K_e

SPRING	ELECTRICAL RESISTOR EQUIV ANALOG	EXAMPLES OF VARIATIONS (Ke)		
m Kb Ks	Rb Rs	Kb 1×10^5 Ks 1×10^6	Kb 1×10^5 Ks 5×10^5	Kb 1×10^5 Ks 1×10^5
$1/Ke = \dfrac{1}{Kb} + \dfrac{1}{Ks}$	$1/Re = \dfrac{1}{Rb} + \dfrac{1}{Rs}$	Ke = 91k lb/in	Ke = 83k lb/in	Ke = 50k lb/in

Kb = STIFFNESS OF BEARING, lb/in (N/m)   N/m = 175 X lb/in
Ks = STIFFNESS OF SUPPORT, lb/in (N/m)
Ke = EQUIVALENT STIFFNESS

NOTE: SPRINGS IN PARALLEL SIMPLY ADD, AS DOES
    ELECTRICAL RESISTANCE IN SERIES: Ke = Kb + Ks

The effective stiffness is

$$K_e = \frac{1}{1/K_b + 1/K_s}$$

If the bearing stiffness K_b is 2.5×10^6 lb/in (4.4×10^8 N/m) and the support stiffness K_s is low, i.e., 7.0×10^5 lb/in (1.2×10^8 N/m), then the effective stiffness is 5.47×10^5 lb/in (9.57×10^7 N/m), which moves the first mode resonance from 4300 cpm to 3550 cpm, which is the running speed of the pump.

The mode shapes of the rotor are shown in Figs. 3 and 4. An animated display is used so that one can better see the rotor gyrations in synchronous whirl. The first modes are shown with a lighter, normal, and correct coupling (15 lb; 6.8 kg). Figure 2 shows the mathematical model of this pump with the impeller at station 2, the radial bearing at station 11 at 2.5×10^6 lb/in (4.4×10^8 N/m) stiffness, the outboard thrust/radial bearing at station 18 at 3.5×10^6 lb/in (6.1×10^8 N/m) and the coupling at station 21.

Figure 3a, obtained from a finite element computer analysis of the mathematical model in Fig. 1, shows the first mode with a rotor weight of 110.4 lb (50 kg), a rotor length of 30.3 in (77.0 cm), and a first mode undamped resonance (critical) of 4717 cpm. This is a pivotal mode, with

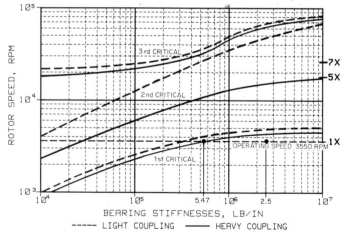

FIG. 1   Undamped critical speed map of single-stage overhung pump, comparing normal and heavy coupling weights (lb/in $\times$ 175 = N/m).

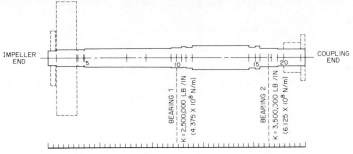

FIG. 2 Rotor cross section of single-stage overhung pump with normal coupling weight. Rotor weight = 110.4 lb (50 kg); rotor length = 30.3 in (77.0 cm); number of stations = 22; number of bearings = 2.

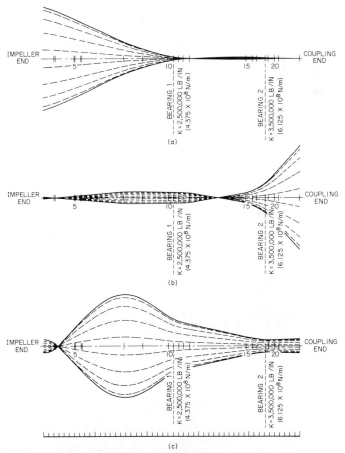

FIG. 3 First, second, and third resonate animated mode shapes of single-stage overhung pump with light coupling of 15 lb (6.8 kg) and rigid foundation. Rotor weight = 110.4 lb (50 kg); rotor length = 30.3 in (77.0 cm); number of stations = 22; number of bearings = 2. (*a*) Mode 1: frequency = 4717 cpm; (*b*) mode 2: frequency = 53,482 cpm; (*c*) mode 3: frequency = 67,522 cpm.

No. UNITS	CRITICAL SPEED rpm	(Hz)	WMODE, lb	ITMODE, lb/in²	KMODE, lb/in	DIM. STRAIN ENERGY USHAFT	UBEARING
1	4717	(79)	57.0	1.84E-01	3.60E+04	91	9
2	53482	(891)	6.0	3.59E-01	4.90E+05	36	64
3	67522	(1125)	9.8	6.42E-01	1.27E+06	42	58

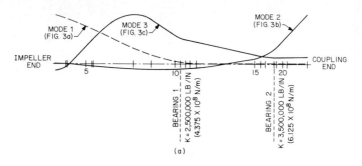

(a)

No. UNITS	CRITICAL SPEED rpm	(Hz)	WMODE, lb	ITMODE, lb/in²	KMODE, lb/in	DIM. STRAIN ENERGY USHAFT	UBEARING
1	4085	(68)	58.0	1.81E-01	2.75E+04	59	41
2	21251	(354)	13.8	3.50E-01	1.76E+05	10	90
3	40036	(667)	17.1	1.34E-00	7.78E+05	37	63

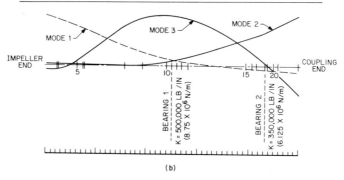

(b)

FIG. 4 Synchronous critical speed summary with first three mode shapes of a single-stage overhung pump with light coupling of 15 lb (6.8 kg) and rigid or flexible foundation. Rotor weight = 110.4 lb (50 kg); rotor length = 30.3 in (77.0 cm); number of stations = 22; number of bearings = 2. (*a*) Rigid support, modes 1, 2, and 3 respectively: 4717, 53,482, and 67,522 cpm; (*b*) flexible support, modes 1, 2, and 3 respectively: 4085, 21,251, and 40,036 cpm. (lb × 0.454 = kg; lb/in × 175 = N/m).

100% of the normalized motion at the impeller end. Any motion at the antifriction bearings is greatly restrained.

Figure 3b shows the second resonance mode, at 53,482 cpm, which is not to be encountered. The coupling motion is now the greatest motion.

Figure 3c shows the third mode, at 67,522 cpm.

Figure 4a shows an overlay of all three modes with a summary of the criticals, the modal mass, and the relative strain energy (91) in the shaft at station 10 (impeller side of radial bearing). A lesser strain energy is at the radial bearing (station 11).

Figure 4b summarizes what happens to critical speed modes if either a more *flexible* bearing or a soft structure is provided intentionally or nonintentionally. Also note that the criticals are lowered significantly and the strain energy is transferred more from the shaft into the bearings, i.e., strain values under the *U-shaft* column are less than under *U-bearing* column. The first critical is 4085 cpm at a pump speed of 3550 rpm (+15%). A 15% margin of separation may be close enough to excite (cause a rise in vibration) the rotor if the resonance response envelope is too wide. However, this is unlikely on antifriction bearings (spiky/narrow response) but possible on sleeve bearings (low/broad response).

Figure 5a is a summary which shows the response of a rigid support and an *excessively heavy* (62 lb = 28 kg) coupling, which is as heavy as the impeller. Note that the first mode is again only

No. UNITS	CRITICAL SPEED rpm	(Hz)	WMODE, lb	ITMODE, lb/in²	KMODE, lb/in	DIM. STRAIN ENERGY USHAFT	UBEARING
1	4279	(71)	65.2	1.80E-01	3.39E+04	91	9
2	15865	(264)	53.3	3.28E-01	3.81E+05	73	27
3	64970	(1083)	8.9	6.54E-01	1.07E+06	51	49

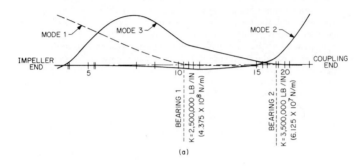

(a)

No. UNITS	CRITICAL SPEED rpm	(Hz)	WMODE, lb	ITMODE, lb/in²	KMODE, lb/in	DIM. STRAIN ENERGY USHAFT	UBEARING
1	3580	(60)	66.6	1.79E-01	2.43E+04	61	39
2	9339	(156)	59.6	3.47E-01	1.48E+05	35	65
3	35365	(589)	22.0	7.01E-01	7.82E+05	25	75

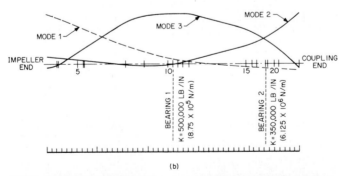

(b)

FIG. 5 Synchronous critical speed summary with first three mode shapes of a single-stage overhung pump with heavy coupling of 62 lb (28 kg) and rigid foundation. Rotor weight = 157.4 lb (71.4 kg); rotor length = 30.3 in (77.0 cm); number of stations = 22; number of bearings = 2. (*a*) Rigid support, modes 1, 2, and 3 respectively: 4279, 15,865, and 64,970 cpm; (*b*) flexible support, modes 1, 2, and 3 respectively: 3580, 9339, and 35,365 cpm. (lb × 0.454 = kg; lb/in × 175 = N/m).

slightly above the operating speed, i.e., 4279 cpm compared with 3550 ($+21\%$). The bearing stiffness is assumed to be the controlling stiffness. Many assume that the structure or base stiffness is one order above the bearing stiffness ($K_s = 10K_b$). This assumption that the bearing stiffness is the controlling stiffness variable is often a very poor assumption. The larger the pump size, the more this is true. That is why an $8 \times 6 \times 13$ pump was used as an example.

Further, the second mode, at 15,865 cpm, is in an area where the blade passing frequency (5 $\times$ 3550 = 17,750 cpm) can easily excite this mode, given little variation in support stiffness.

Figure 5b is a summary sheet which best illustrates the problem:

1. The baseplate was improperly installed and grouted.

2. The elastomeric coupling designed for low-duty, low-speed, and torsional damping was *too* heavy, i.e., too much overhung weight.

Note that the first critical is in sympathy with the pump operating speed, which becomes intolerable with the operating time *limited* to one to two days, due to bearing failures.

The stiffness on antifriction bearings was determined from a program written by M. E. Leader of Monsanto, using values projected by an article written by E. P. Garguilo, DuPont.[2] The correction consisted of converting the 62-lb (28-kg) coupling to a 15-lb (6.8-kg) series dry flex disk-type coupling and stiffening the support by flushing the baseplate cavity with a degreasing fluid and pressure injection of epoxy to fill the baseplate voids.

It should be remembered that the blade passing frequencies will normally be the strongest exciting force. On this pump, the frequency is five times running speed (5 vanes times each cutwater). Since there are two cutwaters, there can also be a frequency at 10 times running speed. The $5\times$ frequency is shown on Fig. 1. Also, this $5\times$ frequency excitation could excite the second mode because the second mode critical could fall anywhere between the solid and dashed lines, depending on baseplate stiffness.

The instruments used in diagnosing this problem were force-effective seismic sensors (velocity or piezoelectric accelerometers). They are preferred for pumps, particularly those with antifriction bearings.

MULTISTAGE PUMP EXAMPLE

To show the mechanical rotor variations, a six-stage boiler-feed pump with a design capacity of 1250 gpm (284 m³/h), 2200 ft (670 m) total head, and driven by a 1000-hp (746-kW), two-pole

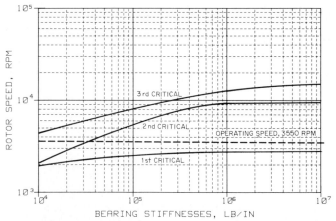

FIG. 6 Undamped critical speed map of multistage boiler-feed pump with plain journal bearings (lb/in $\times$ 175 = N/m).

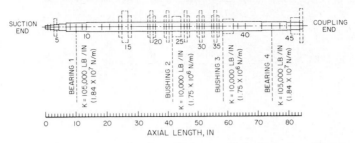

FIG. 7   Rotor cross section of multistage high-pressure boiler-feed pump with plain journal bearings. Rotor weight = 377.7 lb (171.3 kg); rotor length = 84.6 in (215 cm); number of stations = 47; number of bearings or bushings = 4. (in × 2.54 = cm).

motor has been selected. This pump utilizes interstage bushings as support bearings to the rotor. The contribution of these bushings as bearings will probably be less than might be assumed.

Pressure or seal leakage control bushings contribute rotor support if they are long. The hot feedwater has very low viscosity and little damping. The bearing stiffness will be relative to the eccentricity ratio of the shaft in the bushings. An eccentricity ratio of unity (maximum) implies that the shaft is rubbing directly on its bushing.

The impeller weight is increased by the water trapped in each impeller. Many pump manufacturers improperly list pump undamped critical speeds from *dry* pump data or calculations. The bushings, labyrinths, and wear rings all contribute to the *actual* critical speed (see Subsec. 2.2.1). Also, bearing housing resonances are more common than expected.

Figures 6 to 9 illustrate in the same fashion as the previous example the design audit of a steam-turbine-driven, 4 × 8 × 10½, six-stage, boiler-feed pump using hydrodynamic radial and thrust bearings. No problems were experienced with this pump.

A more complete listing of data from analysis is shown with an added breakdown of the rotor model and the output of the first, second, and third resonant modes. The first critical (rotor resonance) is now a cylindrical mode and not a conical mode, as previously seen for the overhung impeller of the single-stage process pump.

ALIGNMENT OF PUMPS AND DRIVERS

Outside of serious unbalance of pump components, there is no single contributor of poor mechanical performance more significant than poor alignment. Incorrect alignment between a pump and its driver can cause:

1. Extreme heat in couplings
2. Extreme wear in gear couplings and fatigue in dry element couplings
3. Cracked shafts and totally failed shafts, with failure due to reverse bending fatigue transverse to the shaft axis initiating at the change of section between the large end of the coupling hub taper and the shaft
4. Preload on bearings (evident by an elliptical and flattened orbit resembling a deflated beach ball); pure asymmetry of vertical and horizontal vibration can be misleading since the bearing spring constants could vary greatly in the k_{yy} (vertical) and the k_{xx} (horizontal) axis.
5. Bearing failures plus thrust transmission through the coupling, which can be totally locked (axial vibration checks across the coupling, i.e., at each adjacent machine, will generally confirm this condition)

Significant changes in the cold nonrunning alignment of a pump and driver can take place if the temperature rise in each machine is different and if the piping imposes forces on the pump.

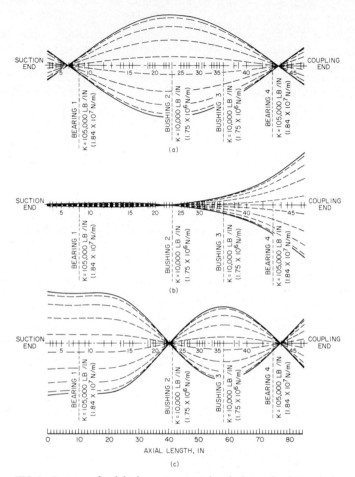

FIG. 8 First, second and third resonate animated mode shapes of multistage high-pressure boiler-feed pump with plain journal bearings. Rotor weight = 377.7 lb (171.3 kg); rotor length = 84.6 in (215 cm); number of stations = 47; number of bearings or bushings = 4. (*a*) Mode 1: frequency = 2614 cpm; (*b*) mode 2: frequency = 5223 cpm; (*c*) mode 3: frequency = 8134 cpm. (in × 2.54 = cm).

Therefore, alignment under actual operating conditions must be predicted or, if unknown, confirmed by instrumentation. In either case, an allowance must be made in the initial cold alignment to compensate for changes in alignment from cold idle to hot running.

There are several techniques for measuring cold and hot alignment. The cold alignment is generally measured by either face and rim (Fig. 10) or reverse dial indicator (Fig. 11) methods.

The face and rim method has a sensitivity advantage when the diameter of a coupling exceeds the indicator span of reverse indicator bracket tooling. This is rare, as the pump will generally have a spacer coupling and the reach of the reverse indicators can be increased by clamping onto the shaft behind each coupling half. The face and rim method would also have an advantage if either the driver or the gear could not be rotated, as it seems unlikely that the pump could not be rotated. In order to compensate for the measuring surface's not being circular or smooth, both shafts should be rotated together when using this method.

No. UNITS	CRITICAL SPEED rpm	(Hz)	WMODE, lb	ITMODE, lb/in²	KMODE, lb/in	DIM. STRAIN ENERGY USHAFT	UBEARING
1	2614	(44)	200.6	8.23E-01	3.89E+04	51	49
2	5223	(87)	51.2	2.40E-01	3.96E+04	17	83
3	8134	(134)	127.8	1.38E-00	2.40E+05	46	54

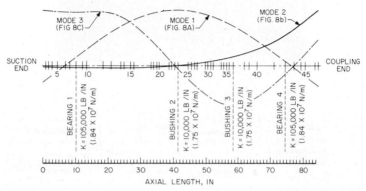

FIG. 9 First three critical speed mode shapes of multistage high-speed boiler-feed pump superimposed (in $\times$ 2.54 = cm).

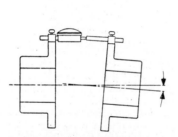

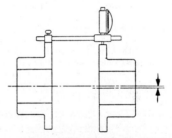

CHECK FOR ANGULAR MISALIGNMENT

Dial indicator measures maximum longitudinal variation in hub spacing through 360° rotation.

1. Attach dial indicator to hub, as with a hose clamp; rotate 360° to locate point of minimum reading on dial; then rotate body or face of indicator so that zero reading lines up with pointer.

2. Rotate both half couplings together 360°. Watch indicator for misalignment reading.

3. Driver and driven units will be lined up when dial indicator reading comes within maximum allowable variation for that coupling style. Refer to specific installation instruction sheet for the coupling being installed. *Note:* If both shafts cannot be rotated together, connect dial indicator to the shaft that is rotated.

CHECK FOR PARALLEL MISALIGNMENT

Dial indicator measures displacement of one shaft center line from the other.

4. Reset pointer to zero and repeat operations 1 and 2 when either driven unit or driver is moved during aligning trials.

5. Check for parallel misalignment as shown. Move or shim units so that parallel misalignment is brought within the maximum allowable variations for the coupling style.

6. Rotate couplings several revolutions to make sure no "end-wise creep" in connected shafts is measured.

7. Tighten all locknuts or capscrews.

8. Recheck and tighten all locknuts or capscrews after several hours of operation.

FIG. 10 Face and rim dial indicator method (Courtesy Rexnord)

WHAT IS MISALIGNMENT

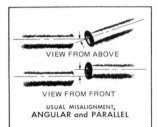
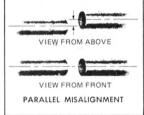
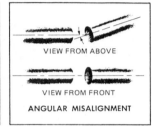

FIG. 11 Model used for training machinist in the use of reverse dial indicator techniques for alignment of machine shafts. Misalignment can be measured as *parallel* or *angular* offset.

Disadvantages of Face and Rim Method

1. Diameters of the rim must be true (circular) and smooth and the face reading surface must be flat and smooth, unless both shafts are rotated together.

2. The driver and pump cannot float axially while reading is being taken or an error will be introduced into the face (angular) reading. (A fixed axial stop will assist in reducing errors.)

Reverse Dial Procedure for Measuring Alignment (Hot or Cold)   Several procedures have been suggested by various people to estimate or actually measure alignment while a pump is running at operating temperature. Some techniques are:

1. Shutdown after temperatures have stabilized for "hot check" by dial indicators

2. Optical measurements cold to hot (A. J. Campbell, Compressor Engineering Corp., Houston)

3. Dodd bars (DynAlign) technique (R. Dodd, Chevron[3])

4. Acculign bench mark gauges (J. Essinger, Shell Oil)

5. Water-cooled probe stands (C. Jackson, Monsanto)

6. Instrumented coupling (e.g., Indikon)

Refer to Ref. 4 for more specific information on the above techniques.

After measurements have been made and actual misalignment is determined for the hot running condition of the pump, the next step is to calculate the alignment correction required between machines to bring them into alignment during operation. Trial-and-error methods should be discouraged. Plotting actual shaft positions to scale allows one to graphically measure the required cold alignment corrections. Other methods—for example, those that can best be carried out with the aid of a programmable calculator—quickly and accurately calculate vertical and horizontal and inboard and outboard changes in machine positions to accomplish the correct cold alignment. See Ref. 11 for a calculator program for both reverse dial and face and rim methods.

Offered here are a graphical procedure and an example for obtaining the desired alignment between a pump and driver for a case where the thermal growth has been estimated and a calculated cold alignment is desired to achieve the final alignment during operation.

The example is illustrated in Fig. 12. Consider a steam-turbine-driven boiler-feed pump which has the following heat rise predictions by the manufacturers: steam end 0.002 in (0.051 mm), exhaust end 0.012 in (0.305 mm), pump inboard support 0.006 in (0.152 mm), and outboard bearing 0.004 in (0.102 mm). The horizontal length is laid on a graph with 1 div. = 1 in (25.4 mm). The vertical movement plots are laid out with 1 div. = 0.001 in (1 mil; 0.0254 mm).

Based on 0.001-in (0.0254-mm) sag (see Fig. 19), the field readings needed to meet the above absolute requirements are shown in Fig. 13.

The actual cold alignment is checked, and the reverse dial readings are as recorded in Fig. 14. There are 0.125-in (3.175-mm) shims under all support feet. It is decided to move the pump rather than the turbine, and therefore the correct cold position of the pump must be calculated.

To correct the turbine dial readings to position zero at the top, simply add −12 to all four turbine readings (Fig. 15). To correct for sag (see Fig. 19 for explanation), *subtract* 1 from the left and right readings and 2 from the bottom reading (Fig. 16). Finally, plot the absolute shaft positions on graph paper, leaving the turbine "in place," so to speak, thereby determining two points across the 16-in (40.64-cm) indicator span to define *where* the pump shaft lies with *respect to* the turbine (Fig. 17).

To plot the horizontal corrections, reduce the final horizontal readings *only* to zero on the least

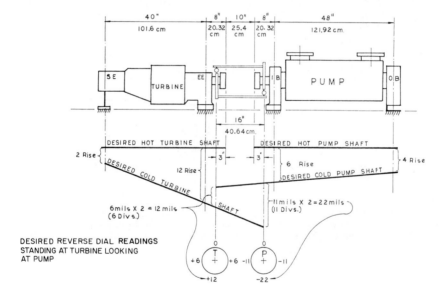

DESIRED REVERSE DIAL READINGS ARE
CALCULATED AS FOLLOWS:

- BOTTOM = 2 X VERTICAL OFFSET
- LEFT = VERTICAL OFFSET − HORIZONTAL OFFSET
- RIGHT = VERTICAL OFFSET + HORIZONTAL OFFSET
- <u>POSITIVE</u> MEANS THE TURBINE SHAFT BEING
  INDICATED ON IS BELOW THE PUMP SHAFT
  EXTENSION VIA REACH BAR
- <u>NEGATIVE</u> MEANS THE PUMP SHAFT BEING
  INDICATED ON IS ABOVE THE TURBINE SHAFT
  EXTENSION VIA REACH BAR

FIG. 12   Alignment example of a turbine-driven boiler-feed pump with heat-rise data from the manufacturers plotted to calculate the desired cold alignment.

FIG. 13 Desired dial indicator readings corrected for indicator bar sag.

FIG. 14 Actual alignment readings.

FIG. 15 Dial readings corrected to position zero at top.

FIG. 16 Actual alignment readings corrected for indicator bar sag.

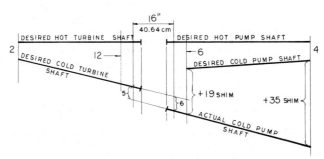

FIG. 17 Plot of absolute vertical shaft positions. It can be seen from the graph that a 0.019-in (0.483-mm) shim is required to raise the inboard end of the pump and a 0.035-in (0.889-mm) shim is required to raise the outboard end of the pump to compensate properly for thermal growth.

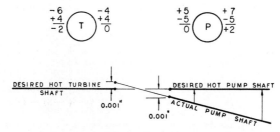

FIG. 18 Required horizontal corrections.

numerical reading, by adding 4 to the turbine readings and −5 to the pump reading, as shown in Fig. 18.

Dial indicator readings can be in metric units and the scale in centimeters rather than inches. A scale of between 500:1 and 1000:1 is suggested. A 1000:1 scale is in use here (horizontal scale equals 1000 times vertical scale).

As shown in Fig. 12, the driver and pump are purposely misaligned so that actual operating temperatures will put the two shafts within acceptable limits. The acceptable limits for pump final alignment are 0.001 in/in (0.025 mm/mm) of coupling flex plane separation. If a spacer coupling is 5 in (13 cm) between flex planes on the flexible coupling, then the shafts must be within 0.005 in (0.127 mm) vertical or horizontal offset. Pure angular misalignment in one plane is not desired, as it reduces the tolerance by 2:1 for gear couplings (dry coupling would have severe fatigue in one flex plane). The above limits should be reduced by one-half.

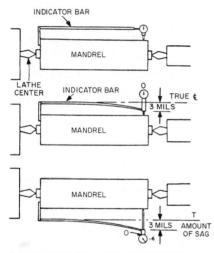

(a) CHECKING INDICATOR BAR FOR SAG

(b) **UNCORRECTED INDICATOR READINGS**

(c) **CORRECTED INDICATOR READINGS**

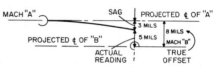

(d) CORRECTING READING FOR SAG

Indicator bar sag can be determined by firmly affixing it to a sag-free shaft mandrel, usually 4-in (10-cm) diameter or larger, dependent on length. The mandrel may be supported between lathe centers, mounted on knife edges, or held and rotated by hand. With the indicator bar positioned on top of the mandrel, the sag of the bar will be down toward the mandrel. Set indicator face to read zero at this position. By zeroing the indicator, you have errored the indicator by the amount of the sag. Rotate the mandrel 180° (indicator at bottom position). The indicator bar will sag away from the mandrel; hence the indicator reading will be twice the actual bar sag and will read negative as shown in (a).

$$\left(\frac{TIR}{2} = Sag \right)$$

Once the indicator bar sag is determined, it should be permanently stamped on the bar. This true sag must be accounted for when determining sweep readings.

To correct your sweep readings for sag, subtract twice the amount of true sag from the bottom reading (B) and correct the side readings (R & L) by subtracting the amount of the sag.

Sweep reading for an 8 mil vertical offset before making correction for a 3 mil indicator bar sag would be as shown in (b)

As shown in (c) and (d), correct sweep readings for indicator bar sag (amount of sag was 3 mils):

$$B = (+10) - (-6) \text{ or } +16$$
$$R = (\ +5) - (-3) \text{ or } +8$$
$$L = (\ +5) - (-3) \text{ or } +8$$

FIG. 19   Illustrative procedure to determine the amount of sag in an indicator bar (bracket). (Courtesy Ref. 3)

INSTALLATION SUGGESTIONS AND USE OF DIAL INDICATORS _____

1. Nonferrous shim packs should be installed under all feet of the pump and driver, in particular when installing a new pump. The amount should be 0.125 to 0.250 in (3.175 to 6.35 mm) in no more than three pieces to start, e.g., one 0.125-in (3.175-mm) and two 0.0625-in (1.59-mm) full shims of stainless steel.

2. Motors have four feet generally, and any "soft foot" should be compensated first. A soft foot is one that is shorter than the other two or three feet, a condition that puts a twist or strain in the equipment. Simply place a dial indicator stem vertically against the motor foot and release the hold-down bolts sequentially around the unit, recording and retightening at each step. If a 0.002-in (0.050-mm) spring-up occurs on three feet, for example, and 0.006 in (0.152 mm) occurs on the fourth foot, then add 0.004 in (0.10 mm) of shim to the fourth foot, eliminating the soft foot.

3. Provide low-sag tooling to reach over the coupling (coupling left in place) for reverse indicator alignment. A 0.001- to 0.0015-in (0.025- to 0.038-mm) sag is easy to accomplish on indicator reach bars.

4. Let the indicator indicate on its own bracket or bracket pin, thus preventing any poor surface condition of shaft or coupling from contributing to poor measurements.

5. Support the dial indicator weight on the motor or pump shaft, so it does not contribute to "reach bar" sag.

6. Do not overlook the fact that many times one can clamp to the shaft behind each coupling hub and obtain more span and therefore better accuracy.

7. Record all data looking the *same* way down the unit, i.e., top east, bottom, west *or* top, north bottom, south *or* top, right bottom, left. It is suggested that the driver-pump always be viewed from the driver end.

8. Turn the shafts in the direction they normally turn and approach the 90° points in a precise manner (do not back up and introduce backlash errors). Turning in the normal direction is good training because, on gear units, it reduces helix angle lift errors.

9. If the motor can be turned down from the end opposite the end from which the measurements are taken, do so. Regardless, always release the strap wrench or spanner bar before recording each ¼-point reading.

10. Obtain center zero dial indicators *or* revolution counter indicators *or* carefully note all indicator movements with a mirror to assure, for example, that 0.090 in was not really −0.010 in. The algebraic sum of horizontal and vertical readings should be near equal.

INSTRUMENTS FOR VIBRATION ANALYSIS _____

One fact about end-suction and between-bearing pumps is that external visual evidence of mechanical problems is very limited. Only three gauges for mechanical trouble exist: temperature, vibration, and sound.

It is normal for a machine to vibrate at some level; such vibrations are caused by manufacturing defects, design limits of the pump, casting irregularities, less than optimum application, and a maintenance/installation problem. When the velocity vibration level starts to increase 0.1 in/s (2.5 mm/s) zero to peak (0–P) above the "as new installed level," the vibration should be analyzed to determine the possible sources of the mechanical and/or hydraulic problem. Several mechanical and/or hydraulic problems may be producing, for instance, the 1× running speed frequency vibration. The key in using vibration to define the mechanical and/or hydraulic problems is to determine the *frequency* at which the vibration occurs. Vibration *amplitude* is also an important factor because it indicates the severity of the vibration. Field vibration data are normally a complex vibration waveform. By using a tunable analyzer, the complex vibration signal, as shown in Fig. 20, can be filtered or tuned into its basic frequency components, i.e., all complex signals are summations of the harmonics and subharmonics 1×, 0.5×, 6×, 30×, etc. By comparing these filtered components of the complexed vibration signal with an analysis chart and some common sense experience, probable causes of the vibration can be listed.

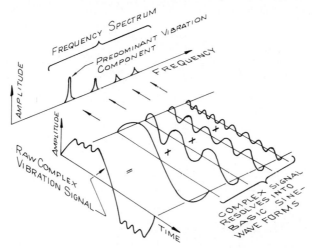

FIG. 20 Complex vibration signal resolves into sine wave spectrum.

The first step toward resolving the vibration problem is to convert the mechanical movement to an equivalent electrical signal so that it can be filtered and measured.

Often used analyzer systems are:

1. A tunable ac-battery-powered analyzer with strobe light, providing amplitude, frequency, and phase. A plotter accessory can also be attached for copies of the data.

2. A small battery-powered, internally driven, tunable analyzer with a built-in plotter using an accelerometer or velocity sensor.

3. A spectrum analyzer, ac powered, that receives the signal directly from a vibration transducer or the recorded signal from a battery-powered four-channel FM/AM cassette tape recorder.

For startup or where the problem is tougher, one can add

1. Eight-channel FM tape recorder.

2. Four-channel oscilloscope with blanking and time display.

3. Tracking filter displaying revolutions per minute, amplitude, and phase, capable of tracking runup/rundown data.

Provisions should be made for the use of all types of sensors, as there are advantages in each. As more complex problems continue to appear, tunable analyzers with a sensor are not just a requirement but a necessity in any maintenance reliability program. The choice of a displacement sensor (eddy current probe), velocity, or seismic sensor, or an accelerometer depends on the frequency range to be analyzed and the type of pumping equipment. There is no one vibration sensor for all jobs!

Of the three types of vibration measurements, acceleration and displacement are dependent on frequency and velocity is independent of frequency. Most engineers and technicians select a measurement that is independent of frequency for a datum to judge the general health of new and used pumps. With the exception of low-speed pumps and motors, 1750 rpm or less, unfiltered velocity and filtered velocity are used for most basic data. Figure 21 shows the frequency relationships (log) versus output (log) of three different measurement sensors with reference to a constant velocity of 0.3 in/s (7.6 mm/s). The figure gives an overview of present sensor limits and shows that each sensor is like a window through which portions of the frequency spectrum may be observed. The figure also shows that the accelerometer is the choice sensor at high frequency because it measures the square of the frequency. The advantage of displacement at low frequen-

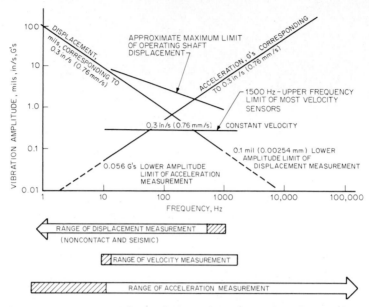

FIG. 21 Limitations on machinery vibration analysis systems and transducers. (mils × 0.0254 = mm; in/s × 25.4 = mm/s.) (Ref. 10)

cies is due to its high output; the disadvantage of displacement at high frequencies is that the output signal will disappear into the background noise of most measuring systems.

One should not confuse the measurement parameters (displacement, velocity, and acceleration) with the sensors (eddy current probes for displacement, velocity sensors, and accelerometers). The basic relationship of these measurement parameters with commonly used units are shown on a simple sine wave in Fig 22.

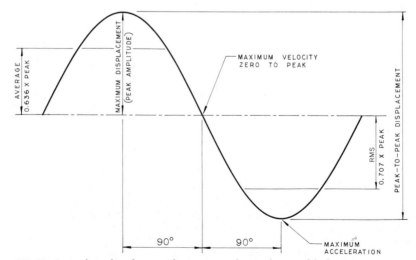

FIG. 22 Basic relationship of measured parameters with a simple sinusoidal vibration.

While the velocity sensor is not necessarily the *best* all-around type of sensor, it does have the advantage of high self-generating output (up to 1000 ft [300 m] of cable), can be mounted in any position, and is influenced only slightly (less than 5%) by transverse sensitivity (side forces). The disadvantages are that the output signal below 600 cpm is significantly nonlinear but can be corrected, the accuracy is limited at $\pm 8\%$ to 1000 Hz, and the sensor will most likely have problems in 1 to 2 years when mounted in field applications where vibration is high, especially vane passing frequencies.

The piezoelectric accelerometer is a very light and compact sensor that measures vibration using a mass mounted on a piezoelectric crystal. Its output is low and requires a charge amplifier in the lead even with very short leads. The accelerometer is small and can be mounted virtually anywhere; it has a 1 to 3% influence factor from transverse side forces. A good rule of thumb on the usable frequency range is one-fifth to one-third of the resonant frequency. The disadvantages are that the sensor is sensitive to mounting torque, although stud mount is the best method to mount accelerometers. A lot of data are produced, of which some may be the data from an excited accelerometer resonance or cable noise. An impedance matching device can be built into the accelerometer for use at temperatures below 250°F (120°C), and cable noise can be greatly reduced with the voltage and charge sensitivity greatly improved, e.g., 100 mV/g and 50 to 100 pC/g (where g = number of accelerations of gravity). For higher temperatures, the accelerometer will need a separate charge amplifier and may need heat insulation, such as MICA wafers (refer to API 678, dated 1981).

Most so-called ultrasonic analyzers use the accelerometer as a structural microphone. Many have chosen a carrier frequency in the megahertz range to improve the signal-to-noise ratio and make characteristic high-frequency patterns. Most of these systems are still in the development stage.

Techniques for Taking Data   The second most important part of a vibration analysis program is the type of data taken and the techniques used to take the data. The purpose of taking vibration data on a pump is to either perform an analysis because someone noticed a noise or increased vibration level, or as a part of a periodic preventative maintenance program. It has been proven from experience that the velocity measurement is the best method for determining acceptable levels of centrifugal pump vibration. This is not to say that displacement and acceleration are not measures of vibration severity; they are, but it is necessary to know the frequency of the vibration. Displacement is preferred by a few for frequencies less than 6000 cpm. Accelerometers matched with analyzers can be purchased with signal integration that will give reliable readings in velocity in the 3000- to 60,000-cpm range. For readings above 60,000 cpm, the sensor would generally be an accelerometer reading in g's, peak or integrated to read velocity zero to peak or 0–P.

Vibration amplitude is an important parameter because it indicates the approximate severity of the dynamic stress levels in the pump. Experience has shown that the shaft bearings and seal will probably fail in a pump with a velocity reading of 0.5 in/s (13 mm/s) 0–P. Also, catastrophic failure will probably occur when a pump is at 1 in/s (2.5 mm/s) 0–P. Pumps with velocity readings of 0.05 to 0.15 in/s (1.3 to 3.8 mm/s) 0–P, will perform well mechanically. Vibration readings are taken in the horizontal and vertical planes on the bearing housing of horizontal-shaft pumps.

To have a worthwhile maintenance reliability program with pumps, vibration readings must be recorded regularly (e.g., monthly). This can range from a trend plotting of unfiltered vibration to a full vibration analysis using a real-time analyzer to generate the frequency spectrum. A standard method used by many companies consists of taping pump vibrations with a battery-powered cassette recorder using a velocity sensor. Readings can then be processed through a real-time analyzer and recorded on an XYY' plotter. The best application of this method is during startup and repair evaluation.

As an alternate method, a spectrum analyzer/plotter that produces a spectrum on a 4 in $\times$ 6 in (10 mm $\times$ 15 mm) card with a frequency plot versus amplitude can be used. This procedure has in some installations detected and corrected 95% of the mechanical problems before failure. Experience has shown that had unfiltered displacement readings been taken, only 60 to 70% of the mechanical problems would have been observed. During these recordings, emphasis should be placed on the change in vibration levels, which is a better indication of a mechanical problem than absolute vibration.

One of the best pieces of data available for the pump's equipment file is a vibration record taken during the manufacturer's test or during water batching or commissioning. It is advisable to request a witness performance test on key or critical pumps. The purpose of this test is to assure mechanical reliability *along with* performance.

The manufacturer should be asked about the availability and type of vibration analysis equipment and sensors. Regardless of the instruments used, the vibration data sheet for the tested pump should have a sketch of where all vibration points were taken. The manufacturer should also supply a complete mechanical description of the number of impeller vanes, number of casing volute cutwaters or diffuser vanes, type of coupling, length of coupling spacer, etc.

There are several different methods for taking periodic vibration data on pumps:

1. Using a handheld battery-powered velocity probe/readout, a machinist or operator logs unfiltered readings taken at one or two points on the bearing housing. When the reading reaches 0.3 to 0.5 in/s (8 to 13 mm/s) 0–P, the pump is pulled for maintenance. Readings are usually taken every two weeks.

2. Vibration points in the vertical, horizontal, and axial directions are recorded on a tabulated chart in unfiltered and filtered velocity at the various peak amplitudes, using a battery-powered tunable analyzer with a velocity sensor.

3. Vibrations in the vertical, horizontal, and axial direction are taken at each bearing, using a velocity sensor. The signal is recorded on a tape recorder, preferably a battery-powered FM/AM cassette. These data are then processed through a real-time analyzer. A spectrum hard copy is made on an XYY' plotter of velocity versus frequency.

4. Key vibration points are fed directly from a velocity sensor or an accelerometer/charge amplifier through a long extension cable to a safe area, where a real-time analyzer processes the signal into a velocity versus frequency spectrum or a g's (acceleration) versus frequency spectrum. Hard copies for records are made on a XYY' plotter. This method requires two technicians with radios.

The most accurate are methods 3 and 4. The most costly to run in workerhours per point is method 2. The least accurate is, of course, method 1, but it is a popular screening technique.

Use of Vibration Sensors   The use of a handheld velocity sensor with an aluminum extension rod or a light-duty vise grip with the probe mounted on the top of the grip has produced some high and misleading vibration readings because of extension resonances. For instance, the vise grip should not be used because of a 5000-cpm resonance. A 9-in (23-cm) long by ⅜-in (0.95-cm) diameter extension to the velocity pickup should not be used above 16,000 cpm. The approximate axial natural frequency in cycles per minute for a rod extension from the probe, in tension and compression, can be expressed as

in USCS units
$$f_n = 188 \sqrt{\frac{AE}{WL}}$$

in SI units
$$f_n = 946 \sqrt{\frac{AE}{WL}}$$

where W = pickup weight (force), lb (N)
  L = length of rod, in (m)
  A = cross-sectional area of rod, in² (m²)
  E = modulus of elasticity of rod,° lb/in² (kPa)

Use of attachements above these listed frequencies will produce a higher amplitude.

The best and simplest method of holding a velocity probe to a pump is a two-bar magnetic holder on the end of a velocity probe. Proper cleaning and some paint removal are generally necessary for good attachment. Periodic wiping of the magnetic bar to remove iron filings is also necessary. After mounting the probe, give it a light twist and a rocking motion; if it twists easily

---

° For aluminum, $E = 10.3 \times 10^6$ lb/in² (71×10^6 kPa). For steel, $E = 30 \times 10^6$ lb/in² (207×10^6 kPa).

or rocks, change locations or reclean the surface. This location should be marked and future readings taken on the same spot; otherwise the trend plots will vary.

If there is a concern about an extension or magnetic holder resonance, test this by holding the sensor, without the extension or magnetic holder, directly to a reasonably flat spot and noting any differences. Holding the sensor on a flat spot is generally safe up to 60,000 cpm.

When measuring vibration on an electric motor, there is always the possibility of false readings at 60 and/or 120 Hz due to electrical induction by the motor. This can be checked by two methods:

1. Hold the sensor by its cord and move it toward the motor, noting any increase in amplitude.

2. Using a two-channel oscilloscope, trigger the filtered signal against line voltage. In-phase signals mean the vibration is electrically induced.

For field use, one usually does not have to contend with temperatures above 250°F (120°C) direct to the sensor; thus, accelerometers with built-in impedance matching devices can be used and 100 mV/g voltage sensitivities can be obtained and transmitted 300 ft (90 m) if necessary. If the frequency range is low, and it is for pumps, then a charge sensitivity in the order of 100 pC/g can also be obtained.

Techniques for Taking Preliminary Vibration Readings   Some key points to remember before you start your analysis of the vibration problem:

1. Do not reach a decision on what the problem is before you record and analyze the data. By deciding too quickly what the problem is, you will most likely neglect other important factors.

2. Before you take the data, take time to review maintenance logs, talk with the area mechanic and/or operator, and make notes on the following:

    a. Are there any unusual sounds (cavitation, bearings, etc.)

    b. Is there any movement in the discharge pressure gage?

    c. What is the direction of rotation?

    d. Are the flush and cooling lines lined up properly?

    e. Is there any movement in the coupling shim pack?

    f. Are there any foundation cracks?

    g. Are pipe supports functioning properly?

    h. Has a suction screen been installed?

    i. What is the magnitude of the liquid velocity in the suction line?

    j. Is the automatic oiler level adjusted correctly?

    k. Where is the pump being operated?

    l. What are the flow, suction, and discharge pressure?

    m. Is the pump's minimum flow bypass system in service?

    n. Has the process changed?

    o. What is the suction valve stem orientation?

    p. Have there been any color changes in the paint?

    q. Are there any loose parts, including the coupling guard?

    r. Determine the color and feel of the oil (if possible)

    s. What is the bearing housing temperature?

    t. How is the coupling guard attached (attachment to the bearing housing is poor practice)?

You will be surprised how much this information aids in an analysis. Example: you record a high 1× in the radial plane, low values of 2× and 5×, and several high-frequency components at about 0.15 in/s (4 mm/s) 0–P. The 1× could be a bent shaft, loose coupling, plugged impeller, bad coupling unbalance, upper and lower case halves misaligned, and a whole list of running frequencies symptoms. During your review of maintenance logs, you noted that the impeller had been replaced because there had been distillation column tray part damage. The next questions you should ask are, "Did maintenance reinstall the suction screen?" (if not, another tray part may have lodged in the impeller) and, "Was the impeller rebalanced after it was trimmed from maximum diameter?"

3. Do not try to interpret partial vibration readings for someone looking over your shoulder before you have even taken all the readings! Sit down in a quiet place with your notes on installation and maintenance, a symptoms list, and a severity chart and then make the analysis. Analysis is not a simple task, but with some experience you will build confidence and it will become second nature.

VIBRATION DIAGNOSTICS

Analysis Symptoms   The vibration severity chart and vibration identification chart are guides, but experience, a set of procedures, and study of the literature will make diagnostics easier. To be effective, one must be thoroughly familiar with the machinery's internal construction, installation, and basic control system. Both mechanical and hydraulic mechanisms can produce symptoms of vibration.

The vibration analysis symptoms, or vibration severity criteria, chart has taken on many forms since the Rathbone chart of 1939. Perhaps the most widely used symptoms chart in the turbomachinery field today is the original paper published by Sohre.[5] A condensation and revision of the original paper is shown in Table 2. Although this chart includes some symptoms that will never appear in pumps, it is one of the better references for vibration analysis. The way the table gives percentages of cases showing the symptoms for the causes listed is unique. As one learns to use the chart and modifies it with experience, a good diagnostic tool will be developed.

A good guide for unfiltered bearing cap velocity limits on field installed pumps is given in Table 3. A guide for shop testing new and rebuilt pumps is given in Ref. 6.

Comments on Table 2   In the following comments, the numbers correspond to "Cause of vibration" in Table 2.

1. Long, high-speed rotors often require field balancing at full speed to make adjustments for rotor deflection and final support conditions. Corrections can be made at balancing rings or at coupling bolts.

2. Bent rotors can sometimes be straightened by the "hot-spot" procedure, but this should be regarded as a temporary solution because bow will come back in time. Several rotor failures have resulted from this practice. If blades or disks have failed, check for corrosion fatigue, stress corrosion, resonance, off-design operation.

3. Straighten bow slowly, running on turning gear or at low speed. If rubbing occurs, trip unit immediately and keep the rotor turning 90° using a shaft wrench every 5 min until the rub clears; then resume slow run. This may take 12 to 24 h.

4. Often requires complete rework or new case, but sometimes a mild distortion corrects itself with time (requires periodic internal and external realignment). Usually caused by excessive piping forces or thermal shock.

5. Usually caused by poor mat under the foundation or thermal stress (hot spots) or unequal shrinkage. May require extensive and costly repairs.

6. Slight rub may clear, but trip the unit immediately if a high-speed rub gets worse. Turn by hand until clear.

7. Unless thrust bearing has failed, this is caused by rapid changes of load and temperature. Machine should be opened and inspected.

8. Usually caused by excessive pipe strain and/or inadequate mounting and foundation, but is sometimes caused by local heat from pipes or the sun's heating the base and foundation.

9. Most trouble is caused by poor pipe supports (should use spring hangers), improperly used

expansion joints, and poor pipe line up at casing connections Foundation settling can also cause severe strain.

10. Bearings may become distorted from heat. Make a hot check, if possible, observing contact.

11. Watch for brown discoloration, which often precedes recurring failures. This indicates very high local oil film temperatures. Check rotor for vibration. Check bearing design and hot clearances. Check condition of oil, especially viscosity.

12. Check clearances and roundness of journal, as well as contact and tight bearing fit in the case. Watch out for vibration transmission from other sources and check the frequency. May require antiwhirl bearings or tilting-shoe bearings. Check especially for resonances at whirl frequency (or multiples) in foundation and piping.

13. This can excite resonances and criticals and combinations thereof at two times running frequency. Usually difficult to field balance because, when horizontal vibration improves, vertical vibration gets worse and vice versa. It may be necessary to increase horizontal bearing support stiffness (or mass) if the problem is severe.

14. Usually the result of slugging the machine with fluid, solids built up on rotor, or off-design operation (especially surging).

15. The frequency at rotor support critical is characteristic. Disks and sleeves may have lost their interference fit by rapid temperature changes. Parts usually are not loose at standstill.

16. Often confused with oil whirl because the characteristics are essentially the same. Before suspecting any whirl, make sure everything in the bearing assembly is absolutely tight with an interference fit.

17. This should always be checked.

18. Usually involves sliding pedestals and casing feet. Check for friction, proper clearance, and piping strains.

19. To obtain frequencies, tape a microphone to the gear case and record noise on magnetic tape.

20. Loose coupling sleeves are notorious troublemakers, especially in conjunction with long, heavy spacers. Check tooth fit by placing indicators on top, then lifting by hand or a jack and noting looseness (should not be more than 1–2 mils [0.025 to 0.05 mm] at standstill, at most). Use hollow coupling spacers. Make sure coupling hubs have at least 1 mil/in (1 mm/m) interference fit on shaft. Loose hubs have caused many shaft failures and serious vibration problems.

21. Try field balancing; more viscous oil (colder); larger, longer bearings with minimum clearance and tight fit; stiffen bearing supports and other structures between bearing and ground. This is basically a design problem. It may require additional stabilizing bearings or a solid coupling. It is difficult to correct in the field. With high-speed machines, adding mass at the bearing case helps considerably.

22. These are criticals of the spacer-teeth-overhang subsystem. Often encountered with long spacers. Make sure of tight-fitting teeth with a slight interference at standstill and make the spacer as light and stiff as possible (tubular). Consider using a solid or membrane coupling if the problem is severe. Check coupling balance.

23. Overhang criticals can be exceedingly troublesome. Long overhangs shift the nodal point of the rotor deflection line (free-free mode) toward the bearing, robbing the bearing of its damping capability. This can make critical speeds so rough that it is impossible to pass through these speeds. Shorten the overhang or put in an outboard bearing for stabilization.

24. Casing resonance is also called case drumming. It can be very persistent but is sometimes harmless. The danger is that parts may come loose and fall into the machine. Also, rotor/casing interaction may be involved. Diaphragm drumming is serious, since it can cause catastrophic failure of the diaphragm.

25. Local drumming is usually harmless, but major resonances, resulting in vibration of the entire case as a unit, are potentially dangerous because of possible rubs and component failures, as well as possible excitation of other vibrations.

26. Similar problems as in 24 and 25 with the added complications of settling, cracking, warping, and misalignment. This cause may also produce piping troubles and possible case warpage. Foundation resonance is serious and greatly reduces unit reliability.

TABLE 2   Vibration Analysis Symptoms

Cause of vibration	Predominant frequencies											Vert.	Hor.
	0–40%	40–50%	50–100%	1× running frequency	2× running frequency	Higher multiples	½ running frequency	¼ running frequency	Lower multiples	Odd frequency	Very high frequency		
1. Initial unbalance	..	..	..	90	5	5	..	..	..	..	..	40	50
2. Permanent bow or lost rotor parts (vanes)	..	..	..	90	5	5	..	..	..	..	..	↓	↓
3. Temporary rotor bow	..	..	..	90	5	5	..	..	..				
4. Casing distortion Temporary		10		80	5	5	..	..	..	..			
Permanent		10		80	5	5	..	..	..	..			
5. Foundation distortion	..	20	..	50	20	..	..	..	..	10	..	↓	↓
6. Seal rub	10	10	10	20	10	10	..	..	10	10	10	30	40
7. Rotor rub, axial		20		30	10	10	..	..	10	10	10	30	40
8. Misalignment	..	..	..	40	50	10	..	..	..	..	..	20	30
9. Piping forces	..	..	..	40	50	10	..	..	..	..	..	20	30
10. Journal & bearing eccentricity	..	..	..	80	20	..	..	..	..	..	..	40	50
11. Bearing damage		20		40	20	..	..	..	..	20	..	30	40
12. Bearing & support excited vibration (oil whirls, etc.)	10	70	..	..	..	..	10	10	..	..	..	40	50
13. Unequal bearing stiffness horizontal-vertical	..	..	..	..	80	20	..	..	..	..	..	40	50
14. Thrust bearing damage	90				..	..	..	..	..	10		20	30
Insufficient tightness in assembly of:	Predominant frequency will show at lowest critical or resonant frequency						..	..	..	..			
15. Rotor (shrink fits)	40	40	10	..	..	..	..	..	..	10	..	40	50
16. Bearing liner	90		..	..	..	..	..	..	..	10	..	40	50
17. Bearing cases	90		..	..	..	..	..	..	..	10	..	40	50
18. Casing & support	50		..	..	..	..	..	..	..	50	..	40	50
19. Gear inaccuracy	..	..	..	..	..	20	..	..	..	20	60	30	50
20. Coupling inaccuracy or damage	10	20	10	20	30	10	..	..	..	..	..	30	40

TABLE 2 continued

| Direction and location of predominant amplitude | | | | | | | Amplitude response to speed variation during vibration-test runs | | | | | | | | | | | |
| | | | | | | | Coming up | | | | | | Slowing down | | | | | |
Axial	Shaft	Bearings	Casing	Foundation	Piping	Coupling	Stays same	Increases	Decreases	Peaks	Comes suddenly	Drops out suddenly	Stays same	Increases	Decreases	Comes suddenly	Drops out suddenly	Cause
10	90	10	..	..	..	..	..	100	..	Peaks at critical	..	..	..	..	100	..	..	1
			..	..	..	..	..	100	..		..	..	..	..	..	..	..	2
			..	..	..	..	30	60	5		5	..	30	5	50	5	10	3
			..	..	..	..	30	50	5		5	10	30	5	50	5	10	4
			..	..	..	..	40	60	..		..	..	40	..	60	..	..	
	40	30	10	10	10	..	20	80	..		..	..	20	..	80	..	..	5
30	80	10	10	..	..	..	10	70	..		10	10	10	..	70	10	10	6
30	70	10	20	..	..	..	10	40	10		20	20	10	..	50	20	20	7
50	80	10	10	..	..	..	20	30	10		20	20	20	..	40	20	20	8
50	80	10	10	..	..	..	20	40	..		20	20	20	..	40	20	20	9
10	90	10	..	..	..	..	40	50	10		..	..	40	10	50	..	..	10
30	70	20	10	..	..	..	10	50	10		20	10	10	10	50	10	20	11
10	50	20	20	20	..	..	..	10	..	..	90	..	..	..	10	..	90	12
10	40	30	30	..	..	..	..	40	..	50	10	..	..	..	40	..	10	13
50	60	20	20	..	..	..	20	50	10	..	10	10	20	10	50	10	10	14
10	60	20	20	..	..	..	..	..	..	..	90	10	..	..	..	10	90	15
10	80	10	10	..	..	..	..	..	..	..	90	10	..	..	..	10	90	16
10	70	20	10	..	..	..	..	..	..	..	90	10	..	..	..	10	90	17
10	50	20	30	..	..	..	..	..	..	..	90	10	..	..	..	10	90	18
20	80	10	10	..	..	..	20	20	20	20	10	10	20	20	20	10	10	19
30	70	20	..	..	..	10	10	20	..	20	Loose sleeve, friction or dirt 40 in teeth 10		10	..	20	10	40	20

TABLE 2 continued

Cause of vibration	Predominant frequencies											Vert.	Hor.
	0-40%	40-50%	50-100%	1× running frequency	2× running frequency	Higher multiples	½ running frequency	¼ running frequency	Lower multiples	Odd frequency	Very high frequency		
21. Rotor & bearing system critical	..	..	..	100	..	..	..	..	..	..	..	40	50
22. Coupling critical	..	..	..	100	Also make sure tooth fit is *tight!*							20	40
23. Overhang critical	..	..	..	100	..	..	..	..	..	..	..	40	50
Structural resonance of: 24. Casing	..	10	..	70	10	..	10	..	..	..	..	40	50
25. Supports	..	10	..	70	10	..	10	..	..	..	..	40	50
26. Foundation	..	20	..	60	10	..	10	..	..	..	..	30	40
27. Pressure pulsations	Most troublesome if combined with resonance									100	..	30	40
28. Electrically excited vibration	..	..	..	..	..			..	..	..	..	30	40
29. Vibration transmission	..	..	..	..	..			..	..	90	..	30	40
30. Valve vibration	..	..	..	..	..			..	..	..	100	30	40
Problem	The section below is meant to identify basic mechanisms												
31. Subharmonic resonance	..	Rare—Look for aerodynamic origin (seals)					100			..	..	30	30
32. Harmonic resonance	..	..	..	..	100		..	..	..	..	..	40	40
33. Friction induced whirl	80	10	10	..	..	..	..	..	..	..	..	40	50
34. Critical speed	..	..	..	100	..	..	..	..	..	..	..	40	50
35. Resonant vibration	..	..	..	100	..	..	..	..	..	..	..	40	40
36. Oil whirl	..	100	Watch for aerodynamic rotor-lift (partial admission, etc.)							..	..	40	50
37. Resonant whirl	..	100	..	..	..	..	..	..	..	..	..	40	50
38. Dry whirl	..	..	..	..	..	..	..	..	..	..	100	30	40
39. Clearance induced vibrations	10	80	10	..	..	..	..	..	..	..	..	40	50
40. Torsional resonance	..	..	..	40	20	20	..	..	..	20	..	Torsional	
41. Transient torsional	..	..	..	50	..	..	..	..	..	50	..	..	↓

Numbers indicate percent of cases showing above symptoms, for causes listed in vertical column at left.

SOURCE: From *The Practical Vibration Primer* by Charles Jackson. Copyright © 1979 by Gulf Publishing Company, Houston, Texas. Used with permission. All rights reserved.

TABLE 2 continued

Direction and location of predominant amplitude							Amplitude response to speed variation during vibration-test runs											
							Coming up						Slowing down					
Axial	Shaft	Bearings	Casing	Foundation	Piping	Coupling	Stays same	Increases	Decreases	Peaks	Comes suddenly	Drops out suddenly	Stays same	Increases	Decreases	Comes suddenly	Drops out suddenly	Cause
10	70	30	..	..	..	..	..	20	..	80	..	..	..	..	20	..	..	21
40	10	10	..	..	..	80	..	20	..	80	..	If loose	..	..	20	50 If loose		22
10	70	10	..	..	..	20	..	30	..	70	..	..	..	..	30	..	..	23
10	..	40	40	10	10	..	..	20	..	80	..	..	..	..	20	..	..	24
10	..	20	50	20	10	..	..	20	..	80	..	..	..	..	20	..	..	25
30	..	10	40	40	10	..	..	20	..	80	..	..	..	..	20	..	..	26
30	Can excite whirls or resonance	30	30	40	..		90			10%—Depending on origin of disturbance			90	10 →→→→				27
30		40	40	20	..		90	..	..		..	..	90		..	..	..	28
30		40	40	20	..		90	..	..		..	..	90		..	..	..	29
30	..	..	80	10	10	..	80	..	..	..	10	10	80		..	10	10	30
40	20 If bearing is excited	80	20	20	20	..	..	20	..	20	30	30	..	..	20	30	30	31
20	20	10	10	30	30	..	20	20	..	60	..	..	20	..	20	..	..	32
10	80	20	..	..	..	..	..	..	..	..	90	10	..	..	..	10	90	33
10	60	40	..	..	..	..	..	20	..	80	..	..	..	..	20	..	..	34
20	20	10	20	30	20	..	..	20	..	80	..	..	..	..	20	..	..	35
10	80	20	..	..	..	..	..	..	..	..	100	..	..	..	..	..	100	36
10	20	20	20	20	20	..	..	..	..	..	80	20	..	..	..	20	80	37
30	40	20	20	10	..	10	..	..	..	..	80	20	80	..	..	..	20	38
10	70	10	10	..	..	10	..	..	..	..	80	20	20	..	..	20	60	39
..	100	Lateral amplitude 40	40	..	..	10	..	20	..	30	30	20	20	..	..	20	30	40
..	Torsion 100	40	40	..	..	10	..	..	..	50	30	20	..	..	..	30	20	41

TABLE 3   Bearing Cap Data—Velocity Unfiltered

Smooth	Acceptable	Marginal	Planned shutdown for repairs	Immediate shutdown
0.1 in/s (p) and less	0.1–0.2 in/s (p)	0.2–0.3 in/s (p)	0.3–0.5 in/s (p)	0.5 in/s (p)

Note: For gearing, add 0.1 in/s to all values.
   p = peak
   mm/s = 25.4 × in/s

27. Pressure pulsations can excite other vibrations with possible serious consequences. Eliminate such vibrations using restraints, flexible pipe supports, sway braces, shock absorbers, etc., plus isolation of the foundation from piping, building, basement, and operating floor.

28. Occurs mostly at two times line frequency (7200 cpm), coming from motor and generator fields. Turn the fields off to verify the source. Usually harmless, but if the foundation or other components (rotor critical or torsional) are resonant, the vibrations may be severe. There is a risk of catastrophic failure if there is a short circuit or other upsets.

29. This can excite serious vibrations or cause bearing failures. Isolate the piping and foundation and use shock absorbers and sway braces.

30. Valve vibration is rare but sometimes very violent. Such vibrations are aerodynamically excited. Change the valve shape to reduce turbulence and increase rigidity in the valve gear. Make sure the valve cannot spin.

31. The vibration is exactly one-half, one-quarter, one-eighth of the exciting frequency. It can be excited only in nonlinear systems; therefore look for such things as looseness and aerodynamic or hydrodynamic excitations. It may involve rotor "shuttling." If so, check the seal system, thrust clearances, couplings, and rotor-stator clearance effects.

32. The vibrations are at two, three, and four times exciting frequency. The treatment is the same as for direct resonance: change the frequency and add damping.

33. If the cause is intermittent, look into temperature variations. Usually the rotor must be rebuilt, but first try to increase stator damping, add larger bearings (tilting-shoe), increase stator mass and stiffness, improve the foundation. This problem is usually caused by maloperation, such as quick temperature changes and fluid slugging. Use membrane-type coupling.

34. This is basically a design problem but is often aggravated by poor balancing and a poor foundation. Try to field-balance the rotor at operating speed, lower oil temperature, and use larger and tighter bearings.

35. Add mass or change stiffness to shift the resonant frequency. Add damping. Reduce excitation, improve system isolation. Reducing mass or stiffness can leave the amplitude the same even if resonant frequency shifts, because of stronger amplification. Check "mobility."

36. Stiffen the foundation or bearing structure. Add mass at the bearing, increase critical speed, or use tilting-shoe bearings (which is the best solution). First, check for loose fit of bearings in bearing case.

37. Same comments as 36 with additional resonance of rotor, stator, foundation, piping, or external excitation; find the resonant members and the sources of excitation. Tilting-shoe bearings are the best. Check for loose bearings.

38. Sometimes you can hear the squeal of a bearing or seal, but frequency is usually ultrasonic. Very destructive. Check for rotor vanes hitting the stator, especially if clearances are smaller than the oil film thickness plus rotor deflection while passing through the critical speed.

39. Usually accompanied by rocking motions and beating within clearances. It is serious especially in the bearing assembly. Frequencies are often below running frequency. Make sure everything is absolutely tight, with some interference. Line-in-line fits are usually not sufficient to positively prevent this type of problem.

40. This problem is very destructive and difficult to find. The symptoms are gear noise, wear on the back side of teeth, strong electrical noise or vibration, loose coupling bolts, and fretting corrosion under the coupling hubs. There is wear on both sides of coupling teeth and possibly torsional-fatigue cracks in keyway ends. The best solution is to install properly tuned torsional vibration dampers.

41. Similar to 40 but encountered only during startup and shutdown because of very strong torsional pulsations. It occurs in reciprocating machinery and synchronous motors. Check for torsional cracks.

Impeller Unbalance Impeller unbalance appears as a $1\times$ running speed frequency vibration approximately 90% of the time and may be mechanical or hydraulic in origin. Impeller mechanical unbalance is a frequent cause of mechanical seal and bearing failures. Many mechanics will never think of checking impeller balance until heavily pitted areas appear. Because of the nonhomogeneous nature of most castings, corrosion is usually more aggressive in one area of the impeller. The degree of etching and/or surface pitting is a judgmental indicator of balance change. Impeller balancing should be part of the shop repair procedure for impellers over 10 in (25 cm) at 3600 rpm. It is good practice, when balancing an impeller, to keep the impeller bore to balance mandrel fit no greater than 0.001 in (0.0254 mm) loose. Installing the impeller with the keyway up on the balance mandrel and pump shaft will help eliminate some of the unbalance due to shaft centerline shift.

EXAMPLE A 38-lb (17.2-kg), 15½-in (39.4-cm) diameter impeller operating at 3600 rpm is balanced on a machine good to 25×10^{-6} in (635×10^{-6} mm) using an expanding mandrel. The impeller is then installed on its shaft, which has a loose fit of 0.0035 in (0.0889 mm). The forces created by this shift in the center of mass is calculated as follows:

in USCS units   Unbalance   = eccentricity of impeller (in) $\times$ impeller wt. force (oz)

$$= \frac{0.0035}{2} \times 38 \times 16 = 1.064 \text{ oz} \cdot \text{in}$$

$$\text{Unbalance force} = 1.77 \left(\frac{\text{rpm}}{1000} \right)^2 \times \text{unbalance (oz} \cdot \text{in)}$$

$$= 1.77 \left(\frac{3600}{1000} \right)^2 \times 1.064 = 24.4 \text{ lb}$$

in SI units   Unbalance   = eccentricity of impeller (mm) $\times$ impeller wt. mass (g)

$$= \frac{0.0889}{2} \times 17.2 \times 1000 = 765 \text{ g} \cdot \text{mm}$$

$$\text{Unbalance force} = 0.01094 \left(\frac{\text{rpm}}{1000} \right)^2 \times \text{unbalance (g} \cdot \text{mm)}$$

$$= 0.01094 \left(\frac{3600}{1000} \right)^2 \times 765 = 108.5 \text{ N}$$

The example also points out that the unbalance force generated by loose fit impellers with keyways mounted in one plane could be quite high. This force could be minimized by staggering keyways or randomly orienting the impellers on the balance mandrel. Shifting of shrink-fitted, well-balanced impellers on multistage and high-speed double-suction pumps after a period of operation can result in unbalance. The shifting of the impeller is due to the relaxation of residual stresses that built up as the impeller cooled and contracted around the shaft. Shaft vibration and flexing tend to relieve the residual stress and cause the impeller to cock or bow the shaft from the original balance centerline.

Standards should be referred to for balancing pumps and their drivers. When balancing, consideration must be given to the need for balancing at rated speed in order to properly evaluate the importance of shaft deflection due to modal components of unbalance. See Refs. 12 and 13.

Hydraulic Unbalance Uneven flow distribution entering the impeller can cause a 1× running speed frequency type of vibration. The unbalance occurs because the flow is not equal in all vane passages. An example of this is a double-suction impeller with a short, straight run to the pump and an elbow in the horizontal plane. Flow from the elbow does not have time to straighten and therefore enters both sides of the impeller unequally. A similar condition results if suction is taken from a tee off the main header. Unequal and unsteady flow into the impeller may cause axial thrust and high axial vibrations. Thus, it is good design practice to install elbows vertically in double-suction pumps. (See Chap. 10.)

In double-suction pumps, the nonsymmetrical positioning of the impeller or the offset of the upper case half to the lower case half will cause a 1× unbalance due to nonsymmetrical flow.

Recirculation forces and pulsation recirculation within a pump (which can occur when flow is less than design) may manifest themselves in the form of a noise and/or vibration with random frequencies, along with pressure pulsation that may be seen on a pressure gage. Recirculation may also appear in the piping system as vibration and noise. Increased NPSHA has helped in a few cases, especially if the recirculation is mainly on the suction side of the impeller. Once a pump has a recirculation problem in a given system and the flow cannot be increased using a bypass, little can be done to the pump itself unless the system characteristics allow an impeller change. This is discussed in more detail in Subsecs. 2.3.1, 2.3.2, and 2.3.4.

Antifriction Bearings Vibrations generated by ball bearings cover a wide range of high frequencies that are not necessarily a multiple of the shaft running speed. The frequency readings obtained during analysis are somewhat unsteady because of the resolution of the filters in a hand-tuned analyzer. The amplitude reading may also be somewhat unsteady.

Experience has shown that hand-tuned field analyzers tend to show the last stage of the bearing failure. Monitoring of stress waves and/or shock pulses (impact energy) on the pump bearing housing will show failure trends which will generally precede an increase in the detectable level of mechanical vibration. This method of failure detection is called acoustic high-frequency monitoring or incipient-failure detection (IFD). A comparison between conventional methods and the acoustic high-frequency method is shown in Fig. 23.

Accurate analysis of pump bearings and other machinery can also be made using a velocity sensor good to 1500 Hz or an accelerometer. Data should be recorded and processed through a real-time analyzer with at least 256-line resolution capability and a band selectable analysis option. Analysis of antifriction bearings using a real-time analyzer and equations for calculating frequencies generated by defective bearings can be found in Ref. 7.

Accurate and extended analysis of pump bearing vibration is generally not needed. A majority of bearing problems are currently identified by an acoustic noise during operation. What is needed by maintenance personnel is a quick and reliable method to monitor bearings and determine when

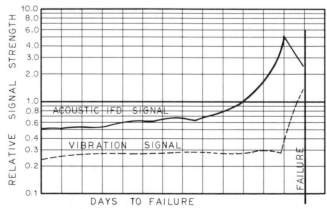

FIG. 23 Relative signal strength versus days to failure for acoustic IFD and conventional vibration monitoring methods.

a bearing is failing and the rate of bearing deterioration. At present, there are some expensive instruments that can be purchased, but none of them meet maintenance requirements.

Currently, the best method for reducing bearings analysis is prevention of the failure. Most antifriction pump bearings fail for one of several reasons: (1) water gets into the oil, (2) automatic oiler is not adjusted properly (this cause is most often overlooked and will continue to produce a short life to failure cycle), (3) product gets into the oil, (4) acidic vapors condense and break down the oil, (5) mounting techniques or fits are improper, (6) new bearing is defective. The solution to high-humidity problems and problems with acidic units is the use of an oil mist lubrication system. If this cannot be economically installed, then an aggressive preventative maintenance program on a monthly to every-other-month basis is required.

Baseplates   With the change from low rotating speeds and cast iron baseplates to the less rigid fabricated steel baseplates and higher rotating speeds, higher operating temperatures, and larger impellers have greatly increased the probability of baseplate vibrations, distortion, and a decreased stiffness for rotor dynamics. Reference 6 has increased coverage for piping loads and the option of a heavy-duty baseplate over the proposed standard baseplate. The reference specifies a standard baseplate and pedestal support that are twice as rigid as specified in the fifth edition of the reference standards.

Baseplate vibration problems can be resolved at the design stage or during construction. Engineering specifications should call for leveling screws, grout filling holes [4 in (10 cm) minimum, 6 in (15 cm) preferred for each bulkhead section], venting holes [4½ in (11 cm) holes to each bulkhead section], and corrosion protection. A check of the outline dimension approval drawing should also be made for proper grout hole placement (bulkhead and cross bracing must be shown on the drawing). Construction specifications should call for proper baseplate preparation before grouting.[8] API pumps should have epoxy grout bonding the baseplate to the concrete foundation.

After proper cure, the baseplate should be tapped for voids, especially between and under the pump centerline supports.

EXAMPLE   A high vertical vibration occurred on the coupling end bearing at vane passing frequency on a multistage volute pump. The vertical vibration was 2.5 times the horizontal; thus, a check of the bearing pedestal and baseplate was in order. The hammer test on the pan of the baseplate under the bearing pedestal showed complete lack of grout. Vertical vibration on the pan was 1.8 times the horizontal. Regrouting eliminated the pan vibration and reduced the vertical vibration to one-half of the horizontal. Lack of proper stiffness from the baseplate can and will lower some pump critical speeds into the operating range.

When a complete analysis is being done on a problem pump, take several pedestal readings (top, middle, and bottom on the side and end) and several readings on the pan of the baseplate. The middle, bottom, and pan readings should show good attenuation.

REFERENCES

1. IRD Mechanalysis: "Methods of Vibration Analysis," technical paper no. 104-1975, IRD, Cleveland.

2. Garguilo, E. P. Jr.: "A Simple Way To Estimate Bearing Stiffness," *Machine Design*, July 24, 1980, p. 107.

3. Dodd, V. R.: *Total Alignment*, Petroleum Publishing, Tulsa, 1975.

4. "Alignment Tutorium," *Proc. Tex. A&M Turbomachinery Sym.*, December 1980.

5. Sohre, J. S.: "Operating Problems with High Speed Turbomachinery: Causes and Corrections," ASME Petroleum Mechanical Engineering Conference, Dallas, September 1968.

6. American Petroleum Institute: Centrifugal Pumps for General Refinery Services, API Standard 610, 6th ed., Washington, D.C., 1981.

7. Taylor, J. I.: "An Update of Determination of Anti-Friction Bearing Condition by Spectral

Analysis," Vibration Institute, Machinery Vibrations Monitoring Analysis Seminar, New Orleans, April 1981.

8. Murray, M. G. Jr.: "Better Pump Grouting," *Hydrocarbon Processing*, February 1974.

9. Jackson, C.: *A Practical Vibration Primer*, Gulf Publishing, Houston, 1979.

10. Mitchell, J. S.: *An Introduction to Machinery Analysis and Monitoring*, 1st ed., PennWell Books, Tulsa, 1981.

11. Messina, J. P., and S. P. D'Alessio: "Align Pump Drives Faster by Well-Planned Procedure," *Power*, March 1983, p. 107.

12. American National Standard: Procedures for Balancing Flexible Rotors, ANSI S2.42, New York, 1982.

13. American National Standard: Balance Quality of Rotating Rigid Bodies, ANSI S2.19, New York, 1975.

FURTHER READING

Bloch, H. P.: "Improve Safety and Reliability of Pumps and Drivers," *Hydrocarbon Processing*, May 1977, p. 213.

Bussemaker, E. J.: "Design Aspects of Baseplates for Oil and Petrochemical Industry Pumps," IMechE Paper C45/81, Netherlands, 1981, p. 135 (English ed.).

Jackson, C.: "Alignment of Rotating Equipment," NPRA, MC-74-7, Houston, 1974.

Jackson, C.: "Alignment of Pumps," Centrifugal Pump Engineering Seminar, ASME South Texas Section Professional Development, Sec. 9, Houston, 1979.

Sprinker, E. K., and F. M. Patterson: "Experimental Investigation of Critical Submergence for Vortexing in a Vertical Cylindrical Tank," ASME Paper 69-FE-49, June 1969.

Von Nimitz, W. W.: "Dynamic Design Criteria for Reciprocating Compressor and Pump Installations," presented at the 27th Annual Petroleum Mechanical Engineering Conference, New Orleans, Sept. 19, 1972.

Wachel, J. C., and C. L. Bates: "Escape Piping Vibrations While Designing," *Hydrocarbon Processing*, October 1976, p. 152.

2.3.4
CENTRIFUGAL PUMP MINIMUM FLOW CONTROL SYSTEMS

WILLIAM O'KEEFE

Although a constant-speed centrifugal pump will operate over a wide range of capacities, it will encounter difficulty at low flow. In general, low-flow problems are worse for large high-energy pumps, for pumps handling hot or abrasives-laden liquids, for pumps designed for high efficiency at best efficiency point (BEP), and for pumps for low net positive suction head (NPSH).

SOURCES OF TROUBLE AT REDUCED FLOW

The sources of the pump's distress are fourfold—thermal, hydraulic, mechanical, and abrasive wear.

Thermal   Inescapable energy conversion loss in the pump warms the liquid. See Subsec. 2.3.1 and Chap. 12 for discussion of the thermal effects.

Hydraulic   When flow decreases far enough, the impeller encounters suction or discharge recirculation, or perhaps both. Flashing, cavitation, and shock occur, often with vibration and serious damage. See Subsecs. 2.3.1 and 2.3.2 for discussion of this.

Mechanical   Both constant and fluctuating loads in the radial and axial directions increase as pump capacity falls. Bearing damage, shaft and impeller breakage, and rubbing wear on casing, impeller, and wear rings can occur. See Subsecs. 2.3.2 and 2.3.3 for discussion of this.
   Axial-flow and mixed-flow pumps with high specific speed give comparatively higher head and take comparatively more power at low flow. A bypass system may be necessary not only to reduce stress but also to prevent motor overload. See Subsec. 2.3.1 and Sec. 8.1 for discussion of this.

Abrasive Wear   Liquids containing a large amount of abrasive particles, such as sand or ash, must flow continuously through the pump. If flow decreases, the particles can circulate inside the pump passages and quickly erode the impeller, casing, and even wear rings and shaft.

METHODS OF PROTECTION

Many systems of piping, valves, and controls are available to prevent damage to centrifugal pumps at low flow. Cost, reliability, and maintainability are criteria for the more highly developed systems. Every pump with a driver of 10 hp (7 kW) or larger should be examined to determine whether low-flow protection is needed. In some pump installations, no formal protection exists. For these pumps, heat loss from the case and perhaps natural circulation in sections of the inlet and discharge lines remove enough heat to prevent harm even at shutoff. Many small low-energy, low-temperature, single-stage pumps can withstand hydraulic effects for long periods of low flow or shutoff.

Larger pumps, with less cooling surface per unit of casing volume, and high-head and/or high-temperature pumps with high power input need specific protection against damage from low-flow operation. Protection methods fall into two classes: pump shutoff and bypass for flow back to an upstream receiver or to waste.

Shutoff Systems

PUMP MOTOR SHUTOFF AT PREDETERMINED DISCHARGE HEAD   This requires only an accurate pressure switch, set for a discharge pressure corresponding to a total head at the pump's minimum required capacity. (For conservative results, calculate on the basis of maximum suction pressure.) The method can serve a pump discharging through a float valve to a tank. Restart can be manual, time-delay, or upon tank level decrease. The pump characteristic should rise continuously to shutoff. Control circuitry may require delays and rejection of spurious pressure surges. The method will not respond to an increase in inlet temperature, and therefore calculations must take maximum design temperature into account if that condition determines the minimum flow required.

PUMP MOTOR SHUTOFF AT PREDETERMINED TEMPERATURE OR DIFFERENTIAL TEMPERATURE   A temperature sensor mounted on the pump casing or in the discharge and close to the pump can signal for a shutdown of all or some of the operating pumps on either a high-temperature sensing or a high-differential-temperature sensing between inlet and discharge. This method can protect against large temperature rises (occurring at perhaps 5% or less of capacity) but is less sensitive to smaller rises at higher percentages of capacity, where adverse hydraulic effects begin for large and high-power pumps.

PUMP MOTOR SHUTOFF AT PREDETERMINED CAPACITY   A low differential pressure across a metering orifice on the discharge causes a signal for shutoff of some or all of the operating pumps. The method requires a primary meter element, piping, and a switch to respond to low differential pressure. The method will not respond to an increase in inlet temperature. Flow can also be sensed by monitoring motor current.

Bypass Systems

In many pump systems, the pump cannot stop but must continue to deliver fractional flow to the system. In these, a branch line on the discharge takes away enough flow to assure the required minimum through the pump. The branch line ordinarily bypasses the liquid back upstream far enough to permit cooling before re-entry to the pump. *Bypass* is the term covering several variations.

CONTINUOUS BYPASS OF MINIMUM FLOW   This is a simple arrangement, with only a pipe and a means for dissipating head to control bypass flow. A partly open globe valve or single orifice can establish bypass flow rate for small pumps and low pressure drops. Series orifices (Fig. 1) are necessary for high pressure drops. The loss of energy in this system is high for any but small pumps with low-percentage bypass flow.

The bypass is sized to maintain required minimum flow at shutoff conditions, dissipating the pump total head, when the pump is operating and the system is completely shut off.

ON-OFF BYPASS OF MINIMUM FLOW   Flow reduction to the selected minimum causes a valve to open in the bypass line, allowing additional flow to maintain the total pump minimum flow required.

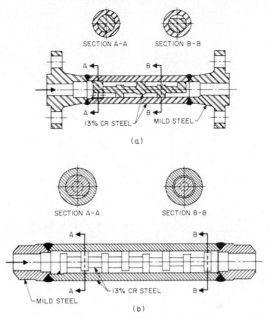

FIG. 1 Compact orifice assemblies break down high pressure: (*a*) orifice assembly for 2000 psig, (*b*) orifice assembly for 3500 psig. (Worthington Pump)

A flow sensor (primary metering element or metering orifice) supplies information to a transmitter and control to signal the valve to open. Orifices downstream of the bypass valve can produce all or part of the pressure drop required to protect the valve. If the pump characteristic rises reasonably steeply to shutoff, the signal for opening of the bypass can be discharge pressure or total head of the pump.

If the minimum flow is a large fraction (say 30% or more) of full capacity, on-off bypass will give large load swings or process changes as flow suddenly doubles and halves near the minimum flow point. A deadband in capacity is often necessary to prevent hunting when flow is near the minimum (Fig. 2).

MODULATED BYPASS   Flow reduction to the selected minimum causes modulated opening of a control valve in the bypass line. As flow demand continues to decrease, the bypass valve opens wider to maintain minimum flow through the pump. This method has become customary on large high-head pumps to reduce harmful forces associated with a sudden change in flow and to conserve power, especially if bypass is used often (Fig. 2).

FACTORS IN DESIGN OF A MINIMUM FLOW CONTROL SYSTEM _____

Pump Size   Higher capacity, brake power, specific speed, and suction specific speed all tend to make a minimum flow control system difficult to design and costly to build and operate.

Discharge Pressure   High discharge pressures mean high head loss in bypass lines. Liquids that can flash and cavitate demand special precautions in valving, orifices, and piping.

Available Heat Sink   Bypass flow must go far enough upstream to prevent progressive temperature buildup or flow disturbance in the pump suction. This may mean a simple discharge

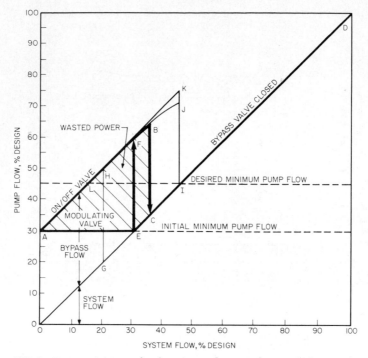

FIG. 2 Various operating modes of a minimum flow control system. In this example, the system is sized to bypass 30% of pump design flow, which was the initial required minimum pump flow. *ABCD:* increasing system flow. On-off valve is open on startup and closes at 35% of system flow to prevent hunting. *DEFA:* decreasing system flow. On-off valve opens at 30% of system flow. *AED:* modulating valve used in place of on-off valve. *AFBCEA:* pump flow is in excess of mimimum required until system flow is shut off. Therefore on-off valve wastes more power than modulating valve. *AHGD:* if operation of on-off valve is set below system flow (required minimum pump flow), pump is not protected during system flow range *E* to *G*. *AJID:* an undersized minimum flow bypass can be partially upgraded if it is desired later to increase pump flow without increasing the capacity of the bypass. Resetting the bypass to open at the higher desired flow is effective only if system flow is in the range *L* and greater. *AK:* theoretical pump flow (assuming pump head curve does not decrease with flow), in contrast to curve *AJ,* which shows the effects of decreasing bypass flow with decreasing pump head. (Courtesy Public Service Electric and Gas)

back to an uninsulated inlet line or discharge to a receiving tank or cooler with enough area and enough inflow of cool liquid to handle the thermal load. Bypass flow can discharge into a deaerator storage tank, a condenser, a flash tank, or a cooling pond. Altitude, distance, and pressure inside the receiving tank are also factors, as is the fact that the interior must be available for inspection and for repair of spray or distribution pipes and orifices.

Design of Pump   A pump's design and materials of construction often affect the minimum allowable percentage of flow. With thermal effects, pumps vary in the length of time that they will tolerate shutoff or low flow. This may be important in design of the bypass system's valves and control. Some large boiler-feed pumps can tolerate zero flow for several seconds in spite of very high power input.

Hydraulic effects at low flow are most apparent in high-energy pumps. The pump manufacturer should state the minimum hydraulically acceptable flow for a given pump. With axial-flow pumps, the power and head increase as flow falls off may be a factor in selecting percentage bypass.

Liquid Pumped   Liquids that flash and cavitate readily call for a high percentage of bypass flow. Liquids near the boiling point and high-pressure feedwater are examples. Abrasives in the liquid may require more bypass than would be needed for thermal reasons alone.

Energy Costs   A high cost for pump drive power suggests close attention to bypass power loss. This loss warrants attempts to reduce flow in all types of bypass systems by substituting on-off and modulating systems for continuous-flow systems. Balancing higher initial cost and perhaps higher maintenance charges against lower energy costs over lifetime operation is a large part of the study.

ESTIMATING MINIMUM PUMP FLOW

The bypass system designer must know the minimum allowable flow through a pump before designing the system. A minimum of 5% of design capacity in a 50-gpm (3.1–l/s) pump is obviously less demanding than a 70% value for a 50,000-gpm (3150 l/s) pump. Several considerations enter choice of minimum pump flow.

Thermal Considerations   See Subsec. 2.3.1 and Chap. 12 for a method of calculating temperature rise and for a discussion of judgmental factors that must be considered in deciding on minimum bypass flow rate.

For pumps continuously recirculating hot water, the bypass flow percentage can be high if the inlet water temperature is close to the flashing point. Calculate the amount of bypass on the basis of what is necessary to prevent the temperature from decreasing the NPSH available to less than that required by the pump.

Hydraulic Considerations   The minimum bypass flow necessary to prevent hydraulic trouble depends on the liquid, pump type, and fundamental parameters of design. Manufacturers are often pressured by competitive factors to show high full-load efficiency, good suction performance, and low minimum or bypass flow. A very low bypass requirement is a competitive help because piping and valves will be smaller and energy losses and flow effects in the receiver will be lower. If, however, the pump cannot run for more than a few minutes at an excessively low flow, the owner may have to pay for decreased system flexibility, heavy repairs, and eventual rework of the entire bypass system.

For bypass flow rates required to prevent hydraulic troubles, see Subsec. 2.3.1, where guidelines based on point of occurrence of suction recirculation in the pump are used. The charts there cover various configurations having specific speeds to 10,000 (6100 in units of rpm, l/s, and m).

Pumps requiring very high minimum flow are often paralleled for operating reasons. If total flow decreases substantially, shutting off one or more pumps will restore flow in the rest, so that the bypass systems may not have to handle full minimum flow for all pumps.

Mechanical Considerations   It is necessary to know how total head, radial thrust, axial thrust, and brake power vary with capacity before deciding on minimum allowable flow. Bearing capacity, motor rating, and stresses in drive and driven components are important influences.

Abrasive Wear Considerations   Where the purpose of bypass is the protection of the pump against abrasives in the liquid, reliance must be on the manufacturer's or user's experience with almost identical pumps and liquid/solid mixtures. Heavy wear can occur at flows as high as 85% of best efficiency point capacity.[2]

BYPASS SYSTEMS

For all but the simplest of low-head, low-percentage bypass piping systems, careful tracing of liquid and vapor pressures and flow effects through all stages of the piping and at all flow conditions is advisable.

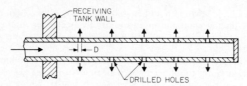

FIG. 3 Simple spray pipe has orifices to distribute flashing liquid and produce back pressure.

Orifices Simple, easily replaced orifices that reduce the pressure are an effective way to reduce bypass head. Several stages may be necessary, however, to break down high-pressure drops without flashing (Fig. 1). For calculation of flow through standard-shaped orifices, see Sec. 8.1 and 8.2.

Coefficients of discharge for oddly shaped multistage orifices are difficult to calculate. Manufacturers of specialties can furnish curves of delivery as a function of pressure, however.

Orifices to maintain back pressure on the main pressure-reducing orifice or control valve may be simply holes in water-distribution headers inside storage tanks or condensers (Fig. 3). These spray pipes have much less energy per jet than the main pressure-reducing orifice. In another version of this spray header, flow from pairs of jets intersects and dissipates jet energy before the jet strikes a wall or vessel internals (Fig. 4). Flow distribution will not be uniform if the sizing of the spray header results in high kinetic energy and momentum force within the inlet fluid compared to friction losses along the header. For these conditions, flow through the holes will increase toward the end of the pipe. A usual rule of thumb is to make the total cross-sectional area of all the orifices in the distribution header no greater than one-quarter of the area of the pipe cross-section.

The practical limit on maximum jet velocity through an antiflash spray header is 100 ft/s (30 m/s). This limits the back pressure. The pressure differential ΔP in lb/ft^2 (N/m^2) across a spray header containing N holes is

$$\Delta P = \frac{\gamma}{2g}\left(\frac{V}{C_D}\right)^2$$

where γ = specific weight of liquid, lb/ft^3 (N/m^3)
$\quad V$ = jet velocity, ft/s (m/s)
$\quad g$ = acceleration due to gravity, ft/s^2 (m/s^2)
$\quad C_D$ = coefficient of discharge

The number of holes drilled in the spray header is

$$N = \frac{4Q}{\pi V D_I^2}$$

where Q = design flow, ft^3/s (m^3/s)
$\quad V$ = jet velocity, ft/s (m/s)
$\quad D_I$ = inner diameter of holes, ft (m)

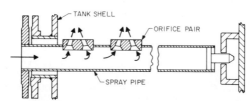

FIG. 4 Advanced design of spray pipe has intersecting jets and is readily removable.

A coefficient of discharge C_D of approximately 0.6 is accurate enough even for flashing flow (as for single-phase liquid), provided the liquid velocity in holes is larger than in the pipe.

A fixed orifice delivers a given flow at a given differential pressure. If upstream head drops and flow decreases, perhaps because of control valve closure in the bypass line, the pressure drop at the downstream fixed orifice may fall enough to permit flashing and cavitation in the control valve. The addition of a variable orifice (Fig. 5) at the end of the bypass pipe can help maintain pressure on the liquid at all flows. The spring-loaded piston skirt in this orifice reduces the port area enough to create back pressure in the bypass line and, under equilibrium conditions, to let the control valve open slightly wider, reducing wear on it. An eroded or damaged variable orifice can be easily removed and replaced.

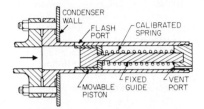

FIG. 5   Variable orifice prevents flashing upstream and adjusts for flow changes. (Yarway Corp)

All orifices in bypass service should be in accessible locations, with provision for inspection and change. Isolating valves and erosion-resistant inserts are common accessories.

Valves   For most pumps, an actuated valve is needed to control or shut off bypass line flow. See Chap. 7 for general information on valves.

For cold water at low pressure, a simple power-actuated globe-type valve is often adequate. In modulating bypass systems, the valve must resist throttling damage, particularly if the water is hot. Staging the pressure drop in the valve is the most common way to reduce or eliminate flashing and cavitation damage to valve trim or body. Elements in the valve may also subdivide and then recombine flow.

Discharge Valve Position   When a flow meter cannot be used, or is not available, to actuate the bypass control valve, the position of the discharge valve can operate the bypass. If pump flow control is solely by discharge valve throttling, then valve position or signal to position can actuate the bypass valve controller. If discharge valve position is a function of pump flow, then at a selected position the bypass valve can open or begin to open and modulate as the discharge valve closes further. Closure in the reverse direction of discharge valve position is also a simple arrangement.

This system will not operate as precisely as a system monitored by a flow meter, and it depends on knowledge of valve characteristics and system head.

Piping and Fittings   Pipe material like that used for the main discharge is adequate for nearly all of the bypass line. Near the orifice and control valve outlets, however, heavier wall or higher chrome content will lengthen life. Welded piping is common for high pressures. To prevent erosion of pipe fittings, they should not immediately follow an orifice.

Flow Meters   Flow rate is the variable most often sensed to obtain a signal for opening and closing the valve. The meter may have any type of primary element that will produce a clear signal at the flow rate requiring bypass and at the flow rate at which bypass is to cease. In addition, if the system is a modulating one, a flow meter before the pump, or after it but before the bypass branchoff, is needed to signal flow rate of bypass.

A simple orifice meter is common and may serve other control and metering purposes. If the meter is for the bypass system alone, then the permanent head loss at full capacity may be worth consideration in evaluation.

Flow meter location can be upstream or downstream of the pump, depending on the bypass system. Located upstream of the pump, the meter is under lower internal pressure but needs larger flanges. Its orifice, smaller in percentage of pipe area, may cause inlet flow problems for the pump or may require an inconvenient piping run to assure accurate metering. Located downstream between pump and bypass branchoff, the primary element often has smaller flanges or end connections, but internal pressure is higher.

In any location upstream of the bypass branchoff, the primary element will measure total flow and therefore indicate actual flow through the pump, which is necessary for modulation bypass control. The flow to boiler or process, however, will not be measured, and so another meter will be needed downstream of the bypass branchoff. For economic reasons, often only one meter is installed if the selected bypass system permits.

Controls   Controls range in complexity from a simple pressure switch that stops a pump drive motor through relays to modulating systems that meter flow to maintain a selected minimum flow through the pump. The control system may have to provide a deadband near the opening point and be sufficiently stable to hold bypass flow nearly constant in spite of erratic flow during upsets and startup. For example, when the primary flow-measuring element is upstream of the bypass branchoff, sufficient decrease in flow will cause the bypass control to open in an on-off bypass system. The main control valve, farther downstream, will then also open to maintain flow to the boiler or process. The primary element will sense the added flow and, in a simple bypass control system, close the bypass valve, initiating hunting in flow and valve action. To prevent this, the set point for closing must be above at least twice the minimum total flow. If minimum flow is in the 30 to 50% range, this wastes energy during bypass. This is avoided if a modulating system (described below) is installed.

EXAMPLES OF SYSTEMS

Continuous Bypass System   Figure 6 illustrates a simple system, with a bypass line branching off the pump discharge upstream of the main line check valve and containing a fixed orifice dimensioned by analysis to provide minimum required pump flow. The bypass receiving tank must be at a lower pressure than pump discharge and may be the pump suction source or some other vessel that will return the liquid to the pump (if it is not to be wasted). Locating the bypass branchoff before the discharge check valve as shown keeps backflow from process or flow from a parallel operating pump from going back to the receiving tank or back through the pump during a pump shutdown.

The choice of size for the bypass pipe depends on flow, equivalent line length, and how close the liquid temperature is to the boiling point. If the pressure breakdown after the orifice results in flashing flow, the orifice should discharge into the larger receiving tank. In that case, eliminate the shutoff valve downstream of the orifice because the valve is likely to erode. Waterhammer and erosion are not problems in well-designed continuous bypass systems. Paralleled pumps may discharge into the main line downstream of the bypass.

On-Off Bypass of Minimum Flow   In Fig. 7, a bypass line runs to a below-surface spray pipe in an overhead pressurized receiver. For low bypass flows and low heads of cool water, the on-off valve may be single-stage. Control is simple, with a deadband of 2 to 5% of design-point delivery (Fig. 2). The meter, downstream of the bypass branchoff, serves system control functions.

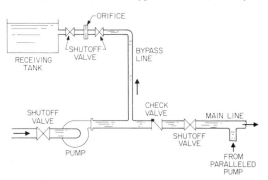

FIG. 6   Continuous bypass is simple and reliable protection.

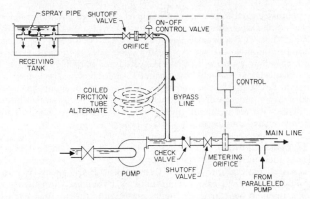

FIG. 7 Energy-conserving on-off bypass has several design variations.

Bypass piping is full of water, eliminating the danger of a sudden impact of a water slug on the orifice.

Should the bypass piping discharge above the water level in the receiving tank, a check valve in this line is advisable. Without a check valve, the water in the bypass piping will drain back through an idle pump to the level in the receiving tank, if this is the suction source. When the pump is started, water slugging could occur in the empty portion of the pipe and fittings.

With high-pressure hot water, the design is more exacting, for the reasons given for the continuous bypass systems of Fig. 6. The control valve may have stages, but a multistage orifice of the Fig. 1 type, in conjunction with a single-stage control valve, can be designed to take the pressure drop.

When there is danger that the bypass water will be close to flashing downstream of the orifice, a spray pipe (Figs. 3 and 4) or variable orifice (Fig. 5) located at the end of the bypass line is advisable. Size the antiflash orifices to provide back pressure to suppress cavitation at the main pressure-reducing orifice. Take care to direct the high-velocity jets from the flashing sprays away from any internal tubes in the receiving tank and rather onto structural members. If the bypass discharges above the receiving tank and there is no check valve, as noted above, water hammer forces will be suppressed somewhat at the bypass discharge if a variable orifice is selected for installation there.

When an internal antiflash device is not desirable, a simple orifice plate or nipple-type orifice at the end of the bypass pipe and just outside the receiving tank wall will serve. Whenever possible, as long as installation limits permit, locate the pressure-reducing valve at the receiver wall if flashing flow is expected but there is no need for an antiflash orifice.

For systems having very high pressure drop, consideration has been given to sizing the bypass pipe to provide most of the head loss. This friction tube system, as it is called, produces smooth throttling. The control valve and, if required, an antiflash orifice located near the receiver will keep the liquid in a liquid state. The pressure breakdown action, over a distance, involves less shock than when concentrated in a valve or orifice. With this system, an inexpensive on-off control valve can be used.

Pipe inside diameter and length are the variables in the design of a friction tube. Pipe velocities for water of good quality, such as boiler feedwater, can be as high as 150 ft/s (46 m/s). For high velocities, entrance and exit losses and velocity head effects must be part of the design data. Chrome-moly alloy pipe, such as ASTM-A335, P5, or better, is recommended. Coiling the pipe (Fig. 7) to increase flow length is a design alternative. The bypass flow must arrive at the control valve at a pressure and temperature within the valve's capabilities as specified by the manufacturer.

A degree of bypass flow control would be possible in a system with two or more friction tubes in parallel, each with its own on-off valve. For example, one branch could open at a selected minimum flow of 25% capacity and bypass 12%, protecting the pump down to 13% main line flow. Below that, opening a second bypass branch, with 13% capacity, would give protection down to zero main flow.

The bypass is usually sized to provide the required minimum flow rate to protect the pump under pressure conditions when all system flow is shut off. As can be seen from Fig. 2, to provide pump protection at all flows, the bypass valve must be set to open when the system flow is equal to required pump minimum flow. If set to open at a lower flow, pump will be underprotected. If set to open at a higher flow, pump will be overprotected and pumping power will be wasted.

If the bypass flow originally selected is later found to be too low, the bypass valve can be set to open at a system flow equal to the higher required minimum pump flow. This resetting is effective, provided minimum system flow is maintained at not less than the difference between the initial and the new desired minimum pump flow. If prolonged system flows are less than the above, a larger flow bypass is required.

Modulated Bypass System

In the system of Fig. 8, for example, the bypass runs to a condenser which is under vacuum. The valve, whose number of stages depends on the maximum pressure drop across it, has high-pressure hot water upstream and relatively cool vapor downstream when the pump delivery is above minimum flow. When flow decreases to the minimum, the valve begins to open on signal from the meter upstream of the pump or bypass branchoff.

In most cases, the water left in the valve body will have cooled somewhat, so that the first release of water through the valve's critical last stage will be less likely to flash under the high initial pressure drop. As soon as the flow through the first stages increases to the point where the stages can participate in pressure breakdown, the last stage's pressure drop becomes much less, so that the stage can better withstand the hot water.

Downstream of the bypass control valve, the pipe rapidly fills with water and flashed vapor. The throttling orifice retards flow, building pressure in the pipe and suppressing vapor formation. Farther on is the spray pipe in the condenser, distributing mixed water and vapor.

This system can be designed to provide suitable pressure drops at valve and orifices and along the piping, so that at full bypass flow there will be no flashing until the final spray pipe orifices. At low bypass flow, however, the pressures and pressure drop between the control valve and throttling orifice are so low that flashing and cavitation will occur at either the valve or the throttling orifice. A bypass valve location near the condenser, if possible, or a variable antiflash orifice (Fig. 5) helps suppress low-flow flashing and cavitation.

Design the bypass system so that, at complete system shutoff, the sum of the pressure drop through the design-maximum open position of the modulating valve plus the pressure drop through the main orifice and the remainder of the bypass system permits the required minimum pump flow to pass. When main flow begins and increases, the control valve must reduce the bypass flow by adding resistance to the changing orifice and bypass piping pressure drop, so that total system head remains equal to that required by the pump at minimum flow.

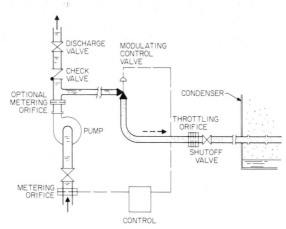

FIG. 8 Modulated bypass is needed when energy loss is to be minimized.

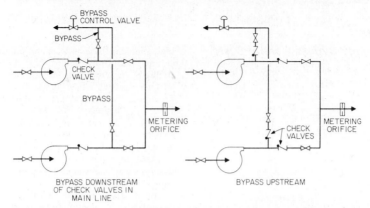

FIG. 9 If a single bypass protects two paralleled and similar pumps, bypass branchoff and check valve location may differ.

Single Bypass for Two or More Pumps   It is possible to provide a single bypass valve for two or perhaps more pumps (Fig. 9). The bypass line and valve must be able to handle simultaneously the total bypass flow of all pumps. The pumps must operate together, and they must have identical head-capacity curves, especially if flat. Otherwise bypass flow will differ from pump to pump, with the pump with the lower head not having minimum flow protection or being completely shut off.

In the system of Fig. 9, left, the main line check valve prevents backflow through an idle pump. If the branchoff were that of Fig. 9, right, an additional check valve in each bypass connection would be necessary to stop the backflow, and therefore this latter branchoff is not recommended.

Bypass Protection for Pumps in Series   A single bypass line and valve can protect two pumps in series (Fig. 10). The designer must provide for the larger of the two separately evaluated bypass flows and must take into account the heating effect of the upstream pump. The pumps cannot operate singly unless additional piping and controls are installed.

Minimum Flow Control Systems for Variable-Speed Pumps   In systems with turbine or other variable-speed drives, capacity control in normal operation can be largely through speed control. The effect of decreased speed and discharge head is to reduce the minimum flow requirement of the pump. For large high-head pumps, the reduction is rarely large enough to make a bypass control system in which minimum flow is a function of pump speed economically attractive. Instead, a single minimum flow can be taken, perhaps reduced by a small amount to allow for the beneficial effect of speed reduction. A constant-speed booster pump, requiring less minimum flow than the boiler-feed pump, can be protected by the bypass flow for the turbine-driven boiler-feed pump. The design study would have to take into account the heating effect in

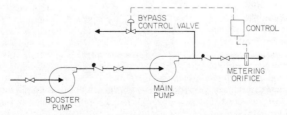

FIG. 10 A single bypass line and valve can protect both the main pump and its booster.

the booster pump and also the extent to which the control valve, rather than turbine speed, controls flow.

Automatic Bypass Control System

A single element that combines the pump check valve, pipe tee, pressure-reducing orifice, flow meter, and control valve can be an on-off or modulating bypass control, as in Figs. 11 and 12. A lift check acts as a flow meter, as in Fig. 13. This opens and closes a pilot valve which operates an integral-control pressure-reducing valve. The element is self-powered, requiring no external source of electricity or air, and is designed to be fail-safe. The highest pressure drops can be handled with commercially available systems. The system design must suit the pump characteristics.

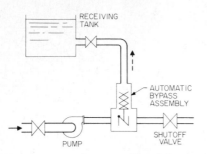

FIG. 11 Automatic bypass assembly includes several essential elements and is self-powered. (Yarway Corp).

As in other bypass systems, protection is necessary to prevent flashing in the piping downstream of the control element. A fixed orifice for on-off control or a variable orifice for modulating control is needed at the end of the bypass line and at the receiving tank wall. No antiflash orifice is required if the automatic control is at the end of the bypass line.

PUMP WARM-UP CIRCUITRY

A pump with close internal clearances, such as a boiler-feed pump for high pressures, requires slow initial warming by water circulating through a separate, relatively low-flow system. If the bypass line were open during warm-up, the flow through the pump and bypass could increase the warm-up flow enough to rotate the rotor prematurely. Therefore the bypass control system must assure closure of the line during warm-up, thereby allowing the orifice in the special warm-up drain line to control flow at a safe limit.

PROCEDURE FOR DETERMINING HEAD REQUIRED OF ON-OFF PRESSURE-REDUCING ORIFICE

First, use the pump performance curve to determine pump total head TH in feet (meters) at the minimum required pump flow. If shutoff head SOH is taken instead, then the actual bypass flow will be less than required by a factor of $(TH/SOH)^{1/2}$.

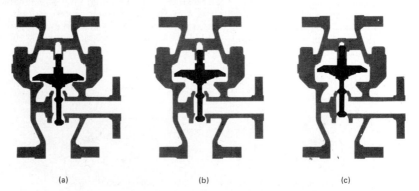

        (a)   (b)   (c)

FIG. 12 Automatic bypass assembly of one variety links lift check valve and control valve directly: (a) no process flow, (b) bypass and process flow, (c) no bypass flow. (Yarway Corp).

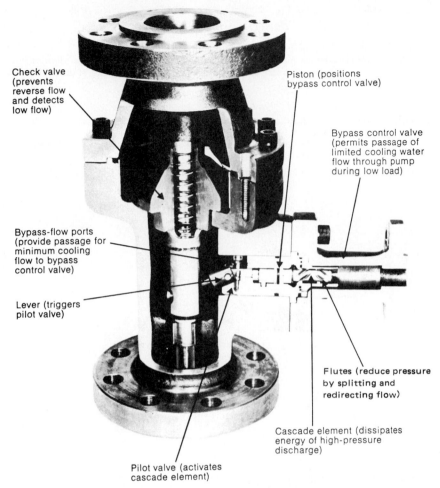

Check valve
(prevents
reverse flow
and detects
low flow)

Piston (positions
bypass control valve)

Bypass control valve
(permits passage of
limited cooling water
flow through pump
during low load)

Bypass-flow ports
(provide passage for
minimum cooling
flow to bypass
control valve)

Lever (triggers
pilot valve)

Flutes (reduce pressure
by splitting and
redirecting flow)

Cascade element (dissipates
energy of high-pressure
discharge)

Pilot valve (activates
cascade element)

FIG. 13 Automatic bypass assembly with staged pressure breakdown in control valve has a pilot valve to position the control valve. (Yarway Corp)

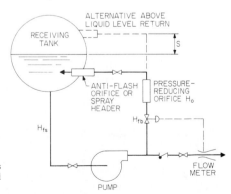

FIG. 14 Head at pressure-reducing orifice depends on pump head and losses in lines ahead of pump and in bypass.

ALTERNATIVE ABOVE
LIQUID LEVEL RETURN

RECEIVING
TANK

S

ANTI-FLASH
ORIFICE OR
SPRAY
HEADER

PRESSURE-
REDUCING
ORIFICE H_o

H_{fs}

H_{fb}

D

FLOW
METER

PUMP

Second, with the notation of Fig. 14, calculate the orifice head H_o in feet (meters):

$$H_o = TH - (H_{fs} + H_{fb}) - S$$

where H_{fs} = sum of all suction pipe losses at bypass design flow, ft (m)
H_{fb} = sum of all bypass pipe and antiflash and control valve losses at bypass flow, ft (m)
S = alternate return elevation above liquid level (usually negligible), ft (m)

REFERENCES

1. Mikasinovic, M., and P. Tung: "Sizing Centrifugal Pumps for Safety Service," *Chem. Eng.*, Feb. 23, 1981, p. 83.

2. McQueen, R.: "Minimum Flow Requirements for Centrifugal Pumps," *Pump World* **6**(2):10 (1980).

SECTION 2.4
CENTRIFUGAL PUMP PRIMING

IGOR J. KARASSIK
C. J. TULLO

A centrifugal pump is primed when the waterways of the pump are filled with the liquid to be pumped. The liquid replaces the air, gas, or vapor in the waterways. This may be done manually or automatically.

When a pump is first put into service, its waterways are filled with air. If the suction supply is above atmospheric pressure, this air will be trapped in the pump and compressed somewhat when the suction valve is opened. Priming is accomplished by venting the entrapped air out of the pump through a valve provided for this purpose.

Unlike a positive displacement pump, a centrifugal pump that takes its suction from a supply located below the pump, which is under atmospheric pressure, cannot start and prime itself (unless designed to be self-priming, as described later in this section). At its rated capacity, a positive displacement pump will develop the necessary pressure to exhaust air from its chambers and from the suction piping. Centrifugal pumps can also pump air at their rated capacity, but only at a pressure equivalent to the rated head of the pump. Since the specific weight of air is approximately $\frac{1}{800}$ that of water, a centrifugal pump can produce only $\frac{1}{800}$ of its rated liquid pressure. For every 1 ft (1 m) water has to be raised to prime a pump, the pump must produce a discharge head of air of approximately 800 ft (m). It is therefore apparent that the head required for a conventional centrifugal pump to be self-priming and to lift a large column of water (and in some cases to discharge against an additional static liquid head) when pumping air is considerably greater than the rating of the pump. Centrifugal pumps that operate with a suction lift can be primed by providing (1) a foot valve in the suction line, (2) a single-chamber priming tank in the suction line or a two-chamber priming tank in the suction and discharge lines, (3) a priming inductor at the inlet of the suction line, or (4) some form of vacuum-producing device.

FOOT VALVES

A foot valve is a form of check valve installed at the bottom, or foot, of a suction line. When the pump stops and the ports of the foot valve close, the water cannot drain back to the suction well if the valve seats tightly. Foot valves were very commonly used in early installations of centrifugal pumps. Except for certain applications, their use is now much less common.

A foot valve does not always seat tightly, and the pump occasionally loses its prime. However, the rate of leakage is generally small, and it is possible to restore the pump to service by filling and starting it promptly. This tendency to malfunction is increased if the water contains small particles of foreign matter, such as sand, and foot valves should not be used for such service. Another disadvantage of foot valves is their unusually high frictional loss.

The pump can be filled through a funnel attached to the priming connection or from an overhead tank or any other source of water. If a check valve is used on the pump and the discharge line remains full of water, a small bypass around the valve permits the water in the discharge line to be used for repriming the pump when the foot valve has leaked. Provision must be made for filling all the waterways and for venting out the air.

PRIMING CHAMBERS

The Single-Chamber Tank   A single-chamber primer is a tank with a bottom outlet that is level with the pump suction nozzle and directly connected to it. An inlet at the top of the tank connects with the suction line (Fig. 1). The size of the tank must be such that the volume contained between the top of the outlet and the bottom of the inlet is approximately three times the volume

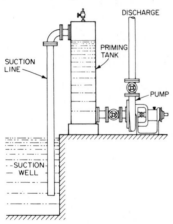

of the suction pipe. When the pump is shut down, the liquid in the suction line may leak out, but the liquid in the tank below the suction inlet cannot run back to the supply. When the pump is started, it will pump this entrapped liquid out of the priming chamber, creating a vacuum in the tank. The atmospheric pressure on the supply will force the liquid up the suction line into the priming chamber.

Each time the pump is stopped and restarted, a quantity of liquid in the priming tank must be removed to create the required vacuum, and because of possible backsiphoning, the liquid volume in the tank is reduced. Unless the priming tank is refilled, it has limited use. Automatic refilling of the priming tank cannot occur if the pump has a discharge check valve unless there is sufficient backflow from the discharge system before the check valve seats.

Because of their size, the use of single-chamber priming tanks is restricted to installations of relatively small pumps, usually to a 12-in (305-mm) suction line (approximately 2000 gpm [450 m³/h]).

FIG. 1   Single-chamber priming tank.

The Two-Chamber Tank   This type of priming tank is an improvement over the single-chamber design. It consists of integral suction and discharge chambers connected by a vacuum breaker or equalizing line. The operation and automatic refilling feature of this device are shown in Fig. 2. Commercial priming chambers are readily available with proper automatic vents and other features.

PRIMING INDUCTORS

If a separate source of water of sufficient capacity and pressure is available, it can be used to fill the suction line of the pump to be primed through the use of an inductor, as shown in Fig. 3. The pressure must be equal to about 4 lb/in² for each foot (90 kPa for each meter) of head necessary to prime the pump, measured from the lowest water level in the sump from which priming must be accomplished to the top of the pump. The amount of liquid necessary depends on the pressure.

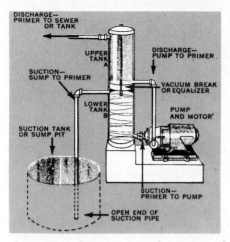 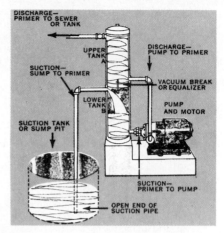

Phantom View of Primer in Use with Pump Stopped (*left*)

When the pump stops, the liquid in the upper chamber runs back into the pump and lower chamber of primer by gravity, thus refilling them with liquid and keeping the pump always ready for starting.

Phantom View of Primer in Use with Pump Running (*right*)

When the pump starts, it draws liquid from the lower chamber and discharges it through the upper chamber. Withdrawal of liquid from the lower chamber creates a partial vacuum in this chamber, which causes the liquid in the sump or well to rise in the suction pipe and flow through the primer to the pump.

FIG. 2 Two-chamber priming tank. (Apco/Valve and Primer)

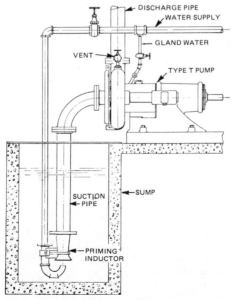

FIG. 3 Priming inductor. (Nagle Pumps)

Since the inductor is a positive-pressure device, leaks in the suction line or at the pump stuffing box are not critical. The service liquid can be left on or turned off after pumping is initiated. If left on, the head developed by the pump (therefore the flow also) will be increased.

An additional feature of the priming inductor is that operation is possible even when the suction line is covered with sediment. When the pump and suction line are filled, backflow results, mixing the solids and the liquid.

TYPES OF VACUUM DEVICES

Almost every commercially made vacuum-producing device can be used with systems in which pumps are primed by evacuation of air. Formerly water-jet, steam-jet, or air-jet primers had wide application, but with the increase in the use of electricity as a power source, motor-driven vacuum pumps have become popular.

Ejectors   Priming ejectors work on the jet principle, using steam, compressed air, or water as the operating medium. A typical installation for priming with an ejector is shown in Fig. 4. Valve V_1 is opened to start the ejector, and then valve V_2 is opened. When all the air has been exhausted from the pump, water will be drawn into and discharged from the ejector. When this occurs, the pump is primed and valves V_2, and V_1 are closed in that order.

An ejector can be used to prime a number of pumps if it is connected to a header through which the individual pumps are vented through isolating valves.

Dry Vacuum Pumps   Dry vacuum pumps, which may be of either the reciprocating or the rotary oil-seal type, cannot accommodate mixtures of air and water. When they are used in priming systems, some protective device must be interposed between the centrifugal pump and the dry vacuum pump to prevent water from entering the vacuum pump. The dry vacuum pump is used extensively for central priming systems.

Wet Vacuum Pumps   Any rotary, rotative, or reciprocating pump that can handle air or a mixture of air and water is classified as a wet vacuum pump. The most common type used in priming systems is the Nash Hytor pump (Fig. 5). This is a centrifugal displacement type of pump consisting of a round, multiblade rotor revolving freely in an elliptical casing partially filled with liquid. The curved rotor blades project radially from the hub and, with the side shrouds, form a series of pockets and buckets around the periphery.

The rotor revolves at a speed high enough to throw the liquid out from the center by centrifugal force. This forms a solid ring of liquid revolving in the casing at the same speed as the rotor, but following the elliptical shape of the casing. It will be readily seen that this forces the liquid to alternately enter and recede from the buckets in the rotor at high velocity.

Referring to Fig. 5 and following through a complete cycle of operation in a given chamber,

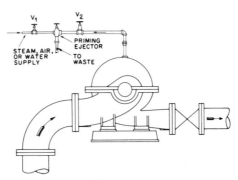

FIG. 4   Arrangement for priming with an ejector.

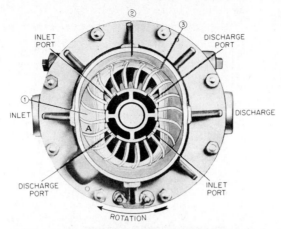

FIG. 5   Operating principle of Nash pump. (Nash Engineering)

we start at point A with the chamber (1) full of liquid. Because of the effect of the centrifugal force, the liquid follows the casing, withdraws from the rotor, and pulls air through the inlet port, which is connected to the pump inlet. At (2) the liquid has been thrown outwardly from the chamber in the rotor and has been replaced with air. As rotation continues, the converging wall of the casing at (3) forces the liquid back into the rotor chamber, compressing the air trapped in the chamber and forcing it out through the discharge port, which is connected with the pump discharge. The rotor chamber is now full of liquid and ready to repeat the cycle. This cycle takes place twice in each revolution.

If a solid stream of water circulates in this pump in place of air or of an air and water mixture, the pump will not be damaged but it will require more power. For this reason, in automatic priming systems using this type of vacuum pump, a separating chamber or trap is provided so that water will not reach the pump. Water needed for sealing a wet vacuum pump can be supplied from a source under pressure, with the shutoff valve operated manually or through a solenoid connected with the motor control. It is, however, preferable to provide an independent sealing water supply by mounting the vacuum pump on a base containing a reservoir. This is particularly desirable in locations where freezing may occur, as a solution of antifreeze can be used in the reservoir.

CENTRAL PRIMING SYSTEMS

If there is more than one centrifugal pump to be primed in an installation, one priming device can be made to serve all the pumps. Such an arrangement is called a central priming system (Fig. 6). If the priming device and the venting of the pumps are automatically controlled, the system is called a central automatic priming system.

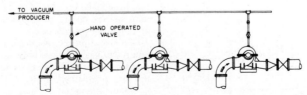

FIG. 6   Connections for a central priming system.

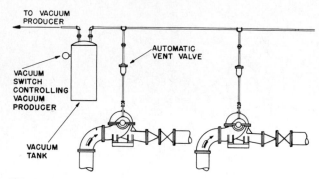

FIG. 7   Vacuum-controlled central automatic priming.

Vacuum-Controlled Automatic Priming System   A vacuum-controlled automatic priming system consists of a vacuum pump exhausting a tank. The pump is controlled by a vacuum switch and maintains a vacuum in the tank of 2 to 6 inHg (50 to 150 mmHg) above the amount needed to prime the pumps with the greatest suction lift. The priming connections on each pump served by the system are connected to the vacuum tank by automatic vent valves and piping (Fig. 7). The vacuum tank is provided with a gage glass and a drain. If water is detected in the vacuum tank as the result of leakage in a vent valve, the tank can be drained. Automatic vent valves consist of a body containing a float which actuates a valve located in the upper part. The bottom of the body is connected to the space being vented. As air is vented out of the valve, water rises in the body until the float is lifted and the valve is closed.

A typical vent valve designed basically for vacuum priming systems is illustrated in Fig. 8. The valve is provided with auxiliary tapped openings on the lower part of the body for connection to any auxiliary vent points on the system—for instance, when air is to be exhausted simultaneously from the high point of the discharge volute and the high point of the suction waterways. When one or more of these vent points are points of higher pressure, such as the top of the volute of the pump, an orifice is used in the vent line to limit the flow of liquid. Otherwise, a relatively high constant flow of water from the discharge back to the suction would take place, causing a constant loss. Where a unit is used more or less constantly, a separate valve should be used for each venting point.

The system shown in Fig. 7 is the most commonly used of central automatic priming systems. It can use either wet or dry motor-driven vacuum pumps. Most central priming systems are provided with two vacuum pumps. The usual practice is to have the control of one vacuum pump switched on at some predetermined vacuum and the control of the second switched on at a slightly lower vacuum.

Combination dry reciprocating-type vacuum pump and vacuum tank units are commercially available in various sizes. The operation of a typical unit is explained in Fig. 9.

Automatic priming systems using ejectors are also feasible, and several such systems are commercially available.

SECTION A - A

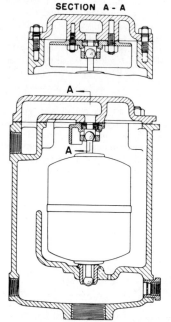

FIG. 8   Automatic vent valve. (Nash Engineering)

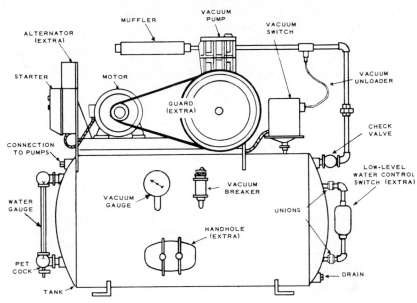

How the Primer Works

The primer automatically stops and starts itself to maintain a minimum vacuum in its tank at all times, regardless of whether the centrifugal pumps it serves are operating or not.

A vacuum header is run from the primer to the centrifugal pumps, and connection is made to the priming valve mounted on each pump. The vacuum in the primer tank removes the air from the pump and suction line via the vacuum header and priming valve. Water rises up the suction line and fills the pump. The water cannot enter the vacuum header because the float mechanism in the priming valve closes when the water reaches it.

If, for any reason, the water level in the pump drops, the priming valve immediately opens and the vacuum restores the water level.

The pumps are thus kept permanently primed.

FIG. 9   Combination dry reciprocating-type vacuum pump and vacuum tank. (Apco/Valve and Primer)

SELF-CONTAINED UNITS

Centrifugal pumps are available with various designs of priming equipment that makes them self-contained units. Some have automatic priming devices, which are basically attachments to the pump and become inactive after the priming is accomplished. Other units, which are self-priming pumps, incorporate a hydraulic device that can function as a wet vacuum pump during the priming period (see below). For stationary use, the automatically primed type is more efficient. The self-priming designs are generally more compact and are preferred for portable or semiportable use.

An automatically primed motor-driven pump uses a wet vacuum pump either directly connected to the pump or driven by a separate motor. In a directly connected unit, as soon as the centrifugal pump is primed, a pressure-operated control opens the vacuum pump suction to atmosphere so that it operates unloaded. With a separately driven vacuum pump, the controls stop the vacuum pump when the centrifugal pump is primed.

SELF-PRIMING PUMPS

The basic requirement for a self-priming centrifugal pump is that the pumped liquid must be able to entrain air in the form of bubbles so that the air will be removed from the suction side of the pump. This air must be allowed to separate from the liquid after the mixture of the two has been discharged by the impeller, and the separated air must be allowed to escape or to be swept out through the pump discharge. Such a self-priming pump therefore requires, on its discharge side, an air-separator, which is a relatively large stilling chamber, or reservoir, either attached to or built into the pump casing.

There are two basic variations of the manner in which the liquid from the discharge reservoir makes the pump self-priming: (1) recirculation from the reservoir back to the suction and (2) recirculation within the discharge and the impeller itself.

Recirculation to Suction

In such a pump, a recirculating port is provided in the discharge reservoir, communicating with the suction side of the impeller. Before the first time the pump is started, the reservoir is filled. As the pump is started, the impeller handles whatever liquid comes to it through the recirculating port plus a certain amount of air from the suction line. This mixture of air and liquid is discharged to the reservoir, where the two elements are separated, the air passing out of the pump discharge and the liquid returning to the suction of the impeller through the recirculation port. This operation continues until all the air has been exhausted from the suction line. The vacuum thus produced draws the liquid from the suction supply up the suction piping and into the impeller. After all the air has been exhausted and liquid is drawn into the pump, the pressure difference between the pump body and the inlet causes the priming valve, which permits communication between discharge and suction passages, to close. It is essential that the reservoir in the suction side remain filled with liquid when the pump is stopped, so that the pump is ready to restart. This is accomplished by incorporating either a valve or some sort of trap between the suction line and the impeller.

Pumps with recirculation to suction are seldom used today, and by far the most common arrangement is that with recirculation at the discharge.

Recirculation at Discharge

This form of priming is distinguished from the preceding method by the fact that the priming liquid is not returned to the suction of the pump but mixes with the air either in the impeller or at its periphery. The principal advantage of this method, therefore, is that it eliminates the complexity of internal valve mechanisms.

One such self-priming pump is illustrated in Fig. 10. An open impeller (A) rotates in a volute casing (B), discharging the pumped liquid through passage C into the reservoir (D). When the pump starts, the trapped liquid carries entrained air bubbles from the suction to the discharge chamber. There, the air separates from the liquid and escapes, while the liquid in the reservoir returns to the impeller through the recirculation port (F), re-enters the impeller and, after mixing once more with air bubbles, is discharged through passage C. This operation is repeated continuously until all the air in the suction line has been expelled. Once the pump is primed, a uniform pressure distribution is established around the impeller, preventing further recirculation. From this moment on, the liquid is discharged into the reservoir both at C and at F.

Figure 11 illustrates another form of self-priming pump with recirculation at the discharge. In this arrangement the return of the liquid to the impeller periphery takes place through a communicating passage (A) located at the bottom of the pump casing. Once the pump is primed, liquid is delivered into the reservoir at the discharge both through its normal volute discharge and through port A. In the pump illustrated in Fig. 11, a check valve at the pump suction (B) acts to prevent the draining of the pump after it has been stopped.

Regenerative Turbine Pumps

Because these pumps can handle relatively large amounts of gas, they are inherently self-priming as long as sufficient liquid remains in the pump to seal the clearance between the suction and discharge passages. This condition is usually met by building a trap in the pump suction.

SPECIAL APPLICATIONS

Systems for Sewage Pumps

A pump handling sewage or similar liquids containing stringy material can be equipped with automatic priming, but special precautions must be used to prevent carry-over of the liquid into the vacuum-producing device. One approach is to use a

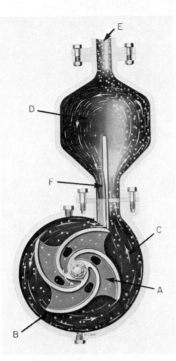

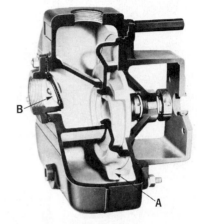

FIG. 10 Self-priming pump with recirculation at discharge. (Worthingon Pump)

FIG. 11 Self-priming pump with recirculation at discharge. (Peabody Barnes)

tee on the suction line immediately adjacent to the pump suction nozzle, with a vertical riser mounted on the top outlet of the tee. This riser is blanked at the top, thus forming a small tank. The top of the tank is vented to a vacuum system through a solenoid valve. The solenoid valve in turn is controlled electrically, through electrodes located at different levels in the tank. The solenoid valve closes if the liquid reaches the top electrode and opens if the liquid level falls below the level of the lower electrode.

Another solution permits the use of an automatic priming system with a separate motor-driven vacuum pump controlled by a discharge-pressure-actuated pressure switch. An inverted vertical loop is incorporated in the vacuum pump suction line to the pump being primed. This prevents the sewage from being carried over into the vacuum pump because this pump shuts down before the liquid reaches the top of the loop.

Systems for Air-Charged Waters

Some types of water, particularly from wells, have considerable dissolved gas that is liberated when the pump handles a suction lift. In such installations, an air-separating tank (also called a *priming tank* or an *air eliminator*) should be used in the suction line. One type (Fig. 12) uses a float-operated vent valve to permit the withdrawal of air or other gas. Another common arrangement uses a float valve mounted on the side of the tank to directly control the starting and stopping of the vacuum pump. Unless the air-separating tank is relatively large and the vacuum pump is not oversized, there is danger of frequent starting and stopping of the vacuum pump in such a system. When sand is present as an impurity in the water, the air-separating tank can be made to also function as a sand trap.

Systems for Units Driven by Gasoline or Diesel Engines

An automatic priming system using motor-driven vacuum pumps can be used for centrifugal pumps driven by diesel engines if a reliable source of electric power is available in the station. An auxiliary vacuum pump

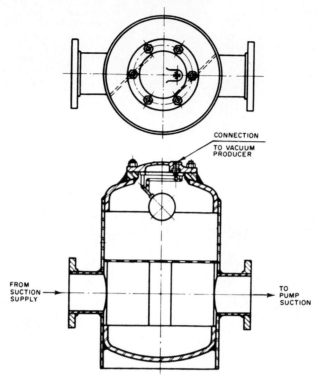

CONNECTION
TO VACUUM
PRODUCER

FROM
SUCTION →
SUPPLY

TO
PUMP
SUCTION

FIG. 12 Air-separating chamber.

driven by a gasoline engine might be desirable for emergency use in case of electric power failure. Alternatively, a direct-connected wet vacuum pump with controls similar to those used in motor-driven automatically primed units is very satisfactory.

The choice of the priming device for a gasoline-engine centrifugal pump depends on the size of the pump, the required frequency of priming, and the portability of the unit. Most portable units are used for relatively low heads and small capacities, for use in pumping out excavations and ditches, for example. Self-priming pumps of various types are most satisfactory for this service and are preferable to regular centrifugal pumps.

It is possible to utilize the vacuum in the intake manifold of a gasoline engine as a means of priming or keeping the pump primed. The rate at which the air can be drawn from the pump in this manner is relatively low, so that many of these units use foot valves. They are initially primed by filling the pump manually. Provision must be made to prevent water from being drawn over into the manifold.

For larger-volume low-head portable units, it is possible to use a wet vacuum pump belted to the main shaft by a tight-and-loose pulley. The vacuum pump can then be stopped when the unit is primed. For permanent installations, a separate wet vacuum pump driven by a small gasoline engine is generally preferred.

TIME REQUIRED FOR PRIMING

The time required to prime a pump with a vacuum-producing device depends on (1) the total volume to be exhausted, (2) the initial and final vacuums, and (3) the capacity of the vacuum-producing device over the range of vacuums that will exist during the priming cycle. The calculations for determining the time necessary to prime a pump are complicated. To permit close approximations, jet primers are usually rated in net capacity for various lifts. It is necessary to divide the volume to be exhausted by the rating to obtain the approximate priming time. Unless such a simplified method is available, the selection of the size of a primer is best left to the vendor of the equipment.

Central automatic priming systems are usually rated for the total volume to be kept primed. The time initially required to prime each unit served by the central system is not usually considered, as the basic function of the system is to keep the pumps primed and in operating condition at all times.

PREVENTION OF UNPRIMED OPERATION

Various controls may be used to prevent the operation of a pump when it is unprimed. These controls depend upon the type of priming system used. For most installations, a form of float switch in a chamber connected with the suction line is used. If the level in the chamber is above the impeller eye of the pump, the float switch control allows the pump to operate. If the liquid falls below a safe level, the float switch acts through the control to stop the pump, to prevent its being started, to sound an alarm, or to light a warning lamp. Such a valve and switch are illustrated in Fig. 13.

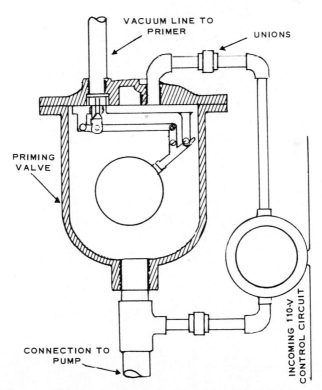

FIG. 13 Priming valve with water level control switch. (Apco/Valve and Primer)

GENERAL

Since a great number of automatic priming devices and systems are available, care should be taken to use the type or variation best suited to the application. The discussion of priming in this section does not, of course, cover all makes and modifications available for every specific application.

An automatic priming system will often allow units to be operated with excessive air leakage into the suction lines. This is poor practice because it requires the operation of the vacuum producer for greater periods of time than normally necessary.

FURTHER READING

Apco/Valve and Primer Corp.: *Automatic Pump Primers*, Technical Bulletin 721, Schaumburg, Ill., 1978.

Karassik, I. J., and R. Carter: Chap. 21 in *Centrifugal Pumps: Selection, Operation, and Maintenance*, McGraw-Hill, New York, 1960.

DISPLACEMENT PUMPS

SECTION 3.1
POWER PUMPS

FRED BUSE

A power pump is a constant-speed, constant-torque, and nearly constant-capacity reciprocating machine whose plungers or pistons are driven through a crankshaft from an external source.

The pump's capacity fluctuates with the number of plungers or pistons. In general, the higher the number, the less capacity variation at a given rpm. The pump is designed for a specific speed, pressure, capacity, and horsepower. The pump can be applied to horsepower conditions less than the specific design point, but at a sacrifice of the most economical operating condition.

Pumps are built in both horizontal and vertical construction. Horizontal construction is used on plunger pumps up to 200 hp (149 kW). This construction is usually below waist level and permits ease of assembly and maintenance. Horizontal pumps are built with three or five plungers. Horizontal piston pumps are rated to 3000 hp (2235 kW) and usually have two or three pistons, which are single- or double-acting. Vertical construction is used on plunger pumps up to 1500 hp (1118 kW), with the fluid end above the power end. This construction eliminates the plunger weight on the bushings, packing, and crosshead and aligns the plunger with the packing. A special sealing arrangement is required to prevent the liquid of the fluid end from mixing with the oil of the power end. There can be three to nine plungers.

Plungers are applied to pumps with pressures from 1000 to 30,000 lb/in² (69 to 2069 bar).° Maximum developed pressure with a piston is around 2000 lb/in² (138 bar). The pressure developed by the pump is proportional to the power available at the crankshaft. This pressure can be greater than the rating of the discharge system or that of the pump. When the pressure developed is greater than these ratings, a mechanical failure can result. To prevent this, a pressure relief device should be installed between the pump discharge flange and the first valve in the discharge system.

DESIGN

Brake Horsepower   The brake horsepower for the pump is

in USCS units
$$bhp = \frac{Q \times Ptd}{1714 \times ME}$$

° 1 bar = 10⁵ Pa. For a discussion of bar, see *SI Units—A Commentary* in the front matter.

in SI units

$$kW = \frac{Q \times Ptd}{36 \times ME}$$

where Q = delivered capacity, gpm (m³/h)
$\quad Ptd$ = developed pressure, lb/in² (bar)
$\quad ME$ = mechanical efficiency, %

Capacity   The capacity Q is the total volume of fluid delivered per unit of time. This fluid includes liquid, entrained gases, and solids at the specified condition.

Displacement   Displacement D, gpm (m³/h), is the calculated capacity of the pump with no slip losses. For single-acting plunger or piston pumps, this is

in USCS units

$$D = \frac{A \times m \times n \times s}{231}$$

in SI units

$$D = A \times m \times n \times s \times 6 \times 10^{-5}$$

where A = cross-sectional area of plunger or piston, in² (mm²)
$\quad m$ = number of plungers or pistons
$\quad n$ = rpm of pump
$\quad s$ = stroke of pump, in (mm) (half the linear distance the plunger or piston moves linearly in one revolution)
$\quad 231$ = in³/gal

For double-acting plunger or piston pumps, this is

in USCS units

$$D = \frac{(2A - a)m \times n \times s}{231}$$

in SI units

$$D = (2A - a)m \times n \times s \times 6 \times 10^{-5}$$

where a is the cross-sectional area of piston rod in in² (mm²).

Pressure   The pressure Ptd used to determine brake horsepower is the differential developed pressure. Because the suction pressure is usually small relative to the discharge pressure, discharge

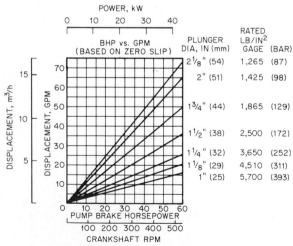

FIG. 1   Typical power pump performance: brake horsepower versus gallons per minute (kilowatts versus cubic meters per hour) based on zero slip. (Ingersoll-Rand)

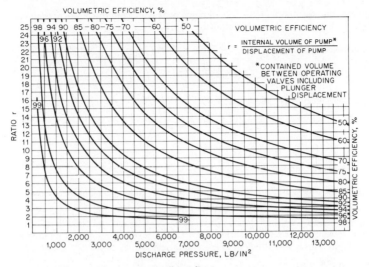

FIG. 2 Volumetric efficiency. (Ingersoll-Rand)

pressure is used in lieu of differential pressure. Figure 1 shows a typical performance curve for a power pump.

Slip   Slip S is the capacity loss as a percentage of the suction capacity. It consists of stuffing box loss B_1 and valve loss V_1:

$$S = B_1 + V_1$$

Volumetric Efficiency   Volumetric efficiency VE is the ratio of the discharge volume to the suction volume, expressed as a percentage plus slip. It is proportional to the ratio r and developed pressure (Fig. 2), where r is the ratio of internal volume of fluid between valves when the plunger or piston is at the top of the peak of its back stroke $(C + D)$ to the plunger or piston displacement (D) (Fig. 3).

Since discharge volume cannot be readily measured at discharge pressure, it is taken at suction pressure. Taking the discharge volume at suction pressure results in a higher volumetric efficiency than using the calculated discharge volume at discharge pressure because of fluid compressibility. Compressibility becomes important when pumping water or other liquids over 6000 lb/in² (414 bar), and it should be taken into consideration to determine the actual delivered capacity into the discharge system.

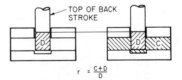

FIG. 3   The ratio r. (Ingersoll-Rand)

Figure 2 shows the approximate volumetric efficiency for water (not including slip).

Based on expansion back to suction capacity,

$$VE = \frac{1 - Ptd \times \beta \times r}{1 - Ptd \times \beta} - S$$

Based on discharge capacity,

$$VE = 1 - Ptd \times \beta \times r - S$$

where β is the compressibility factor of the liquid being pumped. Figure 4 shows approximate values of β for various liquids.

When the compressibility factor is not known but suction and discharge density can be determined in pounds per cubic foot, the following equations can be used for calculating VE:

based on suction capacity

$$VE = r - \frac{\rho_d}{\rho_s}(r - 1) - S$$

based on discharge capacity

$$VE = 1 - r\left(1 - \frac{\rho_s}{\rho_d}\right) - S$$

where ρ_s = suction density
ρ_d = discharge density

Stuffing Box Loss Stuffing box loss B_1 is usually considered negligible when calculating S.

Valve Loss Valve loss V_1 is the flow of liquid going back through the valve while it is closing and/or seated. This is a 2 to 10% loss, depending on valve design and condition.

Slip is affected by the viscosity of the liquid. Table 1 gives slip values for a pump with a plate valve at 150 rpm. Slip is also affected by speed and pressure, as indicated in Table 2.

Mechanical Efficiency The mechanical efficiency ME of a power pump at full load pressure and speed is 90 to 95%, depending on size, speed, and construction. Mechanical efficiency is affected by speed and slightly by developed pressure, as indicated in Tables 3 and 4. When a single built-in gear is part of the power frame, the pump's mechanical efficiency is 80 to 85%.

Speed Design speed n of power pumps is from 300 to 800 rpm, depending on capacity, size, and horsepower. To maintain good packing life, speed is sometimes limited to a plunger speed of 140 to 150 ft/min (0.71 to 0.76 m/s). Pump speed is also limited by valve life and allowable suction conditions.

Speed is the limiting factor of liquid separation from the plunger. Low speed limits with sleeve bearings must be considered to prevent loss of the lubrication film.

Plunger Load In addition to the horsepower limit, a power pump is designed for a plunger load (PL) limit. This is the load in pounds (newtons) that is applied to the plunger or piston and the bearing system. The industry's definition of plunger load is

in USCS units $\qquad PL = Ptd \times A$

in SI units $\qquad PL = 0.1\,Ptd \times A$

where Ptd = developed pressure, lb/in² (bar)
A = cross-sectional area of the plunger or piston, in² (mm²)

With double-acting pistons, the piston-rod area a is subtracted on the forward stroke.

The bearing system is rated for a specific load at design speed. Higher plunger loads will result in shorter bearing life. With sleeve bearings, high plunger loads at low speed will destroy the lubrication film. The load on the bearings from the moment of inertia of unbalanced reciprocating and unbalanced rotating parts is considered in bearing selection and is approximately 10 to 25% of the rated plunger load.

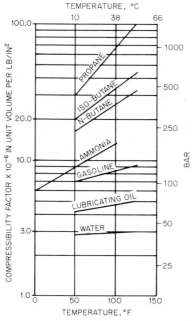

FIG. 4 Compressibility factor. (Ingersoll-Rand)

TABLE 1   Slip in a Pump with a Plate Valve Rotating at 150 rpm

Centistokes	100	1,000	2,000	6,000	10,000	12,000
Slip, %	8	8.5	9.5	20	41	61

TABLE 2   Slip as a Function of Pump Speed and Pressure

Pressure, psi (bar)	Slip, %		
	At 440 rpm	At 390 rpm	At 365 rpm
4000 (275)	11	22	34
3000 (207)	9	20	31
2000 (138)	7	18	30
1000 (69)	7	15	27.5

TABLE 3   Effect of Speed on Mechanical Energy at Constant Developed Pressure

% of full speed	44	50	73	100
ME, %	93.3	92.5	92.5	92.5

TABLE 4   Effect of Pressure on Mechanical Efficiency at Constant Speed

% of full-load developed pressure	20	40	60	80	100
ME, %	82	88	90.5	92	92.5

Unbalanced Reciprocating Parts Force (F_{rec}*)*   Parts are one-third of the connecting rod weight, crosshead, crosshead bearing, wrist pin, pony rod, and plunger. Additional parts on vertical pumps are pull rods, yoke, and plunger nut.

$$F_{rec} = \frac{W}{g}\,\omega^2 R \left(\cos\theta + \frac{R}{L}\cos 2\theta \right)$$

where F_{rec} = reciprocating parts force, lb (N)
W = weight of all reciprocating or rotating parts, lb (N)
g = 32.2 ft/s^2 (9.81 m/s^2)
ω = $(2\pi/60) \times n$ (n = pump rpm)
R = one-half of stroke, ft (m)
L = length of connecting rod ℄ to ℄ , ft (m)
θ = crank angle; usually maximum force is at θ = 0°, cos θ = 1

Unbalanced Rotating Parts Force (F_{rot}*)*   Parts are two-thirds of the connecting rod weight, crank end bearing, and crankpin.

$$F_{rot} = \frac{W}{g}\,\omega^2 R$$

where the variables are as above.

Number of Plungers or Pistons Table 5 lists the industry's terms for the number of plungers or pistons (m) on the crankshaft. They are the same for single- or double-acting pumps.

TABLE 5 Industry Terms for Number of Plungers or Pistons

Number	Term
1	Simplex
2	Duplex
3	Triplex
4	Quadruplex
5	Quintuplex
6	Sextuplex
7	Septuplex
9	Nonuplex

Pulsations The pulsating characteristics of the output of power pumps are extremely important in the pump application. The magnitude of the discharge pulsation is mostly affected by the number of plungers or pistons on the crankshaft. In Fig. 5, r is the radius of the crank in feet (meters), L is the length of the connecting rod in feet (meters), $C = L/r$, and $\omega = (2\pi/60) \times$ rpm. Thus, X, the linear movement of the piston or plunger is

$$X = r\left[\,1 - \cos\theta + L\left(1 - \sqrt{1 - \frac{r^2}{L^2}(\sin\theta)^2}\,\right)\right]$$

The approximate velocity of the plunger or piston is

$$S = \frac{dx}{dt} = r\left(\sin\theta + \frac{\sin 2\theta}{2C}\right)\omega$$

The approximate acceleration of the plunger or piston is

$$G_p = \frac{d^2x}{dt^2} = r\left(\cos\theta + \frac{\cos 2\theta}{C}\right)\omega^2$$

The momentary rate of discharge capacity is the cross-sectional area A of the plunger or piston times velocity:

for single-acting plunger or piston

$$A = 0.785 \times D^2$$

for double-acting piston

$$A - a = 0.785(D^2 - d^2)$$

The total discharge Q is equal to ALS', where S' is the number of effective strokes in a given time. The quantiy Q is the area of the curve, the mean height of which is

$$\frac{\text{Total } Q}{t} = \frac{ALS'}{t} = ALS''$$

where S'' the number of strokes per second and t is time in seconds.

Figures 6 and 7 show discharge rates for two types of pumps. Tables 6 and 7 show how variations from the mean are related to changes in the number

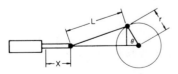

FIG. 5 Crankshaft position. (Ingersoll-Rand)

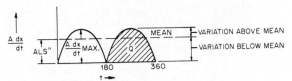

FIG. 6 Discharge rate for single double-acting pump. (Ingersoll-Rand)

of plungers and changes in $C = L/r$. Table 6 shows variation with the number of plungers with a C of approximately 6:1 and shows that pumps with an even number of throws have a higher variation than adjacent pumps with an odd number of throws. Table 7 shows the variations in a triplex with changes in $C = L/R$.

Net Positive Suction Head Required (NPSHR)

The NPSHR is the head of clean, clear liquid required at the suction connection centerline to ensure proper pump suction operating conditions. For any given plunger size, rotating speed, pumping rate, and pressure, there is a specific value of NPSHR. A change in one or more of these variables changes the NPSHR. For a given plunger size, the NPSHR changes approximately as the square of the speed.

It is a good practice to have the NPSH 3 to 5 lb/in² (0.2 to 0.3 bar) greater than the NPSHR. This will prevent release of vapor and entrained gases into the suction system. Released vapors and gases under repeated compression and expansion will cause cavitation damage in the internal passages.

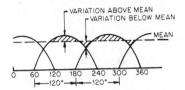

FIG. 7 Discharge rate for triplex single-acting pump. (Ingersoll-Rand)

Figure 8 shows NPSHR for a triplex pump as a function of rotating speed and plunger diameter.

Net Positive Suction Head Available (NPSHA)

The NPSHA is the static head plus atmospheric head minus lift loss, frictional loss, vapor pressure, velocity head, and acceleration head in feet (meters) available at the suction connection centerline.

Acceleration head can be the highest factor of NPSHA. In some cases it is 10 times the total of all the other losses. Data from both the pump and the suction system are required to determine acceleration head; its value cannot be calculated until these data have been established.

Acceleration Head

The flow in the suction line is always fluctuating, continuously accelerating or decelerating. The acceleration head is not a loss because the energy is restored during deceleration. Acceleration head loss is defined as

TABLE 6 Effect of Number of Plungers on Variations in Capacity from the Mean (C Approximately 6:1)

Type	Number of plungers	% above mean	% below mean	Total %	Plunger phase
Duplex (double)	2	24	22	46	180°
Triplex	3	6	17	23	120°
Quadruplex	4	11	22	33	90°
Quintaplex	5	2	5	7	72°
Sextuplex	6	5	9	14	60°
Septuplex	7	1	3	4	51.5°
Nonuplex	9	1	2	3	40°

TABLE 7   Effect of Change in C on Variations in Capacity from Mean for a Triplex Pump

C	% above mean	% below mean	Total %
4:1	8.2	20.0	28.2
5:1	7.6	17.6	25.2
6:1	6.9	16.1	23.0
7:1	6.4	15.2	21.6

$$ha = \frac{L \times V \times n \times C}{g \times K}$$

where ha = acceleration head loss, ft (m)
    L = length of suction pipe, ft (m)
    V = mean velocity in suction pipe, ft/s (m/s):

in USCS units $V = \dfrac{\text{gpm} \times 0.321}{\text{area of suction pipe, in}^2}$

in SI units $V = \dfrac{\text{m}^3/\text{h} \times 277.8}{\text{area of suction pipe, mm}^2}$

    n = rpm of the pump
    C = factor for type of pump:

duplex	0.115		quintuplex	0.04
triplex	0.066		sextuplex	0.055
quadruplex	0.08		septuplex	0.028

    g = acceleration of gravity, 32.2 ft/s^2 (9.81 m/s^2)
    K = factors for various fluids:

water	1.4
petroleum	2.5
liquid with entrained gas	1.0

When the suction system consists of pipes of various sizes, calculate the acceleration head for each section separately. Add the acceleration head of all sections to obtain the total.

If the calculated NPSHA, including acceleration head, is greater than the suction system can provide, the system NPSH should be increased. This can be accomplished by

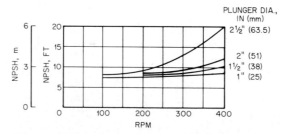

FIG. 8   NPSHR for a triplex pump. (Ingersoll-Rand)

- Increasing the static head
- Increasing the atmospheric pressure
- Adding a booster pump to the system
- Adding a pulsation dampener

The basic definition of acceleration head is

$$G_s = V \times n \times C = \text{acceleration of liquid in suction line, ft/s}^2 \text{ (m/s}^2)$$

$$F_s = \frac{W_s \times G_s}{g} = \text{force to produce acceleration, lb (N)}$$

where W_s is

in USCS units
$$\frac{L \times A_s \times \gamma}{144}$$

in SI units
$$\frac{L \times A_s \times \gamma}{102}$$

The theoretical head is

in USCS units
$$H_t = \frac{144\, F_s}{A_s \times \gamma} = \text{theoretical head, ft of liquid}$$

in SI units
$$H_t = \frac{102\, F_s}{A_s \times \gamma} = \text{theoretical head, m of liquid}$$

where A_s = cross-sectional area of the pipe, in^2 (mm^2)
γ = density, lb/in^2 (kg/l)

Substituting, the theoretical head

in USCS units
$$H_t = \frac{144 \times W_s \times V \times n \times C}{A_s \times \gamma \times g} = \frac{L \times V \times n \times C}{g}$$

in SI units
$$H_t = \frac{102 \times W_s \times V \times n \times C}{A_s \times \gamma \times g} = \frac{L \times V \times n \times C}{g}$$

$$ha = \frac{H_t}{K}$$

where K is the ratio of the theoretical head to the actual head. Therefore, the acceleration head is

$$ha = \frac{L \times V \times n \times C}{g \times K}$$

Liquid Separation from the Plunger   The pump speed at which water will separate from the end of the plunger may be calculated from the following:

in USCS units
$$\text{rpm} = 54.5 \sqrt{\frac{(34 - h_s - h_f)A_s}{LR[l - (l/L/R)]A_p}}$$

in SI units
$$\text{rpm} = 16.6 \sqrt{\frac{(10.36 - h_s - h_f)A_s}{LR[l - (l/L/R)]A_p}}$$

where h_s = suction head, ft (m)
h_f = piping frictional loss, ft (m)
A_s = area of suction pipe, in^2 (mm^2)
L = length of connecting rod, ℄ to ℄, ft (m)
R = crank radius, ft (m)
l = length of pipe where resistance of flow is to be measured, ft (m)
A_p = area of plunger, in^2 (mm^2)

EXAMPLE

$$h_s = 4 \text{ ft (lift)} \ (1.219 \text{ m})$$

$$h_f = 0.146 \text{ ft} \ (0.044 \text{ m})$$

$$A_s = 113 \text{ in}^2 \ (72903 \text{ mm}^2)$$

$$L = 2.5 \text{ ft} \quad (0.762 \text{ m})$$

$$R = 0.41 \text{ ft} \quad (0.125 \text{ m})$$

$$l = 10 \text{ ft} \quad (3.05 \text{ m})$$

$$L/R = 6{:}1$$

$$A_p = 38.48 \text{ in}^2 \ (24826 \text{ mm}^2)$$

in USCS units $\text{rpm} = 54.5 \sqrt{\dfrac{(34 - 4 - 0.146)113}{2.5 \times 0.41[10 - (10/6)]38.48}} = 177$

in SI units $\text{rpm} = 16.6 \sqrt{\dfrac{(10.36 - 1.219 - 0.044)72{,}903}{0.762 \times 0.125[3.05 - (3.05/6)]24{,}826}} = 177$

CONSTRUCTION

The following is a description of the liquid end and power end components of a power pump.

Liquid End The liquid end consists of the cylinder, plunger or piston, valves, stuffing box, manifolds, and cylinder head. Figure 9 shows the liquid end of a horizontal pump, and Fig. 10 shows the same for a vertical pump.

CYLINDER (WORKING BARREL) The cylinder is the body where the pressure is developed. It is continuously under fatigue. Cylinders on many horizontal pumps have the suction and discharge manifolds made integral with the cylinder. Vertical pumps usually have separate manifolds.

A cylinder containing the passages for more than one plunger is referred to as a single cylinder. When the cylinder is used for one plunger, it is called an individual cylinder. Individual cylinders are used where developed stresses are high. The forging may have a 4:1 to 6:1 forging reduction on all sides of the cylinder to obtain a homogeneous forging. Stress at the cylinder's intersecting bore can be based on a double hoop stress.

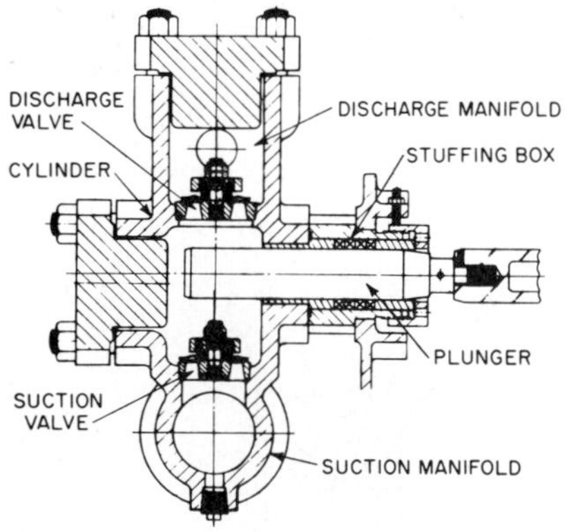

In Figs. 11 and 12, the distances b and d are the internal diameters, a and c are the diameters of the nearest obstruction of the solid material of the cylinder, P is the developed pressure, and S is the resulting stress. Allowable stresses range from 10,000 to 25,000 lb/in² (690 to 1725 bar), depending on the material of the cylinder and the liquid being pumped. The allowable stress is a function of the fatigue stress of the material for the liquid pumped and the life cycles required.

Instantaneous pressure in the cylinder may be two to three times design pressure. When the liquid contains entrained gas that can be realeased because of inadequate suction pressure, the instantaneous pressures may be four to five times the design pressure, resulting in short cylinder life.

FIG. 9 Liquid end, horizontal pump. (Ingersoll-Rand)

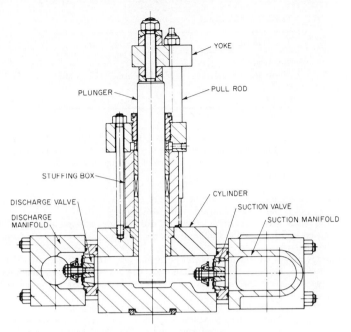

FIG. 10　Liquid end, vertical pump. (Ingersoll-Rand)

Cast cylinders are usually limited to the developed pressures listed in Table 8. Forged cylinders are made from 1020, 4140, 17-4 PH, 15-5 PH, 316 L, 304 and nickel aluminum bronze.

The internal bores usually have a minimum of 63 rms finish for pressures above 3,000 lb/in^2 (207 bar). Internal radii should not be less than ¼ in (1.5 mm).

PLUNGER　The plunger transmits the force that develops the pressure. It is solid up to 5 in (127 mm) in diameter. Above that dimension, it may be made hollow to reduce weight. Small-diameter plungers used for 6000 lb/in^2 (414 bar) and above should be reviewed for possible buckling. Plunger speed ranges from 150 to 350 ft/min (46 to 107 m/s). The finish is 16 rms with a 30 to 58 Rockwell C hardness. Materials of construction are Colmonoy No. 6 on 1020, chrome plate on 1020, 440C, 316, ceramic on 1020 with a 200°F limit, and solid ceramic. Ceramic is used for soft water, crude oil, mild acids, and mild alkalis. The problem with coatings on plungers is that at high pressures the liquid gets through the pores and underneath the coating. The pressure underneath the coating will cause it to flake off the plunger.

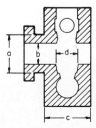

FIG. 11　Stress dimensions, horizontal liquid end. (Ingersoll-Rand)

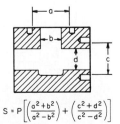

$$ S = P\left[\left(\frac{a^2+b^2}{a^2-b^2}\right)+\left(\frac{c^2+d^2}{c^2-d^2}\right)\right] $$

FIG. 12　Stress dimensions, vertical liquid end. (Ingersoll-Rand)

TABLE 8   Maximum Pressure in Cast Cylinders

Material	lb/in² (bar)
Cast iron	2000 (137)
Aluminum bronze	2500 to 3000 (172 to 207)
Steel	3000 (207)
Ductile iron	3000 (207)

PISTONS   Pistons are used for water pressures up to 2000 lb/in² (138 bar); for higher pressures a plunger is usually used. Pistons are cast iron, bronze, or steel with reinforced elastomer faces (Fig. 13). They are most frequently used on duplex double-acting pumps. The latest trend is to use single-acting pistons on triplex pumps.

STUFFING BOX   The stuffing box of a pump consists of the box, lower and upper bushing, packing, and gland (Fig. 14). It should be removable for maintenance.

The stuffing box bore is machined to a 63 rms finish to ensure packing sealing and life. A single hoop stress is used to determine the stuffing thickness, with an allowance of 10,000 to 20,000 lb/in² (690 to 1380 bar).

The bushings have a 63 rms finish with approximately 0.001 to 0.002 in (0.02 to 0.05 mm) diametrical clearance per inch of plunger diameter. The lower bushing is sometimes secured in an axial position to prevent the working of the packing. Bushings are made of bearing bronze, Ni-resist, or 316.

Packing is V- or chevron-shaped (Fig. 15). Some types use metal backup adapters. Unit packing consists of top and bottom adapter with a seal ring. A stuffing box can use three to five rings of packing or units, depending on the pressure and on the fluid being pumped. Packing or seal ring is reinforced asbestos, Teflon, duck, or neoprene.

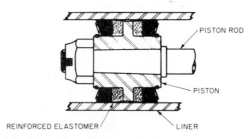

FIG. 13   Elastomer face piston. (FWI)

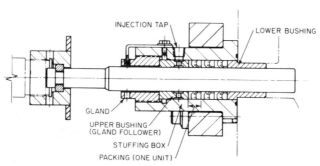

FIG. 14   Stuffing box. (Ingersoll-Rand)

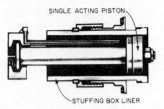

FIG. 15   Chevron packing. (Ingersoll-Rand)

FIG. 16   Single-acting piston stuffing box. (Ingersoll-Rand)

Packing can be made self-adjusting by installing a spring between the bottom of the packing and the lower bushing or bottom of the cylinder. This arrangement eliminates overtightening and alllows for uniform break-in of the packing. The packing is lubricated by injecting grease through a fitting, by gravity oil feed, or by an auxiliary lubricator driven through a take-off on the crankshaft.

For chemical or slurry service a lower injection ring is used for flushing. This prevents concentrated pumped fluid from impinging directly on the packing. This injection can be a continuous flush or can be synchronized to inject only on the suction stroke. Flush glands are employed where toxic vapor or flashing appears after the packing.

The stuffing box of a double-acting piston does not require an upper and lower bushing because the piston is guided by the cylinder liner. Single-acting pistons do not employ a stuffing box. Leakage of the piston goes into the frame extension to mix with the stuffing box continuous-circulating lubricant.

CYLINDER LINER   The cylinder wear liner (Figs. 13 and 16) is usually of Ni-resist material. Its length is slightly longer than the stroke of the pump, to allow for an assembly entrance taper of the piston into the liner. On double-acting pumps, the liner has packing to prevent leakage from the high-pressure side to the low side of the cylinder. Because of the brittleness of the liner, the construction should be such that the liner is not compressed. The finish of the liner is 16 rms.

VALVES   There are many types of valves. Which type is used depends on the application. The main parts of a valve are the seat and the plate. The plate movement is controlled by a spring or retainer. The seat usually uses a taper where it fits into the cylinder or manifold. The taper not only gives a positive fit but also permits easy replacement of the seat.

Some pumps use the same size suction and discharge valves for interchangeability (Fig. 10). Some use longer suction valves than discharge valves for NPSHR reasons. Others use larger discharge valves than suction valves because the discharge valve is on top of the suction valve (Fig. 9).

Because of space considerations, valves are sometimes used in clusters on each side of the plunger to obtain the required total valve area.

Table 9 shows seat and plate hardness for some valve materials. Seats and plates made of 316 material are chrome-plated or Colmonoy No. 6 plated to give them surface hardness.

TABLE 9   Recommended Material Hardness for Valve Plate and Seat

Material	Plate	Seat
	Rockwell C hardness	
329	30 to 35	38 to 43
440	44 to 48	52 to 56
17-4 PH	35 to 40	40 to 45
15-5 PH	35 to 40	40 to 45
	Brinell hardness	
316	150 to 180	150 to 180

TABLE 10 Types of Valves and Their Applications

TYPE	SKETCH A = SEAT AREA B = SPILL AREA	PRESSURE, PSI (BAR)	APPLICATION
PLATE		5,000 (345)	CLEAN FLUID. PLATE IS METAL OR PLASTIC
WING		10,000 (690)	CLEAN FLUIDS. CHEMICALS
BALL		30,000 (2,069)	FLUIDS WITH PARTICLES. CLEAR, CLEAN FLUID AT HIGH PRESSURE. BALL IS CHROME PLATED
PLUG		6,000 (414)	CHEMICALS
SLURRY	INSERT	2,500 (172)	MUD, SLURRY. POT DIMENSIONS TO API-12. POLYURETHANE OR BUNA-N INSERT

SOURCE: Ingersoll-Rand Co.

Seats and plates have a 32 rms finish. Table 10 shows various types of valves and their applications. Spill velocity through the valves is

in USCS units
$$V \text{ (ft/s)} = \frac{\text{gpm through valve} \times 0.642}{\text{spill area of the valve, in}^2}$$

in SI units
$$V \text{ (m/s)} = \frac{\text{m}^3/\text{h through valve} \times 555.6}{\text{spill area of valve, mm}^2}$$

The quantity 0.642 (555.6) is used because all the liquid passes through the valve in half the stroke. Some representative velocities are given in Table 11.

MANIFOLDS   These are the chambers where liquid is dispersed or collected for distribution before or after passing through the cylinder. On horizontal pumps the suction and discharge manifold is usually made integral with the cylinder (Fig. 17). Some horizontal pumps and some vertical pumps have only the discharge manifold integral with the cylinder (Fig. 18). Most vertical pumps have the suction and discharge manifold separate from the cylinder (Fig. 10).

TABLE II Valve Spill Velocities

Valve	Spill velocity, ft/s (m/s)
Clean liquid suction valve	3–8 (0.9–2.4)
Clean liquid discharge valve	6–20 (1.8–6)
Slurry suction and discharge valve	6–12 (1.8–3.7)

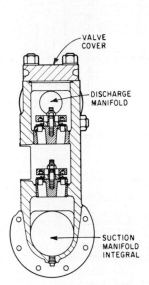

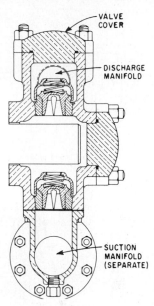

FIG. 17 Integral suction and discharge manifold. (Gardner-Denver)

FIG. 18 Separate suction with integral discharge manifold. (Gardner-Denver)

Suction manifolds are designed to eliminate air pockets from the flange to the valve entrance (Fig. 19). Separate suction manifolds are cast iron or fabricated steel. Discharge manifolds are steel forgings or fabricated steel. The manifolds have a minimum deflection to prevent gasket shift when subjected to the plunger load (Fig. 19).

The velocity through the manifolds of a clean liquid is 3 to 5 ft/s (0.9 to 1.5 m/s) at the suction and 6 to 16 ft/s (1.8 to 4.9 m/s) at the discharge. Suction and discharge manifold velocities on slurry service are 6 to 10 ft/s (1.8 to 3 m/s). Slurry service has a minimum of 6 ft/s (1.8 m/s) to prevent the fallout of slurries.

in USCS units

$$V \text{ (ft/s)} = \frac{\text{gpm of pump} \times 0.321}{\text{cross-sectional area of the manifold, in}^2}$$

in SI units

$$V \text{ (m/s)} = \frac{m^3/\text{h of pump} \times 277.8}{\text{cross-sectional area of manifold, mm}^2}$$

Waterhammer creates an additional pressure (which is added to the rated pump pressure), as does hydraulic shock loading. The discharge manifold rating is then made equal to or greater than the sum of these pressures.

VALVE COVERS   Valve covers are used to provide accessibiity to the valves without disturbing the cylinder or manifolds (Figs. 17 and 18).

PLUNGER COVERS AND/OR CYLINDER HEADS   These are used on horizontal pumps to provide accessibility to the plunger, piston, and cylinder liner (Figs. 17, 18, and 20).

Power End   The power end contains the crankshaft, connecting rod, crosshead, pony rod, bearings, and frame (Figs. 21 and 22). Basic designs are horizontal and vertical with sleeve or antifriction bearings.

FRAME   The frame absorbs the plunger load and torque. On vertical pumps with an outboard stuffing box (Fig. 10), the frame is in compression (Fig. 21). With horizontal single-acting pumps,

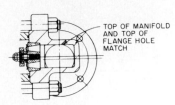

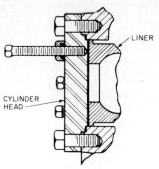

FIG. 19 Suction-manifold construction to eliminate air pockets. (Ingersoll-Rand)

FIG. 20 Plunger cover. (FWI)

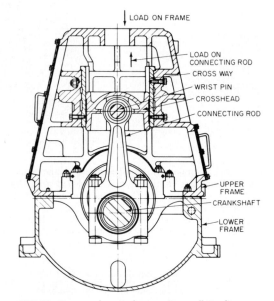

FIG. 21 Power end, vertical pump. (Ingersoll-Rand)

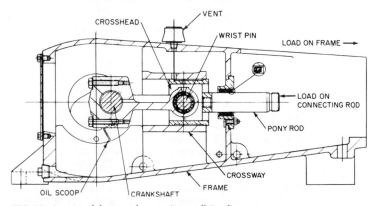

FIG. 22 Power end, horizontal pump. (Ingersoll-Rand)

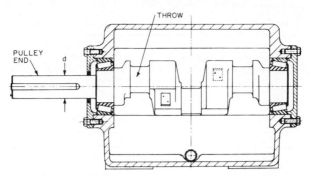

FIG. 23 Cast crankshaft. (Ingersoll-Rand)

the frame is in tension (Fig. 22). Frames are usually close-grain cast iron. Slurry pump frames, designed for mobile service at various sites, are usually fabricated steel. The frame is vented to the atmosphere.

When the working atmosphere is detrimental to the working parts in the frame, such as ammonia attack on bronze bearings, for example, the frame may be purged continuously with nitrogen.

CRANKSHAFT   The crankshaft (Figs. 23, 24, and 25) varies in construction depending on the design and power output of the pump. In horizontal pumps, the crankshafts are usually of nodular iron

FIG. 24   Crankshaft machined from billet. (Ingersoll-Rand)

FIG. 25   Cast crankshaft with integral gear. (Continental-EMSCO)

TABLE 12 Order of pressure Build-Up, or Firing Order

Throw from pulley end	No. of plungers or pistons	Pressure build-up order								
		1	2	3	4	5	6	7	8	9
Duplex	2	1	2							
Triplex	3	1	3	2						
Quintaplex	5	1	3	5	2	4				
Septuplex	7	1	4	7	3	6	2	5		
Nonuplex	9	1	5	9	4	8	3	7	2	6

or cast steel. Vertical pumps use forged steel or machined billet crankshafts. Because the crankshafts have relatively low speeds and mass, counterweights are not used. Except for duplex pumps, the crankshaft usually has an odd number of throws to obtain the best pulsation characteristics. The firing order in a revolution depends on the number of throws on the crankshaft (Table 12).

The pulley end stress is 2000 to 3000 lb/in² (14 to 20.7 mPa), where

in USCS units

$$\text{Stress} = \frac{3.96 \times 10^5 \times bhp}{d^3 \times n}$$

in SI units

$$\text{Stress} = \frac{6 \times 10^7 \text{ kW}}{d^3 \times n}$$

and d is the pulley end diameter in inches (millimeters) and n is the pump speed in revolutions per minute. Bearing surfaces are ground to 16 rms finish.

Many crankshaft designs incorporate rifle drilling between throws to furnish lubrication to the crankpin bearings.

CONNECTING RODS AND ECCENTRIC STRAPS The connecting rods (Fig. 26) transfer the rotating force of the crank pin to an oscillating force on the wrist pin. Connecting rods are split perpendicular to their centerline at the crank pin end for assembly of the rod onto the crankshaft.

The cap and rod are aligned with a close-tolerance bushing or body-bound bolts. The rods are either rifle-drilled or have cast passages for transferring oil from the wrist pin to the crank pin. A connecting rod with a tension load is made of forged steel, cast steel, or fabricated steel. Rods with a compression loading are cast nodular steel or aluminum alloy. Connecting rod finish, where the bearings are mounted, is 32 to 63 rms.

The ratio of the distance between the centerlines of the wrist pin and of the crank pin bearings to half the length of the stroke is referred to as L/R. The ratio directly affects the pressure pulsations, volumetric efficiency, size of pulsation dampener, speed of liquid separation, acceleration head, moment of inertia forces, and size of the frame. Low L/R results in high pulsations. High L/R reduces pulsations but may result in a large and uneconomical power frame. The common industrial L/R range is 4:1 to 6:1.

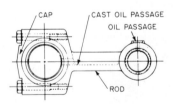

FIG. 26 Connecting rod. (Ingersoll-Rand)

The eccentric strap (Fig. 27) has the same function as a connecting rod except that the former usually is not split. The eccentric strap is furnished with antifriction bearings, whereas connecting rods are furnished with sleeve bearings. Eccentric straps are applied to mud and slurry pumps, which are started up against full load without requiring a bypass line.

WRIST PIN Located in the crosshead, the wrist pin transforms the oscillating motion of the connecting rod to reciprocating motion.

The maximum stress in the wrist pin from deflection should not exceed 10,000 lb/in² (69 mPa):

in USCS units

$$S = \frac{PL \times l}{8 \times 0.098d^3}$$

FIG. 27 Eccentric strap. (Gardner-Denver)   **FIG. 28** Crosshead. (Gardner-Denver)

in SI units
$$S = 1.25 \frac{PL \times l}{d^3}$$

where PL = plunger load, lb (N)
  l = length under load, in (mm)
  d = diameter of pin, in (mm)

Large pins are hollow to reduce the oscillating mass and assembly weight. Depending on the design, pins can have a tight straight fit, a taper fit, or a loose fit in the crosshead. The pin is case-hardened and has a 16 rms finish. When needle or roller bearings are used for the wrist pin bearing, the wrist pin is used as the inner race.

CROSSHEAD The crosshead (Fig. 28) moves in a reciprocating motion and transfers the plunger load to the wrist pin. The crosshead is designed to absorb the side, or radial, load from the plunger as the crosshead moves linearly on the crossway. The side load is approximately 25% of the plunger load. For cast iron crossheads the allowable bearing load is 80 to 125 lb/in^2 (551 to 862 kPa). Crossheads are grooved for oil lubrication with a bearing surface of 63 rms. Crossheads are piston type (full round) or partial-contact type. The piston type should be open end or vented to prevent air compression at the end of the stroke.

On vertical pumps the pull rods go through the crosshead so that the crosshead is under compression when the load is applied (Fig. 21).

CROSSWAYS (CROSSHEAD GUIDE) The crossway (Fig. 29) is the surface on which the crosshead reciprocates. On horizontal pumps it is cast integral with the frame (Fig. 22). On large frames it is usually replaceable and is shimmed to effect proper running clearance (Figs. 21 and 29). The crossway finish is 63 rms.

FIG. 29 Crossways. (Gardner-Denver)

PONY ROD (INTERMEDIATE ROD)   The pony rod is an extension of the crosshead on the horizontal pumps (Figs. 28 and 29). It is screwed or bolted to the crosshead and extends through the frame (Fig. 22). A seal on the frame and against the pony rod prevents oil from leaking out of the frame. A baffle is fixed onto the rod to keep leakage from coming in contact with the frame (Fig. 22).

PULL ROD (TIE ROD)   On vertical pumps two pull rods go through the crosshead. The rods are secured by a shoulder and nut so that the cast iron crosshead is in compression when the load is applied (Fig. 21). The rods extend out of the top of the frame and fasten to a yoke. The plunger is attached to the middle of the yoke with an aligning feature for the plunger (Fig. 10).

Bearings   Both sleeve and antifriction bearings are used in power pumps. Some frames use all sleeve, others use all antifriction, and others use a combination of both.

SLEEVE BEARINGS   When properly installed and lubricated, sleeve bearings are considered to have infinite life. The sleeve bearing is designed to operate within a certain speed range, and too high or too low a speed will upset the film lubrication. Sleeve bearings cannot be operated satisfactorily below 40 rpm with the plunger fully loaded, using standard lubrication. Below this speed, film lubrication is inadequate. The finish on the sleeve bearing is 16 rms. Clearances are approximately 0.001 in per inch (0.001 mm per millimeter) of diameter of the bearing.

Wrist Pin Bearings   These bearings have only oscillating motion. On single-acting pumps with low suction pressure, there is adequate reverse loading on the bearing to permit replenishment of the oil film. High suction pressure on horizontal pumps increases the reverse loading. On vertical pumps, high suction pressure can produce a condition of no reversal loading, and in this case higher oil pressure is required. Allowable projected area loading is 1200 to 1500 lb/in^2 (8.27 to 10.3 mPa) with bearing bronze.

Crank Pin Bearings   The crank pin bearing is a rotating split bearing and has a better oil film than the wrist pin bearing. It is clamped between the connecting rod and cap. The bearing is bronze-backed babbitt metal or a trimetal automotive bearing. Allowable projected area load is 1200 to 1600 lb/in^2 (8.27 to 11.03 mPa).

Main Bearings   The main bearings absorb the plunger load and gear load. The total plunger load varies during the revolution of the crankshaft. The triplex main bearings receive the greatest variations in loading because the crankshaft has the greatest relative span between bearings. Sleeve bearings are flanged to lock them in an axial position, and the flange absorbs residual axial thrust. On large-stroke vertical pumps, there is a main bearing between every connecting rod. Split bearings are bronze-backed babbitt metal with an allowable projected area load of 750 lb/in^2 (5.1 mPa).

ANTIFRICTION BEARINGS   A pump with antifriction bearings can be started under full plunger load without a bypass line. Antifriction bearings allow the pump to operate continuously at low speeds with full plunger load. They are selected for a 30,000- to 50,000-h B-10 bearing life. Slurry pipeline applications may require 100,000 h. Eccentric straps are used in place of connecting rods.

Wrist Pin Bearings   These are needle or roller bearings. The outer race is a tight fit into the strap, and the wrist pin is used as the inner race. This reduces the size of the bearing and strap. The wrist pin is held in the crosshead with a taper or keeper plate (Fig. 31).

Crank Pin Bearings   These are roller bearings. The outer race is mounted separately from the inner race; the inner race is mounted on the crankshaft and secured axially by a shoulder on the shaft and by a keeper plate, or by keeper plates on both sides of the race. The outer race is assembled in the strap in the same way as the inner race mounting. The strap is then slipped over the shaft and assembled to the inner race (Fig. 31).

Main Bearing   The main bearings used in conjunction with the sleeve wrist bearing and the sleeve crank pin bearings are usually tapered roller bearings, as in the case of of a horizontal triplex plunger pump. The main bearings are ordinarily mounted directly into the frame (Fig. 30).

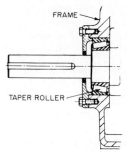

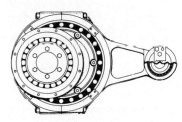

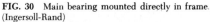

FIG. 30 Main bearing mounted directly in frame.
(Ingersoll-Rand)

FIG. 31 Main bearing mounted in bearing holder.
(Gardner-Denver)

The main bearings used with full antifriction bearing design are self-aligning spherical roller bearings. These bearings compensate for axial and radial movement of the crankshaft and are usually mounted in bearing holders, which in turn are mounted on the frame (Fig. 31). This type of design is used on mud and slurry pumps.

LUBRICATION   SAE 30 to 40 oil is used for bearing lubrication. With splash lubrication, the cheeks of the crankshaft or oil scoops (Fig. 22) throw oil by centrifugal force against the frame wall. This oil is then distributed by gravity to the crosshead, wrist pin, and crank pin. At low speeds there is not enough centrifugal force for proper oil distribution, so that partial force feed is then employed to get proper lubrication. Force-feed lubrication requires ½ to 1 gpm (1.9 to 3.8 l/m) per bearing with an oil pressure of 25 to 40 lb/in^2 (1.7 to 2.8 bar).

APPLICATIONS

Some of the applications for power pumps are

Ammonia service	Liquid petroleum gas
Carbamate service	Liquid pipeline
Chemicals	Power oil
Crude-oil pipeline	Power press
Cryogenic service	Soft-water injection for water flood
Fertilizer plants	Slurry pipeline (70% by weight)
High-pressure water cutting	Slush ash service
Hydro forming	Steel mill descaling
Hydrostatic testing	Water blast service

The following applications should be reviewed with the pump supplier:

Cryogenic service
Highly compressible liquids
Liquids over 250°F (120°C)
Liquids with high percentage of entrained gas
Low-speed operation
Slurry pipeline
Special fluid end materials
Viscosity over 250 SSU

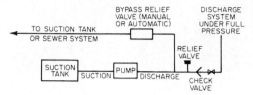

FIG. 32   Bypass relief system. (Ingersoll-Rand)

Duty Service   In making an economical pump selection, the type of full-load service should be specified:

Continuous:	8 to 24 h/day
Light:	3 to 8 h/day
Intermittent:	up to 3 h/day
Cyclic:	30 seconds loaded out of every 3 minutes

Relief Valve   Relief valves should be set at the following percentage above operating pressure:

Duplex double-acting:	25%
Triplex and above:	10%

Bypass Relief   Starting a power pump under full load requires a high starting torque. Also, a pump with sleeve bearing construction, when starting under full load, may not have adequate bearing lubrication to reach operating speed.

The starting torque on the driver and plunger load on the bearings may be reduced by

1. Installing a bypass line from the discharge line back to the suction or to a drain (Fig. 32). This line is located between the pump discharge and a check valve before the discharge piping system. The bypass valve is operated manually or automatically. The bypass line reduces starting torque from mechanical losses and liquid inertia of the suction and bypass systems. After the pump is up to speed, the bypass is closed and the pump goes on stream.
2. Use of suction valve unloaders. The unloader is a mechanism that lifts the suction valve plates off the seats before and during start-up. This stops pumping action by the pump. There are only mechanical losses to be overcome. After the pump is up to speed, the suction valves are closed in unison. On large-flow pumps, a distributor is used to close the suction valves in synchronism with the plunger pressure build-up order.

Pulsation Dampeners   Dampeners are used to reduce suction and discharge pulsations. They are especially useful on pumps with high pressure pulsations, as in duplex and triplex pumps. A properly sized and located suction dampener can reduce the system pulsations to an equivalent pipe length of 5 to 15 pipe diameters. The dampener should be located on the same side of the manifold as the pipe, not at the dead end of the manifold.

An effective suction dampener is a vertical air chamber made from pipe with a connection at the top for recharging. The inside diameter of the chamber should be as close as possible to the inside diameter of the pipe. The height is approximately 8 to 10 pipe diameters, with a minimum of 2 ft (0.6 m). The air cushion should be 1 ft (0.3 m) in height.

Dampeners on the discharge are mostly nitrogen-charged bladder bottles. Charge pressure is approximately 66% of the discharge system pressure.

SECTION 3.2
STEAM PUMPS

ROBERT M. FREEBOROUGH

BASIC THEORY

A reciprocating positive displacement pump is one in which a plunger or piston displaces a given volume of fluid for each stroke. The basic principle of a reciprocating pump is that a solid will displace an equal volume of liquid. For example, when an ice cube is dropped into a glass of water, the volume of water that spills out of the glass is equal to the submerged volume of the ice cube.

In Fig. 1 a cylindrical solid, a plunger, has displaced its volume from the large container to the small container. The volume of the displaced fluid (B) is equal to the plunger volume (A). The volume of the displaced fluid equals the product of the cross-sectional area of the plunger and the depth of submergence.

All reciprocating pumps have a fluid-handling portion, commonly call the *liquid end*, which has

1. A displacing solid called a *plunger* or *piston*.
2. A container to hold the liquid, called the *liquid cylinder*.
3. A suction check valve to admit fluid from the suction pipe into the liquid cylinder.
4. A discharge check valve to admit flow from the liquid cylinder into the discharge pipe.
5. Packing to seal the joint between the plunger and the liquid cylinder tightly to prevent liquid from leaking out of the cylinder and air from leaking into the cylinder.

These basic components are identified on the rudimentary liquid cylinder illustrated in Fig. 2. To pump the liquid through the liquid end, the plunger must be moved. When the plunger is moved out of the liquid cylinder as shown in Fig. 2, the pressure of the fluid in the cylinder is reduced. When the pressure becomes less than that in the suction pipe, the suction check valve opens and liquid flows into the cylinder to fill the volume being vacated by withdrawal of the plunger. During this phase of operation, the discharge check valve is held closed by the higher pressure in the discharge pipe. This portion of the pumping action of a reciprocating positive displacement pump is called the *suction stroke*.

The withdrawal movement must be stopped before the end of the plunger gets to the packing.

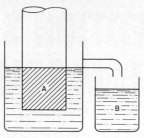

FIG. 1 The volume of liquid displaced by a solid equals the volume of the solid.

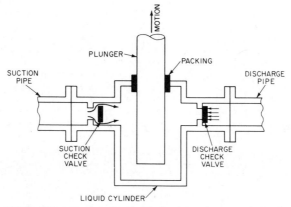

FIG. 2 Schematic of the liquid end of reciprocating pump during the suction stroke.

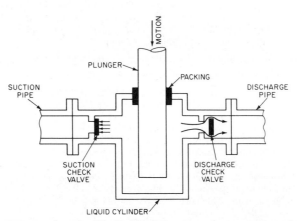

FIG. 3 Schematic of the liquid end of a reciprocating pump during the discharge stroke.

The plunger movement is then reversed, and the *discharge stroke* portion of the pumping action is started, as illustrated in Fig. 3.

Movement of the plunger into the cylinder causes an increase in pressure of the liquid contained therein. This pressure immediately becomes higher than suction pipe pressure and causes the suction check valve to close. With further plunger movement, the liquid pressure continues to rise. When the liquid pressure in the cylinder reaches that in the discharge pipe, the discharge check valve is forced open and liquid flows into the discharge pipe. The volume forced into the discharge pipe is equal to the plunger displacement less very small losses. The plunger displacement is the product of its cross-sectional area and the length of stroke. The plunger must be stopped before it hits the bottom of the cylinder. The motion is then reversed, and the plunger again goes on suction stroke as previously described.

The pumping cycle just described is that of a *single-acting* reciprocating pump. It is called single-acting because it makes only one suction stroke and only one discharge stroke in one reciprocating cycle.

Many reciprocating pumps are *double-acting*, i.e., they make two suction and two discharge strokes for one complete reciprocating cycle. Most double-acting pumps use as the displacing solid a piston which is sealed to a bore in the liquid cylinder or to a liquid cylinder liner by piston packing. Figure 4 is a schematic diagram of a double-acting liquid end. In addition to a piston with packing, it has two suction and two discharge valves, one of each on each side of the piston. The piston is moved by a piston rod. The piston rod packing prevents liquid from leaking out of the cylinder. When the piston rod and piston are moved in the direction shown, the right side of the piston is on a *discharge stroke* and the left side of the piston is simultaneously on a *suction stroke*. The piston packing must seal tightly to the cylinder liner to prevent leakage of liquid from the high-pressure right side to the low-pressure left side.

The piston must be stopped before it hits the right side of the cylinder. The motion of the piston is then reversed, so that the left side of the piston begins its discharge stroke and the right side begins its suction stroke.

A reciprocating pump is not complete with a liquid end only; it must also have a driving mechanism to provide motion and force to the plunger or piston. The two most common driving mechanisms are a reciprocating steam engine and a crank-and-throw device. Those pumps using the steam engine are called *direct-acting steam pumps*. Those pumps using the crank-and-throw device are called *power pumps*. Power pumps must be connected to an external rotating driving force, such as an electric motor, steam turbine, or internal combustion engine.

DIRECT-ACTING STEAM PUMPS

Direct-acting steam pumps are mainly classified by the number of working combinations of cylinders. For example, a duplex pump (Fig. 5) has two steam and two liquid cylinders mounted side by side, and a simplex pump (Fig. 6) has one steam and one liquid cylinder.

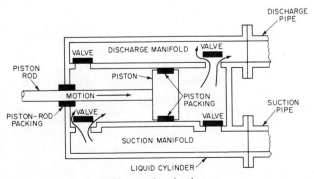

FIG. 4 Schematic of a double-acting liquid end.

FIG. 5   Duplex steam pump. (Worthington Pump)

Additionally, simplex and duplex pumps may be further defined by (1) cylinder arrangement, whether horizontal or vertical; (2) number of steam expansions in the power end; (3) liquid end arrangement, whether piston or plunger; and (4) valve arrangement, i.e., cap and valve plate, side pot, turret type, etc.

Although this section will refer to steam as the driving medium, compressed gases such as air or natural gas can be used to drive a steam pump. These gases should have oil mist added to them prior to entering the pump to prevent wear of the steam end parts.

Steam End Construction and Operation   The driving mechanism, or *steam end,* of a direct-acting steam pump includes the following components, as illustrated in Fig. 7:

1. One or more steam cylinders with suitable steam inlet and exhaust connections
2. Steam piston with rings
3. Steam piston rods directly connected to liquid piston rods
4. Steam valves which direct steam into and exhaust steam from the steam cylinder
5. A steam valve actuating mechanism which moves the steam valve in proper sequence to produce reciprocating motion

The operation of a steam pump is quite simple. The motion of the piston is obtained by admitting steam of sufficient pressure to one side of the steam piston while simultaneously exhausting steam from the other side of the piston. There is very little expansion of the steam since it is

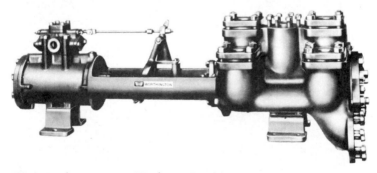

FIG. 6   Simplex steam pump. (Worthington Pump)

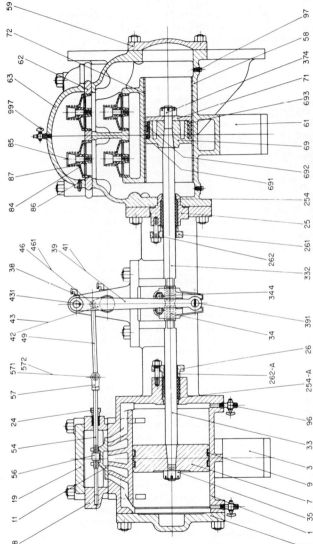

FIG. 7 Typical section of duplex steam pump: (1) Steam cylinder with cradle, (2) steam cylinder head, (3) steam cylinder foot, (7) steam piston, (9) steam piston rings, (11) slide valve, (18) steam chest, (19) steam chest cover, (24) valve rod stuffing box gland, (25) piston rod stuffing box, liquid, (26) piston rod stuffing box gland, steam, (33) steam piston rod, (34) steam piston spool, (35) steam piston nut, (38) cross stand, (39) long lever, (41) short lever, (42) upper rock shaft, long crank, (43) lower rock shaft, short crank, (46) crank pin, (49) valve rod link, (54) valve rod, (56) valve rod nut, (57) valve rod head, (58) liquid cylinder, (59) liquid cylinder head, (61) liquid cylinder foot, (62) valve plate, (63) force chamber, (69) liquid piston body, (71) liquid piston follower, (72) liquid cylinder lining, (84) metal valve, (85) valve guard, (86) valve seat, (87) valve spring, (96) drain valve for steam end, (97) drain plug for liquid end, (254) liquid piston rod stuffing box bushing, (332) liquid piston rod, (344) piston rod spool bolt, (374) liquid piston rod nut, (391) lever pin, (431) lever key, (461) crank pin nut, (571) valve rod head pin, (572) valve rod head pin nut, (691) liquid piston snap rings, (692) liquid piston bull rings, (693) liquid piston fibrous packing rings, (997) air cock, (251) liquid piston rod stuffing box, (254A) steam piston rod stuffing box bushing, (261) piston rod stuffing box gland, liquid, (262) piston rod stuffing box gland lining, liquid, (262A) piston rod stuffing box gland lining, steam. (Worthington Pump)

admitted at a constant rate throughout the stroke. The moving parts, i.e., the steam piston, the liquid piston, and the piston rod or rods, are cushioned and brought to rest by exhaust steam trapped in the end of the steam cylinder at the end of each stroke. After a brief pause at the end of the stroke, steam is admitted to the opposite side of the piston and the pump strokes in the opposite direction.

STEAM VALVES   Since the steam valve and its actuating mechanism control the reciprocating motion, any detailed description of the construction of the direct-acting steam pump should rightfully begin with a discussion of steam valve types, operation, and construction.

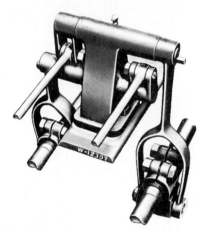

FIG. 8 Steam valve actuating mechanism for a duplex pump (Worthington Pump)

DUPLEX STEAM VALVES   The steam valves in a duplex steam pump are less complicated than those in a simplex pump and will be described first. As previously stated, the duplex steam pump can be considered as two simplex pumps arranged side by side and combined to operate as a single unit. The piston rod of one pump, in making its stroke, actuates the steam valve and thereby controls the admission or exhaust of steam in the second pump. A valve gear cross stand assembly is shown in Fig. 8. The wishbone-shaped piston rod lever of one side is connected by a shaft to the valve rod crank of the opposite side. The steam valve is connected to the valve rod crank by the steam valve rod and steam valve link. Through this assembly, the piston rod of one side moves the steam valve of the opposite side in the same direction. When the first pump has completed its stroke, it must pause until its own steam valve is actuated by the movement of the second pump before it can make its return stroke. Since one or the other steam cylinder port is always open, there is no "dead center" condition; hence the pump is always ready to start when steam is admitted to the steam chest. The movements of both pistons are synchronized to produce a well-regulated flow of liquid free of excessive pulsations and interruptions.

FLAT SLIDE STEAM VALVES   Steam enters the pump from the steam pipe into the steam chest on top of the steam cylinder. Exhaust steam leaves the pump through the center port of five ports, as shown in Fig. 9. Most duplex pumps use a flat slide valve which is held against its seat by steam pressure acting upon its entire top area; this is called an *unbalanced* valve. The flat, or D, valve, as it is often called, is satisfactory for steam pressures up to approximately 250 lb/in^2 (17 bar°) and has reasonable service life, particularly where steam end lubrication is permissible. On large pumps, the force required to move an unbalanced valve is considerable, and so a balanced piston valve, which will be discussed later, is used.

The slide valve shown in Fig. 9 is positioned on dead center over the five valve ports. A movement of the valve to the right uncovers the left-side steam port and the right-side exhaust port, which is connected through the slide valve to the center exhaust port. The main steam piston will be moved from left to right by the admitted steam. Movement of the slide valve from dead center to the left would, of course, cause opposite movement of the steam piston.

The steam valves of a duplex pump are mechanically operated, and their movements are dependent upon the motion of the piston rod and the linkage of the valve gear. In order to ensure that one piston will always be in motion when the other piston is reversing at the end of its stroke, lost motion is introduced into the valve gear. Lost motion is a means by which the piston can move during a portion of its stroke without moving the steam valve. Several lost motion arrangements are shown in Fig. 9.

If the steam valves are out of adjustment, the pump will have a tendency not to operate

°1 bar = 10^5 Pa. For a discussion of bar, see *SI Units—A Commentary* in the front matter.

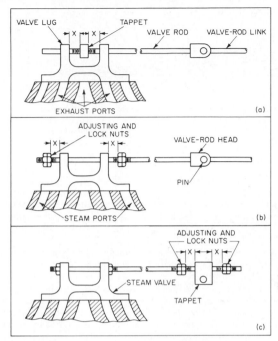

FIG. 9   Duplex pump steam valve lost-motion arrangements.

through its designed stroke. Increasing the lost motion lengthens the stroke; if this is excessive, the piston will strike the cylinder head. Reducing the lost motion shortens the stroke; if this is excessive, the pump will short-stroke, with a resultant loss in capacity.

The first step in adjusting the valves is to have both steam pistons in a central position in the cylinder. To accomplish this, the piston is moved toward the steam end until the piston strikes the cylinder head. With the piston rod in this position, a mark is made on the rod flush with the steam end stuffing box gland. Next, the piston rod is moved toward the liquid end until the piston strikes, and then another mark is placed on the rod halfway between the first mark and the steam end stuffing box gland. After this, the piston rod is returned toward the steam end until the second mark is flush with the stuffing box gland. The steam piston is now in central position. This procedure is repeated for the opposite piston rod assembly.

The next step is to see that both steam valves are in a central position with equal amounts of lost motion on each side, indicated by distance X in Fig. 9.

Most small steam pumps are fitted with a fixed amount of lost motion, as shown in Fig. 9a. With the slide valve centered over the valve ports, a properly adjusted pump will have the tappet exactly centered in the space between the valve lugs. The lost motion (X) on each side of the tappet will be equal.

Larger pumps are fitted with adjustable lost motions, such as are shown in Fig. 9b. The amount of lost motion (X) can be changed by moving the locknuts. Manufacturers provide specific instructions for setting proper lost motion. However, one rule of thumb is to allow half the width of the steam port on each side for lost motion. A method to provide equality of lost motion is to move the valve each way until it strikes the nut and then note if both port openings are the same.

In some cases it is desirable to be able to adjust the steam valves while the pump is in motion. With the arrangements mentioned above, this cannot be done because the steam chest head must be removed. In a pump equipped with a lost-motion mechanism like that shown in Fig. 9c, all adjustments are external and can therefore be made while the pump is in operation.

BALANCED PISTON STEAM VALVE   The balanced piston steam valve (Fig. 10) is used on duplex steam pumps when the slide-type valve cannot be used because of size. The balanced piston valve can

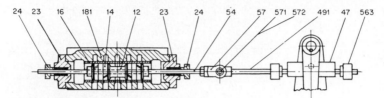

FIG. 10   Balanced piston steam valve: (12) piston valve, (181) steam chest, (23) valve rod stuffing box, (24) valve rod stuffing box gland, (54) valve rod, complete, (57) valve rod head, (14) piston valve ring, (16) piston valve lining, (47) lost-motion block tappet, (491) valve rod link, (563) valve rod collar. (Worthington Pump)

also be used without lubrication at pressures above 250 lb/in^2 (17 bar) and temperatures above 500°F (260°C). At higher pressures, wire drawing or steam cutting can occur as the piston slowly crosses the steam ports. To prevent wear and permanent damage to the steam chest and piston, piston rings are used on the steam valve and a steam chest liner is pressed into the steam chest to protect it.

CUSHION VALVES   Steam cushion valves are usually furnished on larger pumps to act as an added control to prevent the steam piston from striking the cylinder heads when the pump operates at high speeds. As previously shown, the steam end has five ports; the outside ports are for steam admission, and the inside ports are for steam exhaust. As the steam piston approaches the end of the cylinder, it covers the exhaust port, trapping a volume of steam in the end of the cylinder. This steam acts as a cushion and prevents the piston from striking the cylinder head. The cushion valve is simply a bypass valve between the steam and exhaust ports; by opening or closing this valve the amount of cushion steam can be controlled.

If the pump is running at low speed or working under heavy load, the cushion valve should be opened as much as possible without allowing the piston to strike the cylinder head. If the pump is running at high speed or working under light load, then the cushion valve should be closed. The amount of steam cushion and, consequently, the length of stroke can be properly regulated for different operating conditions by the adjustment of this valve.

SIMPLEX STEAM VALVES   The simplex pump steam valve is steam-operated, not mechanically operated as duplex steam valves are. The reason for this is that the piston rod assembly must operate its own steam valve. Consequently the travel of the valve cannot be controlled directly by means of the piston rod motion. Instead, the piston rod operates a pilot valve by means of a linkage similar to that used with a duplex pump. This controls the flow of steam to each end of the main valve, shuttling the steam back and forth. The arrangement illustrated in Fig. 11 is one of the designs available to produce this motion.

With the pilot valve in the position shown in Fig. 11, steam from the live steam space flows through the pilot valve steam port into the steam space at the left-hand end of the main valve (balanced piston type). Simultaneously, the D section of the pilot valve connects the steam space at the right-hand end of the main valve with the exhaust port, thereby releasing the trapped steam. The main valve has moved completely across to the right-hand end of the chest. The main valve in this position permits steam to flow from the chest to the left-hand steam cylinder port and, at the same time, connects the right-hand steam cylinder port with the exhaust port.

The steam piston now moves to the right, and after the lost motion is taken up in the valve gear, the pilot valve moves to the left. In this position the cycle described above now takes place at the opposite end of the steam chest. Since the main valve is steam-operated, it can be in only two positions, either at the left-hand or at the right-hand end of the chest. Hence it is impossible to have it at dead center. In other words, steam can always flow either to one side or to the other of the steam piston, irrespective of the position of the steam piston.

For the valve to operate smoothly and quietly, an arrangement must be provided to create a cushioning effect on the valve travel. The steam piston, as it approaches the end of its travel, cuts off the exhaust port and traps a certain amount of steam, which acts as a cushion and stops the steam piston.

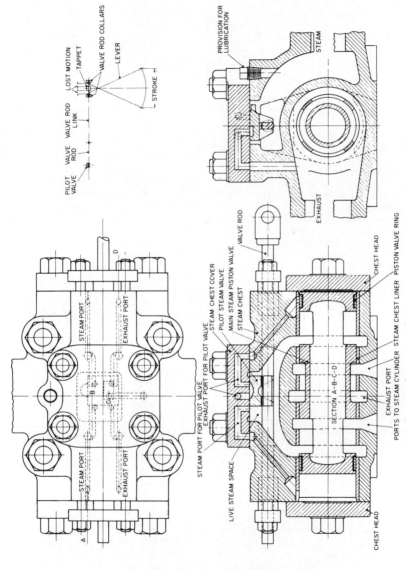

FIG. 11 Simplex-type steam valve. (Worthington Pump)

All valve adjustments are outside of the steam chest, and so it is possible to adjust the valve while the pump is in operation. The effect of decreasing or increasing the lost motion is the same as that described for duplex pumps. The lost-motion arrangement is the same as that shown in Fig. 9c.

STEAM END MATERIALS   For most services, cast iron is an excellent material for the steam cylinder and it is the major element of the steam end. It is readily cast in the complicated shape required to provide the steam porting. It possesses good wearing qualities, largely because of its free graphite content. This is required in the piston bores, which are continuously being rubbed by the piston rings. At high steam temperatures and pressures, ductile iron or steel is used. In the latter case, however, cast iron steam cylinder liners are frequently used because of their better wear resistance.

Counterbores are provided at each end of a steam cylinder so that the leading piston ring can override, for a part of its width, the end of the cylinder bore to prevent the wearing of a shoulder on the bore.

The cylinder heads and steam pistons are also usually made of cast iron. The cylinder head has a pocket cast in it to receive the piston rod nut at the end of the stroke. Most steam pistons are made in one piece, usually with two piston ring slots machined into the outside circumference.

The relatively wide piston rings are usually made from hammered iron. They are split so that they can be expanded to fit over the piston and snapped into the grooves in the piston. They must be compressed slightly to fit into the bores in the cylinder. This ensures a tight seal with the cylinder bores even as the rings wear during operation. In services where steam cylinder lubrication is not permissible, a combination two-piece ring of iron and bronze is used to obtain longer life than is obtained with the hammered iron rings.

The piston and valve rods are generally made from steel, but stainless steel and Monel are also commonly used. Packing for the rods is usually a braided graphited asbestos.

Drain cocks and valves are always provided to permit drainage of condensation, which forms in the cylinder when a pump is stopped and cools down. On each start-up these must be cracked open until all liquid is drained and only steam comes out; then they are closed.

The steam end and liquid end are joined by a cradle. On most small pumps the cradle is cast integral with the steam end. On large pumps it is a separate casting or fabricated weldment.

Liquid End Construction   Steam pumps are equipped with many types of liquid ends, each being designed for a particular service condition. However, they can all be classified into two basic types, the piston, or inside-packed, type and the plunger, or outside-packed, type.

The piston pump (Fig. 7) is generally used for low and moderate pressures. Because the piston packing is located internally, the operator cannot see the leakage past it or make adjustments that could make the difference between good operation and packing failure. Generally, piston pumps can be used at higher pressures with noncorrosive liquids having good lubricating properties, such as oil, than with corrosive liquids, such as water.

Plunger pumps, illustrated in Fig. 12, are usually favored for high-pressure and heavy-duty

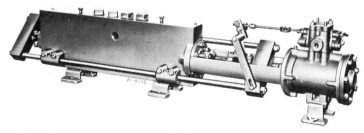

FIG. 12   Simplex-type plunger pump. (Worthington Pump)

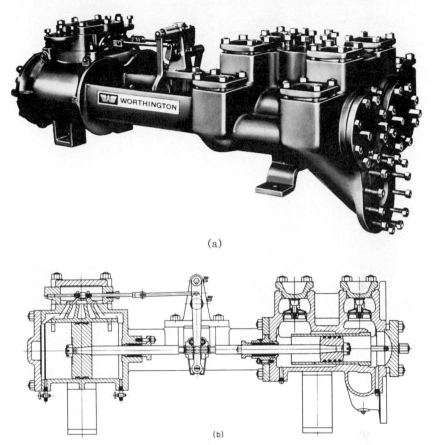

(a)

(b)

FIG. 13 Side-pot piston pump. (Worthington Pump)

service. Plunger pumps have stuffing box packing and glands of the same type as those on the piston rods of piston pumps. All packing leakage is external, where it is a guide to adjustments that control the leakage and extend packing and plunger life. During operation, lubrication can be supplied to the external plunger packing to extend its life. Lubrication cannot be supplied to the piston packing rings on a piston pump.

PISTON-TYPE LIQUID ENDS   The most generally used piston pump is the cap-and-valve-plate design, illustrated in Fig. 7. This is usually built for low pressures and temperatures, although some designs are used up to 350 lb/in² (246 bar) of discharge pressure and 350°F (177°C). The discharge valve units are mounted on a plate separate from the cylinder and have a port leading to the discharge connection. A dome-shaped cap, subject to discharge pressure, covers the discharge valve plate. The suction valve units are mounted in the cylinder directly below their respective discharge valves. A passage in the liquid cylinder leads from below the suction valves down between the cylinders of a duplex pump to the suction connection.

Side-pot liquid ends are used where the operating pressures are beyond the limitations of the cap-and-valve-plate pump. Figure 13 illustrates this design. Suction valves are placed in individual pots on the side of the cylinders and discharge valves in the pots above the cylinders. Each valve can be serviced individually by removing its cover. The small area of the valve covers exposed to discharge pressure makes the sealing much simpler than is the case in the cap-and-

FIG. 14 Close-clearance liquid end pump. (Worthington Pump)

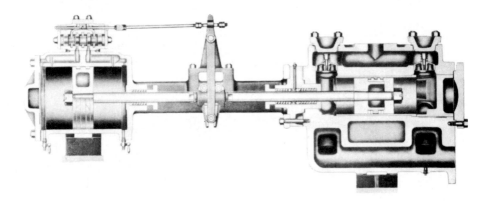

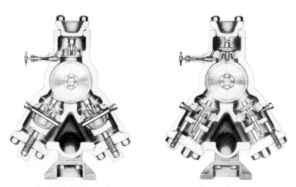

FIG. 15 End views of close-clearance liquid end pump, showing disk valve assembly and wing valve assembly. (Worthington Pump)

valve design. Side-pot liquid ends are widely used in refinery and oil field applications. This design is commonly employed to the maximum pressure practicable for a piston pump.

There are several specially designed piston-type liquid ends which have been developed for specific applications. One of these is the close-clearance design illustrated in Figs. 14 and 15. This pump can handle volatile liquids, such as propane or butane, or a liquid which may contain entrained vapors.

The close-clearance cylinder is designed to minimize the dead space when the piston is at each end of its stroke. The liquid valves are placed as close as possible to the pump chamber to keep clearance to a minimum. The suction valves are positioned below the cylinder at the highest points in the suction manifold to ensure that all the gases are passed into the pump chamber. Although these pumps are of close-clearance design, they are not compressors and can vapor-bind, i.e., a large amount of gas trapped below the discharge valve will compress and absorb the entire displacement of the pump. When this occurs, the discharge valve will not open and this will cause a loss of flow. Hand-operated bypass or priming valves are provided to bypass the discharge valve and permit the trapped gases to escape to the discharge manifold. When the pump is free of vapors, the valves are closed.

There are a number of other special piston pump designs for certain services in addition to the most common types just described. One of these special designs is the wet vacuum pump, which features tight-sealing rubber valves that permit the pump to handle liquid and air or noncondensable vapors. Another special design is made of hard, wear-resistant materials to pump cement grout on construction projects. Another design, shown in Fig. 16, has no suction valves and is made for handling viscous products such as sugarcane pulp, soap, white lead, printer's ink, and tar. The liquid flows into the cylinder from above through a suction port which is cut off as the piston moves back and forth.

PISTON PUMP LIQUID END MATERIALS AND CONSTRUCTION   The materials used for piston pump liquid ends vary widely with the liquids handled. Most of the services to which these pumps are applied use one of the common material combinations listed in Table 1.

The liquid cylinder, the largest liquid end component, is most frequently made from cast iron or bronze. However, other materials are also used. Cast steel cylinders are used in refineries and chemical plants for high-pressure and high-temperature applications. Nickel cast steels are used for low-temperature services. Ni-Resist cast iron, chrome-alloy steels, and stainless steels are occasionally used for certain corrosive and abrasive applications but tend to make pump cost very high. The liquid cylinder heads and valve covers are usually made from the same material as the liquid cylinder.

As was the case in the steam end, a liquid cylinder liner is used to prevent wear and permanent damage to the liquid cylinder. Liners must be replaced periodically when worn by the piston packing to the point that too much fluid leaks from one side of the piston to the other. The liners may be either of a driven-in (or pressed-in) type or of a removable type, which is bolted or clamped in position in the cylinder bore.

The pressed-in type (Fig. 7) derives its entire support from the drive fit in the cylinder bore. As a rule, such a liner is relatively thin and is commonly made from a centrifugal casting or a

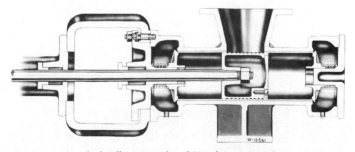

FIG. 16   A pump for handling viscous liquid (Worthington Pump)

TABLE 1 Material and Service Specifications for Pump Liquid Ends

Part	Regular fitted (RF)	Bronze fitted (BF)	Fully bronze fitted (FBF)	All iron fitted (AIF)	All bronze (AB)
Cylinders	Cast iron	Cast iron	Cast iron	Cast iron	Bronze
Cylinder liners	Bronze	Bronze	Bronze	Cast iron	Bronze
Piston	Cast iron	Cast iron	Bronze	Cast iron	Bronze
Piston packing	Fibrous	Fibrous	Fibrous	Cast iron, 3-ring	Fibrous
Stuffing box	Cast iron, bronze bushed	Cast iron, bronze bushed	Cast iron, bronze bushed	Cast iron	Bronze
Piston rod	Steel	Bronze	Bronze	Steel	Bronze
Valve service	Bronze	Bronze	Bronze	Steel	Bronze
Services for which most often used	Cold water; other cold liquids not corrosive to iron and bronze	Same as for RF, with reduced maintenance; continuous hot water	Boiler feed; intermittent hot-water; sodium chloride, brines	Oils and other hydrocarbons not corrosive to iron or steel; caustic solutions	Mild acids which would attack iron cylinders but not acid-resisting bronze

cold-drawn brass tube. After a driven liner is worn to the point where it must be replaced, it is usually removed by chipping a narrow groove along its entire length. This groove is cut as nearly as possible through the liner without damage to the wall of the cylinder bore. After the groove has been cut, the liner can be collapsed inward and removed.

The removable type of liner (Fig. 13) may be removed and replaced without damage. Instead of being a driven fit throughout most of its length, it is located longitudinally by a flange at the cylinder head end and is a driven fit only at the end of the cylinder bore. A flange on the liner fits into a recess at the beginning of the cylinder bore. This flange is held in contact with a shoulder by jack bolts or a spacer between the cylinder head and the end of the liner. Sometimes a packing ring is used between the flange and shoulder for a positive seal. Removable liners are heavier than pressed-in ones.

There are several designs of pistons and piston packings used for various applications. The three most common are as follows:

1. The body-and-follower type of piston with soft fibrous packing or hard-formed composition rings (Fig. 17). The packing is installed in the packing space on the piston with a clearance in both length and depth. This clearance permits fluid pressure to act on one end and the inside of the packing to hold and seal it against the other end of the packing space and the cylinder liner bore.

2. The solid piston or, as shown in Fig. 18, a body and follower with rings of cast iron or other materials. This type is commonly used in pumps handling oil or other hydrocarbons. The metal rings are split with an angle or step-cut joint. Their natural tension keeps them in contact with the cylinder liner, assisted by fluid pressure under the ring.

3. The cup piston (Fig. 19), which consists of a body-and-follower type of piston with molded cups of materials such as rubber reinforced with fabric. Fluid pressure on the inside of the cup presses the lip out against the cylinder bore, forming a tight seal.

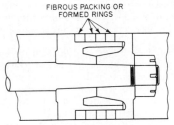

FIG. 17 Body-and-follower piston.

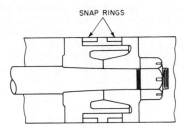

FIG. 18 Body-and-follower piston with snap rings.

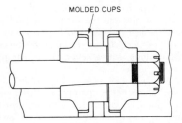

FIG. 19 Body-and-follower piston with molded cups.

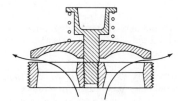

FIG. 20 Stem-guided disk valve.

The piston rod stuffing boxes are usually made separate from, but of the same material as, the liquid cylinder. When handling liquids with good lubrication properties, the stuffing boxes are usually packed full with a soft, square, braided packing which is compatible with the liquid. When the liquid has poor lubricating properties, a lantern ring is installed in the center of the stuffing box with packing rings on both sides of it. A drilled hole is provided through which a lubricant, grease or oil, can be injected into the lantern ring from the outside of the stuffing box. At higher temperatures, approximately 500°F (260°C) or higher, a cooling water jacket is added to the outside of the stuffing box or as a spacer between the stuffing box and the liquid cylinder. The purpose of the cooling water jacket is to extend packing life by keeping the packing cool.

The liquid end valves of all direct-acting steam pumps are self-acting, in contrast to the mechanically operated slide valves in the steam end. The liquid end valves act like check valves; they are opened by the liquid passing through and are closed by a spring plus their weight.

Liquid end valves are roughly divided into three types: the disk valve for general service and thin liquids, the wing-guided valves for high pressures, and either the ball or the semispherical valve for abrasive and viscous liquids.

The valve shown in Fig. 20 is typical of the disk type. This stem-guided design is commonly

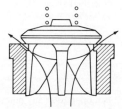

FIG. 21 Wing-guided valve.

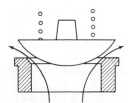

FIG. 22 Semispherical valve.

used in the cap-and-valve-plate design. For hot-water boiler-feed and general service, the disk, seat, and stem are usually made of bronze, although other alloys may also be used. For lower temperatures and pressures, the disk may be made of rubber, which has the advantage of always making a tight seal with the valve seat.

The wing-guided valve shown in Fig. 21 is typical of the design used for high pressures. It derives its name from the wings on the bottom of the valve, which guide it in its seat. The beveled seating surfaces on the valve and seat tend to form a tighter seal than the flat seating surfaces on a disk valve. There is also less danger that a solid foreign particle in the liquid will be trapped between the seat and the valve. This type of valve is commonly made from a heat-treated chrome-alloy-steel forging, although a cast hard bronze and other materials may be used.

The ball valve, as its name suggests, is a ball which acts like a check valve. It is usually not spring-loaded, but guides and lift stops are provided as necessary to control its operation. The ball may be made of rubber, bronze, stainless steel, or other materials as service conditions require. The semispherical valve (Fig. 22) is spring-loaded and can therefore be operated at higher speeds than the ball valve. Both the ball and the semispherical type have the advantage of having no obstructions to flow in the valve seat (the disk valve seat has ribs and the wing-guided valve has vanes which obstruct the flow). The one large opening in the seat and the smooth spherical surface of ball and semispherical valves minimize the resistance to flow of viscous liquids. These types are also used for liquids with suspended solids because their rolling seating action prevents trapping of the solids between the seat and the valve.

PLUNGER-TYPE LIQUID ENDS   As mentioned previously, plunger-type pumps are used where dependability is of prime importance, even when the pump is operated continuously for long periods and where the pressure is very high. Cast liquid end plunger pumps are used for low and moderate pressures. Forged liquid end pumps (Fig. 23), which are the most common plunger types, are used for high pressures and have been built to handle pressures in excess of 10,000 lb/in^2 (69 MPa).

Most of these designs have opposed plungers, i.e., one plunger operating into the inboard end of the liquid cylinder and one into the outboard end. The plungers are solidly secured to inboard and outboard plunger crossheads. The inboard and outboard plunger crossheads are joined by side rods positioned on each side of the cylinder. With this arrangement, each plunger is single-acting, i.e., it makes only one pressure stroke for each complete reciprocating cycle. The pump, however, is double-acting because the plungers are connected by the side rods.

PLUNGER PUMP LIQUID END MATERIALS   The liquid cylinder of a forged liquid end plunger-type pump is most commonly made from forged steel, although bronze, Monel, chrome alloy, and stainless steels are also used. The stuffing boxes and valve chambers are usually integral with the cylinder (Fig. 23), which is desirable for high temperatures and pressures since high-pressure joints are minimized.

The liquid plungers may be made of a number of materials. The plungers must be as hard and smooth as possible to reduce friction and to resist wear by the plunger packing. Hardened chrome-alloy steels and steel coated with hard-metal alloys or ceramics are most commonly used.

The stuffing box packing used will vary widely depending upon service conditions. A soft, square packing cut to size may be used. However, solid molded rings of square, V-lip, or U-lip design are commonly used at higher pressures. Oil or grease is frequently injected into a lantern ring in the center of the stuffing box to reduce friction and reduce packing and plunger wear.

The liquid valves may be of any of the types or materials described above. However, the wing-guided valve with beveled seating surfaces is the most common because it is most suitable for high pressures.

DIRECT-ACTING STEAM PUMP PERFORMANCE _____

The direct-acting steam pump is a very flexible machine. It can operate at any point of pressure and flow within the limitations of the particular design. The speed of, and therefore the flow from, the pump can be controlled from stop to maximum by throttling the steam supply. This can be done by either a manual or an automatically operated valve in the steam supply line. The maxi-

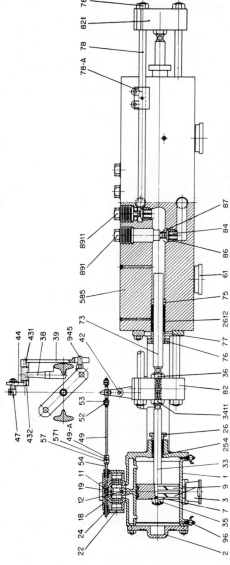

FIG. 23 Simplex plunger pump with forged steel cylinder: (1) steam cylinder with cradle, (2) steam cylinder head, (3) steam cylinder foot, (7) steam piston, (9) steam piston rings, (11) slide valve, (12) piston valve, (18) steam chest, (19) steam chest cover, (22) steam chest head, (24) valve rod stuffing box gland, (26) steam piston rod stuffing box gland, (33) steam piston rod, (35) steam piston rod nut, (36) plunger nut, (38) cross stand, (39) lever, (42) fulcrum pin, (44) crank, (47) tappet, (49) valve rod link, (49A) valve rod link head, (52) lost-motion adjusting nut, (53) lost-motion locknut, (54) valve rod, (57) valve rod head, (61) liquid cylinder foot, (73) plunger, (75) plunger lining, (76) plunger gland flange, (77) plunger gland lining, (78) side rod, (78A) side rod guide, (82) plunger crosshead, (84) metal valve, (86) valve seat, (87) valve spring, (96) steam cylinder drain valve, (254) piston rod stuffing box bushing, (431) lever key, (432) crank key, (571) valve rod head pin, (585) liquid cylinder, (781) side rod nut, (821) plunger crosshead, rear, (891) suction valve plug, (945) crosshead pin, (2612) lantern gland, (8911) discharge valve plug, (3411) steam piston rod jam nut. (Worthington Pump)

3.41

mum speed of a particular design is primarily limited by the frequency with which the liquid valves will open and close smoothly. The pump will operate against any pressure imposed upon it by the system it is serving, from zero to its maximum pressure rating. The maximum pressure rating of a particular design is determined by the strength of the liquid end. In a particular application, the maximum liquid pressure developed may be limited by the available steam pressure and by the ratio of the steam piston and liquid piston areas.

Steam Pump Capacity   The flow to the discharge system is termed the pump *capacity*. The capacity, usually expressed in U.S. gallons per minute (cubic meters per minute), is somewhat less than the theoretical *displacement* of the pump. The difference between displacement and capacity is called *slip*. The displacement is a function of the area of the liquid piston and the speed at which the piston is moving.

The displacement of a single double-acting piston can be calculated from the formula

in USCS units $$D = \frac{12 \times AS}{231} \quad \text{or} \quad 0.0408 d^2 S$$

in SI units $$D = 1 \times 10^{-6} \times AS$$

where D = displacement, gpm (m³/min)
$\quad A$ = area of piston or plunger, in² (mm²)
$\quad S$ = piston speed, ft/min (m/min)
$\quad d$ = diameter of the liquid piston or plunger, in (mm)

For a duplex double-acting pump, D is multiplied by 2. This formula neglects the area of the piston rod. For very accurate calculations, it is necessary to deduct the rod area from the piston area. This is normally not done, and the resultant loss is usually considered part of the slip.

The slip also includes losses due to leakage from the stuffing boxes, leakage across the piston on packed-piston pumps, and leakage back into the cylinder from the discharge side while the discharge valves are closing. Slip for a given pump is determined by test. For a properly packed pump, slip is usually 3 to 5%. As a pump wears, slip will increase, but this can be compensated for by increasing the pump speed to maintain the desired capacity.

Piston Speed   Although piston speed in feet per minute (meters per minute) is the accepted term used to express steam pump speed, it cannot easily be measured directly and is usually calculated by measuring the revolutions per minute of the pump and converting this to piston speed. One revolution of a steam pump is defined as one complete forward and reverse stroke of the piston. The relationship between piston speed and rpm is

in USCS units $$S = \frac{\text{rpm} \times \text{stroke}}{6}$$

in SI units $$S = 0.002 \times \text{rpm} \times \text{stroke}$$

where S = piston speed, ft/min (m/min)
$\quad$ rpm = revolutions per minute
stroke = stroke of pump, in (mm)

A steam pump must fill with liquid from the suction supply on each stroke, or it will not perform properly. If the pump runs too fast, the liquid cannot flow through the suction line, pump passageways, and valves fast enough to follow the piston. On the basis of experience and hydraulic formulas, maximum piston speeds that vary with the length of stroke and the liquid handled can be established.

Table 2 shows general averages of maximum speed ratings for pumps of specified stroke handling various liquids. Some pumps, by reason of exceptionally large valve areas or other design features, may be perfectly suitable for speeds higher than shown. From the table it should be noted that piston speed should be reduced for viscous liquids. Unless the net positive suction head is proportionally high, viscous liquids will not follow the piston at high speeds because frictional resistance in suction lines and in the pump increases with viscosity and rate of flow. Pumps handling hot water are run more slowly to prevent boiling of the liquid as it flows into the low-pressure area behind the piston.

TABLE 2 Average Maximum Speed Ratings

Stroke length, in°	Piston speed, ft/min°					
	Cold water; oil to 250 SSU	Oil, 250–500 SSU	Oil, 500–1000 SSU	Oil, 1000–2500 SSU	Oil, 2500–5000 SSU	Boiler feed 212°F (100°C)
3	37	35	33	29	24	22
4	47	45	42	36	31	28
5	53	51	47	41	35	32
6	60	57	53	46	39	36
7	64	61	57	49	42	39
8	68	65	61	53	45	41
10	75	72	67	58	49	45
12	80	77	71	62	52	48
15	90	86	80	69	57	54
18	95	91	85	73	62	57
24	105	100	94	81	68	63
36	120	115	107	92	78	72

°SI conversion factors: in $\times$ 25.4 = mm; ft/min $\times$ 0.3048 = m/min.

Size of Liquid End   The size of a steam pump is always designated as follows:

Steam piston diameter $\times$ liquid piston diameter $\times$ stroke

For example, a 7½ $\times$ 5 $\times$ 6 steam pump has a 7½-in (191-mm) diameter steam piston, a 5-in (127-mm) diameter liquid piston, and a 6-in (152-mm) stroke.

To determine the liquid piston diameter for a specified capacity, the following procedure is used. First a reasonable stroke length is assumed and the maximum piston speed for this stroke and the type of liquid pumped is selected from Table 2. Then the desired capacity is increased by 3 to 5% to account for slip. The result is the desired displacement. Then for either a simplex or duplex pump the liquid piston diameter can be calculated as follows:

	For simplex pumps	*For duplex pumps*
in USCS units	$d_l = 4.95 \left(\dfrac{D}{S}\right)^{1/2}$	$d_l = 3.5 \left(\dfrac{D}{S}\right)^{1/2}$
in SI units	$d_l = 1128.4 \left(\dfrac{D}{S}\right)^{1/2}$	$d_l = 797.9 \left(\dfrac{D}{S}\right)^{1/2}$

where d_l = liquid piston diameter, in (mm)
  D = displacement, gpm (m³/min)
  S = piston speed, ft/min (m/min)

Using the resultant liquid piston diameter, the next larger standard piston size is selected.

Size of Steam End   To calculate the size of the steam end required for a specific application, the basic principle of steam pump operation should be considered. A simple schematic of a steam pump is shown in Fig. 24.

In order for the pump to move, the force exerted on the steam piston must exceed the force on the liquid piston which is opposing it. The force on the steam piston is the product of the net steam pressure and the steam piston area. The *net* steam pressure is the steam inlet pressure minus the exhaust pressure. The force acting on the liquid piston is the product of the net liquid pressure and the liquid piston area. The *net* liquid pressure is the pump discharge pressure minus the suction pressure or *plus* the suction lift. This may be expressed algebraically as follows:

$$P_s A_s > P_l A_l$$

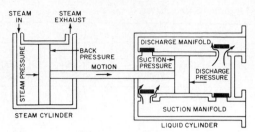

FIG. 24 Schematic showing direction of forces on pistons.

where P_s = net steam pressure, lb/in^2 (bar)
  A_s = steam piston area, in^2 (mm^2)
  P_l = net liquid pressure, lb/in^2 (bar)
  A_l = liquid piston area, in^2 (mm^2)

Since the pistons are circular, the squares of their diameters are directly proportional to their areas, and the above formula can be rewritten as

$$P_s d_s^2 > P_l d_l^2$$

where d_s = steam piston diameter, in (mm)
  d_l = liquid piston diameter, in (mm)

In practice it is necessary for the force on the steam piston to exceed the force opposing it on the liquid piston by a considerable amount. This is because of mechanical losses, which include stuffing box friction, friction between piston rings and cylinder of both liquid and steam ends, and the operation of the valve gear. These losses are determined by test and are accounted for in size calculations by introduction of a mechanical efficiency figure. Mechanical efficiencies are expressed as a percentage, with 100% being a perfect balance of forces acting on the steam and liquid pistons as expressed in the above formula. Since the efficiencies of two identical pumps may vary with stuffing box and piston ring packing tightness, the efficiencies published by manufacturers tend to be conservative.

With the mechanical efficiency factor inserted, the formula of forces becomes

$$P_s d_s^2 E_m = P_l d_l^2$$

where P_s = net steam pressure, lb/in^2 (bar)
  d_s = steam piston diameter, in (mm)
  P_l = net liquid pressure, lb/in^2 (bar)
  d_l = liquid piston diameter, in (mm)
  E_m = mechanical efficiency, expressed as a decimal

This formula is commonly used to determine the *minimum* size of steam piston required when the liquid piston size has already been selected and the net steam and net liquid pressures are known. For this calculation, the formula is rearranged to the form

$$d_s = d_l \left(\frac{P_l}{P_s E_m} \right)^{1/2}$$

The efficiency of a long-stroke pump is greater than that of a short-stroke pump. Although mechanical efficiency varies with stroke length, any two pumps of the same size are capable of the same efficiency.

Table 3 shows typical mechanical efficiencies that can be used to determine required steam end size.

Steam Consumption and Water Horsepower After determining the proper size of other pump types, the next concern usually is to calculate the maximum brake horsepower so that the proper size of driver can be selected. With a steam pump, the next step is usually to determine

TABLE 3   Typical Mechanical Efficiency
Values

Stroke length, in (mm)	Mechanical efficiency, %	
	Piston pump	Plunger pump
3 (76)	50	47
4 (102)	55	52
5 (127)	60	57
6 (152)	65	61
8 (201)	65	61
10 (254)	70	66
12 (305)	70	66
18 (457)	73	69
24 (610)	75	71

the steam consumption. This must be known to ensure that the boiler generating the steam is large enough to supply the steam required by the pump as well as that required for all its other services.

To determine the steam consumption, it is necessary first to calculate the water horsepower as follows:

in USCS units
$$whp = \frac{Q \times P_l}{1715}$$

in SI units
$$kW = \frac{Q \times P_l}{36}$$

where whp = water horsepower
Q = pump capacity, gpm (m^3/h)
P_l = net liquid pressure, lb/in^2 (bar)

A steam consumption chart (Fig. 25) affords a means of quickly obtaining an approximate figure for the steam rate of direct-acting steam pumps. For duplex pumps, divide the water horsepower by 2 before applying it to the curves. These curves were made up on the basis of water horsepower per cylinder; if the above procedure is not followed, inaccurate results will be obtained.

Starting with the water horsepower per cylinder:

1. Move vertically to the curve for steam cylinder size.

2. Move horizontally to the curve for 50-ft/min (15.15-m/min) piston speed. This is the basic curve from which the other curves were plotted.

3. Move vertically to the piston speed at which the pump will run.

4. Move horizontally to the steam rate scale and read it in pounds per water horsepower-hour (kilograms per kilowatt-hour).

5. Multiply the above result by total water horsepower to obtain the steam rate in pounds per hour (kilograms per hour).

For steam cylinders with diameters as shown, but with longer stroke, deduct 1% from the steam rate for each 20% of additional stroke. Thus, a 12 × 24 steam end will have a steam consumption about 5% less than a 12 × 12 steam end. For 5¼ × 5 and 4½ × 4 steam ends, the 6 × 6 curve will give approximate figures. For cylinders of intermediate diameters, interpolate between the curves.

To correct for superheated steam, deduct 1% for each 10° of superheat. To correct for back pressure, multiply the steam rate by a correction factor equal to

$$\left(\frac{P + BP}{P}\right)^{1/2}$$

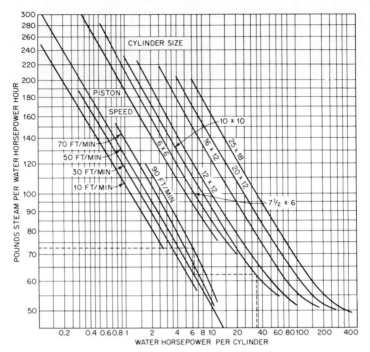

FIG. 25   Approximate steam consumption for steam pumps (pounds of steam per water horsepower-hour = 1.644 kg of steam per water kilowatt-hour; water horsepower per cyliner = 1.341 wkW per cylinder; ft/s = 0.3047 m/s; in = 25.4 mm). (From Hydraulic Institute, "Hydraulic Institute Standards," 12th ed., 1969)

where P = net steam pressure to drive pump, lb/in^2 (bar)
   BP = back pressure, lb/in^2 (bar)

Direct-acting steam pumps have inherently high steam consumption. This is not necessarily a disadvantage, however, when the exhaust steam can be used for heating the boiler-feed water or for building heating or process work. Because these pumps can operate with a considerable range of back pressure, it is possible to recover nearly all the heat in the steam required to operate them. Since they do not use steam expansively, they are actually metering devices rather than heat engines and as such consume heat from the steam only as the heat is lost via radiation from the steam end of the pump. These pumps act, in effect, like a reducing valve to deliver lower pressure steam that contains nearly all its initial heat.

Suction Systems and Net Positive Suction Head
A majority of pump engineers agree that most operating problems with pumps of all types are caused by failure to supply adequate suction pressure to fill the pump properly.

The steam pump industry uses the term *net positive suction head required* (NPSHR) to define the head, or pressure, required by the pump over its datum, usually the discharge valve level. This pressure is needed to (1) overcome frictional losses in the pump, (2) overcome the weight and spring loadings of the suction valves, and (3) create the desired velocity in the suction opening and through the suction valves. The NPSHR of a steam pump will increase as the piston speed and capacity are increased. The average steam pump will have valves designed to limit the NPSHR to 5 lb/in^2 (0.3 bar) (34.5 MPa) or less at maximum piston speed.

If the absolute pressure in the suction system minus the vapor pressure is inadequate to meet or exceed the NPSHR, the pump will cavitate. Cavitation is the change of a portion of the liquid to vapor, and it causes a reduction in delivered capacity, erratic discharge pressure, and noisy operation. Even minor cavitation will require frequent refacing of the valves, and severe cavitation can result in cracked cylinders or pistons or failure of other major parts.

Flow Characteristics   The flow characteristics of duplex and simplex pumps are illustrated in Fig. 26. The flow from a simplex pump is fairly constant except when the pump is at rest. However, because the flow must stop for the valves to close and for the forces on both sides of the steam and liquid pistons to reverse, there is uneven and pulsating flow. This can be compensated for, in part, by installing a pulsation-dampening device on the discharge side of the pump or in the discharge line.

In a duplex pump, one piston starts up just before the other piston completes its stroke, and the overlapping of the two strokes eliminates the sharp capacity drop.

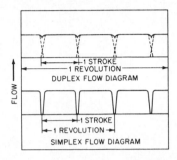

FIG. 26   Flow characteristics of simplex and duplex pumps.

FURTHER READING

Grobholz, R. K.: "Get the Right Steam Pump to Handle the Job," *Mill and Factory*, 1955.

Hydraulic Institute: *Hydraulic Institute Standards*, 12th ed., Cleveland, 1969.

Schaub, D. G.: "Reciprocating Steam Pumps," *South. Power and Ind.*, 1955.

Wright, E. F.: "Direct-Acting Steam Pumps," in *Standard Handbook for Mechanical Engineers* (T. Baumeister and L. Marks, eds.), 7th ed., McGraw-Hill, New York, 1967, pp. 14–9 to 14–14.

Wright, E. F., in *Pump Questions and Answers* (R. Carter, I. J. Karassik, and E. F. Wright, eds.) McGraw-Hill, New York, 1949.

SECTION 3.3
DIAPHRAGM PUMPS

WARREN E. RUPP

Diaphragm pumps are displacement pumps with flexible membranes clamped at their peripheries in sealing engagement with a stationary housing. The central portion moves in a reciprocating manner through mechanical means, such as a crank or an eccentric cam, or by fluid means, such as compressed air or liquid under alternating pressure. An inlet check valve and an outlet check valve control the flow of pumped liquid into and out of the pumping chamber. A distinguishing feature of all diaphragm pumps is that they have no seals or packing and can be used in applications requiring zero leakage. They are also self-priming and can run dry without damage.

MECHANICALLY DRIVEN DIAPHRAGM PUMPS

Mechanically driven diaphragm pumps are usually considered positive displacement pumps similar in many respects to mechanically driven piston pumps, but there are exceptions. Figures 1 and 2 show a popular make of a mechanically driven single-diaphragm pump with a spring on the plunger rod. If the operating pressure exceeds the maximum recommended pumping pressure, the spring compresses and does not move the diaphragm. This type of pump is widely used in the construction industry, where pumps may suck in rocks or other debris. The spring can compress and thus keep a rock from being pushed through the wall of the pumping chamber or causing the cranking mechanism to fail.

In single-diaphragm pumps, the pumped liquid can have a lot of inertia if the suction and discharge lines are relatively long. A pumping cycle consists of an inlet stroke and a discharge stroke. A simple accumulator on the suction (inlet) side of the pump will allow the pump to draw liquid from the accumulator while it simultaneously draws liquid through the suction line. During the discharge stroke, the accumulator can refill with liquid through the suction line. If the discharge line from the pump is relatively long, the inertia of the liquid can be great, as mentioned above, and can impose severe loads on the diaphragm and cranking means as the diaphragm enters the discharge stroke of a complete cycle. The spring on the plunger rod (Fig. 1) can absorb some of the cranking energy early in the discharge stroke and "give it back" during the latter part of the discharge stroke, greatly reducing the inertia loading on the diaphragm and cranking means.

Automotive fuel pumps are usually of the diaphragm type. The diaphragm is moved mechan-

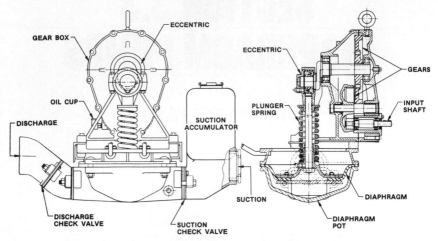

FIG. 1 Cross section of a mechanically driven single-diaphragm pump.

ically by a cam on a suction stroke and is returned by a spring for the discharge stroke. Thus, the spring determines the discharge pressure and provides a nearly constant pump pressure to the carburetor regardless of pump (engine) speed or rate of fuel consumption.

Mechanically driven diaphragm pumps operate by a reciprocating plunger rod usually secured to plates on both sides of the central portion of the diaphragm. The diaphragms are customarily fabric-reinforced elastomers (usually synthetic rubbers) similar in many ways to the fabric-reinforced materials used in pneumatic tires. The diaphragms are normally molded with a convoluted section between the central clamped area and the clamped periphery. This convoluted section permits longer strokes than would be possible otherwise. Mechanically driven diaphragms are unbalanced in operation. The suction and discharge pressures are created by the forces on the central movable portion of the diaphragm. The mechanical force imposed by the plunger rod and central clamping plate is balanced by opposing forces on the unsupported, usually convoluted, section of the diaphragm.

Mechanically driven diaphragm pumps are sometimes duplexed so that the reciprocating means acts alternately on two diaphragms—with one on a suction stroke while the other is on a

FIG. 2 Mechanically driven single-diaphragm pump, engine-powered. (Gorman-Rupp)

discharge stroke and vice versa. A connector called a walking beam is pivoted between two diaphragms. As one diaphragm is pushed down on a discharge stroke, the other diaphragm is simultaneously pulled up on a suction stroke. The pumping chambers with inlet and outlet check valves are manifolded together to a common inlet and a common outlet. The principal advantage of the duplex diaphragm pump is its more nearly constant flow.

Mechanically driven diaphragm pumps are used primarily in the construction industry, for dewatering foundations and cofferdams, and in sewage treatment plants, for pumping lime slurries. They are normally limited to 50 ft (15.2 m) of head—25 ft (7.6 m) of suction lift and 25 ft (7.6 m) of discharge head measured under static conditions. Dynamic forces are so difficult to calculate and measure that the rating of these pumps is simply 50 ft (15.2 m) total static head.

Hand-operated diaphragm pumps are used extensively as bilge pumps on sailboats, where the loss of power is a major concern and hand operation is essential.

HYDRAULICALLY ACTUATED DIAPHRAGM PUMPS

Used extensively as metering pumps, hydraulic diaphragm pumps employ an adjustable-stroke reciprocating piston which forces hydraulic fluid into and out of a chamber behind a diaphragm. The diaphragm is hydraulically balanced and is simply a flexible membrane separating the hydraulic fluid from the pumped liquid. Because the diaphragm is balanced, this type of pump can be used in pumping systems up to 4000 lb/in^2 (275.2 bar°) and at pumping rates from zero to above 2000 gal/h (7.76 m^3/h). Spring-loaded ball check valves are used with a metering accuracy of $\pm$ 1% of maximum flow rate. These pumps are positive displacement pumps and must be provided with a bypass arrangement in case the discharge is blocked. Primary markets for these pumps include water and waste treatment plants and the chemical process, pulp and paper, petrochemical, and petroleum industries. Typical liquids handled are hydrazine, phosphate, acids, filter aids, aluminum chloride, caustic solvents, vinyl acetates, mercaptan, and glycol.

PNEUMATICALLY POWERED DIAPHRAGM PUMPS

Pneumatic diaphragm pumps have unique operating characteristics. The most common types are double-diaphragm pumps, i.e., duplex pumps. Figure 3 is a cross section of an air-operated double-diaphragm pump. Figure 4 is a photograph of the same pump. The principle of operation is quite simple. There are two diaphragm chambers and two flexible diaphragms. Clamped in sandwich fashion at their outer edges, the shaft-connected diaphragms move simultaneously in a parallel path. Compressed air directed to the back side of the left diaphragm moves both diaphragms to the left while air is exhausted from the back side of the right diaphragm. On completion of a stroke, an air distribution valve automatically transfers the compressed air to the back side of the right diaphragm and exhausts air from the left chamber. This continuous reciprocating motion creates an alternating intake and discharge of pumped liquid into and out of each chamber to effect a nearly continuous pumping action from the combined chambers. The pump of Fig. 3 has its inlet at the top and discharges from the bottom. Equipped with flap-type check valves, this pump can handle nearly pipe-sized solids. Because the discharge is from the bottom of the diaphragm chambers, the pump is ideally suited for pumping solids in suspension which may tend to settle out, particularly when the pumping rate is reduced or when the pump is shut down. The bottom outlet enables foreign matter to be easily pumped out of the chambers.

The air distribution valve drawn schematically in Fig. 3 shows an easily understood four-way plug valve. A quarter turn of the valve reverses the flow of compressed air to and from the chambers. In practice, most air-operated double-diaphragm pumps use some type of four-way spool valve.

Figure 5 is a cross section of a popular air-operated double-diaphragm pump with ball-type check valves. Figure 6 is a photograph of the same pump. The inlet is at the bottom of the diaphragm chambers, and the outlet is at the top. Top discharge has the advantage of allowing air

° 1 bar = 10^5 Pa. For a discussion of bar, see *SI Units—A Commentary* in the front matter.

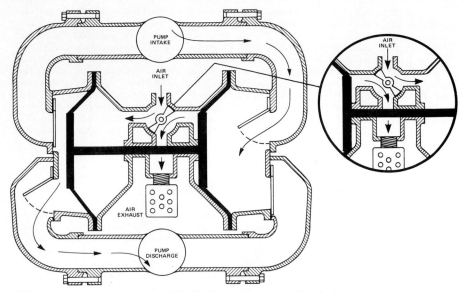

FIG. 3 Cross section of an air-operated double-diaphragm pump with flap check valves. The air distribution valve is schematic for ease of understanding. Circular inset shows reversed air flow to and from chambers behind diaphragms.

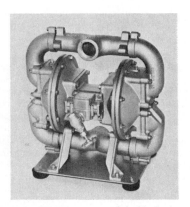

FIG. 4 SandPIPER air-operated double-diaphragm pump with flap check valves. (Warren Rupp)

or vapor to be easily expelled from the chambers. Trapped air or vapor in pumps having bottom outlets can reduce the volumetric displacement of the pumps as the air or vapor is successively compressed and expanded instead of liquid being displaced. This can be of concern in applications requiring relatively high pumping pressure. In applications where the pumping rate is relatively high, there is sufficient turbulence and mixing of air or vapor with the pumped liquid to purge the pumping chambers of the gases.

Figure 7 shows the composite performance chart of a typical 2-inch (51-mm) air-operated double-diaphragm pump. Note that for a constant supply pressure of compressed air, the pump discharge pressure decreases with increasing capacity—similar to a centrifugal pump having a steep characteristic performance curve. An interesting feature of the performance of an air-oper-

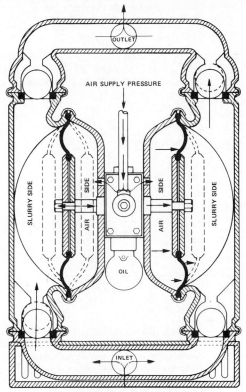

FIG. 5 Cross section of an air-operated double-diaphragm pump with ball check valves.

FIG. 6 CHAMP air-operated double-diaphragm pump with ball check valves. (Wilden Pump & Engineering)

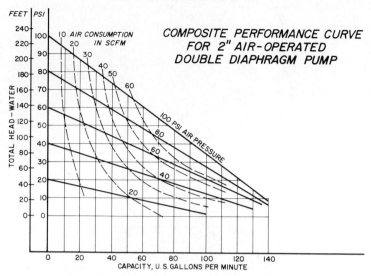

FIG. 7   Performance chart.

ated double-diaphragm pump is that the pump discharge pressure, measured in pounds per square inch (bar), remain the same for a given capacity and air inlet pressure regardless of the specific gravity of the liquid being pumped. For centrifugal pumps, the discharge pressure is directly proportional to the specific gravity of the liquid being pumped.

Figure 7 shows that air consumption decreases with decreased capacity. Air consumption is approximately proportional to pumping rate: zero air consumption at zero pumping rate and maximum air consumption at maximum pumping rate. This is a very important feature. It allows the pumps to be used in applications where the pumping rate must be varied over a wide range or from no flow to the desired flow rate—like turning a water faucet on and off.

MATERIALS OF CONSTRUCTION

Check valves for diaphragm pumps are basically of three types: ball, flap, and poppet (Fig. 8). Flap valves preferably hang in a vertical position with horizontal flow through the valves. Their principal advantage is the ability to handle large objects in suspension. Flap valves with elastomeric hinges are extensively used.

Ball valves usually depend upon gravity to return the ball from the opened to the closed posi-

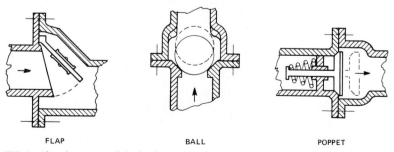

FIG. 8   Three basic types of check valves.

tion. Therefore, the valve seats are horizontal and there is vertical flow through the valve seats. Poppet valves are usually guided by a valve stem and are spring-loaded. They are not position-sensitive and can operate in any position.

Since many diaphragm pumps are used to pump abrasive slurries, the valves usually have elastomeric faces or are elastomeric balls. Large ball check valves may have metal cores covered by a thick wall of synthetic rubber, or they may be of solid rubber.

Diaphragms are customarily made of fabric-reinforced synthetic rubber. Diaphragm materials include most of the synthetic rubbers: neoprene, Buna N, butyl, Viton, and Teflon. Many combinations of pump and diaphragm materials are possible to cover a wide range of pumped products. For some solvents and aggressive acids or alkalis, Teflon diaphragms can be used either directly or as an overlay on the conventional diaphragm.

The diaphragms in air-operated double-diaphragm pumps are essentially balanced and act simply as membranes separating the compressed air from the product being pumped. The only unbalance occurs on the suction stroke of one diaphragm while it is being pulled by the shaft-connected other diaphragm. Where the suction lift is minimal, the unbalance may be negligible.

Pump materials include cast iron, aluminum, stainless steel, Carpenter 20, Hastelloy C, and plastics reinforced with glass fiber.

ADVANTAGES AND LIMITATIONS

Diaphragm pumps are ideally suited for handling abrasive slurries because the liquid velocity through the check valves and pumping chambers does not exceed pipeline velocity and so there is minimal scouring and abrasion from the slurry.

Because there are no close-fit sliding or rubbing parts and velocities are low, these pumps can be used for liquids with viscosities up to 50,000 SSU (11,000 cSt). Because there is minimum turbulence and mixing, they are ideally suited for shear-sensitive materials, such as latex.

While air-operated diaphragm pumps are displacement pumps, they are not positive displacement pumps. The maximum pumping pressure cannot exceed the pressure of the compressed air powering the pumps.

Pumping Dry Powders   Since diaphragm pumps can pump air as well as liquids, they are used successfully to pump dry powders. The air acts as a fluid medium for the powders in suspension, and the pump moves the air containing the suspended powder. Sometimes it is necessary to inject air into the powder to lower the apparent specific gravity and to get the powder into suspension.

Size Limitations for Air-Operated Diaphragm Pumps   There is a practical limit to the size of air-operated diaphragm pumps. For pumping rates above 250 gal/min gpm (58.3 m³/h), the cost of compressed air and the cost of pump components become excessive. For greater pumping rates, it is cheaper to use several pumps in parallel. There is no problem in installing air-operated double-diaphragm pumps in parallel (as there is in paralleling centrifugal pumps).

Pressure Limitations   There is no real limit to pump pressure. Compressed air above 125 lb/in³ (8.6 bar) is not usually available. Pumps operating at higher pressures could be made if the market justified their design and manufacture.

Pump Controls   An air pressure regulator in the compressed air supply line can control the pumping pressure. An air line valve can control the pumping rate. Thus, the pressure and capacity are easily controlled.

Liquids Handled   Pumped liquids and slurries handled by diaphragm pumps include ceramic slurry, paint, cement grout, chemicals, glue, resins, petroleum products, driller's mud, mill scale, ore concentrates, printer's ink, sewage, filter aids, latex, waste oils, wood preservatives, core washes, asphaltic coatings, bilge waste, radioactive waste, lapping compounds, porcelain frit, mine tailings, volatile solvents, coolant with metal fines, varnish, acids, coatings, soapstone slurries, explosives, lime slurries, yeast, chocolate, and wine.

In summary, the limitations of diaphragm pumps are that they are

1. Not practical for pumping rates above 250 gal/min (58.3 m^3/h)
2. Not manufactured for operating pressures above 125 lb/in^2 (8.6 bar)

The advantages of these pumps are as follows:

1. Self-priming from dry start
2. Infinitely variable pumping rate and pressure within pressure and capacity range of pump
3. No seals or packing
4. Can run dry indefinitely
5. Discharge can be throttled to zero flow indefinitely
6. Explosionproof for use in hazardous environments
7. Power used in proportion to pumping rate
8. Can be used in confined areas; no heat build up
9. Can pump abrasive slurries and solids in suspension
10. Can pump viscous liquids up to 50,000 SSU (11,000 cSt)
11. Gentle pumping of shear-sensitive materials
12. Can pump dry powders in air suspension
13. No close-fit, sliding, or rotating parts in contact with liquid
14. No bypass required as in other displacement pumps
15. Zero leakage
16. Simple to maintain and to repair
17. No bedplate or couplings to align
18. Can be used in handling aggressive chemical solutions
19. Can handle a wider range of materials than any other type pump

SECTION 3.4
SCREW PUMPS

J. R. BRENNAN
G. J. CZARNECKI
J. K. LIPPINCOTT

Screw pumps are a special type of rotary positive displacement pump in which the flow through the pumping elements is truly axial. The liquid is carried between screw threads on one or more rotors and is displaced axially as the screws rotate and mesh (Fig. 1). In all other rotary pumps the liquid is forced to travel circumferentially, thus giving the screw pump with its unique axial flow pattern and low internal velocities a number of advantages in many applications where liquid agitation or churning is objectionable.

The applications of screw pumps cover a diversified range of markets: navy, marine, and utilities fuel oil service; marine cargo; industrial oil burners; lubricating oil service; chemical processes; petroleum and crude oil industries; power hydraulics for navy and machine tools; and many others. The screw pump can handle liquids in a range of viscosities, from molasses to gasoline, as well as synthetic liquids in a pressure range from 50 to 5000 lb/in² (3.5 to 350 bar°) and flows up to 5000 gal/min (1135 m³/h).

Because of the relatively low inertia of their rotating parts, screw pumps are capable of operating at higher speeds than other rotary or reciprocating pumps of comparable displacement. Some turbine-attached lubricating oil pumps operate at 10,000 rpm and even higher. Screw pumps, like other rotary positive displacement pumps, are self-priming and have a delivery flow characteristic which is essentially independent of pressure.

According to the Hydraulic Institute Standards, screw pumps are classified into single- or multiple-rotor types. The latter are further divided into timed and untimed categories.

The single-screw pump (Fig. 2) exists only in a limited number of configurations. The rotor thread is eccentric to the axis of rotation and meshes with internal threads of the stator (rotor housing or body); alternatively the stator is made to wobble along the pump centerline.

Multiple-screw pumps are available in a variety of configurations and designs. All employ one driven rotor in mesh with one or more sealing rotors. Several manufacturers have two basic configurations available—single-end (Fig. 3) and double-end (Fig. 4) construction—of which the latter is the better known.

As with every pump type, there are certain advantages and disadvantages characteristic of

° 1 bar = 10⁵ Pa. For a discussion of bar, see *SI Units—A Commentary* in the front matter.

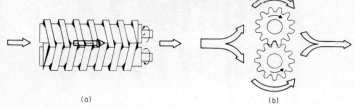

FIG. 1 Diagrams of screw and gear elements, showing (*a*) axial and (*b*) circumferential flow.

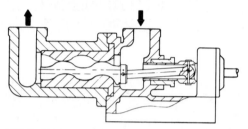

FIG. 2 Single-rotor pump.

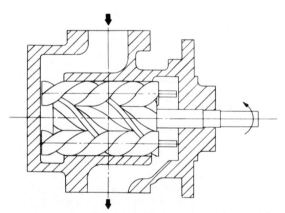

FIG. 3 Multiple-screw single-end arrangement.

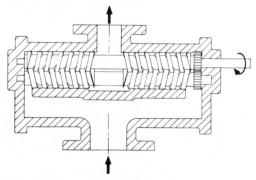

FIG. 4 Multiple-screw double-end arrangement.

screw pump design; these should be recognized in selecting the best pump for a particular application.

Advantages

1. Wide range of flows and pressures
2. Wide range of liquids and viscosities
3. High speed capability, allowing freedom of driver selection
4. Low internal velocities
5. Self-priming, with good suction characteristics
6. High tolerance for entrained air and other gases
7. Minimum churning or foaming
8. Low mechanical vibration, pulsation-free flow, and quiet operation
9. Rugged, compact design—easy to install and maintain
10. High tolerance to contamination in comparison with other rotary pumps

Disadvantages

1. Relatively high cost because of close tolerances and running clearances
2. Performance characteristics sensitive to viscosity change
3. High pressure capability requires long pumping elements

THEORY

In screw pumps, it is the intermeshing of the threads on the rotors and the close fit of the surrounding housing which create one or more sets of moving seals in series between pump inlet and outlet. These sets of seals act as a labyrinth and provide the screw pump with its positive pressure capability. The successive sets of seals form fully enclosed cavities (Fig. 5) which move continuously from inlet to outlet. These cavities trap liquid at the inlet and carry it along to the outlet, providing a smooth flow.

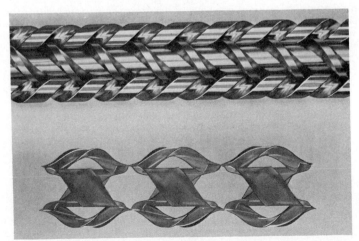

FIG. 5 Axially moving seals and cavities. Alternate cavities filled with oil shown below. (DeLaval-Pyramid)

Delivery   Because the screw pump is a positive displacement device, it will deliver a definite quantity of liquid with every revolution of the rotors. This delivery can be defined in terms of displacement D, which is the theoretical volume displaced per revolution of the rotors and is dependent only upon the physical dimensions of the rotors. It is generally measured in cubic inches (cubic millimeters) per revolution. This delivery can also be defined in terms of theoretical capacity Q_t, measured in U.S. gallons per minute (cubic meters per hour), which is a function of displacement and speed N:

in USCS units
$$Q_t = \frac{DN}{231}$$

in SI units
$$Q_t = 6 \times 10^{-8} DN$$

If no internal clearances existed, the pump's actual delivered capacity, or net capacity, Q would equal the theoretical capacity. Clearances, however, do exist, with the result that, whenever a pressure differential occurs, there will always be internal leakage from outlet to inlet. This leakage, commonly called *slip* S, varies depending upon the pump type or model, the amount of clearance, the liquid viscosity at pumping conditions, and the differential pressure. For any given set of these conditions, the slip for all practical purposes is unaffected by speed. The delivered capacity, or net capacity, therefore, is the theoretical capacity less slip: $Q = Q_t - S$. If the differential pressure is almost zero, the slip may be neglected and $Q = Q_t$.

The theoretical capacity of any pump can readily be calculated if all essential dimensions are known. For any particular thread configuration, assuming geometric similarity, the size of each cavity mentioned earlier is proportional to its length and cross-sectional area. The length is defined by the thread pitch measured in terms of the same nominal diameter which is used in calculating the cross-sectional area (Fig. 6). Therefore the volume of each cavity is proportional to the cube of this nominal diameter, and the pump theoretical capacity is also proportional to the cube of this nominal diameter and the speed of rotation, i.e.,

$$Q_t = KD^3N$$

Thus it can be seen that a relatively small increase in pump size can give a large increase in capacity.

Slip can also be calculated, but usually it depends upon empirical values developed by extensive testing. These test data are the basis of the design parameters used by every pump manufacturer. Slip generally varies approximately as the square of the nominal diameter. The net capacity therefore is

$$Q = KD^3N - S$$

Pressure Capability   As mentioned earlier, screw pumps can be applied over a wide range of pressures, up to 5000 lb/in^2 (345 bar), provided the proper design is selected. Internal leakage must be restricted for high-pressure applications. Close-running clearances and high accuracy of the conjugate rotor threads are requirements. In addition, an increased number of moving seals between inlet and outlet are employed, as in classic labyrinth-seal theory. The additional moving

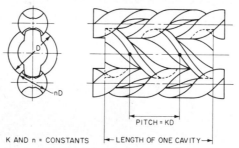

K AND n = CONSTANTS |◄—LENGTH OF ONE CAVITY—►|

FIG. 6   Screw-thread proportions, showing lead, diameter, etc.

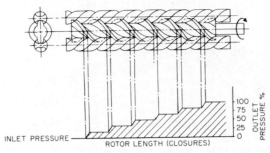

FIG. 7   Pressure gradient along a screw set.

seals are obtained by a significant increase in length of the pumping elements for a given size of rotor and pitch. Here minimum pump length is sacrificed in order to gain pressure capability.

The internal leakage in the pumping elements resulting from the differential pressure between outlet and inlet causes a pressure gradient across the moving cavities. The gradient is approximately linear (Fig. 7) when measured at any instant. Actually the pressure in each moving cavity builds up gradually and uniformly from inlet to outlet pressure as the cavity moves toward the outlet. In effect, the pressure capability of a screw pump is limited by the allowable pressure rise across any one set of moving seals. This pressure rise is sometimes referred to as *pressure per closure* or *pressure per stage* and generally is of the order of 125 to 150 lb/in² (8.6 to 10.3 bar) with normal running clearances, but it can go as high as 500 lb/in² (35 bar) when minimum clearances are employed.

Design Concepts   The pressure gradient in the pump elements of all types of screw pumps produces various hydraulic reaction forces. The mechanical and hydraulic techniques employed for absorbing these reaction forces are among the fundamental differences in the types of screw pumps produced by various manufacturers. Another fundamental difference lies in the method of engaging, or meshing, the rotors and maintaining the running clearances between them. Two basic design approaches are used:

1. *Timed rotors* rely on external means for phasing the mesh of the threads and for supporting the forces acting on the rotors (Fig. 4). In this concept, theoretically, the threads do not come into contact with each other or with the housing bores in which they rotate.

2. *Untimed rotors* rely on the precision and accuracy of the screw forms for proper mesh and transmission of rotation. They utilize the housing bores as journal bearings supporting the pumping reactions along the entire length of the rotors (Fig. 3).

Timed screw pumps require separate timing gears between the rotors and separate support bearings at each end to absorb the reaction forces and maintain proper clearances. Untimed screw pumps do not require gears or external bearings and thus are considerably simpler in design.

CONSTRUCTION

Basic Types   As indicated in the introduction, there are three major types of screw pumps:

1. Single-rotor
2. Multiple-rotor timed
3. Multiple-rotor untimed

The second and third types are available in two basic arrangements—single-end and double-end. The double-end construction (Fig. 8) is probably the best-known version as it was by far the most widely used for many years because of its relative simplicity and compactness of design.

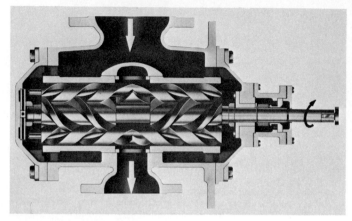

FIG. 8   Double-end pump. Flow path provides axial balance. (DeLaval-Pyramid)

Double-End Screw Pumps   The double-end arrangement is basically two opposed single-end pumps or pump elements of the same size with a common driving rotor of opposed double-helix design within one casing. As can be seen from Fig. 8, the fluid enters a common inlet with a split flow going to the outboard ends of the two pumping elements and is discharged from the middle or center of the pump elements. The two pump elements are, in effect, pumps connected in parallel. The design can also be provided with a reversed flow or low-pressure applications. In either of these arrangements, all axial loads on the rotors are balanced, as the pressure gradients in each end are equal and opposite.

The double-end screw pump construction is usually limited to low- and medium-pressure applications, with 400 lb/in² (28 bar) a good practical limit to be used for planning purposes. However, with special design features, applications to 1400 lb/in² (97 bar) can be handled. Double-end pumps are generally employed where large flows are required or where very viscous liquids are handled.

Single-End Screw Pumps   All three types of screw pumps are offered in the single-end construction. As pressure requirements in many applications were raised, the single-end design came into much wider use because it provided the only practical means for obtaining the greater number of moving seals necessary for high-pressure capability. The only penalty in the use of the single-end pump is the complexity of balancing the axial loads.

The single-end construction is most often employed for handling low-viscosity fluids at medium-to-high pressures or hydraulic fluids at very high pressures.

The single-end design for high pressure is developed by literally stacking a number of medium-pressure, single-end pumping elements in series within one pump casing (Fig. 9). The single-end construction also offers the best design arrangement for quantity manufacture.

Special mention must be made of the single-end, single-rotor design (Fig. 10). The pump elements of this design consist of only a stator and one rotor. The stator has a double-helix internal thread and is constructed of an elastomeric material chemically bonded to a metal tubing. Rotating inside the stator is the rotor, consisting of a single external helical thread. The rotor is constructed of hard chrome stainless- or tool-steel material. One version of this design uses internally developed pressure in the pump to compress the elastomeric stator on the rotor, thus maintaining minimum running clearances.

In most single-end designs, special axial balancing arrangements must be used for each of the rotors, and in this respect the design is more complicated than the double-end construction. For smaller pumps, strictly mechanical thrust bearings can be used for differential pressures up to 150 lb/in² (10.3 bar), while hydraulic balance arrangements are used for higher pressures. For large pumps, hydraulic balance becomes essential at pressures above 50 lb/in² (3.5 bar).

Hydraulic balance is provided through the balance piston (Fig. 11) mounted on the rotors

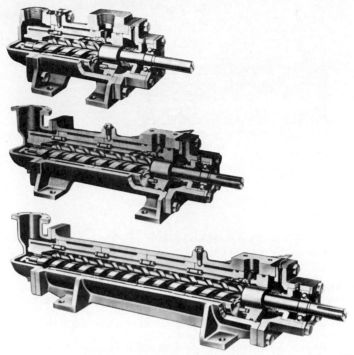

FIG. 9 Increasing pump pressure capability by modular design. (DeLaval-Pyramid)

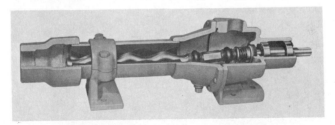

FIG. 10 Single-rotor pump. (Robbins-Meyers)

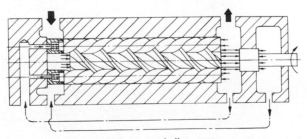

FIG. 11 Axial balancing of power and idler rotors.

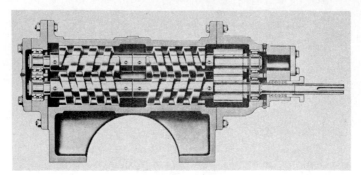

FIG. 12   Double-end internal gear design. (Sier Bath)

between the outlet and seal or bearing chambers, which are at inlet pressure. This piston is exposed to discharge pressure in the outlet chamber and is equal in area to the exposed area of the driven rotor threads; thus the hydraulic forces on the rotor are canceled out.

Timed Design   Timed screw pumps having timing gears and rotor support bearings are furnished in two general arrangements: internal and external. The internal version has both the gears and the bearings located in the pumping chamber and is relatively simple and compact (Fig. 12). This version is generally restricted to the handling of clean lubricating fluids which serve as the only lubrication for the timing gears and bearings.

The external timing arrangement is the most popular and is extensively used. It has both the timing gears and the rotor support bearings located outside the pumping chamber (Fig. 13). This type can handle a complete range of fluids, both lubricating and nonlubricating, and with proper selection of materials has good abrasion resistance. The timing gears and bearings are oil-bath lubricated from an external source. This arrangement requires the use of four stuffing boxes or mechanical seals, as opposed to the internal type, which employs only one shaft seal.

Manufacturers of the timed screw pump claim that the timing gears transmit power to the rotors without the necessity of metallic contact between the screw threads, thus promoting long

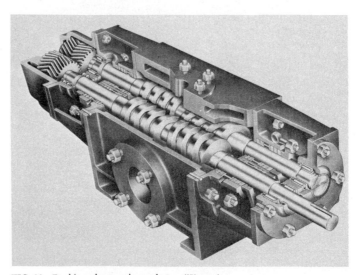

FIG. 13   Double-end external gear design. (Warren)

pump life. The gears are timed at the factory to maintain the proper clearance between the screws. Manufacturers also claim that the bearings at each end of the rotating elements completely support the rotors so that they do not come in contact with the housing; hence no liner is required. One set of these bearings also positions and supports the timing gears.

The timing gears are usually helical, herringbone, hardened-steel gears with tooth profiles designed for efficient, quiet, positive drive of the rotors. Cast iron timing gears are also used. Antifriction radial bearings are usually of the heavy-duty roller type, while the thrust bearings which position the rotors axially are either double-row ball-thrust or spherical-roller types.

The housing can be supplied in a variety of materials, including cast iron, ductile iron, cast steel, stainless steel, and bronze. In addition, the rotor bores of the housing can be lined with industrial hard chrome for abrasion resistance.

Since the rotors are not in metallic contact with the housing or with one another, they can also be supplied in a variety of materials, including cast iron, heat-treated alloy steel, stainless steel, Monel, and nitralloy. The outside of the rotors can also be furnished with a hard coating of tungsten carbide, chrome oxide, or ceramic.

Untimed Design   The untimed type of screw pump has rotors which have generated mating-thread forms that allow any necessary driving force to be transmitted smoothly and continuously between the rotors without the use of timing gears. The rotors can be compared directly with precision-made helical gears with a high helix angle. This design usually employs three rotor screws with the center, or driven, rotor in mesh with two close-fitting sealing, or idler, rotors symmetrically positioned about the central axis (Fig. 14). A close-fitting housing provides the only transverse bearing support for both driven and idler rotors.

The use of the rotor housing as the only means for supporting idler rotors is a unique feature of the untimed screw pump. There are no outboard support bearings required on these rotors. The idler rotors in their related housing bores are in effect partial journal bearings which generate a hydrodynamic fluid film that prevents metal-to-metal contact. The load-supporting capability of this design follows closely the laws of hydrodynamic film theory. The key parameters of rotor size, clearance, surface finish, speed, fluid viscosity, and bearing pressure are related as in a journal bearing.

Since the idler rotors are supported by the bores along their entire length, there are no bending loads applied to them. The central driven rotor is also not subjected to any bending loads because of the symmetrical positioning of the idler rotors and the use of two threads on all the rotors (Fig. 15). This is quite different from the two-rotor design common to the timed type, where the hydraulic forces generated in the pump cause significant bending loads on the intermeshing pairs of screws.

In contrast to timed pumps, the untimed design, with its absence of timing gears and bearings, appears very simple, but its success depends entirely upon the accuracy and finish of the rotor threads and rotor housing bores. Special techniques and machine tools have been developed to manufacture these parts. The combination of design simplicity and manufacturing techniques has

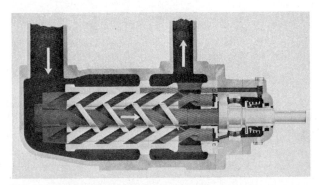

FIG. 14   Single-end design. (Roper)

enabled this design to be used in very long rotor lengths, with a multiplicity of sealing closures, for applications up to 5000 lb/in² (350 bar).

In special applications handling highly aerated oils, a four-rotor design is sometimes employed in the untimed type. Three idler rotors are equally spaced radially at 120°F (49°C) around the driven rotor. This design is not truly a positive displacement pump, and it falls outside the scope of this handbook.

The rotors in untimed pumps are generally made of gray or ductile iron or carbon steel. The thread surfaces are often hardened for high pressure and abrasive resistance. Flame hardening, induction hardening, and nitriding are currently used. Through-hardened tool steel or stainless steel can be used in some critical applications.

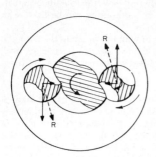

FIG. 15   Force diagram on rotor set.

Rotor housings, or liners, are made of gray pearlitic iron, bronze, or aluminum alloy. In many instances the bores as well as the rotors may be treated by the application of dry lubricant or toughening coatings. The pump casings are made of gray iron or ductile iron or cast steel where shock or other safety requirements demand it.

In many untimed designs an antifriction bearing is employed on the shaft end of the driven rotor (Fig. 9) to provide precise shaft positioning for mechanical seal and coupling alignment. This bearing can be either an external grease-sealed bearing or an internal type with the pumped fluid providing the lubrication. The bearing also supports overhung loads with belt or gear drives.

Seals   As with any rotary pump, the sealing arrangement for the shafts is very important and often is critical. Every type of rotary seal has been used in screw pumps at one time or another. All types of pumps require at least one rotary seal on the drive shaft. The timed screw pumps with external timing and bearings require additional seals at each rotor end to separate the pumped fluid from the lubricating oil necessary for the gears and bearings.

For drive shafts, rotary mechanical seals as well as stuffing boxes or packings are used, depending on the manufacturer and/or customer preference. Double back-to-back arrangements with a flushing liquid are sometimes used for very viscous or corrosive substances.

The mechanical seal is gaining wider use with the advent of new elastomers such as Viton, Butyl, and Nordel. The rotary components are made of carbon, bronze, cast iron, Ni-resist, carbides, or ceramics. The mechanical seal can be designed to be completely or partly independent of the fluid pressure to which it is exposed, and it also has the capability of operating at subatmospheric pressures without drawing in air. In spite of these advantages, the stuffing box is still preferred by some users. The stuffing box requires regular maintenance and tightening, which many users find objectionable, but a mechanical seal failure usually results in a major shutdown.

PERFORMANCE

Performance considerations of screw pumps are very closely related to applications, and so any discussion must cover both viewpoints.

In the application of screw pumps, there are certain basic factors which must be considered to ensure a successful installation. These are fundamentally the same regardless of the liquids to be handled or the pumping conditions.

The pump selection for a specific application is not difficult if all the operating parameters are known. It is often quite difficult, however, to obtain this information, particularly inlet conditions and fluid viscosity, since it is a common feeling that, inasmuch as the screw pump is a positive displacement device, these items are unimportant.

In any screw pump application—regardless of design—suction lift, viscosity, and speed are inseparable. Speed of operation, therefore, is dependent upon viscosity and suction lift. If a true picture of these latter two items can be obtained, the problem of making a proper pump selection becomes simpler and the selection will result in a more efficient unit.

Inlet Conditions The key to obtaining good performance from a screw pump, as with all other positive displacement pumps, lies in a complete understanding and control of inlet conditions and the closely related parameters of speed and viscosity. To ensure quiet, efficient operation, it is necessary to completely fill with liquid the moving cavities between the rotor threads as they open to the inlet, and this becomes more difficult as viscosity, speed, or suction lift increases. Basically it can be said that, if the liquid can be properly introduced into the rotor elements, the pump will perform satisfactorily. The problem is getting it in!

It must be remembered that a pump does not pull or lift liquid into itself. Some external force must be present to push the liquid into the rotor threads initially. Normally atmospheric pressure is the only force present, but there are some applications where a positive inlet pressure is available.

Naturally, the more viscous the liquid, the greater the resistance to flow and therefore the slower the rate of filling the moving cavities of the threads in the inlet. Conversely, low-viscosity liquids flow quite rapidly and will quickly fill the rotor threads. It is obvious that if the rotor elements are moving too fast, the filling will be incomplete and a reduction in output will result. To obtain complete filling, the rate of liquid flow into the pumping elements should always be greater than the rate of cavity travel. Table 1 lists examples of safe internal axial velocity limits found from experience by one screw pump manufacturer for various liquids and pumping viscosities with only atmospheric pressure available at the pump inlet. It is quite apparent from Table 1 that pump speed must be selected to satisfy the viscosity of the liquid being pumped. The pump speed of rotation is directly related to the internal axial velocity, which in turn is a function of the screw thread lead. The lead is the advancement made along one thread during a complete revolution of the driven rotor, as measured along the axis. In other words, it is the distance traveled by the moving cavity in one complete revolution of the driven rotor.

TABLE 1 Internal Axial Velocity Limits

Liquid	Viscosity, SSU	Velocity, ft/s (m/s)
Diesel oil	32	30 (9)
Lubricating oil	1,000	12 (3.7)
#6 fuel oil	7,000	7 (2.1)
Cellulose	60,000	½ (0.15)

Fluids and Vapor Pressure In many cases screw pumps handle a mixture of liquids and gases, and therefore the general term *fluid* is more descriptive. Most of these fluids, especially petroleum products, because of their complex nature, contain certain amounts of entrained and dissolved air or some other gas, which is released as vapor when the fluid is subjected to reduced pressures. If the pressure drop required to overcome entrance losses is sufficient to reduce the static pressure significantly, vapors are released in the rotor cavities and cavitation results.

Vapor pressure is a key fluid property which must always be recognized and considered. This is particularly true of volatile petroleum products, such as gasoline, for example, which has a very low vapor pressure. Crude oil is an example of a volatile fluid where vapor pressure has been overlooked in the past when applying screw pumps. The vapor pressure of a liquid is the absolute pressure at which the liquid will change to vapor at a given temperature. A common example is that the vapor pressure of water at 212°F (100°C) is 14.7 lb/in^2 (1 bar). For petroleum products, the Reid vapor pressure (absolute) is usually the only information available. This is vapor pressure determined by the ASTM Standard D-323 procedure.

In all screw pump applications, the absolute static pressure must never be allowed to drop below the vapor pressure of the fluid; this will prevent boiling and the release of gases, which again will cause cavitation.

Cavitation, as mentioned previously, results when fluid vaporizes in the pump inlet because of incomplete filling of the pump elements and a reduction of pressure. Under these conditions, vapor bubbles, or voids, pass through the pump and collapse as each moving cavity opens to discharge pressure. The result is noisy vibrations, the severity depending on the extent of vaporization or incomplete filling and the magnitude of the discharge pressure. There is also an attendant reduction in output. It is therefore very important to be fully aware of the characteristics of entrained and dissolved air as well as of the vapor pressure of the fluid to be handled. This is particularly true when a suction lift exists.

Suction Lift Suction lift occurs when the total available pressure at the pump inlet is below atmospheric pressure and is normally the result of a change in elevation and pipe friction. Although screw pumps are capable of producing a high vacuum, it is not this vacuum that forces the fluid to flow. As previously explained, it is atmospheric or some other externally applied pressure that forces the fluid into the pump. Since atmospheric pressure at sea level corresponds to 14.7 lb/in^2 (1 bar) absolute, or 30 inHg (762 mmHg), this is the maximum amount of pressure available for moving the fluid, and suction lift cannot exceed these figures. In practice, the available pressure is lower since some of it is used up in overcoming friction in the inlet lines, valves, elbows, etc. It is considered the best practice to keep suction lift as low as possible. The maximum suction lift available is calculated as follows (Fig. 16):

Total head at source = velocity head + elevation head + static head + frictional head loss

in USCS units
$$H = h + \frac{33.9}{w} = h_v + Z + h_s + \Sigma(h_f) = \frac{V^2}{2g} + Z + \frac{144 P_s}{w} + \frac{144 P_f}{w}$$

in SI units
$$H = \frac{V^2}{2g} + Z + \frac{1.02 \times 10^4\, P_s}{w} + \frac{1.02 \times 10^4\, P_f}{w}$$

Static head at pump inlet = net positive suction head + liquid vapor pressure, ft (m) abs

$$h_s = NPSH + h_{vp} \quad \text{or} \quad NPSH = h_s - h_{vp}$$

Maximum suction lift available = $NPSH$ expressed in reference to atm (gage reading)

in USCS units
$$MSLA = 1 \text{ atm} - NPSH$$

in SI units
$$MSLA = 10.36 \text{ m} - NPSH$$

V = velocity, ft/s (m/s)
P_g, h_g = pressure gage readings at pump inlet flange, lb/in^2 (bar) gage and inHg (mmHg) (vac)
P_s = absolute static pressure at pump inlet, lb/in^2 (bar) abs
h_s, h_{sg} = static head at pump inlet, ft (m) of liquid abs or gage
Z = elevation head, ft (m) in reference to datum
h = reservoir liquid level, ft (m) in reference to datum
h_v, h_f = velocity head and frictional head loss, ft (m)
P_{vp}, h_{vp} = liquid vapor pressure, lb/in^2 (bar) abs, or head, ft (m) abs
P_{sv} = net positive inlet pressure, lb/in^2 (bar) abs
$NPSH$ = net positive suction head, ft (m) of liquid abs
P_f = frictional pressure loss, lb/in^2 (bar)
$MSLA$ = maximum suction lift available from pump, ft (m) of liquid or inHg (mmHg) (vac)
w = specific weight (mass) of liquid, lb/ft^3 (kg/m^3)

The majority of screw pumps operate with suction lifts of approximately 5 to 15 inHg (127 to 381 mmHg). Lifts corresponding to 24 to 25 inHg (610 t 635 mmHg) are not uncommon, and there are numerous installations operating continuously and satisfactorily where the absolute suction pressure is within 0.5 in (13 mm) of a perfect vacuum. In the latter cases, however, the pumps are usually taking the fluid from tanks under vacuum, and no entrained or dissolved air or gases are present. Great care must be taken in selecting pumps for these applications since the inlet losses can very easily exceed the net suction head available for moving the fluid into the pumping elements.

The defining of suction requirements by the user and the stating of pump suction capability by the manufacturer have always been complex problems. Some manufacturers are reluctant to publish a single value of minimum inlet pressure required for satisfactory operation for fear that it may not cover all conditions encountered in practice, such as gas entrainment or temperature fluctuations resulting in viscosity changes and hence changes in line losses. One manufacturer refers to maximum suction lift available (MSLA), specified in inches (millimeters) of mercury, which varies with the viscosity of the fluid pumped and the pumping speed. These MSLA ratings are actually determined from operating test data under precisely controlled conditions.

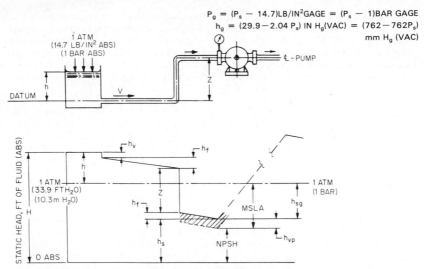

FIG. 16   Relationship between hydraulic gradient, NPSH, and MSLA.

For those engineers accustomed to working with NPSH (net positive suction head), it might prove helpful to point out that the NPSH required by a pump is equal to atmospheric pressure minus MSLA.

To allow the pump manufacturer to offer the most economical selection and also assure a quiet installation, accurate suction conditions should be clearly stated. Specifying a higher suction lift than actually exists results in selection of a pump that operates at a lower speed than necessary. This means not only a larger and more expensive pump, but also a costlier driver. If the suction lift is higher than stated, the outcome could be a noisy pump installation.

There are many known instances of successful installations where screw pumps were properly selected for high suction lift. There are also, unfortunately, many other installation with equally high suction lifts which are not so satisfactory. This is because proper consideration was not given, at the time the pump was specified and selected, to the actual suction conditions at the pump inlet. Frequently suction conditions are given as "flooded" simply because the source feeding the pump is above the inlet. In many cases no consideration is given to outlet losses from the tank or to pipe friction in the inlet lines, and these can be exceptionally high in the case of viscous liquids.

Where it is desired to pump extremely viscous products, such as grease, chilled shortening, cellulose preparations, and the like, care should be taken to use the largest possible size of suction piping, to eliminate all unnecessary fittings and valves, and to place the pump as close as possible to the source of supply. In addition, it may be necessary to supply the liquid to the pump under some pressure, which may be supplied by elevation, air pressure, or mechanical means.

Entrained and Dissolved Air   As mentioned previously, a factor which must be given careful consideration is the possibility of entrained air or other gases in the liquid to be pumped. This is particularly true of installations where recirculation occurs and the fluid is exposed to air either through mechanical agitation, leaks, or improperly located drain lines.

Most liquids will dissolve air or other gases retaining them in solution, the amount being dependent upon the liquid itself and the pressure to which it is subjected. It is known, for instance, that lubricating oils at atmospheric temperature and pressure will dissolve up to 10% air by volume and that gasoline will dissolve up to 20%.

When pressures below atmospheric exist at the pump inlet, dissolved air will come out of solution; both this and entrained air will expand in proportion to the existent absolute pressure. This expanded air will accordingly take up a proportionate part of the available volume of the moving cavities, with a consequent reduction in delivered capacity. See the more detailed study at the end of this section entitled "Effect of Entrained or Dissolved Gas on Performance."

One of the apparent effects of handling liquids containing entrained or dissolved gas is noisy pump operation. When such a condition occurs, it is usually dismissed as cavitation and let go at that; then too, many operators never expect anything but noisy operation from rotary pumps. This should not be the case, particularly with screw pumps. With properly designed systems and pumps, quiet, vibration-free operation can be produced and should be expected. Noisy operation is inefficient; steps should be taken to make corrections until the objectionable conditions are overcome. Proper pump design and size with proper speed selection can go a long way toward overcoming the problem.

Viscosity It is not very often that a screw pump is called upon to handle liquids at a constant viscosity. Normally, because of temperature variations, a wide range of viscosities will be encountered; for example, a pump may be required to handle a viscosity range from 150 to 20,000 SSU, the higher viscosity usually resulting from cold-starting conditions. This is a perfectly satisfactory range for a screw pump, but a better and a more economical selection can be made if information can be obtained concerning such things as the amount of time the pump is required to operate at the higher viscosity, whether the motor can be overloaded temporarily, whether a multispeed motor can be used, and if the discharge pressure will be reduced during the period of high viscosity.

Quite often only the type of liquid is specified, not its viscosity, and assumptions must be made for the operating range. For instance, Bunker C or No. 6 fuel oil is known to have a wide range of viscosity values and usually must be handled over a considerable temperature range. The normal procedure in a case of this type is to assume an operating viscosity range of 20 to 700 SSF. The maximum viscosity, however, might very easily exceed the higher value if extra-heavy oil is used or if exceptionally low temperatures are encountered. If either should occur, the result may be improper filling of the pumping elements, noisy operation, vibration, and overloading of the motor.

Although it is the maximum viscosity and the expected suction lift that determine the size of the pump and set the speed, it is the minimum viscosity that determines the capacity. Screw pumps must always be selected to give the specified capacity when handling the expected minimum viscosity since this is the point at which maximum slip, hence minimum capacity, occurs (Fig. 17).

It should also be noted that the minimum viscosity often determines the selection of the pump model since most manufacturers have special lower-pressure ratings for handling liquids having a viscosity of less than 100 SSU.

Non-Newtonian Liquids The viscosity of most liquids is unaffected by any agitation or shear to which they may be subjected, as long as the temperature remains constant; these liquids are accordingly known as *true* or *Newtonian* liquids because they follow Newton's definition of viscosity. There is another class of liquids, however, such as cellulose compounds, glues, greases, paints, starches, slurries, and candy compounds, which display changes in viscosity as agitation is

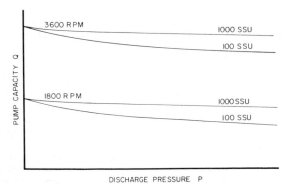

FIG. 17 Head-capacity performance curve with vicosity as parameter, for two speeds.

varied at constant temmperature. The viscosity of these substances will depend upon the shear rate at which it is measured, and these fluids are termed *non-Newtonian*.

If a substance is known to be non-Newtonian, the expected viscosity under actual pumping conditions should be determined, since it can vary quite widely from the viscosity under static conditions. Since a non-Newtonian substance can have an unlimited number of viscosity values (as the shear rate is varied), the term *apparent viscosity* is used to describe its viscous properties. Apparent viscosity is expressed in absolute units and is a measure of the resistance to flow at a given rate of shear. It has meaning only if the shear rate used in the measurement is also given.

The grease-manufacturing industry is very familiar with the non-Newtonian properties of its products, as evidenced by the numerous curves that have been published of apparent viscosity plotted against rate of shear. The occasion is rare, however, when one is able to obtain accurate viscosity information when it is necessary to select a pump for handling these products.

It is practically impossible in most instances to give the viscosity of grease in the terms most familiar to the pump manufacturer, i.e., Saybolt seconds universal or Saybolt seconds Furol, but only a rough approximation would be of great help.

For applications of this type, data taken from similar installations are most helpful. Such information should consist of the type, size, capacity, and speed of installed pumps, suction pressure, and temperature at the pump inlet flange, total working suction head, and, above all, the pressure drop in a specified length of piping. From the latter, a satisfactory approximation of viscosity under operating conditions can be obtained.

Speed   It was previously stated that viscosity and speed are closely tied together and that it is impossible to consider one without the other. Although rotative speed is the ultimate outcome, the basic speed which the manufacturer must consider is the internal axial velocity of the liquid going through the rotors. This is a function of pump type, design, and size.

Rotative speed should be reduced when handling liquids of high viscosity. The reasons for this are not only the difficulty of filling the pumping elements, but also the mechanical losses that result from the shearing action of the rotors on the substance handled. The reduction of these losses is frequently of more importance than relatively high speeds, even though the latter might be possible because of positive inlet conditions.

Capacity   The delivered capacity of any screw pump, as stated earlier, is theoretical capacity less internal leakage, or slip, when handling vapor-free liquids. For a particular speed, $Q = Q_t - S$, where the standard unit of Q and S is the U.S. gallon per minute (cubic meter per minute).

The delivered capacity of any specific rotary pump is reduced by

1. Decreasing speed
2. Decreased viscosity
3. Increased differential pressure

The actual speed must always be known; most often it differs somewhat from the rated, or nameplate, specification. This is the first item to be checked and verified in analyzing any pump performance. It is surprising how often the speed is incorrectly assumed and later found to be in error.

Because of the internal clearances between rotors and their housing, lower viscosities and higher pressures increase slip, which results in a reduced capacity for a given speed. The impact of these characteristics can vary widely for the various types of pumps. The slip, however, is not measurably affected by changes in speed and thus becomes a smaller percentage of the total flow at higher speeds. This is a very significant factor in the handling of low-viscosity fluids at higher pressures, particularly in the case of untimed screw pumps, which favor high speed for best results and best volumetric efficiency. This will not generally be the case with pumps having support-bearing speed limits.

Pump volumetric efficency E_v is calculated as

$$E_v = \frac{Q}{Q_t} = \frac{Q_t - S}{Q_t}$$

with Q_t varying directly with speed. As stated previously, the theoretical capacity of a screw pump varies directly as the cube of the nominal diameter. Slip, however, varies with approximately the square of the nominal diameter. Therefore, for a constant speed and geometry, doubling the rotor size will result in an eightfold increase in theoretical capacity and only a fourfold increase in slip. It follows, therefore, that the volumetric efficiency improves rapidly with increase in rotor size.

On the other hand, viscosity change affects the slip inversely to some power which has been determined empirically. An acceptable approximation for the range 100 to 10,000 SSU is obtained by using the 0.5 power index. Slip varies directly with approximately the square root of differential pressure, and a change from 400 SSU to 100 SSU will double the slip in the same way as will a differential pressure change of 100 to 400 lb/in² (7 to 28 bar):

$$S = K\sqrt{P/\text{viscosity}}$$

Figure 18 shows capacity and volumetric efficiency as functions of pump size.

Pressure   Screw pumps do not in themselves create pressure; they simply transfer a quantity of fluid from the inlet to the outlet side. The pressure developed on the outlet side is solely the result of resistance to flow in the discharge line. The slip characteristic of a particular pump type and model is one of the key factors which determine the acceptable operating range, and it is generally well defined by the pump manufacturer.

Horsepower   The brake horsepower (bhp), or, in SI units, the brake kilowatts, required to drive a screw pump is the sum of the theoretical liquid horsepower (kilowatts) and the internal

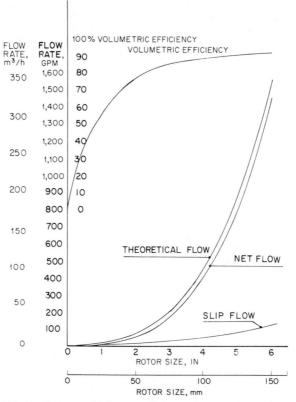

FIG. 18   Capacity and volumetric efficiency as functions of pump size.

power losses. The theoretical liquid horsepower (twhp) is the actual work done in moving the fluid from its inlet pressure condition to the outlet at discharge pressure. Note: This work is done on all the fluid of theoretical capacity, not just delivered capacity, as slip does not exist until a pressure differential ΔP occurs. Screw pump power ratings are expressed in terms of horsepower (550 ft-lb/s) in USCS units and in terms of kilowatts in SI units, and theoretical liquid horsepower (kilowatts) can be calculated: $twph = Q_t \Delta P/1714 \left(tKw = \dfrac{Q_t \, \Delta P}{36} \right)$. It should be noted that the theoretical liquid horsepower (kilowatts) is independent of viscosity and is a function only of the physical dimensions of the pumping elements, the rotative speed, and the differential pressure.

The internal power losses are of two types: mechanical and viscous. The mechanical losses include all the power necessary to overcome the frictional drag of all the moving parts in the pump, including rotors, bearings, gears, and mechanical seals. The viscous losses include all the power lost from the fluid drag effects against all the parts in the pump as well as from the shearing action of the fluid itself. It is probable that mechanical loss is the major component when operating at low viscosities and high speeds and viscous loss is the larger at high-viscosity and slow-speed conditions.

In general, the losses for a given type and size of pump vary with viscosity and rotative speed and may or may not be affected by pressure, depending upon the type and model of pump under consideration. These losses, however, must always be based upon the maximum viscosity to be handled since they will be highest at this point.

The actual pump power output (whp) or wkW, or delivered liquid horsepower (kilowatts), is the power imparted to the liquid by the pump at the outlet. It is computed similarly to theoretical liquid horsepower (kilowatts), using Q in place of Q_t; hence the value will always be less.

The pump efficiency E_p is the ratio of the pump power output to the brake horsepower (Fig. 19).

INSTALLATION AND OPERATION

Rotary pump performance and life can be improved by following the recommendations on installation and operation given below.

Pipe Size   Resistance to flow usually consists of differences in elevation, fixed resistances of restrictions, such as orifices, and pipe friction. Nothing much can be done about the first, since

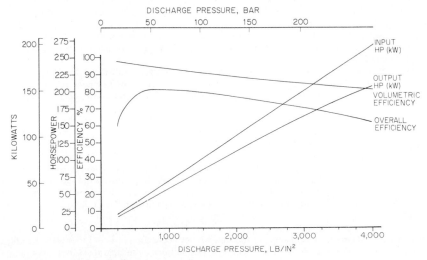

FIG. 19   Typical overall efficiency curves.

this is the basic reason for using a pump. Something, however, can be done about restrictions and pipe friction. Literally millions of dollars are thrown away annually because of the use of piping that is too small for the job. To be sure, all pipe friction cannot be eliminated as long as fluids must be handled in this manner, but every effort should be made to use the largest pipe that is economically feasible. Numerous tables are available from which frictional losses in any combination of piping can be calculated; among the most recent are the tables in the Hydraulic Institute *Pipe Friction Manual.* There are also other similar hydraulic engineering references.

Before any new installation is made, the cost of larger-size piping, which will result in lower pump pressures, should be carefully balanced against the cost of a less expensive pump, smaller motor, and a saving in power over the expected life of the system. The larger piping may cost a little more in the beginning, but the ultimate savings in power will often offset many times the original cost. These facts are particularly true for extremely viscous fluids. Although most engineers dealing with fluids of this type are conscious of what can be done, it is surprising how many installations are encountered where considerable savings could have been made if a little more study had been made initially.

Foundation and Alignment   The pump should be placed on a smooth, solid foundation readily accessible for inspection and repair. It is essential that the power shaft and drive shafts be in perfect alignment. The manufacturer's recommendation of concentricity and parallelism should always be followed and checked occasionally.

The suction pipe should be as short and straight as possible, with all joints airtight. There should be no points at which air or other entrapped gases may collect. If it is not possible to have the fluid flow to the pump under gravity, a foot or check valve should be installed at the end of the suction line, or as far from the pump as possible. All piping should be independently supported to avoid strains on the pump casing.

Start-Up   A priming connection should be provided on the suction side, and a relief valve should be set from 5 to 10% above the maximum working pressure on the discharge side.

Under normal operating conditions with completely tight inlet lines and wetted pumping elements, a screw pump is self-priming. Starting the unit may involve simply opening the pump suction and discharge valves and starting the motor. It is always advisable to prime the unit on initial starting to wet the screws. On new installations, the system may be full of air, which must be removed; if this air is not removed, the performance of the unit will be erratic, and in certain cases air in the system can prevent the unit from pumping. Priming the pump should preferably consist of filling with fluid not only the pump but as much of the suction line as possible.

The discharge side of the pump should be vented at start-up. Venting is especially essential where the suction line is long or where the pump is initially discharging against system pressure.

If the pump does not discharge after being started, the unit should be shut down immediately. The pump should then be primed and tried again. If it still does not pick up fluid promptly, there may be a leak in the suction pipe, or the trouble may be traceable to excessive suction lift from an obstruction, throttled valve, or other causes. Attaching a gage to the suction pipe at the pump will help locate the trouble.

Once the screw pump is in service, it should continue to operate satisfactorily with practically no attention other than an occasional inspection of the mechanical seal or packing for excessive leakage and a periodic check to be certain alignment is maintained within reasonable limits.

Noisy Operation   Should the pump develop noise after satisfactory operation, this is usually indicative of excessive suction lift resulting from cold liquid, air in the liquid, misalignment of the coupling, or, in the case of an old pump, excessive water.

Shutdown   Whenever the unit is shut down, if the operation of the system permits, both suction and discharge valves should be closed. This is particularly important if the shutdown is to be for an extended period because leakage in the foot valve, if the main supply is below the pump elevation, could drain the oil from the unit and necessitate repriming as in the initial starting of the system.

Abrasives   There is one other point that has not yet been discussed, and that is the handling of liquid containing abrasives. Since screw pumps depend upon close clearances for proper pumping action, the handling of abrasive fluids will usually cause rapid wear. Much progress has been made in the use of harder and more abrasive-resistant materials for the pumping elements so that a good job can be done in some instances. It cannot be said, however, that performance is always satisfactory when handling liquids laden excessively with abrasive materials. On the whole, screw pumps should not be used for handling fluids of this character unless shortened pump life and an increased frequency of replacement are acceptable.

EFFECT OF ENTRAINED OR DISSOLVED GAS ON PERFORMANCE

The following data have been taken directly from H. Gartmann, *DeLaval Engineering Handbook*, 3d ed., McGraw-Hill, pages 6–56 to 6–59, by permission of the publisher and Transamerica DeLaval Inc., Pyramid Pump Division.

A very important factor in rotary pump applications is the amount of entrained and dissolved air or other gas in the liquid handled. This is especially true if the suction pressure is below atmospheric. It is generally neglected since rotary pumps are of the displacement type and hence are self-priming. If the entrained or dissolved gases are a large percentage of the volume handled and if their effect is neglected, there may be noise and vibration, loss of liquid capacity, and pressure pulsations.

The amount of entrained gas is extremely variable depending upon the viscosity and type of liquid and how much agitation it may have received.

There is little information available covering the solubility of air and other gases in liquids, especially all those handled by rotary pumps. C. S. Cargoe of the National Bureau of Standards developed the following formula about 1930, based on literature data available at the time, to show the solubility of air at atmospheric pressure in oils, both crude and refined, and other organic liquids:

$$\log_{10} A = \frac{792}{t + 460} - 4 \log_{10} \text{sp. gr.} - 0.4$$

where A = volume of dissolved air, in³/gal (in³/gal = 0.231 cm³/l)
t = temperature, ° F (°F = (°C × ⅘) + 32)
sp. gr. = specific gravity of the liquid

This equation is plotted as Fig. 20, taken from a paper on rotary pumps by Sweeney in the February 1943 issue of the *Journal of the Society of Naval Engineers*. The equation and curve should be considered as approximate only, since some liquids have a higher affinity for gases. For example, gasoline at atmospheric pressure will dissolve as much as 20% of air by volume.

This actual displacement is measured in terms of volume of fluid pumped and will be the same whether the fluid is a liquid, gas, or a mixture of both, as long as the fluid can get to and fill the pump moving voids.

If the fluid contains 5% entrained gas by volume and no dissolved gas and the suction pressure

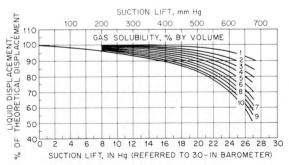

FIG. 20   Effect of dissolved gas on liquid displacement.

is atmospheric, the mixture is then 95% liquid and 5% gas. This mixture fills up the moving voids on the inlet side, but 5% of the space is filled with gas, the remainder with liquid. Therefore, in terms of amount of liquid handled, the output is reduced directly by the amount of gas present, or 5%. The liquid displacement as a function of the theoretical displacement when the suction pressure is atmospheric then becomes

$$D' = D(1 - E)$$

where D = theoretical displacement
  D' = liquid displacement
  E = percent entrained gas by volume at atmospheric pressure, divided by 100

Assume the fluid handled is a mixture containing 95% liquid and 5% entrained gas by volume at atmospheric pressure and no dissolved gas. Also assume that the inlet pressure at the pump is p_i in lb/in^2 abs (bar), which is below atmospheric. The entrained gas will increase in volume as it reaches the pump in direct ratio to the absolute pressures. The new mixture will contain a greater percentage of gas, and the portion of theoretical displacement available to handle liquid becomes

$$D' = \frac{D(1 - E)}{(1 - E) + Ep/p_i}$$

where p = atmospheric pressure, lb/in^2 abs (lb/in^2 = 0.06897 bar)
  p_i = inlet pressure, lb/in^2 abs

Note that p_i depends upon the vapor pressure of the liquid, the static lift, and the frictional and entrance losses to the pump.

In the above equation, if the atmospheric pressure is 14.7 lb/in^2 abs (0.986 bar), the pump inlet pressure 5 lb/in^2 abs (0.345 bar), and the vapor pressure very low, then the liquid displacement is 86.6% of theoretical.

If dissolved gases in liquids are considered, the effect on the liquid displacement reduction is the same as that due to entrained gases, since in the latter case the dissolved gases come out of solution when the pressure is lowered. For example, assume a liquid free of entrained gas but containing gas in solution at atmospheric pressure and the pumping temperature. So long as the inlet pressure at the pump does not go below atmospheric pressure and the temperature does not rise, gas will not come out of solution. If pressure below atmosphperic does exist at the pump inlet, gas will evolve and expand to the pressure existing. This will have the same effect as entrained gas taking up available displacement capacity, and reduce the liquid displacement accordingly. The liquid displacement then will be

$$D' = \frac{D}{1 + y(p - p_i)/p_i}$$

where the symbols have the meanings given above and y is the percent of dissolved gas by volume at pressure p divided by 100. If the operating conditions are 9% dissolved gas at 14.7 lb/in^2 abs (0.986 bar) with a pump inlet pressure of 5 lb/in^2 abs (0.345 bar), the liquid displacement will be 85.2% of theoretical.

If there are both entrained and dissolved gases in the material to be pumped, the liquid displacement becomes

$$D' = \frac{Dp_i(1 - E)}{(1 - E)[p_i + y(p - p_i)] + Ep}$$

where the symbols have the meanings given above. For the operating conditions 5% entrained gas, 9% dissolved gas at 14.7 lb/in^2 abs (0.986 bar), and a pump inlet pressure of 5 lb/in^2 abs (0.345 bar), the liquid displacement is 75.2% of theoretical. Figure 21 shows graphically the reduction in liquid displacement as a function of pump inlet pressure, expressed in terms of suction lift, for different amounts of dissolved gas, neglecting slip.

Figure 22 shows the reduction in liquid displacement as a function of pump inlet pressure, expressed as suction lift, for different amounts of entrained air only, neglecting slip. From this figure it may be noted that a very small air leak can cause a large reduction in liquid displacement, especially if the suction lift is high.

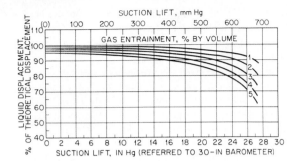

FIG. 21 Effect of entrained gas on liquid displacement.

From these few examples and curves, it would appear that the problem of entrained and dissolved gases could be resolved by providing ample margins in pump capacity. Unfortunately, capacity reductions from the causes mentioned are attended by other and usually more serious difficulties.

The operation of a rotary pump is such that, as rotation progresses, closures are formed which fill and discharge in succession. If the fluid pumped is compressible, such as a mixture of oil and air, the volume within each closure is reduced as it comes in contact with the discharge pressure. This produces pressure pulsations, the intensity and frequency of which depend upon the discharge pressure, the number of closures formed per revolution, and the speed of rotation. Under some conditions the pressure pulsations are of high magnitude and can cause damage to piping and fitting or even the pump, and will almost certainly be accompanied by undesirable noise.

The amount of dissolved gas may be reduced by lowering the suction lift. This may often be controlled by pump location, suction pipe diameter, and piping arrangement.

Many factors are associated with the amount of entrained air that can exist in a given installation. It is prevalent in systems where the liquid is handled repeatedly and during each cycle is exposed to, or mechanically agitated in, air. Unfortunately, in many cases the system is such that air entrainment cannot be entirely eliminated, as in the case of the lubrication system of a reduction gear. Considerable work has been done by oil companies on foam dispersion, and while it has been recommended that special oils be used which are inhibited against oxidation and corrosion, all agree that the best cure is to remove or reduce the cause of foaming, namely, air entrainment.

Even though air entrainment cannot be entirely eliminated, in many cases it is possible, by adhering to the following rules, to reduce it and its ill effects on rotary pump performance:

1. Keep liquid velocity low in the suction pipe to reduce turbulence and pressure loss. Use large and well-rounded suction bell to reduce entrance loss.

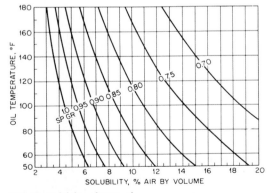

FIG. 22 Solubility of air in oil.

2. Keep suction lift low. If possible, locate pump to provide positive head on the inlet.

3. Put suction piping inside a reservoir to obtain maximum submergence.

4. Submerge all return lines, particularly from bypass and relief valves, and locate them away from the suction.

5. Keep circulation rate low and avoid all unnecessary circulation of the fluid.

6. Do not exceed rated manifold pressures on machinery lubricating systems, as the increased flow through sprays and bearings increases the circulation rate.

7. Heat the fluid where practical to reduce viscosity and to drive off entrained air. Fluids of high viscosity will entrain and retain more air than fluids of low viscosity.

8. Avoid all air leaks, no matter how small.

9. Provide ample vents; exhauster fans to draw off air and vapors have been used with good results.

10. Centrifuging will break a foam and remove suspended foreign matter, which promotes foaming.

11. Use a variable-speed drive for the pump to permit an adjustment of pump capacity to suit the flow requirements of the machinery.

SECTION 3.5
ROTARY PUMPS

C. W. LITTLE, JR.

Rotary pumps are rotary positive displacement pumps in which the main pumping action is caused by relative movement between the rotating and stationary elements of the pump. The rotary motion of these pumps distinguishes them from reciprocating positive displacement pumps, in which the main motion of moving elements is reciprocating. The positive displacement nature of the pumping action of rotary pumps distinguishes them from the general class of centrifugal pumps, in which liquid displacement and pumping action depend in large part on developed liquid velocity.

It is characteristic of rotary pumps, as positive displacement pumps, that the amount of liquid displaced by each revolution is independent of speed. Also, it is characteristic of rotary pumps that a time-continuous liquid seal of sorts is maintained between the inlet and outlet ports by the action and position of the pumping elements and the close running clearances of the pump. Hence rotary pumps generally do not require inlet and outlet valve arrangements, as reciprocating pumps do.

Certain general actions are common to all of the many types of rotary pumps. The following terms are useful in describing these actions. This nomenclature is consistent with that used in *Hydraulic Institute Standards.*°

Rotary pumps are useful in handling both fluids and liquids, where *fluid* is a general term that includes liquids, gases, vapors, and mixtures thereof, and sometimes solids in suspension, and where *liquid* is a more specific term that is limited to true liquids which are relatively incompressible and relatively free of gases, vapors, and solids.

Parts of a Rotary Pump The *pumping chamber* is generally defined as all the space inside the pump that may contain the pumped fluid while the pump is operating. Fluids enter the pumping chamber through one or more *inlet ports* and leave through one or more *outlet ports,* all of which usually include arrangements for liquidtight and airtight connections to external fluid systems. The *body* is that part of the pump which surrounds the boundaries of the pumping chamber and is sometimes called a *casing* or a *housing.* In a very few rotary pumps the body may also be a rotating assembly, but in most types it is stationary and is sometimes called the *stator. Endplates*

° *Hydraulic Institute Standards,* 13th ed., Cleveland, 1975.

are those parts of the body or those separate parts which close the ends of the body to form the pumping chamber. They are sometimes called *pump covers.*

The *rotating assembly* generally includes all the parts of the pump that rotate when the pump is operating, while the *rotor* is the specific part of the rotating assembly which rotates in the pumping chamber. Rotors may be given specific names in specific types of rotary pumps; they may be called gears, screws, etc. Most rotary pumps have *drive shafts* which accept driving torque from a power source. The majority of rotary pumps are mechanically coupled to the driving power source with *couplings* of various types, but a few are coupled to the driving source magnetically or electromagnetically in configurations called *sealless drives.*

Pump *seals* are of two general types, static and moving. *Static seals* provide a liquidtight and airtight seal between demountable stationary parts of the pumping chamber, and *moving seals* are used at pumping chamber boundary locations through which moving elements extend, usually shafts. Moving seals also are formed between pump rotors in some types of rotary pumps. A cavity in the pump body through which a shaft extends is called a *seal chamber,* and leakage through the seal chamber is controlled by either a *radial seal,* which seals on its outside diameter through an interference fit with its mating bore and on the rotating shaft with a radially loaded sliding surface, or a *mechanical seal,* in which two seal faces are opposingly loaded axially and maintained close to each other at all times. Sometimes a *stuffing box* is used instead of a seal chamber. A compressible sealing material called *packing* is compressed in the stuffing box by a part called a *gland* or *gland follower,* which keeps the stuffing in intimate contact with the stationary and rotating surfaces in the stuffing box. Devices called *lantern rings* or *seal cages* are used to allow lubrication or cooling of the stuffing or to control the net pressure on the stuffing in the stuffing box.

Where drive shafts are used, the generally accepted direction of rotation is determined as clockwise or counterclockwise when viewing the pump from the driver end of the drive shaft. In multiple-rotor positive displacement pumps, torque may be transmitted to the rotors, and the angular relationship between them maintained, by the action of *timing gears,* sometimes called *pilot gears.*

A number of auxiliary devices and arrangements may be used with rotary pumps, but two are somewhat characteristic of positive displacement pumps. The pressure at the outlet port and in the outlet portion of the pumping chamber can become damagingly high if the pump discharge is obstructed or blocked, and *relief valves* are used to limit the pressure there by opening an auxiliary passage at a predetermined pressure. The valve may be integral with the body, or integral with an endplate, or attachable. The low velocity (relative to some centrifugal pumps) of the fluid flowing through rotary pump chambers permits some control of pump or fluid temperature by passages or jackets in or on the pump body or endplates through which an auxiliary fluid may be circulated to transfer heat to or from the pump fluid. Such pumps are called *jacketed pumps.*

Pumping Action of Rotary Pumps   The pumping sequence in all rotary positive displacement pumps includes three elementary actions. The rotating and stationary parts of the pump act to define a volume, sealed from the pump outlet and open to the pump inlet, which grows as the pump rotating element rotates. Next, the pump elements establish a seal between the pump inlet and some of this volume, and there is a time, however short, when this volume is not open to either the inlet or outlet parts of the pump chamber. Then the seal to the outlet part of the chamber is opened, and the volume open to the outlet is constricted by the cooperative action of the moving and stationary elements of the pump. In all the many types of rotary pumps the action of the pumping volume elements must include these three conditions: *closed-to-outlet open-to-inlet, closed-to-outlet closed-to-inlet, open-to-outlet closed-to-inlet.* For a good pumping action the open-to-inlet (OTI) volume should *grow* smoothly and continuously with pump rotation, the closed-to-inlet-and-outlet (CTIO) volume should *remain constant* with pump rotation, and the open-to-outlet (OTO) volume should *shrink* smoothly and continuously with pump rotation. At no time should any fluid in the pumping chamber be open to both inlet and outlet ports simultaneously if the pump is truly a positive displacement pump.

ROTARY PUMP TYPES

The major types of rotary pumps are named and described below. Although this is not an exhaustive list, it illustrates all the various rotary pump principles of operation commonly in use.

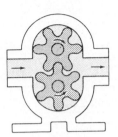

FIG. 1   External gear pump.

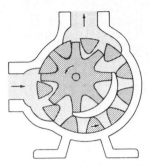

FIG. 2   Internal gear pump.

Gear Pumps   Gear pumps are rotary pumps in which two or more gears mesh to provide the pumping action. It is characteristic that one of the gears be capable of driving the others. The mechanical contacts between the gears form a part of the moving fluid seal between the inlet and the outlet port, and the outer radial tips of the gears and the sides of the gears form a part of the moving fluid seal between inlet and outlet ports. The gear contact locus moves along the tooth surfaces and then jumps *discontinuously* from tooth to tooth as the gears rotate. (These two characteristics distinguish gear pumps from lobe pumps, in which the rotors are *not* capable of driving each other and in which fluid seal contact locus between lobes moves *continuously* across all the radial surfaces of the lobes.)

The two main types of gear pumps are *external* and *internal*. Generally, external gear pumps are arranged so that the center of rotation of each element is external to the major diameter of an adjoining gear and all gears are of the external-tooth type. The center of rotation of at least one gear in an internal gear pump is inside the major diameter of an adjoining gear, and at least one gear is an internal-tooth type or a crown-tooth type.

Figure 1 shows a section through an external gear pump, and Fig. 2 shows a section through an internal gear pump. In these drawings, and in those of following types, the unshaded zones are parts of the body or stator, the lightly shaded zones are areas where liquid may be present in the pump chamber, and the dark zones are parts of the rotating assembly.

The OTI volume of the pump chamber in gear pumps is defined by the body walls and by the gear tooth surfaces between the fluid seal points where the gears mesh and the fluid seal points where each tooth tip meets and seals with the body walls as it leaves the OTI volume. The fluid trapped between the gear teeth and the body walls is sealed from both inlet and outlet chambers and is the CTIO volume. The OTO volume is defined by the body walls and those gear tooth surfaces between the fluid seal points where each tooth tip leaves the body wall and enters the OTO volume and the fluid seal points where the gears mesh.

A part or all of the side (or axial) surfaces of the gears run in small-clearance contact with the axial end faces of the pumping chamber. The gear teeth run in small-clearance contact with each other where they mesh. The tips run in small-clearance contact with the radial surfaces of the pumping chamber in their travel from the OTI volume to the OTO volume. Load-bearing contact between rotors or between rotors and the stator may exist in all three of these zones, and the apertures defined by the running clearances in these zones determine the amount of fluid leakage from the OTO volume to the OTI volume for any given pressure difference between these volumes, for any given effective fluid viscosity. (See discussion of *slip* later in this section.)

Pumping torque is shared by both rotors, and the proportional amount of the total torque felt by each rotor at any instant of time is determined by the locus of the fluid seal point between rotor gear teeth. As this fluid seal point moves *toward* the center of rotation of a rotor gear, the pumping torque on that gear *increases,* and as the seal point moves *away* from the center, the torque *decreases*. Methods of computing this torque are given later. External timing gears are not necessary in gear pumps, but they may be used to transfer torque from one rotating assembly to another to avoid rapid wear between rotor teeth when the pump handles nonlubricating liquids.

A special form of gear pump is illustrated in Fig. 3. It is called a *screw-and-wheel pump*. The driving gear is helical, and the driven gear is a special form of a spur gear. The helical gear always

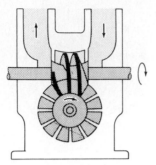

FIG. 3   Screw-and-wheel pump.

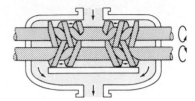

FIG. 4   Multiple-rotor screw pump.

is the driving, or power, rotor in this type of pump, and external timing gears are not used. The pumping torque in the screw-and-wheel pump is felt both by the screw and by the wheel, and the amount of torque felt by each is determined by the fluid seal contact locus points between the two rotors. As in other gear pumps, the running clearances between rotors and between rotors and the body walls determine leakage from the OTO volume to the OTI volume.

Multiple-Rotor Screw Pumps   Figure 4 shows a multiple-rotor screw pump. Usually the screw-shaped rotors in this type of pump cannot drive each other, and timing gears are required. The OTI volume is defined by the body walls and by the screw surfaces between the fluid seal meshing contact between screws and the fluid seal contact between the major diameter of the screw flights and the body wall where the screws adjoin the OTI volume. The CTIO volume is trapped in the screw flights between the fluid seal points adjoining the OTI and the OTO volumes and the screw mesh fluid seal points there. The OTO volume is defined by the body walls and by the screw surfaces between the fluid seal meshing contact between screws and the fluid seal contact between the major diameter of the screw flights and the body wall where the screws adjoin the OTO volume.

The major diameters of the screw rotors run in small-clearance contact with stator walls. The sides of the screw flights run in small-clearance contact with each other where they mesh. Load-bearing contact may exist in both these zones, and the apertures defined by the running clearances determine the amount of fluid leakage from the OTO volume to the OTI volume for any given pressure difference and effective fluid viscosity.

The nature of the contact locus between screws where they mesh determines the pumping torque felt by each rotating assembly. In most two-screw pumps, the pumping torque is divided equally between the screws at all times. The axial-end faces of each screw section feel thrust force generated by the pressure difference between the OTO volume and the OTI volume. Consequently, as illustrated in Fig. 4, opposing sections are usually used in screw pumps designed for high-pressure service, to balance the large net end-thrust forces that otherwise would be imposed on thrust bearings at the rotor ends. Screw pumps are discussed in more detail in Sec. 3.4.

Circumferential Piston Pumps   An external circumferential piston pump is illustrated in Fig. 5. Pistonlike rotor elements, supported from cylindrical hubs inset into the pump endplate, travel in circular paths in mating body bores. This is called *external* because the centers of rotation of the rotors are external to the major diameter of adjoining rotors. The OTI volume is defined by the body walls and by the rotor piston element surfaces between the contact locus points where the rotors form fluid seals with stator walls as they enter (zone between rotors) or leave (top and bottom zones) the OTI volume. The rotors do *not* mesh or touch, and fluid seals exist *only* between the rotor and stator surfaces and not between rotors. The OTO volume is defined by the surfaces of rotor piston elements between their fluid seal contacts with stator walls where they leave (zone between rotors) or enter (top and bottom zones) the OTO volume and by body walls. It is characteristic of circumferential piston pumps either that the rotors do not mesh with, or contact, each other or that they form no fluid seal if they do. This distinguishes them from lobe, gear, and screw pumps.

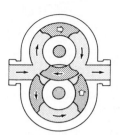

FIG. 5   External circumferential piston pump.

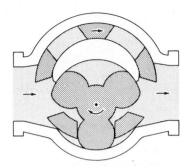

FIG. 6   Internal circumferential piston pump.

The radial surfaces and the axial-end surfaces of the rotor piston elements run in close-clearance contact with body walls, and load-bearing contact may exist in these zones. The apertures defined by the running clearances there determine the amount of fluid leakage from the OTO volume to the OTI volume for a given pressure difference and a given effective viscosity.

Each rotor in an external circumferential piston pump feels the full pumping torque alternately.

Figure 6 shows an internal circumferential piston pump, where the center of rotation of one of the rotors is *inside* the major diameter of the other rotor. Unlike the internal gear pump, which it physically resembles, there is no fluid seal between rotors in the internal circumferential piston pump. In this type of pump the smaller, or *idler*, rotor operates with balanced torque and all the pumping torque is felt continuously by the larger, or power, rotor. With no pumping torque transfer from rotor to rotor, timing gears are usually not needed, even for a pump handling nonlubricating liquids.

Lobe Pumps   The lobe pump receives its name from the rounded shape of the rotor radial surfaces, which permits the rotors to be continuously in contact with each other as they rotate. Figure 7 shows a single-lobe pump, and Fig. 8 shows a multiple-lobe pump.

Unlike gear pumps, neither the number of lobes nor their shape permits one rotor to drive the other, and all true lobe pumps require timing gears. The OTI volume is defined by the body surfaces, the rotor surfaces, the contact between rotors, and the contact between rotor lobe ends and the body. The CTIO volume is defined by the contacts between lobe ends and the body wall and the adjoining body wall and lobe surfaces. The OTO volume is defined by the body walls, the rotor surfaces, the lobe-to-body-wall contacts, and the lobe-to-lobe contacts. In the two-rotor lobe pump the torque is shared by both rotors, with the proportional amount of torque dependent on the position of the rotor-to-rotor contact point on the rotor contact locus. When the contact point is at the major locus radius (maximum lobe radius of one rotor in contact with minimum lobe radius of adjoining rotor), one rotor feels the full pumping torque while the other rotor feels a balanced torque. The transfer of full pumping torque from one rotor to the other takes place as many times in each complete revolution of a rotor as there are lobes on the rotor.

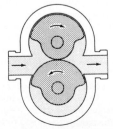

FIG. 7   Lobe pump, single lobe.

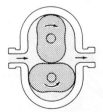

FIG. 8   Lobe pump, multiple lobes.

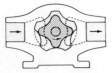

FIG. 9 Internal gear, or internal lobe, pump.

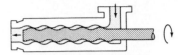

FIG. 10 Single-screw pump.

An *internal lobe pump,* sometimes called an internal gear pump without crescent, is one in which a single rotor with lobelike peripheral shape is moved in a combination of rotation and gyration of center of rotation in a body with internal lobe-shaped contours in such a way that the rotor always touches the body at two or more locations to preserve the fluid seal between OTI and OTO volumes. It is illustrated in Fig. 9. The OTI volume is defined by the outer rotor surface and the inner body surface and the fluid seal points between them. The CTIO volume is defined by the outer rotor surface and the inner body surface between two adjacent fluid seal points, and the OTO volume is defined by the outer rotor surface, the inner body surface, and the rotor-to-body fluid seal points. Most pumps of this type have one fewer rotor lobe than internal body lobe cavity; consequently the term *progressing tooth gear pump* is sometimes used. The full pumping torque is felt by the single rotor, but the torque is cyclic, dependent on the position of the rotor and its sealing arrangement with the body, the number of torque cycles per rotor revolution being equal to the number of lobes on the rotor.

The *single-screw pump,* illustrated in Fig. 10, is similar in principle of operation to the internal lobe pump except that, in the former, the cavity progresses *axially* along a screwlike member with lobular outer surfaces as it rotates in a body with internal matching lobular grooves. A high thrust load at high pressures is felt by the rotor in the single-ended version illustrated in Fig. 10, but this could be balanced by the use of two opposing sections, as in the two-screw pump illustrated in Fig. 4. The full pumping torque is felt by the single rotor.

Rigid Rotor Vane Pumps In the various types of rigid rotor vane pumps, movable sealing elements in the form of rigid blades, rollers, slippers, shoes, buckets, etc., are moved, generally radially inward and outward by cam surfaces, to maintain a fluid seal or seals between the OTI and OTO volumes during the operation of the pump. When the cam surface is internal to the body member and the vanes are mounted in or on the rotor, the pump is called an *internal vane pump* or *vane-in-rotor pump.* The OTI volume is defined by the body walls, the rotor walls, the fluid seal contact between rotor and body, the fluid seal contacts between vanes and rotor, and the fluid seal contacts between vanes and body. The CTIO volume is defined by the body wall surface, the rotor surfaces, and the vane-to-rotor and vane-to-body fluid seal points. The OTO volume is defined by the body surface, the rotor surface, the vane-to-body fluid seal points, and the vane-to-rotor fluid seal points. In internal vane pumps, the volume radially inward behind the vanes needs to be either always of composite constant volume or else vented, because of the pistonlike pumping action of the vanes on fluids trapped there when the vanes are in the form of blades, rollers, etc. However, no such venting is required when the vanes are in the form of rocking slippers. An internal vane pump is illustrated in Fig. 11.

When the cam surface is the external radial surface of the rotor and the vane or vanes are mounted in the body or stator, the pump is called an *external vane pump* or *vane-in-body pump.* The OTI, CTIO, and OTO volumes are defined as for internal vane pumps when multiple external nal vanes are used. Figure 12 shows a single-vane external vane pump. In this case the CTIO volume is defined by the rotor surface, the body surface, and the fluid seal points between them. Vane pumps are single-rotor pumps, and this rotor feels the full pumping torque.

Rotary Piston Pumps Rotary piston pumps are true rotary pumps that require no inlet or outlet valves. They get their name from the pistonlike elements that reciprocate in bores of the rotor as the pump rotor rotates. The pistonlike elements operate off camming surfaces in much the same way that blade and roller vanes do in vane

FIG. 11 Internal (vane-in-rotor) pump.

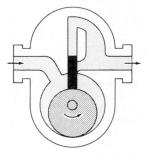

FIG. 12   External (vane-in-body) pump.

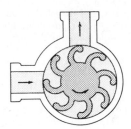

FIG. 13   Flexible vane pump.

pumps, but the pumping action comes directly from the reciprocal movement of the piston elements in their cavities. When the piston elements move axially as the rotating element rotates, the pump is called an *axial piston pump*. The OTI volume is defined by the inlet chamber body walls and by the part of each piston element cylinder exposed to the inlet chamber as it passes. The CTIO volume is defined by the cylinder-to-body-wall and cylinder-to-piston-element volume when the cylinder is sealed from both inlet and outlet, and the OTO volume is defined by the outlet chamber walls and the part of each piston element cylinder exposed by the piston when open to the outlet chamber. The camming action is provided by an inclined plate, adjustable or fixed in angle of incline, that determines the axial position of each piston element as a function of rotor angular position. Since it is a single-rotor pump, the full pumping torque is felt by the rotor.

The pumping action in a *radial piston pump* is the same as in an axial piston pump, but the radially reciprocal piston element movements in the cylinder and bores of the radial pump are generally caused by an eccentric camming surface on the shaft.

Flexible Member Pumps   Several types of rotary positive displacement pumps depend on the elasticity of flexible parts of the pump. The *flexible vane pump*, illustrated in Fig. 13, has a pumping action similar to that of an internal vane pump with the OTI, CTIO, and OTO volumes defined by the rotor surfaces and the body surfaces and the fluid seal contacts between the rotor flexible vanes and the body surfaces.

The *flexible liner pump*, illustrated in Fig. 14, is similar in pumping action to the external vane pump, and all three chamber volumes are defined by the inner surface of the body, the outer surface of the liner, and the liquid seal contact between liner and body bore. Most flexible liner pumps, unlike other rotary pumps, have at least one position of the rotor in which there is no fluid seal between the OTI and OTO volumes. The pump depends only on fluid velocity and inertia to limit backflow during this position.

The *flexible tube pump*, illustrated in Fig. 15, is one in which the three volumes are bounded only by the inner surface of the flexible tube and defined by the locus of the compression points

FIG. 14   Flexible liner pump.

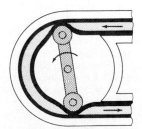

FIG. 15   Flexible tube pump.

on the tube by the rollers or shoes and the body wall. The OTI volume is the tube volume between the inlet and the first compression, or "nip," point; the CTIO volume is that contained between two nip points; and the OTO volume is that contained in the tube volume between the pump outlet and the adjacent nip point. Flexible member pumps, being single-rotor pumps, feel all the pumping torque on the single rotor.

Other Types   There are many other possible types of rotary pumps which have not been included here, but the types listed above and the numerous design variations of each type constitute the bulk of rotary positive displacement pumps used today. Economy of manufacture, special pressure-balancing arrangements or relief valve arrangements, or other such considerations have led to some specific pump designs requiring that the pump be operated in one direction only. However, the principle of operation of almost all rotary pumps permits the pumps to operate in either direction equally well.

INDUSTRY CLASSIFICATIONS _____

Overlying the pump-type classification according to principle of operation are rotary pump classifications according to various industrial, governmental, or military specifications. *It is important that both the manufacturer and the user know of any voluntary or regulatory specifications governing the construction and materials of rotary pumps destined for use in specific industries.* These specifications generally are available either from governmental agencies or from professional societies or trade associations.

For example, the International Association of Milk, Food and Environmental Sanitarians, the United States Public Health Service, and the Dairy Industry Committee have jointly issued specifications called "The Three-A Sanitary Standards for Pumps for Milk and Milk Products." These standards govern the materials of construction of the pump, the surface finish and shape details of fluid contact surfaces, the finish and shape of external surfaces, the method of mounting, even to the details of the legs for the pump base, and restrictions on openings, gaskets, seals, and other pump auxiliary features. Pumps constructed to meet "The Three-A Sanitary Standards" are called *sanitary pumps* and can carry the Three-A seal of approval. Similar specifications have been generated by the International Association of Milk, Food and Environmental Sanitarians, the United States Public Health Service, the United States Department of Agriculture, the Institute of American Poultry Industries, and the Dairy and Food Industries Supply Association to cover design features in pumps used in the handling of cracked eggs.

Standards for pumps used in various processes in the petroleum industry are concerned with the materials and design features that are intended to prevent catastrophic failure when explosive, flammable, or toxic fluids are being handled. Other standards govern the specifications for pumps used on firefighting equipment; military specifications govern the materials and construction of pumps used by the armed services; detailed specifications generated by governmental agencies and professional societies, such as the American Society of Mechanical Engineers, control manufacturing procedures and design limitations of pumps for nuclear power service, etc. In most cases, the descriptors "sanitary," "aseptic," "explosionproof," "N Stamp," and so on, may not be applied casually but can be used only when the manufacturer warrants the pump to meet the specifications. The cost of manufacturing pumps to meet these various special industry specifications generally makes these pumps more expensive than those intended for general service.

Three main industry *classes* of rotary pumps are:

1. The class of rotary pumps used in commercial, industrial, governmental, or military *liquid-handling* tasks. In this class an immense variety of commercially important liquids (well over 1500) are handled. The performance of the pump is judged on its ability to handle the specific liquids in the specific applications. In this class the hydraulic power generated by the pump is usually secondary and the liquid-handling task is primary. Liquid handling may include such factors as liquid transfer and delivery, control of amount transferred, control of flow rates, and control of delivery-point pressures.

2. The class of rotary pumps used in *hydraulic power* applications. In this class, the hydraulic power generated by the pump is of primary importance and hydraulic fluids are carefully

developed and selected to complement pump performance and hydraulic power system requirements. Rotary pump types commonly used in this class include gear and vane pumps, with design features permitting operation at *high pressures* (up to several thousand pounds per square inch [several hundred bar]) on selected hydraulic power fluids.

3. The class of rotary pumps specifically designed for use in a single type of application or on a single type of liquid, in the category of *retail* equipment. This is a miscellaneous class ranging from miniature pumps for home aquariums to fuel-injection pumps for automotive vehicles.

The first class has the widest variety of pump types and industry specifications.

MATERIALS OF CONSTRUCTION

General information on materials of construction is given in Chap. 5. The following paragraphs cover only those particular properties of materials important to rotary pumps.

Special consideration in the selection of materials for use in rotary pumps is needed in five areas: rotary pumps constructed of rigid materials require consideration of the material's *modulus of elasticity, coefficient of friction and nongalling properties* in sliding contact, and *coefficient of thermal expansion;* rotary pumps with flexible members additionally require consideration of the material's *bulk modulus* and *time of recovery* following deformation.

The close running clearances in most rotary pumps constructed of rigid materials impose the requirement that these materials resist deformation and deflection by the various forces present when the pump is operated. Otherwise such deformation or deflection could open the clearances to lower operating efficiency dramatically or close the clearances to cause high mechanical loading or seizing between moving parts and stationary parts. These same considerations require the selection of materials that have compatible coefficients of thermal expansion when the pump must operate at a variety of ambient and process temperatures. Even though care is taken to select materials of high rigidity and similar thermal expansion coefficients, the deflection of rotating assemblies to cause relatively high load-bearing sliding contact between rotating and stationary parts requires that these materials have good bearing characteristics and resistance to galling, up to the point of compressive yield of the mating materials. This is even more important when the pumped liquid has no lubricity. Furthermore, some materials used in centrifugal pumps where surfaces are noncontacting, to provide resistance to corrosion by the pumped fluid, may not be usable in rotary pumps, where sliding contact between rotor and stator may continually wipe or wear away the passivating or protective layer on the materials.

In general, these material selection restrictions become more severe when the pump is to handle low-viscosity fluids at high pressures, particularly those without lubricity or those that contain abrasive fines, and less severe when the pump is to handle lubricating, clean fluids at medium or high viscosities and at medium or low pressures, where larger operating clearances may be used. Even in those pumps where either the mechanical design or the hydraulic pressure balancing design would permit operation of the pump without load-bearing contact between rotary and stationary parts in steady-state dynamic conditions, the high transient forces generated in start-up or shutdown or during unusual operating conditions (such as those accompanying heavy cavitation) may cause sufficient deflection to warrant the same consideration in the selection of materials as for pumps in which load-bearing contact is ordinary.

The performance of flexible member pumps is dependent on the material chosen for the flexible member. The bulk modulus must be high enough to keep the distortion of the flexible member under pressure within functional limits. In many cases, as in flexible vane and flexible tube pumps, the ability of the flexible member to spring back to its original shape (recovery) after the flexing or compression by the pumping action is essential. If the tube in a tube pump, once flattened, were to remain flat and not spring back to its tubular shape, the pump would not operate. If the vanes in a vane pump, once deflected by a cam surface, were to remain deflected, the pump would not operate. Consequently, the materials for flexible members in flexible member rotary pumps should be chosen not only for the initial properties that produce the desired pumping effect but also for resistance to the deterioration of these properties from fatigue, chemical attack by the fluid being pumped, and the temperatures to which the members may be exposed in pumping service.

Frequently, the natural properties of a flexible material may be strengthened and enhanced by lamination or by filling with other materials that may limit deflection or bulging or that may supply additional snapback or recovery after deflection. Other reinforcing materials, such as metal cores molded into flexible vane rotors, may be used to support the flexible member at locations where high stresses may exist.

Where the pump must meet limiting specifications imposed by various industry applications, additional constraints may be imposed on material selection. For example, nonsparking metals not subject to brittle fracture may be required when the pump is to be used to handle flammable or explosive fluids. Only certain materials are approved for use in sanitary pumps for food and dairy service, and the pump bodies of pumps for use in certain petroleum processing applications cannot be constructed of materials such as gray cast iron which may fail in brittle fracture. Many materials commonly used in centrifugal pumps, because of these restraints, cannot be used in rotary pumps.

Most rotary pumps are low-speed pumps. Hence, for the same liquid horsepower, the torque transmitted through the drive shaft to the rotating assembly in rotary pumps is usually much higher than that transmitted to rotating impellers in centrifugal pumps. This is particularly true where the rotary pump is pumping a high-viscosity liquid against a high differential pressure; both the materials and the mechanical design in these high-torque locations sometimes are critical, disqualifying some materials that otherwise would be suitable for use in rotary pumps.

Because of the complexity of these many design considerations, it is not practical to outline a complete procedure for selecting materials for every given application. It should be noted, however, that a pump perfectly suitable for handling a lubricating liquid at medium or low pressures may fail miserably if used to pump a nonlubricating liquid, or a liquid containing abrasive, or a high-temperature liquid, or a high-viscosity liquid at high pressures. Thus there are some general considerations other than corrosion resistance that affect the choice of materials for each application.

OPERATING CHARACTERISTICS

The operating characteristics of rotary pumps are covered in the following paragraphs. For clarity, it is assumed first that the pumped fluid is a true liquid—i.e., it has a Newtonian viscosity and is incompressible—and that its resistance to shear (shear stress) is in direct proportion to the rate of shear (shear strain). The effects of the properties of the pumped fluid on operating characteristics, including viscosity, vapor pressure, temperature, multiphase fluids, liquids with gases, liquids with solids, liquids with abrasives, shear-sensitive liquids, and lubricity, are discussed later. A common liquid used in evaluation tests of basic pump performance is cool water, with or without very small amounts of soluble oil for some lubricity. Hence water may be assumed to be the fluid in the following discussion of basic operating characteristics.

Pump Displacement   The displacement D of a rotary pump is the total net fluid volume transferred from the OTI volume to the OTO volume during one complete revolution of the driving rotor. A standard unit of displacement is cubic inches (cubic millimeters) per revolution. For any given pump, the displacement depends only upon the physical dimensions of the pump elements and the pump geometry and is independent of other operating conditions. In those pumps designed for variable displacement, the pump usually is rated at its maximum displacement.

The displacement of any rotary pump may be computed by the general method of integrating, over one complete revolution of the drive shaft, the differential rate of net volume transfer with respect to angular displacement of the drive shaft through any complete planar segment taken through the pump chamber between the inlet and outlet ports. For any given type of pump, the coordinate system should be selected and the plane located to simplify the computation. Most pump rotors have constant radial dimensions in the axial direction in the body cavity and sweep a right circular cylinder of volume while rotating. Consequently, in single-rotor pumps, or in multiple-rotor pumps in which no sealing contact exists between rotors (all dynamic sealss are formed between rotor elements and body surfaces and none are formed between rotors), the vol-

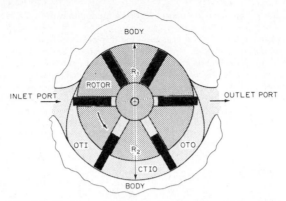

FIG. 16 Displacement calculation dimensions for an internal vane pump.

ume transfer computation may be based on polar coordinates centered on each rotor axis and the contribution to net volume transfer can be computed for each rotor independently.

In general, the axial dimension of the rotor in the body cavity can be most simply expressed if the planar segment is taken through the rotor axis, or at least parallel to the rotor axis. Also, for most types of rotary pumps, the computation is simplified if the intersections of the plane with the body cavity occur in CTIO regions, usually midway between the inlet and outlet ports of the pump. This is particularly true for those rotary pumps which pump equally well in either direction of rotation and are generally symmetric. In many cases the computation can be further simplified by separating the differential statement for volume transfer through the plane from the inlet to the outlet from that for volume transfer through the plane from the outlet to the inlet and expressing the result as a difference. Examples of this method of computation for some commonly used types of rotary pumps follow.

A section through a vane-in-rotor pump is shown in Fig. 16. Let z be the axial distance toward the front endplate from the rotor end surface next to the rear endplate, and let Z be the total axial length of the rotor. Let r be the radial distance from the rotor axis. Let R_1 be the minimum radial dimension of the rotor elements at the intersection of the plane with the minor cam radius of the pump chamber in the CTIO zone, and let R_2 be the maximum radial dimension of the rotor elements at the intersection of the plane with the major cam radius of the pump chamber in the CTIO zone. Let ϕ be the angular displacement of the drive shaft (assumed to be direct-coupled to the rotor; no gear increase or decrease). Then the general equation for D is

$$D = \int_{\phi=0}^{\phi=2\pi} \int_{r=R_1}^{r=R_2} \int_{z=0}^{z=Z} kr \, d\phi \, dr \, dz = k\pi Z(R_2^2 - R_1^2) \qquad (1)$$

where k is a constant used to convert D to desired units ($k = 1$ if z and r are in feet and D is in cubic feet per revolution).

It may be noted that the equation describing the transition of the major radius cam surface to the minor radius cam surface is not used or needed in Eq. 1, because the planar segment is entirely in the CTIO zone of the pump. Also, the integration limits for r were chosen by noting that the net volume transfer for all $r < R_1$ cancels and equals zero. The same result is obtained if the integral is expressed as the difference of the positive contribution of the integration limits 0 to R_2 and the negative contribution of the integration limits 0 to R_1.

The same computations and formula would apply to flexible vane pumps with the vanes on the rotor and to any vane-in-rotor pump where the camming surface creating the pumping action is formed on the interior surface of the body chamber. The equation can be used for most single-rotor positive displacement pumps with appropriate selection of the numerical values for R_1 and

R_2. For example, in the vane-in-body pump, R_1 is the minimum radius of the rotor cam and R_2 is the maximum radius of the rotor cam.

An example extending this method of computing displacement to multiple-rotor pumps that have no rotor-to-rotor sealing contact is the computation on the external circumferential piston pump shown in Fig. 17. A derivation can be made in the same manner used for the vane pump, and the individual equations for the net volume flow for each rotor can be computed and summed.

The pump in Fig. 17 is symmetrical, and the net contribution of each rotor occurs only during a total of half of the drive shaft revolution. In the derivation shown in Eq. 1, this would mean a change in the integration limits of ϕ from 2π to π for the contribution of each rotor. However, when the contributions from each of the two rotors are added, the result is the same as Eq. 1.

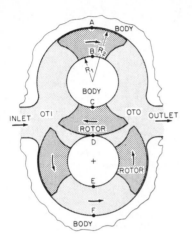

FIG. 17 Displacement calculation dimensions for an external circumferential piston pump.

Another method could be used. Because of the pump symmetry, inspection shows that there are two net volume components continually transferred from the inlet to the outlet chamber by the motion of the rotors in zones A-B and E-F and one equal volume element continually transferred from the outlet to the inlet chamber by the cooperative action of the rotors in zone C-D. Consequently, the net volume being transferred from outlet to inlet in zone C-D is continually canceled by one of the volumes continually being transferred from inlet to outlet at either zone A-B or zone E-F. The entire computation, then, can be made only for zone A-B or zone E-F over the entire revolution. The resulting formula for displacement would be that given in Eq. 1, with R_1 being the minimum radial dimension of the piston element of the rotor and R_2 being the maximum radial dimension of the piston element of the rotor.

The direct computation of displacement for multiple-rotor positive displacement pumps where a moving seal is formed by contact between rotors is much more complex. In such pumps the equation of motion of the locus of the contact point between rotors as a function of angular displacement of the drive shaft is needed for a rigorous solution of the displacement. The differential rate of volume transfer is not constant with angular displacement, but decreases as the contact locus moves toward the inlet chamber and increases as the contact locus moves toward the outlet chamber. In this case, a graph of pump displacement versus angular displacement would not be a straight line (or constant function of angular displacement), as in the prior two examples. In effect, the motion of the contact locus superimposes a ripple on the steady-state component of the differential rate of volume transfer.

Figure 18 shows the locus of the contact line between lobes in a two-lobe pump. The locus is shown on a plane taken perpendicular to the planar segment used for computing displacement, which is taken through the axis of both rotors with the center of rotation of the two rotors being points A and B. Point O is midway between the two rotor centers. The lower-case letters represent key points in the differential rate of volume transfer. Remembering that the velocity of the movement of the locus point toward or away from the outlet port determines the amount of plus or minus deviation from the average differential transfer rate, the following observations can be made.

The *maximum* instantaneous differential volume transfer rate occurs as the locus passes through point O on its way from c to d or from f to a. The *minimum* instantaneous rate of differential transfer occurs when the locus passes through point b or point f. The average differential rate of transfer occurs at points a, c, d, and f. In the symmetric case shown, if R_1 is the distance from O to A or B, if x is the distance from O to b or e, and if R_2 is the maximum radial dimension of a lobe, then derivations similar to those given before will give the maximum and minimum differential rates of volume transfer:

$$\frac{dD}{d\phi_{max}} = kZ(R_2^2 - R_1^2)$$

$$\frac{dD}{d\phi_{min}} = kZ(R_2^2 - R_1^2 - x^2) = kZ[2(R_2R_1 - R_1^2)] \qquad (2)$$

since, for the lobe pump shown, $R_2 = R_1 + x$.

In most pumps of practical design the peak-to-peak amplitude of the ripple is less than 10% of the steady-state component of displacement. Consequently, displacement for multiple-rotor pumps with contacting rotors can be *approximated* by computing the peak displacement (where R_2 is the maximum radial dimension of the rotor and R_1 is one-half the distance between rotors of equal size). The formula is given in Eq. 3:

$$D_{max} = 2\pi kZ(R_2^2 - R_1^2)$$

$$D_{av} = 2\pi kZ\left(\frac{R_2^2}{2} + R_2R_1 - \tfrac{3}{2}R_1^2\right) \qquad (3)$$

$$D_{min} = 4\pi kZ(R_2R_1 - R_1^2)$$

If the rotors are of unequal size, the transfer rate must be computed for each rotor, with R_1 taken as the radius of the pitch circle of the rotor. This also applies to gear pumps, where R_1 is the radius of the pitch circle of the gears. Displacement computed by this simplified method will usually be within 5% of the true average displacement for lobe pumps and within 1% of the true average of

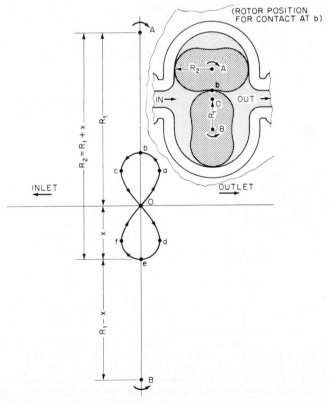

FIG. 18 Rotor contact locus of lobe pump.

displacement for gear pumps. For a closer approximation, Eq. 3 may be used. It is precise only when the ripple waveform has a zero average component and when it is symmetric.

If the displacement cannot be computed precisely from the geometry of the pump, it should be established by test and stated by the manufacturer because displacement is needed to compute the efficiency of the pump under various operating conditions.

Pump Slip   Slip S in a rotary pump is the quantity of fluid which leaks from the OTO volume to the OTI volume per unit of time. Slip depends on the clearances between the rotating and stationary members that define the "leak path orifice," on the differential pressure between the OTO volume and the OTI volume, and on the characteristics of the fluid handled (in particular the viscosity); in those rotary pumps where liquid velocity is high, slip may be secondarily dependent on pump speed. A common unit of slip is U.S. gallons per minute (cubic meters per minute).

Slip in a rotary pump is an important factor of pump performance and applications. A good understanding of the concept of slip and of the effects of pump design, type, manufacturing tolerances, fluid conditions, and system operating conditions on the magnitude of slip is necessary for both pump designers and pump users.

Slip in a rotary pump occurs only when a pressure difference exists between the inlet and outlet chambers of the pump. This pressure difference causes the pumped fluid to flow between the outlet and inlet chambers through clearances between rotors and clearances between rotors and body members. It has the same effect as a shunt, or bypass, around the pump from the outlet port to the inlet port. Most rotary pumps are constructed so that the clearances in the pump generally are of the same nature as those found between two parallel flat plates, with one plate stationary and the other moving. Elements of the clearances have long, narrow, rectangular cross sections. In most pumps these clearances across the narrow dimension range from essentially zero to a few thousandths of an inch, and consequently even minor variations in manufacturing tolerances can cause considerable variations in the percentage change of the aperture volume. Also, the movement, or deflection, of movable elements in the pump when exposed to pressure differences can cause relatively large percentage changes in these clearances in different locations in the pump. Consequently each pump must be tested to determine slip under any given operating condition.

The major paths of the backflow through the pump in the presence of a positive differential pump pressure are the clearances between the end faces of the rotors and the endplates of the pump chamber and those between the outer radial surfaces of the rotors and the inner radial surfaces of the chamber. The width, length, and height of the apertures thus formed vary considerably with different positions of the rotor as the drive shaft turns slowly through a complete revolution. If the differential pressure across the pump remains constant during a revolution, then the instantaneous slip rate usually varies throughout the revolution. This variation in the slip is caused by the same effect that would be produced if the physical dimensions of the equivalent bypass around the pump were varied as a function of angular rotation of the drive shaft. This is one of the common causes of flow pulsation in rotary pumps; it is particularly dominant when large amounts of slip occur when low-viscosity liquids are being pumped at high pressure.

The average slip for any given operating condition can be determined by measuring the flow rate of fluid through the outlet port (incompressible liquid) and subtracting that flow rate from the flow rate Q_d that would be produced by the total displacement of the pump at that speed. Most slip paths are constant in width but may vary in height with runout of the outside diameter of the rotors or with wobble of the end faces of the rotors as they rotate. The paths also vary considerably in length during a revolution because of the relative positions of the mating rotary and body-sealing surfaces during different angular positions of the rotor.

In general, slip increases as the ratio of the height of the slip path to its length increases and decreases as this ratio decreases. During pump design, the clearances or height of the slip path may be increased if the length of the slip path is increased correspondingly, without materially affecting the slip characteristic of the pump.

The effect of pressure on slip is complex. The primary effect is direct: *slip increases in direct proportion to pressure*. However, there are several secondary effects, which can be classified into three general categories.

One category is the effect of pressure differences across the pump on the dimensions of the slip path. This occurs because of the deflection of pump elements as a function of pressure. This effect is relatively small in most rigid element pumps but may become extremely large in such pumps as flexible vane pumps, where the pressure may cause the vanes to flatten out and move away from the body walls. One generalization is that slip may be dramatically increased in flexible member pumps at high pressures, but clearances may be closed by high-pressure deflection of the rotors in rigid rotor pumps and the effective slip may be decreased.

Another category is the indirect effect of pressure on the fluid velocity through the slip paths. At any given viscosity, the flow through these paths may have the characteristics of turbulent flow or the characteristics of viscous flow or the characteristics of slug flow, depending on fluid velocity. The majority of practical applications would require that slip be a minor percentage of the pump displacement. To remain so, the velocity of fluid flow through slip paths would normally be in the viscous-flow region, and slip would then be directly proportional to the pressure difference. A pressure increase could cause a change to turbulent flow and a corresponding change in slip as a function of pressure.

A third category is the indirect effect of pressure on the effective compression ratio of compressible fluids. The compression ratio reduces the amount of net volume flow through the outlet port relative to the displacement of the pump. Although not a true slip in the sense discussed up to this point, the type of "slip" caused by this effect reduces the net volume delivered through the outlet port and consequently affects the volumetric efficiency. This effect is a secondary effect in most liquids but can become a large component of slip in aerated or compressible liquids. An increase in compression ratio caused by an increase in pressure difference causes an increase in slip from this effect.

Pump Capacity   The capacity Q of a rotary pump is the net quantity of fluid delivered by the pump per unit time through its outlet port or ports under any given operating condition. When the fluid is essentially noncompressible, capacity is numerically equal to the total volume of liquid displaced by the pump per unit time minus the slip, all expressed in the same units. The capacity of a rotary pump operating with zero slip is called the *displacement capacity Q_d*. A common unit of capacity is U.S. gallons per minute (cubic meters per minute):

$$Q = kDN - S = Q_d - S \tag{4}$$

where $k = 0.004329$ (.03471), with S and Q in gallons per minute (cubic meters per minute), D in cubic feet (cubic meters), and N in revolutions per minute.

Pump Speed   The speed N of a rotary pump is the number of revolutions of the driving, or main, rotor per unit time. When no gear reduction or increase exists between the drive shaft and the main rotor, the speed may be measured or set at the drive shaft. A common unit of speed is revolutions per minute.

Pump Pressure   The absolute pressure of the fluid at any location in the pump, expressed in pounds per square inch (bar°), is the total pressure there and is the basis for other pressure definitions associated with pump operation. Most of the pressure-associated operating characteristics of a rotary pump involve conditions where the *velocity pressure P_v* caused by fluid velocity is small relative to the total pressure and may be neglected, or where the fluid velocities (at pump locations used to determine pressure differences) are sufficiently alike that the velocity pressures cancel when the total pressure difference is computed. Should this not be so, velocity pressure may be computed as

in USCS units
$$P_v = \frac{wV^2}{288g}$$

in SI units
$$P_v = \frac{wV^2}{20{,}387g}$$

where w = fluid specific weight (mass), lb/ft^3 (kg/m^3)
   V = fluid velocity, ft/s (m/s)
   g = acceleration due to gravity, ft/s^2 (m/s^2)

or as

° 1 bar = 10^5 Pa. For a discussion of bar, see *SI Units—A Commentary* in the front matter.

in USCS units
$$P_v = \frac{0.000357w}{g}\left(\frac{Q}{a}\right)^2$$

in SI units
$$P_v = \frac{13,600w}{g}\left(\frac{Q}{a}\right)^2$$

where Q = flow, gal/min (m³/min)
a = cross-sectional area perpendicular to flow, in² (mm²)

Several pressure terms are of interest. The *outlet pressure P_d* is the total pressure at the outlet of the pump. In pumps with multiple outlets, this pressure is usually defined at the location in the outlet manifold where the pump is mated to the external piping system. Although composed of the sum of system and velocity pressures external to the pump, the outlet pressure is most commonly expressed as the *gage pressure* at the outlet port. Gate pressure (pounds per square inch gage or bar) is the difference between the absolute pressure and atmospheric pressure at the point of measurement.

The *inlet pressure P_s* is the total pressure at the inlet to the pump or, for multiple-inlet pumps, at the manifold location where the pump connects to the external piping system. In common practice the inlet pressure may be variously expressed as absolute pressure (lb/in² abs) (pounds per square inch abs or bar), as positive or negative gage pressure (pounds per square inch gage or bar) or as vacuum (inches or millimeters of mercury).

The pump *differential pressure P_{td}* is the algebraic difference between the outlet pressure and the inlet pressure, with both expressed in the same units. The differential pressure is used in the determination of power input and in evaluating the slip characteristics of the pump:

$$P_{td} = P_d - P_s \tag{5}$$

The *net inlet pressure P_{sv}* of a rotary pump is the difference between the inlet pressure expressed in absolute units and the vapor pressure of the fluid expressed in absolute units:

$$P_{sv} = P_{sa} - P_{\text{vapor}} \tag{6}$$

The *required net inlet pressure P_{svr}* is the minimum net inlet pressure that can exist without the creation of enough vapor in the inlet to interfere with proper operation of the pump. The inlet pressure usually is not the true minimum pressure in the OTI volume. The flow of fluid in its passage through the OTI volume causes fluid friction pressure loss, which causes the pressure to drop below the inlet pressure at some point in the OTI volume or inlet chamber. This inlet pressure loss increases with fluid velocity, and hence with pump speed, and with fluid viscosity and is a function of the pump geometry, which determines the fluid path lengths and local velocities where the fluid changes direction in flowing around curves or corners in the pump OTI volume boundary. In pumps operating at pressures where appreciable slip occurs, the pump differential pressure may have an indirect effect on the inlet loss. Consequently, the required net inlet pressure for a given pump is usually established by the manufacturer for specific speeds, pressures, and fluid characteristics. In practice, the pump user is warned that the net inlet pressure is near or at the required net inlet pressure by noisy and rough operation of the pump caused by incomplete filling of the CTIO volume with an accompanying reduction in pump capacity.

Other pressure ratings for pumps are the *maximum outlet working pressure P_{dr}*, the *maximum inlet working pressure P_{sr}*, and the *maximum differential pressure P_{tdr}*. The maximum outlet working pressure is the maximum gage pressure at the outlet port permitted for safe operation of the pump, and the maximum inlet working pressure is the maximum gage pressure permitted at the inlet port for safe operation. These two ratings are usually determined by the stiffness and strength of the pump body or casing and the type of seals used in the pump. The maximum differential pressure is the maximum allowable difference between the outlet pressure and the inlet pressure, measured in the same units, and is determined by the ability of the rotating assembly and its fluid seal contact zones to withstand pressure difference between the OTO and OTI volumes.

Pump Power The *total power input* to a pump (ehp) is the total power required by the pump driver or the pump prime mover under given operating conditions. Sometimes called *driver power*, total power input is the sum of the power required to overcome losses in the pump driver or prime mover; to overcome mechanical friction, fluid friction, and slip losses in the pump; and to deliver the net power imparted by the pump to the fluid discharged from it.

The *pump power input* (php) is the net power delivered to the pump drive shaft by the pump driver under given operating conditions. It is the net power available after subtraction of the power loss of the driver and associated transmission devices from the total power input.

The *pump power output* (whp) is the power imparted to the fluid delivered by the pump at given operating conditions and is frequently called *liquid horsepower*. It is the power remaining after the amount of slip power loss, mechanical power loss, and fluid friction power loss in the pump has been subtracted from the pump power input. The relationship between these power terms may be expressed as:

$$\text{ehp} = \text{driver and transmission power loss} + \text{php} \tag{7}$$

$$\text{php} = \text{pump power loss} + \text{whp} \tag{8}$$

A common unit used for expressing power ratings is the *horsepower*, equal to 550 ft·lb/s; the SI unit for power is the kilowatt. The pump power output can be computed by the formula

in USCS units
$$\text{whp} = \frac{QP_{td}}{1714}$$

in SI units
$$\text{kW} = \frac{QP_{td}}{36} \tag{9}$$

The constant 1714 (36) gives whp (kW) in horsepower (kilowatts) when Q is in gallons per minute (cubic meters per hour) and P_{td} is in pounds per square inch (bar).

The total volume of liquid handled by the pump is larger than Q when slip is present. The amount of slip actually represents "wasted" power and affects pump efficiency. The difference between pump power input and pump power output actually consists of three power-loss amounts: the amount of power loss represented by slip, the amount of power loss represented by mechanical friction in the pump, and the amount of power loss represented by fluid friction (a function of viscosity of the fluid and pump shear stresses on the fluid). To determine the combined mechanical and fluid friction losses, the *displacement power output* (dph) is first computed by the following formula and then subtracted from php to give the combined frictional losses:

$$\text{dhp} = \frac{(Q + S)P_{td}}{1714} = \frac{DNP_{td}}{395,934} = \frac{Q_d P_{td}}{1714}$$

$$\text{dkW} = \frac{DNP_{td}}{600} = \frac{Q_d P_{td}}{36} \tag{10}$$

where dhp (dkW) is horsepower (kilowatts) when Q, Q_d, and S are in gallons per minute (cubic meters per hour), P_{td} is in pounds per square inch (bar), D is in cubic inches (liters) per revolution, and N is in revolutions per minute.

Pump Efficiency Several efficiencies can be computed for a pump. The *overall unit efficiency* E_o is the percentage of the total power input delivered as pump power output and is computed by the equation

in USCS units
$$E_o = \frac{\text{whp}}{\text{ehp}} \times 100$$

in SI units
$$E_o = \frac{\text{wkW}}{\text{ekW}} \times 100 \tag{11}$$

The *pump efficiency*, or *pump mechanical efficiency*, E_p is the ratio of the pump power output to the pump power input. It may be computed by the equation

in USCS units
$$E_p = \frac{\text{whp}}{\text{php}} \times 100$$

in SI units
$$E_p = \frac{\text{wkW}}{\text{pkW}} \times 100 \tag{12}$$

The *volumetric efficiency* E_v of a pump is the percentage of pump displacement per unit time delivered as pump capacity. The equation for computing the volumetric efficiency is

in USCS units
$$E_v = \frac{231Q}{DN} \times 100 = \frac{Q}{Q_d} \times 100$$

$$(13)$$

in SI units
$$E_v = \frac{1000Q}{DN} \times 100$$

where E_v is a percent when Q and S are gallons per minute (cubic meters/min), D is in cubic inches (liters) per revolution, and N is in revolutions per minute. Equation 13 may be stated in alternative ways as follows:

$$E_v = \frac{Q}{Q+S} \times 100 \tag{14}$$

$$E_v = \frac{kDN - S}{kDN} \times 100 \tag{15}$$

Pump Performance   Figures 19, 20, and 21 show the change in displacement capacity Q_d, capacity Q, and slip S as differential pressure P_{td} across the pump, liquid viscosity ν, and pump speed N are varied. Several assumptions are made in these graphs. It is assumed that inlet conditions are satisfactory and that there is no inlet effect on pump capacity over the charted range. It is assumed that the liquid viscosity is Newtonian and that the liquid is incompressible. In Fig. 19 it is assumed that viscosity is constant and relatively low, approximately that of water, and that the speed is within the normal speed range of the pump. In Fig. 20 it is assumed that the viscosity is constant and relatively low and that the pressure is within the normal rated pressure range of the pump. In Fig. 21 it is assumed that both pressure and speed are within the normal ratings of the pump.

The graph of Fig. 19 is plotted with capacity, slip, and displacement capacity on a linear scale as the ordinate and pressure on a linear scale as the abscissa. A further assumption is that the size of the clearances and the viscosity of the liquid are such that the slip increases proportionately with pressure. The solid lines are the ideal characteristics when secondary effects are neglected. It may be noted that at zero pressure S is zero and Q equals Q_d. As the pressure increases, S

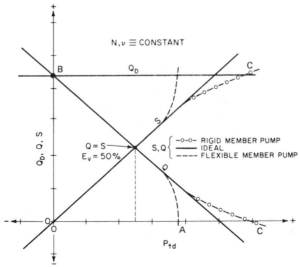

FIG. 19   Variation of Q_D, Q, and S with P_{td}, N, and ν constant.

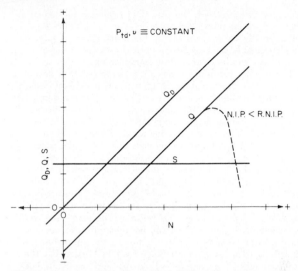

FIG. 20 Variation of Q_D, Q, and S with N, P_{td} and ν constant.

increases until it equals Q_d at pressure B. If the pressure imposed on the pump were to increase past this point, S would exceed Q_d and the flow through the pump would be from outlet to inlet, causing a negative Q. Even though rotary pumps are not normally operated in the higher pressure end of this range, the condition of pressure B may be reached when a valve is closed, blocking the outlet of the pump. Pressure B, then, represents dead-ended pressure developed by a rotary pump when its outlet line is blocked. It may be noted that, should Q_d be numerically equal to

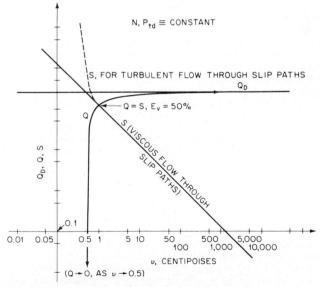

FIG. 21 Variation of Q_D, Q, and S with ν, N, and P_{td} constant.

100 in whatever units are used, then the plot of Q is numerically equal at each point to the volumetric efficiency E_v of the pump and Q becomes zero at pressure B.

Pressure B data usually are not supplied in the rating charts of most pump manufacturers, since these values are usually far beyond the normal pressure rating of the pump for efficient operation. However, they can be estimated by extrapolating the data given. There are cautions, however. A flexible member can quickly reach a pressure at which flexible member deflection becomes excessive, and the pressure at which $Q = 0$ is reached very rapidly. This is illustrated in Fig. 19 by the dashed line breaking away from the solid Q line and intersecting the abscissa at pressure point A. This characteristic of flexible member pumps provides a self-limited maximum pressure developed by the pump should the pump be dead-ended. In rigid rotor pumps, as the pressure increases beyond the normal operating range, deflections of the shafts and rotors bring the rotors into heavy bearing contact with the body chamber walls, reducing the dimensions of the slip clearance path. The result of this action is illustrated by the dashed line leaving the Q and S curves and intersecting the abscissa and Q_d line at pressure point C. Consequently in actuality the zero flow pressure, which is ideally at point B, may be considerably different from this extrapolated value, and it should be measured if it is of importance in the application.

Another characteristic which is normally not in the range of data supplied by manufacturers is the effect of a net negative total differential pressure. This may occur from time to time when system conditions change or if there is a variable positive static pressure on the inlet which sometimes exceeds the discharge or outlet pressure. In this case the slip reverses and adds to the capacity of the pump, causing total flow through the pump to be greater than the pump displacement capacity. This is illustrated by the extensions of the ideal slip line into the negative region in Fig. 19. It should be noted that this characteristic may easily occur when Q_d is equal to zero because the pump is stopped. In other words, most rotary positive displacement pumps are not effective as a valve to stop flow through the pump caused by pressure differences in the system when the pump is stopped. In those applications where it is important that flow stop when the pump stops and where inlet or outlet static pressure heads exist, valving in the system external to the pump must be used to stop the flow. For example, in intermittent deliveries where the pump is "lifting" liquid from a source below its inlet and delivering it to a discharge point with a piping system physically higher than the source of liquid at the inlet, if no valving arrangement is used in connection with the pump when the pump is stopped, the liquid will gradually drain backward through the pump into the liquid source and may create fairly large errors in the amount of liquid transferred if the pump should be used in metering applications.

Under the same assumptions given for Fig. 19, Fig. 20 shows the relative independence of slip with speed when differential pressure is constant. It may be noted that the chart of Q intersects the abscissa at the point where the speed is low enough to reduce the displacement capacity to equal slip at the pressure of operation. The speed at which Q_d equals $2S$ is the speed at which S equals Q, and the volumetric efficiency E_v is 50%. The volumetric efficiency (Eq. 15) increases with speed because the ratio of $kDN - S$ to kDN increases with speed. As the speed increases, the pumping action of the shear stress in the clearances tends to reduce slip below the ideal line. However, for most rotary pumps, the detrimental effects of speed increase above normal operating range are usually caused by inlet losses in the pump. This effect is discussed later in this section.

Figure 21 shows the effect of viscosity on slip and capacity in a rotary pump. The graphs are on a log-log scale. In this particular chart it is assumed that the pressure, speed, and viscosity combine to keep flow through clearances of the pump in the viscous flow region. Slip, then, is directly proportional to the total pressure difference across the pump and inversely proportional to viscosity. This is expressed in Eq. 16, where the constant K is a function of pump geometry and size:

$$S = \frac{KP_{td}}{\nu} \tag{16}$$

The constant K is sometimes called the *coefficient of slip*. It may be designated as K_s. It includes all the constants needed to express slip in the desired units of flow.

As viscosity increases, slip becomes arbitrarily small and the capacity of the pump approaches the displacement capacity. As the viscosity decreases, slip very rapidly approaches the displacement capacity and the capacity of the pump drops very rapidly to zero or to a negative quantity.

For any given pump and any given pump speed N and pressure difference P_{td}, there is a viscosity below which slip flow through clearances of the pump will change to turbulent flow. It is unlikely that this change will occur simultaneously through all slip paths. However, once it begins to occur, slip will increase much more rapidly with further reductions in viscosity because of the turbulent flow relationship of slip, pressure, and viscosity expressed in Eq. 17:

$$S = \frac{KP_{td}^{1/2}}{\nu^{1/x}} \tag{17}$$

where x usually is in the range of 4 to 10.

Pump volumetric efficiency E_v drops very rapidly with viscosity if the viscosity is lower than that required for 50% volumetric efficiency (represented by the crossover point of the slip and capacity curves). This crossover point occurs for almost all rotary pumps operating at rated P_{td} for viscosities between 0.1 and 10 cP, and for the majority of commercially available models this point usually falls in the viscosity range of 0.3 to 3 cP at P_{tdr}.

The effect of inlet pressure on capacity can be seen clearly if secondary effects are eliminated. It is assumed that a pump is operating within its normal pressure rating and within its normal speed rating and that viscosity is sufficiently high to reduce slip to a negligible or zero value. A graph of capacity as a function of inlet pressure is shown in Fig. 22 with these assumptions.

There is no change in capacity as inlet pressure is lowered until the pressure reaches pressure A on the graph. If the inlet pressure were lowered further, the capacity would drop as shown. The cause of this drop is complex in detail but simple in concept. The liquid flow from the inlet port through the inlet chamber of the pump causes a pressure drop, which causes a minimum pressure point somewhere in the inlet chamber. When the pressure in the liquid at this minimum pressure point approaches the vapor pressure of the liquid, vapor begins to form there. When the amount and time persistence of this vapor cause vapor to be swept into the CTIO volume of the pump, the amount of liquid in this volume is reduced. When this volume reaches higher pressure, the vapor condenses, leaving a deficiency of liquid volume. For example, if half the fluid volume swept from the inlet chamber is vapor, then only half the normal-capacity liquid volume is available at the outlet chamber, and the capacity is reduced accordingly.

An increase in speed would mean an increase in capacity, all other things being equal. This would increase the pressure drop between the inlet port and the inlet chamber and correspondingly increase the absolute inlet pressure (which is measured at the inlet port) at which capacity would begin to drop (pressure A). Correspondingly, if the speed and capacity were constant and the viscosity were to increase, the pressure drop between the inlet port and the minimum pressure

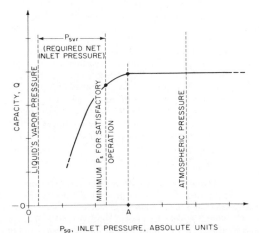

FIG. 22 Variation of Q with P_s, other operating conditions constant.

point in the inlet chamber would increase with viscosity. This also would cause pressure A to move to higher inlet absolute pressures. Operation with absolute inlet pressures below pressure A for any given speed and viscosity is usually unsatisfactory both because of the drop in capacity (and hence in volumetric efficiency) and because of the noisy and rough operation caused by the formation and collapse of vapor. For lower viscosity liquids, where the collapse of the vapor bubbles formed may be quite rapid, cavitation may cause a significant amount of damage to body or rotor surfaces. It is important to understand that cavitation damage may occur even though the absolute inlet pressure is above pressure A. There may be locations in the inlet chamber, particularly where flow direction changes rapidly, as around sharp corners, where cavitation (vapor formation in the form of very fine bubbles) may occur. However, these vapor cavities may be swept into higher pressure regions of the inlet chamber and collapse near body or rotor surfaces to cause cavitation damage even though the capacity of the pump is not affected. For any given viscosity, then, there is an upper limit to the speed at which the pump may be operated.

The inlet pressure may be allowed to drop slightly below pressure A without significant deterioration of pump performance. However, there is a point at which the pressure becomes too low for satisfactory pump operation. This pressure determines the required net inlet pressure P_{svr} for the particular pump and particular set of operating conditions.

For any given set of operating conditions, the satisfaction of required net inlet pressure is a main limitation on pump operating speed.

The other main limit on pump operating speed is the *pump outlet pressure*. In every pump application there is some frictional loss in the outlet system of the pump. Even if the pump outlet is opened to the atmosphere, a pressure drop exists between some maximum pressure point in the pump outlet chamber and the pump outlet itself. However, by far the most common situation is one in which a significant liquid friction pressure is developed in the system external to the outlet of the pump. This liquid pressure usually is a function of pump capacity and hence of pump speed. If the inlet conditions are maintained to keep the inlet pressure above pressure A as the pump speed increases, a speed will be reached at which the pump outlet pressure equals the outlet pressure rating of the pump. Operations at speeds higher than this would cause the pump outlet pressure to exceed the rated pressure, and this could result in possible damage to the pump.

These two limits on pump speed are illustrated in Fig. 23. Three sets of operating conditions are shown. In operating condition 1, the outlet system friction resisting liquid flow (outlet system impedance) is relatively high and the outlet pressure developed by flow is directly proportional to pump capacity (viscous flow). There is no static head in the system, and the outlet pressure is developed only when the pump causes liquid to flow through the outlet system. The liquid viscosity is assumed to be low enough to permit some slip in the pump. The net inlet pressure is assumed to be above pressure A over the range of operation shown. Operating condition 2 is the same as operating condition 1 except that a static head (static outlet pressure) exists in the outlet system. This head is the pressure at which the chart of pump outlet pressure in condition 2 intercepts the zero speed line. Operating condition 3 is one in which the impedance to liquid flow in the outlet system is relatively low but the liquid viscosity is relatively high. In this condition Q is equal to the displacement capacity Q_d as speed is increased, until a speed is reached at which the available net inlet pressure of the pump drops to the required net inlet pressure of the pump. If the speed were increased beyond this point, the capacity of the pump would drop rapidly as an ever-increasing part of the pump fluid becomes vapor instead of liquid. The upper limit of speed for satisfactory operation of the pump is shown as N_1, N_2, and N_3 for the three conditions described. The locations of N_1, N_2, and N_3 on the speed axis are independent of each other because they depend primarily on the operating conditions of the pump and the system in which it operates and on the conditions of the pumped fluid. For example, a negative head in the outlet system could cause N_3 to move to a higher value than N_2, and N_2 to move to a higher value than N_1. Operation with a liquid with lower viscosity (but still sufficient to reduce slip to zero) could cause N_3 to be higher than N_1.

In most pump applications one of the two limits described determines the maximum permitted speed of operation of the pump. However, if neither of these conditions limits the speed and the speed is continuously increased, a speed will be reached at which the peripheral velocity of the rotors will exceed the cavitation velocity of the liquid. A further increase in speed beyond this point would be limited by the cavitation occurring at the rotor outer radial surfaces.

Pump Power Requirements   The *displacement capacity* (Q_d) *liquid horsepower* (dhp; dkW) of any rotary pump of any type and of any size depends only on the displacement capacity

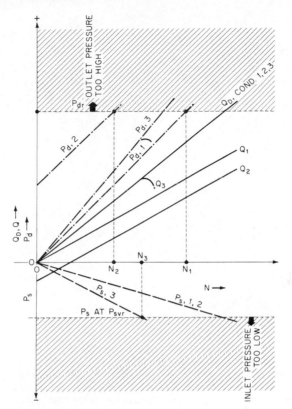

FIG. 23 P_d and P_s limits on pump speed.

at the given pump speed and the total differential pressure P_{td} across the pump. It does not depend on any characteristics of the liquid being pumped. The mechanical friction horsepower (mhp; mkW) in a pump has two components. One is the mechanical friction horsepower required by all elements *external* to the pumping chamber; this part usually is independent of the liquid being pumped and is dependent only on the lubricity of the lubricant used, if any, and on the pump speed and pump differential pressure. The mechanical friction horsepower *inside the pump* depends also on the pump speed and pump differential pressure, but it usually also depends on the lubricity of the liquid or fluid being handled by the pump. The *liquid friction horsepower*, or *viscous horsepower*, vhp (vkW), depends primarily upon the viscosity of the liquid being handled and on the shear rate in the liquid, which is a function of pump design and pump speed. The *pump power input*, php (pkW), which can be expressed as the sum of the displacement capacity horsepower, the mechanical friction horsepower, and the viscous horsepower, is equal to a constant times the required torque times the pump speed:

in USCS units $\qquad$ php $= \text{dhp} + \text{mhp} + \text{vhp} = K_h T_p N$ $\qquad\qquad$ (18)

in SI units $\qquad$ pkW $= \text{dkW} + \text{mkW} + \text{vkW} = K_h T_p N$

Of the three parts of pump power input, only the displacement horsepower is uniquely determined by the displacement capacity of the pump at its elected speed of operation and by the pressure across the pump. This may be expressed as in Eq. 19:

$$\text{dhp} = K_h T_d N = k_d Q_d P_{td} = k_d (kDN) P_{td} \qquad\qquad (19)$$

where 1 hp = 1.3 kW. This is the absolute minimum horsepower required for operation of that particular pump at a given speed against a given pressure difference. It would be the total horse-

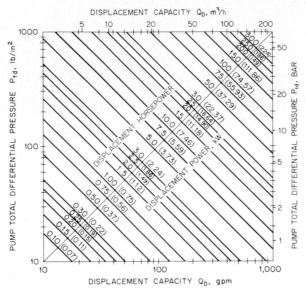

FIG. 24 Pump displacement horsepower variation with Q_d and P_{td}, all rotary pumps, all liquids.

power required only if there were no mechanical or liquid friction horsepower losses and no slip losses in the pump (volumetric efficiency at 100%). Equation 19 can be solved in terms of torque, displacement, and pump differential pressure to give Eq. 20. The constant given is for torque in foot-pounds (kilogram-meters), displacement in cubic inches (liters) per revolution, differential pressure in pounds per square inch (bar), and speed in revolutions per minute:

$$\text{in USCS units } T_d = 0.01326 DP_{td} \tag{20}$$

$$\text{in SI units } \quad T_d = 1.622 DP_{td}$$

Equation 19 also may be used to generate the chart in Fig. 24, which can be used to determine the minimum basic horsepower input to a pump expected to deliver a given flow rate against a given pressure differential. Equation 13 can be used to compute Q_d when the actual capacity Q is known and the pump volumetric efficiency E_v is known. In any case, if the displacement capacity is known for whatever speed is required to produce the desired capacity, the chart of Fig. 24 can be entered from the known displacement capacity and from the known pressure difference to determine the displacement horsepower. To this displacement horsepower must be added the basic mechanical friction horsepower and viscous horsepower of the pump. The significance of displacement horsepower is that it is a measure of all the energy required to impart energy of motion to the external capacity and internal slip flow of the pump against the specified pressure. The remaining horsepower required is consumed in the heat generated by mechanical and liquid viscous friction.

The torque required to overcome the frictional losses in the pump may become large when the pump handles certain kinds of liquids. When the pump handles nonlubricating liquids and pump differential pressure is high, the mechanical friction in the pump chamber may become high and increase the pump torque requirement. When the pump handles high-viscosity liquids, the torque required by the viscous friction grows as viscosity increases, even if pump speed and pressure remain constant. This torque depends on a number of design features of the pump and usually will grow proportionally to the viscosity to the nth power, where n usually ranges between 0.3 and 1.0 for most rotary pumps. Increasing any clearances in a pump reduces the shear stress in the liquid in those clearances at any given speed and consequently reduces the torque to overcome the viscous friction. Too great an increase in clearances can cause slip, which will increase the total power loss for a given capacity by a drop in volumetric efficiency. If a pump is designed for a liquid with a specific viscosity, the minimum power clearance is that at which volumetric efficiency just reaches 100%. A further reduction of clearance beyond that point would not improve volumetric efficiency but would increase viscous horsepower.

Most pump manufacturers publish graphs or tables of total pump horsepower required for any set of operating conditions within the ratings of the pump. Many pump manufacturers also identify the maximum permitted torque on the drive shaft of the pump. Once the horsepower requirements are determined for any specific operating conditions, the required torque may be computed from Eq. 18, in which K_h equals 1.903×10^{-4} (1.026×10^{-3}) for speed in revolutions per minute and torque in foot-pounds (kilogram-meters). The torque limitation usually is of importance on low-speed, high-horsepower applications, such as might occur when the pump handles a liquid of very high viscosity at reasonably high pressures at low speeds.

ROTARY PUMP APPLICATIONS

The immense variety of applications and system conditions, including the large number of different kinds of fluids handled in rotary pump applications, precludes a comprehensive coverage in a handbook of this type. For more detailed information on pump applications, a recent reference is Hicks and Edwards, *Pump Application Engineering*, McGraw-Hill, New York, 1971. Additional detail on pump testing, application, and maintenance can be found in *Hydraulic Institute Standards*, 13th ed., Cleveland, 1975. Good coverage of the complexity of the viscosity of non-Newtonian fluids is contained in a *Chemical Engineering* refresher course prepared by Martin Wohl and published serially in *Chemical Engineering* (January–August 1968). An excellent source of application information for a specific type of pump is a manufacturer of that pump. Many manufacturers have available application, installation, and maintenance manuals.

In the remainder of this section are given some highlights of requirements in mechanical installation, of system considerations, and of the effect of various kinds of pumped fluids in various pump applications.

Mechanical Installation of Pumps   All rotary pumps, particularly rigid rotor pumps, must be installed so that no large mechanical forces of any kind other than those imposed by pump operating pressures can act to warp or distort the pump chamber or rotating assembly. Relatively small distortions of a few thousandths of an inch may cause interference between rotating and stationary parts and generate high wear or pump damage.

To avoid such unwanted distortions, the pump must not be installed with rigid fittings to long, rigid piping systems, either where the weight of the piping system is supported by the pump body or where large mechanical forces on the pump body can be generated by thermal expansion of the piping system. The problem of distortion of operating clearances in the pump chamber is not as severe for flexible member pumps, but large mechanical forces tending to distort the pump body may cause serious distortion of mechanical sealing arrangements and unusually high wear rates on bearing systems. A cardinal rule, then, is that a *rotary pump must not be so rigidly coupled to either the fluid piping system or the driver that it supports the weight of either or that it is exposed to high forces caused by thermal expansion of either.* A poor mechanical installation of a pump might be the sole cause of unsatisfactory operation.

System Considerations   Most rotary pumps will be damaged if operated without liquid flowing through the pump chamber. If the system provides a positive static pressure on fluid present at the inlet, the pump usually will not be damaged by running dry. If a negative inlet gage pressure, or suction lift, is a condition at the pump inlet, the system installation should be checked to ensure airtight seals at all demountable joints in the inlet system between the pump inlet and the body of fluid to be pumped. The sealing arrangement on the pump must be monitored to prevent any leakage of air or other gas through the seal. If these conditions are met, liquid will enter the pump very soon after pump operation starts because of the vacuum generated by pump operation. If they are not met, a sufficient flow of air or other gas through leaks in the inlet system or seal may satisfy the pump capacity requirements and the pump will run dry indefinitely. An alternate way to ensure liquid at the pump inlet is to install a foot or check valve in the liquid-submerged portion of the pump inlet piping system. Once the pump is primed, the foot valve will keep liquid in the pump inlet and prevent its running dry.

If there is a possibility that a valve in the outlet piping system of the pump may close or remain closed to block all flow from the pump, then a pressure relief (bypass) valve should be used with the pump. This valve may be installed externally or may be an integral part of the pump. A pressure relief valve usually is not necessary in flexible member rotary pumps. Pressure relief valves also may be used to limit the pump discharge pressure to a selected predetermined maximum.

The inlet system of the pump must be arranged in such a way that fluid is always present at the pump inlet at a pressure at or above the inlet pressure needed to satisfy the required net inlet pressure.

Effects of Pumped Fluid   The variety of fluids handled by rotary pumps requires that each pump application be considered in terms of the effects of the pumped fluid on pump performance. General summaries of the effects of various fluid characteristics on pump performance are covered in the following paragraphs.

The *temperature* of the pumped fluid affects pump performance in three main ways. If the pump is to handle a fluid at temperatures considerably different from ambient, the materials of construction, both in the pump and in the pump seals, and the operating clearances to which the pump is manufactured must be selected to give the desired operating characteristics at the temperature of operation. The selection process becomes even more stringent when the pump is operated over a wide range of temperatures. Such a use may preclude the use of pump construction materials having high thermal coefficients of expansion. Furthermore, in the actual application it may be necessary to preheat or precool the pump to or near the temperature of operation before the pump is started, in order to avoid thermal shock when the fluid enters the pump chamber. If this is not done, the resulting rapid heating or cooling of pump members from the inside out can cause transient conditions that may damage the pump. The preheating or precooling can be accomplished with a secondary medium (a jacketed pump) or, if the state of the liquid permits, the pump may be started on fluid at ambient temperature and the temperature of the fluid gradually lowered or raised to the desired operating temperature while the pump is operating.

A second effect of fluid temperature is on the vapor pressure of the fluid. For most liquids, vapor pressure increases with increased temperature; consequently the required net inlet pressure would cause a corresponding increase in the necessary inlet pressure at the pump.

A third effect of fluid temperature is on the viscosity of the fluid. For the majority of commercial liquids, the viscosity tends to increase with decreasing temperature and to decrease with increasing temperature. If the pump application requires operation over a wide temperature range, the highest viscosity at this temperature range must be known to determine whether the pump is operating below the upper speed limit imposed by this fluid viscosity.

Another important characteristic of the pumped fluid is its *viscosity*. Earlier parts of this section describe the general effects of viscosity upon pump performance. However, very few commercial liquids are completely Newtonian in viscosity. The effective viscosity of the fluid under the actual projected or operating conditions should be determined by test or computation. If the pump is to operate over a range of speeds and pressures, the maximum and minimum effective viscosities of the liquid over this range should be determined to provide for the additional slip or the additional horsepower that may be required, to ensure satisfactory selection of operating speeds and pump driver horsepower. A brief summary of the effects of time and shear rate on the viscosity of fluids is given in *Hydraulic Institute Standards*, 13th ed., p. 158, and in the reference on viscosity cited earlier. The importance of a good knowledge of the viscous behavior of fluids in pump selection and application is illustrated by the following example. The shear stress in either a pseudoplastic or thixotropic fluid decreases as shear rate increases. In such fluids a high viscosity at rest may decrease rapidly as shear rate increases. In such a case, the most satisfactory choice for the application would be a pump much smaller than usual, operated at much higher speeds than would otherwise be permitted.

Still another important fluid characteristic is *lubricity*. Many rotary pumps are not designed to operate on liquids with no lubricity. Either the wear rate or the pump mechanical friction may be inefficiently high if these pumps are used to pump nonlubricating fluids.

Knowledge of the *corrosiveness* of the pumped fluid on the pump materials in contact with the fluid is important to satisfactory pump application. The clearances in rotary pumps are small, and corrosion rates of only a few thousandths of an inch per year may seriously affect the efficiency of the pump, particularly when it is handling low-viscosity fluids. Also, the general compatibility of the fluid with the materials of pump construction must be considered. For example, a certain solvent to be pumped may soften, dissolve, or destroy the elasticity of flexible members in the pump, including those flexible members used in the sealing arrangements. Hence the effect of the pumped fluid on the physical and chemical state of materials used in pump construction should be considered in choosing a particular pump for a particular application.

Very few commercially handled fluids are pure liquids. In actual systems most fluids have air or other gases dissolved or entrained in them, or the system may operate where liquid vapor can exist in various concentrations in the pumping chamber. In other cases, wanted or unwanted solids may be contained in the liquid. These solids may be abrasive or nonabrasive to the materials of the pump.

If appreciable amounts of gas are entrained or dissolved in the pumped fluid (for example, an aerated ice cream mix or various foamed plastics), it is necessary to select a type of rotary pump in which the action of the pump returns no part of the fluid in the discharge chamber to the inlet chamber during operation. One type of rotary pump usually satisfactory for aerated fluids is the lobe pump. This type of rotary pump has been used as a rough vacuum pump and as a compressor and consequently is capable of handling the gaseous phase of the fluid.

The pump geometry in applications involving nonabrasive solids, particularly those that are fibrous and tend to mat, must be such as to maintain high fluid velocities everywhere in the pump. Abrasive solids represent a formidable problem for almost any rotary pump. Again, because of the small operating clearances, the tendency of the abrasives in the pumped fluid to wear down and open clearances usually will cause a rapid decline in pump efficiency, particularly when the abrasives are carried in low-viscosity liquids. Very few rotary pumps that will handle abrasive fines in a low-viscosity liquid at medium or high pressures are commercially available. Most of the pumps used for this service are constructed of very-high-hardness (usually expensive) materials.

Many other characteristics of the pumped fluid could be of importance in the selection and application of rotary pumps. Some liquids are sensitive to shear. They must be handled in pumps with low shear rates. Some liquids cannot tolerate exposure to the atmosphere either because they may explode or because they may instantly crystallize into hard crystals which would damage the pump seal arrangement or otherwise interfere with satisfactory operation of the pump. Rotary pumps for these applications should have only static seals or multiple-section rotary seals with a protective liquid or gas in the seal zones between the pumped fluid and the outside atmosphere. For example, stationary and moving seals on aseptic pumps may be designed with multiple seals allowing the use of steam under pressure as a sterile barrier between the pump chamber and the outside world. Pumps used to pump human blood, such as the tube pump, may use disposable, sterilizable pump elements. Pumps used in industrial applications where conditions may become dangerous or unsatisfactory if the pump liquid is exposed to atmosphere may use multiple sealing arrangements with an inert liquid under pressure acting as a barrier between the pumping chamber and the outside atmosphere.

For effective application of rotary pumps, then, the temperature, viscosity, lubricity, corrosiveness, and nonliquid content of the pumped fluid must be known, along with any special characteristics of the liquid, such as shear sensitivity or atmospheric sensitivity. When these are known, the proper pump type, seal arrangements, horsepower requirements, and speed can be determined.

The positive displacement nature of rotary pumps makes them suitable for many metering applications. The pump would be a perfect meter with zero error if there were no slip. It would be a good meter if the slip were low or maintained constant over the entire range of operating conditions. A useful ratio in a comparison of different types of rotary pumps for a given metering application is the ratio of minimum capacity in any of the operating conditions to maximum capacity in any of the operating conditions at a given speed. In variable-speed applications the ratio of minimum volumetric efficiency to maximum volumetric efficiency is substituted. The metering effectiveness is highest as this ratio approaches unity. The difference between this ratio and 1×100 is the percentage of change that can be expected either in the flow rate of the pump or in the total amount of fluid displaced for a given number of revolutions, under extremes of operating conditions (pressures, temperatures, viscosities, etc.).

SECTION 3.6

DISPLACEMENT PUMP PERFORMANCE, INSTRUMENTATION, AND DIAGNOSTICS

J. C. WACHEL
FRED R. SZENASI

The most common operational and reliability problems in reciprocating positive displacement pump systems are characterized by

- Low net positive suction head (NPSH), pulsations, pressure surge, cavitation, waterhammer
- Vibrations of pump or piping
- Mechanical failures, wear, erosion, alignment
- High horsepower requirements, high motor current, torsional oscillations
- Temperature extremes, thermal cycling
- Harsh liquids: corrosive, caustic, colloidal suspensions, precipitates (slurries)

While any component in a pump system may be defective, most operational problems are caused by liquid transient interaction of the piping system and pump system or by purely mechanical interaction of the pump, drive system, foundation, etc. This section discusses hydraulic and mechanical problems and suggests measurement and diagnostic procedures for determining the sources of these problems.

HYDRAULIC AND MECHANICAL PUMP PROBLEMS

Inadequate NPSH Net positive suction head available (NPSHA) is the static head plus atmospheric head minus lift loss, frictional loss, vapor pressure, and acceleration head available at the suction connection centerline.

Acceleration head can be the highest factor of NPSHA. In some cases it is 10 times the total of all the other losses. Data from both the pump and the suction system are required to determine acceleration head; its value cannot be calculated until these data have been established. Inadequate NPSH can cause cavitation, the rapid collapse of vapor bubbles, which can result in a variety of pump problems, including noise, vibration, loss of head and capacity, and severe erosion of the valves and surfaces in the adjacent inlet areas. To avoid cavitation of liquid in the pump or piping, the absolute liquid static pressure at pumping temperatures must always exceed the vapor pressure of the liquid. The pressure at the pump suction should include sufficient margin to allow for the presence of pulsations as well as pressure losses due to flow.

Positive Displacement Pump Pulsations The intermittent flow of a liquid through pump internal valves generates liquid pulsations at integral multiples of the pump operating speed. For example, a 120-rpm triplex pump generates pulsations at all multiples of pump speed (2 Hz, 4 Hz, etc.); however, the most significant components will usually be multiples of the number of plungers (6 Hz, 12 Hz, 18 Hz, etc.). Resultant pulsation pressures in the piping system are determined by the interaction of the generated pulsation spectrum from the pump and the acoustic length resonances of liquid in the piping. For variable-speed units, the discrete frequency components change in frequency as a function of operating speed and the measured amplitude of any pulsation harmonic can vary substantially with changes in the location of the measurement point relative to the pressure nodes and antinodes of the standing wave pattern.

Piping System Pulsation Response Since acoustic liquid resonances occur in piping systems of finite length, these resonances will selectively amplify some pulsation frequencies and attenuate others. Resonances of individual piping segments can be described from organ pipe acoustic theory. The resonant frequencies of standing pressure waves depend upon the velocity of sound in the liquid being pumped, pipe length, and end conditions. The equations for calculating these frequencies are shown in Fig. 1. All of the integral multiples (N) of a resonance can occur, and it is desirable to mismatch the excitation frequencies from any acoustical resonances. A 2:1 diameter increase or greater would represent an open end for the smaller pipe. Closed valves, pumps, or a 2:1 diameter reduction represent closed ends. For example, a 2-in (51-mm) diameter pipe which connects radially into two 8-in (203-mm) diameter volumes would respond acoustically as an open-end pipe.

Complex piping system responses depend upon the termination impedances and interaction

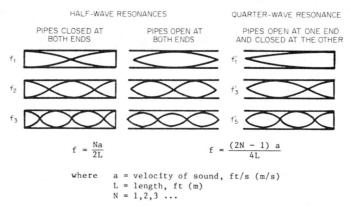

$$f = \frac{Na}{2L} \qquad f = \frac{(2N - 1)\,a}{4L}$$

where a = velocity of sound, ft/s (m/s)
L = length, ft (m)
N = 1,2,3 ...

FIG. 1 Organ pipe resonant mode shapes.

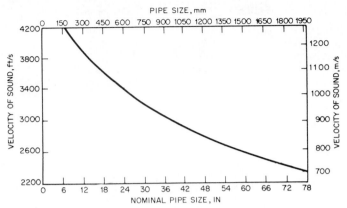

FIG. 2   Velocity of sound in water at 14.6 lb/in² (1 bar), 60°F (15.6°C) versus nominal pipe size with 0.25-in (6.35-mm) wall thickness.

of acoustical resonances and cannot be handled with simplified equations. An electoacoustic analog[1] or digital computer can be used for the more complex systems.

Velocity of Sound in Liquid Piping Systems   The acoustic velocity of liquids can be determined by the following equation:

$$a = C_1 \sqrt{\frac{K_s}{\text{sp. gr.}}} \tag{1}$$

where a = velocity of sound, ft/s (m/s)
$\quad C_1$ = 8.615 for USCS units, 1.0 for SI units
$\quad K_s$ = isentropic bulk modulus, lb/in² (kPa)
sp. gr. = specific gravity

In liquid piping systems, the acoustic velocity can be significantly affected by pipe wall flexibility. The acoustic velocity can be adjusted by the following equation:

$$a_{\text{adjusted}} = a \sqrt{\frac{1}{1 + \dfrac{DK_s}{tE}}} \tag{2}$$

where D = pipe diameter, in (mm)
$\quad t$ = pipe wall thickness, in (mm)
$\quad E$ = elastic modulus of pipe material, lb/in² (kPa)

Pipe wall radial compliance can reduce the velocity of sound in liquid in a pipe as shown in Fig. 2.[2]

The bulk modulus of water can be calculated with the following equation[3] for temperatures from 0 to 212°F (0 to 100°C) and pressures from 0 to 4.4 × 10⁴ lb/in² (0 to 3 kbar°):

$$K_s = K_0 + 3.4P \tag{3}$$

where K_s = isentropic bulk modulus, kbar
$\quad K_0$ = constant from Table 1, kbar
$\quad P$ = pressure, kbar
(1 kbar = 10⁵ kPa = 14,700 lb/in²)

°1 bar = 10⁵ Pa. For a discussion of bar, see *SI Units—A Commentary* in the front matter.

TABLE 1 Constant K_0 for Evaluation of Isentropic Bulk Modulus of Water from 0 to 3 kbar

$$K_s = K_0 + 3.4P$$

Temperature, °C (°F)	Isentropic constant K_0, kbar[a]
0 (32)	19.7
10 (50)	21.0
20 (68)	22.0
30 (86)	22.7
40 (104)	23.2
50 (122)	23.5
60 (140)	23.7
70 (158)	23.7
80 (176)	23.5
90 (194)	23.3
100 (212)	22.9

[a]1 kbar = 14,700 lb/in².

SOURCE: Ref. 3.

The calculation of the isentropic bulk modulus of water is accurate to ±0.5% at 68°F (20°C) and lower pressures.[4] At elevated pressures (greater than 3 kbar) and temperatures (greater than 100°C), the error should not exceed ±3%.

The bulk modulus for petroleum oils (hydraulic fluids) can be obtained at various temperatures and pressures by using Figs. 3 and 4, which were developed by the American Petroleum Institute

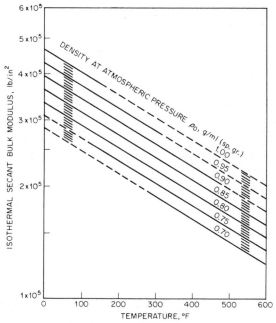

FIG. 3 Isothermal secant bulk modulus at 20,000 lb/in² gage for petroleum oils. (1 lb/in² = 6.895 kPa; °C = (°F − 32)/1.8)

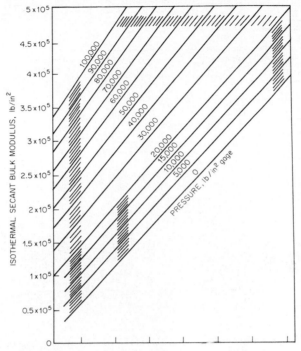

FIG. 4 Pressure correction for isothermal secant bulk modulus for petroleum oils. ($1 \text{lb/in}^2 = 6.895$ kPa)

(API).[4-7] Figure 3 relates density (mass per unit volume) and temperature to the isothermal secant bulk modulus at 20,000 lb/in^2 (137,900 kPa), and Fig. 4 corrects for various pressures.

The isentropic tangent bulk modulus is needed to calculate the speed of sound in hydraulic fluids and can be readily obtained from Figs. 3 and 4 as follows:

1. Read the isothermal secant bulk modulus at the desired temperature from Fig. 3.

2. Using the value for isothermal secant bulk modulus obtained from Fig. 3, go to Fig. 4 and locate the intersection of the pressure line with that value. Move vertically to the pressure line representing twice the normal pressure. Read the adjusted isothermal secant bulk modulus for the double value of pressure.

3. Multiply the adjusted isothermal secant bulk modulus by 1.15 to obtain the value of the isentropic tangent bulk modulus (compensation for the ratio of specific heats).

The isothermal tangent bulk modulus has been shown to be approximately equal to the secant bulk modulus at twice the pressure[5] within ±1%. The relationship between isothermal bulk modulus K_t and isentropic bulk modulus K_s is

$$K_s = K_t \, c_p/c_v \qquad (4)$$

The value of c_p/c_v for most hydraulic fluids is approximately 1.15.

PULSATION CONTROL

Pulsation control can be achieved by judicious use of acoustic filters and side branch accumulators. Acoustic filters are liquid-filled devices consisting of volumes and chokes which use reactive fil-

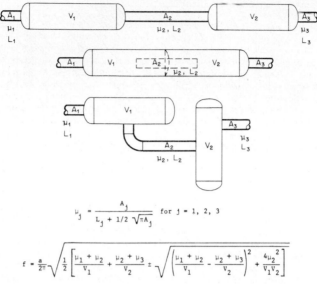

$$\mu_j = \frac{A_j}{L_j + 1/2 \sqrt{\pi A_j}} \quad \text{for } j = 1, 2, 3$$

$$f = \frac{a}{2\pi} \sqrt{\frac{1}{2} \left[\frac{\mu_1 + \mu_2}{V_1} + \frac{\mu_2 + \mu_3}{V_2} \pm \sqrt{\left(\frac{\mu_1 + \mu_2}{V_1} - \frac{\mu_2 + \mu_3}{V_2} \right)^2 + \frac{4\mu_2^2}{V_1 V_2}} \right]}$$

FOR EQUAL VOLUMES, THE RESONANT FREQUENCY IS APPROXIMATELY:

$$f = \frac{a}{\sqrt{2}\,\pi} \sqrt{\frac{\mu_2}{V_1}}$$

FIG. 5 Two-chamber resonator system with both ends open. V = volume, ft³ (m³); f = resonant frequency, Hz; L = choke tube length, ft (m); A = choke tube area, ft² (m²); a = acoustic velocity, ft/s (m/s); μ = acoustic parameter.

tering techniques to attenuate pulsations. Side branch resonators are of two types: quarter-wave-length resonant stubs and gas-charged accumulators. (The terms *accumulators, dampeners,* and *dampers* are used interchangeably in the liquid filter industry.)

Acoustic Filters An acoustic filter consisting of two volumes connected by a small-diameter choke can significantly reduce the transmission of pulsations from the pump into the suction and discharge piping systems. The equations given in Fig. 5 can be used to calculate the resonant frequency for a simple volume-choke-volume filter. The filter should be designed to have a resonant frequency no more than one-half the lowest frequency desired to be reduced, referred to as the cutoff frequency. Such a filter is called a low-pass filter since it attenuates frequencies above the cutoff frequency.

A special case for symmetric liquid-filled filters can be obtained by choosing equal chamber and choke lengths. This reduces the equation in Fig. 5 to

$$f = \frac{ad}{\pi \sqrt{2LD}} \tag{5}$$

where d = choke diameter, in (mm)
$\quad L$ = chamber and choke length, ft (m)
$\quad D$ = chamber diameter, in (mm)

Normally, a good filter design will have a resonant frequency less than one-half the plunger frequency and will have a minimal pressure drop. For example, a triplex pump running at 600 rpm generates pulsations at all multiples of 10 Hz. The largest amplitudes would normally be at 30 Hz, 60 Hz, 90 Hz, etc. The filter resonant frequency should be set at 15 Hz or lower. For water, in which the velocity of sound is 3200 ft/s (976 m/s), and a volume bottle size inner diameter of 19 in (482.6 mm),

in USCS units
$$\frac{L}{d} = \frac{3200}{\pi\sqrt{2}\,(15)(19)} = 2.53\,\frac{\text{ft}}{\text{in}}$$

(6)

in SI units
$$\frac{L}{d} = \frac{976}{\pi\sqrt{2}\,(15)(482.6)} = 0.03\,\frac{\text{m}}{\text{mm}}$$

If the choke diameter is selected to be 1.049 in (26.64 mm), then the length of each volume bottle and choke tube is 2.65 ft (0.81 m). See also Ref. 8.

Side Branch Accumulators   Liquid-filled, quarter-wavelength, side branch accumulators reduce pulsations in a narrow frequency band and can be effective on constant-speed positive displacement pumps. However, in variable-speed systems, accumulators with or without a bladder can be made more effective by partially charging them with a gas (nitrogen or air) since the gas charge cushions hydraulic shocks and pulsations. If properly selected, located, tuned, and charged, a wide variety of accumulators (weight- or spring-loaded; gas-charged) can be used in positive displacement pump systems to prevent cavitation and waterhammer, damp pulsations, and reduce pressure surges.[9,10] Improper sizing or location can aggravate existing problems or cause additional ones. Typically, the best location for accumulators is as close to the pump as possible.

Gas-charged dampeners, or accumulators, such as those depicted in Fig. 6, are most commonly used and can be quite effective in controlling pulsations. These devices are commercially available from several sources. Important to their effectiveness are their location and volume and the pressure of the charge. When gas-charged dampeners are used, the gas pressure must be monitored and maintained since the gas can be absorbed into the liquid. The system pressure can sometimes be lower than the gas charging pressure, such as on start-up; therefore, a valve should be installed to shut off the accumulator during start-up to eliminate gas leakage to the primary liquid. When the valve is closed, the accumulator is decoupled from the system and is not effective. Accumulators with integral check valves should be adjusted so that pressure transients do not close the check valve and render the accumulator ineffective. Accumulators which have bladders (Figs. 6b, d, and e) to separate the gas charge from the liquid have some distinct advantages, particularly if gas absorption is a problem. Accumulators with flexible bladders must be carefully maintained since failure of a bladder could release gas into the liquid system and could compromise the effectiveness of the dampener.

The in-line gas dampener (Fig. 6e) has a cylinder around the pipe containing a gas volume and bladder. The liquid enters the dampener through small holes in the circumference of the pipe and impinges upon the bladder, which produces the same acoustic effect as a side branch configuration.

It is not always possible to design effective pulsation control systems using simplified techniques. For complicated piping systems with multiple pumps, an electroacoustic analog[1] is recommended for designing optimum filters or accumulators. This tool has become widely accepted for designing reliable piping systems for reciprocating liquid pumps and gas compressor units. The reduction in pulsations in a liquid pump system in a nuclear plant is shown in Fig. 7. These results were obtained with a two-volume acoustic filter system designed with the electroacoustic analog. The volume diameter was 19.3 in (49 cm), and the length was 4 ft (1.2 m). The choke tube diameter was 0.8 in (2 cm), and its length was 7 ft (2.1 m). The speed of the triplex (2-cm) pump was 360 rpm, and the filter resonant frequency was set at 8.1 Hz for a velocity of sound of 4550 ft/s (1390 m/s).

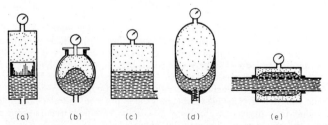

FIG. 6   Types of accumulators: (*a*) piston, (*b*) diaphragm, (*c*) gas-charged, (*d*) bladder, (*e*) in-line.

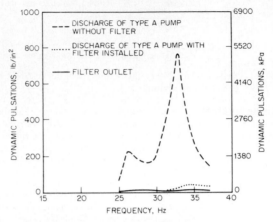

FIG. 7 Effect of acoustic filter on pulsations.

PIPING VIBRATIONS

When mechanical resonances are excited by pulsations, vibrations in the pump and piping can sometimes be 20 times higher than under off-resonant conditions. When the mechanical resonances coincide with the acoustic resonances, an additional amplification factor as high as 300 can be encountered.

Piping system mechanical natural frequencies can be calculated using simplified design procedures to provide effective detuning from known excitation sources. A nomogram (Fig. 8) for calculating the lowest natural frequency of uniform steel piping spans[11] can be used in designing piping systems and in diagnosing and solving vibration problems. For example, welded between two bottles, a 4-in (102-mm) pipe that is 10 ft (3 m) long and has an inner diameter of 3.826 in (97.18 mm) would have a mechanical frequency of 74 Hz. If the pipe was an equal-leg L bend ($L = 5$), the natural frequency would be 50 Hz.

To minimize piping vibration problems, all unnecessary bends (considering routing and thermal flexibility) should be eliminated since they provide a strong coupling point between pulsation excitation forces and the mechanical system. When bends are needed, use the largest enclosed angle possible and locate restraints near each bend. Piping should also have supports near all piping size reductions and at large masses (valves, accumulators, flanges, etc.). Small auxiliary piping connections (vents, drains, pressure test connections, etc.) should be designed such that the mass of the valve and flange is effectively tied back to the main piping, thus eliminating relative vibration.

DIAGNOSTICS AND INSTRUMENTATION

The diagnosis of vibrations in positive displacement pumps should usually include dynamic pressure measurements in the cylinders and piping near the pump. These measurements can be obtained by the use of piezoelectric or strain gage pressure transducers. If cavitation or flashing is suspected, a pressure transducer capable of measuring static and large dynamic pressures should be used; even then, cavitation-produced pressure shocks may damage the transducer.

Accelerometers with low frequency characteristics may be used with electronic integrators to obtain accurate vibration displacement data from the pump case, cylinders, or piping. Similarly, velocity or seismic pickups may be employed. Maximum vibrations usually occur at the middle of piping spans and at unsupported elbows (out of plane).

Real-time analyzers and oscilloscopes may be used to display the resulting signals. A field

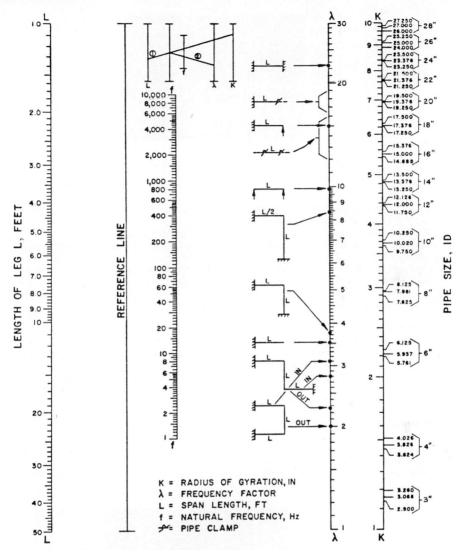

FIG. 8 Natural frequency of uniform steel piping spans. (1 ft = 0.3048 m; 1 in = 2.54 cm)

example showing the diagnosis of cavitation at the pump suction using a strain gage diaphragm pressure transducer is given in the oscilloscope trace of Fig. 9. The vapor pressure (gage) for this system was 25 lb/in² (172 kPa). Note that the negative half of the cycle is flattened when vapor pressure is reached and that very high amplitude pressure spikes are apparent. For liquids with dissolved gases, lower pulsation amplitudes can produce cavitation; however, the cavitation is usually less severe.

Typical field pulsation data obtained on a three-plunger pump is shown in Fig. 10, which is a frequency spectrum of the pulsations made by a real-time analyzer. Each spike represents a frequency multiple of running speed. Note that the third spike, representing the plunger frequency, has the largest amplitude; however, the components at one and two times pump speed are also

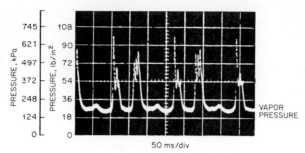

FIG. 9   Complex wave data showing cavitation effects of pressure wave in a liquid piping system.

significant, which means that these frequency components should be considered for pulsation control.

The vibration frequencies of the piping should be compared with pulsation frequencies to evaluate potential pulsation excitation of mechanical resonances. A check of the piping mechanical natural frequencies from the nomogram (Fig. 8) should be made to evaluate the possibility of a mechanical resonance. High vibrations produced by low-level pulsations at a particular frequency are indicative of a mechanical resonance, which can usually be corrected by additional piping restraints or snubbers. Once the causes of the pulsations and vibrations are diagnosed, the techniques presented above can be used to develop solutions.

Thermal Problems   In pump systems having high thermal gradients, large forces and moments on the pump case can cause misalignment of the pump and its driver as well as pump case distortion resulting in vibrations, rubbing (wear), bearing failure, seal leakage, etc. High stresses can be imposed on the piping, resulting in local yielding or damage to the piping restraints, snubbers, or support system. Misalignment problems commonly exhibit a high second-order component of the shaft vibrations. Proximity probes can be used at the bearings to measure movement of the shaft relative to the bearing centerline.

Diagnosis of Shaft Failures   Pump and driver shafting can experience high stresses during start-up and normal operation because of the uneven torque loading of the positive displacement pumping action. Shaft failures are strongly influenced by the torsional resonances of the system, which are the angular natural frequencies of the system.

Torsional vibrations can be measured using velocity-type torsional transducers which mount

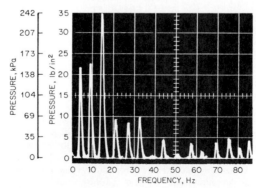

FIG. 10   Typical field data recorded on a three-plunger pump.

on a stub shaft or by measuring the gear tooth passing frequency with a magnetic transducer or proximity probe and using frequency-to-voltage converters to give the change in tooth passing frequency (the torsional vibrational velocity). Spectral analysis of these signals defines the torsional amplitudes and natural frequencies. The stresses can be calculated by using the mode shape of the specific resonant natural frequency and combining all the torsional loads. Torsional natural frequencies, mode shapes, and stresses can be calculated by using either the Holzer technique or digital computer programs.[12]

Torsional problems can usually be solved by changing the coupling stiffness between the driver and pump or by using a flywheel in an effective location. The addition of a flywheel will tend to smooth the torque oscillations. Pumps with a greater number of cylinders and equal cylinder phasing usually operate more smoothly with lower shaft stresses.

REFERENCES

1. Von Nimitz, W. W.: "Reliability and Performance Assurance in the Design of Reciprocating Compressor and Pump Installation," 1974 Purdue Compressor Technology Conference.

2. Sparks, C. R., and J. C. Wachel: "Pulsation in Centrifugal Pump and Piping Systems," *Hydrocarbon Processing*, July 1977, p. 183.

3. Hayward, A. T. J.: "How to Estimate the Bulk Modulus of Hydraulic Fluids," *Hydraulic Pneumatic Power*, January 1970, p. 28.

4. Wright, W. A.: "Prediction of Bulk Moduli and Pressure-Volume-Temperature Data for Petroleum Oils, *ASLE Transactions* **10**:349 (1967).

5. API Technical Data Book, *Petroleum Refining*, 2d ed., Washington, D.C., 1972.

6. Klaus, E. E., and J. A. O'Brien: "Precise Measurement and Prediction of Bulk-Modulus Values for Fluids and Lubricants," *Journal of Basic Engineering*, September 1964, p. 469.

7. Noonan, J. W., "Ultrasonic Determination of the Bulk Modulus of Hydraulic Fluids," *Materials Research and Standards*, December 1965, p. 615.

8. Hicks, E. J., and T. R. Grant: "Acoustic Filter Controls Reciprocating Pump Pulsations," *Oil and Gas Journal*, January 15, 1979, p. 67.

9. *Hydraulic Institute Standards for Centrifugal, Rotary and Reciprocating Pumps*, 14th ed., 1983.

10. *Machine Design*, Fluid Power Reference Issue, vol. 52, no. 21, Penton Publications, Cleveland, 1980.

11. Wachel, J. C., and C. L. Bates: "Techniques for Controlling Piping Vibration and Failures," ASME paper 76-Pet-18, 1976.

12. Szenasi, F. R., and L. E. Blodgett: "Isolation of Torsional Vibrations in Rotating Machinery," *Proceedings of the National Conference on Power Transmission*, vol. II, Illinois Institute of Technology, 1975.

FURTHER READING

Positive Displacement Pumps: Reciprocating, API Standard 674, 1st ed., Washington, D.C., 1980.

Positive Displacement Pumps: Controlled Volume, API Standard 675, 1st ed., Washington, D.C., 1980.

SECTION 3.7
DISPLACEMENT PUMP FLOW CONTROL

WILL SMITH

FLOW CONTROL IN INDIVIDUAL PUMPS

An inherent characteristic of positive displacement pumping of relatively incompressible liquids is that flow rate is proportional to displacement rate and independent of pressure levels. The capacity of a centrifugal pump operating at constant speed varies from a maximum flow at no developed pressure to zero flow at a definite limiting pressure known as shutoff head. The *average* capacity of positive displacement pumps at constant speed is, within design limits of pressure, practically constant, even though flow rate pulsations do occur as individual displacements are forced into the discharge pipe.

Flow control of positive displacement pumps is accomplished by

1. Changing the displacement rate
2. Changing the displacement volume
3. Changing the proportion of the displacement delivered into the piping system

Throttle Control in Direct-Acting Steam Pumps   Direct-acting pumps are controlled by speed change, which is effected by throttling the flow of steam (or other motive gas) to the drive cylinder. The magnitude of excess force on the drive piston over that required for the driven, or pumping, piston to force pumpage into the piping system dictates the stroke rate and therefore the capacity of the pump. A sensor which detects the desired result of pumping (pressure, level, flow rate, etc.) may be used to modulate the steam throttle valve.

Speed Control in Power-Driven Pumps   Speed modulation is the most common means of flow control for power-driven pumps.

The most rudimentary speed control is intermittent (start-stop) operation. The average capacity over relatively long time periods depends upon the percentage of time the pump operates at 100% versus the percentage of time it operates at zero flow. Of course, consideration must be given to the frequency of starts since electric motors may overheat if there is insufficient time for cooling after the inrush of starting current.

When continuous modulation of flow is desired, means of varying the driver speed or the ratio of speed reduction between the driver and pump are employed.

DRIVER SPEED CONTROL

1. *Multispeed constant-torque motors* offer limited steps in speeds.
2. *Variable-frequency motor* controls permit a wide range of fully modulated speeds but, at present, are limited in maximum horsepower (Subsec. 6.2.2).
3. *Direct current and wound rotor alternating current motors* permit modulation of speeds *within* definite limits of torque capability (Subsec. 6.2.2).
4. *Gasoline and diesel engines* provides speed variation capabilities within the limits where driver torque is adequate to satisfy the constant-torque requirements of the pump (Subsec. 6.1.3).
5. *Steam or gas turbines* can operate over a limited speed range within their particular output torque limits (Subsecs. 6.1.2 and 6.1.5).

SPEED CHANGER CONTROL   Since *most* positive displacement pumps operate at significantly lower speed than *most* drivers, a speed reduction unit is coupled between driver and pump. Capacity control by varying the ratio of the speed reduction is quite convenient. A number of methods are used.

1. *Variable-ratio belt drives* change the speed by changing the pitch diameter of driver and driven sheaves in response to capacity requirements (Subsec. 6.2.5).
2. *Hydraulic torque converters* (hydrodynamic drives) vary the speed by regulating the *amount* of active drive fluid in the coupling. Since these converters relate torque to speed, they are sensitive to the driven torque requirement change that may derive from system pressure changes (Subsec. 6.2.3).
3. *Hydroviscous drives* vary the slip between input and output shafts by adjusting the *distance* between elements in a viscous fluid environment (Subsec. 6.2.3).
4. *Eddy-current couplings* also vary the slip between driver and driven elements but do so by varying the strength of a magnetic field (Subsec. 6.2.1).

Variable Displacement Flow Control   Rather than by speed control (changes in the number of displacements in a given period of time), flow may be varied by changing the *volume* displaced per stroke.

Variable strokes are generally limited to small metering or "controlled volume" pumps. Although large-capacity variable-stroke pumps have been built, the complexity of the mechanism involved and the development of other effective and economical means of flow control have precluded their use (Sec. 9.16).

Changing Delivery to the System   Another means of capacity control is *not* to vary the pump capacity but to alter the amount of pumpage delivered to the system. There are two popular means of doing this.

SUCTION VALVE UNLOADING   Suction valve unloaders cause the displacement to be ineffective during the strokes for which the unloaders are activated. They command either full or zero delivery while the pump is kept running. Thus current inrush problems from frequent motor starts are avoided. Since discharge pressure is not developed when suction valves are unloaded, energy consumption is held to a minimum.

A pump valve is unloaded by mechanically preventing the valve plate from returning to its seat. If the suction valve is held away from its seat, liquid will ebb and flow through the valve from suction header to pump cylinder and from cylinder to header during the stroking of the plunger or piston.

The unloader mechanism does not move the valve. The valve is lifted off its seat by the pres-

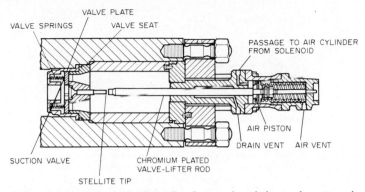

FIG. 1 Synchronized unloading keeps the valve open through the use of a suction valve unloader.

sure differential created by normal pump operation. In synchronized unloading, the suction valve unloader (Fig. 1) is retracted only when it is certain that the incoming flow will keep the valve open.

The unloader never resists pressure in the cylinder; it merely holds the valve plate against valve springs that had already been compressed and moves away from the valve while the springs are still compressed. This permits the valve to close when the plunger begins its discharge stroke.

For smooth transition between full flow and zero flow, the unloading and loading are accomplished in an immediate and identical sequence of the pumping order of the plungers. It is the function of the synchronized suction valve unloader distributor (Fig. 2) to control unloading events in that proper and immediate sequence. Similarity can be drawn to the gasoline engine distributor in its function of assuring the proper and immediate sequence of firing of the spark plugs.

The distributor is driven directly by the crankshaft so that its position is positively indexed to the stroking of the plungers. It provides an electrical signal to a solenoid valve which admits motive fluid pressure to the unloader chamber when the suction valve to the particular cylinder has been opened. Then, in sequence, each of the remaining suction valves is retained so that, in the first revolution of the pump after a control signals the distributor to initiate unloading, all suction valves are unloaded and the pump capacity is reduced to zero. When the control again signals for capacity, each suction valve is released, in sequence, during a single revolution of the crank.

While unloaded, the discharge valves do not function; they remain closed and serve as check valves against the pressurized discharge system. The suction valves remain open. Liquid enters from and is dispelled back into the suction manifold through the open suction valves. Therefore, no work is applied to the liquid, not even that required to open the suction and discharge valves.

When the control calls for loading, the load is applied in steps equal to the number of plungers.

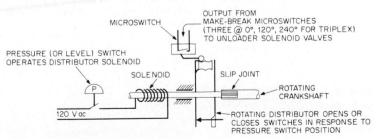

FIG. 2 Electromechanical distributor system.

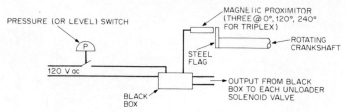

FIG. 3 Magnetic distributor system.

Thus the pumping horsepower, as well as the liquid inertia in the pumping system, is changed in a number of steps equal to the number of plungers over the time span of one crank revolution.

The unloader consists of a spring-loaded pneumatic cylinder attached to the suction valve cover. The spring advances holding fingers which keep the valve away from its seat. Pneumatic pressure, applied through a solenoid valve in response to the distributor signal, creates a force on a piston which opposes the spring force and allows the suction valve to load (close). The system is normally arranged to fail in the unloaded position on either electrical or pneumatic pressure failure.

Two distributor systems have been successfully used. One system (Fig. 2) employs an electromechanical distributor with cam, solenoid-operated clutch, and microswitches. The control signal activates the clutch. This draws a cam which is rotating with the crankshaft into a position where it sequentially activates microswitches which ernergize the individual unloader solenoids.

The second system (Fig. 3) employs a magnetic proximitor which constantly senses the position of the crankshaft, on which a rotating flag is attached. A solid-state circuit integrates the control signal with the proximitor position so that an output signal is delivered to the solenoids only when the controller and the proximitor combine to indicate that it is the proper instant for a change.

BYPASS CONTROL   Ultimate delivery of pumpage to the system may be controlled by a discharge-to-suction bypass valve. Pump operation is continuous with the possibility of modulating the flow to the system by regulating the portion "bypassed" to suction. A serious disadvantage is the full power utilization and full-load wear and tear even when at zero system capacity.

Evaluation of Flow Control Options   Proper selection of the means of capacity control requires consideration of system control requirements and then evaluation of the options available to satisfy those requirements.

Precision, responsiveness, and degree of modulation required may limit options. First cost, maintenance requirements, and the effect on pump operating cost should then be weighed for viable options. To evaluate the effect of operating costs, it is necessary to determine (or reasonably assume) the expected pattern of operation throughout the capacity range. For example, if practically all operation except infrequent start-ups is to be at full pump capacity, little penalty should be assigned to the full power usage of a bypass valve or to the part-load losses inherent with hydroviscous or eddy-current drives.

If smooth modulation of flow is not essential, start-stop or suction valve unloading may be the preference. If smooth modulation *and* considerable operation at part capacity are necessary, variable-frequency motors or variable-pitch V-belt drives may be indicated.

Table 1 and Fig. 4 provide some general guidelines. Each application should be evaluated on the merits of requirements and options.

FLOW CONTROL IN COMBINED DISPLACEMENT AND CENTRIFUGAL PUMPS

Systems which involve both centrifugal and reciprocating positive displacement pumps deserve some special consideration. Centrifugal pumps are often used as suction boosters to overcome

TABLE 1   Comparison of Several Capacity Control Schemes for Positive Displacement Pumps

Control method	Degree of modulation	First cost	Operating cost	Comments
		Directing-acting pumps		
Steam throttle	Full; zero to 100%	Low	Low	Steam pressure required to balance liquid piston force and overcome breakaway friction. Steam volume throttled to produce desired capacity.
		Power pumps		
Start-stop	Zero *or* 100%	Low	Low	Limited in frequency of starts because of temperature rise from inrush.
Multispeed motors	Steps dependent on motor winding	Medium	Low	Cost of motor controller switch gear must be assessed.
Variable frequency	Full: zero to 100% +	High	Low	Limited by current-handling capacity of solid-state controller.
Direct current	Full: zero to 100%	High	Medium	Drive is torque-speed sensitive, pump is torque-pressure sensitive at all speeds. Check drive for required torque at minimum and maximum speeds.
Wound rotor motors	Full: zero to 100%	High	Low	Drive is torque-speed sensitive. Pump is torque-pressure sensitive at all speeds. Check drive for required torque at minimum and maximum speeds.
Combustion engines	Variable for limited range	High	Low	Torsional analysis required to avoid high torsional stresses.
Steam or gas turbines	Variable for limited range	Medium	Medium	Drive is torque-speed sensitive. Pump is torque-pressure sensitive at all speeds. Check drive for required torque at minimum and maximum speeds.
Hydraulic torque converter	Full: zero to 100%	High	High	Low full speed efficiency.
Hydroviscous speed control	Full: zero to 100%	High	High at part load	Efficiency porportional to capacity. Loss to heat exchanger.
Fluid coupling	Full: zero to 100%	Medium	High	Full-speed slip loss greater than hydroviscous. Loss to heat exchanger,
Magnetic (eddy) coupling	Full: zero to 100%	High	High	Full-speed slip loss.

TABLE 1　(*cont.*)

Control method	Degree of modulation	First cost	Operating cost	Comments
		Power pumps		
Suction valve unloaders	Zero *or* 100%	Low	Low	Synchronization with suction stroke avoids start-stop and shock problems.
Bypass valve	Limited	Low	High	Uses full power at zero capacity with full valve wear.
Variable-pitch belt	Limited	Low	Low	Limited to belt drive horsepower.

acceleration head requirements peculiar to reciprocating pumps but are rarely used to supplement flow. Some unique characteristics of each type of pump which affect the other type must be considered in the design, operation, and control of the interrelated system.

Pertinent Positive Displacement Pump Characteristics

1. The discharge pressure product is a function of the system requirement only and is independent of pump capacity.
2. The flow rate pulses between maximum and minimum values for each revolution of the crankshaft.
3. The pulsating flow imposes an acceleration head which adds to the net positive inlet pressure required (Sec. 3.1).
4. Being more energy-efficient, the positive displacement pump is normally the lead pump and the centrifugal pump is the supplement.

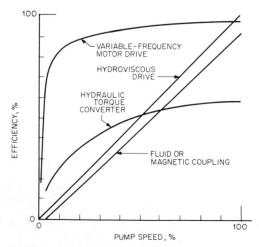

FIG. 4　Relative efficiencies of speed control options.

Pertinent Centrifugal Pump Characteristics

1. The dynamic head (or pressure rise) produced is a function of pump capacity as well as of system requirements.
2. For satisfactory operation, the flow must be kept within a limited range of the best efficiency capacity (Subsec. 2.3.1).
3. When used as a suction booster, the centrifugal pump must be designed so that it cannot introduce air into the gas-intolerant reciprocating pump. (Gas in positive displacement pumps, like liquids in positive displacement compressors, may cause severe hydraulic and mechanical shock).
4. Centrifugal pumps do not generate acceleration heads that impose on net positive suction head required. However, if a centrifugal pump is connected in parallel to a common suction line with a reciprocating pump, some of the *system* acceleration head loss from the positive displacement pump may affect the centrifugal pump.

There are only two cases that need be considered: centrifugal pumps feeding (1) in series into reciprocating pumps to increase suction NPSHA to the reciprocating pump or (2) in parallel to augment the delivered capacity. It would be most unusual to encounter a positive displacement pump feeding into the suction of a centrifugal pump because of the high pressure that could be imposed on the centrifugal pump suction and because of the amplification of flow pulsations resulting from interaction of the characteristics of the two pumps, which could be deleterious to both pumps.

Series Operation, Suction Boost
The flow rate is always determined by the positive displacement pump capacity and is independent of the centrifugal booster pump. The capacity is therefore controlled as with any positive displacement pump.

The purpose of the centrifugal pump is to supply sufficient pressure to satisfy the suction requirements of the positive displacement pump. The centrifugal pump must be sized to provide the maximum rate of flow of the positive displacement pump at a pressure high enough to meet the NPSH requirement of the latter.

With a flow-controlled positive displacement pump, speed control of the booster might be considered, as illustrated in Fig. 5. The system pressure requirement can be satisfied by the pos-

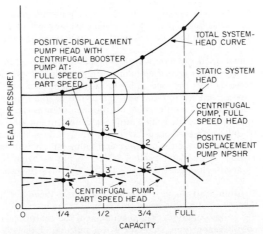

FIG. 5 A centrifugal booster pump can satisfy the NPSH requirement of a positive displacement pump by operating either on its characteristic full speed curve 1-2-3-4 or with speed reduction along curve 1-2'-3'-4'.

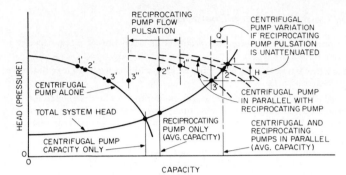

FIG. 6 Parallel operation of undampened reciprocating and centrifugal pumps.
1. The reciprocating pump will operate at a capacity corresponding to its speed regardless of its discharge pressure.
2. If the system head is only static (constant head), the centrifugal pump capacity is constant and the system flow variation corresponds to the reciprocating pump variation.
3. If the system head is partially dynamic (as illustrated), the centrifugal pump will experience flow variations.
   a. The flatter the centrifugal pump curve, the greater the capacity and smaller the head variations on the centrifugal pump.
   b. The flatter the system curve, the lower the capacity and greater the head variations on the centrifugal pump.

itive displacement pump at all flow rates; therefore total energy requirements may be reduced by using the less efficient centrifugal pump to develop no more than the required suction head.

If the centrifugal booster pump is run at constant speed, its share of the total head requirement will increase as system capacity is reduced.

Parallel Operation To increase the capacity of an existing pumping system, it is sometimes convenient to consider paralleling a positive displacement pump with a centrifugal pump. The interactions of the two pumps *and* the system must be considered and controlled for satisfactory operation.

Suction stabilizers and discharge dampeners, desirable for most reciprocating pump installations, become mandatory when a centrifugal pump and a positive displacement pump are connected in parallel. These devices are used to isolate, as much as possible, the effects of positive displacement pump flow pulsations from the sensitive centrifugal pump operating characteristics.

Centrifugal pumps in parallel with positive displacement pumps can supplement the system capacity only so long as they generate sufficient discharge pressure to satisfy the system requirement at the higher total flow. Undampened flow pulsations shift the centrifugal shutoff flow point relative to system flow at the frequency of the reciprocating pulsation. From the *system standpoint*, then, the total flow varies along curve 1-2-3-2-1 in Fig. 6. From the *centrifugal pump standpoint*, it varies along its characteristic curve, 1'-2'-3'-2'-1'. The reciprocating pump operates along curve 1"-2"-3"-2"-1".

For purely static system-head curves, the centrifugal pump, theoretically, would not "sense" the pulsating flow. For dynamic system heads, the centrifugal pump flow varies contracyclically with the reciprocating pump pulsations. The magnitude of the centrifugal pump flow variation due to the positive displacement pump pulsations is a function of the system head curve, the centrifugal pump curve, the degree of flow pulse imposed, and the liquid bulk modulus (or compressibility).

FLOW CONTROL IN POSITIVE DISPLACEMENT PUMPS IN SERIES _____

Whenever positive displacement pumps are installed in series (such as for multiple-station pipeline service), some variable-capacity capability is mandatory. Otherwise any variation in capacity between series pumps or series groups of pumps would result in disastrous interstation pressures.

The flow is determined by the first pump in the line. Subsequent pumps should be equipped with capacity controls responsive to their *inlet* pressures. The preferred control is a continuous-modulation type, such as variable speed, unless large reservoir capacity is available between pumps.

Recirculation from discharge to suction through a throttling bypass valve may be used, at the expense of energy, to match the delivery from downstream pumps to upstream pumps. Excessive bypass flow in relation to through flow can result in significant temperature rise of the liquid.

When downstream pump capacity is less than the upstream delivery, the interpump pressure will rise, indicating a need for increased capacity. When the downstream pump capacity exceeds the upstream delivery, the downstream pump will "draw down" the interpump pressure to the point where net positive inlet pressure difficulties are ensured.

Reservoir accumulators between pumps permit a limited degree of mismatch in capacities, allowing the use of incremental capacity controls, such as suction valve unloading and intermittent pump bypass.

In any event, effective discharge dampeners on the upstream pump and suction stabilizers on the downstream pump are necessary to prevent the inherent pressure pulsations and control modulation variances from adversely affecting operation.

FLOW CONTROL IN POSITIVE DISPLACEMENT PUMPS IN PARALLEL _____

Since the capacity of positive displacement pumps is virtually independent of head, there is no problem from the head capacity characteristics in parallel operation—even for pumps of different sizes.

Particular attention to suction acceleration head requirements is necessary when more than one pump is taking suction from a common suction pipe. Since it cannot be assured that two pumps will not, at some point, have simultaneous suction strokes, the acceleration head requirements of each pump must be added for that portion of the suction line that furnishes the total flow. This suggests extralarge suction lines, individual ones from the liquid source, or suction stabilizers sized for the total flow with individual outlet connections for each pump.

JET PUMPS

Alex M. Jumpeter

The term *jet pump*, or *ejector*, describes a pump having no moving parts and utilizing fluids in motion under controlled conditions. Specifically, motive power is provided by a high-pressure stream of fluid directed through a nozzle designed to produce the highest possible velocity. The resultant jet of high-velocity fluid creates a low-pressure area in the mixing chamber, causing the suction fluid to flow into this chamber. Ideally, there is at this point an exchange of momentum that produces a uniformly mixed stream traveling at a velocity intermediate between the motive and suction velocities. The diffuser is shaped to reduce the velocity gradually and convert the energy to pressure at the discharge with as little loss as possible. The three basic parts of any ejector are the nozzle, the diffuser, and the suction chamber, or body (Fig. 1).

DEFINITION OF TERMS

A definition of standard ejector terminology is as follows:

Ejector General name used to describe all types of jet pumps which discharge at a pressure intermediate between motive and suction pressures

Eductor A liquid jet pump using a liquid as motive fluid

Injector A particular type of jet pump which uses a condensable gas to entrain a liquid and discharge against a pressure higher than either motive or suction pressure; principally, a boiler injector

Jet compressor A gas jet pump used to boost pressure of gases

Siphon A liquid jet pump utilizing a condensable vapor, normally steam, as the motive fluid

Of concern to this text are jet pumps used to pump liquids. Eductors, being the most common, will receive the principal treatment herein. Sizing parameters for siphons will also be presented.

EDUCTORS

Theory and Design Eductor theory is developed from the Bernoulli equation. Static pressure at the entrance to the nozzle is converted to kinetic energy by permitting the fluid to flow freely through a converging-type nozzle. The resulting high-velocity stream entrains the suction fluid in the suction chamber, resulting in a flow of mixed fluids at an intermediate velocity. The diffuser section then converts the velocity pressure back to static pressure at the discharge of the eductor.

The Bernoulli equation for the motive fluid across the nozzle of an eductor is

$$\frac{P_1}{\gamma_1} + \frac{V_1^2}{2g} = \frac{P_s}{\gamma_1} + \frac{V_N^2}{2g} \tag{1}$$

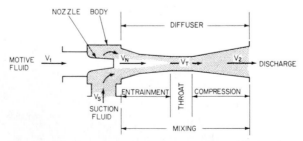

FIG. 1 Jet nozzles convert pressure energy to velocity; diffusers entrain and mix the fluids and convert velocity back to pressure.

where P_1 = static pressure upstream, lb/ft² (N/m²)
   P_s = static pressure at suction (nozzle tip), lb/ft² (N/m²)
   V_1 = velocity upstream of nozzle, ft/s (m/s)
   V_N = velocity at nozzle orifice, ft/s (m/s)
   γ_1 = specific weight (force) of motive fluid, lb/ft³ (N/m³)

Upstream of the nozzle, all the energy is considered static head, so that the velocity term V_1 drops out, yielding

$$\frac{V_N^2}{2g} = \frac{P_1 - P_s}{\gamma_1}$$

(2)

This term is called the *operating head*.

Across the diffuser, the same principle applies for the mixed fluid stream, except that the effect is the reverse of a nozzle; hence,

$$\frac{P_s}{\gamma_2} + \frac{V_T^2}{2g} = \frac{P_2}{\gamma_2} + \frac{V_2^2}{2g}$$

where P_s = static pressure at suction, lb/ft² (N/m²)
   P_2 = static pressure at discharge, lb/ft² (N/m²)
   V_T = velocity at diffuser throat, ft/s (m/s)
   V_2 = velocity downstream, ft/s (m/s)
   γ_2 = specific weight (force) of mixed fluids, lb/ft³ (N/m³)

At the discharge, it is assumed that all velocity head has been converted to static head; hence $V_2 = 0$ and

$$\frac{V_T^2}{2g} = \frac{P_2 - P_s}{\gamma_2}$$

(3)

This term is called the *discharge head*. The head ratio R_H is then defined as the ratio of the operating head to the discharge head:

$$R_H = \frac{V_N^2/2g}{V_T^2/2g} = \frac{V_N^2}{V_T^2} = \frac{(P_1 - P_s)/\gamma_1}{(P_2 - P_s)/\gamma_2} = \frac{(P_1 - P_s)\gamma_2}{(P_2 - P_s)\gamma_1}$$

(4)

Since ratios are involved, it is convenient to replace specific weight with specific gravity:

$$R_H = \frac{(P_1 - P_s)(\text{sp.gr.}_2)}{(P_2 - P_s)(\text{sp.gr.}_1)}$$

(5)

When the suction and motive fluids are the same, no gravity correction is required and Eq. 5 becomes

$$R_H = \frac{H_1 - H_s}{H_2 - H_s}$$

(6)

where $H_1 - H_s$ = operating head, ft (m)
   $H_2 - H_s$ = discharge head, ft (m)

Entrainment conditions are defined by the basic momentum equation:

$$M_1 V_N + M_s V_s = (M_1 + M_s)V_T$$

where M_1 = mass of motive fluid, slugs (kg)
   M_s = mass of suction fluid, slugs (kg)
   V_N = velocity at nozzle discharge, ft/s (m/s)
   V_s = velocity at suction inlet, ft/s (m/s)
   V_T = velocity at diffuser throat, ft/s (m/s)

The velocity of approach at the suction inlet is zero; therefore rearranging yields

$$M_s = M_1 \left(\frac{V_N}{V_T} - 1 \right)$$

and the term below is defined as the *weight operating ratio:*

$$R_w = \frac{M_s}{M_1} = \frac{V_N}{V_T} - 1 \tag{7}$$

Observe that the term V_N^2/V_T^2 has previously been defined as the head ratio R_H; therefore

$$R_w = \sqrt{R_H} - 1 \tag{8}$$

The volume ratio R_q is then simply

$$\frac{Q_s}{Q_1} = R_w \frac{\text{sp.gr.}_1}{\text{sp.gr.}_2} \tag{9}$$

where Q_s = suction flow in volumetric units
      Q_1 = motive flow in volumetric units

The maximum theoretical performance of eductors is calculated from the above relationships. In practice, there are energy losses associated with the mixing of two fluids and frictional losses in the diffuser. These losses are accounted for by the use of an empirical factor to reduce the theoretical maximum performance. Figure 2 shows this factor plotted against NPSH (net positive suction head) for a single-nozzle and annular-nozzle eductor. In an annular-nozzle eductor, the motive fluid is introduced around the periphery of the suction fluid, either by a ring of nozzles (Fig. 15) or by an annulus created between the inner wall of the diffuser and the outer wall of the suction nozzle (Fig. 14). The NPSH is the head available at the centerline of the eductor to move and accelerate suction fluid entering the eductor mixing chamber. NPSH is the total head in feet (meters) of fluid flowing and is defined as atmospheric pressure minus suction pressure minus vapor pressure of suction or motive fluid, whichever is higher.

The efficiency factor is introduced into Eq. 8 as

$$R_w = \epsilon \sqrt{R_H} - 1$$

This equation is used to calculate the motive quantity or pressure from the operating parameters. The nozzle and diffuser diameters are calculated from the equation $Q = wAV$, using suitable nozzle and diffuser entrance coefficients. The principal problems in design concern the size and proportions of the mixing chamber, the distance between nozzle and diffuser, and the length of the diffuser. Eductor designs are based on theory and empirical constants for length and shape. The most efficient units are developed from calculated designs which are then further modified by prototype testing.

Increased viscosity of motive or suction fluid increases the frictional and momentum losses and therefore reduces the efficiency factor of Fig. 2. Below 20 cP, the effect is minimal (approximately 5% lowering of ϵ). Above this value, the loss of performance is more noticeable and empirical data or pilot testing is used to determine sizing parameters.

Figure 3 shows the operating ratio R_w versus the head ratio R_H for various lift conditions. The efficiency factor has been incorporated into this curve.

The final size of the eductor is determined by the discharge line and is based on normal pipeline velocities, which are usually 3 to 10 ft/s (0.9 to 3 m/s). Figure 4a and 4b are used for estimating eductor size. To illustrate the use of Figs. 3 and 4, consider the following example.

EXAMPLE 1   It is desired to remove 100 gpm (22.7 m³/h) of water at 100°F (38°C)

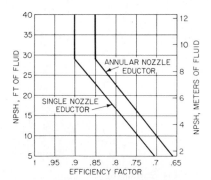

FIG. 2   NPSH versus efficiency factor. (Schutte and Koerting)

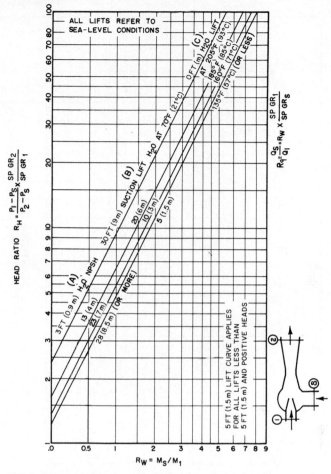

FIG. 3 Estimating operating ratios for liquid jet eductors. An eductor can be designed for only one head point. (Schutte and Koerting)

from a pit 20 ft (6.1 m) deep. Discharge pressure is 10 lb/in² (0.69 bar°) gage. Motive water is available at 60 lb/in² (4.1 bar) gage and 80°F (26.6°C). The eductor is to be located above the pit. Find the eductor size and motive water quantity required.

Solution To use Fig. 3, it is necessary to determine the NPSH and the head ratio R_H. The centerline of the eductor is chosen as the datum plane, and NPSH is taken to be atmospheric pressure minus suction lift minus vapor pressure at 100°F (38°C):

in USCS units $$NPSH = 34 \text{ ft} - 20 \text{ ft} - 1.933 \text{ inHg} \left(\frac{13.6}{12} \right) = 11.81 \text{ ft}$$

in SI units $$NPSH = 10.36 \text{ m} - 6.1 \text{ m} - 49 \text{ mmHg} \left(\frac{13.6}{1000} \right) = 3.59 \text{ m}$$

° 1 bar = 10⁵ Pa. For a discussion of bar, see *SI Units—A Commentary* in the front matter.

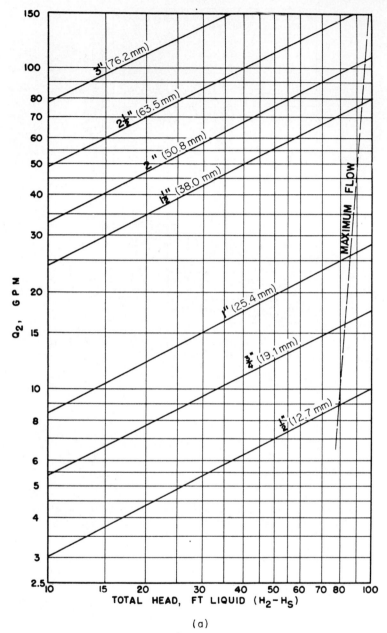

(a)

FIG. 4 Sizing curve (gpm × 0.227 = m³/h; ft × 0.3048 = m).

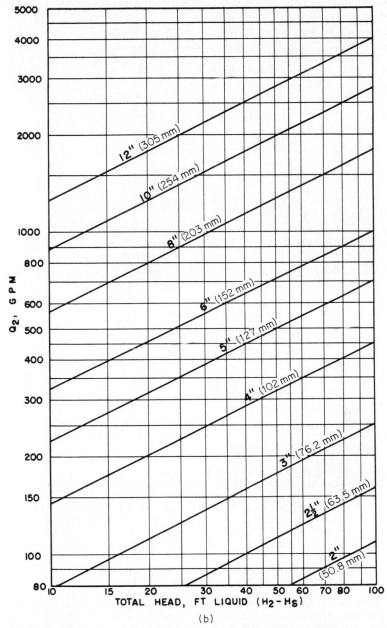

FIG. 4 *(continued)*

4.7

Since motive and suction are the same fluid, it is convenient to work in feet (meters) rather than pounds per square inch (bar), and

$$P_1 = 60 \text{ lb/in}^2 \text{ gage} = 138.6 \text{ ft } H_2O \text{ (4.1 bar} = 42 \text{ m)}$$

$$P_2 = 10 \text{ lb/in}^2 \text{ gage} = 23.1 \text{ ft } H_2O \text{ (0.69 bar} = 7 \text{ m)}$$

$$P_3 = -20 \text{ ft } (-6.1 \text{ m})$$

Then

in USCS units $\qquad R_H = \dfrac{138.6 - (-20)}{23.1 - (-20)} = \dfrac{158.6}{43.1} = 3.68$

in SI units $\qquad R_H = \dfrac{42 - (-6.1)}{7 - (-6.1)} = 3.68 \text{ m}$

Enter Fig. 3 at $R_H = 3.68$ and $NPSH = 11.81$ (3.59 m); read $R_w = 0.48$. Since there is no gravity correction,

in USCS units $\qquad R_w = R_q = \dfrac{0.48 \text{ gal suction}}{\text{gal motive}}$

in SI units $\qquad R_w = R_q = \dfrac{0.48 \text{ m}^3 \text{ suction}}{\text{m}^3 \text{ motive}}$

The same result can be obtained by using the efficiency factor from Fig. 2. Then R_w is $0.77\sqrt{R_H} - 1 = 0.48$ and the required motive fluid is

in USCS units $\qquad \dfrac{100 \text{ gpm suction}}{0.48} = 208 \text{ gpm at 60 lb/in}^2 \text{ gage}$

in SI units $\qquad \dfrac{22.7 \text{ m}^3/\text{h}}{0.48} = 47.3 \text{ m}^3/\text{h at 4.1 bar gage}$

Discharge flow is

in USCS units $\qquad\qquad 208 + 100 = 308 \text{ gpm}$

in SI units $\qquad\qquad 47.3 + 22.7 = 70.0 \text{ m}^3/\text{h}$

The size is obtained from Fig. 4. Enter Fig. 4b at $Q_2 = 308$ gpm (70 m³/h) and discharge head $(H_2 - H_3) = 23.1 - (-20) = 43.1$ ft [$7 - (-6.1) = 13.1$ m]; read eductor size of 4 in (102 mm) based on the discharge connection.

Note If there were any appreciable length of run on the discharge line, it would be necessary to calculate the pressure drop in this line and recalculate the eductor size after adding the line loss to the discharge head required. Frictional losses on the suction side must also be included. In the example chosen, however, 100 gpm (22.7 m³/h) in a 4-in (102-mm) suction line 20 ft (6.1 m) long will have negligible frictional loss, less than 0.25 ft (0.08 m) H_2O.

Performance Characteristics Figure 5 illustrates the performance characteristics of eductors. Note the sharp break in capacity below the design point. For this reason all eductors are not designed for a peak efficiency. It is often advantageous to have a wide span of performance with lower efficiencies rather than a peak performance with very limited range.

Applications Beside the obvious advantages of being self-priming, having no moving parts, and requiring no lubrication, eductors can be made from any machinable material in addition to special materials, such as stoneware, Teflon,®° heat-resistant glass, and fiberglass. The applications throughout industry are too numerous to mention, but some of the more common will be discussed here. The type of eductor is determined by the service intended.

° Teflon is a registered trademark of E. I. DuPont de Nemours and Co., Inc.

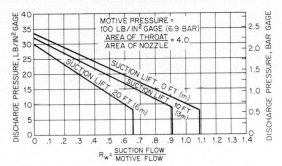

FIG. 5 Characteristic performance of an eductor.

GENERAL PURPOSE EDUCTORS Table 1 is a capacity table for a general purpose eductor used for pumping and blending. This type of eductor, illustrated in Fig. 6, has a broad performance span rather than a high peak efficiency point. Standard construction materials for this type of eductor are cast iron, bronze, stainless steel, and PVC. Typical uses include cesspool pumping, deep-well pumping, bilge pumping aboard ship, and condensate removal.

The following problem illustrates the use of Table 1.

EXAMPLE 2 Pump 30 gpm (6.81 m³/h) of water from a sump 5 ft (0.61 m) below ground. Discharge to drain at atmospheric pressure. Motive water available is 40 lb/in² (2.8 bar) gage.

Solution Enter left side of Table 1 at 5 ft (1.5 m) suction lift and 0 lb/in² (bar) gage discharge pressure. Read horizontally across to 40 lb/in² (2.8 bar) gage operating water pressure. Read 9.6 gpm (2.18 m³/h) suction and 7.3 gpm (1.66 m³/h) operating fluid. These values are obtained in a 1-in (25.4-mm) eductor with a capacity ratio of 1.0.

To determine the capacity ratio of the required unit, divide the required suction by the quantity handled in 1-in (25.4-mm) eductor:

in USCS units $$\text{Capacity ratio} = \frac{30}{9.6} = 3.13$$

in SI units $$\text{Capacity ratio} = \frac{6.81}{2.18} = 3.13$$

Referring to the bottom of Table 1, a 2-in (50.8-mm) eductor with a capacity ratio of 4.0 is obtained. The required motive flow is then

in USCS units $\qquad\qquad$ 4(7.3) = 29.2 gpm

in SI units $\qquad\qquad$ 4(1.66) = 6.64 m³/h

and the suction capacity is

in USCS units $\qquad\qquad$ 4(9.6) = 38.4 gpm

in SI units $\qquad\qquad$ 4(2.18) = 8.72 m³/h

A 1½-in (38-mm) unit can handle 2.89 times the values in Table 1, or 27.7 gpm (6.3 m³/h) suction when using 21 gpm (4.8m³/h) motive water at 40 lb/in² (2.8 bar) gage. If suction flow rate is not critical, some capacity can be sacrificed in order to use a smaller and therefore lower-cost eductor. If optimum performance is desired, it is necessary to size a special eductor using Figs. 3 and 4.

Figure 7 illustrates more streamlined versions for higher suction lifts or applications involving the handling of slurries. This type of eductor is often used to remove condensate from vessels

TABLE 1 Capacity Table of Standard 1-in (25.4-mm) Water-Jet Eductors, gpm[a]

Suction lift, ft (m)	Discharge pressure, lb/in² (bar) gage	Function	Operating water pressure, lb/in² (bar) gage							
			10 (0.69)	20 (1.4)	30 (2.1)	40 (2.8)	50 (3.4)	60 (4.1)	80 (5.5)	100 (6.9)
0 (0)	0 (0)	Suction	5.85	8.1	9.5	10.0	12.0	12.0	12.0	12.0
		Operating	3.55	5.0	6.1	7.1	7.9	8.7	10.0	11.0
	5 (0.34)	Suction	...	1.4	4.1	6.0	8.0	10.0	11.0	12.0
		Operating	...	4.9	6.1	7.0	7.9	8.6	10.0	11.0
	10 (0.69)	Suction	...	..	0.28	2.3	4.8	6.4	8.8	11.0
		Operating	...	..	5.9	6.8	7.8	8.5	9.8	11.0
	15 (1.0)	Suction	...	..	...	...	1.2	3.4	5.9	8.6
		Operating	...	..	...	...	7.7	8.4	9.8	11.0
	20 (1.4)	Suction	...	..	...	...	...	0.3	3.5	5.9
		Operating	...	..	...	...	...	8.2	9.7	11.0
	25 (1.7)	Suction	...	..	...	...	...	...	0.83	3.9
		Operating	...	..	...	...	...	...	9.6	11.0
	30 (2.1)	Suction	...	..	...	...	...	...	...	1.7
		Operating	...	..	...	...	...	...	...	11.0
5 (1.5)	0 (0)	Suction	4.4	6.8	8.6	9.6	11.0	11.0	12.0	12.0
		Operating	3.9	5.3	6.4	7.3	8.1	8.8	10.0	11.0
	5 (0.34)	Suction	...	1.5	3.2	5.0	7.0	9.0	11.0	11.0
		Operating	...	5.2	6.3	7.2	8.0	8.7	10.0	11.0
	10 (0.69)	Suction	...	..	...	1.9	3.6	5.6	8.6	10.0
		Operating	...	..	...	7.1	7.9	8.6	10.0	11.0
	15 (1.0)	Suction	...	..	...	...	1.1	2.6	5.8	8.3
		Operating	...	..	...	...	7.8	8.6	9.9	11.0
	20 (1.4)	Suction	...	..	...	...	...	...	3.3	5.6
		Operating	...	..	...	...	...	...	9.8	11.0
	25 (1.7)	Suction	...	..	...	...	...	...	0.47	3.6
		Operating	...	..	...	...	...	...	9.8	11.0
	30 (2.1)	Suction	...	..	...	...	...	...	...	1.5
		Operating	...	..	...	...	...	...	...	11.0
10 (3.0)	0 (0)	Suction	2.0	4.6	6.7	8.3	9.0	10.0	10.0	10.0
		Operating	4.2	5.5	6.6	7.4	8.2	9.0	10.0	11.0
	5 (0.34)	Suction	...	..	2.0	4.3	5.9	7.7	9.9	10.0
		Operating	...	..	6.5	7.4	8.2	8.9	10.0	11.0
	10 (0.69)	Suction	...	..	...	1.1	3.0	4.5	8.1	9.6
		Operating	...	..	...	7.3	8.1	8.8	10.0	11.0
	15 (1.0)	Suction	...	..	...	...	1.1	2.1	5.6	7.3
		Operating	...	..	...	...	8.0	8.7	10.0	11.0
	20 (1.4)	Suction	...	..	...	...	...	...	2.8	5.3
		Operating	...	°	...	...	...	...	9.9	11.0
	25 (1.7)	Suction	...	..	...	...	...	...	...	2.8
		Operating	\...	..	...	...	...	...	...	11.0
	30 (2.1)	Suction	...	..	...	...	...	...	...	1.1
		Operating	...	..	...	...	...	...	...	11.0
15 (4.6)	0 (0)	Suction	...	3.3	5.3	7.9	8.4	8.9	8.9	9.1
		Operating	...	5.7	6.8	7.6	8.4	9.1	10.0	12.0
	5 (0.34)	Suction	...	..	...	4.0	4.9	7.3	8.6	9.1
		Operating	...	..	...	7.6	8.3	9.0	10.0	11.0
	10 (0.69)	Suction	...	..	...	...	2.4	4.0	6.4	8.6
		Operating	...	..	...	...	8.2	9.0	10.0	11.0
	15 (1.0)	Suction	...	..	...	...	...	...	4.2	6.8
		Operating	...	..	...	...	...	...	10.0	11.0
	20 (1.4)	Suction	...	..	...	...	...	...	2.1	4.5

Suction lift, ft (m)	Discharge pressure, lb/in² (bar) gage	Function	Operating water pressure, lb/in² (bar) gage							
			10 (0.69)	20 (1.4)	30 (2.1)	40 (2.8)	50 (3.4)	60 (4.1)	80 (5.5)	100 (6.9)
	25 (1.7)	Operating	...	..	...	...	...	...	10.0	11.0
		Suction	...	..	...	...	...	...	...	1.9
		Operating	...	..	...	...	...	...	...	11.0
	0 (0)	Suction	...	2.0	4.0	6.4	7.8	7.8	7.8	7.8
		Operating	...	6.0	7.0	7.8	8.6	9.3	11.0	12.0
	5 (0.34)	Suction	...	..	...	2.8	3.9	6.3	7.8	7.8
		Operating	...	..	...	7.7	8.5	9.2	10.0	12.0
	10 (0.69)	Suction	...	..	...	...	1.2	3.1	5.7	7.1
20 (6.1)		Operating	...	..	...	...	8.3	9.1	10.0	12.0
	15 (1.0)	Suction	...	..	...	...	...	...	3.6	5.4
		Operating	...	..	...	...	...	...	10.0	11.0
	20 (1.4)	Suction	...	..	...	...	...	...	1.4	3.8
		Operating	...	..	...	...	...	...	10.0	11.0
	25 (1.7)	Suction	...	..	...	...	...	...	...	1.5
		Operating	...	..	...	...	...	...	...	11.0

Relative capacities of standard sizes

Size eductor, in (mm)	½ (12.7)	¾ (19.1)	1 (25.4)	1½ (38.1)	2 (50.8)	2½ (63.5)	3 (76.2)	4 (102)	6 (152)
Capacity ratio	0.36	0.64	1.00	2.89	4.00	6.25	9.00	16.00	36.00

[a]gpm × 0.227 = m³/h.
SOURCE: Schutte and Koerting.

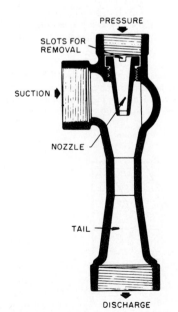

PRESSURE

SLOTS FOR REMOVAL

SUCTION

NOZZLE

TAIL

DISCHARGE

FIG. 6 General purpose eductor. (Schutte and Koerting)

vessels under vacuum. The advantage is that eductors require only 2 ft (0.61 m) NPSH and, being smaller than mechanical pumps, save considerable space. Further, a partial vapor load is much less likely to vapor-lock a jet pump because the venturi tube minimizes the expansion effect of flashing vapor. Sizing is done in the manner illustrated in Example 1, using Figures 3 and 4.

MIXING EDUCTORS   While any eductor is inherently a mixing device, some are specifically designed as mixers. They are used to replace mechanical agitators and are located inside the tank containing the fluid to be agitated. Figure 8 illustrates the simplest type of eductor, the *Sparger nozzle*. These units entrain approximately three times the volume of suction fluid to motive fluid. A 20-lb/in² (1.4-bar) drop across the nozzle is recommended for proper mixing. Figure 9 shows the motive capacities for this type of eductor. Sparger nozzles are normally used for shallow tanks, whereas the tank mixer described below is preferred for deeper vessels.

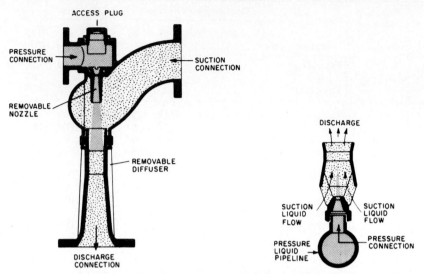

FIG. 7   Streamlined eductor. (Schutte and Koerting)   **FIG. 8**   Sparger nozzle. (Schutte and Koerting)

Figure 10 illustrates a type of eductor called a *tank mixer*. It is installed under the tank containing the fluid to be agitated. Motive capacities are shown in Table 2. The units are usually custom designed for a specific entrainment ratio, the required capacity being determined by the quantity of tank fluid, the ratio of mixture desired, and the depth of the tank being agitated.

EXAMPLE 3   It is desired to blend recycled tank fluid into a tank 20 ft (6.1 m) deep in a volume ratio of 1 motive to 1.5 suction. The tank contains 7500 gal (28.4 m^3), and it is desired to turn over the tank in 30 min. The motive pump will deliver 60 lb/in^2 (4.14 bar) gage at the eductor nozzle. What size mixing eductor is needed?

Solution   The 500 gal (28.4 m^3) turned over in 30 min is equivalent to 250 gpm (56.8 m^3/h). Since the motive fluid in this case is recycled from the tank, both motive and suction fluid contribute to the tank turnover. In the ratio of 1.5 suction to 1 motive fluid, the motive quantity required to attain a circulation rate of 250 gpm (56.8 m^3/h) is 100 gpm (22.7 m^3/h). To

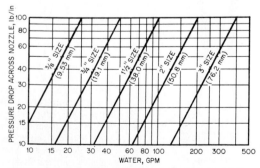

FIG. 9   Motive capacity of Sparger nozzles (gpm × 0.227 = m^3/h; lb/in^2 × 0.0689 = bar). (Schutte and Koerting)

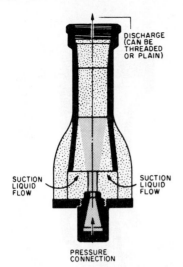

DISCHARGE
(CAN BE
THREADED
OR PLAIN)

SUCTION
LIQUID
FLOW

SUCTION
LIQUID
FLOW

PRESSURE
CONNECTION

FIG. 10 Tank mixing eductor. (Schutte and Koerting)

select the size, it is necessary to obtain the differential pressure across the nozzle orifice of the eductor. Since the eductor is below the tank, the net driving head is 60 lb/in^2 gage − 20/2.31 = 51.35 lb/in^2 (4.14 − 6.1/10.2 = 3.54 bar) gage across the nozzle. Enter Table 2 and interpolate between 50 and 60 lb/in^2 (3.4 and 4.1 bar) gage. A 1½-in (38-mm) eductor will pass only 73 gpm (16.6 m^3/h), whereas a 2-in (51-mm) eductor will pass 129 gpm (29.3 m^3/h). The selection would then be a 2-in (51-mm) mixing eductor.

SPINDLE PROPORTIONING EDUCTORS Another type of mixing eductor is illustrated in Fig. 11. Typical applications of this type include mixing hydrocarbons with caustic, oxygen, or copper chloride slurries; producing emulsions; and proportioning liquids in chemical process industries. In critical applications the regulating spindle is sometimes fitted with a diaphragm operator to achieve close control. Table 3 shows capacities on several typical applications for units of this type.

SAND AND MUD EDUCTORS Figure 12 illustrates a sand and mud eductor used for pumping out wells, pits, tanks, sumps, and similar containers where there is an accumulation of sand, mud, slime, or other material not easily handled by other eductors. With this type of eductor the bottom of the pressure chamber is fitted with a ring of agitating nozzles which stir the material in which the jet is submerged to allow maximum entrainment. Relative capacities for this type of eductor are shown in Table 4, which is used in the same manner as Table 1. The required suction flow is divided by the suction capacity selected from Table 4 under the appropriate motive pressure. This value is the capacity ratio. From the table select the eductor by choosing the next highest capacity ratio. Actual capacities are then determined by multiplying the values in the table by the capacity ratio of the eductor selected. Maximum discharge head is read from the table.

TABLE 2 Motive Capacities of Tank Mixing Eductors, gpm[a]

Size, in (mm)	Pressure difference, inlet to tank, lb/in^2 (bar) gage							
	10 (0.69)	20 (1.4)	30 (2.1)	40 (2.8)	50 (3.4)	60 (4.1)	80 (5.5)	100 (6.9)
½ (12.7)	3.5	5.0	6.0	7.0	8.0	8.5	10.0	11.0
¾ (19.1)	10.0	14.5	17.5	20.0	23.0	24.5	29.0	32.0
1 (25.4)	14.2	20.0	25.0	28.0	30.0	34.5	40.0	44.5
1¼ (31.8)	22.0	31.0	37.5	44.0	50.0	53.0	62.5	69.0
1½ (38.1)	31.5	45.0	54.0	63.0	72.0	76.5	90.0	99.0
2 (50.8)	56.0	80.0	96.0	112.0	128.0	136.0	160.0	176.0
3 (76.2)	126.0	180.0	216.0	252.0	288.0	306.0	360.0	396.0
4 (102)	224.0	320.0	384.0	448.0	512.0	544.0	640.0	704.0
5 (127)	350.0	500.0	600.0	700.0	800.0	850.0	1000.0	1100.0
6 (152)	494.0	720.0	864.0	1008.0	1152.0	1224.0	1440.0	1584.0

[a]gpm × 0.227 = m^3/h.

SOURCE: Schutte and Koerting Co.

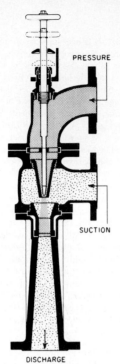

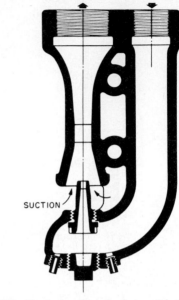

FIG. 11 Proportioning eductor. (Schutte and Koerting)

FIG. 12 Sand and mud eductor. (Schutte and Koerting)

TABLE 3 Capacities—Proportioning Eductor

Motive liquid/ suction fluid	Naphtha/ copper chloride slurry	Hydrocarbon/ hydrocarbon	Gasoline/ slurry	Gasoline/ water	Sour kerosene/ kerosene slurry
Pressure, lb/in^2 (bar) gage					
Motive	165 (11.4)	295 (20.3)	170 (117.1)	75 (5.2)	146 (31.7)
Suction	40 (2.8)	5 (0.3)	75 (5.2)	50 (3.4)	60 (41.3)
Discharge	75 (5.2)	10 (0.7)	100 (68.9)	50 (3.4)	70 (48.2)
Flow, gpm (m^3/h)					
Motive	30 (2.1)	10 (0.7)	90 (62.0)	170 (117.1)	482 (332.1)
Suction	20 (1.4)	58 (4.0)	74 (51.0)	42 (30.0)	700 (482.3)
Discharge	50 (3.4)	68 (4.7)	164 (113.0)	212 (146.1)	1182 (814.4)
Eductor size, in (mm)	1½ (38.1)	3 (76.2)	4 (102)	4 (102)	6 (152)

SOURCE: Schutte and Koerting.

TABLE 4 Relative Capacities of Sand and Mud Eductors

Capacity of standard 3-in (76.2-mm) eductor			
Operating water pressure, lb/in² (bar) gage	40.0 (2.8)	50.0 (3.4)	60.0 (41.3)
Total motive fluid, gpm (m³/h)	69.5 (15.8)	77.5 (17.6)	85.0 (19.3)
Net suction fluid, gpm (m³/h)	30.0 (6.8)	34.5 (7.8)	38.5 (8.7)
Maximum discharge head, ft (m)	22.0 (6.7)	26.0 (7.9)	32.0 (9.8)

Relative capacities of standard sizes					
Size eductor, in (mm)	1½ (38.1)	2½ (63.5)	3 (76.2)	4 (102)	5 (127)	6 (152)
Capacity ratio	0.29	0.62	1.00	1.85	2.80	3.80

SOURCE: Schutte and Koerting.

SOLIDS-HANDLING EDUCTORS   Figure 13 illustrates a specific type of eductor called a *hopper eductor*, made for handling slurries or dry solids in granular form and used for ejecting sludges from tank bottoms, pumping sand from filter beds, and washing or conveying granular materials. Typical construction is cast iron with hardened steel nozzle and throat bushings. In operation, the washdown nozzles are adjusted to provide smooth flow down the hopper sides, thus preventing bridging of the material being handled and also sealing the eductor suction against excess quantities of air. Without this seal, the capacities shown in Table 5 should be divided by approximately 3. Table 6 shows typical materials handled by this eductor and their bulk density. Use of the capacity table for hopper eductors is similar to use of Tables 1 and 4, except the suction quantities required are expressed in cubic feet (cubic meters). Capacity ratio is determined by dividing the value in the table into the required suction flow, and the next largest size eductor is selected.

Another type of solids-handling eductor is illustrated in Fig. 14. This *annular-orifice eductor* is used where the material being handled tends to agglomerate and gum up when wetted and has been used successfully for handling and mixing hard-to-wet solids. In this unit intimate mixing occurs in the throat, and the device is virtually clogproof. Normally this unit is installed directly over the tank into which the mixture is discharged. Table 7 shows capacities for this type of unit.

Capacity Table 7 is similar to Table 5, and the selection method is the same as discussed previously.

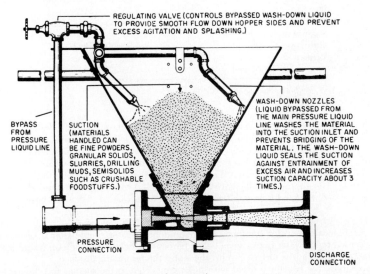

FIG. 13   Hopper eductor. (Schutte and Koerting)

TABLE 5 Relative Capacities of Hopper Eductors

Capacity of standard 1½-in (38-mm) eductor				
Operating water pressure, lb/in² (bar) gage	30 (2.1)	40 (2.8)	50 (3.4)	60 (4.1)
Suction capacity,, ft³/h (m³/h)	13 (0.34)	36 (1.0)	72 (2.0)	90 (2.5)
Maximum discharge pressure, lb/in² (bar) gage	14 (1.0)	17 (1.1)	18 (1.2)	20 (1.4)
Motive water consumption, gpm (m³/h)[a]	35 (7.9)	40 (9.1)	45 (10.2)	50 (11.4)

Relative capacities of standard sizes				
Size, in (mm)	1½ (38.1)	2 (50.8)	3 (76.2)	4 (102)	6 (152)
Capacity ratio	1.00	1.60	3.50	6.00	18.00

[a]Based on using approximately 10% motive water through washdown nozzles.

Source: Schutte and Koerting

TABLE 6 Typical Materials Handled by Hopper Eductors

Material	Approx. bulk density, lb/ft³ (kg/m³)
Borax	50–55 (800–880)
Charcoal	18–28 (290–450)
Diatomaceous earth	10–20 (160–320)
Lime, pebble	56 (900)
Lime, powdered	32–40 (510–640)
Fly ash	35–40 (560–640)
Mash	60–65 (960–1040)
Rosin	67 (1070)
Salt, granulated	45–51 (720–820)
Salt, rock	70–80 (1120–1280)
Sand, damp	75–85 (1200–1360)
Sand, dry	90–100 (1440–1600)
Sawdust, dry	13 (210)
Soda ash, light	20–35 (320–560)
Sodium nitrate, dry	80 (1280)
Sulfur, powdered	50–60 (800–960)
Wheat	48 (770)
Zinc oxide, powdered, dry	10–35 (160–560)

SOURCE: Schutte and Koerting.

MULTINOZZLE EDUCTORS   Figure 15 illustrates an annular multinozzle eductor designed for special applications where the suction fluid contains solids or semisolids. It is used primarily for large flows at low discharge heads. Because these units have relatively large air-handling capacities, they are well suited for priming large pumps, such as dredging pumps, where air pockets can cause these pumps to lose their prime. These eductors are designed by using the basic equations for head ratio. The appropriate efficiency factor is selected from Fig. 2, and the volumetric flow ratio is calculated. Figure 4 is used to size the eductor after discharge flow has been determined.

DEEP-WELL EDUCTORS   The eductor illustrated in Fig. 16 is typical of those used in conjunction with a mechanical pump for commercial and residential water supply from a deep well. The eductor is used to lift water from a level below barometric height up to a level where the suction of the motive pump at the surface can lift the water the remaining distance.

In operation, the eductor is fitted with hoses connected to the suction and discharge of the motive pump and dropped into the well casing. An initial prime is required, which is maintained by the foot valve at the suction of the eductor. When the surface pump is activated, pressure water through the eductor entrains water from the well, lifting it high enough to enable the mechanical pump to carry it to the surface. A bypass valve at the surface diverts the suction quantity to a receiving tank.

Capacities of these units are dependent upon the depth of the well and the centrifugal pump. The standard commercial unit has 1-in (25-mm) pressure and 1¼-in (32-mm) discharge connections and is available with a variety of nozzle and diffuser combinations for use with standard centrifugal pumps at varying depths. The following example illustrates how to calculate this type unit.

EXAMPLE 4   A centrifugal pump with a capacity of 100 gpm (22.7 m³/h) at a total discharge head of 150 ft (45.7 m) and requiring 10 ft (3.05 m) NPSH is available to operate an eductor to pump water at 50°F (10°C) from a level 50 ft (15.2 m) below grade. Find the quantity of water that can be delivered at 60 lb/in² (4.1 bar) gage (see Fig. 17).

Solution   The available operating head is 138.6 ft + 50 ft − frictional loss (42.2 + 15.2 − frictional loss). As a first assumption, the frictional loss is ignored and the head ratio is 188.6/40.83 = 4.62 (57.4/12.38 = 4.62). From Fig. 2 at NPSH 33.6 ft (10.24 m), ϵ = 0.9 and

$$R_w = R_q = 0.9\sqrt{4.62} - 1 = 0.934$$

(from Eqs. 8 and 9)

With Q_R fixed at 100 gpm (22.7 m³/h),

in USCS units
$$Q_1 = \frac{100}{1 + R_q} = \frac{100}{1.934} = 51.7 \text{ gpm}$$

in SI units
$$Q_1 = \frac{22.7}{1.934} = 11.7 \text{ m}^3/\text{h}$$

PRESSURE

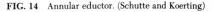

FIG. 14   Annular eductor. (Schutte and Koerting)

The motive line size is now chosen by selecting a reasonable velocity and frictional loss. Choosing a 2-in (51-mm) pipe size, velocity is 5.57 ft/s (1.70 m/s) and frictional loss is 6.1 ftH₂O (1.86 mH₂O). The revised operating head becomes 188.6 − 6.1 = 182.5 ft (57.4 − 1.86 = 55.6 m) and

in USCS units $R_q = 0.9\sqrt{182.5/40.83} - 1 = 0.9$
in SI units $R_q = 0.9\sqrt{55.6/12.38} - 1 = 0.9$

then Q_1 = 100/1.9 = 52.6 gpm (22.7/1.9 = 11.9 m³/h) (This value is close enough so that a third trial is not necessary.) The suction flow that can be delivered is then

in USCS units   100 − 52.6 = 47.4 gpm
in SI units   22.7 − 11.9 = 10.8 m³/h

TABLE 7 Relative Capacities of Annular Eductors

	Capacity of standard 1½-in (38.1-mm) mixing eductor, 5 lb/in² (0.34 bar) gage discharge pressure				
Motive pressure, lb/in² (bar) gage	30 (2.1)	40 (2.8)	60 (4.1)	80 (5.5)	100 (6.9)
Entrainment, ft³/h (m³/h)	2.6 (0.07)	7.1 (0.20)	17.9 (0.51)	22.0 (0.62)	23.8 (0.67)
Motive flow, gpm (m³/h)	12.7 (2.88)	14.6 (3.31)	17.9 (4.06)	20.7 (4.70)	23.1 (5.24)

	Relative capacities of standard sizes					
Size, in (mm)	1¼ (31.8)	1½ (38.1)	2 (50.8)	2½ (63.5)	3 (76.2)	4 (102)
Capacity ratio	0.62	1.00	1.43	2.86	4.76	8.80

SOURCE: Schutte and Koerting.

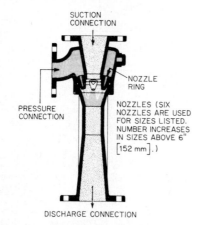

FIG. 15 Multinozzle eductor. (Schutte and Koerting)

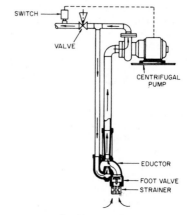

FIG. 16 Centrifugal-jet pump combination.

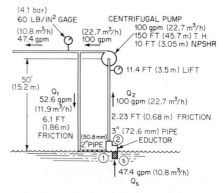

FIG. 17 Centrifugal-jet pump for Example 4.

PRIMING EDUCTORS—WATER-JET EXHAUSTERS   Educators are often used as priming devices for mechanical pumps. In this application the eductor is used to remove air rather than water. Liquid jets are not well suited for pumping noncondensables; therefore the capacities are low. However, the volume being primed is usually small, and so the low capacity is not a factor. When larger volumes are involved, such as condenser water boxes, it is more feasible to use an exhauster. The water-jet eductor of Fig. 6 is converted to a water-jet exhauster by replacing the jet nozzle with a solid-cone spray nozzle. Evacuating rates and capacity tables for such a unit are shown in Fig. 18 and Table 8. Eductors have approximately one-fifth the air-handling capacities of water-jet exhausters when supplied with similar motive quantities and pressures.

TABLE 8   Approximate Water Consumption of Water-Jet Exhausters, gpm[a,b]

Size, in (mm)	Water pressure, lb/in² (bar) gage				
	30 (2.1)	40 (2.8)	60 (4.1)	80 (5.5)	100 (6.9)
½ (12.7)	2.6	2.9	3.4	3.8	4.2
¾ (19.1)	4.6	5.3	6.4	7.4	8.3
1 (25.4)	6.2	6.8	8.1	9	10
1½ (38.1)	20	23	27	30	32
2 (50.8)	28	31	36	41	45
2½ (63.5)	46	51	60	67	73
3 (76.2)	66	73	86	96	106

[a]gpm $\times$ 0.227 = m³/h

[b]All flows at 15 in (381 mm) Hg abs.

SOURCE: Schutte and Koerting.

EXAMPLE 5   From Fig. 18 and Table 8, determine size and water consumption to exhaust 15 standard ft³/min (0.42 m³/min) of air at 20 inHg (508 mmHg) abs discharging to atmosphere using 60 lb/in² (4.14 bar) gage motive water at 80°F (27°C).

Solution   Enter Fig. 18 at 80°F (27°C) (1); read horizontally to the suction pressure 20 inHg (508 mmHg) abs (2); project vertical line to 60 lb/in² (4.14 bar) gage motive pressure (3); project a horizontal line for the capacity of a 1-in (25-mm) exhauster (4); divide desired flow by the capacity of a 1-in (25-mm) unit, which is 1.9 standard ft³/min, (0.054 m³/min), to find capacity ratio: 15/1.9 = 7.9 (0.42/0.054 = 7.9).

The capacity ratio table shows that a 3-in (76-mm) exhauster with a capacity ratio of 9.0 is required. The motive water quantity from Table 8 is 86 gpm (19.5 m³/h). *Note:* Table 8 gives water consumption at 15 inHg (381 mmHg) abs; since flow varies as the square of pressure differential across the nozzle, the exact flow is obtained as follows:

Nozzle upstream pressure:

in USCS units $60 + 14.7 = 74.7$ lb/in² abs

in SI units $4.14 + 1.01 = 5.15$ bar abs

Nozzle downstream pressure:

in USCS units $20 \text{ inHg abs} \left(\dfrac{14.7 \text{ lb/in}^2}{30 \text{ inHg}} \right) = 9.8$ lb/in² abs

in SI units $508 \text{ mmHg abs} \left(\dfrac{1.01 \text{ bar}}{762 \text{ mmHg}} \right) = 0.67$ bar abs

Operating differential:

in USCS units $\qquad\qquad\qquad\qquad$ $74.7 - 9.8 = 64.9 \text{ lb/in}^2$

in SI units $\qquad\qquad\qquad\qquad\;\;$ $5.15 - 0.67 = 4.48 \text{ bar}$

Table differential:

in USCS units $\qquad\qquad\qquad$ $74.7 - 15\left(\dfrac{14.7}{30}\right) = 67.35 \text{ lb/in}^2$

in SI units $\qquad\qquad\qquad\;\;$ $5.15 - 381\left(\dfrac{1.01}{762}\right) = 4.64 \text{ bar}$

Actual flow:

in USCS units $\qquad\qquad\qquad$ $86\left(\dfrac{64.9}{67.35}\right)^{1/2} = 84.4 \text{ gpm}$

in SI units $\qquad\qquad\qquad\;\;$ $19.5\left(\dfrac{4.48}{4.64}\right)^{1/2} = 19.2 \text{ m}^3/\text{h}$

SIPHONS

Operation   As previously defined, the term *siphon* refers to a jet pump utilizing a condensable vapor to entrain a liquid and discharge to a pressure intermediate between motive and suction pressure. The principal motive fluid is steam.

In an eductor the high-pressure motive fluid enters through a nozzle and creates a vacuum by jet action, which causes suction fluid to enter the mixing chamber. The siphon of Fig. 19 is identical to the eductor of Fig. 6 except that, unlike the eductor, the siphon motive nozzle is a converging-diverging nozzle to achieve maximum velocity at the nozzle tip. The velocity is supersonic at this point. The motive fluid is condensed into the suction fluid on contact and imparts its energy to the liquid, thus impelling it through the diffuser. The diffuser section is the same as an eductor diffuser, and it converts the velocity energy to pressure at the discharge. To achieve maximum performance, the siphon nozzle must be expanded to the desired suction pressure in order to achieve the highest possible velocity. Since negligible radiation losses are encountered, the siphon is 100% thermally efficient in that the heat in the incoming water plus the heat in the operating steam must equal the heat of the mixture plus its mechanical energy. Furthermore, the momentum of the incoming water plus the momentum of the expanded steam is equal to the momentum of the discharge mixture less impact and frictional losses.

It is important that the motive steam be condensed in the suction liquid prior to the throat for proper operation. If condensation does not occur, full available energy is not transferred. Furthermore, energy must he expended to recompress the uncondensed steam. For this reason, discharge temperature cannot exceed the boiling point at the discharge pressure. The fact that energy is required to recompress any uncondensed steam explains why air is a very poor motive fluid for liquid pumping.

STANDARD SIPHONS   Tables 9 and 10 illustrate the capacity and operating characteristics of standard siphons. These tables are similar to the eductor capacity tables. To size a unit, read the suction capacity of a 1½-in (38-mm) unit from Table 9 at the appropriate motive steam pressure, suction lift, and discharge head. Divide the desired suction flow by this capacity to find the capacity ratio. From Table 10 find the unit with the next largest capacity ratio. Read the motive steam required under the proper motive pressure.

Standard materials of construction are cast iron, bronze, stainless steel, and Pyrex®. If desired, special capacity ratios can be achieved by using a custom-designed unit. Sizes over 6 in (152 mm) can be fabricated of any suitable material.

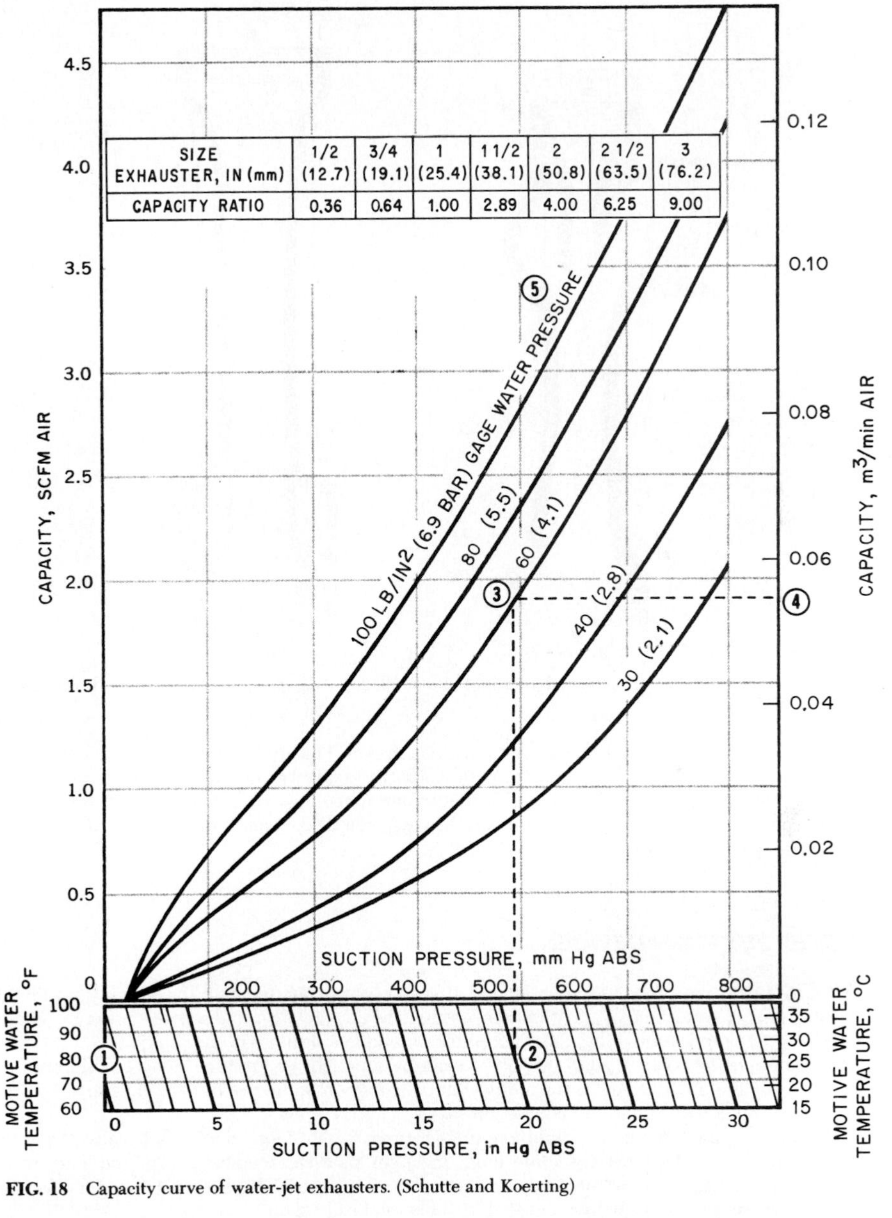

SIZE EXHAUSTER, IN (mm)	1/2 (12.7)	3/4 (19.1)	1 (25.4)	1 1/2 (38.1)	2 (50.8)	2 1/2 (63.5)	3 (76.2)
CAPACITY RATIO	0.36	0.64	1.00	2.89	4.00	6.25	9.00

FIG. 18 Capacity curve of water-jet exhausters. (Schutte and Koerting)

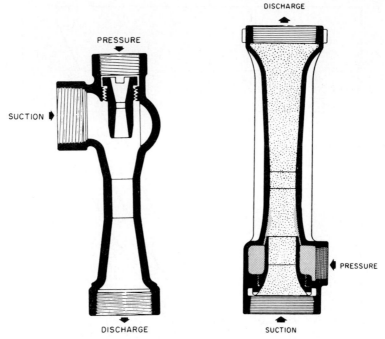

FIG. 19   Standard siphon. (Schutte and Koerting)   **FIG. 20**   Annular siphon. (Schutte and Koerting)

ANNULAR SIPHONS   Figure 20 illustrates an annular siphon. This unit is identical to the eductor of Fig. 14 except that steam is the motive fluid. Capacity Table 11 is used in the same manner as previous examples. This type of siphon is used when inline flow is desired or when the suction liquid contains some solids. Units are available in cast iron through an 8-in (203-mm) size. Special materials or fabricated designs are also available.

OTHER JET PUMP DEVICES

Air Siphons   As previously mentioned, air is a very poor motive fluid for entraining a liquid because energy must be expended in compressing the air back to the discharge pressure. There are, however, applications where it is necessary to sample a liquid with no dilution.

   Small units (less than 1 gpm, 0.23 m³/h) can be supplied for limited discharge pressures, as indicated by Fig. 21. With air as the motive fluid, the suction liquid can be very close to its boiling point and only a very slight NPSH is required.

   When air is used as a motive fluid, the smaller sizes operate more efficiently because the air is more intimately mixed with a suction fluid. In larger sizes the tendency is for the fluid to be discharged in slugs, since intimate mixing does not readily occur. This has a detrimental effect on the performance and especially on available discharge head.

Air-Lift Eductors   Air-lift pumps are frequently used for difficult pumping operations. Compressed air is forced into the bottom of a pipe submerged in the liquid to be pumped. The expanding air, as it rises up the pipe, entrains the suction fluid.

   If compressed air is not available, it is possible to lift water higher than 34 ft (10 m) with the use of an eductor-air-lift combination. Figure 22 illustrates the suction capacity of a 2-in (51-mm) eductor drawing water form a 50-ft (15-m) depth and discharging it to the atmosphere. In oper-

TABLE 9 Relative Capacities of Steam-Jet Siphons, gpm[a]

Suction lift, ft (m)	Suction temp., °F[b]	Operating steam pressure, lb/in² (bar) gage								Operating steam pressure, lb/in² (bar) gage							
		0-ft (m) discharge head								20-ft (6.1-m) discharge head							
		40 (2.8)	50 (3.4)	60 (4.1)	80 (5.5)	100 (6.9)	120 (8.3)	160 (11)	240 (16)	40 (2.8)	50 (3.4)	60 (4.1)	80 (5.5)	100 (6.9)	120 (8.3)	160 (11)	240 (16)
1 (0.3)	70	52	51	51	49	46	43	37	30	36	47	48	48	45	41	35	29
	90	45	44	43	42	40	37	33	26	37	44	43	43	41	38	33	26
	110	40	38	36	36	35	33	28	22	38	39	37	37	36	34	30	23
	130	35	32	30	30	29	29	25	…	35	33	31	31	30	29	25	…
	150	26	25	24	24	24	24	21	…	26	25	24	24	23	22	20	…
	165	17	17	17	18	18	17	17	…	17	17	17	17	17	17	16	…
10 (3.0)	70	38	38	37	35	30	28	25	20	27	36	37	35	31	29	25	19
	90	34	34	33	30	27	25	21	17	26	32	32	29	26	23	20	17
	110	28	27	26	25	23	21	18	…	26	27	27	25	23	21	18	…
	130	21	21	21	20	18	16	14	…	21	22	22	21	18	16	14	…
	145	16	16	16	16	14	12	…	…	16	16	16	16	14	12	…	…
15 (4.6)	70	34	32	30	26	23	21	18	14	24	33	32	27	24	23	19	15
	90	29	28	26	23	20	18	16	12	23	28	27	23	20	19	16	12
	110	24	23	22	19	17	15	13	…	23	23	22	19	17	15	13	…
	130	17	17	17	15	13	11	…	…	17	17	17	14	13	…	…	…
	145	10	12	11	9	…	…	…	…	11	11	10	10	…	…	…	…
20 (6.1)	70	26	23	21	18	16	15	13	…	24	24	22	19	17	15	12	…
	90	22	19	17	15	14	12	11	…	19	20	18	15	14	12	11	…
	110	18	16	14	12	11	10	…	…	17	16	14	12	11	…	…	…
	125	13	12	11	…	…	…	…	…	12	11	10	…	…	…	…	…

4.23

TABLE 9 Relative Capacities of Steam-Jet Siphons, gpm[a] (continued)

Suction lift, ft (m)	Suction temp, °F[b]	Operating steam pressure, lb/in² (bar) gage 40-ft (12.2-m) discharge head								Operating steam pressure, lb/in² (bar) gage 50-ft (15.2-m) discharge head							
		40 (2.8)	50 (3.4)	60 (4.1)	80 (5.5)	100 (6.9)	120 (8.3)	160 (11)	240 (16)	40 (2.8)	50 (3.4)	60 (4.1)	80 (5.5)	100 (6.9)	120 (8.3)	160 (11)	240 (16)
1 (0.3)	70	...	20	33	47	44	41	36	29	...	...	18	44	44	41	36	28
	90	...	18	36	43	42	39	34	27	...	...	21	42	42	39	34	27
	110	...	20	36	37	37	35	30	24	...	...	24	37	36	34	30	24
	130	...	23	31	31	30	29	25	...	...	...	26	30	30	29	25	...
	150	...	24	24	24	24	24	20	...	...	...	24	24	24	24	20	...
	165	...	17	18	18	18	18	16	...	...	...	18	18	18	18	16	...
10 (3.0)	70	...	...	24	34	30	27	24	18	...	...	...	35	30	27	23	18
	90	...	...	26	29	26	24	20	17	...	...	...	29	27	24	21	17
	110	...	...	27	26	23	20	18	...	...	...	...	25	23	20	18	...
	130	...	...	22	21	19	16	14	...	...	...	...	21	19	17	14	...
	145	...	...	16	16	14	12	...	...	...	...	...	16	14	13	...	...
15 (4.6)	70	...	...	23	28	24	22	19	15	...	...	...	27	24	21	18	14
	90	...	...	20	24	20	18	16	...	...	...	...	24	21	18	16	...
	110	...	...	21	19	17	15	...	...	...	...	...	19	17	15	...	...
	130	...	...	17	14	12	...	...	...	...	...	...	14	...	...	...	...
	145	...	...	11	...	...	...	...	...	...	...	...	...	...	...	...	...
20 (6.1)	70	...	...	21	19	16	15	12	...	...	...	...	...	16	15	...	...
	90	...	...	18	16	14	13	11	...	...	...	...	...	15[c]	13[c]	...	...
	110	...	...	...	11	10	...	...	...	...	...	...	...	...	...	...	...

[a] gpm = 0.227 m³/h.   [b] °C = (°F − 32)/1.8.   [c] Suction temperature 85°F (29.4°C).

SOURCE: Schutte and Koerting.

TABLE 10 Steam Consumption of Steam-Jet Siphons, lb/h[a]

Siphon size, in (mm)	Capacity ratio	Operating steam pressure, lb/in² (bar) gage							
		40 (2.8)	50 (3.4)	60 (4.1)	80 (5.5)	100 (6.9)	120 (8.3)	160 (11.0)	240 (16.5)
½ (12.7)	0.125	40	47	54	69	83	97	126	184
¾ (19.1)	0.222	70	83	96	122	147	173	222	322
1 (25.4)	0.346	110	130	150	190	230	270	350	510
1½ (38.1)	1.000	318	376	434	550	665	780	1,012	1,475
2 (50.8)	1.38	440	520	600	761	920	1,080	1,400	2,040
2½ (63.5)	2.0	635	750	865	1,100	1,329	1,558	2,020	2,940
3 (76.2)	3.11	990	1,170	1,350	1,710	2,065	2,425	3,145	4,590
4 (102)	5.54	1,760	2,085	2,400	3,045	3,685	4,320	5,500	8,170
6 (152)	12.45	3,960	4,680	5,400	6,850	8,280	9,710	12,600	18,360

[a]Lb/h × 0.454 = kg/h.

SOURCE: Schutte and Koerting.

TABLE 11 Relative Capacities of Annular Siphons

Capacity of standard 3-in (76.2-mm) siphon, water temperature 100°F (37.8°C), 0 suction lift				
Steam pressure, lb/in² (bar) gage	50 (3.45)	75 (5.17)	100 (6.90)	125 (8.62)
Steam consumption, lb/h (kg/h)	1180 (535)	1620 (735)	2060 (934)	2490 (1130)
Max back pressure, lb/in² (bar) gage at zero flow	12 (0.83)	18 (1.24)	22 (1.52)	35 (2.41)
Suction capacity, gpm (m³/h)	140 (31.8)	130 (29.5)	120 (27.2)	110 (25)
Discharge pressure, lb/in² (bar) gage	5 (0.34)	8 (0.55)	12 (0.83)	30 (2.07)

Relative capacities of standard sizes							
Size, in (mm)	1¼ (31.8)	1½ (38.1)	2 (50.8)	2½ (63.5)	3 (76.2)	4 (102)	6 (152)	8 (203)
Capacity ratio	0.13	0.21	0.30	0.60	1.00	1.85	4.0	7.1

SOURCE: Schutte and Koerting.

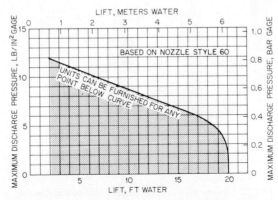

FIG. 21 Air pumping liquid. (Schutte and Koerting)

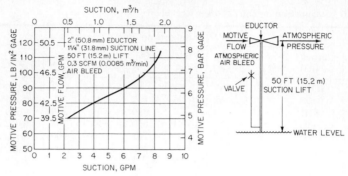

FIG. 22 Suction capactiy of eductor–air-lift combination (gpm $\times$ 0.227 = m^3/h).

ation, an air line from the atmosphere enters the suction pipe near the water level. As the eductor creates a vacuum in this line, atmospheric pressure forces air into the suction pipe. Once in the line, the rising air carries the suction fluid to the surface and both fluids are discharged to the atmosphere through the eductor.

No sizing data are presented since this type of pump is best specified according to specific conditions.

Boiler Injectors   The boiler injector is a jet pump utilizing steam as a motive fluid to entrain water, and it is used as a boiler feedwater heater and pump. It differs from a siphon in that the discharge pressure is higher than either motive or suction pressure. This is achieved by the double-tube design shown in Fig. 23. In operation, the lower nozzle is activated by pulling the handle partway back. The lower jet creates a vacuum in the chamber, causing water to be induced into the unit. When water is spilling out the overflow, the handle is drawn back all the way. This closes the overflow and simultaneously admits motive steam to the upper jet. This second jet, which is of the straight or forcing type, then picks up the discharge from the first jet and imparts a velocity to the water through the discharge tube. The energy contained is sufficient to open the check valve and discharge against the boiler pressure.

The now-obsolete steam locomotives were the largest users of this type of injector. Principal use at present is as a backup to a regular boiler-feed pump. Capacity Table 12 illustrates the range of capacities available for the double-tube injector.

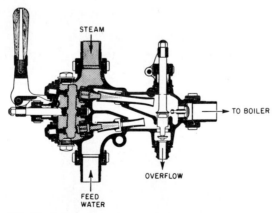

FIG. 23   Boiler injector, starting position. (Schutte and Koerting)

TABLE 12   Capacities of Boiler Injectors, gph[a]

Size no.	Size iron pipe conn., in (mm)	Size copper pipe OD, in (mm)	Size overflow (drip funnel) pipe, in (mm)	Steam pressure, lb/in² (bar)				
				50 (3.45)	100 (6.90)	150 (10.34)	200 (13.79)	250 (17.24)
0	¼ (12.7)	⅜ (9.53)	¼ (12.7)	80	100	110	100	90
1	⅜ (9.53)	½ (12.7)	¼ (12.7)	110	140	180	140	130
2	½ (12.7)	⅝ (15.9)	⅜ (9.33)	170	210	230	200	190
3	¾ (19.1)	⅞ (22.2)	½ (12.7)	280	340	400	340	320
3½	¾ (19.1)	⅞ (22.2)	½ (12.7)	400	470	550	470	440
4	1 (25.4)	1⅛ (28.6)	¾ (19.1)	530	620	720	620	590
5	1¼ (31.8)	1½ (38.1)	1 (25.4)	680	800	920	800	750
6	1¼ (31.8)	1½ (38.1)	1 (25.4)	820	990	1130	990	930
7	1½ (38.1)	1¾ (44.4)	1¼ (31.8)	1070	1370	1610	1370	1290
8	1½ (38.1)	1¾ (44.4)	1¼ (31.8)	1400	1800	2100	1800	1700
9	2 (50.8)	2¼ (57.2)	1½ (38.1)	1700	2100	2500	2100	2000
10	2 (50.8)	2¼ (57.2)	1½ (38.1)	2000	2500	2900	2500	2300
11	2½ (63.5)	2¾ (69.8)	2 (50.8)	2500	3000	3500	3000	2800
12	2½ (63.5)	2¾ (69.8)	2 (30.8)	3000	3600	4300	3600	3400
14	3 (76.2)	3¼ (82.6)	2½ (63.5)	3900	4600	5500	4600	4400
16	3 (76.2)	3¼ (82.6)	2½ (63.5)	5000	6000	7000	6000	5700

[a]gph × 0.00379 = m³/h

SOURCE: Schutte and Koerting.

C · H · A · P · T · E · R · 5

MATERIALS
OF
CONSTRUCTION

SECTION 5.1
MATERIALS OF CONSTRUCTION OF METALLIC PUMPS

WARREN H. FRASER

The essential requirements for a successful pump installation are performance and life. *Performance* is the rating of the pump—head, capacity, and efficiency. *Life* is the total number of hours of operation before one or more pump components must be replaced to maintain an acceptable performance. Initial performance is the responsibility of the pump manufacturer and is inherent in the hydraulic design. Life is primarily a measure of the resistance of the materials of construction to corrosion, erosion, or a combination of both under operating conditions.

The selection of the most economical material for any particular service, however, requires a knowledge not only of the pump design and manufacture but also of the erosion-corrosion properties of the materials under consideration when subjected to the velocities encountered in the pump. Very little corrosion information exists on the effects of the velocities encountered in centrifugal pumps for liquids other than seawater.

Despite this limitation, experience has provided the designer of pumping systems with a specialized branch of metallurgical and corrosion engineering adequate for most pumping problems. Factors that lead to a long pump life are

1. Neutral liquids at low temperatures
2. Absence of abrasive particles
3. Continuous operation at or near maximum efficiency capacity of pump
4. Adequate margin of available NPSH over NPSH required as stated on manufacturer's rating curve

Any pumping installation that satisfies all these criteria will have a long life. A typical example would be a waterworks pump. Some waterworks pumps with bronze impellers and cast iron casings have a life of 50 years or more. At the other extreme would be a chemical pump handling a hot, corrosive liquid with abrasive particles carried in suspension. Here the life might be measured in months rather than in years despite the fact that the construction was based upon the most resistant materials available.

Most pumping applications lie somewhere between these two extremes. Aside from straight corrosion or erosion from abrasive particles in the fluid, the greatest single factor that reduces pump life is operation at flows other than the maximum efficiency or rated capacity of the pump. The vane angles of the impeller are designed to match the fluid angles at the maximum efficiency capacity. At flows other than the rated capacity, the fluid angles no longer match the vane angles and separation occurs with increasing intensity as the operating point moves away from the maximum efficiency capacity. The destruction of the impeller vanes is particularly severe at the inlet to the impeller, as this is the point of lowest pressure in the pump. In addition to surface damage to the inlet vanes from separation, localized cavitation damage may occur during sustained operation at capacities below 50 or above 125% of the maximum efficiency capacity. This is not to say that many pumps do not operate under these adverse operating conditions, but their life is considerably less than that of the same pump operating at or near its maximum efficiency capacity.

TYPES OF CORROSION

Erosion Corrosion   The rate of corrosion of most metals is increased when there is a flow of liquid relative to the component. The rate of corrosion is also dependent upon the angle of attack of the piece to the direction of flow. As a general rule, the rate of corrosion increases as the angle of the attack increases, that is, as the separation of the fluid from the metal surface becomes larger. The greater the separation, the more intense the turbulence and therefore the greater the rate of metal removal.

The corrosion rate of most metals and alloys in any liquid environment under static conditions depends upon the resistance of the film that forms on the surface and protects the base metal from further attack. Damage to or removal of this film by erosion exposes the unprotected base metal to the corrosive environment, and metal removal continues unabated.

In centrifugal pumps the impeller is particularly susceptible to erosion corrosion. Although the casing can be damaged by erosion corrosion, the problem is usually secondary to that of the impeller. The diffuser-type casing, with its many vanes, is more susceptible to erosion corrosion than is the volute-type casing, with only one vane—the casing tongue—as an obstruction to the line of flow.

Wearing rings are also susceptible to erosion corrosion and should receive special consideration in material selection. The high fluid velocities through the small clearance annulus can result in high rate of wear unless the proper material is selected.

When considering any material for a pump service, the erosion-corrosion properties of the material must be evaluated. This can be a difficult task, as many of the properties and characteristics of both the liquid to be pumped and the materials under consideration are not known in sufficient detail to allow an optimum solution. Most materials used in pump construction have been corrosion-tested under static conditions in a wide variety of liquids, but few have been tested in an environment of high velocity. For example, the corrosion rate of the 300 series of austenitic stainless steels remains virtually unchanged over a wide range of seawater velocities, whereas the corrosion rate for the bronzes increases rapidly over the same range. The rate of corrosion, however, is only part of the problem in evaluating erosion corrosion. The erosion resistance of a metal or alloy is dependent primarily on the hardness of the material. The hardness of a material, however, is not necessarily compatible with its corrosion resistance. The 300 series of austenitic stainless steels are not hardenable but exhibit low rates of corrosion in most liquids. The 400 series of stainless steels, on the other hand, are hardenable to 500 Brinell but exhibit higher corrosion rates in many more liquids than do the austenitic stainless steels.

Most pump components of standard designs are limited to a hardness of approximately 350 Brinell. Above 350 Brinell the standard machining operations of turning, boring, drilling, and tapping become uneconomical. Stuffing box sleeves and wearing rings of higher hardness can be provided, and the finishing operation of cylindrical components can be performed by grinding.

As a general rule, the material should be selected primarily on the basis of its velocity-corrosion resistance to the liquid being pumped, provided the liquid is free of abrasive solids. If the liquid contains abrasive solids, then the material of construction should be selected primarily for abrasive-wear resistance, provided the velocity-corrosion-resistance characteristics are acceptable.

Corrosion Fatigue When evaluating material for a pump component that is subjected to a cyclic stress, the endurance limit of the material must be considered. The *endurance limit* is the maximum cyclic stress that the material can be subjected to and still not fail after an infinite number of cyclic stress reversals. The endurance limit of steel, for example, is approximately 50% of its tensile strength. A 100,000-lb/in^2 (690 mPa) tensile steel could be subjected to a static load in tension to produce 100,000 lb/in^2 (690 mPa) stress, but the same steel subjected to a cyclic stress of 100,000 lb/in^2 (690 mPa) would fail in a short period of time. If the stress were reduced to 50,000 lb/in^2 (345 mPa), however, the endurance limit would not be exceeded and the metal could be subjected to 50,000 lb/in^2 (345 mPa) stress reversals and not fail. If the same steel were subjected to a cyclic stress of 50,000 lb/in^2 (345 mPa) in a corrosive environment, however, failure would occur quite rapidly. The failure is caused by corrosion fatigue. As a result of the cyclic stressing of the piece, minute cracks develop at the surface. In a corrosive environment the surface of the metal exposed by the cracks corrodes rapidly. The crack then penetrates deeper, corrosion develops further, and the piece will ultimately fail.

Experience has shown that the maximum combined stress in a pump shaft should not exceed 7500 lb/in^2 (52 mPa) for those portions exposed to the liquid pumped. Above this value the premature failure of shafts increases sharply. Since the corrosion-fatigue strength of any metal is dependent more on the corrosion resistance of the metal than on its tensile strength, the life of any pump component subjected to cyclic loading can only be estimated. A pump shaft, for example, is subjected to a complete stress reversal for each revolution and will have some definite life based upon the corrosion-fatigue strength of the material in the particular application and the rotational speed of the pump. The best way to guard against short shaft life is to protect the shaft against exposure to the liquid by means of sleeves.

Intergranular Corrosion Intergranular corrosion is the corrosion of the grain boundaries in the body of the material. Unlike direct or galvanic corrosion of the metal surface, intergranular corrosion is insidious in that the mechanical properties of the material can be drastically reduced with little apparent surface damage. As the grain boundaries alone are affected, the material appears sound from a surface inspection. Progressive intergranular corrosion can proceed, however, to the point where the material literally disintegrates.

Intergranular corrosion of the austenitic stainless steels occurs as a result of carbides precipitating out at the grain boundaries during the slow cooling of the casting. When exposed to a corrosive environment, the carbides are preferentially attacked, and the strongly bonded matrix of the metal grains is destroyed. The precipitation of carbides can be controlled by heating the casting to 2000°F (1100°C) and then quenching. At 2000°F (1100°C) the carbides are held in solution, and the rapid quench prevents their precipitation.

Susceptibility to intergranular corrosion in the austenitic stainless steels can also be reduced by controlling the carbon content of the alloy. The standard austenitic stainless steels of the 300 series have a carbon content in excess of 0.08%. Without proper heat treatment, these steels are susceptible to intergranular corrosion. Extra-low-carbon steels are available in the 300 series and are identified by the suffix letter *L*. These steels have a carbon content of less than 0.03% and are much less susceptible to intergranular corrosion.

The possibility of intergranular corrosion must be considered when castings of austenitic stainless steels are indicated for impellers and casings of centrifugal pumps or for the liquid ends of reciprocating pumps. For moderate-size castings, the low-carbon grades are adequate when properly heat treated, provided post-heat treatment is not necessary. The use of the more costly 0.03% carbon steels is not justified in this case. In the case of larger, more intricate castings, however, the 0.03% carbon steels should be considered. This is particularly true for large open or semiopen impellers used in mixed-flow and propeller pumps. Here the large, unshrouded vanes can be severely distorted during quenching. In this case a 0.03% carbon steel with forced air cooling is the preferred selection.

If impeller vane tips damaged as a result of erosion or other corrosion are to be restored during the life of the pump, the impeller casting must be specified with a maximum carbon content of 0.03%. If not, subsequent welding will precipitate carbides at the grain boundaries, and rapid failure may occur from intergranular corrosion at the welded areas.

Cavitation Erosion   Cavitation erosion is the removal of metal as a result of high localized stresses produced in the metal surface from the collapse of cavitation vapor bubbles. In a corrosive environment the rate of damage is accelerated as the corrosion products are continuously removed, and the corrosion proceeds unabated.

While every effort should be made in the design and application of centrifugal pumps to prevent cavitation, it is not always possible to do so at capacities less than the rated maximum efficiency capacity of the pump. It must be recognized that at low-flow operation the stated NPSH required curve is not usually sufficient to suppress all cavitation damage. The stated NPSH required is that required to produce the head, capacity, and efficiency shown on the rating curve. At low flows it must be expected that some cavitation damage will occur. It would be impractical to supply an NPSH that would suppress all cavitation at these low flows, as it would be many times that required at the best efficiency point. Therefore, it should be anticipated that some cavitation will occur at low-flow operation and should be considered in any evaluation of the material for impellers.

Open-type mixed-flow impellers that produce heads in excess of 35 ft (10.7 m) are particularly susceptible to cavitation erosion in the clearance space between the rotating vanes and the stationary housing. This is usually referred to as *vane-tip erosion* and is caused by a cavitating vortex in the clearance space between the vane and the housing. Here again it would be impractical to provide sufficient NPSH to eliminate the cavitation. Any evaluation of the impeller and housing material for a pump of this type should include the possibility of vane-tip cavitation.

Extensive laboratory tests of the resistance of a wide range of materials to cavitation erosion have produced data for all the materials used in centrifugal pump construction. Despite the complexity of the cavitation process and the mass of laboratory data available, it is possible to correlate the laboratory data and field experience to develop the following tabulation of the cavitation-resistance properties of pump materials, listed in order of increasing cavitation resistance:

1. Cast iron
2. Bronze
3. Cast steel
4. Manganese bronze
5. Monel
6. 400 series stainless steel
7. 300 series stainless steel
8. Nickel-aluminum bronze

Abrasive Wear   Abrasive wear is the mechanical removal of metal from the cutting or abrading action of solids carried in suspension in the pumped liquid. An undulating matte-finish wear pattern can usually be identified as abrasive wear. The rate of wear for any material is dependent upon the following characteristics of the suspended solids:

1. Solids concentration
2. Solids size and mass
3. Solids shape, that is, spherical, angular, or sharp fractured surfaces
4. Solids hardness
5. Relative velocity between solids and metal surface

The rate of wear is also dependent upon the materials selected for the rotating and stationary components of a centrifugal pump. Although metal hardness is not the sole criterion of resistance

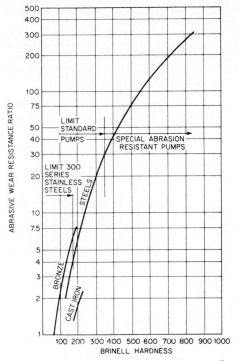

FIG. 1 Abrasive-wear-resistance ratio versus Brinell hardness.

to abrasive wear, hardness does provide a convenient index in the selection of ductile materials usually available for centrifugal pumps. Such an index is shown in Fig. 1, where the abrasive-wear-resistance ratio is shown as a function of Brinell hardness for various materials. It should be noted that a brittle material, such as cast iron, exhibits a much lower ratio than either the steels or bronzes of the same hardness. There is some evidence that, while the rate of abrasive wear for a ductile material is proportional to the square of the velocity of the abrasive particles, the rate of abrasive wear of a brittle material may be as high as the sixth power of the particle velocity.

While it is impossible to establish any direct relation between the life of pump components and the quantity and characteristics of abrasive particles pumped, the following tabulation can be used as a guide in material selection, listed in order of increasing abrasive-wear resistance:

1. Cast iron
2. Bronze
3. Manganese bronze
4. Nickel-aluminum bronze
5. Cast steel
6. 300 series stainless steel
7. 400 series stainless steel

Graphitization　Gray iron consists of a matrix of iron and graphite. The graphite exists as flakes and produces the characteristic gray appearance of cast iron. The presence of the graphite also provides a lubricant during machining. This property, plus the fragility of the chips, accounts for the excellent machining qualities of cast iron.

These characteristics, in addition to low foundry costs, combine to make gray iron the most widely used metal in the pump industry. Aside from the low tensile strength and ductility of cast iron relative to the steels, the corrosion-resistance properties of cast iron must be carefully considered. The presence of graphite in the matrix of cast iron produces the unique corrosion effect known as graphitic corrosion or, more simply, graphitization.

In the presence of an electrolyte, a galvanic cell exists between the iron and the graphite particles. In the combination of iron and graphite, iron becomes the anode and the graphite becomes the cathode. A galvanic current flows from the iron to the graphite, and the iron goes into solution. The result is a gradual depletion of the iron in the matrix until only the graphite remains. The original casting in iron has now been reduced to a porous graphite structure interspersed with the corrosion products of iron. The physical properties of the graphite structure are greatly inferior to those of cast iron, and the structure fails rapidly. In fact, the effect is so dramatic that, while the casting appears sound from outward appearances, pieces may be broken off with the fingers.

The effect of graphitization on a cast iron impeller pumping seawater has been observed many times. The result is a rapid deterioration of the impeller vanes as the soft graphite structure is scoured away, progressively exposing new metal to further attack. The same impeller pumping a nonelectrolyte, such as fresh water, shows no effect of graphitization. Experience has shown that a cast iron impeller should never be used on brackish water or seawater; the result is inevitably destruction by graphitization.

Experience has also shown that cast iron casings are much more resistant to destructive graphitization than are cast iron impellers. While it is true that examination of the inside surface of the casing may reveal a layer of graphite, the velocities encountered in the casing very often are not sufficient to scour away the graphite, and the base material is protected against further attack. This is true, however, only for pumps producing 100 ft (30 m) of head or less. For heads in excess of 100 feet (30 m), alternate casing materials should be considered for brackish water or seawater services.

MATERIALS OF CONSTRUCTION

Impellers　　The following criteria should be considered in the selection of the material for the impeller:

1. Corrosion resistance
2. Abrasive-wear resistance
3. Cavitation resistance
4. Casting and machining properties
5. Cost

For most water and other noncorrosive services, bronze satisfies these criteria and as a result is the most widely used impeller material for these services. Bronze impellers, however, should not be used for pumping temperatures in excess of 250°F (120°C). This is a limitation imposed primarily because of the differential rate of expansion between the bronze impeller and the steel shaft. Above 250°F (120°C) the differential rate of expansion between bronze and steel will produce an unacceptable clearance between impeller and shaft. The result will be a loose impeller on the shaft.

Cast iron impellers are used to a limited extent in small, low-cost pumps. As cast iron is inferior to bronze in corrosion, erosion, and cavitation resistance, low initial cost would be the only justification for a cast iron impeller.

Stainless steel impellers of the 400 series are widely used where bronze would not satisfy the requirements for corrosion, erosion, or cavitation resistance. The 400 series of stainless steels are not used for seawater, however, as pitting will limit their performance life. The 400 series of stainless steels should be used where the pumping temperatures exceed 250°F (120°C), as the differential expansion problem no longer exists with a steel impeller on a steel shaft.

The austenitic stainless steels of the 300 series are the next step up on the corrosion- and cavitation-resistance scale. Initial cost is a factor here that should be evaluated against the increased life.

Casings   The following criteria should be considered in the selection of material for centrifugal pump casings:

1. Strength
2. Corrosion resistance
3. Abrasive-wear resistance
4. Casting and machining properties
5. Cost

For most pumping applications, cast iron is the preferred material for pump casings when evaluated against initial cost. For single-stage pumps, cast iron is usually of sufficient strength for the pressures developed. For corrosive and volatile petroleum products, it may be necessary to specify cast steel or cast stainless steels of the 400 or 300 series.

Cast iron casings for multistage pumps are limited to approximately 1000-lb/in^2 (6.9-mPa) discharge pressure and 350°F (177°C). For temperatures above 350°F (177°C) and pressures up to 2000-lb/in^2 (13.8-mPa) discharge pressure, a cast steel is usually specified for split-casing multistage pumps. For pressures higher than 2000 lb/in^2 (13.8 mPa), a cast or forged steel barrel-type casing is required.

In any evaluation of cast iron versus steel casings, consideration should be given to the probability of casing erosion during operation. Erosion can occur either from abrasive particles in the fluid or from wire drawing across the flange of a split-case pump. While the initial cost of a steel casing is higher than that of a cast iron casing, a steel casing can often be salvaged by welding the eroded portions and then remachining. Salvaging a cast iron casing by welding is not practical, and the casing usually must be replaced.

The ductile irons are useful casing materials for pressure and temperature ratings between cast iron and the steels. While the modulus of elasticity for the ductile irons is essentially the same as that for cast iron, the tensile strength of the former is approximately double that of the latter. In any evaluation of the ductile irons as a substitute for the steels in the intermediate pressure and temperature range, it must be remembered that ductile iron casings cannot be effectively repair-welded in the field.

Shafts   The following criteria should be considered in the selection of the material for a centrifugal pump shaft:

1. Endurance limit
2. Corrosion resistance
3. Notch sensitivity

The endurance limit is the stress below which the shaft will withstand an infinite number of stress reversals without failure. Since on stress reversal occurs for each revolution of the shaft, this means that, ideally at least, the shaft will never fail if the maximum bending stress in the shaft is less than the endurance limit of the shaft material.

In practice, however, the endurance limit is substantially reduced because of corrosion and stress raisers, such as threads, keyways, and shoulders on the shaft. In selecting the shaft material, consideration must be given to the corrosion resistance of the material in the fluid being pumped as well as to the notch sensitivity. A more detailed discussion of the effect of endurance limit on shaft design is given above under the heading Corrosion Fatigue.

TABLE 1 Material Selection Chart for Volute Casing Pumps

Part	Fresh water, 40–250°F (4–120°C)	Seawater		Boiler feed			Sewage, 40–90°F (4–32°C)	Mine water, 40–90°F (4–32°C)	Condensate 90–120°F (32–49°C)
		40–80°F (4–27°C)	>80°F (>27°C)	250°F (120°C)	350°F (177°C)	>350°F (>177°C)			
Impeller									
Cast iron	...	...	...	...	...	...	1	...	...
Bronze	1	1	...	1	...	...	...	...	1
400 series stainless steel	...	2	1	2	1	1	2	2	2
300 series stainless steel	...	1	...	...	...	...	...	1	...
Ni resist	...	1	...	...	...	...	...	...	...
Casing									
Cast iron	1	1	...	1	...	...	1	...	1
Bronze	...	2	2	...	...	...	...	...	...
400 series stainless steel	...	...	...	2	1	1	1	2	2
300 series stainless steel	...	...	1	...	...	...	...	1	...
Ni resist	...	2	1	...	...	...	...	...	...
Wearing rings									
Bronze	1	1	...	1	...	...	...	...	1
400 series stainless steel	2	...	1	2	1	1	1	1	2
300 series stainless steel	...	2	2	...	...	...	...	2	...
Monel	...	...	...	...	...	...	...	...	...
Shaft									
Steel	1	...	...	1	...	...	1	...	1
400 series stainless steel	...	1	1	2	1	1	2	1	2
300 series stainless steel	...	2	2	...	...	...	...	2	...
Monel	...	...	...	...	...	...	...	...	...

Ratings: 1 = most economical to give acceptable service; 2 = extended life at additional cost.

Wearing Rings   The following criteria should be considered in the selection of the material for the wearing rings:

1. Corrosion resistance
2. Abrasive-water resistance
3. Galling characteristics
4. Casting and machining properties

The wearing rings of centrifugal pump consist of the impeller ring rotating in the bore of the stationary, or casing, ring. As the purpose of the wearing rings is to provide a close running clearance to minimize leakage from the discharge to the suction of the impeller, an increase in leakage as a result of wear in the rings will have a direct effect on the head, capacity, and efficiency of the pump.

To reduce the rate of wear of the wearing rings, and thereby increase the life of the pump, special consideration must be given to the corrosion and abrasive-wear characteristics of the ring material in any evaluation of wearing-ring selection.

Bronze is the most widely used material for wearing rings because it exhibits good corrosion resistance for a wide range of water services. In addition, bronze exhibits good wear characteristics in clear liquids but wears rapidly when abrasive particles are present. The bronzes also are low on the galling scale if metal-to-metal contact occurs. This is especially true for the 8 to 12% leaded bronzes, and the leaded bronzes should be used whenever possible. The casting and machining properties of most grades of bronze are excellent

In applications where bronze is not suitable because of either corrosion or abrasive-wear limitations, or where pumping temperatures exceed 250°F (120°C), stainless steel rings are required. Unlike bronze, the stainless steels of the 300 and 400 series exhibit a great tendency to gall. The risk of wearing-ring seizure can be minimized by increasing the clearance between the rings,

TABLE 2   Material Selection Chart for Wet-Pit Diffuser Pumps

Part	Fresh water, 40–250°F (4–120°C)	Seawater 40–80°F (4–27°C)	Seawater >80°F (>27°C)	Mine water 40–90°F (4–32°C)	Condensate 90–120°F (32–49°C)
Diffuser					
Cast iron	1	. . .	. . .	. . .	1
Bronze	. . .	2	. . .	1	. . .
400 series stainless steel	. . .	. . .	. . .	. . .	2
300 series stainless steel	. . .	2	2	2	. . .
Ni resist	2	1	1	1	. . .
Elbow and column					
Steel	1	. . .	. . .	. . .	1
Cast iron	. . .	1	. . .	. . .	. . .
Ni resist	. . .	2	1	1	. . .
Wearing rings					
Bronze	1	1	. . .	. . .	1
400 series stainless steel	2	. . .	. . .	. . .	2
300 series stainless steel	. . .	2	1	1	. . .
Monel	. . .	. . .	2	2	. . .
Shaft					
400 series stainless steel	1	. . .	. . .	. . .	1
300 series stainless steel	. . .	1	1	1	. . .
Monel	. . .	2	2	2	. . .

Ratings: 1 = most economical to give acceptable service; 2 = extended life at additional cost. Impeller selection same as in Table 1.

TABLE 3   Material Selection Chart for Reciprocating Pumps

Part	Fresh water, 250°F (120°C)	Seawater 40–90°F (4–32°C)	Seawater >90°F (>32°)	Water flood, 90°F (32°C)	Brines, 40–90° (4–32°C)	Oil, 250°F (120°C)	Boiler feed, 300°F (149°C)
Valve service							
Steel	1	. . .	. . .	1	. . .	1	1
Bronze	1	1	1	1	1	1	1
400 series stainless steel	2	. . .	. . .	2	. . .	2	2
300 series stainless steel	. . .	2	2	2	2	. . .	. . .
Monel	. . .	2	2	2	2	. . .	. . .
Liquid cylinder							
Cast iron	1	1	. . .	1	1	1	1
Bronze	1	1	1	1	1	. . .	1
Steel	2	. . .	. . .	. . .	. . .	2	1
400 series stainless steel	. . .	. . .	. . .	2	. . .	. . .	2
300 series stainless steel	. . .	2	2	2	2	. . .	. . .
Aluminum-bronze	. . .	2	2	2	2	. . .	. . .
Plungers							
Steel	. . .	. . .	. . .	. . .	. . .	1	. . .
400 series stainless steel	1	. . .	. . .	1	. . .	1	1
300 series stainless steel	. . .	1	1	1	1	. . .	. . .
Hard-faced stainless steel	2	2	2	2	2	2	2

Ratings: 1 = most economical to give acceptable service; 2 = extended life at additional cost.

specifying a difference in hardness between the two rings of 125 to 150 Brinell, or providing serrations in one of the wearing-ring surfaces.

It is apparent that increasing the clearance between the wearing rings is the least costly procedure to reduce the risk of galling or seizure, but increasing the wearing-ring clearance will reduce the output and efficiency of the pump. In large, low-head pumps the loss in efficiency is less than 1%, but in small, high-head units the loss in efficiency can be significant. The serrated

TABLE 4   Combinations to Be Avoided When Area of Metal Considered Is Small Relative to Area of Coupled Metal

Metal considered	Coupled metal					
	Steel	Cast iron	Bronze	400 series stainless steel	300 series stainless steel	Monel
Steel	. . .	X	X	X	X	X
Cast iron	. . .	. . .	X	X	X	X
Bronze	. . .	. . .	. . .	X	X	X
400 series stainless steel	. . .	. . .	X	. . .	X	X
300 series stainless steel	. . .	. . .	X	. . .	. . .	X
Monel	. . .	. . .	. . .	. . .	. . .	. . .

TABLE 5 Combinations to Be Avoided When Area of Metal Considered Equal to Area of Coupled Metal

Metal considered	Coupled metal					
	Steel	Cast iron	Bronze	400 series stainless steel	300 series stainless steel	Monel
Steel	. . .	. . .	X	X	X	X
Cast iron	. . .	. . .	X	X	X	X
Bronze	. . .	. . .	. . .	. . .	. . .	. . .
400 series stainless steel	. . .	. . .	X	. . .	X	X
300 series stainless steel	. . .	. . .	. . .	. . .	. . .	. . .
Monel	. . .	. . .	. . .	. . .	. . .	. . .

rings can be used on smaller pumps to maintain the efficiency, but only at an increase in manufacturing costs.

SELECTION OF MATERIALS OF CONSTRUCTION

The selection of the materials for pumps is at best a compromise between the cost of manufacture and the anticipated maintenance costs. Many pump installations start out with a low service factor and, through operating experience, are gradually upgraded in materials until an acceptable and scheduled replacement program is achieved. It must be anticipated that for the more corrosive services modification and replacement of the wetted parts will be necessary during the life of the pump. In the initial selection of the most economical materials for centrifugal and reciprocating pump applications, Tables 1, 2, and 3 will prove helpful.

When the liquid being pumped is an electrolyte, particular attention must also be directed at an evaluation of the probability of an unacceptable level of galvanic corrosion. The electrolytes most commonly encountered in pump applications are seawater, brines, and mine waters. For these services the use of incompatible materials can lead to rapid failure, and this is particularly true when a small area of a less noble metal is in contact with a larger area of a more noble metal. Bronze wearing rings, for example, should never be used with a stainless steel impeller when pumping an electrolyte because the unequal area between the wearing ring and the impeller would result in a rapid preferential corrosion rate of the ring. An even more destructive combination would be the use of bronze screws or bolts in a stainless steel impeller or casing. Tables 4 and 5 list combinations of metals most commonly used in pumps that should be avoided when the application requires the pumping of an electrolyte.

SECTION 5.2
MATERIALS OF CONSTRUCTION OF NONMETALLIC PUMPS

ROBERT D. NORTON

REASONS FOR USING NONMETALLIC PUMPS

Three major reasons for using nonmetallic, rather than metallic, centrifugal pumps are

1. Cost savings for equal corrosion resistance
2. No contamination of liquid pumped
3. Lighter weight

The first reason, cost savings, is the major factor. As a general rule, the cost of a heavy-duty filament-reinforced plastic (FRP) centrifugal pump is about equal to that of a 316 (18-8) stainless steel horizontal dry-pit pump and 20 to 40% less than that of a vertical wet-pit pump. The cost of the liquid end of an industrial-size horizontal pump, which represents approximately 80% of the total pump cost, is about equal in both materials. In a vertical wet-pit pump, the nonmetallic column pipe, discharge pipe, and miscellaneous flanges and fittings, which cost much less than equivalent components made of 316 stainless steel, are generally supplied by mass producers of pipe and fittings. Therefore, a vertical pump made of nonmetallic components is considerably less expensive than an equivalent metal pump.

Nonmetallic pumps can be used for pumping deionized and demineralized water. They provide ideal corrosion resistance, excellent nongalling characteristics for nonlubricating liquids, and do not add any impurities to the liquid being pumped. They are useful in fish farming, large aquariums, and desalination plants, to name just a few applications, and are less noisy than metallic pumps. Their nonmagnetic and nonsparking properties make them very useful when the liquid being pumped is flammable.

Because of the lighter weight of nonmetallic pumps, they are easier to handle and repair, particularly where mobility is important, as in mining or marine applications.

TABLE 1 Comparative Properties of Typical Protective Coatings and Elastomers

ANSI/ASTM D1418-77	NR/IR	AU/EU	CR	NBR
Elastomer Common Name	Gum/Natural	Urethane	Neoprene	Nitrile/Buna-N
ASTM D-2000; SAE J-200 Military: MIL STD 417 Protective Coatings Code	AA RN G•R	BG SB U	BC-BE SC F•N	BF-BG-BK-CH SB J•P
Chemical Name Definition	Polyisoprene	Polyester/ Polyether Urethane	Poly- Chloroprene	Butadiene Acrylo- Nitrile
Hardness Range: Duro A Specific Gravity Of Base Low Temp.-Min. Service °F * High Temp. - Max. Service °F *	30-90 0.93 -20 to -60 185	60-99 1.06 -10 to -30 185	40-95 1.23 -10 to -50 220	40-95 1.00 +30 to -40 240

RESISTANCE AND CHEMICAL PROPERTIES COMPARISON

	NR/IR	AU/EU	CR	NBR
Abrasion	Excellent	Outstanding	Excellent	Good
Absorption, Water	Very Good	Good	Good	Good
Acid-Concentrated	Fair To Good	Poor	Good	Good
Acid-Dilute	Fair To Good	Fair	Excellent	Good
Adhesion To Fabrics	Excellent	V. Good to Excel.	Excellent	Good
Adhesion To Metals	Excellent	Excellent	Excellent	Excellent
Chemicals	Fair To Good	Fair	Fair To Good	Fair To Good
Cold	Excellent	Excellent	Good	Fair To Good
Dielectric Strength	Excellent	Excellent	Good	Poor
Dynamic Properties	Excellent	Excellent	Fair	Good To Excel.
Electrical Insulation	Good To Excel.	Fair To Good	Fair To Good	Poor
Flame	Poor	Fair	Good	Poor
Heat	Good	Good	Very Good	Good
Heat Aging	Fair	Good	Good	Good
Hydrocarbons-Aliphatic	Poor	Good To Excel.	Fair To Good	Excellent
Hydrocarbons-Aromatic	Poor	Fair To Good	Fair	Good
Hydrocarbons-Oxygenated	Fair To Good	Poor	Poor	Poor
Impermeability	Fairly Low	Fairly Low	Low	Low
Oil-Animal & Vegetable	Poor To Good	Good To Excel.	Good	Very Good
Oil And Gasoline	Poor	Good To Excel.	Good	Excellent
Oxidation	Good	Excellent	V. Good To Excel.	Good
Ozone	Poor To Fair	Excellent	V. Good To Excel.	Fair
Radiation	Excellent	Very Good	Very Good	Very Good
Rebound-Cold	Excellent	Good	Very Good	Good
Rebound-Hot	Excellent	Good	Very Good	Good
Set, Compression	Good	Fair	Fair To Good	Good
Solvents, Lacquer	Poor	Poor	Poor	Fair
Steam	Fair To Good	Poor	Fair	Fair To Good
Sunlight Aging	Poor	Good	Very Good	Poor
Swelling In Oil	Poor	Good To Excel.	Good	Very Good
Tear	Good To V. Good	Excellent	Good	Fair
Tensile Strength	Excellent	Excellent	Good	Good To Excel.
Water	Fair To Good	Poor	Fair	Fair To Good
Weather	Fair	Excellent	Excellent	Fair
Generally Resistant To:	Water, Air And Average Concentration Acids, Bases, Alcohols, Salts, Ketones, Best Abrasion Resistance	Moderate Chemicals, Oils, Fats, Greases And Many Hydrocarbons.	Moderate Acids And Chemicals. Ozone, Oils, Fats, And Many Solvents. Oily Abrasive Applications.	Most Hydrocarbons, Fats, Oils, Greases, Hydraulic Fluids, Chemicals And Solvents.
Generally Affected Or Attacked By:	Not For: Ozone. Strong Acids. Bases, Oils. Solvents, Most Hydrocarbons.	Not For: Concentrated Acids. Ketones, Esters. Chlorinated And Nitro Hydrocarbons.	Not For: Oxidizing Acids, Esters And Ketones. Aromatic. Chlorinated And Nitro Hydrocarbons.	Not For: Ozone. Ketones, Esters. Aldehydes, Nitro And Chlorinated Hydrocarbons. Polar Solvents. MEK.

CIIR	CSM	EPDM	FKM	AFMU	SI	SBR
Chlorobutyl	Hypalon*	EPDM/EPT	Viton*/Fluorel*	Teflon*	Silicone	GRS/Buna-S
AA-BA RS B•D•O	CE SC H	BA-CA-DA RS E•Q	HK — V	— — T	FC-FE-GE TA K	AA-BA RS S
Chloro-Isobutylene Isoprene	Chloro-Sulfonated Polyethylene	Ethylene Propylene Polymer	Fluorinated Hydrocarbon	Tetrafluoro-Ethylene Resin	Poly-Siloxane	Styrene Butadiene
40-75 0.92 -10 to -60 250 to 300	40-95 1.12-1.28 -30 to -60 275	40-90 0.86 -20 to -60 300	55-95 1.85 +10 to -10 400 to 600	— — -120 450	40-85 1.14-2.05 -60 to -150 460	40-90 0.94 0 to -50 195
Good Very Good Good Excellent Good	Excellent Very Good Very Good Excellent Good	Good To Excel. V. Good To Excel. Excellent Excellent Good	Good Very Good Excellent Excellent Good To Excel.	Good — Excellent Excellent Fair To Good	Poor Excellent Fair Excellent Excellent	Good To Excel. Good To V. Good Fair To Good Fair To Good Good
Good Excellent Good Excellent Fair	Excellent Excellent Good V. Good To Excel. Fair	Good To Excel. Excellent Excellent Excellent Good To Excel.	Fair To Good Excellent Good Good Good To Excel.	Fair To Good Excellent Excellent — —	Excellent Good To Excel. Outstanding Good Poor	Excellent Fair To Good Very Good Excellent Good
Good To Excel. Poor Very Good Very Good Poor	Good Good Excellent Very Good Fair To Good	Excellent Poor Excellent Excellent Poor	Fair To Good Excellent Outstanding Outstanding Excellent	— Excellent Excellent Excellent Excellent	Excellent Fair To Good Outstanding Outstanding Poor	Good To Excel. Poor Fair To Good Fair To Good Poor
Poor Good Very Good Very Good Poor	Fair Poor To Fair Low To Very Low Good Good	Poor Good To V. Good Fairly Low Good Poor	Excellent Poor Very Low Excellent Excellent	Excellent Excellent — Excellent Excellent	Poor Fair Fairly Low Good To Excel. Fair	Poor Good Fairly Low Poor To Good Poor
Excellent Excellent Good Poor Very Good	Excellent Outstanding Very Good Fair To Good Good	Excellent Outstanding Outstanding Very Good Very Good	Outstanding Outstanding Very Good Fair To Good Good	Excellent Excellent Fair To Good — —	Excellent Excellent Very Good Excellent Excellent	Fair Poor Excellent Good Good
Fair Fair To Good Good Very Good Poor	Fair Poor Fair Outstanding Good To Excel.	Good Poor To Fair Excellent Outstanding Poor	Fair To Good Poor To Fair Fair To Good Outstanding Excellent	— Excellent — Excellent —	Fair Poor Fair Excellent Fair	Good Poor Fair To Good Poor Poor
Good Good Good Good To Excel.	Fair Fair Fair Excellent	Fair To Good Good To Excel. Excellent Excellent	Fair Good To Excel. Fair To Good Excellent	— — Excellent Excellent	Poor Poor Fair Excellent	Fair Good to Excel. Fair To Good Fair
Animal And Vegetable: Oils. Fats. Greases. Air, Gas, Water. Many Oxidizing Chemicals And Ozone.	Strong Acids And Bases. Freons. Hydroxides. Ozone. Alcohols. Etching. Alkaline And Hypochlorite Solutions.	Vegetable And Animal Fats. Oils: Ozone. Many Strong And Oxidizing Chemicals. Ketones. Alcohols.	All Aromatic, Aliphatic And Halogentated Hydrocarbons. Many Acids. Animal And Vegetable Oils.	Most Known Fluid Chemicals.	Moderate Or Oxidizing Chemicals. Ozone. Concentrated Sodium Hydroxide.	Water, Air. Anti-Freeze. Detergents. Salt Solutions. Bases. Alcohols And Some Acids.
Not For: Oils. Solvents. Aromatic Hydrocarbons.	Not For: Ketones. Esters. Certain Chlorinated Oxidizing Acids. Chlorinated. Nitro And Aromatic Hydrocarbons.	Not For: Mineral Oils. Solvents. Aromatic Hydrocarbons.	Not For: Ketones. Esters. And Nitro Containing Compounds.	Not For: Molten Alkali Metals. Fluorine And Related Compounds.	Not For: Many Solvents. Oils. Concentrated Acids. Sulfurs.	Not For: Oils. Greases. Solvents And Hydrocarbons. Ozone And Strong Acids.

®Fluorel is a Registered Trademark of 3M Companies. Viton, Hypalon, and Teflon are Registered Trademarks of E. I. Du Pont de Nemours & Co., Inc.

*Temperature rating depends upon operating pressure and if service is continuous.

SOURCE: Protective Coatings, Inc.

NONMETALLIC MATERIALS

Useful nonmetallic materials for pumps are plastics, ceramics, glass, rubberlike elastomers, and carbon.

Names of some of the materials used for major and minor pump components are: Teflon (polytetrafluoroethylene), Neoprene (polychloroprene), Buna N (butadiene acrylonitrile), Viton (fluorinated hydrocarbon), EPDM (ethylene propylene polymer), Hypalon (chlor-sulfanated polyethylene), Kynar (polyvinylidene fluoride), Noryl (propylene oxide), Ryton (polyphenylene sulfide), PVC (polyvinyl chloride), and PP (polypropylene).

Plastics   Various compositions of plastics are available. Generally, they are of two classes, *thermosets* and *thermoplastics*. *Thermoset* materials are shaped by initial heating and cannot be reshaped by further heating. *Thermoplastic* materials can be reshaped by subsequent reheating.

THERMOSETS   These materials generally have greater tensile strength, heat deflection temperatures, impact strength, and hardness than thermoplastics. The most commonly used thermosets are vinyl esters, polyesters, and epoxies. Any of these resins, when reinforced with glass filament, provide pump housings which may be as strong as cast iron. Theomoset plastics are regularly used in the manufacture of centrifugal pumps as large as 200 hp (150 kW) and 4000 gpm (900 m^3/h), and it is expected that these size ranges will increase in the future.

THERMOPLASTICS   Thermoplastic materials provide wider range of corrosion resistance than thermosets but lack sufficient tensile strength in the larger sizes to be used without an enclosure or housing for support. Commonly used thermoplastics are PVC and PP.

Ceramics and Glass   Ceramics, glass, and stoneware provide a very wide range of corrosion resistance as well as better high-temperature and abrasion resistance. However, they are weak in tensile strength and must therefore be used as liners or coatings inside metal housings on large pumps; when used thusly, they are subject to cracking from impact shock because of their inherent brittleness. Their complete resistance to all solvents is an advantage over plastics in some cases.

Rubberlike Elastomers   These materials are suitable where impact-type abrasion from rounded particles is a problem; the elasticity absorbs the impact without permanent deformation. The corrosion resistance of synthetic rubbers varies widely from one type to another but is generally good, and it can be used where there is a severe corrosive atmosphere. A summary of the physical and chemical properties of 11 commonly used elastomers is given in Table 1.

Elastomers are generally recommended for gaskets and seal rings in nonmetallic pumps. Elastomers are also used as coatings and linings; as the basic housing material, as, for example, in a peristaltic "hosepump"; and as line-shaft bearings for vertical wet-pit pumps.

COATINGS AND LININGS

Nonmetallic materials are applied as coatings and liners as well as in a solid piece. In general practice, coating or lining material is selected for corrosion resistance and housing or support material is chosen for strength and economy only. It follows naturally therefore that many pump designs are offered with rubberlike elastomeric coatings or linings as well as with graphite, ceramic, or Teflon coatings and linings, all backed by a cast iron or other similar low-cost, high-strength housing or support that is not resistant to corrosion. The reason is almost always cost, i.e., a less expensive alternate on initial price.

Limitations   Coatings, because they are applied to a support as particles, are prone to permeation by the liquid being pumped. If the liquid permeates the coating and is corrosive to the housing or support, then failure will eventually occur. It is simply a matter of time.

The thermal coefficient of expansion of most plastics is several times that of cast iron, the typical base material. Therefore, with either coatings or linings, any large temperature rise may cause the coating or lining to separate from the housing metal, destroying the coating or lining and exposing the support metal structure to corrosion.

Coatings and liner materials in plastic, and rubberlike materials particularly, generally require much larger dimensional tolerance than metal parts. It becomes more difficult therefore, on pump repair overhauls, to maintain the perfect sealing integrity that the noncorrosive housing requires. This fact may justify the normal industry acceptance of rubber-lined pumps for abrasive slurries that are noncorrosive along with an industry dissatisfaction if the abrasive liquid is also corrosive.

If the abrasive particles pumped are very sharp, elastomer liners may tear. Particle size is usually limited to ⅜ in (10 mm). The tensile modulus of soft rubber is so low that, because of centrifugal force, use of lined impellers is generally limited to tip speeds of 80 to 90 ft/s (24 to 27 m/s) or about 150 ft (46m) head and moderate temperatures.

Surface-Hardened Solid-Section Nonmetallic Parts   An additional category of surface improvement which provides a composite, solid material without the disadvantages of coatings and linings is well known in metal castings as surface heat treating. This process improves various properties, such as hardness, at the solid metal surface without causing any loss of the advantages of the solid metal section.

A similar surface-hardness improvement has recently been provided by a thermoset resin transfer process that makes possible a full ¼-in-thick (6.4-mm) composite, abrasion-resistant surface for all wetted parts. A typical part so treated is the backplate (rear volute piece) of a small vinyl ester FRP centrifugal pump. The metal mold cavity is filled to the required depth with catalyzed resin mixed with hard silicon carbide and aluminum oxide spheres in various small sizes. (Such material provides abrasion resistance in the same way that hard silica sand particles provide abrasion resistance to a cement road surface.) Before this section hardens, the remaining volume of the mold is filled under moderate pressure with the usual mixture of glass reinforcement layers and catalyzed pure resin. After oven curing, the mold is opened. The backplate casting is now a single piece with a highly abrasion- and corrosion-resistant thick surface, equivalent to a hard heavy liner yet bonded thoroughly to its FRP corrosion-resistant housing. The base resin is the same throughout, and thus the basic design problems that arise when the mechanical properties of coatings and linings are different from those of the base material are eliminated.

NONMETALLIC CENTRIFUGAL PUMPS

Wet-Pit Pumps   A useful application of a nonmetallic wet-pit pump is for handling chemical or general industrial liquid waste containing waste chemicals from neutralization processes, acid spills, acid cleaning waste from boiler tubes, and so forth. A typical section cut of a nonmetallic 100-hp (75-kW) heavy duty pump is shown in Fig. 1. Figure 2 is a photograph of a similar pump partly cut open.

The pump in Fig. 1 is typically 10 ft (3 m) in overall length below the support plate and remains submerged constantly in liquid with a pH range of as much as 2 to 14. The submerged shaft and metal fasteners for a pump as large as this are made of a corrosion-resistant alloy, such as a Carpenter number 20 stainless steel containing at least 30% nickel and 20% chromium (ADI CN-7M). Other nonmetallic parts are manufactured of glass-reinforced vinyl ester, epoxy, or polyester thermoset resins. The pump liquid-end parts, as well as the column support pipe and discharge pipe, have a minimum tensile strength of 15,000 lb/in^2 (103 MPa).

The wall thicknesses chosen are heavy enough to provide a tensile strength at least equal to that of cast iron. Since these nonmetallic materials lack the hardness of metal parts, threaded connections are avoided and all connectors are therefore flanged. Impeller radial side thrust loads and intermediate shafting support loads of pumps of this type are carried on journal sleeve bearings of Teflon, preferably reinforced with relatively inert materials which improve the mechanical properties. The casing may be single or double volute, and the impeller may be open, enclosed, or of the vortex type.

Teflon is used as the bearing surface because of its wide range of chemical inertness and its good low-friction qualities. Since most waste pits contain solids in suspension, the Teflon sleeve bearings are generally flushed with clear water from an external source (item no. 93 in Fig. 1). In this design the Teflon sleeve bearings are enclosed in filament-reinforced thermoset bearing holders which are in turn securely clamped between the rabbet-fitted flanges approximately, and no

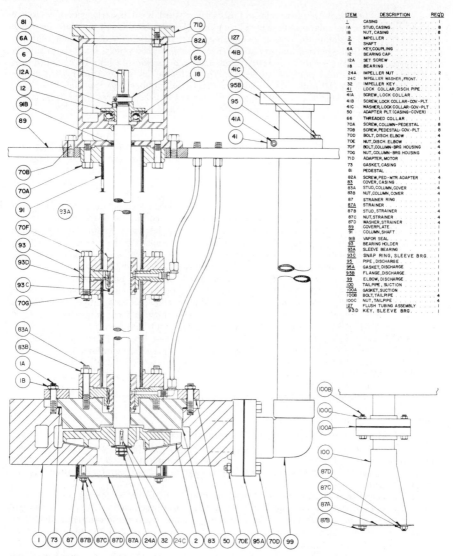

ITEM	DESCRIPTION	REQ'D
1	CASING	1
1A	STUD, CASING	8
1B	NUT, CASING	8
2	IMPELLER	1
6	SHAFT	1
6A	KEY, COUPLING	1
12	BEARING CAP	1
12A	SET SCREW	1
18	BEARING	1
24A	IMPELLER NUT	2
24C	IMPELLER WASHER, FRONT	1
32	IMPELLER KEY	1
41	LOCK COLLAR, DISCH. PIPE	1
41A	SCREW, LOCK COLLAR	1
41B	SCREW, LOCK COLLAR-COV-PLT	1
41C	WASHER, LOCK COLLAR-COV-PLT	1
50	ADAPTER PLT. (CASING-COVER)	1
66	THREADED COLLAR	1
70A	SCREW, COLUMN-PEDESTAL	8
70B	SCREW, PEDESTAL-COV-PLT	8
70D	BOLT, DISCH. ELBOW	4
70E	NUT, DISCH. ELBOW	4
70F	BOLT, COLUMN-BRG. HOUSING	4
70G	NUT, COLUMN-BRG. HOUSING	4
71D	ADAPTER, MOTOR	1
73	GASKET, CASING	1
81	PEDESTAL	1
82A	SCREW, PED-MTR. ADAPTER	4
83	COVER, CASING	1
83A	STUD, COLUMN, COVER	4
83B	NUT, COLUMN, COVER	4
87	STRAINER RING	1
87A	STRAINER	1
87B	STUD, STRAINER	4
87C	NUT, STRAINER	4
87D	WASHER, STRAINER	4
89	COVERPLATE	1
91	COLUMN, SHAFT	1
91B	VAPOR SEAL	1
93	BEARING HOLDER	1
93A	SLEEVE BEARING	1
93C	SNAP RING, SLEEVE BRG.	1
95	PIPE, DISCHARGE	1
95A	GASKET, DISCHARGE	1
95B	FLANGE, DISCHARGE	1
99	ELBOW, DISCHARGE	1
100	TAILPIPE, SUCTION	1
100A	GASKET, SUCTION	1
100B	BOLT, TAILPIPE	4
100C	NUT, TAILPIPE	4
127	FLUSH TUBING ASSEMBLY	1
93D	KEY, SLEEVE BRG.	1

FIG. 1 Typical section of a wet-pit, nonmetallic vertical centrifugal pump. Underlined numbers indicate non-metallic parts. (Fybroc Division, Met Pro)

more than, every 30 in (76 cm) of shaft length. The Teflon sleeves may have refinements, such as spiral grooves, to better distribute the cooling liquid and provisions to keep the sleeve from moving up or down or rotating in the bearing holders.

The column pipe housing surrounding the shaft allows cooling water to exit only from weep holes at the top of the top column section, so that the space around the shaft is always pressurized with the clean flushing liquid up to only a short distance from the mounting plate at the top of the pump. This pressurizing prevents floating or suspended grit and solids from entering the column pipe and getting into the sleeve bearing areas. No stuffing box is required because the clean liquid level in the column pipe is maintained at approximately weep hole elevation. Unless the sump liquid is toxic or highly flammable, there is no need for a stuffing box in this design.

A pump of this type generally costs at least 30% less than a 316 stainless steel pump or 50% less than a Carpenter number 20 stainless pump. This pump also weighs less and so is easier to remove and repair. Other design features of this type of pump are generally similar to those of vertical wet-pit metal pumps, which are covered in other sections of this handbook. A nonmetallic pump does differ in one respect: because of limitations in manufacturing, no renewable casing or impeller wearing rings are provided. Instead of a radial clearance at the running joint between these parts, an axial adjustable clearance is utilized.

Dry-Pit Pumps   A typical section of a non-metallic horizontal dry-pit pump is shown in Fig. 3. The casing may be single or double vol-ute, and the impeller may be open, enclosed, or of the vortex type. Although a single mechanical seal is shown, double seals are often used in chemical plant service. The double seal (Subsec. 2.2.3) is particularly advantageous on nonmetallic pumps because dissipation of heat is aided by the "in-and-out" water circulation through the mechanical seal. If the water is cir-culating whenever the pump is running, the pump can run dry for extended periods with-out damage. Additionally, the circulating water barrier fluid in a double seal will carry away contamination in the pumpage if the inner seal element should fail. The water bar-rier fluid also acts as a safety zone, separating a dangerous corrosive liquid from the atmosphere.

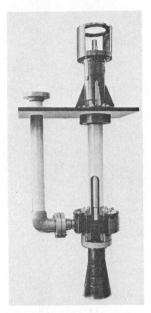

FIG. 2   A typical wet-pit, nonmetallic vertical cen-trifugal pump partly cut open. (Fybroc Division, Met Pro)

Pumps of this horizontal dry-pit type are available in solid casings of thermoset, ther-moplastic, and carbon, and there is also a wide variety of ceramic, elastomer, and thermoplas-tic coatings and liners for standard metal hous-ings. When such coatings and liners are used, however, problems may arise because of a substantial difference between the thermal expansion coefficient of the coating or lining and that of the metal support or because of the dimensional changes that occur when the coating or liner swells. Glass coatings are usually applied from 0.040 to 0.080 in (1 to 2 mm) thick. Brittleness, the principal disadvantage of glass coatings, may cause failure from solid impact or thermal shock. These disadvantages can cause severe problems with high-speed, close-clearance centrifugal pumps.

COATED OR LINED PUMPS   Graphite-coated wet-end parts of centrifugal pumps are sometimes used where plastic materials are not satisfactory. This is particularly true when corrosion resistance to acids containing chlorinated organic solvents is required. One such example is shown in Fig. 4. The volute casing, impeller, cover, and shaft sleeve of this pump are graphite, impregnated and bonded with chemical-resistant resins. The impervious graphite is a chemically inert, usually machinable, lightweight material resistant to a wide range of chemicals, including all organic acids, most inorganic acids, and all organic solvents. The bonding resin, however, can be attacked by strong oxidizing agents, such as concentrated sulfuric acid, nitric acid, and sodium chlorite.

The characteristics of the resin binder for the graphite particles is sometimes a factor in select-ing this type of pump. The tensile and bending strength of these carbon-based materials is rela-tively low, and the parts must be supported by clamping.

Typical specifications for the AVS-designed pump shown in Fig. 4 are as follows: maximum operating temperature, 338°F (170°C); maximum operating pressure, 90 lb/in^2 (620 kPa); heads to 120 ft (37m); capacities to 1400 gpm (320 m^3/h).

Graphite wet-end parts may be treated with silicon carbide for added abrasion resistance.

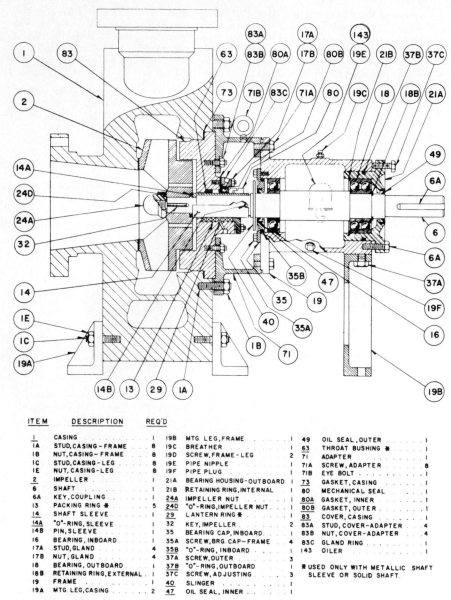

ITEM	DESCRIPTION	REQ'D					
<u>1</u>	CASING	1	19B	MTG. LEG, FRAME	1	49	OIL SEAL, OUTER	1
1A	STUD, CASING – FRAME	8	19C	BREATHER	1	<u>63</u>	THROAT BUSHING ✳	1
1B	NUT, CASING – FRAME	8	19D	SCREW, FRAME – LEG	2	71	ADAPTER	1
1C	STUD, CASING – LEG	8	19E	PIPE NIPPLE	1	71A	SCREW, ADAPTER	8
1E	NUT, CASING – LEG	8	19F	PIPE PLUG	1	71B	EYE BOLT	1
<u>2</u>	IMPELLER	1	21A	BEARING HOUSING – OUTBOARD	1	<u>73</u>	GASKET, CASING	1
6	SHAFT	1	21B	RETAINING RING, INTERNAL	1	80	MECHANICAL SEAL	1
6A	KEY, COUPLING	1	<u>24A</u>	IMPELLER NUT	1	<u>80A</u>	GASKET, INNER	1
13	PACKING RING ✳	5	<u>24D</u>	"O"-RING, IMPELLER NUT	1	<u>80B</u>	GASKET, OUTER	1
<u>14</u>	SHAFT SLEEVE	1	<u>29</u>	LANTERN RING ✳	1	<u>83</u>	COVER, CASING	1
<u>14A</u>	"O"-RING, SLEEVE	1	32	KEY, IMPELLER	2	83A	STUD, COVER – ADAPTER	4
14B	PIN, SLEEVE	1	35	BEARING CAP, INBOARD	1	83B	NUT, COVER – ADAPTER	4
16	BEARING, INBOARD	1	35A	SCREW, BRG. CAP – FRAME	4	83C	GLAND RING	1
17A	STUD, GLAND	4	<u>35B</u>	"O"-RING, INBOARD	1	143	OILER	1
17B	NUT, GLAND	4	<u>37A</u>	SCREW, OUTER	3			
18	BEARING, OUTBOARD	1	<u>37B</u>	"O"-RING, OUTBOARD	1	✳ USED ONLY WITH METALLIC SHAFT		
18B	RETAINING RING, EXTERNAL	1	37C	SCREW, ADJUSTING	3	SLEEVE OR SOLID SHAFT.		
19	FRAME	1	40	SLINGER	1			
19A	MTG. LEG, CASING	2	<u>47</u>	OIL SEAL, INNER	1			

FIG. 3 Typical section of a dry-pit, nonmetallic horizontal centrifugal pump. Underlined numbers indicate nonmetallic parts. (Fybroc Division, Met Pro)

BASEPLATES   Because of leakage from the packing seal, both cast iron and steel baseplates present a severe maintenance problem when strong acids (such as muriatic) are pumped. Since a metal baseplate might not get the critical maintenance attention given to the pump, the baseplate often deteriorates to where it practically disappears in a typical chemical plant.

Figure 5 shows a nonmetallic horizontal dry-pit pump and molded fiberglass baseplate in

FIG. 4   Centrifugal pump with wet end made of graphite impregnated with chemical-resistant resins. (Union Carbide)

chemical-resistant resins. Baseplate sizes are available conforming to the ANSI B73.1 standard dimensions. Most baseplates incorporate a catch basin under the pump stuffing box area.

In the ANSI chemical pump sizes, fiberglass baseplates are twice the cost of steel, and in the larger sizes, those accommodating motor frames through 256T, 10 to 20 hp (7.5 to 15kW), they cost approximately 50% more than channel steel. However, this higher price for fiberglass baseplates is only a small fraction of the total pump cost. Where corrosion is a problem, they offer maintenance savings.

NONMETALLIC POSITIVE DISPLACEMENT PUMPS

Diaphragm, gear, screw, and peristaltic pumps can be made completely or partially of nonmetallic materials.

One example of the advantages of the very broad corrosion resistance of nonmetallic materials is the solid Teflon diaphragm pump shown in Fig. 6. This pump has 2-in (50.8-mm) ports and will deliver 70 gpm (16 m³/h) at 100 ft (30 m) head. Materials are all Teflon for liquid-end parts, with Teflon valve balls, diaphragms, seal rings, and main housing. The unit is satisfactory to 225°F (107°C) pumping temperature. This air-driven pump is inherently self-priming, will run dry without damage for short periods, and possess the overall wide chemical resistance of Teflon. This design is also available in polypropylene at approximately half the price of the all-Teflon model.

Another example of a nonmetallic positive displacement pump is the peristaltic pump shown in Fig. 7. The pumping action of this "hosepump" results from the alternate compression and relaxing of the specifically designed, resilient, nylon-reinforced Buna or natural rubber hose. A

FIG. 5   Nonmetallic horizontal pump and molded fiberglass chemical-resistant resin baseplate. (Fybroc Division, Met Pro)

FIG. 6   Solid Teflon diaphragm pump. (American Pump)

FIG. 7   Peristaltic pump uses buna or natural rubber hose reinforced with braided nylon cord. (Waukesha-Bredel Pump, Abex)

liquid lubricant in the housing minimizes sliding friction. The fluid being pumped is in contact with the inner wall of the hose only. Abrasive particles in the fluid are cushioned in the thick inner hose wall during compression and returned to the fluid stream after compression. The pump has no seals or valves and is completely self-priming. It is used for handling such fluids as waste sludges, lime and cement mortar, shear-sensitive latex emulsions, glue and adhesives, and thickener underflow at flow rates to 300 gpm (70 m³/h) and pressures to 220 lb/in² (1500 kPa). Two advantages this pump has over progressive cavity pumps (see following paragraph) are (1) it can run dry and (2) it requires no seals or packings. The advantage it has over diaphragm sludge pumps is that it requires no valves.

Another example is the progressive cavity pump, a widely used type of nonmetallic positive displacement pump (Sec. 3.4, Fig. 10). A key factor of this design is the nonmetallic elastomer stator. Almost all applications of this type of pump require a stator (pump casing) made of a nonmetallic elastomer (a material which can be stretched 100% at normal temperature and recover to within 10% of normal size) for proper functioning.

INSTALLATION AND OPERATION

Important factors affecting the installation and operation of nonmetallic pumps are due to differences between the mechanical properties of nonmetallic units and those of their metallic counterparts.

Tensile Strength When compared with cast iron, the tensile strength of nonmetallic components can vary from almost equality in the case of reinforced thermosets down to about 10% that of cast iron in the case of unreinforced PVC and sintered carbon. Casing materials of very low tensile strength, such as Teflon, Kynar, and rubber, should be supported in a metal housing.

Modulus of Elasticity Whereas steel has a tensile modulus of about 30×10^6 lb/in^2 (207 MPa), most nonmetallic materials have a tensile modulus of not more than 10 or 20% of this. For this reason, it is generally recommended that a nonmetallic pump connected to metal piping (high modulus) should have flexible expansion joints on both the suction and the discharge. If the non-metallic pump is connected to plastic (usually PVC) piping, the stress from the low-tensile-modulus plastic pipe is relatively low and thus the pump may not require the protection of expansion joints.

The low-modulus characteristic of nonmetallic materials is not always a disadvantage. With a high modulus pump material, such as iron or steel, severe piping expansion may distort the pump casing where there are no expansion joints. This in turn may cause misalignment that results in repeated seal, bearing, coupling, and shaft failures.

Thermal Conductivity The thermal conductivity of plastics is several times lower than that of metal. Plastic is almost an insulator. However, the opposite is true of carbon nonmetallics.

In a pump with a metal packing box, the heat generated when the packing rubs the shaft is carried away by the metal. Since this dissipation of heat is not possible with a plastic stuffing-box housing, much more attention must be given to cooling the packing in a plastic pump. This can generally be done quite readily by an external water flush to the packing area.

An advantage of the low thermal conductivity of plastic is that the insulation provided by the plastic offers protection against freeze-up in cold weather.

Coefficient of Thermal Expansion These values are much higher in plastic pumps than in metal pumps. This is generally not a serious problem but must be considered in design, particularly on a vertical submerged pump with a long shaft. The plastic column pipe on a long-shaft pump may have an expansion coefficient much greater than that of the metal pump shaft. Proper attention to running clearances will generally take care of this problem.

Since plastics are limited to moderate temperatures, 150 to 250°F (66 to 121°C), their high coefficient of expansion is rarely a key factor. Excessive temperatures may be reached through malfunction, however. This occurs in plastic-lined pumps if solids jam clearances, in which case the high differential expansions between the plastic liner and its metal housing may cause the plastic to peel. With line-shaft bearings in a vertical sump pump, a failure of bearing external cooling lubrication will cause the Teflon bearing to heat up rapidly. Its high coefficient of expansion may cause rapid filling up of the bearing clearance and consequent seizing of the shaft.

Nonmetallic Flanges Typical nonmetallic pump flanges have the same bolt sizes as the comparable metallic parts. This is done to conform to standard metal piping systems. For best design, the bolts should be proportioned to the tensile and compressive strengths of the flanged material, but for convenience and for accommodation to the metal industries, nonmetallic plastic pumps have the same bolt sizes as metal pumps. Because of this, there is a tendency for inexperienced pipe fitters to apply the same torque to these bolts that they would to metal joints. Most nonmetallic flange bolting should not be stressed above 30 to 40 ft·lb (40 to 55 N·m) of torque. Metal washers should be used under all nut and bolt heads. Flange gaskets should be 1/8 in (3 mm) minimum thickness, with a Shore A or B durometer hardness of 40 to 70, i.e., a soft gasket.

FIG. 8 Centrifugal pump casing and vortex (recessed) impeller made of fiberglass-reinforced vinyl (or epoxy). (Fybroc Division, Met Pro)

MANUFACTURING PROCESS

Wet Lay-Up or Contact Molding   Layers of fiberglass reinforcement (usually mat or cloth) are placed in a mold. A catalyzed thermosetting resin is applied over the reinforcement. A roller is used to remove air bubbles and to contour the composites to the mold form. After air curing, parts are often cured in an oven to accelerate the development of the final physical properties of the material.

An automated variation of this method, called spray-up, has a mixture of catalyzed resin and chopped fiberglass sprayed from a gun against a mold surface, followed by the usual roller procedure.

This method can be used to produce large parts, such as cores for large pump casings. Uniformity of resin distribution is difficult to control, and the method generally applies best to simple shapes. Low-volume production is usually best served by this method, except in certain cases involving mechanized spray-up lines.

Part cost range: high. Tool cost range: low.

Filament Winding   Continuous fiberglass filament is saturated with resin and wound under tension onto a mandrel with the shape of the finished part. When winding is completed and the part cured, the mandrel is removed through the end of the part. Accurate control of filament tension and resin is required. A high percentage of reinforcement produces an outstanding strength-to-weight ratio. Filament winding does not permit fast production rates and is limited to shapes of rotation. Openings in the finished laminate reduce the strength of the part. The method is used for large pump circular casings.

Part cost range: medium. Tool cost range: low.

Compression Molding   A measured amount of partially cured thermosetting compound is placed in a heated mold cavity. The mold is closed, and heat and pressure are applied, causing the compound to flow and fill the cavity as heat completes the curing. Low finishing cost and little material waste are achieved in this molding method, but it is not recommended for complex configurations, such as undercuts and side draws.

Part cost range: low. Tool cost range: high.

Resin Transfer Molding   Fiberglass reinforcement is placed in a two-piece mold. The mold is closed, and catalyzed thermosetting resin is injected into the cavity under low pressure. As the resin saturates the reinforcement material, air is expelled through vents in the mold. High strengths are obtained from the use of continuous filament reinforcements. This method is used for medium-size industrial pump volutes and major parts.

Part cost range: medium. Tool cost range: medium.

Injection Molding   A thermoforming molding compound is fed from a hopper into a transfer chamber. The material is melted in this chamber and then forced by a plunger through the sprues, runners, and gates of a closed mold into the mold cavity. After the part solidifies, it is removed from the mold. Good dimensional accuracy is achieved, but part size is limited. This method is used for small mass-produced items for such designs as washing machine pumps and insecticide spray pumps.

Part cost range: low. Tool cost range: high.

REPAIRING NONMETALLIC PUMPS

Very little field rebuilding of corroded or abraded pump parts has been attempted in the field to date. Flanges pulled off by piping stress can be cemented back into place with adhesive repair kits. For other repairs, however, it is safest to completely replace the worn or damaged part. Unlike metallic pumps, nonmetallic pumps are not provided with replaceable wearing rings. Therefore, casing and impellers must be replaced entirely when wear becomes excessive.

CORROSION RESISTANCE TABLE

For a general corrosion table for nonmetallic pumps, refer to Table 9 in the appendix. This table covers corrosion in four key polymers and four key metals widely used for shafting and fasteners in the wetted parts of production-size nonmetallic pumps in corrosive services.

Teflon, glass, carbon, and graphite were excluded from the table because their corrosion resistance is satisfactory for almost all of the liquids tabulated. Some exceptions: graphite is attacked by strong oxidizing agents, such as concentrated sulfuric or nitric acid, and glass is attacked by caustics.

Table 9 was compiled by the author from private correspondence and publication excerpts from Fybroc Division of Met Pro, ICI, Union Carbide, Pennwalt, and Dow Chemical.

These corrosion data should be used only for general guidance. No recommendations are intended as guarantees by either the author or his correspondent sources. Pump manufacturers and material suppliers will generally provide test coupons for specific corrosion testing.

C·H·A·P·T·E·R·6

PUMP DRIVERS

SECTION 6.1
PRIME MOVERS

6.1.1
ELECTRIC MOTORS AND MOTOR CONTROLS

A. A. DIVONA
A. J. DOLAN

TYPES OF MOTORS

Alternating-Current Motors

SQUIRREL-CAGE INDUCTION MOTOR   By far the most common motor used to drive pumps is the squirrel-cage induction motor (Fig. 1). This motor consists of a conventional stator wound with a specific number of poles and phases, and a rotor which has either cast bars or brazed bars imbedded in it.

The squirrel-cage induction motor operates at a speed below synchronous speed by a specific slip or revolutions per minute. The synchronous speed is defined as

$$N = \frac{f \times 60 \times 2}{p}$$

where N = speed, rpm
$\quad f$ = line-power frequency, Hz
$\quad p$ = number of poles

The percent slip is defined as

$$\% \text{ slip} = \frac{(N - s) \times 100}{N}$$

where s = slip, rpm

When the stator winding of a squirrel-cage induction motor is connected to a suitable source of power, a magnetic flux is generated in the air gap between the stator and rotor of the motor. This flux revolves around the perimeter of the air gap and induces a voltage in the rotor bars. Since the rotor bars are short-circuited to each other at their ends (end rings), a current circulates in the rotor bars. This current and the air-gap flux interact, causing the motor to produce a torque.

FIG. 1 Cross section of typical squirrel-cage induction motor. (Westinghouse Electric)

The squirrel-cage induction motor exhibits a characteristic speed-torque relationship that is determined by the resistance of the rotor bars. Thus, the desired speed-torque characteristics are obtained by selecting a metal of suitable resistance when designing the rotor bars.

Figure 2 suggests several typical speed-torque characteristics which have been standardized by NEMA (National Electrical Manufacturers Association), covering motor frames 143T through 449T. Motors larger than 449T may not have these same values but generally have the same characteristic curves. Also, single-phase motors may not exhibit these characteristics and are defined specifically by NEMA with different values.

Most pumps are driven by NEMA B characteristic motors when operated from three-phase power sources.

WOUND-ROTOR INDUCTION MOTOR   The wound-rotor induction motor is in every respect similar to the squirrel-cage version except that the rotor is wound with insulated wire turns and this winding is terminated at a set of slip rings on the rotor shaft. Connections are made to the slip rings through brushes and in turn to an external resistor, which can be adjusted in ohmic value to cause the motor speed-torque characteristics to be changed.

Figure 3 demonstrates the speed-torque characteristics of a wound-rotor induction motor for several resistor values. It will be noticed that increasing the external resistance of the control will cause the peak torque of the motor to be developed at lower speeds until the peak torque occurs at zero speed. Increasing the resistance beyond this value will cause the motor to have a limited torque as, for example, curves 4, 5, and 6. This motor can be used where torque control is required or where variable speed is necessary. In the variable-speed application, the rotor resistance is adjusted to produce a motor torque that matches the load torque at the specific speed desired.

SYNCHRONOUS MOTOR   The synchronous motor is also similar to the squirrel-cage induction motor except that it operates at synchronous speed and its rotor is constructed with definite salient poles on which a field coil is wound and connected to a source of direct current for excitation. The most common synchronous motor is constructed with slip rings on the rotor shaft to connect the dc excitation to the field coils.

There are various means of providing the dc power to the slip rings:

1. *Static excitation*   The power to be connected to the slip ring brushes on the motor shaft is obtained from a transformer and rectifier package external to the motor.

2. *Direct-connected exciter*   This arrangement has a dc generator directly connected to the

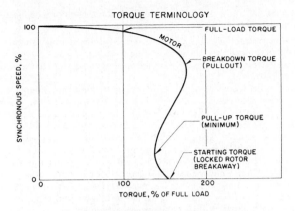

TORQUE TERMINOLOGY

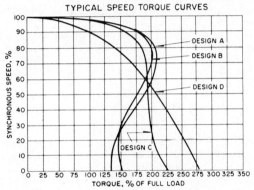

TYPICAL SPEED TORQUE CURVES

FIG. 2 These curves are characteristic of NEMA frame-size motors through frame size 449T. Motors larger than NEMA sizes may have considerably less starting torque. Very large motors may have only 30 to 50% of the starting torques shown here. (Westinghouse Electric)

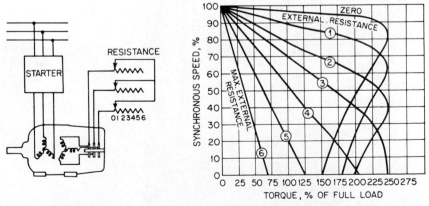

FIG. 3 Typical speed-torque characteristics of a wound-rotor induction motor. (Westinghouse Electric)

synchronous motor shaft (Fig. 4). The dc power from this generator is connected to the brushes of the synchronous motor slip rings.

3. *Motor-generator exciter*   The dc power for exciting the synchronous motor is generated by means of a remote motor-generator set operating from normal ac power, and the dc voltage from this motor-generator set is connected to the brushes of slip rings of the synchronous motor.

Another form of synchronous motor is known as the *brushless synchronous motor* (Fig. 5). As the name implies, this motor has its rotating field excited without the use of slip rings for connecting the external direct current to the motor field. The construction of this motor incorporates a shaft-connected ac generator. The field of the ac generator is physically stationary and connected to a source of dc voltage. The rotor of this ac generator is connected through a solid-state controlled rectifier mounted on the synchronous motor rotor and in turn connected to the synchronous motor field. This arrangement facilitates a connection between external excitation power and the rotating field of the synchronous motor through the air gap of the shaft-connected ac generator. The brushless synchronous motor has many advantages over the conventional slip-ring synchronous motor. Among these are the elimination of brushes and slip rings, which are high-maintenance items; the elimination of sparking devices, which are not permissible in certain atmospheres; and the use of static devices for field control, which are more reliable than conventional electromagnetic controls.

Synchronous motors are used for pump applications requiring larger horsepower ratings at lower speed conditions, as illustrated in Fig. 6. Also, they are used on applications where it is desired to have a high power factor or a power-factor-correction capability. Of less importance is the characteristic of the synchronous motor that it will always operate at synchronous speed (does not have a slip) regardless of load. Synchronous motors are started on their damper windings (the same as squirrel-cage induction motors), and when they have accelerated to within 5% of synchronous speed, the field is applied and the motor accelerates to synchronous speed (Fig. 7). Typical characteristic curves are shown in Fig. 8.

Direct-Current Motors   Dc motors are only occasionally used to drive pumps. This is the case when direct current is the only power available, as in ship, railway, aircraft, and emergency-battery operation.

There are three types of dc motors available (Fig. 9): shunt, series, and compound-connected.

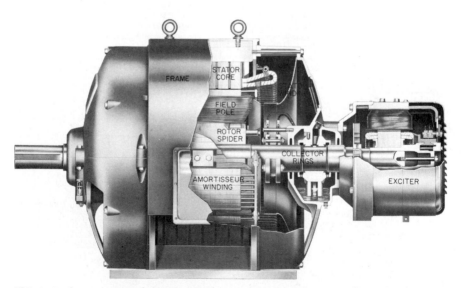

FIG. 4   Synchronous motor with direct-connected exciter. (Electric Machinery Manufacturing)

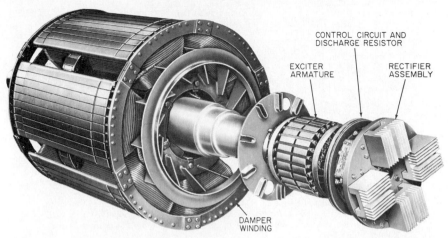

FIG. 5 Brushless synchronous motor with ac generator mounted on the shaft with rectifier and control devices. (Electric Machinery Manufacturing)

Larger horsepower ratings of shunt-wound dc motors are frequently qualified as "stabilized shunt-wound" motors and incorporate a series field similar to that of a compound wound motor. This is necessary to adjust the regulation of the shunt motor so as not to exhibit a rising speed-torque characteristic. It is important to be aware of the speed at which a dc motor will operate on pump applications because of pump performance guarantees.

When operating from constant dc voltage, dc motors can be made to provide up to a 4:1 speed range with a field rheostat for special applications. Many adjustable speed drives utilize a dc motor

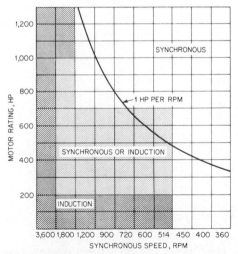

FIG. 6 Below 514 rpm, or powers greater than approximately 1 hp/rpm (0.746 kW/rpm), synchronous motors are a better selection than squirrel-cage induction motors because higher cost can generally be offset by higher power factor and efficiency. (From *Power* special report, "Motors," June 1969)

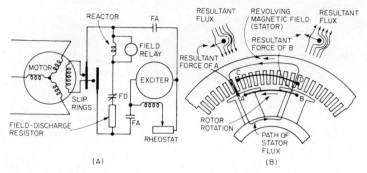

FIG. 7 Method of starting synchronous motor. (*A*) Typical field control (brush type) energizes the dc field to pull the rotor into synchronism as it comes up to speed. Field relay senses change in induced frequency as motor speeds up. (*B*) Damper winding is similar to a squirrel-cage rotor winding. It produces most of the starting torque, and the synchronous motor starts with essentially induction-motor characteristics. (From *Power* special report, "Motors," June 1969)

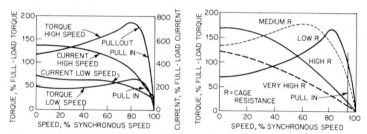

FIG. 8 Characteristics of synchronous motors depend on rotor design. Torque and current relations are influenced by synchronous speed. High-resistance cage produces high starting torque but low pull-in torque. (From *Power* special report, "Motors," June 1969)

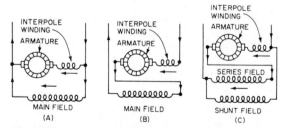

FIG. 9 Types of dc motors. (*A*) Shunt motor has field winding of many turns of fine wire connected in parallel with the armature circuit. The interpole winding aids commutation. (*B*) Series motor has field in series with the armature. Field has a few turns of heavy wire carrying full-motor current flowing in the armature. (*C*) Compound motor has both a shunt and a series field to combine characteristics of both shunt- and series-type motors in the same machine. (From *Power*, special report, "Motors," June 1969)

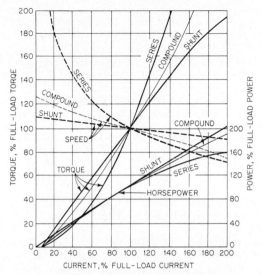

FIG. 10 Speed, torque, and power characteristics of dc motors. (From *Power* special report, "Motors," June 1969)

and different means to produce a varying dc voltage to operate this motor. These include variable-voltage motor-generator sets, solid-state silicon-controlled rectifier packages, and variable-ratio transformers with rectifiers. Figure 10 illustrates typical characteristics of the dc motors discussed.

MOTOR ENCLOSURES

Electric motors are manufactured with a variety of mechanical enclosure features to provide protection to the working parts for specific environmental conditions. Although these special enclosures are classified, they are not, for reasons of economy, available for all size motors. The electric motor industry has incorporated a number of specific enclosure classifications on standard designs of small- and medium-size motors.

The specific enclosure classifications provided for electric motors are as follows:

Open This enclosure permits passage of external cooling air over and around the windings and rotor of a motor and normally includes no special restrictions to ventilation other than that which is inherent in the mechanical parts of the motor.

Open dripproof This is an open machine with ventilating openings designed to permit satisfactory operation when liquids or solids fall on the machine at any angle up to 15° from the vertical. This construction includes mechanical baffling to prevent materials from entering the machine within these limits.

Splashproof This is an open machine with ventilating openings designed to permit satisfactory operation when liquids or solids fall *directly* on the machine or come toward the machine in a straight line at any angle up to 100° from the vertical.

Guarded enclosure This provision limits the size of the ventilating opening to prevent accidental contact with the operating parts of the motor other than the shaft.

Semiguarded This is an open construction where some of the ventilating openings are guarded and the remaining openings are left open.

Open, externally ventilated. This motor is ventilated by a separate motor-driven blower mounted on the motor enclosure (piggyback construction).

Open, pipe-ventilated   This motor is equipped to accommodate an air-inlet duct or pipe for accepting cooling air from a location remote from the motor. Air is circulated in the motor either by its own internal-blower parts or by an external blower, in which case the motor is said to be *forced-ventilated.*

Weather-protected type I   This is an open-construction motor with ventilating openings to minimize the entrance of rain, snow, and airborne particles to the motor electrical parts, and with the openings arranged to prevent the passage of a ¾-in (19-mm) round rod.

Weather-protected type II   This construction has the same features as the type I machine, but in addition the intake and discharge ventilating passages are designed so that high-velocity air and airborne particles blown into the machine by storms can be discharged without entering the internal ventilating passages of the motor leading to the electrical parts. The ventilating passages leading to the electrical parts of the motor are provided with baffles or other features to allow at least three abrupt changes of at least 90° for the ventilating air. The intake air path and openings are proportioned to maintain a maximum of 600 ft/min (3 m/s) velocity of the entering air.

Totally enclosed   This motor is designed without air openings, and so there is no free exchange of air between the inside and outside of the motor frame; the construction is not liquid- or airtight.

Totally enclosed, fan-cooled   This totally enclosed motor is equipped with an external fan operating on the motor shaft to circulate external air over the outside of the motor.

Explosionproof   This totally enclosed motor is designed to withstand an internal explosion of gas or vapor and constructed to prevent ignition by the internal explosion of gases or vapors outside the motor.

Totally enclosed, pipe-ventilated   This motor enclosure is similar to the open, pipe-ventilated motor except that it is equipped to accept outlet ducts or pipes in addition to inlet ducts or pipes.

Totally enclosed, water-cooled   This is a totally enclosed motor cooled by water passages or conductors internal to the motor frame.

Totally enclosed, water-to-air heat exchange   This is a totally enclosed motor equipped with a water-to-air heat exchanger in a closed, recirculating air loop through the motor. Air is circulated through the heat exchanger and motor by integral fans or fans separate from the rotor shaft and powered by a separate motor.

Totally enclosed, air-to-air heat exchange   This motor is similar to the water-to-air heat exchanger motor, except external air is used to remove the heat from the heat exchanger instead of water.

Submersible   This totally enclosed motor is equipped with sealing features to permit operation while submerged in a specified medium at a specified depth.

Environmental Factors   The environmental conditions of the pump application dictate the type of motor enclosures to be used. A brief set of rules for selecting motor enclosures follows.

Dripproof   For installation in nonhazardous, reasonably clean surroundings free of any abrasive or conducting dust and chemical fumes. Moderate amounts of moisture or dust and falling particles or liquids can be tolerated.

Mill and chemical motor (to frame 449T)   For installation in nonhazardous, high-humidity, or chemical applications free of clogging materials, metal dust, or chips and/or where hosing down or severe splashing is encountered.

Totally enclosed, nonventilated or fan-cooled   For installation in nonhazardous atmospheres containing abrasive or conducting dusts, high concentrations of chemical or oil vapors, and/or where hosing down or severe splashing is encountered.

Totally enclosed, explosionproof   For installation in hazardous atmospheres containing:

> *Class I, Group D*   Acetone, acrylonitrile, alcohol, ammonia, benzine, benzol, butane,

dichloride, ethylene, gasoline, hexane, lacquer-solvent vapors, naphtha, natural gas, propane, propylene, styrene, vinyl acetate, vinyl chloride, or xylenes

Class II, Group G   Flour, starch, or grain dust

Class II, Group F   Carbon black, coal, or coke dust

Class II, Group E   Metal dust including magnesium and aluminum or their commercial alloys.

Note   Under Class 1 only, there are two divisions which allow some latitude on motor selection. Generally, Class 1, Division 1 locations are those in which the atmosphere is or may be hazardous under normal operating conditions, including locations which can become hazardous during normal maintenance. An explosionproof motor is mandatory for Division 1 locations. Class 1, Division 2 refers to locations where the atmosphere may become hazardous only under abnormal or unusual conditions (breaking of a pipe, for example). In general, motor in a standard enclosure can be installed in Division 2 locations if the motor has no normally sparking parts. Thus, open or standard totally enclosed squirrel-cage motors are acceptable, but motors with open slip rings or commutators (wound rotor, synchronous or dc) are not allowed unless the commutators or slip rings are in an explosionproof enclosure.

BEARINGS AND LUBRICATION

Motors are generally available with oil-lubricated sleeve bearings or antifriction bearings.

Sleeve bearings in which the bearings are lubricated with an oil-impregnated wick are used on motors rated to approximately 1 hp (0.75 kW). Sleeve-bearing motors larger than 1 hp (0.75 kW) are ring-lubricated. Lubricating oil is drawn up from the bearing sump to the bearing by a ring that rolls over the top of the motor shaft as the shaft rotates. Larger motors having bearing heat losses that cannot be dissipated directly may require the use of a forced-lubrication system wherein oil is pumped into the bearing and allowed to recirculate through a heat exchanger pump. The oil delivered to each bearing is metered to provide only the required amount. A lubrication system composed of heat exchanger, sump, and pump is normally common to a number of bearings, rather than having a single lubricating pump for each bearing. Other types of bearings must be used in place of or in addition to sleeve bearings when thrust loads are present. Smaller sleeve bearings are in the form of a cylindrical shell and are made of bronze or babbitt metal. Larger sleeve bearings are usually split on a horizontal centerline, allowing easy assembly and disassembly. The bearing housing is also split on the horizontal centerline and held together with bolts between the top and bottom halves.

Ball-type antifriction bearings are used on motors where thrust forces are present and where motors will not operate in a horizontal position. Ball bearings are usually grease-lubricated and can be operated for prolonged periods of time between lubrications. Ball-type antifriction bearings are standard on NEMA size motors. More highly loaded ball bearing applications may require oil lubrication with a pump system similar to that used with sleeve bearings.

Roller-type antifriction bearings are used where very high radial loads are present. These bearings are not usually used on motors and cannot accept thrust loads.

Kingsbury thrust bearings are used where very high thrust loads must be accommodated by the motor. The Kingsbury bearing consists of a series of rotating segments that ride on a stationary bearing ring. This type of bearing is oil-lubricated with either a sump or a pump system, depending on size and rating.

MOTOR INSULATION

Classes   Insulation is used in a motor to electrically insulate the windings from the mechanical parts of the motor as well as to insulate the spaces between the turns of the coil winding. Also considered to be insulation are those materials used to secure the windings and to make them rigid and impervious to ambient conditions. Consequently, motor insulation is a complex system utilizing many different materials and parts to effectively insulate the windings. The parts considered in an insulation system include slot cells, phase barriers, conductor insulation, slot wedges, end-turn supports, tie material, and winding impregnation material.

There are four basic classes of insulating materials currently recognized by the motor industry. Each differs according to its physical properties and can withstand a certain maximum operating temperature (frequently termed *total temperature* or *hot-spot temperature*) and provide a practical and useful insulation life. The insulation classes and their maximum operating temperatures are

Class A	90°C
Class B	130°C
Class F	155°C
Class H	180°C

Those factors which contribute to the maximum operating temperature of a motor insulation system are the ambient temperature, the temperature rise in the motor winding caused by motor losses, and any overload allowance designed into the motor (service factor).

The current standard for motors within the range of NEMA ratings (frames 140T to 449T) requires nameplate marking for the maximum allowance ambient temperature, the power rating, the associated line current needed to develop this power, the class of insulation used, and the service factor provided. Motors larger and smaller than NEMA frame sizes have nameplate marking for maximum allowable ambient temperature, temperature rise in degrees Celsius either by thermometer or resistance measurement, power rating, line current needed to develop this rating, and any service factor provided.

In addition to ambient temperature, there are several additional environmental conditions that must be considered when applying an electric motor. In applications where chemical fume or moisture levels are abnormal and can cause decomposition of an insulation system, standard insulation will be inadequate. These applications require a motor with a premium insulation system that will incorporate highly resistant components and may include special impregnation techniques. Chemical fumes and moisture can also be destructive to the mechanical parts of a motor, and special protective treatment should be provided to these parts. NEMA frame-size motors have a special motor for chemical industry applications which has standard features to resist these environmental factors.

Applications with excessive vibration can destroy a winding and damage the mechanical parts of a motor. In such cases, it is advisable to provide (1) extra treatment for the winding to ensure that it is rigid and will not vibrate and chafe the insulating materials and (2) a mechanical construction that will have the strength to withstand the above.

If abrasive dust is present, the motor insulation should be protected with a resilient surface coating to withstand the impact of the abrasive particles.

Since all insulation systems employ components that can in some degree support fungus growth in tropical locations, motors applied in such areas should incorporate fungus-proofing treatment on the insulation.

Obviously, applications exhibiting a combination of any or all of the environmental conditions discussed should have special protection for each condition.

COUPLING METHODS FOR PUMP APPLICATIONS

Direct Coupling   Pumps are usually directly coupled to motors, and where the pump is not close-coupled, it is usually coupled by means of a flexible coupling. The use of flexible coupling permits minor misalignment (angular and parallel) between motor and pump shafts. However, most flexible couplings tend to become rigid when they are transmitting torque and therefore can impart thrust to the motor bearing if this thrust should prevail when the motor and pump are started. Ball bearing motors have reasonable thrust capacity and are therefore capable of accepting these thrust components.

Sleeve bearings, on the other hand, have a negligible thrust capacity unless a thrust bearing is specifically provided in the motor (ball thrust, Kingsbury bearing, etc.), and thus sleeve bearing motors must be protected from thrust components. Since a pump normally has a negligible end-float motion, it is customary to position the motor rotor halfway in its end-float capability and couple the motor and the pump with a limited-end-float flexible coupling that will permit the sum of the coupling end float and pump end float to be approximately 50% of the motor end float. This will ensure that all factors involved will never permit the motor shaft to reach either extreme shaft position and cause thrust on the motor bearings. See also Subsec. 6.3.1.

Larger pumps and motors can require torque values that make it necessary to use solid couplings. Extreme care must be exercised in these applications to ensure a good, stable alignment that establishes the motor in its mid-float position. If the motor has antifriction bearings (ball bearings), it will be necessary to have the outer races free-floating in their housings to prevent preloading the bearings.

Close-Coupling Close-coupled pumps have become very popular. In this arrangement, no coupling is provided between the pump and motor shafts and the pump housing is flange-mounted between close-tolerance fits on both the motor and the pump flanges. The pump impeller is mounted directly on the motor shaft. Care must be taken in this arrangement to ensure that the motor shaft runout or axial movement plus machine tolerances do not cause interference between the pump housing and its rotor. The motor shaft material must be compatible with the fluid being pumped, and if the pump impeller is held in place by a nut, the threat must respect the rotation of the motor. High-pressure close-coupled pumps of a nonbalanced design can cause excessive shaft thrust, which may be incorporated in the motor bearing capacity. It is always good practice with close-coupled pumps to provide some form of flinger on the motor shaft to prevent liquids that leak past the pump seal from entering the motor bearing.

Flanged Motors Flanged motors allow an easy means of aligning pump housings with motors. This construction is usually in the form of a vertical mounting in which the motor is set on top of the pump and the pump supports the motor weight. The pump and motor shafts are normally coupled, and those comments made under the subject of direct-coupling methods are applicable. Also, as in the case of coupled pumps, this construction permits thrust forces that must be considered when selecting a motor if the pump does not have a thrust bearing.

A further extension of flange motors includes the vertical hollow-shaft motor, and with this design a variable length of shafting connects the pump and motor. The pump shaft passes through the center of the motor bore, and the motor torque is imparted to the pump shaft by a suitable coupling at the top of the motor. The weight of the shaft and the pump impeller and the force of the hydraulic thrust are assumed by the motor bearings.

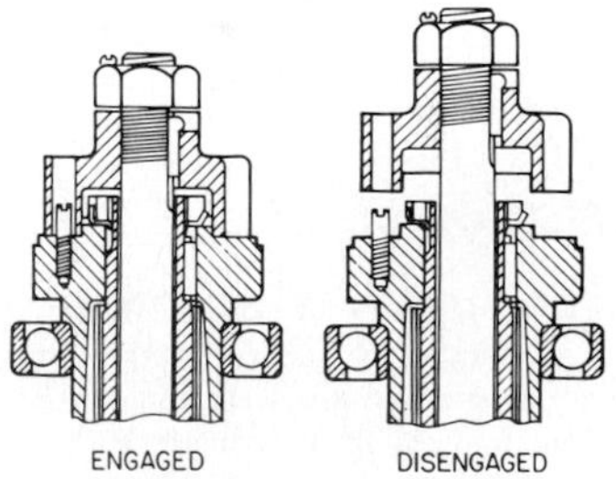

The coupling on the motor can be made a

ENGAGED   DISENGAGED

FIG. 11 Self-release coupling connecting pump head shaft to hollow shaft of vertical motor disengages as a result of pump shaft couplings unscrewing. (U.S. Electrical Motors)

self-release coupling (Fig. 11) to prevent the motor from delivering torque to the shaft in the event the motor is started in the wrong direction and to prevent reversed motor rotation from unscrewing the threaded joints between lengths of pump shafting.

Another modification to a coupling is a *nonreverse ratchet* (Fig. 12), which prevents the remaining head of liquid in a pump from rotating the pump in the reverse direction when the pump is stopped. This prevents possible overspeeding of the pump and motor when the pump is connected to a large reservoir and most of the total pump head is static. This also prevents a pump with a long discharge column pipe from running in reverse with no liquid in the upper portion of the pipe to lubricate the line shaft bearings. Starting a pump capable of back spinning is also prevented.

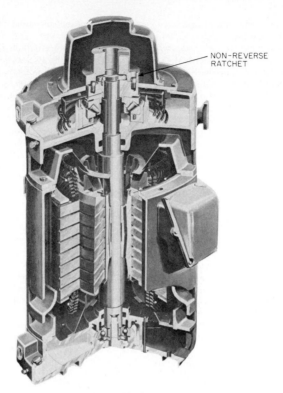

NON-REVERSE
RATCHET

FIG. 12   Section of vertical hollow shaft motor showing nonreverse ratchet. Spring-loaded pins ride on ratchet plate in one direction only. (General Electric)

PERFORMANCE

Motors are designed to produce their rated power, torque, and speed at a specific line voltage, line frequency, and ambient temperature. The motor will also operate at a specific efficiency and power factor when all of these conditions are met. The normal operating conditions of a motor are stipulated on its nameplate with values for power, speed, ambient temperature, and frequency. If the operating conditions are different from the nameplate ratings, the motor performance will be altered.

Voltage   Ac motors are designed to operate satisfactorily at their nameplate voltage rating with a ±10% variation from nameplate voltage when operating at rated nameplate frequency. This dictates that the motor will develop rated power and speed to a pump and will operate at a safe insulation temperature over the range of voltage. The motor torque will vary directly with the square of the applied voltage divided by the nameplate voltage. This affects the peak torque of the motor and will cause the motor speed-torque curve as shown in Fig. 2 to be altered. Within the ±10% voltage band, a motor can be expected to accelerate and operate a pump safely and continuously. A motor should never be expected to operate continuously beyond the ±10% band. If the voltage varies more than ±10%, the pump and motor may not operate satisfactorily.

For example, assume a pump is operated by a NEMA design B motor (Fig. 2) that will produce 200% pull-out torque at rated voltage. If the line voltage were to fall to 70% of the rated nameplate voltage, the motor would produce only 49% of its peak torque value. The pull-out torque of the motor would then become 0.49 × 200 = 98% of rated torque. It then becomes questionable whether the motor will be able to sustain the pump load, and the motor can be expected to lose speed, stall, or become overloaded.

In a similar sense, a motor may be unable to accelerate a pump if low line voltage exists. In the example previously discussed, this same motor develops 150% of rated torque when started at zero speed and rated voltage. If the line voltage is again 70% of nameplate voltage, the motor will develop $0.49 \times 150 = 73\%$ of rated torque. This may be a problem with certain types of pumps, such as a constant displacement pump. It is conceivable that this would not be a problem in starting a centrifugal pump because of its square-law speed-torque characteristics. If the motor voltage never increased beyond 70%, the centrifugal pump would not reach normal operating speed. An exception to this rule is the commutating ac motor, for which a $\pm 6\%$ voltage variation is allowable.

Varying motor voltage from nameplate rated voltage will also affect the motor operating speed, power factor, and efficiency established for rated voltage and load. An induction motor will operate at several rpm faster than nameplate speed at 10% over voltage and several rpm below nameplate speed at 10% under voltage. Synchronous motors, on the other hand, will not change speed over a $\pm 10\%$ voltage range.

Dc motors can also be operated over a $\pm 10\%$ voltage range from nameplate rated voltage. However, it should be recognized that different types of dc motors will have different speed and torque characteristics over the voltage range. This should be taken into account when meeting pump performance requirements.

Frequency   Ac motors will operate satisfactorily at rated load and voltage with a frequency variation up to $\pm 5\%$ from rated nameplate frequency. However, the speed of the motor will vary with the applied frequency. Synchronous motor speed will vary directly with applied frequency, and induction motors will vary almost directly.

When an ac motor must operate with a varying voltage and frequency, the *combined* variation of voltage and frequency must not be more than $\pm 10\%$ from rated nameplate voltage and frequency, provided the frequency variation does not exceed $\pm 5\%$ from rated nameplate frequency.

Frequency variation from the motor nameplate frequency will cause motors to operate at a power factor and efficiency other than those established for rated frequency.

Speed and Speed Range   Synchronous and induction motors are intended basically to operate at one specific speed. Some special applications have been designed using both synchronous and induction motors with adjustable-frequency power sources to develop an adjustable-speed characteristic, but these are unique and special. Adjustable-speed applications are usually satisfied by using one of the adjustable-speed drive packages that are available at this time. These drives are described in Section 6.2.

Pumps requiring discrete speed adjustment can be operated with motors that are designed for multiple-speed operation, and these motors may be selected from a variety of squirrel-cage induction motors. Speed ranges that are even multiples can be obtained with a squirrel-cage induction motor equipped with a reconnectable single winding. Applications requiring a speed range of an uneven ratio normally require a two-winding squirrel-cage induction motor. However, larger horsepower motors requiring speeds of an uneven ratio can be designed with a single-winding incorporated pole-amplitude modulation. This design depends upon a concept which establishes a modulation of the frequency of the motor flux.

Acceleration   A motor must be capable of accelerating as well as driving a pump at rated speed and power. The acceleration can be analyzed by examining a typical pump-motor combination involving a centrifugal pump driven by a six-pole squirrel-cage induction motor rated 10 hp (7.46 kW) with NEMA design B torque characteristics. The curve in Fig. 13 demonstrates this combination when the pump is loaded and the motor is operating at nameplate frequency and voltage. It will be noted that the torque produced by the motor at any speed is greater than the torque required by the pump at that speed. The excess torque at any speed is available to accelerate the mass (WK^2) of the entire motor and pump-rotating parts. The time to accelerate the pump is equal to

in USCS units
$$t = \frac{WK^2 \times \Delta\text{rpm}}{308T}$$

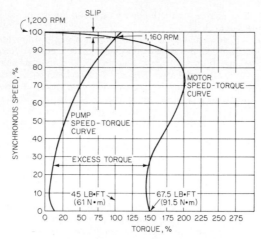

FIG. 13 Typical speed-torque characteristic curves for a centrifugal pump and a squirrel-cage induction, NEMA design B motor. Excess torque accelerates pump.

in SI units
$$t = \frac{MK^2 \times \Delta \mathrm{rpm}}{9.55\,T}$$

where t = time, s
WK^2 = total weight (force) moment of inertia, lb·ft^2
Δrpm = change in speed
T = torque, lb·ft (N·m)
MK^2 = total mass moment of inertia, kg·m^2

Because the difference in torque is not uniform over the speed range, the curve can be analyzed by assuming discrete changes in speed and an average torque over this change in speed. A time period can be calculated for each discrete speed change, and all time values can be totaled to obtain the complete acceleration time (see also Subsec. 2.3.1 and Sec. 8.1).

Increasing the inertia of the pump, the operating speed of the pump and motor, or the torque required by the pump at any speed will result in a longer acceleration time, which may not be possible for the motor. Each motor can operate at a reduced speed and at a torque in excess of rated torque for a given time. Beyond this time, the windings and/or rotor can be damaged.

When an application is analyzed and found to present an acceleration-time problem, consideration should be given to

1. Unloading the pump during acceleration
2. Reducing the inertia of the rotating parts
3. Applying a motor which has greater capability for acceleration

When larger pumps are started and a reduction in line voltage is realized because of the high starting current, the motor torque is reduced by the square of the voltage ratio. Naturally, the acceleration time is greater because of the reduced torque produced by the motor. This will not hamper the ability of the motor to accelerate the pump as long as the motor develops more torque than is required to drive the pump at any speed over the accelerating range.

A similar analysis can be made for synchronous motors once the accelerating speed–torque curve for the motor is known. It should be recognized that a synchronous motor operates as a squirrel-cage induction motor up to the moment of synchronization. At that time the synchronous motor must have an additional capability of synchronizing torque, frequently called *pull-in torque*, to accelerate the motor from subsynchronous to synchronous speed. If the motor cannot develop this synchronous torque, it will "pull out" and usually be shut down by a control function. In the case of pump applications, an economic advantage can be realized by unloading the pump during acceleration to reduce the acceleration and synchronizing torque required by the motor.

Attention should always be given to the breakaway torque that is required to start the pump from zero speed. This is particularly important with constant displacement pumps where the pump will operate at a constant torque over the entire accelerating speed range.

Service Factor Motors are available with service factor ratings which range from 1.0 to as high as 1.5. A service factor implies that a motor has a built-in thermal capacity to operate at the nameplate power times the service factor stamped on the nameplate. It should be noted, however, that when the motor is operated at the service factor power, the motor will operate at what is termed a *safe temperature*. This means that the motor will operate at a total temperature that is greater than the temperature for a motor designed for the same power with a 1.0 service factor. Consequently, it is not advisable to apply a motor with a service factor larger than 1.0 where the continuous power requirements will be greater than the normal power.

A service factor rating on a motor is to provide an increased power capacity beyond nominal nameplate capacity for occasional overload conditions. Also, the speed-torque characteristics are related to the nominal power rating and not the service factor power.

Efficiency Motors are designed to operate with an efficiency expressed in percent at rated voltage, frequency, and power. Efficiency is defined as

$$\text{Efficiency } \% = \frac{\text{shaft output power} \times 100}{\text{electrical input power}}$$

The efficiency of a given motor design will vary slightly from unit to unit because of manufacturing tolerances and variations in materials. For this reason, guaranteed efficiencies are usually lower than actual efficiencies. When motors are operated at reduced powers, the tendency is for the efficiency to decrease.

Several other factors have an effect on motor efficiency. Increasing the applied voltage and operating a motor at its rated power will increase efficiency very slightly, while decreasing applied voltage will decrease the efficiency noticeably. Also, increasing the frequency will cause a very slight increase in efficiency, and decreasing the applied frequency will cause a slight decrease in efficiency.

Dynamometers are used to determine efficiency for small motor ratings—up to approximately 500 hp (373 kW)—and standard methods and formulas are used for calculating efficiency for large motors. For the latter, efficiency values can vary, depending upon which "standard" procedure is used. NEMA has established one method, but some suppliers use different methods, depending upon their national standards or established practices.

Higher efficiencies can be designed into a motor by selecting materials and proportions that reduce losses, but such design choices could make the motor more costly. The current energy crisis has caused many manufacturers to offer high-efficiency motors that are 0.5 to 1% more efficient than standard motors.

Motor operating efficiency in a typical application often may not be that described by the manufacturer. Quoted efficiencies are always at rated power output, rated frequency, and rated voltage. However, it is almost universal in application that a motor is oversized for its applied load and frequently operates at other-than-normal voltage. Both of these variables can greatly reduce the efficiency of a motor.

Power Factor The power factor of a motor is expressed as

$$PF = 100 \cos \theta$$

where θ is the angle between voltage and current at motor terminals (leading or lagging).

The power factor at which a motor operates is dependent on the design of the motor and is established at rated voltage, frequency, and power output.

For induction motors, the power factor can never be 100% leading. A number of factors will influence the power factor of an induction motor:

Condition	Effect on power factor
Increase applied voltage	Decrease
Decrease applied voltage	Increase
Increase load	Increase
Decrease load	Decrease
Increase applied frequency	Slight increase
Decrease applied frequency	Slight increase

In synchronous motors, it is usual to use two varieties of motors, the 100 (unity) and 80% leading motors. The power factor of a synchronous motor operated at rated voltage and frequency is fixed by its field excitation and its power output. At a given power output, the power factor can be adjusted over a range by adjusting the field excitation. Increasing the field excitation will cause the motor to operate at a more leading power factor, and, conversely, reducing the field excitation will make the power factor lag.

Varying the power output of a synchronous motor with a constant field excitation will vary the operating power factor. A decrease in power output will cause a more leading power factor; conversely, increasing the power output will induce operation at a lagging power factor. Consequently, to operate the motor at rated power factor with a varying output power, it is necessary to adjust the field excitation. However, this is not normally done because a synchronous motor is frequently used for improving the power factor and the more leading power factor is used to accommodate power factor improvement. When a synchronous motor is overloaded and operates at a more lagging power factor, it is not usual to increase the excitation beyond its rated excitation because of the extra heating this will develop in the motor. In this case, the more lagging power factor is simply accepted.

TYPES OF CONTROLS

Alternating-Current Motor Starters

MANUAL STARTERS   Manual motor starters are designed to provide positive overload protection and start and stop control of single-phase and polyphase motors. A single manual operating handle provides the control and indication of "on," "off," and "tripped" states.

MAGNETIC STARTERS   Magnetic motor starters are designed to control a motor by incorporating a magnetically operated contactor to apply power to the motor terminals. An overload relay is incorporated to protect the motor from overloading. Magnetic starters are available for reversing and nonreversing service and are also made as noncombination or combination types. The noncombination starter combines only the motor contactor and overload relay. The combination starter combines these parts along with either a circuit breaker or a fused switch to provide short-circuit protection.

REDUCED-VOLTAGE STARTERS   Reduced-voltage starters are available in several types and are basically magnetic starters with additional features to provide reduced voltage, which in turn provides for reduced motor-starting current and/or torque. These starters include the following types:

Primary-resistor Starters, sometimes known as *cushion-type* starters, will reduce the motor torque and starting inrush current to produce a smooth, cushioned acceleration with closed transition. Although not as efficient as other methods of reduced-voltage starting, primary-resistor starters are ideally suited to applications where reduction of starting torque is of prime consideration. A typical diagram for this type of starter is shown in Fig. 14.

Autotransformer Starters are the most widely used reduced-voltage starters because of their efficiency and flexibility. All power taken from the line, except transformer losses, is transmitted to

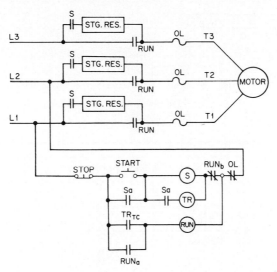

FIG. 14 Typical primary-resistor, reduced-voltage starter wiring diagram. (Westinghouse Electric)

the motor to accelerate the load. Taps on the transformer allow adjustment of the starting torque and inrush to meet the requirements of most applications. The following characteristics are produced by the three voltage taps:

Tap, %	Starting torque, % locked torque	Line inrush, % locked current
50	25	28
65	42	45
80	64	67

A typical diagram for this type of starter is shown in Fig. 15.

Part-winding Starting provides convenient, economical, one-step acceleration at reduced current where the power company specifies a maximum or limits the increments of current drawn from the line. These starters can be used with standard dual-voltage motors on the lower voltage and with special part-winding motors designed for any voltage. When used with standard dual-voltage motors, it should be established that the torque produced by the first half-winding will accelerate the load sufficiently so as not to produce a second undesirable inrush when the second half-winding is connected to the line. Most motors will produce a starting torque equal to between one-half and two-thirds of NEMA standard values with half the winding energized and draw about two-thirds of normal line-current inrush. A typical diagram is shown in Fig. 16.

Star-delta Starters have been applied extensively to starting motors driving high-inertia loads with resulting long acceleration times. They are not, however, limited to this application. When 6 or 12 lead delta-connected motors are started star-connected, approximately 58% of full-line voltage is applied to each winding and the motor develops 33% of full-voltage starting torque and draws 33% of normal locked rotor current from the line. When the motor has accelerated, it is reconnected for normal delta operation. A typical diagram is shown in Fig. 17.

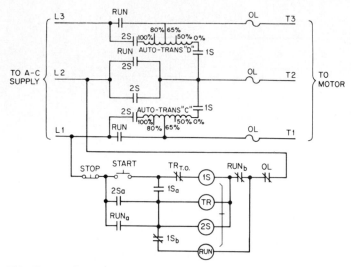

FIG. 15 Typical autotransformer, reduced voltage starter wiring diagram. (Westinghouse Electric)

WOUND-ROTOR MOTOR STARTERS   These magnetic motor starters are used for starting, accelerating, and controlling the speed of wound-rotor motors. The primary control includes overload protection and low-voltage protection or low-voltage release, depending on the type of pilot device. Disconnect switches, circuit breakers, and reversing can be added to the primary circuit when required. Reversing starters are not designed for plugging.

The secondary circuit contains the NEMA recommended number of accelerating or running contactors and resistors to allow approximately 150% of motor full-load torque on first point of acceleration. Additional accelerating points can be added for high-inertia loads or exceptionally

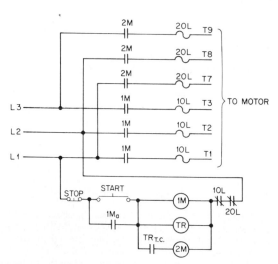

FIG. 16 Typical part-winding starter wiring diagram. (Westinghouse Electric)

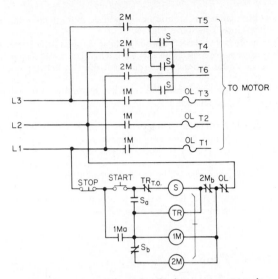

FIG. 17   Typical star-delta, reduced-voltage starter wiring diagram. (Westinghouse Electric)

smooth starts. Adjustable timing relays permit field adjustment. Standard starting duty NEMA 135 resistors allow 10 s starting out of every 80 s. A typical diagram is shown in Fig. 18.

SYNCHRONOUS-MOTOR STARTER   Synchronous, magnetic, full-voltage starters provide reliable automatic starting of synchronous motors. They can be used whenever full-voltage starting is permissible. Automatic synchronization is provided by field relay, which assures application of the field at the proper motor speed and at a favorable angular position of stator and rotor poles. As a result,

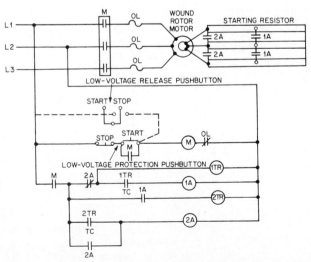

FIG. 18   Typical wound-rotor motor starter wiring diagram. (Westinghouse Electric)

line disturbance resulting from synchronization is reduced and effective motor pull-in torque is increased. A typical diagram is shown in Fig. 19.

Brushless synchronous-motor starters require special consideration inasmuch as all brushless synchronous motors are not constructed in the same way. The usual starter incorporates a low-power adjustable dc excitation source to energize and control the output of an integral shaft-connected exciter.

In addition, a pull-out relay and a timing relay are incorporated to initiate synchronization and stop the motor in event of pull-out.

Direct-Current Motor Starters
Direct-current motor starters are designed to apply normal voltage to the motor field and, by means of a resistor, reduce voltage to the armature. Timing relays and contractors progressively short out the resistor until full voltage is on the armature.

If the motor is equipped with a field rheostat for field-range speed adjustment, the starter will apply the preset field voltage as adjusted by the field rheostat after full voltage is applied to the armature.

These starters incorporate a field failure relay to deenergize in the event of field failure and overload relays to protect the motor against overspeed.

SEALLESS PUMP MOTORS

Various centrifugal pump designs are available which require no shaft sealing, i.e., no packing or mechanical seal. These pumps are completely leakproof, and some are submersible. Such pumps are used when leakage cannot be tolerated or when pumping conditions such as pressure and/or temperature make conventional sealing difficult, if not impossible.

The shaft seal is eliminated by joining the pump and motor housings together to create a single leakproof unit. The impeller and motor rotor are mounted on a single shaft. Most designs permit the pumped liquid to circulate through the motor rotor and motor bearings. However, there are

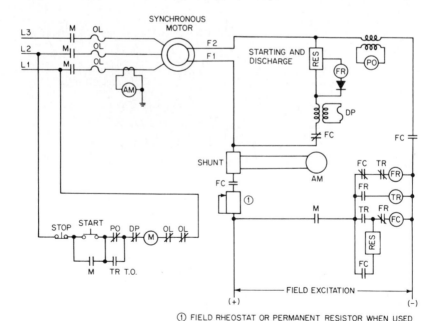

① FIELD RHEOSTAT OR PERMANENT RESISTOR WHEN USED

FIG. 19 Typical synchronous-motor starter wiring diagram. (Westinghouse Electric)

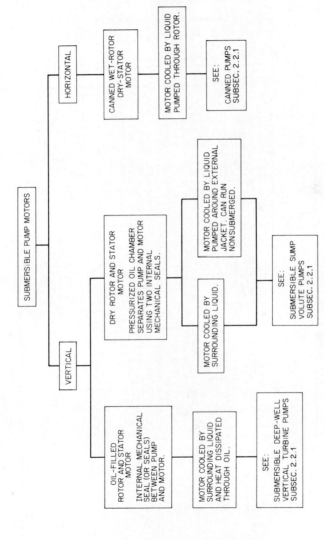

FIG. 20 Integral pump motors for sealless centrifugal pumps.

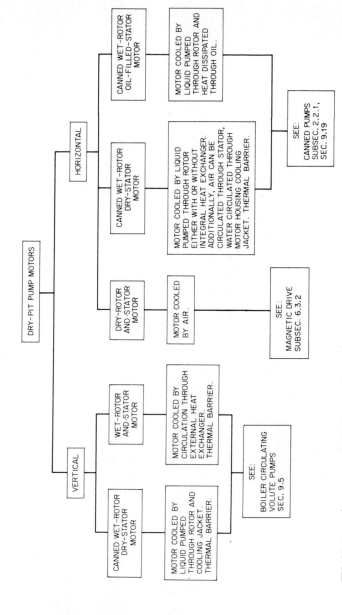

FIG. 20 Integral pump motors for sealless centrifugal pumps. (*cont.*).

6.24

designs where either circulation is also through the stator or the stator is sealed and filled with dielectric oil.

Another popular method used for smaller pumps is to separate the pump shaft completely from the motor shaft. A permanent magnet coupling is provided to transmit the motor torque to the pump shaft through a nonmagnetic section of the pump volute. The pump and motor are integral in these designs. See Subsec. 6.3.2.

Figure 20 classifies various motor designs and references other places in the handbook where the special features and applications of these units are described in more detail.

6.1.2
STEAM TURBINES

WALLACE L. BERGERON

DEFINITIONS

"A steam turbine may be defined as a form of heat engine in which the energy of the steam is transformed into kinetic energy by means of expansion through nozzles, and the kinetic energy of the resulting jet is in turn converted into force doing work on rings of blading mounted on a rotating part."[1]

This definition may be restated:

"A steam turbine is a prime mover which converts the thermal energy of steam directly into mechanical energy of rotation."[2]

REASONS FOR USING STEAM TURBINES

Steam turbines are used to drive pumps for a variety of reasons:

1. The economical generation of steam often requires boiler steam pressures and temperatures that are considerably in excess of those at which the steam is utilized. Steam may also be used at two or more pressure levels in the same plant. Pressure reduction can be accomplished through valves, pressure-reducing stations, or use of a steam turbine.

Pressure reduction by using a steam turbine and thereby developing power to drive a pump permits lower utility costs. The incremental increase in steam flow and consequently in fuel costs for the same lower pressure steam heating load is in most instances less than the cost of purchased power for a motor-driven pump.

2. A pump driven by a steam turbine may be operated over a wide speed range, utilizing the turbine governor system or a separately controlled valve in the turbine or in the steam line to the turbine. Operation at variable speeds is an inherent characteristic of steam turbines and does not require the use of special speed-changing devices, as is the case with other prime movers.

The overall efficiency of the turbine and pump unit can be optimized by operating at reduced speeds and at the resultant reduced power ratings. Pump performance can be controlled by reducing the speed of the pump rather than throttling it. Although the turbine efficiency normally declines when operating at a reduced speed, the steam flow will still be less than when the pump is throttled.

Operation at reduced power but at constant speed is also permitted by the speed governor, which throttles the steam to the nozzles as the power is reduced. Efficiency may be improved by equipping the turbine with auxiliary steam valves that are closed for reduced power operation. Closing these valves reduces the available nozzle area and reduces the pressure drop across the governor valve.

When the turbine is operated with the auxiliary steam valve closed, the steam flow will approximate that for the same turbine designed for the reduced rating.

3. The use of a steam turbine driver permits the driven pump to operate essentially independent of the electric power or distribution system. The steam turbine is not affected by electric power stoppages or interruptions and is therefore ideal for critical pumping operations.

4. A turbine may be used as a secondary driver for a pump; it may also drive an independent standby or emergency pump. The particular plant design may not afford sufficient steam for the pump to be normally driven by the steam turbine. However, in the event of an electric power failure or power system disturbance, a steam turbine may be employed as a dual drive or to drive a separate pump to assure continued operation of the plant until the electric power system is again operable.

5. The steam turbine controls—governor system and overspeed trip system—are inherently sparkproof. Consequently steam turbines can be readily applied to drive centrifugal pumps in a wide variety of hazardous atmospheres without entailing additional cost for explosionproof or sparkproof construction.

6. Steam turbines can normally be readily altered to accommodate an increase in rating for increased pump output or for new pump applications. This inherent flexibility of a steam turbine also permits it to be readily altered to accommodate changes in the initial steam pressure and temperature and in the exhaust steam pressure at which the turbine operates.

7. Steam turbines have a starting, or breakaway, torque of approximately 150 to 180% of the rated torque. Additional starting torque can be readily furnished by designing the turbine for the additional required steam flow—and without reducing the efficiency at the normal operating rating by using an auxiliary steam valve. The additional starting torque can often be obtained without increasing the turbine frame size.

8. Steam turbines can be used to drive all types of pumps.

9. Steam turbines are inherently self-limiting with respect to the power developed. Special protective devices do not have to be furnished to prevent damage to the turbine because of overload conditions. The maximum power that can be developed by a turbine is a function of the flow areas provided in the design of the nozzle ring and governor valve. Application of a load greater than that which can be developed by the turbine causes the turbine to slow down to a speed at which the torque generated by the turbine matches that required by the pump.

10. When the pump application requires the driver to be designed with excess power or to permit operation of the pump "at the end of the curve," the steam turbine can be designed for the corresponding rating without reducing the turbine efficiency when operating at the normal rating. Closing an auxiliary steam valve furnished for operation at the normal rating preserves efficiency because the turbine governor valve is not throttling to obtain the lower rating.

11. With respect to the operation of the various types of pump drivers and their supporting systems, steam turbines afford minimum maintenance, low vibration, and a quiet installation.

TYPES OF STEAM TURBINES

Single-Stage Turbines   A single-stage steam turbine is one in which the conversion of the kinetic energy to mechanical work occurs with a single expansion of the steam in the turbine—from inlet steam pressure to exhaust steam pressure.

A single-stage turbine may have one or more rows of rotating buckets which absorb the velocity energy of the steam resulting from the single expansion of the steam.

Single-stage turbines are available in wheel diameters of 9 to 28 in (22 to 71 cm). The overall efficiency of a turbine for a particular operating speed and steam conditions is normally dependent on the wheel diameter. The efficiency will generally increase with an increase in wheel size, and therefore the steam rate will be less (for the more usual speeds and steam conditions).

The larger-wheel-diameter steam turbines can be furnished with more nozzles to provide increased steam flow capacity and consequently greater power capabilities. The larger-wheel-diameter turbines are therefore furnished with larger steam connections, valves, shafts, bearings, etc. Consequently the size of the turbine will generally increase with increases in power rating.

Multiple-Stage Turbines   A multiple-stage turbine is one in which the conversion of the energy occurs with two or more expansions of the steam in the turbine. The number of stages (steam expansions) is a function of three basic parameters: thermodynamics, mechanical design, and cost. The thermodynamic considerations include the available energy and speed. The mechanical considerations include speed, steam pressure, steam temperature, etc., most of which are material limits. Cost considerations include the number, type, and size of the stages; the number of governor-controlled valves; the cost of steam; and the number of years used as a basis for the cost evaluation.

The two factors generally used in selecting multistage turbines are initial cost and steam rate. Since these two factors are a function of the total number of stages, the application becomes a factor of stage selection. The initial cost increases with the number of stages, but the steam rate generally improves.

Multiple-stage turbines are normally used to drive pumps when the cost of steam or the available supply of steam requires turbine efficiencies greater than these available with a single-stage turbine, or when the steam flow required to develop the desired rating exceeds the capability of single-stage turbines.

Multiple-stage turbines can be furnished with a single or multiple governor valves. A single governor valve is often of the same design whether used in a single-stage or multiple-stage turbine and generally has the same maximum steam flow, pressure, and temperature parameters. Multiple valves are used when the parameters for a single valve are exceeded or to obtain improved efficiency, particularly at reduced power outputs.

Shaft Orientation   Some steam turbines, particularly single-stage turbines, can be furnished with vertical downward shaft extensions. The application of such turbines can require considerable coordination between the pump and the turbine manufacturer to assure an adequate thrust bearing in the turbine, shaft length and details, mounting flange dimensions, and even shaft runout.

Vertical shaft pumps are frequently driven by horizontal turbines through a right-angle speed-reduction gear unit.

Direct-Connected and Geared Turbines   Steam turbines can be directly connected to the pump shaft so that the turbine operatess at the pump speed or can drive the pump through a speed-reduction (and even speed-increase) gear unit, in order to permit the turbine to operate at a more efficient speed.

Turbine Stages   The two types of turbine stages are impulse and reaction. The turbines discussed in this subsection employ impulse stages because steam turbines driving pumps normally have impulse-type stages.

In the ideal impulse stage, the steam expands only in the fixed nozzles and the kinetic energy is transferred to the rotating buckets as the steam impinges on the buckets while flowing through the passages between them. The steam pressure is constant, and the steam velocity relative to the bucket decreases in the bucket passages.

In a reaction stage, the steam expands in both the fixed nozzles and the rotating buckets. The kinetic energy is transferred to the rotating buckets by the expansion of the steam in the passages between the buckets. The steam pressure decreases as the steam velocity relative to the buckets increases in the bucket passages.

In an impulse stage, the steam can exert an axial force on the buckets as it flows through the blade passages. While this force is usually referred to as a reaction, the use of the term does not imply a reaction-type stage.

The larger buckets used in the last stages of an impulse-type multistage turbine can be of a free-vortex design—twisted and tapered. Such a bucket is ideally subjected to a nearly pure impulse force at its root and a nearly pure reaction force at its tip, but, in reality, this bucket is a high reaction-design bucket compared with a normal impulse-stage bucket. A steam turbine stage with such a bucket design is still referred to as an impulse stage because the primary conversion of kinetic energy is by a reduction rather than an increase in relative steam velocity.

A reaction turbine has more stages than an impulse turbine for the same application because of the small amount of kinetic energy absorbed per stage, and requires a larger thrust bearing and/or a balancing piston because of the pressure drop across the moving blades. The small pressure drop per stage and the pressure drop across the moving blades require that the steam-leakage losses be minimized by elaborate sealing between the tips of the nozzle blades and the rotor, and the tips of the moving blades and the casing.

The small pressure differential across the rotating blades of an impulse stage results in smaller thrust bearings and no close blade-tip clearances. Consequently impulse turbines can be started more quickly without thermal-expansion damage, and their stage efficiencies remain relatively constant over the life of the turbine.

CONSTRUCTION DETAILS

Component Parts   The main components of a single-stage steam turbine are shown in Fig. 1.

Function and Operation—Single-Stage Turbine   The *steam chest* and the *casing* contain the steam furnished to the turbine, being connected to the higher-pressure steam supply line and the lower-pressure steam exhaust line, respectively. The steam chest, which is connected to the casing, houses the governor valve and the overspeed trip valve. The casing contains the rotor and the nozzles through which the steam is expanded and directed against the rotating buckets.

The *rotor* consists of the shaft and disk assemblies with buckets. The shaft extends beyond the casing and through the bearing cases. One end of the shaft is used for coupling to the driven pump. The other end serves the speed governor and the overspeed trip systems.

The *bearing cases* support the rotor and the assembled casing and steam chest. The bearing cases contain the journal bearings and the rotating oil seals, which prevent outward oil leakage

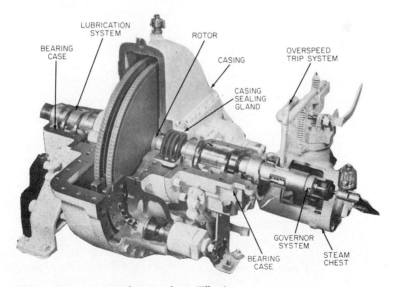

FIG. 1   Component parts of steam turbines. (Elliott)

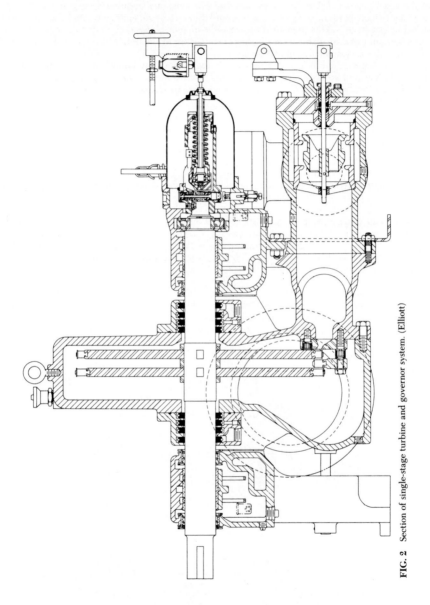

FIG. 2 Section of single-stage turbine and governor system. (Elliott)

and the entrance of water, dust, and steam. The steam end bearing case also contains the rotor positioning bearing and the rotating components of the overspeed trip system. An extension of the steam end bearing housing encloses the rotating components of the speed governor system.

The *casing sealing glands* seal the casing and the shaft with spring-backed segmental carbon rings (supplemented by a spring-backed labyrinth section for the higher exhaust steam pressures).

The *governor system* commonly consists of spring-opposed rotating weights, a steam valve, and an interconnecting linkage or servomotor system. Changes in the turbine inlet and exhaust steam conditions, and the power required by the pump will cause the turbine speed to change. The change in speed results in a repositioning of the rotating governor weights and subsequently of the governor valve.

The *overspeed trip system* usually consists of a spring-loaded pin or weight mounted in the turbine shaft or on a collar, a quick-closing valve which is separate from the governor valve, and interconnecting linkage. The centrifugal force created by rotation of the pin in the turbine shaft exceeds the spring loading at a preset speed. The resultant movement of the trip pin causes knife edges in the linkage to separate and permit the spring-loaded trip valve to close.

The trip valve may be closed by disengaging the knife edges manually, by an electric or pneumatic signal, by low oil pressure, or by high turbine exhaust steam pressure.

The two usual types of lubrication systems are oil-ring and pressure. The *oil-ring* lubrication system employs an oil ring(s) which rotates on the shaft with the lower portion submerged in the oil contained in the bearing case. The rotating ring(s) transfers oil from the oil reservoir to the turbine shaft journal bearing and rotor-locating bearing. The oil in the bearing case reservoirs is cooled by water flowing in cooling water chambers or tubular heat exchangers.

A *pressure* lubrication system consists of an oil pump driven from the turbine shaft, an oil reservoir, a tubular oil cooler, an oil filter, and interconnecting piping. Oil is supplied to the bearing cases under pressure. The oil rings may be retained in this system to provide oil to the bearings during startup and shutdown when the operating speed and bearing design permit.

Typical sectional drawings are shown in Figs. 2, 3, and 4.

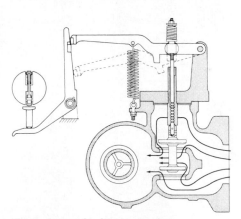

FIG. 3   Section of turbine overspeed trip system.

GOVERNORS AND CONTROLS

Governor systems are speed-sensitive control systems that are integral with the steam turbine. The turbine speed is controlled by varying the steam flow through the turbine by positioning the governor valve. Variations in the power required by the pump and changes in steam inlet or exhaust conditions alter the speed of the turbine, causing the governor system to respond to correct the operating speed.

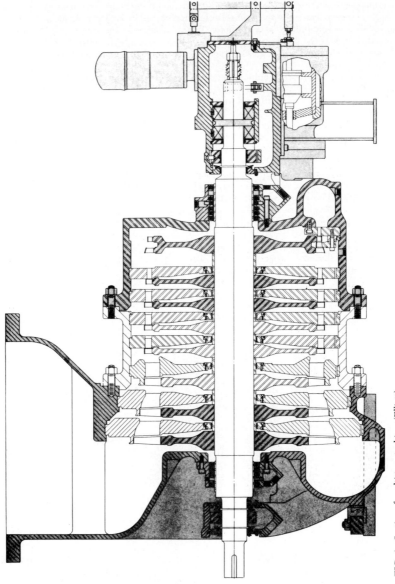

FIG. 4 Section of multistage turbine. (Elliott)

Control systems, unlike governor systems, are not directly speed-sensitive but respond to changes in pump or pump-system pressures and then reposition the turbine "governor" valve to maintain the preset pressure. Consequently, changes in turbine steam conditions or in the power required by the pump result in a repositioning of the turbine governor valve or of a separate steam valve only after the pressure being sensed by the controller has changed.

Even when a control system is furnished, a speed governor is also normally furnished. The speed governor is set for a speed slightly higher than the desired operating speed in order to function as a pre-emergency governor, that is, prevent the turbine from reaching the trip speed when the controller causes the turbine to operate at a speed above rated speed.

Governor systems are defined by their performance as follows:[2]

Class of governor system	Speed range, % (as specified)	Maximum speed regulation, %	Maximum speed variation, %, $\pm$	Maximum speed rise, %
A	10–65	10	0.75	13
B	10–80	6	0.50	7
C	10–80	4	0.25	7
D	10–90	0.50	0.25	7

Speed range is the percentage below rated speed for which the governor speed setting may be adjusted. For example, a turbine with 4000 rpm rated speed and a governor system having a 30% range can be operated at a minimum speed of 2800 rpm:

$$4000 - \frac{30 \times 4000}{100} = 2800$$

If the speed range had been specified as plus 5% and minus 25%, for example, the maximum and minimum speeds would be 4200 and 3000 rpm.

Steady-state *speed regulation* is the change in speed required for the governor system to close the governor valve when the load is *gradually* reduced from rated load to no load with turbine steam conditions constant. Regulation is always expressed as a percentage of rated speed and calculated as follows:

$$\frac{\text{No-load speed} - \text{rated speed}}{\text{Rated speed}} \times 100 = \% \text{ regulation}$$

Therefore

$$\text{No-load speed} = \text{rated speed} \left(1 + \frac{\% \text{ regulation}}{100} \right)$$

A turbine with 4000 rpm rated speed and equipped with a NEMA A speed governor system would have 4400 rpm maximum no-load speed:

$$4000 \times \left(1 + \frac{10}{100} \right) = 4400 \text{ rpm}$$

Consequently, whenever the turbine is developing less than rated power, the operating speed will be greater than rated speed, as required for the governor system to reposition the governor valve.

Speed variation, expressed as a percentage, is the total magnitude of the fluctuations from the set speed permitted by the governor system when the turbine is normally operating at rated speed, power, and steam conditions. This is the "insensitivity" of the governor system. The speed variation equation is

$$\frac{\text{Maximum speed} - \text{minimum speed}}{\text{Rated speed} \times 2} \times 100 = \pm \% \text{ speed variation}$$

A turbine with 4000 rpm rated speed and NEMA A governor system would have ± 30 rpm maximum speed variation:

$$4000 \times \frac{\pm 0.75}{100} = \pm 30 \text{ rpm}$$

Maximum speed rise represents the momentary increase in speed when the load is suddenly reduced from rated power to no-load power with the pump still coupled to the turbine shaft. Shortly after the sudden loss of load, the governor system will cause the turbine speed to be reduced to the no-load speed.

Maximum speed rise is also expressed as a percentage of rated speed and calculated as follows:

$$\frac{\text{Maximum speed} - \text{rated speed}}{\text{Rated speed}} \times 100 = \% \text{ speed rise}$$

Therefore

$$\text{Maximum speed} = \text{rated speed} + \frac{\text{rated speed} \times \% \text{ speed rise}}{100}$$

A turbine with 4000 rpm rated speed and equipped with a NEMA A governor will have a maximum speed rise to

$$4000 + \frac{4000 \times 13}{100} = 4520 \text{ rpm}$$

The setting of the *overspeed trip* is a function of the maximum speed rise of the governor system—it must be higher than the maximum speed rise. The recommended settings are

Class of governor system	Overspeed trip setting, (% of rated speed)
A	115
B	110
C	110
D	110

The overspeed trip system for a turbine with 4000 rpm rated speed and equipped with a NEMA A governor is set for operation at

$$4000 \times \frac{115}{100} = 4600 \text{ rpm}$$

The speed-sensitive portion of the speed governor system is usually a set of spring-loaded rotating weights. Movement of the weights caused by a change in turbine speed positions the governor valve through a suitable linkage.

The speed-sensitive element can also be other devices that are speed-responsive, such as a positive displacement oil pump, electric generator, or a magnetic impulse signal generator.

The rotating-weight governor system is a direct-acting type and is classified as a NEMA A governor. *Direct-acting* designates a governor system in which the speed-sensitive element also provides the power for positioning the governor valve.

The NEMA B, C, and D governor systems have speed-sensitive elements which position the governor valve through a relay or servomotor system instead of actuating the valve directly. The speed-sensitive element can therefore be more precise and sensitive, as required for the improved governor system performance.

THEORY

A steam turbine develops mechanical work by converting to work the available heat energy in the steam expansion. Heat and mechanical work, being two forms of energy, can be converted from one to the other.

The heat energy is converted in two steps. The steam expands in nozzles and discharges at a high velocity, converting the available heat energy to velocity (kinetic) energy. The high-velocity steam strikes moving blades, converting the velocity energy to work. Since the total heat energy available in the steam is converted to velocity (kinetic) energy, the magnitude of the steam velocity is dependent upon the available energy.

The mechanical work that is developed in the turbine by the high-velocity steam striking the buckets is a function of the speed of the buckets. Maximum work occurs when the bucket velocity is approximately one-half the steam jet velocity for an impulse stage and one-fourth the steam jet velocity for a velocity-compounded impulse stage. While the steam jet velocity is fixed by the available heat energy, the bucket velocity is fixed by the speed of the turbine and the diameter of the turbine wheel on which the buckets are mounted. The work developed, or the efficiency of the turbine, ignoring losses in the turbine, is therefore determined by the size of the turbine and the turbine (pump) speed for a fixed amount of available heat energy.

The most common single-stage turbine is the velocity-compounded (Curtis) type. The complete expansion from inlet to exhaust pressure occurs in one step. The Curtis stage, with two rows of rotating buckets and two re-entry-type velocity-compounded stages, is illustrated in Fig. 5.

Single-stage turbines are available with a wide range of efficiencies since they are manufactured with a variety of wheel diameters: 9 to 28 in. (22 to 71 cm).

Multistage turbines are manufactured with a more limited variety of wheel sizes. The efficiency of multistage turbines is varied primarily by varying the number of stages. When the total available energy of the steam results in a steam velocity which is greater than twice the bucket velocity (using convenient wheel sizes), then a multistage turbine will be more efficient. In a multistage turbine, the total steam expansion is divided among the various impulse stages to produce the desired steam velocity for each row of buckets.

A steam turbine is normally evaluated using *steam rate*—the amount of steam required by the turbine to produce the specified power per hour at the specified speed—rather than *efficiency*. The steam rate is a direct function of the turbine efficiency.

The steam consumption can be expressed either as steam rate, pounds of steam per horsepower-hour (kilograms per kilowatt-hour) or as steam flow, pounds of steam per hour (kilograms per hour). The higher the efficiency, the lower the steam rate or steam flow, and vice versa.

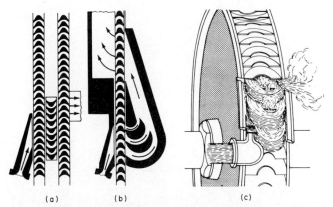

(a) (b) (c)

FIG. 5 Velocity-compounded stages: (*a*) Curtis stage; steam flows once through moving buckets; (*b*) re-entry stage; steam flows twice through moving buckets; (*c*) re-entry stage; steam flows three times through moving buckets. (Reprinted with permission from *Power*, June 1962)

The total available energy of the steam is that available from an isentropic expansion. For given initial steam pressure and temperature and exhaust pressure, the available energy in British thermal units per pound (kilojoules per kilogram) of steam can be obtained from the tables or the Mollier chart in Ref. 3.

The available energy can be converted to power units and expressed as the theoretical steam rate—pounds per horsepower-hour or pounds per kilowatt-hour (kilograms per kilowatt-hour). The theoretical steam rate is the steam rate for a 100% efficient turbine and therefore can be used more conveniently than energy in British thermal units per pound (kilojoules per kilogram) for the calculation of turbine steam rates. Theoretical steam rates can be obtained directly from theoretical steam rate tables (ASME) or from the polar Mollier chart (Elliott Company).

The actual steam rate for a turbine is greater than the theoretical steam rate because of the losses that occur in the turbine when the available energy is converted to mechanical work and because of the ratio of the steam velocity to bucket velocity. The energy remaining in the steam exhausting from the turbine is greater than that after an isentropic expansion, as illustrated in Mollier diagram shown in Fig. 6, where

1 = energy in steam at initial steam pressure and temperature

2 = energy in steam at exhaust pressure for an isentropic expansion

3 = actual energy in steam at exhaust pressure

The efficiency of the turbine is

$$\frac{h_1 - h_3}{h_1 - h_2} \quad \text{or} \quad \frac{\text{theoretical steam rate}}{\text{actual steam rate}}$$

The governor and trip valve pressure drop losses are a function of the sizes of these two valves and the steam flow. The governor valve pressure drop will vary more than the trip valve pressure drop because of changes in valve position with power required by the driven pump, speed variations, and changes in inlet and/or exhaust steam conditions.

The nozzle loss is due to friction in the nozzles as the steam expands. The efficiency of the nozzles is a function of the ratio of the actual and ideal exit steam velocities squared. The efficiency is usually between 95 and 99%.

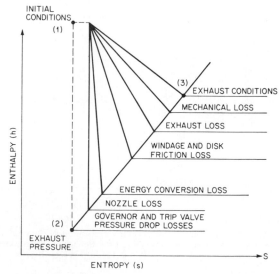

FIG. 6 Mollier diagram with energy losses.

The windage and disk friction losses are due to the friction between the steam and the disks and the blades fanning the steam. This loss varies inversely with the specific volume of the steam, increases with exhaust pressure, and increases with the diameter of the wheel and the length of the blades.

The use of a larger-diameter wheel may increase the efficiency, but the windage and disk friction losses will reduce the improvement and may even cause a net loss in overall efficiency.

The exhaust losses represent the kinetic energy remaining in the steam as a result of the velocity of the steam leaving the bucket and the pressure drop in the steam as it passes out the exhaust connection.

The energy conversion loss is due to the nonideal conversion of the steam velocity energy to mechanical work in the buckets as a function of the steam velocity and bucket velocity, plus nonideal nozzle and bucket angles, friction in the system, etc.

The performance that can be expected from a single-stage Curtis-type turbine may be obtained from Figs. 7, 8, and 9, and Table 1 after determining the theoretical steam rate:

$$\text{Steam rate} = \frac{\text{base steam rate}}{\text{superheat correction factor}} \times \frac{\text{power} + \text{power loss}}{\text{power}}$$

TABLE 1   Temperature of Dry and Saturated Steam
To obtain superheat, subtract temperature given below from total initial temperature. Pressure in kPa = 6.895 × lb/in.² Temperature in °C = (°F −32) ÷ 1.8.

Lb/in² gage	Saturation temp., °F	Lb/in² gage	Saturation temp., °F	Lb/in² gage	Saturation temp., °F	Lb/in² gage	Saturation temp., °F
0	213	150	366	300	422	450	460
5	228	155	368	305	423	455	461
10	240	160	371	310	425	460	462
15	250	165	373	315	426	465	463
20	259	170	375	320	428	470	464
25	267	175	378	325	429	475	465
30	274	180	380	330	431	480	466
35	281	185	382	335	432	485	467
40	287	190	384	340	433	490	468
45	293	195	386	345	434	495	469
50	298	200	388	350	436	500	470
55	303	205	390	355	437	510	472
60	308	210	392	360	438	520	474
65	312	215	394	365	440	530	476
70	316	220	396	370	441	540	478
75	320	225	397	375	442	550	480
80	328	230	399	380	444	560	482
85	328	235	401	385	445	570	483
90	331	240	403	390	446	580	485
95	335	245	404	395	447	590	487
100	338	250	406	400	448	600	489
105	341	255	408	405	449	610	491
110	344	260	410	410	451	620	492
115	347	265	411	415	452	630	494
120	350	270	413	420	453	640	496
125	353	275	414	425	454	650	497
130	356	280	416	430	455	660	499
135	358	285	417	435	456	670	501
140	361	290	419	440	457	680	502
145	364	295	420	445	458	690	504

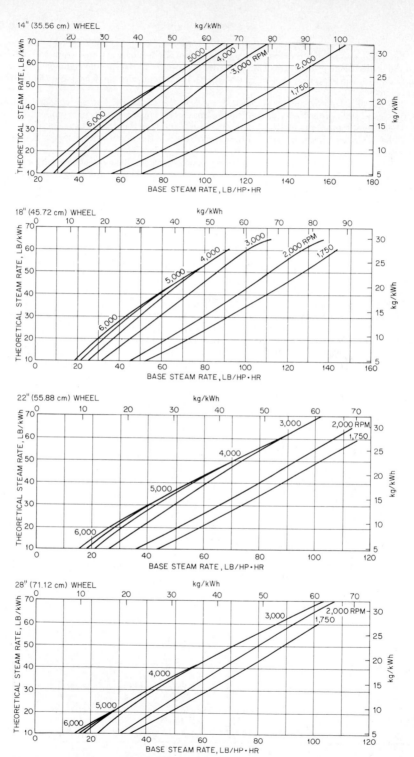

FIG. 7 Base steam rates. (Elliott)

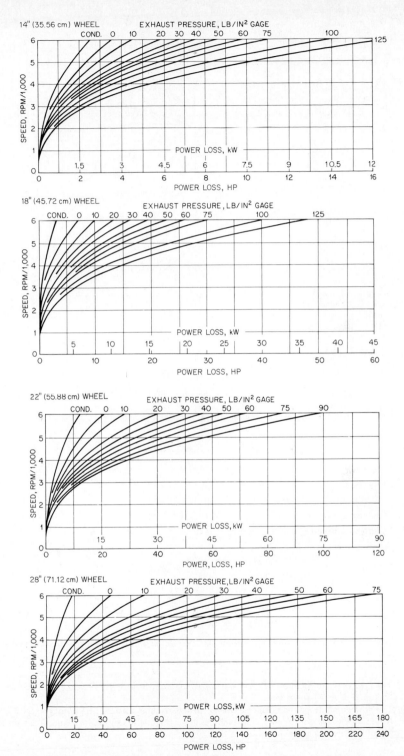

FIG. 8 Power loss (lb/in² × 6.895 = kPa). (Elliott)

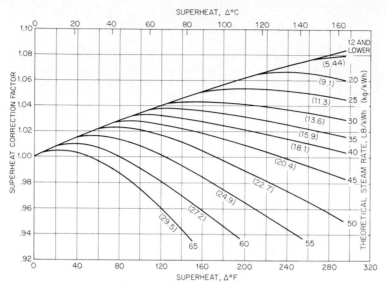

FIG. 9   Superheat correction factor. (Elliott)

SAMPLE CALCULATION

Steam conditions: 250 lb/in² (1724 kPa) gage inlet, 575°F (302°C) inlet, and 50 lb/in² (345 kPa) gage exhaust
Design conditions: turbine to develop 500 hp (373 kW) at 4000 rpm

1. Theoretical steam rate = 26.07 lb/kWh (11.82 kg/kWh)
2. Base steam rate for 28-in (71.12-cm) wheel (Fig. 7) = 36 lb/hp·h (22 kg/kWh)
3. Power loss for 28-in (71.12-cm) wheel (Fig. 8) = 55 hp (41 kW)
4. Temperature of dry and saturated inlet steam (Table 1) = 406°F (208°C)
5. Superheat (Table 1) = 575 − 406 = 169Δ°F (302 − 208 = 94Δ°C)
6. Superheat correction factor (Fig. 9) = 1052
7. Steam rate =

in USCS units
$$\frac{36}{1.052} \times \frac{500 + 55}{500} = 38.0 \text{ lb/hp·h}$$

in SI units
$$\frac{21.89}{1.052} \times \frac{373 + 41}{373} = 23.1 \text{ kg/kWh}$$

The ability of a particular size turbine to develop the required power is determined primarily by:

1. The flow capacity of the inlet and exhaust connection from Figs. 10 and 11, where steam flow = power × steam rate
2. The flow capacity of the nozzles available in a particular turbine (the number and size of the nozzles vary considerably with each design of turbine manufactured, and thus a meaningful plot cannot be included here).

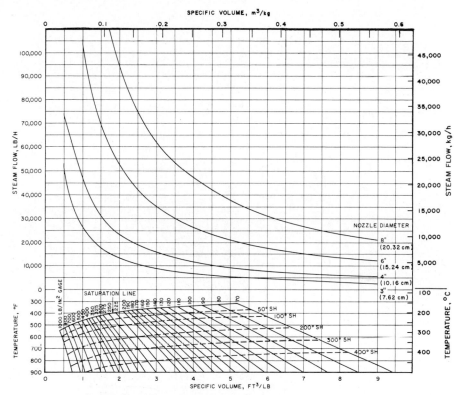

FIG. 10 Nominal inlet flow capacity. Read inlet nozzle size required to pass maximum flow, based on 150-ft/s (45.7-m/s) steam velocity (lb/in^2 × 6.895 = kPa; °F$_{SH}$ × 0.555 = °C$_{SH}$).

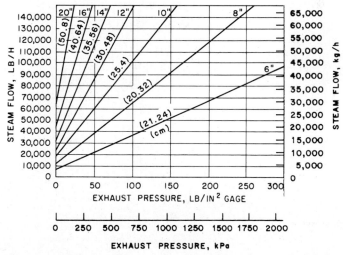

FIG. 11 Nominal exhaust flow capacity. Noncondensing exhaust nozzles, based on 200-ft/s (61 m/s) steam velocity.

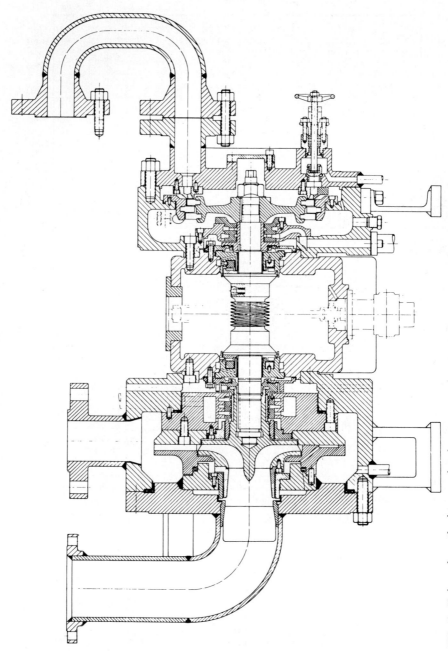

FIG. 12 High-speed centrifugal pump and single-stage steam turbine mounted on a common shaft. Package is designed for boiler feed in marine, industrial, and process steam plants. (Pacific Pumps, Division of Dresser Industries)

EVALUATION OF COSTS

The economical selection of a turbine considers the initial cost of the turbine and the operating costs. A lower-steam-rate (more efficient) turbine generally has a higher initial cost than a turbine with a higher steam rate. The operating costs are the cost of the steam for the number of years upon which the evaluation is to be based:

Total cost =

initial cost + (power × steam rate × steam cost × operating hours per year × number of years)

The cost of installation—foundation, steam piping, and cooling water service (also electric service if required)—is not normally considered unless there are significant differences between the turbines: single-stage versus multistage, direct-connected versus turbine and gear, vertical turbine versus horizontal turbine with a right-angle gear, etc.

The economical selection of a steam turbine versus an electric motor, diesel engine, gas engine, gas turbine, etc., must include an evaluation of all installation costs, costs of supporting equipment (starters, switch gear, fuel-supply system, etc.), operating costs, and the initial costs of the various drivers being considered.

REFERENCES

1. Church, E. F.: *Steam Turbines*, McGraw-Hill, New York, 1950.
2. "Single-Stage Steam Turbines for Mechanical Drive Service," NEMA pub. SM22-1970, 1970.
3. Keenan, J. H., and F. G. Keyes: "Thermodynamic Properties of Steam," John Wiley & Sons, Inc., New York, 1936.

FURTHER READING

W. L. Bergeron: "Governors and Controls: Each Has Specific Operation," *Sugar J.*, March 1965.

"Multistage Steam Turbines for Mechanical Drive Service," NEMA pub. SM21-1970, 1970.

Polar Mollier Chart, Elliott Company

Skrotzki, B. G. A.: "Steam Turbines," *Power* special report, 1962.

Steen-Johnsen, H.: "Selecting the Right Steam Turbine for Industrial Use," *Power Eng.*, April 1965.

Steen-Johnsen, H.: "Here's What You Should Know Before You Specify Control Systems for Mechanical Drive Steam Turbines," *Power Eng.*, February, March 1966.

Steen-Johnsen, H.: "Mechanical Drive Turbines" *Oil Gas Equip.*, March, April, May, 1967.

Steen-Johnsen, H.: "How to Estimate the Size and Cost of Mechanical Drive Steam Turbines," *Hydrocarbon Process.*, October 1967.

"Theoretical Steam Rate Tables," ASME, New York, 1969.

6.1.3
ENGINES

F. J. GUNTHER

ENGINE SELECTION AND APPLICATION

The internal combustion engine is used extensively as a driver for centrifugal and displacement pumps. Depending on the application, the engine fuel may be gasoline, natural gas, liquid petroleum gas (LPG), sewage gas, or diesel fuel. It may be either liquid- or air-cooled.

Basic Design Variations   The basic design of the engine may vary. The cylinder block construction is vertical or horizontal, in-line or V-type. The number of cylinders ranges from 1 to 20. The engine cycle is either four (one power stroke in two revolutions of the crankshaft) or two (one power stroke in one revolution of the crankshaft). The combustion chamber and cylinder head design are classed as L-head (valves in the cylinder block) or valve-in head. In the diesel engine, the combustion chamber may be of the precombustion chamber design, where an antechamber is used to initiate combustion, or of the direct injection design, where the fuel is injected directly into the cylinder. The piston design action may be vertical, horizontal, at an angle as in the V-type engine, or opposed piston (two pistons operating in the same cylinder).

Power Ratings   The power range of engines in current production, depending on displacement, number of cylinders, and speed, is as follows:

1. Air-cooled gasoline, natural gas, and diesel: 1.0 to 75 hp (0.75 to 56 kW)
2. Liquid-cooled gasoline: 10 to 300 hp (75 to 224 kW)
3. Liquid-cooled natural gas, LPG, and sewage gas: 10 to 15,000 hp (75 to 11,200 kW)
4. Liquid-cooled diesel: 10 to 50,000 hp (75 to 37,300 kW)
5. Dual fuel, natural gas, LPG, and diesel: 150 to 25,000 hp (112 to 18,700 kW)

   Typical engine driver applications are represented in Figs. 1 and 2.
   The rating of the internal combustion engine is the most important consideration in making the proper selection. The general practice is to rate engines according to the severity of the duty to be performed. The most common rating classifications are maximum, standby or intermittent, and continuous.
   The maximum output is based on dynamometer tests that are corrected to standard atmo-

FIG. 1 Natural gas engine driving a horizontal water pump for Winnipex, Canada, water utility. (Waukesha Motor)

spheric conditions for temperature and barometric pressure. In applications, this power rating is reduced by accessories such as cooling fans, air cleaners, and starting systems.

Standby, or intermittent, and continuous ratings are arrived at by applying a percentage factor to the net maximum power rating. For example, 75 to 80% is used for continuous and 90% for intermittent.

Duty cycle is a term used to describe the load pattern imposed on the engine. If the load factor (ratio of average load to maximum capabilities) is low, we call the duty cycle "light," but if it is high we classify the cycle "heavy." Continuous, or heavy-duty, service is generally considered to be 24 h/day with little variation in load or speed. Intermittent service is classified as duty where an engine is called upon to operate in emergencies or at reduced loads at frequent intervals.

FIG. 2 Diesel engine driving pumps for flood-control station in Seattle. (Waukesha Motor)

In analyzing power problems when selecting a proper engine, certain terms are used in the industry:

DISPLACEMENT   The volume in cubic inches (cubic centimeters) of an engine cylinder is

in USCS units $D = \text{bore (in}^2) \times 0.7854 \times \text{stroke (in)} \times \text{no. of cylinders}$

in SI units $D = \text{bore (cm}^2) \times 0.7854 \times \text{stroke (cm)} \times \text{no. of cylinders}$

TORQUE   The twisting effort of the engine in pound-feet (Newton-meters) is

in USCS units $T = 5252 \times \dfrac{\text{bhp}}{\text{rpm}}$

in SI units $T = 9545 \dfrac{\text{bkW}}{\text{rpm}}$

ENGINE POWER   This is a measure of the theoretical characteristics of an engine. Brake horsepower (bkW) is the measurable power after the deduction for frictional losses:

in USCS units $\text{bhp} = T \times \dfrac{\text{rpm}}{5252}$

in SI units, $\text{bkW} = T \times \dfrac{\text{rpm}}{9545}$

BRAKE MEAN EFFECTIVE PRESSURE   The average cylinder pressure to give a resultant torque at the flywheel in pounds per square inch (kilopascals)

in USCS units $\text{bmep} = \dfrac{792,000 \times \text{bhp}}{\text{rpm} \times D}$ (four-cycle)

                         $\text{bmep} = \dfrac{396,000 \times \text{bhp}}{\text{rpm} \times D}$ (two cycle)

in SI units $\text{bmep} = \dfrac{120 \times 10^6 \times \text{bkW}}{\text{rpm} \times D}$ (four-cycle)

                         $\text{bmep} = \dfrac{60 \times 10^6 \times \text{bkW}}{\text{rpm} \times D}$ (two cycle)

PISTON SPEED   At a given speed, the average velocity of piston in feet per minute (centimeters per minute) is

in USCS units $\text{Piston speed} = \text{stroke (in)} \times 2 \times \dfrac{\text{rpm}}{12}$

in SI units $\text{Piston speed} = \text{stroke (cm)} \times 2 \times \dfrac{\text{rpm}}{12}$

In selecting an engine for a particular application, the following variables should be considered:

Altitude
Ambient air temperature
Rotation and speed
Bmep and piston speed
Maintenance
Type of fuel

Operating atmosphere (dust and dirt)

Vibrations and torsionals

The observed power is that which an engine will produce at the existing altitude and temperature. All engine manufacturers publish power ratings corrected to certain conditions; for conditions other than these it is necessary to correct by applying a percentage factor for altitude and temperature. Generally this is 3½% per thousand feet (305 m) above sea level and 1% for every 10°F (5.6°C) above 60°F (50.4°C). In a turbocharged engine, there is no established standard and the engine manufacturer should be consulted.

The basic rotation of engines in current production is counterclockwise when viewed from the flywheel end of the engine, although many of the larger engines are available in both counterclockwise and clockwise rotation. The speed of the engine is generally fixed by the equipment being driven. Through the use of speed-increasing or -reducing gear boxes, the proper engine for a given application may be selected. A gear box may also be used to correct a rotation problem.

Speed ranges for engines generally fall into three categories:

High—above 1500 rpm

Medium—700 to 1500 rpm

Low—below 700 rpm

High-speed engines generally offer weight and size advantages as well as cost savings and thus are used for standby applications. On the other hand, medium- or low-speed engines, although heavier and larger, offer a gain in service life and lower maintenance costs.

The speed flexibility of an engine drive is important when the engine is to be used to drive a pump that must move variable quantities of liquid. The engine speed may be changed very simply either manually or through the use of liquid or pressure controls.

Bmep is generally a measure of load, and piston speed a measure of potential wear and maintenance. Although the introduction of the turbocharged and intercooled engine has somewhat changed the consideration given these factors, it is still important to consider them in selecting engines where long life is a factor.

Maintenance of engines has been considered by some as objectionable and more costly than electric power. A recent innovation of engine manufacturers, in the form of a service contract for installations where trained personnel are not available or desirable for economic reasons, can eliminate these objections and costs. The complete maintenance of the engine is done on a fixed-fee basis for a designated period of time.

The remaining conditions listed above will be discussed in detail later.

FUEL SYSTEMS

Gasoline   Gasoline is used primarily with standby pumping units. Inasmuch as the spark ignition system first introduced the internal combustion engine to power applications, gasoline was used as the primary fuel. Commercial gasoline has an average heating value of 19,000 Btu/lb (44.2 MJ/kg). It is easy to transport and handle and, unlike gaseous fuels, does not require pressure storage and regulating equipment. The starting capabilities of a gasoline engine are satisfactory, provided the engine is in good operating condition. With the high-power engine of today, refinery control can produce a fuel matched to the operating conditions.

Gasoline does have some disadvantages that are reducing its use as an engine fuel. In small-volume usage, it is safe and easily handled. In larger volumes, it becomes expensive and hazardous. Because it is not entirely stable, it will deteriorate when exposed to gums and resins in storage over a period of time. There is also the possibility of condensation of water in the fuel, which is detrimental to good operation. The danger of fire is always present because of leaks in the system. Finally, the increased production and distribution of gasoline have made it a target of increasing taxation, making it economically prohibitive in many installations.

Gas A gaseous fuel system using natural gas, LPG, or sewage gas may be a simple manually controlled system, such as a gasoline engine, or a carefully engineered automatic system. The basic gas carburetion system consists of a carburetor and pressure regulator mounted on the engine. A gas distribution system, like a water supply system, must be at some designated pressure and flow, and so a field pressure regulator is required. The characteristics of this regulator will depend upon the gas analysis, the displacement of the engine, the speed range, and local regulations. A typical schematic of a gas fuel system is shown in Fig. 3. The location of the regulator is generally under the jurisdiction of the gas utility that supplies it. In most cases a single field regulator is all that is required, but at times this can cause problems. For example, subsequent installation of gas-burning equipment used intermittently may cause gas pressure regulation not compatible with the small amount required for pilot lighting. A single regulator installed some distance from the engine could result in hard starting because the engine vacuum is not sufficient for a full gas flow. To eliminate this problem, the initial regulator is set at a higher pressure in order to give a readily available supply of fuel for all devices.

Natural gas has an average heating value of 800 to 1000 Btu/ft^3 (29.8 to 37.3 MJ/m^3). Commercial butane has a value of 2950 Btu/ft^3 (110 MJ/m^3), and propane a value of 3370 Btu/ft^3 (126 MJ/m^3). Commercial LPG fuel, which is a mixture of butane and propane, varies in both amount and corresponding heating value.

LPG fuel is produced by mechanical and compression processes, and the methods of distribution and handling must meet regulations. Natural gas is usually supplied under moderate pressures, seldom exceeding 50 lb/in^2 (345 kPa), while LPG is supplied as a liquid under pressures as high as 200 lb/in^2 (1380 kPa) in warm weather. As a result, natural gas will readily mix with air and burn, whereas LPG fuel must be changed from a liquid to a vapor with the addition of heat, as shown in Fig. 4.

As a natural process in the modern waste-treatment plant, sewage gas may be produced in the sewage digester. This gas has the same basic qualities as natural gas, being about 65 to 70% methane. It has a heating value of 550 to 700 Btu/ft^3 (20.5 to 26.1 MJ/m^3). The same basic carburetion system used for natural gas is used. Sewage gas contains inert substances, particularly hydrogen sulfide or free sulfur, which in the presence of free moisture or moisture resulting from combustion will form sulfurous acid, which is corrosive and thus damaging to the valves, pistons, and cylinder walls of an engine. An engine can tolerate from 10 to 30 g of sulfur per 100 ft^3 (350 to 1100 g per 100 m^3). Beyond this, a filtering system to remove the sulfur and moisture is advisable.

Diesel The diesel engine over the years has been used for larger power systems. The initial cost of the system is justified to an extent by lower cost of the fuel. The better fuel economy of the diesel engine and its torque characteristics also are important factors in the selection of this fuel for many applications. Diesel fuel has one distinct advantage: it does not form the dangerous fuel vapors that other fuels do. It does require, however, a good fuel-filtering system because of contaminants in the fuel which can create problems for the precision design of the fuel-injection system.

Commercial diesel fuels are the residue that remains after the more volatile fractions of crude oil have been removed. The heating value is generally about 19,000 Btu/lb (4.4 MJ/kg). In the

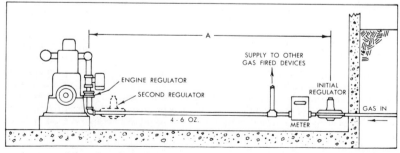

FIG. 3 Typical natural gas fuel system. (Waukesha Motor)

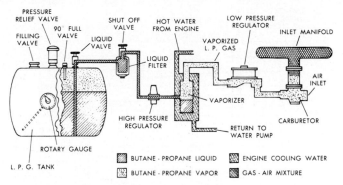

FIG. 4 Typical LPG fuel system. (Waukesha Motor)

diesel engine the fuel is injected into the cylinder at the end of the compression stroke in an atomized form. The compression stroke results in a temperature sufficient to ignite the fuel without the use of any ignition device. Although fuel systems from different engine manufacturers vary, the basic components are the same.

The larger diesel engines may be designed to operate on a dual fuel system. The engine operates on a gaseous fuel with a pilot injection of diesel fuel for ignition. In case of a loss of the gaseous fuel supply, the engine will convert to 100% diesel fuel.

Relative Performance Curves Typical performance curves comparing power, torque, and part-load fuel economy are shown in Fig. 5.

COOLING SYSTEMS

Cooling is essential in all internal combustion engines because only a small portion of the total heat energy of any fuel is converted to useful energy. The remainder is dissipated into the coolant, exhaust and lubricating oil and by radiation. A hypothetical heat balance is shown in Fig. 6.

Specific data and recommendations on cooling requirements vary from one manufacturer to another. In general, the heat rejection to an engine cooling system will range between 30 to 60 Btu/hp·min (25 to 51 MJ/kWh) for diesel engines and up to 70 Btu/hp·min (59 MJ/kWh) for natural gas and gasoline engines. This heat must be transferred to some form of heat-exchange medium.

In designing any cooling system, certain factors must be considered:

1. Additional heat from driven equipment, such as the cooling of speed-reducing or -increasing gears where the engine coolant is the medium

2. Water-cooled exhaust manifolds on the engines or water-cooled exhaust turbochargers or after coolers

3. High ambient temperatures or heat from nearby equipment

4. Variation in the heat-exchanger coolant temperature

5. Entrapment of substantial quantities of air in the coolant water

6. Inability to maintain a clean cooling system

With any cooling system, one of the most important factors of design is the temperature drop across the engine. Most engine manufacturers desire a temperature differential of no more than 10 to 12°F (5.6 to 6.7°C), and closer values are desirable. A jacket-water temperature across the engine of 170°F (77°C) is preferred, and in high-temperature or waste-heat-recovery systems, temperatures of 200°F (93°C) or more are common and not harmful.

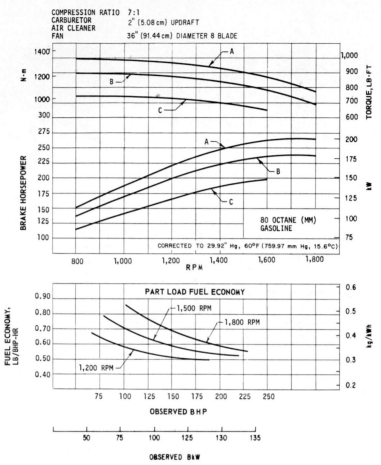

FIG. 5a   Gasoline engine performance curves: (A) maximum, (B) intermittent, (C) continuous ratings of engines with accessories. (Waukesha Motor)

Radiator   The radiator cooling system is perhaps the most common and best understood method of cooling. It is based on a closed system of tubes through which the jacket water passes. The heat is dissipated by a fan, creating a stream of moving air passing through the tubes. The fan is driven either by the engine or by an auxiliary source of power (Fig. 7).

An engine in a fixed outside installation can be cooled without much difficulty. Certain factors must be considered, such as ambient temperature, direction of the prevailing wind, and presence of foreign airborne materials. In high temperatures (usually above 110°F [43°C]), a larger radiator is required. If the prevailing winds are extremely high, the unit can be located to offset normal fan flow. Screening can be used to prevent the clogging of the air passes in the radiator where the atmosphere tends to contain foreign airborne material, such as dust.

Radiator cooling may be used in an inside installation, but there are certain problems which, unless properly anticipated, limit this system. The recirculation of cooling air and the radiation of exhaust heat from the engine create a problem. As was previously pointed out, every 10°F (5.6°C) rise above 60°F (15.6°C) results in a 1% loss in power. When 5 to 10% of the total heat put into an engine is radiated, some means of power ventilation must be provided to remove this heat through ducts, louvers, or forced ventilation with a separate fan. An installation of this type is illustrated in Fig. 8.

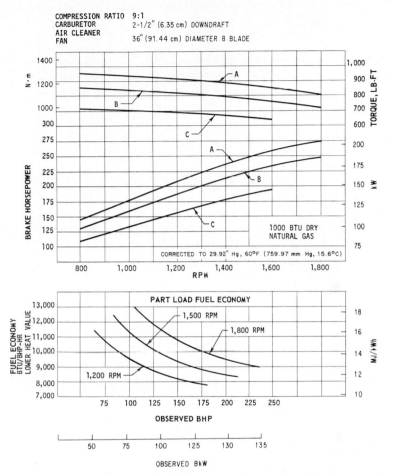

COMPRESSION RATIO 9:1
CARBURETOR 2-1/2" (6.35 cm) DOWNDRAFT
AIR CLEANER
FAN 36" (91.44 cm) DIAMETER 8 BLADE

FIG. 5b Gas engine performance curves: (A) maximum, (B) intermittent, (C) continuous ratings of engines with accessories. (Waukesha Motor)

Heat Exchanger The heat exchanger cooling system (Fig. 9) is the best system for a stationary engine installation. Using a tube bundle in a closed shell, the cooling exchange medium is water (often called raw water). This water may be plant or process water; it may be recirculated or, in standby installations, allowed to pass to waste. The system has the advantage of the radiator cooling system in that it is self-contained: the quantity and quality of the water in the engine can be controlled. It has the further advantage of not being affected by the flow of heat to air movement if the heat of radiation is taken into account in the design of the system. On the other hand, the cooling medium, unless used in a plant system, is a disadvantage because it is costly. A separate pump is required to provide the necessary water for cooling unless city water or process water is under sufficient pressure.

City Water and Standpipe City water cooling is designed to take water directly from the city main or from the pump the engine is driving. It is used on some emergency or standby installation. It is simple and inexpensive, gives unlimited cooling for moderate-size engines, is easily understood, and will operate instantly in an emergency. On the negative side, the cooling water is wasted, corrosive elements may be introduced into the engine jacket water system, and it may create excessive temperature changes across the engine jacket.

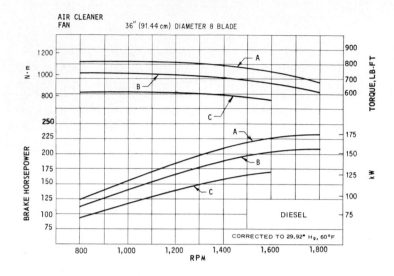

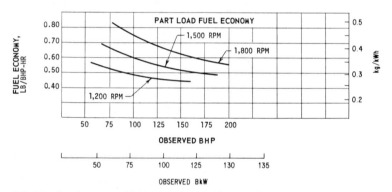

FIG. 5c   Diesel engine performance curves: (A) maximum, (B) intermittent, (C) continuous ratings of engines with accessories. (Waukesha Motor)

Standpipe cooling (Fig. 10) is basically the same as city water cooling except that a thermostatic valve is employed to admit makeup water as required. The vertical pipe is a blending tank into which city water is introduced only in the amount necessary for makeup. The standpipe system is inexpensive and simple to operate.

Ebullition   In installations where heat is required for process equipment, a method of high-temperature, or ebullition, cooling is being used as a very economical method, particularly with larger installations. This system has been termed *steam cooling, high-temperature,* or *Vapor-phase* (a registered trademark). In this system the coolant leaves the engine at a temperature equal to or above its atmospheric boiling point and under sufficient pressure to remain liquid until discharged from the engine into a flash chamber, where a drop in pressure causes the formation of steam, which is condensed and returned to the engine at very near the discharge temperature. A schematic of this system is shown in Fig. 11. This system has the advantages of a very small temperature differential across the engine, which minimizes distortion of all working parts, and a constant working temperature regardless of load. Because of the higher working temperatures,

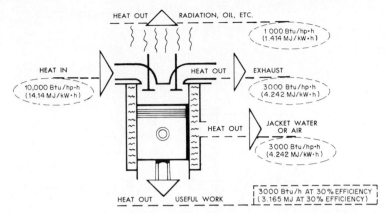

FIG. 6 Hypothetical heat balance. (Waukesha Motor)

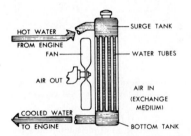

FIG. 7 Radiator cooling system. (Waukesha Motor)

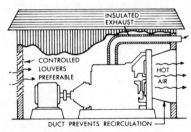

FIG. 8 Radiator cooling system for an inside installation. (Waukesha Motor)

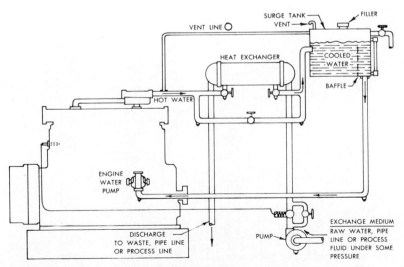

FIG. 9 Heat exchanger cooling system. (Waukesha Motor)

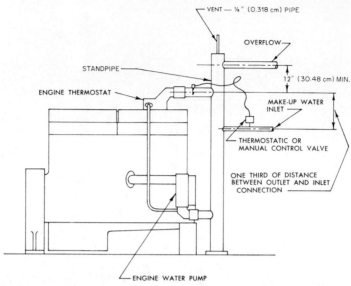

FIG. 10 Standpipe cooling system. (Waukesha Motor)

the combustion area and crankcase of the engine have fewer liquid by-products of combustion and corrosive materials. Of prime importance is the waste heat that can be recovered for plant process with a very small amount of makeup water for cooling.

Cooling Tower   Cooling towers are used in some large or multiple-engine installations. Through the use of a current of air, produced either by a natural draft or by mechanical means, a tower causes a sensible heat flow from the cooling water to the air. Atmospheric, or natural, draft towers depend upon natural wind velocities and thus can vary widely. Mechanical draft

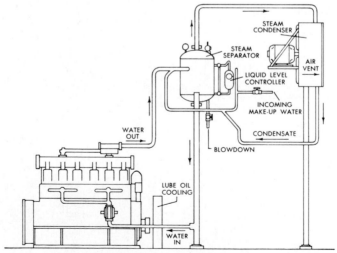

FIG. 11   High-temperature cooling system. (Waukesha Motor)

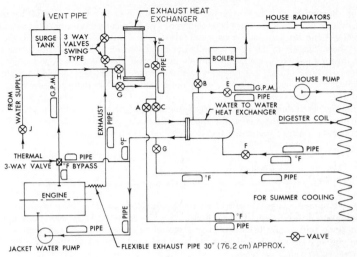

FIG. 12   Complete cooling system, including waste-heat recovery. (Waukesha Motor)

towers, where the air supply can be controlled, can be put in any area, but the limit to their cooling capacity is the power required to operate them. As the water volume increases, the volume of air required and the pressure the fan has to operate against increase. A point is reached where the cost of installation and operation becomes prohibitive. A diagram of a system that combines components of the cooling systems mentioned plus the waste-heat recovery system silencer, to be mentioned later, is shown in Fig. 12.

AIR-INTAKE SYSTEMS

A most important consideration in the application of an engine to any pump drive is the engine's ability to "breathe." As air is required for combustion, the design engineers of any project must take into account the necessary provisions for this air. The environment, the service, the speed range of the engine, the duty cycle, and the location from which the combustion air is to be taken are of vital importance.

The first step in the selection of any intake system is to determine the volume of air required for combustion, which may be calculated by the formula

$$\text{Required air} = \frac{B^2 \times S \times \text{rpm} \times N}{2200 \times K}$$

where B = cylinder bore, in (cm)
S = piston stroke, in (cm)
rpm = engine speed, rpm
N = number of cylinders
K = constant: two-cycle, 1; four-cycle, 4

Volumetric efficiency will vary with engine design, but an average of 80% may be used to determine the air cleaner size.

The two basic cleaners available are wet and dry. The wet, or oil-bath, cleaner consists of either an oil wire mesh or an oil bath through which the air must pass. A dry cleaner uses a paper or cloth filter that traps dust, lint, etc. but allow the air to pass through.

The installation of a suitable air cleaner is important. In cases where adequate air can be supplied through proper ventilation of the area surrounding the engine, it is best to mount the cleaner on the engine. If it is necessary to bring air to the engine from outside the area of the building, certain design factors must be considered. The pipe connections from an outside cleaner should be tight and mechanically strong, and fabric hose should not be used unless the length is relatively short. The air to the engine should not be heated by close proximity to the engine or any other heating device because, as previously mentioned, there is a loss in power with air temperatures above 60°F (15.6°C). To avoid restrictions in the system, there should be no sharp bends in the piping. Finally, any outside air cleaner must be designed so that moisture, such as rain or snow, cannot enter the system.

The air cleaner system on a turbocharged engine must be able to remove any impurities in the air that would be detrimental to the efficiency of the turbocharger. Because of the increased air requirements, a larger air-cleaner system must be used on such units.

EXHAUST SYSTEMS

An engine consumes a large volume of air for combustion, and that volume must be removed after combustion. It is necessary that the exhaust back pressure be kept at a minimum while this volume is being removed. The exhaust piping should be properly sized, and long sweep elbows should be employed if necessary. Unless the back pressure is kept low, the following conditions can result:

1. Loss of power
2. Poor fuel economy
3. High combustion temperatures with increased maintenance
4. High jacket water temperatures
5. Crankcase sludging with resulting corrosion and bearing wear

All internal combustion engines create noise. Depending upon the location of the engine, this noise can be a problem. Normally an unmuffled engine when measured from a distance of about 10 ft (3 m) creates a decibel noise level ranging from 100 for the small- and medium-size engine to 125 for the larger engine. Thus most engine installations incorporate some form of exhaust silencer, or muffler. Depending upon the degree of silencing, a muffler will reduce the unmuffled decibel reading by 30 to 35 dB. Various types are manufactured to meet the required conditions and are classified as follows:

1. Standard or industrial
2. Semiresidential (high-degree)
3. Residential or hospital (supercritical)

The basic exhaust silencer is designed as either a dry type or wet type. The latter is used in installations such as sewage or water treatment plants, where the recovery of heat for plant processes is important. Basically this type may be classified as a low-pressure boiler. Water is admitted to the silencer through tubes or coils to pick up the heat from the exhaust. As shown in Fig. 6, approximately 30% of the total heat input into the engine is exhausted. The wet-type silencer is designed to regain about 60 to 70% of this heat. This silencer has an advantage in its ability to operate either wet or dry. Thus when heat is not required, it may be operated dry and vice versa.

STARTING SYSTEMS

Starting methods for engines fall into two broad categories, direct and auxiliary.

Direct   The direct system, used primarily with large engines, employs some means of exerting a rotating force on the crankshaft, such as the introduction of high-pressure air directly into the cylinders of the engine. A direct system on small air- or water-cooled engines using either a rope or a hand crank has to a large extent been discarded in favor of an auxiliary method.

Auxiliary   The auxiliary system employs a small gear which meshes with a larger gear (ring gear) on the engine flywheel. The ratio of the number of gear teeth on the large gear to those on the small gear is called the *cranking ratio*. Generally, the larger the ratio, the better the cranking performance. The auxiliary system uses several means to drive the small gear:

1. Electric motor

2. Air motor

3. Hydraulic motor

4. Auxiliary engine

ELECTRIC MOTOR   The electric motor may be either dc or ac. The dc motor most commonly used is available in 6, 12, 24, or 32 V. The voltage size will depend upon the size of the engine, the ambient temperature, and the desired cranking speed of the engine. The dc system requires a source of outside power, usually in the form of a storage battery. To assure prompt starting, a charging system for the battery is required in the form of either a charging generator driven by the engine or an ac-powered battery trickle charger. The latter is recommended for standby installations where an engine-driven charging generator functions only when the engine is operating. During idle periods, the battery will lose its charge unless maintained by a trickle charger. In recent years there has been a trend toward the use of an ac generator, or alternator, which has the advantages of small size, higher voltage and amperage, competitive price, and good charging ability under idle speed conditions.

Another type of electric motor is a line voltage starter available in 110, 220, or 440 V ac. It has the advantages of faster and more powerful cranking, the elimination of the battery and charging system, less maintenance, and sustained cranking through unlimited available electric power. Its disadvantages are a higher initial cost, the requirement of high line voltage at the site, the danger to personnel due to the high voltage, and the requirement to conform to existing wiring and installation codes.

AIR MOTOR   The air motor, which is usually of the rotary-vane type, uses high-pressure air in the range of 50 to 150 lb/in^2 (340 to 1030 kPa) to turn it in starting the engine. It is mounted on the engine flywheel housing to mesh with the gear on the flywheel in the same manner as the electric motor. An outside source of air from an air compressor, usually with a 250-lb/in^2 (1720-kPa) capacity, is required. A pressure-reducing valve is installed in the line to the engine. The high-pressure air stored in an adequate receiver is sufficient for several starting cycles. This starting system has the advantages of faster cranking, sustained cranking as long as the air supply lasts, suitability in hazardous locations where an electric system might be dangerous, and ability to operate on either compressed air or high-pressure natural gas. Its disadvantages include a higher initial cost, the requirement of an air-compressor system, and, finally, a shutdown condition if the air supply is depleted before the engine starts.

HYDRAULIC MOTOR   The hydraulic motor system consists of the motor, an oil reservoir, an accumulator, and some means of charging the accumulator. The accumulator, which is a simple cylinder with a piston, is charged on one side with nitrogen gas. As the hydraulic fluid, usually oil, is pumped into the other side, the gas is compressed to a very high pressure. When released, the fluid turns the motor, which in turn rotates the engine. The system can be charged by hand, with an engine-driven pump, or with an electric-motor-driven pump. Generally an engine-driven or electric-motor-driven pump is used in conjunction with the hand pump in case of an engine or electrical failure. This system has the same basic advantages of the air motor except that there is no prolonged starting. If the engine is in good operating condition, the cranking is fast and a start is instantaneous, but, if not, it is necessary to recharge the system before another start can be made.

AUXILIARY ENGINE   A small auxiliary air-cooled or water-cooled engine is sometimes employed for starting. It may be mounted on the engine in the same manner as the other systems, or a belt drive may be employed. Some form of speed reduction is required to reduce the higher speed of the auxiliary engine to that required for proper cranking. The principal advantage of such a system is a complete independence from outside sources of power, such as batteries, air, or pumps, but this is offset by a higher initial cost and the required regular maintenance of the engine.

IGNITION SYSTEMS

The internal combustion engine requires some means of igniting the combustible charge in the cylinder at the proper time. Today's high-compression gasoline and gas engines demand a system which will produce a high-tension spark across a short gap in the combustion chamber for the ignition of the charge. It is obvious that the design of the combustion chamber must be such that this combustible mixture of fuel and air be present between this discharge gap when the spark occurs or ignition will not take place.

Ignition systems for gasoline or gas engines are considered as high or low tension. The energy system for the high-tension system is either an electric generator and battery or a magneto. The generator and battery produce a direct current at 6 to 12 V potential, and the magneto produces an alternating current with higher peak voltages. With either energy source, this system has a primary circuit for the low-voltage current and a secondary circuit for the high-voltage current.

The primary circuit consists of the battery, an ammeter, an ignition switch, a primary coil, and breaker points and a condenser in the distributor. When the ignition switch and the breaker points close, a current flows through the circuit and builds up a magnetic field in the primary coil. Opening the breaker points breaks this circuit, causing the magnetic field to start collapsing. As the field collapses, it produces a current that flows in the same direction in the primary circuit and charges the condenser plates. The condenser builds up a potential opposing flow which discharges back through the current. This results in a sudden collapse of the remaining magnetic field and the induction of a high voltage into the secondary winding of the coil. The breaker points are opened and closed by a cam which is engine-driven, usually at half engine speed.

The secondary circuit consists of the secondary coil winding, the lead to the distributor rotor, the distributor, the spark plug leads or wires, and the spark plug. A magneto eliminates the battery but includes in its construction the balance of both primary and secondary circuits. It may have either a rotating coil and stationary permanent magnets or a stationary coil and rotating magnets. The relative movement of the primary coil winding and the magnets induces an alternating current in the primary circuit, the breaking of which induces a high-voltage current in the secondary circuit.

The development of the modern engine has required many refinements in spark plug design, but basically a spark plug consists of two electrodes, one grounded through the shell of the plug and the other insulated with porcelain or mica. The insulated electrode is exposed to the combustion. The heat flow occurs from this electrode to the spark plug shell through the grounded electrode.

Recent developments in ignition systems have produced the low-tension and breakerless systems. The breakerless system has eliminated most of the moving parts in the distributor system. The breaker points in the distributor system are actually a switch which opens and closes the primary circuit of an ignition coil. In the breakerless system, the use of solid-state devices provides a switch with no moving parts to wear or require adjustment.

The low-tension magneto system has been developed to reduce electrical stresses in the ignition circuit. The secondary coil has been removed from the magneto proper and relocated near each spark plug. The low voltage generated by the magneto is transmitted through the wiring harness to secondary coils, which then step up to the voltage to be transmitted through short leads to the spark plug. These leads may be insulated to withstand the stresses imposed upon them. This results in a minimum of electrical stresses with a resulting longer life for all components of the system.

As was mentioned previously under fuel systems, the diesel engine uses the heat of compression for ignition; thus no auxiliary systems, such as the systems mentioned above, are required. The fuel is injected into the combustion chamber under relatively high pressure through the use of a fuel pump and injection nozzle. This system may be either an individual pump and nozzle for each cylinder, commonly called a *unit injection*, or a multicylinder pump which maintains a high pressure in a common fuel line connected to each injection nozzle. The latter is normally called the *common rail system*.

ENGINE INSTALLATIONS

Foundation The correct foundation, mounting, vibration isolation, and alignment are most important to the success of any engine installation. All stationary engines require a foundation, or mounting base. There are many variations, but all basically serve to isolate the engine from the surrounding structures and absorb or inhibit vibrations. Such a base also provides a permanent and accurate surface upon which the engine and usually the pump may be mounted.

To meet these requirements, the foundation must be suitable in size and mass, rest on an adequate bearing surface, provide an accurately finished mounting surface, and be equipped with the necessary anchor bolts.

The size and mass of the foundation will depend upon the dimensions and weight of the engine and the pump (if a common base is considered). The following minimum standards should be followed:

1. Width should exceed the equipment width and length by a minimum of 1 ft (0.3 m).

2. The depth should be sufficient to provide a weight equal to 1.3 to 1.5 times the weight of the equipment. This depth may be determined by the following formula

in USCS units
$$H = \frac{(1.3 - 1.5)W}{L \times B \times 135}$$

in SI units
$$H = \frac{(1.3 - 1.5)W}{L \times B \times 2162}$$

where H = depth of foundation, ft (m)
W = weight (mass) of equipment, lb (kg)
L = length of foundation, ft (m)
B = width of foundation, ft (m)
135 = density of concrete, lb/ft^3 (2162 kg/m^3)

The soil-bearing load in pounds per square foot (kilograms per square meter) should not exceed the building standard codes. It may be calculated by the formula

$$\text{Bearing load} = \frac{(2.3 - 2.5)W}{B \times L}$$

Foundation or anchor bolts used to hold the equipment in place should be of SAE grade No. 5 bolt material or equivalent. The diameter, of course, is determined by the mounting holes of the equipment. The length should be equivalent to a minimum embedded length of 30 times the diameter plus the necessary length for either a J or an L hook. An additional 5 to 6 in (13 to 15 cm) should be provided above the top surface of the foundation for grout, sole plate, chocks, shims, equipment base washers and nuts, plus small variations in the surface level. Around the bolts, it is a good practice to place a sleeve of iron pipe or plastic tubing to allow some bending of the bolts to conform with the mounting hole locations. This sleeve should be about two-thirds the length of the bolt, with its top slightly above the top surface of the foundation to prevent concrete from spilling into the sleeve.

Sole plates running the length of the equipment are recommended for mounting directly to the foundation. Made of at least ¾-in (19-mm) hot- or cold-rolled steel and a width equivalent to the base-foot mounting of the equipment, they will provide a level means of mounting and avoid variations in the level of the concrete. These plates should be drilled for the mounting holes and drilled and tapped for leveling screws, which will permit the plates to be leveled and held during the pouring of grout.

Alignment Although the alignment will vary with the type of engine and the pumping equipment, the basic objective remains the same. The driven shaft should be concentric with the driver shaft, and the centerlines of the two shafts should be parallel to each other. Rough alignment should be made through the use of chocks and shims. A dial indicator should be used to check deflection by loosening or tightening the anchor bolt nuts until there is less than a 0.005-in (0.13-μm) reading at each bolt. Shims should be added or removed to arrive at this point. A final check should be made with all the conditions "hot," as the engine and its driven equipment expand at the rate of 0.000006 in/°F (0.27 μm/°C) above ambient hot to cold. Although the coupling, or driving, member between the engine and the pump is not discussed in this section, it must be considered in the final alignment.

Vibration Isolation   It is desirable to isolate the engine, and at times the pumping equipment, from the building structure because of vibrations. Cork (Fig. 13) is used in the larger and heavier installations. A combination of cork and rubber pads may be used at each mounting hole

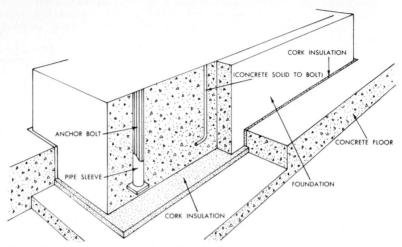

FIG. 13   Engine foundation installation. (Waukesha Motor)

on small- or medium-size installations, and spring isolators may be used on a complete installation if flexible hoses are used for fuel, water, and air connections where required. The manufacturer of the engine and the pumping equipment should be consulted in the use of any isolating material or device.

Vibrations are closely associated with the driving and driven equipment, couplings, and other connections. These linear vibrations may be caused by improper supports of the unbalanced parts, which produce a *torsional* vibration. An understanding of this vibration is important since its elimination is the responsibility of the engine and the driven-equipment manufacturer. It is complex and cannot be detected without the use of calculations and special instruments.

The basic concept involves an elastic element, such as an engine crankshaft, which tends to twist when any firing impulses are applied. When these forces are removed, the elastic body will try to return to its original position. The driven mass and the connecting elements tend to resist these external impulses. The natural elasticity of the crankshaft and its connecting system allows a small amount of torsional deflection and tends to reduce the deflection as the external impulses reduce in force. Other reciprocating forces in the engine and external forces in the driven equipment may excite vibrations in the entire system. When all these forces come into resonance with the natural frequency of the entire system, torsional vibration will occur. This vibration may or may not be serious but, being complex, cannot be solved hastily.

The engine and pump manufacturer designs and constructs the product so that critical harmonic vibrations will not be present under normal speeds and loads. However, there is no way to control the combination. An analysis of the complete system should be made. This requires a study of the mass elastic system of the combination, involving the mass weight and radius of gyration of all rotating parts. This study should be made either by the engine or pump manufacturer or by a torsional-analysis specialist.

FURTHER READING

Gunther, F. J.: "Gas Engine Power for Water and Wastewater Facilities," *Water & Sewerage Works* **112** and **113** (November 1965 to July 1966).

6.1.4
HYDRAULIC TURBINES

WARREN G. WHIPPEN

PUMPING APPLICATIONS _____

In many pumping applications the liquid remains at a high pressure after it has completed its cycle. It is often economically desirable to recover some of the energy, which will otherwise have to be dissipated when the liquid is brought back to a lower pressure. Instead of running the liquid through a pressure-reducing valve and therefore destroying the energy, a hydraulic turbine may be installed, and this turbine can therefore assist in driving the process pumps. There are many such applications in use. Among these are the use of turbine-driven pumps in gas-cleaning operations. Here the solutions from scrubber towers are passed through turbines which in turn drive pumps. Some of the solutions involved in such a process are water saturated with carbon dioxide at a temperature of 40 to 70°F (4 to 21°C) and potassium carbonate containing dissolved carbon dioxide at a specific gravity of 1.31. A liquid at the much lower specific gravity of 0.84 has been used in power-recovery turbines in glycol-ethylene hydration processes.

One of the oldest applications of turbine-driven pumps was for fire-fighting equipment in mills having a natural head of water. The water was routed through a hydraulic turbine which drove the pump to pressure a sprinkler system or provide water to the fire-hose connections. Today hydraulic turbines are used to drive pumps that generate fire-fighting foam.

At thermal power plants, cooling water returning from the cooling towers has been used to drive a hydraulic turbine directly connected to a pump that provides part of the cooling water. Naturally in such an application the power from the turbine alone is not enough to maintain the pumping system and therefore an auxiliary power source is also required.

Turbines using oil pressure have been employed at thermal power plants. Large steam turbines obtain bearing oil and control oil pressure from a main feed pump that is directly connected to the steam turbine shaft. A small amount of the oil at this high pressure is used for the relay control, with the majority of the oil going to the bearings at a lower pressure. To reduce the oil pressure from that required by the relays to that needed by the bearings, the oil is passed through a turbine connected to a booster pump, which pressurizes the oil at the intake of the main pump. Since it is preferable to locate pumps and electrical equipment above flood level, some recent steam power plant installations have used the return from the condenser to the river (normally an appreciable drop in head) to drive a hydraulic turbine which in turn may assist in driving the condensate pumps.

It is often economical to use a small volume of water at a high head to move a large volume of water at a low head. Of course, the reverse application can also be accomplished. The former procedure is employed at hydroelectric projects where it is necessary to operate fishways to enable migrating fish to continue traveling upstream over the dam. The large volume of water is used both to attract the fish to the fish flume and to transport the fish to the fish ladders. An example of such an installation can be seen in Fig. 1.

Turbines have also been used to start large pumping units when the hydraulic conditions are suitable. The impulse turbine, which develops maximum torque at zero speed, is especially useful for this application. The power required on such a starting turbine would be less than the power required on a starting motor to do the same job. Many electric controls can be eliminated when starting with a turbine. An illustration of this application is shown in Fig. 2.

Desalinization plants have been proposed wherein salt is removed from seawater by pumping it through a membrane at high pressure. Only part of the water goes through the membrane, with most of the water being used to carry away the salt. The excess high-pressure water is then put through a power-recovery turbine which helps to drive the high-pressure pumps.

It is often desirable to have the turbine and pump running at different speeds, and this has been accomplished by means of a variable-speed coupling between turbine and pump. Excellent low-speed gear increasers or reducers of up to 35,000 hp (26,000 kW) are being built by many manufacturers. In applications such as starting a pump by means of a turbine, a mechanism may be needed to disengage the turbine after the pump has been started, in order to minimize windage and friction. Disengaging couplings may be hydraulic or mechanical and may be actuated when the unit is stationary or rotating, depending on the application. However, it is possible to allow the turbine to spin in air with the pump after the pump has been started. Figure 3 is an illustration of a turbine-drive pump arrangement.

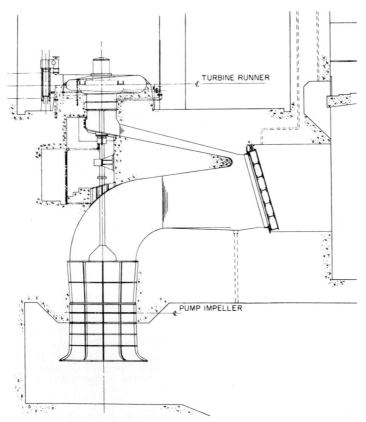

TURBINE RUNNER

PUMP IMPELLER

FIG. 1   Turbine-driven fishway pumps.

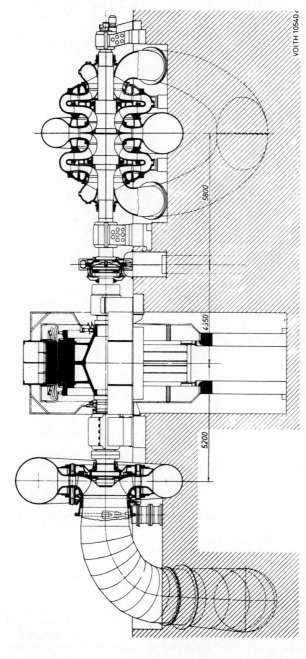

FIG. 2 Impulse runner used to start a large pump. Left to right: Francis turbine, generator motor, starting impulse turbine, and two-stage double-suction storage pump. (Adapted from Voith Publication 1429, J. M. Voith GmbH, Heidenheim, Germany)

VOITH 10540 c

5800

4350

5200

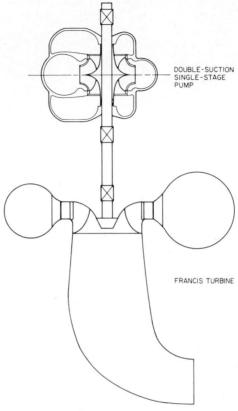

FIG. 3   Double-suction single-stage pump driven by a Francis turbine. (Allis-Chalmers)

TYPES OF TURBINES

There will be three types of turbines discussed: propeller (fixed and adjustable blade), Francis, and impulse. Figures 4 to 7 are illustrations of these turbine runners. Figure 8 shows a complete adjustable-blade unit. The fixed- and adjustable-blade propellers are essentially the same type with the exception that the adjustable is suited to a much wider range of loading conditions. This versatility is reflected in a higher manufacturing cost for the adjustable blade. The Francis runner is much like a centrifugal pump running backward. The impulse runner (or Pelton wheel) is for high-head applications. The water is first channeled through a nozzle which directs a jet of water into the atmosphere surrounding the runner. This jet then strikes the bowl-shaped buckets of the impulse wheel, discharging into the atmosphere. When a large back pressure is present in the system, the impulse wheel cannot discharge properly and the performance drops off rapidly as the back pressure is increased. However, it is sometimes possible to admit air pressure into the impulse runner housing and thereby lower the level of the liquid to below the level of the runner. When the back pressure cannot be eliminated, a Francis machine will perform much better than an impulse machine. Figure 9 illustrates the relative efficiencies of the different types of runners. The fixed-blade propeller and Francis runners are designed to operate most efficiently at a point near full load, and no adjustment can be made when the load drops. The impulse and adjustable-blade runners, however, are designed for high efficiencies over a large load range. The adjustable-blade propeller accomplishes this by changing the angle of its blades by means of linkage in the hub of the runner. With the impulse runner, the size of the jet is controlled by the nozzle, which enables the runner to maintain high efficiencies at low loads.

FIG. 4　Fixed-blade propeller runner.

Specific Speed　This is the speed at which a unit will run if the runner diameter is such that, running at 1 ft (1 m) net head, it will develop 1 hp (1 kW):

in USCS units

$$N_s = \frac{\text{rpm} \times \text{hp}^{1/2}}{\text{ft}^{5/4}}$$

in SI units

$$N_s = \frac{\text{rpm} \times \text{kW}^{1/2}}{\text{m}^{5/4}}$$

The specific speed is an important factor governing the selection of the type of runner best suited for a given operating range. The impulse wheels have very low specific speeds relative to propellers, and the specific speed of a Francis turbine lies between the impulse and propeller. Table 1 shows some statistics on existing units.

FIG. 5　Adjustable-blade propeller runner.

FIG. 6　Francis runner. (Allis-Chalmers)

FIG. 7 Impulse (or Pelton) runner. (S. Morgan Smith)

SIZES AND RATINGS

Although power-recovery turbines are relatively small, there are much larger turbines in existence which could be used to drive pumps. Turbines of the following sizes do exist: propeller, 405.5-in (10.3-m) diameter; Francis, 288-in (7.3-m) diameter; and impulse, 176-in (4.5-m) diameter. The extreme range of power, head, and speed is shown in Table 2.

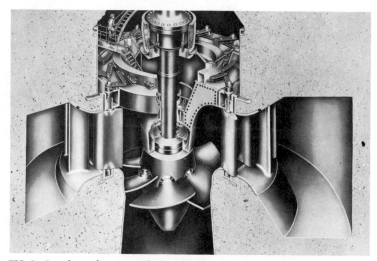

FIG. 8 Complete turbine unit with adjustable-blade propeller. (Allis-Chalmers)

TABLE 1 Turbine Statistics of Existing Units

Speed, rpm	Power, hp[a]	Head, ft[b]	N_s, USCS[c]	Type[d]	Use[e]	Supplier[f]
450	2000	925	3.94	I	SUP	BLH
450	450	485	4.19	I	SUP	BLH
750	170	575	3.47	I	SUP	BLH
340	570	485	3.57	I	SUP	BLH
1775	264	1085	4.63	I	SUP	BLH
437.5	425	405	4.97	I	SUP	BLH
1770	640	960	8.38	I	SUP	BLH
1775	230	1085	4.32	I	SUP	BLH
720	885	460	10.05	I	SUP	BLH
900	640	866	4.85	I	SUP	BLH
3550	85	1920	2.58	I	SUP	BLH
690	535	86.5	60.50	F	FW	BLH
126	670	80	13.63	F	FW	BLH
450	2600	118	58.90	F	SUP	LEF
525	1050	83.5	67.39	F	SUP	LEF
900	750	123	60.18	F	SUP	LEF
1000	375	96	64.43	F	SUP	LEF
700	53	150	9.71	I	SUP	LEF
750	160	30	135.13	P	SUP	LEF
882	128	70	49.26	F	SUP	LEF
1750	145	135	45.78	F	SUP	LEF
1750	220	135	56.39	F	SUP	LEF
280	200	14	146.20	P	SUP	A-C
550	400	35	129.21	P	IR	A-C
3450	7.5	196	12.89	I	PP	A-C
1775	353	1085	5.36	I	ST	A-C
1185	280	231	22.01	I	ST	A-C
1775	300	1510	3.27	I	CH	A-C
1300	1.6	150	3.13	I	PP	A-C
700	1760	798	6.92	I	ST	A-C
600	392	460	5.58	I	SUP	A-C
122	973	65	20.61	F	FW	A-C
108	1200	75	16.95	F	FW	A-C

[a]kW = 0.746 hp.
[b]m = 0.3048 ft.
[c]N_s (SI) = 3.814 N_s (USCS).
[d]I = impulse, P = propeller, F = Francis.
[e]SUP = supplement electric motor to drive pump, FW = drives fishway pumps, ST = scrubber-tower application, PP = turbine drives petroleum pumps, CH = power recovery in chemical plant, IR = irrigation project.
[f]BLH = Baldwin-Lima-Hamilton Corp., LEF = The James Leffel Co., A-C = Allis-Chalmers Corporation.

TABLE 2 Range of Power, Head, Discharge, and Speed of Existing Units of One Manufacturer

	Propeller		Francis		Impulse	
	Low	High	Low	High	Low	High
Power, hp[a]	82.5	268,000	1.2	820,000	1.6	330,000
Head, ft[b]	6.0	180	4.0	2,204	75.0	5,790
Speed, rpm	50	750	56.4	3,500	180	3,600

[a]kW = 0.746 hp
[b]m = 0.3048 ft

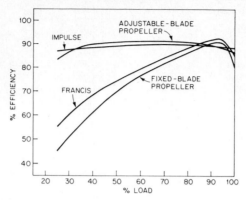

FIG. 9   Efficiency versus load for different runner types.

GENERAL CHARACTERISTICS

Sigma   As in pumps, the problem of cavitation also exists in turbines. Sigma is defined as follows:

$$\sigma = \frac{H_a - H_s}{H}$$

where H_a = feet (meters) of atmospheric pressure minus vapor pressure
  H_s = distance in feet (meters) the centerline of the blades is above tailwater for **vertical** units or distance in feet (meters) the highest point of the blade is above **tailwater for** horizontal units
  H = feet (meters) of elevation between inlet head and tailwater

  Critical sigma is defined as that point at which cavitation begins to affect the performance of the turbine. Figure 10 shows the relationship between critical sigma and specific speed **and also** shows which type of runner is best for a given specific speed. The impulse wheel is not **affected** by sigma since it is a free jet action and therefore not subject to low-pressure areas.

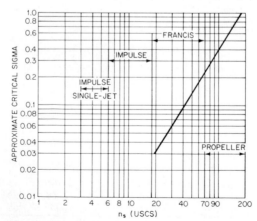

FIG. 10   Critical sigma versus N_s. Specific speed N_s (SI) = $3.814N_s$ (USCS).

Affinity Laws The relationships between head, discharge, speed, power, and diameter can be seen in the following equations, where Q = rate of discharge, H = head, N = speed, P = power, D = diameter, and subscripts denote two geometrically similar units with the same specific speed:

$$\frac{Q_1}{N_1 D_1^3} = \frac{Q_2}{N_2 D_2^3}$$

$$\frac{Q_1^2}{H_1 D_1^4} = \frac{Q_2^2}{H_2 D_2^4}$$

$$\frac{N_1^2 D_1^2}{H_1} = \frac{N_2^2 D_2^2}{H_2}$$

$$\frac{P_1}{N_1^3 D_1^5} = \frac{P_2}{N_2^3 D_2^5}$$

Most designs used are tested as exact homologous models, and performance is stepped up from the model by the normal affinity laws given above. Because of difficulties in measuring large flows at the field installation, only approximate or relative metering is normally done.

Speed Characteristic This is defined as the peripheral speed of the runner divided by the spouting velocity of the water:

$$\phi = \frac{u}{(2gH)^{1/2}}$$

where ϕ = speed characteristic
  u = peripheral speed of runner, ft/s (m/s)
  H = head, ft (m)

A plot of ϕ versus speed can be seen in Fig. 11, which shows that ϕ becomes constant at 0.46 after the specific speed has dropped into the impulse-runner region. Theoretically, for maximum energy conversion, ϕ should equal 0.5 for impulse runners. However, because of small losses in the runner, this value is set at approximately 0.46. The control of the hydraulic turbine can be accomplished by means of a governor or a flow or load control system.

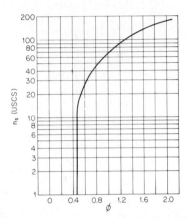

FIG. 11 Specific speed versus ϕ. Specific speed N_s (SI) = $3.814 N_s$ (USCS).

Torque All hydraulic turbines have maximum torque at zero speed; therefore they have ideal starting characteristics. The pump can be accelerated to design speed by gradually opening the turbine gates while keeping within the limitations of hydraulic transients.

DATA REQUIRED FOR TURBINE SELECTION

The turbine supplier should have the following information in order to select the best combination of size and type of runner:

1. Head available and head range.
2. Power and speed required to drive pump.
3. Description of fluid to be handled, including chemical composition and specific gravity.
4. Possibility of adding air to system. Turbine will operate satisfactorily without air, but air may be added to system to reduce pressure fluctuations (normally one-third of rpm in frequency) at part load and possibly to smooth unit operation at full load.

5. If corrosive fluid is to be handled, description of the materials required.

6. Controls from pumping process that will affect turbine operation. If speed is a controlling factor, the possibility of using a speed-varying device should be considered.

7. Back pressure on the turbine.

8. Noise-level limitations for the turbine. Figure 12 illustrates noise levels which have been recorded on some rather large turbine units.

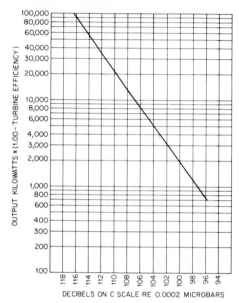

FIG. 12   Sound level readings taken 3 ft (0.9 m) from main shaft. Hydraulic losses are approximated output in kilowatts × (1.00 − turbine efficiency). (From L. F. Henry: "Selection of Reversible Pump/Turbine Specific Speeds," paper presented at meeting on Pumped-Storage Development and Its Environmental Effects, University of Wisconsin, September 1971)

FURTHER READING

Daugherty, R. L., and A. C. Ingersoll: *Fluid Mechanics*, McGraw-Hill, New York, 1954.

Henry, L. F.: "Selection of Reversible Pump/Turbine Specific Speeds," paper presented at meeting on Pumped-Storage Development and Its Environmental Effects, University of Wisconsin, September 1971.

Roth, H. H., and T. F. Armbruster: "Starting Large Pumping Units," *Allis-Chalmers Eng. Rev.* 31 (3)(1966).

Schlichting, H.: *Boundary Layer Theory*, McGraw-Hill, New York, 1960.

Stepanoff, A. J.: *Centrifugal and Axial Flow Pumps*, Wiley, New York, 1957.

Wilson, P. N.: "Turbines for Unusual Duties," *Water Power*, June 1971.

6.1.5
GAS TURBINES

RICHARD G. OLSON

THERMODYNAMIC PRINCIPLE AND CLASSIFICATIONS _____

The gas turbine is an internal combustion engine differing in many respects from the standard reciprocating model. In the first place, the process by which a gas turbine operates involves steady flow; hence pistons and cylinders are eliminated. Secondly, each part of the thermodynamic cycle is carried out in a separate apparatus. The basic process involves compression of air in a compressor, introduction of the compressed air and fuel into the combustion chamber(s), and finally expansion of the gaseous combustion products in a power turbine. Figure 1 illustrates a simple gas turbine.

The Brayton, or Joule, Cycle   Figure 2 shows an ideal Brayton, or Joule, cycle illustrated in PV and TS diagrams. This cycle is commonly used in the analysis of gas turbines.

The inlet air is compressed isentropically from point 1 to point 2, heat is added at an assumed constant pressure from point 2 to point 3, the air is then expanded isentropically in the power turbine from point 3 to point 4, and finally heat is rejected at an assumed constant pressure from point 4 to point 1. From Fig. 2, the compressor work is $h_2 - h_1$, the turbine work is $h_3 - h_4$, and the difference is the net work output. The heat input is $h_3 - h_2$. One can then derive an expression for thermal efficiency as follows:

$$\eta = \frac{(h_3 - h_4) - (h_2 - h_1)}{h_3 - h_2}$$

In reality, the cycle is irreversible and the efficiency of the compression, combustion, and expansion must be taken into account. However, an examination of the thermal efficiency equation shows the need of high compressor and turbine efficiencies in order to produce an acceptable amount of work output. The importance of h_3 is also readily apparent. In noting that h_3 is directly proportional to temperature, one can appreciate why continuous emphasis is being placed on the development of materials and techniques that permit higher combustion temperatures and correspondingly higher turbine inlet temperatures.

Classifications   In an *open-cycle* gas turbine the inlet air mixes directly with the combustion products and is exhausted to the atmosphere after passing through the power turbine. The *closed-cycle* gas turbine uses a heat exchanger to transfer heat to the working fluid, which is continuously recirculated in a closed loop. A *combined cycle* uses the principles of both the open and closed cycles. In current practice, the combined cycle uses an open cycle to provide shaft work while the heat from the exhaust is partially recovered in a waste-heat boiler. The heat recovered then proceeds through a standard steam power cycle until heat is rejected to the most readily available low-temperature reservoir.

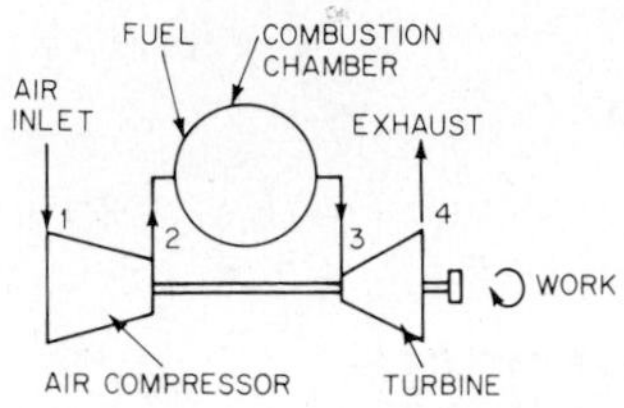

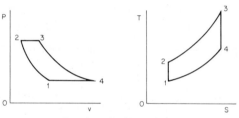

FIG. 1   Components of a simple gas turbine.

Of the above configurations, the open-cycle gas turbine is most extensively used today for driving centrifugal pumps. This is probably due to the important consideration of minimum capital investment for each power output.

Two distinct types of open-cycle gas turbines have evolved, the *single-shaft* and *split-shaft* versions. The single-shaft gas turbine was developed primarily for the electric power industry and uses a compressor and a power turbine integrated on a common shaft. As the unit is used continuously at a single rotational speed, the compressor and power turbine efficiency can be optimized.

The split-shaft gas turbine was developed primarily for mechanical drive applications where output power and speed might be expected to be variable. Figure 3 illustrates such a turbine. A typical curve of power output versus shaft speed is illustrated in Fig. 4.

FIG. 2   *PV* and *TS* diagrams for an ideal Brayton, or Joule, cycle.

Split-shaft gas turbines are available in *conventional* and *aircraft-derivative* versions. The conventional gas turbine evolved from steam turbine technology and is illustrated in Fig. 5. Figure 6 shows a modified jet engine used as a source of hot gas to a power turbine. Note that the jet engine combines the compression, combustion, and power turbine necessary to drive the compressor.

RATINGS

In the evolution of the gas turbine as a prime mover, various organizations have put forth standard conditions of inlet temperature and elevation to allow direct comparison of various gas turbines.

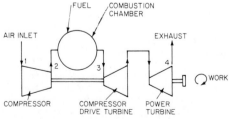

FIG. 3   Components of a split-shaft gas turbine.

Four common standards exist:

ISO (International Standards Organization): sea level and 59°F (15°C)

NEMA (National Electrical Manufacturers Association): 1000 ft (304.8 m) above sea level and 80°F (27°C)

CIMAC (Congrès International des Machines à Combustion): sea level and 59°F (15°C)

Site: actual elevation and design temperature at installation site

With the evolution of higher combustion temperatures and with the greater need for power over relatively short daily periods, new ratings have developed: emergency (maximum intermittent), peaking (intermittent), and base load. These classifications are based on the number of hours per unit of time that a gas turbine is operated and are related to the material used in the power turbine blading. A common standard is to use materials suitable for 100,000 h of continuous operation. Higher temperatures are permitted, but at the sacrifice of the life of the material and an increase in maintenance costs.

In a pump-driving application, the cycle of operations should be considered in specifying a gas turbine driver. A typical curve of gas turbine output as a function of inlet temperature (Fig. 7) clearly indicates the necessity of specifying an accurate design temperature. Figure 8 is a typical correction curve for altitude.

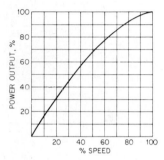

FIG. 4   Typical curve of power output versus shaft speed for a split-shaft gas turbine.

FUELS

A wide range of fuels, from natural gas to the bunker oils, may be burned in simple-cycle gas turbines. In most cases units can operate on gas or liquid fuels, and some turbines have automatic switchover capability while under load.

Although it would appear advantageous to choose the cheapest fuel available, there may be disadvantages in such a choice: impurities such as vanadium, sulfur, and sodium definitely result in high-temperature corrosion; solid impurities and a high ash content will lead to erosion problems. One is therefore often faced with a trade-off between the cost of treating a fuel and the higher original cost.

With ever-increasing firing temperatures, research is continuing in the field of high-temperature corrosion with or without accompanying erosion problems. Present approaches to the problem include the use of inhibitors in the fuel as well as coatings for power turbine blades. Another solution which has been used on some crude oil pipelines is the topping unit. This is a small still which removes a specific fraction of the crude oil for burning as gas turbine fuel. It is good practice to include a detailed fuel analysis as part of any request for bids.

ENVIRONMENTAL CONSIDERATIONS

Legislation at both the federal and state levels has been concerned with the protection of our environment. An engineer must now consider the effects of the project on the environment.

Noise   Any piece of dynamic mechanical equipment will emit airborne sounds which, depending on frequency and level, may be classified as noise.

The primary sources of noise in the gas turbine are the turbine inlet and exhaust and the accessories and support systems, such as fin fan lubrication oil coolers, auxiliary air blowers and fans, starting devices, and auxiliary lubrication oil pumps. It is common practice for manufacturers to provide inlet and exhaust silencers as well as some form of acoustic treatment for auxiliaries.

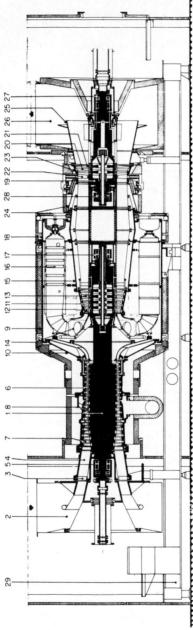

FIG. 5 Gas turbine cross section: (1) gas generator turbine rotor, (2) inlet air casing, (3) compressor inlet guide vanes, (4) conical inlet casing with bearing support, (5) compressor radial-axial bearing, (6) compressor stator, (7) compressor stator blades, (8) compressor rotor blades, (9) combustion chamber main casing, (1) intermediate conical casing with diffuser, (11) gas generator turbine stator blade carrier, (12) gas generator turbine stator blades, (13) gas generator turbine rotor blades, (14) hot gas casing, (15) intermediate casing, (16) nine incorporated combustion chambers, (17) compressor radial bearing, (18) central casing, (19) power turbine rotor, (20) power turbine stator blades, (21) power turbine rotor blades, (22) power turbine adjustable rotor blades, (23) power turbine stator blade carrier, (24) conical intermediate piece, (25) outlet diffuser, (26) outlet casing, (27) power turbine radial-axial bearing, (28) power turbine radial bearing, (29) metallic foundation. (Turbodyne)

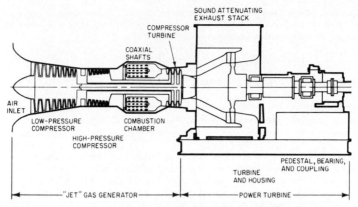

FIG. 6 Cross section of aircraft-derivative gas turbine driver. (Turbodyne)

In considering the degree of acoustic treatment necessary for each installation, the practicing engineer must consider the following parameters:

1. Federal law (Walsh-Healy Act) limiting the time a worker may spend in a noisy environment

2. Local codes and their interpretation

3. Plant location and existing noise level at site

4. Site topography, including any noise-reflective surfaces

5. Applicable ASME standards and other relevant standards

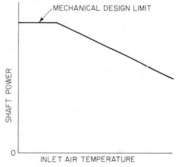

FIG. 7 Typical curve of gas turbine output versus inlet temperature.

Emissions The federal air quality acts require each state to develop a plan for achieving satisfactory air quality. Specifically, goals have been put forth to limit suspended particulate matter and oxides of sulfur and nitrogen. Therefore the engineer must investigate existing regulations during the planning stage of a pump installation.

In general, it can be stated that gas turbines have low particulate emissions. The amount of sulfur oxides exhausted to the atmosphere is in direct proportion to the content of sulfur in the

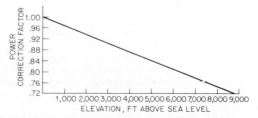

FIG. 8 Typical curve of gas turbine power correction versus altitude. (ft × 0.3048 = m)

fuel, and current practice calls for elimination at the source. The formation of oxides of nitrogen is a direct result of combustion. Manufacturers are currently committing a considerable amount of resources to the investigation and solution of this problem.

GAS TURBINE SUPPORT SYSTEMS

Starting Systems   A form of mechanical cranking is necessary to bring a gas turbine up to its self-sustaining speed. The amount of energy necessary will depend on each manufacturer's design. Available systems include electric motors, diesel engines, and gas-expander turbines.

Lubrication   The gas turbine manufacturer normally provides a combined pump-turbine lubrication oil system. A main lubrication oil pump of sufficient capacity for the combined system is necessary, in addition to a standby pump in the event of failure of the main pump. A reservoir sized for a retention time of at least 4 min is usually specified. Filtration to 10-μm particle size should be adequate for most gas turbines and pumps.

Lubrication oil can be cooled by various means, the selection of which depends upon local conditions. The simplest and cheapest method uses a shell-and-tube heat exchanger with water as the cooling medium. In arid regions a fin fan cooler with direct air-to-water cooling is commonly used.

Inlet Air Filtration   The degree of inlet air filtration needed is primarily a function of the size and number of particles in the atmosphere surrounding the installation. In most cases a simple tortuous-path precipitator will suffice. For dirtier atmospheres or in arid regions where sandstorms occur, inertial separators followed by a rolling-media filter should be considered.

Any filtration will result in a loss in performance because of the pressure drop across the inlet filter. Conservative design practice calls for a face velocity of 500 ft/min (150 m/min) for a rolling-media filter and 1700 ft/min (520 m/min) for a tortuous-path precipitator.

Control   Most manufacturers offer control packages which provide proper sequencing for automatic start-up, operation, and shutdown. During automatic start-up, the sequencer receives signals from various transmitters to ensure that auxiliaries are functioning properly and with the aid of timers brings the unit on line through a planned sequence of events.

Key operating parameters—for example, output speed, gas turbine compressor speed, turbine inlet temperature, lubrication oil temperature and pressure—are continually monitored during normal operation. Signals from a pump discharge-pressure transmitter can be fed into a speed and fuel controller to cause the unit to respond to changes in speed and/or output.

As with start-up, normal and emergency shutdowns are accomplished through the sequencer. Most standard control packages are easily adaptable to remote control by means of cable or microwave.

APPLICATION TO PUMPS

Gas turbines are available to drive centrifugal pumps in a wide range of speeds and sizes, from 40 hp (30 kW) to over 20,000 hp (15,000 kW). It is not practical to list here all the available units since they are too numerous and are continually being upgraded and added to. A listing can be found in *Sawyer's Gas Turbine Catalog*, which is published annually.

Pipeline Service   Crude oil pipeline service has seen a tremendous growth in the use of gas turbine drivers since the mid 1950s, and this trend will continue as crude oil production becomes more and more remote from its markets. The advantages of the gas turbine for this application are as follows:

1. Installed cost is usually lower than that of a corresponding reciprocating engine.
2. Variable-speed operation allows maintenance of a specific discharge pressure under a wide range of operating conditions, thus achieving maximum flexibility.

FIG. 9 A 3300-bhp (2460-bkW) gas-turbine-driven waterflood package. (Solar Division of International Harvester)

3. Normal gas turbine control system is easily adapted to unattended operation and remote control.

4. Operating experience has proved the gas turbine to have a high degree of reliability.

5. Gas turbines can be packaged into modules for ease of transportation and erection.

Water Flooding   An important aspect of oil production is the use of secondary recovery methods to increase the output of crude oil reservoirs whose pressure does not allow the crude to flow freely to the surface. Flooding the reservoir with high-pressure water has been a primary technique for years. The development of high-pressure centrifugal pumps has allowed increased water flow into oil fields.

As most oil production is located in remote areas, the packaged gas turbine driver has become increasingly popular. Figure 9 shows a typical 3300-bhp (2460-brake kW) split-shaft gas-turbine-driven centrifugal water flood pump package. A cutaway view of a typical 1100-bhp (820-brake kW) gas turbine is shown in Fig. 10; note the reduction gearing at the exhaust end for direct

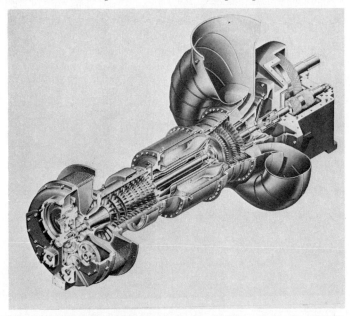

FIG. 10   Cutaway view of 1100-bhp (820-bkW) split-shaft gas turbine driver. (Solar Division of International Harvester)

TABLE 1 Typical Data Sheet for Gas Turbine Driven Centrifugal Pump

Information from purchaser

Pumping requirements
- Service______
- Liquid______
- Pumping temp.______
- Capacity (total) normal/max______/______
- No. pumps operating______
- Specific gravity at pump temp.______
- Viscosity at pumping temp.______
- Total head______
- NPSH available______

Pump type, materials, and accessories

Site conditions
- Elevation, ft(m)______
- Range of site ambient temperature

	Dry bulb	Wet bulb
Design, °F (°C)	______	______
Maximum, °F (°C)	______	______
Minimum, °F (°C)	______	______

Atmospheric air
- Dust below 10 μm, ppm______
- 10 μm and above, ppm______
- Corrosive constituents:
  - Sulfur, ammonia, ammonium salts, salt or seacoast, other______

Noise specifications
- City, state, federal, other______

Emission specifications
- City, state, federal, other______

Utilities available at site
- Steam: Pressure, lb/in^2 (kPa) gage______
  - Temp., °F (°C)______
  - Quantity, lb/h (kg/h)______
- Electricity

	V	Phase	Cycles
ac	______	______	______
dc	______	______	______

- Cooling water
  - Source______quality______
  - Supply temp.______min______max
  - Supply press, lb/in^2 (kPa) gage______
  - Max return, °F (°C)______

Information from manufacturer

Pumping requirements
- Service______
- Liquid______
- Pumping temp.______
- Capacity (total) normal/max______/______
- No. pumps operating______
- Specific gravity at pump temp.______
- Viscosity at pumping temp.______
- Total head______
- NPSH available______

Bhp (brake W)		
required	____normal	____max
Speed, rpm	____normal	____max
Efficiency	____normal	____max

Pump type, materials, and accessories

Gas turbine excluding gear

	Design	Max	Min
Dry bulb temp., °F (°C)	____	____	____
Output, hp (kW)	____	____	____
Heat rate, Btu/hp·h (J/kWh)(LHV)	____	____	____
Output shaft speed, rpm	____	____	____
Turbine inlet temp., °F (°C)	____	____	____
Gear loss	____	____	____
Net hp (kW)	____	____	____

Total utility consumption
- Cooling water, gpm (l/s)______
- Electric power, kW ac/dc______
- Steam, lb/h (kg/h)______
- Compressed air, standard ft^3/min (m^3/min)______

Shipping data

	Turbine	Aux. items
Shipping wt, tons (kg)	______	______
Max erection wt, tons (kg)	______	______
Max maint. wt, tons (kg)	______	______
Length, ft-in (m)	______	______
Width, ft-in (m)	______	______
Height, ft-in (m)	______	______

Accessories included (as required by purchaser)

Inlet air filter	___yes	___no
Inlet air silencer	___	___
Exhaust silencer	___	___
Exhaust duct	___	___

TABLE 1 Typical Data Sheet for Gas Turbine Driven Centrifugal Pump

Information from purchaser	Information from manufacturer		
Fuel	Starting equipment	___yes	___no
Gas_____liquid_____	Load gear (if required)	___	___
Analysis attached	Driven pump	___	___
	Coupling	___	___
Accessory items required (see list in	Fire protection system	___	___
right-hand column)	Equipment enclosure	___	___
	Baseplate or soleplates	___	___
	Combined turbine-pump	___	___
	Lub oil system	___	___
	Main lube pump	___	___
	Auxiliary lub. pump	___	___
	Lub. reservoir	___	___
	Lub. filter	___	___
	Lub. oil cooler	___	___
	Unit control panel	___	___
	Auxiliary motor	___	___
	Control center	___	___

SOURCE: Adapted from API Standard 616.

driving of a pump. Supporting systems, such as starter, lubrication oil pumps, governor, and fuel oil pumps, are driven off the accessory pods located at the air inlet end. Offshore platform installations require drivers with a minimum vibration as well as small, unbalanced inertial forces. The gas turbine fits both of these descriptions very well.

Cargo Loading   Another interesting application of this prime mover is in the field of cargo loading, where units are currently in operation charging tankers with crude oil. Selection of a gas turbine pumping unit with critical speeds above the normal operating range allows great flexibility of operation, particularly during final topping operations.

Application Considerations   The application of gas turbine drivers can vary from a simple driver for one pump operating at constant flow and discharge pressure to a multiplicity of units operating at variable speed on a pipeline. For the purposes of this discussion, it is assumed that the pumping system has been analyzed, a pump selection has been made, and all possible operating conditions have been analyzed so that brake horsepower (brake kilowatt) and speed requirements are known.

The brake horsepower (brake kilowatt) output of the gas turbine must equal or exceed that required by the pump. This output can be determined by the use of specific performance curves similar to Fig. 7 as corrected for elevation (Fig. 8). Gear losses as necessary are added to the brake horsepower (brake kilowatt) required by the pump. Intermediate brake horsepower (brake kilowatt) and speed requirements should then be checked against a gas turbine output versus speed curve (Fig. 4).

A torsional analysis of the combined unit is made (usually by the gas turbine manufacturer) to ensure the absence of any critical speeds in the operating range. Table 1 is a typical data sheet recommended for use when purchasing a gas turbine to drive a centrifugal pump.

FURTHER READING _____

Baumeister, T., and L. Marks: *Standard Handbook for Mechanical Engineers*, 8 ed., McGraw-Hill, New York, 1978.

Combustion Gas Turbines for General Refinery Services, API Standard 616, 1982.

Dacy, J. R.: "Gas Turbines Used on Arctic Pipelines," *Oil Gas J.* **200**:26 (1973).

Proposed Gas Turbine Procurement Standard, ASME.

Sawyer, J. W. (ed.): *Sawyer's Gas Turbine Engineering Handbook,* 3 ed., Gas Turbine Publications, Stamford, Conn., 1985.

Schiefer, R. B.: "The Combustion of Heavy Distillate Fuels in Heavy Duty Gas Turbines," *ASME* 71-GT-56, 1971.

Shepherd, D. G.: *Introduction to the Gas Turbine,* 2 ed., Van Nostrand, Princeton, N.J., 1960.

SECTION 6.2
SPEED-VARYING DEVICES

6.2.1
EDDY-CURRENT COUPLINGS

W. J. BIRGEL

DESCRIPTION

The eddy-current coupling is an electromechanical torque-transmitting device installed between a constant-speed prime mover and a load to obtain adjustable-speed operation. Generally ac motors are the most commonly used pump drives, and they inherently operate at a fixed speed. Insertion of an eddy-current slip coupling into the drive train will allow desired adjustments of load speeds.

In most cases, the eddy-current coupling has an appearance very similar to that of its driving motor except that the coupling has two shaft extensions. One shaft, which operates at constant speed, is connected to the motor, while the other shaft, providing the adjustable-speed output, is connected to the load. A typical self-contained air-cooled eddy-current slip coupling is shown in Fig. 1.

The input and output members are mechanically independent, with the output magnet member revolving freely within the input ring or drum member. An air gap separates the two members, and a pair of antifriction bearings maintain their proper relative position. The magnet member has a field winding which is excited by direct current, usually from a static power supply. Application of this field current to the magnet induces eddy currents in the ring. The interaction between these currents and magnetic flux develops a tangential force tending to turn the magnet in the same direction as the rotating ring. The net result is a torque available at the output shaft for driving a load. An increase or decrease in field current will change the value of torque developed, thereby allowing adjustment of the load speed. Field current, output torque, and load speed are usually not proportional. Therefore load torque characteristics and speed range must be known for proper adjustable-speed drive size selection.

Most eddy-current couplings have load speed control. An integral part of this system is a tachometer generator or magnetic pickup which provides an indication of exact output speed and enables control of load speeds within relatively close tolerances regardless of reasonable variations in load torque requirements. Slip-type adjustable-speed drives without load speed control will experience fluctuations in load speed when load torque requirements vary.

FUNDAMENTALS

The eddy-current coupling, like many other adjustable-speed drives, operates on the slip principle and is classified as a torque transmitter. This means that the input and output torques are essen-

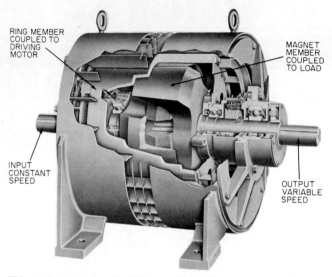

RING MEMBER
COUPLED TO
DRIVING
MOTOR

MAGNET
MEMBER
COUPLED
TO LOAD

INPUT
CONSTANT
SPEED

OUTPUT
VARIABLE
SPEED

FIG. 1 Cutaway of an Ampli-Speed eddy-current slip coupling. (Electric Machinery Mfg.)

tially equal, disregarding frictional and windage losses. The motor input power is equal to the sum of load power and what is known as slip loss. This slip loss is the product of slip speed, which is the difference between motor and load speed, and the transmitted torque.

The various relationships may be expressed as follows:

in USCS units $\text{Motor hp} = \dfrac{\text{rpm}_1 \times T}{5250}$

in SI units $\text{Motor kW} = \dfrac{\text{rpm}_1 \times T}{9545}$

in USCS units $\text{Load hp} = \dfrac{\text{rpm}_2 \times T}{5250}$

in SI units $\text{Load kW} = \dfrac{\text{rpm}_2 \times T}{9545}$

in USCS units $\text{Slip loss (hp)} = \dfrac{(\text{rpm}_1 - \text{rpm}_2) \times T}{5250} = \dfrac{\text{rpm}_3 \times T}{5250}$

in SI units $\text{Slip loss (kW)} = \dfrac{\text{rpm}_3 \times T}{9545}$

This may be further expressed as follows:

in USCS units $\text{Slip loss (hp)} = \dfrac{\text{rpm}_3}{\text{rpm}_2} \times \text{load hp}$

in SI units $\text{Slip loss (kW)} = \dfrac{\text{rpm}_3}{\text{rpm}_2} \times \text{load kW}$

or

in USCS units $\text{Slip loss (hp)} = \dfrac{\text{rpm}_3}{\text{rpm}_1} \times \text{motor hp}$

in SI units
$$\text{Slip loss (kW)} = \frac{\text{rpm}_3}{\text{rpm}_1} \times \text{motor kW}$$

where rpm$_1$ = motor speed at designated load
    rpm$_2$ = load speed at designated load
    rpm$_3$ = slip speed at designated load
      T = load torque, ft·lb (N·m)

An eddy-current coupling must slip in order to transmit torque. The normal minimum value of slip for a centrifugal pump application is usually 3%, but values from 2 to 5% are common. The above formulas hold true regardless of the type of load involved.

The efficiency of an eddy-current coupling can never be numerically greater than the percentage of output speed. This effectively takes into consideration only the slip losses, and a true efficiency value must also include frictional, windage, and plus excitation losses. The frictional and windage losses are constant for a fixed motor speed and therefore increase in significance with speed reduction. Excitation losses, on the other hand, decrease with reduction in output speed. The overall effect of these losses is an efficiency versus speed relationship which is somewhat linear with efficiency values anywhere from 2 to 5 points less than the percentage of output speed.

LOAD CHARACTERISTICS

In the application of slip couplings for continuous pump loads, both variable- and constant-torque requirements are encountered.

Variable-Torque Loads   Variable-torque loads are those where the torque increases with the speed and varies approximately as the square of the speed while the load power varies approximately as the speed cubed. The centrifugal pump fits into this classification. To be specific, the above torque-power-speed relationship exists only where the friction head is the total system head, as would be the case if a centrifugal pump were pumping from and to reservoirs having the same liquid levels or in a closed loop.

The various relationships of load power, motor power, and slip loss applying to a typical variable-torque load are shown in Fig. 2, and again frictional and windage losses have been disregarded.

Static heads, which usually exist in centrifugal pump systems, do not significantly affect the selection of a suitable slip coupling for a specific requirement. Pump efficiency, on the other hand, can be quite an important factor when it decreases significantly with speed reduction. Pumps with relatively flat efficiency curves are most desirable for adjustable-speed duty.

Centrifugal pumps usually operate against some static head. This causes the pump torque to follow a closed discharge characteristic until sufficient speed is reached to cause the resultant discharge head to equal or exceed the static head. This is equivalent to a closed discharge or normally unloaded condition. The pump load then follows a different curve to the full-load condition at minimum slip. These characteristics are illustrated in Fig. 3. Curve OAB indicates the closed discharge power. At point A the static head is overcome, and the pump load then rises along curve AC; this causes a significant difference in slip loss as shown. Under conditions as indicated, where static head is not exceeded and water does not start to flow until 80% of full speed is reached, the slip loss never exceeds 10% of the full-load rating. In general, this is true of a pump with high static head.

Constant-Torque Loads   Constant-torque loads are those requiring essentially constant torque input regardless of operating speed. Positive displacement pumps generally are of this type. The load characteristics showing the division between slip loss and load power are illustrated in Fig. 4. Note that the driving motor output does not change, regardless of load speed and power.

The slip loss characteristics make slip couplings undesirable for large power loads if any appreciable speed range is required. However, in relatively small units, the simplicity and ease of speed control will frequently justify the use of slip couplings instead of more efficient but more complex speed-control systems.

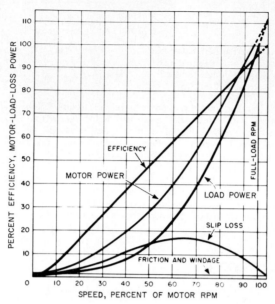

FIG. 2 Load, power, and slip characteristics for a friction-only pumping system.

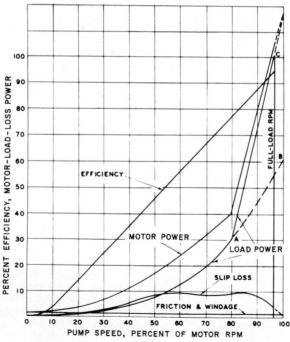

FIG. 3 Percent efficiency, motor load loss power versus pump speed in percent of motor speed for a static head and friction pumping system.

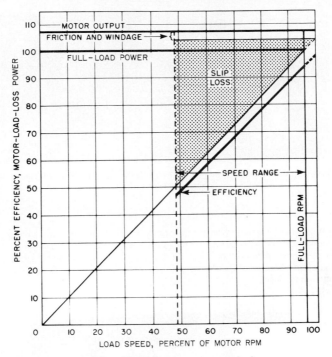

FIG. 4 Power, torque, and losses for a constant-torque load.

The constant-torque-load capacity of a slip coupling is largely limited by slip loss and, to a lesser degree, by breakaway torque and minimum slip. Since slip loss is directly proportional to slip, the desired speed range will definitely limit the torque, and therefore the power, that can be transmitted. In some cases breakaway torque is important as the static friction may be quite high. It is usually recommended that at least 150% of starting torque be available for any constant-torque load.

CONSTRUCTION

Eddy-current couplings are available in both horizontal (Fig. 5) and vertical (Fig. 6) configurations as might be required for any pump mechanical arrangement. Horizontal machines in smaller sizes are frequently close-coupled to the drive motor in what is known as *integral construction*. Larger sizes are usually flexibly coupled to the drive motor and pump load.

Vertical motors and slip couplings are close-coupled to limit overall height and to prevent vibration problems. Pump hydraulic thrust requirements can be accommodated, when necessary, in much the same way as in constant-speed motor applications. Because of mechanical limitations, thrust bearings are frequently located in the bottom of the adjustable-speed drive.

Enclosures available will vary depending on the type of cooling involved. Obviously water-cooled types need little if any enclosure adaptation for virtually any installation. However, caution should be exercised in outdoor use where freezing can occur. Also, vertical installations utilizing water cooling may require special considerations.

Air-cooled couplings are more universally used for centrifugal pump loads but do require attention on enclosure design. Indoor installations in clean atmospheres need only open or drip-

FIG. 5   Horizontal centrifugal pump driven by a 1000-hp (746-kW), 1780-rpm induction motor through a stepup gear and an eddy-current coupling. (Electric Machinery Mfg.)

proof enclosures where heat rejection is not a problem. In some cases, intake and/or discharge covers for connection to ductwork may be necessary for environmental isolation.

Weather-protected enclosures are available for outdoor installations. A NEMA Type I rating is normally adequate because of the slip coupling's inherent mechanical design and relatively low field winding voltage levels.

FIG. 6   Vertical centrifugal pumps driven by 40-hp (30-kW), 1750-rpm induction motors and eddy-current couplings. (Electric Machinery Mfg.)

CONTROLS

Precise speed control of an eddy-current coupling is obtained by the use of feedback circuitry. This consists of an output speed sensing device which produces a voltage proportional to speed, a speed selector which establishes a command voltage, and a resultant voltage which is impressed on an amplifier circuit to control the rectifier supplying excitation to the coupling field. A simplified diagram is shown in Fig. 7.

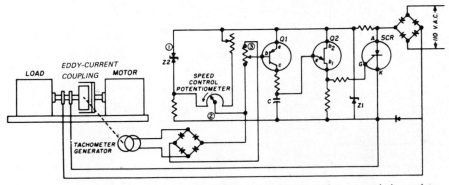

FIG. 7 Simplified diagram of an eddy-current coupling control. Semiconductor devices provide the rectifying and control elements for adjusting the speed of the eddy-current coupling and for maintaining output speed at the desired point. (Electric Machinery Mfg.)

Modern semiconductor technology involving transistors, silicon-controlled rectifiers, and integrated circuits is used extensively in circuit design. Repositioning of the speed control potentiometer creates an imbalance in the control circuitry. This results in an excitation change to regain balance by modification of the speed feedback signal to a suitable value at the new operation point. The response speed of the control is extremely high. In all cases of significant speed change, the excitation is maximum or minimum, resulting in a net torque differential and maximum rate of speed change.

The speed control potentiometer may be positioned by mechanical or electrical means which are responsive to liquid level, pressure, flow, or other parameters. Electrical signals can, in some cases, be applied directly to the speed control circuitry or may be conditioned for such use through amplification and isolation.

A typical two-pump liquid-level control system schematic is shown in Fig. 8. One unit is programmed for lead pump operation with the output speed proportional to a range of liquid-level variations. The two units operate in parallel when additional pumping capacity is required. Circuitry is usually provided for lead pump alternation, and alarms for high or low levels, for loss of air, or for other malfunctions can be readily added.

Constant-pressure system controls are also frequently encountered in eddy-current coupling applications. Pressure transmitters and two-mode controllers make up the bulk of the instrumentation package. The interface with the coupling control is otherwise similar to that in the liquid-level control.

APPLICATIONS

Slip couplings are applied to centrifugal pumps for water and waste-water pumping in municipal installations, for boiler-feed pumping, for circulating water and condensate pumping in power plants, for fan and stock pumping in paper mills, and for reciprocating pumping in a multitude of applications and industries.

In the water and waste-water fields, slip couplings are used extensively for raw- and finished-

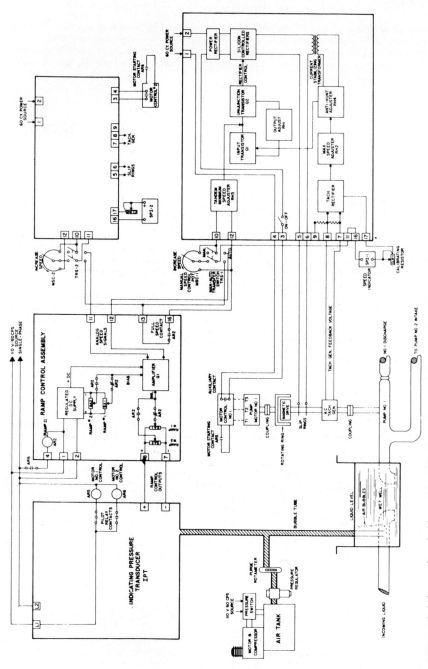

FIG. 8 Block diagram for a double-ramp liquid-level control system. (Electric Machinery Mfg.)

water pumping, lift-station pumping, raw-sewage pumping, and effluent and sludge pumping. Almost any pumping problem, where cyclic constant-speed pumping or throttling or other means of flow control are alternate considerations, can be conveniently solved with the use of an eddy-current slip coupling as the adjustable-speed flow controlling device.

Potable water treatment and distribution facilities are continually confronted with substantial fluctuations in demand through daily, weekly, and even seasonal periods. Distribution systems which depend on direct pumping usually must utilize total or partial adjustable-speed operation for high-service and booster requirements. The quick response of eddy-current slip couplings makes them extremely well suited for this duty.

Waste-water collection systems, where inflow conditions to lift stations and treatment plants vary widely throughout the day, can realize many advantages when designs are based on adjustable speed with eddy-current slip couplings.

RATINGS AND SIZES

Eddy-current couplings are available in a wide range of ratings and sizes, from fractional power units up through 10,000 hp (7500 kW) and beyond. The type of cooling employed is an important consideration; some manufacturers use either water or air exclusively and others use a combination of the two. In addition, the type of load is a factor since thermal capability will vary significantly between water- and air-cooled units.

Centrifugal pump variable-torque loads are usually best handled by air-cooled couplings having high-torque capabilities at low slip values and limited heat-dissipating capabilities. The selection chart shown in Table 1 is representative of one manufacturer's line of couplings designed specifically for centrifugal pump loads. Where full-torque capabilities are realized at 3% slip below motor speed, thermal loads at two-thirds of motor speed will be 16.2% of rated speed load power. Sizes starting at 3 to 5 hp (2.2 to 3.7 kW) and extending up through 3000 to 5000 hp (2240 to 3730 kW) are generally available.

Eddy-current couplings for pump loads requiring constant-torque drives have rather limited usage. In small sizes where thermal capabilities are proportionately greater than the 6:1 ratio encountered on large units, constant-torque loads can be adequately handled by air-cooled couplings. Beyond 100 hp (75 kW), air-cooled units can be impractical because of thermal and starting-torque requirements.

Water-cooled eddy-current couplings have somewhat different characteristics, making them more suitable for constant-torque and large-power variable-torque loading. High starting torque, high minimum slip, and high thermal capabilities all tend to lead to those conclusions. Thermal capabilities are frequently equal to or greater than full-load power ratings. The starting torque is usually the maximum torque, and as a result low slip values are limited.

Figure 9 illustrates the differences in the speed-torque curves of variable-torque air-cooled and constant-torque water-cooled couplings.

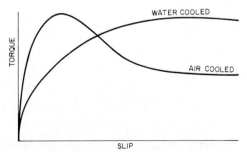

FIG. 9 Speed-torque curves for variable-torque air-cooled and constant-torque motor-cooled couplings.

TABLE 1 Eddy-Current Coupling Selection Chart for Centrifugal Pumps, hp output[a]

Drive motor input speed, rpm	Unit size	Percent of motor full-load speed			
		98	97	96	95
1750	M110	10	16	18	22
	M111	15	22	27	32
	M132	26	37	45	50
	M133	37	51	62	68
	M154	58	74	86	91
	M155	78	97	109	114
	M186	120	148	164	170
	M187	160	200	218	225
	M188	199	248	272	280
	S189	265	320	326	335
	S208	330	385	380	380
	S209	420	405	395	380
	S238	610	597	579	561
	S276	800	775	750	725
	S326	1100	1065	1030	1000
1150	M110	4	6	8	10
	M111	6	9	12	15
	M132	11	17	21	25
	M133	16	24	30	35
	M154	26	37	45	51
	M155	36	50	59	66
	M186	57	75	90	100
	M187	78	102	120	135
	M188	98	127	151	166
	S189	132	168	193	210
	S208	190	210	235	245
	S209	290	305	295	285
	S238	355	380	400	400
1150	S239	470	455	442	428
	S276	615	600	580	565
	S326	870	844	818	793
870	M111	3	5	7	8
	M132	6	9	12	15
	M133	9	13	17	21
	M154	15	21	27	31
	M155	21	29	36	41
	M186	33	45	56	63
	M187	45	62	75	85
	M188	58	80	97	108
	S189	75	110	128	140
	S208	120	145	160	170
	S209	170	225	240	235
	S238	241	265	280	285
	S239	336	375	364	353
	S276	355	490	480	465
	S326	718	696	675	654
700	M154	8	14	18	21
	M155	13	20	25	29
	M186	20	28	35	42
	M187	25	36	45	56
	M188	30	45	55	70
	S189		73	85	95
	S208		110	115	120
	S209		160	185	200
	S238		200	210	217
700	S239		316	316	306
	S276		365	375	405
	S326		597	579	561
585	S189		49	64	75
	S208		80	90	100
	S209		110	135	150
	S238		140	160	170
	S239		223	270	272
	S276		270	280	290
	S326		520	507	492
495	S208			55	65
	S209			100	115
	S238			135	138
	S239			200	220
	S276		205	235	250
	S326		410	460	445
435	S208			55	60
	S209			80	90
	S328			115	120
	S239			167	188
	S276			190	210
	S326		330	390	405
385	S276		275	150	170
	S326			320	340

[a] 1 hp = 0.746 kW.

TABLE 2 Typical Eddy-Current Coupling Downthrust
Capabilities, lb[a]

Unit size	Motor speed, rpm					
	1750	1150	870	700	585	495
M110 & M111	1665	1900	2150			
M132 & M133	1555	1790	2040			
M154 & M155	2615	3065	3465	3825	4005	
M186, 7 & 8	2200	2660	3060	3420	3600	
S189	2090	2550	2950	3310	3490	
S208 & S209	2920	3530	4080	4580	4860	5180
S238 & S239	2240	2860	3410	3910	4190	4500
S276		3900	4605	5170	5670	6150
S326		2545	3260	3810	4290	4780

[a] 1 lb = 4.45 N.

THRUST CAPABILITIES FOR VERTICAL UNITS

Most vertical eddy-current couplings have limited external downthrust load capabilities when provided with standard bearing arrangements. Typical values are listed in Table 2 for the sizes listed in Table 1. Bearing life is an important factor, and the values shown are based on a minimum life of 5 years in accordance with manufacturer's standards at an average output speed of 85% of input speed. Bearings are angular-contact ball type with grease lubrication. Where higher thrust values are encountered or where longer bearing life or adherence to AFBMA life standards must be met, spherical roller bearings with oil lubrication can be furnished.

6.2.2
SINGLE-UNIT ADJUSTABLE-SPEED ELECTRIC DRIVES

E. O. POTTHOFF

Electric drives are available featuring a single rotating element and some associated control for performing the adjustable-speed driving function. The use of a single rotating element differentiates this type of drive from the eddy-current coupling motor, hydraulic coupling motor combinations, and adjustable-speed belt drives, all of which are tandem drives.

The speed of the single-unit adjustable-speed drive is controlled through the interaction of the control and the motor. For this reason, one must consider these two elements as a drive and not consider the motor alone.

Subsection 6.1.1 described motors that will be further discussed in this subsection. Previous discussions of control were limited to that required for starting and protection; this subsection will also describe specific controls required for speed control. The discussion will cover the following types of drives:

- Ac adjustable-voltage drives
- Wound-rotor induction motors with several different types of secondary controls
- Adjustable-frequency drives
- Modified Kraemer drives
- Dc motors with silicon-controlled rectifier (SCR) power supplies

All have physical features, operating characteristics, or prices that make them particularly valuable in some specific segment of the pump driver spectrum.

ALTERNATING-CURRENT ADJUSTABLE-VOLTAGE DRIVES _____

This drive consists primarily of an adjustable voltage, constant frequency control, and a motor (M) conveniently configured as shown in Fig. 1. The motor must possess high slip characteristics and other characteristics that allow it to work successfully with the associated control. The motor is designed for operation and tested with its associated control. Because the motor has high slip characteristics, insulation is of Class F rating.

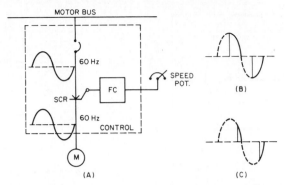

FIG. 1 Block diagram for ac adjustable-voltage drive. (General Electric)

The fundamental parts of the control are a circuit protective device, such as a circuit breaker, an SCR assembly, and firing circuitry, identified as FC in Fig. 1. Other parts are required to complete the control but serve as auxiliaries. The main function of this control is to provide a voltage to the motor at a level that will ensure a desired motor speed. The control also protects the motor under abnormal operating conditions and the motor cable under short-circuit conditions and provides a current-limit function so that the motor draws a maximum of some preselected value, such as 150% of normal, under all operating conditions.

The SCR receives impressed voltage, usually at 60 Hz. The SCR can be turned on (or become conducting) by means of current pulses received from the firing circuits that energize the circuits at different points on the sinusoidal wave. Once conductance starts, it continues until voltage disappears at the end of the half-cycle. The SCR must be pulsed again in order to become conducting. This occurs while the voltage is in the negative phase of its sinusoidal generation and is performed by an SCR connected in parallel to the first and having reverse polarity. The negative loop conductance continues until the voltage again returns to zero. The firing controls are designed to turn the SCR on repetitively sometime during each voltage half-cycle. Figure 1A shows the shape of the applied voltage between the SCR and the motor for the wave illustration shown. In this case, the SCR is turned on at the beginning of each half-cycle.

By delaying the firing, or turning on, of the SCR until later in voltage generation, shorter intervals of applied voltage and lower levels of voltage appear across the motor. The solid line of Fig. 1B illustrates applied motor voltage when the voltage half-wave is half-completed. Figure 1C illustrates the motor-applied voltage when the half-waves are approximately 75% completed. Notice in these illustrations that the frequency of the voltage applied to the motor does not vary; only the voltage magnitude does.

The exact point in the half-wave when firing occurs is controlled by a low-energy electronic signal that may come from a potentiometer, as in Fig. 1A, or from some process instrument signal. By increasing the signal level, voltage applied to the motor increases; on the other hand, a decrease in signal level decreases the voltage level.

Figure 1A shows the SCR on the utility side of the motor as a convenience in illustration. In reality, the SCR is generally placed at the motor neutral point. This reduces voltage from the SCR to ground and allows the motor impedance to protect the SCR somewhat from damage caused by transient voltage spikes entering from the electric utility line. However, placing the SCR at the motor neutral point does require the use of six motor conductors instead of the conventional three.

Figure 2 illustrates representative motor speed-torque curves at rated and other voltages as well as a pump speed-torque curve varying as the square of the speed. The motor possesses approximately 10% slip under rated torque and voltage conditions. Note that motor torque does not break down at speeds lower than breakdown torque speed. These various characteristics help to identify the motor as one designed to operate specifically for this application. Note that by reducing applied motor voltage, motor torque decreases, causing pump operation at lower speed.

Ratings of these drives are limited essentially to low power levels. Figure 3 provides a guide to those available in open construction only. All ratings with maximum rated speeds existing

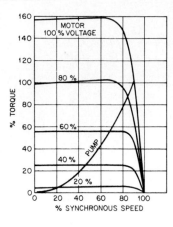

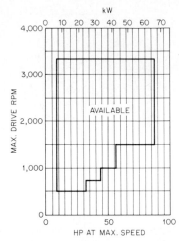

FIG. 2 Torque versus speed for ac adjustable-voltage drive. (General Electric)

FIG. 3 Availability of ac adjustable-voltage drives.

within the solid envelope are available. Some, but very few, are available outside of this envelope. Totally enclosed motors are more restricted in supply than open motors and are limited to approximately 40 hp (30 kW) maximum. In addition to these restrictions, load torques must not exceed values varying as the square of the speed; thus this drive is not a candidate for driving a constant-torque load. Table 1 gives some pertinent application information.

WOUND-ROTOR INDUCTION MOTORS

Liquid Rheostat Controls This form of drive consists primarily of a full-voltage, nonreversing (FVNR) starter, a wound-rotor induction motor, and a liquid rheostat, all of which integrate

TABLE 1 Alternating-Current Adjustable-Voltage Drive Data

Drive element	Power rating[a]	Voltage rating	Max rated speed, rpm	Speed range, %	Enclosure	Mounting	No. of speed points
Motor	See Fig. 3 for open frame ratings; limited availability outside these limits	200 230 460	1640 1095 820 655 545 & others	100	Vertical: shielded dripproof, TEFC not generally available	Vert. or horiz.	Infinite
	Totally enclosed ratings limited as shown under Enclosure				Horizontal: dripproof, TEFC to 40 hp (30 kW) approx.		
Control	5–150 hp (3.7–112 kW)	200 230 460	...	...	NEMA 1 NEMA 12	Wall or floor	Infinite

[a]Drive is capable of operating with power varying as the cube of speed maximum. Specifically, these ratings are not suitable for constant-torque applications.

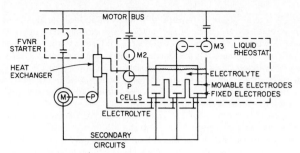

FIG. 4   Block diagram for wound-rotor induction motor with liquid rheostat secondary power. (General Electric)

into a configuration as illustrated in Fig. 4. The FVNR starter switches power to and from the motor stator as well as provides generally accepted protection to the motor from short circuits, overloads, etc. Secondary cables connect the motor rotor and fixed electrodes in the liquid rheostat. The fixed electrodes exist in separate cells of the rheostat, one for each phase. Movable electrodes, one located in each cell, are suspended from a horizontal bar, and this bar, the vertical-electrode suspension bars, the movable electrodes, and the electrolyte filling the cells complete the Y, or common point in the external motor rotor circuit. When the upper electrodes are moved up or down, the secondary circuit resistance of the drive varies and this causes a change in the motor speed-torque characteristic.

As already pointed out, the motor starter provides the normal protective functions for the motor. Control circuitry allows the motor to be started with maximum secondary resistance in the rotor circuit, thus drawing minimum motor current from the power supply.

Figure 4 illustrates a motor (M3) operating a pulley that changes the position of the movable electrode. This motor could be controlled manually by a push button or automatically through a controller position providing either a raise-lower or modulated voltage signal. If desired, a pneumatic cylinder may substitute for motor M3 to provide power to move the electrodes.

The electrolyte receives the drive slip losses dissipated in each of the cells, and of course this heat must be removed into some heat sink capable of dissipating it. In some installations, an electrolyte pump driven by motor M2 circulates the electrolyte through a heat exchanger before returning it to the individual cells. Returning cooled electrolyte re-enters the cells under the fixed electrodes and passes through holes in the electrodes before passing vertically through the cells. The heat exchanger primary coolant can be tap water or mill water, which provides a good means of passing drive slip losses as heat directly out of the station, or it can be the pump discharge water, which conveys heat directly away from the building, as illustrated in Fig. 4.

The liquid rheostat is factory-assembled with the rheostat, electrolyte pump, and its motor and electrode-positioning assembly packaged in a single steel enclosure suitable for control lineups except for ratings above 700 hp (522 kW). The heat exchanger may be of the sleeve (or wrap-around) type, as illustrated in Fig. 4, and comes as a separate device to be attached to station effluent piping. Other forms of exchangers, such as liquid-to-air or shell-and-tube heat exchangers, are available.

Changing the position of the movable electrode results in a change in motor speed-torque relations, as shown in Fig. 5. Each curve shown (except the pump curve) represents a motor characteristic for a discrete secondary resistance. Notice the number associated with each curve; it represents the percentage of secondary resistance external to the motor rotor; 100%

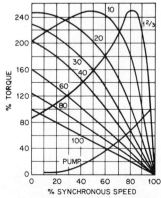

FIG. 5   Torque versus speed for wound-rotor induction motor with variable secondary resistance. (General Electric)

ohms provide 100% motor torque at zero speed. In examining individual curves, note that as the percent resistance increases, the slope of the motor speed-torque curve decreases, thus allowing the motor to slide down the pump speed-torque curve and assume a lower speed.

This drive utilizes resistance only in the motor secondary circuit as a means of controlling speed. Reactance is available as a substitute for resistance in other forms of secondary controllers, but its use reduces the drive power factor and thereby increases motor current, reduces efficiency, and increases motor heating. As a further effect, the additional motor current may require the use of a larger motor frame to accommodate it.

Figures 6a, b, and d provide clues to the availability of enclosed vertical and horizontal wound-rotor induction motors. Figure 6a illustrates the point that vertical, totally enclosed wound-rotor induction motors are not generally available but may be available in random powers and speeds. Totally enclosed water-to-air-cooled (TEWAC) construction becomes available at some minimum point as illustrated. In large ratings, shown on the right, TEWAC construction becomes very important as it provides a convenient way of capturing motor losses and expelling them in cooling water external to the building. Both curves of Fig. 6a terminate at 1800 rpm synchronous speed because higher speed ratings are generally unavailable.

Figure 6b delineates the left-hand areas, where totally enclosed fan-cooled (TEFC) construction should be used for horizontal wound-rotor induction motors because it is the only one available, the middle area, where TEFC construction should be used because it is less expensive than TEWAC construction, and, finally, the right-hand area, where TEWAC construction is less expensive than TEFC construction. Again, motor speeds are limited to 1800 rpm synchronous.

Vertical wound-rotor induction motors are not available in explosionproof construction, as their omission from Fig. 6c implies.

Finally, the areas of availability and unavailability are delineated in Fig. 6d for horizontal wound-rotor induction motors in explosionproof construction. Again, the curve terminates at 1800 rpm synchronous speed for reasons already expressed.

Delineation of motor availability is an important criterion in drive selection. Table 2 presents significant data useful in drive selection.

Tirastat II* Secondary Controls   This drive utilizes the same wound-rotor induction motor and FVNR starter as the preceding but substitutes a Tirastat II controller for the liquid rheostat. Figure 7A shows the configuration with the wound-rotor induction motor and starter identical to those of Fig. 4.

The Tirastat controller in its simplest form consists of the components shown in Fig. 7A. Resistors identified as R1 are permanently connected across the motor secondary phases. Additional resistors (R2) are connected across motor phases through SCR and diodes. The SCR are individually turned on (become conducting) by minute current pulses generated by the firing circuit. As

---

*Trademark of General Electric Co.

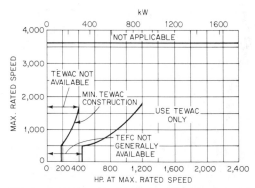

FIG. 6a   Approximate availability of totally enclosed vertical wound-rotor induction motors. (General Electric)

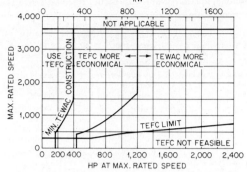

FIG. 6b Approximate availability of totally enclosed horizontal and vertical squirrel-cage induction motors and horizontal wound-rotor induction motors. Wound-rotor induction motor speeds limited to 1800 rpm maximum. (General Electric)

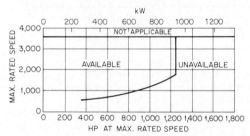

FIG. 6c Approximate availability of explosionproof vertical squirrel-cage induction motors. (General Electric)

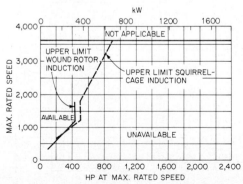

FIG. 6d Approximate availability of explosionproof horizontal squirrel-cage and wound-rotor induction motors. (General Electric)

TABLE 2   Wound-Rotor Induction Motor with Liquid Rheostat Drive Data

Drive element	Power rating[a]	Voltage rating	Max rated speed, rpm	Speed range, %	Enclosure	Mounting	No. of speed points
Motor	No limits for open motors; see Fig. 6a, b, d for restrictions by enclosure	Any NEMA standard	All established by number of motor poles and supply frequency	60	Vertical: shielded dripproof, TEFC generally not available, TEWAC available per Fig. 6a Horizontal: dripproof, TEFC, explosionproof per Fig. 6d	Vert. and horiz.	Infinite
Control	25 hp (18.7 kW) min; no stated max	For any motor secondary voltage	. . .	60	NEMA 1	Lineup or singly Floor only	Infinite

[a]Drive may be designed for constant torque or for torque varying as the square of the speed.

the voltage loop across any SCR drops to zero, the SCR shuts off and does not turn on again until the firing circuit refires it. The diodes allow return of the current to the motor but block out current reversals to the SCR.

Drive speed varies as a function of changing controller average resistance. Thus, resistance can be varied by adjusting time on–time off ratios in the SCR to give the desired average resistance. Figure 7B shows controller secondary resistance with the SCR not firing (conducting). In this condition, the controller resistance value equals the resistance of R1. Figure 7C shows controller resistance with all SCR firing continuously. The decrease in resistance is due to additional

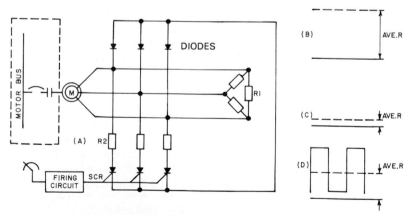

FIG. 7   Block diagram for wound-rotor induction motor with Tirastat II secondary control power. (General Electric)

TABLE 3 Wound-Rotor Induction Motor with Tirastat II Controller Drive Data

Drive element	Power rating[a]	Voltage rating	Max rated speed, rpm	Speed range, %	Enclosure	Mounting	No. of speed points
Motor	No limits for open motors; see Fig. 6a, b, d for restrictions by enclosure	Any NEMA standard	All established by number of motor poles and supply frequency	60	Vertical: shielded dripproof, TEFC generally not available, TEWAC available per Fig. 6a Horizontal: dripproof, TEFC, explosionproof per Fig. 6d	Vert. and horiz.	Infinite
Control	5 hp (3.7 kW) min. 600 hp (448 kW) max.	For any motor secondary voltage	. . .	50 70	NEMA 1 NEMA 12	Lineup or singly	Infinite

[a]Drive may be designed for constant torque or for torque varying as the square of the speed.

resistance placed in parallel with resistor R1. Figure 7D shows controller resistance varying from maximum to minimum in approximately equal time periods. The average resistance value then approximates half of the sum of R1 and R2. The time on–time off of the SCR can be varied automatically by the controller to provide the average resistance in the motor circuit necessary to give the desired motor speed.

The motor starter provides normal motor protective functions as well as short-circuit protection of the motor cables and starter. The Tirastat II controller monitors motor current and limits it to some preselected value, such as 150% of normal, under all conditions.

Packaging of the secondary controller is of particular interest. Both packaging and configuration lend themselves to mounting the resistors on or near the control enclosure or removing them some distance from the enclosure. By placing the resistors outdoors or at some indoor location away from the operations, heat losses can be released to the environment with impunity to operators.

The speed-torque ability of this drive would be very close to that illustrated by Fig. 5. The comments contained in the text describing that figure apply equally well here. Table 3 presents significant data useful in drive selection.

Contact Secondary Controls Many pump drives utilize a very simple combination of FVNR starter, wound-rotor induction motor, and a form of contact making secondary control. Figure 8 shows the configuration with a motor and FVNR starter identical with those of Figs. 4 and 7A.

The main difference between this drive and the two described previously lies in the construction of the secondary controller and the characteristics of the drive. Resistor R in

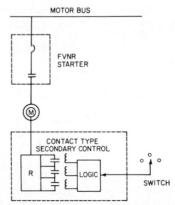

FIG. 8 Block diagram for wound-rotor induction motor with magnetic secondary control power. (General Electric)

TABLE 4 Wound-Rotor Induction Motor with Contact-Type Secondary Control Drive Data

Drive element	Power rating[a]	Voltage rating	Max rated speed, rpm	Speed range, %	Enclosure	Mounting	No. of speed points
Motor	No limits for open motors; see Fig. 6a, b, d for restrictions by enclosure	Any NEMA standard	All established by number of motor poles and supply frequency	60	Vertical: shielded dripproof, TEFC generally not available, TEWAC, available per Fig. 6a Horizontal: dripproof, TEFC, explosionproof per Fig. 6d	Vert. and horiz.	
Control	No limits	For any motor secondary voltage	. . .	75	NEMA 1	Lineup or singly	Discrete only

[a]Drive may be designed for constant torque or for torque varying as the square of the speed.

Fig. 8 is a three-phase resistor connected across the slip rings of the motor. Its resistance rating is selected to provide minimum motor torque at standstill or adequate torque at minimum speed. Contacts in the form of magnetic contactors, drum, cam, or dial switches are closed by a logic device (generally magnetic) to short-circuit part or all of the resistor. Thus, instead of a modulated or infinitely variable resistor as encountered with the other two drives, this drive modifies secondary resistance in discrete steps, providing discrete speed-torque curves instead of an infinite family. The number of speed-torque curves the control generates will depend directly on the number of contacts provided in the secondary-control circuit.

Either a manual switch or some automatic contact-making device actuated by the process can be utilized to actuate the control logic.

Secondary controls are normally packaged in the factory with resistors and contact-making secondary controller integrally assembled.

The speed-torque capability of this drive would be very close to that illustrated in Fig. 5, with the exact number of curves determined by the number of contacts of the contact-making secondary controller and the shape of the curves determined by the resistance selected. Table 4 presents significant data useful in drive selection.

ADJUSTABLE-FREQUENCY DRIVES

This drive consists of an adjustable-frequency inverter control and a constant-speed motor, as shown in Fig. 9. The inverter control is generally built in either pulse width modulation (PWM) or square wave construction. The following description is of a PWM inverter. The control consists fundamentally of a circuit-protective device, a diode bridge, and an inverter section consisting of an SCR and a firing-control section (FC) for the SCR. All are packaged in a steel enclosure. Required auxiliaries, such as special power supplies, motor protective devices, and electric protective devices, complete the package.

Only low voltages at 60 Hz are used to energize the motor bus. A diode bridge rectifies voltage. In turn an inverter inverts the direct current into an adjustable-frequency adjustable voltage which varies linearly with frequency. In one design, this voltage is not sinusoidal but rather consists of a number of dc pulses of positive or negative polarity, as shown between inverter and motor in Fig. 9. The firing controls modulate the width of the pulses and the number of pulses per half-cycle to vary the apparent frequency and to maintain motor voltage at a constant volts-per-cycle value. The firing controls automatically introduce additional pulses or withdraw pulses as bandwidths reach their limits. Different manufacturers feature different designs; however, all designs have the same objective, namely, varying output frequency to obtain different motor speeds.

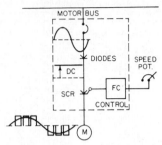

Firing circuits composed of solid-state devices trigger the SCR in accordance with logic controlled by the setting adjustment of a speed potentiometer, as indicated in Fig. 9, or the signal from the process. The control provides normal protection of the motor as well as short-circuit protection of control, motor, and cables. In addition, the control monitors and limits current drawn by the motor under all conditions to a preadjusted value, such as 150% of normal.

FIG. 9 Block diagram for ac adjustable-frequency drive. (General Electric)

Figure 10 displays some representative speed-torque characteristics of the drive utilizing a squirrel-cage induction motor and a representative centrifugal pump. Note that each characteristic intersects the zero-torque point at a different speed value. This characteristic differs from those of other drives, illustrated in Figs. 2 and 5. The steepness of the slope of each characteristic as it rises from its zero-torque value indicates its low-speed regulation and provides a clue to its ability to maintain speed with little fluctuation as load torque varies slightly. It should be obvious that motor speed is a function of frequency adjustment; the voltage is adjusted only to accommodate changes in motor impedance. Table 5 lists significant application data.

MODIFIED KRAEMER DRIVES

Kraemer drives have been used as heavy industrial drives for many years. Although functioning very successfully, each has three rotating units requiring maintenance, and each has significant losses.

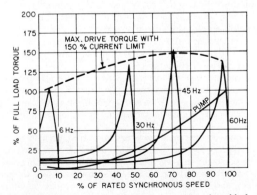

FIG. 10 Torque versus synchronous speed for adjustable-frequency drive. (General Electric)

TABLE 5   Adjustable-Frequency Drive Data

Drive element	Power rating[a]	Voltage rating	Max rated speed, rpm	Speed range, %	Enclosure	Mounting	No. of speed points
Motor	All NEMA ratings for open motors; see Fig. 6b, c, d for restrictions by enclosures	460 or 230	All established by number of motor poles and supply frequency	100 plus limited over-speed	Vertical shielded dripproof, TEFC per Fig. 6b, explosion-proof per Fig. 6c Horizontal: dripproof, TEFC per Fig. 6b explosion-proof per Fig. 6d	Vert. or horiz.	Infinite
Control	Up through 800 hp (597 kW); higher ratings available as custom-built units	460 or 230	Capable of producing 150% of rated frequency but at reduced torque	150	NEMA 1 or NEMA 12	Floor	Infinite

[a]Drive may be designed for constant torque or for torque varying as the square of the speed.

In recent years the advance of semiconductors has simplified the drive configuration to that of Fig. 11, which involves an FVNR starter, a wound-rotor induction motor, a solid-state converter, and an acceleration section. The acceleration section and contactors 1 and 2 can be omitted if the converter rating is large. The FVNR starter switches power to the motor and protects both the motor and the converter. Contactor C2, if provided, serves as a synchronizing contactor between converter and power line and is closed only when converter and line frequencies and voltages are compatible. This contactor may be placed on either side of the converter unit.

Under running conditions, rotor slip energy of low voltage and frequency (Fig. 11B) flows to the low-frequency side of the converter. The converter unit inverts this voltage to a fixed line frequency. Thus all motor rotor slip energy except converter unit losses are returned to the power source, thereby improving drive efficiency.

Under starting conditions, acceleration occurs with contactor C2 open and C1 closed. Accelerating contactors progressively and automatically short-circuit the resistor R to allow the motor to accelerate to some preselected speed. When converter output voltage and frequency match line values, contactor C2 closes and C1 opens automatically. The drive then operates as already described.

The converter generally consists of a diode bridge and SCRs with their firing circuitry for inverting. Required auxiliaries, such as special power supplies, complete the package. Generally all control materials are packaged in a single lineup to facilitate building and installation.

The motor starter provides normal protection of the motor as well as short-circuit protection of control, motor, cables, and converter. In addition, the converter monitors and limits current drawn from the line under all conditions to a preadjusted value, such as 150% of normal.

Although the modified Kraemer drive is familiar to many, its usefulness as a pump drive will be ascertained only after additional operational experience has been gathered.

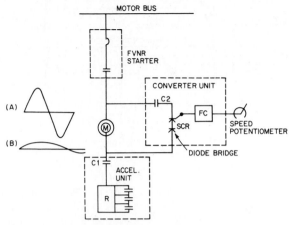

FIG. 11 Block diagram for modified Kraemer drive.

DIRECT-CURRENT MOTOR WITH SCR POWER-SUPPLY DRIVES

General industries use many dc motors with SCR power supplies. These drives perform very well in the exacting circumstances they normally face. Some of these drives are used to drive pumps. They are configured as shown in Fig. 12 and consist of a dc shunt wound motor and an SCR power supply.

The dc power supply unit rectifies motor bus voltage to an adjustable dc voltage level which in turn energizes the armature of the motor. The second function of the power supply unit is to rectify motor bus voltage and apply this constant dc voltage to the motor shunt field. By adjusting the dc voltage to the motor armature, the speed of the motor can be adjusted to a desired value.

The SCR power supply unit consists primarily of a short-circuit protective device, a switching contactor, SCR power modules, firing circuitry, a speed regulator, and a shunt field rectifier. The firing circuitry responds to a low-energy-level speed potentiometer or some process variable. The SCRs in turn respond to the firing circuitry and translate those signals into the correct dc voltage to be applied to the armature of the motor. Control circuitry is included to monitor motor current and limit it to a preselected value, such as 150% of normal.

The SCR power supply unit comes packaged for easy installation in the field.

Speed-torque curves for the drive and for a representative centrifugal pump are shown in Fig. 13. The steepness of the motor torque provides a clue to its stiffness, that is, its ability to resist

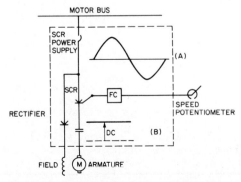

FIG. 12 Block diagram for dc motor with SCR power supply.

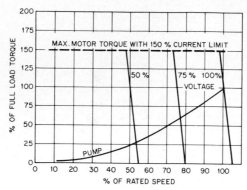

FIG. 13 Torque versus speed for dc motor with SCR power supply.

TABLE 6 Direct-Current Motor with SCR Power-Supply Drive Data

Drive element	Power rating[a]	Voltage rating	Max rated speed, rpm	Speed range, %	Enclosure	Mounting	No. of speed points
Motor	Open horizontal machines can be built at least up to 1500 hp (1120 kW); see Enclosure for other limits	240 500 550	Rating increases inversely with power rating; can match almost all pump speed requirements	At least 95	Vertical: open/ dripproof up through 300 hp, (224kw) totally enclosed up through 200 hp (149kw) Horizontal: open/ dripproof, totally enclosed up through 200 hp (149kw)	Vert. and horiz.	Infinite
SCR power supply	Can be built at least up to 1500 hp (1120 kW)	Any NEMA ac input voltage; will match motor voltage	. . .	At least 95	NEMA 1; NEMA 12	Floor	Infinite

[a]Drive can power either constant-torque loads or those varying as the square of the speed.

speed change because of a change in load torque. Motor torques are limited to some ceiling, such as 150% of normal, by the current limit circuitry of the drive. Table 6 provides significant data useful for drive selection.

DRIVE COMPARISONS

The remaining part of this subsection will deal with drive costs, features, and performances, factors normally factored into criteria involved in drive selection. The relative importance of these factors seldom remains stable but changes as circumstances change. These comparisons cover all drives discussed in this subsection except the modified Kraemer drive and the wound-rotor induction motor with contact secondary control. The former is omitted because its costs have not as yet stabilized, the latter because its costs are highly dependent upon the number of secondary speed steps involved and motor secondary current values.

Drive Costs   Columns 1 and 2 of Table 7 compare the initial costs of the five drives discussed. Prices are shown in terms of percentages, with the wound-rotor induction motor with liquid rheostat taken as 100. Percentages are used instead of monetary costs because the cost relationships of these drives remain fairly stable although prices increase and decrease with time. Column 1 indicates that the ac adjustable-voltage drive has the lowest price for power ratings up through 50 hp (37 kW). This is true at 60 hp (45 kW), but difference between ac adjustable-voltage drive and wound-rotor induction motor and liquid rheostat narrows. At 75 hp (56 kW), the two cost about the same. At 100 hp (75 kW) and above, the ac adjustable-voltage drive prices exceed those of the wound-rotor induction motor with a liquid rheostat, and the availability of the ac adjustable voltage motor becomes irregular. As a result, the ac adjustable-voltage drive becomes generally unattractive at power ratings above 75 hp (56 kW).

At 100 hp (75 kW), the wound-rotor induction motor with liquid rheostat has the lowest initial cost and maintains this advantage through the maximum pump ratings of interest, as indicated in col. 2. The remaining drives cost more.

Of course, initial cost is not the only economic factor making up total cost. Both drive efficiency and power factor may affect the total cost of the drive during its lifetime. Figure 14 is a guide to drive efficiencies for the various drives under consideration. The comparison covers a speed range of 30% with torque loads varying as the square of the speed. A 50-hp (37-kW) 1200 rpm drive is used as a basis for comparison because it represents a probable norm for all pumps in use today. The relative spread between the various drive efficiencies shown in the figure will be maintained even with a change in power or speed, although the curves will be adjusted upward slightly for higher powers and downward slightly for lower powers.

Figure 15 illustrates representative power factors of these drives for a 50-hp (37-kW) drive operating through a 30% speed range with torque varying as the square of the speed. Power factor may be of critical importance in influencing total costs highly dependent on the terms of the rate structure applied by the electric utility furnishing power.

Operating Characteristics   Operating characteristics provide another criterion for drive comparisons. In general, a comparison can be made by reviewing the following characteristics of each drive:

Characteristic	Reference
Accelerating currents	Col. 3 of Table 7
Accelerating torques	Figs. 2, 5, 10, 13
Maximum rated speed	Tables 1–6
Speed range	Tables 1–6
Number of speed points in speed range	Tables 1–6

TABLE 7 Comparisons of Drive Types

Type	(1)[a] Relative initial uninstalled approx. cost 5–50 hp (3.7–37 kW), %	(2)[a] Relative initial uninstalled approx. cost 100–500 hp (75–373 kW), %	(3) Approx. max current during starting, %	(4) Wear points in rotating equipment	Wear points in control equipment
Ac adjustable-voltage drive	80	. . . .	100–150	2 motor bearings	None
Wound-rotor induction motor with liquid rheostat	100	100	100–150	2 motor bearings, 1 set of collector rings, 1 set of brushes	Bearings in electrolyte pump and motor Electrolyte pump seals
Wound-rotor induction motor with Tirastat II controller	120	120	100–150	2 motor bearings, 1 set of collector rings, 1 set of brushes	None
Adjustable-frequency drive	120–170	140–180	100–150	2 motor bearings	None
DC motor with SCR power supply: Custom-designed[b]	112–155	140–170	100–150	2 motor bearings, 1 commutator, 1 set of brushes	None
Pre-engineered[b]	90–120	Not applicable	100–150	2 motor bearings, 1 commutator, 1 set of brushes	None

[a]Prices cover open vertical motors only.

[b]Use custom-designed prices for drives allowing physical amendment in factory to provide special drive sequencing and speed-control feature. For applications involving manual starting and stopping and manual speed adjustment as well as a willingness to accept factory standard ratings, use pre-engineered prices.

All the drives examined compare favorably for each characteristic and are considered generally suitable to perform functionally as pump drives.

Mechanical Features The multifaceted pump supply industry places demands on its drives requiring certain features that may serve as criteria as follows:

Mechanical feature	Reference
Enclosure	Tables 1–6 and Fig. 6a, b, c, d
Mounting	Same as above
Bearing capabilities	All have generally adequate radial and thrust capabilities
Hollow shafts for verticals	All are in supply except dc motor (available with solid shafts only)

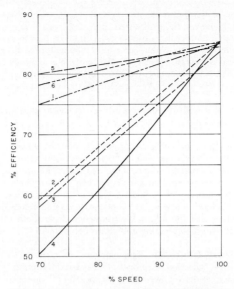

FIG. 14 Efficiency comparison: 50 hp (37 kW), 1200 rpm synchronous speed. (1) PWM adjustable-frequency drive, (2) wound-rotor induction motor with liquid rheostat, (3) wound-rotor induction motor with Tirastat II control, (4) ac adjustable voltage drive, (5) dc motor with SCR power supply, (6) square-wave adjustable-frequency drive. (General Electric)

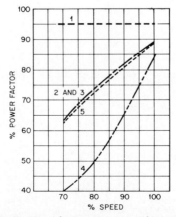

FIG. 15 Power factor comparison: 5 hp (37 kW), 1200 rpm synchronous speed. (1) PWM adjustable-frequency drive, (2 *and* 3) wound-rotor induction motor with liquid rheostat or Tirastat II control, (4) ac adjustable voltage drive, (5) dc motor with SCR power package or square-wave adjustable-frequency drive. (General Electric)

6.107

The reference column shows that the mounting and bearing capabilities of all drives compared are generally adequate. However, the enclosure criterion shows a major variation in availability. The squirrel-cage induction motor obviously possesses all needed enclosure features for both horizontal and vertical mountings. The wound-rotor induction motor has limited availability in vertical, TEFC enclosure and is unavailable in vertical or horizontal explosionproof construction. The ac adjustable-voltage drive is unavailable in vertical TEFC or explosionproof construction and is limited to approximately 40 hp (30 kW) maximum in horizontal TEFC or explosionproof construction.

TEWAC construction is useful, as indicated in Figs. 6a and b. In usual practice, this construction finds use in large drives only.

Mechanical Simplicity This category is generally one of the most important in the minds of operations personnel particularly and one of the best judgment criteria. It consists of counting the number of wear points (or parts) in the rotating and control equipment. The rationale behind this approach is based on the concept that parts subject to wear are the eventual causes of failure in the drive: the fewer the number of such parts, the less cause for failure. Column 4 of Table 7 provides the number and identity of wear points of the drives under consideration. Obviously the ac adjustable-voltage drive and the adjustable-frequency drive (both utilizing a squirrel-cage induction motor) have the simplest rotating and control equipment.

Removal of Heat from Building This category has come to the foreground as a very important criterion in drive selection. Removal of heat from the pump building has become a steadily increasing problem as drive powers have increased. Both operator comfort and successful drive operation are at stake as the area around drives becomes hotter and building temperatures increase.

This general problem is affected in a major way by the amount of heat emitted by the rotating equipment alone and the drive in general. The latter may or may not be of consequence, depending on whether all or some of the drive losses are emitted in the building—and if so, where. Other factors of consequence are the relation of drive heat losses to building size and the geographic location.

Figure 16 shows maximum heat loss distribution of 50-hp (37-kW) drives operating through a 30% speed range with torques varying as the square of the speed. Maximum motor heat losses are shown separately to indicate the amount of heat emitted around the motor. These occur at maximum speed. The maximum drive heat losses show losses emitted by the motor and its control at minimum or maximum speed. The crosshatched areas depict the control losses, and the non-

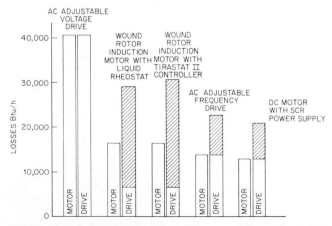

FIG. 16 Maximum loss comparison allocated to motor and drive: 50 hp (37 kW), 1200 rpm. Torque varies as the speed squared; 30% range. Crosshatched areas show control losses. (1 Btu/h = 1055 J/h) (General Electric)

crosshatched areas show associated motor losses. It is important that the destiny of the control heat losses be kept in mind, as some of these are dissipated in the pump building. Others may be dissipated in cooling water or outdoors. Total losses dissipated in the building may vary greatly, depending upon the physical handling of these losses.

Losses of drives at different ratings but operating through a 30% range may be approximated by extrapolation.

FINAL SELECTION OF DRIVES

In the final analysis, no simple routine can be suggested for drive selection. Each drive must be reviewed in the light of the application. By reviewing the capabilities of the various drives in each of the criteria used in this comparison and the requirements of the application, one has a starting point that should lead to successful drive selection.

6.2.3
FLUID COUPLINGS

CONRAD L. ARNOLD

TYPES OF FLUID COUPLINGS

The term *fluid coupling* can be loosely used to describe any device utilizing a fluid to transmit power. The fluid is invariably a natural or synthetic oil because oil is capable of transmitting power, is a lubricant, and is able to absorb and dissipate heat. Manufacturers have tried water as the fluid in fluid couplings, but sealing problems (keeping water out of the bearings and oil out of the water) and corrosion have prevented its use in any standard catalog drive.

All fluid couplings may be broken down in four categories:

1. Hydrokinetic
2. Hydrodynamic
3. Hydroviscous
4. Hydrostatic

HYDROKINETIC DRIVES

Although all types of fluid couplings are used in starting and controlling pumps, the one most commonly used is the hydrokinetic machine (Fig. 1).

Basic Principle In the hydrokinetic drive, commonly known as a fluid drive or hydraulic coupling, oil fluid particles are accelerated in the impeller (driving member) and then decelerated as they impinge on the blades of the runner (driven member). Thus power is delivered in accordance with the basic law of kinetic energy: $E = \frac{1}{2}M(V_1^2 - V_2^2)$, where E represents energy, M is the mass of the working fluid, V_1 is the velocity of the oil particles before impingement, and V_2 is the velocity after impingement on the runner blades. This principle is used in traction units and, with modification, in torque converters. Neither of these offers controlled variable speed.

In variable-speed units, the mass of the working fluid can be changed while the machine is operating and infinitely variable output speed is achieved. Variation of oil quantity can be accomplished in four ways: scoop-trimming couplings, leakoff couplings, scoop-control couplings, and put-and-take couplings (Figs. 2 to 5).

IMPELLER   OIL VORTEX   RUNNER

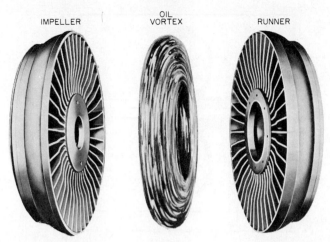

FIG. 1   Power-transmitting elements of a hydrokinetic coupling. (American Davidson)

Components   The following components are common to all the above types with few exceptions.

The *housing* of the fluid drive serves four purposes—as a reservoir for the nonworking oil, as a support for the bearings and scoop tube, as a guard to surround the moving parts, and as a container for oil particles and vapors that prevents their escape to the atmosphere. It also supports the oil pump when an internal pump is used. On small units, the housing is of end-bell construction; all others are split on the horizontal centerline to facilitate inspection and maintenance.

Bearings are used to support the shafts radially and axially. In the case of smaller industrial units, ball or roller bearings are usually used; larger machines utilize babbitt radial bearings and Kingsbury thrust bearings. Input sleeve bearing pillow blocks often support the internal oil pump driving and driven gears. In most cases the thrust bearings are designed to handle only the internal thrust of the fluid drive. Thrust developed by sleeve bearing driving motors can be accepted by

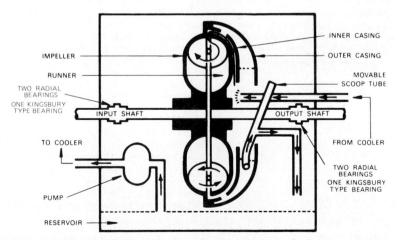

FIG. 2   Hydrokinetic coupling, scoop-trimming type. (American Davidson)

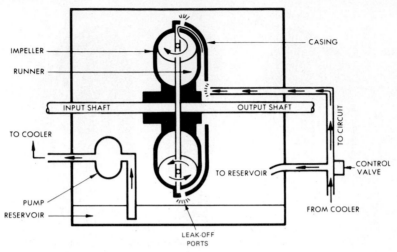

FIG. 3   Hydrokinetic coupling, leakoff type. (American Davidson)

the hydraulic couplings, but driven machines must usually have provisions to absorb their own thrust.

Shafts support the rotors and transmit driving torque to and from them. In some cases shafts are hollow and are used to supply oil to the bearings and to the working circuit (Figs. 2 to 5).

Rotors are often compared to halves of grapefruit after the meat has been removed and may be fabricated in three ways. The lightest-duty units are equipped with die-cast rotors of SAE 356 aluminum. Heavier-duty units have rotors which are machined out of 4130 or 4340 aircraft-quality steel forgings.

Inner and outer *casings* are bowl-shaped members which bolt to the front of the impeller to contain the oil in two connected areas known as the *working circuit* (Fig. 2). One chamber is formed by the impeller and inner casings. The other is formed between the inner and outer casings and can be called the scoop-tube chamber. Ports in the inner casing permit oil to flow from one chamber to the other.

The *scoop tube* (Fig. 2) can be moved radially or rotated inside the scoop tube chamber and is supported by sleeve or antifriction bearings. The pickup end of the tube is between the two casings facing the direction of oil rotation. Linkages permit the tube to be moved from outside the housing, and seals prevent the leakage of oil or vapors at this penetration.

An *oil pump* is provided which may be an internally mounted gear pump driven from the input shaft or an externally mounted positive displacement motor-driven pump. In cases where extreme reliability is required, emergency standby ac and/or dc driven pumps may be furnished. These pumps furnish light turbine oil to lubricate, transmit power, and remove heat from the fluid drive. In many cases they supply lubrication to the driver, the intermediate gear boxes, and the driven machine.

Oil coolers are required on all drives rated above 3 hp (2.2 kW). These coolers remove heat dissipated by the fluid drive and other machines for which they furnish lubrication. Shell-and-tube water-to-oil exchangers are normally supplied, although finned-tube air-to-oil exchangers are utilized where water is not available or economical. On pipeline work, it is common to use in-line coolers and the product of the pipeline is put through one side of the cooler to remove heat from the fluid-drive oil system.

Manifolds are usually used on scoop-controlled couplings in lieu of housings. They provide passages to permit oil flow to and from the working circuit and support the scoop tube and, sometimes, one bearing on the output shaft.

Operation   The flow of oil in the *scoop-trimming* fluid drive is begun by the circulating pump, which is driven at constant speed by the input shaft, or external motor driver. The circu-

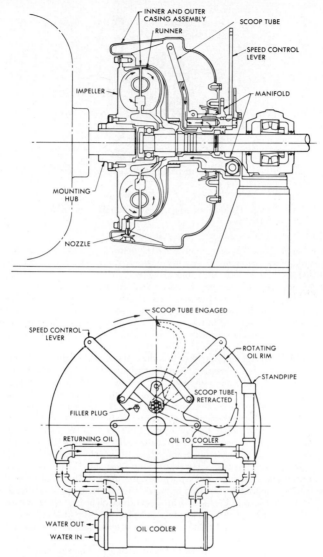

FIG. 4 Hydrokinetic coupling, scoop-control type.

lating pump moves the oil from the reservoir at the bottom of the housing to an external oil cooler (if used) and then to the rotating elements. Oil entering the rotating casing is acted upon by centrifugal force caused by the casing's rotating at the input speed. This centrifugal force throws the oil outward against the side of the casing and into the impeller and runner, or working circuit, where it takes the form of an annular ring. Communication ports in the inner casing permit the oil level to equalize in the two chambers.

    The amount of oil in the working circuit is regulated by the scoop tube acting as a sliding weir. The scoop tube removes the oil from the casing and empties it into the oil reservoir at the bottom of the housing, where it is ready to begin the cycle once more.

    By either manual or automatic control, the scoop tube is moved in the casing. This, in turn,

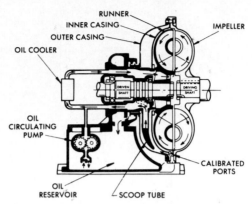

FIG. 5 Hydrokinetic coupling, put-and-take type. (American Davidson)

sets the level of the oil in the working circuit since the oil tends to seek the same level in the entire assembly. The scoop tube is designed to give fast response for both increase and decrease of output speed as required. In the *leakoff* type of fluid drive (Fig. 3), the scoop tube and outer casings are not used. Oil flow is initiated by a pump, usually of the viscous or centrifugal type, driven by the input shaft, through a heat exchanger (if required) and to a two-way control valve. This control valve modulates between the two extreme positions: all oil to the working circuit and all oil dumped back to the reservoir. Oil in the working circuit is thrown out through orifices called *leakoff ports.* Flow is created by centrifugal head, which varies with the depth of oil in the coupling.

If oil is added to the coupling faster than it is thrown out of the orifices, the quantity of oil in the unit and the output speed increase. Obviously the converse is true, and if oil is put into the working circuit at *exactly* the same rate that it is "leaked off," the unit runs at constant speed. This type of unit lends itself well to closed-loop automatic control which compensates for the differential flow through the leakoff ports. Manual control is questionable since oil must be added at *exactly* the rate it is discharged or output speed will drift.

In the *scoop-control* fluid drive, the communication ports in the inner casing are closed to form orifices and the scoop tube casing is sealed at the shaft. This breaks the unit into two separate chambers, the working circuit between the impeller and inner casing and the rotating reservoir between the inner and outer casings. The two are connected only by the orifices, or leakoff ports. Usually the housing, three bearings, and input shaft are omitted. In this configuration, the input rotor and casings are supported by the driving motor. In some cases the mounting is accomplished through a solid hub as shown in Fig. 4, and in others through a disk which is capable of flexing to absorb slight misalignment. The runner and output shaft are supported either by a pilot bearing and an outboard bearing or by a pilot bearing and the driven machine through a piloted flexible coupling.

Oil flow is initiated by the scoop in the reservoir acting as a pump. This flow is directed out through the manifold to the oil cooler, back to the manifold, and into the working circuit.

A portion of the oil constantly flows through calibrated nozzles in the inner casing to the outer casing, where it is held in an annular ring against the outer casing by centrifugal force. The fluid drive is initially charged with just enough oil to fill the impeller and runner and the cooler circuit so that the idle oil in the outer casing is a subtraction from the working circuit. The movable scoop tube adjusts the oil quantity in the outer casing and thus regulates the oil quantity in the working circuit. The scoop tube can be fully engaged, where it skims off all the oil in the casing and thus fills the working circuit, or it can be retracted completely, where all the oil lies idle in the outer casing and the unit is "declutched." Intermediate positions regulate torque and speed of the drive.

Put-and-take couplings (Fig. 5) have not been manufacturered for the past 30 years. There was, in the design of such couplings, a variation of the scoop control coupling wherein the position of the scoop tube was fixed; thus the tube provided circulation only between working circuit and cooler. The amount of oil in the coupling itself was regulated by a gear-type pump which was operated in one direction to pump oil from a reservoir into a unit, stopped to maintain constant coupling speed, or reversed to remove oil from the drive and pump it into the reservoir. This created very unwieldy control systems having very poor response characteristics with some hunting, and the design became obsolete.

Reversibility can be obtained by reversing the driving motor, provided that the unit incorporates oil pumps which are not affected by input shaft rotation. In addition, units utilizing scoop tubes must have dual tips which can accept the flow of oil from either side.

In all fluid drives, the same fluid is utilized to transmit power, to remove absorbed heat, and to lubricate. Thus there is no requirement for internal seals, or slingers, and positive lubrication is assured. Since the power-transmitting medium is the heat-absorbing medium, there are no problems of heat transfer encountered in units utilizing oil pumps. This type of unit can be selected with the capability of dissipating 100% or more of the driving-motor rated power.

HYDRODYNAMIC DRIVES

This type of fluid coupling is occasionally used to drive pumping equipment, usually in the portable pump field (Fig. 6).

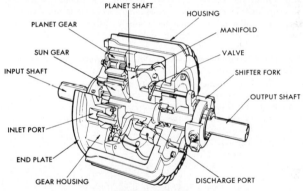

FIG. 6   Hydrodynamic coupling. (American Davidson)

Basic Principle   In the most common forms of hydrodynamic drives, planetary gear trains utilize some components as oil pumps. Throttling the discharge of these pumps creates back pressure and increases drive torque.

Components   The input shaft, supported by a bearing either on an independent bearing pedestal or on a packaged subbase assembly, drives the housing, endplate, manifold, and planetary gear shafts. These planet gears are partially surrounded by the manifold, which forms a pump cavity. The sun gear drives the output shaft. The control yoke moves the internal valve.

Operation   When the driver is started, oil flow is initiated by planet gears rotating against the sun gear. With the control yoke in the low-speed position, the mixing valve is positioned to admit air mixed with oil, the pump discharge valve ports are wide open, and the pump-developed head is approaching zero. The force on the pump gear teeth approaches zero, and the output speed is minimum. As the control yoke is moved, the pump discharge valves begin to close, less air is admitted, and the discharge pressure rises. This develops resistance to pump rotation and imparts a force on the sun gear, and the output shafts begins to rotate. If the oil discharge ports are closed, theoretically the pump pressure will rise until the pump gear is locked to the sun gear. This would rotate the output shaft at exactly input speed. In practice, leakage permits the pump to rotate and the output shaft turns at a slightly lower speed than the input shaft. Reversibility can be achieved simply by reversing the driving motor.

HYDROVISCOUS DRIVES

Hydroviscous drives are relatively new in commercial use. There are several manufacturers in the United States who are marketing this type of drive for a wide range of pump applications (Fig. 7).

Basic Principle   Hydroviscous drives operate on the basic principle that oil has viscosity and energy is required to shear it. More energy is required to shear a thin film than a thick one. The hydroviscous drive varies its torque capability by varying the film thickness between driving and driven members.

Components   The following components are common to all hydroviscous drives. The primary variations from one manufacturer to another are in the mechanics of control, the numbers of disks, and the support of the rotors.

The *housing* serves the same purpose as in other fluid drives, supporting bearings, guarding moving parts, and containing oil and vapors. In addition, one manufacturer uses cored passages in the housing to introduce water to remove heat from the working fluid. Bearings are usually of the antifriction type in small machines with sleeve and Kingsbury bearings available in large units. Shafts support rotors, transmit driving torque, and, in most cases, are hollow to supply cooling oil and control oil.

The *rotors* have the driving hub keyed on the inside to driving disks and the driven hub keyed at its inner diameter to the driven disks.

The *disks* are made of various materials and are usually grooved with some type of pattern to direct cooling oil flow.

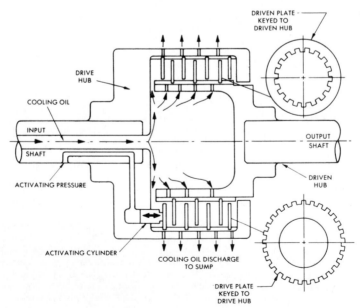

FIG. 7   Hydroviscous coupling. (American Davidson)

Pistons are hydraulic. When moved by control hydraulic oil, they force the disk stack closer together.

Oil pumps are usually motor-driven but sometimes are driven by the input shaft of the coupling. It is not uncommon to have two separate pumping systems, one providing high-pressure control oil and the other lower-pressure cooling oil.

Oil coolers are usually shell-and-tube water-to-oil heat exchangers, although air-to-oil exchangers can be furnished and, as mentioned earlier, cored housings can sometimes be used.

Operation   Oil flow is initiated by the oil pumps, which force cooling oil through the disk stack, draining into the sump. With the control set at minimum speed, the disks are at maximum spacing and the coupling transmits minimum torque. As pressure is applied to the piston, the disks are forced together. This decrease in film thickness between disks increases the force transmitted from one plate to the next. At maximum piston pressure, the spacing between plates is zero and the output shaft is driven at input shaft speed. In the full-speed condition, this device is actually a lockup mechanical clutch; at reduced speeds, it is an oil shear coupling; and in a narrow band between these two points of operation, it must be looked upon as an oil-cooled mechanical clutch. Reversibility can be accomplished by reversing the driving motor if oil pumps are driven by separate motors.

HYDROSTATIC DRIVES

Basic Principle   There are many variations of hydrostatic variable-speed drives, but in one form or another they invariably use positive displacement hydraulic pumps in conjunction with positive displacement hydraulic motors.

In some cases, varying amounts of fluid are bypassed from the pump discharge back to the pump suction. This provides a controllable, variable flow to the positive displacement motor and therefore a variable output speed. This system has no particular advantages over the more common variable-speed drives. The higher-than-average first costs and above-average maintenance required explain why this type of hydrostatic system is seldom used.

In other cases, the hydrostatic drive system uses variable-flow positive displacement pumps which may be of the sliding vane type or axial piston type (Fig. 8). Reducing the discharge flow on the hydraulic pump reduces output speed; increasing pump flow increases output speed. This type of variable-speed drive is offered in package form with pump, piping, and motor mounted in a common housing. It offers the capability of torque multiplication, maintains a relatively constant efficiency regardless of speed, has excellent control characteristics, and is widely used in the

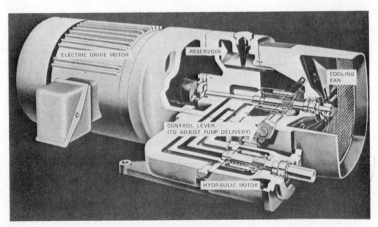

FIG. 8   Typical package hydrostatic drive. (Sperry Vickers)

machine tool and other industries. The output shaft can be reversed by valving (without changing motor rotation). This design has inherently high first cost and maintenance requirements, precluding significant use as a pump driver.

CAPACITY

Hydrokinetic Drive Being centrifugal machines, fluid drives follow very familiar laws: power varies as speed raised to the third power (Fig. 9), as diameter to the second power, and directly as the density of the working fluid.

Thus hydraulic capacity is governed by speed, diameter, and operating fluid; mechanical capability is governed by the structural design of housing, bearings, shafts, rotors, and casings; thermal capacity is limited by the capacity of the oil pumps, the specific heat and thermal conductivity of the oil, and the ability of the heat exchanger to dissipate heat.

It should be noted that in scoop-trimming couplings the oil pumps are usually sized solely for heat dissipation. In the leakoff coupling, the orifices are sized to permit enough oil flow to dissipate heat, and the pumps must handle this plus enough to fill the coupling in a reasonable time. The scoop-control coupling has limited flow and pressure since both are generated by the scoop tube. This may preclude its uses in certain positive displacement pumping applications or where installation of coolers at a remote location is required.

Hydrodynamic Drive This type of coupling varies so much in configuration that it is impossible to establish similar laws. Since any given machine has a definable torque limitation, we can state that power varies directly with speed.

Since standard units have no provision for removal and replacement of the working fluid, all cooling must be provided on the exterior surfaces of the rotating housing. This becomes a decided limitation if the unit is to be used with constant-torque loads and has limited the available sizes to some degree.

Hydroviscous Drive As would be expected, this device also follows the centrifugal laws: power varies as speed raised to the third power and as diameter raised to the second. However,

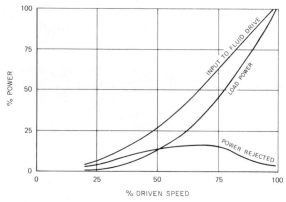

FIG. 9 Load power varies as speed cubed. This figure represents a pump operating in a system where all the head is frictional, or where head varies as flow squared. In this system, the driven pump operates at only one point on its characteristic curve, and therefore the shape of the pump curve is academic. Although the rapid loss of hydraulic coupling efficiency at reduced speeds is quite obvious, the dramatic decrease in pump power requirements results in a total system power which is most acceptable. Note that the heat rejection requirements in the fluid drive are maximum at about 18% of the total load power.

the density of the working fluid has little or no effect; instead capacity varies directly with viscosity. Thus hydraulic capacity varies with speed, diameter, and viscosity. Mechanical capability is a relatively simple matter of structural design. However, thermal design is critical. Since power-transmitting capability varies with viscosity, which in turn varies with temperature, disk design is most important. Free area available for cooling oil flow varies with disk spacing, output speed, and heat load.

REGULATION

The output speed of all fluid couplings (hydrostatic drives are not being considered) is affected, to some degree, by changes in load. While this may be significant in cases of single-cylinder, low-speed reciprocating pumps, it is insignificant on multicylinder reciprocating and all centrifugal pumps. In these cases, ±1% speed regulation is considered normal. In special cases, regulation has been guaranteed at ±0.3%.

TURNDOWN

Standard catalog hydrokinetic and hydroviscous units offer the regulation described above over a 5-to-1 turndown on centrifugal machines and 4-to-1 turndown when driving positive displace-

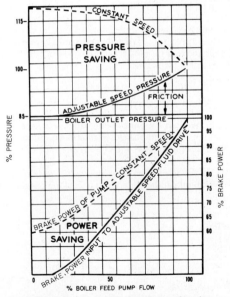

FIG. 10 Comparison of adjustable-speed and constant-speed pressure and power characteristics for a typical centrifugal boiler-feed pump. This figure is typical of a boiler-feed pump where a high percentage of the developed head is relatively constant. In this case, this fixed head is the boiler pressure. The figure demonstrates the savings in pressure and power realized when this system is used rather than a feedwater regulating valve.

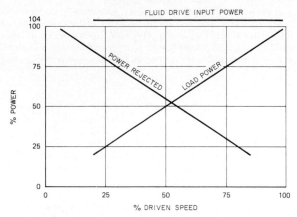

FIG. 11 Positive displacement pump with constant discharge head. This figure assumes that a positive displacement pump is working on a system where pressure is constant. Although this type of system is seldom found, it is shown here to demonstrate that the rapid reduction in fluid drive efficiency does not require overmotoring the pump. Although system efficiency is much poorer than that of a bypass valve, fluid drives are used to provide no-load starting, isolation of torsional vibrations in reciprocating pumps, and elimination of the bypass valve in slurry systems where erosion is severe.

ment pumps on constant-pressure systems. Specially designed fluid drives have been sold which give stable control at 10-to-1 turndown. Hydrodynamic units are limited primarily by heat dissipation capabilities and range from turndown values of 100 to 1 to 1.2 to 1.

RESPONSE

It must be recognized that all fluid couplings being discussed here are slip devices. Thus any demand speed change cannot be accomplished in microseconds or milliseconds. However, the time required to change the torque applied varies from one type of unit to another.

Hydrokinetic Drive   In the scoop-trimming fluid drive, response speed is affected by many factors. The speed with which oil can be added to the working circuit (a factor of the size of the

FIG. 12   Thirteen fluid drives driving reciprocating pumps on a coal pipeline. The fluid drive absorbs a large percentage of the pulsations created by the reciprocating pumps and controls their speed to provide proper pipeline flow. (American Davidson)

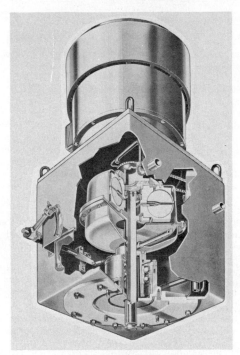

FIG. 13 Cutaway view of vertical fluid drive suitable for operation with vertical pumps. Note NEMA P pump flange and output shaft. (American Davidson)

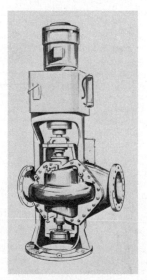

FIG. 14 Fluid drive of Fig. 13 mounted on a vertical-shaft pump. (American Davidson)

oil pumps) or removed from it (a factor of the size of the scoop tube) influences response capability.

In the leakoff unit, the size of the leakoff ports determines how quickly the unit will empty. However, the oil pumps must be sized to replace this oil and have additional capacity to fill the coupling in a reasonably short time.

Scoop-control units are limited by the ability of the scoop tube to pump oil from the reservoir into the working circuit and by the ability of the leakoff ports to return it to the reservoir. Some special marine couplings utilize quick-dumping valves, but these are seldom, if ever, used with pump drives.

Obviously the speed at which the scoop tube is moved is also significant. Large polar moments of inertia (WK^2 values) of the driven equipment will increase response time.

The scoop-trimming coupling offers the best overall response characteristics of the hydrokinetic drives, and standard catalog machines have normal fill times ranging from 10 to 15 s. They will accomplish 90% of a 10% step speed change in the 40 to 100% speed range in 7 to 20 s if coupled to a "normal load inertia." Special units are in operation where this change is accomplished in 2 to 6 s.

Hydrodynamic and Hydroviscous Drive Both hydrodynamic and hydroviscous couplings respond very quickly to a change in demand for torque output. Both require a mechanical motion (change in valve position or change in spacing between disks) followed immediately by a change in pressure or in film thickness.

In most cases the torque available for speed change and the WK^2 values involved are of such a magnitude that the major portion of the response time is caused by inertial effects rather than by the time required to change torque. This is particularly true in the deceleration of centrifugal pumps. Unless auxiliary brakes are built in, none of the hydrokinetic, hydrodynamic, or hydroviscous drives can provide dynamic braking. On a demand to decrease speed, they can at best reduce driving torque to zero. Under these circumstances, the only retarding force to slow the inertia of the driven machine is the load it developed. In the case of centrifugal pumps on fixed systems, this load would fall off as the cube of speed, and below 40% of full speed, such pumps have an almost insignificant braking effect.

EFFICIENCY

There are two kinds of losses present in hydrokinetic, hydrodynamic, and hydroviscous couplings. First, we will consider what are termed *circulation losses*. They are made up of bearing friction, windage, and the power required to accelerate the oil in the rotor. On internal pump units, the power required to drive the oil pump is included. As an average, these losses represent approximately 1.5% of the unit rating, and for most purposes these losses may be considered as being constant, regardless of output speed.

Second are *slip losses*. As is the case on similar slip machines, such as mechanical clutches and eddy-current couplings, the torque on the input shaft equals the torque on the output shaft. Therefore any reduction in the speed of the output shaft has a directly related power loss inside the machine. In other words,

$$\text{Slip efficiency} = \frac{\text{output speed}}{\text{input speed}} \times 100$$

The total fluid drive losses are the sum of the two inefficiencies. The complete energy formula is

in USCS units $\text{Fluid drive input horsepower} = \dfrac{\text{output horsepower}}{\text{output speed/input speed}} + \left(\begin{array}{c}\text{circulation horse-}\\ \text{power losses}\end{array}\right)$

in SI units $\text{Fluid drive input kilowatts} = \dfrac{\text{output kilowatts}}{\text{output speed/input speed}} + \left(\begin{array}{c}\text{circulation}\\ \text{kilowatt losses}\end{array}\right)$

At maximum designed operating speed (which is usually about 98% of driven speed), the total coupling efficiency is approximately 96.5%, with 1.5% of the losses being circulation losses and

2% being slip losses. The hydroviscous unit can be operated at 100% driven speed, but under these conditions it is not a fluid coupling.

Since the circulation losses become relatively insignificant at reduced speeds, approximate calculations may be made using the formula

$$\text{Efficiency} = \frac{\text{output speed}}{\text{input speed}}$$

CONTROLLERS

Of the hydrokinetic devices, the scoop-trimming and scoop-control units require that a mechanical motion be imparted to the scoop tubes for control, and some device must be furnished to provide this motion. This may be a hand crank on a manual control system. Simple mechanical systems are often used—a typical example is a weighted float with a rope connected to the scoop tube, controlling level in a tank. However, most installations utilize electric, electrohydraulic, hydraulic, or pneumatic actuating devices. It is not surprising that the pipeline and refinery industries use electrohydraulic actuators similar to those used on valves. The electric utility industry prefers pneumatic or electric damper operators. The only criterion for actuator selection is compatibility with the other elements of the control system.

The leakoff devices require a signal to the control valve. At present, this is standardized as a hydraulic-pressure signal, although special transducers would permit the use of other types of signals. The hydrodynamic devices are available with manual level control (which could be adapted to actuators) or with closed-loop constant-pressure or constant-temperature systems. All hydroviscous drives utilize oil pressure applied to a piston to "clamp" or vary the spacing of the disks. This hydraulic pressure may be varied by almost any type of signal, provided that the proper servos are utilized. Thus the signal may be electric, hydraulic, or pneumatic.

CAPACITIES AVAILABLE

Standard catalog variable-speed fluid couplings are available from one or more manufacturers in the speeds and powers shown in Table 1. Special designs are available for higher power ratings.

TABLE 1 Variable-Speed Coupling Capacities

| | Hydrokinetic | | | |
Input speed, rpm	Min/max input hp for horizontal-shaft units[a]	Max input hp for vertical-shaft units[a]	Hydrodynamic min/max input hp[b]	Hydroviscous min/max input hp[b]
720	40,000–45,000	. . . . . .	. . . .	3,000– 8,000
900	1,000–7,500	6,000	1–25	3,000–10,000
1,200	1,000–14,000	17,000	1–30	3,000–15,000
1,800	1,000–14,000	18,000	1–60	3,000–20,000
3,600	1,000–30,000	29,000	. . . .	3,000–20,000

[a]1 hp = 0.746 kW.
[b]Horsepowers apply to either horizontal or vertical units.

DIMENSIONS

To give some idea of physical dimensions, Table 2 lists approximate dimensions for hydrokinetic drives of one U.S. manufacturer. These feature a scoop-trimming coupling and are probably the largest unit for a given speed and power.

TABLE 2 Hydrokinetic Drive Dimensions

Power at 1800 rpm, input hp (kW)	Length, in (cm)	Width, in (cm)	Shaft height, in (cm)
5 (3.73)	24 (61.0)	15 (38.1)	11½ (29.2)
20 (14.9)	39 (99.1)	18 (45.7)	12½ (31.7)
50 (37.3)	38 (96.5)	28 (71.1)	19 (48.3)
100 (74.6)	38 (96.5)	28 (71.1)	19 (48.3)
1000 (746)	74 (188)	50 (127)	30 (76.2)
4000 (2980)	102 (259)	62 (157)	42 (107)

SELECTION AND PRICING

Because of variations in fluid coupling design for different sizes and speeds, it is virtually impossible to develop rule-of-thumb methods of estimating costs. The price list of one major fluid drive manufacturer indicates that prices can range from $60 per horsepower for sophisticated machines down to $25 per horsepower for others (in 1984 dollars). Fluid couplings prices can be increased dramatically by specific requirements for exotic controls, backup pumps and heat exchanger equipment, and other accessories. Because of this fact, it is recommended that the manufacturers be contacted to obtain even budget prices.

The basic information required by the manufacturer for selection and pricing is as follows:

1. Speed and type of driver
2. Power required by the driven machine at, or at least, one operating point
3. Character of driven machine—smooth or pulsating load; how torque requirements change with speed
4. Cooling medium available and temperature of medium
5. Control type
6. Accessories
7. Special specification requirements

Reasonable budget figures can usually be obtained with items 1 to 3 only.

CONCLUSION

Fluid couplings are utilized to drive pumps in virtually all pump applications requiring variable flow or pressure. They are used primarily to improve efficiency and controllability, to permit no-load starting, and to reduce pump and system wear. They are standardized to the degree that units are available to handle most pumping applications. Most manufacturers stand ready to develop new designs as the requirements of the marketplace change.

6.2.4
GEARS

H. O. KRON

H. O. KRON

USE OF GEARS WITH PUMP DRIVES

The main use of gearing in pump drives is to reduce the speed of the prime mover (a motor or an engine) to the level applicable to the pump. In some cases, however, the gears are employed to step up the speed because the pump must operate at a speed higher than that of the prime mover.

There are other jobs that gear drives must perform with pump units. For instance, a vertical pump may be combined with a prime mover that must operate horizontally (as in the case of a diesel engine). A right-angle gear set (Figs. 1 and 4) can be incorporated into the drive of such a combination to transmit the power "around the corner" even if the gears are of a 1:1 ratio and no speed change is involved.

At times, too, a gear drive must combine the power shafts of two prime movers, say, an electric motor and a diesel engine. This combination is desirable where there is a need for emergency power in cases of electric power failure. The motor is used ordinarily and the diesel is reserved for emergencies. In such an arrangement, the motor may be mounted vertically on top of the gear drive to drive right through the shaft, whereas the diesel is geared at right angles to the motor.

Gear drives are also used to vary the speed of a pump. Change gear arrangements (Fig. 2) may be used to vary the gear ratio. Also, numerous types of variable-speed devices, such as variable-speed pulley belts, friction rollers, hydraulic couplings, hydrostatic and hydroviscous drives, eddy-current and electric drives, are often utilized.

TYPES OF GEARS

The gears generally used for pump applications are parallel-shaft helical or herringbone gears and right-angle spiral-bevel gears. Spur gears are used on occasion, particularly in low-power, low-speed pump drives. Straight bevel and hypoid gears are also occasionally used in right-angle drives. As in the case of spur gears, straight-bevel gears are limited in power and speed. Hypoid gears are employed only infrequently for pumps because they are generally more costly than the other right-angle types. Worm gears, too, are employed only on occasion, in cases where an overall package requires a compact gear arrangement or when a high ratio of speeds is called for. Worm gears are limited in power capacity, and the efficiency of this type of drive is lower than that of other types. Figure 3 shows the various types of gears.

FIG. 1   Spiral-bevel gear, right-angle vertical pump drive.

Parallel-Shaft Gearing   A high-speed parallel-shaft gear drive is shown in Fig. 5.

HELICAL VERSUS SPUR GEARS   Spur gears transmit power between parallel shafts without end thrust. They are simple and economical to manufacture and do not require thrust bearings—but they are generally used only on moderate-speed drives.

One of the first decisions that must be made when considering a parallel-shaft gear reducer or power transmission drive is whether the gears should be spur or helical and, if helical, whether they should be single or double. It is generally acknowledged that helical gears offer better performance characteristics than do spur gears, but since helical gears, size for size, often are somewhat more expensive, some users have shied away from them to keep the cost of the gear drive to a minimum. However, cost studies that compared similar size spur and helical gears found that helical gears are actually a better buy.

The geometries of spur and helical gears are both the involute tooth form. Slice a helical gear at right angles to its shaft axis and you have the typical spur gear profile. Typical of a spur gear, however, is that, when driving another spur, its teeth make contact with the teeth of the mating gear along the full length of the face. The load is transferred in sequence from tooth to tooth.

A more gradual contact between mating gears is obtained by slanting the teeth in a way to form helices that make a constant angle—a helix angle with the shaft axis. Tooth contact

FIG. 2   Four-speed change gear, variable-speed drive.

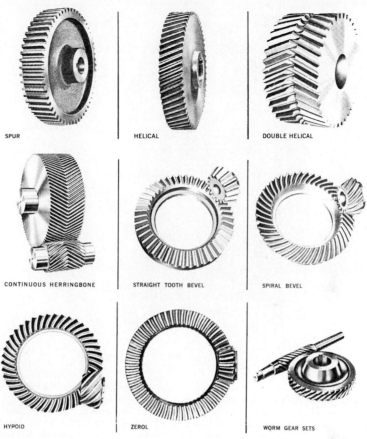

SPUR | HELICAL | DOUBLE HELICAL

CONTINUOUS HERRINGBONE | STRAIGHT TOOTH BEVEL | SPIRAL BEVEL

HYPOID | ZEROL | WORM GEAR SETS

FIG. 3 Types of gears.

between the teeth of mating helical gears is gradual, starting at one end and moving along the teeth so that, at any instant, the line of contact runs diagonally across the teeth. The effect of the tooth helix is to give multiple tooth contact at any time. Gear geometry can be arranged to give from two to six or more teeth at any time.

Because of the greater number of teeth in contact, a helical gear has a greater effective face width (up to 75% more) than an equivalent spur gear. Also, the effect of the tooth helix on the profile geometry in the plane of rotation is to make the pinion equivalent to a pinion with a greater number of teeth, thereby increasing its power capacity. A helical gear is capable of transmitting up to 100% more power than an equivalent spur gear. Furthermore, a helical gear set will give smoother, quieter operation.

The recommended upper limit of pitch line velocity for commercial spur gears is around 1000 ft/min (300 m/min). The upper limit for equivalent helical gears is about five times that, or 5000 ft/min (1500 m/min). Of course, as precision goes up, so do the permissible operating speeds for both spur and helical gears. Velocities in the 30,000-ft/min (9100-m/min) range are not uncommon for helical gears.

In addition to the normal radial loads produced by spur gears, helical gearing also produces an end thrust along the axis of rotation. The end thrust is a function of the helix angle: the larger the helix angle, the greater the thrust produced. Mounting assemblies and bearings for helical gearing must be designed to receive this thrust load.

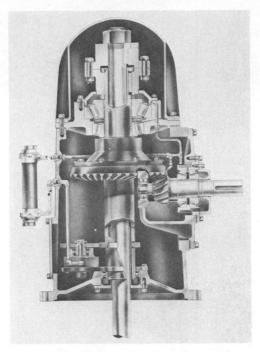

FIG. 4　Cross section of spiral-bevel vertical pump drive.

DOUBLE HELICAL GEARS　　Gears of this type have two sets of opposed helical teeth. Each set of teeth has the same helix angle and pitch, but the helices have opposing hands of cut. Thus, the thrust loads in two sets of teeth counterbalance each other and no thrust is transmitted to shaft and bearings. Also, because end thrust is eliminated, it is possible to cut the teeth with greater helix angles than is generally used in helical gears. Tooth overlap is greater, producing a stronger and smoother tooth action.

The advantages attributed to helical gears are also applicable to double helical gears. Double helical gearing finds application in high-speed pump applications where a large helix angle must be combined with tooth sharing and elimination of end thrust for extremely smooth gear action.

Single helical gears, however, have some attractive advantages over double helical gears, the most significant being that, in the former, the external thrust loads do not affect gear tooth action. With a double helical gearing, a thrust load tends to unload one of the helices and overload the opposite one.

Furthermore, the gear face for a single helical gear can be made narrower than for a double helical gear because the need for a groove between the two helices is eliminated. This leads to the use of a narrower, stiffer pinion with less tooth deflection and torsional windup and, generally, to a more favorable critical speed condition.

In addition, the accuracy of a single helical gear is greater because it is not subject to apex runout. With a double helical gear, the tolerances in tooth position tend to unload one helix cyclically and induce axial vibrations in the pinion.

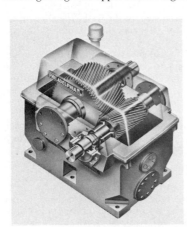

FIG. 5　High-speed parallel-shaft gear drive.

CONTINUOUS-TOOTH HERRINGBONE GEARS   Gears of this type are double helical gears cut without a groove separating the two rows of teeth. Because of the arched construction of these gears, they are often known as "the gears with a backbone." Continuous-tooth herringbone gears are used for the transmission of heavy loads at moderate speeds where continuous service is required, where shock and vibration are present, or where a high reduction ratio is necessary in a single train. Because of the absence of a groove between opposing teeth, a herringbone gear has greater active face width than the hobbed double helical gear and therefore is stronger. There is also no end thrust, as the opposing helices counterbalance one another.

Much of the success of the continuous-tooth herringbone gear is due to the greater number of teeth in contact and to the continuity of tooth action, which is an outgrowth of the larger helix angle. These larger helix angles can be fully utilized without creating bearing thrust loads. Continuous-tooth herringbone gears normally are furnished with a helix angle of 30°.

Crossed-Axis Gearing

STRAIGHT-BEVEL GEARS   Gears of this type transmit power between two shafts usually at right angles to each other. However, shafts other than 90° can be used. The speed ratio between shafts can be decreased or increased by varying the number of teeth on pinion and gear. These gears are designed to operate at speeds up to 1000 ft/min (300 m/min) and are more economical than spiral-bevel gears for right-angle power transmission where operating conditions do not warrant the superior characteristics of spiral-bevel gearing. When shafts are at right angles and both shafts turn at the same speed, the two bevel gears can be alike and are called *miter gears*.

SPIRAL-BEVEL GEARS   Spiral-bevel gear teeth and straight-bevel gear teeth are both cut on cones. They differ in that the cutters for straight-bevel teeth travel in a straight line, resulting in straight teeth, whereas the cutters for spiral-bevel gear teeth travel in the arc of a circle, resulting in teeth which are curved and are called spiral. Figure 4 shows a cross section of a spiral-bevel vertical pump drive.

Spiral-bevel gearing is superior to straight-bevel in that, in the former, loading is always distributed over two or more teeth in any given instant. Recommended maximum pitch line velocity for spiral bevels is about 8000 ft/min (2400 m/min). Spiral bevels are also smoother and quieter in action because the teeth mesh gradually. Because of the curved teeth, spiral-bevel pinions may be designed with fewer teeth than straight-bevel pinions of comparable size. Thrust loads are greater for spiral-bevel gearing than for straight-tooth bevels, however, and vary in axial direction with the direction of rotation and hand of cut of the pinion and gear. Where possible, the hand of spiral should be selected such that the pinion tends to move out of mesh.

ZEROL GEARS   Zerol-bevel gears are cut on conical gear blanks and have curved teeth similar to the spiral bevels, but the teeth are cut with a circular cutter which does not pass through the cone apex. Thus, these are spiral-bevel gears with zero spiral angle (hence the name). Furthermore, the tooth bearing is localized as in spiral-bevel gearing; thus stress concentration at the tips of the gear teeth is eliminated. Zerol-bevel gears are replacing straight-bevel gearing in many installations because their operation is smoother and quieter as a result of their curvature and their operating life is longer. Like straight-bevel gears, zerol gears have the advantage of no inward axial thrust under any conditions. The zero spiral angle produces thrust loads equivalent to those in straight-bevel gears.

HYPOID GEARS   Hypoid-bevel gears have the general appearance of spiral-bevel gears but differ in that the shafts supporting the gears are not intersecting. The pinion shaft is offset to pass the gear shaft. The pinion and gear are cut on a hyperboloid of revolution, the name being shortened to *hypoid*. Hypoid gears can be made to provide higher ratios than spiral-bevel gears. They are also stronger and operate even more smoothly and quietly. The fact that two supporting shafts can pass each other, with bearings mounted on opposite sides of the gear, provides the ultimate rigidity in mounting.

WORM GEARS   In operation, the teeth on the worm of a worm gear set (Fig. 6) slide against the gear teeth and at the same time produce a rolling action similar to that of a rack against a spur

constant output speed completely free of pulsations. Worm gearing is particularly adaptable to service where heavy shock loading is encountered.

Worm gearing is extremely compact, considering load-carrying capacity. Much higher reduction ratios can be attained through a worm gear set on a given center distance than through any other type of gearing. Thus the number of moving parts in a speed-reduction set is reduced to the absolute minimum. However, worm gearing is limited in power capacity and has lower efficiencies than parallel-shaft and bevel-gear types.

GEAR MATERIALS AND HEAT TREATMENT

Two of the most important factors dictating the success or failure of a gear set are the choice of material and heat treatment. This is especially true for gears designed for higher power. American Gear Manufacturers Association (AGMA) ratings for gear strength and durability are dependent on the choice of material and heat treatment. It is often possible to reduce the size of the gear box substantially by simply changing from low-hardened or medium-hardened gears (about 300 Brinell) to full-hardened gears (about 55 to 60 Rockwell C). Generally used methods for hardening gear sets are

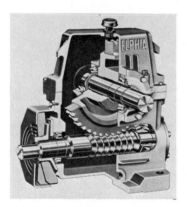

1. Through hardening
2. Nitriding
3. Induction hardening
4. Carburizing and hardening
5. Flame hardening

The use of high-hardness, heat-treated steels permits smaller gears for given loads. Also, hardening can increase service life up to 10 times without increasing size or weight. After hardening, however, the gear must have at least the accuracies associated with softer gears and, for maximum service life, even greater precision.

FIG. 6   Fan-cooled worm gear drive.

Through Hardening   Suitable steels for medium to deep hardening are 4140 and 4340. These steels, as well as other alloy steels with proper hardenability characteristics and carbon content of 0.35 to 0.50, are suitable for gears requiring maximum wear resistance and high load-carrying capacity. Relatively shallow-hardening carbon steel gear materials, types 1040, 1050, 1137, and 1340, cannot be deep hardened and are suitable for gears requiring only a moderate degree of strength and impact resistance. A 4140 steel will produce a hardness of 300 to 350 Brinell. For heavy sections and applications requiring greater hardness, a 4340 steel will provide 350 to 400 Brinell. Cutting of gears in the 380 to 400 Brinell range, although practical, is generally difficult and slow.

Nitriding   Nitriding is especially valuable when distortion must be held to a minimum. It is done at a low temperature (975 to 1050°F, 524 to 566°C) and without quenching—eliminating the causes of distortion common to other methods of hardening and often necessary for finish machining after hardening. Nitrided case depths are relatively shallow so that nitriding is generally restricted to finer pitch gears (four-diameter pitch or finer). However, double-nitriding procedures have been developed for nitriding gears with as coarse as two diametral pitch.

Any of the steel alloys that contain nitride-forming elements, such as chromium, vanadium, or molybdenum, can be nitrided. Steels commonly nitrided are 4140, 4340, 6140, and 8740. It is possible with these steels to obtain core hardnesses of 300 to 340 Brinell and case hardnesses of 47 to 52 Rockwell C. Where harder cases are required, one of the Nitralloy steels may be used. These steels develop a case hardness of 65 to 70 Rockwell C with a core hardness of 300 to 340 Brinell. The depth of case in a 4140 or 4340 steel varies as the length of time in the nitriding furnace. A single nitride cycle will produce a case depth of 0.025 to 0.030 in (0.64 to 0.76 mm) in 72 h. Doubling the time will produce a case depth of 0.045 to 0.050 in (1.14 to 1.27 mm). For the majority of applications, the case depth obtained from a single cycle is ample.

Case depth for Nitralloy steels is somewhat less than the depths obtainable for other alloy steels. In general, alloy steels 4140 and 4340 give up to 50% deeper case than Nitralloy steels for the same furnace time. These cases are tougher but less hard.

Induction Hardening   Two basic types of induction hardening are used by gear manufacturers: coil and tooth-to-tooth. The coil method consists of rotating the work piece inside a coil producing high-frequency electric current. The current causes the work piece to be heated. It is then immediately quenched in oil or water to produce the desired surface hardness. Hardnesses produced by this method range from 50 to 58 Rockwell C, depending on the material. The coil method hardens the entire tooth area to below the root.

Tooth-to-tooth full-contour induction hardening is an economical and effective method for surface hardening larger spur, helical, and herringbone gearing. In this process an inductor passes along the contour of the tooth, producing a continuous hardened area from one tooth flank around the root and up the adjacent flank. The extremely high localized heat allows small sections to come to hardening temperature while the balance of the gear dissipates heat. Thus major distortions are eliminated. The 4140 and 4340 alloy steels are widely used for tooth-to-tooth induction hardening. The hardness of the case produced by this method ranges from 50 to 58 Rockwell C, and the flanks may be hardened to a depth of 0.160 in (4.06 mm). These steels are air-quenched in the hardening process. Plain carbon steels, such as 1040 and 1045, may be used for induction hardening, but they must be water-quenched.

Carburizing   Carburizing with subsequent surface hardening offers the best way to obtain the very high hardness needed for optimum gear life. It also produces the strongest gear, one that has excellent bending strength and high resistance to wear, pitting, and fatigue. The residual compressive stresses inherent in the carburized case substantially improve the fatigue characteristics of this heat-treated material. Normal case depths range from approximately 0.030 to 0.250 in (0.76 to 6.3 mm). Case hardnesses range from 55 to 62 Rockwell C, and core hardnesses from 250 to 320 Brinell. Recommended carburizing-grade steels are 4620, 4320, 3310, and 9310. The main limitation to carburizing and hardening is that the process tends to distort the gear. Techniques have been developed to minimize this distortion, but generally after carburizing and hardening it is necessary to grind or lap the gear to maintain the required tooth tolerances.

Flame Hardening   In tooth-to-tooth progressive flame hardening, an oxyacetylene flame is applied to the flanks of the gear teeth. After the surface has been heated to the proper temperature, it is air- or water-quenched. This method has some limitations; since the case does not extend into the root of the tooth, the durability is improved but the overall strength of the gear is not necessarily improved. In fact, stresses built up at the junction of a hardened soft material may actually weaken the tooth.

Many times it is desirable to use different heat treatments for the pinion and gears. Heat-treatment combinations used for pinions and gears are here listed in order of preference for optimum gear design.

1. Carburized pinion, carburized gear
2. Carburized pinion, through-hardened gear
3. Carburized pinion/nitrided gear
4. Nitrided pinion/nitrided gear
5. Nitrided pinion/through-hardened gear
6. Induction-hardened pinion/through-hardened gear
7. Carburized pinion/induction-hardened gear
8. Induction-hardened pinion/induction-hardened gear
9. Through-hardened pinion/through-hardened gear

OPTIMIZING THE GEARING

The most important factor influencing the durability of a gear set, and hence the gear size, is the hardness of the gear teeth. It is often possible to reduce considerably—sometimes by as much as half—the overall dimensions of a gear set by changing from medium-hardened gears (about 300 Brinell) to full-hardened gears (about 55 to 60 Rockwell C).

There are other factors that play a role in minimizing the dimensions of a gear set, for example, the ratio between face width and pitch diameter and the proper pressure angle and pitch of teeth. Thus, when it comes to deciding between a set of standard catalog gears and gears designed specifically to meet the requirements of the application, the question of cost versus optimizing comes to bear. In general, where there is only a limited number of units to be made, the catalog gears are much less expensive and also much more readily available. There are many applications which call for critical power, speed, or space requirements, however, and it may pay in these applications to select gears that are designed for that application.

Minimizing Gear Noise   Specifying or designing a gear set to produce low noise and vibration levels frequently leads to choices that are the opposite of those for optimizing the gears for strength and size. Generally, parallel-shaft gearing rather than right-angle gearing is preferred for quiet operation because of greater geometric control, inherent ability to maintain tight manufacturing tolerances, and minimum friction during tooth contact. Helical gears, in particular, can have more than one tooth in contact (helical overlap), and some experience has shown as much as a 12-dB reduction in noise using helical instead of spur gears. Double helical or herringbone gearing creates the problem of correctly manufacturing the two helices with the exact same phase and accuracy. Therefore the optimum type of gear throughout all speed ranges becomes the single helical gear. The thrust loads and slight overturning moments of single helical contact are easily handled with modern design techniques.

For quiet, smooth operation, the gears should be designed with some or all of the following properties:

1. Select the finest pitch allowable under load considerations.

2. Employ the lowest pressure angle: 14½ and 20° are most commonly used.

3. Modify the involute profile to include tip and root relief with a crowned flank to ensure smooth sliding into and out of contact without knocking and to compensate for small misalignments.

4. Allow adequate backlash (clearance) for thermal and centrifugal expansion, but not so much as to prevent proper contact.

5. Specify the higher AGMA quality levels, which will reduce the total dynamic load. Generally, AGMA quality 12 or better is required for smooth, quiet operation.

6. Maintain surface finishes of at least 20 rms.

7. Maintain rotor alignments and runouts accurately.

8. Limit rotor unbalance to less than $3W/N$ oz·in ($4760W/N$ g·mm), where W = weight, lb (kg, mass), and N = speed, rpm.

9. Provide a nonintegral ratio ("hunting tooth") to prevent a tooth on the pinion from periodically contacting the same teeth on the mating gear.

10. Have resonances of rotating system members (critical speeds) at least 30% away from operating speed, multiples of rotating speeds, and tooth-mesh frequencies.

11. Have resonances of gear cases and other supporting members 20% away from operating speeds, multiples, and tooth-mesh frequencies.

12. Specify the highest-viscosity lubricant consistent with design and application.

13. Select rolling element bearings to minimize noise generation. Generally, hydrodynamic sleeve bearings are quieter than antifriction types but are more difficult to apply.

14. Since housing design is another area where noise and vibration reductions can be obtained, select an acoustically absorbant material for the housing or design the housing with built-in isolation mounts to cut down any vibration attenuation.

PACKAGED GEAR DRIVES

In many cases it is preferable to select a packaged gear drive rather than a set of open gears that must be mounted and housed.

The relative merits of a packaged drive, or "gear reducer," versus open gearing are many. The packaged drive consists essentially of gears, housing, bearings, shafts, oil seals, and a positive means of lubrication. Frequently, reducers also include any or all of the following: electric motor and accessories, bedplates or motor supports, outboard bearings, a mechanical or electric device providing overload protection, a means of preventing reverse rotation, and other special features as specified.

The advantages of packaged gear drives have been well established and should be given consideration when selecting the type of gear drive for the application:

1. *Power conservation.* Because of accurate gear design, quality construction, proper bearings, and adequate lubrication, minimum loss between applied and delivered power is assured.

2. *Low maintenance.* If the correct design and power capacity for the requirements are selected and the recommended operating instructions are followed, low maintenance costs will result.

3. *Operating safety.* All gears, bearings, and shafts are enclosed in oil-tight, strongly built cast iron or steel housings.

4. *Low noise and vibration level.* Precision gearing is carefully balanced and mounted on accurate bearings. The transmitting motion is uniform and shock-free. The entire mechanism is tightly sealed in sound-damping rigid housing. Noise and vibration are reduced to a minimum.

5. *Space conservation.* Units are entirely self-contained and extremely compact; therefore, they require a small space. This also enables them to be installed in out-of-the-way locations.

6. *Adverse operating conditions.* Enclosure designs have been developed to protect the mechanism from dirt, dust, soot, abrasive substances, moisture, or acid fumes.

7. *Economy.* Units permit the use of high-speed prime movers directly connected to low applied speeds.

8. *Life expectancy.* The life of a unit can be predetermined by design and made unlimited if it is correctly aligned and properly maintained.

9. *Power and ratios.* Units are available in almost all desired ratios and for all practical power requirements.

10. *Cooling systems.* Greater attention to sump capacity for oil and the use of fan air cooling have allowed higher powers to be transmitted through smaller units without overheating.

11. *Appearance.* Housings have been streamlined for eye appeal as well as for reduction of weight and space.

12. *Rugged capabilities.* The ruggedness of steel-constructed welded housings and modern housings produces higher reliability and service life.

Types of Gear Packages Gear packages are used in multiple combinations to produce the ratio desired in the unit. Units are available in single-, double-, and triple-reduction configurations. Three stages of reduction are generally the maximum number used in standard reducers, although it is possible to use four or even more stages. Units for increasing the output speed generally have only one gear set, although at times two-stage units have been used successfully as speed increasers. Gear packages may be assembled with shaft arrangements that are right-handed or left-handed (Fig. 7).

ALLOWABLE SPEEDS OF GEAR REDUCERS The maximum speed of a gear reducer is limited by the accuracy of the machined gear teeth, the balance of the rotating parts, the allowable noise and vibration, the allowable maximum speed of the bearings, the pumping and churning of the lubricating oil, the friction of the oil seals, and the heat generated in the unit.

At high speeds it is possible for inaccuracies in the gear teeth to produce failure even though

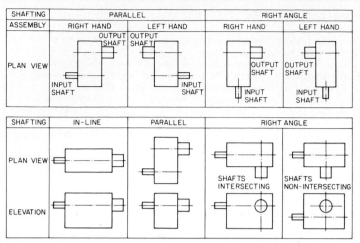

FIG. 7   Shaft arrangements for gear drives.

no power is being transmitted. Gear reducers built in accordance with AGMA specifications are recommended to operate at speeds given in Table 1.

POWER RANGE OF GEAR REDUCERS   The power-transmitting capacity of a reduction gear unit is a function of the output torque and the speed of the reducers. Some types of reducers, such as worm gear reducers, are more satisfactory for high torques and low speeds, whereas others, such as helical herringbone (or double helical), are suitable for high torques and also high speeds. Therefore the range of powers suitable for various units is considerable. A listing of this range obtainable in standard types of reducers is given in Table 1, with a brief explanation of why the range indicated is maintained. These powers are not fixed at the values given since they are continually changing. Although the values given are general, there are many special reducers available outside this range.

RATIOS AND EFFICIENCIES   The ratio of a gear reducer is defined as the ratio of the input shaft speed to the output shaft speed. Different types of gearing allow different ratios per gear stage. Spur gears usually are used with a ratio range of 1:1 to 6:1; helical, double helical, and herringbone with ratios of 1:1 to 10:1; straight-bevel with ratios of 1:1 to 4:1; spiral-bevel (also zerols and hypoids) with ratios of 1:1 to 9:1; and worm gears with ratios of 3¼:1 to 90:1. Planetary gear arrangements allow ratios of 4:1 to 10:1 per gear stage.

  Factors influencing the efficiency of a gear reducer are

1. Frictional loss in bearings
2. Losses due to pumping lubricating oil
3. Windage losses due to rotation of reducer parts
4. Frictional losses in gear tooth action

  It is not uncommon in many types of reducers to have the combined losses due to items 1, 2, and 3 greater than the loss due to item 4. For this reason in some cases the power lost in the reducer remains practically constant regardless of the power transmitted. Therefore it must be realized that the efficiency specified for a reducer applies only when the unit is transmitting its rated power, since when no power is being transmitted through the reducer, all the input power (small as it may be) is used in friction and the efficiency is zero.

  Ratios available in standard reducers and efficiencies to be expected when units are transmitting rated power are given in Table 1.

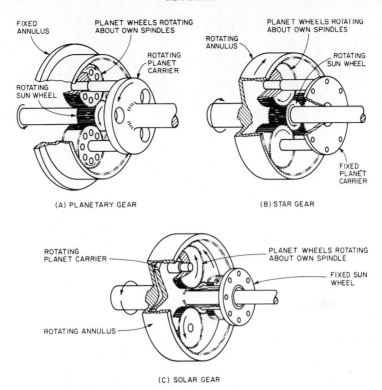

FIG. 8 Single epicyclic gear drives.

Epicyclic Gear Units Epicyclic gear units are sometimes used for pump drive applications. The important advantages are compact configuration, coaxial shafts, and light weight.

The most common types of epicyclic arrangements are planetary, star, and solar (Fig. 8). The planetary configuration (with the planet carriers integral with the output shaft) is the most commonly used arrangement. It is simple and rugged and gives the maximum ratio for the size of gears. The star arrangement (where the planet carrier is fixed and the internal gear rotates integrally with the output shaft) is used for higher-speed applications since centrifugal loads of the planet gears are eliminated with nonrotating planet carrier. The solar arrangement (where the pinion is fixed and the planet carrier is integral with the output shaft) has the input through the internal gear. This arrangement gives epicyclic advantages but allows a low ratio, generally less than 2:1. Higher ratios (in the range of 8:1 to 60:1) can be obtained utilizing double-reduction or compound-planetary arrangements.

INSTALLATION

The basic gear unit is generally shipped from the factory completely assembled. Mating gears and pinions are carefully assembled at the factory to provide proper tooth contact. Nothing should be done to disturb this setting.

Solid Foundation The reducer foundation should be rigid enough to maintain correct alignment with connected machinery. The foundation should have a flat mounting surface in order to assure uniform support for the unit. If the unit is mounted on a surface which is other than hor-

TABLE 1 Guide for Gear Selection and Design (Reflecting Generally Accepted Design Criteria)

	Spur gearing		Helical gearing	
	External	Internal	External	Internal
	Parallel axis	Parallel axis	Parallel axis	Parallel axis
Shaft arrangement	Parallel axis	Parallel axis	Parallel axis	Parallel axis
Ratio range	1:1 to 10:1	1½:1 to 10:1	1:1 to 15:1	2:1 to 15:1 (generally feasible). Ratio depends on pinion gear tooth combination because of clearance requirements
Size availability (including maximum face widths)	Up to 150-in (381-cm) OD, 30-in (76.2-cm) face width. Larger segmental gears can be produced with special processing and tooling	Up to 100-in (254-cm) OD, 16-in (406-cm) face width	Up to 150-in (381-cm) OD, 30-in (76.2-cm) face width	Up to 100 in (254 cm), depending on blank configuration; 16-in (406-cm) maximum face width
Gear tolerances (quality requirements)	See footnote.	See footnote.	See footnote.	See footnote.
Finishing methods (singly or in combination)	Cast; rotary cut, shaped; hobbed: shaved, ground	Same as external spurs	Shaped, hobbed, shaved, ground	Shaped, hobbed, shaved, honed, lapped, ground
Power range, hp (kW)	Commercial: less than 1000 (750)		Commercial: generally up to 50,000 (37,000) However, power limited only by maximum size capacity of design	Same as external helical gearing
Speed range, pitch line velocity, ft/min (m/min)	Commercial: normal up to 1000 (300); special precision up to 20,000 (6100)	Commercial, standard manufacture: up to 1000 (300); precision manufacture: up to 20,000 (6100)	To 30,000 (9100)	

Gear efficiency, %	Commercial: 95 to 98	97 to 99	
Quietness of operation	Commercial: quiet under 500 ft/min (150 m/min); noise increases with increasing pitch-line velocity	Noise level depends on quality of gear. Higher pitch line velocity (above 5000 ft/min (1500 m/min)) requires higher precision gear. Gears operating 30,000 ft/min (9100 m/min) have been made with overall noise level below 90 dB. Quieter than spur gears.	
Load imposed on bearings	Radial only	Radial and thrust	Radial and thrust

	External double helical gearing	Continuous-tooth herringbone gearing
Shaft arrangement	Parallel axis	Parallel axis
Ratio range	1:1 to 15:1	1:1 to 10:1
Size availability (including maximum face widths)	Up to 150-in (381-cm) OD, 30-in (76.2-cm) face width	Up to 150-in (381-cm) OD, 30-in (76.2-cm) face width
Gear tolerances (quality requirements)	See footnote.	See footnote.
Finishing methods (singly or in combination)	Same as helical gearing	Shaped, shaved
Power range, hp (kW)	Same as helical gearing	Up to 2000 (1500)
Speed range pitch line velocity, ft/min (m/min)	Same as helical gearing	Commercial: up to 5000 (1500)
Gear efficiency, %	Same as helical gearing	96 to 98
Quietness of operation	Same as helical gearing	Quiet operation up to 5000 ft/min (1500 m/min). Not generally used at extremely high pitch line velocity, over 20,000 ft/min (6100 m/min).
Load imposed on bearings	Radial only	Radial only

TABLE 1 Guide for Gear Selection and Design (Reflecting Generally Accepted Design Criteria) (*continued*)

	Straight-bevel gearing	Spiral-bevel gearing	Zerol-bevel gearing	Hypoid gearing
Shaft arrangement	Intersecting axis	Intersecting axis	Intersecting axis	Nonintersecting, nonparallel axis
Ratio range	1:1 to 6:1	1:1 to 10:1	1:1 to 10:1	1:1 to 10:1
Size availability (including maximum face widths)	Up to 102-in (55-cm) OD, 12-in (30.5-cm) face width	Up to 102-in (55-cm) OD 12-in (30.5-cm) face width	102-in (55-cm) OD, 12-in (30.5-cm) face	102-in (55-cm) OD, 12-in (30.5-cm) face
Gear tolerances (quality requirements)	See footnote.	See footnote.	See footnote.	See footnote.
Finishing methods (singly or in combination)	Cast, generated, planed	Generated, planed, ground	Generated, planed, ground	Generated, planed, ground
Power range, hp (kW)	Up to 1,500 (1120)	Up to 20,000 (15,000), depending on speed	Same as straight bevel gears	Same as spiral bevel gears, with use of EP lubricants
Speed range pitch line velocity, ft/min (m/min)	Same as spur gearing	Commercial, normal: up to 5,000 (1500); special precision: up to 15,000 (4500)	Up to 15,00 (4500) for ground gears	6,000 (1800) to 10,000 (3000), depending on offset
Gear efficiency, %	Same as spur gearing	96 to 98 (commercial)	94 to 98	85 to 98, depending on offset
Quietness of operation	Same as spur gearing	Noise level depends on quality of gear. Higher pitch line velocity— above 5,000 ft/min (1500 m/min)—requires higher precision gear.	Quieter than straight-bevel gears	Quiet
Load imposed on bearings	Radial and thrust	Radial and thrust	Radial and thrust	Radial and thrust

	Worm gearing	
	Cylindrical	Double-enveloping cone-drive gears
Shaft arrangement	Nonintersecting, nonparallel axis	Right angle in single-reduction units
Ratio range	3½:1 to 100:1	5:1 to 70:1
Size availability (including maximum face widths)	300-in (762-cm) OD, 4-in (10.2-cm) circular pitch	2- to 24-in (5.1- to 61-cm) center distance. For special requirements larger and smaller sizes are available.
Gear tolerances (quality requirements)	Worm gear tolerances given in AGMA Standard: Inspection of Course-Pitch Cylindrical Worms and Worm Gears, Standard No. 234.01	Standard AGMA commercial tolerances. Closer tolerances available for special applications
Finishing methods (singly or in combination)	Worm gears: hobbed, worm-milled, and ground	Worms: threads generated with cutter and finished by polishing. Gears: hobbed. Both members then lapped and matched together
Power range, hp (kW)	Up to 400 (300)	Fractional to 1,430 (1070) dependent on ratio, center distance, and speed
Speed range pitch line velocity, ft/min (m/min)	Up to 6000 (1830)	0 to 2400 rpm, or 2000 (610) rubbing speed with splash lubrication. Higher speeds permissible with special combinations.
Gear efficiency, %	From 25 to 95, depending on ratio	52 to 94, depending on ratio and speed
Quietness of operation	Relatively quiet operation up to 6000 ft/min (1830 m/min)	Smooth and quiet up to 2000 ft/min (610 m/min); can run quietly at higher speeds with special attention to lubrication, mounting, materials, balancing, etc.
Load imposed on bearings	Radial and thrust	Radial and thrust

Gear tolerances are dependent on method of manufacture, application, load requirements, and speeds. For spur, helical, and herringbone gears, AGMA Gear Classification Manual 390.03 lists quality numbers from 3 to 15 for coarse-pitch gears and 5 to 16 for fine-pitch gears, quality increasing as quality number increases. The general range of quality for gears now being manufactured is from quality numbers 5 to 14. Quality numbers relate to runout, tooth-to-tooth, spacing, profile, total composite, and lead tolerances. Bevel and hypoid gear tolerances range from 3 to 13.

izontal, consult the factory to ensure that the design provides for proper tooth contact and adequate lubrication.

The design of fabricated pedestals or baseplates for mounting speed reducers should be carefully analyzed to determine that they are sufficiently rigid to withstand operating vibrations. Vibration dampening materials may be used under the baseplate to minimize the effect of vibrations.

When mounting a drive on structural steel, the use of a rigid baseplate is strongly recommended. Bolt unit and baseplate securely to steel supports with proper shimming to ensure a level surface.

If a drive is mounted on a concrete foundation, allow the concrete to set firmly before bolting down the unit. For the best mounting, grout structural steel mounting pads into the concrete base rather than grouting the gear unit directly into the concrete.

Leveling   If shims are employed to level or align the unit, they should be distributed evenly around the base under all mounting pads to equalize the support load and to avoid distortion of the housing and highly localized stresses. All pads must be squarely supported to prevent distortion of the housing when the unit is bolted down.

Alignment   If the equipment is received mounted on a bedplate, it has been aligned at the factory. However, it may have become misaligned in transit. During field mounting of the complete assembly, it is always necessary to check alignment by breaking the coupling connection and shimming the bedplate under the mounting pads until the equipment is properly aligned. All bolting to the bedplate and foundation must be pulled up tight. After satisfactory alignment is obtained, close up the coupling.

Couplings   Drive shafts should be connected with flexible couplings. The couplings should be aligned as closely as possible following the manufacturer's instructions.

Alignment and Bolting   The gear unit, together with the prime mover and the driven machine, should be correctly aligned. After precise alignment, each member must be securely bolted and dowelled in place. Coupling alignment instructions should be carefully followed.

LUBRICATION

Types of Lubricant   The recommended types of oil for use in gear units are either straight mineral oil or extreme-pressure (EP) oil. In general, the straight mineral oil should be a high grade, well-refined petroleum oil within the recommended viscosity range. It must be neutral in reaction and not corrosive to gears and ball or roller bearings. It should have good defoaming properties and good resistance to oxidation for high operating temperatures.

Gear drives that are subject to heavy shock, impact loading, or extremely heavy duty should use an EP lubricant. EP gear lubricants are petroleum-based lubricants containing special chemical additives. The ones most recommended contain sulfur-phosphorous additives. Sulfur-phosphorous EP oils may be used to a maximum sump temperature of 180°F (82°C).

In general, if units are subjected to unusually high ambient temperatures (100°F, 38°C or higher), extreme humidity, or atmospheric contaminants, use the straight mineral oil recommended.

Grease Lubrication   The lubricant should be high-grade, nonseparating, ball bearing grease suitable for operating temperatures to 180°F (82°C). Grease should be N.L.G.I. No. 2 consistency.

The grease lubricant must be noncorrosive to ball or roller bearings and must be neutral in reaction. It should contain no grit, abrasive, or fillers; it should not precipitate sediment; it should not separate at temperatures up to 300°F (149°C); and it should have moisture-resistant characteristics and good resistance to oxidation.

TABLE 2 Troubleshooting Chart

Trouble	What to inspect	Action
Overheating	1. Unit overload	Reduce loading or replace with drive of sufficient capacity.
	2. Oil-cooler operation	Check coolant and oil flow. Vent system of air. Oil temperatures into unit should be approximately 110°F (43°C). Check cooler internally for build-up of deposits from coolant water.
	3. Oil level	Check oil level indicator to see that housing is accurately filled with lubricant to the specified level.
	4. Bearings adjustment	Bearings must not be pinched. Adjustable tapered bearings must be set at proper bearing lateral clearance. All shafts should spin freely when disconnected from load.
	5. Oil seals or stuffing box	Oil seals should be greased on those units having grease fitting for this purpose. Otherwise, apply small quantity of oil externally at the lip until seal is run in. Stuffing box should be gradually tightened to avoid overheating. Packing should be self-lubricating braided-asbestos type.
	6. Breather	Breather should be open and clean. Clean breather regularly in a solvent.
	7. Grade of oil	Oil must be of grade specified in lubrication instructions. If not, clean unit and refill with correct grade.
	8. Condition of oil	Check to see if oil is oxidized, dirty, or of high sludge content; change oil and clean filter.
	9. Forced-feed lubrication system	Make sure oil pump is functioning. Check that oil passages are clear and permit free flow of lubricant. Inspect oil line pressure regulators, nozzles, and filters to be sure they are free of obstructions. Make sure pump suction is not sucking air.
	10. Coupling alignment	Disconnect couplings and check alignment. Realign as required.
	11. Coupling lateral float	Adjust spacing between drive motor, etc., to eliminate end pressure on shafts. Replace flexible coupling with type allowing required lateral float.
	12. Speed of unit	Reduce speed or replace with drive suitable for speed.
Shaft failure	1. Type of coupling used	Rigid couplings can cause shaft failure. Replace with coupling that provides required flexibility and lateral float.
	2. Coupling alignment	Realign equipment as required.
	3. Overhung load	Reduce overhung load. Use outboard bearing or replace with unit having sufficient capacity.

TABLE 2   Troubleshooting Chart (*continued*)

Trouble	What to inspect	Action
	4. Unit overload	Reduce loading or replace with drive of sufficient capacity.
	5. Presence of high-energy loads or extreme repetitive shocks	Apply couplings capable of absorbing shocks and, if necessary, replace with drive of sufficient capacity to withstand shock loads.
	6. Torsional or lateral vibration condition	These vibrations can occur through a particular speed range. Reduce speed to at least 25% below critical speed. System mass elastic characteristics can be adjusted to control critical speed location. If necessary, adjust coupling weight, as well as shaft stiffness, length, and diameter. For specific recommendations, contact factory.
	7. Alignment of outboard bearing	Realign bearing as required.
Bearing failure	1. Unit overload	See Overheating (item 1). Abnormal loading results in flaking, cracks, and fractures of the bearing.
	2. Overhung load	See Shaft failures (item 3).
	3. Bearing speed	See Overheating (item 12).
	4. Coupling alignment	See Overheating (item 10).
	5. Coupling lateral float	See Overheating (item 11).
	6. Bearings adjustment	See Overheating (item 4). If bearing is too free or not square with axis, erratic wear pattern will appear in bearing races.
	7. Bearings lubrication	See Overheating (items 2, 3, 7, 8, 9). Improper lubrication causes excessive wear and discoloration of bearing.
	8. Rust formation due to entrance of water or humidity	Make necessary provisions to prevent entrance of water. Use lubricant with good rust-inhibiting properties. Make sure bearings are covered with sufficient lubricant. Turn over gear unit more frequently during prolonged shutdown periods.
	9. Bearing exposure to abrasive substance	Abrasive substance will cause excessive wear, evidenced by dulled balls, rollers, and raceways. Make necessary provision to prevent entrance of abrasive substance. Clean and flush drive thoroughly and add new oil.
	10. Damage due to improper storage or prolonged shutdown	Prolonged periods of storage in moist air and at ambient temperatures will cause destructive rusting of bearings and gears. When these conditions are found to have existed, the unit must be disassembled and inspected and damaged parts either thoroughly cleaned of rust or replaced.

TABLE 2 Troubleshooting Chart (*continued*)

Trouble	What to inspect	Action
Oil leakage	1. Oil	Check through level indicator that oil level is precisely at level indicated on housing.
	2. Open breather	Breather should be open and clean.
	3. Open oil drains	Check that all oil drain locations are clean and permit free flow. Drains are normally drilled in the housing between bearings and bearing cap where shafts extend through caps.
	4. Oil seals	Check oil seals and replace if worn. Check condition of shaft under seal and polish if necessary. Slight leakage normal, required to minimize friction and heat.
	5. Stuffing boxes	Adjust or replace packing. Tighten packing gradually to break in. Check condition of shaft and polish if necessary.
	6. Force-feed lubrication to bearing	Reduce flow of lubricant to bearing by adjusting orifices. Refer to factory.
	7. Plugs at drains, levels, etc., and standard pipe fittings	Apply pipe joint sealant and tighten fittings.
	8. Compression-type pipe fittings	Tighten fitting or disassemble and check that collar is properly gripping tube.
	9. Housing and caps	Tighten cap screws or bolts. If not entirely effective, remove housing cover and caps. Clean mating surfaces and apply new sealing compound (Permatex #2 or equal). Reassemble. Check compression joints by tightening fasteners firmly.
Gear wear	1. Backlash	Gear set must be adjusted to give proper backlash. Refer to factory.
	2. Misalignment of gears	Make sure that contact pattern is above approximately 75% of face, preferably in center area. Check condition of bearings.
	3. Twisted or distorted housing	Check shimming and stiffness of foundation.
	4. Unit overload	See Overheating (item 1).
	5. Oil level	See Overheating (item 3).
	6. Bearings adjustment	See Overheating (item 4).
	7. Grade of oil	See Overheating (item 7).
	8. Condition of oil	See Overheating (item 8).
	9. Forced-feed lubrication	See Overheating (item 9).
	10. Coupling alignment	See Overheating (item 10).
	11. Coupling lateral float	See Overheating (item 11).
	12. Excessive speeds	See Overheating (item 12).
	13. Torsional or lateral vibration	See Shaft failure (item 6).
	14. Rust formation due to entrance of water or humidity	See Bearing failure (item 8).
	15. Gears exposure to abrasive substance	See Bearing failure (item 9).

Grease Lubrication of Bearings   Pressure fittings are often supplied in gear units for the application of grease to bearings that are shielded from the oil. Although a film of grease over the rollers and races of the bearing is sufficient lubrication, drives are generally designed with ample reservoirs at each grease point.

Greased bearings should be lubricated at definite intervals. Usually one-month intervals are satisfactory unless experience indicates that regreasing should occur at shorter or longer intervals.

Oil Seals   Oil seals require a small amount of lubricant to prevent frictional heat and subsequent destruction when the shaft is rotating. Normally when a single seal is utilized, sufficient lubricant is provided by spray or splash. Certain design or application requirements dictate that double seals be used at some sealing points. When this is the case, a grease fitting and relief plug are located in the seal retainer to provide lubricant to the outer seal. Grease must periodically be applied betwen the seals by pumping through the fitting until overflow is noted by the relief plug. The greases recommended for bearings may also be used for seals.

TROUBLESHOOTING TIPS

Improper lubrication causes a high percentage of gear reduction unit failures. Too frequently, speed reducers are started up without any lubricant at all. Conversely, units are sometimes filled to a higher oil level than specified in the mistaken belief that better lubrication is obtained. This higher oil level usually results in more of the input power going into churning the oil, creating excessive temperatures with detrimental results to the bearings and gearing. Insufficient lubrication gives the same results.

Gear failure due to overload is a broad and varied area of misapplication. The nature of the load (input torque, output torque, duration of operating cycle, shocks, speed, acceleration, etc.) determines the gear unit size and other design criteria. Frequently, a gear drive must be larger than the torque output capability of the severity of application conditions by providing a higher nominal power which in effect increases the size of the gear unit. If there is any question in the user's mind that the actual service conditions may be more severe than originally anticipated, it is recommended that this information be communicated to the gear manufacturer before start-up. Often there are remedies that can be suggested before a gear unit is damaged by overload, but none are effective after severe damage.

Motors and other prime movers should be analyzed while driving the gear unit under fully loaded conditions to determine that the prime mover is not overloaded and thus putting out more than the rated torque. If it is determined that overload does exist, the unit should be stopped and steps taken either to remove the overload or to contact the manufacturer to determine suitability of the gear drive under the observed conditions.

Table 2 is an extensive troubleshooting chart that should be consulted whenever necessary.

FURTHER READING

AGMA: *AGMA Standards and Technical Publications Index*, AGMA 000-67, Arlington, VA., March 1974.

Dudley, D. W.: *Handbook of Practical Gear Design*, McGraw-Hill, New York, 1984.

"Gear Drives," *Design News*, Jan. 19, 1968.

Hamilton, J. M.: "Are You Paying Too Much for Gears?" *Mach. Design*, Oct. 19, 1972.

Kron, H. O.: "Optimum Design of Parallel Shaft Gearing," *Trans. ASME*, 72PTG-17, October 1972.

Philadelphia Gear Corporation: *Philadelphia Application Engineered Gearing Catalog*, G-76, King of Prussia, Pa., 1976.

Rosaler, R. C., and J. O. Rice: *Standard Handbook of Plant Engineering*, McGraw-Hill, New York, 1983.

6.2.5
ADJUSTABLE-SPEED BELT DRIVES

MILTON B. SNYDER

Mechanical adjustable-speed drives for pump applications are generally of the compound adjustable-pitch-sheave and rubber-belt variety, as illustrated in Figs. 1 and 2. The integrated drive package converts constant input speed to an output that is steplessly variable within a certain range. The drive packages are usually driven by constant-speed ac induction motors and usually contain built-in gear reducers to obtain low output speeds.

Drive packages may be mounted horizontally, vertically, or on a 45° angle and are available in standard open or totally enclosed designs. Some of the possible mounting arrangements are illustrated in Fig. 3. Speed ranges of 10:1 to 2:1 can be obtained with most units. A typical distribution of available output speeds is 4850 to 1.4 rpm, including drives with and without reducer gearing. Increaser gearing (offered as an integrally mounted package) provides speeds to 16,000 rpm.

Alternating-current induction-drive motors used with mechanical adjustable-speed drives usually operate at a speed of 1750 or 1160 rpm. The electric design characteristics of these motors comply with the standards of the National Electrical Manufacturers Association (NEMA). The NEMA design B motor with normal torque and normal slip characteristics is standardly furnished. NEMA design C (high torque, low starting current) and NEMA design D (high torque, high slip) motors may be used when their specific characteristics are dictated by the application.

The mechanical mounting characteristics of the drive motors for mechanical adjustable-speed drives vary from one manufacturer to another. Motors of round body, footless, NEMA C-face construction are commonly used. Sometimes standard foot-mounted motors are supplied as an option in conjunction with a scoop support by some manufacturers or as standard practice by others. Still others may supply mechanically special partial motor frames with their drives.

For a given power rating, many manufacturers will supply drive motors with increased service-factor power. This increased power of the drive motor compensates for and overcomes the inherent mechanical losses of the drive. This in turn causes full-rated power to be developed at the output shaft of the mechanical adjustable-speed drive. Drive output power ratings are more fully discussed below under Rating Basis.

The input speed of the mechanical adjustable-speed drive is typically 1750 or 1160 rpm, as defined by the speed of the drive motor. The output speed of the internal adjustable-speed belt section goes above and below the drive motor speed. A maximum speed of 4200 rpm or greater is not uncommon for fractional and small-integral power drives. The need for stages of output gearing to obtain final output speeds that are usable for pump or other applications is therefore readily apparent.

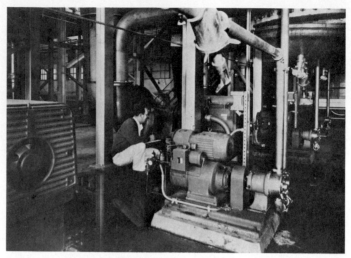

FIG. 1   Pump application utilizing typical mechanical adjustable-speed belt drives.
(Reliance Electric)

Parallel-shaft or right-angle-shaft reducer gearing, at the option of the design engineer or user, may be incorporated as an integral part of the mechanical adjustable-speed drive package. Generally speaking, gear reduction is required when drive maximum output speed must be lower than 1750 rpm or drive minimum output speed lower than 583 rpm.

The American Gear Manufacturers Association (AGMA) does not define standards for reducers used in adjustable-speed applications. However, most all drive manufacturers produce reducers for these drives in accordance with accepted AGMA standards for constant-speed reducers.

FIG. 2   Cutaway view of typical mechanical adjustable-speed drive.
(Reliance Electric)

C—POWER FLOW MOUNTINGS

Vertical   45°   Horizontal   Trunnion

Z—POWER FLOW MOUNTINGS

Vertical   45°   Horizontal   Trunnion

SPECIAL MOUNTINGS and CASE ENCLOSURES

Scoop Mounted   Flange Mounted   Corrosion-Proof   Sanitary-type

FIG. 3   Mechanical adjustable-speed drive mounting and special enclosures.

Where infinite, or stepless, adjustment over a specific finite speed range is necessary, stepless mechanical adjustable-speed drives are generally most economical for standard pump application requirements. Initial costs are usually lower than for comparable electric or hydraulic systems, and the mechanical systems are easier to operate and maintain.

Reliability and accuracy of speed control are advantages of the mechanical adjustable-speed belt drive package. Construction details, size, and mounting dimensions are not standardized and vary with the manufacturer, but all employ dual adjustable-pitch sheaves mounted on parallel shafts at a fixed center distance and a special wide-section rubber V belt to provide a compact assembly. Most of the designs utilize spring-loaded sheaves for control of belt tension.

A high degree of control and flexibility in application is offered by mechanical adjustable-speed belt drive power packages. Capacities range from fractional to 100 hp (75 kW), with maximum speed ratios of 10:1 in the fractional power sizes, decreasing to 6:1 at 30 hp (22 kW) and 3:1 in the largest sizes. Multiple-belt drive arrangements are usually required for capacities over 50 hp (37 kW). Again, depending on capacity, speeds ranging from a maximum of 16,000 to a minimum of 1.4 rpm are possible, although most of the standard units are within the 2 to 5000 rpm range.

OPERATING PRINCIPLE

In Fig. 4, the upper disk assembly is the input assembly driven directly by the shaft of the ac induction motor at a constant speed. This constant-speed disk assembly has one stationary disk member on the *left* and one movable, or sliding, disk member on the *right*. The sliding member is mechanically attached to a positive shifting linkage arrangement consisting of a thrust bearing, bearing housing, and shifting yoke. This linkage is actuated by a control handwheel or some other speed-changing device.

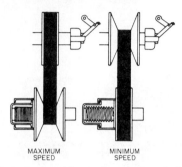

MAXIMUM SPEED   MINIMUM SPEED

FIG. 4   Functional diagram for a mechanical adjustable-speed drive belt section. (Reliance Electric)

In the same view, the lower disk assembly is the output assembly, whose speed is adjustable. In this adjustable-speed disk assembly, the sliding member is next to the spring cartridge on the *left* and the stationary member is on the *right*. Note that this arrangement is just the opposite of that for the upper, or constant-speed, disk assembly. This adjustable-speed disk assembly is mounted on the adjustable-speed output shaft of the drive. A flexible, wide-section, rubber V belt connects the two disk assemblies.

Assuming the minimum speed belt position as a starting point, the positive shifting linkage on the sliding member of the constant-speed disk assembly is moved to the left toward the fixed member. This positive change forces the wide-section V belt to a larger diameter in the constant-speed disk assembly. Simultaneously the belt forces the sliding member on the adjustable-speed disk assembly against its spring so that the belt assumes a smaller running, or pitch, diameter in this disk assembly. The speed of the output shaft increases in stepless increments while the drive motor speed remains constant. Reversing the above procedures reduces the output shaft speed.

When the V belt is at this maximum diameter in the constant-speed disk assembly, it is at a minimum diameter in the adjustable-speed disk assembly and the output shaft speed is at maximum. Conversely, when the V belt is at its minimum diameter in the constant-speed disk assembly, it is at a maximum diameter in the adjustable-speed disk assembly and the output shaft speed is at minimum.

Since one member of each disk assembly is fixed on its shaft, the V belt must move axially as well as along the inclined surface of the disk to assume different diameters and cause the speed to change. Centrifugal force makes this composite movement of the belt effortless when the drive is in operation, but the speed setting of the drive must never be changed while the drive is not operating and motionless. If speed-setting change is attempted while the drive is motionless, destructive, crushing forces are imposed on the belt. These same forces may also damage the positive shifting linkage of the otherwise rugged mechanical adjustable-speed drive.

OTHER BELT DRIVES

Other drives of the mechanical adjustable-speed type, such as the single adjustable-motor-sheave or pulley and the wood block or metal-chain-belt type transmission, are only mentioned briefly here because of their rather limited application as pump drives.

The motor sheave, or pulley, illustrated in Fig. 5 is a simple, single, adjustable-pitch device mounted on a motor shaft. It drives by way of a standard or wide-section V belt to a companion fixed-diameter, flat-face pulley or V sheave. Driving and driven shafts are parallel but must be

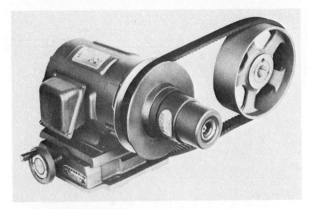

FIG. 5 Typical motor pulley belt drive with adjustable center distance between driving and driven shafts. (Reliance Electric)

arranged so that their center distance is adjustable; this is generally accomplished by means of a sliding motor base. The entire drive assembly is usually operated as an open belt drive.

The wood-block or metal-chain belt transmission uses the original mechanical adjustable-speed belt drive operating principle. This drive utilizes a special wide-section wood-block belt or laminated-metal-chain belt driven by two pairs of positively controlled variable-pitch sheaves. Movement of the sheave flanges is synchronized by a positive linkage arrangement.

The transmission-type drive is generally thought of as a low-speed, high-torque device that can withstand severe overload and abuse for long periods of time.

RATING BASIS

The mechanical adjustable-speed drive package is usually rated as a constant-torque, variable-power device. The power rating is based upon the capacity of the whole unit at maximum speed setting. When operating at any output speed below this maximum, the power capacity is reduced in direct proportion.

A drive unit is made up of three major components, each of which has its own individual torque or power output characteristics. For example, the ac induction-drive motor or prime mover develops constant power at a constant rotational speed.

The adjustable-speed belt section has an output torque characteristic that is a mixture of both constant torque and variable torque over its speed range. This section has a constant-power, variable-torque characteristic when operating above a 1:1 belt position and a constant-torque variable-power characteristic when operating below a 1:1 belt position. Finally, a parallel-shaft or right-angle gear reducer section has a constant-torque, variable-power characteristic. Figure 6 graphically illustrates these relationships.

Since the gear reducer section has a *constant-torque* characteristic, *this section* defines the output characteristic for the entire mechanical adjustable-speed drive.

It should be noted that virtually all manufacturers rate parallel-shaft drives using output shaft power as a base, whereas right-angle output shaft drives are rated on the basis of input power to the reducer minus reducer efficiency.

Because of the relatively low efficiency of right-angle worm gear reducers and right-angle combination worm and helical gear reducers, the power transmission industry follows the practice of rating these units in terms of power at the input shaft. From the earlier discussion on operating principle, you will note that the adjustable-speed output shaft of the belt section of the drive becomes the input shaft of the reducer section of the same drive.

One or more sections of a drive, usually the belt and reducer sections, may be service-factored by frame or case oversizing to permit the rating of these sections for constant power over all or a

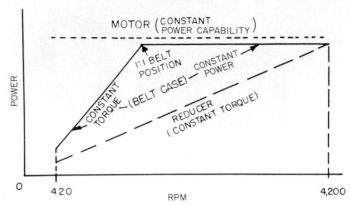

FIG. 6 Power versus output speed relationship for the motor, belt section, and reducer section of a mechanical adjustable-speed belt drive.

portion of their speed range. Since the drive motor is a constant-power device, as already mentioned, oversizing of its frame is unnecessary.

SERVICE FACTORING

Service factoring of a mechanical adjustable-speed belt drive is common practice when the normal operating requirements for steady constant-torque loads running 8 h/day, 5 days/week are to be exceeded.

Drives to be used for other-than-normal service as described above must be selected by use of modifying factors that will provide correct service capacity. Some unusual service requirements are moderate to heavy shock loads, 24 h/day continuous operation, and constant-power demands over a wide speed range.

Other unusual service conditions may be suggested by the following information check list. This list itemizes *required information* data that should be furnished to the variable-speed drive manufacturer for those applications calling for unusual service:

1. Speed and torque required for the application
2. Value and frequency of peak-load condition
3. Hours of operation per day or week
4. Frequency of starts and stops
5. Inertia (WK^2) of the load
6. Frequency of reversals of rotation direction
7. Electric and mechanical overload protection provisions
8. Method used to connect drive output shaft to driven load
9. Any unusual environment or other operating conditions

METHODS OF CONTROL

A variety of control systems have been developed for use with mechanical adjustable-speed drives. For the majority of pump applications, speed is controlled manually through a lever, handwheel, or knob attachment. Remote semiautomatic and automatic control methods in mechanical, pneumatic, or electric forms are also being used.

For manual operation, vernier attachments are often useful to increase the accuracy of speed adjustment. Cams are occasionally employed, mounted externally or internally, to assure a prescribed pattern of output characteristics.

Remote control is usually obtained by means of a positioning motor, which is a fractional-power motor connected to the drive control shifting screw through reduction gearing. The output speed of the drive is then adjusted from a station at a remote location.

For semiautomatic or automatic operation, control systems usually consist of three elements: a sensing unit, a receiver, and a positioning actuator. The sensing unit detects changes in the process being controlled and transmits a signal to the receiver. At the receiver, the signal is analyzed, amplified, and transmitted to the positioning actuator, which adjusts the speed of the mechanical adjustable-speed drive accordingly.

If the process and/or load requirements can be adapted to produce a signal, there should be a suitable control system that can be used for speed adjustment. The only limitation is that the load requirements must follow a specific pattern of some type, regardless of whether the pattern is based on direct, inverse, or proportional relationships.

The pneumatic actuators used for speed adjustment are usually responsive to a 3- to 15-lb/in^2 (20- to 100-kPa) air signal pressure. These pneumatic positioning devices are actually analog piloted servovalve positioning devices that can be used in a great variety of open- and closed-loop process control applications. They can be used to cause a mechanical adjustable-speed drive to either follow or maintain a given process signal from variables such as liquid level, pressure, flow rate, or any other measurable value. The adjustable-speed drive thus becomes the final control element in a closed-loop process control system. Figure 7 is a typical drive equipped with a pneumatic actuator for speed changing.

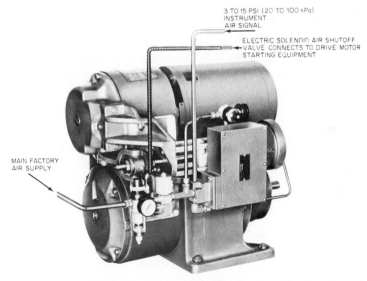

FIG. 7 Mechanical adjustable-speed drive with pneumatic actuator for automatic speed changing. (Reliance Electric)

APPLICATION GUIDELINES

The following is a listing of the more common items to consider when specifying a mechanical adjustable-speed belt drive for a specific application:

1. Manufacturer's size designation of the drive
2. Range of speed variation and actual output speeds

3. Motor specifications: power rating, electric current (single or polyphase, frequency, and voltage), type of enclosure, and other special electric or mechanical modifications

4. Special drive output shaft extension

5. Type of control: handwheel, electric remote, mechanical automatic, pneumatic, etc.

6. Manufacturer's assembly configuration designation and type of mounting, whether standard floor type, trunnion, ceiling, sidewall, or flange

7. Accessory equipment, such as tachometer, magnetic brake

8. Power rating based on constant torque and maximum output speed

9. Type of case enclosure

SECTION 6.3
POWER TRANSMISSION DEVICES

6.3.1
PUMP COUPLINGS AND INTERMEDIATE SHAFTING

FRED K. LANDON

COUPLING TYPES USED IN PUMP DRIVE SYSTEMS

A coupling is used wherever there is a need to connect a prime mover to a piece of driven machinery. The principal purpose of a coupling is to transmit rotary motion and torque from one piece of equipment to another. Couplings may perform other, secondary functions, such as accommodating misalignment between shafts, transmitting axial thrust loads from one machine to another, permitting adjustment of shafts to compensate for wear, and maintaining precise alignment between connected shafts.

Rigid Couplings Rigid couplings are used to connect machines where it is desired to maintain shafts in precise alignment. They are also used where the rotor of one machine is used to support and position the other rotor in a drive train. Since a rigid coupling cannot accommodate misalignment between shafts, precise alignment of machinery is necessary when one is used.

TYPES There are two commonly used types of rigid couplings. One type consists of two flanged rigid members, each mounted on one of the connected shafts (Fig. 1). The flanges are provided with a number of bolt holes for the purpose of connecting the two half-couplings. Through proper design and installation of the coupling, it is possible to transmit the torque load entirely through friction from one flange to the other, which assures that the flange bolts do not experience a shearing stress. This type of arrangement is especially desirable for driving systems where torque oscillations occur, as it avoids subjecting the flange bolts to a shearing stress.

A second type of rigid coupling, known as the *split rigid*, is split along its horizontal centerline (Fig. 2). The two halves are clamped together by a series of bolts arranged axially along the coupling. The rigid coupling and machine shafts may be equipped with conventional keyways, which are in turn fitted with keys to transmit the torque load, or in certain cases the frictional clamping force may be sufficient to permit transmitting the torque by means of friction between shaft and rigid coupling. This type of coupling is commonly used to connect sections of line shafting in a drive train.

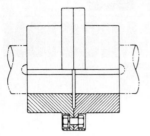

FIG. 1   Flanged rigid coupling. (Koppers)

FIG. 2   Split rigid coupling. (Dodge Manufacturing Division, Reliance Electric)

A variation to the flanged rigid coupling is known as the adjustable rigid coupling (Fig. 3). This coupling is designed along the lines of conventional rigid couplings, except that a threaded adjusting ring is placed between the two flanges. This ring engages a threaded extension on one of the shaft ends. By means of this ring, it is possible to position the pump shaft axially with respect to the driver.

APPLICATIONS   A common application for rigid couplings in the pump industry is in vertical drives, where the prime mover (generally an electric motor) is positioned above the pump. In such cases, both machines can employ a common thrust bearing, which is generally located in the motor. The coupling flange bolts must be capable of transmitting any down thrust from the pump to the motor. In applications where the thrust from the pump is *toward* the motor, it is common practice to provide shoulders on the shafts to transmit the axial force.

Many pump drive systems require a rigid coupling which is capable of providing axial adjustment to compensate for wear in the pump impeller or impellers. The adjustable rigid coupling is used for this purpose. The threaded adjusting ring attached to a mating threaded extension of the pump shaft permits vertical positioning of the impeller or impellers. The hub, which is mounted on the pump shaft, is equipped with a clearance fit and feathered key that permit the hub to slide with respect to the shaft. The load capacity of a coupling of this type is generally limited by the pressure on the pump shaft key since there is no possibility of load being transmitted by interference fit.

A few words of caution should be noted about the use of rigid couplings. First, precise alignment of machine bearings is absolute necessary since there is no flexibility in the coupling to accommodate misalignment between shafts. Secondly, accuracy of manufacture is extremely

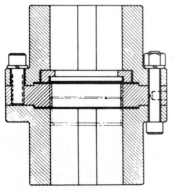

FIG. 3   Adjustable flanged rigid coupling. (Koppers)

FIG. 4   Gear mechanically flexible coupling. (Koppers)

important. The coupling surfaces which interface between driving and driven shafts must be manufactured with high degrees of concentricity and squareness, to avoid the transmittal of eccentric motion from one machine to the other.

Flexible Couplings   Flexible couplings accomplish the primary purpose of any coupling, that is, to transmit a driving torque between prime mover and driven machine. In addtion, they perform a second important function: they accommodate unavoidable misalignment between shafts. A proliferation of designs exists for flexible couplings, which may be classified into two types: mechanically flexible and materially flexible.

MECHANICALLY FLEXIBLE COUPLINGS   Mechanically flexible couplings compensate for misalignment between two connected shafts by means of clearances incorporated in the design of the coupling. The most commonly used type of mechanically flexible coupling is the gear, or dental, coupling (Fig. 4). This coupling essentially consists of two pair of clearance fit splines. In the most common configuration, the two machine shafts are equipped with hub members having external splines cut integrally on the hubs. The two hubs are connected by a sleeve member having mating internal gear teeth. Backlash is intentionally built into the spline connection, and it is this backlash which compensates for shaft misalignment. Sliding motion occurs in a coupling of this type, and so a supply of clean lubricant (grease or oil, depending on the design) is necessary to prevent wear of the rubbing surfaces.

If operation cannot be interrupted to lubricate the couplings, constantly lubricated couplings are used, as shown in Fig. 5. These consist of an oiltight enclosure bolted at one end to the stationary portion of either the driving or driven piece of equipment. The other end of the enclosure has a slip fit inside a cover that is bolted to the other piece of equipment. Some form of packing is used to prevent loss of lubricant at the slip joint. Oil under pressure is brought through the enclosure and impinges upon the meshing gear teeth of the coupling, the excess being collected at the bottom of the enclosure and returned to the oil reservoir.

A second type of mechanically flexible coupling which sees wide usage, especially in low-cost drive systems, is known as the roller-chain flexible coupling (Fig. 6). This coupling employs two sprocket-like members mounted one on each of the two machine shafts and connected by an annulus of roller chain. The clearance between sprocket and roller and, in some cases, the crowning of the rollers provide mechanical flexibility for misalignment. This type of coupling is generally limited to low-speed machinery.

MATERIAL-FLEXIBLE COUPLINGS   These couplings rely on flexing of the coupling element to compensate for shaft misalignment. The flexing element may be of any suitable material (metal, elastomer, or plastic) which has sufficient resistance to fatigue failure to provide acceptable life. Some

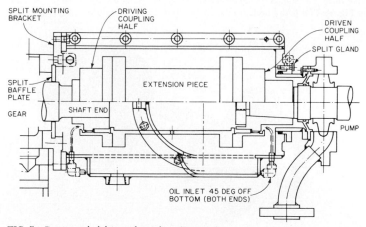

FIG. 5 ʹ Continuously lubricated coupling. (Koppers)

FIG. 6 Roller-chain mechanically flexible coupling.
(Dodge Manufacturing Division, Reliance Electric)

FIG. 7 Metal-disk material-flexible coupling.
(Formsprag)

materials, such as steel, have a finite fatigue limit, and a coupling made of such material must be
operated under conditions of load and misalignment which assure that the stress developed in the
coupling element is within that limit. Other materials, such as elastomers, generally do not have
a well-defined fatigue limit. In these cases, however, heat developed in the material when the
coupling flexes can cause failure if excessive.

One type of material-flexible coupling is the metal-disk coupling (Fig. 7). This coupling consists of two sets of thin sheet-metal disks bolted to the driving and driven hub members. Each set
of disks is made up of a number of thin laminations which are individually flexible and compensate for shaft misalignment by means of this flexibility. These disks may be stacked together as
required to obtain the desired torque transmission capability. This type of coupling requires no
lubrication; however, alignment of the equipment must be maintained within acceptable limits
so as not to exceed the fatigue limit of the material.

Another example of an all-metal material-flexible coupling is the flexible diaphragm coupling,
shown in Fig. 8. This coupling is similar in function to the metal-disk coupling in that the disk

FIG. 8 Diaphragm material-flexible coupling. (Fluid Power Division,
Bendix)

flexes to accommodate misalignment. However, the diaphragm type consists of a single element with a hyperbolic contour which is designed to produce uniform stress in the member from inner to outer diameter. By more efficiently utilizing the material, the weight is reduced correspondingly, thus making this coupling suitable for high-speed applications.

Material-flexible couplings employing elastomer materials are numerous and their designs are varied. By definition, an elastomer is a material which has a high degree of elasticity and resiliency and will return to its original shape after undergoing large-amplitude deformations. One example of an elastomer coupling is the pin-and-bushing coupling (Fig. 9). This design comprises two flanged hub members, one mounted on each machine shaft. The flange of one hub is fitted with pins which extend axially toward the adjacent shaft. The other flange is equipped with rubber bushings, which generally have a metal sleeve at the center. The pins fit into these sleeves and provide transmission of torque through the bushings. Since the bushings are made of flexible material, they can accept slight angularity or offset conditions between the two flanges.

A second group of elastomer couplings employs a sleevelike element which is connected to a hub member on each shaft and transmits torque through shearing of the flexible element. The flexible element may be attached to the machine hubs by a number of different means: it may be chemically bonded, it may be mechanically connected by means of loose-fitting splines (Fig. 10), or it may be clamped to the hubs and held in place by friction (Fig. 11). Misalignment between shafts is accommodated through flexing of the elastomer sleeve.

A third group of couplings utilizes an elastomer member which is loaded in compression to transmit load from one shaft to the other. The elastomer material may be bonded onto the machine shafts (or onto hub members attached to the shafts, Fig. 12) or may be placed loosely into cavities which are formed by members rigidly mounted onto the two shafts (Fig. 13). Again, the elastomer material deflects to compensate for shaft misalignment.

A type of elastomer coupling which is commonly used on low-power drive systems is the rubber jaw coupling (Fig. 14). The heart of this coupling is a "spider" member, having a plurality (usually three) of segments extending radially from a central section. The hubs, which are mounted on driving and driven shafts, each have a set of jaws corresponding to the number of spiders on the flexible element. The spider fits between the two sets of jaws and provides a flexible "cushion" between them. This cushion transmits the torque load as well as compensating for misalignment.

SPRING-GRID COUPLING   There is a type of commercially available flexible coupling which combines the characteristics of mechanically flexible and material-flexible couplings. This is the spring-grid coupling (Fig. 15). This design has two hubs, one mounted on each machine shaft. Each hub has a raised portion on which toothlike slots are cut. A spring steel grid member is fitted, or woven, between the slots on the two hubs. The grid element can slide in the slots to accommodate shaft misalignment and flexes like a leaf spring to transmit torque from one machine to

FIG. 9 Pin-and-bushing elastomer coupling. (Ajax Flexible Coupling)

FIG. 10 Sleeve-type elastomer coupling. (T. B. Woods' Sons)

FIG. 11   Sleeve-type clamped elastomer coupling. (Dodge Manufacturing Division, Reliance Electric)

FIG. 12   Compression-loaded, bonded elastomer coupling. (Koppers)

the other. Unlike most material-flexible couplings, this design requires periodic lubrication to prevent excessive wear of the grid member.

APPLICATIONS   The type of flexible coupling most suitable for a particular application depends upon a number of factors, including power, speed of rotation, shaft separation, amount of misalignment, cost, and reliability. In the design of a system, it is the goal of the designer to use the least expensive coupling which will do the job. In low-cost systems, cost alone may be the most important criterion, and the least expensive coupling which transmits the rated power and accepts some small degree of misalignment is generally the choice, albeit at some sacrifice of reliability and durability. On the other hand, high-power, high-speed machinery generally represents a critical piece of equipment for a power station, sewage plant, or other vital process, and in these cases a coupling should be selected which will not compromise the overall reliability of the system.

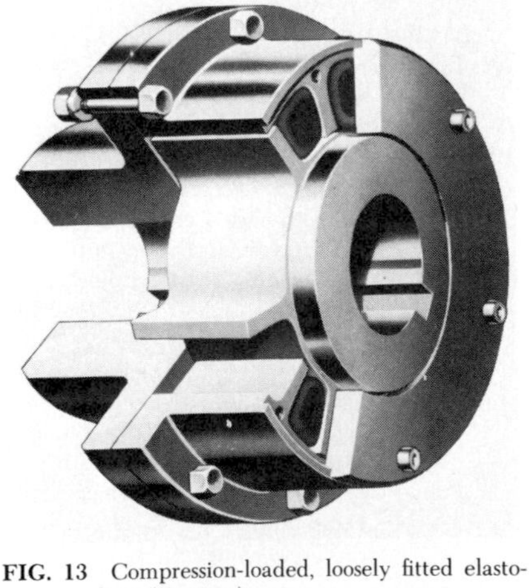

Low-power pumps (up to about 200 hp, 150 kW) driven by electric motors can usually be coupled successfully by any of the couplings described here. Selection procedures vary from manufacturer to manufacturer, but generally the following data are required: power rating, speed, anticipated misalignment, and type of pump (reciprocating, vane, centrifugal, etc.).

Pumps of the same power range are very often driven by reciprocating engines (diesel, gasoline, natural gas). This is quite common in remote areas, such as at pipeline pumping stations, where a source of electric power is not

FIG. 13   Compression-loaded, loosely fitted elastomer coupling. (Koppers)

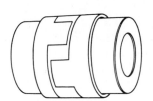

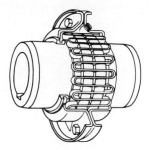

FIG. 14 Rubber jaw coupling. (Lovejoy)

FIG. 15 Spring-grid coupling. (Falk Division, Sundstrand)

available. Since this type of prime mover produces a pulsating type of power, it is often necessary to perform a torsional vibration analysis of the drive system to ensure that the normal operating speed is well removed from a speed which may produce a torsional resonant vibration. Such an analysis requires that the torsional stiffness of the coupling be known. It is quite often possible to tune the drive system to avoid operating at a resonant condition by selecting the proper coupling stiffness. The selection data required for a system of this type are the same as listed above. Most coupling manufacturers will, however, assign a higher service factor to an application involving a reciprocating prime mover, to compensate for fatigue effects due to torque fluctuations. In addition, the remote location of many engine-driven pumps indicates a special need to ensure a high degree of reliability of the system.

Another commonly employed prime mover is a steam turbine. These machines, which range from about 100 hp (75 kW) to well over 50,000 hp (37,000 kW) for pump drives, operate very efficiently and economically, providing there is a source of steam available at the installation. Any flexible coupling employed on a machine driven by a steam turbine must be capable of accepting the thermal gradient at the turbine shaft and must also accommodate the axial growth of the turbine shaft as it warms up to operating speed.

Steam turbines are generally high-speed machines (4,000 to as high as 10,000 to 15,000 rpm in some cases) and as such require a relatively high degree of system balance to avoid critical vibration. Elastomer couplings have occasionally been applied successfully to steam turbine drives, but because of the high speeds involved, all-metal couplings are usually employed. The metal coupling most commonly used on this type of drive is the gear (mechanically flexible) design. High-speed machinery requires that the weight of rotating components be minimized so as to decrease shaft deflections and hence increase the lateral critical speed of the system. The gear coupling is the most efficient design yet devised for transmitting large amounts of power at high speeds and with minimum weight. Where coupling weight and torsional stiffness are critical values to the overall system, special designs may be created which provide the specific values required for satisfactory system operation.

LIMITED END FLOAT Many horizontal motor-driven pump systems utilize motors which are equipped with journal, or sleeve-type, bearings. These bearings are intended only to absorb the transient thrust created by the motor rotor during acceleration and deceleration. The coupling for this type of drive should be equipped with suitable provisions for limiting the axial float of the motor rotor to some fraction of its total float. This may be done by positioning the motor in the center of its axial travel and then employing a coupling having a total float which is *less* than the float of the motor. Any motor thrust is taken by the pump bearing. Typical design practice calls for using a coupling with ⅛-in (0.32-cm) float on motors having a ¼-in (0.64-cm) total end float. Motors having ½-in (1.3-cm) (or greater) rotor float generally use couplings having ³⁄₁₆-in (0.48-cm) total float. Coupling total float can be limited by inserting a button between shaft ends, as shown in Fig. 16. This type of coupling prevents the motor rotor from ever contacting the thrust shoulders on the shaft bearings. Certain types of elastomer and disk couplings having inherent float-restricting characteristics provide centering without any additional modifications (Fig. 17).

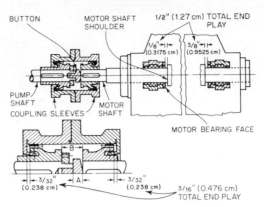

FIG. 16 Motor-driven coupling end float limited by insertion of button between shaft ends. (Worthington Pump)

VERTICAL OPERATIONS   As previously noted, rigid couplings are commonly used on vertical-drive systems where the system characteristics warrant such a coupling. However, many vertical-drive systems require a flexible coupling to accommodate shaft misalignment. It is generally possible to use a nonlubricated coupling, such as one of the many elastomer designs, in a vertical position without modification, provided the shafts are supported in their own bearings and the coupling does not have to transmit a thrust force. Lubricated designs, such as the gear and spring-grid types, usually require some modification to make certain that lubricant is retained in both halves of the coupling. One design which provides the advantages of a gear coupling and is commonly used on pump drives is the vertical double-engagement coupling (Fig. 18). This coupling provides two gear meshes in one half-coupling (thus permitting angular *and* offset shaft misalignment), avoiding the potential leakage problem which exists when a conventional coupling is operated vertically.

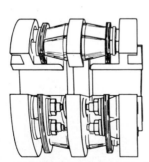

FIG. 17   Metal-disk coupling can be used to limit motor end float. (Thomas Coupling Division, Rexnord)

FIG. 18   Vertical double-engagement gear coupling. (Koppers)

PUMP DRIVE SHAFT SYSTEMS

Pump drive system arrangements may be classified in one of two categories: those which are close-coupled, having a shaft separation of a fraction of an inch (not to be confused with integral motor pumps not having a flexible coupling) and those which, for one or more reasons, have the prime mover located a substantial distance from the pump. These latter types of systems require a modification to the basic flexible coupling designs previously described.

Spacers   One means of accommodating shaft separation in excess of the normal amount provided in a standard coupling is to employ a flanged tubular spacer between the two coupling halves (Fig. 19). These components are lightweight and are commonly used with end-suction pumps, where it is possible to remove the pump impeller while the pump housing and piping, as well as the prime mover, remain in place (Fig. 20). For high-speed drive sys-

FIG. 19   Tubular spacer coupling. (Koppers)

tems, a lightweight spacer is usually the only practical means of achieving the goal. This spacer may be used to tune the system torsionally (as previously discussed) by varying the body diameter and wall thickness. Spacers can be manufactured in lengths up to several feet, but the cost generally restricts its usage to shorter lengths, except where no other means is suitable.

Floating Shafts   Floating shaft couplings have a purpose similar to that of spacer couplings, namely, to connect two widely separated shafts. The basic difference is in the construction of the component. While spacers are normally made with the body having an integrally formed flange to connect the two coupling halves, a floating shaft usually is made by attaching a flange to a piece

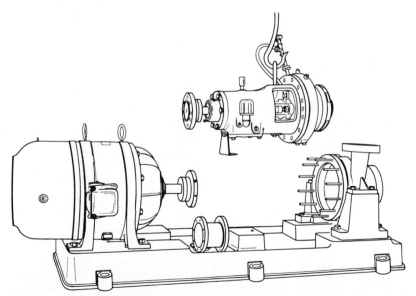

FIG. 20   Spacer coupling enables end-suction pump to be dismantled without moving piping, pump casing, or driver. (Worthington Pump)

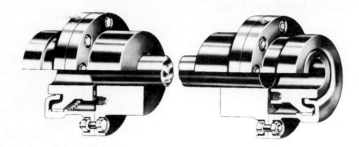

FIG. 21   Section of floating shaft and couplings. (Koppers)

of solid or tubular shafting by means of mechanical keys or by welding (Fig. 21). This type of construction is generally less expensive than a one-piece spacer, especially when very large shaft separations must be spanned. Floating shaft arrangements are widely used on horizontal pump applications of all types and are especially common on vertical pump applications, such as in water pumping and sewage treatment stations. In these latter applications, the pump may be submerged in a pit usually 30 to 50 ft (9 to 15 m) below ground level, while the motor (to prevent damage during flooding) is mounted at ground level. Settings as deep as 100 ft (30 m) have been constructed. Such systems require long floating shafts, which are generally made in sections and supported by line bearings at intermediate supports or floors.

Rigid Shafts   Vertical centrifugal pumps can be designed to contain their own thrust and line bearings. These pumps employ a flexible coupling at the pump shaft. When vertical intermediate shafting is used, the shaft need be designed only to transmit torque. Some pumps are designed to contain only a single line bearing. These pumps require a rigid coupling at the pump shaft so that axial thrust can be carried by the thrust bearing in the driver or gear located above. The construction of a single oil-lubricated line-bearing pump is shown in Fig. 109 of Subsec. 2.2.1. In order to carry thrust, the drive shaft, if made in sections, must use rigid intermediate couplings and the coupling at the driver or gear must also be rigid. Intermediate bearings used with rigid shafting must be designed to provide only lateral support, thereby assuring that all the axial thrust is carried by the driver or gear. An intermediate oil-lubricated sleeve-type guide bearing is shown in Fig. 108 of Subsec. 2.2.1. Pump and intermediate shaft sleeve guide bearings can be either grease- or oil-lubricated, and they can also be antifriction if preferred.

A rigid intermediate shaft connected to a pump provided with a rigid coupling must be designed to carry its own weight, the axial thrust from the pump, and a bending load. Volute-type centrifugal pumps produce a radial reaction on the pump shaft which in turn is transmitted to the vertical shafting through the rigid pump coupling. Intermediate bearings then act as line bearings. The bearing

FIG. 22   Tubular intermediate shafting and universal joints. (H. S. Watson)

supports must also be designed to resist this bending force. The pump shaft and bearing, intermediate shafts and line bearings, and the driver or gear shaft and bearings must be treated as a single shaft supported at each bearing when analyzing the critical speed of the entire rotor system. If the intermediate line bearing supports are assumed to be nodes in the critical speed calculation, then these supports must be absolutely rigid. The supplier of the shafting should, however, confirm what flexibility will be allowed at these supports and give the design forces. The support for the guide bearings must also be designed not to have a natural frequency of vibration within the operating speed range of the pump.

Flexible Drive Shafts   Universal joints with tubular shafting (Fig. 22) can be substituted for flexible couplings whenever it is necessary to (*a*) eliminate the need for critical alignment, (*b*) provide wider latitude in placement of pump and driver, and (*c*) permit large amounts of relative motion between pump and driver. This type of shafting can be used horizontally as well as vertically to provide a short spacer arrangement or a large separation between pump and driver, such as that required for deep settings (Fig. 23).

Flanges are furnished to fit pump and driver shafts to the universal joints, which are splined to allow movement of the shaft. An intermediate steady bearing is required at each joint, which must also support the weight of a section of shafting (Fig. 24). Because of the spline, pump thrust cannot be taken by the driver, and it is necessary to use a combination pump thrust and line bearing. Because of the universal joint, the intermediate bearing or bearings take no radial load from the pumps and therefore act only as a steady bearing or bearings for the shaft.

The shaft must be selected to transmit the required torque and be of a length and diameter that will have a critical speed well removed from the operating speed range of the pump. The

FIG. 23   Sections of flexible shafts and intermediate guide bearings used to transmit torque from motor located above flood elevation to pump located below ground. (Worthington Pump)

support for the guide bearing must not have a natural frequency of vibration within the speed range. Shafts of this type are often pre-engineered and stocked. Selection charts are available from the manufacturer to make a proper size and length selection. Standard tubular, flexible drive shafts have limited torque-carrying capacity.

Design Criteria

TORSIONAL STRESS   The torsional stress in the shafting may be calculated by the following equations:

$$S_s = \frac{16T}{\pi D_o^3} \qquad \text{for solid shafting}$$

$$S_s = \frac{16T}{\pi D_o^3 \left(1 - \dfrac{D^4}{D_o^4}\right)} \qquad \text{for tubular shafting}$$

where S_s = torsional shear stress, lb/in^2
         (N/m^2)
      T = transmitted torque, in·lb (N·m)
     D_o = shaft outer diameter, in (m)
      D = shaft inner diameter (tubular
         shaft only), in (m)

The allowable shear stress depends upon the material being used and whether it is subjected to other loads, such as bending or compression. The design safety factor on the shafting should be equal to or greater than those of the other components in the drive train.

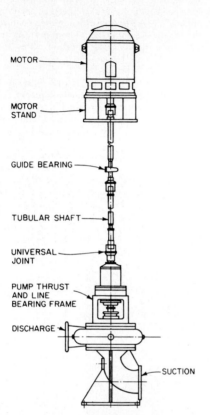

FIG. 24   Vertical pump with flexible shafting. (Worthington Pump)

MOTOR

MOTOR STAND

GUIDE BEARING

TUBULAR SHAFT

UNIVERSAL JOINT

PUMP THRUST AND LINE BEARING FRAME

DISCHARGE

SUCTION

CRITICAL SPEED   The critical speed of a drive shaft is determined by the deflection, or "sag," of the shaft in a horizontal position under its own weight. The less the sag, the higher the critical speed. In practical terms, a long, slender shaft will have a low critical speed and a short, large-diameter shaft will have a very high critical speed. The deflection of a simply supported shaft is calculated as follows:

$$y = \frac{5wL^4}{384EI} = \text{shaft deflection, in (m)}$$

noting that

$$I = \frac{\pi D_o^4}{64} \qquad \text{for solid shafting}$$

$$I = \frac{\pi(D_o^4 - D^4)}{64} \qquad \text{for tubular shafting}$$

where w = weight (force) of shaft per unit length, lb/in (N/m)
     L = length between bearing supports, in (m)
     E = Young's modulus, lb/in^2 (N/m^2)
     I = moment of inertia, in^4 (m^4)

Knowing the natural deflection of the shaft, it is possible to calculate the first critical speed from the equation:

$$N_{crit} = 187.7 \sqrt{\frac{1}{y \text{ (in inches)}}} = 946 \sqrt{\frac{1}{y \text{ (in millimeters)}}}$$

which expresses critical speed directly in revolutions per minute of the rotating shaft.

In practice, the critical speed should be placed well away from the operating speed of the shaft. Actual operating conditions, such as imbalance and clearance in bearings and couplings, tend to reduce the critical speed from its theoretical value. The designer has several options from which to choose to attain the desired critical speed: (1) vary the shaft diameter, (2) use tubular shafting to increase stiffness and reduce deflection, and (3) vary the bearing span.

Varying the diameter of the shaft has a dramatic effect on shaft deflection, since the deflection decreases inversely with the cube of the diameter. The weight of the shaft will increase proportionately to the square of the diameter and may consequently impose excessive loads on bearings. When this is the case, tubular shafting may be substituted for solid shafting. The ratio I/w is much higher for a tubular shaft than for a solid shaft of the same diameter, so that the shaft deflection is again reduced and critical speed increased.

The bearing span may be varied when the size and weight of a single length is broken into sections, with a single engagement coupling used to connect each section and acting as a hinge joint. A self-aligning bearing is placed on each section immediately adjacent to the coupling. Each section may then be treated as an individual shaft, which will result in a substantial reduction in shaft size. The rigidity of the support at each bearing must be taken into consideration when calculating the critical speed of the shaft.

DYNAMIC BALANCE   Drive systems which operate at high speeds and connect two widely separated shafts must be given special consideration with regard to dynamic imbalance. Care must be taken to ensure that the long, slender shaft used in such systems is not bent, either in manufacture or in installation. These systems quite often require shafting which has been dynamically balanced after manufacture to compensate for these normal errors.

6.3.2
MAGNETIC DRIVES

JAN A. HATCH

APPLICATIONS

Magnetic-drive centrifugal pumps were developed to eliminate the dangers of pumping hazardous liquids by the use of a hermetically sealed wet end. Such pumps fall into two general categories: (1) small plastic pumps for acid transfer applications and (2) full-size metal pumps for hostile liquid applications, for operation at high temperatures and pressure, and for reductions in maintenance costs.

The small plastic units employ low-cost ceramic magnets and economically handle corrosive acidic or alkaline liquids on a laboratory or small transfer scale. They are generally limited to flow rates of less than 100 gpm (23 m³/h) and total heads of less than 100 ft (30 m). Temperatures and pressures are limited by the fluoroplastic construction materials used, usually to 300°F (150°C) and 150 lb/in² (1000 kPa).

The full-size metal units contain alnico magnets and are used to eliminate the problems associated with packing and mechanical seals in full-scale process systems where leakage is either dangerous or costly. All metal magnetic-drive pumps may achieve capacities to 4000 gpm (910 m³/h) and single-stage heads to 400 ft (120 m). Temperatures to 850°F (450°C) and systems pressures to 5000 lb/in² (34,500 kPa) can be achieved.

CONSTRUCTION

Magnetic-drive centrifugal pumps eliminate the necessity of packing glands or mechanical seals by interrupting the continuity of pump shaft to motor shaft and enclosing the former in a leak-proof containment shell. As shown in Fig. 1, the pump shaft/impeller assembly is driven by a cylinder of permanent magnets external to the containment shell, which is mechanically coupled to the separate (Fig. 1a) or close-coupled (Fig. 1b) motor shaft. Within the containment shell, a matching set of permanent magnets (synchronous coupling) or a torque ring (eddy-current coupling) is mounted on the impeller shaft. The shaft bearings, as shown in Fig. 2, are lubricated by the liquid being pumped, about 3 to 5% of which is directed through a recirculation path beginning directly below the discharge nozzle (point B), entering the bearing area, and returning to the casing behind the impeller as close as possible to its focus. This recirculated liquid is driven by the pressure/velocity differential of its entrance and exit points and serves not only to lubricate

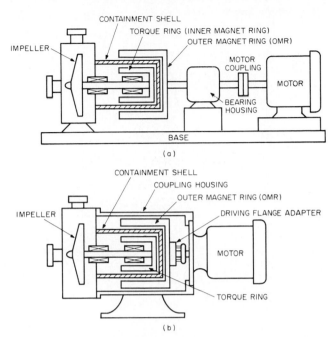

FIG. 1 Construction of magnet-driven sealless pump: (*a*) separate motor drive, (*b*) close-coupled motor drive. (Kontro Pump)

the bearings but also to flush away particles and dissipate the heat generated by hydraulic and bearing friction. In a nonsynchronous (eddy-current) magnetic coupling, this recirculation also carries away the heat generated by magnetic slip. If the recirculated fluid must be filtered, heated, or cooled prior to recirculation, the starting point would be a tapped hole in the flange of the discharge nozzle (point *A*).

The allowable gap between inner and outer magnetic rings is 0.120 to 0.180 in (3.1 to 4.6 mm), enabling a containment shell thickness of 0.060 to 0.120 in (1.5 to 3.1 mm) with 0.030 in (0.76 mm) of clearnace on either side. This is three to four times the allowable gap between the rotor and stator in a canned motor pump, which are therefore limited in can thickness and inside and outside clearances to 0.012 to 0.015 in (0.31 to 0.38 mm) each.

As there are no mechanical seals or motor windings requiring cooling, magnetic-drive pumps are temperature limited only by their construction materials. Steel or alloy magnetic-drive pumps can handle fluids to 750°F (400°C) without cooling water. High pressure as well as high temperature is more easily provided for with magnetic-drive pumps, since high-strength materials and bolting alone (see Fig. 3) can increase pressure handling capability to 5000 lb/in² (34,500 kPa) or more without the need of ancillary equipment to provide back pressure against either a seal or a can.

The hermetically sealed design and high-pressure capability which allow magnetic-drive pumps to be totally leakproof also allow them to operate in full vacuum systems without product air entrainment or creation of a drain on the vacuum system.

Eddy-Current Coupling Figure 4 shows a schematic representation of an eddy-current coupling. The outer magnetic assembly is driven by a separately mounted motor. The outer magnetic ring (OMR) consists of a number of permanent magnets securely bolted to a cylindrical frame and evenly distributed to provide a uniform magnetic field. The torque ring is made of mild steel with an outer facing of copper. The copper facing is usually covered by a nonmagnetic sheath which prevents chemical attack of the copper by an aggressive pumpage. The containment

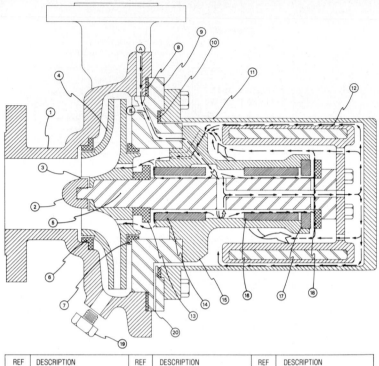

REF	DESCRIPTION	REF	DESCRIPTION	REF	DESCRIPTION
1	PUMP CASING	8	GASKET	15	BEARING HOLDER
2	IMPELLER NUT	9	BACK PLATE	16	BACK BEARING
3	IMPELLER TAB WASHER	10	GASKET	17	THRUST PAD
4	IMPELLER	11	CONTAINMENT SHELL	18	BACK THRUST WASHER
5	PUMP SHAFT	12	TORQUE RING	19	CASING DRAIN
6	FRONT WEAR RING	13	FRONT THRUST WASHER	20	CONTAINMENT SHELL DRAIN
7	BACK WEAR RING	14	FRONT BEARING		(OPTIONAL)

FIG. 2 Wet end of typical centrifugal pump. (Kontro Pump)

FIG. 3 High-pressure magnetic-drive centrifugal pump. (Kontro Pump)

shell is an extension of the pump housing, and thus the torque ring operates in the process pumpage while the OMR operates in the ambient atmosphere surrounding the pump. When the OMR revolves, the magnetic field passes through the containment shell, through the copper facing of the torque ring, and through the mild steel beneath the torque ring and then returns to the OMR to complete the circle. The rotating magnetic field produces eddy currents in the copper facing, and these eddy currents create electromagnets which tend to follow the rotating magnetic field which created them.

The torque ring in Fig. 4 is shown as a series of parallel copper strips laid parallel to the pump shaft. In practice, these strips might be separated and tied together at the ends, much like the squirrel cage of an ac motor. This is called a rodded torque ring. In some cases, the torque ring may be a solid copper cylinder. In any case, the copper conducting paths in the torque ring are firmly connected to the mild steel cylinder, which in turn is solidly bolted to the pump shaft.

The greater the slip in torque ring speed, the greater the eddy current flowing and the greater the torque. If a pumpage has high viscosity when it is cold, the eddy-current drive will provide high starting torques and will also provide greater heating of the pumpage (copper losses are higher at higher slip levels). This heating of the pumpage will lower its viscosity and get the pumping operation under way more quickly than would be the case for a straight magnet-to-magnet drive.

Magnet-to-Magnet (Synchronous) Coupling   Figure 5 shows a schematic representation of a magnet-to-magnet synchronous drive coupling. Just as with an eddy-current coupling, the outer magnetic assembly is driven by a separately mounted motor. The differences between an eddy-current coupling and a magnet-to-magnet coupling occur in the inner magnetic ring (IMR). In the magnet-to-magnet coupling, the IMR contains the same number of magnets as the OMR, the number of magnets being determined by the torque which must be transmitted. Thus, when the OMR rotates, the IMR rotates in synchronism. While the absence of slip means that, rating for rating, the magnet-to-magnet coupling drive has a higher speed than the eddy-current coupling drive, the difference is too slight to offset the lighter weight, lower cost, and higher starting torque advantages of the eddy-current drive. Thus, the eddy-current coupling drive is advantageous in all but a few applications. Preference for the magnet-to-magnet drive is generally associated either with low-speed drives or where the pumpages must be kept at extremely low temperatures and/or are very sensitive to heat.

The IMR is mounted on the same shaft as the pump impeller. The containment shell is an extension of the pump housing and thus the IMR operates in the process pumpage while the OMR operates in the ambient atmosphere surrounding the pump. The OMR is enclosed by a coupling housing to protect it from dirt and to shield it from operating personnel.

Configurations   In addition to standard closed-impeller centrifugal designs, the full-size metal magnetic-drive pumps may be configured as self-primers as shown in Fig. 6, internally-jacketed pumps for molten solids as shown in Fig. 7, or in full compliance with American Petroleum Institute (API) standard 610, sixth edition, as shown in Fig. 8.

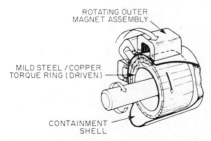

FIG. 4   Schematic of eddy-current magnetic-drive coupling. (Kontro Pump)

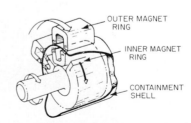

FIG. 5   Schematic of magnet-to-magnet (synchronous) drive coupling. (Kontro Pump)

FIG. 6 Self-priming magnetic-drive centrifugal pump. (Kontro Pump)

FIG. 7 Internally-jacketed magnetic-drive centrifugal pump for molten solids. (Kontro Pump)

FIG. 8 API standard 610 magnetic-drive centrifugal pump. (Kontro Pump)

COST

Magnetic-drive centrifugal pumps are generally more expensive to manufacture than packed or sealed centrifugal pumps, and their use is usually cost-justified by one of the following system problems:

1. The liquid pumped is toxic, carcinogenic, pyroforic, or detrimental to the environment.
2. The liquid is hot (e.g., synthetic heat-transfer oils) and would require a cooling water system if seals were employed.
3. The liquid is expensive, so that the value of the amount lost through seals or packing approaches the value of the pump amortization.
4. The system is under high pressure (e.g., superheated water).
5. Lack of air entrainment is desirable in the end product of a vacuum system (e.g., vegetable oil deodorization).
6. The system has consistently high seal maintenance costs.
7. The pump is run unattended for long periods.
8. The pump is subject to extremes of ambient temperature or environment which would limit seal life.
9. Any combination of these conditions.

6.3.3
HYDRAULIC PUMP AND MOTOR POWER TRANSMISSION SYSTEMS

DAVID ELLER

The use of fluid (hydraulic pump and motor) power transmission systems to drive water pumps is a relatively recent development. The first units of any size were first marketed in the late 1960s and early 1970s and were utilized primarily for construction dewatering purposes since they can be driven by diesel engines or electric motors located well away from construction site cave-ins, without the inconveniences of priming and suction hoses and with no limitation on suction lift or capacity (Fig. 1).

COMPONENTS

A typical unit consists of a submersible axial-flow water pump with a hydraulic motor mounted in the pump bowl (Fig. 2). The thrust bearings, shaft, and hydraulic motor are all sealed to operate under water. The hydraulic motor is driven by oil fed from a hydraulic pump under a pressure of approximately 2500 lb/in^2 (170 bar).° Water flowing past the hydraulic motor acts as a heat exchanger and keeps the system cool. Plumbing extends from the hydraulic motor out through the pump bowl and connects to the hydraulic oil conduits leading to the hydraulic pump driver. Quick-coupling hose connections are generally supplied on each end of the hydraulic hoses. The quick couplings contain spring-loaded ball valves which prevent the oil from leaking from either side when the couplings are disconnected (Fig. 3).

The hydraulic pump is driven by the prime mover, which is generally a diesel engine or electric motor. The hydraulic conduits may be steel-reinforced rubber hoses or steel pipe or a combination of both (Fig. 4). Sometimes it is desirable to utilize a combination of steel pipe and flexible hose as the conduits. This is particularly true when the drive units are to be located a considerable distance from the water pump. When steel pipe is used, it is highly desirable to weld all joints. After welding, the pipes should be thoroughly cleaned and tested up to maximum expected operating pressures and inspected for leaks. The pipes should then be painted with an asphalt-base enamel or epoxy, depending upon the surrounding environment.

Hydraulic pump and motor drives can be adapted to centrifugal volute pumps. Figure 5 illustrates a portable nonclogging-impeller trash or sewage pump so driven.

° 1 bar = 10^5 Pa. For a discussion of bar, see *SI Units—A Commentary* in the front matter.

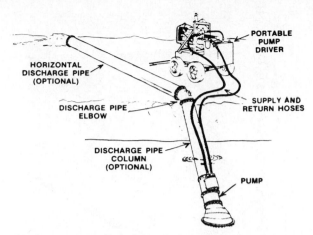

FIG. 1 Typical portable hydraulically driven water pump with diesel prime mover. (M&W Pump)

ADVANTAGES

Although the original users of hydraulically driven water pumps were contractors, farmers, and open pit miners using mostly portable pumps, the system described here offers the advantage of permanent pump stations, such as might be used for municipal storm drainage (Fig. 6) or massive irrigation or drainage works.

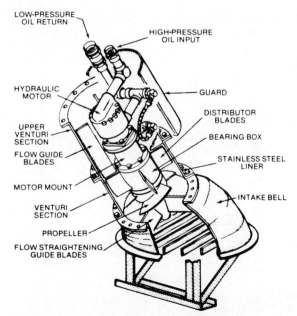

FIG. 2 Interior of a hydraulically driven axial-flow water pump. (M&W Pump U.S. Patent NO. 3,907,463. Other patents pending)

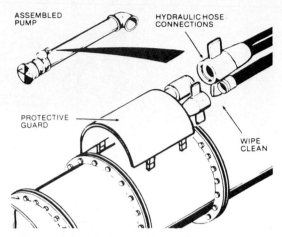

FIG. 3   Quick couplings are used to fasten hydraulic hoses to the water pump.

The system is very simple. The power source can be located close to the pump or in a more accessible or protected area. Other advantages include the ability to vary the speed of the water pump by regulating the amount of hydraulic oil sent to the motor, the ease of automation for automatic or remote control, and safety since there is no high-voltage electricity in the water (Figs. 7 and 8).

FIXED VERSUS VARIABLE FLOWS

A fixed-displacement hydraulic pump is used to drive the fixed-displacement hydraulic motor when a fixed water pump flow is desired. The drive shaft of a fixed-displacement vane pump (see

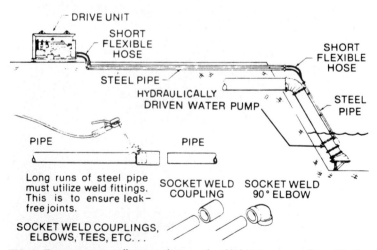

FIG. 4   For permanent installation, steel pipes with welded joints are generally used for the hydraulic oil conduits instead of hoses.

FIG. 5 Portable centrifugal volute trash pump driven by hydraulic motor. (M&W Pump)

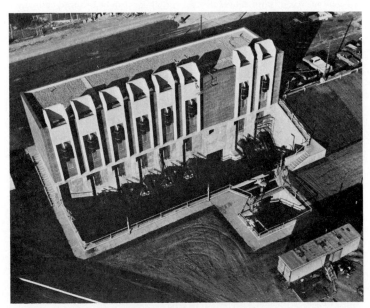

FIG. 6 A 350,000-gpm (79,500-m³/h) municipal pump station for combination storm drainage and final sewage effluent. (M&W Pump)

FIG. 7 Five 42-in (107-cm) double-staged hydraulically driven water pumps with design flows automatically varying from 0 to 50,000 gpm (0 to 11,350 m³/h). (M&W Pump)

Fig. 9) is keyed to a rotor and revolves with it. Rectangular vanes fit into slots in the rotor. As the rotor turns, the vanes are forced out by centrifugal force to make continuous contact with the cam surface. Because of the kidney shape of the cam surface, the space between a vane passing top dead center and the vane ahead of it starts closing down, increasing pressure and forcing oil out through the outlet port. Simultaneously, the vanes on the pump inlet side are passing through the cam kidney-shaped area, which is increasing in size, thus creating a vacuum which pulls oil in through the pump inlet.

When it is desirable (as for final sewage effluent pumping), water pump flow rate can be easily varied through a large range by regulating the amount of oil going to the hydraulic motor in the water pump bowl. Since the hydraulic motor has a fixed displacement, its speed is a direct function of the amount of oil supplied to it by the hydraulic pump connected to the prime mover. Assuming a constant-speed prime mover, the rate of oil flow can be changed easily by utilizing a

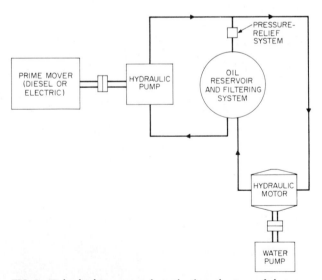

FIG. 8 Hydraulic drives are much simpler than other types of electromechanical equipment and provide a wide speed range, variable-speed and reversing capability, and shock resistance at a relatively low cost.

variable-displacement piston hydraulic pump (Fig. 10). In this type of pump, the pistons are moved into and out of the cylinder block by a swashplate rotating cam whose angle can be changed either manually or automatically.

When the cam is set at zero angle, the piston does not move and no oil is pumped. As the cam angle is increased, the pistons begin to reciprocate and oil is pumped. The amount of oil pumped is directly related to the angle of the swashplate. For automatic systems, the swashplate angle can be regulated by a servo-control valve attached to the hydraulic pump. This valve is simply an apparatus which converts an electric signal to a mechanical force. A typical system would utilize a 0- to 20-mA signal coming from a sensor in the pump sump. The sensor could be a sonic, air bubbler, or float device measuring the wet-well sump level and sending a proportional electric signal back to the servo-control valve on the hydraulic pump.

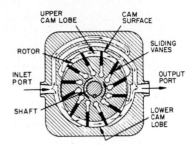

FIG. 9 Principle of fixed-displacement balanced-vane pump.

If the sump level is rising, for instance, the electric signal may increase, which would cause the servo to stroke the hydraulic pump swashplate to a greater angle, increasing the amount of oil being pumped. This in turn would speed up the hydraulic motor and the water pump propeller, and the water pump flow would increase. Thus a relatively small water-pump sump can be utilized and pumped water outflows can be regulated to match highly varying gravity inflows.

EFFICIENCY

The main criticism of hydraulic power transmission systems for water pumps is that power is lost through the hydraulic system. There is obviously a power loss in the hydraulic pump, the hydraulic motor, and the plumbing. However, when evaluating efficiencies against those of other types of power transmission systems, such as gears, belts and pulleys, and direct-connected shafts, it is important to make meaningful comparisons.

Water pump manufacturers typically publish performance curves for their wet-pit pump bowls *only*, and these curves do not include power losses for *any* type of power transmission system. This is done because the pump manufacturer does not know the shaft lengths for all possible extended-shaft pumps. Therefore, there is no way the manufacturer can include the losses for the shaft, bushing and bearing supports, couplings, and other transmission parts. Consequently, a user must add to the evaluation the losses resulting from the extended shafting and other transmission parts.

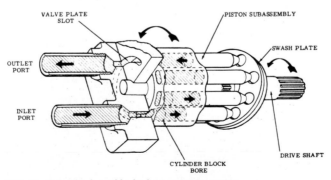

FIG. 10 Principle of variable-displacement piston pump.

TABLE 1   Standard Pump Sizes Readily Available with Hydraulic Drives

Discharge diameter, in (cm)	Capacity range, gpm (m³/h)		Total head range, ft (m)[a]
4 (10)	475–1,025	(110–230)	8–10 (2–12)
6 (14)	700–1,450	(160–330)	18–50 (5–15)
8 (20)	1,400–2,300	(320–520)	5–23 (1.5–7)
12 (39)	2,500–4,000	(570–910)	5–45 (1.5–14)
16 (41)	4,000–8,000	(910–1,800)	5–23 (1.5–7)
18 (46)	5,000–9,000	(1,100–2,000)	18–45 (5–14)
20 (51)	7,000–12,000	(1,600–2,700)	5–20 (1.5–6)
24 (61)	12,000–17,000	(2,700–3,900)	5–22 (1.5–7)
30 (76)	23,000–27,000	(5,200–6,100)	5–22 (1.5–7)
36 (91)	24,000–35,000	(5,400–7,900)	5–19 (1.5–6)
42 (107)	45,000–53,000	(10,000–12,000)	5–16 (1.5–5)
60 (152)	100,000–120,000	(23,000–27,000)	5–16 (1.5–5)

[a]Most units can be double-staged to accomplish twice the heads shown.

For example, a typical extended-shaft pump may have a column shaft 30 ft (9 m) long, and an additional power requirement of 10 to 15% above that of the pump alone would not be unusual for an extended-shaft pump with a gear or belt drive. By comparison, a hydraulically powered pump would typically have a power transmission loss of 20 to 25%—or in other words would require 10 to 15% more power than an equivalent extended-shaft pump. Thus a hydraulically powered pump may require, for example, a 30-hp (22-kW) motor rather than a 25-hp (19-kW) motor or diesel prime mover. However, the additional cost of this larger motor should be weighed against the savings in civil works costs and engineering and installation time and the savings resulting from the versatility and automatic operation possible with the hydraulic system.

AVAILABLE SIZES

Hydraulic pump and motor transmission systems are available for almost any speed output, from 100 to 3000 rpm for power outputs up to 500 hp (370 kW). For larger power drives, speed selection is more limited.

Table 1 illustrates readily available standard pump and hydraulic drive sizes from one manufacturer.

PUMP CONTROLS

AND VALVES

W. O'Keefe

CONTROLS

Pump control in the broadest sense gives the pump user (1) the flow rate, pressure, or liquid level desired, (2) protection for the pump and system against damage from the pumped liquid, and (3) administrative freedom in decisions on operations and maintenance.

Control System Types Pump control systems range in complexity from single hand-operated valves to highly advanced, automatic flow control or pump speed control systems. Pump type and drive type are factors in control system choice. For centrifugal pumps, either change of speed or change of valve setting can control the desired variable. For positive displacement pumps, whether reciprocating, rotary, screw, or other type, control is by change in speed, change in setting of bypass valve, or change in displacement. The last-mentioned method is found in metering and hydraulic drive pumps. Although this chapter considers only control systems having valves as final control elements, the sensing elements discussed also serve in pump speed control systems.

Pump control systems divide readily into two types: on-off and modulating. The on-off system provides only two conditions: a given flow (or pressure) value or a zero value. A valve is therefore either open or closed, and a pump driver is running or not. The modulating system, on the other hand, adjusts valve setting or speed to the needs of the moment. Either type of system can be automatic or manual.

System Essentials All control systems have

1. A sensing or measuring element
2. A means of comparing the measured value with a desired value
3. A final control element (a valve) to produce the needed change in the measured variable
4. An actuator to move the final control element to its desired position
5. Relaying or force-building means to enable a weak sensing signal to release enough force to power the actuator

The sensing or measuring element is often physically separated from the comparison and relaying means, which are usually housed together and called the *controller*. The actuator and valve are physically connected and may be at a distance from the controller.

In a very simple control action, such as one based on an administrative decision to shut down temporarily one of several small parallel pumps in service, some of the five essentials may be supplied by the operator who turns the valve handwheels and pushes the motor stop button. Nevertheless, the essentials must always be present in some form.

EFFECT OF RATE OF CHANGE The nature of the rate of change of the measured variable or desired value with time gives a convenient guideline in pump control. The chief types of change are

1. Slow change (practically steady state)
2. Sudden change from one steady state to another (either a nearly instantaneous step change or a high-rate ramp change)
3. Fluctuation at varying rates and in varying amounts

Slow change involves questions of the ability of the control system to hold the desired value accurately and not lag unnecessarily during the change. Equal accuracy when approaching the new value from above or below is also desired.

Sudden change involves additional questions of how long the system will take to reach the new value, whether it will overshoot and then fluctuate back and forth, and for how long it will do this.

Fluctuation introduces serious questions of whether the system will be excited into amplification of some types of fluctuations and go totally out of control. Systems involving fluctuation changes are the most difficult to design and operate; such factors as the inertia of the control elements, amount of liquid in the system, and dynamic behavior of each element and of the elements together, must be considered.

OPEN-LOOP CONTROL   The simplest mode of automatic control is open-loop control, in which the pump speed (or displacement in some pump types) or the control valve setting is adjusted to and held at a desired value calculated or calibrated to produce the required output of flow, level, or pressure. The calculation can result in a cam for the controller or positioner or a particular characterization of a valve plug. In operation, only the deviation of the input variable from its desired value is measured and the control system adjusts the input variable to eliminate the deviation. Because the output variable is not measured, a change in the conditions on which calculation or calibration was based will introduce output errors. Change of input variable can be done manually or by another control system. For example, a pump may be speeded up by a rheostat, or the air pressure to a valve actuator may be changed by changing a pneumatic pressure control valve setting. Open-loop systems are also called feedforward systems, in contrast to feedback, or closed-loop, systems. Open-loop systems are stable, simple, and quick in response, but they tend to err as downstream conditions change.

CLOSED-LOOP CONTROL   A closed-loop control system eliminates much of the error of the open-loop system. In the basic closed-loop, or feedback, system, the output variable is measured and the value compared with an arbitrary desired or set value. If the comparison reveals an error, the pump speed or control valve setting is changed to correct the error. Large-capacity water tanks or lag in the control system can introduce delays in establishment of the new output value, and the system can therefore overcorrect and oscillate back and forth unless design prevents this.

On-Off Control   The simplest closed-loop systems operate on-off between fixed limits, such as water level or pressure. The on-off action is at the extremes of a wide or narrow band that can be set at any point in the range. For example, a tank level control may work in an on-off band of 1 in (2.54 cm) or 10 in (25.4 cm) at any level in a tank that is 5 ft (1.5 m) deep.

Proportional Control   This is the basic type of closed-loop control. Within a wide or narrow band of output variable values, the controller input, such as actuator air pressure, is proportional to the deviation from the set point, or desired value, at the band center. If the band is very narrow, say 1 in (2.54 cm) of level in a 60-in (1.5-m) tank, then the controller will apply full air pressure to the valve actuator at a ½-in (1.27-cm) deviation from the set level in one direction and minimum air pressure at a ½-in (1.27-cm) deviation in the other direction. This is close to the effect of an on-off control. If the band is wider, say 20 in (50.8 cm) of level in the 60-in (1.5-m) tank, the air pressure will vary from minimum to full pressure over the 20-in (50.8-cm) band and the system will be less sensitive and apply less correction for a given small change in output variable. The lower sensitivity can make the system less likely to overshoot or hunt. Because a given controller output corresponds to every value of deviation from the set point, the simple proportional system will not come back to its set point if the output variable changes as a result of changed demand, such as for more water from the tank. The difference between set point and actual new equilibrium value of level is called *offset*. Narrowing the band will reduce the offset but may cause intolerable oscillations or hunting.

To improve response and stability and to achieve very high accuracy, however, several refinements may be needed. Addition of reset to a simple proportional controller will eliminate offset. This is the proportional-plus-reset or proportional-plus-integral system. In terms of the proportional band, reset means that the band is shifted in such a way as to produce slightly more correction and return the output variable back to what is desired. The reset feature may impair stability, however, because of the added control action.

Derivative action is an added refinement to improve stability and response. In this, the rate of change of the measured output variable is what determines the controller output. A step or sudden change in measured output variable will cause a momentary large increase in controller output which will initiate response. When the derivative action fades, the basic proportional-plus-reset action takes over to restore conditions.

The open-loop system, sensing a change in input variable and therefore giving rapid response, is exploited by adding it to the closed-loop system. An example of a feedforward-feedback system in pump flow control is the three-element boiler feedwater regulator.

Sensing and Measuring Elements   In automatic control of a pump, these elements detect values of and changes in liquid level, pressure, flow rate, chemical concentration, and tem-

perature. The signal emitted by the element often needs amplification or conversion to another medium, which is done in a transducer. Air pressure to electric voltage or current and rotary motion to electric voltage are common transformations.

LIQUID-LEVEL SENSORS   The simplest of several types of sensors is the float in the main tank (or boiler drum) or in a separate float chamber connected at top and bottom to the tank or drum (Fig. 1). The float can be a pivoted type, with motion transmitted outside the chamber by a small-diameter rotating shaft or translational rod attached to the lever arm near its pivot to obtain mechanical advantage. A rod of the latter type can actuate the stem of a balanced valve to control liquid flow and thus liquid level in the supplied tank.

Floats on vertical rods can actuate switches outside and above the float chamber (Fig. 2). Depths can vary from less than 1 to more than 50 ft (0.3 to 15 m), with rod guides often necessary at the greater depths. In some cases the floats slide on the vertical rod between adjustable stops. The floats then push upward or downward on the rod at the desired control levels and trip the switch above. In a displacer-type arrangement for open tanks, a ceramic displacer is suspended from one end of a stainless steel tape that passes over a pulley and down again to a counterweight. The counterweight compensates for part of the ceramic displacer weight, so that it floats in the liquid. The extended pulley shaft drives through a reducing gear to a shaft which carries mercury switches controlling as many as four circuits (Fig. 3). The gearing allows the displacer to travel as far as 30 ft (9 m), with level adjustment between 2 and 27 ft (0.6 to 8 m). A weighted overcenter mechanism in the switches gives quick make and break.

In other applications of the displacer, porcelain bodies on a cable are suspended from the armature of a magnetic head control. In one form, a spring partly supports the weight of the

FIG. 1   Low-water cutoff and alarm are purposes of this liquid level sensor. (McDonnell & Miller)

FIG. 2   Adjustable stops on rod actuate lever arm to tilt mercury switches as float moves. (Autocon Industries)

FIG. 3 Mercury switches in liquid level sensor head are tripped by adjustable cams. (Autocon Industries)

displacers. As liquid rises to the displacers in succession, their apparent weight decreases and the spring can move the cable and armature upward to actuate snap-action switches. The displacers can be moved up and down the cable to initiate action at the desired levels. Three displacers can be mounted on a cable for such applications as one pump actuated by the center displacer, a second pump by either the top or bottom displacer, and an alarm by the third displacer. Displacers are advantageous for dirty or viscous liquids which are still. Levels covered are from 1 to more than 10 ft (0.3 to 3 m).

The pulley shaft of the tape suspended displacer can also drive a potentiometer. The potentiometer output can be applied to solid-state control equipment handling recording, pump start and stop, and alarms.

The connection between float and switch need not be mechanical. An armature can be attached to the top of the float rod sliding in a tube. Outside the nonmagnetic tube is mounted the control switch, with a permanent magnet attached to it and set close to the tube. When the level rises, the armature passes the permanent magnet and attracts it, so that the switch is actuated. A spring retracts the magnet when the levels falls enough, and the switch is reactuated. A float arrangement of this kind, although limited in range, finds use as a low-water cutoff for boilers to 600 lb/in^2 (40 bar°) gage pressure. Switches are dry-contact or mercury type. Two of these units can be placed at different levels to give both high- and low-limit control.

Liquid level control systems without floats operate on several principles. If the liquid is at all conductive, probes can be used. The electrode probes are fixed, usually mounted in the same holder, and extend down into the tank (Fig. 4). Two or three electrodes are most common. Inductive or electronic relays are also part of the control system and actuate pumps or valves.

In a tank filled by a pump, a drop in level below the lower electrode breaks the circuit to allow a relay to start a pump or open a valve. When the liquid rises to the high-level electrode,

°1 bar = 10^5 Pa. For a discussion of bar, see *SI Units—A Commentary* in the front matter.

the direct electric circuit between electrodes is established and a relay stops the pump or closes the valve. In an induction relay (Fig. 5), the line voltage is separated from the control circuit by a primary and secondary coil arrangement. The relay design depends on the specific resistance of the liquid, which can vary from that of metallic circuits to that of demineralized water. Electronic relays have low potential and low electrode current.

Fixed probes usually do not exceed 6 ft (1.8 m) in length, but suspension electrodes are available for deeper tanks or higher level differences. Pressuretight electrode holders capable of operating at 10,000 lb/in² (690 bar) are available. Temperatures are generally limited to 450°F (232°C). For tanks where icing is a problem, a pipe sleeve in which the probes are suspended can be supplied (Fig. 4). An immersion heater near the sleeve bottom warms the water when the pump is not in operation.

Bubble sensors measure liquid level by determining the air pressure required to force a small stream of air bubbles through the lower end of a tube extending to the bottom of an open tank. The air flow tends to keep the tube and tube end clear in liquids that contain solids. Floats and probes are eliminated in this method, and only the air flow regulator and pressure switch are exposed to corrosive effects. The air stream flow rate can vary over a range without affecting air pressure. The specific gravity of the liquid must be known to allow the instrument to be calibrated. Because the measured variable is air pressure, the other instrumentation can be set at distances of 250 ft (76 m) horizontally or vertically. The range of liquid level is 6 in (15.2 cm) to about 32 ft (9.8 m). Sewage, industrial processes, and water supply are some applications.

FIG. 4 Electrodes suspended on cables sense tank water level below ice. (B/W Controls)

Notwithstanding the low air pressure involved, the differential sensitivity of pressure switches for bubbler sensors is about 0.5% of maximum operating range. Air consumption is about 1½ ft³/hr (0.042 m³/h) when the air flow regulator is set for 60 to 80 bubbles/min. The effects of air pressure failure can be prevented by providing a cylinder of carbon dioxide gas; a pressure switch and solenoid valve will introduce the gas to the system if the compressor fails.

PRESSURE SENSORS   Pressure controllers of the simple on-off variety may have a single-pole double-throw mercury switch actuated by a bourdon tube. A typical differential value is 2% of maximum scale reading. Adjustment to desired cutin (low) pressure is made by a knob on the case. Pressure ratings go to 5000 lb/in² (345 bar) for these devices. Proportional control can be added to controllers of this type by incorporating a slidewire potentiometer (Fig. 6).

In other types of pressure sensors, one sensor is provided for pump start and one for pump stop. Adjustable time delay prevents surging or waterhammer from giving a spurious start or stop signal. Increasing liquid pressure transmitted through tubing to an air chamber acts on a bellows and, overcoming adjustable spring tension, trips a mercury switch. The differential sensitivity of the bellows-type sensors is 0.5% of maximum operating range. Maximum pressure is about 175 lb/in² (12 bar) gage, since the systems are intended for use on open tanks. Sensors, timers, and relays can be mounted in a cabinet located near the tank or even near the pump.

Air trapped in a bell and pressurized by rising water is the actuating mechanism for another alarm switch (Figs. 7 and 8). A synthetic rubber diaphragm in the switch body mounted above

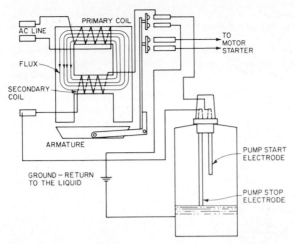

FIG. 5 Induction relay: when liquid reaches pump start electrode, current flows in secondary coil and diverts flux to lift armature and close motor contacts. (B/W Controls)

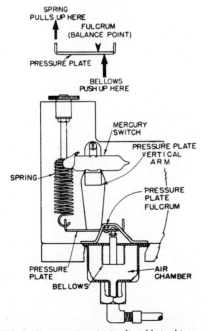

FIG. 6 Pressure trip point is adjustable in this pressure switch. (Autocon Industries)

FIG. 7 Alarm for level rise has bell connected to switch mechanism by 1-ft (0.3-m) pipe. (Autocon Industries)

the bell and connected to it by a small pipe is caused to tilt a mercury switch and thus give the alarm. A rise in level of about 1¼ in (3.2 cm) above the bell mouth will activate the switch.

ALTERNATORS   An alternator may be installed to achieve regular use and equal wear of each pump in multipump installations. The simplest versions serve on 2-pump systems, but more advanced

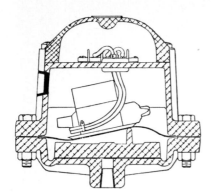

FIG. 8   Switch assembly for level rise alarm. (Autocon Industries)

designs can rotate starting sequence of as many as 12 pumps. In one version of the two-pump alternator, a solenoid plunger picks up and causes a four-pole double-throw switch to take alternate positions, maintaining a position after the solenoid is de-energized. If one pump leads with the other coming into service only to augment it, the switching compensates accordingly.

If starting sequence is to be rotated for more than two pumps, a motor-driven rotor can be advanced a given number of degrees each time a pump motor operates. The rotor contacts are connected together in pairs to provide circuits between pairs of stator contacts. Other control variations available in this regard are a change in sequence after a timed interval and an option of starting the pump that has been idle longest and stopping the pump that has run longest. Although many alternators operate on the same voltage as the loads, variants are available for operation from the low voltage and current ratings of control equipment.

TRANSDUCERS AND TRANSMITTERS   The variable that is most convenient or advantageous to measure is rarely the one best suited for direct use in the control system or for actuation of the final control element. A small differential pressure in a liquid level or flow control system can scarcely open a large valve. Conversion of measured variable values to another signal medium is therefore necessary and is the task of transducers and transmitters. These two terms are used interchangeably to some extent, although the transducer usually converts a signal to an electric current and the output of the transmitter is usually an air pressure.

A pressure-to-voltage transducer is especially useful in pump control. In one design, a bellows subjected to the pressure of the liquid transmits force to a pivoted beam that moves a core in a differential transformer or motion transducer to produce a 3- to 5-V output. The beam is balanced by a spring, and another spring allows the minimum pressure for voltage output to be set, which is equivalent to zero suppression. Zero suppression here can go as high as 95%.

Differential pressure tranducers may operate on a force balance principle with a very small motion of the bellows. The motion is converted to rotary motion in a ram or shaft, which then moves a differential transformer core to give an output signal that can vary from -2.5 to $+2.5$ V dc.

A transmitter is a device which can sense pressure, temperature, flow, liquid level, or differential pressure and convert the signal to a pneumatic pressure for transmission to receiving instruments several hundred feet distant. The pressure-sensing transmitter can span ranges to 80,000 lb/in^2 (5500 bar). The differential pressure type allows low differences in air or liquid pressures, such as a flow orifice develops, to be amplified through linkage and force balance mechanisms. An air pressure of 3 to 15 lb/in^2 (0.2 to 1 bar) in the air line from the transmitter is the result. Differential pressure transmitter designs are available to withstand primary system pressures to 6000 lb/in^2 gage (400 bar), whereas the differential pressures span ranges between 5 to 25 and 200 to 850 in (13 to 64 and 508 to 2160 cm) water.

TELEMETRY SYSTEMS   When pump control must be exercised over distances greater than the few hundred feet over which most pneumatic control equipment can operate, telemetry systems find application. In some of these systems, the sensor's output is converted to a proportionally variable 3- to 15-V dc voltage at the transmitter input. The transmitter then converts the dc voltage to a square-wave pulse with duration varying in proportion to the input signal. The resultant pulse width modulation (PWM) signal is sent directly over transmission lines or via tone carrier in microwave, VHF, or UHF systems. To transmit the bipolar PWM signal alone requires direct wiring with less than 5000 ohms resistance. This means as much as 24 miles (37 km) on direct telephone lines. The PWM signal eliminates loss of information from signal amplitude variations, and the bipolarity of pulses reduces line capacitance effects.

With tone transmission, either amplitude modulation or frequency shift is used.

The receiver converts the pulse signal back to a variable 3- to 15-V dc signal identical to the transmitter input signal and capable of use for indication or control. If transmission is not received for a certain time period, an alarm can be actuated and the pumps started or stopped.

Constant Speed Control   Where pump speed control is economically justified, it is a preferred method of obtaining the desired output parameter, such as flow rate, head, or liquid level. If the pump operates at constant speed, then there are four common control means:

1. On-off
2. Throttling by valve
3. Bypass by valve
4. Submergence

The on-off method with a single pump largely focuses on liquid level or temperature range control. Centrifugal and positive displacement pumps can be controlled by this method. If an accumulator is installed downstream, the method can be extended to head control. With multiple pumps in parallel, flexibility is slightly greater and a coarse control over flow rate is possible.

The simplest mechanism for on-off control of constant-speed pumps is the push button switch and starter for across-the-line start of small pumps. For large pumps, reduced-voltage starting is customary. Number of starts per hour is restricted in all cases to prevent overheating. The electric impulse to start or stop the motor can originate in any of the sensors or switches described above.

Throttling by valve is very common and can provide refined control under difficult conditions where rapid response and outstanding dynamic stability are sought, as in boiler feedwater control. Positive displacement pumps cannot use this method.

Bypass by valve is an occasional variation of valve control based on the bleedoff of discharge liquid to reduce the flow rate at a downstream point or to allow a cooling flow to pass through a constant-speed pump when its discharge has been blocked. The method can serve both centrifugal and positive displacement pumps.

Submergence control, for centrifugal pumps, relies on a temporary decrease in available NPSH to reduce the pump flow rate to the value at which liquid is entering the sump (Fig 9). The method serves for condensate, and design precautions prevent rapid cavitation damage.

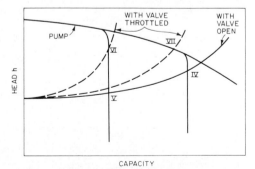

FIG. 9   In submergence control, operation can be at points where restricted-capacity curve intersects piping characteristic (IV, V, VI) or on regular pump characteristic (VII).

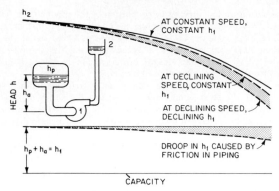

FIG. 10 Centrifugal pump characteristic changes depending on speed and head at inlet.

VALVE-THROTTLING CONTROL   The chief elements of centrifugal pump performance are shown in Fig. 10. At any given flow rate (capacity), a centrifugal pump produces a discharge head consisting of the static head on the pump inlet and the dynamic head imparted by the pump. At higher flow rates, speed usually declines slightly, lowering the characteristic curve as shown. In addition, the higher flow rates produce more frictional head loss in the inlet piping, so that the pump senses an inlet head slightly less than the static head developed by the weight of the liquid column and the effect of compressed gas or upstream pumps.

The pump can deliver any flow rate along the curve. What determines the actual flow rate at any instant is the characteristic curve of the downstream piping (system curve), as shown in Fig. 11. Under zero-flow conditions, there is a gravity head of liquid and perhaps a pressure in a container, such as a boiler drum. When liquid flows, piping friction head is added. Piping friction causes the system curve to turn upward, roughly parabolically. If a downstream control valve, previously wide open, is throttled, a new and more rapidly rising system curve is established. The intersection of a pump curve and system curve plotted on a single chart (Fig. 12) indicates conditions at the pump discharge. The combined plot also shows that the flow rate or discharge head will be modified by a change in other parameters besides throttle valve setting. For example, an increase in pump speed will lift the entire pump curve up and move the intersection point to higher flow rate and head. Decreased pressure in a boiler drum will lower the entire system curve and move the intersection point to higher flow rate and lower head. Throttling the inlet line to the pump will reduce the inlet head and cause the pump curve to start at the same point but slope

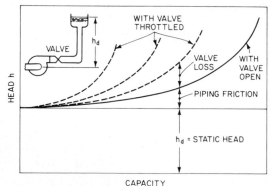

FIG. 11 Static head, piping friction, and valve loss determine the piping characteristic downstream of pump.

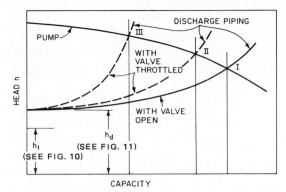

FIG. 12 Intersection of combined pump and piping characteristics is operating point (I, II, III.)

downward at a faster rate. The pump curve will then intersect the unchanged (downstream) system curve at lower head and flow rate.

BOILER FEEDWATER CONTROL　　In its original and simplest form, this control maintained the water level in a boiler drum (Fig. 13). Although this is still the primary objective in many boilers, there are other applications in which balance of steam flow rate against feedwater flow rate is the primary objective, with level maintenance a secondary factor unless it exceeds preset limits. In steam generators operating above the critical pressure of 3206 lb/in^2 (221 bar) abs, the feedwater turns to steam without a water level being visible, so that temperature and flow rates are the variables to be controlled.

Both on-off and modulating control are used in feedwater control systems. One classification of boiler feedwater control systems is based on whether the system is electric or pneumatic. Another classification gives the number of variables sensed to determine control valve position: a single-element regulator senses water level alone, a two-element regulator also senses steam flow, and a three-element regulator adds feedwater flow sensing.

For low-pressure boilers operating on moderate loads at no higher than 600 lb/in^2 (41 bar)

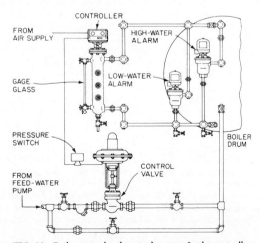

FIG. 13 Boiler water level control system. In the controller, an external magnet senses position of displacer. (Magnetrol)

gage and usually far below, on-off control of constant-speed pumps is sometimes used. The control can be similar to a low-water cutoff device actuated by float, but it has two switches: the switch at the higher water level is for level control, and the one below is for low-water cutoff of fuel and for alarm. Level differences of 1 to 3 in (2.5 to 7.6 cm) start the pump. This type of control can have a third switch, installed to give a separate alarm for low water before the fuel cutoff. With fire-tube boilers, the third switch can give a separate high-water alarm if the pump does not stop when the pump cutout switch is actuated.

Regulators directly actuated by float are also in use. Valves for these regulators are usually two-seated to reduce the thrust required of the float and linkage mechanism. Boiler pressures are low for this method, below 250 lb/in^2 (17 bar), although the regulators can be built to withstand 600 lb/in^2 gage (41 bar). Capacities go to more than 400,000 lb/h (180,000 kg/h) at pressure drops of 100 lb/in^2 (7 bar) across the valve.

For more demanding service, modulating control by means of an amplified signal applied to a valve actuator is necessary. The simplest type of modulating control of this kind is a single-element regulator, serving for fairly constant loads and pressures (Figs. 13 and 14). In the pneumatic system, a change of water level in the boiler drum provides a pneumatic output signal

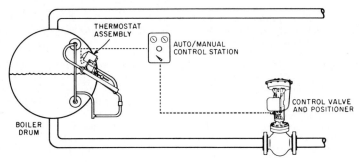

FIG. 14 Single-element boiler feedwater regulator system. (Copes-Vulcan)

which is transmitted to a controller supplying air pressure to the diaphragm or piston actuator of a control valve in the discharge line from the feedwater pump. A sensing thermostat for drum water level may be designed as a proportional controller, with gain changing in proportion to deviation from the set point. Linkage transfers the elongation of the sensing element to the pneumatic transmitter. Torque tubes or magnetic couplings (Fig. 13) may also convert the water level sensing of a float or displacer to a pneumatic signal, in any of the standard pressure ranges. The air pressure required for the feedwater control valves must usually be at least 50 lb/in^2 gage (3.4 bar), and up to 125 lb/in^2 (8.6 bar) gage may be necessary. Balanced-trim valves do not require as high air pressures as do the unbalanced-plug type, and the small amount of leakage is not harmful.

In some single-element systems, the sensing element directly actuates the valve. In one system of this type, a high enough vapor pressure is produced in an enclosed tube to operate the feedwater control valve directly. The vapor pressure generator consists of a slanting inner tube mounted beside the boiler drum, with ends connected to top and bottom of the drum. A finned outer tube, filled with a liquid and connected only to the control valve actuator, envelops the inner tube. A decrease in water level brings more heating steam into the inner tube to warm the liquid in the outer tube and increase its vapor pressure. The vapor pressure is transmitted to the valve actuator to open the valve.

In an electric control system, the level sensor can emit a signal modified by a slidewire potentiometer. The valve operator is an electric motor. In the larger sizes, the valve actuating speed will be low, so that the system cannot respond quickly to rapid change in steaming rate and water level.

Some single-element systems employ a pressure control valve directly upstream of the main feedwater control valve. The upstream valve, called a differential valve, maintains a constant

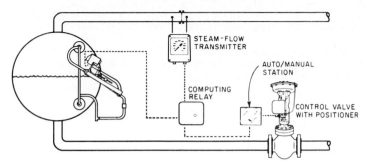

FIG. 15   Two-element regulator system with steam flow measurement. (Copes-Vulcan)

pressure on the feedwater control valve, improving its performance. Pressure control valves of this type have been used on some more advanced systems, too.

Single-element systems are inadequate for a boiler whose steaming rate changes suddenly because of the anomalous behavior of the water level during the change. A sudden demand for steam will reduce pressure in the drum, and steam bubble formation will increase, temporarily raising the water level at precisely the time when a falling level is required to signal for increased feedwater flow. The anomaly is called a rising water level characteristic. If the steaming rate change is gradual, then the characteristic can be constant or even lowering.

The two-element regulator (Fig. 15) solves this problem by sensing steam flow through an orifice in the steam main. The flow rate signal goes to the controller, and a sudden increase in steam flow will temporarily override the spurious water level signal.

Three-element control (Fig. 16) offers a further refinement—it senses feedwater flow rate in addition to water level and steam flow rate. During a rapid and large load swing, the feedwater flow rate can then be adjusted to the steam flow rate, while the water level is simultaneously noted. The feedwater flow rate signal is converted to a linear signal for transmission to a computing relay which can be adjusted for the relative influences of the three variables. A balanced signal then goes to the drum level controller, whose air output actuates the control valve. Three-element control systems can eliminate the effects of pressure variations upstream of the regulating valve.

BUILDING-WATER PRESSURE CONTROL   In tall buildings or large industrial, commercial, or housing water systems, water pressure can be maintained in several ways. If elevated or pressure tanks are not desired, a multiple-pump system may be considered. The pumps can be constant-speed or variable-speed. For constant-speed pumps, pressure sensors bring in individual pumps as required

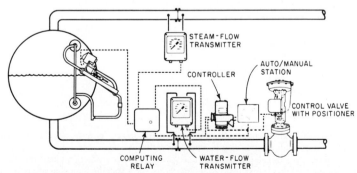

FIG. 16   Three-element regulator system action depends on water level, steam flow, and feedwater flow. (Copes-Vulcan)

to maintain pressure. If a large number of pumps are necessary, then means must be provided to prevent all of them from starting simultaneously on restoration of power after a failure. (See Sec. 9.20.)

Hydropneumatic systems rely on air pressure in the top of a tank into which pumps deliver water intermittently. As water is drawn off, the air expands, reducing its pressure and eventually requiring another pump start. Both pressure and water level are sensed. If the pump liquid does not bring in enough air to the tank to make up for losses, an air compressor or the compressed-air system must supply air. Tank pressure after each level-controlled pumping cycle indicates whether more air is needed or whether air should be bled off. Float or probe sensors can determine liquid level, and various pressure sensors are available.

VALVES

For the final control element, conventional and traditional valves serve for on-off control in pump systems. They also cover much of the modulation need. In recent years, modulating control in demanding services has required development of special control valves, actuators, and accessories. System dynamic characteristics, corrosion, erosion, noise, and costs have influenced the development.

Operation and Valve Types   The on-off operation of pump valves serves for

1. Isolation of a pump: protection, maintenance, removal, administrative reasons
2. Bypass or partial isolation: inlet or outlet block for protection, improved flow control, administrative reasons
3. Pressure relief: protection
4. Venting: removal of gases and vapors from the casing
5. Draining: removal of liquids from the casing

The modulating mode of operation serves for

1. Control of flow rate to pump or of pressure at inlet
2. Control of delivered flow rate or pressure
3. Control of bypass flow rate

Auxiliary flows in lines to packing boxes, seals, and sensing or measuring elements are controlled by either on-off or modulating valves.

The principal types of valves for on-off and much modulating service are

Gate (rare for modulating)

Globe (and angle)

Butterfly

Ball

Eccentric butterfly

Plug

Diaphragm

Check valves and relief valves, although possessing design features peculiar to their nature, make use of the essential features of globe and butterfly valves.

Control valves have developed as a special group for demanding services that require wide modulation range, stability, wear resistance, low noise, or a specific flow characteristic. Ingenuity and experience have combined to evolve many unusual but effective designs, each with advantages and drawbacks.

Control Valves A control valve is a valve that modulates the flow through it to provide the desired downstream (or upstream) pressure, flow rate, or temperature. Although most types of valves can be partly closed and thus give a degree of control that may be acceptable for many purposes, the term *control valve* has come to mean a specialized type of power-actuated valve designed for good performance under steady-state or dynamic flow conditions.

Before examining various control valves and their reasons for existence, some basic practical concepts must be reviewed. A control valve includes

1. A body to contain the pressure, direct the liquid flow, and resist loads from piping and actuation
2. A variable orifice or orifices
3. A stem for positive connection of orifice elements to actuator
4. A piercement through the body wall to allow the stem to pass
5. An actuator to adjust the orifice size

Adaptations of globe valves and angle valves are common. These forms inherently give tight shutoff. High-quality trim helps resist erosion and wear at low flow rates when the orifice is nearly closed. Support is often provided for the stem to prevent vibration and flutter (Fig. 17). The plug can be characterized (shaped to give certain rates of flow for given percentages of opening) as desired. The support of the stem is usually on the bonnet side of the orifice rather than opposite, to keep orifice size down. With flow upward (through the orifice, then past the plug), the actuator must overcome upstream pressure to close the valve. With flow downward (past the plug, then through the orifice), valve motion becomes unstable when the valve is nearly closed.

The basic globe valve is often modified to put two orifices and plugs on the same stem, with the upstream fluid entering the space between and passing in two opposite flows through the two orifices. This is the double-seated valve (Fig. 18). Actuator force is greatly reduced because fluid pressure tends to open one plug and close the other. The balance is not complete, however,

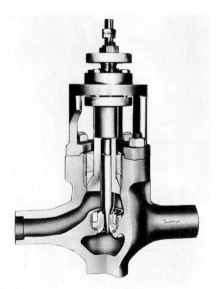

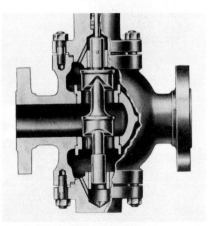

FIG. 17 This control valve is basically a globe type, with reduced trim and guided tapered plug. (Copes-Vulcan)

FIG. 18 Double-seated control valve with characteristized plugs. (Masoneilan International)

because one orifice is usually larger than the other to permit assembly and because there is a difference in head conversion effects in the orifices at low flow rates.

Addition of an internal diaphragm and a port in the body near one orifice makes the valve a three-way type, able to divide flow between two outlet lines or, with reversed flow direction, combine two flows in a desired ratio. The double-seated valve cannot seal tightly because of manufacturing tolerances and thermal and pressure effects on the valve body.

Butterfly valves can modulate flow (Fig. 19). Special vane shapes have been introduced to improve performance. Elastomer or plastomer linings give a tight shutoff on liquids within the temperature range of the materials.

Ball valves as control valves may take conventional form, with an actuator and positioner atop the valve. In other designs, the ball may be merely a fraction of a spherical shell, adequate for sealing on the customary tetrafluoroethylene seat ring but with its edge shaped to develop the required characterized flow as the shell rotates and exposes the orifice. Convex, V-notch, and parabolic edge shapes find use. The conventional ball valve has two variable orifices in series, of course, with a small chamber between them in which some head recovery occurs as fluid momentarily slows.

A specialized form of gate valve can serve as a control valve. This type contains a multiple-orifice plate mounted permanently as a diaphragm perpendicular to the line of flow (Fig. 20). The "disk," a plate which also contains two or more slotted orifices, is mounted to slide vertically across the upstream side of the stationary plate. The degree of orifice coincidence determines the flow rate. Actuation is by a pin mounted on the stem and protruding through the stationary plate into a pocket on the sliding plate. Low vibration and straight-through flow are characteristic of this valve. The actuating force is low at all flow rates because of the sliding action of the lapped disk and plate and because of the disk support. Both disk and plate are made of stainless steel or other alloys.

CAGE VALVES   This type has developed into an entire group of control valves (Figs. 21 to 25). The valve body closely resembles that of the globe valve, with a large orifice in a horizontal area of the central diaphragm. The cage is a hollow cylinder which is held between the bonnet and the

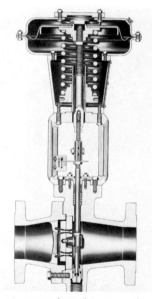

FIG. 19   Butterfly control valve with linkage connecting spindle to diaphragm actuator and positioner. (Masoneilan International)

FIG. 20   Movement of one plate past another opens or closes flow orifices in this control valve. (Jordan Valve Division, Richards Industries)

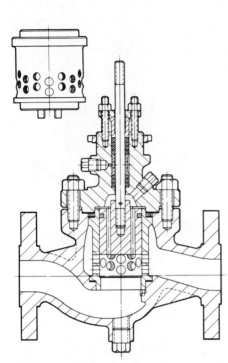

FIG. 21 Hole pattern in cage can reduce noise and cavitation. (ITT Hammel Dahl Conoflow)

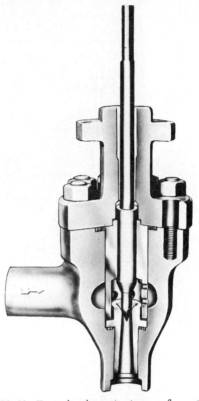

FIG. 22 Tapered outlet section improves flow pattern in this cage valve. (Copes-Vulcan)

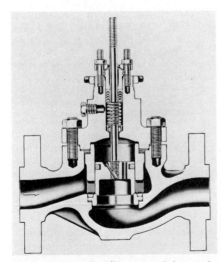

FIG. 23 Holes in plug allow pressure balance in this cage valve. (Fisher Controls)

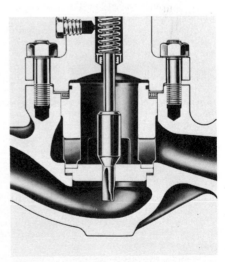

FIG. 24 An extreme in reduced trim for a cage valve. (Fisher Controls)

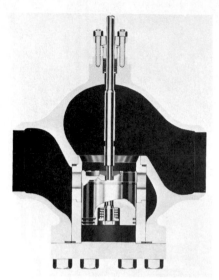

FIG. 25 Cage valve with bottom access. (Copes-Vulcan)

FIG. 26 Control valve with flow through labyrinthine orifices in built-up cage. (Control Components)

edge of a hole in the diaphragm. The disk or plug, sliding up and down inside the cage, is guided by it.

In most cage valves, the lower zone of the annular cage is available for flow control orifices, alternately exposed and covered as the plug moves up and down (Fig. 21). Tight shutoff at these orifices is impossible, of course, and so the bottom of the cage or a separate seat ring is machined and finished to match a seating surface on the plug.

In other cage valves, where the plug may be much smaller than the cage inside diameter (Fig. 22), all control action is at the lower seat ring. The plug guiding is then on the stem or in the seat ring orifice, and the holes in the cage are merely passage holes to distribute the liquid evenly around the perimeter.

The seat ring in cage valves is retained by bonnet bolting forces acting through the cage. Gaskets take up tolerances in dimensions and finish, so that the seat ring need not be pressed or screwed into the valve diaphragm.

The ordinary cage valve has several advantages. Suitable machining of cage holes can give the valve the desired characteristic. Removal and replacement of internals, such as cage and seat ring, are quick and simple. A vertical hole through the plug makes the valve nearly balanced (Fig. 23), although considerable leakage can occur between plug and cage wall if the plug is balanced in this way. Cage wall orifices vary not only in number, size, and cross-sectional form, but also in path and surface roughness.

In one advanced variation of the cage valve, the cage wall is comparatively thick and the many pathways through it are labyrinthine, with several right-angle turns and several orifices and expansion chambers in each pathway (Fig. 26). To make the cage practical from a manufacturing standpoint, it consists of a series of thin disks each carrying a pattern of labyrinthine paths in one surface. The opposite surface is flat, so that when the disks are stacked into a cage, the paths are sealed from one another. Flow in valves of this type can be from inside or outside the cage. Characterization is possible by such means as a change in the number of orifices per disk at various heights in the stack. As in conventional cage valves, the trim and characterization can be quickly changed after bonnet removal. The bonnet bolting, outside the liquid, holds the cage elements in place.

In other cage valves, the passage walls may be wavy, resembling screw threads. This assists in noise reduction in gas valves.

MULTIPLE ORIFICES IN SERIES   Several valves have orifices in series rather than in parallel; the principle is called *cascading*. In one group, a tapered plug with a series of circumferential serrations moves in a tapered seat which may be either conical or stepped (Figs. 27 and 28). In either case the serrations or steps produce a series of small annular chambers alternating with annular restrictions that serve as orifices. Some alternating change in flow direction also occurs to create the desired head loss.

In another group, annular chambers in the wall of a cylindrical cage are separated from one another by ridges that are a close fit with annular ridges on a sliding plug. The orifices narrow as the ridge sets approach one another while the expansion chambers remain nearly constant in size. Repeated change in flow direction and speed produces head loss. In a variation of this type (Fig. 29), the chambers on the cylindrical plug are short, steeply angled helical cuts, so that the liquid takes a helical path. The purpose is to fling the liquid against the walls of the cage and displace cavitation bubbles toward the center and away from wall contact. Shutoff in these valves cannot rely on the multiple-orifice systems but instead depends on a separate conventional seat and plug surface, either upstream or downstream of the orifice system.

Many of these special valves are very expensive because of the multiplicity of complicated parts and because of the reduction in capacity caused by the advanced design. Larger bodies and overall sizes are required for a given flow rate. In pump control, the valves see service on high-pressure feedwater pump minimum-flow recirculating lines, where pressures go as high as 6000 lb/in² (400 bar) and water temperatures range to 500°F (260°C). Tight shutoff over thousands of operating cycles is the goal.

FLOW CHARACTERISTICS   An important parameter for valves in modulating control is the flow characteristic of the valve, often called simply the characteristic. The flow characteristic expresses the way in which the flow through the valve depends on percentage of valve stem travel. The latter may be translatory or rotary motion, of course. A plot of percentage of maximum flow at various percentages of stem travel is the usual quantitative way of showing a characteristic (Fig. 30). Several types of characteristics have become common, either because of inherent desirability or because familiar and traditional types of valves have them.

FIG. 27   Pressure breakdown occurs across several annular orifices and direction changes in this valve. (Masoneilan International)

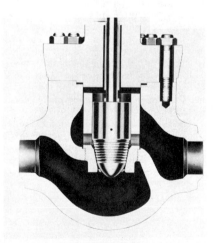

FIG. 28   Serrations on large plug give high pressure drop at low flow. (Copes-Vulcan)

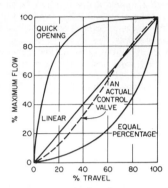

FIG. 29 Pump bypass flow at high pressure drop results when line flow at top ceases and mechanism opens valve at right. (Yarway)

FIG. 30 Basic control valve characteristics. Valves are designed to approach these.

The linear characteristic is a straight line, with flow percentage always equal to stem travel percentage. A quick-opening characteristic, on the other hand, produces proportionately more flow in the early stages of stem travel. An equal-percentage characteristic gives a change that, for a given percentage of lift, is a constant percentage of the flow before the change. A change of 16% of total stem travel will double the flow, so that at a stem travel of about 84%, flow will be 50% of maximum.

Although the linear characteristic would seem best because the rate of flow change is uniform for a given stem travel change, incorporation of the valve into a piping system affects the decision. Since resistance to flow in a given piping system is roughly proportional to the square of the flow rate, the curve of the piping system head loss plotted against flow rate will be a parabola, with resistance increasing at a faster rate than flow. If the piping system and valve are considered together and the flow rate in the system plotted against percentage of valve stem travel, the overall system characteristic will differ from the valve characteristic. The overall system characteristic is displaced upward toward the quick-opening valve characteristic but can have points of flexure. The amount of displacement depends on what part of the total system pressure drop is taken by the valve. Only with a very short outlet pipe would the valve take all the pressure drop, and then its characteristic would be that of the system.

In many systems the pressure drop across the valve is designed to be from one-tenth to one-third of the total system drop. If the valve in such a system has an equal-percentage characteristic, then the characteristic of the overall system will be close to linear as far as the actuator of the valve is concerned (Fig. 31).

The equal-percentage characteristic is obtained by such measures as contouring the valve plug, contouring slots in plug skirt or cage, or suitably spacing holes in the cage.

Valves with a characteristic between linear and equal-percentage are also useful in modulating control. Ball, plug, and butterfly valves are examples. Characterized ball and plug valves are examples of modifications for control characteristic purposes.

RANGEABILITY   Control valve rangeability (Fig. 32) can be important in some cases; it is defined as the ratio of maximum flow to minimum flow at which the valve characteristic is still evident and control is possible. A high rangeability value means that a single valve can handle low as well as high flows, so that auxiliary valves are unnecessary. The best performance in this regard is about 100:1 for special designs under favorable circumstances, and 25:1 is common for conventional valves and ordinary circumstances.

Connected with rangeability is *valve gain*, which is the slope of the flow characteristic curve at any point. In practical terms, valve gain is the change in flow rate per unit of change in stem

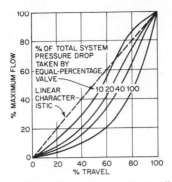

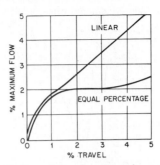

FIG. 31 For a valve in a system, the overall characteristic depends on valve characteristic and pressure breakdown.

FIG. 32 Rangeability of a valve is determined by the point at which valve characteristic is still evident, between 1 and 2% of maximum flow for the two valve characteristics shown here.

travel. A high gain means that a slight movement of the stem causes a large change in flow rate, so that instability occurs more readily. This sets a limit on a valve's rangeability. The quick-opening valve, with a high gain in the nearly closed position, is unsuitable for many modulating tasks.

Valves with approximately an equal-percentage flow characteristic are considered most suitable for the majority of flow control tasks; valves with a linear characteristic are preferred for some applications.

SIZE   The *size* of a valve is an indefinite concept. In the past, a valve's size was understood to be the pipe size of the line connected to it. Venturi valves with tapered end passages leading to reduced-diameter orifices and valves in which the orifice area is reduced for reasons such as cost cutting, characterization, and special advantages have forced users to rate valve size in other terms. A statement of maximum orifice area might be a way to express size, but this too would be ambiguous because the head losses in partial recovery after the orifice are not the only losses in the valve. In addition, the valve may have two or more orifices in series, and the geometry of the orifice itself may affect results.

FLOW COEFFICIENT   The valve flow coefficient C_v (K_v in SI units) is now a frequently used parameter for valve size. It is the number of gallons per minute (cubic meters per hour) of 60°F (15.6°C) water that will flow through a valve at a 1-lb/in² (1-bar) pressure drop across the valve. The upstream test pressure is also stated. The maximum C_v, (K_v), found with the valve fully open, is widely accepted as a measure of valve size. To find the maximum liquid flow rate of a valve at any pressure drop and with a liquid of any specific gravity, the equation is

in USCS units
$$Q = \frac{C_v}{(G_t/\Delta P)^{1/2}}$$

in SI units
$$Q = \frac{K_v}{(G_t/\Delta P)^{1/2}}$$

where Q = flow rate, gpm (m³/h)
C_v = valve flow coefficient in USCS units
K_v = valve coefficient in SI units
G_t = specific gravity of the liquid
ΔP = pressure drop across valve, lb/in² (bar)

PRESSURE RECOVERY AND CAVITATION   A drop in liquid pressure upon passing through a valve is recovered to varying extent downstream (Fig. 33). The degree of recovery depends on valve type: ball and butterfly valves have higher recovery percentages than do globe and angle valves. To avoid cavitation, which is the formation of vapor bubbles near the *vena contracta* of the valve, followed by a sudden damaging collapse near the metal, the static pressure at the *vena contracta* must be above the liquid vapor pressure. This is easier to do with a low-recovery valve because

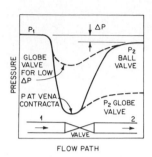

the initial pressure drop need not be as high for a given downstream pressure. Several factors have been devised to indicate pressure recovery. One, C_f, the critical flow factor, is the ratio of pressure recovery, varying for different valve openings. Another, K_m, the valve recovery coefficient, is the ratio of pressure drop across the valve to pressure drop between valve inlet and *vena contracta* at that instant when flow begins to be choked by bubble formation. Both of these factors will be higher for globe valves than for ball and butterfly valves, and the factors serve to indicate valve suitability for marginal cavitation service.

FIG. 33   For a given pressure drop ΔP across a valve, the globe will show a higher pressure at the *vena contracta*, making it more likely that cavitation difficulties will be avoided when the vapor pressure is high.

ACTUATORS   The motion needed to change the valve orifice area and to close the valve tightly is produced by an actuator. The types of motion of the valve plug or disk are either linear or rotary (Fig. 34), the latter being usually 90° but occasionally as low as 70°. These motions can be effected in several ways. The linear translating motion can result from a cylinder or diaphragm actuator working directly or through linkage (Fig. 35). A screw thread at the stem top can convert a rotary motion to linear stem motion, or threads at the stem bottom can engage threads in the valve disk so that rotation of the stem moves the disk. Geared electric motor drives (Fig. 36), cylinders, (Fig. 37), and diaphragm-and-spring actuators (Fig. 38) are common with ball, plug, and butterfly valves. The solenoid valve (Fig. 39) relies on an electromagnetic force to move a disk directly or to initiate the piloting action that allows line fluid to open the valve. The piloted solenoid valve (Fig. 40) relies on fluid pressures to open the main orifice.

The simplest actuator is the manually powered operator, which is a gear box. It provides enough mechanical advantage to overcome starting friction and to seal the valve tightly. Provision for an impact blow to initiate opening is found in some operators.

The choice of actuator depends first on whether the service is on-off or modulating. For on-off service, the actuator need have only enough force to overcome breakaway force or torque and sufficient stroke to open the valve fully. Speed of operation is rarely critical, and motion limits can be designed into valve or actuator. Pneumatically (Fig. 41) and hydraulically powered actuators usually stroke rapidly but can be slowed in either direction by auxiliary valving or controls. On some pneumatic actuators, times to 5 min are possible. Electric-motor-driven actuators are

FIG. 34   Rotary actuator with sealed blade. (Xomox)

slower than pneumatic or hydraulic types and require limit switches to stop the motor at the end of travel.

In modulating service, where the actuator must hold a control valve setting, demands are more severe. The speed of movement, expressed as stroking speed, is sometimes an important factor, especially in emergency shutdown or bypass. The stability of an actuator is partly its ability to hold the valve setting under fluctuating or buffeting loads from the fluid. Damping and high spring rate can help with this. The relation of the natural frequency of the actuator and its adjacent elements to the frequencies encountered in controlling the flow or those experienced from fluid buffeting can also be important.

Stroke length is also a factor. Although the disk in a globe valve or similar type need lift only one-quarter of the seat diameter to give adequate area for full flow, this distance in large valves will exceed the 2-in (5-cm) stroke of most diaphragm actuators. If linkage with its lever advantage is needed to increase thrust, the problem becomes more acute. A cylinder or electric actuator is then necessary.

The source of power for the actuator influences choices too. One standard may be 3- to 25-lb/in^2 (0.2- to 1.7-bar) instrument air pressure, while in other cases much higher air or oil pressure is available.

The diaphragm-and-spring actuator (Fig. 38) is a very common type and has several important advantages. The spring can be preloaded to cause the valve to either close or open fully (be fail-safe) if control air fails. The spring also opposes the force generated by the

FIG. 35   Linkage connects actuator and valve stem. (Masoneilan International)

control signal, doing so in a manner giving proportional control. Of course, the spring's opposition negates much of the force available on the diaphragm, but the simplicity and low friction of this actuator have made it very popular. Most modern types are reversing: the fail-safe action can easily be changed from open to close by turning the diaphragm enclosure upside down and reassembling.

POSITIONERS   With actuators that lack an internal spring, a positioner is needed to adjust the valve position to the desired value. A positioner is a small feedback system that receives an input signal (usually air pressure but sometimes an electric signal) from a controller and adjusts a valve stem position to a prearranged corresponding value. The valve stem position, which is the output, need not vary linearly with input pressure; cams in the mechanism can give a wide range of stem position functions and thus apparently change the characteristic of the valve.

The positioner is a necessity for actuators in which the valve stem position is not a function of the actuator fluid pressure or electric current magnitude. Examples are pneumatic and hydraulic cylinders and electric motors. Even though positioners are not inherently necessary on the diaphragm-and-spring actuator, they are sometimes applied. The reasons for the application hold for other types of actuators, too.

Friction in the actuator diaphragm cylinder, or valve stem packing is one reason. The positioner can cut the dead-band from values such as 5 to 15% to less than 0.5% and can give repeatability of 0.1% of full span.

Need for more force to close a single-seat valve tightly is another reason for using a positioner. If loading pressure must be increased above a standard 15 lb/in^2 gage (1-bar) value, the positioner can control air at a higher pressure and thus greatly increase the stem force.

Split-range operation, in which different valves operate over different parts of the controller output pressure range, calls for positioners. Reversal of valve action, too, is easily achieved with

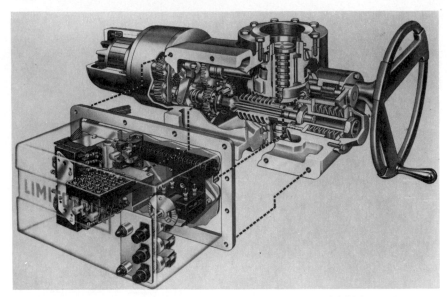

FIG. 36 Electric-motor-driven actuator with mechanism for limiting torque. (Philadelphia Gear)

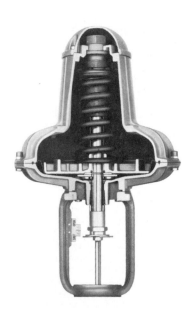

FIG. 37 Cylinder actuator and linkage for ball valve. (Jamesbury)

FIG. 38 Diaphragm-and-spring actuator, reversible type. (Foxboro)

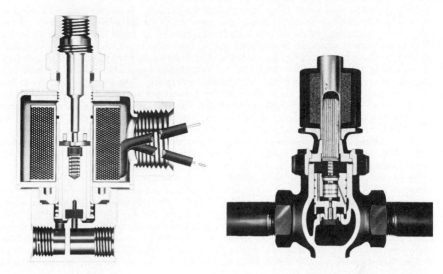

FIG. 39　Solenoid valve for three-way operation is direct-acting. (Skinner Precision Industries)

FIG. 40　Piloted solenoid valve relies on fluid pressure to open main orifice. (Magnetrol Valve)

positioners. A positioner can also speed up valve response because the low-volume positioner will act faster than the high-volume valve actuator and can open a larger air supply than that in the controller. A pneumatic amplifier or booster is an alternative way to do this. Finally, change in control valve characteristic, such as from linear to equal-percentage, is also possible through a positioner cam.

　　Because a positioner is another control loop added to a system, its effect under dynamic conditions may worsen overall performance. If changes or oscillations are slow, the positioner and actuator will follow them accurately and correct for them. For rapid changes, however, the effect of the positioner can be harmful. Evidence shows that if the natural frequency of the complete process control loop is more than 20% of the frequency at which the gain in the positioner-actuator system is attenuated 3 dB, then the positioner will impair system performance. Liquid level control systems are more likely to benefit from positioners than are flow or pressure control systems.

　　Pneumatic positioners may be classified as force-balance or motion-balance types. In the force-

FIG. 41　Opposed pistons drive rack-and-gear mechanism for 90° rotation in this pneumatic actuator. (Worcester Controls)

balance type, the force in the range spring inside the positioner is balanced against control air pressure inside a bellows or double-diaphragm assembly. In the force-balance positioner of Fig. 42, the feedback spring, which can be adjusted for range of pressure and initial actuation pressure, is attached to the actuator diaphragm plate at the bottom and to a double-diaphragm assembly at the top. The upper diaphragm has twice the area of the lower; introduction of signal air from the controller into the space between the two diaphragms forces the assembly upward very slightly but enough to lift a pilot valve at the positioner top and allow supply air pressure to flow through and downward past the feedback spring to press the actuator diaphragm down until forces balance. A reduction in signal pressure allows the double-diaphragm assembly to move downward, first closing the pilot valve and then exposing a hole through the pilot valve stem. Air then bleeds out from the actuator to atmosphere until forces are again in balance.

In the force-balance positioner of Fig. 43, flexure strips and a bell crank convert the vertical actuator motion to a horizontal motion in the double-diaphragm assembly at the top left and the supply valve at the right.

A motion-balance positioner showing the application of a cam to impart a characteristic is shown in Fig. 44. The cam at the lower right is pivoted and caused to rotate by the actuator stem motion. Supply pressure enters the valve assembly block at the left and goes to both booster valves. It also bleeds through a restrictor and nozzle at the bottom of the valve assembly block. The position of the flapper before the nozzle determines the pressure in the diaphragm comparator at the right of the valve assembly block. An increase in signal air pressure to the bottom bellows moves one end of a balance beam and pushes the flapper closer to the nozzle. This builds pressure in the diaphragm comparator and moves it to the right. A linkage transforms this motion into a

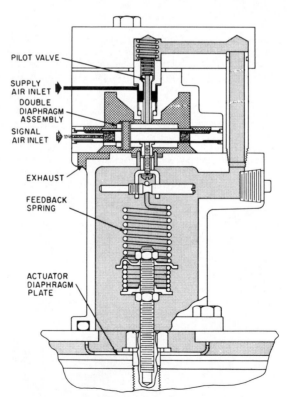

PILOT VALVE

SUPPLY
AIR INLET

DOUBLE
DIAPHRAGM
ASSEMBLY

SIGNAL
AIR INLET

EXHAUST

FEEDBACK
SPRING

ACTUATOR
DIAPHRAGM
PLATE

FIG. 42   Force-balance positioner with spool and sleeve pilot valve. (Masoneilan International)

FIG. 43 Force-balance positioner with flexure linkage. (ITT Hammel Dahl Conoflow)

motion which opens the booster valve to supply air to the cylinder actuator top and permits air to exhaust from the cylinder bottom. The actuator stem moves down until the feedback cam, aided by the comparator linkage, has repositioned the flapper in front of the nozzle.

In another motion-balance positioner (Fig. 45), signal air pressure from 3 to 15 lb/in^2 (0.2 to 1 bar) gage in a bellows opposes a flexure assembly on a shaft which is rotated by the valve stem motion. An increase in signal air pressure to the bellows expands it and moves the lower end of the flexure away from a flapper, permitting the flapper to move toward a nozzle. The resultant

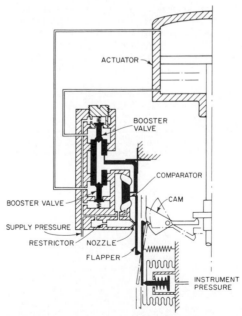

FIG. 44 Motion-balance positioner with cam and diaphragm comparator. (ITT Hammell Dahl Conoflow)

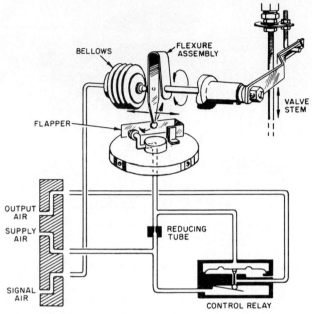

FIG. 45   Motion-balance positioner with flexure assembly. (Foxboro)

buildup of air pressure on the diaphragm of the control relay at the lower right closes the exhaust port and opens the supply port to allow air at full supply pressure to pass to the actuator. The valve stem motion rotates the flexure and thereby shifts the tip touching the flapper. The flapper assumes an equilibrium position proportional to the signal air pressure.

The pneumatic amplifier or booster is a special kind of regulator valve which develops an output air pressure proportional to the input signal pressure. It can be used to boost pressure on an actuator for faster action in cases where the instrument tubing is small-bore and long and the actuator volume is large.

FURTHER READING

ISA *Handbook of Control Valves*, Instrument Society of America, Pittsburgh, 1971.

PUMP SYSTEMS

SECTION 8.1

GENERAL CHARACTERISTICS OF PUMPING SYSTEMS AND SYSTEM-HEAD CURVES

J. P. MESSINA

SYSTEM CHARACTERISTICS AND PUMP HEAD

A pump is used to deliver a specified rate of flow through a particular system. When a pump is to be purchased, this required capacity must be specified along with the total head necessary to overcome resistance to flow and to meet the pressure requirements of the system components. The total head rating of a centrifugal pump is usually measured in feet (meters), and the differential pressure rating of a positive displacement pump is usually measured in pounds per square inch (kilopascals or bar°). Both express, in equivalent terms, the work in foot-pounds (newton-meters) the pump is capable of doing on each unit weight (force) of liquid pumped at the rated flow.† It is the responsibility of the purchaser to determine the required pump total head so that the supplier can make a proper pump selection. Underestimating the total head required will result in a centrifugal pump's delivering less than the desired flow through the system. An underestimate of the differential pressure required will result in a positive displacement pump's using more power than estimated, and the design pressure limit of the pump could be exceeded. Therefore, system pressure and resistance to flow, which are dependent on system characteristics, dictate the required pump head rating.

° 1 bar $= 10^5$ Pa. For a discussion of bar, see *SI Units—A Commentary* in the front matter.

† Work per unit weight mass, rather than weight force, is sometimes called *specific delivery work;* it has the units of newton-meters per kilogram and is equal to total head multiplied by g, the gravitation constant.

THE PUMPING SYSTEM

The piping and equipment through which the liquid flows to and from the pump constitute the pumping system. Only the length of the piping containing liquid controlled by the action of the pump is considered part of the system. A pump and the limit of its system length are shown in Fig. 1.

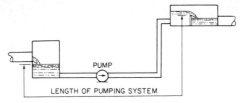

FIG. 1   Length of system controlled by pump.

The pump suction and discharge piping can consist of branch lines, as shown in Fig. 2. There can be more than one pump in a pumping system. Several pumps can be piped together in series or in parallel or both, as shown in Fig. 3. When there is more than one pump, the flow through the system is determined by the combined performance of all the pumps.

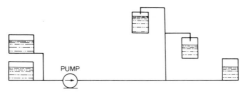

FIG. 2   Branch-line pumping system.

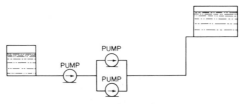

FIG. 3   Pumps in series and in parallel.

The system through which the liquid is pumped offers resistance to flow for several reasons. Flow through pipes is impeded by friction. If the liquid discharges to a higher elevation and/or a higher pressure, additional resistance is encountered. The pump must therefore overcome the total system resistance due to friction and, as required, produce an increase in elevation and/or pressure at the desired rate of flow. System requirements may be such that the pump discharges to a lower elevation and/or pressure but additional pump head is still required to overcome pipe friction and obtain the desired rate of flow.

ENERGY IN AN INCOMPRESSIBLE LIQUID

The work done by a pump is the difference between the energy level at the point where the liquid leaves the pump and the energy level at the point where the liquid enters the pump. Work is also the amount of energy added to the liquid in the system. The total energy at any point in a pumping system is a relative term and is measured relative to some arbitrarily selected datum plane.

An incompressible liquid can have energy in the form of velocity, pressure, or elevation. Energy in various forms is added to the liquid as it passes through the pump, and the total energy added is constantly increasing with flow. It is appropriate then to speak of the energy added by a pump as the energy added per unit of weight (force) of the liquid pumped, and the units of energy expressed this way are foot-pounds per pound (newton-meters per newton) or just feet (meters). Therefore, when adding together the energies in their various forms at some point, it is necessary to express each quantity in common equivalent units of feet (meters) of *head*.

Liquid flowing in a conduit can undergo changes in energy form. Bernoulli's theorem for an incompressible liquid states that in steady flow, without losses, the energy at any point in the conduit is the sum of the *velocity head, pressure head,* and *elevation head* and that this sum is constant along a streamline in the conduit. Therefore, the energy at any point in the system relative to a selected datum plane is

$$H = \frac{V^2}{2g} + \frac{p}{\gamma} + Z \qquad (1)$$

where H = energy (total head) of system, ft·lb/lb or ft (N·m/N or m)
V = velocity, ft/s (m/s)
g = acceleration of gravity, 32.17 ft/s² (9.807 m/s²)
p = pressure, lb/ft² (N/m²)
γ = specific weight (force) of liquid, lb/ft³ (N/m³)
Z = elevation above (+) or below (−) datum plane, ft (m)

The velocity and pressure at the point of energy measurement are expressed in units of *equivalent head* and are added to the distance Z that this point is above or below the selected datum plane. If pressure is measured as gage (relative to atmosphere), total head H is gage; if pressure is measured as absolute, total head H is absolute. Equation 1 can also be applied to liquid at rest in a vertical column or in a large vessel (or to liquids of various densities) to account for changes in pressure with changes in elevation or vice versa.

The equivalent of velocity and pressure energy heads in feet (meters) can be thought of as the height to which a vessel of liquid of constant density has to be filled, above the point of measurement, to create this same velocity or pressure. This is illustrated in Fig. 4 and further explained below.

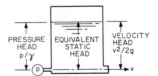

FIG. 4   Equivalent static head.

Velocity Head   The kinetic energy in a mass of flowing liquid is $\frac{1}{2}mV^2$ or $\frac{1}{2}(W/g)V^2$. The kinetic energy per unit weight (force) of this liquid is $\frac{1}{2}(WV^2/Wg)$, or $V^2/2g$, measured in feet (meters). This quantity is theoretically equal to the equivalent static head of liquid that is required in a vessel above an opening if the discharge is to have a velocity equal to V. This is also the theoretical height the jet of liquid would rise if it were discharging vertically upward from an orifice.

A free-falling particle in a vacuum acquires the velocity V starting from rest after falling a distance H. Also $V = \sqrt{2gH}$. All liquid particles moving with the same velocity have the same velocity head, regardless of specific weight. The velocity of liquid in pipes and open channels invariably varies across any one section of the conduit. However, when calculating system resistance it is sufficiently accurate to use the average velocity, computed by dividing the flow rate by the cross-sectional area of the conduit, when substituting in the term $V^2/2g$.

Pressure Head   The pressure head, or flow work, in a liquid is p/γ, with units in feet (meters). A liquid, having pressure, is capable of doing work, for example, on a piston having an area A and stroke L. The quantity of liquid required to complete one stroke is γAL. The work (force × stroke) per unit weight (force) is $pAL/\gamma AL$, or p/γ. The work a pump must do to produce pressure intensity in liquids having different specific weights varies inversely with the specific weight or specific gravity of the liquid. Figure 5 illustrates this point for liquids having specific gravities of 1.0 and 0.5. The less dense liquid must be raised to a higher column height to produce the same pressure as the denser liquid. The pressure at the bottom of each liquid column *H* is the weight of the liquid above the point of pressure measurement divided by the cross-sectional area *A* at the same point, $AH\gamma/A$, which is simply $H\gamma$. Note that in this discussion *A* is in square feet (square meters) and *L* and *H* are in feet (meters).

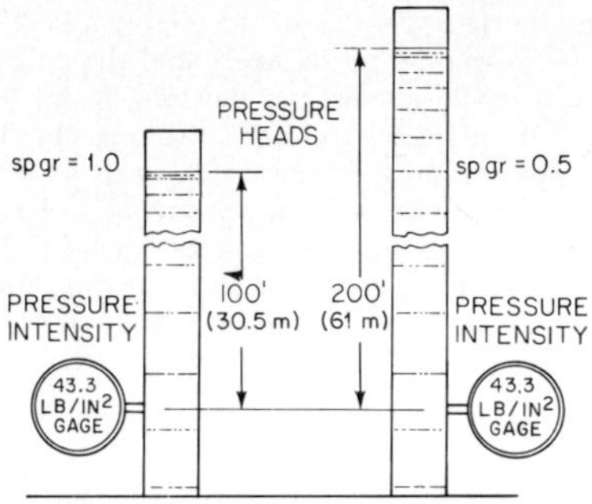

The height of the liquid column in feet (meters) above the point of pressure measurement, if the column is of constant density, is the equivalent pressure static head. When pressure intensity $AH\gamma/A$ is substituted for *p* in p/γ, it can be seen that the pressure head *H* is the liquid column height. Therefore, at the base of equal columns containing different liquids (with equal surface pressures), the pressure heads in feet (meters) are the same but the pressure intensities in pounds per square foot (newtons per square meter) are different. For this reason, it is necessary to identify the liquids when comparing pressure heads.

FIG. 5 Liquids of different specific weights (also specific gravities) require different column heights or pressure heads to produce the same pressure intensity. $(43.3 \text{ lb/in}^2 = 298.5 \text{ kPa})$

Elevation Head The elevation energy, or potential energy, in a liquid is the distance *Z* in feet (meters) measured vertically above or below an arbitrarily selected horizontal datum plane. Liquid above a reference datum plane has positive potential energy since it can fall a distance *Z* and acquire kinetic energy or vertical head equal to *Z*. Also, it requires *WZ* ft·lb (N·m) of work to raise *W* lb (N) of liquid above the datum plane. The work per unit weight (force) of liquid is therefore *WZ/W*, or *Z* ft (m). In a pumping system, the energy required to raise a liquid above a reference datum plane can be thought of as being provided by a pump located at the datum elevation and producing a pressure that will support the total weight of the liquid in a pipe between the pump discharge and the point in the pipe to which the liquid is to be raised. This pressure is $AZ\gamma/A$, or simply $Z\gamma$ lb/ft² (N/m²) or $Z\gamma/144$ lb/in². Since head is equal to pressure divided by specific weight, elevation head is $Z\gamma/\gamma$, or *Z* ft.

Liquid below the reference datum plane has negative elevation head.

Total Head Figure 6 illustrates a liquid under pressure in a pipe. To determine the total head at the pressure gage connection and relative to the datum plane, Eq. 1 may be used (assume that

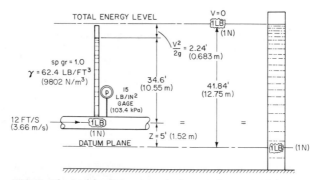

FIG. 6 The total head, or energy, in foot-pounds per pound (newton-meters per newton) is equal to the sum of the velocity, pressure, and elevation heads relative to a datum plane. A unit weight (force) of the same liquid or any liquid raised to rest at the height shown, or under a column of liquid of this same height, has the same head as the unit weight (force) of liquid shown flowing in the pipe.

the gage is at the pipe centerline):

$$H = \frac{V^2}{2g} + \frac{p}{\gamma} + Z$$

in USCS units
$$H = \frac{12^2}{2 \times 32.17} + \frac{144 \times 15}{62.4} + 5$$

$$= 2.24 + 34.6 + 5$$

$$= 41.84 \text{ ft} \cdot \text{lb/lb, or ft}$$

in SI units
$$H = \frac{3.66^2}{2 \times 9.807} + \frac{103.4 \times 1000}{9802} + 1.52$$

$$= 0.683 + 10.55 + 1.52$$

$$= 12.75 \text{ N} \cdot \text{m/N, or m}$$

The total head may also be calculated using the expression

$$H = \frac{V^2}{2g} + \text{manometer height} + Z$$

in USCS units
$$H = 2.24 + 34.6 + 5$$

$$= 41.84 \text{ ft} \cdot \text{lb/lb, or ft}$$

in SI units
$$H = 0.683 + 10.55 + 1.52$$

$$= 12.75 \text{ N} \cdot \text{m/N, or m}$$

The total head of 41.84 ft (12.75 m) is equivalent to 1 lb (N) of the liquid raised 41.48 ft (12.75 m) above the datum plane (zero velocity) or the pressure head of 1 lb (N) of the liquid under a column height of 41.84 ft (12.75 m) measured at the datum plane.

The gage pressure p, and consequently the pressure head, are measured relative to atmospheric pressure. Gage pressure head can therefore be a positive or a negative quantity. The pressure may also be expressed as an absolute pressure (measured from complete vacuum). Therefore, when velocity, pressure, and elevation heads are combined to obtain the total energy at a point, it should be clearly stated that the total head is either feet (meters) gage or feet (meters) absolute with respect to the datum plane.

The pressure and/or velocity of a liquid may at times be given as a pressure head of a liquid having a density different from the density of the liquid being pumped. In the total head, that is, the sum of the pressure, velocity, and elevation heads, the components must be corrected to be equal to the head of the liquid being pumped. For example, if the pressure is measured by a manometer to be 24 in (61 cm) of mercury (sp. gr. = 13.6) absolute, the pressure head, or energy, in foot-pounds per pound (newton-meters per newton) of water pumped at 60°F (15.6°C) is found as follows.

Let subscripts 1 and 2 denote different liquids or, in this example, mercury and water, respectively:

$$h_1 = \frac{p_1}{\gamma_1}$$

$$p_1 = h_1\gamma_1 = p_2$$

$$h_2 = \frac{p_2}{\gamma_2} = \frac{h_1\gamma_2}{\gamma_2}$$

$$= \frac{\gamma_1}{\gamma_2} h_1 \tag{2}$$

$$= \frac{\text{sp. gr.}_1}{\text{sp. gr.}_2} h_1 \tag{3}$$

Therefore

in USCS units
$$h_2 = \frac{13.6}{1} \times \frac{24}{12} = 27.2 \text{ ft abs}$$

in SI units
$$h_2 = \frac{13.6}{1} \times \frac{61}{100} = 8.3 \text{ m abs}$$

Also, $h_2 = (27.2 - 13.6)(^{30}\!\!/_{12}) = -6.8$ ft $[(8.3 - 13.6)(^{76}\!\!/_{200}) = -2.04$ m] gage if corrected to a standard barometer of 30 in (76 cm) of mercury.

PUMP TOTAL HEAD

The total head of a pump is the difference between the energy level at the pump discharge (point 2) and that at the pump suction (point 1), as shown in Figs. 7 and 8. Applying Bernoulli's equation (Eq. 1) at each point, the pump total head TH in feet (meters) becomes

$$TH = H_d - H_s = \left(\frac{V_d^2}{2g} + \frac{p_d}{\gamma_d} + Z_d \right) - \left(\frac{V_s^2}{2g} + \frac{p_s}{\gamma_s} + Z_s \right) \qquad (4)$$

The equation for pump differential pressure P_Δ in pounds per square foot (newtons per square meter) is

$$P_\Delta = P_d - P_s = \left[p_d + \gamma_d \left(Z_d + \frac{V_d^2}{2g} \right) \right] - \left[p_s + \gamma_s \left(Z_s + \frac{V_s^2}{2g} \right) \right] \qquad (5)$$

$$(P_\Delta \text{ in lb/in}^2 = P_\Delta \text{ in (lb/ft}^2) \div 144)$$

where the subscripts d and s denote discharge and suction, respectively, and

H = total head of system, $(+)$ or $(-)$ ft (m) gage or $(+)$ ft (m) abs
P = total pressure of system, $(+)$ or $(-)$ lb/ft^2 (N/m^2) gage or $(+)$ lb/ft^2 (N/m^2) abs
V = velocity, ft/s (m/s)
p = pressure, $(+)$ or $(-)$ lb/ft^2 (N/m^2) gage or $(+)$ lb/ft^2 (N/m^2) abs
Z = elevation above $(+)$ or below $(-)$ datum plane, ft (m)
γ = specific weight (force) of liquid, lb/ft^3 (N/m^3)
g = acceleration of gravity, 32.17 ft/s^2 (9.807 m/s^2)

Pump total head TH and pump differential pressure P_Δ are always absolute quantities since either gage pressures or absolute pressures but not both are used at the discharge and suction connections of the pump and a common datum plane is selected.

Pump total head in feet (meters) and pump differential pressure in pounds per square foot (newtons per square meter) are related to each other as

$$TH = \frac{P_\Delta}{\gamma} \qquad (6)$$

It is very important to note that, if the rated total head of a centrifugal pump is given in *feet (meters)*, this head can be imparted to *all* individual liquids pumped at the rated capacity and speed, regardless of the specific weight (force) of the liquids as long as their viscosities are approximately the same. A pump handling different liquids of approximately the same viscosity will generate the same total head but will not produce the same differential pressure, nor will the power required to drive the pump be the same. On the other hand, a centrifugal pump rated in pressure units would have to have a different pressure rating for each liquid of different specific weight (force). In this section, pump total head will be

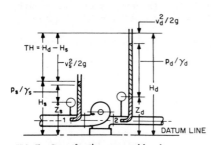

FIG. 7 Centrifugal pump total head.

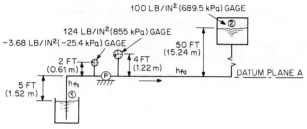

FIG. 8 Example 1.

expressed in feet (meters), the usual way of rating centrifugal pumps. For an explanation of positive displacement pump differential pressure, its use and relationship to pump total head, see Chap. 3.

Pump total head can be measured by installing gages at the pump suction and discharge connections and then substituting these gage readings into Eq. 4. Pump total head may also be found by measuring the energy difference between any two points in the pumping system, one on each side of the pump, providing all losses (other than pump losses) between these points are credited to the pump and added to the energy head difference. Therefore, between any two points in a pumping system where the energy is added only by the pump and the specific weight (force) of the liquid does not change (for example, as a result of temperature), the following general equation for determining pump total head applies:

$$TH = (H_2 - H_1) + \Sigma h_{f(1-2)} \tag{7}$$
$$= \left(\frac{V_2^2}{2g} + \frac{p_2}{\gamma} + Z_2\right) - \left(\frac{V_1^2}{2g} + \frac{p_1}{\gamma} + Z_1\right) + \Sigma h_{f(1-2)}$$

where the subscripts 1 and 2 denote points in the pumping system anyplace upstream and downstream from the pump, respectively, and

H = total head of system, $(+)$ or $(-)$ ft (m) gage or $(+)$ ft (m) abs
V = velocity, ft/s (m/s)
p = pressure, $(+)$ or $(-)$ lb/in^2 (N/m^2) gage or $(+)$ lb/in^2 (N/m^2) abs
Z = elevation above $(+)$ or below $(-)$ datum plane, ft (m)
γ = specific weight (force) of liquid (assumed the same between points), lb/ft^3 (N/m^3)
g = acceleration of gravity, 32.17 ft/s^2 (9.807 m/s^2)
Σh_f = sum of piping losses between points, ft (m)

When the specific gravity of the liquid is known, the pressure head may be calculated from the following relationships:

in feet
$$\frac{p}{\gamma} = \frac{0.016 \text{ lb/ft}^2}{\text{sp. gr.}} \text{ or } \frac{2.3 \text{ lb/in}^2}{\text{sp. gr.}} \tag{8a}$$

in meters
$$\frac{p}{\gamma} = \frac{0.102 \text{ kPa}}{\text{sp. gr.}} \text{ or } \frac{1.02 \times 10^{-3} \text{ bar}}{\text{sp. gr.}} \tag{8b}$$

The velocity in a pipe may be calculated as follows:

in feet per second
$$V = \frac{(\text{gm})(0.408)}{(\text{pipe ID in inches})^2} \tag{9a}$$

in meters per second
$$V = \frac{(\text{m}^3/\text{h})(3.54)}{(\text{pipe ID in cm})^2} \text{ or } \frac{(\text{liters/s})(12.7)}{(\text{pipe ID in cm})^2} \tag{9b}$$

The following example illustrates the use of Eqs. 4 and 7 for determining pump total head.

EXAMPLE 1   A centrifugal pump delivers 1000 gpm (227 m^3/s) of liquid of specific gravity 0.8 from the suction tank to the discharge tank through the piping shown in Fig. 8. (a) Calculate pump total head using gages and the datum plane selected. (b) Calculate total head using the pressures at points 1 and 2 and the same datum plane as (a).

Given: Suction pipe ID $= 8$ in (203 mm), discharge pipe ID $= 6$ in (152 mm), $h_f =$ pipe, valve, and fitting losses, $h_{f_s} = 3$ ft (0.91 mm), $h_{fd} = 25$ ft (7.62 m).

in USCS units   Calculated pipe velocity $= \dfrac{(gpm)(0.408)}{(ID \text{ in inches})^2}$

$$V_s = \frac{1000 \times 0.408}{8^2} = 6.38 \text{ ft/s}$$

$$V_d = \frac{1000 \times 0.408}{6^2} = 11.33 \text{ ft/s}$$

in SI units   Calculated pipe velocity $= \dfrac{(m^3/h)(3.54)}{(ID \text{ in cm})^2}$

$$V_s = \frac{227 \times 3.54}{20.3^2} = 1.95 \text{ m/s}$$

$$V_d = \frac{227 \times 3.54}{15.2^2} = 3.48 \text{ m/s}$$

(a) From Eq. 4,

$$TH = \left(\frac{V_d^2}{2g} + \frac{p_d}{\gamma_d} + Z_d\right) - \left(\frac{V_s^2}{2g} + \frac{p_s}{\gamma_s} + Z_s\right)$$

and Eq. 8,

in USCS units $\dfrac{p}{\gamma} = \dfrac{2.31 \text{ lb/in}^2}{sp. \ gr.}$

in SI units $\dfrac{p}{\gamma} = \dfrac{0.102 \text{ kPa}}{sp. \ gr.}$

Therefore,

in USCS units

$$TH = \left(\frac{11.33^2}{2 \times 32.3} + \frac{2.31 \times 124}{0.8} + 4\right) - \left(\frac{6.38^2}{2 \times 32.2} + \frac{2.31(-3.68)}{0.8} + 2\right)$$
$$= 364 - (-8) = 372 \text{ ft} \cdot \text{lb/lb, or ft}$$

in SI units

$$TH = \left(\frac{3.48^2}{2 \times 9.807} + \frac{0.102 \times 855}{0.8} + 1.22\right) - \left(\frac{1.95^2}{2 \times 9.807} + \frac{0.102(-25.4)}{0.8} + 2\right)$$
$$= 110.9 - (-2.43) = 113.3 \text{ N} \cdot \text{m/N, or m}$$

(b) from Eq. 7,

$$TH = \left(\frac{V_2^2}{2g} + \frac{p_2}{\gamma} + Z_2\right) - \left(\frac{V_1^2}{2g} + \frac{p_1}{\gamma} + Z_1\right) + \Sigma h_{f(1-2)}$$

and Eq. 8,

in USCS units $\dfrac{p}{\gamma} = \dfrac{2.31 \text{ lb/in}^2}{sp. \ gr.}$

in SI units $\dfrac{p}{\gamma} = \dfrac{0.102 \text{ kPa}}{sp. \ gr.}$

Therefore

$$\text{in USCS units} \quad TH = \left(0 + \frac{2.31 \times 100}{0.8} + 50\right) - (0 + 0 - 5) + (3 + 25)$$
$$= 399 - (-5) + 28 = 372 \text{ ft}\cdot\text{lb/lb, or ft}$$

$$\text{in SI units} \quad TH = \left(0 + \frac{0.102 \times 689.5}{0.8} + 15.24\right) - (0 + 0 - 1.52) + (0.91 + 7.62)$$
$$= 103.2 - (-1.52) + 8.53 = 113.3 \text{ N}\cdot\text{m/N, or m}$$

ENERGY AND HYDRAULIC GRADIENTS

The total energy at any point in a pumping system may be calculated for a particular rate of flow using Bernoulli's equation (Eq. 1). If some convenient datum plane is selected and the total energy, or head, at various locations along the system is plotted to scale, the line drawn through these points is called the *energy gradient*. Figure 9 shows the variation in total energy H measured in feet (meters) from the suction liquid surface point 3 to the discharge liquid surface point 4. A horizontal energy gradient indicates no loss of head.

The line drawn through the sum of the pressure and elevation heads at various points represents the pressure variation in flow measured above the datum plane. It also represents the height the liquid would rise in vertical columns relative to the datum plane when the columns are placed at various locations along pipes having positive pressure anywhere in the system. This line, shown dotted in Fig. 9, is called the *hydraulic gradient*. The difference between the energy gradient line and the hydraulic gradient line is the velocity head in the pipe at that point.

The pump total head is the algebraic difference between the total energy at the pump discharge (point 2) and the total energy at the pump suction (point 1).

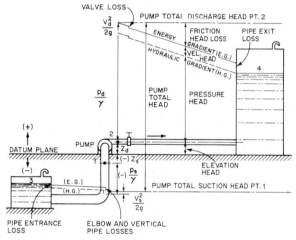

FIG. 9   Energy and hydraulic gradients.

SYSTEM-HEAD CURVES

A pumping system may consist of piping, valves, fittings, open channels, vessels, nozzles, weirs, meters, process equipment, and other liquid-handling conduits through which flow is required for various reasons. When a particular system is being analyzed for the purpose of selecting a pump or pumps, the resistance to flow of the liquid through these various components must be calculated. It will be explained in more detail later in this section that the resistance increases with flow at a rate approximately equal to the square of the flow through the system. In addition to overcoming flow resistance, it may be necessary to add head to raise the liquid from suction level to a higher discharge level. In some systems the pressure at the discharge liquid surface may be higher than the pressure at the suction liquid surface, a condition that requires more pumping head. The latter two heads are *fixed system heads,* as they do not vary with rate of flow. Fixed system heads can also be negative, as would be the case if the discharge level elevation or the pressure above that level were lower than suction elevation or pressure. Fixed system heads are also called *static heads.*

A system-head curve is a plot of total system resistance, variable plus fixed, for various flow rates. It has many uses in centrifugal pump applications. It is preferable to express system head in feet (meters) rather than in pressure units since centrifugal pumps are rated in feet (meters), as previously explained. System-head curves usually show flow in gallons per minute, but when large quantities are involved, the units of cubic feet per second or million gallons per day are used. Although the standard SI units for volumetric flow are cubic meters per second, the units of cubic meters per hour are more common.

When the system head is required for several flows or when the pump flow is to be determined, a system-head curve is constructed using the following procedure. Define the pumping system and its length. Calculate (or measure) the fixed system head, which is the net change in total energy from the beginning to the end of the system due to elevation and/or pressure head differences. An increase in head in the direction of flow is a positive quantity. Next, calculate, for several flow rates, the variable system total head loss through all piping, valves fittings, and equipment in the system. As an example, see Fig. 10, in which the pumping system is defined as starting

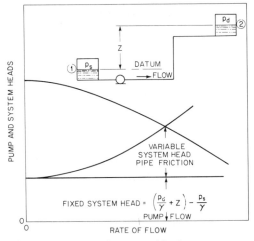

FIG. 10 Construction of system total-head curve.

at point 1 and ending at point 2. The fixed system head is the net change in total energy. The total head at point 1 is p_s/γ, and that at point 2 is $p_d/\gamma + Z$. The pressure and liquid levels do not vary with flow. The variable system head is pipe friction (including valves and fittings). The fixed head and variable heads for several flow rates are added together, resulting in a curve of total system head versus flow.

The flow produced by a centrifugal pump varies with the system head, while the flow of a positive displacement pump is independent of the system head. By superimposing the head-capacity characteristic curve of a centrifugal pump on a system-head curve, as shown in Fig. 10, the flow of a pump can be determined. The curves will intersect at the flow rate of the pump, as this is the point at which the pump head is equal to the required system head for the same flow. When a pump is being purchased, it should be specified that the pump head-capacity curve intersect the system-head curve at the desired flow rate. This intersection should be at the pump's best efficiency capacity or very close to it.

The system-head curve for Example 1 is shown in Fig. 11. This assumes that the suction and

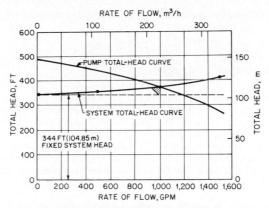

FIG. 11   System total-head curve for Example 1.

discharge liquid levels are 5 ft (1.5 m) below and 50 ft (15 m) above the datum plane, respectively, and do not vary with flow. The pressure in the discharge tank is also independent of flow and is 100 lb/in² (689.5 kPa) gage. These values are therefore fixed system heads. The pipe and fitting losses are assumed to vary with flow as a square function. The length of the pumping system is from point 1 to point 2. The difference in heads at these points plus the frictional losses at various flow rates are the total system head and the head required by a pump for the different flows. It is necessary to calculate the total system head for only one flow rate—say, design—which in this example is 1000 gpm (227 m³/h). The total head at other flow conditions is the fixed system head plus the variable system head multiplied by $(gpm/1000)^2$ $(m^3/h \div 227)^2$. If Example 1 is an existing system, the total head may be calculated by using gages at the pump suction and discharge connections. The total head measured will then be the head at the intersection of the pump and system curves, as shown in Fig. 11. In this example, a correctly purchased pump would produce a total head of 372 ft (113 m) at the design flow of 1000 gpm (227 m³/h).

In systems that are open-ended and in which there is a decrease in elevation from inlet to outlet, a portion of the system-head curve will be negative (Fig. 12). In this example, the pump is used to increase gravity flow. Without a pump in the system, the negative resistance, or static head, is the driving head which moves the liquid through the system. Steady-state gravity flow is sustained at the flow rate corresponding to zero total system head (negative static head plus system resistance equals zero). If a flow is required at any rate greater than that which gravity can produce, a pump is required to overcome the additional system resistance.

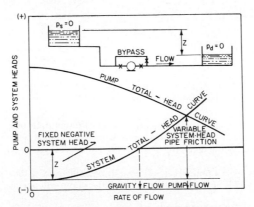

FIG. 12   Construction of system total-head curve to determine gravity flow and centrifugal pump flow.

For additional information concerning the construction of system-head curves for flow in branch lines, refer to Sec. 8.2.

VARIANTS IN PUMPING SYSTEMS

For a fixed set of conditions in a pumping system, there is just one total head for each flow rate. Consequently a centrifugal pump operating at a constant speed can deliver just one flow. In practice, however, conditions in a system vary as a result of either controllable or uncontrollable changes. Changes in the valve opening in the pump discharge or bypass line, changes in the suction or discharge liquid level, changes in the pressures at these levels, the aging of pipes, changes in the process, changes in the number of pumps pumping into a common header, changes in the size, length, or number of pipes are all examples of either controllable or uncontrollable system changes. These changes in system conditions alter the shape of the system-head curve and, in turn, affect pump flow.

Methods of constructing system-head curves and determining the resultant pump flows for two of the more common of these variants are explained here.

Variable Static Head   In a system where a pump is taking suction from one reservoir and filling another, the capacity of a centrifugal pump will decrease with an increase in static head. The system-head curve is constructed by plotting the variable system friction head versus flow for the piping. To this is added the anticipated minimum and maximum static heads (difference in discharge and suction levels). The resulting two curves are the total system heads for each condition. The flow rate of the pump is the point of intersection of the pump head-capacity curve with either one of the latter two system-head curves or with any intermediate system-head curve for other level conditions. A typical head versus flow curve for a varying static head system is shown in Fig. 13.

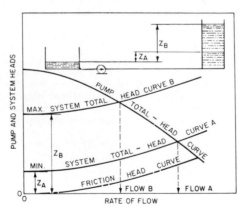

FIG. 13   Construction of system total-head curves for a pumping system having variable static head.

If it is desired to maintain a constant pump flow for different static head conditions, the pump speed can be varied to adjust for an increase or decrease in the total system head. A typical variable-speed centrifugal pump operating in a varying static head system can have a constant flow, as shown in Fig. 14.

It is important to select a pump that will have its best efficiency within the operating range of the system and preferably at the condition at which the pump will operate most often.

Variable System Resistance   A valve or valves in the discharge line of a centrifugal pump alter the variable frictional head portion of the total system-head curve and consequently the pump flow. Figure 15, for example, illustrates the use of a discharge valve to change the

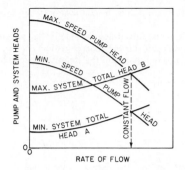

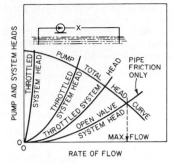

FIG. 14   Varying centrifugal pump speed to maintain constant flow for the different reservoir levels shown in Fig. 13.

FIG. 15   Construction of system total-head curves for various valve openings.

system head for the purpose of varying pump flow during a shop performance test. The maximum flow is obtained with a completely open valve, and the only resistance to flow is the friction in the piping, fittings, and flowmeter. A closed valve results in the pump's operating at shutoff conditions and produces maximum head. Any flow between maximum and shutoff can be obtained by proper adjustment of the valve opening.

DIVIDING TOTAL HEAD AND SYSTEM-HEAD CURVES FOR CENTRIFUGAL PUMPS IN SERIES

Pump limitations or system component requirements may determine that two or more pumps must be used in series. There are practical limitations as to the maximum head that can be developed in a single pump, even if multistaged. When pumping through several system components, there may be pressure limitations which prevent using a single pump to develop all of the head required at the beginning of the system. If several pumps are to be used in series, how should the total head be divided among them?

The sum of the total heads of the pumps must be equal to the required total system head at the design flow. Although mathematically any division of the total head among the pumps to be used is possible (as long as the sum of the pump heads is equal to the total system head), the actual pressure required at various locations along the system flow path determines how the pumping heads are to be divided. An energy or pressure gradient should be drawn for the system. The number of pumps, their locations, and their total heads should be selected to produce the desired pressures (or range of pressures) at critical locations along the system. In addition to considering the pressure loss through components to overcome resistance to flow, consideration should also be given to the minimum pressures required to prevent flashing in piping, cavitation at pump inlets, etc., as well as the maximum working pressures for different parts of the system.

If preferred, the total system can be divided into subsystems, one for each pump (or group of pumps). The end of one subsystem and the beginning of another can be selected anywhere between pumps in series since the pump total head will be unaffected by the division line. Consequently, several system-head curves can be drawn for specification and purchasing purposes—for example, primary condensate pump system, secondary condensate pump system, feed pump system, in a total power plant system.

TRANSIENTS IN SYSTEM HEADS

During the starting of a centrifugal pump and prior to the time normal flow is reached, certain transient conditions can produce or require heads and consequently torques much higher than design. In some cases the selection of the driver and the pump must be based on starting rather than normal flow conditions.

Low- and medium-specific-speed pumps of the radial- and mixed-flow types (less than approximately 5000 specific speed, rpm, gpm, ft units) have favorable starting characteristics. The pump head at shutoff is not significantly higher than that at normal flow, and the shutoff torque is less

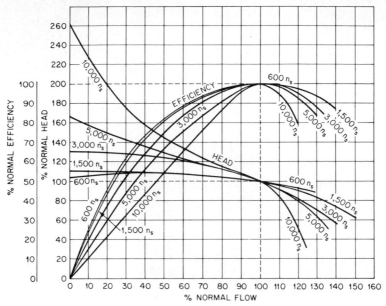

FIG. 16a Approximate comparison of head and efficiency versus flow for impellers of different specific speeds in single-stage volute pumps. Specific speed n_s = rpm [$\sqrt{\text{gpm}}$ per impeller eye$/(TH$ in ft$)^{3/4}$].

To convert to other units:
  rpm, m³/s, m: multiply by 0.01936
  rpm, m³/h, m: multiply by 1.163
  rpm, L/s, m: multiply by 0.6123

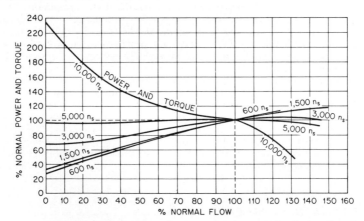

FIG. 16b Approximate comparison of power and torque versus flow for impellers of different specific speeds in single-stage volute pumps.

than that at normal flow. High-specific-speed pumps of the mixed- and axial-flow types (greater than approximately 5000 specific speed) develop relatively high shutoff heads, and their shutoff torque is greater than that at normal flow. These characteristics of high-specific-speed pumps require special attention during the starting period. Characteristics of pumps of different specific speeds are shown in Figs. 16a and 16b.

Starting Against a Closed Valve   When any centrifugal pump is started against a closed discharge valve, the pump head will be higher than normal. The shutoff head will vary with pump specific speed. As shown in Fig. 16a, the higher the specific speed, the higher the shutoff head in percent of normal pump head. As a pump is accelerated from rest to full speed against a closed valve, the head on the pump at any speed is equal to the square of the ratio of the speed to the full speed times the shutoff head at full speed. Therefore, during starting, the head will vary from point A to point E in Fig. 17. Points B, C, and D represent intermediate heads at intermediate

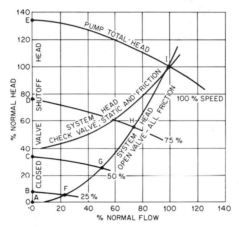

FIG. 17   Variation in head when a centrifugal pump is started with a closed valve, an open valve, and a check valve.

speeds. The pump, the discharge valve, and any intermediate piping must be designed for maximum head at point E.

Pumps requiring less shutoff power and torque than at normal flow condition are usually started against a closed discharge valve. To prevent backflow from a static discharge head prior to starting, either a discharge shutoff valve, a check valve, or a broken siphon is required. When pumps are operated in parallel and are connected to a common discharge header that would permit flow from an operating pump to circulate back through an idle pump, a discharge valve or check valve must be used.

Figure 18 is a typical characteristic curve for a low-specific-speed pump. Figure 19 illustrates the variation of torque with pump speed when the pump is started against a closed discharge valve. The torque under shutoff conditions varies as the square of the ratio of speeds, similar to the variation in shutoff head, and is shown as curve ABC. At zero speed, the pump torque is not zero as a result of static friction in the pump bearings and stuffing box or boxes. This static friction is greater than the sum of running friction and power input to the impeller at very low speeds, which explains the dip in the pump torque curve between 0 and 10% speed. Also shown in Fig. 19 is the speed-torque curve of a typical squirrel-cage induction motor. Note that the difference between motor and pump torque is the excess torque available to accelerate the pump from rest to full speed. During acceleration, the pump shaft must transmit not only the pump torque (curve ABC) but also the excess torque available in the motor. Therefore pump shaft torque follows the motor speed-torque curve less the torque required to accelerate the mass inertia (WK^2) of the motor's rotor.

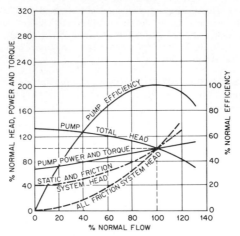

FIG. 18 Typical constant-speed characteristic curves for a low-specific-speed pump.

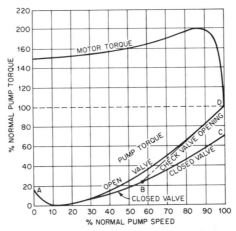

FIG. 19 Variation of torque during start-up of a low-specific-speed pump with a closed valve, an open valve, and a check valve. See Fig. 18 for pump characteristics.

High-specific-speed pumps, especially propeller pumps, requiring more than normal torque at shutoff are not normally started with a closed discharge valve because larger and more expensive drivers would be required. These pumps will also produce relatively high pressures in the pump and in the system between pump and discharge valve. Figure 20 is a typical characteristic curve for a high-specific-speed pump. Curve *ABC* of Fig. 21 illustrates the variation of torque with speed when this pump is started against a closed discharge valve. A typical speed-torque curve of a squirrel-cage induction motor sized for normal pump torque is also shown. Note that the motor has insufficient torque to accelerate to full speed and would remain overloaded at point *C* until the discharge valve on the pump was opened. To avoid this situation, the discharge valve should be timed to open sufficiently to keep the motor from overloading when the pump reaches full speed. To accomplish this timing, it may be necessary to start opening the valve in advance of energizing the motor. Care should be taken not to start opening the discharge valve too soon because this would cause excessive reverse flow through the pump and require the motor to start

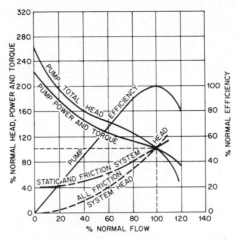

FIG. 20 Typical constant-speed characteristic curves for a high-specific-speed pump.

under adverse reverse-speed conditions. If the driver is a synchronous motor, additional torque at pull-in speed is required, over and above that needed to overcome system head and accelerate the pump and driver rotors from rest. At the critical pull-in point, sufficient torque must be available to pull the load into synchronism in the prescribed time. A synchronous motor is started on low-torque squirrel-cage windings prior to excitation of the field windings at the pull-in speed. The low starting torque, the torque required at pull-in, and possible voltage drop, which will lower the motor torque (varies as the square of the voltage), must all be taken into consideration when selecting a synchronous motor to start a high-specific-speed pump against a closed valve.

If a high-specific-speed pump is to be started against a closed discharge valve, high starting torques can also be avoided by the use of a bypass valve (see Sec. 8.2) or by an adjustable-blade pump.[1]

Starting Against a Check Valve　A check valve can be used to prevent reverse flow from static head and/or head from other pumps in the system. The check valve will open automatically when the head from the pump exceeds system head. When a centrifugal pump is started against a check valve, pump head and torque follow shutoff values until a speed is reached at which shutoff head exceeds system head. As the valve opens, the pump head continues to increase, and, at any flow, the head will be that necessary to overcome system static head or head from other pumps, frictional head, valve head loss, and the inertia of the liquid being pumped.

Figure 19, curve *ABD*, illustrates speed-torque variation when a low-specific-speed pump is started against a check valve with static head and system friction as shown in Fig. 18. Figure 21, curve *ABD*, illustrates speed-torque variation for a high-specific-speed pump started against a check valve with static head and system friction as shown in Fig. 20. The use of a quick-opening check valve with high-specific-speed pumps eliminates starting against higher than full-open-valve shutoff heads and torques.

The speed-torque curves shown for the period during the acceleration of the liquid in the system have been drawn with the assumption that the head required to accelerate the liquid and overcome inertia is insignificant. Acceleration head is discussed in more detail later.

Starting Against an Open Valve　If a centrifugal pump is to take suction from a reservoir and discharge to another reservoir having the same liquid elevation or the same equivalent total pressure, it can be started without a shutoff discharge valve or check valve. The system-head curve is essentially all frictional plus the head required to accelerate the liquid in the system during the starting period. Neglecting liquid inertia, the pump head would not be greater than normal at any speed during the starting period, as shown in curve *AFGHI* of Fig. 17. Pump torque would

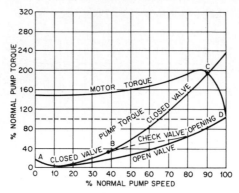

FIG. 21 Variation of torque during start-up of a high-specific-speed pump with a closed valve, an open valve, and a check valve. See Fig. 20 for pump characteristics.

not be greater than normal at any speed during the starting period, as shown in Figs. 19 and 21, curves *AD*. Pump head and torque at any speed are equal to their values at normal condition times the square of the ratio of the speed to full speed, while the capacity varies directly with this ratio.

Starting a Pump Running in Reverse

When a centrifugal pump discharges against a static head or into a common discharge header with other pumps and is then stopped, the flow will reverse through the pump unless the discharge valve is closed or unless there is a check valve in the system or a broken siphon in a siphon system. If the pump does not have a nonreversing device, it will turn in the reverse direction. A pump that discharges against a static head through a siphon system without a valve will have reverse flow and speed when the siphon is being primed prior to starting.

Figures 22 and 23 illustrate typical reverse-speed–torque characteristics for a low- and a high-specific-speed pump. When flow reverses through a pump and the driver offers very little or no torque resistance, the pump will reach higher than normal forward speed in the reverse direction. This runaway speed will increase with specific speed and system head. Shown in Figs. 22 and 23 are speed, torque, head, and flow, all expressed as a percentage of the pump design conditions for the normal forward speed. When a pump is running in reverse as a turbine under no load, the

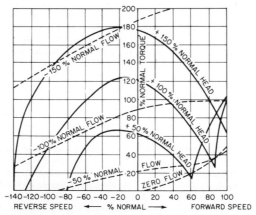

FIG. 22 Typical reverse-speed–torque characteristics of a low-specific-speed, radial-flow, double-suction pump.

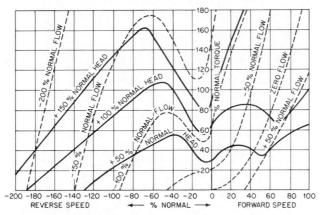

FIG. 23   Typical reverse-speed–torque characteristics of a high-specific-speed, axial-flow, diffuser pump.

head on the pump will be static head (or head from other pumps) minus head loss as a result of friction due to the reverse flow.

If an attempt is made to start the pump while it is running in reverse, an electric motor must apply positive torque to the pump while the motor is initially running in a negative direction. Figures 22 and 23 show, for the two pumps, the torques required to decelerate, momentarily stop, and then accelerate the pump to normal speed. If either of these pumps were pumping into an all-static-head system, starting it in reverse would require overcoming 100% normal head, and it can be seen that a torque in excess of normal would be required by the driver while the driver is running in reverse. In addition to overcoming positive head, the driver must add additional torque to the pump to change the direction of the liquid. This could result in a prolonged starting time under higher than normal current demand. Characteristics of the motor, pump, and system must be analyzed together to determine actual operating conditions during this transient period. Starting torques requiring running in reverse become less severe when the system head is partly or all friction head.

Inertial Head   If the system contains an appreciable amount of liquid, the inertia of the liquid mass could offer a significant resistance to any sudden change in velocity. Upon starting a primed pump and system without a valve, all the liquid in the system must accelerate from rest to a final condition of steady flow. Figure 24 illustrates a typical system head resistance that could be produced by a propeller pump pumping through a friction system when accelerated from rest to full speed. If the pump were accelerated very slowly, it would produce an all-friction resistance with zero liquid acceleration varying with flow approaching curve *OABCDEF*. Individual points on this curve represent system resistance at various constant pump speeds. If the pump were accelerated very rapidly, it would produce a system resistance approaching curve *OGHIJKL*. This is a shutoff condition which cannot be realized unless infinite driver torque is available. Individual points on this curve represent a system resistance at various constant speeds with no flow, which is the same as operating with a closed discharge valve. Curve *LF* represents the maximum total head the pump can produce as a result of both system friction and inertia at 100% speed. Head variation from *L* to *F* is a result of flow through the system, increasing at a decreasing rate of acceleration and increasing friction to the normal operating point *F*, where all the head is frictional. The actual total system-head resistance curve for any flow condition will be, therefore, the sum of the frictional resistance in feet (meters) for that steady-flow rate plus the inertial resistance, also expressed in feet (meters). The inertial resistance at any flow is dependent on the mass of liquid and the instantaneous rate of change of velocity at the flow condition. A typical total system resistance for a motor driven propeller pump is shown as *OMNPQRF* in Fig. 24. For a particular pump and system, different driver-speed–torque characteristics will result in a family of curves in the area *OLF*.

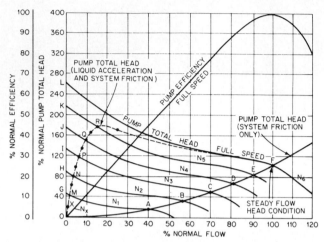

FIG. 24   Transient system head as a result of liquid acceleration and system friction when a propeller pump is started with an open valve.

The added inertial system head produced momentarily when high-specific-speed pumps are started is important when considering the duration of high driver torques and currents, the pressure rise in the system, and the effect on the pump of operation at high heads and low flows. The approximate acceleration head h_a required to change the velocity of a mass of liquid at a uniform rate and cross section is

$$h_a = \frac{L \, \Delta V}{g \, \Delta t} \tag{10}$$

where L = length of constant-cross-section conduit, ft (m)
    ΔV = velocity change, ft/s (m/s)
     g = acceleration of gravity, 32.17 ft/s^2 (9.807 m/s^2)
    Δt = time interval, s

To calculate the time to accelerate a centrifugal pump from rest or from some initial speed to a final speed, and to estimate the pump head variation during this interim, a trial-and-error solution may be used. Divide the speed change into several increments of equal no-flow heads, such as $OGHIKL$ in Fig. 24. For the first incremental speed change, point O to point N_1, estimate the total system head, point M, between G and A. Next estimate the total system head, point X, at the average speed for this first incremental speed change. These points are shown in Fig. 24. Calculate the time in seconds for this incremental speed change to take place using the equation

in USCS units $\Delta t = \dfrac{\Sigma W K^2 \, \Delta N}{307(T_D - T_P)}$ (11a)

where $\Sigma W K^2$ = total pump and motor rotor weight moment of inertia, W = weight (force), K
                 = radius of gyration, lb·ft^2
    ΔN = incremental speed change, rpm
    T_D = motor torque at average speed, ft·lb
    T_P = pump torque at average speed, ft·lb

in SI units $\Delta t = \dfrac{\Sigma M K^2 \, \Delta N}{9.55(T_D - T_P)}$ (11b)

where $\Sigma M K^2$ = total pump and motor rotor mass moment of inertia, M = weight (mass), K =
                 radius of gyration, kg·m^2
    ΔN = incremental speed change, rpm

T_D = motor torque at average speed, N·m

T_P = pump torque at average speed, N·m

In SI units, when diameter of gyration D is used rather than radius of gyration K, and $MD^2 = 4MK^2$, then

$$\Delta t = \frac{\Sigma MD^2 \,\Delta N}{38.2(T_D - T_P)} \qquad (11c)$$

Calculate the acceleration head required to change the flow in the system from point O to point M using Eq. 10 and time from Eq. 11. Add acceleration head to frictional head at the assumed average flow, and if this value is correct, it will fall on the average pump head-capacity curve, point X. Adjust points M and X until these assumed flows result in the total acceleration and frictional heads agreeing with flow X at the average speed. Repeat this procedure for other increments of speed change, adding incremental times to get total accelerating time to bring the pump up to its final speed. Plot system head versus average flow for each incremental speed change during this transient period, as shown in Fig. 24.

Figures 25 and 26 illustrate how driver and pump torques can be determined from their

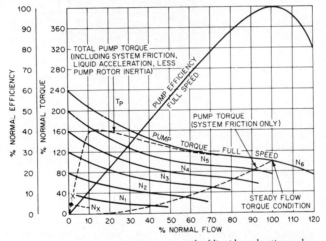

FIG. 25 Transient total pump torque as a result of liquid acceleration and system friction when a propeller pump is started with an open valve. Pump head characteristics are shown in Fig. 24.

respective speed-torque curves. Figure 25 is a family of curves which represent the torques required to produce flow against different heads without acceleration of the liquid or pump for the various speeds selected. Pump torque for any reduced speed can be calculated from the full-speed curve using the relation that torque varies as the second power and flow varies as the first power of the speed ratio. Point X is the torque at the average speed and the trial average flow during the first incremental speed change, which is adjusted for different assumed conditions. Figure 26 shows a typical squirrel-cage induction motor speed-torque curve, and, for the purpose of illustration, it has been selected to have the same torque rating as the pump requires at full speed (approximately 97% synchronous speed). Point X in this figure is the motor torque at the average speed during the first incremental speed change. In Fig. 26, the developed torque curve for the different speeds shown in Fig. 25 is redrawn as the total frictional and inertial pump torque T_P. For the conditions used in this example, and as shown in Fig. 26, after approximately 88% synchronous speed very little excess torque $(T_D - T_P)$ is available for accelerating the pump and motor rotor inertia. However, as long as an induction motor has adequate torque to drive the pump at the normal condition, full speed will be reached providing the time-current demand can be tolerated. A synchronous motor, on the other hand, may not have sufficient torque to pull into

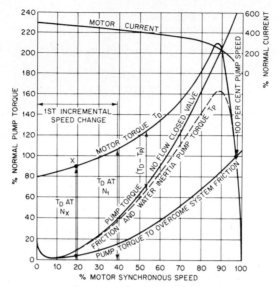

FIG. 26 Propeller pump and motor speed-torque curves, showing effect of accelerating the liquid in the system during the starting period.

step. Liquid acceleration decreases as motor torque decreases, adjusting to the available excess motor torque. The pump speed changes very slowly during the final period. Figure 26 also shows the motor current for the different speeds and torques.

A motor having a higher than normal pump torque rating or a motor having high starting torque characteristics will reduce the starting time, but higher heads will be produced during the acceleration period. Note, however, that heads produced cannot exceed shutoff at any speed. The total torque input to the pump shaft is equal to the sum of the torques required to overcome system friction, liquid inertia, and pump rotor inertia during acceleration. Torque at the pump shaft will therefore follow the motor speed-torque curve less the torque required to overcome the motor's rotor inertia. To reduce the starting inertial pump head to an acceptable amount, if desired, other alternative starting schemes can be used. A short bypass line from the pump discharge back to the suction can be provided to divert flow from the main system. The bypass valve is closed slowly after the motor reaches full speed. A variable-speed or a two-speed motor will reduce the inertial head by controlling motor torque and speed, thereby increasing the accelerating time.

This procedure for developing the actual system head and pump torque, including liquid inertia, becomes more complex if the pump must employ a discharge valve. To avoid high pump starting heads and torques, the discharge valve must be partially open on starting and then opened a sufficient amount before full speed is reached. The valve resistance must be added to the system friction and inertia curves if an exact solution is required.

Siphon Head Between any two points having the same elevation in a pumping system, no head is lost because of piping elevation changes since the net change in elevation is zero. If the net change in elevation between two points is not zero, additional pump head is required if there is an increase in elevation and less head is required if there is a decrease in elevation.

When piping is laid over and under obstacles with no net change in elevation, no pumping head is required to sustain flow other than that needed to overcome frictional and minor losses. As the piping rises, the liquid pressure head is transformed to elevation head, and the reverse takes place as the piping falls. A pipe or other closed conduit that rises and falls is called a *siphon*, and one that falls and rises is called an *inverted siphon*. The siphon principle is valid provided the conduit flows full and free of liquid vapor and air so that the densities of the liquid columns are alike. It is this requirement that determines the limiting height of a siphon for complete recovery because the liquid can vaporize under certain conditions.

Pressure in a siphon is minimum at the summit, or just downstream from it, and Bernoulli's equation can be used to determine if the liquid pressure is above or below vapor pressure. Referring to Fig. 27, observe the following. The absolute pressure head H_S in feet (meters) at the top of the siphon is

$$H_S = H_B - Z_S + h_{f(S-2)} - \frac{V_S^2}{2g} \qquad (12)$$

where H_B = barometric pressure head of liquid pumped, ft (m)
Z_S = siphon height to top of conduit ($= Z_1$ if no seal well is used), ft (m)
$h_{f(S-2)}$ = frictional and minor losses from S to 2 (or 3 if no seal well is used), including exit velocity head loss at 2 (or 3), ft (m)
$V_S^2/2g$ = velocity head at summit, ft (m)

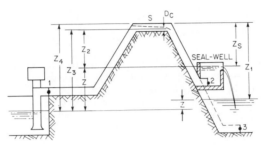

FIG. 27   Pumping system using a siphon for head recovery.

The absolute pressure head at the summit can also be calculated using conditions in the up leg by adding the barometric pressure head to the pump head (TH) and deducting the distance from suction level to the top of the conduit (Z_4), the frictional loss in the up leg ($h_{f(1-S)}$), and the velocity head at the summit. If the suction level is higher than the discharge level and flow is by gravity, the absolute pressure head at the summit is found as above and $TH = 0$.

Whenever Z_1 in Fig. 27 is so high that it exceeds the maximum siphon capability, a seal well is necessary to increase the pressure at the top of the siphon above vapor pressure. Note $Z_1 - Z_s$ represents an unrecoverable head and increases the pumping head. Water has a vapor pressure of 0.77 ft (23.5 cm) at 68°F (20°C) and theoretically a 33.23-ft-high (10.13-cm) siphon is possible with a 34-ft (10.36-m) water barometer. In practice higher water temperatures and lower barometric pressures limit the height of siphons used in condenser cooling water systems to 26 to 28 ft (8 to 8.5 m). The siphon height can be found by using Eq. 12 and letting H_S equal the vapor pressure in feet (meters).

In addition to recovering head in systems such as condenser cooling water, thermal dilution, and levees, siphons are also used to prevent reverse flow after pumping is stopped by use of an automatic vacuum breaker located in the summit. Often siphons are used solely to eliminate the need for valves or flap gates.

In open-ended pumping systems, siphons can be primed by external means of air removal. Unless the siphon is primed initially upon starting, a pump must fill the system and provide a minimum flow to induce siphon action. During this filling period and until the siphon is primed, the siphon head curve must include this additional siphon filling head, which must be provided by the pump. Pumps in siphon systems are usually low-head, and they may not be capable of filling the system to the top of the siphon or of filling it with adequate flow. Low-head pumps are high-specific-speed and require more power at reduced flows than during normal pumping. Figure 28 illustrates the performance of a typical propeller pump when priming a siphon system and during normal operation.

When a pump and driver are to be selected to prime a siphon system, it is necessary to estimate the pump head and the power required to produce the minimum flow needed to start the siphon.

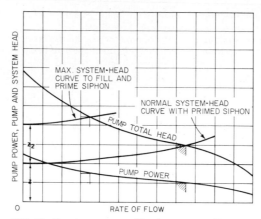

FIG. 28   Transient system total head priming a siphon.

The minimum flow required increases with the length and the diameter and decreases with the slope of the down-leg pipe.[2] Prior to the removal of all the air in the system, the pump is required to provide head to raise the liquid up to and over the siphon crest. Head above the crest is required to produce a minimum flow similar to flow over a broad-crested weir. This weir head may be an appreciable part of the total pump head if the pump is low-head and large-capacity. A conservative estimate of the pump head would include a full conduit aboved the siphon crest. In reality, the down leg must flow partially empty before it can flow full, and it is accurate enough to estimate that the depth of liquid above the siphon crest is at critical depth for the cross section. Table 1 can be used to estimate critical depth in circular pipes, and Fig. 29 can be used to calculate the cross-sectional area of the filled pipe to determine the velocity at the siphon crest.

Until all the air is removed and all the piping becomes filled, the down leg is not part of the pumping system, and its frictional and minor losses are not to be added to the maximum system-head curve to fill and prime the siphon shown in Fig. 28. The total head TH in feet (meters) to be produced by the pump in Fig. 27 until the siphon is primed is

$$TH = Z_3 + h_{f(1-S)} + \frac{V_C^2}{2g} \tag{13}$$

TABLE 1   Values for Determining Pipe-Diameter Ratio vs. $(ft^3/s)/d^{5/2}$ in Circular Pipes

$\dfrac{D_{crit}}{d}$	0.00	0.01	0.02	0.03	0.04	0.05	0.06	0.07	0.08	0.09
0.0	.....	0.0006	0.0025	0.0055	0.0098	0.0153	0.0220	0.0298	0.0389	0.0491
0.1	0.0605	0.0731	0.0868	0.1016	0.1176	0.1347	0.1530	0.1724	0.1928	0.2144
0.2	0.2371	0.2609	0.2857	0.3116	0.3386	0.3666	0.3957	0.4259	0.4571	0.4893
0.3	0.523	0.557	0.592	0.628	0.666	0.704	0.743	0.784	0.825	0.867
0.4	0.910	0.955	1.000	1.046	1.093	1.141	1.190	1.240	1.291	1.343
0.5	1.396	1.449	1.504	1.560	1.616	1.674	1.733	1.792	1.853	1.915
0.6	1.977	2.041	2.106	2.172	2.239	2.307	2.376	2.446	2.518	2.591
0.7	2.666	2.741	2.819	2.898	2.978	3.061	3.145	3.231	3.320	3.411
0.8	3.505	3.602	3.702	3.806	3.914	4.023	4.147	4.272	4.406	4.549
0.9	4.70	4.87	5.06	5.27	5.52	5.81	6.18	6.67	7.41	8.83

All tabulated values are in units of $(ft^3/s)/d^{5/2}$; d = diameter, ft; D_{crit} = critical depth, ft. For SI units, multiply $(m^3/h)/m^{5/2}$ by 5.03×10^{-4} to obtain $(ft^3/s)/ft^{5/2}$ Examples 2, 3, and 4 illustrate the use of this table.

SOURCE: Ref 13.

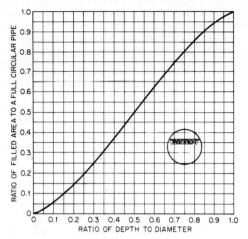

FIG. 29 Area versus depth for a circular pipe.

In SI units:

$$R = \frac{\rho VD}{\mu} \quad (\rho \text{ in kg/m}^3, \ V \text{ in m/s}, \ D \text{ in m}, \ \mu \text{ in N·s/m}^2)$$

1 meter = 3.28 feet

1 VD (V in m/s, D in m) = 129.2 VD'' (V in ft/s, D'' in inches)

where Z_3 = distance between suction level and centerline liquid at siphon crest, ft (m)
$h_{f(1-S)}$ = frictional and minor losses from 1 to S, ft (m)
$V_C^2/2g$ = velocity head at crest using actual liquid depth, approx. critical depth, ft (m)

Use of Eq. 13 permits plotting the maximum system-head curve to fill and prime the siphon for different flow rates. The pump priming flow is the intersection of the pump total head curve and this system-head curve. The pump selected must have a driver with power as shown in Fig. 28 to prime the system during this transient condition.

For the pumping system shown in Fig. 27, after the system is primed, the pump total head reduces to

$$TH = Z + h_{f(1-2)} \tag{14}$$

where Z = distance between suction and seal well levels (or discharge pool level if no seal well is used), ft (m)
$h_{f(1-2)}$ = frictional and minor losses from 1 to 2 (or 3 if no seal well is used), including exit velocity head loss at 2 (or 3), ft (m)

Use of Eq. 14 permits plotting of the normal system-head curve with a primed siphon, shown in Fig. 28, for different flow rates. The normal pump flow is the intersection of the pump total head curve and the normal system-head curve.

If the pump cannot provide sufficient flow to prime the siphon, or if the driver does not have adequate power, the system must be primed externally by a vacuum or jet pump. An auxiliary priming pump can also be used to continuously vent the system, as it is necessary that this be done to maintain full siphon recovery. In some systems the water pumped is saturated with air, and as the liquid flows through the system, the pressure is reduced (and in cooling systems the temperature is increased). Both these conditions cause the release of some of the entrained air. Air will accumulate at the top of the siphon and in the upper parts of the down leg. The siphon works on the principle that an increase in elevation in the up leg produces a decrease in pressure and an equal decrease in elevation in the down leg results in recovery of this pressure. This cannot occur if the density of the liquid in the down leg is decreased as a result of the formation of air pockets. These air pockets also restrict the flow area. A release of entrained air and air leakage into the system through pipe joints and fittings will result in a centrifugal pump's delivering less than design flow, as the head will be higher than estimated. Also, high-specific-speed pumps with rising power curves toward shutoff can become overloaded. In order to maintain full head recovery, it is necessary to continuously vent the siphon at the top and at several points along the down leg, especially at the beginning of a change in slope.[3] These venting points can be manifolded together and connected to a single downward venting system.

Some piping systems may contain several up and down legs, i.e., several siphons in series. Each down leg, as in a single siphon, is vulnerable to air and/or vapor binding. The likelihood of flow reduction, and conceivably in some cases complete flow shutoff, is increased in a multiple siphon system.[3] As shown in Fig. 30, system static head is increased if proper venting is not provided at

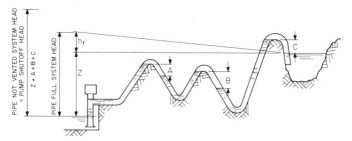

FIG. 30 Multiple siphon system (Z = normal system static head when pipe is flowing full). Pump flow is stopped when normal static head plus sum of air pocket heights equals pump shutoff head.

the top of each siphon. Normally the static head is the difference between outlet and inlet elevations. If air pockets exist, head cannot be recovered and the normal static head is increased by the sum of the heights of all the intermediate liquidless pockets. Flow will stop when the total static head equals the pump shutoff head.

The following examples illustrate the use of Eqs. 3, 9, 12, 13, and 14, Table 1, and Fig. 29.

EXAMPLE 2   A pump is required to produce a flow of 70,000 gpm (15,900 m³/h) through the system shown in Fig. 27. Calculate the system total head from point 1 to point 3 (no seal well) under the following conditions:

Specific gravity = 0.998 for 80°F (26.7°C) water

Barometric pressure = 29 in (73.7 cm) mercury abs (sp. gr. 13.6)

Suction and discharge water levels are equal, $Z = 0$

$Z_1 = 40$ ft (12.2 m)

$h_{f(1-S)} = 3$ ft (0.91 m) up-leg frictional head

$h_{f(S-3)} = 3.3$ ft (1 m) down-leg frictional head, including exit loss

Pipe diameter = 48 in (121.9 cm) ID

Water vapor pressure = 0.507 lb/in² (3.5 kPa) abs at 80°F (26.7°C)

The maximum siphon height may be found from Eq. 12:

$$Z_S = H_B + h_{f(S-3)} - H_S - \frac{V_S^2}{2g}$$

From Eq. 3

$$H_B = \text{barometric pressure head in feet (meters) of liquid pumped}$$

in USCS units
$$\frac{\text{sp. gr.}_1}{\text{sp. gr.}_2} h_1 = \frac{13.6}{0.998} \times \frac{29}{12} = 32.9 \text{ ft abs}$$

$$H_S = \frac{p}{\gamma} = \frac{144 \times 0.507}{62.19} = 1.17 \text{ ft abs}$$

in SI units
$$\frac{\text{sp. gr.}_1}{\text{sp. gr.}_2} h_1 = \frac{13.6}{0.998} \times \frac{73.7}{100} = 10 \text{ m abs}$$

$$H_S = \frac{p}{\gamma} = \frac{3.5 \times 1000}{9769} = 0.358 \text{ m abs}$$

From Eqs. 9 and 12,

in USCS units
$$V_S = \frac{\text{gpm}}{(\text{pipe ID in inches})^2} \times 0.408 = \frac{70,000}{48^2} \times 0.408 = 12.4 \text{ ft/s}$$

$$Z_S = 32.9 + 3.3 - 1.17 - \frac{12.4^2}{2 \times 32.17} = 32.63 \text{ ft}$$

in SI units
$$V_S = \frac{\text{m}^3/\text{h}}{(\text{pipe ID in cm})^2} \times 3.54 = \frac{15,900}{121.9^2} \times 3.54 = 3.79 \text{ m/s}$$

$$Z_S = 10 + 1 - 0.358 - \frac{3.79^2}{2 \times 9.807} = 9.1 \text{ m}$$

Since the maximum height is exceeded ($Z_1 > Z_S$), siphon recovery is not possible. The system total head is therefore found from Eq. 13:

$$TH = Z_3 + h_{f(1-S)} + \frac{V_c^2}{2g}$$

The critical depth D_{crit} is found using Table 1:

in USCS units
$$\text{ft}^3/\text{s} = \frac{70,000}{7.481 \times 60} = 156$$

$$\frac{\text{ft}^3/\text{s}}{d^{5/2}} = \frac{156}{4^{5/2}} = 4.88$$

$$\frac{D_{\text{crit}}}{d} = 0.901$$

$$D_{\text{crit}} = 0.901 \times 4 = 3.6 \text{ ft}$$

in SI units
$$\frac{\text{m}^3/\text{h}}{d^{5/2}} = \frac{15,900}{1.219^{5/2}} = 9695$$

Convert to USCS units (see footnote to Table 1):

$$\frac{\text{ft}^3/\text{s}}{\text{ft}^{5/2}} = 9695 \times 5.03 \times 10^{-4} = 4.88$$

$$\frac{D_{\text{crit}}}{d} = 0.901$$

$$D_{\text{crit}} = 0.901 \times 1.219 = 1.1 \text{ m}$$

To calculate the water velocity at the siphon crest, determine the area of the filled pipe. From Fig. 29, ratio of filled area to area of a full pipe is 0.95 for a depth-to-diameter ratio of 0.901:

in USCS units
$$V_c = \frac{\text{gpm}}{0.95(\text{ID in inches})^2} \times 0.408 = \frac{70,000}{0.95 \times 48^2} \times 0.408 = 13.0 \text{ ft/s}$$

$$Z_3 = 40 - 4 + \frac{3.6}{2} = 37.8 \text{ ft}$$

From Eq. 13
$$TH = 37.8 + 3 + \frac{13.0^2}{2 \times 32.17} = 43.43 \text{ ft}$$

in SI units
$$V_c = \frac{\text{m}^3/\text{h}}{0.95(\text{ID in cm})^2} \times 3.54 = \frac{15,900}{0.95 \times 121.9^2} \times 3.54 = 3.99 \text{ m/s}$$

$$Z_3 = 12.2 - 1.219 + \frac{1.1}{2} = 11.53 \text{ m}$$

From Eq. 13
$$TH = 11.53 + 0.91 + \frac{3.99^2}{2 \times 9.807} = 13.25 \text{ m}$$

EXAMPLE 3 Calculate the minimum total system head using conditions in Example 2 and a seal well, as shown in Fig. 27. Use 2.8 ft (0.853 m) for the frictional head loss $h_{f(S-2)}$.

The maximum siphon height Z_S in Example 2 was found to be 32.63 ft (9.95 m). Therefore from Eq. 14 the total system head after priming is

$$TH = Z + h_{f(1-2)}$$

in USCS units
$$h_{f(1-2)} = h_{f(1-S)} + h_{f(S-2)} = 3 + 2.8 = 5.8 \text{ ft}$$

$$Z = Z_1 - Z_S = 40 - 32.63 = 7.37 \text{ ft}$$

in SI units
$$h_{f(1-2)} = h_{f(1-S)} + h_{f(S-2)} = 0.91 + 0.853 = 1.763 \text{ m}$$

$$Z = Z_1 - Z_S = 12.2 - 9.95 = 2.25 \text{ m}$$

Note that the seal well elevation is above discharge level. Therefore

in USCS units
$$TH = 7.37 + 5.8 = 13.17 \text{ ft}$$

in SI units
$$TH = 2.25 + 1.763 = 4.01 \text{ m}$$

EXAMPLE 4 The dimensions of the down leg in Example 3 require a minimum velocity of 5 ft/s (1.52 m/s) flowing full to purge air from the system and start the siphon. Calculate the system head the pump must overcome to prime the siphon.

in USCS units
$$\text{gpm} = \frac{V(\text{pipe ID in inches})^2}{0.408} = \frac{5 \times 48^2}{0.408} = 28,200$$

$$\text{ft}^3/\text{s} = 28,200 \div 449 = 62.8$$

in SI units
$$\text{m}^3/\text{h} = \frac{V(\text{pipe ID in cm})^2}{3.54} = \frac{1.52 \times 121.9^2}{3.54} = 6400$$

The critical depth is found from Table 1:

in USCS units
$$\frac{\text{ft}^3/\text{s}}{d^{5/2}} = \frac{62.8}{4^{5/2}} = 1.97$$

$$\frac{D_{\text{crit}}}{d} = 0.6$$

$$D_{\text{crit}} = 0.6 \times 4 = 2.4 \text{ ft}$$

in SI units
$$\frac{\text{m}^3/\text{h}}{d^{5/2}} = \frac{6400}{1.219^{5/2}} = 3902$$

Convert to USCS units (see footnote to Table 1):

$$\frac{\text{ft}^3/\text{s}}{d^{5/2}} = 3902 \times 5.03 \times 10^{-4} = 1.97$$

$$\frac{D_{crit}}{d} = 0.6$$

$$D_{crit} = 0.6 \times 1.219 = 0.73 \text{ m}$$

From Fig. 29, the ratio of the filled area to the area of a full pipe is 0.625 for a depth-to-diameter ratio of 0.60:

in USCS units
$$V_c = \frac{5}{0.625} = 8.0 \text{ ft/s}$$

in SI units
$$V_c = \frac{1.52}{0.625} = 2.4 \text{ m/s}$$

From Eq. 13
$$TH = Z_3 + h_{f(1-S)} + \frac{V_c^2}{2g}$$

in USCS units
$$Z_2 = Z_S - 4 + \frac{D_{crit}}{2} = 32.63 - 4 + \frac{2.4}{2} = 29.83 \text{ ft}$$

$$Z = 7.37 \text{ ft} \quad \text{(from Example 3)}$$

$$h_{f(1-S)} \text{ at } 28,200 \text{ gpm} = \left(\frac{28,200}{70,000}\right)^2 \times 3 = 0.49 \text{ ft}$$

$$TH = (7.37 + 29.83) + 0.49 + \frac{8^2}{2 \times 32.17} = 38.68 \text{ ft}$$

in SI units
$$Z_2 = Z_S - 1.219 + \frac{D_{crit}}{2} = 9.95 - 1.219 + \frac{0.73}{2} = 9.1 \text{ m}$$

$$Z = 2.25 \text{ m} \quad \text{(from Example 3)}$$

$$h_{f(1-S)} \text{ at } 6400 \text{ m}^3/\text{h} = \left(\frac{6400}{15,900}\right)^2 \times 0.91 = 0.15 \text{ m}$$

$$TH = (2.25 + 9.1) + 0.15 + \frac{2.4^2}{2 \times 9.807} = 11.79 \text{ m}$$

If the system is not externally primed, the centrifugal pump selected must be able to deliver at least 28,000 gpm (6400 m³/h) at 38.68 ft (11.79 m) total head and must be provided with a driver having adequate power for this condition. After the system is primed, the pump must be capable of delivering at least 70,000 gpm (15,900 m³/h) at 13.17 ft (4.01 m) (see Fig. 28).

HEAD LOSSES IN SYSTEM COMPONENTS

Pressure Pipes   Resistance to flow through a pipe is caused by viscous shear stresses in the liquid and by turbulence at the pipe walls. *Laminar* flow occurs in a pipe when the average velocity is relatively low and the energy head is lost mainly as a result of viscosity. In laminar flow, liquid particles have no motion next to the pipe walls and flow occurs as a result of the movement of particles in parallel lines with velocity increasing toward the center. The movement of concentric cylinders past each other causes viscous shear stresses, more commonly called *friction*. As flow increases, the flow pattern changes, the average velocity becomes more uniform, and there is less viscous shear. As the laminar film decreases in thickness at the pipe walls and as the flow increases, the pipe roughness becomes important since it causes turbulence. *Turbulent* flow occurs when average pipe velocity is relatively high and energy head is lost predominantly because of turbulence caused by the wall roughness. The average velocity at which the flow changes from laminar to turbulent is not definite, and there is a critical zone in which either laminar or turbulent flow can occur.

Viscosity can be visualized as follows. If the space between two planar surfaces is filled with a liquid, a force will be required to move one surface at a constant velocity relative to the other. The velocity of the liquid will vary linearly between the surfaces. The ratio of the force per unit area, called *shear stress*, to the velocity per unit distance between surfaces, called *shear* or *deformation rate*, is a measure of a liquid's *dynamic* or *absolute viscosity*.

Liquids such as water and mineral oil, which exhibit shear stresses proportional to shear rates, have a constant viscosity for a particular temperature and pressure and are called *Newtonian* or *true liquids*. In the normal pumping range, however, the viscosity of true liquids may be considered independent of pressure. For these liquids the viscosity remains constant since the rate of deformation is directly proportional to the shearing stress. The viscosity and resistance to flow, however, increase with decreasing temperature.

Liquids such as molasses, grease, starch, paint, asphalt, and tar behave differently from Newtonian liquids. The viscosity of the former does not remain constant and their shear, or deformation, rate increases more than the stress increases. These liquids, called *thixotropic*, exhibit lower viscosity as they are agitated at a constant temperature.

Still other liquids, such as mineral slurries, show an increase in viscosity as the shear rate is increased and are called *dilatant*.

In USCS units, dynamic (absolute) viscosity is measured in pound-seconds per square foot or slugs per foot-second. In SI measure, the units are newton-seconds per square meter or pascal-seconds. Usually dynamic viscosity is measured in *poises* (1 P = 0.1 Pa·s) or in *centipoises* (1 cP = $\frac{1}{100}$ P):

$$1 \text{ lb·s/ft}^2 = 47.8801 \text{ Pa·s} = 47,880.1 \text{ cP}$$

The viscous property of a liquid is also sometimes expressed as *kinematic viscosity*. This is the dynamic viscosity divided by the mass density (specific weight/g). In USCS units, kinematic viscosity is measured in square feet per second. In SI measure, the units are square meters per second. Usually kinematic viscosity is measured in *stokes* (1 St = 0.0001 m²/s) or in *centistokes* (1 cSt = $\frac{1}{100}$ St):

$$1 \text{ ft}^2/\text{s} = 0.0929034 \text{ m}^2/\text{s} = 92,903.4 \text{ cSt}$$

A common unit of kinematic viscosity in the United States is Saybolt seconds univeral (SSU) for liquids of medium viscosity and Saybolt seconds Furol (SSF) for liquids of high viscosity. Viscosities measured in these units are determined by using an instrument which measures the length of time needed to discharge a standard volume of the sample. Water at 60°F (15.6°C) has a kinematic viscosity of approximately 31 SSU (1.0 cSt). For values of 70 cSt and above,

$$\text{cSt} = 0.216 \text{ SSU}$$

$$\text{SSU} = 10 \text{ SSF}$$

The dimensionless Reynolds number *Re* is used to describe the type of flow in a pipe flowing full and can be expressed as follows:

$$Re = \frac{VD}{\nu} = \frac{\rho VD}{\mu} \tag{15}$$

where V = average pipe velocity, ft/s (m/s)
$\quad\quad D$ = inside pipe diameter, ft (m)
$\quad\quad \nu$ = liquid kinematic viscosity, ft²/s (m²/s)
$\quad\quad \rho$ = liquid density, slugs/ft³ (kg/m³)
$\quad\quad \mu$ = liquid dynamic (or absolute) viscosity slug/ft·s (N·s/m²)

Note: The dimensionless Reynolds number is the same in both USCS and SI units.

When the Reynolds number is 2000 or less, the flow is generally laminar, and when it is greater than 4000, the flow is generally turbulent. The Reynolds number for the flow of water in pipes is usually well above 4000, and therefore the flow is almost always turbulent.

The Darcy-Weisbach formula is the one most often used to calculate pipe friction. This formula recognizes that friction increases with pipe wall roughness, with wetted surface area, with velocity to a power, and with viscosity and decreases with pipe diameter to a power and with density. Specifically, the frictional head loss h_f in feet (meters) is

$$h_f = f \frac{L}{D} \frac{V^2}{2g} \tag{16}$$

where f = friction factor
  L = pipe length, ft (m)
  D = inside pipe diameter, ft (m)
  V = average pipe velocity, ft/s (m/s)
  g = acceleration of gravity, 32.17 ft/s^2 (9.807 m/s^2)

  For laminar flow, the friction factor f is equal to $64/Re$ and is independent of pipe wall roughness. For turbulent flow, f for all incompressible fluids can be determined from the well-known Moody diagram, shown in Fig. 31. To determine f, it is required that the Reynolds number and the relative pipe roughness be known. Values of relative roughness ϵ/D, where ϵ is a measure of pipe wall roughness height in feet (meters), can be obtained from Fig. 32 for different pipe diameters and materials. Figure 32 also gives values for f for the flow of 60°F (15.6°C) water in rough pipes with complete turbulence. Values of kinematic viscosity and Reynolds numbers for a number of different liquids at various temperatures are given in Fig. 33. The Reynolds numbers of 60°F (15.6°C) water for various velocities and pipe diameters may be found by using the VD'' scale in Fig. 31.
  There are many empirical formulas for calculating pipe friction for water flowing under turbulent conditions. The most widely used is the Hazen-Williams formula:

in USCS units $V = 1.318Cr^{0.63}S^{0.54}$ $\tag{17a}$

in SI units $V = 0.8492Cr^{0.63}S^{0.54}$ $\tag{17b}$

where V = average pipe velocity, ft/s (m/s)
  C = friction factor for this formula, which depends on roughness only
  r = hydraulic radius (liquid area divided by wetted perimeter) or $D/4$ for a full pipe, ft (m)
  S = hydraulic gradient or frictional head loss per unit length of pipe, ft/ft (m/m)

  The effect of age on a pipe should be taken into consideration when estimating the frictional loss. A lower C value should be used, depending on the expected life of the system. Table 2 gives recommended friction factors for new and old pipes. A value of C of 150 may be used for plastic pipe. Figure 34 is a nomogram which can be used in conjunction with Table 2 for a solution to the Hazen-Williams formula.
  The frictional head loss in pressure pipes can be found by using either the Darcy-Weisbach formula (Eq. 16) or the Hazen-Williams formula (Eq. 17). Tables in the appendix give Darcy-Weisbach friction values for Schedule 40 new steel pipe carrying water. Tables are also provided for losses in old cast iron piping based on the Hazen-Williams formula with $C = 100$. In addition, values of C for various pipe materials, conditions, and years of service can also be found in the appendix.
  The following examples illustrate how Figs. 31, 32, and 33 and Table 2 may be used.

EXAMPLE 5   Calculate the Reynolds number for 175°F (79.4°C) kerosene flowing through 4-in (10.16-cm), Schedule 40, 3.426-in (8.70-cm) ID, seamless steel pipe at a velocity of 14.6 ft/s (4.45 m/s).

in USCS units $VD'' = 14.6 \times 3.426 = 50$ ft/s $\times$ in

in SI units $VD = 4.45 \times 0.087 = 0.387$ m/s $\times$ m $= 0.387 \times 129.2 = 50$ ft/s $\times$ in

Follow the tracer lines in Fig. 33 and read directly:

$$Re = 3.5 \times 10^5$$

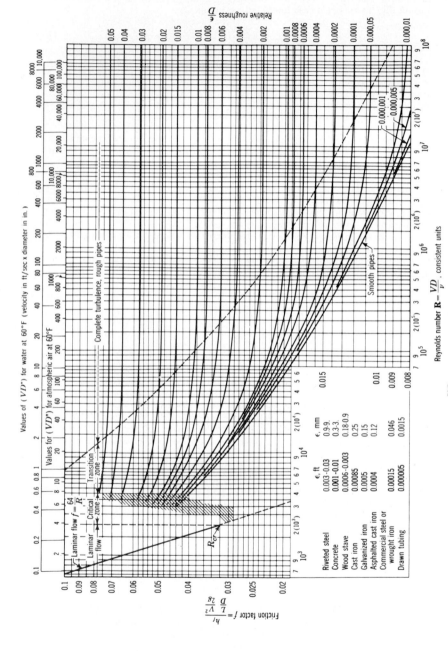

FIG. 31 Moody diagram. (Ref. 14) In SI units: $\mathbf{R} = \dfrac{\rho\,VD}{\mu}$ (ρ in kg/m³, V in m/s, D in m, μ in N·s/m²).

1 meter = 3.28 ft; VD (V in m/s, D in m) = 129.2 VD'' (V in ft/s, D'' in inches).

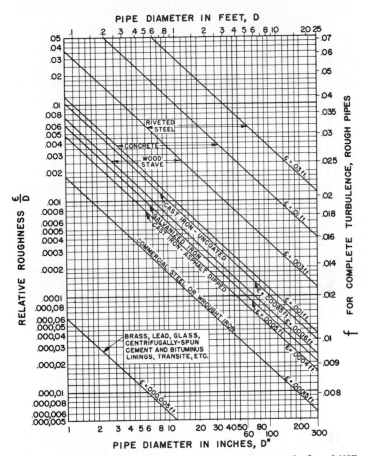

FIG. 32   Relative roughness and friction factors for new, clean pipes for flow of 60°F (15.6°C) water. (Ref. 5) (1 meter = 39.37 in = 3.28 ft)

EXAMPLE 6   Calculate the frictional head loss for 100 ft (30.48 m) of 20-in (50.8-cm), Schedule 20, 19.350-in (49.15-cm) ID, seamless steel pipe for 109°F (42.8°C) water flowing at a rate of 11,500 gpm (2612 m³/h). Use the Darcy-Weisbach formula.

in USCS units $V = \dfrac{\text{gpm}}{(\text{pipe ID in inches})^2} \times 0.408 = \dfrac{11{,}500}{19.35^2} \times 0.408 = 12.53 \text{ ft/s}$

$VD'' = 12.53 \times 19.35 = 242 \text{ ft/s} \times \text{in}$

in SI units $V = \dfrac{\text{m}^3/\text{h}}{(\text{pipe ID in cm})^2} \times 3.54 = \dfrac{2612}{49.15^2} \times 3.54 = 3.83 \text{ m/s}$

$VD = 3.83 \times 0.4915 = 1.88 \text{ m/s} \times \text{m} = 1.88 \times 129.2 = 242 \text{ ft/s} \times \text{in}$

From Fig. 33 $Re = 3 \times 10^6$

From Fig. 32 $\dfrac{\epsilon}{D} = 0.00009$

From Fig. 31 $f = 0.012$

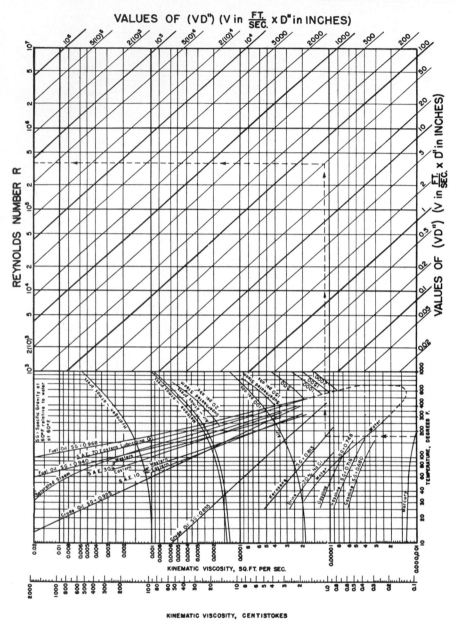

FIG. 33 Kinematic viscosity and Reynolds number. (Ref. 5) [1 ft²/s = 0.0929 m²/s; 1 cSt = 1.0 × 10⁻⁶ m²/s; 1VD (m/s × m) = 129.2 VD″ (ft/s × in); °F = (°C + 18) 1.8]

TABLE 2 Values of Friction Factor C to Be Used with the Hazen-Williams Formula in Fig. 34.

Type of pipe	Age	Size, in[a]	C
Cast iron	New	All sizes	130
	5 years old	12 and over	120
		8	119
		4	118
	10 years old	24 and over	113
		12	111
		4	107
	20 years old	24 and over	100
		12	96
		4	89
	30 years old	30 and over	90
		16	87
		4	75
	40 years old	30 and over	83
		16	80
		4	64
		40 and over	77
		24	74
		4	55
Welded steel	Any age, any size		Same as for cast iron pipe 5 years older
Riveted steel	Any age, any size		Same as for cast iron pipe 10 years older
Wood-stave	Average value, regardless of age and size		120
Concrete or concrete-lined	Large sizes, goodworkmanship, steel forms		140
	Large sizes, good workmanship, wooden forms		120
	Centrifugally spun		135
Vitrified	In good condition		110

[a]In $\times$ 25.4 = mm.
SOURCE: Adapted from Ref. 15.

Using Eq. 16,

in USCS units
$$D = \frac{19.35}{12} = 1.61 \text{ ft}$$

$$h_f = f\frac{L}{d}\frac{V^2}{2g} = 0.012\frac{100}{1.61} \times \frac{12.53^2}{2 \times 32.17} = 1.82 \text{ ft}$$

in SI units

$$h_f = f\frac{L}{D}\frac{V^2}{2g} = 0.012\frac{30.48}{0.4915} \times \frac{3.83^2}{2 \times 9.807} = 0.556 \text{ m}$$

EXAMPLE 7   The flow in Example 6 is increased until complete turbulence results. Determine the friction factor f and flow.

From Fig. 31, follow the relative roughness curve $\epsilon/D = 0.00009$ to the beginning of the zone marked "complete turbulence, rough pipes" and read

$$f = 0.0119 \qquad \text{at } Re = 2 \times 10^7$$

The problem may also be solved using Fig. 32. Enter relative roughness $\epsilon/D = 0.00009$ and read directly across to

$$f = 0.0119$$

An increase in Re from 3×10^6 to 2×10^7 would require an increase in flow to

in USCS units
$$\frac{2 \times 10^7}{3 \times 10^6} \times 11,500 = 76,700 \text{ gpm}$$

in SI units
$$\frac{2 \times 10^7}{3 \times 10^6} \times 2612 = 17,413 \text{ m}^3/\text{h}$$

EXAMPLE 8   The liquid in Example 6 is changed to water at 60°F (15.6°C). Determine Re, f, and the frictional head loss per 100 ft (100 m) of pipe.

$$VD'' = 242 \text{ ft/s} \times \text{in} \qquad \text{(as in Example 6)}$$

Since the liquid is 60°F (15.6°C) water, enter Fig. 31 and read directly downward from VD'' to

$$Re = 1.8 \times 10^6$$

Where the line VD'' to Re crosses $\epsilon/D = 0.00009$ in Fig. 31, read

$$f = 0.013$$

Water at 60°F (15.6°C) is more viscous than 109°F (42.8°C) water, and this accounts for the fact that Re decreases and f increases. Using Eq. 16, it can be calculated that the frictional head loss increases to

in USCS units
$$h_f = f\frac{L}{D}\frac{V^2}{2g} = 0.013\frac{100}{1.61} \times \frac{12.53^2}{2 \times 32.17} = 1.97 \text{ ft}$$

in SI units
$$h_f = f\frac{L}{D}\frac{V^2}{2g} = 0.013\frac{100}{0.4915} \times \frac{3.83^2}{2 \times 9.807} = 1.97 \text{ m}$$

EXAMPLE 9   A 102-in (259-cm) ID welded steel pipe is to be used to convey water at a velocity of 11.9 ft/s (3.63 m/s). Calculate the expected loss of head due to friction per 1000 ft and per 1000 m of pipe after 20 years. Use the empirical Hazen-Williams formula.

From Table 2, $C = 100$.

in USCS units
in SI units
$$r = D/4 = 102/(4 \times 12) = 2.13 \text{ ft}$$
$$r = D/4 = 2.59/4 = 0.648 \text{ m}$$

Substituting in Eq. 17,

in USCS units
$$S^{0.54} = \frac{V}{1.318Cr^{0.63}} = \frac{11.9}{1.318 \times 100 \times 2.13^{0.63}} = 0.0557$$

$$S = (0.0557)^{1/0.54} = 0.0048 \text{ ft/ft}$$

$$h_f = 1000 \times 0.0048 = 4.8 \text{ ft}$$

in SI units
$$S^{0.54} = \frac{V}{0.8492Cr^{0.63}} = \frac{3.63}{0.8492 \times 100 \times 0.648^{0.63}} = 0.0562$$

$$S = (0.0562)^{1/0.54} = 0.0048 \text{ m/m}$$

$$h_f = 1000 \times 0.0048 = 4.8 \text{ m}$$

The problem may also be solved by using Fig. 34, following the trace lines:

$$h_f \approx 5 \text{ ft (m)}$$

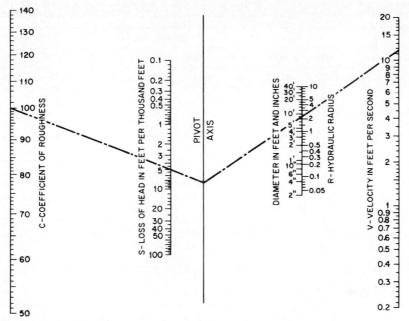

FIG. 34 Nomogram for the solution of the Hazen-Williams formula. Obtain values for C from Table 2. (Ref. 15) (1 m/s = 3.28 ft/s; 1 m = 3.28 ft; 1 m = 39.37 in)

Frictional Loss for Viscous Liquids   Table 3 gives the frictional loss for viscous liquids flowing in new Schedule 40 steel pipe. Values of pressure loss are given for both laminar and turbulent flows.

For laminar flow, the pressure loss is directly proportional to the viscosity and the velocity of flow and inversely proportional to the pipe diameter to the fourth power. Therefore, for intermediate values of viscosity and flow, obtain the pressure loss by direct interpolation. For pipe sizes not shown, multiply the fourth power of the ratio of any tabulated diameter to the pipe diameter wanted by the tabulated loss shown. The flow rate and viscosity must be the same for both diameters.

For turbulent flow and for rates of flow and pipe sizes not tabulated, the following procedures may be followed. For the viscosity and pipe size required, an intermediate flow loss is found by selecting the pressure loss for the next lower flow and multiplying by the square of the ratio of actual to tabulated flow rates. For the viscosity and flow required, an intermediate pipe diameter flow loss is found by selecting the pressure loss for the next smaller diameter and multiplying by the fifth power of the ratio of tabulated to actual inside diameters.

The viscosity of various common liquids can be found in tables in the appendix.

Partially Full Pipes and Open Channels   Another popular empirical equation applicable to the flow of water in pipes flowing full or partially full or in open channels is the Manning formula:

in USCS units
$$V = \frac{1.486}{n} r^{2/3} S^{1/2} \tag{18a}$$

in SI units
$$V = \frac{r^{2/3} S^{1/2}}{n} \tag{18b}$$

where V = average velocity, ft/s (m/s)
n = friction factor for this formula, which depends on roughness only

TABLE 3 Frictional Loss for Viscous Liquids (Copyright 1965 by the Hydraulic Institute)

Loss in pounds per square inch per 100 ft of new Schedule 40 steel pipe based on specific gravity of 1.00. of that liquid. For commercial installations, it is recommended that 15% be added to the values in this table.

gpm	Pipe size	Viscosity, SSU							
		100	200	300	400	500	1000	1500	2000
3	½	11.2	23.6	35.3	47.1	59	118	177	236
	¾	3.7	7.6	11.5	15.3	19.1	38.2	57	76
	1	1.4	2.9	4.4	5.8	7.3	14.5	21.8	29.1
5	¾	6.1	12.7	19.1	25.5	31.9	61	96	127
	1	2.3	4.9	7.3	9.7	12 .1	21.2	36.3	48.5
	1¼	0.77	1.6	2.4	3.3	4.1	8.1	12.2	16.2
7	¾	8.5	17.9	26.8	35.7	44.6	89	134	178
	1	3.2	6.8	10.2	13.6	17	33.9	51	68
	1¼	1.1	2.3	3.4	4.5	5.7	11.4	17	22.7
10	1	4.9	9.7	14.5	19.4	24.2	48.5	73	97
	1¼	1.6	3.3	4.9	6.5	8.1	16.2	24.3	32.5
	1½	0.84	1.8	2.6	3.5	4.4	8.8	13.1	17.5
15	1	11	14.5	21.8	29.1	36.3	73	109	145
	1¼	2.8	4.9	7.3	9.7	12.2	24.3	36.5	48.7
	1½	1.3	2.6	3.9	5.3	6.6	13.1	197	26.3
20	1	18	18	29.1	38.8	48.5	97	145	194
	1¼	4.9	6.4	9.7	13	16.2	32.5	48.7	65
	1½	2.3	3.5	5.3	7	8.8	17.5	26.3	35
	2	0.64	1.3	1.9	2.6	3.2	6.4	9.6	12.9
25	1½	3.5	4.4	6.6	8.8	11	21.9	32.8	43.8
	2	1	1.6	2.4	3.2	4	8	12.1	16.1
	2½	0.4	0.79	1.2	1.6	2	4	5.9	7.9
30	1½	5	5.3	7.9	10.5	13.1	26.3	39.4	53
	2	1.4	1.9	2.9	3.9	4.8	9.6	14.5	19.3
	2½	0.6	0.95	1.4	1.9	2.4	4.7	7.1	9.5
40	1½	8.5	9	10.5	14	17.5	35	53	70
	2	2.5	2.5	3.9	5.1	6.4	12.9	19.3	25.7
	2½	1.1	1.3	1.9	2.5	3.2	6.3	9.5	12.6
50	1½	12.5	14	14	17.5	21.9	43.8	66	88
	2	3.7	4	4.8	6.4	8	16.1	24.1	32.1
	2½	1.6	1.7	2.4	3.2	4	7.9	11.8	15.8
60	2	5	5.8	5.8	7.7	9.6	19.3	28.9	38.5
	2½	2.2	2.4	2.8	3.8	4.7	9.5	14.2	19
	3	0.8	0.8	1.2	1.6	2	4	6	8
70	2½	2.8	3.2	3.4	4.4	5.5	11.1	16.6	22.1
	3	1	1.1	1.4	1.9	2.3	4.6	7	9.3
	4	0.27	0.31	0.47	0.63	0.78	1.6	2.4	3.1
80	2½	3.6	4.2	4.2	5.1	6.3	12.6	19	25.3
	3	1.3	1.4	1.6	2.1	2.7	5.3	8	10.6
	4	0.36	0.36	0.54	0.72	0.89	1.8	2.7	3.6
100	2½	5.3	6.1	6.4	6.4	8	15.8	23.7	31.6
	3	1.9	2.2	2.2	2.7	3.3	6.6	9.9	13.3
	4	0.52	0.57	0.67	0.89	1.1	2.2	3.4	4.5

←— TURBULENT FLOW —→ ←————— LAMINAR FLOW ————→

For a liquid having a specific gravity other than 1.00, multiply the value from the table by the specific gravity
No allowance for aging of pipe is included. For conversions to SI units, see Table A.10, p. A.36.

Viscosity, SSU									
2500	3000	4000	5000	6000	7000	8000	9000	10,000	15,000
294	353	471	589	706	824	942	.....	.....	.....
96	115	153	191	229	268	306	344	382	573
36.3	43.6	58	73	87	101	116	131	145	218
159	191	255	319	382	446	510	573	637	956
61	73	97	121	145	170	194	218	242	363
20.3	24.3	32.5	40.6	48.7	57	65	73	81	122
223	268	357	416	535	624	713	803	892	.....
85	102	136	170	203	237	271	305	339	509
28.4	34.1	45.4	57	68	80	91	102	114	170
121	145	194	242	291	339	388	436	485	727
40.6	48.7	65	81	97	114	130	146	162	243
21.9	26.3	35	43.8	53	61	70	79	88	131
182	218	291	363	436	509	581	654	727	.....
61	73	97	122	146	170	195	219	243	365
32.8	39.4	53	66	79	92	105	118	131	197
242	291	388	485	581	678	775	872	.....	.....
81	97	130	162	195	227	260	292	325	487
43.8	53	70	88	105	123	140	158	175	263
16.1	19.3	25.7	32.1	38.5	45	51	58	64	96
55	66	88	110	131	153	176	197	219	328
20.1	24.1	32.1	40.2	48.2	56	61	72	80	121
9.9	11.8	15.8	19.7	23.7	27.6	31.6	35.5	39.5	59
66	79	105	131	158	184	210	237	263	394
24.1	28.9	38.5	48.2	58	67	77	87	96	145
11.8	14.2	19	23.7	28.4	33.2	37.9	42.6	47.4	71
88	105	140	175	210	245	280	315	350	526
32.1	38.5	51	64	77	90	103	116	129	193
15.8	19	25.3	31.6	37.9	44.2	51	57	63	95
110	131	175	219	263	307	350	394	438	657
40.2	48.2	64	80	96	112	129	145	161	241
19.7	23.7	31.6	39.5	47.4	55	63	71	79	118
48.2	58	77	96	116	135	154	173	193	289
23.7	28.4	37.9	47.4	57	66	76	85	95	142
9.9	11.9	15.9	19.9	23.9	27.9	31.8	35.8	39.8	60
27.6	33.2	44.2	55	66	77	88	100	111	166
11.6	13.9	18.6	23.2	27.8	32.5	37.1	41.7	46.4	70
3.9	4.7	6.3	7.8	9.4	11	12.5	14.1	15.6	23.5
31.6	37.9	51	63	76	88	101	114	126	190
13.3	15.9	21.2	26.5	31.8	37.1	42.4	47.7	53	80
4.5	5.4	7.2	8.9	10.7	12.5	14.3	16.1	17.9	26.8
39.5	47.4	63	79	95	111	127	142	158	237
16.6	19.9	26.5	33.1	39.8	46.4	53	60	66	99
5.6	6.7	8.9	11.2	13.4	15.6	17.9	20.1	22.3	33.5

← ———————— LAMINAR FLOW ———————— →

TABLE 3 Frictional Loss for Viscous Liquids (Continued)

gpm	Pipe size	Viscosity, SSU 100	200	300	400	500	1000	1500	2000
120	3	2.7	3.1	3.2	3.2	4	8	11.9	15.9
	4	0.73	0.81	0.81	1.1	1.3	2.7	4	5.4
	6	.098	0.11	0.16	0.21	0.26	0.52	0.78	1.0
140	3	3.4	4	4.3	4.3	4.6	9.3	13.9	18.6
	4	0.95	1.1	1.1	1.3	1.6	3.1	4.7	6.3
	6	0.13	0.15	0.18	0.24	0.30	0.61	0.91	1.2
160	3	4.4	5	5.7	5.7	5.7	10.6	15.9	21.2
	4	1.2	1.4	1.4	1.4	1.8	3.6	5.4	7.2
	6	0.17	0.18	0.21	0.28	0.35	0.69	1.0	1.4
180	3	5.3	6.3	7	7	7	11.9	17.9	23.9
	4	1.5	1.8	1.8	1.8	2	4	6	8
	6	0.2	0.24	0.24	0.31	0.39	0.78	1.2	1.6
200	3	6.5	7.7	8.8	8.8	8.8	13.3	19.9	26.5
	4	1.8	2.2	2.2	2.2	2.2	4.5	6.7	8.9
	6	0.25	0.3	0.3	0.35	0.43	0.87	1.3	1.7
250	4	2.6	3.2	3.5	3.5	3.5	5.6	8.4	11.2
	6	0.36	0.43	0.45	0.45	0.51	1.1	1.6	2.2
	8	.095	0.12	0.12	0.15	0.18	0.36	0.54	0.72
300	4	3.7	4.3	5	5	5	6.7	10.1	13.4
	6	0.5	0.6	0.65	0.65	0.65	1.3	2	2.6
	8	0.13	0.17	0.17	0.18	0.22	0.43	0.65	0.87
400	6	0.82	1	1.1	1.2	1.2	1.7	2.6	3.5
	8	0.23	0.27	0.29	0.29	0.29	0.58	0.87	1.2
	10	0.08	0.09	0.1	0.1	0.12	0.23	0.35	0.47
500	6	1.2	1.5	1.6	1.8	1.8	2.2	3.2	4.3
	8	0.33	0.39	0.44	0.47	0.47	0.72	1.1	1.5
	10	0.11	0.14	0.15	0.15	0.15	0.29	0.44	0.58
600	6	1.8	2.2	2.3	2.4	2.6	2.7	3.9	5.2
	8	0.47	0.57	0.62	0.67	0.67	0.87	1.3	1.7
	10	0.16	0.18	0.2	0.22	0.22	0.35	0.52	0.7
700	6	2.3	2.7	3	3.2	3.5	3.6	4.6	6.1
	8	0.6	0.74	0.82	0.89	0.93	1	1.5	2
	10	0.2	0.25	0.27	0.3	0.3	0.41	0.61	0.82
800	6	2.8	3.5	3.7	4	4.2	4.8	5.2	6.9
	8	0.78	0.94	1	1.1	1.2	1.2	1.7	2.3
	10	0.26	0.3	0.34	0.38	0.4	0.47	0.7	0.92
900	6	3.5	4.3	4.6	5.0	5.2	6	6	7.8
	8	0.95	1.1	1.3	1.4	1.5	1.5	2	2.6
	10	0.32	0.37	0.43	0.46	0.5	0.52	0.79	1.1
1000	8	1.1	1.4	1.5	1.6	1.8	1.9	2.2	2.9
	10	0.38	0.45	0.5	0.55	0.6	0.6	0.87	1.2
	12	0.17	0.2	0.22	0.24	025	0.29	0.43	0.58

←────────── TURBULENT FLOW ──────── ✕ ───────

				Viscosity, SSU					
2500	3000	4000	5000	6000	7000	8000	9000	10,000	15,000
19.9	23.9	31.8	39.8	47.7	56	61	72	80	119
6.7	8	10.7	13.4	16.1	18.8	21.4	24.1	26.8	40.2
1.3	1.6	2.1	2.6	3.1	3.6	1.2	4.7	5.2	7.8
23.2	27.8	37.1	46.4	56	65	74	84	93	139
7.8	9.4	12.5	15.6	18.8	21.9	25	28.2	31.3	46.9
1.5	1.8	2.4	3.0	3.6	4.2	4.9	5.5	6.1	9.1
26.5	31.8	42.4	53	64	74	85	95	106	159
8.9	10.7	14.3	17.9	21.5	25	28.6	32.2	35.7	54
1.7	2.1	2.8	3.5	4.2	4.9	5.5	6.2	6.9	10.4
29.8	35.8	47.7	60	72	84	95	107	119	179
10.1	12.1	16.1	20.1	24.1	28.1	32.2	36.2	40.2	60
2	2.3	3.1	3.9	4.7	5.5	6.2	7	7.8	11.7
33.1	39.8	53	66	80	93	106	119	133	199
11.2	13.4	17.9	22.3	26.8	31.3	35.7	40.2	44.7	67
2.2	2.6	3.5	4.3	5.2	6.1	6.9	7.8	8.7	13
14	16.8	22.3	27.9	33.5	39.1	44.7	50	56	84
2.7	3.3	4.3	5.4	6.5	7.6	8.7	9.8	10.8	16.3
0.9	1.1	1.5	1.8	2.2	2.5	2.9	3.3	3.6	5.4
16.8	20.1	26.8	33.5	40.2	47	54	60	67	101
3.3	3.9	5.2	6.5	7.8	9.1	10.4	11.7	13	19.5
1.1	1.3	1.7	2.2	2.6	3	3.5	3.9	4.3	6.5
4.3	5.2	6.9	8.7	10.4	12.1	13.9	15.6	17.3	26
1.5	1.7	2.3	2.9	3.5	4.1	4.6	5.2	5.8	8.7
0.58	0.7	0.93	1.2	1.4	1.6	1.9	2.1	2.3	3.5
5.4	6.5	8.7	10.8	13	15.2	17.3	19.5	21.7	32.5
1.8	2.2	2.9	3.6	4.3	5.1	5.8	6.5	7.2	10.8
0.73	0.87	1.2	1.5	1.8	2	2.3	2.6	2.9	4.4
6.5	7.8	10.4	13	16	18.2	20.8	23.4	26	39
2.2	2.6	3.5	4.3	5.2	6.1	6.9	7.8	8.7	13
0.87	1.1	1.4	1.8	2.1	2.4	2.8	3.3	3.5	5.2
7.6	9.1	12.1	15.2	18.4	21.2	24.3	27.3	30.3	45.5
2.5	3	4.1	5.1	6.1	7.1	8.1	9.1	10.1	15.2
1	1.2	1.6	2	2.4	2.9	3.3	3.7	4.1	6.1
8.7	10.4	13.9	17.3	20.8	24.3	27.7	31.2	34.7	52
2.9	3.5	4.6	5.8	6.9	8.1	9.3	10.4	11.6	17.3
1.2	1.4	1.9	2.3	2.8	3.3	3.7	4.2	4.7	7
9.8	11.7	15.6	19.5	23.4	27.3	31.2	35.1	39	58.5
3.3	3.9	5.2	6.5	7.8	9.1	10.4	11.7	13	19.5
1.3	1.6	2.1	2.6	3.1	3.7	4.2	4.7	5.2	7.9
3.6	4.3	5.8	7.2	8.7	10.1	11.6	13	14.5	21.7
1.5	1.8	2.3	2.9	3.5	4.1	4.7	5.2	5.8	8.7
0.72	0.87	1.2	1.5	1.7	2	2.3	2.6	2.9	4.3

————— LAMINAR FLOW —————→

TABLE 3 Frictional Loss for Viscous Liquids (Continued)

gpm	Pipe size	Viscosity, SSU 20,000	25,000	30,000	40,000	50,000	60,000
3	2	19.3	24.1	28.9	38.5	48.2	58
	2½	9.5	11.8	14.2	19	23.7	28.4
	3	4	5	6	8	9.9	11.9
5	2	32	40	48.2	64	80	96
	2½	15.8	19.7	23.7	31.6	39.5	47.4
	3	6.6	8.3	9.9	13.3	16.6	9.9
7	2	45	56	67	90	112	135
	2½	22.1	27.6	33.2	44.2	55	66
	3	9.3	11.6	13.9	18.6	23.2	27.8
10	2½	31.6	39.5	47.4	63	79	95
	3	13.3	16.6	19.9	26.5	33.1	39.8
	4	4.5	5.6	6.7	8.9	11.2	13.4
15	2½	47.4	59	71	95	118	142
	3	19.9	24.9	29.8	39.8	49.7	60
	4	6.7	8.4	10.1	13.4	16.8	20.1
20	3	26.5	33.1	39.8	53	66	80
	4	8.9	11.2	13.4	17.9	22.3	26.8
	6	1.7	2.2	2.6	3.5	4.3	5.2
25	3	33.1	41.4	49.7	66	83	99
	4	11.2	14	16.8	22.3	27.9	33.5
	6	2.2	2.7	3.3	4.3	5.4	6.5
30	3	39.8	49.7	60	80	99	119
	4	13.4	16.8	20.1	26.8	33.5	40.2
	6	2.6	3.3	3.9	5.2	6.5	7.8
40	3	53	66	80	106	133	160
	4	17.9	22.3	26.8	35.7	44.7	54
	6	3.5	4.3	5.2	6.9	8.7	10.4
50	4	22.3	27.9	33.5	44.7	56	67
	6	4.3	5.4	6.5	8.7	10.8	13
	8	1.5	1.8	2.7	2.9	3.6	4.3
60	4	26.8	33.5	40.2	54	67	80
	6	5.2	6.5	7.8	10.4	13	16
	8	1.7	2.2	2.6	3.5	4.3	5.2
70	4	31.3	39.1	46.9	63	78	94
	6	6.1	7.6	9.1	12.1	15.2	18.4
	8	2	2.5	3	4.1	5.1	6.1
80	6	6.9	8.7	10.4	13.9	17.3	20.8
	8	2.3	2.9	3.5	4.6	5.8	6.9
	10	0.93	1.2	1.4	1.9	2.3	2.8
90	6	7.8	9.8	11.7	15.6	19.5	23.4
	8	2.6	3.3	3.9	5.2	6.5	7.8
	10	1.1	1.3	1.6	2.1	2.6	3.1

← ——————————— LAMINAR FLOW ——————————— →

			Viscosity, SSU					
70,000	80,000	90,000	100,000	125,000	150,000	175,000	200,000	500,000
67	77	87	96	120	145	169	193	482
332	37.9	42.6	47.4	59	71	83	95	237
13.9	15.9	17.9	19.9	24.9	29.8	34.8	39.8	99
112	129	145	161	201	241	281	321	803
55	63	71	79	99	118	138	158	395
23.2	26.5	29.8	33	41.4	49.7	58	66	166
157	180	202	225	281	337	393	450	. . . . .
77	88	100	111	138	166	194	221	553
32.5	37.1	41.7	46.4	58	70	81	93	232
111	126	142	158	197	237	276	316	790
46.4	53	60	66	83	99	116	133	331
15.6	17.9	20.1	22.3	27.9	33.5	39.1	44.7	112
166	190	213	237	296	355	415	474	. . . . .
70	80	89	99	124	149	174	199	497
23.5	26.8	30.2	33.5	41.9	50	59	67	168
93	106	119	133	166	199	232	265	663
31.3	35.7	40.2	44.7	56	67	78	89	223
6.1	6.9	7.8	8.7	10.8	13	15.2	17.3	43.3
116	133	149	166	207	49	290	331	828
39.1	44.7	50	56	70	84	98	112	279
7.6	8.7	9.8	10.8	13.5	16.3	19	21.7	54
139	159	179	199	249	298	348	398	. . . . .
46.9	54	60	67	84	101	117	134	335
9.1	10.4	11.7	13	16.3	19.5	22.7	26	65
186	212	239	265	331	398	464	532	. . . . .
63	72	80	89	112	134	156	179	447
12.1	13.9	15.6	17.3	21.7	26	30.3	34.7	87
78	89	101	112	140	168	196	223	559
15.2	17.3	19.5	21.7	27.1	32.5	37.9	43.3	107
5.1	5.8	6.5	7.2	9	10.8	12.6	14.5	36.1
94	107	121	134	168	201	235	268	670
18.2	20.8	23.4	26	32.5	39	45.5	52	130
6.1	6.9	7.8	8.7	10.8	13	15.2	17.3	43.4
110	125	141	156	196	235	274	313	782
21.2	24.3	27.3	30.3	37.9	45.5	53	61	152
7.1	8.1	9.1	10.1	12.6	15.2	17.7	20.2	51
24.3	27.7	31.2	34.7	43.3	52	61	69	173
8.1	9.3	10.4	11.6	14.5	17.3	20.2	23.1	58
3.3	3.7	4.2	4.7	5.8	7	8.2	9.3	23.3
27.3	31.2	35.1	39	48.7	59	68	78	195
9.1	10.4	11.7	13	16.3	19.5	22.8	26	65
3.7	4.2	4.7	5.2	6.6	7.9	9.2	10.5	26.2

← ———————————— LAMINAR FLOW ———————————— →

TABLE 3 Frictional Loss for Viscous Liquids (Continued)

gpm	Pipe size	Viscosity, SSU					
		20,000	25,000	30,000	40,000	50,000	60,000
100	6	8.7	10.8	13	17.3	21.7	26
	8	2.9	3.6	4.3	5.8	7.2	8.7
	10	1.2	1.5	1.8	2.3	2.9	3.5
120	6	10.4	13	15.6	20.8	26	31.2
	8	3.5	4.3	5.2	6.9	8.7	10.4
	10	1.4	1.8	2.1	2.8	3.5	4.2
140	6	12.1	15.2	18.2	24.3	30.3	36.4
	8	4	5.1	6.1	8.1	10.1	12.1
	10	1.7	2	2.4	3.3	4.1	4.9
160	6	13.9	17.3	20.8	27.7	34.7	41.6
	8	4.6	5.8	6.9	9.3	11.6	13.8
	10	1.9	2.3	2.8	3.7	4.7	5.6
180	6	15.6	19.5	23.4	31.2	39	46.8
	8	5.2	6.5	7.8	10.4	13	15.6
	10	2.1	2.6	3.1	4.2	5.2	6.3
200	8	5.8	7.2	8.7	11.6	14.5	17.3
	10	2.3	2.9	3.5	4.7	5.8	7
	12	1.2	1.5	1.7	2.3	2.9	3.5
250	8	7.2	9	10.8	14.5	18.1	21.7
	10	2.9	3.6	4.4	5.8	7.3	8.7
	12	1.5	1.8	2.2	2.9	3.6	4.3
300	8	8.7	10.8	13	17.3	21.7	26
	10	3.5	4.4	5.2	7	8.7	10.5
	12	1.7	2.2	2.6	3.5	4.3	5.2
400	8	11.6	14.5	17.3	23	28.9	34.7
	10	4.7	5.8	7	9.3	11.6	14
	12	2.3	2.9	3.5	4.6	5.8	7
500	8	14.5	18.1	21.7	28.9	36.1	43.4
	10	5.8	7.3	8.7	11.6	14.6	17.5
	12	2.9	3.6	4.3	5.8	7.2	8.7
600	8	17.3	21.7	26	34.7	43.4	52
	10	7	8.7	10.5	14	17.5	21
	12	3.5	4.3	5.2	7	8.7	10.4
700	8	20.2	25.3	30.3	40.5	51	61
	10	8.2	10.2	12.2	16.3	20.4	24.4
	12	4.1	5.1	6.1	8.1	10.1	12.2
800	8	23.1	28.9	34.7	46.2	58	69
	10	9.3	11.6	14	18.6	23.3	27.9
	12	4.6	5.8	7	9.3	11.6	13.9
900	8	26	32.5	39	52	65	78
	10	10.5	13.1	15.7	21	26.2	31.4
	12	5.2	6.5	7.8	10.4	13	15.6
1000	8	28.9	36.1	43.4	58	72	87
	10	11.6	14.6	17.5	23.3	29.1	34.9
	12	5.8	7.2	8.7	11.6	14.5	17.4

←————————————— LAMINAR FLOW —————————————→

				Viscosity, SSU				
70,000	80,000	90,000	100,000	125,000	150,000	175,000	200,000	500,000
30.3	34.7	39	43.3	54	65	76	87	217
10.1	11.6	13	14.5	18.1	21.7	25.3	28.9	72
4.2	4.7	5.2	5.8	7.3	8.7	10.2	11.6	29.1
36.4	41.6	46.8	52	65	78	91	104	260
12.1	13.9	15.6	17.3	21.7	26	30.4	34.7	87
4.9	5.6	6.3	7	8.7	10.5	12.2	14	34.9
42.5	48.5	55	61	76	91	106	121	303
14.2	16.2	18.2	20.2	25.3	30.4	35.4	40.5	101
5.7	6.5	7.3	8.1	10.2	12.2	14.3	16.3	40.7
48.5	56	62	69	87	104	121	139	347
16.2	18.5	20.8	23.1	28.9	34.7	40.5	46.2	116
6.5	7.5	8.4	9.3	11.6	14	16.3	18.6	46.6
55	62	70	78	98	117	137	156	390
18.2	20.8	23.4	26	32.5	39	45.5	52	130
7.3	8.4	9.4	10.5	13.1	15.7	18.3	21	52
20.2	23.1	26	28.9	36.1	43.4	51	58	145
8.2	9.3	10.5	11.6	14.6	17.5	20.4	23.3	58
4.1	4.6	5.2	5.8	7.2	8.7	10.1	11.6	28.9
25.3	28.9	32.5	36.1	45.2	54	63	72	181
10.2	11.6	13.1	14.6	18.2	21.8	25.5	29.1	73
5.1	5.8	6.5	7.2	9	10.9	12.7	14.5	36.2
30.4	34.7	39	43.4	54	65	76	87	217
12.2	14	15.7	17.5	21.8	26.2	30.6	34.9	87
6.1	7	7.8	8.7	10.9	13	15.2	17.4	43.4
40.5	46.2	52	58	72	87	101	116	289
16.3	18.6	21	23.3	29.6	34.9	40.7	46.6	116
8.1	9.3	10.4	11.6	14.5	17.4	20.3	23.2	58
51	58	65	72	90	108	126	145	361
20.4	23.3	26.2	29.1	36.4	43.7	51	58	146
10.1	11.6	13	14.5	18.1	21.7	25.3	28.9	72
61	69	78	87	107	130	152	173	434
24.4	27.9	31.4	34.9	43.7	52	61	70	175
12.2	13.9	15.6	17.4	21.7	26.1	30.4	34.7	87
71	81	91	101	126	152	177	202	506
28.5	32.6	36.7	40.7	51	61	71	82	204
14.2	16.2	18.2	20.3	25.3	30.4	35.5	40.5	101
81	93	104	116	145	173	202	231	578
32.6	37.3	41.9	46.6	58	70	82	93	233
16.2	18.5	20.8	23.1	28.9	34.7	40.5	46.3	116
91	104	117	130	163	195	228	260	650
36.7	41.9	47.1	52	66	79	92	105	262
18.2	20.8	23.4	26.1	32.6	39.1	45.6	52	130
101	116	130	145	181	217	253	289	723
40.7	46.6	52	58	72	87	102	116	291
20.3	23.2	26.1	28.9	36.2	43.4	51	58	145

← ———————— LAMINAR FLOW ———————— →

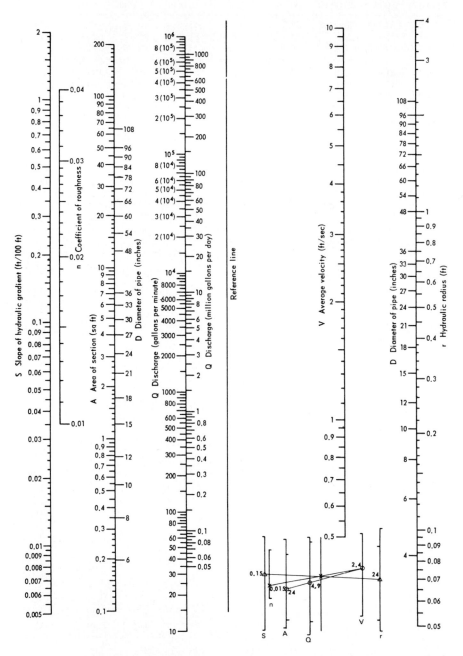

FIG. 35 Nomogram for the solution of the Manning formula. Obtain values of n from Table 4. To solve for S, align A with Q and read V, align V and n intersecting reference line, align r with reference line intersection, and read S. (1 m = 3.28 ft; 1 m = 39.37 in; 1 m^2 = 1550 in^2; 1 m^3/h = 4.4 gpm; 1 m/s = 3.28 ft/s; 1 m/m = 1 ft/ft)

8.48

r = hydraulic radius (liquid area divided by wetted perimeter), ft (m)
S = hydraulic gradient or frictional head loss per unit length of conduit, ft/ft (m/m)

The Manning formula nomogram shown in Fig. 35 can be used to determine the flow or frictional head loss in open or closed conduits. Note that the hydraulic gradient S in Fig. 35 is plotted in feet per 100 ft of conduit length. Values of friction factor n are given in Table 4.

TABLE 4 Values of Friction Factor n to Be Used with the Manning Formula in Fig. 35

Surface	Best	Good	Fair	Bad
Uncoated cast iron pipe	0.012	0.013	0.014	0.015
Coated cast iron pipe	0.011	0.012[a]	0.013[a]	
Commercial wrought iron pipe, black	0.012	0.013	0.014	0.015
Commercial wrought iron pipe, galvanized	0.013	0.014	0.015	0.017
Smooth brass and glass pipe	0.009	0.010	0.011	0.013
Smooth lockbar and welded OD pipe	0.010	0.011[a]	0.013[a]	
Riveted and spiral steel pipe	0.013	0.015[a]	0.017[a]	
Vitrified sewer pipe	0.010 } 0.011	0.013[a]	0.015	0.017
Common clay drainage tile	0.011	0.012[a]	0.014[a]	0.017
Glazed brickwork	0.011	0.012	0.013[a]	0.015
Brick in cement mortar; brick sewers	0.012	0.013	0.015[a]	0.017
Neat cement surfaces	0.010	0.011	0.012	0.013
Cement mortar surfaces	0.011	0.012	0.013[a]	0.015
Concrete pipe	0.012	0.013	0.015[a]	0.016
Wood-stave pipe	0.010	0.011	0.012	0.013
Plank flumes:				
Planed	0.010	0.012[a]	0.013	0.014
Unplaned	0.011	0.013[a]	0.014	0.015
With battens	0.012	0.015[a]	0.016	
Concrete-lined channels	0.012	0.014[a]	0.016[a]	0.018
Cement-rubble surface	0.017	0.020	0.025	0.030
Dry-rubble surface	0.025	0.030	0.033	0.035
Dressed-ashlar surface	0.013	0.014	0.015	0.017
Semicircular metal flumes, smooth	0.011	0.012	0.013	0.015
Semicircular metal flumes, corrugated	0.0225	0.025	0.0275	0.030
Canals and ditches:				
Earth, straight and uniform	0.017	0.020	0.0225[a]	0.025
Rock cuts, smooth and uniform	0.025	0.030	0.033[a]	0.035
Rock cuts, jagged and irregular	0.035	0.040	0.045	
Winding sluggish canals	0.0225	0.025[a]	0.0275	0.030
Dredged earth channels	0.025	0.0275[a]	0.030	0.033
Canals with rough stony beds, weeds on earth banks	0.025	0.030	0.035[a]	0.040
Earth bottom, rubble sides	0.028	0.030[a]	0.033[a]	0.035
Natural stream channels:				
(1) Clean, straight bank, full stage, no rifts or deep pools	0.025	0.0275	0.030	0.033
(2) Same as (1), but some weeds and stones	0.030	0.033	0.035	0.040
(3) Winding, some pools and shoals, clean	0.033	0.035	0.040	0.045
(4) Same as (3), lower stages, more ineffective slope and sections	0.040	0.045	0.050	0.055
(5) Same as (3), some weeds and stones	0.035	0.040	0.045	0.050
(6) Same as (4), stony sections	0.045	0.050	0.055	0.060
(7) Sluggish river reaches, rather weedy or with very deep pools	0.050	0.060	0.070	0.080
(8) Very weedy reaches	0.075	0.100	0.125	0.150

[a]Values commonly used in designing.

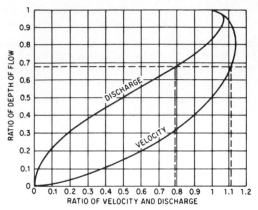

FIG. 36   Discharge velocity of a partially full circular pipe versus that of a full pipe.

If the conduit is flowing partially full, computing the hydraulic radius is sometimes difficult. When the problem to be solved deals with a pipe which is not flowing full, Fig. 36 may be used to obtain multipliers for correcting the flow and velocity of a full pipe to the values needed for the actual fill condition. If the flow in a partially full pipe is known and the frictional head loss is to be determined, Fig. 36 is first used to correct the flow to what it would be if the pipe were full. Then Eq. 18 or Fig. 35 is used to determine the frictional head loss (which is also the hydraulic gradient and the slope of the pipe). The problem is solved in reverse if the hydraulic gradient is known and the flow is to be determined.

For full or partially full flow in conduits which are not circular in cross section, an alternate solution to using Eq. 18 is to calculate an equivalent diameter equal to four times the hydraulic radius. If the conduit is extremely narrow and so width is small relative to length (annular or elongated sections), the hydraulic radius is one-half the width of the section.[4] After the equivalent diameter has been determined, the problem may be solved by using the Darcy-Weisbach formula (Eq. 16).

The hydraulic gradient in a uniform open channel is synonymous with frictional head loss in a pressure pipe. The hydraulic gradient of an open channel or of a pipe flowing partially full is the slope of the free liquid surface. In the reach of the channel where the flow is uniform, the hydraulic gradient is parallel to the slope of the channel bottom. Figure 37 shows that, in a pressure pipe of uniform cross section, the slope of both the energy and hydraulic gradients is a measure of the frictional head loss per foot (meter) of pipe between points 1 and 2. Figure 38 illustrates the flow in an open channel of varying slope. Between points 1 and 2 the flow is uniform and the liquid surface (hydraulic gradient) and channel bottom are both parallel and their slope is the frictional head loss per foot (meter) of channel length.

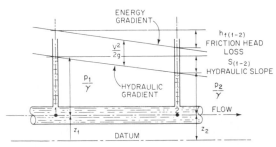

FIG. 37   Slopes of energy and hydraulic gradients measure frictional head loss ft/ft (m/m) of pipe length.

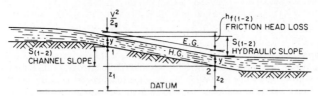

FIG. 38 In an open channel with uniform flow, slopes of channel bottom, energy, and hydraulic gradients are the same as frictional head loss ft/ft (m/m) of channel length.

A rearrangement of Eq. 18 gives

in USCS units
$$S = h_f = \left(\frac{Vn}{1.486r^{2/3}}\right)^2 \qquad (19a)$$

in SI units
$$S = h_f = \left(\frac{Vn}{r^{2/3}}\right)^2 \qquad (19b)$$

The following examples illustrate the application of the Manning formula (Eq. 19), Figs. 35 and 36, and Table 4 to the solution of problems involving the flow of water in open channels.

EXAMPLE 10   The flow through a 24-in (61.0-cm) ID commercial wrought iron pipe in fair condition is 4.9 mgd (772.7 m^3/h). Determine the loss of head as a result of friction in feet per 100 ft (meters per 100 m) of pipe and the slope required to maintain a full, uniformly flowing pipe.

in USCS units $\quad V = \dfrac{\text{gpm}}{(\text{pipe ID in inches})^2} \times 0.408 = \dfrac{4.9 \times 10^6 \times 0.408}{24 \times 60 \times 24^2} = 2.41 \text{ ft/s}$

in SI units $\quad V = \dfrac{m^3/h}{(\text{pipe ID in cm})^2} \times 3.54 = \dfrac{772.7}{60.96^2} \times 3.54 = 0.736 \text{ m/s}$

From Table 4, $n = 0.015$.

in USCS units $\qquad r = D/4 = \text{¾} = 0.5 \text{ ft} \qquad$ (hydraulic radius)

$$S = h_f = \left(\frac{Vn}{1.486r^{2/3}}\right)^2 = \left(\frac{2.41 \times 0.015}{1.486 \times 0.5^{2/3}}\right)^2 \qquad \text{(from Eq. 19a)}$$

$$= 0.0015 \text{ ft/ft}$$

$$100 \times 0.0015 = 0.15 \text{ ft/100 ft} \qquad \text{(slope and frictional head)}$$

in SI units $\qquad r = D/4 = 0.6096/4 = 0.1524 \text{ m} \qquad$ (hydraulic radius)

$$S = h_f = \left(\frac{Vn}{r^{2/3}}\right)^2 = \left(\frac{0.736 \times 0.015}{0.1524^{2/3}}\right)^2 \qquad \text{(from Eq. 19b)}$$

$$= 0.0015 \text{ m/m}$$

$$100 \times 0.0015 = 0.15 \text{ m/100 m} \qquad \text{(slope and frictional head)}$$

The problem may also be solved by using Fig. 35 and following the trace lines:

$$S = h_f = 0.15 \text{ ft/100 ft (m/100 m)}$$

EXAMPLE 11   Determine what the flow and velocity would be if the pipe in Example 10 were flowing two-thirds full and were laid on the same slope.

Follow the trace lines in Fig. 36 and note that the multipliers for discharge and velocity are 0.79 and 1.11, respectively. Therefore

in USCS units $\qquad\qquad\qquad$ Flow $= 0.79 \times 4.9 = 3.87$ mgd

$\qquad\qquad\qquad\qquad\quad$ Velocity $= 1.11 \times 2.41 = 2.68$ ft/s

in SI units
$$\text{Flow} = 0.79 \times 772.7 = 610 \text{ m}^3/\text{h}$$
$$\text{Velocity} = 1.11 \times 0.736 = 0.817 \text{ m/s}$$

Pipe Fittings Invariably a system containing piping will have connections which change the size and/or direction of the conduit. These fittings add frictional losses, called *minor losses*, to the system head. Fitting losses are generally the result of changes in velocity and/or direction. A decreasing velocity results in more loss in head than an increasing velocity, as the former causes energy-dissipating eddies. Experimental results have indicated that minor losses vary approximately as the square of the velocity through the fittings.

VALVES AND STANDARD FITTINGS The resistance to flow through valves and fittings may be found in Refs. 4 and 5 and other sources. Losses are usually expressed in terms of a *resistance coefficient* K and the average velocity head in a pipe having the same diameter as the valve or fitting. The frictional resistance h in feet (meters) is found from the equation

$$h = K \frac{V^2}{2g} \tag{20}$$

where K = resistance coefficient, which depends on design and size of valve or fitting
  V = average velocity in pipe of corresponding internal diameter, ft/s (m/s)
  g = acceleration of gravity, 32.17 ft/s² (9.807 m/s²)

A comparison of the Darcy-Weisbach equation (Eq. 16) with Eq. 20 suggests that $K = f(L/D)$, where L is the equivalent length of pipe in feet (meters) and D is the inside pipe diameter in feet (meters), to produce the same head loss in a straight pipe as through a valve or fitting. The friction in an "equivalent length of pipe" has been another method used to estimate head loss through values and fittings. Values of the ratio L/D have been experimentally determined. This ratio multiplied by the inside diameter of a pipe of specified schedule for the valve or fitting being considered gives the equivalent length of pipe to use to calculate the head lost.

Loss of head in straight pipe depends on the friction factor or Reynolds number. However, with valves and fittings, head is lost primarily because of change in direction of flow, change in cross section, and obstructions in the flow path. For this reason the resistance coefficient is practically constant for a particular shape of valve or fitting for all flow conditions, including laminar flow. The resistance coefficient would theoretically be constant for all sizes of a particular design of valve or fitting except for the fact that all sizes are not geometrically similar. The Crane Company has reported the results of tests that show that the resistance coefficient for a number of lines of valves and fittings decreases with increasing size at flow conditions of equal friction factor and that the equivalent length L/D tends to be constant for the various sizes at the same flow conditions.

When available, the K factor furnished by the valve or fitting manufacturer should always be used rather than the value from a general listing. The Hydraulic Institute lists losses in terms of K through valves and fittings (Tables 5a to 5c); these losses vary with size of the valve or fitting but are independent of friction factor. The Crane Company provides a similar listing of K values (Tables 6a to 6e). The latter listing of flow coefficients is associated with the velocity head $V^2/2g$ that would occur through the internal diameter of the schedule pipes for the various ANSI classes of valves and fittings shown in Table 6e. If the connecting pipe is of a different size and/or schedule, either use the velocity for the pipe shown in Table 6a or use the actual pipe velocity head and correct the resistance coefficient obtained from this table by the multiplier

$$\left(\frac{\text{Actual pipe ID}}{\text{Standard pipe ID}} \right)^4$$

Tables 6 are based on the use of an equivalent length constant for complete turbulent flow for each valve or fitting shown. This constant is shown as the multiplier of the friction factor f_T for the corresponding clean commercial steel pipe with completely turbulent flow. The product $(L/D)(f_T)$ is the coefficient K. The friction factors are given in Table 6a for different pipe sizes, or they can be obtained from Fig. 31 or 32. If the valve or fitting has a sudden or gradual contraction, enlargement, and/or change in direction of flow, appropriate formulas for these conditions are given for the determination of K. If flow is laminar, valve and fitting resistance coefficients are

TABLE 5a   Resistance Coefficients K for Valves and Fittings

BELL – MOUTH
INLET OR REDUCER
K = 0.05

REGULAR
SCREWED
45° ELL

SQUARE EDGED INLET
K = 0.5

LONG
RADIUS
FLANGED
45° ELL

INWARD PROJECTING PIPE
K = 1.0

SCREWED
RETURN
BEND

NOTE: K DECREASES WITH
INCREASING WALL THICKNESS OF
PIPE AND ROUNDING OF EDGES

FLANGED
RETURN
BEND

REG.
LONG
RADIUS

REGULAR
SCREWED
90° ELL

LINE
FLOW

LONG
RADIUS
SCREWED
90° ELL

SCREWED
TEE

BRANCH
FLOW

REGULAR
FLANGED
90° ELL

LINE
FLOW

LONG
RADIUS
FLANGED
90° ELL

FLANGED
TEE

BRANCH
FLOW

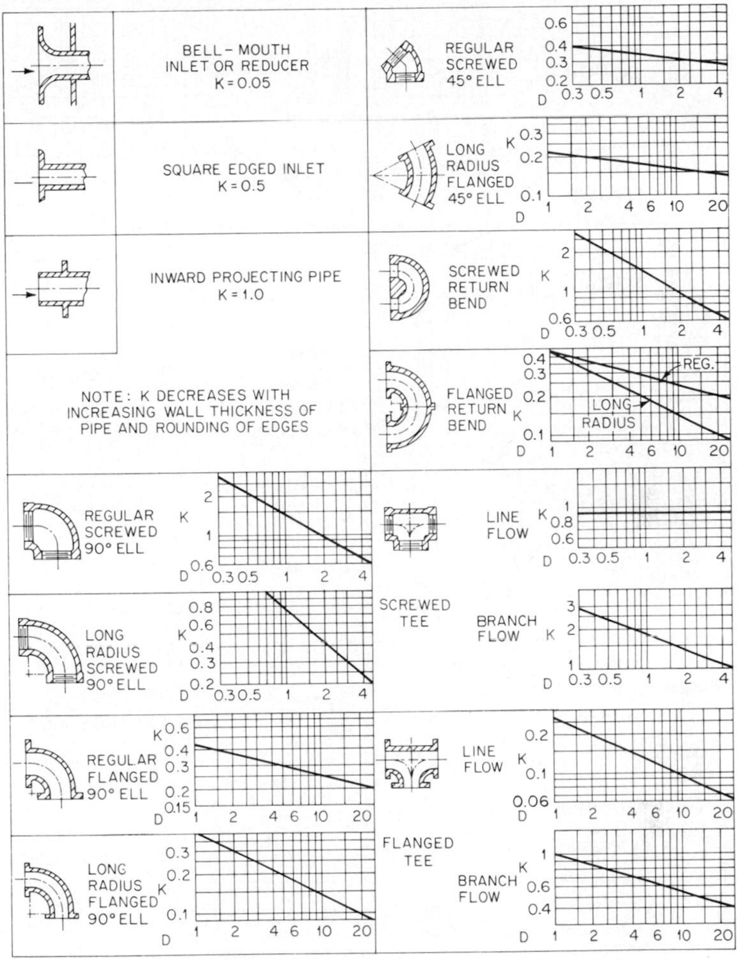

$$h = K\,\frac{V^2}{2g}\ \text{FEET (METERS) OF FLUID}$$

NOTE: D = nominal iron pipe size in inches (in $\times$ 25.4 = mm).

SOURCE: Ref. 5.

obtained from Table 6a based on completely turbulent flow but the pipe frictional loss is calculated using the laminar friction factor $f = 64/Re$ instead of f_T. Also shown in Tables 6 for check valves is the minimum pipe velocity required for full disk lift for the coefficient of resistance listed ($\overline{V}$ = liquid specific volume in cubic feet per pound).

Prior to the 15th printing (1976) of the Crane Company Technical Paper 410, and as shown in the first edition of this text, valve and fitting losses were calculated using an equivalent length of pipe rather than the coefficient K. The Crane Company states that this conceptual change regarding the values of equivalent length L/D and resistance coefficient K for valves and fittings relative to the friction factor in pipes has a relatively minor effect on most problems dealing with turbulent flow and avoids a significant overstatement of pressure drop in the laminar zone.

TABLE 5b Resistance Coefficients K for Valves and Fittings

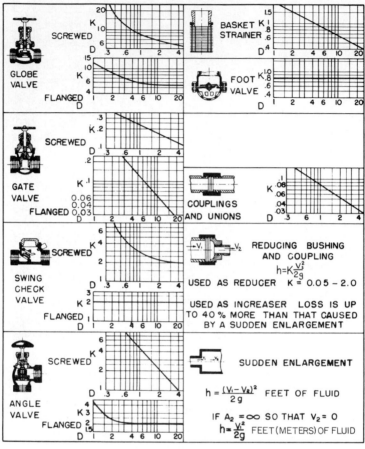

$$h = K\frac{V^2}{2g} \text{ FEET (METERS) OF FLUID}$$

NOTE: D = nominal iron pipe size in inches (in $\times$ 25.4 = mm). For velocities below 15 ft/s (4.6 m/s), check valves and foot valves will be only partially open and will exhibit higher values of K than shown.

SOURCE: Ref. 5.

Valve Flow Coefficient The loss of head through valves, particularly control valves, is often expressed in terms of a *flow coefficient* C_v in USCS units (K_v in SI units). The flow of water in gallons per minute (cubic meters per hour) at 60°F (15.6°C) that will pass through a valve with a 1-lb/in^2 (1-bar) pressure drop is defined as the flow coefficient for a particular valve opening. Since loss of head h is a measure of energy loss per unit weight (force) and since $h = p/\gamma$ and head loss varies directly with the square of the flow through a certain fixed opening, the formulas for flow coefficient are

in USCS units

$$C_v = \text{gpm} \sqrt{\frac{\text{sp. gr.}}{\text{lb/in}^2}} \qquad (21a)$$

also

$$C_v = 29.9d^2/\sqrt{K} \qquad (22a)$$

TABLE 5c Approximate Variation for K Listed in Tables 5a and 5b

	Fitting	Range of variation, %
90° elbow	Regular screwed	±20 above 2-in size[a]
	Regular screwed	±40 below 2-in size
	Long radius, screwed	±25
	Regular flanged	±35
	Long radius, flanged	±30
45° elbow	Regular screwed	±10
	Long radius, flanged	±10
180° bend	Regular screwed	±25
	Regular flanged	±35
	Long radius, flanged	±30
T	Screwed, line or branch flow	±25
	Flanged, line or branch flow	±35
Globe valve	Screwed	±25
	Flanged	±25
Gate valve	Screwed	±25
	Flanged	±50
Check valve[b]	Screwed	±30
	Flanged	$\left\{\begin{array}{l}+200\\-80\end{array}\right.$
Sleeve check valve	—	Multiply flanged values by 0.2 to 0.5
Tilting check valve	—	Multiply flanged values by 0.13 to 0.19
Drainage gate check	—	Multiply flanged values by 0.03 to 0.07
Angle valve	Screwed	±20
	Flanged	±50
Basket strainer	—	±50
Foot valve[a]	—	±50
Couplings	—	±50
Unions	—	±50
Reducers	—	±50

[a]In × 25.4 = mm.
[b]For velocities below 15 ft/s (4.6 m/s), check valves and foot valves will be only partially open and will exhibit higher values of K than shown.

SOURCE: Ref. 5.

in SI units
$$K_v = m^3/h \sqrt{\frac{sp.\ gr.}{bar}} \qquad (21b)$$

also
$$K_v = 0.04d^2/\sqrt{K} \qquad (22b)$$

where d = internal diameter of pipe corresponding to K and as shown in Table 6e, in (mm) and 1 bar = 100 kPa. The conversion from the SI flow coefficient to the USCS flow coefficient is

$$C_v = 1.156\ K_v.$$

EXAMPLE 12   A pumping system consists of 20 ft (6.1 m) of 2-in (51-mm) suction pipe and 300 ft (91.5 m) of 1½-in (38-mm) discharge pipe, both Schedule 40 new steel. Also included are a bell mouth inlet, a 90° short radius (SR) suction elbow, a full port suction gate valve of Class 150 steel, a full port discharge gate valve of Class 400 steel, and a swing check valve. The valves and fittings are screw-connected and the same size as the connecting pipe.

Determine the pipe, valve, and fitting losses when 60°F (15.6°C) oil having a specific gravity of 0.855 is pumped at a rate of 60 gpm (13.6 m³/h). Use resistance coefficients from Tables 5.

TABLE 6a Resistance Coefficient K for Valves and Fittings

PIPE FRICTION DATA FOR CLEAN COMMERCIAL STEEL PIPE WITH FLOW IN ZONE OF COMPLETE TURBULENCE

Nominal Size	½"	¾"	1"	1¼"	1½"	2"	2½, 3"	4"	5"	6"	8-10"	12-16"	18-24"
Friction Factor (f_T)	.027	.025	.023	.022	.021	.019	.018	.017	.016	.015	.014	.013	.012

FORMULAS FOR CALCULATING "K" FACTORS FOR VALVES AND FITTINGS WITH REDUCED PORT

- **Formula 1**

$$K_2 = \frac{0.8 \sin\frac{\theta}{2}(1 - \beta^2)}{\beta^4} = \frac{K_1}{\beta^4}$$

- **Formula 2**

$$K_2 = \frac{0.5 (1 - \beta^2) \sqrt{\sin\frac{\theta}{2}}}{\beta^4} = \frac{K_1}{\beta^4}$$

- **Formula 3**

$$K_2 = \frac{2.6 \sin\frac{\theta}{2}(1 - \beta^2)^2}{\beta^4} = \frac{K_1}{\beta^4}$$

- **Formula 4**

$$K_2 = \frac{(1 - \beta^2)^2}{\beta^4} = \frac{K_1}{\beta^4}$$

- **Formula 5**

$$K_2 = \frac{K_1}{\beta^4} + \text{Formula 1} + \text{Formula 3}$$

$$K_2 = \frac{K_1 + \sin\frac{\theta}{2}[0.8 (1 - \beta^2) + 2.6 (1 - \beta^2)^2]}{\beta^4}$$

- **Formula 6**

$$K_2 = \frac{K_1}{\beta^4} + \text{Formula 2} - \text{Formula 4}$$

$$K_2 = \frac{K_1 + 0.5 \sqrt{\sin\frac{\theta}{2}}(1 - \beta^2) + (1 - \beta^2)^2}{\beta^4}$$

- **Formula 7**

$$K_2 = \frac{K_1}{\beta^4} + \beta \, (\text{Formula 2} + \text{Formula 4}) \quad \text{when } \theta = 180°$$

$$K_2 = \frac{K_1 + \beta \left[0.5 (1 - \beta^2) + (1 - \beta^2)^2 \right]}{\beta^4}$$

NOTES:

$$\beta = \frac{d_1}{d_2}$$

$$\beta^2 = \left(\frac{d_1}{d_2}\right)^2 = \frac{a_1}{a_2}$$

Subscript 1 defines dimensions and coefficients with reference to the smaller diameter.
Subscript 2 refers to the larger diameter.

SUDDEN AND GRADUAL CONTRACTION	**SUDDEN AND GRADUAL ENLARGEMENT**
	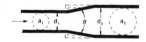
If: $\theta \lessgtr 45°$ .... K_2 = Formula 1	If: $\theta \lessgtr 45°$ .... K_2 = Formula 3
$45° < \theta \lessgtr 180°$ ... K_2 = Formula 2	$45° < \theta \lessgtr 180°$ ... K_2 = Formula 4

NOTE: K is based on use of schedule pipe as listed in Table 6e. In $\times$ 25.4 = mm.

SOURCE: Ref. 4.

The inner diameter of the suction pipe is 2.067 in (52.5 mm), and from Fig. 32, ϵ/D = 0.00087. From Eq. 9,

in USCS units, $V = \dfrac{\text{gpm}}{(\text{pipe ID in inches})^2} \times 0.408 = \dfrac{60}{2.067^2} \times 0.408 = 5.73$ ft

$$VD'' = 5.73 \times 2.067 = 11.8 \text{ ft/s} \times \text{in}$$

TABLE 6b

GATE VALVES
Wedge Disc, Double Disc, or Plug Type

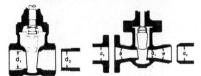

If: $\beta = 1$, $\theta = 0$ . . . . $K_1 = 8\,f_T$
   $\beta < 1$ and $\theta \gtrless 45°$ . . . . $K_2 =$ Formula 5
   $\beta < 1$ and $45° < \theta \gtrless 180°$ . . . $K_2 =$ Formula 6

SWING CHECK VALVES

$K = 100\,f_T$ $K = 50\,f_T$

Minimum pipe velocity   Minimum pipe velocity
(fps) for full disc lift   (fps) for full disc lift
$= 35\,\sqrt{\overline{V}}$ $= 48\,\sqrt{\overline{V}}$

GLOBE AND ANGLE VALVES

If: $\beta = 1$, $K_1 = 340\,f_T$

If: $\beta = 1$ . . . $K_1 = 55\,f_T$

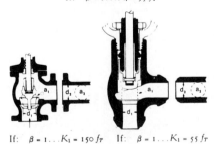

If: $\beta = 1$ . . . $K_1 = 150\,f_T$ If: $\beta = 1$ . . . $K_1 = 55\,f_T$

All globe and angle valves,
whether reduced seat or throttled,
If: $\beta < 1$ . . . $K_2 =$ Formula 7

LIFT CHECK VALVES

If: $\beta = 1$ . . . $K_1 = 600\,f_T$
   $\beta < 1$ . . . $K_2 =$ Formula 7
Minimum pipe velocity (fps) for full disc lift
$= 40\,\beta^2\,\sqrt{\overline{V}}$

If: $\beta = 1$ . . . $K_1 = 55\,f_T$
   $\beta < 1$ . . . $K_2 =$ Formula 7
Minimum pipe velocity (fps) for full disc lift
$= 140\,\beta^2\,\sqrt{\overline{V}}$

TILTING DISC CHECK VALVES

	$\alpha = 5°$	$\alpha = 15°$
Sizes 2 to 8″ . . . $K =$	$40\,f_T$	$120\,f_T$
Sizes 10 to 14″ . . . $K =$	$30\,f_T$	$90\,f_T$
Sizes 16 to 48″ . . . $K =$	$20\,f_T$	$60\,f_T$
Minimum pipe velocity (fps) for full disc lift $=$	$80\,\sqrt{\overline{V}}$	$30\,\sqrt{\overline{V}}$

in SI units $V = \dfrac{\text{m}^3/\text{h}}{(\text{pipe ID in cm})^2} \times 3.54 = \dfrac{13.6}{5.25^2} \times 3.54 = 1.75 \text{ m/s}$

$VD = 1.75 \times 0.0525 = 0.0919 \text{ m/s} \times \text{m}$

$(VD'' = 0.0919 \times 129.2 = 11.8 \text{ ft/s} \times \text{in})$

From Fig. 33, $Re = 1 \times 10^4$, and from Fig. 31, $f = 0.031$. From Eq. 16

in USCS units $h_{fs} = 0.031\,\dfrac{20 \times 12}{2.067} \times \dfrac{5.73^2}{2 \times 32.17} = 1.84 \text{ ft}$

TABLE 6c

STOP-CHECK VALVES (Globe and Angle Types)	FOOT VALVES WITH STRAINER

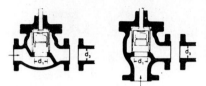

STOP-CHECK VALVES (Globe and Angle Types)

If:
$\beta = 1 \ldots K_1 = 400 f_T$
$\beta < 1 \ldots K_2 =$ Formula 7

Minimum pipe velocity for full disc lift
$= 55 \, \beta^2 \sqrt{\overline{V}}$

If:
$\beta = 1 \ldots K_1 = 200 f_T$
$\beta < 1 \ldots K_2 =$ Formula 7

Minimum pipe velocity for full disc lift
$= 75 \, \beta^2 \sqrt{\overline{V}}$

FOOT VALVES WITH STRAINER

Poppet Disc   **Hinged Disc**

$K = 420 f_T$

$K = 75 f_T$

Minimum pipe velocity (fps) for full disc lift
$= 15 \sqrt{\overline{V}}$

Minimum pipe velocity (fps) for full disc lift
$= 35 \sqrt{\overline{V}}$

If:
$\beta = 1 \ldots K_1 = 300 f_T$
$\beta < 1 \ldots K_2 =$ Formula 7

If:
$\beta = 1 \ldots K_1 = 350 f_T$
$\beta < 1 \ldots K_2 =$ Formula 7

Minimum pipe velocity (fps) for full disc lift
$= 60 \, \beta^2 \sqrt{\overline{V}}$

BALL VALVES

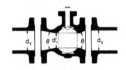

If: $\beta = 1, \theta = 0 \ldots\ldots\ldots\ldots\ldots K_1 = 3 f_T$
$\beta < 1$ and $\theta \gtrsim 45° \ldots\ldots\ldots K_2 =$ Formula 5
$\beta < 1$ and $45° < \theta \gtrsim 180° \ldots K_2 =$ Formula 6

If:
$\beta = 1 \ldots K_1 = 55 f_T$
$\beta < 1 \ldots K_2 =$ Formula 7

If:
$\beta = 1 \ldots K_1 = 55 f_T$
$\beta < 1 \ldots K_2 =$ Formula 7

Minimum pipe velocity (fps) for full disc lift
$= 140 \, \beta^2 \sqrt{\overline{V}}$

BUTTERFLY VALVES

Sizes 2 to 8″ $\ldots K = 45 f_T$
Sizes 10 to 14″ $\ldots K = 35 f_T$
Sizes 16 to 24″ $\ldots K = 25 f_T$

in SI units
$$h_{fs} = 0.031 \, \frac{6.1}{0.0525} \times \frac{1.75^2}{2 \times 9.807} = 0.56 \text{ m}$$

The inner diameter of the discharge pipe is 1.610 in (40.89 mm), and from Fig. 32, $\epsilon/D = 0.0011$. From Eq. 9

in USCS units
$$V = \frac{60}{1.601^2} \times 0.408 = 9.44 \text{ ft/s}$$

$$VD'' = 9.44 \times 1.601 = 15.2 \text{ ft/s} \times \text{in}$$

TABLE 6d

PLUG VALVES AND COCKS

Straight-Way

3-Way

If: $\beta = 1$,
$K_1 = 18 f_T$

If: $\beta = 1$,
$K_1 = 30 f_T$

If: $\beta = 1$,
$K_1 = 90 f_T$

If: $\beta < 1 \ldots K_2 = $ Formula 6

STANDARD ELBOWS

90°

$K = 30 f_T$

45°

$K = 16 f_T$

STANDARD TEES

Flow thru run . . . . $K = 20 f_T$
Flow thru branch . . . . $K = 60 f_T$

MITRE BENDS

∢	K
0°	2 f_T
15°	4 f_T
30°	8 f_T
45°	15 f_T
60°	25 f_T
75°	40 f_T
90°	60 f_T

90° PIPE BENDS AND FLANGED OR BUTT-WELDING 90° ELBOWS

r/d	K	r/d	K
1	20 f_T	8	24 f_T
1.5	14 f_T	10	30 f_T
2	12 f_T	12	34 f_T
3	12 f_T	14	38 f_T
4	14 f_T	16	42 f_T
6	17 f_T	20	50 f_T

The resistance coefficient, K_B, for pipe bends other than 90° may be determined as follows:

$$K_B = (n - 1)\left(0.25\ \pi\ f_T\frac{r}{d} + 0.5\ K\right) + K$$

n = number of 90° bends
K = resistance coefficient for one 90° bend (per table)

CLOSE PATTERN RETURN BENDS

$K = 50 f_T$

PIPE ENTRANCE

Inward Projecting

$K = 0.78$

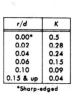

r/d	K
0.00*	0.5
0.02	0.28
0.04	0.24
0.06	0.15
0.10	0.09
0.15 & up	0.04

*Sharp-edged

Flush

For K, see table

PIPE EXIT

Projecting   **Sharp-Edged**   **Rounded**

$K = 1.0$ $K = 1.0$ $K = 1.0$

TABLE 6e Pipe Schedule for Different Classes of Valves and Fittings Associated with K Factors used in Tables 6a to 6d

Class	Schedule
300 and lower	40
400 and 600	80
900	120
1500	160
2500 (sizes ½ to 6 in)[a]	XXS
2500 (sizes 8 in and up)[a]	160

[a]In × 25.4 = mm.

SOURCE: Ref. 4.

in SI units
$$V = \frac{13.6}{4.089^2} \times 3.54 = 2.88 \text{ m/s}$$

$$VD = 2.88 \times 0.04089 = 0.118 \text{ m/s} \times \text{m}$$

$$(VD'' = 0.118 \times 129.2 = 15.2 \text{ ft/s} \times \text{in})$$

From Fig. 33, $Re = 1.5 \times 10^4$, and from Fig. 31, $f = 0.030$. From Eq. 16

in USCS units
$$h_{fd} = 0.030 \frac{300 \times 12}{1.610} \times \frac{9.44^2}{2 \times 32.17} = 92.9 \text{ ft}$$

in SI units
$$h_{fd} = 0.030 \frac{91.5}{0.04089} \times \frac{2.88^2}{2 \times 9.807} = 28.39 \text{ m}$$

The valve and fitting losses from Tables 5 and Eq. 20 are 2-in (51-mm) bellmouth, $K = 0.05$:

USCS units
$$h_{f1} = 0.05 \frac{5.73^2}{2 \times 32.17} = 0.026 \text{ ft}$$

in SI units
$$h_{f1} = 0.05 \frac{1.75^2}{2 \times 9.807} = 0.0078 \text{ m}$$

2-in (51-mm) SR 90° elbow, $K = 0.95 \pm 30\%$

in USCS units
$$h_{f2} = 0.95 \frac{5.73^2}{2 \times 32.17} = 0.48 \pm 0.14 \text{ ft}$$

in SI units
$$h_{f2} = 0.95 \frac{1.75^2}{2 \times 9.807} = 0.15 \pm 0.044 \text{ m}$$

2-in (51-mm) gate valve, $K = 0.16 \pm 25\%$

in USCS units
$$h_{f3} = 0.16 \frac{5.73^2}{2 \times 32.17} = 0.082 \pm 0.021 \text{ ft}$$

in SI units
$$h_{f3} = 0.16 \frac{1.75^2}{2 \times 9.807} = 0.0250 \pm 0.0062 \text{ m}$$

1½-in (38-mm) gate valve, $K = 0.19 \pm 25\%$

in USCS units
$$h_{f4} = 0.19 \frac{9.44^2}{2 \times 32.17} = 0.263 \pm 0.066 \text{ ft}$$

in SI units
$$h_{f4} = 0.19 \frac{2.88^2}{2 \times 9.807} = 0.0803 \pm 0.02 \text{ m}$$

1½-in (38-mm) swing check valve, $K = 2.5 \pm 30\%$

in USCS units
$$h_{f5} = 2.5 \frac{9.44^2}{2 \times 32.17} = 3.46 \pm 1.0 \text{ ft}$$

in SI units
$$h_{f5} = 2.5 \frac{2.88^2}{2 \times 9.807} = 1.06 \pm 0.32 \text{ m}$$

The total pipe, valve, and fitting losses are

in USCS units
$$\Sigma h_f = h_{fs} + h_{fd} + h_{f1} + h_{f2} + h_{f3} + h_{f4} + h_{f5}$$
$$= 1.84 + 92.9 + 0.026 + 0.48 + 0.082 + 0.263$$
$$+ 3.46 = 99.05 \text{ ft}$$

Total variation $= \pm(0.14 + 0.021 + 0.066 + 1.0) = \pm 1.23 \text{ ft}$

in SI units
$$\Sigma h_f = h_{fs} + h_{fd} + h_{f1} + h_{f2} + h_{f3} + h_{f4} + h_{f5}$$
$$= 0.56 + 28.39 + 0.0078 + 0.044 + 0.0250 + 0.0803$$
$$+ 1.06 = 30.17 \text{ m}$$
$$\text{Total variation} = \pm(0.044 + 0.0062 + 0.02 + 0.32) = \pm 0.39 \text{ m}$$

EXAMPLE 13   Solve Example 12 using resistance coefficients from Tables 6. Suction pipe:

in USCS units $h_{fs} = 1.84$ ft   (same as in Example 12)

in SI units $h_{fs} = 0.56$ m   (same as in Example 12)

Discharge pipe:

in USCS units $h_{fd} = 92.9$ ft   (same as in Example 12)

in SI units $h_{fd} = 28.39$ m   (same as in Example 12)

Valve and fitting losses from Tables 6 and Eq. 20: 2-in (51-mm) bellmouth, $K = 0.04$

in USCS units $h_{f1} = 0.04 \dfrac{5.73^2}{2 \times 32.17} = 0.020$ ft

in SI units $h_{f1} = 0.04 \dfrac{1.75^2}{2 \times 9.807} = 0.0062$ m

2-in (51-mm) SR 90° elbow, $K = 30 f_T$

$$f_T = 0.019 \qquad \text{(from Table 5a)}$$
$$K = 30 \times 0.019 = 0.57$$

in USCS units $h_{f2} = 0.57 \dfrac{5.73^2}{2 \times 32.17} = 0.29$ ft

in SI units $h_{f2} = 0.57 \dfrac{1.57^2}{2 \times 9.807} = 0.089$ m

2-in (51-mm) gate valve, $\beta = 1$, $\theta = 0$, $K = 8f_T$

$$K = 8 \times 0.019 = 0.15$$

in USCS units $h_{f3} = 0.15 \dfrac{5.73^2}{2 \times 32.17} = 0.077$ ft

in SI units $h_{f3} = 0.15 \dfrac{1.75^2}{2 \times 9.807} = 0.023$ m

1½-in (38-mm) gate valve, $\beta = 1$, $\theta = 0$, $K = 8f_T$ for Schedule 80

$$K = 8 f_T(1.10/1.500)^4 = 10.62 f_T \text{ for Schedule 40}$$
$$f_T = 0.021 \qquad \text{(from Table 6a)}$$
$$K = 10.62 \times 0.021 = 0.22$$

in USCS units $h_{f4} = 0.22 \dfrac{9.44^2}{2 \times 32.17} = 0.30$ ft

in SI units $h_{f4} = 0.22 \dfrac{2.88^2}{2 \times 9.807} = 0.093$ m

1½-in (38-mm) swing check valve, $K = 100f_T$ (from Table 6b)

Minimum pipe velocity for full disk lift $= 35\sqrt{\overline{V}} = 35\sqrt{0.0189} = 4.81$ ft/s < 9.44 ft/s

$$K = 100 \times 0.021 = 2.1$$

in USCS units $\qquad\qquad\qquad\qquad h_{f5} = 2.1\dfrac{9.44^2}{2 \times 32.17} = 2.91$ ft

in SI units $\qquad\qquad\qquad\qquad\quad h_{f5} = 2.1\dfrac{2.88^2}{2 \times 9.807} = 0.89$ m

Total pipe, valve, and fitting losses:

in USCS units $\qquad \Sigma h_f = h_{fs} + h_{fd} + h_{f1} + h_{f2} + h_{f3} + h_{f4} + h_{f5}$

$\qquad\qquad\qquad\qquad\quad = 1.84 + 92.9 + 0.020 + 0.29 + 0.077 + 0.30 + 2.91$

$\qquad\qquad\qquad\qquad\quad = 98.34$ ft

in SI units $\qquad\quad\; \Sigma h_f = h_{fs} + h_{fd} + h_{f1} + h_{f2} + h_{f3} + h_{f4} + h_{f5}$

$\qquad\qquad\qquad\qquad\quad = 0.56 + 28.39 + 0.062 + 0.089 + 0.023 + 0.093 + 0.89$

$\qquad\qquad\qquad\qquad\quad = 30.05$ m

INCREASERS   The head lost when there is a sudden increase in pipe diameter, with velocity changing from V_1 to V_2 in the direction of flow, can be calculated analytically. Computed results have been confirmed experimentally to be true to within $\pm 3\%$. The head loss is expressed as shown below, with K computed to be equal to unity:

$$h = K\frac{(V_1 - V_2)^2}{2g} = K\left[1 - \left(\frac{D_1}{D_2}\right)^2\right]^2\frac{V_1^2}{2g} = K\left[\left(\frac{D_2}{D_1}\right)^2 - 1\right]^2\frac{V_2^2}{2g} \qquad (23)$$

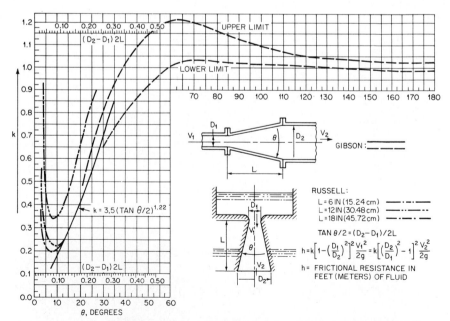

FIG. 39   Resistance coefficients for increasers and diffusers. ($D = $ in or ft [m]; $V = $ ft/s [m/s]). (Ref. 5)

The value of K is also approximately equal to unity if a pipe discharges into a relatively large reservoir. This indicates that all the kinetic energy $V_1^2/2g$ is lost and V_2 equals zero.

The loss of head for a gradual increase in pipe diameter when the flow is through a diffuser can be found from Fig. 39 and Table 6a. The diffuser converts some of the kinetic energy to pressure. Values for the coefficient used with Eq. 23 for calculating head loss are shown in Fig. 39. The optimum total angle appears to be 7.5°. Angles greater than this result in shorter diffusers and less friction, but separation and turbulence occur. For angles greater than 50°, it is preferable to use a sudden enlargement.

REDUCERS   Figure 40 and Table 6a give values of the resistance coefficient to be used for sudden reducers.

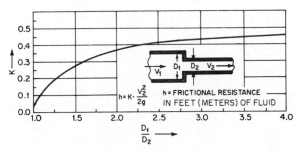

FIG. 40   Resistance coefficients for reducers. (Ref. 5)

BENDS   Figure 41 may be used to determine the resistance coefficient for 90° pipe bends of uniform diameter. Figure 42 gives resistance coefficients for bends that are less than 90° and can be used for surfaces having moderate roughness, such as clean steel and cast iron. Figures 41 and 42 are not recommended for elbows with R/D below 1. Tables 6d and 7 give values of resistance coefficients for miter bends.

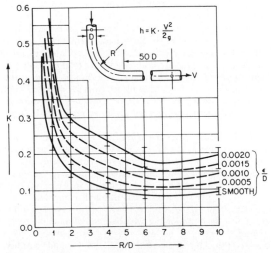

FIG. 41   Resistance coefficients for 90° bends of uniform diameter. (Ref. 5)

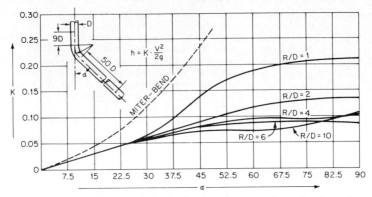

FIG. 42   Resistance coefficients for bends of uniform diameter and smooth surface at Reynolds number $\approx 2.25 \times 10^5$. (Ref. 5)

PUMP SUCTION ELBOWS   Figures 43 and 44 illustrate two typical rectangular to round reducing suction elbows. Elbows of this configuration are sometimes used under dry-pit vertical volute pumps. These elbows are formed in concrete and are designed to require a minimum height, thus permitting a higher pump setting with reduced excavation. Figure 43 shows a long-radius elbow, and Fig. 44 a short-radius elbow. The resulting velocity distribution into the impeller eye and the loss of head are shown for these two designs.

METERS   Orifice, nozzle, and venturi meters (Figs. 45–47) are used to measure rate of flow. These meters, however, introduce additional loss of head into the pumping system. Each of these meters is designed to create a pressure differential through the primary element. The magnitude of the pressure differential depends on the velocity and density of the liquid and the design of the element. The primary element restricts the area of flow, increases the velocity, and decreases the pressure. An expanding section following the primary element provides pressure head recovery and determines the meter efficiency. The pressure differential between inlet and throat taps measure rate of flow; the pressure differential between inlet and outlet taps measures the meter head loss (an outlet tap is not usually provided). Of the three types, venturi meters offer the least resistance to flow, and orifice meters the most.

When meters are designed and pressure taps located as recommended,[6] Figs. 48–50 may be used to estimate the overall pressure loss. In these figures, the loss of pressure is expressed as a percentage of the differential pressure measured at the appropriate taps and values are given for various sizes of meters. This loss of pressure is also the meter total head, or energy loss, since there is no change in velocity head if the pipe inside diameters are the same at the various measuring points. The meter loss of head should be in units of feet (meters) of liquid pumped if other system losses are expressed this way. Reference 6 should be consulted for information concerning formulas and coefficients for calculating differential pressure versus rate of flow.

Screens, Perforated Plates, and Bar Racks

Obstructions to the flow of liquid in the form of multiple orifices uniformly distributed across an open or closed conduit may be used to remove solids, throttle flow, and produce or reduce turbulence. They may be used upstream or downstream from a pump, depending on their purpose, and they therefore introduce a loss of head which must be accounted for. When an obstruction is placed upstream from a pump, a significant reduction in suction pressure and NPSH available can result.

The loss of head results from an increase in velocity at the entrance to the openings, friction, and the sudden decrease in velocity following the expansion of the numerous liquid jets. The total head loss is a function of the ratio of the total area of the openings to the area of the conduit before the obstruction, the thickness of the obstruction, the Reynolds number, and the velocities. Various investigators have determined values for resistance coefficients which can be multiplied

TABLE 7 Resistance Coefficients for Miter Bends at Reynolds Number $\approx 2.25 \times 10^5$

Top row of miter bend diagrams:

Angle	K_s	K_r
5°	0.016	0.024
10°	0.034	0.044
15°	0.042	0.062
22.5°	0.065	0.154
30°	0.130	0.165
45°	0.236	0.320
60°	0.471	0.684
90°	1.129	1.265

Second row of bend diagrams:

Description	K_s	K_r
1.17D, 22.5°	0.112	0.204
1.23D, 30°	0.150	0.268
2.37D, 30°	0.143	0.227
1.06D, 20°	0.108	0.236
1.23D 2.37D, 30°	0.188	0.320
2.37D 1.23D, 30°	0.202	0.323
1.44D, 30°/60°	0.400	0.534
1.44D, 60°/30°	0.400	0.601

Bottom row data tables:

45° / 22.5°:

a/D	K_s	K_r
0.71	0.507	0.510
0.943	0.230	0.415
1.174	0.333	0.384
1.42	0.261	0.377
1.50*	0.280	0.376
1.88	0.269	0.390
2.58	0.338	0.429
3.14	0.346	0.426
3.72	0.356	0.490
4.89	0.389	0.455
5.59	0.392	0.444
6.29	0.399	0.444

22.5°:

a/D	K_s	K_r
1.186	0.120	0.294
1.40	0.125	0.252
1.50*	—	0.250
1.63	0.124	0.266
1.86	0.117	0.272
2.325	0.096	0.317
2.40*	0.095	—
2.91	0.108	0.317
3.49	0.150	0.318
4.65	0.148	0.310
6.05	0.142	0.313

30°:

a/D	K_s	K_r
1.23	0.195	0.347
1.44	0.196	0.320
1.67	0.150	0.300
1.70*	0.149	0.299
1.91	0.154	0.312
2.37	0.167	0.337
2.96	0.172	0.342
4.11	0.190	0.354
4.70	0.192	0.360
6.10	0.201	0.360

30°:

a/D	K_s	K_r
1.23	0.157	0.300
1.67	0.156	0.378
2.37	0.143	0.264
3.77	0.160	0.242

*OPTIMUM VALUE OF α INTERPOLATED

K_s = RESISTANCE COEFFICIENT FOR SMOOTH SURFACE
K_r = RESISTANCE COEFFICIENT FOR ROUGH SURFACE, $\frac{\varepsilon}{D} \cong 0.0022$

SOURCE: Ref. 5.

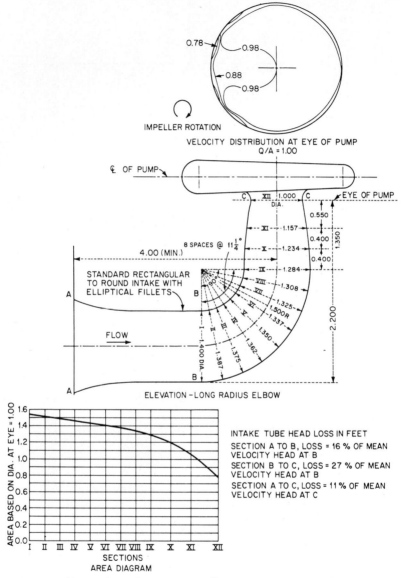

FIG. 43 Head loss in a long-radius pump suction elbow. (in × 25.4 = mm) (Ref. 16)

by the approach velocity head to obtain the loss through these obstructions. According to Idel'chik,[7] loss of head h in feet (meters) may be calculated from the equation

$$h = K_1 \frac{V_1^2}{2g} \tag{24}$$

where K_1 = resistance coefficient
V_1 = average velocity in the conduit approaching the obstruction, ft/s (m/s)
g = acceleration of gravity, 32.17 ft/s² (9.807 m/s²)

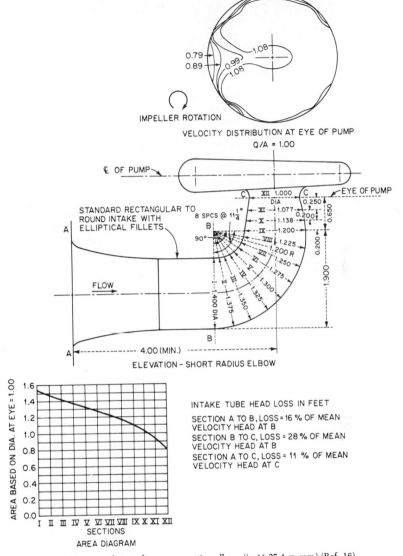

FIG. 44 Head loss in a short-radius pump suction elbow. (in × 25.4 = mm) (Ref. 16)

ROUND-WIRE MESH SCREENS   For flow having Reynolds numbers equal to or greater than 400, the resistance coefficient for flow through a round-wire, plain square mesh screen (Fig. 51a) may be estimated as a function of percentage of open area using the equations

in USCS units
$$Re = \frac{V_o W_d}{\nu 12} \geq 400 \qquad (25a)$$

in SI units
$$Re = \frac{V_o W_d}{\nu 1000} \geq 400 \qquad (25b)$$

$$K_1 = k_0 \left(1 - \frac{A_r}{100} \right) + \left(\frac{100}{A_r} - 1 \right)^2 \tag{26}$$

$$A_r = 100 \left(\frac{S}{S + W_d} \right)^2 \tag{27}$$

$$A_r = 100(1 - MW_d)^2 \tag{28}$$

where Re = screen Reynolds number referred to wire diameter
  V_o = velocity through area of rectangular opening = $100V_1/A_r$, ft/s (m/s)
  W_d = screen wire diameter, in (mm)
  ν = kinematic viscosity, ft²/s (m²/s)
  k_0 = 1.0 for new, perfectly clean screens to 1.3 for normal screens
  A_r = percentage of open area
  S = square space between wires, in (mm)
  M = mesh of screen or number of wires per in (mm)

Using water in a flow range of Reynolds numbers of approximately 60 to 800, Padmanabhan and Vigander[8] experimented with 1.3- to 12.0-mesh screens, 51.4 to 56% open area, and found

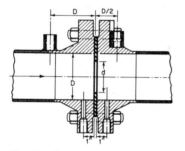

their results to be comparable with those of other investigators. Values of the coefficient of resistance K_1, from the Padmanabhan-Vigander tests and others, vary from 2 decreasing asymptotically to 1 with increase in Reynolds number (Eq. 25) up to approximately 1000 for 47 to 56% open area. Smaller percentage open area values have larger coefficients.

Armour and Cannon,[9] using nitrogen as the fluid, tested various types of fine woven wire screens and derived equations for pressure drop in terms of a friction factor and the screen Reynolds number. Approach velocities ranged from 0.1 to 30 ft/s (0.03 to 9 m/s), resulting in Reynolds numbers from 350 to 275,000. To simplify calculations, screen geometry constants C_1 = (surface area to unit volume ratio)²

FIG. 45 Thin-plate, square-edged orifice meter, showing alternate locations of pressure taps. (Ref. 6)

$\times$ (pore diameter) and C_2 = (screen thickness) $\div$ (void fraction)² $\times$ (pore diameter) have been added to the reference authors' equations to obtain the following expression for screen head loss h in feet (meters):

$$h = C_2 \frac{V_1}{g} (8.61\nu C_1 + 0.52V_1) \tag{29}$$

where V_1 = average velocity in the conduit approaching the screen, ft/s (m/s)
  Table 8 lists C_1 and C_2 values for a sample of plain square screens tested by Armour and Cannon.

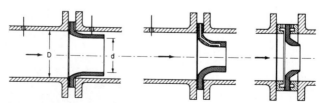

FIG. 46 Shapes of flow nozzle meters and locations of pressure taps. (Ref. 6)

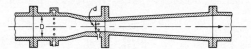

FIG. 47 Herschel-type venturi meter, showing locations of pressure taps. (Ref. 6)

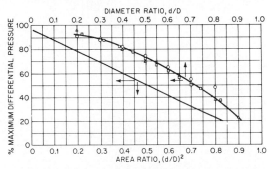

FIG. 48 Overall pressure loss across thin-plate orifices. (Ref. 6)

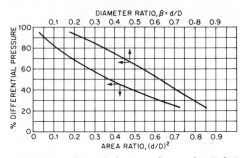

FIG. 49 Overall pressure loss across flow nozzles. (Ref. 6)

PERFORATED PLATES AND BAR RACKS   For flow having Reynolds numbers equal to or greater than 10^5, the resistance coefficients for flow through thick, square-edge perforated plates with round (Fig. 51b) or rectangular (Fig. 51c) openings and racks with rectangular cross-section bars (length = 5 times thickness; Fig. 51g) may be calculated using Eq. 24, Table 9, and the following equations:

$$Re = \frac{V_o D_h}{\nu} \geq 10^5 \qquad (30)$$

in USCS units

$$D_h = \frac{a}{3p} \quad \text{or} \quad D_h = \frac{d_o}{12} \qquad (31a)$$

in SI units

$$D_h = 0.004\,\frac{a}{p} \quad \text{or} \quad D_h = \frac{d_o}{1000} \qquad (31b)$$

FIG. 50 Overall pressure loss across venturi tubes. (Ref. 6)

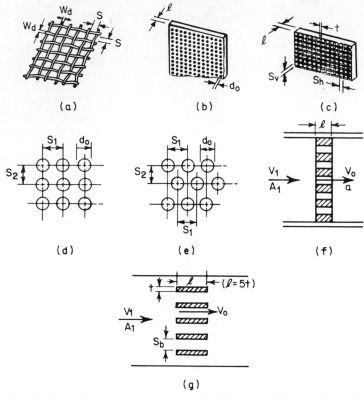

FIG. 51 Explanation of terms used in Eq. 25–28 and 30–37 and Tables 8 and 9 for calculating resistance coefficient K_1. (a) Round-wire, plain square, mesh screen. (b) Round-hole perforated plate. (c) Rectangular-grid perforated plate. (d) Perforated plate, holes in vertical columns. (e) Perforated plate, staggered holes. (f) Perforated plate, cross section. (g) Rectangular bar rack, cross section.

for plates with round holes

$$A_r = 100 \left(\frac{0.785 d_o^2}{S_1 S_2} \right) \tag{32}$$

for plates with single hole in center

$$A_r = 100 \left(\frac{d_o}{d_1} \right)^2 \tag{33}$$

TABLE 8   Plain Square Screen Geometry

Screen mesh size M, in^{-1} (mm^{-1})	Wire diameter W_d, in $\times 10^{-4}$ (mm $\times 10^{-4}$)	Open area A_r, %	Screen constant C_1, ft$^{-1} \times 10^3$ (m$^{-1} \times 10^3$)	Screen constant C_2
30×30 (1.18×1.18)	9.45 (240)	94.4	2.553 (8.376)	1.553
150×150 (5.91×5.91)	2.36 (59.9)	93.0	8.267 (27.12)	2.594
250×250 (9.84×9.84)	1.69 (42.9)	91.7	16.21 (53.18)	2.934
400×400 (15.7×15.7)	1.00 (25.4)	92.2	30.68 (100.7)	2.680

SOURCE: Ref. 9.

TABLE 9 Values of Resistance Coefficient K_1 for Perforated Plates and Bar Racks

$\dfrac{l}{Dh}$	$A_r/100$ or A_o/A_1															
	0.02	0.04	0.06	0.08	0.10	0.15	0.20	0.25	0.30	0.40	0.50	0.60	0.70	0.80	0.90	1.0
0	7000	1670	730	400	245	96.0	51.5	30.0	18.2	8.25	4.00	2.00	0.97	0.42	0.13	0
0.2	6600	1600	687	374	230	94.0	48.0	28.0	17.4	7.70	3.75	1.87	0.91	0.40	0.13	0.01
0.4	6310	1530	660	356	221	89.0	46.0	26.5	16.6	7.40	3.60	1.80	0.88	0.39	0.13	0.01
0.6	5700	1380	590	322	199	81.0	42.0	24.0	15.0	6.60	3.20	1.60	0.80	0.36	0.13	0.01
0.8	4680	1130	486	264	164	66.0	34.0	19.6	12.2	5.50	2.70	1.34	0.66	0.31	0.12	0.02
1.0	4260	1030	443	240	149	60.0	31.0	17.8	11.1	5.00	2.40	1.20	0.61	0.29	0.11	0.02
1.4	3930	950	408	221	137	55.6	28.4	16.4	10.3	4.60	2.25	1.15	0.58	0.28	0.11	0.03
2.0	3770	910	391	212	134	53.0	27.4	15.8	9.90	4.40	2.20	1.13	0.58	0.28	0.12	0.04
3.0	3765	913	392	214	132	53.5	27.5	15.9	10.0	4.50	2.24	1.17	0.61	0.31	0.15	0.06
4.0	3775	930	400	215	132	53.8	27.7	16.2	10.0	4.60	2.25	1.20	0.64	0.35	0.16	0.08
5.0	3850	936	400	220	133	55.5	28.5	16.5	10.5	4.75	2.40	1.28	0.69	0.37	0.19	0.10
6.0	3870	940	400	222	133	55.8	28.5	16.6	10.5	4.80	2.42	1.32	0.70	0.40	0.21	0.12
7.0	4000	950	405	230	135	55.9	29.0	17.0	10.9	5.00	2.50	1.38	0.74	0.43	0.23	0.14
8.0	4000	965	410	236	137	56.0	30.0	17.2	11.1	5.10	2.58	1.45	0.80	0.45	0.25	0.16
9.0	4080	985	420	240	140	57.0	30.0	17.4	11.4	5.30	2.62	1.50	0.82	0.50	0.28	0.18
10	4110	1000	430	245	146	59.7	31.0	18.2	11.5	5.40	2.80	1.57	0.89	0.53	0.32	0.20

NOTE: l = thickness of perforated plate or length of bars, ft (m); D_h, A_r, A_o, and A_1 are as defined following Eq. 37.

SOURCE: Ref. 7.

for plates with square openings ($S = S_h = S_v$)

$$A_r = 100 \left(\frac{S}{S + t} \right)^2 \tag{34}$$

for plates with rectangular openings

$$A_r = 100 \left[\frac{S_h S_v}{(S_h + t)(S_v + t)} \right] \tag{35}$$

for any plate or bar rack

$$A_r = 100 \frac{A_o}{A_1} \tag{36}$$

for single vertical or hosizontal bar racks

$$A_r = 100 \left(\frac{S_b}{S_b + t} \right) \tag{37}$$

where Re = Reynolds number referred to hydraulic diameter
$\quad V_o$ = velocity through area of opening, ft/s (m/s)
$\quad D_h$ = hydraulic diameter (= diameter if openings are round holes), ft (m)
$\quad \nu$ = kinematic viscosity, ft^2/s (m^2/s)
$\quad a$ = area of single opening, in^2 (mm^2)
$\quad p$ = perimeter of single opening, in (mm)
$\quad d_o$ = diameter of hole, in (mm)
$\quad A_r$ = percentage of open area
$\quad S_1$ = horizontal spacing of holes, in (mm)
$\quad S_2$ = vertical spacing of holes, in (mm)
$\quad d_1$ = diameter of approach, in (mm)
$\quad S_h$ = horizontal space between vertical bars, in (mm)
$\quad S_v$ = vertical space between horizontal bars, in (mm)
$\quad t$ = thickness of plate laths or bars, in (mm)
$\quad A_o$ = total area of openings, ft^2 (m^2)
$\quad A_1$ = total area of approach, ft^2 (m^2)
$\quad S_b$ = space between single vertical or horizontal bars, in (mm)

The loss of head through perforated plates may also be calculated by using an *orifice coefficient C*, as suggested by Smith and Van Winkle[10] and Kolodzie and Van Winkle.[11] Test results using air and other gases with equilateral-triangle pitch perforated plates are shown in Table 10 for Reynolds numbers 400 to 20,000. Using single-orifice relations, the following expression equates flow rate to pressure drop:

$$\omega = C A_o \sqrt{ \frac{2g\, \Delta P}{\gamma \left[1 - \left(\dfrac{A_r}{100} \right)^2 \right]} } \tag{38}$$

TABLE 10   Orifice Coefficients C for Equilateral-Triangle Pitch Perforated Plates

Pitch/ D_h	Re	l/D_h								
		0.33	0.43	0.50	0.52	0.65	0.75	0.80	1.0	2.0
2.0	400–4000	0.74–0.69	0.75–0.71	0.76–0.74	—	0.79–0.85	0.82–0.86	0.83–0.87	0.84–0.89	0.77–0.92
2.0	4000–20,000	0.69	0.71	0.74	0.77	0.85	0.86	0.87	0.89	0.92
3.0	400–4000	0.70–0.67	0.72–0.68	0.73–0.72	—	0.75–0.81	0.76–0.82	0.77–0.84	0.78–0.85	0.70–0.87
3.0	4000–20,000	0.67	0.68	0.72	0.74	0.81	0.82	0.84	0.85	0.87
4.0	400–4000	0.71–0.66	0.72–0.67	0.72–0.69	—	0.73–0.77	0.74–0.79	0.75–0.81	0.76–0.83	0.69–0.84
4.0	4000–20,000	0.66	0.67	0.69	0.72	0.77	0.79	0.81	0.83	0.84
5.0	400–4000	0.68–0.65	0.69–0.66	0.70–0.68	—	0.72–0.76	0.73–0.77	0.74–0.78	0.75–0.81	0.65–0.82
5.0	4000–20,000	0.65	0.66	0.68	0.71	0.76	0.77	0.78	0.81	0.82

NOTE: Pitch/D_h = pitch-to-hole-diameter ratio; l/D_h = plate-thickness-to-hole diameter ratio; Re = Reynolds number.

SOURCE: Refs. 10 and 11.

where ω = flow rate, ft³/s (m³/s)
  C = orifice coefficient from Table 10
  A_o = total area of openings, ft² (m²)
  g = acceleration of gravity, 32.17 ft/s² (9.807 m²)
  ΔP = pressure drop, lb/ft² (N/m²)
  γ = specific weight (force) of liquid, lb/ft³ (N/m³)
  A_r = percentage of open area

The resistance coefficient K_1 and the orifice coefficient C may be interchanged in Eqs. 24 and 38 for loss through perforated plates:

$$C = \sqrt{\frac{1 - \left(\dfrac{A_r}{100}\right)^2}{K_1 - \left(\dfrac{A_r}{100}\right)^2}} \tag{39}$$

$$K_1 = \frac{1 - \left(\dfrac{A_r}{100}\right)^2}{C^2 \left(\dfrac{A_r}{100}\right)^2} \tag{40}$$

Throttling Orifices In addition to measuring flow, orifices can be used to (a) reduce flow by adding artificial resistance to increase system head, (b) dissipate energy to provide a desired pressure reduction, and (c) create a high-velocity jet. Orifices for these purposes are called *throttling orifices*. In order to maintain a desired minimum flow to prevent damage to a centrifugal pump, a throttling orifice, or a series of throttling orifices, can be used in the bypass system. The orifice provides additional bypass resistance to maintain the required bypass flow. (See Subsec. 2.3.4.)

A throttling orifice can be fabricated by drilling a hole in a metal plate (or through bar stock) which, when inserted between flanges in a pipe (or threaded to pipe), will create the desired loss of head at the design flow. The resistance coefficient K_1 may be calculated as if the throttling device were a single hole in a perforated plate, using Table 9 and Eqs. 30, 31, and 33 and the loss of head calculated using Eq. 24.

Energy is dissipated through a throttling orifice because pressure head is converted to velocity head and this conversion is followed by an inefficient pressure head recovery. Conditions may exist at the orifice *vena contracta* which could cause vaporization of the high-velocity, low-pressure liquid jet. Care must be taken in selecting the orifice size to avoid excessive cavitation noise and/or choke flow. An orifice cavitation index used to check the orifice selection is described by Tung and Mikasinovic,[12] who also discuss the use of orifices in series to avoid cavitation.

Although orifices are used to meter flow, accuracy requires that they be fabricated to standard proportions and that pressure taps be precisely located (Ref. 6 and Fig. 45). A distinction should be made between *meter differential pressure*, which is used to measure flow and is the difference in pressure at the upstream and downstream *vena contracta* taps, and *meter loss of head* calculated using the resistance coefficient K_1, which is the total overall loss of energy as measured at the upstream tap and past the downstream *vena contracta* tap. The loss of head through a standard orifice meter should be calculated as discussed previously under Meters.

PUMP FLOW, HEAD, AND POWER IN VARYING TEMPERATURE SYSTEMS

In a pumping system where the weight of liquid pumped is constant, the volumetric flow rate will vary through system components having different temperatures. An example would be the condensate and feedwater system in a steam power plant. The following equation may be used to calculate volumetric flow rate using the specific gravity corresponding to the temperature of the liquid at the location in the system where flow is required:

in USCS units,
$$\text{gpm} = \frac{\text{lb/h}}{500\,(\text{sp. gr.})} \tag{41a}$$

in SI units,
$$\text{m}^3/\text{h} = \frac{\text{kg/h}}{998\,(\text{sp. gr.})} \tag{41b}$$

When calculating the total head required of a pump or pumps to overcome total system component losses, the actual volumetric flow rate and temperature through each component must be used, since head loss is a function of velocity and viscosity. Information provided in this chapter permits computing pipe, valve, and fitting losses in ft·lb/lb or ft (N·m/N or m) of liquid passing through the component. If the pump is at a location in the system where the temperature is different than at the locations where the head losses are calculated, the total head to be produced by the pump cannot be found by simply adding the individual component heads.

The pump head required to produce a specific increase in pressure varies inversely with specific gravity. Therefore, to calculate pump total head, either the component total heads must be converted to pump equivalent heads and added together, or all component losses must be expressed in pressure units so that the total of these pressures can be converted to an equivalent pump head at the pump temperature. From Eqs. 2 and 3, the pump equivalent head is

$$h_2 = \frac{\gamma_1}{\gamma_2}\,h_1$$

or

$$h_2 = \frac{\text{sp. gr.}_1}{\text{sp. gr.}_2}\,h_1$$

where the subscript 1 denotes the component and the subscript 2 denotes the pump. From Eq. 6, individual component head losses can be converted to lb/ft^2 (N/m^2) using

$$p_1 = h_1\gamma_1$$

Also for Eq. 6,

$$TH = \frac{P_\Delta}{\gamma_2} = \frac{\Sigma p_1}{\gamma_2}$$

from which total component losses in lb/ft^2 (N/m^2) can be converted to an equivalent total pump head in feet (meters) of liquid to be produced at the pumping temperature.

In a varying temperature system, the positive or negative static head required to raise or lower the liquid pumped is not simply a difference in elevation. A pump must produce pressure in a pipe to raise liquid; the pressure required is proportional to the specific weight (force) of the liquid. The static head required at the pump should be found by expressing the suction and discharge elevation heads Z as pressures at the pump suction and discharge connections (corrected to the reference datum plane, if it is not at the pump centerline elevation) and using actual specific weights (forces) along the pipe. This differential pressure, in lb/ft^2 (N/m^2), is then converted to an equivalent static head using the specific weight (force) or specific gravity at the pump in the above appropriate equations.

When designing a pumping system, there may be several locations for placing a pump to produce a specified flow rate in lb/h (kg/h) and an increase in pressure in lb/ft^2 (N/m^2). If the temperature of the liquid varies at the different pump locations being considered (for example, before or after a feedwater heater in a steam power plant), the pump total head in feet (meters) and the volumetric flow rate in gpm (m^3/h) will vary. While it is true that pump power is proportional to the product of volumetric flow × head × specific gravity, higher pumping temperature (lower specific gravity) will nevertheless result in higher pumping power. For the same conditions of weight (or mass) flow and differential pressure, pump power varies inversely with specific gravity because of the following relationships:

$$\text{Pump power} \propto \text{volumetric flow} \times \text{total head} \times \text{sp. gr.}$$

$$\text{Volumetric flow} \propto \frac{\text{weight or mass flow}}{\text{sp. gr.}}$$

$$\text{Total head} \propto \frac{\text{pressure}}{\text{sp. gr.}}$$

therefore,

$$\text{Pump power} \propto \frac{\text{weight or mass flow}}{\text{sp. gr.}} \times \frac{\text{pressure}}{\text{sp. gr.}} \times \text{sp. gr.}$$

then,

$$\text{Pump power} \propto \frac{(\text{weight or mass flow}) \times \text{pressure}}{\text{sp. gr.}}$$

Following are formulas for calculating pump input power in brake horsepower or brake kilowatts:

in USCS units,
$$bhp = \frac{\text{gpm} \times TH \times \text{sp. gr.}}{3960 \times \text{pump eff.}} \tag{42a}$$

$$= \frac{\text{lb/h} \times \text{lb/in}^2}{858,600 \times \text{pump eff.} \times \text{sp. gr.}} \tag{43a}$$

in SI units,
$$bkW = \frac{\text{m}^3/\text{h} \times \text{m} \times \text{sp. gr.}}{367.7 \times \text{pump eff.}} \tag{42b}$$

$$= \frac{\text{kg/h} \times \text{kPa}}{3,593,000 \times \text{pump eff.} \times \text{sp. gr.}} \tag{43b}$$

REFERENCES

1. Stepanoff, A. J.: *Centrifugal and Axial Flow Pumps*, 2d ed., Wiley, New York, 1948.

2. Moors, J. A.: "Criteria for the Development of Siphonic Action in Pumping Plant Discharge Lines," paper presented at Hydraulic Division, ASCE Annual Convention, New York, October 1957.

3. Richards, R. T.: "Air Binding in Water Pipelines," *AWWA J.* **53** (1962).

4. "Flow of Fluids through Valves, Fittings, and Pipe," Technical Paper 410, 15th printing, Crane, New York, 1980.

5. *Pipe Friction Manual*, 3d ed., Hydraulic Institute, Cleveland, 1961. See also *Engineering Data Book*, 1st ed., Hydraulic Institute, Cleveland, 1979.

6. *Fluid Meters*, 6th ed., American Society of Mechanical Engineers, New York, 1971.

7. Idel'chik, I. E.: *Handbook of Hydraulic Resistance* (translated from Russian), Israel Program for Scientific Translations, catalog no. 1505, 1966 (available from U.S. Department of Commerce, Washington, D.C.).

8. Padmanabhan, M., and S. Vigander: "Pressure Drop Due to Flow through Fine Mesh Screens," *ASCE J. Hydraulics Div.*, 1978, p. 1191.

9. Armour, J. C., and J. N. Cannon: "Fluid Flow through Woven Screens," *AIChE J.* **14**:415 (1968).

10. Smith, P. L., Jr., and M. Van Winkle: "Discharge Coefficients through Perforated Plates at Reynolds Numbers of 400 to 3000," *AIChE J.* **4**(3):266 (1958).

11. Kolodzie, P. A., Jr., and M. Van Winkle: "Discharge Coefficients through Perforated Plates," *AIChE J.* **3**(3):305 (1957).

12. Tung, P. C., and M. Mikasinovic: "Eliminating Cavitation from Pressure-Reducing Orifices," *Chem. Eng.*, December 1983, p. 69.

13. King, H. W., and E. Brater: *Handbook of Hydraulics*, 6th ed., McGraw-Hill, New York, 1976.

14. Streeter, V. L.: *Fluid Mechanics*, 5th ed., McGraw-Hill, New York, 1971.

15. Davis, C. V., and K. E. Sorensen: *Handbook of Applied Hydraulics*, 3d ed., McGraw-Hill, New York, 1969.

16. U.S. Interior Dep., Bureau of Reclamation: Turbines and Pumps, Design Standard no. 6, Washington, D.C., 1956.

FURTHER READING

Engineering Data Book, 1st ed., Hydraulic Institute, Cleveland, 1979.

Flow Meter Engineering Handbook, 4th ed., Minneapolis-Honeywell Regulator Company, Brown Instrument Division, Philadelphia, 1968.

Karassik, I. J., and R. Carter: *Centrifugal Pumps*, McGraw-Hill, New York, 1960.

McNown, J. S.: "Mechanics of Manifold Flow," *Proc. Amer. Soc. Civil Engl.* **79**(258), August 1953.

Simpson, L.L.: "Sizing Piping for Process Plants," *Chem. Eng.*, June 17, 1968, p. 192.

SECTION 8.2
BRANCH-LINE PUMPING SYSTEMS

J. P. MESSINA

In some systems the liquid leaving the pump or pumps will divide into a network of pipes. If the pump is of the centrifugal type, the total pump flow is dependent on the combined system resistance. The total pump flow and flow through each branch can be determined by the following methods. (Review Pump Total Head and System-Head Curves in Sec 8.1.)

BRANCHES IN CLOSED-LOOP SYSTEMS

Figure 1 illustrates a pump and network of piping consisting of three parallel branches in series with common supply and return headers. Junction points 1 and 2 need not be at the same elevation (provided the liquid density remains constant and the pipes flow full and free of vapor) because, in a closed-loop system, the net change in elevation is zero. Figure 2 shows the system total-head curves for each branch line and header considered independent of the others. These curves are constructed for several flow rates by adding the frictional resistances of the pipes, fittings, and head losses through the equipment serviced from point 1 to point 2. Curves A, B, C, and D therefore represent the variation in system resistance in feet (meters) versus flow through each branch and header.

If the valves are open in all branches, the total system resistance, total pump flow, and individual branch flows are found by the following method. First observe that (a) the total flow must be equal to the sum of the branch flows, (b) the head loss or pressure drop across each branch from junction 1 to junction 2 is identical, and (c) the flow divides to produce these identical head losses. Therefore, at several head points, add together the flow through each branch and obtain curve $A + B + C$. Header D is in series with branches A, B, and C, and their system heads are added together for several flow conditions to obtain curve $(A + B + C) + D$. On curve E, the head-capacity characteristics of a centrifugal pump, point X represents the pump flow since at this point the system total head and pump total head are equal. Point Y_{1-2} represents the total head across points 1 and 2, and this head determines the flow through each branch; consequently points a, b, and c give individual branch flows. Curve F represents the head-capacity character-

8.77

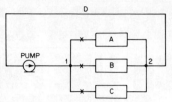

FIG. 1 Closed-loop pumping system with branch lines.

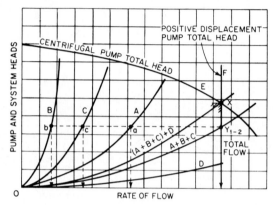

FIG. 2 System-head curves for pump and branch lines shown in Fig. 1 with all valves open.

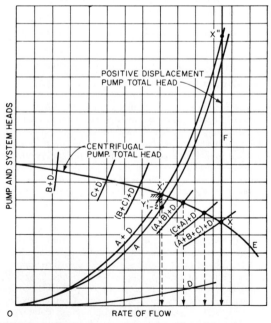

FIG. 3 System-head curves for pump and branch line shown in Fig. 1 with different combinations of open valves.

istics of a positive displacement pump (constant capacity) which would produce the same flow conditions.

If valve A is open and valves B and C are closed, Fig. 3 shows the construction of the curves required to determine pump flow point X'. Obviously the pump flow and branch A flow are the same. Note that the total flow at point X is less than when all valves are open as a result of an increase in system head. If all valves were open and the total flow were obtained by a positive displacement pump having a constant capacity curve F, closing valves B and C would not change the flow. The system head would, however, increase to point X'' and the head would be greater than for a centrifugal pump having curve E.

Also shown in Fig. 3 are the system total-head curves for different combinations of open valves A, B, and C and the resulting flow caused by a pump having characteristic curve E. For these various valve combinations, the head differential across the junction points is found by subtracting the head of curve D from the system total head for the condition investigated, e.g., point Y'_{1-2} for only valve A open. The intersection of a horizontal line through point Y'_{1-2} and the individual branch curves gives the branch flow, as illustrated in Fig. 2.

BRANCHES IN OPEN-ENDED SYSTEMS

Figure 4 illustrates a pump supplying three branch lines which are open-ended and terminate at different elevations. Figure 5 shows the system total-head curves for each branch line and main supply line considered independently of each other. These curves are constructed by starting at elevation heads Z_A, Z_B, Z_C, and Z_D at zero flow. To each of these heads is added the frictional resistances in each line for several flow rates. Frictional losses from the suction tank to junction 1 are included in curve D. Curves A, B, C, and D therefore represent the variation in system resistance in feet (meters) versus flow through each branch and supply line. Note that Z_D is negative because in line D there is a decrease in elevation to point 1.

The total head at the junction is the head Z_D in the suction tank measured above point 1 plus the pump total head less the frictional head loss h_{fD} in line D, and it varies with flow, as illustrated by curve F.

The total system resistance, total pump flow, and individual branch flows are found by the following method. First observe that (a) the total flow must be equal to the sum of the branch flows, (b) the frictional resistance plus the elevation head measured relative to junction 1 for each branch is identical and (c) the flow divides to produce these identical total branch heads. Therefore, at several head points, add together the flow of each branch to obtain curve $A + B + C$. Supply line D is in series with branches A, B, and C, and their system heads are added algebraically for several flow conditions to obtain curve $(A + B + C) + D$. If curve E is the head-capacity characteristics of a centrifugal pump, point X represents the pump flow since at this point

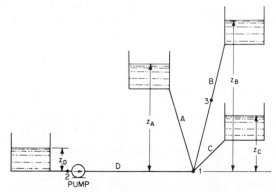

FIG. 4 Open-ended pumping system with branch lines.

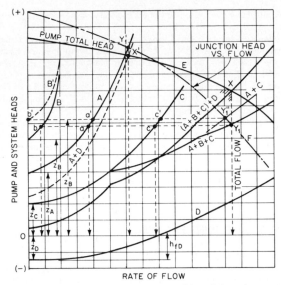

FIG. 5 System-head curves for pump and branch lines shown in
Fig. 4.

the system total head and pump total head are equal. Point Y_1 represents the total head at junction
1, and this head determines the flow through each branch; consequently points a, b, and c give
individual branch flows.

In this example the pump discharges to all tanks, but this should not be assumed. There is a
limiting liquid level elevation for each tank, and, if this level is exceeded, flow will be from the
tank into the junction. Therefore it is possible for the lower-level tanks to be fed by the higher-
level tank and the pump. The limit for the liquid elevation in tank B is Z_B', and it is found from
the intersection of curve $A + C$ with curve F, point Y_1''. The flow in branches A and C is at rates
a' and c' when there is no flow in branch B. This is also a condition similar to closing a valve in
branch B.

If elevation Z_B is greater than the previously found limiting height Z_B', flow in branches A and
C is determined in the following manner. Construct a curve for junction head versus flow by
adding heads and flows that result when the pump and suction tank are in series with each other
and tank B (less line losses) is in parallel with the pump and suction tank. The intersection of this
curve with curve $A + C$ will give the junction head required to determine the individual flows
from the pump and tank and the flows to tanks A and C (not illustrated).

If flow to branches B and C is shut off, Fig. 5 illustrates the construction of the curves required
to determine the pump flow point X' and junction head point Y_1'.

CENTRIFUGAL PUMP BYPASS _____

Bypass orifices around centrifugal pumps are often used to maintain a minimum flow recommended by the pump manufacturer because of one or more of the following reasons:

> Limit the temperature rise to prevent seizing and/or cavitation
> Reduce shaft and bearing loads
> Prevent excessive recirculation in the impeller and casing
> Prevent overloading of driver if pump power increases with decrease in flow

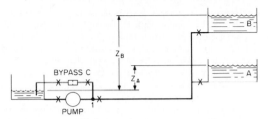

FIG. 6   Pump with bypass to maintain minimum flow.

Figure 6 illustrates a system which under certain conditions reduces pump flow below the recommended minimum. The pump delivers its flow to either tank A or tank B. Figure 7 shows the separate system-head curves for flow to tank A and for flow to tank B. Curve E is the head-capacity characteristics of the centrifugal pump. Individual flow rates to each tank are shown as Q_A and Q_B. The recommended minimum flow is Q_R, which is greater than Q_B by the amount shown. In order to maintain the minimum flow, a bypass orifice with necessary pipe, valves, and fittings is required to pass flow Q_C at total head H_R when the pump discharges to tank B only.

Figure 8, curve C, shows the construction necessary to determine the required bypass head versus flow characteristics of the orifice and pipe. The bypass system-head curve C includes the pipe, valve, and fitting losses from the pump connection between the suction tank and the end of the bypass piping below the suction water level. These losses must be deducted from the total bypass losses to determine the required orifice head.

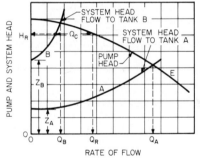

FIG. 7   System-head curves for pump and tanks shown in Fig. 6 with bypass valve closed.

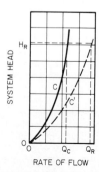

FIG. 8   Bypass orifice system requirements.

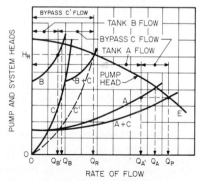

FIG. 9   System-head curves for pump and tanks shown in Fig. 6 with bypass valve open.

Figure 9 illustrates the resultant pump flow with the bypass in operation. Curve C is added to curve B to obtain curve $B + C$ by combining flows through each system at the same heads. Note that flow through the piping from the suction tank to junction 1 is the total from both systems. Therefore, the combined system-head curve $B + C$ should take this into consideration. Similarly, curve C is added to curve A to obtain curve $A + C$. Note that when the flow is directed to tank B with the bypass open, pump flow is increased from Q_B to Q_R and tank flow is decreased from Q_B to Q_B'. When the flow is directed to tank A with the bypass open, pump flow is increased from Q_A to Q_P and tank flow is decreased from Q_A to Q_A'.

If it is desired that there be no reduction in flow and/or that there be no waste of pumping power when flow is to tank A, the bypass can be closed either manually or automatically. If pump flow is monitored, this measurement can be used to open, close, or modulate the bypass valve automatically to maintain desired flow. Refer to Subsec. 2.3.4 for more detailed information.

If operating procedures require that the pump occasionally be run with a closed valve (at the pump discharge or at tanks A and B), the bypass line must be designed to recirculate all of the minimum required pump flow Q_R, dissipating head H_R, shown as curve C', Figs. 8 and 9.

FLOW CONTROL THROUGH BRANCHES

The flow through branches A, B, and C in Figs. 1 and 4 is dependent on the individual branch characteristics. When parallel branches are connected to a pump, the resulting division of flow may not satisfy the requirements of the individual lines. If it is desired that the flow to each branch meet or exceed specified individual line requirements, then it is necessary only to select a pump to provide the maximum head required by any one branch. In those branches where this head is more than required, the flow will be greater than the desired amount. A throttling valve or other flow-restricting device may be used to reduce the flow in these branches to the desired quantity. If the flow is controlled in this manner, the pump need be selected to produce only the minimum total system flow at a total head required to satisfy the branch needing the highest head at junction 1 in Figs. 1 and 4.

The flow through a branch is sometimes dictated by the requirement of a component elsewhere in the system. For example, in the system shown in Fig. 10, the required flow through component D could be greater than the sum of the required flows through components A, B, and C. An additional branch line and control valve F around components A, B, and C may be used to bypass the additional flow needed by component D. Individual component throttling valves may also be used, if needed, to adjust the flow in each branch.

The following example illustrates how flow through branches may be controlled and how pump total head is calculated.

EXAMPLE   A pump is required to circulate water at a rate of 3600 gpm (817 m³/h) through the system shown in Fig. 10. The head versus flow characteristics of the system components A, B, C, D, and E (system pipe and fittings) are shown in Fig. 11. The branch pipe and fitting losses from point 1 to point 2 are included in the total heads for components A, B, and C. Determine

1. The pump total head required and the individual flows through components A, B, and C.

2. The pump total head required if the flow through components A, B, and C need be only 800, 700, and 1000 gpm (182, 159, and 227 m³/h), respectively, and a bypass F is installed. Calculate the individual throttling valve head drops to achieve a controlled branch flow system.

Solution   1. Head versus flow curves A, B, and C of Fig. 11 are added together in parallel, giving curve $(A + B + C)$. Curves $(A + B + C)$, D, and E are added together in series, resulting in curve $(A + B + C) + D + E$. This latter curve indicates that 450 ft (137 m) total pump head is required at 3600 gpm (817 m³/h), point X. Curve $(A + B + C)$ crosses point Y at 3600 gpm (817 m³/h) flow through the branches, and this condition requires 310 ft (94.5 m) total head, which is the head across branch points 1 and 2. From each individual

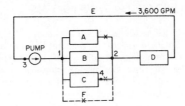

FIG. 10   Example of a branch-flow pumping system.

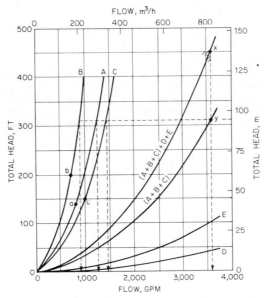

FIG. 11   System-head curves required for solutions to example problems.

component curve, the flow through branches A, B, and C can be read as 1250, 900, and 1450 gpm (284, 204, and 329 m³/h), respectively.

2. Since the total flow through components A, B, and C need be only 2500 gpm (800 + 700 + 1000) [568 m³/h (182 + 159 + 227)], the bypass should be designed to pass 1100 gpm (249 m³/h). Component B requires the maximum head, 200 ft (61 m) differential (point b) across points 1 and 2. Throttling valves are needed in components A and C and bypass F to increase the head in each branch to 200 ft (61 m) at the required flows. Branch A (point a) requires only 140 ft (42.7 m) total head to pass 800 gpm (182 m³/h); therefore the throttling valve must be designed for a 60-ft (42.7-m) head loss. Branch C (point c) requires only 150 ft (45.7 m) total head to pass 1000 gpm (227 m³/h), requiring a throttling valve for a 50-ft (15.3-m) head loss. The bypass control valve and piping should be designed to produce a 200-ft (61-m) head drop at 1100 gpm (249 m³/h).

At 3600 gpm (817 m³/h), the pump is now required to overcome 200 ft (61 m) total head across points 1 and 2, 40 ft (12.2 m) total head through component D, and 100 ft (30.1 m) total head through the system pipe and fittings, component E, for a total of 340 ft (103.3 m). The reduction in pumping head from 450 to 340 ft (137.1 to 103.6 m), a saving of 110 ft (33.5 m), or 24.4% water power, is the result of decreasing the branch head from 310 to 200 ft (94.5 to 61 m) by bypassing the excess flow.

PUMP TOTAL HEAD IN BRANCH-LINE SYSTEMS _____

The total head produced is the difference in total heads measured across the suction and discharge connections of a pump. As explained in Sec. 8.1, the total head is also the difference between the heads at any two points in the pumping system, one on each side of the pump, plus the sum of the head losses between these two points. Confusion sometimes results when the flow through the pump divides into branches in either closed-loop or open-ended systems. The points of head measurement can be in any branch line, upstream or downstream from the pump, regardless of the flow rate in these lines.

In part 2 of the above example, the pump or system total head could be measured using points 3 and 4 in Fig. 10. If the total head measured at the pump suction, point 3, were 25 ft (7.62 m) gage, the head measured at point 4 would be 205 ft (62.48 m) gage above the same reference datum plane, assuming 10 ft (3.05 m) of friction between the pump discharge and point 1. The difference between the head at point 3 and that at point 4 is 180 ft (54.86 m). The loss of head due to friction and the head drop through component C is $10 + 150 = 160$ ft ($3.05 + 47.5 = 48.75$ m). The pump and system total head at 3600 gpm (817 m³/h) is therefore $180 + 160 = 340$ ft ($54.86 + 48.75 = 103.63$ m).

Similarly, the pump and system total head for an open-ended system, such as the one shown in Fig. 4, could be found by measuring, for example, the difference between the head at point 2 and that at point 3. Each head measurement would be referred to a common datum plane. The total head loss from 2 to 1 plus the head loss from 1 to 3, at the rate of flow in their respective lines, added to the difference between the head at 2 and that at 3 is the pump and system total head.

SECTION 8.3
WATERHAMMER

JOHN PARMAKIAN

Waterhammer is a very destructive force which exists in any pumping installation where the rate of flow changes abruptly for various reasons. Most engineers recognize the existence of waterhammer, but few realize its destructive force. Much time and expense have been spent repairing pipelines and pumps damaged by waterhammer. It is thus essential for an engineer to be able to know when to expect waterhammer, how to estimate the possible maximum pressure rise, and if possible how to provide means to reduce the maximum pressure rise to a safe limit.

The computational procedures used for the analysis of waterhammer in pump discharge lines with electric-motor-driven pumps have been known for many years, beginning with the basic waterhammer contributions by Joukousky and Allievi more than 70 years ago. This work was followed in later years by many applications of numeric, graphic, and computer techniques. Although the theory and mechanics of computing waterhammer in pump discharge lines have advanced rapidly in recent years, there are many practical aspects of this subject which are still confusing to engineers. It is the purpose of this section to bring these to the reader's attention. The first and major portion of the section contains a discussion of some practical aspects of waterhammer control devices used in pumping plants; the second indicates the source of various charts which provide ready waterhammer solutions for a variety of these control devices.

NOMENCLATURE

The following is a list of variables commonly used in waterhammer computations. The SI conversion factors for these terms are to be found in Table 1.

a = velocity of pressure wave, ft/s
D = inside diameter of conduit, ft
e = thickness of pipe wall, ft
E = Young's modulus for pipe material, lb/ft^2
g = acceleration of gravity, ft/s^2
H_0 = pumping head for initial steady pumping conditions, ft
H_R = rated pumping head, ft
K_1 = 91,600 $H_R Q_R / W R_{\eta R}^2 N_R^2$, s^{-1}
K = volume modulus of liquid, lb/ft^2

L = total length of conduit, ft
$2L/a$ = round-trip wave travel time, s
N_R = rated pump speed, rpm
η_R = pump efficiency at rated speed and head, decimal form
ρ = pipe line constant = $a\bar{V}_0/gH_0$
Q_0 = initial flow through pump, ft³/s
Q_R = rated flow through pump, ft³/s
μ = Poisson's ratio of pipe material
V_0 = velocity in conduit for initial steady conditions, ft/s
w = specific weight of water, lb/ft³
WR^2 = flywheel effect of rotating parts of motor, pump, and entrained water, lb-ft²

TABLE 1 SI Conversions

To convert	To	Multiply by
$GD^2(\text{kg}\cdot\text{m}^2)$	$WR^2(\text{lb-ft}^2)$	23.73
kg/m²	lb/ft²	0.2048
kg/m³	lb/ft³	0.06243
m	ft	3.281
m/s	ft/s	3.281
m/s²	ft/s²	3.281
m³/s	ft³/s	35.32
mm	ft	3.281×10^{-3}

BASIC ASSUMPTIONS

A considerable number of assumptions were made in the derivation of the fundamental water-hammer equations and in the solution of the various hydraulic transients in pumping systems. These assumptions are often overlooked and involve the physical properties of the fluid and pipeline, the kinematics of the flow, and the transient response of the pump as follows:

1. The fluid in the pipe system is elastic, of homogeneous density, and always in the liquid state.
2. The pipe wall material or conduit is homogeneous, isotropic, and elastic.
3. The velocities and pressures in the pipeline, which is always flowing full, are uniformly distributed over any transverse cross section of the pipe.
4. The velocity head in the pipeline is negligible relative to the pressure changes.
5. At any time during the pump transient, when operation is in the the zones of pump operation, energy dissipation, and turbine operation, there is an instantaneous agreement at the pump, as defined by the steady-state complete pump characteristics, of the pump speed and torque corresponding to the transient head and flow which exist at that moment at the pump.
6. The length between the inlet and outlet of the pump is so short that waterhammer waves propagate between these two points instantly.
7. Windage effects of the rotating elements of the pump and motor during the transients are negligible.
8. Water levels at the intake and discharge reservoirs do not change during the transient period.

FACTORS AFFECTING WATERHAMMER

High- and Low-Head Pumping Systems Waterhammer is of greater significance in low-head pumping systems than in high-head systems. The normal steady water velocities in high-head and low-head pumping systems are usually of about the same order of magnitude. However, the pressure changes are proportional to the rate of change in the velocity of the water in the line. For a given rate of velocity change, however, the pressure changes in the high- and low-head pumping systems are of about the same order of magnitude. Therefore a given head rise would be a larger proportion of the pumping head in a low-head pumping system than a high-head system.

Discharge Line Profile   The pump discharge line profile is usually based on economic, topographic, and land right-of-way considerations. However, in selecting the alignment along which a pump discharge line is to be located, there are other considerations which often make one pipline profile and alignment more favorable than another. For example, upon a power failure at the pump motors, the envelope of the maximum downsurge gradient along the length of the pipeline is a concave curve. Therefore it may be possible to avoid the use of expensive pressure control devices at a pumping plant if the pipeline profile is also concave and is not located above the downsurge gradient curve. In some cases, it may even be economical to lower the profile of the discharge line at the critical locations by deeper excavation. If a surge tank at the pumping plant is definitely required, the most favorable pipeline profile is one with high ground near the pumping plant where the surge tank structure can be placed so that its height above the natural ground line will be much less than what would be necessary if there were no high ground near the plant.

Rigid Water Column Theory   The question is often raised as to whether the rigid water column theory is sufficiently accurate for the computation of waterhammer in pump discharge lines. In the rigid water column theory the water is assumed to be incompressible and the pipe walls rigid. In the author's experience, the accuracy and limitations of the rigid water column theory are often questionable for most waterhammer problems that occur in pump discharge lines.

Waterhammer Wave Velocity   From a practical viewpoint a difference of 15 to 20% in the magnitude of the computed waterhammer wave velocity usually has very little effect on the waterhammer in pump discharge lines. The effect on the waterhammer due to a possible error in the wave velocity can be verified by first computing the wave velocity as accurately as possible and then recomputing the transients for the critical cases with a wave velocity about 20% higher or lower. At installations where alternative materials for the pipeline are being investigated, one waterhammer wave velocity and solution for waterhammer for either alternative will usually suffice regardless of the pipe material finally selected.

Pipeline Size   The diameter of the pipeline is usually determined from economic considerations based on steady-state pumping conditions. However, the waterhammer effects in a pump discharge line can be reduced by increasing the size of the discharge line since the velocity changes in the larger pipeline will be less. This is usually an expensive method of reducing waterhammer in pump discharge lines, but there are sometimes occasions where an increase in pipe size may be justified to avoid the use of more expensive waterhammer control devices.

Number of Pumps   The number of pumps connected to each pump discharge line is usually determined from the operational requirements of the installation, availability of pumps, and other economic considerations. However, the number and size of pumps connected to each discharge line have some effect on the waterhammer transients. For pump start-up with pumps equipped with check valves, the greater the number of pumps on each discharge line, the smaller the pressure rise. Moreover, if there is a malfunction at one of the pumps or check valves, a multiple pump installation on each discharge line would be preferable to a single pump installation because the flow changes in the discharge line due to such a malfunction would be less with multiple pumps. When a simultaneous power failure occurs at all of the pump motors, the fewer the number of pumps on a discharge line, the smaller the pressure changes and other hydraulic transients. For a given total flow in the discharge line, a large number of small pumps and motors will have considerably less total kinetic energy in the rotating parts to sustain the flow than a small number of pumps. Consequently, for the same total flow, the velocity changes and waterhammer effects due to a power failure are a minimum when there is only one pump connected to each discharge line.

Flywheel Effect (WR^2) Another method for reducing the waterhammer effects in pump discharge lines is to provide additional flywheel effect (WR^2) in the rotating element of the motor. As an average, the motor usually provides about 90% of the combined flywheel effect of the rotating elements of the pump and motor. Upon a power failure at the motor, an increase in the kinetic energy of the rotating parts will reduce the rate of change in the flow of water in the discharge line. In most cases an increase of 100% in the WR^2 of large motors can usually be obtained at an increased cost of about 20% of the original cost of the motor. Ordinarily an increase in WR^2 is not an economical method for reducing waterhammer, but it is possible in some marginal cases to eliminate other, more expensive pressure control devices.

Specific Speed of Pumps   For a given pipeline and initial steady-flow conditions, the maximum head rise which can occur in a discharge line subsequent to a power failure where the reverse flow passes through the pump depends first on the magnitude of the maximum reverse flow which can pass through the pump during the energy-dissipation and turbine-operation zones and then on the flow which can pass through the pump at runaway speed in reverse. Upon a power failure, the radial-flow (low-specific-speed) pump will produce slightly more downsurge than the axial-flow (high-specific-speed) and mixed-flow pumps.[1] The radial-flow pump will also produce the highest head rise upon a power failure if the reverse flow is permitted to pass through the pump. There is usually very little head rise at mixed-flow and axial-flow pumps when a power failure occurs and if a water column separation does not occur at some other location in the line.

During a power failure with no valves, the highest reverse speed is reached by the axial-flow pump and the lowest by the radial-flow pump. Care must therefore be taken to prevent damage to the motors with the higher-specific-speed pumps because of these higher reverse speeds. Upon pump start-up against an initially closed check valve, the axial-flow pump will produce the highest head rise in the discharge line since it also has the highest shutoff head. On pump start-up, a radial-flow pump will produce a nominal head rise but an axial-flow pump can produce a head rise of several times the static head.

Complete Pump Characteristics   In order to determine the transient conditions due to a power failure at the pump motors, the waterhammer wave phenomena in the pipeline, the rotating inertia of the pump and motor, and the complete pump characteristics as well as other boundary conditions and head losses must be known. In the solution of waterhammer problems with computers, the complete pump characteristics are sometimes approximated by polynomial expressions in which the coefficients of the polynomial are obtained by fitting a representative curve through several points at specific locations on the pump characteristics diagram. Pump manufacturers sometimes provide limited information to determine such coefficients. However, a comparison between the polynomial values and the complete pump characteristics diagram indicates serious discrepancies in some cases, especially in the zone of energy dissipation. To ensure that a serious error does not result in the computation of the hydraulic transients, therefore, care must be exercised in the use of an approximate polynomial expression as a substitute for the correct complete pump characteristics.

Complex Piping Systems   As noted above in the basic assumptions, the waterhammer theory is strictly applicable for a pipeline of uniform characteristics. However, for waterhammer purposes a complex piping system can be reduced to a satisfactory equivalent uniform pipe system. The approximations are made by neglecting the wave transmission effects at the junctions and points of discontinuity and by utilizing the rigid water column theory. The pertinent waterhammer equations are then found to be analogous to those used in electric circuits. In practice the waterhammer analysis with these approximations will usually give more conservative results than those obtained experimentally from the pipe system.[2]

Available Waterhammer Solutions   The waterhammer solutions for pumping systems with various surge control devices are given in convenient chart form in the references. These include the following:

1. Hydraulic transients at the pump and midlength of the pump discharge lines for radial-flow, mixed-flow, and axial-flow pumps with reverse flow passing through the pumps
2. Surge tanks

3. Air chambers

4. Surge suppressors

5. One-way surge tanks

6. Water column separation

The operation of these surge control devices is described below.

Power Failure at Pump Motors

PUMPS WITH NO VALVES AT THE PUMP   When the power supply to the pump motors is suddenly cut off, the only energy that is left to drive the pump in the forward direction is the kinetic energy of the rotating elements of the pump and motor. Since this energy is usually relatively small relative to that required to maintain the flow of water against the discharge head, the reduction in the pump speed is very rapid. As the pump speed is reduced, the flow of water in the discharge line is also reduced. As a result of these rapid flow changes, waterhammer waves of increasing subnormal pressure are formed in the discharge line at the pump. These subnormal pressure waves move rapidly up the discharge line to the discharge outlet, where complete wave reflections occur. Soon the speed of the pump is reduced to a point where no water can be delivered against the existing head. If there is no control valve at the pump, the flow through the pump reverses even though the pump may still be rotating in the forward direction. The speed of the pump now drops more rapidly and passes through zero speed. Soon the maximum reverse flow passes through the pump. A short time later the pump, acting as a turbine, reaches runaway speed in reverse. As the pump approaches runaway speed, the reverse flow through the pump is reduced. For radial-flow pumps, this rapid reduction in reverse flow produces a pressure rise at the pump and along the length of the discharge line. The results of a large number of waterhammer solutions for a given set of radial-flow (low-specific-speed) pump characteristics are given in chart form in Figs. 1 to 8. These charts furnish a convenient method for obtaining the hydraulic transients at the pump and midlength of the discharge line when no control valves are present at the pump. Although the charts are theoretically applicable to one particular set of radial-flow pump characteristics, they are useful for estimating the waterhammer effects in any pump discharge line equipped with radial-flow pumps. The charts are based on two independent parameters: ρ, the pipeline constant,

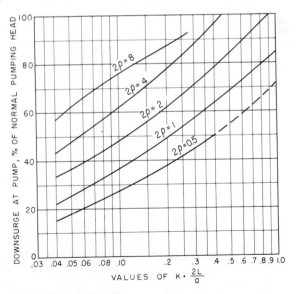

FIG. 1   Downsurge at pump.

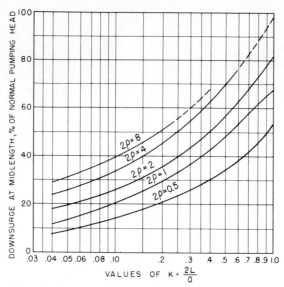

FIG. 2 Downsurge at midlength.

and $K(2L/a)$, a constant which includes the effect of pump and motor inertia and the waterhammer wave travel time of the discharge line. If the frictional head in the discharge line during normal operation is more than 25% of the total pumping head and if water column separation does not occur at any point in the line, the maximum head at the pump with reverse flow passing through the pumps will usually not exceed the initial pumping head.[3]

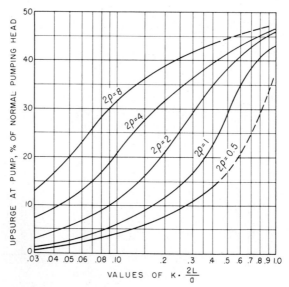

FIG. 3 Upsurge at pump.

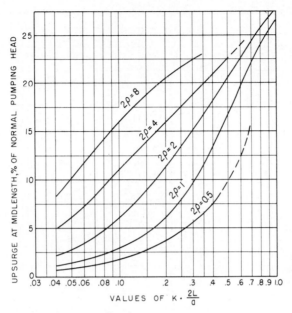

FIG. 4   Upsurge at midlength.

PUMPS EQUIPPED WITH CHECK VALVES   There are a number of problems associated with the use of check valves in pump discharge lines. Under steady flow conditions, the pump discharge keeps the check valve open. However, when the flow through the pump reverses subsequent to a power failure, the check valve closes very rapidly under the action of the reverse flow and the resulting dynamic forces on the check valve disk. Under these conditions, neglecting pipeline friction, the head rise in the discharge line at the check valve is about equal to the head drop which existed at the moment of flow reversal. However, if the check valve closure upon flow reversal is momentarily delayed because of hinge friction, malfunction, or the inertial characteristics of the valve, the maximum head rise in the discharge line at the check valve can be considerably higher. On the other hand, if the check valve closure can be accomplished slightly in advance of flow

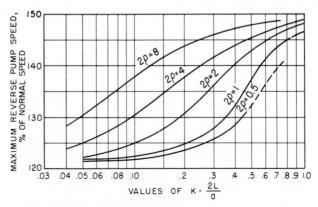

FIG. 5   Maximum reverse speed.

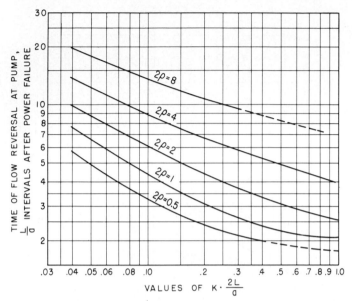

FIG. 6 Time of flow reversal at pump.

reversal, the head rise in the pump discharge line at the valve is even lower than that obtained with a check valve which closes at the moment of flow reversal. This feature is utilized by a number of check valve manufacturers, who provide spring-loaded or lever-arm-weighted devices on the check valve hinge pins to assist in closing the valve disk before the flow reverses. With these devices, the hydraulic forces on the valve disk under normal flow conditions must be suffi-

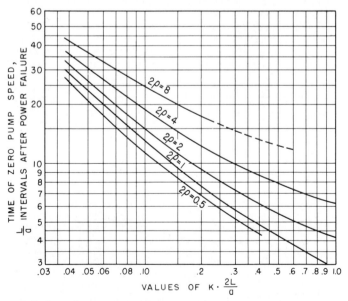

FIG. 7 Time of zero pump speed.

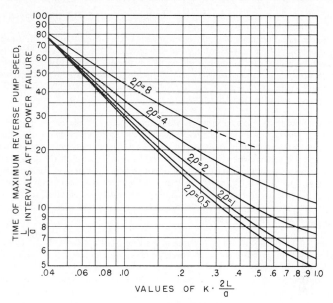

FIG. 8 Time of maximum reverse pump speed.

cient to overcome the spring or lever-arm-weight forces in order to keep the check valve disk wide open so that the head losses at the valve under steady flow conditions will be a minimum.

Check valves in pump discharge lines may be grouped into two general classes, rapid-closing and slow-closing. From the considerations noted above, the primary requirement for a check valve upon a power failure is that it should close quickly, before a substantial reverse flow has been established. When this primary requirement for a fast-closing check valve cannot be met because of the flow characteristics of the system and the design of the check valve, an alternative is to provide a device such as a dashpot, which will slow down or cushion the last portion of the check valve closure. This feature has been utilized by a number of check valve manufacturers.

CONTROLLED VALVE CLOSURE   At most large pumping plant installations, the use of a single-speed discharge valve closure subsequent to a power failure will usually limit the head rise in the discharge line to an acceptable value. However, it will be found that there, with the optimum single-speed closure, there will be some reverse rotation below the maximum runaway speed of the unit in reverse. If it is desired from other considerations to prevent or to limit the reverse speed of the unit, a two-speed valve closure can be used. In such cases the discharge valve should close the major portion of its stroke very rapidly up to the moment that the flow reverses at the pump. It should then complete the remainder of its stroke at a lower rate in order to limit the pressure rise in the discharge line to an acceptable valve. At pumping plants where there is more than one pump on the same discharge line, a compromise must be obtained on the optimum single-speed and two-speed closure rates for the various combination of pumps which might be in operation at the time of a power failure.

SURGE SUPPRESSORS   Surge suppressors are sometimes used in pumping plants to control the pressure rise that occurs in pump discharge lines subsequent to power interruptions. A typical surge suppressor consists of a pilot-operated valve which opens quickly after a power interruption either through the loss of power to a solenoid or by a sudden large pressure reduction or pressure increase at the surge suppressor. This valve provides an opening for releasing water from the pump discharge line. The valve is later closed at a lower rate by the action of a dashpot to control the pressure rise as the flow of water is shut off. A properly sized and field-adjusted surge suppressor can reduce the pressure rise in the discharge line to any desired value, provided that water column separation does not occur at other locations in the discharge line. The charts given in Ref. 4 can be used to determine the required flow capacity of the surge suppressor.

The proper field adjustment of a surge suppressor is very important. If the suppressor opens too rapidly subsequent to a power failure, the downsurge at the pump and along the discharge line profile will be more than if no surge suppressor is present. As a result, a water column separation may be produced at some locations in the discharge line by the premature opening of the surge suppressor. If the suppressor closes too rapidly after the maximum reverse flow has been established, a large pressure rise will occur.

WATER COLUMN SEPARATION   Water column separation in a pump discharge line subsequent to a power failure at the pump motors occurs whenever the momentary hydraulic gradient at any location reduces the pressure in the discharge line to the vapor pressure of water. Whenever this condition occurs, the normal waterhammer solution is no longer valid. If the subatmospheric pressure condition inside the pipe persists for a sufficient period, the water in the discharge line parts and is separated by a section of water and vapor. Whenever possible, water column separation should be avoided because of the potentially high pressure rise which often results when the two water columns rejoin. An approximate waterhammer solution for water column separation in pump discharge lines is given in Ref. 5.

QUICK-OPENING, SLOW-CLOSING VALVES   A quick-opening, dashpot-controlled, slow-closing valve can be used to limit the pressure rise at the high points in the discharge line, where water column separation frequently occurs. When the pressure in the pipeline at the point of water column separation drops below a predetermined value for which the valve is set, the valve opens quickly and a small amount of air is admitted in the pipeline. After the upper water column in the pipeline stops, reverses, and returns to the point of separation near the valve, the valve should be wide open. At first the air and water mixture, and then the clear water discharges through the valve. The open valve then provides a point of relief to reduce the pressure rise caused by the rejoining of the water columns. The valve is later closed slowly under the action of a dashpot so that the head rise that occurs in the discharge line at the valve location when the reverse flow is shut off is not objectionable. Whenever these valves are used, precautions should be taken to ensure that they are properly sized, field-adjusted to the proper opening and closing times, and adequately protected against freezing.

ONE-WAY SURGE TANKS   The one-way surge tank, which was introduced by the writer about 24 years ago[6] is an effective and economical pressure control device for use at locations where water column separation occurs. A one-way surge tank is a relatively small tank filled with water to a level far below the hydraulic gradient. It is connected to the main pipeline with check valves which are held closed by the discharge line pressure. Upon a power failure, when the pressure in the discharge line at the one-way surge tank drops below the head corresponding to the water level in the tank, the check valve opens quickly and the tank starts to drain, filling the void formed by the separation of the water columns. When the flow in the upper column starts to reverse, the check valves at the one-way surge tank close before any appreciable reverse flow is established in the discharge line. Thus there is no pressure rise when the water columns rejoin. The initial level of water in the one-way surge tank is usually maintained automatically with float control or altitude valves. It should be noted that the one-way surge tank does not act during the start-up cycle of the pump discharge line and that it must also be protected against freezing.

AIR CHAMBERS   An effective device for controlling the pressure surges in a long pump discharge line is a hydropneumatic tank or air chamber. The air chamber is usually located at or near the pumping plant. It can be of any desired configuration and may be placed in a vertical, horizontal, or sloping position. The lower portion of the chamber contains water, and the upper portion contains compressed air. The desired air and water levels are maintained with float level controls and an air compressor. When the power failure occurs at the pump motor, the head and flow developed by the pump decrease rapidly. The compressed air in the air chamber then expands and forces water out of the bottom of the chamber into the discharge line, thus minimizing the velocity changes and waterhammer effects in the line. When the pump speed is reduced to the point where the pump cannot deliver water against the existing head, which is usually a fraction of a second after a power failure, the check valve at the discharge side of the pump closes rapidly and the pump then slows down to a stop. A short time later, the water in the discharge line slows down to a stop, reverses, and flows back into the air chamber. As the reverse flow enters the chamber, usually through a throttling orifice, the air volume in the chamber decreases and a head rise above the pumping head occurs in the discharge line. The magnitude of this head rise depends on the throttling orifice and on the initial volume of air in the air chamber.

The results of a large number of graphic waterhammer–air chamber solutions are given in Ref. 7. Another presentation of air chamber charts using the rigid water column theory is given in Ref. 8.

SURGE TANKS   Because it has no moving parts which can malfunction, a surge tank is one of the most dependable devices that can be used at a pumping plant to reduce waterhammer resulting from rapid changes of flow in the discharge line subsequent to a power failure at the pump motor. Following a power failure, the water in the surge tank provides a nearby source of potential energy which will effectively reduce the rate of change of flow and the waterhammer in the discharge line. The charts given in Ref. 7 provide a ready means for calculating the surges in the pipeline due to the sudden starting or stopping of a pump.

One of the disadvantages of a conventional surge tank is that, since the top of the tank must extend above the normal hydraulic gradient to avoid spilling, the tank must be quite tall and expensive at high-head pumping installations. In order to obtain the most economical surge tank design, care should be given to the proper sizing of the throttling device at the base of the tank.

NONREVERSE RATCHETS   Another device occasionally used for reducing waterhammer in a pump discharge line upon a power failure is a nonreverse ratchet on the pump and motor shaft and motor shaft which prevents the reverse rotation of the pump. This device is effective for controlling waterhammer when there is a power failure because of the large reverse flow which can pass through the stationary impeller. Except on small pumps, experience to date with nonreverse ratchet mechanisms has been very disappointing. At a number of moderate-size pump installations where these devices were used, the shock to the pump and motor shaft system caused by the sudden shaft stoppage created other serious mechanical difficulties.

AUTOMATIC RESTART OF MOTORS   At small unattended pumping plants, it is often desirable after a power failure to automatically return the pumps to service as soon as the power is restored. However, it was found that occasionally, subsequent to a very short power outage, an induction motor could restart and come quickly up to forward speed while a reverse flow was still passing through the pump. Under these conditions, waterhammer in the discharge line is very objectionable. If the pump motor has the capability of restarting under such transient conditions, a time delay or similar device should be installed at the motor controls so that the pump can be restarted only when it is safe to do so.

Normal Pump Start-Up

WITH CONTROLLED VALVE OPENING   At some pumping plants, the pump is brought up to speed against a closed valve on the discharge side of the pump. The valve is then opened slowly, and there is very little waterhammer in the discharge line. However, it will be found that nearly all of the pump flow in the discharge line is established with only a relatively small valve opening, since the head losses across the valve decrease very rapidly during the opening stroke. For long discharge lines, the head loss and flow characteristics of the valve during the opening stroke must be considered in determining the optimum rate of opening.

WITH CHECK VALVES   At pumping plants where the pipeline is held full with pump check valves, waterhammer in the discharge line due to a pump start-up can be objectionable in some cases. If the motor comes up to speed very rapidly, the pump will develop a pressure rise in the discharge line as the sudden increase in flow moves into the line. As noted above and in Ref. 1, this pressure rise is lower for radial-flow (low-specific-speed) pumps than for axial-flow (high-specific-speed) pumps.

WITH CASING UNWATERED   At pumping plants equipped with large pumps, normal starting of a pump is often performed with the pump casing unwatered. This is accomplished by depressing the water level below the pump impeller by means of compressed air, which is admitted into the pump casing with the pump discharge valve closed and the discharge line full. After the motor has been synchronized on the line, the compressed air in the pump casing is released, allowing water to re-enter the pump from the suction elbow, after which the discharge valve is slowly opened. This type of operation has been satisfactory with most large pumping units, and there are normally no significant waterhammer effects on the discharge line. However, there have been some difficulties with this type of operation at a few large pump units. In the latter case, when the rising water level in the suction elbow first reaches the pump impeller, a very fast pumping action occurs within a few seconds and a severe uplift of the pump and motor from the thrust bearing could occur at this moment. If the discharge valve is still closed when this fast pumping action occurs, there is no waterhammer effect in the discharge line.

WITH SURGE TANKS OR AIR CHAMBERS   With a surge tank or air chamber at the pumping plant, it makes very little difference whether the increased pump flow is sudden or gradual, inasmuch as the major portion of the sudden increased flow will enter the surge tank or air chamber. With these devices the steep front of the pressure rise at the pump is transformed into smaller pressure rise in the discharge line and a subsequent slow oscillating movement in the surge tank or air chamber.

Normal Pump Shutdown   The pumping installation which produces the least waterhammer effect in a pump discharge line during a normal pump shutdown is one in which the control valve on the discharge side of the pump is first closed and the power to the pump motor then is shut off. If only check valves are in operation on the discharge side of the pumps and the power to one of several pump motors connected to the same discharge line is cut off, the flow at the pump which has been shut down will reverse rapidly and the check valve will close rapidly. The use of antislam or slow-closing features at the check valves will reduce the waterhammer effect in the discharge line.

CONCLUSIONS

A variety of waterhammer control devices for pumping plants are available to the designer. In most cases the experienced designer can narrow the choice of the most suitable device to a few practical alternatives. A prior knowledge of the available waterhammer solutions for these devices will reduce the amount of detailed computational work which must be done to determine the critical hydraulic transient effects.

EXAMPLE   Consider a power failure at the pumping plant installation shown in Fig. 9. This installation consists of three pumps which discharge into a steel pipeline. Aside from isolation valves, there are no check valves in the system. The basic data for this installation are as follows:

$$D = 32 \text{ in}; e = \tfrac{3}{16} \text{ in}; Q_0 = Q_R = 33.7 \text{ ft}^3/\text{s (for three pumps)}$$
$$V_0 = 6.03 \text{ ft/s (for three pumps)}; H_0 = H_R = 220 \text{ ft}$$

400-hp motors at each pump
WR^2 of each pump and motor = 385 lb-ft^2
$N_R = 1760 \text{ rpm}; \eta_R = 0.847$

$$\frac{D}{e} = \frac{32}{0.1875} = 171$$

$a = 3000$ fts from Fig. 10

$$\frac{2L}{a} = \frac{2(3940)}{3000} = 2.63 \text{ s}$$

$$2\rho = \frac{aV_0}{gH_0} = \frac{(3000)(6.03)}{(32.2)(220)} = 2.55$$

$$K = \frac{(91,600)(H_R Q_R)}{WR^2 \eta_R N_R^2} = \frac{(91,600)(220)(33.7)}{3(385)(0.847)(1760)^2} = 0.224$$

$$K\frac{2L}{a} = 0.59$$

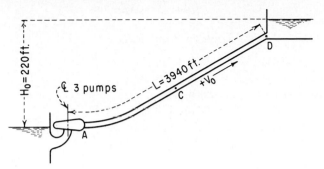

FIG. 9 Pipeline profile.

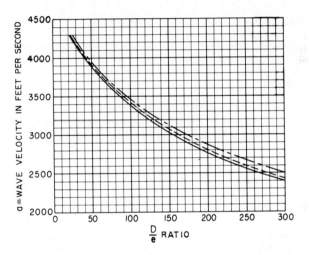

$$a = \sqrt{\dfrac{1}{\dfrac{w}{g}\left(\dfrac{1}{K} + \dfrac{Dc_1}{Ee}\right)}}$$

where:

a = wave velocity (ft. per sec.)

g = acceleration of gravity (ft. per sec.^{2})

$\dfrac{D}{e} = \dfrac{\text{diameter of pipe}}{\text{thickness of pipe}}$

E = Young's modulus for steel pipes = 4.32×10^9 (lb. per ft.^{2})

K = volume modulus of water = 43.2×10^6 (lb. per ft.^{2})

w = 62.4 = specific weight of water (lb. per ft.^{3})

μ = 0.3

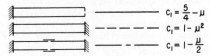

$c_1 = \dfrac{5}{4} - \mu$

$c_1 = 1 - \mu^2$

$c_1 = 1 - \dfrac{\mu}{2}$

FIG. 10 Pressure wave velocity in steel pipes.

8.97

From Figs. 1 to 8 the following results are obtained:

1. Downsurge at pump = $(0.92)(220)$ = 202 ft
2. Downsurge at midlength = $(0.64)(220)$ = 141 ft
3. Upsurge at pump = $(0.42)(220)$ = 92 ft
4. Upsurge at midlength = $(0.23)(220)$ = 51 ft
5. Maximum reverse speed = $(1.45)(1760)$ = 2550 rpm
6. Time of flow reversal at pump = $3.5\,L/a$ = 4.6 s
7. Time of zero pump speed = $5.8\,L/a$ = 7.6 s
8. Time of maximum reverse speed = $10.0\,L/a$ = 13.1 s

As noted in the discussion on pumps equipped with check valves, the upsurge, or head rise, at the pump above the normal head would have been about 202 ft if there were check valves at the pumps which closed at the time of flow reversal.

PRESSURE PULSATIONS IN PUMPING SYSTEMS

Motor-driven centrifugal pumps often produce objectionable pressure pulsations in pump discharge lines. The frequency of these pulsations results from the rotating and stationary components of the pump. The following pressure pulsation frequencies have been observed at a number of major pump installations:

1. Fundamental shaft frequency, once-per-revolution pressure pulsation
2. Impeller vane frequency, product of the shaft frequency and the number of impeller vanes
3. Scroll case frequency, product of shaft frequency and the number of guide vanes at either the inlet or discharge side of the pump

Pressure Pulsations at Fundamental Shaft Frequency In pumping systems, objectionable pressure pulsations with a frequency corresponding to the shaft frequency result primarily from a hydraulic unbalance in the pump impeller. The source of such unbalance is usually an eccentricity of the flow passage in the impeller, but in some cases it may be the eccentricity of the outer periphery of the impeller and the pumping action exerted by this eccentricity.

Considerable care must be taken when dynamically balancing a pump impeller to avoid pressure pulsations. In accomplishing this, balance weights are often welded to the top surface or an eccentric machine cut is taken on the outer periphery. Although such surfaces are out of sight after the unit is assembled, objectionable pressure pulsations with a frequency corresponding to the rotational speed of the unit can occur. Such sources of pressure pulsations are very difficult to correct after the unit has been installed. In order to avoid such difficulties, any welding at the top surface or metal removal at the outer periphery of the impeller should be done in such a manner that these surfaces remain smooth and concentric with the axis of rotation. Balance weights on surfaces normal to the axis of rotation should be applied or covered in such a manner that the surface remains flat, smooth, and normal to the axis of rotation.

Pressure Pulsations at Impeller Vane Frequency The tongue of the pump is the source of the pressure pulsations which are transmitted to the discharge line with a frequency equal to the product of the shaft frequency and the number of impeller vanes. An explanation of the phenomenon is as follows. The velocity distribution at the exit of the pump impeller is not uniform. As this nonuniform flow passes the tongue of the pump casing, an abrupt change in the direction of the impeller exit velocity vector occurs at the proximity of the tongue. This produces a positive pressure wave at the pressure face and a negative pressure wave at the back face of the tongue. From this location, the positive pressure wave travels directly up the discharge line and the negative pressure wave travels completely around the pump casing and is attenuated before reaching the discharge line. These positive pressure pulsations have a frequency equal to the prod-

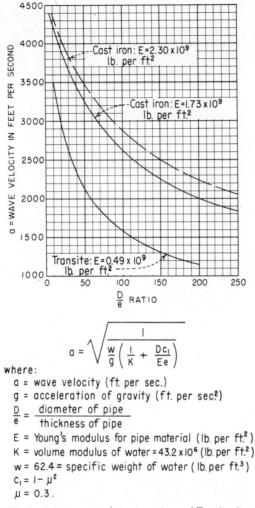

$$a = \sqrt{\dfrac{1}{\dfrac{w}{g}\left(\dfrac{1}{K} + \dfrac{Dc_1}{Ee}\right)}}$$

where:

a = wave velocity (ft. per sec.)

g = acceleration of gravity (ft. per sec.²)

$\dfrac{D}{e} = \dfrac{\text{diameter of pipe}}{\text{thickness of pipe}}$

E = Young's modulus for pipe material (lb. per ft.²)

K = volume modulus of water = 43.2×10^6 (lb. per ft.²)

w = 62.4 = specific weight of water (lb. per ft.³)

$c_1 = 1 - \mu^2$

$\mu = 0.3$.

FIG. 11 Pressure wave velocity in cast iron and Transite pipes.

uct of the pump speed and the number of impeller vanes. The most effective method for reducing the magnitude of these pressure pulsations is to provide large radial clearance between the outer diameter of the impeller and all guide vanes, consistent with the head discharge requirements. Several manufacturers adopt a minimum radial clearance of 5% of the impeller diameter.

Pressure Pulsations at Scroll Case Frequency Objectionable pressure pulsations in pumping systems have been observed at a frequency corresponding to the product of the rotating speed and the number of guide vanes on either the inlet or discharge side of the pump. These pulsations have been observed only on those pumping units where the number of guide vanes is an exact multiple of the number of impeller blades. It does not exist where this multiple relation does not exist.

An explanation of the source of these pressure changes is as follows. As the pump impeller rotates, all of the impeller vanes simultaneously cross the flow lines between a corresponding number of guide vanes. This disturbance in the flow pattern by all of the runner blades simultaneously produces periodic pressure changes inside the unit which correspond to the product of the rotational speed and the number of guide vanes. The remedy for eliminating these periodic pressure changes is relatively simple. It is necessary only to avoid having the number of vanes on the suction and discharge sides of the pumps equal to an exact multiple of each other or of the number of impeller vanes.

REFERENCES

1. Donsky, B.: "Complete Pump Characteristics and the Effects of Specific Speeds on Hydraulic Transients," *ASME J. Basic Eng.*, December 1961.
2. Jones, S. S.: "Water-Hammer in a Complex Piping System: Comparison of Theory and Experiment," *ASME* Paper 64—WA/FE-23, New York, 1964.
3. Kinno, H., and J. F. Kennedy: "Waterhammer Charts for Centrifugal Pump Systems," *Proc. ASCE, J. Hydraulics Div.*, May 1965.
4. Lundgren, C. W.: "Charts for Determining Size of Surge Suppressors for Pump Discharge Lines," *ASME J. Eng. Power*, January 1967.
5. Kephart, J. T., and K. Davis: "Pressure Surges Following Water-Column Separation," *Trans. ASME J. Basic Eng.*, September 1961.
6. Parmakian, J.: "One-Way Surge Tanks for Pumping Plants," *Trans. ASME* **80** (October 1958).
7. Parmakian, J.: *Waterhammer Analysis*, Dover, New York, 1963.
8. Combes, G., and R. Borot: "New Graph for the Calculation of Air Reservoirs, Account Being Taken of the Losses of Head," *La Houille Blanche*, October–November 1952.

FURTHER READING

Martin, C. S.: "Represenation of Pump Characteristics for Transient Analysis: Performance Characteristics of Hydraulic Turbines and Pumps," FED vol. 6 *ASME*, 1983.

Parmakian, J.: "Pressure Surges at Large Pump Installations," *Trans. ASME*, August 1953.

Parmakian, J.: "Pressure Surges in Pump Installations," *Trans. ASCE* **120** (1955).

Parmakian, J.: "Unusual Aspects of Hydraulic Transients in Pumping Plants," *J. Boston Soc. Civil Eng.*, January 1968.

Proceedings of the 2nd International Conference on Pressure Surges, London, 1976. BHRA Fluid Engineering, Cranford, Bedford MK430AJ, England.

SECTION 8.4
PUMP NOISE

FRED R. SZENASI
CECIL R. SPARKS
J. C. WACHEL

The major concern regarding pump noise falls into two categories:

1. Noise levels which do not meet applicable environmental criteria. Examples range from personnel noise exposure criteria to overside noise criteria for submarines.

2. Noise signatures which can be used to diagnose faulty pump operation or incipient failure.

The proliferation of industrial noise regulations in recent years has taken much of the guesswork out of allowable noise levels insofar as personnel and community exposure is concerned, and various noise standards have specified noise measurement techniques. Several organizations have developed test procedures and codes for machinery-generated noise levels.[1,2] The Hydraulic Institute code was specifically developed for the measurement of airborne sound generated by pumps (Ref. 3, p. 297).

The most common approach for controlling airborne noise levels from pumps is to interrupt the paths by which noise reaches the listener. When noise is an indicator of abnormal pump operation, modification of pump internals or operating conditions is normally required.

The measurement of noise for diagnostic purposes is not well prescribed, either for instrumentation or for interpretation. Even a well-designed and properly operated pump will of course produce noise. Variations in noise amplitude and frequency which result from malfunction or improper operating conditions will depend upon the type and design of the pump and the type of problem causing the noise. Measurement and analysis techniques for interpreting these signatures will depend upon whether the noise is solid-, liquid-, or airborne and upon the nature of coexisting noise from other sources.

Determining the source and cause of noise is the first step in evaluating whether noise is normal or an indicator of possible problems. Noise in pumping systems can be generated both by the mechanical motion of pump components and by the liquid motion in the pump and piping systems. Liquid noise sources can result from vortex formation in high-velocity (shear) flow, from pulsating flow, and from cavitation and flashing.

Noise from internal mechanical and liquid sources can be propagated to the environment by several paths, including the pump and support structure, attached piping, the liquid in the piping, and ultimately the surrounding air itself.

This section discusses various pump noise-generating mechanisms (sources) and common noise conduction paths as a basis for both effective diagnostics and treatment.

SOURCES OF PUMP NOISE

Effective control of pump noise requires knowledge of the liquid and mechanical noise-generation mechanisms and the paths by which noise can be transmitted to a listener.

Mechanical Noise Sources   Mechanical sources are vibrating components or surfaces which produce acoustic pressure fluctuations in an adjacent medium. Examples are pistons, rotating unbalance vibrations, and vibrating pipe walls.

In positive displacement pumps, noise is generally associated with the speed of the pump and the number of pump plungers. Liquid pulsations are the primary mechanically induced noise, and these in turn can excite mechanical vibrations in components of both pump and piping system. Incorrect crankshaft counterweights will also cause shaking at running speed, which may loosen anchor bolts and produce rattling of the foundation or skid. Other mechanical noises are associated with worn bearings on the connecting rods, worn wrist pins, or slapping of the pistons or plungers.

In centrifugal machines, improper installation of couplings often causes mechanical noise at twice pump speed (misalignment). If pump speed is near or passes through the lateral critical speed, noise can be generated by high vibrations resulting from imbalance or by the rubbing of bearings, seals, or impellers. If rubbing occurs, it may be characterized by a high-pitched squeal. Windage noises may be generated by motor fans, shaft keys, and coupling bolts.

Liquid Noise Sources   When pressure fluctuations are produced directly by liquid motion, the sources are fluid dynamic in character. Potential fluid dynamic sources include turbulence, flow separation (vorticity), cavitation, waterhammer, flashing, and impeller interaction with the pump cutwater. The resulting pressure and flow pulsations may be either periodic or broad-band in frequency and generally excite either the piping or the pump itself into mechanical vibration. These mechanical vibrations can then radiate acoustic noise into their environment.

In general, pulsation sources are of four types in liquid pumps:

1. Discrete-frequency components generated by the pump impeller or plungers

2. Broad-band turbulent energy resulting from high flow velocities

3. Impact noise consisting of intermittent bursts of broad-band noise caused by cavitation, flashing, and waterhammer

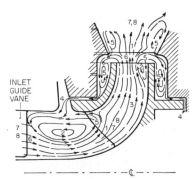

FIG. 1   Secondary flows in and around a pump impeller stage. (Ref. 11)

4. Flow-induced pulsations caused by periodic vortex formation when flow is past obstructions and side branches in the piping system

A variety of secondary flow patterns which produce pressure fluctuations are possible in centrifugal pumps, as shown in Fig. 1, particularly for operation at off-design flow. The numbers shown in the flow stream are the locations of the following flow mechanisms:

1. Stall

2. Recirculation (secondary flow)

3. Circulation

4. Leakage

5. Unsteady flow fluctuations

6. Wake (vortices)

7. Turbulence

8. Cavitation

Most of these unstable flow patterns produce vortices by boundary layer interaction between a high-velocity and low-velocity region in a fluid field, e.g., by flow around obstructions or past deadwater regions or by bidirectional flow. The vortices, or eddies, are converted to pressure perturbations as they impinge on the sidewall and may result in localized vibration excitation of the piping or pump components. The acoustic response of the piping system can strongly influence the frequency and amplitude of this vortex shedding. Experimental work has shown that vortex flow is most severe when a system acoustic resonance coincides with the natural or preferred generation frequency of the source. This source frequency has been found to correspond to a Strouhal number (S_n) from 0.2 to 0.5, where

$$S_n = \frac{f_e D}{V}$$

where f_e = vortex frequency, Hz
  D = a characteristic dimension of the generation source, ft (m)
  V = flow velocity in the pipe, ft/s (m/s)

For flow past tubes D is the tube diameter, and for branch piping excitation D is the diameter of the branch pipe. The basic Strouhal equation is further defined in Table 1, items 4A and B. As an example, flow at 100 ft/s (30 m/s) past a 12-in diameter (0.3-m) stub line would exhibit instability tendencies at a frequency of approximately 50 Hz. If the stub is acoustically resonant at a frequency near 50 Hz, rather large pulsation amplitudes can result.

When a centrifugal pump is operated at flows less than or greater than best efficiency capacity, noise is usually heard around the pump casing. The magnitude and frequency of this noise vary from pump to pump and are dependent on the magnitude of the pump head being generated, the ratio of NPSH required to NPSH available, and the amount by which pump flow deviates from ideal flow. Noise is often generated when the vane angles of the inlet guide, impeller, and casing (or diffuser) are incorrect for the actual flow rate. Another major source contributing to this noise is referred to as *recirculation*.[4,5]

Before the pressure of the liquid flowing through a centrifugal pump is increased, the liquid

TABLE 1 Piping Vibration Excitation Forces

Generation mechanism	Excitation frequency f_e, Hz
1. Reciprocating compressors	nf
2. Reciprocating pumps	nf, nPf
3. Centrifugal compressors and pumps	nf, nBf, nvf
4. Flow excitation	
A. Flow through restrictions	$\dfrac{0.2V}{D}$ to $\dfrac{0.5V}{D}$
B. Flow past stubs	$\dfrac{0.5V}{D}$
C. Flow turbulence due to quasi-steady flow	0–30 Hz (typically)
D. Cavitation and flashing	Broad band

n = 1, 2, 3, . . .
f = running speed, Hz
P = number of pump plungers
B = number of blades
v = number of volutes or diffuser vanes
V = flow velocity, ft/s (m/s)
D = restriction diameter, ft (m)

must pass through a region where its pressure is less than that existing in the suction pipe. This is due in part to acceleration of the liquid into the eye of the impeller. It is also due to flow separation from the impeller inlet vanes. If flow is in excess of design and the incident vane angle is incorrect, high-velocity, low-pressure eddies will form. If the liquid pressure is reduced to the vaporization pressure, the liquid will flash. Later in the flow path the pressure will increase. The implosion which follows causes what is usually referred to as *cavitation noise*. The collapse of the vapor pockets, usually on the nonpressure side of the impeller blades, causes severe damage (blade erosion) in addition to noise.

Sound levels measured at the casing of an 8000-hp (5970-kW) pump and near the suction piping during cavitation are shown in Fig. 2. The cavitation produced a wide-band shock that excited many frequencies; however, in this case, the vane passing frequency (number of impeller blades times revolutions per second) and multiples of it predominated. Cavitation noise of this type usually produces very-high-frequency noise, best described as "crackling."

Cavitation-like noise can also be heard at flows less than design, even when available inlet NPSH is in excess of pump required NPSH, and this has been a puzzling problem. A recent explanation offered by Fraser[4,5] suggests that noise of a very low, random frequency but very high intensity results from backflow at the impeller eye or at the impeller discharge, or both, and every centrifugal pump has this recirculation under certain conditions of flow reduction. Operation in a recirculating condition can be damaging to the pressure side of the inlet and/or discharge impeller blades (and also to casing vanes). Recirculation is evidenced by an increase in loudness of a banging type, random noise, and an increase in suction and/or discharge pressure pulsations as flow is decreased. Refer to Subsecs. 2.3.1 and 2.3.2 for further information.

Pressure regulators or flow control valves may produce noise associated with both turbulence and flow separation. These valves, when operating with a severe pressure drop, have high flow velocities which generate significant turbulence. Although the generated noise spectrum is very broad-band, it is characteristically centered around a frequency corresponding to a Strouhal number of approximately 0.2.

CAVITATION AND FLASHING   For many liquid pump piping systems, it is common to have some degree of flashing and cavitation associated with the pump or with the pressure control valves in the piping system. High flow rates produce more severe cavitation because of greater flow losses through restrictions.

In the suction piping of positive displacement pumps, high-amplitude pulsations can be generated by the plungers and amplified by the system acoustics and cause the dynamic pressure to periodically reach the vapor pressure of the liquid even though the suction static pressure may be above this pressure. As the cyclic pressure increases, the vapor bubbles collapse, producing noise and shock to the system, and this can result in erosion as well as undesirable noise. (See Fig. 9, Sec. 3.6.)

Flashing is particularly common in hot water systems (feedwater pump systems) when the hot, pressurized water experiences a decrease in pressure through a restriction (e.g., flow control valve). This reduction of pressure allows the liquid to suddenly vaporize, or flash, which results in

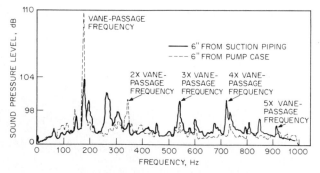

FIG. 2   Noise spectra of cavitation and vane passage on a centrifugal pump.

a noise similar to cavitation. To avoid flashing after a restriction, sufficient back pressure should be provided. Alternately, the restriction could be located at the end of the line so that the flashing energy can dissipate into a larger volume. Refer to Subsec. 2.3.4 for additional information.

NOISE CONTROL TECHNIQUES

Environmental noise usually does not emanate directly from the energy source; rather, it is transmitted along mechanical and/or liquid paths before it finally radiates from some vibrating surface into the surrounding environment.

The approaches to treating pump noise generally include the following:

1. *Source Modification.* Modify the basic pump design or operating condition to minimize the generation of acoustic energy.

2. *Interruption of Transmission.* Prevent sources from generating airborne (or overside) noise by interrupting the path between the energy source and the listener. This approach may range from isolation mounts at the source to physically removing the listener.

Source Modification Equipment modification to eliminate the noise source is usually quite specific to each particular situation; therefore, only general guidelines can be given. Many technical papers relate pump configurational modifications to the degree of noise reduction achieved; however, noise reduction depends on many parameters, and hence a particular modification may or may not help in a specific case. On the other hand, if the noise results from a resonant condition in the pump system, almost any reasonably conceived modification can destroy the resonance with varying degrees of improvement.

Some of the source modification approaches for pump applications are:

1. Increase or decrease pump speed to avoid system resonances of the mechanical or liquid systems.

2. Increase liquid pressures (NPSH, etc.) to avoid cavitation or flashing; decrease suction lift.

3. Balance rotating and/or oscillating components.

4. Change drive system to eliminate noisy components.

5. Correct acoustic resonance to minimize liquid-borne energy.

6. Modify centrifugal pump casing vanes so that clearance between impeller diameter and casing cutwater (tongue) or diffuser vanes is increased.

7. Modify centrifugal pump impeller discharge blade configuration by pointing, slanting, grooving, adding holes or staggering one-half pitch (if double suction).

8. Modify centrifugal pump casing cutwater (tongue) by slanting or adding holes.

9. Replace pump with different model or type to permit operation at reduced speed and the least number required.

10. If noise is due to operation of a centrifugal pump at flows less than design and recirculation is the problem, install minimum flow recirculation system bypass to increase total pump flow; if several pumps are operating in parallel, operate all pumps at the same speed and the least number required.

11. Use heavier bearing lubricant and/or increase number of bearing rolling elements.

12. Inject small quantity of air into the suction of a centrifugal pump to reduce cavitation noises.

The degree of improvement which can be achieved by any of the source modification approaches obviously depends upon the particulars of each installation, i.e., the basic causes of excess noise. Modification of the pump internals, for example, is extremely difficult in existing installations and can produce undesirable side effects unless the pump was poorly designed or selected initially.

As described earlier, the major sources of internal noise in reciprocating pumps are usually associated with piston-induced pulsations, piston mechanical reactions, turbulence, vortex for-

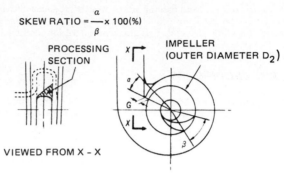

FIG. 3 Cutwater of a spiral casing and impeller. (Ref. 9)

mation from separated flow around obstructions, and cavitation. In centrifugal pumps, in addition to recirculation noise, interaction of the impeller flow with the pump case (especially the cutwater), high-velocity and pressure gradients at the impeller blade tip, and flow separation can make significant contributions to pulsation levels and noise. Internal modifications to the pump can ameliorate any or all of these conditions if they are severe initially, and the techniques are well known to most pump designers: use adequate valve sizes, avoid high velocities and obstructed flows, keep pressures above the vapor pressure of the fluid being pumped, degasify the fluid, provide adequate pulsation control equipment, and maintain proper angles of attack in centrifugal machines.

Many references give examples of how changing pump design parameters affects noise.[6-10] While such examples are valuable, they are of interest more in suggesting approaches than in predicting the degree of noise reduction which can be achieved in other pump applications.

Sudo, Komatsu, and Kondo[9] investigated pressure pulsations (and noise) generated in a centrifugal pump as a result of interference between the impeller discharge vanes and the receiving spiral casing single cutwater vane. The geometry of the pump (Fig. 3) defines the gap G between the impeller outer diameter D_2 and the casing cutwater. How varying the gap G/D_2 and the skew ratio (inclination of the cutwater or impeller vanes) affected discharge pressure pulsations is shown

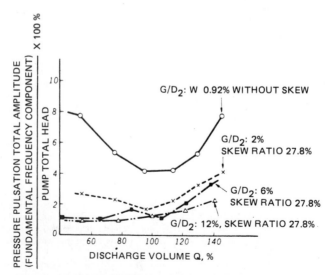

FIG. 4 Effects of gap width between cutwater of a spiral casing and trailing edge of the impeller blade. (Ref. 9)

in Fig. 4. Increasing G and the skew ratio decreases pulsation and noise amplitudes. Some investigators suggest that the ratio of cutwater diameter to impeller outer diameter be as large as 2:1 for optimum operation. When the cutwater diameter (or gap) is unknown, a rule of thumb suggested to optimize impeller diameter selection is not to use an impeller larger than 85% of maximum diameter.

To reduce the effect of impeller/casing vane passing pulsations, and consequently noise, double-suction impellers should have staggered vanes, i.e., the discharge vane tips should be shifted one-half pitch. This allows the fluctuation of the flow from each half of the impeller to interfere and thus reduce pressure pulsations at the pump discharge.

Florjančič, Schöffler, and Zogg[10] have reported that centrifugal pump impeller blade and casing tongue configuration can affect sound pressure levels and alter pump head and efficiency (Fig. 5).

Control of Noise Paths   These approaches consist of system modifications and treatments to disrupt the conduction of sound, whether borne by the structure, liquid, or air. They include approaches other than those which directly affect the originating source of oscillating energy (Fig. 6).

LIQUID PATH   Noise generated in the region of the pump is conducted both upstream and downstream by the liquid in the piping system. Such paths are most effectively reduced by acoustic

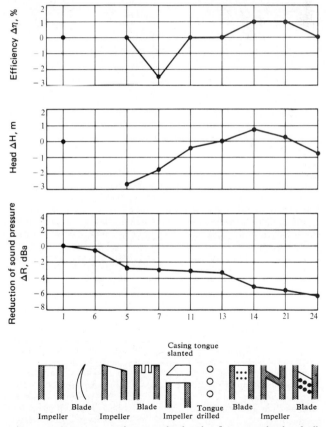

FIG. 5 Reduction in sound pressure level with influences on head and efficiency at 100% flow for various impeller blade and casing tongue configurations. (Ref. 10)

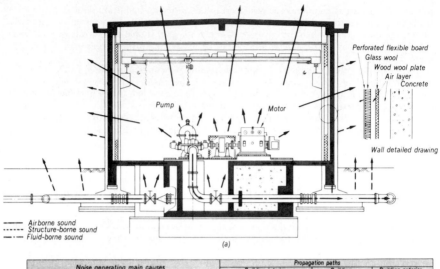

	Noise generating main causes		Propagation paths		
			Building interior	Building	Building exterior
Pump	Hydraulic causes	Cavitation			
		Impeller and casing interference		Air	→ Building wall	Air
		Vortex · Surging			
	Mechanical causes	Unbalance of rotor		Foundation pipe wall	Pipe wall	Ground
		Exciting vibration of bearing			
Motor · Auxiliary machinery		Cooling fan, etc.		Water		Pipe wall
		Unbalance of rotor			
Valve		Cavitation · Vortex			

(b)

FIG. 6   Noise propagation paths in a pumping plant. (Ref. 9)

(pulsation) filters or other pulsation control equipment, such as side branch accumulators (see Sec. 3.6).

STRUCTURE-BORNE NOISE   Obvious paths for solid-borne noise are the attached piping and pump support systems. Oscillatory energy generated near the pump by pulsation, cavitation, turbulence, etc., can be conducted as solid-borne noise for substantial distances before it is radiated as acoustic noise into the atmosphere. Some success in controlling solid-borne noise propagation can be achieved by the use of flexible couplings in the piping systems, mechanical isolation (vibration mounts, etc.) for the pump and drive systems, and resilient pipe hangers and supports.

Techniques for the vibration isolation of mechanical equipment are well known and available from many suppliers of vibration isolators or mounts (Table 2). These may consist of resilient supports at each mounting point of the machine, although it is often necessary to mount the pump and drive system on a single rigid skid (to assure alignment) and then isolate the skid from its

TABLE 2   Manufacturers of Isolators

Company	Product
Aeroflex Laboratories, Inc.	Helical isolators
Consolidated Kinetics Corporation	Airmount isolators
Firestone Company	Airmount isolators
Hamilton Kent Manufacturing Company	Tylox vibration mounts
Robintech, Inc.	Met-L-Flex isolators
Royal Products, Inc.	Machine mounts

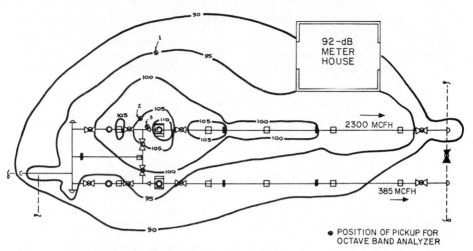

FIG. 7 Contours of equal sound level (decibels). (Ref. 12)

support system. For best isolation, the lowest resonant frequency of the supported system should be well below the minimum operating frequency, and none of the higher resonant modes should be coincident with running speed or multiples thereof.

The application of elastomeric coatings to the exterior surface of the pump or piping to damp pipe wall vibrations is normally ineffective except on very thin conduit. However, such coatings may have a small acoustic effect in confining or absorbing high-frequency noise which would otherwise be radiated by the pipe wall vibrations. (See the following discussion on pipe wraps.)

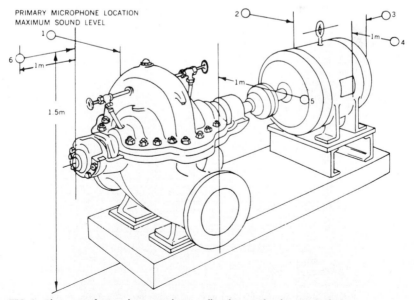

FIG. 8 Placement of microphones on a horizontally split centrifugal pump. (Ref. 3)

AIRBORNE NOISE   Although acoustic energy can be generated in the pumped liquid by purely fluid processes (turbulences, etc.), most noise radiated to the surrounding air is the direct result of mechanical vibrations in the pump case, the pipe wall, or other structures to which the pump system is coupled by liquid or mechanical attachment. (The exceptions to purely mechanical sources are windage noise produced by the rotating coupling, by cooling air in the drive motors, etc.)

When excessive noise is encountered, the source or sources can sometimes be identified by making sound measurements at points in a grid around the suspect equipment and plotting sound level contours. A sound level contour of a typical flow control valve installation is shown in Fig. 7. Octave band analyses or spectral analyses and contours can also provide clues by identifying the source of various frequency components of the noise. These components can also be compared with the information in Table 1 to aid in determining the source and location of the noise-generating mechanism.

The Hydraulic Institute code[3] gives specific details of the procedures for noise measurement around pumps, including microphone locations (Fig. 8), measurement procedures, and a data sheet (Fig. 9).

The approaches for reducing noise from pumps and piping systems after it is airborne generally consist of either interrupting the transmission path (barriers) or controlling the reverberation characteristics of the pump room.

A highly reflective (reverberant) pump room or enclosure can increase pump noise levels several decibels by reflections of the noise back and forth in the enclosure. The maximum reduction which can be achieved by the application of acoustic absorption material to the interior surfaces is normally about 10 dB for a highly reverberant enclosure. At most, such a treatment can reduce

---

AIRBORNE SOUND LEVEL TEST REPORT FOR PUMPING EQUIPMENT   REPORT FORM

SUBJECT:

Model:_____ Manufacturer:_____ Serial:_____
Rated Pump Speed:_____ Capacity:_____ Total Head:_____
Type of Driver:_____ Speed:_____
Auxiliaries such as Gears:_____
Applicable Figure No:_____
Description:_____

TEST CONDITIONS:

Distance from Subject to Microphone:_____Meters
Operating Speed as Tested:_____
Height of Microphone Above Reflecting Plane:_____Meters
Reflecting Plane Composition:_____
Primary Microphone Location No:_____
Remarks:_____

INSTRUMENTATION:

Microphone:_____ No._____
Sound Level Meter:_____ No._____
Octave Band Analyzer_____ No._____
Calibrator:_____ No._____
Other:_____ No._____

DATA:

db Ref. 2 × 10⁻ N/m²	BACK-GROUND	LOCATION *										AV.
		1	2	3	4	5	6	7	8	9	10	
dB A												
63												
125												
250												
500												
1k												
2k												
4k												
8k												

(Left axis label: MIDBAND FREQ. Hz)

*Corrected for background sound. Readings having 3 dB corrections must be reported in brackets. Only octave bands of interest as defined on page 299 need be reported.

TESTED BY:_____ DATE:_____

REPORTED BY:_____ DATE:_____

FIG. 9 Hydraulic Institute data sheet for measurement of airborne sound from pumping equipment. (Ref. 3)

noise levels to those which would exist if the pump were operating in the open, totally free of reflecting surfaces. Quantitatively, the noise reduction NR in decibels which can be achieved is

$$NR = 10 \log \frac{\overline{\alpha}_a}{\overline{\alpha}_b}$$

where $\overline{\alpha}_a$ = average absorption coefficient of the surfaces after treatment
$\overline{\alpha}_b$ = average absorption coefficient of the surfaces before treatment

The average absorption coefficient is defined as follows:

$$\overline{\alpha} = \frac{\alpha_1 A_1 + \alpha_2 A_2 + \alpha_3 A_3 + \cdots \alpha_n A_n}{A_1 + A_2 + A_3 + \cdots A_n}$$

where $\alpha_1, \alpha_2, \alpha_3, \ldots$ are absorption coefficients of various surface areas within the enclosure and $A_1, A_2, A_3, \ldots$ are the corresponding surface areas.

Absorption coefficients for typical building materials are given in Table 3. Note that the absorption coefficients vary with frequency, and hence calculated values of NR can be quite frequency-sensitive. Absorption data for various absorbing materials can best be obtained from the suppliers. One of the most common and most effective of such materials is fiberglass matting or the more rigid fiberglass board.

The noise reduction that can typically be attained with fiberglass piping insulation is shown

TABLE 3   Sound Absorption Coefficients of Common Construction Materials

Material	Frequency, Hz					
	125	250	500	1000	2000	4000
Brick						
Unglazed	0.03	0.03	0.03	0.04	0.04	0.05
Painted	0.01	0.01	0.02	0.02	0.02	0.02
Concrete block, painted	0.10	0.05	0.06	0.07	0.09	0.08
Concrete	0.01	0.01	0.015	0.02	0.02	0.02
Wood	0.15	0.11	0.10	0.07	0.06	0.07
Glass	0.35	0.25	0.18	0.12	0.08	0.04
Gypsum board	0.29	0.10	0.05	0.04	0.07	0.09
Plywood	0.28	0.22	0.17	0.09	0.10	0.11
Soundblox concrete block						
Type A (slotted), 6 in (15 cm)	0.62	0.84	0.36	0.43	0.27	0.50
Type B, 6 in (15 cm)	0.31	0.97	0.56	0.47	0.51	0.53
Carpet	0.02	0.06	0.14	0.37	0.60	0.65
Fiberglass typically 4 lb/ft³, (64 kg/m³), hard backing						
1 in (2.5 cm) thick	0.07	0.23	0.48	0.83	0.88	0.80
2 in (5 cm) thick	0.20	0.55	0.89	0.97	0.83	0.79
4 in (10 cm) thick	0.39	0.91	0.99	0.97	0.94	0.89
Polyurethane foam, open-cell						
¼ in (0.6 cm) thick	0.05	0.07	0.10	0.20	0.45	0.81
½ in (1.3 cm) thick	0.05	0.12	0.25	0.57	0.89	0.98
1 in (2.5 cm) thick	0.14	0.30	0.63	0.91	0.98	0.91
2 in (5 cm) thick	0.35	0.51	0.82	0.98	0.97	0.95
Hairfelt						
½ in (1.3 cm) thick	0.05	0.07	0.29	0.63	0.83	0.87
1 in (2.5 cm) thick	0.06	0.31	0.80	0.88	0.87	0.87

[a]For specific grades, see manufacturer's data; note that the term NCR, when used, is a single-term rating that is the arithmetic average of the absorption coefficients at 250, 500, 1000, and 2000 Hz.

SOURCE: Ref. 8.

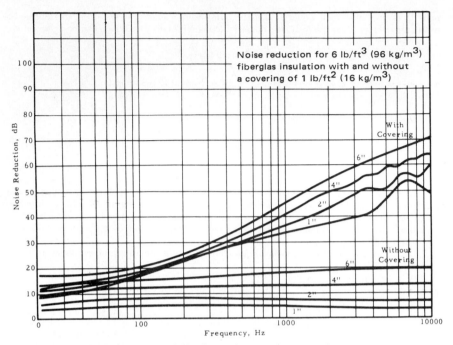

FIG. 10   Noise reduction for piping with fiberglass insulation. (1 in = 2.54 cm)

in Fig. 10. Note that noise reduction varies with noise frequency and with the thickness and density of the fiberglass matting. Densities of 5 to 6 lb/ft^3 (80 to 96 kg/m^3), found in the rigid fiberglass board, are normally more effective than the 1 to 2 lb/ft^3 (16 to 32 kg/m^3) found in rolled matting.

Normally, the average absorption in a room must be increased by a factor of at least 3 before noise improvement is discernible to the ear. Unless the absorption coefficient of the untreated room is less than about 0.3 in the frequency range of maximum noise, the addition of absorbing material to the walls and/or ceiling will likely be ineffective. Nevertheless, the noise reduction could amount to 5 dB and thus could mean the difference between compliance and noncompliance with existing criteria.

The use of barriers near or enclosing the pump can be much more effective than environmental absorption treatments, particularly if the barriers totally confine the noise-producing components (Fig. 11). Acoustic barriers simply interrupt the airborne path from the source to the listener. Such barriers work best (have the highest transmission loss) when

1. Their area density is high

2. They are total enclosures and all joints are well sealed against air leaks

3. They are not mechanically tied to the vibrating surfaces of the pump (i.e., they are mechanically isolated)

For total enclosures, the barrier area density controls transmission loss for noise frequencies above the lowest mechanical flexural frequency of the barrier panels. Loosely sprung but well-sealed and heavy panels therefore serve this function well. Treating the interior of the enclosure with sound-absorbing material tends to reduce the apparent source strength by preventing reverberant buildups in the enclosure. Note that the use of a total enclosure around a pump may cause other operational problems which must be dealt with, e.g., overheating, formation of explosive mixtures, or difficulty in getting to the pump for maintenance.

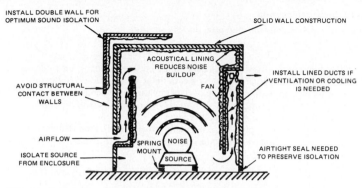

FIG. 11 Design of an effective sound-isolating enclosure requiring air circulation for cooling. (Ref. 8)

Although partial barriers around a pump usually cannot provide the degree of noise reduction afforded by total enclosures, they may be quite useful. Such barriers normally redirect the noise away from the listener and work best when they are located very near either the source or the listener. They also work better in an open or acoustically dead room than in a highly reverberant one. Their performance can often be enhanced by treating the interior (pump-side) surface with absorbing material or by reflecting the noise into an absorbing surface. The transmission loss of partial barriers is usually not critical so long as it is 20 dB or greater, as reverberation and flanking noise paths will normally limit overall reduction to less than 20 dB.

Exterior control techniques for pipe noise usually consist of pipe wrapping materials in addition to the resilient pipe supports and hangers already mentioned. The function of such materials is identical to that of total enclosures; hence, pipe wraps should be

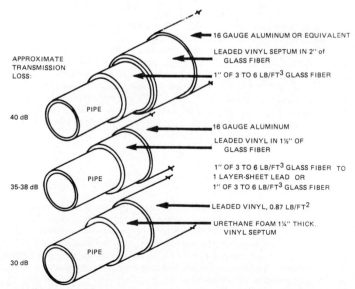

FIG. 12 Three possible pipe wrap configurations. (1 in = 2.54 cm; 1 lb/ft^3 = 16 kg/m^3) (Ref. 12)

1. Massive (high area density)
2. Mechanically decoupled from the pipe (loosely sprung)
3. Airtight

For all the very-low-frequency noise, decoupling between a heavy, airtight outer coating and the vibrating pipe wall can be achieved by a compliant acoustic absorbing material, such as preformed fiberglass pipe insulation. This material provides some acoustic absorption of noise as it "passes through," but, more important, it provides vibratory decoupling between the vibrating pipe wall and the external pipe wrap shell or layer.

Multiple-layered panels or pipe wrap normally provide more attenuation than single homogenous layers, so long as the layers are mechanically decoupled. Examples of several pipe wrap configurations are given in Fig. 12.

NOISE MEASUREMENT

In very general terms, sound consists of cyclic modulations of ambient pressure at frequencies which are detectable to the human ear, normally between 30 Hz and 17 kHz. For our purposes, noise is simply unwanted sound. The complications in measuring noise or in assessing its effects on personnel arise chiefly because of the remarkable complexity of the human ear and its very nonlinear acoustic response with both frequency and amplitude. The dynamic range of the human ear is 160 dB, from a minimum detectable level of about 3.7×10^{-9} lb/in^2 to about 0.37 lb/in^2 (2.6×10^{-10} to 0.026 bar°). In order to compensate for the ear's dynamic range, the decibel scale was adapted to roughly simulate the ear's nonlinear sensitivity and reduce numerical sound level values to a manageable range. The strength of familiar sounds relative to their typical decibel level is given in Fig. 13.

The decibel is a unit of sound measurement relating the dynamic pressure variations produced by the source of interest to a reference sound pressure (0.0002 μbar). The sound pressure level in decibels is equal to 20 times the common logarithm of this ratio:

$$SPL = 20 \log \left(\frac{P}{0.0002} \right)$$

where P is the rms sound pressure in microbars. For example, if one sound pressure level is twice another, the measured noise level will be 6 dB greater.

The basic instrument used to measure sound intensity is the sound level meter, which provides a numerical reading of decibel values by integrating sound throughout the audible frequency spectrum. Since the frequency response of the ear varies with sound intensity, the sound level meter has three internal filters that can be used to approximate the ear's frequency response at 45, 75, and 95 dB. These are called the A, B, and C filters, respectively. In recent years, use of the C weighing filter has become popular in prescribing noise criteria. The dBC scale has been thus adapted, partially because it is easy to use and partially because it attenuates very-low-frequency noise, which is not as potentially injurious to the ear.

CRITERIA

In order to decide whether noise treatment is necessary, the sound levels must be measured and compared with applicable criteria. Most protective noise criteria have been developed to protect a specific statistical sample of people from prescribed typical noise conditions and exposure patterns. While such criteria are admittedly approximate, they are usually sufficiently conservative to protect a large majority of those who may be exposed, whether they are written to prevent hearing loss, community annoyance, speech and communication masking, or any of the other physiological or psychological effects of noise.

° 1 bar = 10^5 Pa. For a discussion of bar, see SI Units—A Commentary in the front matter.

AT A GIVEN DISTANCE FROM NOISE SOURCE

ENVIRONMENTAL

DECIBELS
RE 0.0002 MICROBAR

140

50-HP SIREN (100')

F-84 AT TAKEOFF (80' FROM TAIL)
HYDRAULIC PRESS (3')
LARGE PNEUMATIC RIVETER (4') **130**

BOILER SHOP (MAXIMUM LEVEL)

PNEUMATIC CHIPPER (5')

120

MULTIPLE SANDBLAST UNIT (4')   ENGINE ROOM OF SUBMARINE (FULL SPEED)
TRUMPET AUTO HORN (3')   JET ENGINE TEST CONTROL ROOM
AUTOMATIC PUNCH PRESS (3') **110**

CHIPPING HAMMER (3')   WOODWORKING SHOP

CUT-OFF SAW (2')   INSIDE DC-6 AIRLINER
ANNEALING FURNACE (4')   WEAVING ROOM
AUTOMATIC LATHE (3') **100**

SUBWAY TRAIN (20')   CAN MANUFACTURING PLANT
HEAVY TRUCKS (20')   INSIDE CHICAGO SUBWAY CAR
TRAIN WHISTLES (500')   INSIDE MOTOR BUS
10-HP OUTBOARD (50') **90**   INSIDE SEDAN IN CITY TRAFFIC

SMALL TRUCKS ACCELERATING (30')

80
LIGHT TRUCKS IN CITY (20')   OFFICE WITH TABULATING MACHINES

HEAVY TRAFFIC (25' TO 50')
AUTOS (20')

70
AVERAGE TRAFFIC (100')

ACCOUNTING OFFICE
CONVERSATIONAL SPEECH (3')   CHICAGO INDUSTRIAL AREAS
60

15,000-kVA, 115-kV TRANSFORMER (200')

50   PRIVATE BUSINESS OFFICE

LIGHT TRAFFIC (100')
AVERAGE RESIDENCE

40

MINIMUM LEVELS FOR RESIDENTIAL
AREAS IN CHICAGO AT NIGHT
30   BROADCASTING STUDIO (SPEECH)

BROADCASTING STUDIO (MUSIC)

20   STUDIO FOR SOUND PICTURES

10

THRESHOLD OF HEARING - YOUNG MEN { **0**
1000 TO 4000 CPS

FIG. 13 Typical overall sound levels. (1 ft = 0.3048 m) (Ref. 13)

When noise criteria are legislated into noise codes or regulations, they normally specify noise measurement locations, instruments, and sampling times as well as allowable levels and exposure durations. Thus they have taken much of the guesswork out of evaluating machinery noise. The main criterion for pump and other machinery installations can be obtained from the U.S. Department of Labor, OSHA Bulletin 334. OSHA guidelines for allowable daily noise exposure as a function of time and sound pressure level for personnel near the installations are given in Table 4. The allowable exposure time depends upon the maximum sound level in the work space. When the noise exposure for personnel is intermittent, the allowable exposure may be calculated from Table 4 by computing the fractions of actual exposure time at a given noise level to the allowable

TABLE 4   OSHA Noise Exposure Limits

Sound level, dBA	Allowable exposure time	
	Minutes	Hours
90	480	8.00
91	420	7.00
92	360	6.00
93	320	5.33
94	280	4.67
95	240	4.00
96	210	3.50
97	180	3.00
98	160	2.67
99	140	2.33
100	120	2.00
101	105	1.75
102	90	1.50
103	80	1.33
104	70	1.16
105	60	1.00
106	54	0.90
107	48	0.80
108	42	0.70
109	36	0.60
110	30	0.50
111	27	0.45
112	24	0.40
113	21	0.35
114	18	0.30
115 (maximum allowable level)	15	0.25

exposure time at that level. If the sum of these fractions is less than 1, the noise level can be considered safe. It is important to know one underlying implication in this criterion. If a reduction of the noise level is not economically feasible, it may be reasonable to schedule operators' work so that the criterion levels will not be exceeded. This rescheduling may permit a wider latitude in the final solution of a plant noise problem.

MODEL TESTING

It is possible to conduct tests on reduced-size pumps and pumping systems in order to predict and adjust pressure pulsations and noise from the pump and piping. This has been successfully done by Sudo, Komatsu, and Kondo[9] using a model variable-speed, single-stage, double-suction pump in addition to mathematical computer confirmation. A comparison of performance with nonstaggered and staggered impeller vanes revealed the good pulsation suppression characteristics of the latter impeller configuration (Fig. 14).

These investigators, in their modeling, used the following relationships to provide the required similarity between model and prototype:

1. Equal ratios of pipe length to wavelength

2. Equal ratios of pipe cross-sectional area

3. Model pump speed adjusted to make the product of pulsation wavelength and frequency equal to the speed of sound in water

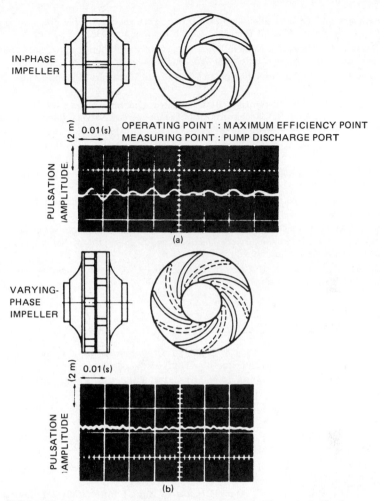

FIG. 14 Comparison of pressure pulsation waveforms: (*a*) a double-suction in-phase centrifugal pump impeller and (*b*) a varying-phase (staggered) centrifugal pump impeller. (Ref. 9)

REFERENCES

1. American National Standards Institute: Acoustical Terminology (Including Mechanical Shock and Vibration), ANSI Standard S1.1, New York, 1976.

2. American National Standards Institute: Specification for Sound Level Meters, ANSI Standard S1.4, New York, 1971.

3. *Hydraulic Institute Standards for Centrifugal, Rotary and Reciprocating Pumps*, 13th ed., Cleveland, 1975.

4. Fraser, W. H.: "Flow Recirculation in Centrifugal Pumps," in *Proceedings of the Tenth Turbomachinery Symposium*, Texas A & M, College Station, Texas, 1981.

5. Fraser, W. H.: "Recirculation in Centrifugal Pumps: Materials of Construction of Fluid

Machinery and Their Relationship to Design and Performance," paper presented at winter annual meeting of ASME, Washington, D.C., November 16, 1982.

6. Meyerson, N.: "Sources of Noise in Power Plant Centrifugal Pumps with Considerations for Noise Reduction," *Noise Control Eng.* **2**(2):74 (1974).

7. Evans, L. M.: "Control of Vibration and Noise from Centrifugal Pumps," *Noise Control*, January 1958, p. 28.

8. Berendt, R. D., E. L. R. Corliss, and M. S. Ojalvo: *Quieting: A Practical Guide to Noise Control*, Handbook 119, U.S. Department of Commerce and National Bureau of Standards, Washington, D.C., July 1976.

9. Sudo, S., T. Komatsu, and M. Kondo: "Pumping Plant Noise Reduction," *Hitachi Rev.* **29**(5):217 (1980).

10. Florjančič, D., W. Schöffler, and H. Zogg: "Primary Noise Abatement on Centrifugal Pumps," *Sulzer Tech. Rev.* **1**:24 (1980).

11. *Centrifugal Pump Hydraulic Instability*, CS-1445 Research Project 1266-18, Electric Power Research Institute, 3412 Hillview Ave., Palo Alto, Calif., 1980.

12. Sharp, J. M., G. Damewood, C. R. Sparks, M. T. Hanchett: *Noise Abatement at Gas Pipeline Installations*, vol. I: *Physiological, Psychological and Legal Aspects of Noise*, Project No. 23, American Gas Association, New York, 1959.

13. Peterson, A. P. G., and E. E. Gross, Jr.: *Handbook of Noise Measurement*, 5th ed., General Radio, West Concord, Mass., 1963.

FURTHER READING

Beranek, L. L.: "Noise Reduction," in *Noise and Vibration Control*, paper for program at MIT, Cambridge, Mass., 1960; published by McGraw-Hill, New York, 1971.

Edison Electric Institute, *Electric Power Plant Environmental Noise Guide*, no. 78-51 chap. 4, Washington, D.C., 1978.

Kamis, P. A.: "Techniques for Reducing Noise in Industrial Hydraulic Systems," *Pollution Eng.*, May 1975, p. 46.

Kondo, M., et al.: "Pressure Fluctuations in Pump Pipeline," in *Proceedings of the 17th Congress IAHR*, 1977–78. International Association for Hydraulic Research, Delft, Netherlands.

Matley, J., and staff of *Chemical Engineering: Fluid Movers, Pumps, Compressors, Fans and Blowers*, McGraw-Hill, New York, 1979.

Salmon, V., J. S. Mills, and A. C. Petersen: *Industrial Noise Control Manual*, Industrial Noise Services, contract no. HSM 99-73-82, Palo Alto, Calif., June 1975, Tables 4.1 and 4.2.

Sparks, C. R., and J. C. Wachel: "Pulsation in Centrifugal Pump and Piping Systems," *Hydrocarbon Processing*, July 1977, p. 183.

Sperry Rand: *More Sound Advice: How to Quiet Hydraulically Powered Equipment*, bulletin 74-154, Troy, Mich., 1974.

Thompson, A. R., and R. M. Hoover: "Reduction of Boiler Feed Pump Noise in an 800 MW Coal-Fired Generating Unit," IEEE paper F78-819-5 delivered at Joint Power Generation Conference, Dallas, September 1978.

Tung, P. C., and M. Mikasinovic: "Eliminating Cavitation from Pressure-Reducing Orifices," *Chem. Eng.*, December 1983, p. 69.

Wojda, H. W.: "First Aid for Hydraulic System Noises," *Pollution Eng.*, April 1975, p. 38.

PUMP SERVICES

SECTION 9.1
WATER SUPPLY

F. G. HONEYCUTT, JR.
D. E. CLOPTON

SOURCES OF WATER

Surface Water Surface water supplies are obtained from streams, rivers, lakes, and reservoirs. The quantity of water available from a surface supply can be determined with reasonable accuracy from yield studies which take into account the local effects of rainfall, runoff, evaporation and sedimentation rates, and other hydrological factors. Development of a surface supply usually requires pumps to transport raw water from the source to a treatment plant and to provide the head necessary for proper hydraulic operation of the treating facilities. Pumps utilized for this purpose are classified as low-lift pumps because relatively low discharge heads are required.

Selection of a specific type of pump for low-lift service is dependent on intake conditions. Since surface water supplies vary significantly in temperature, bacteria count, and turbidity at varying depths and since the water level may fluctuate considerably, it is necessary to provide some type of intake structure which will permit withdrawal of water at several elevations. Multiple intake ports equipped with trash racks and water screens provide this capability and provide protection from fish and debris. The design and location of the intake structure influence the selection of either a horizontal or vertical pump for low-lift service.

Groundwater In many areas of the United States where rainfall and runoff are sparse, significant supplies of water are available from underground sources. The groundwater table is formed when rainfall percolates through the soil and reaches a zone of saturation, the depth of which is governed by soil characteristics and subsurface conditions.

Groundwater can be developed as a source of supply through utilization of wells or springs. Shallow wells generally utilize the water table as a source, while deep wells utilize water held in a pervious subsurface stratum (aquifer). Deep wells generally provide more constant and more prolific supplies than do shallow wells.

Artesian wells may be developed when an aquifer outcrops at a surface elevation significantly higher than the ground elevation at the well site. The aquifer is thus pressurized, and water flows from the well without pumpage. A natural artesian well may occur if a fault extends from the aquifer to the ground surface.

Springs occur where the groundwater table outcrops at the surface of the earth. However, the supply available from springs is seldom great enough to serve more than one or two homesteads and is thus usually insignificant in terms of supplying communities. Surface streams may also be fed from the groundwater table, and vice versa. Figure 1 depicts these various conditions.

Well water is usually pumped to a treatment site or into a distribution system by either vertical turbine line shaft pumps or submersible pumps. Economics dictates the selection of one type over

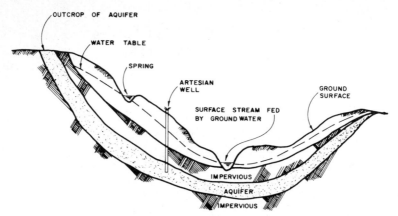

FIG. 1 Subsurface conditions for development of groundwater supplies.

the other. In general, at depths greater than 500 ft (150 m) submersible pumps become economically competitive with line shaft pumps. For shallower depths line shaft pumps are used almost exclusively.

Effects of Source on Water Quality

Although the quality of water obtained from both groundwater and surface supplies varies greatly according to local climatological, hydrological, and geological conditions, some general comparisons can be made between groundwater and surface supplies. Surface supplies generally contain more bacteria, algae, and suspended solids than do groundwater sources and thus require specific treatment to eliminate the resulting turbidity, colors, odors, and tastes.

Groundwater supplies generally contain few bacteria and in some instances are pure enough for domestic use without chemical treatment. Frequently, however, groundwater supplies contain significant amounts of dissolved minerals. Depending upon the mineral type, the resulting supply may exhibit such characteristics as extreme hardness, toxicity, or undesirable color, odor, or taste. Special treatment, such as aeration, lime softening, and disinfection, may be required to remove such objectionable characteristics.

USES OF WATER

Classifications History has shown that significant concentrations of population have always occurred where a water supply was readily available. Accordingly, technology has been developed to make full use of the available supply, not only for domestic use but also for public, commercial, industrial, and agricultural consumption.

Domestic usage consists of water used for household purposes. The amounts used for such purposes vary with the standard of living of the consumers, the quality of the water, whether or not metering devices are used, and other factors. Consumption for domestic purposes is generally in the range of 50 to 60 gal per capita per day (gpcd) [190 to 230 liters (lpcd)], which is defined as the total quantity of water used in one calendar year divided by (365 × average number of persons supplied).

Public usage includes the water used for such purposes as street cleaning, water for public parks, and supply to public buildings. Such consumption generally amounts to about 10 to 15 gpcd (38 to 57 lpcd).

Commercial usage varies according to the number and size of shops and stores in the area served. Attempts have been made to assign commercial consumption on the basis of gallons per day per square foot of floor space. However, the nature of business conducted in commercial installations varies widely, and accurate allocation of specific demands for commercial use must be based on a thorough investigation of the individual establishments.

Industrial usage can play a major role in the design of water supply, treatment, and distribution systems. In heavily industrialized areas, industrial usage can account for 30% or more of the total water consumption.

Agricultural usage includes the water used for irrigation and for watering livestock. In most localities, irrigation water will not be a factor in system design. However, in semiarid locations where crops rely heavily on irrigation water, consideration must be given to such needs, and a thorough analysis must be made.

Water Consumption   A necessary element in the design and selection of pumping equipment for water supply projects is the determination of the amount of water required. Population forecasts and estimates of future usage serve as a basis for the design of municipal and urban systems.

The rate of consumption is usually expressed as the average annual usage in gallons per capita per day. Actual average use varies throughout the country, generally from 120 to 200 gpcd (450 to 760 lpcd). In metropolitan areas an average value of 175 gpcd (660 lpcd) is often used for design purposes.

Demand and Daily Fluctuations   Water supply systems are subject to wide fluctuations in demand. The rate of consumption varies seasonally, monthly, daily, and hourly. For design purposes it is essential that a reasonably dependable relationship between certain water use rates be determined from past experience records. The usage and demand rates are as follows:

1. *Average daily demand* may be expressed as gallons (liters) per capita per day or million gallons (liters) per day (the total year's usage divided by 365). The average daily demand rate is generally used as the yardstick by which all other demand rates are measured.

2. *Maximum monthly demand* is the average daily use during the month of greatest consumption. It is determined by dividing that month's total usage by the number of days in the month.

3. *Maximum daily demand* is the total amount of water used during the day of heaviest consumption in the year. Experience shows this demand rate to occur on from 3 to 5 consecutive days during the year.

4. *Maximum hourly demand* is the rate of use during the hour of peak demand on the day of maximum demand. This demand rate normally establishes the highest rate of design for distribution systems and "peaking pumpage."

To record the hourly variations in water consumption over a 24-h period is no simple undertaking. It involves taking synchronized hourly readings of all pump discharge rates and storage levels and, from these, computing the hourly rate of consumption. Such recordings of hourly consumption rates made for Dallas are shown in Fig. 2 and can be used to establish system design parameters.

Water Consumption Rates as a Basis of Design   The variations in demand, expressed as a ratio of the average daily demand, must be considered in basic designs of water supply systems.

Figure 3 illustrates graphically the relationship that is characteristic of many water supply demand rates. It also serves to establish percentage ratios for design purposes as follows:

Maximum monthly rate = 155% × average daily rate

Maximum daily rate = 186% × average daily rate

Maximum hourly rate = 343% × average daily rate

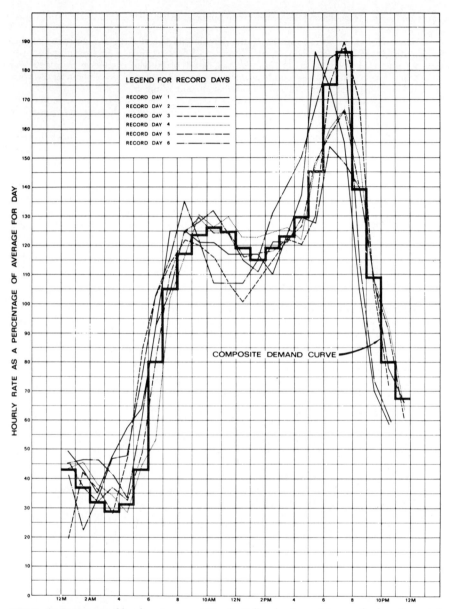

FIG. 2 Percentage rate of hourly water consumption for Dallas, showing variations on the days of recorded maximum demand and composite demand curve. This information can be used as a basis of system design. (URS/Forrest and Cotton)

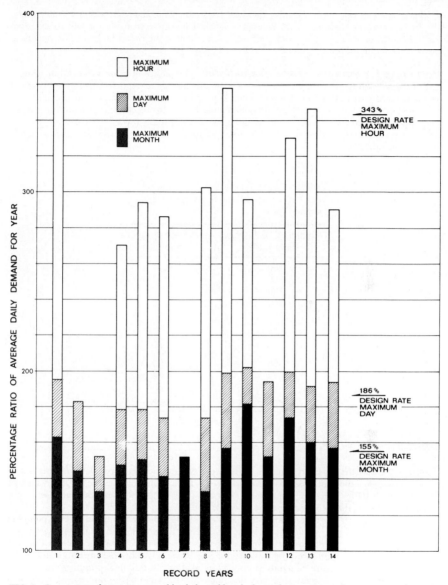

FIG. 3 Comparison of maximum monthly, daily, and hourly demand rates, expressed as percentage of average daily rate for the year. (URS/Forrest and Cotton)

Alternately, the design demand rates may be expressed in terms of per capita consumption:

Average daily demand = 175 gpcd (662 lpcd)

Maximum monthly demand = 270 gpcd (1022 lpcd)

Maximum daily demand = 325 gpcd (1230 lpcd)

Maximum hourly demand = 600 gpcd (2270 lpcd)

These demand rates are indicative of water requirements in the southwestern United States and will vary in other regions. Nevertheless, the values shown are reasonable and will assist designers in establishing local criteria. Above all, such values depict the broad range of pumping rates for which equipment must be chosen.

Variations for Pumping-Time Evaluations The pumps of most water supply systems must operate continuously throughout the year. Studies are often required for determination of operating costs and evaluation of equipment. Thus, pumping time in days, which can also be expressed as a percentage of average daily demands, is a useful piece of information in various pumping design problems. Based upon studies made of municipally owned waterworks of various sizes in the southwestern United States, the total pumpage for the year will vary approximately as shown in Table 1.

TABLE 1 Number of Days of Pumpage in Year Expressed as Ratio of Average Daily Demand Rate

Demand, % of average daily rate	Days of pumpage per year
190	11
185	14
180	12
165	11
155	13
145	17
135	12
115	27
85	95
75	64
65	89
	365

PUMPING STATIONS

Pumping Capacity Determination of the capacity required at a particular pumping station must be based on a thorough analysis of the proposed system. Such factors as the projected average and maximum daily demands, the safe yield of the available supply, and the function of the pumping station in the total system must be considered.

Accurate forecasts of future demands will often establish the design criteria for a pumping station, which should be capable of supplying demands for the area served for many years. Common practice involves sizing the station for anticipated demands for 25 years or more, with initial installation of only enough pumping capacity for 5 to 10 years. Additional capacity can then be added, as directed by future demands, simply by installation of additional pumping units.

Limitations on pumping capacity may be imposed by the yield of the raw water supply. It is sometimes desirable to size pumps to deliver only the safe yield of the source. More often, however, it is practical to impose severe overdrafts on the supply source for short periods of time and to allow the source to refill during sustained periods of low demand. In such cases, pumping capacity may exceed the safe yield of 200% or more, depending on the magnitude of anticipated peak demands and the length of time for which such demands must be satisfied.

The function of the pumping station in overall system operation can also affect the determination of required pumping capacities. It is sometimes practical to provide constant-rate pumpage from the source by constructing a balancing reservoir near the treatment site. The balancing reservoir must have ample capacity to permit withdrawal at varying rates (in accordance with treated water demands) without overflowing or draining. Since the balancing reservoir is sized for a constant input, the transmission line from the source requires no peaking capacity and the size of the transmission line is thus minimized. This type of operation is thus economically feasible when the savings realized from the reduced size of the transmission line are greater than the cost of constructing a balancing reservoir and variable-rate pumping and transmission facilities for the short distance from the balancing reservoir to the treatment facilities. Accordingly, this operational scheme becomes desirable when the source of supply is located a great distance from the treatment site.

Selection of Pump Type   The location and configuration of the pumping station and intake structure and the anticipated heads and capacities are the major factors influencing the selection of a specific type of pump. If the pumping station and intake structure are to be located within a surface reservoir, vertical turbine pumps with columns extending down into a suction well are a logical choice.

In many cases, however, the pumping station is located downstream from the dam, with connecting suction piping from the intake structure. In such instances, a horizontal centrifugal pump represents a more logical selection. Horizontal centrifugal pumps of split-case design are commonly used in waterworks because the rotating element can be removed without disturbing suction and discharge piping. Selection of a bottom suction pump (in lieu of side suction) should be considered when possible because the former requires less space on the station floor.

Effect of Source on Selection of Pump Materials   Although many service conditions are involved in the selection of pump materials, the primary factors which can be related to source of supply are alkalinity and abrasiveness. The alkalinity (or acidity) of a raw water source is reflected by the raw water pH. In general, a pH above 8.5 or below 6.0 precludes the use of a standard bronze-fitted pump (cast iron casing, steel shaft, bronze impeller, wearing rings, and shaft sleeve). The high pH values often associated with groundwater sources then dictate the use of all-iron or stainless steel–fitted pumps.

Abrasiveness, which may result from sand and other suspended matter in a surface water supply, may dictate the selection of stainless steel or nickel–cast iron casing; cast iron, nickel–cast iron, or chrome steel impellers; and stainless steel, phosphor bronze, or Monel wearing rings, shafts, sleeves, and packing glands.

Suction and Discharge Piping   In order to minimize head loss and turbulence, the use of long-radius bends in both suction and discharge piping is strongly recommended. American Water Works Assocation (AWWA) approved double-disk gate valves with outside screw and yoke or AWWA-approved butterfly valves of the proper classification are recommended for use as isolation valves in the pump suction and discharge piping. Additionally, a check valve should be provided on the discharge side of the pump to prevent backflow through the pump upon shutdown or power failure. Many types of check valves have been used satisfactorily in such applications. However, the regulated opening and closing times afforded by cone valves, combined with excellent throttling characteristics, have proved effective in minimizing surges and should be considered for pump check service if economically feasible.

Manifolding of suction and discharge headers is common practice in the design of pumping stations because parallel operation can be readily achieved with such an arrangement. Suction headers may be located directly below the pumps (Fig. 4) or along the outside wall of the pump station (Fig. 5), depending on the location of the intake structure and the configuration of the suction piping. Interconnection of discharge headers (Fig. 6) provides additional system flexibility and added protection in the event of line breakage.

Pump Drivers   The waterworks industry has evolved to the point of almost exclusive use of electric motors as pump drivers. Diesel or gasoline engines may be used as emergency drivers or as the primary drivers where reliable electric power is not available. However, the high costs of continuous operation and the limitations of rotative speed preclude the use of diesel or gasoline drivers in most installations.

Although some dc motors are used in the waterworks industry, the great majority of electric motors used as pump drivers are ac motors of the squirrel-cage-induction, wound-rotor, or syn-

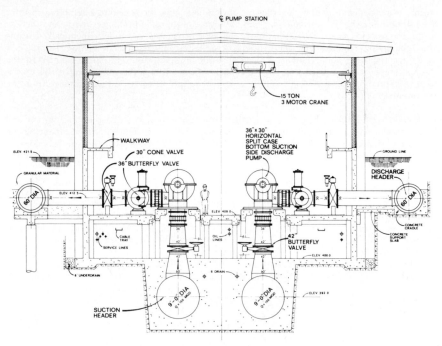

FIG. 4 Typical section of the Forney raw water pumping station in Dallas, showing pump suction and discharge piping. (1 in = 2.54 cm; 1 ft = 0.3048 m) (URS/Forrest and Cotton)

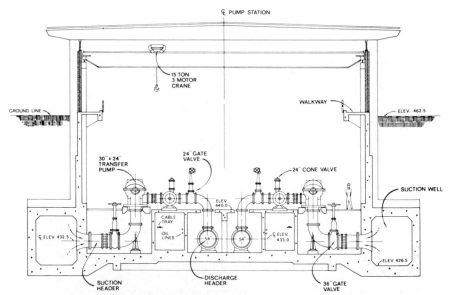

FIG. 5 Typical section of the high-service pumping station of the East Side water treatment plant in Dallas. (1 in = 2.54 cm; 1 ft = 0.3048 m) (URS/Forrest and Cotton)

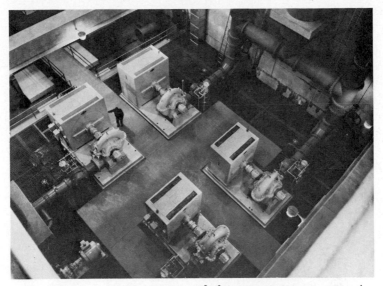

FIG. 6 Dallas Iron Bridge 100-mgd (3.78 × 10^5 m^3/day) raw water pumping station, showing layout with bottom-suction, side-discharge pumping units. (URS/Forrest and Cotton)

chronous type. Hydraulic or magnetic variable-speed devices can be used in conjunction with the pump driver to vary pumping rates in accordance with demand.

System-Head Curves The capacity of a pumping installation cannot be determined without an accurate determination of the head requirements of the system. Consequently, a system-head curve must be derived, depicting calculated losses through the system for various pumping rates. The construction of a system-head curve must be based upon a logical sequence of determinations of the various components of system losses.

A preliminary sketch or schematic should be derived, showing the configuration and size of suction and discharge piping, including all pipe, valves, and fittings.

FRICTIONAL LOSSES Frictional loss in the discharge piping can be determined for various pumping rates from the formula:

in US units
$$h_f = KQ^{1.85} \tag{1a}$$

in SI units
$$h_f = 2.618 \times 10^{-5} KQ^{1.85} \tag{1b}$$

where h_f = head loss, ft (m)
K = a constant depending on pipe size and friction factor C
Q = flow, mgd (m^3/h)

Equation 1 is a modification of the basic Hazen-Williams formula for the special case of circular pipes flowing full. Values of 10^5K for various pipe sizes and friction factors are listed in Table 2. Common practice has been to design systems on the basis of $C = 100$. However, investigations for the Dallas Water Utilities have indicated that a regulated cleaning program for pipelines can result in sustained values of C as high as 135 or 140. Accordingly, consideration should be given to providing for periodic cleaning of pipelines so that the initial design capacity of the pumping units can be sustained.

To demonstrate the use of Eq. 1, assume that it is desired to compute the head losses due to pipe friction as a result of pumping at a rate of 30 mgd (4730 m^3/h) through 1050 ft (320 m) of

TABLE 2 Computed Valued of 10^5K for Use in Eq. 1

Pipe diameter, in.	Hazen-Williams C value								
	70	80	90	100	110	120	130	135	140
6	11,800	9,240	7.420	6,100	5,110	4,350	3,750	3,590	3,280
8	2,900	2,270	1,823	1,500	1,240	1,070	924	881	805
10	982	767	617	507	425	362	312	298	272
12	404	315	253	209	175	149	128.5	122.6	112.1
14	190	149	119.7	98.4	82.5	70.4	60.5	57.8	52.9
16	99.6	77.8	62.6	51.6	43.3	36.8	31.6	30.3	27.7
18	56.1	43.9	35.2	29.0	24.3	20.7	17.84	17.04	15.6
20	33.6	26.2	21.1	17.33	14.53	12.4	10.67	10.17	9.30
21	26.4	20.7	16.6	13.67	11.47	9.77	8.43	8.05	7.35
24	13.8	10.3	8.68	7.13	5.99	5.09	4.39	4.19	3.83
27	12.4	6.08	4.89	4.02	3.37	2.87	2.47	2.36	2.16
30	4.66	3.64	2.91	2.41	2.02	1.717	1.48	1.412	1.291
33	2.94	2.29	1.964	1.516	1.269	1.081	0.933	0.890	0.814
36	1.92	1.50	1.206	0.993	0.832	0.708	0.610	0.583	0.534
39	1.298	1.016	0.816	0.670	0.563	0.480	0.413	0.395	0.361
42	0.906	0.706	0.570	0.469	0.392	0.334	0.287	0.276	0.251
45	0.646	0.504	0.406	0.334	0.280	0.238	0.206	0.1964	0.1796
48	0.462	0.379	0.290	0.239	0.200	0.1702	0.1470	0.1401	0.1280
51	0.352	0.275	0.221	0.1816	0.1522	0.1296	0.1119	0.1067	0.0975
54	0.266	0.208	0.1673	0.1377	0.1152	0.0982	0.0848	0.0807	0.0738
57	0.204	0.160	0.1285	0.1056	0.0886	0.0753	0.0650	0.0621	0.0567
60	0.1594	0.1244	0.1000	0.0823	0.0691	0.0587	0.0507	0.0484	0.0442
63	0.1256	0.0982	0.0789	0.0650	0.0545	0.0464	0.0400	0.0382	0.0349
66	0.1000	0.0785	0.0630	0.0518	0.0434	0.0370	0.0319	0.0304	0.0278
69	0.0805	0.0630	0.0507	0.0417	0.0349	0.0298	0.0256	0.0245	0.0224
72	0.0655	0.0511	0.0412	0.0339	0.0284	0.0242	0.0209	0.01991	0.01820
75	0.0538	0.0414	0.0339	0.0278	0.0233	0.01982	0.01710	0.01632	0.01493
78	0.0444	0.0347	0.0279	0.0229	0.01924	0.01637	0.01413	0.01348	0.01231
81	0.0314	0.0289	0.0232	0.01906	0.01600	0.01360	0.01173	0.01121	0.01024
84	0.0309	0.0242	0.01942	0.01600	0.01340	0.01141	0.00984	0.00939	0.00859
90	0.0222	0.01730	0.01390	0.01143	0.00957	0.00816	0.00704	0.00672	0.00614
96	0.01614	0.01262	0.01013	0.00834	0.00700	0.00596	0.00513	0.00491	0.00448
102	0.01200	0.00939	0.00756	0.00621	0.00520	0.00444	0.00381	0.00365	0.00334
108	0.00910	0.00711	0.00572	0.00471	0.00395	0.00336	0.00289	0.00276	0.00253
120	0.00545	0.00426	0.00343	0.00283	0.00235	0.00200	0.00173	0.00165	0.00151

NOTE: In Eq. 1a,

$K = (1594/C)^{1.85}(L/d^{4.87})(1/0.446)$

C = Hazen-Williams coefficient of pipe friction

D = pipe length, ft; 10^5K values are for $L = 1.0$ ft and must be multiplied by true line length to determine line segmental K

d = pipe diameter, in

1 in = 25.4 mn

36-in diameter (914.4-mm) pipe having a friction factor of 130. From Table 2, $10^5K = 0.610$ or unit $K = 0.00000610$:

$$\text{Segmental } K = 0.00000610 \times 1050 = 0.006405$$

in USCS units
$$Q^{1.85} = 30^{1.85} = 540.3$$

$$h_f = 0.006405 \times 540.3 = 3.46 \text{ ft}$$

in SI units
$$Q^{1.85} = 4730^{1.85} = 6.29 \times 10^6$$

$$h_f = 2.618 \times 10^{-5} \times 0.006405 \times 6.29 \times 10^6 = 1.06 \text{ m}$$

Entrance and exit losses and losses through valves and fittings can be calculated from the data in Sec. 8.1.

STATIC HEAD AND PRESSURE DIFFERENTIAL   In computing total system head, consideration must be given to the effects of static head and pressure differential. Static head can be calculated simply from the difference in elevation between supply level and discharge level, and differential pressure can be calculated from the difference between terminal pressure and suction pressure.

PUMP CHARACTERISTIC CURVES   The shape of a centrifugal pump curve must be considered in selection of a specific pump. If, for example, the water level at the source of supply or at the point of discharge is subject to wide variation, a pump with a steep characteristic curve near the design point should be selected to minimize the effects of head variations on pump capacity. Additionally, the pump shutoff head (or head of impending delivery) must exceed the static head to ensure that the pump will operate upon opening of the discharge valve.

NET POSITIVE SUCTION HEAD   As a final step in selection of a specific pump for a particular application, suction conditions should be investigated to determine the net positive suction head (NPSH) available. Failure to meet the NPSH requirements of the pump selected will result in cavitation of the pump impeller and very low pumping efficiency.

TOTAL SYSTEM HEAD   The summation of all components of system head, as calculated for various flows, results in a graphical plot of the system-head curve (Fig. 7). To determine the capability of a specific centrifugal pump operating under system conditions, the pump characteristic curve should be superimposed on the system-head curve. The intersection of the two curves then represents the capacity that the specific pump can deliver.

Parallel and Series Operation   The installation of multiple pumping units operating in parallel is common practice in the waterworks industry because, with proper design and regulation, it permits the most efficient use of pumping facilities and allows smooth transitions in pumping rates as demand fluctuates. This type of operation is particularly adaptable to pumpage from treatment facilities into a distribution system but is also applicable to raw water pumpage, which is generally required to match treated water pumpage. Special care must be taken in sizing the individual pumping units to ensure efficient operation and to prevent units from operating significantly above or below design rates during parallel operation. The principle upon which design

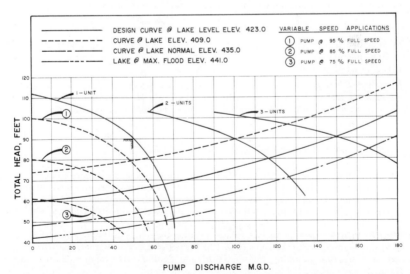

FIG. 7   System-head curve for the Forney raw water pumping station of Dallas. (1 ft = 0.3048 m; 1 mgd = 157.7 m³/h; elev. in ft × 0.3048 = elev. in m) (URS/Forrest and Cotton)

must be based is that *total station discharge may be determined by adding the individual pump discharges associated with any particular head.* Obviously then, as additional pumps are placed in operation, the station flow will increase. However, since the system-head curve rises with increasing flow, *the operation of two identical pumps in parallel will not produce a discharge equal to twice the capacity of one pump.* As more pumps are placed in operation, the incremental increase in pumping capacity becomes smaller.

It should be noted that, when only a portion of the pumps are in operation (total station flow is less than design capacity), the total system head is reduced and the individual pumping units are thus operating at a rate exceeding the design capacity.

If flows are increased substantially past the design point, the NPSH available may become inadequate and cavitation may occur. Additionally, the possibility of overloading the pump driver is introduced, particularly in pumps designed for high heads. In selecting a pumping unit, it is therefore necessary to check conditions at runout capacity (the maximum discharge and lowest head anticipated) in order to ensure proper pump operation.

The undesirable effects of operating a pump at capacities lower than design flow are similar to those resulting from overpumping, but for different reasons. Low discharge rates result in recirculation through the pump, causing cavitation, vibration, and noise. Moreover, the radial force on the impeller increases substantially, thus increasing the stress on the shaft and bearings (to an even greater extent than would result from overpumping). Drivers of pumps designed for low heads may be subjected to overload at low capacities (and thus high heads).

In determining pumping capacities for series operation, *heads are added.* Thus two identical pumps with capacity of 30 mgd (4730 m³/h) at 100 ft (30 m) of system head would, if placed in series, discharge 30 mgd (4730 m³/h) at 200 ft (60 m) of system head. This type of operation is frequently employed in raw water pumping stations in the form of multistage, vertical turbine pumps and is frequently utilized to boost pressures in a distribution system.

Variable-Speed Applications   Installation of two or more variable-speed pumping units will allow gradual increase or decrease of station discharge as dictated by demand. When the required discharge is less than the capacity of two pumps, the variable-speed units may be operated in parallel. Moreover, the installation of two variable-speed units permits operation at flows in excess of the minimum allowable flow and ensures satisfactory efficiency under all conditions. Figure 8 typifies the relationships between head, capacity, and efficiency at varying speeds. Typical procedure for an installation with two variable-speed units is as follows:

1. If station flow is less than the capacity of one pump, one variable-speed unit is operated.

2. If station flow is between one and two times the capacity of a single unit, both variable-speed units are operated (in lieu of operating one unit at full speed and the other at a speed less than that required to produce minimum allowable flow).

3. If station flow is more than two times the capacity of a single unit, the two variable-speed units are operated, along with as many constant-speed units as are required to ensure operation of all pumps at speeds which will produce flows exceeding minimum requirements.

USES OF PUMPS AT WATER TREATMENT PLANTS

Pumps are an integral part of virtually every water treatment plant in existence and have a variety of uses, including low-lift service, coagulant feed, carbon slurry transfer and feed, fluoride feed, delivery of samples from selected points to the chemical laboratory, plant water supply, wash water supply, surface wash supply, and high service pumpage to the distribution system (Fig. 9).

Low-Lift Pumps   Low-lift pumps, located at either the source or the treatment site, are often outdoor, weatherproof vertical pumps (Fig. 10). They may also be horizontal centrifugal pumps housed in a separate pumping station. Head and capacity requirements are as discussed previously.

Coagulant Feed Pumps   Pumps used in chemical feed applications are usually not centrifugal because of the kinds of materials handled. More often, chemicals are fed by a jet pump

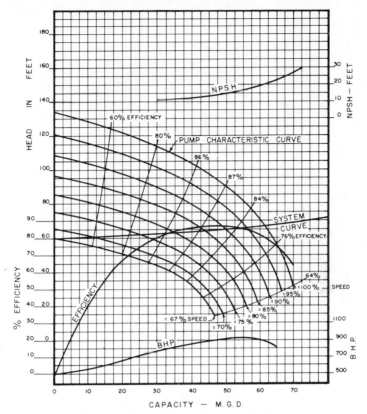

FIG. 8 System-head curves for variable-speed operation for the Forney raw water pumping station in Dallas. (1 ft = 0.3048 m; 1 bhp = 0.746 bkW; 1 mgd = 157.7 m³/ h) (URS/Forrest and Cotton)

or hydraulic eductor which utilizes a moving stream of water to create a vacuum at the suction port of the eductor. This type of pump has no moving parts and is thus adaptable to the feeding of coagulants such as alum or ferric sulfate in solution. A constant-level solution tank is often used to maintain a constant head on the eductor suction. The eductor capacity is then a function of water supply pressure and frictional loss in the feed lines. Eductors are normally rated according to suction capacity, which must be added to the amount of flow contributed by the motive fluid in calculating head losses through the feed system. The manufacturer's literature should be consulted to determine the water supply requirements for various sizes of eductors.

Carbon Slurry Pumps Treatment plants which receive raw water from surface sources often require the addition of activated carbon to combat taste and odor problems caused by suspended material and microorganisms carried into the supply by floodwaters. Consequently carbon must be stored in ample quantities to supply immediate needs. The activated carbon is often transferred from storage tanks to a constant-level solution tank in the form of a slurry. Progressing cavity-type pumps equipped with variable-speed drives are suitable for this transfer service, provided the proper construction materials are selected. The range of capacities required is equal to the difference between minimum and maximum carbon feed rates, and discharge head can be calculated as for any pumping system. However, selection of operating speed and power requirements must be based on estimates of the abrasive and viscous characteristics of the slurry.

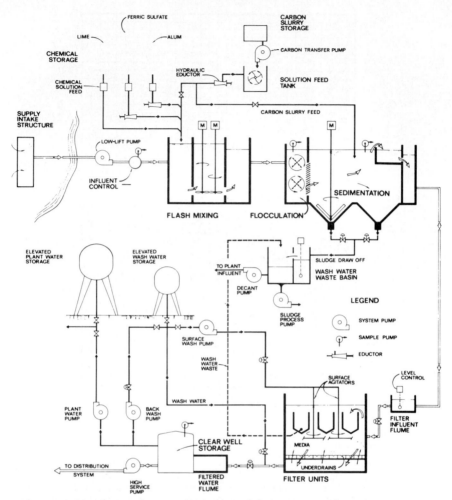

FIG. 9   Typical flow diagram of water treatment plant, including raw water supply.

Fluoride Pumps   The addition of hydrofluosilicic acid to drinking water is achieving growing acceptance in the United States as an effective preventative of tooth decay. Pumps used for addition of this material should be positive displacement, mechanical diaphragm, metering pumps constructed of special materials, such as Kralastic, Penton, and Teflon. A relief valve should be installed on the discharge line to protect the pump in the event of a line blockage.

Diaphragm pumps operate on the principle of the linear motion of a flexible diaphragm, which pulls the solution through an intake port during the backward stroke and forces solution out the discharge port on the forward stroke. Feed rate can be varied by adjusting the stroke length or by a multistep pulley belt-connected to the motor driver. Hydrofluosilicic acid is usually supplied as a slurry that is 20 to 30% hydrofluosilicic acid. Of this acid, 79% is in the form of fluoride ion. In sizing the fluoride feed pumps, consideration must be given to the following factors:

1. Difference between the fluoride ion content of the raw water and that desired in the finished water

FIG. 10 Raw water pumping station for North Texas Municipal Water District, showing vertical turbine pumps with variable-speed drives. (Byron Jackson Pump Division, Borg-Warner)

2. Strength of the slurry

3. Specific gravity of the slurry

For example, assume that the fluoride ion content of the raw water is 0.3 parts per million (ppm) and the desired concentration in the treated water is 0.8 ppm, that the slurry is 20% hydrofluosilicic acid with a specific gravity of 1.22, and that the plant raw water inflow is 100 mgd (15,800 m³/h). The required dosage is then

$$\frac{0.8 - 0.3}{(0.79)(0.20)(1.22)} = 0.192 \text{ ppm of slurry}$$

and the amount of slurry needed is

in USCS units $\qquad$ 0.192 ppm $\times$ 100 mgd $\times$ 8.34 lb/gal = 160 lb/day

in SI units $\quad$ 0.192 ppm $\times$ 15,800 m³/h $\times$ 10⁻⁶ $\times$ 24 h/day $\times$ 998.3 kg/m³ = 72.7 kg/day

Head losses to the point of application can be calculated (with allowances for the viscosity of the slurry) and the proper feed rate accomplished by adjustment of the stroke length. A diaphragm pump is capable of providing a repeatable accuracy of ±1%, a desirable characteristic in light of the fact that excess fluoride in drinking water can produce harmful rather than beneficial effects on the teeth.

Sampling Pumps Small-capacity centrifugal pumps are generally used for delivery of samples from various points in the plant to the chemical laboratory for analysis. Capacities required are generally in the vicinity of 5 to 10 gpm (1.1 to 2.3 m³/h). The required discharge head can be determined from a schematic layout of the suction and discharge piping, with provision for some 10 to 15 lb/in² (70 to 100 kPa) pressure at the chemical laboratory faucet. Head losses through the system are computed as in any pumping system, and such computations are simplified somewhat by the fact that no interconnections are involved, i.e., each sample pump must have a separate discharge line to the chemical laboratory.

Plant Water Pumps Water used for various purposes throughout the treatment plant is usually taken from the treated water at the end of the process. Such water may be pumped back directly into the plant water system or to an elevated storage tank which supplies the head required for adequate pressures throughout the plant. In either case a thorough study of the plant water system is necessary to determine the amount of water required. The plant water tank can be provided with automatic start and stop controls, based on the water level in the tank, to prevent complete draining or overflowing. Allowances should be made to ensure that the tank is sufficiently elevated to provide ample pressures at water-consuming devices. Plant water pumps will generally be of horizontal split-case construction and must be capable of filling the elevated tank at rates determined from analysis of the plant water system.

Wash Water Pumps   The procedure for selecting wash water pumps is much the same as that for plant water pumps. An elevated wash water tank should be sized large enough to permit backwashing of one filter at a time and should be elevated as necessary to provide the required flows through the wash water piping system. Required backwash rates vary from about 15 to 22.5 gpm/ft^2 (37 to 55 m^3/h/m^2) of filter surface area. In determining head losses from the elevated tank through the filters, consideration must be given not only to the head losses in piping, valves, and fittings but also to the head losses in the filter underdrain system [which may range from 3 to 8 ft (0.9 to 2.4 m) at a backwash rate of 15 gpm/ft^2, (37 m^3/h/m^2), depending on the underdrain system installed] and the filter bed. The wash water tank should be sized to allow backwashing of one filter for approximately 10 min and should be equipped with controls for automatically starting and stopping the wash water pumps at predetermined levels. The required pumping capacity depends on the estimated frequency and rate of backwashing.

Wash water pumps may also be used to supply water directly to the backwash piping system. The head required in such cases is that needed to overcome all losses in the wash water piping, underdrain system, and filter bed, and the capacity should equal the maximum backwash rate. An air release valve, check valve, and throttling valve should be provided on the discharge side of the wash water pump, and standby service is highly recommended.

Since the wash water pumps generally use treated water from clear well storage, horizontal centrifugal pumps should be used only when positive suction head is available. Otherwise a vertical pump unit may be suspended in the clear well.

Surface Wash Pumps   Manufacturers of rotary agitators for surface wash systems generally specify a minimum pressure for proper operation of their equipment. This pressure [generally from 40 to 100 lb/in^2 (280 to 690 kPa)] must be added to the system piping losses in determining head requirements for surface wash pumps. The required discharge may vary from 0.2 to 1.5 gpm/ft^2 (0.5 to 3.7 m^3/h/m^2) of surface area, depending on the supply pressure, size, and type of agitator supplied.

High-Service Pumps   High-service pumps at a water treatment plant are those pumps which deliver water to the distribution system. The term *water distribution system* as used herein is defined as embodying all elements of the municipal waterworks between the treatment facilities and the consumer. The function of the distribution system is to provide an efficient means of delivering water under reasonable pressure in volumes adequate to meet peak consumer demands in all parts of the area served.

High-service pumps may be of either vertical or horizontal construction, depending on the required capacity and the design and configuration of treated water storage facilities from which the pumps take suction. The high-service pumping station often houses the plant water and wash water pumps in addition to the high-service pumps because all such pumps utilize treated water to perform their required services (Fig. 11).

Operating conditions in the distribution system play an important role in the determination of high-service pump capacities. In small municipalities, for example, it may be possible to pump from the treatment plant at a constant rate equal to the average daily demand and to supply peak demands from elevated storage tanks throughout the system. However, as systems become larger, the need for variable-rate pumping from the treatment plant increases. If sufficient storage is available in the distribution system for supplying peak hourly demands, it may be possible to provide variable-speed high-service pumps with total capacity equal to the maximum capacity of the treatment plant and to supply hourly peaks from storage. However, inasmuch as peak hourly demands may be two or more times as great as the peak daily demand and may be sustained for several hours, the required system storage for such operation can exceed 30% of the maximum

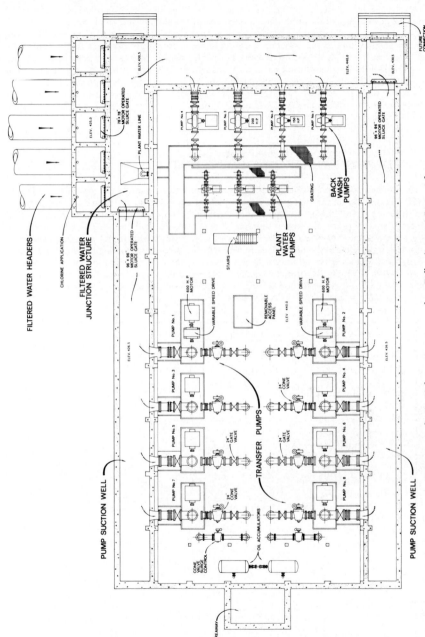

FIG. 11 Plan of the high-service pumping station in the East Side water treatment plant, Dallas. (1 in = 2.54 cm; 1 ft = 0.3048 m) (URS/Forrest and Cotton)

daily demand. In very large systems, provision for this amount of storage is simply impractical. It may then become necessary to design treatment facilities for capacities in excess of the maximum daily demands and to supply only a portion of the peaking water from storage.

In any event, as distribution systems become larger, variable-speed pumping becomes more desirable and development of a system-head curve becomes more complex. In many cases pumpage from treatment plants during off-peak hours exceeds pumpage during peak hours because storage tanks must be filled during off-peak hours. System-head curves must be derived for each of the above conditions and also for a third condition, which represents those periods when storage tanks are full but are not being utilized to supply demands. Typical system-head curves for all three conditions are shown on Fig. 12. Only a thorough analysis of the entire distribution system can provide the data necessary for the proper selection of high-service pumps.

BOOSTER PUMPS IN DISTRIBUTION SYSTEMS _____

In order to maintain distribution system pressures within the desirable 40- to 90-lb/in² (275- to 620-kPa) range, booster pumps may be required at various locations, as dictated by topographical conditions and system layout. Booster stations may include in-line vertical or horizontal pumps connected directly to the pipeline or separate storage reservoirs and pumps. The latter method is preferred but is not always practicable because of land requirements for the storage reservoir.

In-line booster pumps should be sized to match the capacities of incoming pipelines and should be capable of supplying the additional head required to increase pressures to desirable ranges in the affected area of the distribution system. Determination of discharge head must take into account the pressure on the suction side of the pump as well as static head, frictional losses, and desired residual pressure on the discharge side.

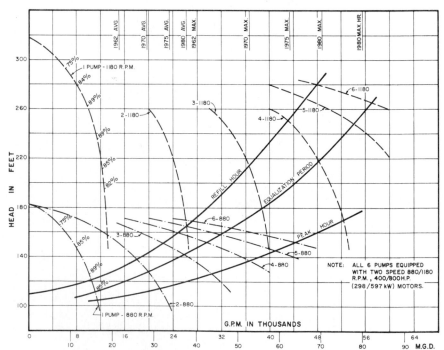

FIG. 12 Typical system-head curves for high-service pumpage to water distribution system. (1 ft = 0.3048 m; 1 gpm = 0.227 m³/h)

Many water utilities refuse to accept in-line pumps as satisfactory for booster service and require construction of a pressure-equalizing storage reservoir. Horizontal or vertical booster pumps then take suction from the storage reservoir and return water at higher pressures to the mains. Pressure switches or floats automatically start and stop the booster pumps and thus eliminate the need for continuous attendance.

GROUNDWATER WELLS

Site Selection   Foremost among the factors which must be considered in selection of a site for a proposed well are local regulations governing proximity to such potential contaminants as cesspools, privies, animal pens, or abandoned wells. Also to be considered is the effect on adjacent wells, because if wells are located too close together excessive head losses may result.

Types of Pumps Used   The two basic types of pump used in groundwater wells are vertical turbine, line shaft pumps and submersible pumps. The vertical turbine pump consists of three basic components: the driving head, the column-pipe assembly, and the bowl assembly. The driving head is mounted above ground and consists of the pump discharge elbow, the motor support, and the stuffing box. The column-pipe assembly (consisting of shaft, bearings, and bearing retainers) and the bowl assembly (consisting of a suction head, impeller or impellers, discharge bowl and intermediate bowl or bowls) are suspended from the driver head. Use of multiple bowls and impellers results in a form of series operation and permits pumpage against very high heads.

The submersible pump utilizes a waterproof electric motor located below the static level of the well to drive a series of impellers and produce a series operation similar to that of a line shaft pump. However, the length of shafting required is greatly reduced, and thus the shaft losses and total thrust are minimized. As a result, the submersible pump becomes economically competitive with the line shaft pump at great depths.

Air-lift pumps, which operate on the principle that a mixture of air and water will rise in a pipe surrounded by water, may be used in some cases. Such pumps are easy to maintain and operate and can be used in a crooked well or with sandy water. However, they are relatively inefficient (usually 30 to 50%) and allow very little system flexibility.

Reciprocating pumps are also used in some cases where small capacities are required from deep wells. Such pumps can be driven by electric motors or windmills, but they are generally noisy and are more expensive than centrifugal pumps.

Determining Pump Capacity   The capacity of any well is dependent on such factors as screen size, well development, aquifer permeability, recharge of groundwater supply from rainfall and streams, and available head. The basic procedure used in sizing a pump for well service involves drilling the well and performing a test operation. First the static head, or elevation of the groundwater table prior to pumping, is determined. Water is then pumped at various rates and the drawdown associated with each pumping rate determined. A plot of drawdown versus pumping rate can then be derived. Pumping rate is usually measured by a weir, orifice, or pitot tube, and drawdown is determined with a detector line and gage or with an electric sounder.

From the test data and from a preliminary layout of discharge piping, a system-head curve can be derived, with drawdown added to frictional losses for each pumping rate. The pump characteristic curve can then be superimposed on the system-head curve to determine the capacity that can be attained with a specific pump. It should be noted that pump curves for line shaft pumps are based on the results of shop tests, which do not allow for column frictional or line shaft and thrust losses. Consequently the laboratory characteristic curve for any line shaft pump must be adjusted to field conditions.

Field pumping head can be determined by subtracting column frictional losses from the laboratory head. Field brake horsepower (brake kilowatts) is determined by adding shaft brake horsepower (brake kW), which depends on shaft diameter and length, and on rotative speed, to laboratory brake horsepower (brake kW). Field efficiency is determined from the formula

in USCS units $\qquad$ Field efficiency $= \dfrac{\text{gpm} \times \text{field head in feet}}{3960 \times \text{field brake horsepower}}$

in SI units $\qquad$ Field efficiency $= \dfrac{m^3/h \times \text{field head in meters}}{367.5 \times \text{field brake kW}}$

Since thrust loads cause additional losses in the motor bearing, it is necessary to determine the additional power required to overcome thrust losses. Total thrust load is equal to the sum of the shaft weight and the hydraulic thrust (which varies with laboratory head for any particular impeller), and losses due to thrust amount to approximately 0.0075 hp/100 rpm/1000 lb (0.00126 kW/100rpm/1000 N) of thrust. Motor efficiency is then calculated by dividing the motor's full load power input (without thrust load) by the sum of full load power input and loss due to thrust. Overall efficiency then equals the product of field efficiency and motor efficiency.

As a result of the efficiency losses produced by shaft weight and length in line shaft pumps, it is usually more economical to use a submersible pump at depths of more than about 500 ft (150 m). Sizing a submersible pump requires calculations similar to those for a line shaft pump. However, the submersible pump installation requires a check valve in the column pipe, which must be considered in the determination of frictional losses. Moreover, the efficiency losses resulting from the motor cable (expressed as a percentage of input electric power) must be considered in determining overall efficiency, which can be calculated from the formula:

$$\text{Overall efficiency} = \frac{\text{water power} \times (\% \text{ motor efficiency} - \% \text{ cable loss})}{\text{shop brake power} \times 100}$$

where

in USCS units $\qquad$ Water power in hp $= \dfrac{\text{gpm} \times \text{field head in feet}}{3960}$

in SI units $\qquad$ Water power in kW $= \dfrac{m^3/h \times \text{field head in meters}}{367.5}$

Cable size must be selected on the basis of motor power and motor input amperes, voltage, and cable length.

Well Stations A typical well station generally includes a small building for housing the pump, pump controls, metering and surge-control facilities, and chemical feed equipment. Submersible pumps do not require a pump house for protection, but if pump controls or chemical feed equipment are provided, an enclosure of some type is required. Since well water supplies are often pumped directly into the distribution system, a differential producer is usually installed for metering. Moreover, chemicals may be added at the well station to minimize corrosion, control bacteria, decrease hardness, and inject fluorides into the water supply. Surges are usually controlled by installation of a surge valve in the pump discharge line. Controls may also be installed to permit starting and stopping of the pump from remote central locations and to provide for measurement and control of well drawdown.

FURTHER READING

Fair, Gordon Maskew, John Charles Geyer, and Daniel Alexander Okun: *Water and Wastewater Engineering*, Wiley, New York, 1968.

Karassik, Igor J., and R. Carter: *Centrifugal Pumps: Selection, Operation, and Maintenance*, McGraw-Hill, New York, 1960.

Messina, J. P.: "Operating Limits of Centrifugal Pumps in Parallel," *Water and Sewage Works*, 1969, p. R-79.

Singley, J. E., and A. P. Black: "Water Quality and Treatment: Past, Present and Future," *Journal AWWA.* **64**(1):6 (1972).

Texas Water Utilities Association, *Manual of Water Utilities Operations*, 5th ed., Lancaster Press, Lancaster, Pa., 1969.

URS/Forrest and Cotton, Inc., Consulting Engineers, "Report to the City of Dallas on Distribution System Analysis," January 1958.

SECTION 9.2
SEWAGE TREATMENT

H. H. BENJES, SR.
W. E. FOSTER

Sewage is defined as the spent water of a community. Although it is more than 99.9% pure water, it contains wastes of almost every form and description. Raw sewage, when fresh, is gray and looks something like dirty dishwater containing bits of floating paper, garbage, rags, sticks, and numerous other items. If allowed to go stale, it turns black and becomes very malodorous. About 25% of the waste matter of normal domestic sewage is in suspension; the remainder is in solution.

Sewage contains many complex organic and mineral compounds. The organic portion of sewage is biochemically degradable and, as such, is responsible for the offensive characteristics usually associated with sewage. Sewage contains large numbers of microorganisms, most of which are bacteria. Fungi, viruses, and protozoa are also found in sewage, but to a lesser extent. Although most of the microorganisms are harmless and can be used to advantage in treating the sewage, the viruses and some of the bacteria are pathogenic and can cause disease.

Sewage flow generally averages between 50 and 200 gal per capita per day (gpcd) (190 and 760 lpcd). In the absence of better information, an average figure of 100 gpcd (380 lpcd) is generally used for design purposes. The rate of flow usually varies from minimum in the early morning to maximum in the later afternoon. Minimum flow ranges from 50 to 80% and maximum dry-weather flow from 140 to 180% of average flow. The extent of variation decreases as the size of the system increases. Wet-weather flows can be 600 gpcd (2270 lpcd) or more because of the extraneous water entering sewers from roof drains, areaway drains, footing drains, etc.

SEWAGE SYSTEMS

In most instances, sewage systems are divided into two parts: collection systems and treatment systems.

Collection Collection systems consist of a network of sewers which collect and convey sewage from individual residences, commercial establishments, and industrial plants to one or more points of disposal. Pumping stations are often needed at various points in the system to pump from one drainage area to another or to the treatment plant. The judicious location of pumping stations enhances the economy of the overall design by eliminating the need for extremely deep sewers.

FIG. 1 Factory-built conventional lift station. (Smith & Loveless Division, Ecodyne)

Small-system pumping stations (Figs. 1 and 2) are frequently built underground and may be factory-built. For larger stations, superstructures should be in keeping with surrounding development. It has been said that people smell with their eyes and their ideas as well as their noses, and for this reason aboveground structures should be attractive, with landscaped grounds, to overcome the popular prejudices against sewage works. Stations can be and have been designed and constructed in residential areas where the neighbors apparently are not aware that the stations are not homes.

Treatment Treatment facilities can be many and varied, with the extent and nature of the treatment determined to a large degree by the proposed use of the receiving stream and its ability to assimilate pollutants. Most conventional treatment plants being built today can be classified as either primary, biological, or advanced waste treatment. Other alternatives, such as physical-chemical or chemical-biological treatment, are also used on occasion but on a lesser scale. The treatment needs of smaller communities are sometimes satisfied by package treatment plants or by waste stabilization lagoons.

Primary treatment involves removal of a substantial amount of the suspended solids but little or no colloidal or dissolved matter. Primary treatment facilities normally include screening, grit removal, and primary sedimentation. The sewage is often chlorinated during primary treatment in order to sterilize the wastes.

Biological treatment uses bacteria and other microorganisms to break down and stabilize the organic matter. Trickling filters and the many variations of the activated sludge process are the most popular biological treatment concepts presently in use. Biological treatment is generally followed by final sedimentation of the solids produced by the microorganisms.

Advanced waste treatment is a very complex subject, and it can range from a limited objective, such as phosphate removal, to whatever additional treatment is necessary for water reuse pur-

FIG. 2 Factory-built pneumatic ejector lift station. (Smith & Loveless
Division, Ecodyne)

poses. Advanced waste treatment usually follows conventional primary and biological treatment
and can include phosphate removal, nitrate removal, multimedia filtration, carbon adsorption,
and ion exchange. Where zero discharge is required, it may be necessary to follow advanced waste
treatment with spray irrigation of the plant effluent or other methods of disposal.

Combined primary and biological treatment using the activated sludge process is perhaps the
most commonly used treatment concept currently in use. A schematic drawing of a typical activated sludge treatment plant is shown in Fig. 3. In the example, liquid treatment is accomplished
by coarse screening, grit removal, fine screening (or comminution), and primary settling, followed
by aeration, final settling, and chlorination. Sludge processing includes thickening, dewatering,
incineration, and liquid disposal of ash. There are many variations to this layout, but the one
shown includes most of the pumping applications normally encountered in treatment plant
design. Pumping requirements will of course vary from plant to plant, depending on the process
used, the site size and topography, and the relative location of the various structures and
equipment.

PUMP APPLICATIONS

Most of the pumping applications associated with the collection and treatment of sewage can be
classified according to the general nature of the liquid to be handled. The primary classifications
are (1) raw sewage, (2) settled sewage, (3) service water, and (4) sludge. There are also, however,
a number of specialized applications involving the handling of abrasive materials, such as grit and
ash. The types of pumps recommended for sewage applications are indicated in Table 1. Although
included in the table, chemical pumps are not discussed in this section. They are covered in Sec.
9.6.

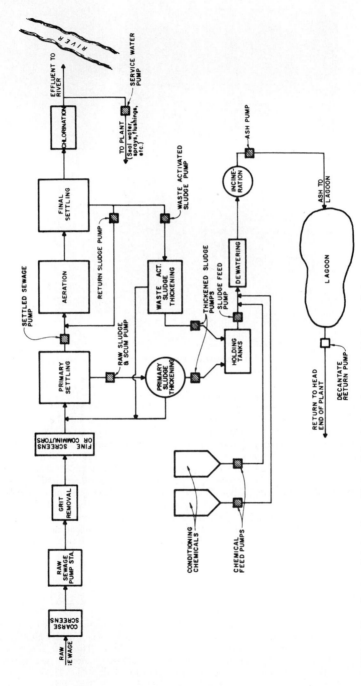

FIG. 3 Typical activated sludge plant, with dewatering, incineration, and liquid disposal of ash.

TABLE 1 Types of Pumps Generally Used in Sewage Applications

Application	Conventional sewage	Diffuser	Torque flow	Clear-water volute	Ash	Screw	Pneumatic ejector	Air lift	Positive displacement	Chemical
Raw sewage	X	...	...	...	...	X	X	...	...	...
Grit	...	...	X	...	...	...	...	X	...	...
Primary sludge										
Less than 2% solids	X	...	X	...	...	...	...	...	...	...
More than 2% solids	...	...	...	...	...	...	...	...	X	...
Normal primary scum	X	...	...	...	...	...	...	...	X	...
Diluted scum	X	...	...	...	...	...	...	...	...	...
Biological sludge	X	X	...	...	...	X	...	X	...	...
Thickened biological sludge	...	...	X	...	...	...	...	...	X	...
Digested sludge, recirculation	X	...	...	...	...	...	...	...	...	...
Settled sewage	X	X	...	...	...	X	...	...	...	...
Plant effluent	X	X	...	...	...	X	...	...	...	...
Service or nonpotable water	X	X	...	X	...	...	...	...	...	...
Ash sluice	...	...	...	...	X	...	...	...	...	...
Decantate or supernatant liquor	X	X	...	...	...	...	...	...	...	...
Chemical solution	...	...	...	...	...	...	...	...	...	X

Raw Sewage Raw sewage pumps are used to lift liquid wastes from one level of the collection system to another or to the treatment plant for processing. Regardless of where the pumps are located, the basic design considerations remain the same.

Even though the sewage is normally screened at larger installations before entering the suction wet well, it still contains a large quantity of problem material, such as grit, rags, stringy trash, and miscellaneous solids small enough to pass through the coarse screens. Screens are often omitted from smaller installations because large objects are not as much of a problem because of the smaller size of the incoming sewers.

Raw sewage pumping installations are usually sized so that their firm capacity either is equal to the maximum total inflow rate of the incoming sewers or can be expanded to accommodate this level. *Firm capacity* is defined as total station capacity with one or more of the largest units out of service.

Pneumatic ejectors (Fig. 2) are sometimes used where the required capacity is less than that provided by the smallest conventional sewage pump. This type of unit, however, should not be used where more than 50 connections are expected.

Conventional sewage pumps are, by far, the most common pumps used for the handling of raw sewage. A conventional sewage pump is more specifically described as an end-suction, volute-type centrifugal with an overhung impeller of either the nonclog (Fig. 4) or the radial- or mixed-flow type (Fig 5), depending on capacity and head.

Nonclog pumps are all based on an original development by Wood at New Orleans. Actually, no pump has been developed that cannot clog, either in the pump or at its appurtenances. Experience shows that rope, long stringy rags, sticks, cans, rubber and plastic goods, and grease are objects most conducive to clogging.

Nonclog impellers are used almost exclusively today for pumps smaller than 10 in (25 cm). These pumps differ from clear-water pumps in that they are designed to pass the largest solids possible for the pump size. The conventional nonclog impeller contains two blades, although some manufacturers are now offering a single-blade ("bladeless") impeller. The two-blade impeller has thick vanes with large fillets between the vanes and the shroud at the vane entrance. The bladeless impeller has no vane tips to catch trash. On the other hand, it is inherently out of balance because of its lack of symmetry.

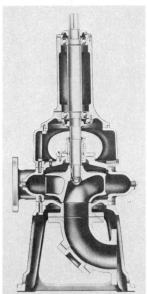

FIG. 4 Nonclog sewage pump. (Fairbanks-Morse Pump Division, Colt Industries)

The larger raw sewage pumps are equipped with either mixed-flow or radial-flow impellers, depending on head conditions. Both have two or more vanes, depending on pump size and the size of solids to be handled. The vane tips are sharper than for the nonclog impeller, resulting in a higher operating efficiency. The heavier vanes are not necessary because the vane openings can be larger than on the smaller pumps. Experience indicates that stringy trash will not clog an impeller with vane openings larger than 4 in (102 mm) in diameter.

Volute-type pumps should be used in dry-pit applications only. Units are available for wet-pit installation, but their use should be limited to temporary applications with low load factors . Although nonclog pumps 8 in (203 mm) and smaller are available with self-priming (Sec. 2.4), most conventional sewage pumps are located so that the impeller is always below low water level in the suction wet well. This eliminates the need for specialized priming systems. Submersible nonclog pumps are also available, but their use generally is not recommended.

Self-priming pumps have been used successfully to pump raw unscreened sewage, particularly in the southern part of the United States. The self-priming feature eliminates the dry-pit cost and

FIG. 5   Vertical sewage pumping units at the Lemay Pump Station, Metropolitan St. Louis Sewer District, showing 36-in (914.4-mm), 40,000-gpm (9080-m³/h) pumps close-coupled to 1250-hp (932-kW) motors.

gives the centrifugal pump the gas-handling advantage of positive displacement pumps. Operating costs are higher, though, because the design efficiencies generally run about 10 to 15% lower than for the conventional nonclog units.

Archimedean screw pumps (Fig. 6) are occasionally used for raw sewage pumping applications. These units are advantageous in that they do not require a conventional wet well, and they are self-compensating in that they automatically pump the liquid received regardless of quantity as long as it does not exceed the design capacity of the pump. This is done without benefit of special drive equipment. Also, as shown by Fig. 6, the total operating head of a screw pump installation is less than for those pumps which require conventional suction and discharge piping. Screw pumps, however, have a practical limitation as to pumping head. Generally speaking, they are not used for lifts in excess of 25 ft (7.6 m).

Settled Sewage   Settled sewage pumps are used to lift partially or completely treated waste from one part of the plant to another or to the receiving stream. In Fig. 3, these pumps are the settled sewage pump, service water pump, and decantate return pump.

The liquid to be handled usually contains some solids, but grit and most of the rags and other stringy material have already been removed. Sufficient firm capacity should be provided to meet peak flow requirements. In no case should fewer than two units be provided.

Diffuser pumps (Subsec. 2.2.1) are commonly used for the pumping of settled sewage. Depending on head conditions, they may be of either the propeller or mixed-flow design. Although normally installed in wet-pit applications, these units are sometimes mounted on suction piping and installed in a dry pit. Either type of application is acceptable, although economics usually dictates a wet-pit installation. Head and capacity conditions will determine which type of unit is applicable.

A conventional sewage pump may be used to pump settled sewage where a dry-pit installation

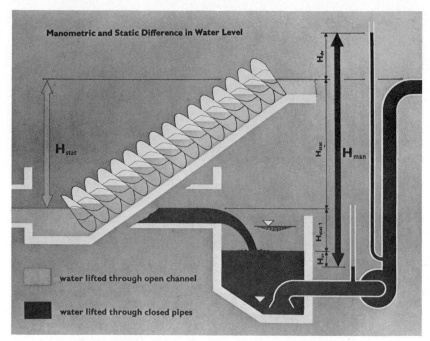

FIG. 6   Archimedean screw pump. (Beloit-Passavant)

is desirable. This is not usually an economical application, but it is acceptable as far as suitability of equipment is concerned. As previously indicated, wet-pit application of volute-type pumps is discouraged, primarily because of the high maintenance factor on the submerged bearings and the relative inaccessibility of the pumping unit for repair.

Archimedean screw pumps can be used to pump settled sewage, provided the lift is not excessive. As previously noted, this type of pump has certain inherent advantages.

Service Water   Plant effluent water is frequently used for flushing, gland seal, foam control sprays, chlorine injector operation, lawn sprinkling, fire protection, and various other services in a waste-water treatment plant. Except for the fact that some solids have to be contended with, this application is much the same as that found in building-water supply and small distribution systems.

Screening of solids is normally required; this can be accomplished either before or after the pumps, depending upon various circumstances. Pipeline-type strainers are recommended as they are not only economical but require a minimum of space, can be automatically backflushed, and are much easier to operate than alternative equipment.

Any type of conventional volute or diffuser clear-water pump can be used on service water applications, provided the effluent water is screened prior to entering the pump. Pumps capable of handling some solids should be used in those instances where prescreening is not practical.

Sludge and Scum   This classification is divided into two separate categories, based on the concentration of solids in the liquid to be handled. Specialized pumping equipment is required for more concentrated sludges, whereas pumping of dilute sludge and scum is somewhat comparable to the handling of settled sewage.

DILUTE SLUDGE OR SCUM   For the purposes of this manual, *dilute sludge and scum* is defined as having less than 2% solids. An exception is digested sludge recirculation, which generally exceeds the 2% limit. This is included along with the more dilute sludges since the same type of pumping equipment is used.

Normally, the handling of dilute sludge is limited to the transfer of biological sludge back to the treatment process or to some other point for further concentration and/or dewatering and disposal. When digesters are used as part of the treatment facilities, sludge is often recirculated through external heat exchangers in order to maintain temperatures conducive to anaerobic bacterial action. This recirculation also helps keep the contents of the digester mixed. Occasionally, primary sludge and scum are handled in diluted form.

The firm capacity of dilute sludge pumping facilities should be equal to anticipated peak loading. Biological sludge return pumps should have a capacity range from 25 to 100% of average design raw sewage flow to the plant. Digested sludge recirculation pumps should be sized to turn over the contents of the digester frequently enough to maintain the desired temperature. Diluted primary and waste biological sludge pumps should have sufficient capacity to handle peak sludge loading at conservative solids concentrations.

Conventional sewage pumps are suitable for handling dilute sludge and scum. Either the nonclog or mixed-flow impeller may be used, depending upon capacity requirements. Dry-pit installations are recommended.

Diffuser pumps are particularly suitable for handling biological sludge that does not contain any appreciable amount of trash or stringy material. They are not recommended, however, for handling diluted scum or for recirculating digested sludge. Depending on capacity requirements, diffuser pumps may be of either the mixed-flow or propeller design. Wet-pit applications are most common, although dry-pit installations are occasionally used.

Torque flow (or vortex) pumps (Fig. 7) are often used to handle dilute sludges which contain some grit. These units are particularly suitable for this type of service because their design is such that close running tolerances are not required; this allows the use of specially hardened materials, such as high-nickel iron, which are not easily machined. The most common applications of torque flow pumps are for the pumping of nondegritted dilute primary sludge to gravity thickening and the recirculation of digested sludge.

Screw pumps can be used in certain instances for handling biological sludge. Use of screw pumps is generally limtied to low to medium lifts and to those instances where the point of discharge is close to the sludge source.

Air-lift pumps are suitable for transferring biological sludge where the lift is small and the point of discharge nearby. A typical air-lift pump installation is shown in Fig. 8. Total head should not exceed 4 to 5 ft (1.2 to 1.5 m). The ability of an air-lift pump to vary capacity is somewhat limited, ranging from about 60 to 100% of the rated amount. These pumps are inexpensive in first cost but have an operating efficiency of only about 30%. They are very easy to install, and maintenance is minimal since there are no moving parts. Air-lift pumps are commonly used to transfer sludge at package treatment plants.

FIG. 7   Torque flow pump. (Wemco Division, Envirotech)

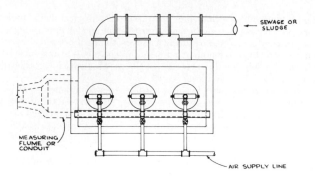

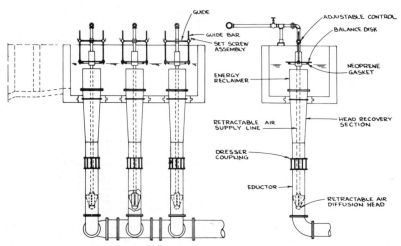

FIG. 8   Typical air-lift pump installation. (Walker Process Equipment Division, Chicago Bridge & Iron)

CONCENTRATED SLUDGE OR SCUM   Concentrated sludge or scum is defined as having more than 2% solids. The single exception is in the case of the recirculation of digested sludge. As previously discussed, this has been included in the dilute sludge classification.

Each pumping installation should have enough firm capacity to handle peak design sludge quantities while operating part time. The proportion of operating time at peak loading should vary from about 25% for primary sludge pumps to close to 80% for pumps feeding dewatering equipment.

Only positive displacement pumps are recommended for handling concentrated sludge and scum, mainly because they can pump viscous liquids containing entrained gas without losing prime. Also, these materials are thixotropic, and conventional formulas for frictional losses are not always valid. An arbitrary allowance of at least 25 lb/in² (170 kPa) should be added to the pumping head calculated by conventional methods to allow for changes in viscosity and partial clogging of pipelines. Positive displacement pumps are able to maintain a relatively constant capacity regardless of variations in discharge head.

For most applications, positive displacement pumps may be of either the plunger (Fig. 9) or the progressing cavity design (Fig. 10). The performance of both depends upon close running clearances; consequently they have a high incidence of maintenance, especially where gritty substances are encountered. Even so, they represent the best pumping equipment currently available,

FIG. 9   Plunger-type sludge pump. (ITT Marlow Pumps)

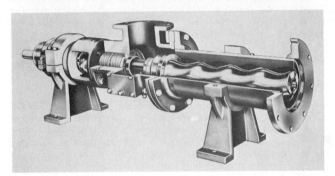

FIG. 10   Progressing cavity sludge pump. (Moyno Pump Division, Robbins & Myers)

and both designs have been used with success. Lobe-type gear pumps have been used for specialized applications. These are to be avoided, however, where there is any possibility that the material to be pumped will contain even a small amount of grit.

Plunger pumps should be of the heaviest design available and should be rated for capacity at about one half of full stroke. The shorter the stroke, the more stable the operation and the less maintenance required. Heads as high as 80 to 100 lb/in² (550 to 690 kPa) are available and should be specified in order to give as much flexibility as possible.

Specially designed progressing cavity pumps are available for handling sewage sludges. Wear increases along with pump speed, and so excessive speed should be avoided. Ideally, the maximum speed of a progressing cavity pump should not exceed 350 rpm. These units are readily available with head capabilities up to 50 lb/in² (345 kPa) and should be so specified.

Certain of the newer sludge conditioning and dewatering processes, such as heat treatment and pressure filtration, require pumps having a head capability in excess of 500 lb/in² (3450 kPa). This is extremely difficult service, and special care should be taken in selecting the type of equipment to be used. So far, this area of application has received very little consideration from pump manufacturers.

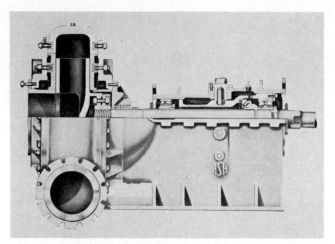

FIG. 11 Ash pump. (Allen-Sherman-Hoff Pump, subsidiary of Envirotech)

Other Uses   Grit may be handled with reasonable success with either a torque or an air-lift pump. Considerable flushing water is required with a torque flow pump, and to a lesser extent with an air-lift unit. A special ash pump, (Fig. 11) is required where it is necessary to dispose of incinerator residue in a liquid form. These units are especially designed for ash sluicing service and are made of special hardened metals. No other pump should be considered for this service.

PUMP SELECTION

Various factors should be considered when selecting pumping equipment. These include the number of units to be installed, operating frequency, and station reliability requirements. Once these factors have been fully evaluated, head-capacity curves should be prepared in order to match the pumps properly with system requirements. This is necessary because the capacity of most pumps varies with the total head at which the unit operates. When a pump is referred to as having a certain capacity, this capacity applies to only one point on the characteristic curve.

Number of Pumps   The number of pumps to be provided at a particular installation depends largely on the required capacity and range of flow. In considering capacity, it is customary to provide a total pumping capability equal to the maximum expected inflow with at least one of the largest pumping units out of service. A minimum of two pumps should be installed in any installation except where pneumatic ejectors are used to serve fewer than 50 houses. Two pumps are customarily installed where the maximum inflow is less than 1.0 mgd (160 m^3/h). At larger installations, the size and number of units should be such that the range of inflow can be met without starting and stopping pumps too frequently and without requiring excessive wet-well storage capacity. Variable-capacity pumps can be used to match pumping rate with inflow rate.

Where variable-capacity pumps are used, a minimum of two units should be installed. In those cases where more than one variable-capacity unit is required to handle peak flow, three units should be installed. In this manner it is possible to maintain a reasonable rate of flow through each pump. Operation of a single variable-capacity pump in parallel with a constant-capacity pump requires the variable-speed unit to operate at almost no capacity whenever total inflow barely exceeds the rating of the constant-capacity unit. This is extremely difficult service and should be avoided. As a general rule, pumping rates of less than 20% of the rated capacity for which a pump is designed will result in excessive internal recirculation and unstable operation. Recirculation can occur in some pumps at more than 50% of rated capacity. See Subsec. 2.3.1.

Operating Frequency Pump size should be coordinated with wet-well design in order to avoid frequent on-off cycling of pumps. Excessive starting will cause undue wear on the starting equipment. Also, standard motors should not be started more than six times an hour. Where more frequent starting is required, special motors should be provided. Inflow into the wet well without pumping should not exceed about 30 min if septicity is to be prevented.

Cycle time is defined as the total time between starts of an individual pump. It can be determined by comparing the volume between the on and the off levels in the wet well with the pump capacity. Cycle time is computed as follows:

in USCS units
$$CT = \frac{V}{D - Q} + \frac{V}{Q}$$

in SI units
$$CT = 60 \left(\frac{V}{D - Q} + \frac{V}{Q} \right)$$

where CT = cycle time, min
V = wet-well volume between on and off levels, gal (m³)
D = rated pump capacity, gpm (m³/h)
Q = wet-well inflow, gpm (m³/h)

With a given wet-well volume and pumps having a uniform pumping rate, minimum cycle time will occur when the rate of inflow is equal to one-half the discharge rate of the individual pump under consideration. The formula for cycle time simplifies to

in USCS units $\qquad CT = 2V/Q$
in SI units $\qquad CT = 120\ V/Q$

An effective wet-well volume of at least 2.5 times the discharge rate of the pump under consideration is required in order not to exceed the six starts per hour recommended above for pumps having a uniform pumping rate.

Reliability With its increased awareness of and concern for environmental matters, the public has little tolerance for the bypassing of sewage equipment because of power outages, equipment failure, insufficient pumping capacity, or whatever. Reliability is of extreme importance, and the design of pumping facilities should be premised on providing continuous service. Where electric motors are used, two incoming power lines from separate sources with automatic switching from the preferred source to the standby source are the minimum required for reliability. Standby engine-driven pumps, engine-driven right-angle gear drives, or standby engine-driven generators should be provided where dual electric service cannot be obtained or where the degree of reliability provided by two feeds is not considered adequate. Raw sewage pumping installations are particularly critical. Plant pumping installations usually can be out of service for as long as four hours without adversely affecting the treatment process, provided the liquid will flow by gravity through the plant.

Speed The maximum speed at which a pump should operate is determined by the net positive suction head available at the pump, the quantity of liquid being pumped, and the total head. When specifying pumps, especially those which are to operate with a suction lift, the speed at which the pumps will operate should be checked against limiting suction requirements as set forth by the Hydraulic Institute.

In general, it is not good practice to operate sewage pumping units at speeds in excess of 1750 rpm. This speed is applicable only to smaller units. Larger pumps should operate at lower speeds.

Preparation of Head-Capacity Curves Pump selection generally involves preparation of a system head-capacity curve showing all conditions of head and capacity under which the pumps will be required to operate. Frictional losses can be expected to increase with time, materially affecting the capacity of the pumping units and their operation. For this reason, system curves should reflect the extreme maximum and minimum frictional losses to be expected during the lifetime of the pumping units as well as high and low wet-well levels.

Where two or more pumps discharge into a common header, it is usually advantageous to omit the head losses in individual suction and discharge lines from the system head-capacity curves. This is advisable because the pumping capacity of each unit will vary depending upon which units are in operation. In order to obtain a true picture of the output from a multiple-pump installation, it is better to deduct the individual suction and discharge losses from the pump characteristic curve. This provides a modified curve which represents pump performance at the point of connection to the discharge header. Multiple-pump performance can be determined by adding the capacity for points of equal head from the modified curve. Figure 12 shows a typical set of system curves, together with representative individual pump characteristic curves, modified pump curves, and combined modified curves for multiple-pump operation. Intersection of the modified individual and combined pump curves with the system curves shows total discharge capacity for each of the several possible pumping combinations. A typical set of system curves consists of two curves with a Williams-Hazen coefficient of $C = 100$ (one for maximum and one for minimum static head) and two curves with a Williams-Hazen coefficient of $C = 140$ (for maximum and minimum static head). These coefficients represent the extremes normally found in sewage applications.

Pumps should be selected so that the total required capacity of the installation can be delivered with maximum water level in the wet well and maximum friction in the discharge line. Pump efficiency should be maximum at average operating conditions. In the case of Fig. 12, assuming that the total capacity of the installation is to be obtained by operating pumps 1, 2, and 3 in parallel, the total head required at the discharge header would be approxiamtely 51 ft (15.5 m). Projecting this point horizontally to the individual modified pump curves and thence vertically to the pump characteristic curves, the required head for pumps 1 and 2 should be 54 ft (16.5 m) and for pump 3 approximately 57 ft (17.4 m). The difference between the head obtained from the pump characteristic curve and the modified curve is the head loss in the suction and discharge piping for the individual pumping units.

Figure 12 also shows the minimum head at which each pump has to operate, approximately 39 ft (11.9 m) for pumps 1 and 2, and about 42 ft (12.8 m) for pump 3. These minimum heads are important and should be made known to the pump manufacturer since they will usually deter-

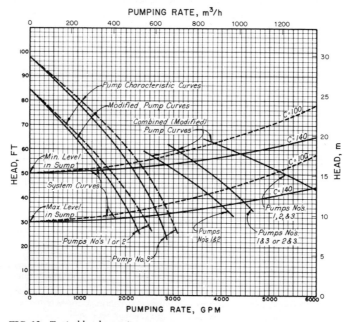

FIG. 12 Typical head-capacity curves.

mine the maximum brake power required to drive the pump and the maximum speed at which the pump may operate without cavitation.

PUMP DRIVERS

In the majority of cases, pumps are driven by electric motors. Sometimes, however, they are driven by gasoline, gas, or diesel units where firm power is not available or where pumping is required only at infrequent intervals. Variable-speed drivers are used extensively in sewage applications. These units generally consist of variable-speed motors or constant-speed motors with adjustable slip couplings of either the eddy-current or the fluid coupling type. Selection of the type of variable-speed drive to be used is usually based on initial cost and space considerations, as there is little difference in operating efficiency. See Sec 6.2 for these and other types of speed-varying devices.

Variable-speed drivers are particularly appropriate for raw sewage installations which discharge to a treatment plant. Use of this equipment allows the treatment facilities to operate continuously instead of intermittently surging the plant at incremental pumping rates. Variable-speed drivers are used to pump settled sewage and biological sludge where intermittent surging would adversely affect the process. Also, sludge pumps used to feed dewatering equipment are often equipped with variable-speed drives since it is necessary to vary the rate of discharge with the dewatering characteristics of the sludge.

The choice between horizontal- and vertical-drive motors depends considerably upon station arrangement and the availability of dry-pit space. Horizontal motors are usually preferred, provided there is space and no potential flooding problem. Horizontal pumps are more easily maintained, and they are generally less expensive in first cost. Vertical drivers are generally used, however, for the pumping of raw sewage because of their smaller space requirements. Also, vertical units are advantageous in that the motor is located higher and is less susceptible to flooding.

Intermediate shafting is preferred with smaller vertical dry-pit pumps. This allows the drivers to be located at ground level, out of flood range. However, intermediate shafting is not practical for large pumps because of the size of the shafting and the intermediate bearings. Where vertical-drive motors are used without intermediate shafting, the motors are set directly above the pumps and are connected by means of a flexible or a rigid coupling. Separate support of the drive motors is sometimes required.

PUMP CONTROLS

Some means of controlling pump operation is required at most pumping installations. This is usually done from either wet-well level or flow.

Level Control   With level control, each pump is turned on and off at specific water levels in the suction wet well; in the case of variable-capacity pumps, the level control attempts to maintain a preset level, once started. Pumps turn on with a rising level and off as the level lowers. Level control is generally used in raw sewage pumping applications. In this manner it is possible to match discharge with incoming flow.

Flow Control   Flow control is used sometimes where there is no limitation on the availability of flow to the pump suction and where it is desirable to maintain a predetermined rate of discharge. Where flow control is used, a flowmeter is used as the primary instrument to measure flow and to serve as a basis for varying pump speed, which in turn controls capacity. The speed can be changed either manually or automatically through closed-loop instrumentation.

Additional Level Control   Low-water pump cutoff and high-level alarm are provided on most pumping installations. The low-level cutoff is required to prevent the pumps from running dry, and the high-level alarm notifies the operator in the event the pump should fail to operate.

When the pump stops because of low-water cutoff, there is usually some means of indicating this to the operator.

MISCELLANEOUS DESIGN CONSIDERATIONS _____

In addition to the matters discussed above, there are certain other items which should be given consideration in the design of pumping installations.

Piping and Valves   Suction and discharge piping should normally be sized so that the maximum velocities do not exceed 5 and 8 ft/s (1.5 and 2.4 m/s), respectively. Higher velocities, however, may be justified by economic analysis for particular installations. Lines less than 4 in (102 mm) in diameter should not be used for raw sewage. Preferably, sludge lines should be at least 6 in (152 mm) in diameter; 4-in (102-mm) lines are sometimes used for dilute biological sludge.

Valves should be installed on the suction and discharge sides of each pump to allow removal and maintenance of individual pumping units without disturbing the function of the remainder of the installation. It is customary to use either ball or plug valves on raw sewage and concentrated sludge applications. Either plug or butterfly valves can be used for settled sewage or for dilute sludge.

Piping should be designed with sufficient flexibility to avoid stress on the pump flanges. Flange-coupling adapters are sometimes used for this purpose on both the suction and discharge sides of the pump.

Surge Control   Careful attention should be given to surge control wherever a pump discharges into a force main of appreciable length. Generally this is a problem only in the design of raw sewage pumping stations located within the collection system. Changes in fluid motion caused by starting or stopping of pumps or by power failure can create surge conditions.

Surges caused by normal starting and stopping of pumps driven by electric motors may be controlled (1) by selecting individual pump capacities such that the change in velocity in the system when a single pump starts or stops will not result in excessive surges, (2) by using variable-speed drives to bring pumps gradually on or off line, or (3) by using power-operated valves which are controlled so that the pumps are started and stopped against a closed valve.

Surges caused by power failure can be controlled by devices designed to open on an increase in pressure, by devices which will exhaust sewage from the system upon sudden pressure drop in anticipation of surge, or by a surge tank.

Pump Seals   Most sewage and sludge pumps can be obtained with either mechanical seals or stuffing boxes. Mechanical seals have the disadvantage of requiring the pump to be dismantled so that the seal can be repaired. Often it is easier to replace the seal rather than repair it, and it is desirable to keep a spare on hand for this purpose. Water-seal stuffing boxes are recommended for most sludge pumps and for larger sewage pumps. Grease seals are sometimes used for some of the smaller sewage pumps which do not run continuously.

Water serves multiple purposes as a sealing medium: it seals, lubricates, and flushes. Flushing is particularly important where abrasive material is involved in that it helps prevent this material from entering the seal. Grit and ash are very abrasive, and either will cut the shaft sleeves in a relatively short time. Where pumps are controlled automatically, a solenoid valve interlock with the pump starting circuit should be provided in the seal water connection to each pump. A manual shutoff valve and strainer should be provided on each side of each solenoid valve, and a bypass line should be provided around it.

Mechanical seals can be lubricated by the sewage being pumped, provided it is filtered. When mechanical seals are used, a connection is normally provided between the pump discharge and the seal with a 20- to 40-μm in-line strainer to prevent foreign material from entering the seal.

Pump Bearings   Pump bearings must be adequate for the service and should be designed on the basis of not less than a minimum life of five years in accordance with the Anti-Friction Bearings Manufacturers Association life and thrust values. The larger sewage pumps are usually equipped with both case and impeller rings of bronze or chrome steel.

Cleanout Ports Pumps should be provided, where possible, with cleanout ports on both the suction and discharge sides of the impeller. These are desirable for inspection and maintenance purposes.

Wet-Well Design Raw sewage wet wells should not be so large that sewage is retained long enough to go septic. It is usually desirable to limit storage to a maximum of 30 min. Shorter retention time is desirable. With the variable-speed controls now available, many stations can be designed so that the pumping rate matches the inflow rate and the inherent difficulties of frequent pump cycling or long retention times in wet wells can be avoided.

FURTHER READING

American Society of Civil Engineers: *Design and Construction of Sanitary and Storm Sewers*, Manual and Report on Engineering Practice No. 37; WPCF Manual of Practice No. 9, New York, 1969, pp. 287–331.

Benjes, H. H.: "Design of Sewage Pumping Stations," *Public Works*, August 1960.

Benjes, H. H.: "Sewage Pumping," *J. Sanit. Eng. Div., Proc. ASCE*, June 1958.

Great Lakes–Upper Mississippi River Board of State Sanitary Engineers: *Recommended Standards for Sewage Works*, Health Education Service, Albany, N.Y., 1978.

Parmakian, J.: *Water Hammer Analysis*, Prentice-Hall, Englewood Cliffs, N.J., 1955.

Rich, G. R.: *Hydraulic Transients*, 2d rev. ed., McGraw-Hill, New York, 1963.

Water Pollution Control Federation: *Safety and Health in Wastewater Systems*, WPCF Manual of Practice No. 1, Washington, D.C., 1983.

Water Pollution Control Federation: *Design of Wastewater and Stormwater Pumping Stations*, WPCF Manual of Practice No. FD-4, Washington, D.C., 1980.

Water Pollution Control Federation: *Wastewater Treatment Design*, WPCF Manual of Practice No. 8, Washington, D.C., 1977.

SECTION 9.3
DRAINAGE AND IRRIGATION

JOHN S. ROBERTSON

DRAINAGE PUMPS

Drainage pumps are used to control the level of water trapped in a protected area. Entrapment occurs when high lake levels, stream stages, and tides preclude the normal discharge of streams, storm runoff, and seepage from the protected area. These high-water conditions are created by floods and hurricanes or impoundment.

Floods and hurricanes occur infrequently and are relatively short-lived. However, as normal drainage is not possible at such times, all interior drainage that cannot be ponded must be pumped over the protective works if the protective works is a levee or through it if it is a concrete flood wall. Pumps for this purpose are strategically located at the edge of ponding areas and streams, in sewer systems, or in the protective works. These pumps are operated continuously or are cycled on and off as necessary to maintain the water level in the protected area below the elevation at which damage would be experienced. As their use is required only during emergency situations and usually under adverse weather conditions, reliability of operation is essential.

Where valuable low-lying areas must be protected against inundation due to backwater, gravity drainage from the area is generally not possible. In such situations all seepage and storm-water runoff entering the protected area must be pumped. The pumping of seepage is often a continuous operation, whereas the pumping of storm-water runoff is intermittent. Both can occur simultaneously because inflow due to seepage does not stop during a rainstorm. It therefore is necessary to provide pumps that can handle seepage and storm water simultaneously.

IRRIGATION PUMPS

Irrigation pumps play an important part in making vast areas of arid and semiarid land agriculturally productive. These pumps take water from surface sources, from subsurface sources, and in ever-increasing amounts from sewage treatment facilities and pump it to the point of application.

Water from surface sources, such as streams, lakes, and ponds, is pumped directly into the distribution system or into a conveyance system. If pumped into a conveyance system, it flows either by gravity or under pressure to the distribution point or to a booster pumping station. Most of the pumps used in these intallations are mounted permanently and are arranged to operate as necessary without constant attendance during the growing season.

Water from subsurface sources is usually pumped directly into the distribution system. In such installations a well is drilled in the vicinity of the area to be irrigated and is fitted with the proper size pump. In many instances more than one well is needed to provide the required capacity.

Effluent waters from different types of sewage treatment facilities are sometimes used for irrigation on a year-round basis in land disposal systems. In these systems, the treated effluent is temporarily stored in holding tanks or ponds and is applied to the disposal area at predetermined rates via sprinkler systems. As the disposal area may be a considerable distance from the storage area and as pressure is needed for operation of the sprinkler system, pumping is required.

PUMP TYPES

Centrifugal pumps are used almost exclusively in drainage and irrigation installations. Because of the large selection of propeller, volute, turbine, and portable pumps manufactured today, there is little difficulty in finding a pump that will meet the conditions encountered in these fields. Pumps in the sizes needed to meet the requirements of the majority of installations are available as standard items. Specialized pumps and extremely large pumps are designed and built to meet the needs of individual projects.

Propeller Pumps　These units are used for low-head pumping. As most of the pumping in drainage and irrigation is low-head, the propeller pump is the most widely used type. In general, vertical single-stage axial- and mixed-flow pumps are used; however, there are instances where two-stage axial-flow pumps should be considered for economic reasons.

Horizontal axial-flow pumps are used for pumping large volumes against low heads and usually employ siphonic action when not of the submersible type. When higher heads are involved, these pumps can be arranged to operate with siphonic action until the back pressure places the hydraulic gradient above the pump.

Variable-pitch propeller pumps rotating at constant speed can be operated efficiently over a wide range of head-capacity conditions by varying the pitch of the propeller blades. These pumps are used when the head-capacity conditions cannot be met with the more economical fixed-blade pumps and where the pumps will be operated often enough and long enough to warrant the expense. The blade-control system needed for such pumps is more sophisticated than in any of the systems usually provided on drainage and irrigation projects. As a consequence, operation and maintenance must be performed by organizations employing competent and experienced personnel.

Volute Pumps　Volute pumps are used when pumping from surface sources and in general when the total head exceeds approximately 45 ft (14 m). Such pumps are available in many types, such as vertical and horizontal shaft, end suction, bottom suction, and double suction with semi-open or closed impellers, etc. These pumps are mounted in dry pits when located below grade and on slabs or floors when located above grade. The particular type used depends on the capacity and kind of service to be performed.

Deep-Well Turbine Pumps　Deep-well turbine pumps (vertical-shaft, single-suction pumps having one or more stages) are used in irrigation primarily to pump water from a subsurface source into a distribution system. The head against which the pump will operate determines the number of stages that must be provided. For large capacities, more than one pump will be needed.

Submersible Turbine Pumps　In these deep-well units, the motor is close-coupled to the pump and submerged in the well. This type of pump is used for high-head applications where long intermediate shafts are undesirable.

Portable Pumps　In drainage work, portable pumps are used as emergency equipment to control ponding elevations where mobile equipment, such as tractors and trucks, having power takeoffs are available to drive them. In irrigation work they are usually used to irrigate from a surface source. Such pumps are small-capacity, low-head, economical pumps that must be submerged in order to be used.

PUMP SELECTION

A pump selection study should always be made, and its importance cannot be emphasized enough. Such studies permit selection of the type of pump and discharge system best suited to the project and provide the information needed to proceed with the design of the installation.

In many cases, the type of pump required will be obvious. If more than one kind of pump would be satisfactory, the specifications should permit the pump manufacturer to make the choice. This is particularly advantageous when competitive bidding is involved. The practice of the Corps of Engineers in this regard is to write a performance specification and allow the pump manufacturer to determine the type, size, and speed of the pump.

Before initiating such a study, all previous studies made to determine the total pumping requirement or station capacity, pertinent water-surface elevations, terrain, utility locations, proposed station or well locations, points of discharge, and the proposed method of operation should be reviewed. Also to be considered is the experience of the personnel that will be responsible for the operation and maintenance of the installation.

Number of Pumps First costs are generally of more concern than operating costs in drainage and irrigation work becaue the operating period for the majority of installations is relatively short and occurs only once a year. Costs can be minimized by using as few pumps as possible. However, one-pump installations are seldom used except in the case of wells. For reliability, a minimum of two pumps should be installed in drainage pumping stations, where the loss of even one pump during an emergency situation could result in considerable damage. Three or more pumps are preferred. Standby units are provided only in those installations where continuous operation precludes taking a pump out of service for maintenance.

The number of pumps ultimately used should be consistent with the demands of the project. For instance, when the installation is located in an agricultural area or is a part of an urban sewer system, a standby pump should be provided, since these installations must always be capable of discharging project requirements during periods of blocked drainage. In this way considerable damage may be avoided. When the installation is used to pump storm water from pondage or irrigation water from a lake, the loss of a pump is not critical and so a standby pump is not needed.

If during pump selection it is found that the rated power of the prime mover exceeds the maximum power requirements of the pump by a considerable amount, the contemplated number of pumps should be increased or decreased, provided the change results in a better power match without increasing the overall cost of the installation. By increasing the number of pumps and thereby reducing the required power, there is a possibility that the size of the prime mover can be reduced and that the pump power will approach the rated power of the prime mover. On the other hand, decreasing the number of pumps will increase the power requirements. This increase may be sufficient to either utilize most of the excess capacity in the prime mover or require the use of a larger one.

For installations requiring the use of large pumps, foundation conditions become important. To prevent the installation from being relocated to a less desirable site or the necessity of providing a more expensive pile foundation because the bearing pressures at the selected site exceed the allowable limit, the number of pumps should be increased, provided the loading can be reduced to an acceptable amount and the resulting installation continues to be the most economical.

Inadequate depth on the suction side of the pumps may necessitate the use of more pumps. Should the water not be deep enough to provide the submergence needed by the contemplated pump, more pumps of a smaller size may have to be used or the sump and approach channel may have to be excavated to the needed depth. The latter alternative could cause operational and maintenance problems and might be the more expensive solution.

Capacity The capacity of a pump is a function of the total pumping requirement, the number of pumps, and, in the case of wells, the capacity of the well. Whenever possible, all pumps in a multiple-pump installation should be of the same capacity. This is advantageous from a cost standpoint as well as from a maintenance standpoint. In drainage installations, three pumps are generally provided in order to have the capability of pumping not less than two-thirds of the project requirement with one pump inoperative. Each pump would then have a capacity equal to one-third of the total capacity. When more than three are used, the capacity should be the total required capacity divided by the number of pumps being used. If two pumps are used, each pump should be sized to pump not less than two-thirds of the total capacity.

In irrigation installations utilizing multiple pumps, the capacity of each pump should be the same. In single-pump installations, the pump should be sized to meet project requirements. When wells are involved, the capacity of the pump will be determined by the capacity of the well. Often several wells are needed to give the capacity to satisfy the established requirements.

When a standby pump is to be provided, its capacity should be equal to that of the largest pump being furnished.

Head   Total head is the algebraic difference between the total discharge head and the total suction head. In drainage and irrigation work total suction head can usually be determined, but this is not always the case for the total discharge head. The losses in the discharge systems often must be determined by hydraulic model test rather than by calculation and therefore must be estimated for preliminary selection purposes. The head specified will have to be some head other than total head; generally pool-to-pool head is used.

Total discharge head is defined in the Hydraulic Institute Standards, and its value is to a great extent determined by the type of discharge system used. A number of the many possible discharge systems used are shown in Fig. 1. The losses in systems A, B, C, D, and E can be calculated. Thus the performance for pumps discharging into these systems can be put on a total head basis.

System A is an "over the levee" siphonic discharge line, the use of which can seldom be justified. However, continuous operation over long periods could effect a savings that would be sufficient to justify the additional cost of the installation and the taking on of the operational hazards usually encountered with such lines. The total discharge head for this system is equal to the height of the discharge pool, stream, or lake above the impeller plus the exit loss and the discharge line losses from the pump discharge nozzle to the line terminus. The absolute pressure at the high point of the line should be not less than 9 ft (2.7 m). Lower values have been used successfully, but this should be the exception rather than the rule. For additional information on the determination of heads in a siphonic system, refer to Siphon Head in Sec. 8.1.

Flap valves should be installed on the discharge end of all lines subjected to a cycling operation. The closing of these valves following pump shutdown prevents reverse flow into the protected area and permits development of the pressures needed to keep the lines full (primed) during those short periods when pumps are idle. These pressures will be less than atmospheric pressure, with the minimum pressure (absolute) being at the high point. Should there be significant leakage at the joints or in the valves, the pressures will rise and the water level at the high point will drop. If the water level drops below the invert of the line at the high point, the two legs of the line will be separated by an air space and priming will be necessary when the pump is started.

Air-release valves should be installed at the high point of all siphonic discharge lines to provide an escape for the air being compressed as the water rises in the line during priming, and vacuum-break valves should be used to prevent reverse flow into the protected area. For lines equipped with flap valves, the vacuum-break valve should be manually operated and used when the flap gate fails to seat properly or to provide a rapid means of draining the line when pumping is no longer required. Units for almost any size line can be obtained from manufacturers who specialize in such equipment or they can be assembled, using swing check and angle valves.

System B in Fig. 1 is an "over the levee" nonsiphonic discharge line. To preclude siphonic action, this line should be vented at the high point with a vent having a diameter that is approximately one-fourth that of the line. The invert of this line at the high point should be placed at the same elevation as the top of the protective works so that the pumped flow can discharge from the down leg of the line under gravity without backwater effects for all discharge pool elevations up to maximum. Thus the total discharge head will have a constant value because it will not be affected by changes in the level of the discharge pool. To ensure adequate prime mover capacity when using this system, it is the practice in all cases to use the top of the line at the high point in lieu of the hydraulic gradient when determining the total discharge head. Therefore total discharge head is the height of the top of the line at the high point above the impeller plus the

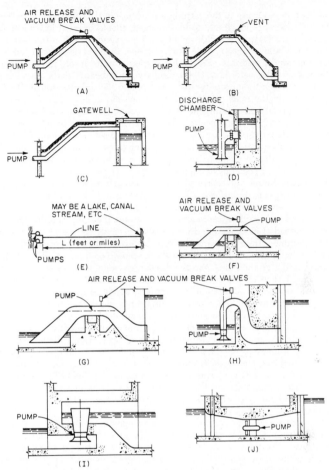

FIG. 1 Discharge systems. (Adapted from Dept. of the Army: "Mechanical and Electrical Design of Pumping Stations," EM 1110-2-3105, Washington, D.C., 1962)

velocity head in the line at the high point based on a full pipe and the losses in the line from the pump discharge nozzle to the beginning of the down leg of the line.

System C is used when there is a conduit carrying the normal gravity discharge under the levee adjacent to the station that must be valved off against reverse flow into the protected area during periods of high water. The closure gate is located in a gate well constructed on the stream or lake side of the levee to prevent subjecting the gravity conduit to high-water conditions. The pump discharge lines go over the levee and terminate in the gate well above the maximum water level. This shortens the lines and reduces the cost. The total discharge head for this system is equal to the height of the top of the line at the terminal end above the impeller plus the exit loss and the losses in the line between the pump discharge nozzle and the terminal point. The total discharge head for this system, as in system B, is independent of the discharge pool and therefore constant. Neither flap valves nor vents are required on these lines.

System D is used when the pumping station is constructed as an integral part of the levee or flood wall. The invert of the pump discharge line is placed at an elevation that is above the stream or lake level that will prevail approximately 70% of the time or as is dictated by the physical dimensions of the pump. Owing to the extreme turbulence in the discharge chamber, gates with multiple shutters, which are less likely to be damaged, should be used instead of flap gates on discharge lines that are larger than 36 in (914 mm) in diameter. When the water level in the discharge chamber is below the top of the discharge line, the total discharge head is determined in the same manner as for system C. For higher discharge water levels, the total discharge head is equal to the height of the water level in the discharge chamber above the impeller plus the exit loss and the losses between the pump discharge nozzle and the chamber side of the flap valve.

System E is perhaps the most common discharge system in use today. It is used to connect one pump or several manifolded pumps with a lake, canal, stream, ditch, reservoir, or sprinkler system. For short lines and low static heads, valve and fitting losses, frictional losses, and exit losses are very important, whereas in long lines or very high static head installations only frictional losses are given consideration. In manifolded installations using propeller pumps, a check valve and gate valve are installed immediately downstream of the pump. The gate valve should always be opened before the pump is started because the motors provided are not usually sized to operate against shutoff head. Positive shutoff valves are placed immediately downstream of volute or turbine pumps since these pumps are usually started and stopped against a closed valve. They also prevent reverse flow into the sump when one of the pumps is inoperative.

Pool-to-pool head is the difference in elevation between the sump and discharged-water surfaces and is used instead of total head in drainage work because the losses in the discharge system are not easily determined. Installations of this type are exemplified by systems F, G, H, I, and J in Fig. 1. For such installations it is best to specify the pumps on a pool-to-pool basis, to have the pump manufacturer design the pump and the discharge system, and to verify the predicted performance by model test. It should be noted that in such installations the discharge systems are usually constructed within the confines of the pumping station structure.

System F is operated as a siphon with the pump supplying energy equivalent to the pool-to-pool head plus the system losses. The invert of the pump discharge pipe at the highest point is located above the maximum river stage, and vacuum pumps are generally used to aid in priming the pump. This system is used for pool-to-pool heads of up to approximately 6 ft (1.8 m) and where the physical dimensions of a vertical pump would make it necessary to operate against higher heads. In estimating the losses, entrance losses, which are small [approximately 0.14 ft (4 cm)], should be neglected. The centerline of the suction piping should be assumed to make an angle of 45° with the horizontal, and the diameter of the discharge piping as measured at the discharge flange of the discharge elbow should be such that the velocity at maximum discharge is approximately 12 ft/s (0.037 m/s) or less.

System G is used for pool-to-pool heads of up to approximately 15 ft (4.6 m). The water passages change in cross section from round at the pump bowl to rectangular at each end. The width at the suction end is the same as or less than that of the suction bay, and the height is such that the entrance to the suction passage is always submerged when the pump is in operation. The discharge velocity should be kept to approximately 6 ft/s (1.8 m/s). Multiple-shutter gates (Fig. 2) arranged to be raised when the pump is in operation are provided to prevent prime mover overload when the pump is started and to prevent reverse flow when the pump is stopped or inoperative. These pumps are usually large and slow and require substantial prime movers. Vacuum priming equipment is used so that additional power will not be required for priming.

System H is essentially the same as system G, except that the former is used for pool-to-pool heads up to approximately 26 ft (7.9 m). Also, a splitter may be required in the discharge water passage for structural purposes as well as for keeping the multiple-shutter gates to a reasonable size. The pumps used with this system are vertical and in general smaller than those used with system G.

System I can be constructed with the lip of the pump column either above or below the design flood elevation or the maximum surge, if it is being provided for hurricane protection. If the lip of the pump column is above and the chances of reverse flow through the pumps are extremely remote, then gating of either the pump column or the discharge water passage is unnecessary. If the lip of the pump column is below, then a decision as to whether or not to gate both the pump column opening and the discharge water passage or just the water passage must be made. In general, if the pumps will be in operation continuously during high-water conditions and if reverse flow through an inoperative pump will not have substantial detrimental effects, only the multiple-shutter gate at the end of the discharge water passage need be provided. Reverse flow could occur should the flap for some reason fail to close. Like system G, splitter walls may be

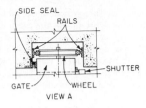

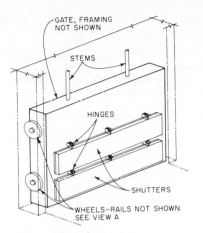

FIG. 2 Multiple-shutter gates.

required if pumps are large. Also, the clearance between the lip of the pump column and the ceiling of the discharge water passage is critical and should be determined by model test. The shape of the pump column lip will also affect pump efficiency and should be determined by test.

System J could have many configurations but, regardless of arrangement, would have to be gated on the suction side with a positive shutoff gate, such as a pressure-seating slide gate, and on the discharge side with a positive shutoff gate and a multiple-shutter gate or a flap gate if the installation is small and a pipe is used in place of formed water passages. When formed water passages are used, transition sections between the pump and the gates sections will be needed and should be designed by the pump manufacturer.

TOTAL SUCTION HEAD   The practice for propeller pump installations is to dimension the station sump or sump bays in accordance with Hydraulic Institute standards and to make the distance between the sump floor and the lip of the suction bowl conform to the standards or the recommendations of the pump manufacturer. The approach and entrance velocities resulting therefrom are small enough to be disregarded when calculating total suction head. Total suction head in vertical propeller pump installations is the height from the centerline or eye of the propeller to the water surface in the sump. For drainage installations using vertical propeller pumps, a total suction head of zero is not uncommon. In submerged horizontal propeller pump installations, total suction head is the height from the centerline of the propeller shaft to the water surface in the sump, and the minimum value should be not less than $1.2D$, where D represents the diameter of the propeller.

Volute pumps equipped with a formed suction or suction piping may operate with either a suction lift or a suction head, depending on the suction water level. In either case, all the losses between the entrance and the eye of the impeller should be included in any calculation. Approach velocity is not a consideration in these installations, but entrance losses are.

TOTAL SUCTION LIFT   Approach and entrance velocities can be disregarded in suction lift installations by proper dimensioning of the station sump or sump bays and by proper setting of the suction bell. The sump or sump bay dimensions used, as in suction head installations, should conform to the Hydraulic Institute standards or to the recommendations of the pump manufacturer. The suction bell for both horizontal propeller pump and vertical volute pump installations should be set with the lip located approximately $0.5D$ above the sump floor. For horizontal propeller pump installations, the minimum submergence of the suction bell should be approximately $0.25D$, where D is the diameter of the suction pipe; for volute pump installations it should be $1.5D$, where D is the diameter of the suction bell. In determining the total suction lift for these installations, it is assumed that the suction piping is a part of the pump and that the approach and entrance velocities can be disregarded. Total suction lift therefore is the height from the water surface in the sump to the centerline of the propeller shaft or to the eye of the impeller.

Setting   The setting of the pump or the locating of the centerline or eye of the propeller or impeller with respect to the water surface should be given careful consideration when selecting the pump to be used. Some installations offer few if any problems in this regard, while in others setting may have a considerable effect on the size and number of pumps selected.

TURBINE PUMPS   Drawdown is a consideration in any well installation. In setting the pump to prevent cavitation, sufficient NPSH at the eye of the first-stage impeller and/or sufficient depth over the suction bell lip or tailpipe to prevent vortexing should be maintained when maximum drawdown is being experienced.

VOLUTE PUMPS   Volute pumps may operate with either a suction head or a suction lift. If with a suction lift and of the horizontal type, the pump should be set above the maximum anticipated elevation of the suction water source in order to avoid inundation. As priming will be necessary when starting, a suction lift would be considered a satisfactory arrangement when long periods of continuous operation are anticipated.

Suction head installation is the preferred practice since priming is unnecessary and should be used whenever conditions permit. In drainage work vertical volute pumps are used for pumping small amounts of rainfall runoff and seepage flows. These pumps are usually located in a dry sump adjacent to the storm water pump sump, with motors and valve operators located on the operating floor above. The submergence of these pumps should be such that when discharging at maximum capacity, the total suction head is zero or above. In many instances water surface fluctuations of just a few feet are experienced. When small volumes are involved, a cycling operation occurs. For this type of operation, the pumps are usually started at the maximum water level and stopped at the minimum level.

PROPELLER PUMPS   Sufficient water depth does not always exist or cannot always be provided to give the submergence needed to permit the smallest and perhaps the most efficient pump to be used. Excavation is one answer. However, depending on the soil type, the location of the installation, the kind of construction used, the silt-carrying characteristics of the stream, and the frequency of operation, excavation may be an operational and maintenance headache and it is impractical when sumps are to be made self-draining. Another alternative—and the one most frequently used in drainage work—is to set the centerline or eye of the propeller at or slightly below the minimum sump elevation and select a pump that will operate at this setting with little or no cavitation damage. This means that a larger pump operating at a lower speed should be used.

Prime Movers   The prime movers used in drainage and irrigation installations are electric motors and diesel and gas engines. The one to be used in any particular situation must be determined before the pump can be selected.

Electric motors are the most economical installations; they should be used when a reliable source of electric power is available and when the cost of bringing it into the pumping station is not unreasonable. A reliable source of electric power is a source that historically has not suffered outages under the climatic conditions that will prevail during the time the pumps will be required to operate. Two feeders having separate origins and not subject to simultaneous outages are sometimes provided to ensure the reliability needed for drainage installations. Such an arrangement has been satisfactorily employed in urban areas but would not be a practical solution in remote areas not yet electrified. The cost of constructing and maintaining even one transmission line in such an area could be prohibitive.

The motors in all but the largest installations should be of the squirrel-cage induction type. In those installations where the motor rating is numerically larger than the speed, the type of motor used should be the one having the lowest overall first cost. It may be either a squirrel-cage-induction or a synchronous motor.

All motors should be full-voltage starting except in those instances where the local power company indicates that reduced-voltage starting is necessary. For unattended operation and for drainage stations pumping seepage or pumping from a sewer system where frequent cycling is usually necessary, control devices set to start and stop the motors automatically at predetermined sump or discharge pool levels should be provided. In drainage pumping stations not subject to a cycling operation, motors are started manually by the operator and stopped automatically by a control device.

Engines are used to drive pumps when it is not feasible to use electric motors. They are more expensive than motors but reliable if properly maintained and serviced. They are also variable-speed drives that should be operated at constant speed whenever possible. The requirements of most installations can be met with constant-speed operation. However, for those that cannot, the number of speeds used should be held to a minimum.

Engines should not be cycled on and off but should be operated on a continuous basis. For those installations where the inflow is not sufficient, continuous operation can be obtained by returning a part of the pumped discharge back to the sump. This is accomplished by connecting the pump discharge line and the sump with a valved line.

Gas engines are seldom used, but their use should be considered when the installation is close to a natural gas main.

Right-angle reduction gears are used to transmit the power from the engine to the pump shaft of vertical propeller pumps. For horizontal pump installations where the engine shaft parallels the pump shaft but is at a different elevation and off to one side, silent chain drives are used. For other horizontal installations parallel-shaft gear units may be used. A service of 1.50 should be used when determining the equivalent power of these units. Right-angle units should be of the hollow-shaft type only if vertical adjustment of the pump impeller is required.

Adequate fuel storage in addition to the day tanks should be provided. Storage facilities should be sufficient to provide fuel at the maximum rate of consumption for a period of 48 hours or less, as demanded by the anticipated pumping requirements. Larger fuel storage installations may be needed if replenishment supplies are not readily available. The design of the facilities should be in accordance with the standards of the National Board of Fire Underwriters and local agencies having jurisdiction.

PUMPING STATION

Sump   The sump is perhaps the most important element in the structure of the pumping station. Unless it is properly located, designed, and sized, the flow conditions within could have an adverse effect on the operation of the pump. There are many variations in sump arrangements that are acceptable; however, best results are obtained when the sumps or sump bays are oriented parallel to the line of flow. Flows approaching from an angle create dead spots and high local velocities, which result in the formation of vortices, nonuniform entrance velocities, and an increase in entrance losses. The flow to any pump should not be required to pass another pump before reaching the pump it is meant for. When sumps or sump bays are normal to the direction of flow, such as in sewer systems, the distance between the sump or sump bay entrance and the pump must be sufficient for the flow to straighten itself out before reaching the pump. For additional information relative to sump design and sizing, refer to Chap. 10. If the installation is large enough to warrant it, modeling of the sump to permit the best design to be determined is advocated.

Sumps in drainage installations pumping storm water should be either located above normal water levels, in which case they would be self-draining, or isolated from normal flows by gates. In small installations, motorized pressure-seating gates should be used; in large installations, roller gates raised and lowered with a crane or by some other suitable system should be used. These gates should be sized so that the velocity through them will not exceed 5 ft/s (1.5 m/s) for any condition of flow. One gate should be located directly opposite each pump when all pumps are installed in a common sump or should be located at the entrance to each sump bay when pumps are separated.

Frequent cycling of pumps is encountered in installations which pump from sewer systems and which pump seepage. In such installations, the sump should be sized so that the volume stored in it and in the ponding area or the sewer lines, as the case may be, within the limits of the operating range, will be sufficient to prevent the starting of the pumps oftener than once every 4 min.

Superstructure A superstructure is provided on practically all drainage pumping stations but not on all irrigation pumping stations. The type of superstructure provided should be consistent with the surrounding area and should have a minimum number of windows and openings. In rural areas, corrugated sheet metal structures, which are inexpensive, have been used extensively. These provide adequate protection from the elements and from vandalism. To eliminate the need for an indoor crane, hatches in the roof over each pump can be provided to permit a truck crane to remove motors and pumps as a unit. For engine-driven pumps and larger pump installations, indoor cranes of the appropriate size and type should be provided for installation, removal, and maintenance.

Corrosion The corrosion of electrical and mechanical equipment in housed and unhoused stations can be controlled by painting exterior surfaces with a good paint and by installing strip heaters in all electrical enclosures. In large housed installations, heating of the operating room area in addition to the use of strip heaters should be considered. All equipment below the operating room floor level should be coated with a paint that is suitable for the exposure. In dry sump stations, enamels should be satisfactory. In wet sumps that are kept dry during inoperative periods, cold-applied coal tar enamel, which is easily repaired, is preferred. In installations where the equipment is continuously immersed, coal tar epoxy paint or vinyl paint should be used. If the water is extremely corrosive, consideration should be given to mounting galvanic anodes on the pumps in addition to painting.

FURTHER READING

Department of the Army, Office of the Chief of Engineers: "Interior Drainage of Leveed Urban Areas: Hydrology," EM 1110-2-1410, Washington, D.C., 1965.

Department of the Army, Office of the Chief of Engineers: "General Principles of Pumping Station Design and Layout," EM 1110-2-3102, Washington, D.C., 1962.

Department of the Army, Office of the Chief of Engineers: "Mechanical and Electrical Design of Pumping Stations," EM 1110-2-3105, Washington, D.C., 1962.

Hicks, T. G.: *Pump Selection and Application*, McGraw-Hill, New York, 1957.

Houk, I. E.: *Irrigation Engineering*, Wiley, New York, 1951.

Hydraulic Institute Standards for Centrifugal, Rotary, and Reciprocating Pumps, 14th ed., Hydraulic Institute, Cleveland, 1983.

Israelson, O. W.: *Irrigation Principles and Practices*, 2d ed., Wiley, New York, 1950.

U.S. Department of the Interior, Bureau of Reclamation: Turbines and Pumps, Design Standard 6, Denver, 1960.

SECTION 9.4
FIRE PUMPS

MARIO Di MASI

Compelling reasons dictate the installation of fire protection systems driven by stationary fire pumps. Foremost among these reasons is protection—protection of lives, of equipment, possessions, and inventories, and of major assets, such as hospitals, hotels, office and residential buildings, and warehouses. There are two additional, less obvious but equally compelling, reasons: the reduction of costs and the protection of income-generating operations.

Costs are reduced in a simple manner. Over the projected life of a facility, the total of construction costs plus fire protection equipment costs plus fire insurance costs is lower than the total of construction costs plus fire insurance costs without fire protection equipment.

Virtually every American city has fire hydrants. Each has its head-flow discharge capabilities. Most cities have mobile fire-fighting equipment which includes "pumpers"—trucks equipped with engine-driven boost pumps. The hydrant-pumper combination can extinguish many kinds of fires—grass fires, single-family-dwelling fires, fires in low-rise business and manufacturing establishments, and so forth—but there comes a point beyond which the head-flow capabilities of the hydrant-pumper combination can no longer provide adequate protection. That point might be reached when a facility becomes too high, too big, too remote, too combustible . . . The "rules" by which one can determine whether a stationary fire pump system is required are contained in the following National Fire Protection Association publications, available from National Fire Protection Association, Batterymarch Park, Quincy, Mass. 02269. Only the most recent edition should be used:

NFPA Pamphlet 13, "Sprinkler Systems"

NFPA Pamphlet 14, "Standpipe, Hose Systems"

NFPA Pamphlet 16, "Deluge Foam-Water Systems"

THE COGNIZANT AUTHORITY

A facility can be self-insured. If so, users can write the fire pump specifications strictly in accordance with their perception of the application's requirements. Even such self-insured users are advised to conform to two specifications insofar as possible because a demonstrable effort to comply might lessen liability if someone is subsequently injured or killed. These two specifications are NFPA Pamphlet 20, "Centifugal Fire Pumps," and "Factory Mutual Loss Prevention Data 3-7N." Also, American Standards Institute (ANSI) standard B73.1 might apply.

In most applications, pumps are *required* to conform to NFPA Pamphlet 20. When the user arranges for fire insurance, the insurer makes compliance a condition of the insurance rate and will require the pumping equipment to be either Factory Mutual Research Corporation (FM) approved or Underwriters Laboratories (UL) listed. In both cases, the requirements of NFPA Pamphlet 20 will apply, but the FM and UL organizations interpret the requirements of the pamphlet somewhat differently.

FIRE PUMP CRITERIA

The following are some of the more important requirements applicable to approved/listed fire pumps.

The pumps (excluding jockey pumps), controls, and accessories must all be ordered on a single purchase contract, stipulating compliance with NFPA Pamphlet 20 and satisfactory performance of the equipment when installed. The pump manufacturer must be responsible for the proper operation of the installed assembly as demonstrated by field tests.

Each fire pump, driver, and controller and some of the accessories must have the appropriate UL listing and/or FM nameplate. (Equipment bearing such a nameplate is said to be "labeled.") For example, the pump nameplate will indicate the approval/listing agency, pump model, serial number, design capacity, design pressure, rated speed, maximum power, and shutoff or churn pressure.

If an adequate, reliable supply of water is available and if a positive pressure will always be available at the pump suction, the pump may be of the horizontal end-suction or of the horizontal split-case design. Multistage case pumps are available for high-head applications. A horizontal fire pump driven by an electric motor is shown in Fig. 1, and Fig. 2 depicts a horizontal fire pump driven by a diesel engine. A horizontal fire pump requires a reservoir located above the pump suction if the suction piping is inadequate.

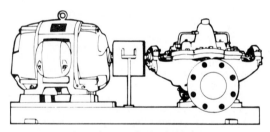

FIG. 1   Motor-driven horizontal split-case fire pump.

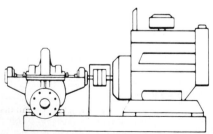

FIG. 2   Diesel-driven horizontal split-case fire pump.

If the pump must lift the water from a well, pond, river, etc., a vertical turbine pump must be used. A vertical turbine fire pump driven by an electric motor is shown in Fig. 3. When driven by a diesel engine, a universal drive shaft and a right-angle gear drive are included, as shown in Fig. 4. The vertical pump must be installed so that the static water level is never below the bottom impeller.

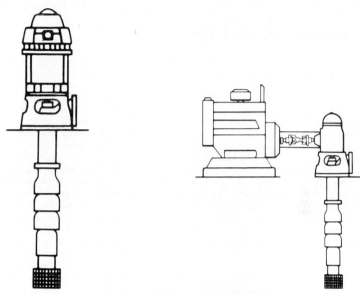

FIG. 3   Motor-driven vertical turbine fire pump.   **FIG. 4**   Vertical turbine fire pump driven by diesel engine through right-angle gear.

The inner column (drive shafting) for vertical pumps is available in either of two configurations: (1) open line shaft (OLS) construction, in which steel shafting rotates in water-lubricated rubber bearings which are centered and stabilized by rigid bearing retainers, is used for static water levels 50 ft (15 m) or less below the pump discharge flange and (2) enclosed line shaft (ELS) construction, in which steel shafts rotate in oil-lubricated bronze sleeve bearings, is used for static water levels of more than 50 ft (15 m). In ELS construction, the outside of the sleeve bearings are threaded and tubes enclosing and supporting the bearings and shafts isolate the shafting from the water being pumped.

Horizontal split-case fire pumps must have shutoff (no flow) pressures no higher than 120% of design (operating) pressures. End-suction horizontal pumps and vertical turbine pumps are required to have shutoff pressures no higher than 140% of design pressures. When the pressure against the pump discharge is reduced so that the pump produces 150% of its design capacity, the discharge pressure must be no lower than 65% of the design pressure.

Static pressure tests are required. The pumping unit must withstand 250 lb/in^2 gage (17.2 bar°) or 150% of its operating head, whichever is greater. For example, suppose that the design conditions for a horizontal split-case pump require 3000 gpm (700 m^3/h) at a discharge head of 300 ft (8.9 bar). The maximum shutoff head is 1.2 × 300 = 360 ft (10.7 bar). At 1.5 × 3000 = 4500 gpm 1050 m^3/h), the discharge pressure must be at least 0.65 × 300 = 195 ft (5.8 bar). The pump must withstand a static pressure of 1.5 × 3000/2.31 = 195 lb/in^2 gage (13.4 bar) or 250 lb/in^2 gage (17.2 bar), whichever is greater.

Certified performance test curves are required for most fire pumps. The curves show pump head, power, and efficiency at various pump capacities. The pump manufacturer might perform the tests using a calibrated laboratory motor, a dynamometer, or the job diesel engine.

°1 bar = 10^5 Pa. For a discussion of bar, see *SI Units—A Commentary* in the front matter.

The major determinants of pump size are the pump head and capacity required, the nature of the supply source, and the pressure (if any) at which the water is delivered to the pump suction. UL listed pumps are available for all commonly encountered head-capacity requirements. If application requirements exceed the capabilities of UL listed pumps, manufacturers can furnish unlabeled pumps made in accordance with the spirit of NFPA Pamphlet 20.

MATERIALS OF CONSTRUCTION

Since most fire pumps handle clear water and since wear (from heavy-duty use) is not usually a problem, UL listed pumps are ordinarily made of commonly used industrial materials.

Horizontal pumps (Fig. 5) have casings made of cast iron. Enclosed impellers, made of bronze, and bronze case wear rings are used. Stuffing boxes are equipped with bronze lantern rings and bronze split-type packing glands; shafts are sealed with graphite asbestos packing. (As a reminder, all pump shaft packings must have a small amount of leakage when the pump is running to prevent the packing from being burned.) The impeller shaft, usually steel, will have a bronze sleeve through the packing; the sleeve will be sealed to the shaft with either an O ring or a gasket. Ball bearings will be grease-lubricated. In horizontal split-case pumps, the inboard bearing will be of the single-row, radial type and the outboard bearing will be of the duplex, angular-contact type.

Vertical turbine pumps (Fig. 6) will have cast iron bowls, enclosed bronze impellers, and steel or stainless steel impeller shafts. The vertical drive line will be either OLS construction, which includes steel shafting, rubber sleeve bearings, and bronze bearing retainers, or ELS construction, which includes steel shafting, steel enclosing tubes, and bronze sleeve bearings, all arranged for oil lubrication. Vertical turbine pumps usually have cast iron aboveground discharge heads. The column pipe is furnished in 10-ft (3-m) (nominal) joints made of steel, with either threaded or flanged joints. Flanged column pipe requires accurately machined flanges with rabbet (registered) fits for accurate alignment.

Since some vertical turbine fire pumps are installed in either brackish or salt water, UL listed pumps made of appropriate materials are available.

Stationary fire pumps can be purchased as packaged systems, which save installation time and money. Single or multiple units of horizontal or vertical pumps can be packaged. Characteristically, each packaged unit includes the pump(s), driver(s), controller(s), headers, accessories, and piping mounted on a common base. To the extent possible, all wiring and piping connections are made and the unit is factory-tested. Installation consists simply of positioning and leveling the package and making external piping and electrical connections.

Most modern fire protection systems are designed to operate automatically. The starting sequence is initiated by a pressure drop caused by activation of an automatic sprinkler or a manual hose valve. In order to keep the system filled and pressurized without activating the main fire pump(s), a jockey, or pressure maintenance, pump is used. The jockey pump is small so that it can restore small system pressure losses but cannot supply the large flows and pressures required when the fire protection system is activated. Any pump suitable for the application may be used for jockey pump service; jockey pumps are not covered by UL listing or FM approval requirements.

FIRE PUMP DRIVERS

NFPA Pamphlet 20 describes the requirements for fire pump drivers, which must be UL or FM approved. Approved drivers include various types of electric motors and diesel engines. The driver must be selected to provide the maximum power required by the pump anywhere along its performance curve. (Spark ignition internal combustion engines and dual drives—a motor and an engine as alternate drivers on the same pump—have not had NFPA approval since the 1974 edition of Pamphlet 20.)

Most stationary fire pumps are driven by electric motors. To justify the use of an electric motor, a single power station or substation which can guarantee virtually uninterrupted power is

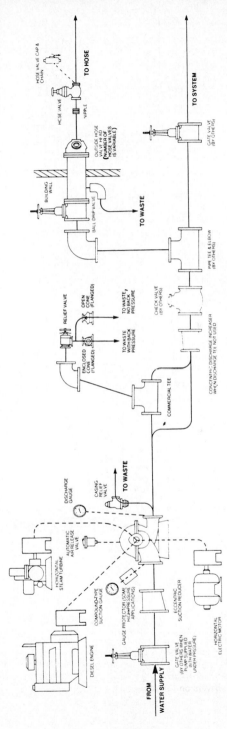

FIG. 5 Typical horizontal fire pump system.

9.55

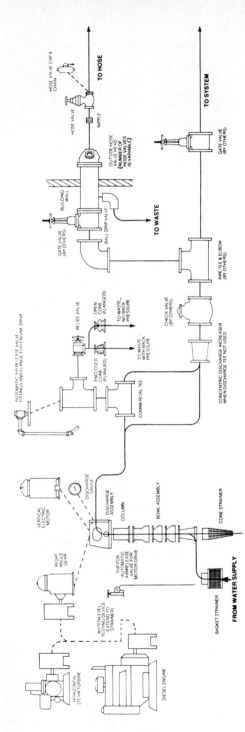

FIG. 6 Typical vertical turbine fire pump system.

preferred. If the source cannot guarantee continuous power, the power must be supplied by two or more stations located and equipped so that an accident or a fire at one will not cause an interruption of the power supplied by the other.

An acceptable alternative for motor-driven pumps is to use commercially available electric power and add a diesel-driven emergency generator set. When alternative sources of electric power are used, all on-site electrical switching equipment must be of the automatic transfer type.

Ordinarily, there is no reason not to use the least expensive motor—an open, dripproof (ODP), squirrel-cage induction motor with across-the-line starting. However, if required, weather-protected I or II (WPI or WPII) or totally enclosed, fan-cooled (TEFC) motors may be used. Depending upon city codes and the current available for starting, wound rotor, star delta, wye delta, primary resistance, or part-winding start motors may be used.

UL listed or FM approved diesel engines are frequently used to drive stationary fire pumps. Equipped with battery packs and automatic controls, they rival electric motors for reliability and eliminate concern over the dependability of the source of electric power. However, they do require additional considerations.

Diesel engines are rated to produce their maximum power at 500 ft (152 m) above sea level and 85°F (29.5°C) ambient air temperature. As these conditions change, the following deductions apply:

1. Reduce power by 3% for each additional 1000 ft (305 m) of elevation.
2. Reduce power by 1% for each 10 °F (5.6 °C) rise above 85°F (29.5°C) ambient air temperature.

If the listed engine with the required power is not available, the engine chosen must produce at least 10 hp (7.457 kW) more than the maximum brake power required by the pump anywhere along its performance curve.

If a right-angle gear drive is used (as with vertical turbine pump), the engine power requirement must be increased to compensate for the gear loss.

The minimum temperature of the room or pump house in which the engine is installed must be as specified by the engine manufacturer, usually 70°F (21.1°C). An automatic engine water preheater is required to maintain the water jacket temperature at a minimum of 120°F (48.9°C). A shell-and-tube heat exchanger is required. Pressurized water from the fire pump passes through a valve into the cooling water loop, which must include the following:

Indicating manual shutoff valve at entrance

Approved flushing-type strainer

Pressure regulator, which might include an integral strainer

Automatic electric solenoid valve

Indicating manual shutoff valve

Pressure gage

Bypass line, also equipped with two manual shutoff valves and flushing-type strainer

Each UL listed or FM approved engine must also be equipped with a governor capable of regulating the engine speed through that part of the pump operating range between maximum load and shutoff with an accuracy of ±10%. The engine must also be equipped with a shutdown device which will limit the engine to approximately 120% of its maximum rated speed; once actuated, this device must be manually reset.

An instrument panel secured to the engine at an appropriate place must include a tachometer, an hour meter, an oil pressure gage, and a water temperature gage.

Occasionally, one encounters an application for a fire pump driven by a steam turbine. Although no approved listed steam turbine is available, its use is acceptable if a dependable source of steam is available.

FIRE PUMP CONTROLLERS

Most modern fire pumps are started automatically by a pressure signal from the pump discharge line. Each fire pump must have its own approved/listed controller. Each jockey pump normally has its own nonlisted controller.

Depending upon the requirements of the application, fire pump controllers for electric motors are available in NEMA 2, 3, or 4 enclosures. Each controller must be arranged to match the starting characteristics of its motor and must include

Manual disconnect switch

Circuit breaker

Starter without heaters or contactors

Mercoid pressure switch

Minimum-run timer to prevent motor cycling

The diesel engine controller is arranged to permit either automatic or manual start. The manual start mode is required to permit periodic run tests. The power source for the controller and

---

1. **Common to horizontal and vertical pumps:**
    a. Casing relief valve—to prevent no-flow condition when system is running at shutoff:
    b. Automatic air release valve—to vent entrapped air in automatic systems:
    c. Umbrella cock—to vent air in manually operated systems:
    d. Hose valve head with hose valves, caps, and chains—to permit pump flow (capacity) tests (most modern systems use flowmeters in lieu of hose valve heads):
    e. Ball drip valve—installed upstream to prevent freeze damage to hose valve head installed outside:
    f. Overflow cone—to show whether relief valve is open:
    g. Commercial discharge tee with (if required) 90° elbow—used when main relief valve is required:
2. **Horizontal pumps only:**
    a. Eccentric suction reducer—required when size of pump suction does not match size of suction pipe:
    b. Concentric discharge increaser—required when size of discharge pipe does not match discharge size of pump:
    c. Main relief valve—required when pump shutoff pressure plus suction pressure exceeds system design pressure and when engine or other variable-speed driver is used:
    d. Splash partition—used for motor-driven units where hose head valves are mounted indoors near pump:
3. **Vertical pumps only:**
    a. Main relief valve—required when engine or other variable-speed driver is used:
    b. Water level testing device—to determine distance to surface of water; required for well pump installations:
4. **All engine-driven pumps:**
    a. Dual set of lead-acid or nickel-cadmium batteries with rack and cables:
    b. Fuel tank with gage and fittings (sized to meet requirements of authority exercising approval):
    c. Muffler to permit engine noise to escape (since engines run only during fire emergencies and during periodic tests, engine noise is not usually a problem):
    d. Flexible connector for engine exhaust

FIG. 7 Available fire pump fittings.

for starting the engine is a dual set of batteries. The controller is arranged to show the following engine conditions:

Low lubricating oil pressure

High water jacket temperature

Failure to start automatically

Shutdown due to overspeed

Battery failure (on the panel, each battery has a light to indicate battery failure)

The engine controller must also provide the means of relaying the following information to remote indicators:

Engine running

Engine switch in off or manual position

Trouble signal (activated by one or any combination of the engine or controller signals described above)

Engine controllers are now available which include built-in battery chargers.

The jockey pump nonlabeled controller is a combination starter with a Mercoid pressure switch, a minimum-run timer, and either a fusible disconnect or a circuit breaker. The jockey pump's function is to keep the system pressurized so that the main fire pump(s) will not start in response to extraneous low-pressure signals.

FIRE PUMP FITTINGS

Various fittings peculiar to the industry are available from fire pump manufacturers. These are shown in Fig. 7, opposite page.

A typical horizontal fire pump system is shown in Fig. 5, and a typical vertical turbine pump fire system in Fig. 6.

SECTION 9.5
STEAM POWER PLANT

IGOR J. KARASSIK

STEAM POWER PLANT CYCLES

Power is produced in a steam power plant by supplying heat energy to the feedwater, changing it into steam under pressure, and then transforming part of this energy into mechanical energy in a heat engine to do useful work. The feedwater therefore acts merely as a conveyor of energy. The basic elements of a steam power plant are the heat engine, the boiler, and a means of getting water in the boiler. Modern power plants use steam turbines as heat engines; except for very small plants, centrifugal boiler-feed pumps are used.

This basic cycle is improved by connecting a condenser to the steam turbine exhaust and by heating the feedwater with steam extracted from an intermediate stage of the main turbine. This results in an improvement of the cycle efficiency, provides deaeration of the feedwater, and eliminates the introduction of cold water into the boiler and the resulting temperature strains on the latter. The combination of the condensing and feedwater heating cycle (Fig. 1) requires a minimum of three pumps: the condensate pump, which transfers the condensate from the condenser hot well into the direct-contact heater; the boiler-feed pump; and a circulating pump, which forces cold water through the condenser tubes to condense the exhaust steam. This cycle is very common and is used in most small steam power plants. A number of auxiliary services not illustrated in Fig. 1 are normally used, such as service water pumps, cooling pumps, ash-sluicing pumps, oil-circulating pumps, and the like.

The desire for improvements in operating economy dictated further refinements in the steam cycle, and these have created new or altered services for power plant centrifugal pumping equipment. Some of these refinements involved a steady increase in operating pressures until 2400-lb/in² (165-bar°) steam turbines became quite common and many plants are operating at supercritical steam pressures of 3500 lb/in² (240 bar).

Other refinements were directed toward a greater utilization of heat through increased feedwater heating, introducing a need for heater drain pumps—equipment with definite problems of its own. Finally, the introduction of forced or controlled circulation as opposed to natural circulation in boilers created a demand for pumping equipment of again an entirely special character.

While direct-contact heaters would have thermodynamic advantages, a separate pump would be required after each such heater. The use of a group of closed heaters permits a single boiler-

°1 bar = 10^5 Pa. For a discussion of bar, see *SI Units—A Commentary* in the front matter.

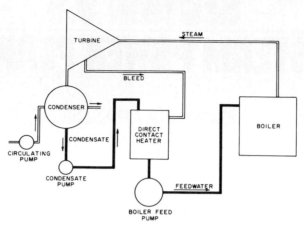

FIG. 1 Simple steam power cycle.

feed pump to discharge through these heaters and into the boiler. The average power plant is based on a compromise system: one direct-contact heater is used for feedwater deaeration while several additional heaters of the closed type are located upstream as well as downstream of the direct-contact heater and of the boiler-feed pump (Fig. 2). Such a cycle is termed an *open cycle*. The major variation is the *closed cycle*, where the deaeration is accomplished in the condenser hot well and all heaters are of the closed type (Fig. 3).

STEAM POWER PLANT PUMPING SERVICES

Pumps are very important components of a steam electric power plant. The major applications which immediately come to mind involve the condensate, boiler-feed, heater drain, and condenser circulating pumps. The all-inclusive category of "miscellaneous pumps" includes such a variety of services that it merits being broken down into its components and included in a representative listing. Table I provides such a listing for conventional (fossil fuel) steam power plants. The list is not necessarily complete but reasonably representative.

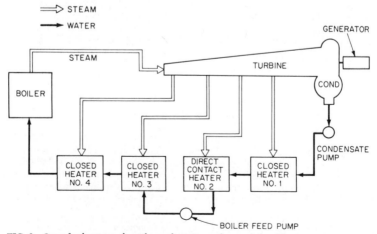

FIG. 2 Open feedwater cycle with one deaerator and several closed heaters.

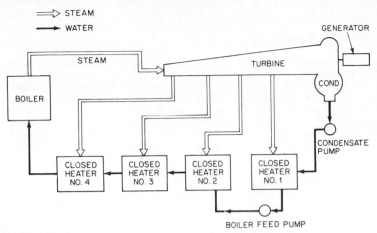

FIG. 3   Closed feedwater cycle.

TABLE I   Pump Services in Conventional Steam Power Plants

Turbogenerator and auxiliaries	Fuel oil system (*cont.*)
Condenser circulating pumps	Distillate oil unloading pumps
Screen wash-water pumps	Fuel oil additive unloading pumps
Cooling tower make-up pumps	Fuel oil additive transfer pumps
Steam generator equipment	Fuel oil additive metering pumps
Condensate pumps	Fuel oil hose drain pumps
Condensate booster pumps	Lubricating oil system
Boiler-feed pumps	Lubricating transfer pumps
Boiler-feed booster pumps	Starting oil pumps
Deaerator make-up pumps	Main oil pumps
Heater drain pumps (low and high pressure)	Emergency oil pumps
Chemical feed system	Centrifuge feed pumps
Amine pumps	Fire protection system
Hydrazine pumps	Fire pumps
Phosphate pumps	Jockey pumps
Caustic feed pumps	Foam proportioning pumps
Acid feed pumps	Heating, ventilating, and air-conditioning
Ammonia pumps	system
Regeneration waste pumps	Hot-water circulating pumps
Demineralizer pumps	Chilled-water pumps
Neutralizing metering pumps	Service water system
Neutralizing tank sump pumps	Service water pumps
Acid bulk-transfer pumps	Air preheater wash pumps
Caustic bulk-transfer pumps	Cooling water booster pumps
Inlet and effluent demineralizer waste tank	Primary air heating coil condensate return
pumps	pumps
Fuel oil system	Heating drain tank return pumps
Fuel oil transfer pumps	Sump pumps
Secondary fuel oil pumps	Closed cooling water system pumps
Secondary fuel oil heater drip pumps	Miscellaneous
Ignitor oil pumps	Ash sluice pumps
Auxiliary boiler fuel pumps	Slurry pumps
Warm-up oil pumps	Acid cleaning pumps
High-temperature oil-circulating pumps	Hydrostatic pressure test pumps
Low-temperature oil-circulating pumps	

BOILER-FEED PUMPS

Under the term *conditions of service* are included not only the pump capacity, discharge pressure, suction conditions, and feedwater temperature but also the chemical analysis of the feedwater, the pH at pumping temperature, and other pertinent data which may reflect upon the hydraulic and mechanical design of the boiler-feed pumps. Preferably, a complete layout of the feedwater system and of the heat balance diagram should be supplied to the boiler-feed pump manufacturer. The study of this layout will often permit the manufacturer to suggest an alternate arrangement of the equipment which would result in a more economical operation, in a lower installation cost, or even in longer equipment life so as to reduce the eventual maintenance expense.

Boiler-Feed Pump Capacity

The total boiler-feed pump capacity is established by adding to the maximum boiler flow a margin to cover boiler swings and the eventual reduction in effective capacity from wear. This margin varies from as much as 20% in small plants to as little as 5% in the larger central stations. The total required capacity must be either handled by a single pump or subdivided between several duplicate pumps operating in parallel. Industrial power plants generally use several pumps. Central stations tend to use single full-capacity pumps to serve turbogenerators up to a rating of 100 or even 200 MW and two pumps in parallel for larger installations. There are obviously exceptions to this practice: some engineers prefer the use of multiple pumps even for small installations, while steam-turbine-driven boiler-feed pumps designed for full capacity are being applied for units as large as 700 MW. A spare boiler-feed pump is generally included in industrial plants. There is a trend, however, in central stations to eliminate spare pumps when two half-capacity pumps are used and, in a few cases, even if a single full-capacity boiler-feed pump is installed.

Suction Conditions

The net positive suction head (NPSH) represents the net suction head at the pump suction, referred to the pump centerline, *over and above* the vapor pressure of the feedwater. If the pump takes its suction from a deaerating heater, as in Fig. 2, the feedwater in the storage space is under a pressure equivalent to the vapor pressure corresponding to its temperature. Therefore the NPSH is equal to the static submergence between the water level in the storage space and the pump centerline less the frictional losses in the intervening piping. Theoretically, the required NPSH is independent of operating temperature. Practically, this temperature must be taken into account when establishing the recommended submergence from the deaerator to the boiler-feed pump. A margin of safety must be added to the theoretical required NPSH to protect the boiler-feed pumps against the transient conditions which follow a sudden reduction in load for the main turbogenerator.

Whereas the previous discussion applies primarily to the majority of installations, where the boiler-feed pump takes its suction from a deaerating heater, it holds as well in the closed feed cycle (Fig. 3). The discharge pressure of the condensate pump must be carefully established so that the suction pressure of the boiler-feed pump cannot fall below the sum of the vapor pressure at pumping temperature and the required NPSH.

Transient Conditions Following Load Reduction

Following a sudden load reduction, the turbine governor reduces the steam flow in order to maintain the proper relation between turbine and generator power and to hold the unit at synchronous speed. The consequence of this reduction is a proportionate pressure reduction at all successive turbine stages, including the bleed stage which supplies steam to the direct-contact heater. The check valve in the extraction line closes and isolates the heater from the turbine. As hot feedwater continues to be withdrawn from the heater and cold condensate to be admitted to the heater, the pressure in the direct-contact heater starts to drop rapidly. The check valve reopens when the heater pressure has been reduced to the prevailing extraction pressure and stable conditions are re-established.

It should be noted that, even though the feedwater system in a drum boiler may be provided with a three-element feedwater regulator, the feedwater flow will not instantaneously follow the steam flow as soon as the steam demand is reduced by a reduction in unit load. Because of the time lag between the reduction in steam demand and that of the fuel-burning rate and because of the heat retention in the steam generator, there is a momentary rise in the boiler pressure, with the resultant collapse of some of the steam and water bubbles in the boiler drum. This lowers the

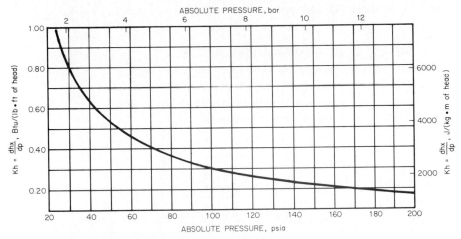

FIG. 4 Enthalpy change with change of vapor pressure, for water.

apparent boiler drum level, causing the level control to override to some degree the impulse from the change in steam flow. Therefore, there will generally be a definite lack of correlation between feedwater and steam flow following a sudden drop in load. The exact degree of the difference between these two flows will depend upon the particular type and setting of the feedwater controls. In some extreme cases, the feedwater flow after a sudden drop in load can actually exceed the feedwater flow at maximum design conditions. Thus, it is a safer practice to assume that the feedwater flow will not be reduced and to assume that the NPSH required will in turn correspond to at least its value under flow conditions preceding the drop in load.

In the interval, however, the pressure at the boiler-feed pump suction is reduced correspondingly. Unfortunately, until the suction piping has been completely voided of the feedwater it contained prior to the load reduction, its temperature and vapor pressure will not be reduced. As a consequence, the available NPSH will diminish and may become insufficient to provide adequate pump operation. In such a case, the pump will flash and serious damage may be incurred.

The factor which establishes the adequacy of an installation from the point of view of suction conditions after a load drop is the ratio between the direct-contact heater storage capacity and the suction piping volume. Based on a number of simplifying assumptions, a formula has been developed for the minimum value of this ratio:

$$\text{Minimum } \frac{Q_h}{Q_s} = \frac{h_{x0} - h_{c2}}{K_h H_x} \tag{1}$$

where Q_h = volume of feedwater in heater storage, gal (m³)
Q_s = volume of feedwater in suction piping, gal (m³)
h_{x0} = enthalpy of feedwater under initial conditions, Btu/lb (J/kg)
h_{c2} = enthalpy of condensate to heater under final conditions, Btu/lb (J/kg)
K_h = change in enthalpy with pressure at steam conditions prior to load reduction, Btu/lb·ft absolute pressure (J/kg·m) (Fig. 4)
H_x = available excess NPSH = NPSH available − NPSH required, ft (m)

This relationship is somewhat conservative and does not take into account the residence time of the condensate in the piping and the closed heaters between the condenser hot well and the direct-contact heater. A slightly less conservative formula which takes some account of this residence time is

$$\text{Minimum } \frac{Q_h}{Q_s} = \frac{h_{x0} - [(h_{c0} + h_{c2})/2]}{K_h H_x} \tag{2}$$

where h_{c0} = enthalpy of condensate to heater under initial conditions, Btu/lb (J/kg)

For example, let

Initial heater pressure = 153 lb/in² (10.5 bar)

Initial feedwater temperature = 360°F (182°C)

Initial feedwater enthalpy = 331.4 Btu/lb (770.8 kJ/kg)

Final condensate enthalpy = 82.95 Btu/lb (192.9 kJ/kg)

K_h (from Fig. 4) = 0.22 Btu/lb/ft (1679 J/kg/m)

H_x (available excess NPSH) = 15 ft (4.57 m)

Then

in USCS units $\qquad$ Minimum $\dfrac{Q_h}{Q_s} = \dfrac{331.4 - 82.95}{0.22 \times 15} = 75.3$

in SI units $\qquad$ Minimum $\dfrac{Q_h}{Q_s} = \dfrac{770{,}800 - 192{,}900}{1679 \times 4.57} = 75.3$

This means that for safe operation after a sudden load drop in this particular case, the heater storage volume must be at least 75.3 times the volume of the suction piping.

More complex but more rigorous calculations of the minimum ratio of heater storage volume to suction piping volume are provided in Ref. 1.

Even where analysis indicates that the boiler-feed pumps are assured of their required NPSH during a reduction in turbine load, there is no guarantee that their operation will not be interrupted by flashing at some point in the suction piping. The criterion in determining the probability of flashing in the suction piping is to consider that the water which left the heater outlet at a saturated condition must pick up static pressure, by means of the vertical drop, at a rate at least equal to the pressure decay rate of the heater, or it will flash.

The most adverse conditions are those introduced by locating a horizontal run of piping too close to the heater outlet. A typical case is illustrated in Fig. 5. (Since this example is used merely to illustrate the unfavorable effect of such a piping layout, the unit system used is immaterial and the example has been expressed in USCS units.) In the comparison of the two installations, we will stipulate that the total length and the sizes of the piping are the same for both arrangements and that the volumes of the suction piping between the heater outlet and points C and E of the two arrangements are the same. To simplify the comparison, the vertical distances between A and B, B and D, and A and E have been expressed in pounds per square inch instead of feet.

If a time interval x is selected such that water having left the heater outlet at the start of the transient conditions will have reached points C and E, respectively, at the end of the time interval, it becomes apparent that in the case illustrated on the left side of Fig. 5, the pressure gain at point C is only 3 lb/in² by virtue of the vertical drop. Thus, x seconds after a pressure drop in the direct-contact heater from 52 to 48 lb/in² gage, the pressure at point C will be 51 lb/in², which is below the vapor pressure at the new temperature (296°F), and so flashing will occur. On the other hand, in the case of a straight vertical drop (right side of Fig. 5), after the same time interval x, the pressure will be 60 lb/in², which exceeds the vapor pressure, and so no flashing will occur. Formulas 1 and 2 can be used to determine the adequacy of the piping layout by selecting the critical point in the piping (in this case, point C) and substituting in the formulas so that Q_s = volume in suction piping to point C and H_x = static head to point C less frictional losses to point C.

In the event that circumstances do not permit the provision of sufficient NPSH margin to provide adequate protection to the boiler-feed pumps during a sudden turbine load reduction, two alternate means are available to compensate for these circumstances:

1. A small amount of steam from the boiler can be admitted to the direct-contact heater through a pressure-reducing valve, to reduce the rate of pressure decay in the heater.

2. A small amount of cold condensate from the discharge of the condensate pumps can be made to bypass all or some of the closed heaters and be injected at the boiler-feed pump suction to subcool the feedwater, thus providing additional NPSH margin during load reduction.

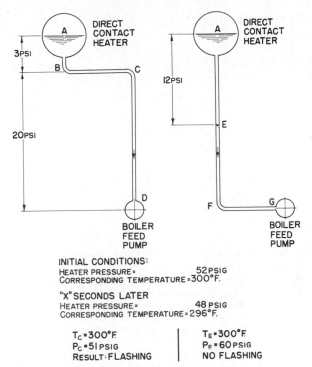

FIG. 5 Comparison of suction piping arrangements.

Figure 6 illustrates the effect of subcooling (or temperature depression) on the available NPSH at various initial feedwater temperatures. Figure 7 shows, for varying ratios of injection flows, the temperature depression resulting from cold water injection plotted against the difference in temperature between the feedwater and the injection stream. For instance, if it were desired to provide 20 ft (6.1 m) additional NPSH to a boiler-feed pump which handles 325°F (163°C) water,

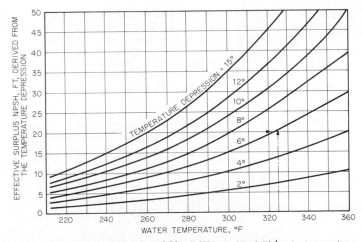

FIG. 6 Effect of subcooling on available NPSH at various initial water temperatures. [°C = (°F − 32)%; 1 ft = 0.3048 m].

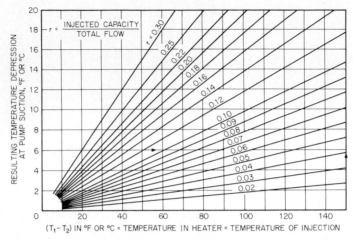

FIG. 7 Required amount of cold water injection for a given temperature depression.

the required temperature depression is 6°F (3.3°C). If the injection water temperature is 190°F (87.8°C), the difference between feedwater and injection water temperature is 135°F (75.2°C). From Fig. 7 we can see that the injection flow must be 4.5% of the total feedwater flow.

An analysis of the relative merits of the two methods of protecting boiler-feed pumps against the unfavorable effects of transient conditions is presented in Ref. 2. Either corrective action can be initiated automatically. This involves constant monitoring of the suction pressure and of the vapor pressure of the feedwater at the pump suction. The difference between the two is then constantly compared with a pre-established minimum NPSH. Any transient condition that causes the available NPSH to fall below this desired minimum initiates corrective action, be it admission of cold condensate at the pump suction or admission of auxiliary steam to the direct-contact heater.

BOOSTER PUMPS

The increasing sizes of modern boiler-feed pumps coupled with the practice of operating these pumps at speeds considerably higher than 3600 rpm have led to NPSH requirements as high as 150 to 250 ft (46 to 76 m). In most cases, it is not practical to install the direct-contact heaters from which the feed pumps take their suction high enough to meet such requirements. In such cases, it has become the practice to use boiler-feed booster pumps operating at lower speeds, such as 1750 rpm, to provide a greater available NPSH to the boiler-feed pumps than can be made available from strictly static elevation differences. Such booster pumps are generally of the single-stage, double-suction design.

Discharge Pressure and Total Head   The discharge pressure is the sum of the maximum boiler drum pressure and the frictional and control losses between the boiler-feed pump and the boiler drum inlet. The required discharge pressure will generally vary from 115 to 125% of the boiler drum pressure. The net pressure to be generated by the boiler-feed pump is the difference between the required discharge pressure and the available suction pressure. This must be converted to a total head, using the formula

in USCS units $$\text{Total head, ft} = \frac{\text{net pressure, lb/in}^2 \times 2.31}{\text{sp. gr.}}$$

in SI units $$\text{Total head, m} = \frac{\text{net pressure, bar} \times 10.2}{\text{sp. gr.}}$$

Slope of the Head-Capacity Curve   In the range of specific speeds normally encountered in multistage centrifugal boiler-feed pumps, the rise of head from the point of best efficiency to shutoff will vary from 10 to 25%. Furthermore, the shape of the head-capacity curve for these pumps is such that the drop in head is very slow at low capacities and accelerates as the capacity is increased.

If the pump is operated at constant speed, the difference in pressure between the pump head-capacity curve and the system-head curve must be throttled by the feedwater regulator. Thus the higher the rise of head toward shutoff, the more pressure must be throttled off and, theoretically, wasted. Also, the higher the rise, the greater the pressure to which the discharge piping and the closed heaters will be subjected. However, it is not advisable to select too low a rise to shutoff because too flat a curve is not conducive to stable control; a small change in pressure corresponds to a relatively great change in capacity, and a design that gives a very low rise to shutoff may result in an unstable head-capacity curve, difficult to use for parallel operation. When several boiler-feed pumps are to be operated in parallel, they must have stable curves and equal shutoff heads. Otherwise, the total flow will be divided unevenly and one of the pumps may actually be backed off the line after a change in required capacity occurs at light flows.

Driver Power   A boiler-feed pump will generally not operate at any capacity beyond the design condition. In other words, a boiler-feed pump has a very definite maximum capacity because it operates on a system-head curve made up of the boiler drum pressure plus the frictional losses in the discharge. If, as it should be, the design capacity of the pump is chosen as the maximum capacity that can be expected under emergency conditions, there can be no further increase under any operating conditions since the pressure requirement corresponding to an increased capacity would exceed the design pressure of the pump. Even when the design pressure includes a safety margin, the boiler demand does not exceed the design capacity, and the feedwater regulator will impart additional artificial frictional losses to increase the required pressure up to the pressure available at the pump.

When two pumps are operated in parallel, feeding a single boiler, the situation is somewhat different. If one of the pumps is taken off the line at part load, the remaining pump could easily operate at capacities in excess of its design since its head-capacity curve would intersect the system-head curve at a head lower than the design head (Fig. 8). In such a case, it is necessary to determine the pump capacity at the intersection point; the power corresponding to this capacity will be the maximum expected. It is not always necessary to select a driver that will not be overloaded at any point on the boiler-feed pump operating curve. But while electric motors used on boiler-feed service generally have an overload capacity of 15%, it is usually the practice to reserve this overload capacity as a safety margin and to select a motor that will not be overloaded at the design capacity. Exceptions occur in the case of very large motors. For instance, if the pump brake horsepower is 3100, it is logical to apply a 3000-hp motor which will be overloaded by about 3% rather than a considerably more expensive 3500-hp motor. Because steam turbines are not built in definite standard sizes but can be designed for any intermediate rating, they are generally selected with about 5% excess power over the maximum expected pump power.

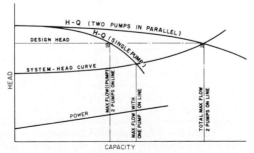

FIG. 8   Method of determining maximum pump power for two boiler-feed pumps operating in parallel.

General Structural Features   Boiler-feed pumps designed for discharge pressures of less than 1500 lb/in² (103 bar) are generally of the axially split casing type (Subsec. 2.2.1, Figs. 18, 19, and 20). Double-casing, radially split pumps (Subsec. 2.2.1, Figs. 21, 22 and 24) are used for higher discharge pressures. The selection of materials for boiler-feed pump casings and internal parts is discussed in Sec. 5.1.

High-Speed, High-Pressure Boiler-Feed Pumps   As steam pressures rose from 1250 to 1800, then to 2400 and even to 3500 lb/in² (from 85 to 240 bar), the total head that had to be developed by the pump rose from somewhere around 4000 ft (1220 m) to as high as 7000 and 12,000 ft (2140 and 3660 m). The only means available of achieving these higher heads at 3600 rpm was to increase the number of stages. The pumps had to have longer and longer shafts. This threatened to interfere with the long uninterrupted life between pump overhauls to which steam power plant operators were beginning to become accustomed. The logical solution was to reduce the shaft span by reducing the number of stages. Experience had indicated by 1953 that stage pressures could be increased from the 250- to 350-lb/in² (17- to 24-bar) range commonly used at 3600 rpm to as high as 800 or 1000 lb/in² (55 or 70 bar). In turn, these higher heads per stage could better be achieved by increasing the speed of rotation than by increasing impeller diameters. As a result, boiler-feed pumps in large central stations today generally operate at speeds from 5000 to 9000 rpm.

Boiler-Feed Pump Drives   The majority of boiler-feed pumps in small and medium-size steam plants are driven by electric motors. It used to be the practice to install steam-turbine-driven standby pumps as a protection against the interruption of electric power, but this practice has disappeared in central steam stations and is encountered rarely even in industrial plants.

Recent years have seen a trend away from electric motor drive to steam turbine drive in large central steam stations for units in excess of 200 MW, because

1. The use of an independent steam turbine increases plant capability by eliminating the auxiliary power required for boiler feeding.
2. Proper utilization of the exhaust steam in the feedwater heaters can improve cycle efficiency.
3. In many cases, the elimination of the boiler-feed pump motors may permit a reduction in the station auxiliary voltage.
4. Driver speed can be matched ideally to the pump optimum speed.
5. A steam turbine provides variable-speed operation without an additional component, such as a hydraulic coupling.

Operation of Boiler-Feed Pumps at Reduced Flows   Operation of centrifugal pumps at shutoff or even at certain reduced flows can lead to very undesirable results. This subject is covered in detail in Subsecs. 2.3.1 to 2.3.4, Sec. 8.1, and Chap. 12, where methods for calculating minimum permissible flows and means for providing the necessary protection against operation below these flows are discussed.

CONDENSATE PUMPS

Condensate pumps take their suction from the condenser hot well and discharge either to the deaerating heater in open feedwater systems (Fig. 2) or to the suction of the boiler-feed pumps in closed systems (Fig. 3). These pumps, therefore, operate with a very low pressure at their suction—from 1 to 3 inHg (25 to 75 mmHg) absolute. The available NPSH is obtained by the submergence between the water level in the condenser hot well and the centerline of the condensate pump first-stage impeller. Because it is desirable to locate the condenser hot well as low as possible and avoid the use of a condensate pump pit, the available NPSH is generally extremely low, on the order of 2 to 4 ft (0.6 to 1.2 m). The only exception to this occurs when vertical-can condensate pumps are used, since these can be installed below ground and higher values of submergence can be obtained. Frictional losses on the suction side must be kept to an absolute minimum. The piping connection from the hot well to the pump should therefore be as direct as possible and of ample size and should have a minimum of fittings.

Because of the low available NPSH, condensate pumps operate at relatively low speeds, ranging from 1750 rpm in the low range of capacities to 880 rpm or even less for larger flows.

It is customary to provide a liberal excess capacity margin above the full-load steam condensing flow to take care of the heater drains that may be dumped into the condenser hot well if the heater drain pumps are taken out of service for any reason.

Types of Condensate Pumps Both horizontal and vertical condensate pumps are used. Depending on the total head required, horizontal pumps may be either single-stage or multistage.

Figure 9 shows a single-suction, single-stage pump with an axially split casing used for heads up to about 100 ft (30 m). It is designed to have discharge pressure on the stuffing box. The suction opening in the lower half of the casing keeps the suction line at floor level. An oversize vent at the highest point of the suction chamber permits the escape of all entrained vapors, which will be vented back to the condenser and removed by the air-removal apparatus.

Multistage pumps are used for higher heads. A two-stage pump is shown in Fig. 10, with the impellers facing in opposite directions for axial balance. By turning the impeller suctions toward the center, both boxes are kept under positive pressure to prevent leakage of air into the pump. For higher heads and larger capacities, a three-stage pump, as in Fig. 11, may be used. The first-stage impeller is of the double-suction type and is located centrally in the pump. The remaining impellers are of the single-suction type and are also arranged so that both stuffing boxes are under pressure. Two liberal vents connecting with the suction volute on each side of the first-stage double-suction impeller permit the escape of vapor.

Recently a trend has developed toward the use of vertical-can condensate pumps (Subsec. 2.2.1, Fig. 111). The chief advantage of these pumps is that ample submergence can be provided without the necessity of building a dry pit. The first stage of this pump is located at the bottom of the pumping element, and the available NPSH is the distance between the water level in the hot well and the centerline of the first-stage impeller.

Condensate pumps are located very close to the condenser hot well, and the suction piping is generally so short that the frictional losses in this piping are not very significant. However, strainers should normally be installed in this piping, and a great deal of attention must be paid to the frictional losses through them and to their frequent cleaning. Cases have been reported on occa-

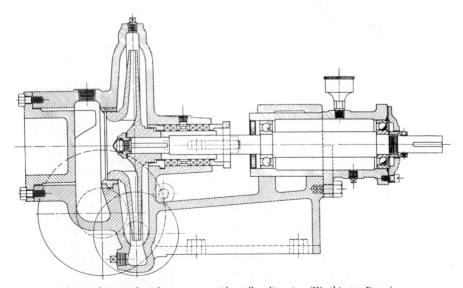

FIG. 9 Single-stage horizontal condensate pump with axially split casing. (Worthington Pump)

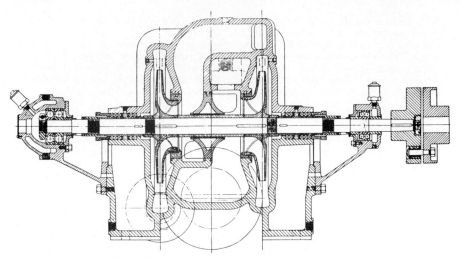

FIG. 10 Two-stage horizontal condensate pump with axially split casing. (Worthington Pump)

sion where the pressure drop across these strainers was sufficient to cause flashing at the suction nozzle of the condensate pumps.

The increasing use of full-flow demineralizers in condensate systems and, in general, the increasing discharge pressures required from the condensate pumps have resulted in the frequent need to split condensate pumping into two parts. The condensate pumps proper thus develop only a small portion of the total head required. The balance of the required head is provided by separate condensate booster pumps, which have generally been of the conventional horizontal, axially split casing type.

To prevent air leakage at the stuffing boxes of condensate pumps, these are always provided with seal cages. The water used for gland sealing must be taken from the condensate pump discharge manifold beyond all the check valves.

Normally, condensate pumps have been standard-fitted, with cast iron casings and bronze internal parts, but the emergence of once-through boilers has created the need to eliminate all copper alloys in the condensate system to avoid deposition of copper on the boiler tubes, and stainless steel–fitted pumps are becoming the rule for this service.

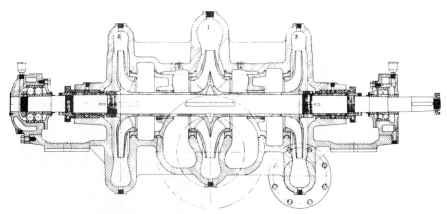

FIG. 11 Three-stage horizontal condensate pump with axially split casing. (Worthington Pump)

Condensate Pump Regulation When a condensate pump operates in a closed cycle ahead of the boiler-feed pump, the two pumps can be considered as a combined unit insofar as their head-capacity curve is concerned. Variation in flow is accomplished either by throttling the boiler-feed pump discharge or by varying the speed of the boiler-feed pump.

In an open feedwater system, several means can be used to vary the condensate pump capacity with the load:

1. The condensate pump head-capacity curve can be changed by varying the pump speed.
2. The condensate pump head-capacity curve can be altered by allowing the pump to operate in the "break" (Figs. 12 and 13).
3. The system-head curve can be artificially changed by throttling the pump discharge by means of a float control.
4. The pump can operate at the intersection of its head-capacity curve and the normal system-head curve. The net discharge is controlled by bypassing all excess condensate back to the condenser hot well.
5. Methods 3 and 4 can be combined so that the discharge is throttled back to a predetermined minimum, but if the load, and consequently the flow of condensate to the hot well, are reduced below this minimum, the excess condensate handled by the pump is bypassed back to the hot well.

The impulse for the controls used in methods 1, 3, and 4 is taken from the deaerator level.

Operating in the break, or "submergence control" as it has often been called, has been applied very successfully in a great many installations. Condensate pumps designed for submergence control require specialized hydraulic design, correct selection of operating speeds, and limitation of stage pressures. The pump is operating in the break (i.e., cavitates) at all capacities. However, this cavitation is not severely destructive because the energy level of the fluid at the point where the vapor bubbles collapse is insufficient to create a shock wave of a high enough intensity to inflict physical damage on the pump parts. If, however, higher values of NPSH were required—as for instance with vertical-can condensate pumps because of their generally higher operating speed—operation in the break would result in a rapid deterioration of the impellers. It is for this reason that submergence control is not applicable to can-type condensate pumps.

The main advantages of submergence control are its simplicity and the fact that the power required for any operating condition is less than with any other system. Disadvantages occur when the pump is operating at very light loads, however, because the system head may require as little

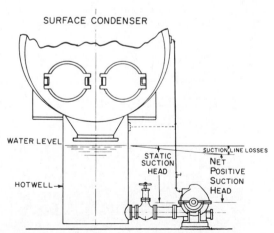

FIG. 12 Typical hookup for submergence-controlled condensate pump.

as one-half of the total head produced in the normal head-capacity curve. In this case, the first stage of a two-stage pump produces no head whatsoever and, if the axial balance was achieved by opposing the two impellers, a definite thrust is imposed on the thrust bearing, which must be selected with sufficient capacity to withstand this condition. In addition, no control is available to provide the minimum flow that may be required through the auxiliaries, such as the ejector condenser.

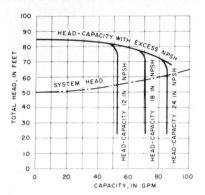

FIG. 13   Characteristic of a condensate pump operating on a submergence-controlled system.

The condensate pump discharge can be throttled by a float control arranged to position a valve that increases the system-head curve as the level in the hot well is drawn down. This eliminates the cavitation in the condensate pump, but at the cost of a slight power increase. Furthermore, the float necessarily operates over a narrow range, and the mechanism tends to be somewhat sluggish in following rapid load changes, often resulting in capacity and pressure surges.

When condensate delivery is controlled through bypassing, the hot well float controls a valve in a bypass line connecting the pump discharge back to the hot well. At maximum condensate flow, the float is at its upper limit with the bypass closed and all the condensate is delivered to the system. As the condensate flow to the hot well decreases, the hot well level falls, carrying the float down and opening the bypass. Sluggish float action can create the same problems of system instability in bypass control as in throttling control, however, and the power consumption is excessive because the pump always operates at full capacity.

A combination of throttling and bypassing control eliminates the shortcoming of excessive power consumption. The minimum flow at which bypassing begins is selected to provide sufficient flow through the ejector condenser.

A modification of the bypassing control for minimum flow is illustrated in Fig. 14, which shows a thermostatic control for condensate recirculation. With practically constant steam flow through the ejector, the rise in condensate temperature between the inlet and outlet of the ejector condenser is a close indication of condensate flow rate through the ejector condenser tubes. Therefore an automatic device to regulate the condensate flow rate can be controlled by this temperature differential. A small pipe is connected from the condensate outlet on the ejector condenser back into the main condenser shell. An automatic valve is installed in this line and is actuated and controlled by the temperature rise of the condensate. Whenever the temperature rises to a certain

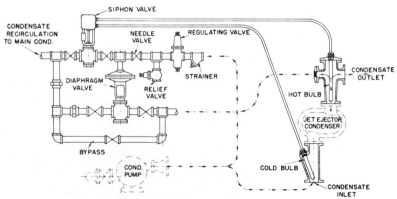

FIG. 14   Thermostatic control for condensate recirculation.

predetermined figure, indicating a low flow of condensate, the automatic valve begins to open, allowing some of the condensate to return to the condenser and then to the condensate pump, which supplies it to the ejector at the increased rate. When the temperature rise through the ejector condenser is less than the limiting amount, indicating that ample condensate is flowing through the ejector condenser, the automatic valve remains closed.

Just as in the case of boiler-feed pumps—or, as a matter of fact, of any pumps—condensate pumps should not be operated at shutoff or even at certain reduced flows. This subject is covered in detail in Subsecs. 2.3.1 and 2.3.4.

HEATER DRAIN PUMPS

Service Conditions   Condensate drains from closed heaters can be flashed to the steam space of a lower-pressure heater or pumped into the feedwater cycle at some higher-pressure point. Piping each heater drain to the heater having the next lower pressure is the simpler mechanical arrangement and requires no power-driven equipment. This "cascading" is accomplished by an appropriate trap in each heater drain. A series of heaters can thus be drained by cascading from heater to heater in the order of descending pressure, the lowest being drained directly to the condenser.

This arrangement, however, introduces a loss of heat since the heat content of the drains from the lowest-pressure heater is dissipated in the condenser by transfer to the circulating water. It is generally the practice, therefore, to cascade only down to the lowest-pressure heater and pump the drains from that heater back into the feedwater cycle, as shown in Fig. 15. Because the pressure in that heater hot well is low (frequently below atmospheric even at full load), heater drain pumps on that service are commonly described as on "low-pressure heater drain service."

In an open cycle, drains from heaters located beyond the deaerator are cascaded to the deaerator. Although the deaerator is generally located above the closed heaters, the difference in pressure is sufficient to overcome both the static and the frictional losses. This difference in pressure decreases with a reduction in load, however, and at some partial main turbine load it becomes insufficient to evacuate the heater drains and they have to be switched to a lower-pressure heater or even to the condenser, with a subsequent loss of heat. To avoid these complications, a "high-pressure heater drain pump" is generally used to transfer these drains to the deaerator. Actually this pump has a "reverse" system head to work against; at full load, the required total head may be negative, whereas at light loads the required head is at its maximum.

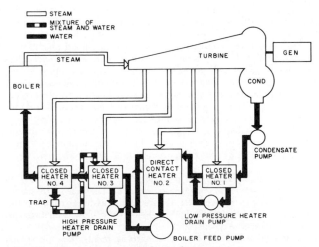

FIG. 15   Typical arrangement for heater drain pumps.

High-pressure drain pumps are subject to more severe conditions than boiler-feed pumps encounter:

1. Their suction pressure and temperature are higher.
2. The available NPSH is generally extremely limited.
3. They are subject to all the transient conditions to which the feed pump is exposed during sudden load fluctuations, and these transients are more severe than those at the feed pump suction.

Types of Heater Drain Pumps　In the past, heater drain pumps were always horizontal, either single-stage or multistage, depending upon total head requirements. In the single-stage type, end-suction pumps of the heavier "process pump" type (Subsec. 2.2.1, Fig. 101) were preferred for both low- and high-pressure service. Recently, however, just as it has invaded the field of condensate service, the vertical-can pump (Subsec. 2.2.1, Fig. 111) has been frequently applied on heater drain service. As previously described, the advantages of the vertical-can pump are lower first cost and a built-in additional NPSH because the first-stage impeller is lowered below floor level in the can. Against these advantages, one must weigh certain shortcomings. A horizontal heater drain pump is more easily inspected than a can pump. The external grease- or oil-lubricated bearings of the horizontal pump are less vulnerable to the severe operating conditions during swinging loads than the water-lubricated internal bearings of the can pump. If vertical-can heater drain pumps have a bearing in the suction bell, consideration must be given to the fact that the water in the immediate location of that bearing is at near saturated pressure and temperature conditions (high temperature and low pressure). To keep the water in the bearing from flashing, additional water should be piped back to the bearing from a higher stage.

Heater drain pumps should be adequately vented to the steam space of the heater. Because heater drain pumps and especially those on low-pressure service may operate with suction pressures below atmospheric, it is necessary to provide a liquid supply to the seal cages in the stuffing boxes. Low-pressure heater drain pumps can use cast iron casings and bronze fittings if no evidence of corrosion erosion has been uncovered. On high-pressure services, stainless steel fittings are generally mandatory and 5% chrome stainless steel casings should preferably be used.

CONDENSER CIRCULATING PUMPS

Types of Pumps　Condenser circulating pumps may be of either horizontal or vertical construction. For many years the low-speed, horizontal, double-suction volute centifugal pump (Subsec. 2.2.1, Fig. 3) was the preferred type. This pump has a simple but rugged design which allows ready access to the interior for examination and rapid dismantling if repairs are required.

Although many installations of horizontal circulating pumps are still encountered, many larger central stations have switched over to vertical pumps, which fall into two categories: dry pit, which operate surrounded by air, and wet pit, which are either fully or partially submerged in the water pumped. The choice between these two types is somewhat controversial; strong personal preferences exist in favor of one or the other construction. A number of factors must be considered when choosing between dry- and wet-pit pumps for condenser circulating service. Some of the factors involved lend themselves to a straight economic evaluation, their advantages and disadvantages being readily expressed in dollars. Other factors, no less important, are intangible, and only experience and sound judgement can give them their proper and deserved weight.

Mechanical Considerations　The most popular type of pump during the past 35 years for condenser circulating service in dry-pit installations has been the single-suction, medium-specific-speed, mixed-flow pump (Subsec. 2.2.1, Figure 109). This design combines the high efficiency and low maintenance of the horizontal double-suction radial-flow centrifugal pump with lower cost and slightly higher rotative speeds.

Because of their suction and discharge nozzle arrangements, these pumps are ideally suited for vertical mounting in a dry pit, preferably at the lowest water level, so that they are self-priming on starting. They are directly connected to solid-shaft induction or synchronous motors, either close-coupled or with intermediate shafting between the pump and the motor, which is then mounted well above the pump pit floor.

Like the horizontal double-suction pump, the vertical dry-pit mixed-flow pump is a compact and sturdy piece of equipment. Its rotor is supported by external oil-lubricated bearings of optimum design. This construction requires the least attention, for the oil level can be easily inspected by means of an oil sight glass mounted at the side of the bearing or oil reservoir.

Because the rotor is readily removed through the top of the casing, facilitating maintenance and replacement, the pump does not have to be removed from its mounting and the suction and discharge connections do not have to be broken to make periodic inspections or repairs.

In recent years power plant designers have shown a preference for the wet-pit column-type condenser circulating pump, although most recently this trend has appreciably slowed down, and several major utilities are reverting to the dry-pit type.

Unless qualified, the term *wet pit* normally implies a diffuser-type pump. There have been instances when a conventional volute-type pump has been submerged into a pit and operated as a wet-pit pump, but this is the exception to the rule. The volute pump is essentially a dry-pit pump.

The wet-pit pump (Subsec. 2.2.1, Figs. 112, 113, and 114) employs a long column pipe that supports the submerged pumping element. It is available with open main shaft bearings lubricated by the water handled, or preferably with enclosed shafting and bearings, lubricated by clean, fresh, filtered water from an external source. Even in the latter case there is some danger of contamination of the lubricating water from seepage into the shaft enclosure tube during shutdowns. Even with the best design and the best care, a water-lubricated bearing is not the equal of an oil-lubricated bearing. Consequently, higher maintenance costs may be expected with a wet-pit pump.

Larger power plants are generally located near centers of population and as a result often have to use badly contaminated water—either fresh or salt—as a condenser cooling medium. With such water, a fabricated steel column pipe and elbow would give short life, so cast iron, bronze, or even a more corrosion-resistant cast metal must be used. This results in a very heavy pump when large capacities are involved. Pulling up the column in a long pump requires special and expensive facilities, and, in addition, the discharge flange must be disconnected when withdrawing the pump and column from the pit. To avoid the necessity of lifting the entire pump when the internal parts require maintenance, some units (Subsec. 2.2.1, Fig. 114) are built so that the impeller, diffuser, and shaft assembly can be removed from the top without disturbing the column pipe assembly. (The driving motor must still be removed.) These designs are commonly designated "pull-out."

The selection of materials for either dry- or wet-pit pumps can vary considerably, depending on the character of the circulating water. The dry-pit pump requires the least amount of changes in material since the impeller and rings are the parts that would be primarily affected. The wet-pit pump, however, requires a careful selection of all submerged parts, particularly when seawater is to be pumped. Special alloys are usually required for the impellers and rings. Because of the danger of electrolytic attack, material changes may also often be required for such parts as the shaft, diffuser, column pipe, and shaft enclosure tube.

Performance Characteristics Condenser circulating pumps are normally required to work against low or moderate heads. Extreme care should be exercised in calculating the system frictional losses, which include losses from friction in the condenser. If more total head is specified than is rquired, the resulting driver size may be unnecessarily increased. For instance, an excess of 1 or 2 ft (0.3 or 0.6 m) in an installation requiring only 20 ft (6 m) of head represents an increase of 5 to 10% in excess power costs.

The range of suction lift for dry-pit pumps must be determined very accurately and checked with the manufacturer to ensure that cavitation will be avoided in the installation. Priming facilities must be provided, or the pump must be installed in a dry pit at such an elevation that the water in the suction channel leading to the pump will be maintained at the level recommended by the manufacturer. This presents no problem in a wet-pit installation because the pump column can be made long enough to provide adequate submergence, even with minimum water levels in the suction well or pit. The dry-pit pump will generally have 3 to 4% higher efficiency than the wet-pit type and therefore 3 to 4% lower power consumption. The two types are available for the same specific speed range. When pumping total head is 25 ft (7.6 m) or less, an axial-flow propeller (approximately 10,000 specific speed in USCS units) can be used in either type of pump.

The low-specific-speed, double-suction pump has a very moderate rise in head with reducing capacities and a nonoverloading power curve with a reduction in head. The mixed-flow impeller with a higher specific speed has a steeper head-capacity curve and a reasonably flat power curve that is also nonoverloading. As the specific speed increases, the steepness of the head-capacity curve increases and the curvature of the power curve reverses itself, hitting a maximum at the lowest flow. Finally, the curve of a high-specific-speed propeller pump has the highest rise in both head-capacity and power-capacity curves toward zero flow. The head range developed by the mixed-flow pump is ideal for condenser service; this pump is usually furnished with an enclosed impeller, which produces a relatively flat head-capacity curve and a flat power characteristic.

System Hydraulics The dry-pit pump is not too sensitive to the suction well design since the inlet piping and the formed design of the suction passages into the pump normally ensure a uniform flow into the eye of the impeller. On the other hand, the higher-speed wet-pit pumps are more sensitive to departures from ideal inlet conditions than the low-speed centrifugal volute pump or the medium-speed mixed-flow pump. A discussion of the arrangements recommended for wet- and dry-pit pump installations is presented in Sec. 10.1.

Drivers Whether a dry-pit or a wet-pit pump is used, the axial thrust and weight of the pump rotor are normally carried by a thrust bearing in the motor, and the driver and driven shafts are connected through a rigid coupling. The higher rotative speeds of the wet-pit pumps reduce the cost of the electric motors somewhat. This difference may be offset considerably, however, by the fact that the thrust load of the wet-pit pump is considerably higher than that of the dry-pit pump.

BOILER CIRCULATING PUMPS

The forced circulation, or controlled circulation, boiler requires the use of circulating pumps which take their suction from a header connected to several downcomers, which originate from the bottom of the boiler drum and discharge through the various tube circuits operating in parallel (Fig. 16). The circulating pumps therefore must develop a pressure equivalent to the frictional losses through these tube circuits. Thus, in the case of a boiler operating at 1900 lb/in^2 gage (131 bar), the boiler circulating pump must handle feedwater at approximately 630°F (332°C) under a suction pressure of 1900 lb/in^2 gage (131 bar). Such a combination of high suction pressure and high water temperature at saturation imposes very severe conditions on the circulating pump stuffing boxes, making it necessary to develop special designs for this part of the pump.

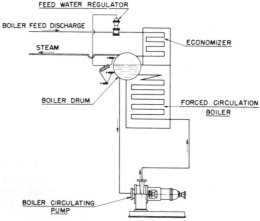

FIG. 16 Forced, or controlled, circulation cycle.

The net pressure to be developed by these pumps is relatively low, ranging from 50 to 150 lb/in^2 (3.4 to 10.3 bar). Hence these are single-stage pumps with single-suction impellers and a single stuffing box. This imposes an extremely severe axial thrust on the pump, placing a load of several tons on the thrust bearing. In many cases a thrust-relieving device must be incorporated to permit pump start-up. Two general types of construction are used for this service: (1) the conventional centrifugal pump with various stuffing box modifications (Fig. 17) and (2) the submersible motor pump of either the wet- or dry-stator type (Fig. 18). In the lower boiler pressure range—up to 500 or 600 lb/in^2 gage (34 or 41 bar) the construction shown in Fig. 19 may be used. The pump is of the same general type as is used on high-pressure heater drain service. The packed stuffing box is preceded by a pressure-reducing bushing. Feedwater from the boiler-feed pump discharge, at a temperature lower than in the boiler drum and at a pressure somewhat higher than pump internal pressure, is injected into the middle of this bushing. Part of this injected feedwater proceeds toward the pump interior, making a barrier against the outflow of high-temperature water. The rest proceeds outward to a bleed portion of the bushing, from where it is bled to a lower pressure. The packing need withstand only the lower boiler-feed pump temperature and a much lower pressure than boiler pressure.

More sophisticated designs are required for pressures from 1800 to 2800 lb/in^2 gage (124 to 193 bar) (Fig. 17). The shaft is sealed by two floating ring pressure breakdowns and a water-jacketed stuffing box. Boiler feedwater is injected at a point between the lower and upper stacks of floating ring seals at a pressure about 50 lb/in^2 (3.5 bar) above the pump internal pressure. Here again, part of this injection leaks into the pump interior and the rest leaks past the upper

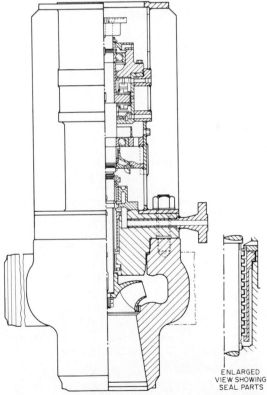

ENLARGED
VIEW SHOWING
SEAL PARTS

FIG. 17   Vertical injection-type boiler circulating pump for high pressures. (Ingersoll-Rand)

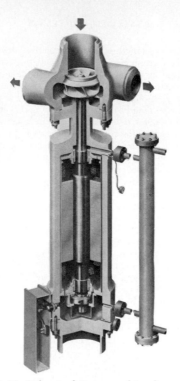

FIG. 18   Boiler circulating pump driven by a wet-stator motor. (Hayward Tyler)

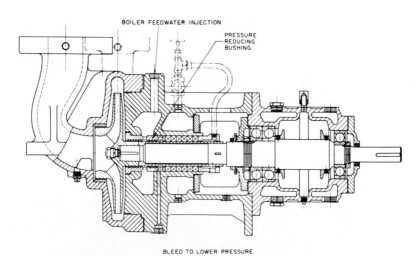

FIG. 19   End-suction boiler circulating pump for low-pressure range. (Worthington Pump)

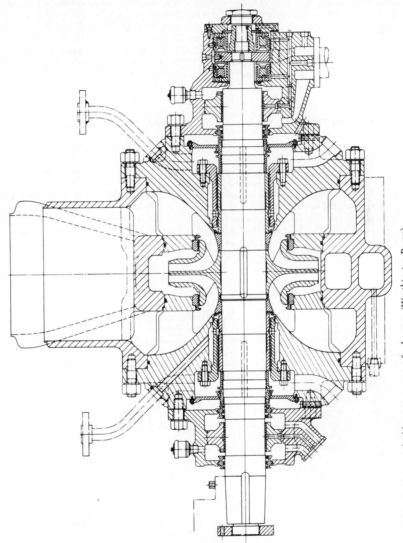

FIG. 20 Single-stage double-suction reactor feed pump, (Worthington Pump)

stack of seals to a region of low pressure in the feed cycle. Leakage to atmosphere is controlled by the conventional stuffing box located above the upper stack.

The available NPSH may not be sufficient at start-up, when the water in the boiler is cold and the pressure is low. Therefore certain installations include two-speed motors so that a lower NPSH is required at start-up. There is an added advantage to this arrangement: under normal operating conditions, the feedwater will be at boiler saturation temperature and therefore will have a specific gravity of as low as 0.60; at start-up, however, the specific gravity will be 1.0. The power consumption on cold water would therefore be some 65% higher than in normal operation if the pump operates at the same speed, and a much larger motor would be required. If a two-speed motor is used, however, the pump is operated at lower speed when the water is cold, and the motor need be only large enough to supply the maximum power required under normal operating conditions.

ASH-HANDLING PUMPS

Boilers that use solid fossil fuels produce refuse that is generally classified as ash and includes bottom ash or slag, fly ash, and mill rejects such as pyrites, etc. This refuse has to be removed and disposed of, either by hydraulic or pneumatic conveying. The latter, of course, does not involve the use of pumps and need not be discussed here.

Hydraulic conveying is generally restricted to bottom ash and mill rejects. There are a great number of different systems in use today, and the reader should consult boiler design and operation literature to become acquainted with this subject. What is common to most hydraulic conveying systems is that they use centrifugal pumps to handle concentrated ash slurries that may contain considerable amounts of coarse, heavy pieces of stone, slate, and iron pyrites. Pumps suitable for this service are described in detail in Subsec. 9.17.2.

NUCLEAR POWER PLANTS

In oversimplified form, the nuclear energy steam power plant differs from the conventional power plant only in that it uses a different fuel. Thus what is called the secondary cycle (consisting of turbogenerator, condenser and auxiliaries, and boiler-feed pumps) is not very different from its counterpart in the conventional steam power plant. The main differences are a desire for even greater equipment reliability and a preference for an absence or minimum of leakage to avoid any possibility of contamination with radioactive material. One other difference distinguishes most nuclear power plants today from their fossil fuel counterparts: their operating steam pressures and temperatures are much lower. Consequently, in most cases, reactor feed pumps are single-stage pumps; a typical section is shown in Fig. 20. The lower operating conditions result in higher heat rates, and the flows—both of the feedwater and of the condenser circulation—are about one-third higher than for fossil fuel power plants of equal megawatt rating.

More detailed information on other nuclear power plant pumping services is given in Subsec. 9.15.1.

REFERENCES

1. Liao, G. S., and P. Leung: "Analysis of Feedwater Pump Suction Pressure Decay," *ASME J. Eng. Power*, April 1972, p. 83.

2. Liao, G. S.: "Protection of Boiler Feed Pump against Transient Suction Pressure Decay," *ASME J. Eng. Power*, July 1974, p. 247.

FURTHER READING

Karassik, I. J.: "Steam Power Plant Clinics," a series of articles in *Combustion Engineering*, 1958 through 1976.

SECTION 9.6
CHEMICAL INDUSTRY

JOHN R. BIRK
JAMES H. PEACOCK

SOLUTION CHARACTERISTICS _____

Chemical industries can be broadly defined as those which make, use, or dispose of chemicals. The pumps used in these industries are different from those used in other industries primarily in the materials from which they are made. Although cast iron, ductile iron, carbon steeel, and aluminum- or copper-base alloys will handle a few chemical solutions, most chemical pumps are made of stainless steel, the nickel-base alloys, or the more exotic metals, such as titanium and zirconium. Pumps are also available in carbon, glass, porcelain, rubber, lead, and whole families of plastics, including the phenolics, epoxies, and fluorocarbons. Each of these materials has been incorporated into pump design for just one reason—to eliminate or reduce the destructive effect of the chemical liquid on the pump parts.

Since what type of corrosive liquid is to be pumped determines which of these materials is most suitable, a careful analysis of the chemical solution to be handled is the first step in selecting the proper materials for pump construction.

Major and Minor Constituents Foremost in importance in a study of the characteristics of any solution are the constituents of the solution. This means not only the major constituents but the minor ones as well, for in many instances the minor constituents will be the more important. They can drastically alter corrosion rates, and therefore a full and detailed analysis is most critical.

Concentration Closely allied to what the constituents are is the concentration of each. Merely stating "concentrated," "dilute," or "trace quantities" is basically meaningless because of the broad scope of interpretation of these terms. For instance, some interpret *concentrated* as meaning any constituent having a concentration of greater than 50% by weight, whereas others interpret any concentration above 5% in a like manner. Hence it is always desirable to cite the percentage by weight of each and every constituent in a given solution. This eliminates multiple interpretations and permits a more accurate evaluation. It is also recommended that the percentage by weight of any trace quantities be cited, even if this involves only parts per million. For example, high-silicon iron might be completely suitable in a given environment in the absence of fluorides. If, however, the same environment contained even a few parts per million of fluorides, the high-silicon iron would suffer a catastrophic corrosion failure.

Temperature Generalized terms such as *hot, cold,* or even *ambient* are ambiguous. The preferred terminology would be the maximum, minimum, and normal operating temperature. In general, the rate of a chemical reaction increases approximately two to three times with each 18°F (10°C) increase in temperature. Since corrosion can be considered a chemical reaction, the importance of temperature or temperature range is obvious.

A weather-exposed pump installation is a good illustration of the ambiguity of the term *ambient*. There could be as much as a 150°F (83°C) difference between an extremely cold climate and an extremely warm climate. If temperature cannot be cited accurately, the ambient temperature should be qualified by stating the geographic location of the pump. This is particularly important for materials that are subject to thermal shock in addition to increased corrosion rate at higher temperatures.

Acidity and Alkalinity More often than not, little consideration is given to the pH of process solutions. This may be a critical and well-controlled factor during production processing, and it can be equally revealing in evaluating solution characteristics for material selection. One reason the pH may be overlooked is that it generally is obvious whether the corrosive substance is acidic or alkaline. However, this is not always true, particularly with process solutions in which the pH is adjusted so that the solutions will always be either alkaline or acidic. When this situation exists, the precise details should be known so that a more thorough evaluation can be made. It is also quite important to know when a solution alternates between acidic and alkaline conditions because this can have a pronounced effect on materials selection. Some materials, while entirely suitable for handling a given alkaline or acidic solution, may not be suitable for handling a solution whose pH is changing.

Solids in Suspension Erosion-corrosion, velocity, and solids in suspension are closely allied in chemical industry pump services. Pump design is a very critical factor when the solution to be pumped contains solids. It is not uncommon for a given alloy to range from satisfactory to completely unsatisfactory in a given chemical application when hydraulic design is the only variable. Failure to cite the presence of solids on a solution data sheet is not an uncommon occurrence. This undoubtedly is the reason for many catastrophic erosion-corrosion failures.

Aerated or Nonaerated The presence of air in a solution can be quite significant. In some instances, it is the difference between success and failure in that it can conceivably render a reducing solution oxidizing and require an altogether different material for pump construction. A good example of this would be a self-priming nickel-molybdenum-alloy pump for handling commercially pure hydrochloric acid. This alloy is excellent for the commercially pure form of this acid, but any condition that can induce even slightly oxidizing tendencies renders this same alloy completely unsuitable. The very fact that the pump is a self-primer means that aeration is a factor to contend with, and extreme caution must be exercised in using an alloy that is not suitable for an oxidizing environment.

Transferring or Recirculating This item is important because of the possible buildup of corrosion product or contaminants, which can influence the service life of the pump. Such a buildup of contaminants can have a beneficial or deleterious effect, and for this reason it should be an integral part of evaluating solution characteristics.

Inhibitors or Accelerators Both inhibitors and accelerators can be intentionally or unintentionally added to the solution. Inhibitors reduce corrosivity, whereas accelerators increase corrosivity. Obviously, no one would add an accelerator to increase the corrosion rate on a piece of equipment, but a minor constituent added as a necessary part of a process may serve as an accelerator; thus the importance of knowing the presence of such constituents.

Purity of Product Where purity of product is of absolute importance, particular note should be made of any element that may cause contamination problems, whether it be discoloration of product or solution breakdown. In some environments, pickup of only a few parts per billion of certain elements can create severe problems. This effect is particularly important in pump applications where velocity effects and the presence of solids can alter the end result, as contrasted with other types of process equipment where the velocity and/or solids may have little or no effect.

When a material is basically suitable for a given environment, purity of product should not be a problem. However, this cannot be an ironclad rule, particularly with chemical pumps.

Continuous or Intermittent Duty   Depending upon the solution, continuous or intermittent contact can have a bearing on service life. Intermittent duty in some environments can be more destructive than continuous duty if the pump remains half full of corrosive during periods of downtime and accelerated corrosion occurs at the air-liquid interface. Perhaps of equal importance is whether or not the pump is flushed and/or drained when not in service.

CORROSIVES AND MATERIALS

Metallic or Nonmetallic   Materials for chemical industry pump applications can, in general, be divided into two very broad categories: metallic and nonmetallic. The metallic category can be further divided into ferrous and nonferrous alloys, both of which have extensive application in the chemical industry. The nonmetallic category can be further divided into natural and synthetic rubbers, plastics, ceramics and glass, carbon and graphite, and wood. Of these nonmetallic materials, wood, of course, has little or no application for pump services. The other materials have definite application in the handling of heavy corrosives. In particular, plastics in recent years have gained widespread acclaim for their corrosion resistance and therefore are much used in chemical environments. For a given application, a thorough evaluation of not only the solution characteristics but also the materials available should be made to ensure the most economical selection.

Source of Data   To evaluate material for chemical pump services, various sources of data are available. The best source is previous practical experience within one's own organization. It is not unusual, particularly in large organizations, to have a materials group or corrosion group whose basic responsibility is to collect and compile corrosion data pertaining to process equipment in service. These sources should be consulted whenever a materials evaluation program is being conducted. A second source of data is laboratory and pilot-plant experience. Though the information from this source cannot be as valuable and detailed as plant experience, it certainly can be very indicative and serve as an important guide. The experience of suppliers can be a third source of information. Though suppliers cannot hope to provide data on the specific details of a given process and the constituents involved, they normally can provide assistance and materials for test to facilitate a decision. Technical journals and periodicals are a fourth source of information. A wealth of information is contained in these publications, but if an excellent information retrieval system is not available, it can be very difficult to locate the information desired.

Reams of information have been published in books, tables, charts, periodicals, bulletins, and reports pertaining to materials selection for various environments. It is not the intent of this section of the handbook to make materials recommendations. However, it is deemed advisable to provide some general comments and to point out a few applications having unusual characteristics.

Sulfuric Acid   This is the most widely used chemical in industrial applications today, and much time is spent in evaluating and selecting materials for applications involving sulfuric acid with and without constituents. The following are some of the applications that merit special consideration.

DILUTION OF COMMERCIALLY PURE SULFURIC ACID   When sulfuric acid is diluted with water, there is considerable evolution of heat. At times, the mixing of the acid and water takes place not in the mixing tank but in the pump transferring the acid. This means that heat is evolved as the solution is passing through the pump. Temperatures of 200°F (93°C) or higher are reached, depending upon the degree of dilution and the amount of mixing taking place in the pump. Thus, the heat evolved in the dilution would restrict material selection. Very few metallics or nonmetallics are resistant to 70% sulfuric acid at temperatures approaching 200°F (93°C).

SULFURIC ACID SATURATED WITH CHLORINE   It is a well-known fact that any solution involving wet chlorine is extremely corrosive. In a solution containing sulfuric acid and chlorine, the specific weight percentage of sulfuric acid determines whether the solution will accelerate corrosion. Because of the hygroscopic nature of concentrated sulfuric acid, it will absorb moisture from the chlorine. Thus, when a sulfuric acid–chlorine solution contains at least 80% sulfuric acid, there need be little concern for the chlorine because dry chlorine is essentially noncorrosive. In such a case a material selection can be made as if sulfuric acid were the only constituent. If the solution is saturated with chlorine but contains less than approximately 80% sulfuric acid, however, the material selection must be based not only on the sulfuric acid but also on the wet chlorine. This, of course, is a very corrosive solution, and extreme caution must be exercised in selecting the material to be used.

RAYON SPIN BATHS   Rayon spin baths can contain up to 20% sulfuric acid. In addition, they contain significant quantities of zinc sulfate, sodium sulfate, carbon disulfide, and hydrogen sulfide. Temperatures up to 200°F (93°C) are often encountered. In solutions of this type, it is not unusual to base a material selection strictly on the concentration of sulfuric acid and the temperature because the other constituents are thought to exert a very slight influence. On the contrary, the presence of carbon disulfide and hydrogen sulfide makes rayon spin baths strongly reducing and can render an otherwise suitable material completely unsuitable. Even trace quantities of these sulfides exert a significant influence at any temperature and must be considered in selecting a material for rayon spin baths.

SULFURIC ACID CONTAINING SODIUM CHLORIDE   It is quite apparent that the addition of sodium chloride to sulfuric acid will result in the formation of hydrochloric acid and thus necessitate a material that will resist the corrosive action of hydrochloric acid also. Though this may seem obvious, it is amazing how often it is ignored. This is particularly true in 10 to 15% sulfuric acid pickling solutions to which sodium chloride has been added to increase the rate of pickling, with little or no consideration being given to the destructive effect of the salt addition on the process equipment handling the pickling solution.

PIGMENT MANUFACTURE   A slurry of titanium dioxide in sulfuric acid is one of the processing stations in the manufacture of pigment. A variety of metallics and nonmetallics would be suitable for this application in the absence of the titanium dioxide solids, but the presence of the solids circulating in a pump renders practically all of the normal sulfuric acid-resistant pump materials unsuitable. Special consideration must be given to materials that will resist the severe erosion-corrosion encountered in this type of service.

SULFURIC ACID CONTAINING NITRIC ACID, FERRIC SULFATE, OR CUPRIC SULFATE   The presence of these compounds in sulfuric acid solutions will drastically alter the suitability of materials that can be used. Their presence in quantities of 1% or less can make a sulfuric acid solution oxidizing, whereas it would normally be reducing. Their presence, singly or in combination, could serve as a corrosion inhibitor, thus in certain instances allowing a stainless steel, such as type 316, to be used. On the other hand, the same compounds could serve as a corrosion accelerator for a nonchromium bearing alloy, such as nickel-molybdenum alloys, and thus render it completely unsuitable.

Nitric Acid   In the concentrations normally encountered in chemical applications, nitric acid presents fewer problems than sulfuric acid. The choice of metallic materials for various nitric applications is somewhat broader than the choice of nonmetallic materials. Nitric acid, being a strongly oxidizing acid, permits the use of stainless steel quite extensively, but its oxidizing characteristics restrict the application of nonmetallics in general and plastics in particular. Requiring special evaluation are such aggressive solutions as fuming nitric acid; nitric-hydrofluoric; nitric-hydrochloric (some of which fall into the aqua regia category); nitric-adipic combinations; and practically any environment consisting of nitric acid in combination with other constituents. Invariably, additional constituents in nitric acid result in more aggressive corrosion; hence material selection becomes quite critical.

Hydrochloric Acid   Both commercially pure and contaminated hydrochloric acid present difficult situations in selecting pump materials. The most common contaminant that creates problems is ferric chloride, the presence of which can render this otherwise reducing solution oxidizing and thus completely change the material of construction that can be used. Addition of a very few parts per million of iron to commercially pure hydrochloric acid can result in the formation of enough ferric chloride to cause materials such as nickel-molybdenum, nickel-copper, and zirconium to be completely unsuitable. Conversely, the presence of ferric chloride can make titanium completely suitable. Nonmetallics find extensive application in many hydrochloric acid environments. Often the limiting factors for the nonmetallics are temperature, mechanical properties, and suitability for producing pump parts in the design desired. With the nonmetallics, the near-complete immunity from corrosion in such environments subordinates corrosion resistance to other factors.

Phosphoric Acid   The increasing use and demand for all types of fertilizers have made phosphoric acid a very important commodity. In the wet process of producing phosphoric acid, the phosphate rock normally contains fluorides. In addition, at various stages of the operation the solution will also contain sulfuric, hydrofluoric, fluosilicic, and phosphoric acids as well as solids. In some instances the water used in these solutions may have an exceptionally high chloride content, which can result in the formation of hydrochloric acid, which further aggravates the corrosion problem. It is also typical for certain of these solutions to contain solids, which of course create an erosion-corrosion problem. Pure phosphoric and superphosphoric acids are relatively easy to cope with from a material standpoint, but when the solution contains all or some of the aforementioned constituents, a very careful materials evaluation must be conducted. Such environments are severely corrosive in the absence of solids and cause severe erosion-corrosion and a drastically reduced service life when solids are present. This is particularly significant with any type of chemical pump.

Chlorine   Little need be said about the corrosivity of chlorine. Wet chlorine, in addition to being extremely hazardous, is among the most corrosive environments known. Dry chlorine is not corrosive, but there are those who contend that dry chlorine does not exist. Chlorine vapor combined with the moisture in the atmosphere, for instance, can create severe corrosion problems. In any case, selecting the most suitable material for any type of chlorine environment requires very careful evaluation.

Alkaline Solutions   With some exceptions, alkaline solutions, such as sodium hydroxide or potassium hydroxide, do not present serious corrosion problems at temperatures below 200°F (93°C). However, in certain applications, purity of product is of utmost concern, necessitating selection of a material that will have essentially a nil corrosion rate. Among the exceptions to the rule that alkaline solutions are relatively noncorrosive are bleaches, alkaline brines, and other solutions containing chlorine in some form.

Organic Acids   Organic acids are much less corrosive than inorganic acids. This does not mean, however, that they can be taken lightly. For instance, acetic, lactic, formic, and maleic acids all have their corrosive characteristics and must be treated accordingly when evaluating metallics and nonmetallics.

Salt Solutions   Normally considered neutral, salt solutions do not present a serious corrosion problem. In some instances, process streams adjust pH to maintain a slightly alkaline environment, and such solutions are even less corrosive than when they are neutral. On the other hand, when a process stream has a pH adjustment to maintain a slightly acidic environment, the liquid becomes considerably more corrosive than neutral salt solutions. This condition requires that more effort be expended in evaluating the solution before making a material selection.

Organic Compounds   Most organic compounds do not present corrosion problems of the same magnitude as inorganic compounds. This does not mean that any material arbitrarily selected will be a suitable choice. It does mean that there will be more materials available to choose from, but each application should be considered on its own merits. Of particular concern in this area are chlorinated organic compounds and those that will produce hydrochloric acid when moisture is present. Plastics, categorically, possess excellent corrosion resistance to inorganic compounds within their temperature limitations, but they do exhibit some weaknesses in their corrosion resistance to organic compounds.

Water   Water is less corrosive than most of the other mediums encountered in the chemical and allied industries. For the term *water* to be meaningful, however, it is extremely important to know the specific kind of water: demineralized, fresh, brackish, salt, boiler feed, mine. These waters and the various constituents in them can demand a variety of materials, indicated, for example, by the spectrum of materials being studied and used in desalination programs. Because they are likely to have a very pronounced effect on our total economy, precise materials evaluation and selection are integral parts of these programs.

TYPES OF PUMP CORROSION

The types of corrosion encountered in chemical pumps may at first appear to be unusual compared with those found in other process equipment. Nevertheless pumps, like any other type of chemical process equipment, experience basically only eight forms of corrosion, of which some are more predominant in pumps than in other types of equipment. It is not the intent here to describe in detail these eight forms of corrosion, but it is desirable to enumerate them and provide a brief description of each so that they can be recognized when they occur.

General, or Uniform, Corrosion   This is the most common type, and it is characterized by essentially the same rate of deterioration over the entire wetted or exposed surface. General corrosion may be very slow or very rapid, but it is of less concern than the other forms of corrosion because of its predictability. However, predictability of general corrosion in a pump can be a difficult task because of the varying velocities of the solution in the pump.

Concentration Cell, or Crevice, Corrosion   This is a localized form of corrosion resulting from small quantities of stagnant solution in areas such as threads, gasket surfaces, holes, crevices, surface deposits, and the underside of bolt and rivet heads. When concentration cell corrosion occurs, the concentration of metal ions or oxygen in the stagnant area is different from the concentration in the main body of the liquid. This causes electric current to flow between the two areas, resulting in severe localized attack in the stagnant area. Usually this form of corrosion does not occur in chemical pumps except in misapplications or in designs where the factors known to contribute to concentration cell corrosion have been ignored.

Pitting Corrosion   This is the most insidious form of corrosion, and it is very difficult to predict. It is extremely localized and manifested by small or large holes (usually small), and the weight loss due to the pits will be only a small percentage of the total weight of the equipment. Chlorides in particular are notorious for inducing pitting, which can occur in practically all types of equipment. In some instances, this form of corrosion can be closely allied to concentration cell corrosion, since pits may initiate in the same areas where concentration cell corrosion appears. Pitting is common in areas other than stagnant areas, wheres concentration cell corrosion is basically confined to areas of stagnation.

Stress Corrosion Cracking   This is localized failure caused by a combination of tensile stresses in a medium. Undoubtedly more research and development have been conducted on this form of corrosion than on any of the others. Nevertheless, the exact mechanism of stress corrosion cracking is still not well understood. Fortunately, castings, because of their basic overdesign, seldom experience stress corrosion cracking. Corrosion fatigue, which can be classified as stress corrosion cracking, is of concern in chemical pump shafts because of the repeated cyclic stressing. Failures of this type occur at stress levels below the yield point as a result of the cyclic application of the stress.

Intergranular Corrosion   This is a selective form of corrosion at and adjacent to grain boundaries. It is associated primarily with stainless steels but can also occur with other alloy systems. In stainless steels, it occurs when the material is subjected to heat in the 800 to 1600°F (427 to 871°C) temperature range. Unless other alloy adjustments are made, this form of corrosion can be prevented only by heat-treating. It is easily detectable in castings because the grains are quite large relative to those in wrought material of equivalent composition. In some instances, uniform corrosion is misinterpreted as intergranular corrosion because of the etched appearance of the surfaces exposed to the environment. Even in ideally heat-treated stainless steels, very slight accelerated attack can be noticed at the grain boundaries because these areas are more reactive than the grains themselves. The untrained eye should be careful to avoid confusing general and intergranular corrosion. Stainless steel castings will never encounter intergranular corrosion if they are properly heat-treated after being exposed to temperatures in the 800 to 1600°F range (427 to 871°C).

Galvanic Corrosion   This occurs when dissimilar metals are in contact or are otherwise electrically connected in a corrosive medium. Corrosion of the less noble metal is accelerated, and corrosion of the more corrosion-resistant metal is decreased. The farther apart the metals or alloys are in the electromotive series, the greater the possibility of galvanic corrosion. When it is necessary to have two dissimilar metals in contact, the total surface area of the less resistant metal should far exceed that of the more resistant material. This tends to prevent premature failure by simply providing a substantially greater area of the more corrosion-prone material. This form of corrosion is not common in chemical pumps but may be of some concern with accessory items in contact with pump parts and exposed to the environment.

Erosion-Corrosion   This type of failure is characterized by accelerated attack resulting from the combination of corrosion and mechanical wear. It may involve solids in suspension and/ or high velocity. it is quite common with pumps where the erosive effects prevent the formation of a passive surface on alloys that require passivity to be corrosion-resistant. The ideal material for avoiding erosion-corrosion in pumps would possess the characteristics of corrosion resistance, strength, ductility, and extreme hardness. Few materials possess such a combination of properties.

Selective Leaching Corrosion   This, in essence, involves removal of one element from a solid alloy in a corrosive medium. Specifically, it is typified by dezincification, dealuminumification, and graphitization. This form of attack is not common in chemical pump applications because the alloys in which it occurs are not commonly used in heavy chemical applications.

TYPES OF CHEMICAL PUMPS

The second step in selecting a chemical pump is to determine which type of pump is required, based on the characteristics of the liquid and on the desired head and capacity. It should also be noted that not all types are available in every material of construction, and the final selection of pump type may depend on the availability of designs in the proper material.

Centrifugal Pumps   Centrifugal pumps are used extensively in the chemical industry because of their suitability in practically any service. They are available in an almost unending array of corrosion-resistant materials. While not built in extremely large sizes, pumps with capacity ranges of 5000 to 6000 gpm (1100 to 1400 m³/h) are commonplace. Heads range as high as 500 to 600 ft (150 to 180 m) at standard electric motor speeds. Centrifugal pumps are normally mounted in the horizontal position, but they may also be installed vertically, suspended in a tank, or hung in a pipeline similar to a valve. They are simple, economical, dependable, and efficient. Disadvantages include reduced performance when handling liquids of more than 500 SSU (108 cSt) viscosity and the tendency to lose prime when comparatively small amounts of air or vapor are present in the liquid.

Rotary Pumps   The gear, screw, deforming-vane, sliding-vane, axial-piston, and cam types are used for high-pressure service. They are particularly adept at pumping liquids of high viscosity or low vapor pressure. Their constant displacement at a set speed makes them ideal for use in metering small quantities of liquid. Since they operate on the positive displacement principle, they are inherently self-priming. When built of materials that tend to gall or seize on rubbing contact, the clearances between mating parts must be increased, with the result of decreased efficiency. The gear, sliding-vane, and cam units are generally limited to use on clear, nonabrasive liquids.

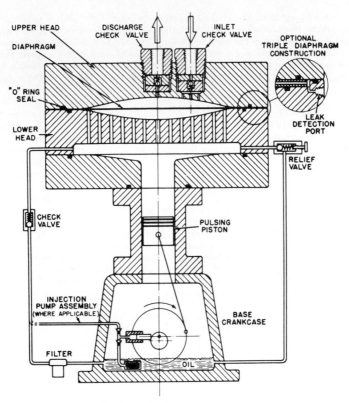

FIG. 1   Construction details of a diaphragm pump.

Diaphragm Pumps   These units are also classed as positive displacement pumps since the diaphragm acts as a limited displacement piston. Pumping action is obtained when the diaphragm is forced into reciprocating motion by mechanical linkage, compressed air, or oil from a pulsating external source. This type of construction eliminates any connection between the liquid being pumped and the source of energy and thereby eliminates the possibility of leakage. This characteristic is of great importance when toxic or very expensive liquids are being handled. Disadvantages include a limited selection of corrosion-resistant materials, limited head and capacity range, and the necessity of using check valves in the suction and discharge nozzles. Construction details are shown in Fig. 1.

Regenerative Turbine Pumps   Flow rates up to 100 gpm (23 m³/h) and heads up to 700 ft (210 m) are easily handled with this type of pump. When it is used for chemical service, the internal clearances must be increased to prevent rubbing contact, which results in decreased efficiency. These pumps are generally unsuitable for solid-liquid mixtures of any concentration.

CHEMICAL PUMP DESIGN CONSIDERATIONS

Casting Integrity   Practically all the major components of chemical pumps are castings. Needless to say, it is fruitless to evaluate thoroughly the detailed characteristics of the solution to be pumped and the construction material to be used if the component castings are not of high enough quality to provide good service life. This is probably of more concern in chemical pump applications than in any other type of service because leakage, loss of product, and downtime can be extremely costly, as well as very dangerous. Quality products in any industry are no accident, but are the result of concentrated effort. When producing pumps for chemical applications, whether of solid metallics, solid nonmetallics, or lined, there is no excuse for sacrificing quality. Being acquainted with materials of construction and how they are utilized to produce chemical pumps can be extremely beneficial in securing the best material and design for a given application.

Mechanical Properties   It has been mentioned that there are several factors which determine whether or not a certain material can be utilized for a particular design. Of these, the mechanical properties are the most important. Materials may possess outstanding corrosion resistance but may be completely impossible to produce in the form of a chemical pump because of their limited mechanical properties. Accordingly, it is advisable to be aware of the mechanical properties of any material being considered in a corrosion evaluation program. This gives a relatively good indication of whether or not a particular design may be available. Since most materials are covered by ASTM or other specifications, such sources can be used for reference. Tables of mechanical properties and other characteristics and of proprietary materials not included in any standard specification should be readily available from manufacturers.

Weldments   Welded construction should impose no limitation, provided the weldment is as good as or better than the base material. If the weldment meets this requirement, there is no reason for welded construction to be objectionable. Materials requiring heat treatment to achieve maximum corrosion resistance must be treated after welding, or other adjustments must be made, to make certain corrosion resistance is not sacrificed.

Section Thickness   Wall sections are generally made thicker than required for handling a noncorrosive liquid so that full pumping capability will be maintained even after the loss of some material to the corrosive medium. Parts that are subject to corrosion from two or three sides, such as impellers, must be made considerably heavier than their counterparts in water or oil pumps. Pressure-containing parts are also made thicker so that they will remain serviceable after a specified amount of corrosive deterioration. Areas subject to high velocities, such as the cutwater of a centrifugal pump casing, are reinforced to allow for the accelerated corrosion caused by the high velocities.

Threads   Threaded construction of any type in the wetted parts must be avoided whenever possible. The thin thread form is subject to attack from two sides, and a small amount of corrosive deterioration will eliminate the holding power of the threaded joint. Pipe threads are also to be avoided because of their susceptibility to attack.

Gaskets   Gasket materials must resist being corroded by the chemical being handled. Compressed asbestos, lead, and certain synthetic rubbers have been used extensively for corrosion services. In recent years the fluorocarbon resins have come into widespread use because of their almost complete corrosion resistance.

Power End   This assembly, consisting of the bearing housing, bearings, oil or grease seals, and bearing lubrication system, is normally made of iron or steel components; thus it must be designed to withstand a severe chemical plant environment. For example, where venting of the bearing housing is required, special means of preventing the entrance of water, chemical fumes, or dirt must be incorporated into the vent design.

The bearing which controls axial shaft movement is usually selected to limit shaft movement to 0.002 in (0.051 mm) or less. Endplay values above this limit have been found detrimental to mechanical seal operation.

Water jacketing of the bearing housing may be necessary under certain conditions to maintain bearing temperatures below 225°F (107°C), the upper limit for standard bearings.

Maintenance   Maintenance of a chemical pump in a corrosive environment can be a very costly and time-consuming item. It can be divided into two categories—preventive and emergency. When evaluating materials and design factors, maintenance aspects should be high on the priority list. The ease and frequency of maintenance are two very critical items and should be considered part of a preventive maintenance program. Such a program can be the most effective way of eliminating emergency shutdowns caused by pump failure. Furthermore, the knowledge gained in a routine preventive maintenance progam can be of unlimited value when a breakdown does occur, since repair personnel will have acquired a thorough knowledge of the construction details of the pump and this can minimize downtime.

STUFFING BOX DESIGN

The area around the stuffing box probably causes more chemical pump failures than all other parts combined. The problem of establishing a seal between a rotating shaft and the stationary pump parts is one of the most intricate and vexing problems facing the pump designer.

Packings   Braided asbestos, lead, fluorocarbon resins, aluminum, graphite, and many other materials or combinations of materials have been used to establish a seal (discussed in more detail in Subsec. 2.2.2). Inconsistent as it seems, a small amount of liquid must be allowed to seep through the packing and lubricate the surface between packing and shaft. This leakage rate is hard to control, and usually the packing is overtightened and the leakage is stopped. The unfortunate results of this condition is rapid scoring of the shaft surface, making it much harder to adjust the packing to the proper tension. Recommendations as to the type of packing to be used for various chemical services should come from the packing manufacturer.

Mechanical Seals   Mechanical shaft seals as described in Subsec. 2.2.3 are used extensively on chemical pumps. Once again the primary consideration is selection of the proper materials for the type of corrosive being pumped. Stainless steels, ceramics, graphite, and fluorocarbon resins are used to make most seal parts. Several of the large manufacturers of this type of equipment have developed very complete files on seal designs for various chemical services; they are therefore in the best position to make recommendations.

Temperature   One of the most important factors affecting the stuffing box sealing medium is the operating temperature. Most packings are impregnated with a lubricating grease that breaks down at temperatures above 250°F (120°C), causing further temperature increases. The friction of the packing rubbing on the shaft can easily generate temperatures of this magnitude. One of the not-so-obvious results of this temperature increase is corrosive attack on the pump parts in the heat zone. Many materials that are suitable at the pumping temperature will be completely unsuitable in the presence of the corrosive at elevated stuffing box temperatures. Another source of heat is, of course, the chemical solution itself. These liquids often are in the 300°F (150°C) range, and some go as high as 700°F (370°C).

The best answer to the heat problem is removal of the heat by means of a water jacket around the stuffing box. While the heat conductivity is rather low for most chemical pump materials, the stuffing box area can generally be maintained in the 200 to 250°F (93 to 120°C) range. This cooling is of further benefit in that it prevents the transfer of heat along the shaft to the bearing housing, thereby eliminating other problems around the bearings.

Pressure   Stuffing box pressure varies with suction or discharge pressure depending on its location in the pump, impeller design, and the degree of maintenance of close-fitting seal rings. Variations in impeller design include those using vertical or horizontal seal rings in combination with balance ports, as opposed to those using back vanes or pump-out vanes. All impeller designs depend upon a close running clearance between the impeller and the stationary pump parts. This clearance must be kept as small as possible to prevent excessive recirculation of the liquid and the resulting loss of efficiency. Unfortunately, most chemical pump materials tend to seize when subjected to rubbing contact, and the running clearances must therefore be increased considerably above those used in pumps for other industries.

At pressures above 100 lb/in² (690 kPa), packing is generally unsatisfactory unless the stuffing box is very deep and the operator is especially adept at maintaining the proper gland pressure on the packing. Mechanical seals incorporating a balancing feature to relieve the high face pressure are the best means of sealing at pressures above 100 lb/in² (690 kPa).

Shaft The pump shaft can create additional stuffing box problems. Obviously, a shaft that is out of round or bent will form a larger hole in the packing than a perfect shaft will, thereby allowing liquid to escape. Lack of static or hydraulic balance in the impeller produces a dynamic bend in the shaft, resulting in the same condition. Undersize shafts, or those made of materials that bend readily, will deflect from their true center in response to radial thrust on the impeller. This action produces a secondary hole in the packing and again allows the liquid to escape.

Mechanical seal operation is also impaired when the shaft is bent or deflected during operation. Since the flexible member of the seal must adjust with each revolution of the shaft, excessive deflection results in shortened seal life. If the deflection is of more than nominal value, the flexible seal member will be unable to react with sufficient speed to keep the seal faces together, allowing leakage at the mating faces.

An arbitrary limit of 0.002 in (0.051 mm) has been established as the maximum deflection or shaft runout at the face of the stuffing box consistent with good pump design.

Shaft Surface In the stuffing box region, the shaft surface must have corrosion resistance at least equal to and preferably better than that of the wetted parts of the pump. In addition, this surface must be hard enough to resist the tendency to wear under the packing or mechanical seal parts. Further, it must be capable of withstanding the sudden temperature changes often encountered in operation.

Since it is not economically feasible to make the entire pump shaft of stainless alloys, and physically impossible to make functional carbon, glass, or plastic shafts, chemical pumps often have carbon steel shafts with a protective coating or sleeve over the steel in the stuffing box area. Cylindrical sleeves are sometimes made so that they may be removed and replaced when they become worn. Other designs utilize sleeves that are permanently bonded to the shaft to obtain lower runout and deflection values.

Another method of obtaining a hard surface in this region is the welded overlay or spray coating of hard metals onto the base shaft. These materials are generally lacking in corrosion resistance, however, and they have not been widely accepted for severe chemical service. Ceramic materials applied by the plasma spray technique possess excellent corrosion resistance but cannot achieve the complete density required to protect the underlying shaft.

Composite shafts utilizing carbon steel for the power end and a higher alloy for the wet end have been used extensively where the high-alloy end has acceptable corrosion resistance. The two ends are joined by various welding techniques, and the combination of metals is therefore limited to those that can be easily welded together. On such assemblies, the weld joint and the heat-affected zone must be outside the wetted area of the shaft.

Other Designs Elimination of the stuffing box and its associated problems has been the object of several chemical pump designs other than the diaphragm pumps previously mentioned.

Vertical submerged pumps utilize a sleeve bearing in the area immediately above the impeller to limit the flow of liquid up along the shaft. For chemical service, the problem of materials associated with this bearing, and its lubrication, have been major disadvantages.

Canned pumps, wherein the motor windings are hermetically sealed in stainless steel "cans," also avoid the use of a stuffing box. The pumped liquid is circulated through the motor section, lubricating the sleeve bearings which support the entire rotating assembly. Disadvantages have again centered around the selection of bearing materials compatible with the corrosive liquid, the lubrication of these bearings when nonlubricating liquids are being handled, and the probability of the liquid paths through the motor section becoming clogged in the presence of solid-liquid mixtures.

DESIGNING WITH SPECIAL MATERIALS

A number of low-mechanical-strength materials have been used extensively in chemical pump construction. While breakage problems are inherently associated with these materials, their excellent corrosion resistance has allowed them to remain competitive with stronger alloys. Their low tensile strength and brittleness make them sensitive to tensile or bending stresses, requiring special pump designs. The parts are held together by clamps and are braced to prevent bending. The unit must also be protected from sudden temperature changes and from mechanical impact from outside sources.

High-Silicon Iron Although produced by very few manufacturers, high-silicon (14.2 to 14.7%) iron is the most completely corrosion-resistant metallic available at an economical price. This resistance, coupled with a hardness of approximately 520 Brinell, makes high-silicon iron an excellent material for handling abrasive chemical slurries. The same hardness, however, precludes normal machining operations, and the parts must be designed for machine grinding. It also eliminates the possibility of using drilled or tapped holes for connecting piping to the pump parts. Special designs are therefore required for process piping, stuffing box lubrication, and drain connections.

Ceramics and Glass These materials are similar to high-silicon iron in hardness, brittleness, and susceptibility to thermal or mechanical shock. Pump designs must therefore incorporate the same special considerations.

Glass linings or coatings on iron or steel parts are sometimes used to eliminate some of the undesirable characteristics of solid glass. While this method provides means for connecting process piping and protects the glass from mechanical shock, the dissimilar expansion characteristics of the two materials generate small cracks in the glass, allowing the corrosive liquid to attack the armor material.

Plastics Both thermosetting and thermoplastic substances are used extensively in services where chlorides are present. Their primary disadvantage is loss in strength at higher pumping temperatures. Phenolic and epoxy parts are subject to a gradual loss of dimensional integrity because of the "creep" characteristics of these materials. The low tensile strength of the unfilled resins again dictates a design that will place these parts in compression and will eliminate bending stresses. Typical construction details are shown in Fig. 2.

Fluorocarbon Resins Both polytetrafluoroethylene and hexafluoropropylene possess excellent corrosion resistance. These resins have been used for gaskets, packing, mechanical seal parts,

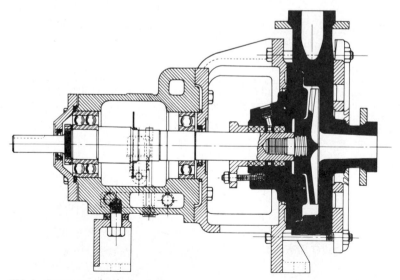

FIG. 2 Construction details of a plastic pump.

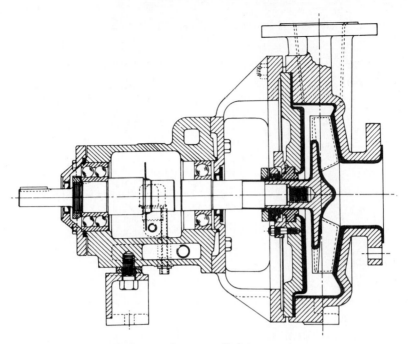

FIG. 3 Metallic pump with plastic or other nonmetallic lining.

and flexible piping connectors. Several pumps made of these materials have reached the market in recent years, and undoubtedly they will be followed by many more. Problems associated with these materials have centered around their tendency to cold-flow under pressure, as well as their high coefficient of expansion relative to the metallic components of the unit. Pumps may be made of heavy solid sections, as illustrated in Fig. 2, or may use more conventional metallic components lined with the fluorocarbon material, as shown in Fig. 3.

CHEMICAL PUMP STANDARDS

In 1962 a committee of the Manufacturing Chemists Association (MCA) reached agreement with a special committee of the Hydraulic Institute on a proposed American Standards Association (ASA) standard for chemical process pumps. This document was referred to as the American Voluntary Standard or the Manufacturing Chemists Association Standard. In 1971 it was accepted by the American National Standards Institute (ANSI) and issued as ANSI Standard B123.1 (renumbered ANSI B73.1 in 1974). Many pump manufacturers in the United States and a few in foreign countries are building pumps which meet these dimensional and design criteria.

It is the intent of this standard that pumps of similar size from all sources of supply shall be dimensionally interchangeable with respect to mounting dimensions, size and location of suction and discharge nozzles, input shafts, base plates, and foundation bolts. Table 1 lists the pump dimensions that have been standardized, and a cross-sectional assembly of a pump meeting these criteria is shown in Fig. 4.

It is also the intent of this standard to outline certain design features which will minimize maintenance problems. The standard states, for instance, that the pump shaft should be sized so that the maximum shaft deflection, measured at the face of the stuffing box when the pump is operating under its most adverse conditions, will not exceed 0.002 in (0.051 mm). It does not

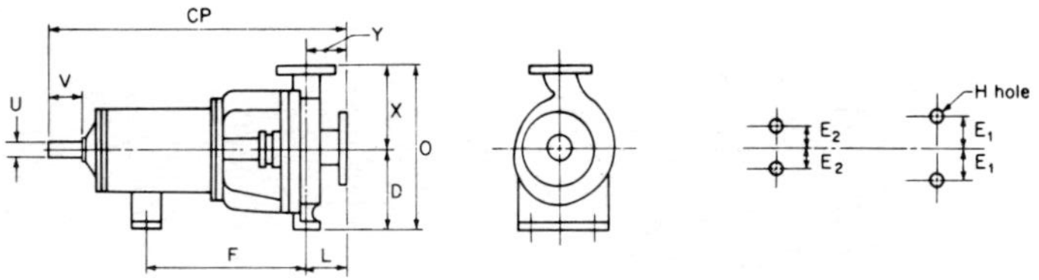

TABLE 1a Standard ANSI B73.1 Dimensions in Inches for Horizontal Pump Shown in Fig. 4

Dimension designation	Suction × discharge × nominal impeller diameter	*CP*	*D*	$2E_1$	$2E_2$	*F*	*H*	*O*	*U* Diam.	*U* Keyway	*V*, min	*X*	*Y*
AA	1½ × 1 × 6	17½	5¼	6	0	7¼	⅝	11¾	⅞	3/16 × 3/32	2	6½	4
AB	3 × 1½ × 6	17½	5¼	6	0	7¼	⅝	11¾	⅞	3/16 × 3/32	2	6½	4
A10	3 × 2 × 6	23½	8¼	9¾	7¼	12½	⅝	16½	1⅛	¼ × ⅛	2⅝	8¼	4
AA	1½ × 1 × 8	17½	5¼	6	0	7¼	⅝	11¾	⅞	3/16 × 3/32	2	6½	4
A50	3 × 1½ × 8	23½	8¼	9¾	7¼	12½	⅝	16¾	1⅛	¼ × ⅛	2⅝	8½	4
A60	3 × 2 × 8	23½	8¼	9¾	7¼	12½	⅝	17¾	1⅛	¼ × ⅛	2⅝	9½	4
A70	4 × 3 × 8	23½	8¼	9¾	7¼	12½	⅝	19¼	1⅛	¼ × ⅛	2⅝	11	4
A05	2 × 1 × 10	23½	8¼	9¾	7¼	12½	⅝	16¾	1⅛	¼ × ⅛	2⅝	8½	4
A50	3 × 1½ × 10	23½	8¼	9¾	7¼	12½	⅝	16¾	1⅛	¼ × ⅛	2⅝	8½	4
A60	3 × 2 × 10	23½	8¼	9¾	7¼	12½	⅝	17¾	1⅛	¼ × ⅛	2⅝	9½	4
A70	4 × 3 × 10	23½	8¼	9¾	7¼	12½	⅝	19¼	1⅛	¼ × ⅛	2⅝	11	4
A20	3 × 1½ × 13	23½	10	9¾	7¼	12½	⅝	20½	1⅛	¼ × ⅛	2⅝	10½	4
A30	3 × 2 × 13	23½	10	9¾	7¼	12½	⅝	21½	1⅛	¼ × ⅛	2⅝	11½	4
A40	4 × 3 × 13	23½	10	9¾	7¼	12½	⅝	22½	1⅛	¼ × ⅛	2⅝	12½	4
A80[a]	6 × 4 × 13	23½	10	9¾	7¼	12½	⅝	23½	1⅛	¼ × ⅛	2⅝	13½	4
A90[a]	8 × 6 × 13	33⅞	14½	16	9	18¾	⅞	30½	2⅜	⅝ × 5/16	4	16	6
A100[a]	10 × 8 × 13	33⅞	14½	16	9	18¾	⅞	32½	2⅜	⅝ × 5/16	4	18	6
A110[a]	8 × 6 × 15	33⅞	14½	16	9	18¾	⅞	32½	2⅜	⅝ × 5/16	4	18	6
A120[a]	10 × 8 × 15	33⅞	14½	16	9	18¾	⅞	33½	2⅜	⅝ × 5/16	4	19	6

[a]Suction connections may have tapped bolt holes.

TABLE 1b Standard ANSI B73.1 Dimensions in Millimeters for Horizontal Pump Shown in Fig. 4

Dimension designation	Suction × discharge × nominal impeller diameter	CP	D	2E₁	2E₂	F	H	O	U Diam.	U Keyway	V, min	X	Y
AA	40 × 25 × 150	445	133	152	0	184	16	298	22.23	4.76 × 2.38	51	165	102
AB	80 × 40 × 150	445	133	152	0	184	16	298	22.23	4.76 × 2.38	51	165	102
A10	80 × 50 × 150	597	210	248	184	318	16	420	28.58	6.35 × 3.18	67	210	102
AA	40 × 25 × 200	445	133	152	0	184	16	298	22.23	4.76 × 2.38	51	165	102
A50	80 × 40 × 200	597	210	248	184	318	16	425	28.58	6.35 × 3.18	67	216	102
A60	80 × 50 × 200	597	210	248	184	318	16	450	28.58	6.35 × 3.18	67	242	102
A70	100 × 80 × 200	597	210	248	184	318	16	490	28.58	6.35 × 3.18	67	280	102
A05	50 × 25 × 250	597	210	248	184	318	16	425	28.58	6.35 × 3.18	67	216	102
A50	80 × 40 × 250	597	210	248	184	318	16	425	28.58	6.35 × 3.18	67	216	102
A60	80 × 50 × 250	597	210	248	184	318	16	450	28.58	6.35 × 3.18	67	242	102
A70	100 × 80 × 250	597	210	248	184	318	16	490	28.58	6.35 × 3.18	67	280	102
A20	80 × 40 × 330	597	254	248	184	318	16	520	28.58	6.35 × 3.18	67	266	102
A30	80 × 50 × 330	597	254	248	184	318	16	546	28.58	6.35 × 3.18	67	292	102
A40	100 × 80 × 330	597	254	248	184	318	16	572	28.58	6.35 × 3.18	67	318	102
A80[a]	150 × 100 × 330	597	254	248	184	318	16	597	28.58	6.35 × 3.18	67	343	102
A90[a]	200 × 150 × 330	860	368	406	229	476	22	775	60.33	15.88 × 7.94	102	406	152
A100[a]	250 × 200 × 330	860	368	406	229	476	22	826	60.33	15.88 × 7.94	102	457	152
A110[a]	200 × 150 × 380	860	368	406	229	476	22	826	60.33	15.88 × 7.94	102	457	152
A120[a]	250 × 200 × 380	860	368	406	229	476	22	851	60.33	15.88 × 7.94	102	483	152

[a]Suction connections may have tapped bolt holes.

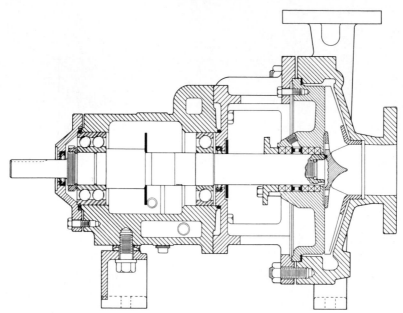

FIG. 4 Cross-sectional assembly of a typical pump made to the ANSI B73.1 Standard.

specify shaft diameter since impeller diameter, shaft length, and provision for operation with liquids of high specific gravity would determine the proper diameter.

The standard also states that the minimum bearing life, again under the most adverse operating conditions, should be not less than two years. Bearing size is not specified but is to be determined by the individual manufacturer and will be dependent upon the load to be carried.

Additional specifications in the standard include hydrostatic test pressure, shaft finish at rubbing points, and packing space.

A similar document, ANSI B73.2, covers vertical in-line centrifugal pumps for chemical process. Dimensional criteria are shown in Table 2.

Other dimensional standards for both horizontal and vertical pumps are used in foreign countries. In 1971 the International Organization for Standardization (ISO) reached agreement on a set of dimensional standards for horizontal end-suction centrifugal pumps. This document, ISO 2858, in SI units, describes a series of pumps of slightly lower capacity than described in ANSI B73.1. Technical specifications are covered in ISO 5199. Codes for acceptance tests are given in ISO 2548 and ISO 3555.

The British Standards Institution issued British Standard 4082 in 1966 to describe a series of vertical in-line centrifugal pumps. While dimensional interchangeability was the primary reason for this standard, it also includes requirements for hydrostatic testing of the pump parts. It is made up of two sections: Part 1 covers pumps having the suction and discharge nozzles in a horizontal line (the I configuration), and Part 2 covers pumps in which the nozzles are on the same side of the pump parallel to each other (the U configuration).

CONSTRUCTION OF NONMETALLIC PUMPS

For more detailed information regarding the construction of nonmetallic pumps, refer to Sec. 5.2.

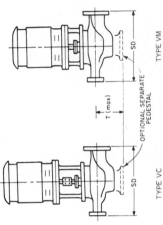

TYPE VC

TYPE VM

TABLE 2a Standard ANSI B73.2 Dimensions for Vertical In-Line Pumps—USCS Units

Pump designation, VC (or VM)[a]	ANSI 125, 150, 250, or 300 flange size, in		SD (+0.10, −0.08), in	T (max), in
	Suction	Discharge		
2015/15	2	1½	14.96	
2015/17	2	1½	16.93	6.89
2015/19	2	1½	18.90	
3015/15	3	1½	14.96	
3015/19	3	1½	18.90	7.87
3015/24	3	1½	24.02	
3020/17	3	2	16.93	
3020/20	3	2	20.08	7.87
3020/24	3	2	24.02	
4030/22	4	3	22.05	
4030/25	4	3	25.00	8.86
4030/28	4	3	27.95	
6040/24	6	4	24.02	
6040/28	6	4	27.95	9.84
6040/30	6	4	29.92	

[a]Sequence defines design, suction flange size, discharge flange size, and SD dimension.

TABLE 2b Standard ANSI B73.2 Dimensions for Vertical In-Line Pumps—SI Units

Pump designation[a], VC (or VM)	Flange size, mm		SD (+2.5, −2.0), mm	T (max), mm
	Suction	Discharge		
50-40-380	50	40	380	175
50-40-430	50	40	430	175
50-40-480	50	40	480	175
80-40-380	80	40	380	200
80-40-480	80	40	480	200
80-40-610	80	40	610	200
80-50-430	80	50	430	200
80-50-510	80	50	510	200
80-50-610	80	50	610	200
100-80-560	100	80	560	225
100-80-635	100	80	635	225
100-80-710	100	80	710	225
150-100-610	150	100	610	250
150-100-710	150	100	710	250
150-100-760	150	100	760	250

[a]Sequence defines design, suction flange size, discharge flange size, and SD dimension.

FURTHER READING

Fontana, M. G., and N. D. Greene: *Corrosion Engineering*, McGraw-Hill, New York, 1978.

Lee, J. A.: *Materials of Construction for Chemical Process Industries*, McGraw-Hill, New York, 1950.

National Association of Corrosion Engineers: *Corrosion Data Survey*, NACE, Houston, 1967.

National Association of Corrosion Engineers: *Proceedings, Short Course on Process Industry Corrosion*, NACE, Houston, 1960.

SECTION 9.7
PETROLEUM INDUSTRY

A. W. ELVITSKY

USE OF PUMPS

Pumps of all types are used in every phase of petroleum production, transportation, and refining.

Production pumps include reciprocating units for mud circulation during drilling and sucker-rod, hydraulic rodless, or motor-driven submersible centrifugal units for lifting crude to the surface. The largest use of centrifugal pumps in production is for water-flooding (secondary recovery, subsidence prevention, or pressure maintenance).

Transportation pumps include units for gathering, for on and offshore production, for pipelining crude and refined products, for loading and unloading tankers, tank cars, barges, or tank trucks, and for servicing airport fueling terminals. The majority of the units are centrifugal.

Refining units vary from single-stage centrifugal units to horizontal and vertical multistage barrel-type pumps handling a variety of products over a full range of temperatures and pressures. Centrifugal pumps are also used for auxiliary services, such as cooling towers and cooling water.

CENTRIFUGAL PUMPS

Except for some comments about the use of displacement pumps for handling viscous liquids, this section is restricted to centrifugal pumps—the type most frequently used in the petroleum industry—and includes a detailed analysis of the principal types.

Refinery Pumps   Major refinery processes that utilize centrifugal pumps are crude distillation, vacuum tower separation, catalytic conversion, alkylation, hydrocracking, catalytic reforming, coking, and hydrotreatment for the removal of sulfur and nitrogen. The products resulting from these processes include motor gasoline, commercial jet fuel and kerosene, distillate fuel oil, residual fuel oil, and lubricating oils. The American Petroleum Institute Standard 610, which is periodically revised, has established minimal specifications for the design features required for centrifugal pumps used for general refinery service.

CONSTRUCTION   Figure 1 illustrates the details of a refinery process pump meeting API 610 requirements. The stuffing box is equipped with a mechanical seal. The suction nozzle may be

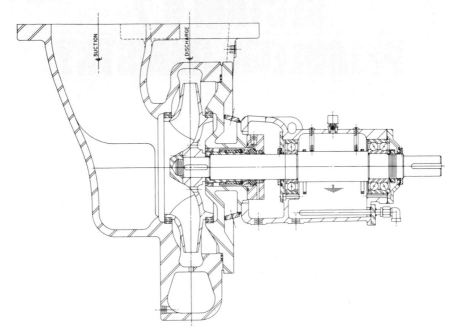

FIG. 1 Top-suction, single-stage, double-volute, centerline-mounted, overhung, radially split case refinery process pump with mechanical seal. (United Centrifugal Pumps)

located either at the end or at the top. If a spacer coupling is used between the pump shaft and driver shaft, the bearing bracket and cover may be removed without disconnecting the suction or discharge piping. This disassembly feature is common to all present-day refinery pumps. Because of the overhung shaft design, there is only one stuffing box to seal from leakage to the atmosphere. The thrust bearing for this arrangement, however, is subject to a load caused by exposure of one end of the shaft to suction pressure. To meet API Standard 610, the bearings must be designed for a minimum life of 16,000 h.

For suction pressures in excess of 250 lb/in² (1724 kPa) gage, a common practice is to make the diameter of the back wearing rings smaller than that of the suction side wearing rings, or eliminate them completely, in order to produce an equalizing thrust in the opposite direction and decrease the net load on the thrust bearing. Where casting limitations permit, double-volute pumps are used to limit the radial load imposed on the impeller by uneven pressure distributions in the casing. To comply with API 610, the shaft deflection at the wearing rings for one- and two-stage pumps, under most severe dynamic conditions, must be limited to one-half the minimum diametrical clearance specified. For operating temperatures above 500°F (260°C), the API recommends that consideration be given to increasing the minimum specified clearances. This is necessary if dissimilar metals are used or if a rapid warm-up procedure is to be employed. To aid in maintaining alignment at various temperatures, pump mounting feet are located on the case on the same centerline as the shaft. For pumping temperatures above 400°F (200°C), bearing cooling or stuffing box cooling is used to limit heat transfer along the shaft and/or lower the environmental temperature of the mechanical seal or packing. For NPSH requirements lower than the capability of a single-suction design, double-suction pumps are available for capacities above 600 gpm (136 m³/h).

Overhung shaft construction is nominally limited by most manufacturers to pumps with drivers rated below 500 hp (373 kW). Units with greater power are designed with bearings on both ends of the shaft, with the impeller or impellers in between (commonly designated as two-bearing or inboard-outboard bearing design). Ball bearing construction in compliance with API 610 is used up to a limit of DN factor of 300,000. The DN factor is the product of the bearing bore (D) in

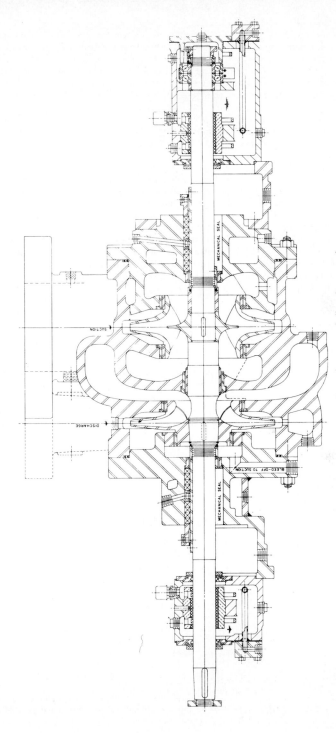

FIG. 2 Top-suction, double-suction, first-stage, two-stage, centerline mounted, radially split case refinery process pump. (United Centrifugal Pumps)

9.103

millimeters and the operating speed (N) in revolutions per minute. For values above 300,000, double-bearing construction is used with sleeve radial bearings and ball or hydrodynamic thrust bearing. The double-suction, two-stage, two-bearing features may be combined into one design, as shown in Fig. 2. Note that two stuffing boxes are required, both under suction pressure through the use of a bleedoff arrangement from behind the second stage. Two-stage units are available in either overhung or two-bearing construction up to 500 hp (373 kW) and in two-bearing construction above that value.

An innovation which has challenged the use of the standard process unit is the vertical in-line pump (Fig. 3). This unit has the advantage of relatively low installation cost, since it is simply mounted in the line as a valve would be, and it is not subject to pipe strains, which are a common problem on standard units.

PERFORMANCE   Figure 4 illustrates the wide range of capacity and head requirements that can be met by refinery pumps. The range shown varies slightly with the manufacturer. The range of in-line pumps is almost identical to that of horizontal units.

MATERIALS   Refinery pumps handle a variety of products with specific gravities from 0.3 to 1.3 and viscosities from values below that of water to as high as 15,000 SSU (3240 cSt) for centrifugal pumps and even higher for rotary pumps, over a wide range of temperatures. The product may be as inert as a lubricating oil or extremely corrosive. Many materials are utilized to satisfy these many requirements; the most common are given in Table 1. Caution must be exercised in selecting material for rotating parts because the physical properties of cast iron, bronze, and Ni-Resist severely limit the allowable peripheral speed for these materials. Material specifications are given in Table 2. Stuffing boxes are normally equipped with balanced mechanical seals. In special cases, double or tandem mechanical seals are used. One of the sealing rings is made of carbon, and the mating ring is either stellite, Ni-Resist, or tungsten carbide. Mechanical seals have proved extremely reliable and are not subject to the erratic performance sometimes encountered with packing.

Drivers for refinery pumps are electric motors or steam turbines. Centrifugal pumps run backward as hydraulic turbines are also used. To ensure safety and reliability, it is common practice

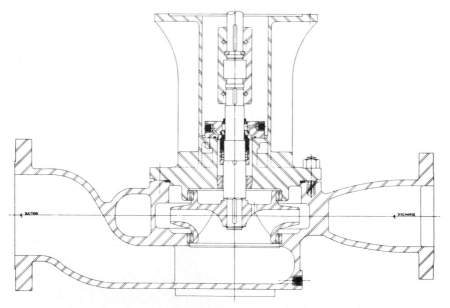

FIG. 3   Vertical, in-line, single-stage, double-volute, radially split case refinery process pump. (United Centrifugal Pumps)

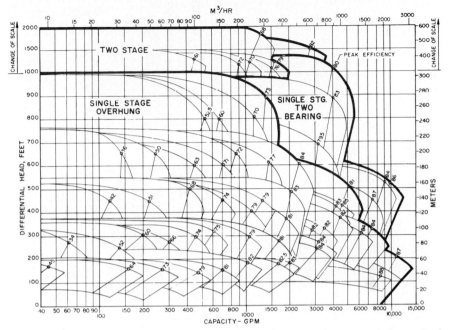

FIG. 4 Performance coverage for single-stage and two-stage refinery process pumps. (United Centrifugal Pumps)

to subject all refinery pumps to a hydrostatic test equivalent to that required by the discharge flange rating. In addition, a performance test in accordance with the Hydraulic Institute standards is normally specified. If critical suction conditions warrant it, an NPSH test may also be conducted.

Pipeline Pumps Centrifugal pumps are used on every major liquid pipeline in the world to transport a variety of fluids, including crude oil, motor gasoline, fuel oil, jet fuel, liquefied petroleum gases, and anhydrous ammonia. Pump efficiency is of primary importance because of the power required to transport the liquid. Most pipeline systems install pumps in series because the differential head is primarily made up of energy loss due to friction and an outage of one of the units would result in only a partial loss in the through-put capacity. For pipelines where the differential head is mostly static, such as the Trans-Alpine line, pumps are installed in parallel. Series arrangement in a static system would be unsuitable because the differential head required could not be obtained unless all of the units were operating.

CONSTRUCTION A single-stage double-suction pump, a two-stage double-suction pump, and a four-stage pipeline pump are shown in Figs. 5, 6, and 7, respectively. All the units are double-volute and axially split. The single-stage double-suction unit is typical of those used in series operation on large-diameter pipelines with capacities as high as 2 million barrels per day. One pipeline alone has installed more than one hundred 5000-hp (3730-kW) pumps of this construction. The two-stage double suction pump is similar to Trans-Alpine pumps. The unit illustrated in Fig. 6 is a 12,000-hp (8950-kW) pump operating at 3000 rpm in parallel. For lateral and smaller-diameter pipeline pumping stations, multistage units (Fig. 7) are used. The number of stages chosen corresponds to the differential head required. Multistage units may be destaged initially to produce lower heads, with a subsequent power saving, or the nozzling may be arranged for series or parallel operation of a portion of the stages. The exact arrangement depends on the system characteristics and initial and ultimate capacities to be pumped.

TABLE 1 Refinery Process Pump Materials

Trim part	Steel case, cast iron trim	Steel case, bronze trim	Steel case, Ni-Resist trim	Steel case, 11-13 chrome trim	11-13 chrome case and trim	18-8 S.S. case and trim	316 S.S. case and trim
Cover	Cast steel	Cast steel	Cast steel	Cast steel	11-13 chrome	18-8 S.S.	316 S.S
Case stud	ASTM-A193 GR B7	ASTM-A193 GR B7	ASTM-A193 GR B7	ASTM-A193 GR B7	ASTM-A193 GR B7	ASTM-A193 GR B7	ASTM-A193 GR B7
Case nut	ASTM-A194 GR 2H	ASTM-A194 GR 2H	ASTM-A194 GR 2H	ASTM-A194 GR 2H	ASTM-A194 GR 2H	ASTM-A194 GR 2H	ASTM-A194 GR 2H
Shaft	SAE 4140 HT	SAE 4140 HT	SAE 4140 HT	SAE 4140 HT	SAE 4140 HT	18-8 S.S.	316 S.S.
Impeller	Cast iron	Bronze	Ni-Resist	Cast steel	11-13 chrome	18-8 S.S.	316 S.S.
Impeller wear ring	Cast iron	Bronze	Ni-Resist	11-13 chrome hardened	11-13 chrome hardened	18-8 S.S. hard-faced	316 S.S. hard-faced
Case wear ring	Cast iron	Bronze	Ni-Resist	11-13 chrome hardened	11-13 chrome hardened	18-8 S.S. hard-faced	316 S.S. hard-faced
Shaft sleeve, packed pump	11-13 chrome hardened	11-13 chrome hardened	11-13 chrome hardened	1020 hard-faced	11-13 chrome hardened	18-8 S.S. hard-faced	316 S.S. hard-faced
Shaft sleeve, mechanical seal	11-13 chrome	11-13 chrome	11-13 chrome	11-13 chrome	11-13 chrome	18-8 S.S.	316 S.S.

Gland	Steel	Steel	Steel	Steel	Steel	18-8 S.S.	316 S.S.
Gland stud	ASTM-A193 GR B7	ASTM-A193 GR B7	ASTM-A193 GR B7	ASTM-A193 GR B7	ASTM-A193 GR B7	ASTM-A193 GR B8	ASTM-A193 GR B8M
Lantern ring	Cast iron	Cast iron	Cast iron	Cast iron	Cast iron	18-8 S.S.	316 S.S.
Throat bushing	Cast iron	Bronze	Ni-Resist	11-13 chrome hardened	11-13 chrome hardened	18-8 S.S.	316 S.S.
Throttle bushing	Cast iron	Bronze	Ni-Resist	11-13 chrome hardened	11-13 chrome hardened	18-8 S.S. hard-faced	316 S.S. hard-faced
Gasket, sleeve	18-8 S.S. annealed	18-8 S.S. annealed	18-8 S.S. annealed	18-8 S.S. annealed	18-8 S.S. annealed	18-8 S.S. annealed	316 S.S. annealed
Gasket, case	18-8 S.S. with asbestos	18.8 S.S. with asbestos	18-8 S.S. with asbestos	18-8 S.S. with asbestos	18-8 S.S. with asbestos	18-8 S.S. with asbestos	316 S.S. with asbestos
Impeller nut	Steel	Steel	Steel	Steel	Steel	18-8 S.S.	316 S.S.
Bearing shield	Bronze	Bronze	Bronze	Bronze	Bronze	Bronze	Bronze
Oil rings	Brass	Brass	Brass	Brass	Brass	Brass	Brass
Bearings bracket	Cast iron	Cast iron	Cast iron	Cast iron	Cast iron	Cast iron	Cast iron
Heat exchanger assembly	Steel	Steel	Steel	Steel	Steel	Steel	Steel
Coupling guard	Fab steel	Fab steel	Fab steel	Fab steel	Fab steel	Fab steel	Fab steel
Base plate	Fab steel	Fab steel	Fab steel	Fab steel	Fab steel	Fab steel	Fab steel

TABLE 2 Material Specifications

Material	Castings	Forgings	Bars	Studs	Nuts
Cast iron	A-48[a]				
Ni-Resist	A-436, Types 1 and 2				
Bronze	B-143 Alloy 1A B-143 Alloy 2B B-144 Alloy 3B B-145 Alloy 4A		B-139 Alloy 510	B-124 Alloy 655	
Carbon steel	A-216 GR WCB	A-266, Class 1 A-266, Class 2	A-108 GR 1018 A-575 GR 1020		A-108 GR 1018
Alloy steel (SAE 4140) 11–13 chrome steel	A-296 GR CA-15	A-182 GR F-6 A-336 CL F-6	A-434 Class BC or BD A-276 Type 410, 416, or 420	A-193 GR B-7 A-193 GR B6	A-194 GR 2H A-194 GR 6
18–8 stainless steel	A-296 GR CF-8	A-182 GR F-304 A-366 CL F-8	A-276 Type 304	A-193 GR B8	A-194 GR 8
316 stainless steel	A-296 GR CF-8M	A-182 GR F-316 A-366 CL F-8M	A-276 Type 316	A-193 GR B8M	A-194 GR 8

[a]All entries are ASTM numbers.

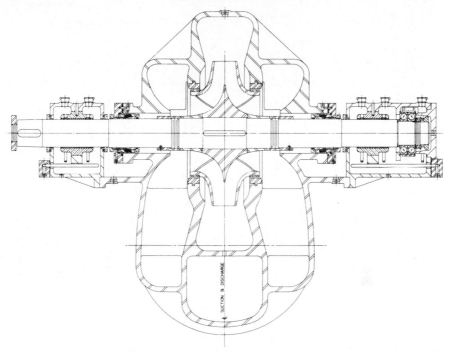

FIG. 5 Double-suction, double-volute, axially split, single-stage pipeline pump with mechanical seals. (United Centrifugal Pumps)

All units are equipped with sleeve radial bearings and either ball or hydrodynamic thrust bearings, if rotating speeds dictate. The single- and two-stage double-suction pumps are inherently balanced axially. The multistage units utilize opposed impeller design to obtain axial balance. All modern units are equipped with mechanical seals in the stuffing box. Since most pipeline stations are now designed for unattended outdoor service, a number of safeguard controls are used. These include warnings for low suction pressure, high discharge pressure, high bearing or case temperature, vibration, and excessive seal leakage. Tank farms utilize single- and double-suction in-line vertical pumps or, if the tank is to be pumped dry, vertical canned pumps.

PERFORMANCE   Figure 8 indicates the typical available range for single- and two-stage pipeline pumps, and Fig. 9 shows the range for multistage units. Pipeline units are custom-designed, and the range is being extended daily. Electric motors are normally used as drivers. With the utilization of gas turbines, the maximum power employed has increased dramatically in the last decade. Another use of gas turbines on existing lines is for peak-capacity booster station service, in which case pumps are direct-driven at operating speeds as high as 6000 rpm.

MATERIALS   Table 3 lists pump part materials commonly specified for crude and product pipeline service. Also included is a list of materials for waterflood pumps, to be discussed later.

The typical piping of a main-line pump station in which the units are arranged in series is shown in Fig. 10. As pipeline capacities have increased, one of the major problems has been the pressure loss at each station. For the station shown, three discharge valves, one 16-in (406-mm) ball valve manifolded in parallel with two 24-in (610-mm) control valves, were used. This arrangement limited the calculated pressure drop to that of a length of 36-in (914-mm) pipe equal to the distance across the manifold and still allowed the use of control valves of proven size. When line conditions require throttling, the 16-in (406-mm) ball valve is first completely closed. If additional throttling is needed, the two 24-in (610-mm) control valves are closed to produce the required

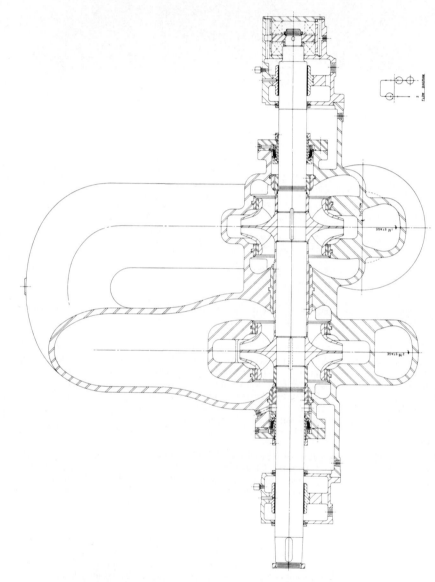

FIG. 6 Double-suction, double-volute, axially split, two-stage pipeline pump with mechanical seals. (United Centrifugal Pumps)

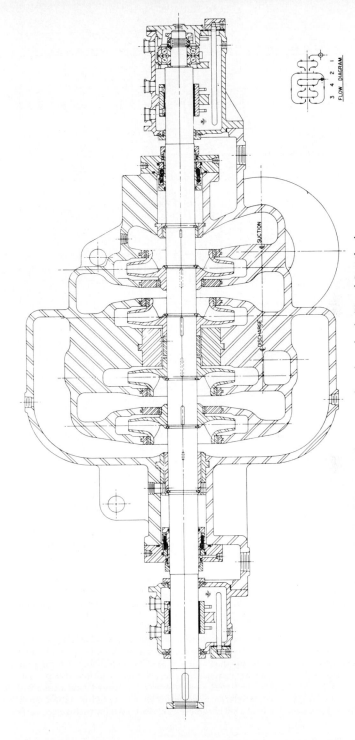

FIG. 7 Single-suction, double-volute, axially split, four-stage pipeline pump with mechanical seals. (United Centrifugal Pumps)

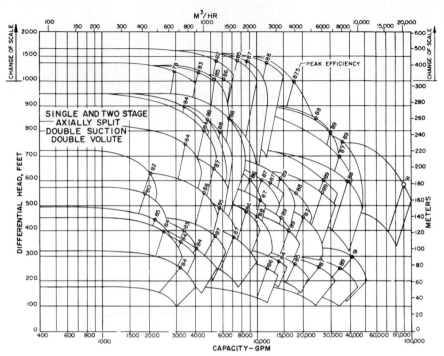

FIG. 8 Performance coverage for double-suction, double-volute, axially split single-stage and two-stage pumps. (United Centrifugal Pumps)

pressure drop. In order to move scrapers or batch separators through each station without interrupting flow through the pumps, signals (PIG SIG), hydraulically operated 24-in (610-mm) valves, and sequence control wiring are used. The distance between PIG SIG 1 and PIG SIG 2 represents the volume of station loop to be displaced. The tripping of PIG SIG 3 and PIG SIG 4 controls the valve opening and closing required to divert flow from the station discharge piping to behind the scraper or batch separator and force it to leave the station in the same relative position as it entered. Elbows at pump suction are arranged so as to avoid uneven flow distribution in the inlet of the double-suction impeller.

Pipeline pumps are tested in accordance with Hydraulic Institute standards. Reduced-speed tests may be required because of the power limitations of manufacturing plants and available drivers, and such tests have proved to be extremely accurate representations of the full-speed performance.

Special Services Because of the extreme range in heads and capacities, operating pressures, and temperatures, the petroleum industry utilizes a great number of pumps designed specifically for a given service or process. A few of these special services are as follows.

WATERFLOOD PUMPS Centrifugal pumps are used for water injection when the capacity required is above 10,000 barrels per day. Injection pressures vary considerably. One of the highest is obtained through the use of the pump shown in Fig. 11. Two of these nine-stage units, when operated in series and driven by a 6000-rpm gas turbine, are capable of injection pressures as high as 8000 lb/in² (55,200 kPa) gage. The barrel was constructed of forged steel with a 316 stainless steel welded overlay applied to the interior and hydrostatically tested at 12,000 lb/in² (82,700 kPa) gage. Other parts were made of 18-8 chrome-nickel steel overlaid with a hard-surface mate-

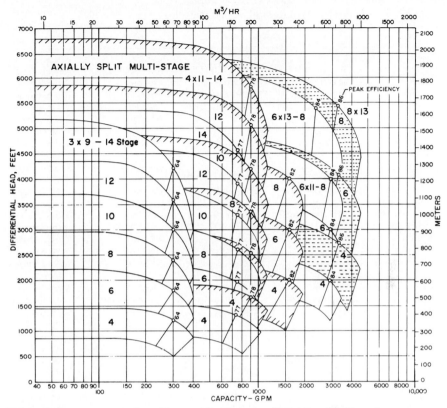

FIG. 9 Performance coverage for multistage axially split pumps. (United Centrifugal Pumps)

rial at all close-clearance parts. The shaft was of K-500 Monel. The stuffing box contained a limited-leakage breakdown bushing for increased reliability at the high operating pressures and speeds. The majority of waterflood injection pressures are such that axially split multistage pumps of the design shown in Fig. 7 may be used. As many as 14 stages in one pump are utilized to obtain the required differential head; if still greater pressures are required, the units may be piped in series. Axially split pumps have been hydrostatically tested to 6000 lb/in² (41,400 kPa) gage for a working pressure of 4000 lb/in² (27,600 kPa) gage. Materials are given in Table 3.

REACTOR FEED PUMPS   One of the most difficult services encountered in the petroleum industry is the high-temperature, high-pressure reactor feed, or charge, pump. The construction is similar to that of the waterflood pump shown in Fig. 11, but the reactor feed unit is subject to operating temperatures in the range of 600 to 700°F (316 to 371°C) with discharge pressures of 2000 lb/in² (13,800 kPa) gage or more. The corrosiveness of the pumped fluid requires the use of a number of dissimilar metals, and compensation must be made for the different rates of expansion at the operating temperatures. Extreme care is required in the start-up and operation of these units.

LPG   Multistage axially split centrifugal pumps have been installed in pipelines transporting liquefied petroleum gases with specific gravities as low as 0.35. The low lubricity of the pumped fluid requires careful selection of wearing-part materials. Stuffing boxes utilize single or double mechanical seals. At ambient temperatures, the suction pressures are as high as 1000 lb/in² (6900 kPa) gage. These fluids experience a large temperature rise due to compression.

TABLE 3 Crude Pipeline, Product Pipeline, and Waterflood Pump Materials

Part	Crude		Product	Waterflood
Case	Cast steel	Cast steel	Cast steel	316 S.S.
Shaft	SAE 4140 HT	SAE 4140 HT	SAE 4140 HT	Monel
Impeller	Cast steel	Bronze	Cast steel	316 S.S.
Impeller wear ring	Type 410 350-375 BHN	Type 410 350-375 BHN	Type 410 350-375 BHN	316 S.S. hard-faced
Case wear ring	Ni-Resist	Bronze	Ni-Resist	316 S.S. hard-faced
Case separation ring	Ni-Resist	Cast iron	Ni-Resist	316 S.S. hard-faced
Shaft sleeve, high pressure	Type 410 375-400 BHN	Type 410 375-400 BHN	Carbon steel	316 S.S. hard-faced
Shaft sleeve, low pressure	Type 410	Type 410	Carbon steel	316 S.S. hard-faced
Intermediate shaft sleeve	SAE 1020 hard-faced	Type 410 375-400 BHN	Cast iron	316 S.S. hard-faced
Throttle bushing	Ni-Resist	Bronze	Ni-Resist	316 S.S. hard-faced
Throat bushing	Ni-Resist	Bronze	Ni-Resist	316 S.S. hard-faced
Gasket, case, split	Asbestos composition	Asbestos composition	Asbestos composition	Asbestos composition
Bearing bracket	Cast iron	Cast iron	Cast iron	Cast iron
Sleeve bearing	Bronze, babbitt-lined	Bronze, babbitt-lined	Bronze, babbitt-lined	Bronze, babbitt-lined
Oil-ring carrier	Cast iron	Cast iron	Cast iron	Cast iron
Bearing shield	Bronze	Bronze	Bronze	Bronze
Oil ring	Red brass	Red brass	Red brass	Red brass
Splitters	18-8 S.S.	18-8 S.S.	18-8 S.S.	316 S.S.
Impeller locating rings	18-8 S.S.	18-8 S.S.	18-8 S.S.	316 S.S.

Many additional special designs could be mentioned. The wide range of operating conditions and products in the petroleum industry frequently requires pumps specifically engineered for a certain service. The pump designer is continually challenged to provide a safe, reliable, and economical design.

PUMPING OF VISCOUS LIQUIDS

In many petroleum industry applications, an important factor in the selection of a pump is the viscosity of the liquid. As liquid viscosity increases, the pump is faced with a more difficult task to perform, and an understanding of the relationship between viscosity and pump performance becomes essential to proper sizing of both the pump and its driver.

Centrifugal pumps are routinely applied on services having viscosities below 3000 SSU (660 cSt) and may be used up to 15,000 SSU (3300 cSt). (For background information on viscosity, refer to Sec. 8.1.) They are sensitive, however, to changes in viscosity and will exhibit significant reductions in capacity and head, and rather drastic reductions in efficiency, at moderate to high

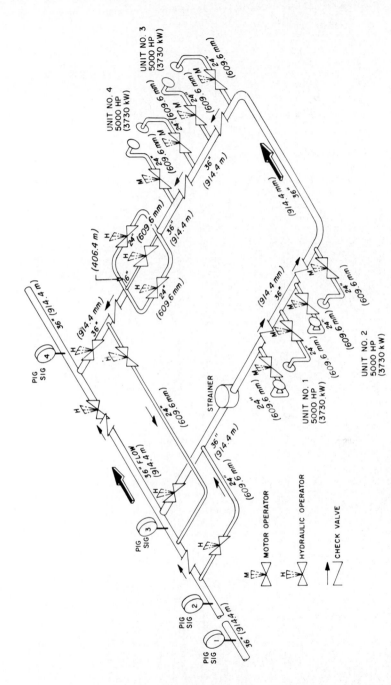

FIG. 10 Main-line pumping station piping. (United Centrifugal Pumps)

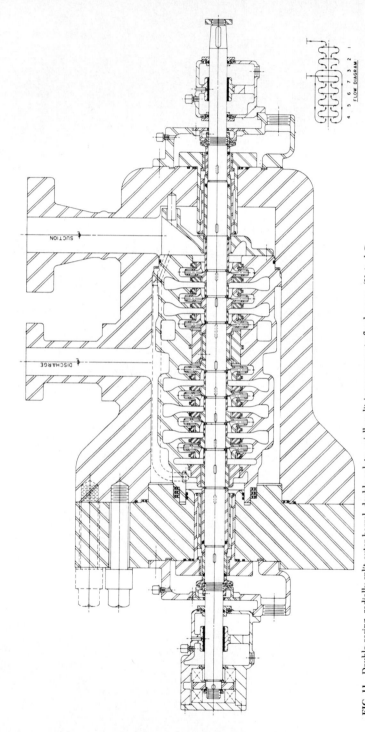

FIG. 11 Double-casing, radially split outer barrel, double-volute, axially split inner case waterflood pump. (United Centrifugal Pumps)

TABLE 4 Effect of Viscosity on Performance of a Typical Centrifugal Pump Operating at Best Efficiency Point

Viscosity, SSU (cSt)	Capacity, gpm (m³/h)	Total head, ft (m)	Efficiency, %	Brake power, bhp (kW)[a]
Nil (nil)	3000 (681)	300 (91)	85	241 (180)
500 (110)	3000 (681)	291 (89)	71	279 (208)
2,000 (440)	2900 (658)	279 (85)	59	312 (233)
5,000 (1110)	2670 (606)	264 (80)	43	373 (278)
10,000 (2200)	2340 (531)	243 (74)	31	417 (311)
15,000 (3300)	2100 (477)	228 (69)	23	473 (353)

[a]All values of brake power based on liquid having a specific gravity of 0.90.

values of viscosity. The extent of these effects may be seen in Table 4, constructed with the aid of Fig. 12, which provides a convenient means for determining the viscous performance of a centrifugal pump when its water performance is known. To use Fig. 12, enter at the bottom with the pump capacity, proceed vertically upward to the total head (head per stage for multistage pumps), then horizontally right or left to the viscosity, and vertically upward again to the curves for correction factors. The values thus obtained for the respective correction factors are multiplied by the water performance values for capacity, total head, and efficiency to obtain the equivalent values for viscous performance. By using the individual correction factors for total head, it is even possible to approximate the shape of the head-capacity characteristic when the viscous liquid is being handled, at least to 120% of the best efficiency point (Q_N). The total head at shutoff will remain essentially the same for viscous and nonviscous liquids.

Centrifugal pump performance is almost invariably specified by the manufacturer on the basis of handling clean cold water, even where the pump has been specifically designed for petroleum industry applications. Pump selections, however, must necessarily be made to satisfy viscous conditions of service and require application of these correction factors in the reverse direction. In this case, Fig. 12 provides an approximation of equivalent water performance that is probably within the limits of accuracy of the graph, for viscosities in Saybolt seconds universal numerically equal to pump capacity in gallons per minute. For higher viscosities, the initial solution for equivalent water performance, determined as indicated in the following paragraph, may need to be adjusted and then checked by conversion of water performance to viscous performance.

To determine approximate equivalent water performance when viscous performance is known, enter Fig. 12 at the bottom with the viscous capacity, proceed vertically upward to the desired viscous head (head per stage for multistage pumps), then horizontally right or left to the viscosity, and vertically upward to the correction factor curves for capacity and head. In this case, divide the viscous performance values by the correction factors to obtain the equivalent water performance values. The pump selection can then be made on the basis of ratings established for water, and efficiency can be calculated for the viscous liquid using the efficiency correction factor applied to the pump efficiency for water.

EXAMPLE To select a pump to handle 500 gpm (114 m³/h) of 3000-SSU (660-cSt) liquid against a head of 150 ft (46 m), proceed as follows:

From Fig. 12, determine $C_Q = 0.80$ and $C_H = 0.81$.

The water capacity is

in USCS units	$Q_W = 500/0.80 = 625$ gpm
in SI units	$Q_W = 114/0.80 = 142$ m³/h

and the water head is

in USCS units	$Q_H = 150/0.81 = 185$ ft
in SI units	$Q_H = 46/0.81 = 57$ m

For 625 gpm (142 m³/h), 185 ft (57 m), 3000 SSU (660 cSt), Fig. 12 indicates $C_Q = 0.83$,

10-106   Petroleum Industry

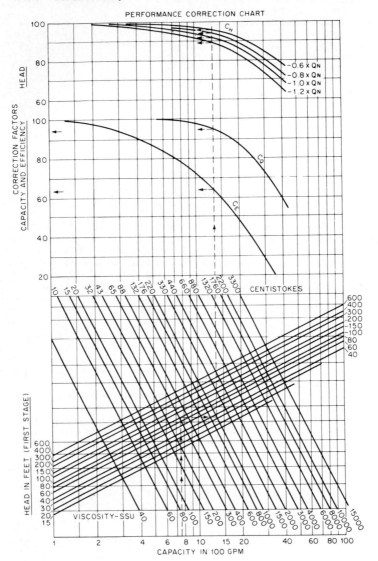

FIG. 12   Viscosity correction curves for centrifugal pumps. (1 gpm = 0.227 m³/h; 1 ft = 0.3048 m) (Reprinted from *Hydraulic Institute Standards*, 14th ed., Hydraulic Institute, Cleveland, 1983)

$C_H = 0.85$, and $C_E = 0.42$. If the values for C_Q and C_H obtained from these calculations are roughly within 2% of those taken directly from Fig. 12, the pump may be selected on the basis of the water capacity and water head already obtained. Since, in this case, these values differ from the first approximation by more than 3%, the water performance should be adjusted as follows.

Adjust the water capacity:

$$Q_W \times \frac{0.80}{0.83} = 602 \text{ gpm } (137 \text{ m}^3/\text{h})$$

Adjust the water head:

$$H_W \times \frac{0.81}{0.84} = 178 \text{ ft } (55 \text{ m})$$

Select the pump for 602 gpm (137 m^3/h) and 178 ft (55 m) and determine water efficiency from the manufacturer's rating.

Assuming the water efficiency in this case is 75%, then the viscous efficiency $E_V = 0.75 \times 0.42 = 0.315$, or 31.5%.

For a specific gravity of 0.90:

in USCS units $$\text{bhp} = \frac{\text{gpm} \times \text{ft of head} \times \text{sp. gr.}}{3960 \times \text{efficiency}} = \frac{500 \times 150 \times 0.9}{3960 \times 0.315} = 54$$

in SI units $$\text{brake kW} = \frac{\text{m}^3/\text{h} \times \text{m} \times \text{sp. gr.}}{376.5 \times \text{efficiency}} = \frac{114 \times 46 \times 0.90}{376.5 \times 0.315} = 40$$

The Hydraulic Institute standards limit the use of Fig. 12 to radial-flow centrifugal pumps with open or closed impellers, handling Newtonian liquids within the pump's normal operating range and within the capacity limits indicated on the chart. Failure to heed these limitations may result in misleading expectations of performance with viscous liquids. Tests on vertical turbine pumps, for example, which normally have mixed-flow rather than radial-flow impellers, have shown that, while the viscous efficiency predicted from this chart is quite accurate, the viscous head and viscous capacity predictions are substantially above the demonstrable values. In other words, the performance reductions determined from the chart would not be severe enough for that type of pump. For centrifugal pumps which do not fall within the limits of applicability of Fig. 12, viscous performance can be accurately established only by test.

Displacement pumps, both rotary and reciprocating, are also frequently used for pumping viscous fluids, and some types are well suited for use where viscosities are beyond the limit that can be handled with centrifugal pumps. In fact, many common designs are suitable only for use with liquids which are at least moderately viscous since they depend on the viscosity to maintain the lubricating and sealing films between the various internal parts of the pumps. Most gear pumps and screw pumps are in this category.

As with centrifugal pumps, the performance of displacement pumps may be significantly affected by changes in liquid viscosity, but the nature of these changes may be quite different. At constant speed, changes in viscosity are likely to have little or no effect on pump capacity. Total head, or differential pressure across the pump, would probably increase with increasing viscosity because of higher system resistance. Thus, the brake power required would increase even though pump efficiency would not suffer nearly so drastically, if at all, as for a centrifugal pump.

The influence of pump design on how performance varies with viscosity is greater in displacement pumps than in centrifugal pumps. Since there are so many designs, it is not practical to attempt to provide here a general means of determining their viscous performance. This should be done only on the basis of information provided by the pump manufacturer.

SECTION 9.8
PULP AND PAPER MILLS

J. F. GIDDINGS

Additive Any material such as clay, filler, or color added to stock to contribute specific properties.

Alkali *Total alkali:* Chemical used in cooking for soda process $Na_2CO_3 + Na_2SO_4$, all expressed as Na_2. *Active alkali:* Soda process NaOH only (as Na_2O); sulfate process $NaOH + Na_2S$ (as Na_2O). *Effective alkali:* Sulfate; $NaOH + \frac{1}{2} Na_2S$.

Chest Vessel for storing pulp.

Cooking Action of the chemical used to break down lignin bond between cellulose fiber in wood and other organic matter.

Digester Pressure vessel used to contain chemical action of cooking chemical and raw cellulose material.

Fourdrinier Continuous wire upon which pulp or paper sheet is produced.

Freeness A measure of degree of refinement of stock and, hence, of its ability to drain water.

Groundwood A mechanical pulp formed by simple grinding to break down wood structure.

Kraft Process Method of separating cellulose fiber from lignins by using caustic soda in presence of sulfur radical.

Lignin A generic term used to refer to the complex organic matter present in wood which acts as binding agent for cellulose fibers.

Liquor *Black liquor:* Solution of water plus residual organic matter (or lignins) in the wood after washing of raw stock. *Green liquor:* solution of smelt from recovery furnace when dissolved in either water or weak washed liquor. *White liquor:* Solution of caustic soda (or other alkali), sometimes in the presence of a sulfur radical; this is the liquor charged to the digester for cooking.

Neutral Sulfite A cooking process using a solution of about 10% sodium sulfite and 5% caustic soda mixed for cooking wood or agricultural fiber to produce high-yield pulp.

Refining A mechanical process carried out on stock to improve the ability of the fiber to form paper sheet. Different techniques are employed—some designed to shorten the fiber, others to increase the amount of fibrils (or "whisker") on the fiber.

Soda Process   Similar to kraft process but uses caustic soda without presence of sulfur.

Stock   A generic term for the suspension of cellulose fiber in water or chemicals. *Bleached stock:* The same after bleaching. *Brownstock:* The same before bleaching. *Raw stock:* The product discharged from digester(s) before any washing or other treatment.

Sulfidity

$$\frac{Na_2S}{NaOH + Ha_2S}$$

Sulfite Process   Method of separating cellulose fiber from lignin with acid.

Vat   Semicylindrical mold or container for holding stock during washing or sheet information.

Yield   Percentage of cellulose fiber in the form of pulp produced from a given weight of wood or other raw material. *High yield:* The same when some lignins are allowed to remain in the finished product.

GENERAL

Apart from the petrochemical industry, there are few continuous-process plants where the pumps are so much in evidence and where reliability is so essential as in the pulp and paper industry.

Before paper leaves the machine room, 100 to 200 tons of water will have been pumped to the mill for every ton of paper produced. These figures represent only the basic amount of water taken from a river and rejected as effluent; the amount of liquid circulated is several times greater.

Although many attempts have been made to utilize a dry process for papermaking, both pulping (the separating of crude fibers from the raw material) and papermaking (actual treatment of the fiber and mechanical formation of the sheet) still require water as the medium to convey fiber. Throughout the mill, pumps are used to transfer or circulate fibers suspended in water (stock), chemicals or solids in solution (liquors), or residues and waste matter as slurries, as well as to supply water for general services.

There are at least 150 pumps installed in a modern pulp mill and another 50 or so in a paper mill. About 1000 kW·h is required for the production of one ton of pulp, and a further 500 kW·h for the finished paper. Of this total, about 25% is used in pumping. This means that even in a medium-size mill the installed power required for pumps alone is approximately 10,000 hp (7457 kW), and often it is much higher.

PULPING PROCESSES

Raw Materials   In the past the traditional fibrous raw materials for the manufacture of paper were cloth and agricultural residues; today the vast bulk of cellulose pulp is made from wood. Although over half of this is produced from softwoods (long-fibered), an increasing amount is being now produced from hardwoods (short-fibered). Traditional raw materials, such as linen, cotton waste, straw, and agricultural residues, are still used in small quantities, particularly where the paper sheet requires special properties. These distinctions are important in the selection of pumps, for the liquors have different characteristics. For example, straw black liquor is much more viscous than wood black liquor, and the proper corrections must be made in calculating the pump performance and pipe frictional losses.

Many grades of paper include waste paper, and specialized plants exist for the processing of waste paper before it is used for papermaking. Special care is necessary in the selection of pumps to handle waste-paper stock because of the large amount of foreign matter present—rope, string, metal, and synthetic fiber—all of which can cause problems in the process and pumping.

Groundwood Pulp   This type of pulp is produced by simply grinding away wood by mechanical action. Almost all of the wood is used in the pulp, including many of the resins and other complex organic compounds. The fibers are bruised so that the pulp has inferior strength.

Large amounts of water are required for cooling and for carrying away the groundwood pulp; the latter is usually acidic (pH 4 to 5), and so corrosion-resistant materials must be used. The pulp is used primarily for newsprint and magazines. Depending on the end use, some mechanical treatment (refining) may be required to alter the characteristics of the pulp, particularly the viscosity. In some cases a mild bleach may be used to improve the color.

Semichemical, or Chip Groundwood, Pulp   To increase pulp yield, the wood can be treated with a mild chemical before it is subjected to mechanical treatment. The object is to break down to some extent the lignin bond, and this is achieved most efficiently when the wood is first chipped. The mechanical treatment is frequently effected by two roughened rotating disks. The chips are fed between them, and the action separates the fibers and produces a pulp. This type of pulp is often used as a filler for other products and for the manufacture of corrugating medium.

Chemical Pulping   Wood is a complex, nonuniform material containing about 50% by weight cellulose fiber, 30% lignins, and 18 to 20% carbohydrate. The remainder is proteins, resins, and other complex organic compounds, which vary from one species to another. Cellulose resists attack from most chemicals, whereas the carbohydrates and other organic materials generally form compounds with the chemical cooking liquor. Some paper products can use the carbohydrate fraction to contribute bulk to the sheet, and for such papers groundwood and semichemical pulps are used. Where high strength is required, cooking is necessary to separate the fibers completely from the remainder of the wood.

Most cooking of wood is done in a pressure vessel at high temperature and pressure in the presence of an acid or alkali.

There is considerable tradition in chemical pulping, and a number of different processes are used. For many years the traditional method of producing pulp for high-grade papers was the acid sulfite process. This has been largely superseded in recent years by an alkaline process using sodium-based liquors in the presence of a sulfur radical; this is known as the sulfate or kraft process. The main reasons for the change to the sulfate process have been lower corrosion rates, ease of chemical recovery, and a stronger pulp. The properties of the liquids pumped in the two processes are different, and the pumps require different materials of construction.

Typical Sulfate Process   Pulpwood logs are first chipped to about ¾ by ⅛ in (19 by 3 mm) and then charged into either a continuous digester (Kamyr type) or a batch digester, each having a volume of about 5000 ft³ (141 m³). Cooking liquor (NaOH plus up to 30% Na₂S) is then allowed to react with the wood chips for 2 to 2½ h at a temperature up to 350°F (177°C) and a pressure in the digester of 80 to 100 lb/in² (551 to 689 kPa). In many mills the heating of the chips and cooking liquor is by direct steam injection to the digester. In others some form of indirect heating is used with a closed recirculation system. In the latter case the digester circulating pumps are a critical item since they must handle hot caustic solutions and entrained solid matter in a closed, pressurized circuit. After cooking, the contents of the digester are discharged to atmospheric pressure into a vessel called the blow tank, where the sudden expansion causes the fibers to separate from the liquid, which is now known as black liquor.

At this point, the process splits into two streams—one for fiber processing and the other for chemical recovery. The fiber is washed and screened and then formed into a pulp or paper sheet. The black liquor is washed from the pulp and treated for chemical recovery. Because the most troublesome liquors are to be found in the recovery process and bleach plant, the selection of these pumps is critical for the successful operation of the process.

The chemistry of the recovery process is as follows. After concentration of the black liquor in multiple-effect evaporators to about 50% total solids, the final concentration to 60 to 65% is done by direct contact with hot flue gas from the waste heat or recovery boiler. The 65% concentration black liquor is mixed with salt cake (Na₂SO₄) before being sprayed into the furnace under pressure generated by high-pressure pumps. The furnace atmosphere is maintained with a minimum of excess air so that the Na₂SO₄ is reduced to Na₂S, and sodium carbonate (Na₂CO₃) is formed in the process. These molten chemicals run out as a smelt and are dissolved in a tank to form green liquor. This liquor is then causticized with lime to form caustic soda (NaOH), with the Na₂S still present along with other residual chemicals, thus forming the regenerated cooking liquor known as white liquor. The calcium carbonate (CaCO₃) formed is burned in a lime kiln for reuse in

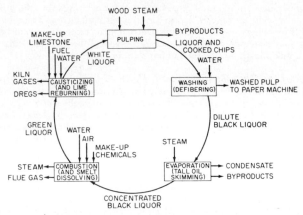

FIG. 1 The recovery cycle in the sulfate process.

causticizing. Various lime slurries and residues are formed during this process. The white liquor is then clarified and reused in the digesters, completing the cycle, as shown in Fig. 1.

There are a variety of other pulping processes in use, but the sulfate process offers so many advantages that almost all recent installations have been of this type.

Bleaching   Bleaching may be considered an extension of the cooking process, the object being to remove the coloring matter, carbohydrate, and lignins so that the remaining pulp contains a maximum percentage of alpha cellulose, which is the purest cellulose form and the one most resistant to attack from normal chemicals. Because of this resistance, special highly reactive chemicals, such as chlorine, must be used for bleaching; these produce corrosive liquors in the bleach plant.

LIQUIDS PUMPED IN A MILL

There are, broadly, three categories of liquids to be pumped in a paper mill:

1. *Water* and similar fluids
2. *Liquors* and *slurries*—mainly chemicals and solids in solution or suspension
3. *Stock*—suspension of cellulose fibers in water

Water   Apart from the quantities involved, there are no special requirements concerning the water in pulp and paper mills, for operating conditions are well within normal limits. However, iron and carbon steel piping should not be used in bleach pulp mills because of iron pickup.

Process water treatment is frequently used to purify process water for the mill and to remove undesirable elements such as iron. Higher-quality water is required for chemical reparation in the bleach plant and for boiler feedwater where demineralizer plants are used. Rubber- or epoxy-resin-lined pumps are used for those components in contact with the demineralized water.

PUMPS FOR MILL WATER   For the majority of pumps, standard cast iron, bronze, or stainless steel fittings are used except as noted for demineralized water. In many mills, however, stainless-steel-fitted pumps are standard because this permits a minimum number of spares to be held in stock for other duties.

In the paper mill, water that has been removed from the sheet on the paper machine has a very low fiber content—less than 0.05%—and is known as white water. In some installations, fiber recovery plants are used, but the fiber content does not usually cause any pumping problems.

Much of this water is recirculated, and where bleached products are produced, pumps must be constructed of a low-grade stainless steel. All bronze is also suitable for service with bleached products.

Liquors and Slurries   Depending on the process and the particular point in that process, the liquor characteristics may require special pumps or special materials. Although the liquor cycle is a difficult one as far as the pumps are concerned, standard designs should be used whenever possible since this reduces the number of different types of pumps in the mill. In some cases it may be necessary to use a higher material specification than necessary to achieve interchangeability.

Liquor and slurry pumps may be grouped as follows:

Group A   Standard designs suitable for most process uses where corrosion is not a major factor. Pumps may be 304 stainless steel with casings of Meehanite or 2 to 5% nickel iron.

Group B   Standard end-suction designs suitable for corrosive liquors. Pumps may be 316 stainless steel fitted with casing of 316 stainless steel.

Group C   Standard or nonstandard designs suitable for special services. Pumps are similar to group B for most applications but are of 317, or 317L stainless steel. For most corrosive services, glass-reinforced epoxy resin, rubber, titanium, and stainless steels similar to alloy 20 are used for both impeller and casing. Mechanical seals in place of packed boxes are usually fitted to these pumps.

Recommendations for liquor and slurry pumps are

1. All liquor pumps should be classified as slurry type with open nonshrouded impellers of the end-suction and back pull-out type.
2. Where stainless-steel-fitted pumps are used, different metals or different harnesses should be used for materials in close contact. If this is not feasible because of the stainless steel grades required, clearances should not be less than 0.040 in (1 mm).
3. For group C pumps, in particular, it may be necessary to depart from a standard design or type of centrifugal pump. For example, if a positive displacement characteristic is required, then a screw-type pump may be used with confidence. In addition, all pumps handling stock with consistency above 2 to 2½% must be regarded as nonstandard types.

Once the pumps are grouped as above, it becomes necessary to decide which pump may be used for specific liquors. Requirements for individual mills will differ in detail, but the following may be taken as an indication of current practice, particularly in modern sulfate mills. In every case, manufacturers should be made aware of the liquor characteristics and of the location of the pump in the process.

COOKING LIQUOR (WHITE LIQUOR—SULFATE PROCESS)   This is essentially an alkaline solution made by causticizing green liquor. The liquor is prepared at concentrations over the range of 50 to 100 g/liter depending on the wood species, and the amount of active alkali (expressed as Na_2O) may be from 14 to 30% of the dry wood weight. White liquor is mainly sodium hydroxide, with a small percentage of sodium sulfide which depends on the mill sulfidity. Higher values of active chemical are used in bleached pulp mills. The term *sulfidity* is used to denote the ratio of chemicals present; it is frequently expressed as Na_2O and calculated from the expression

$$\frac{Na_2S}{NaOH + Na_2S}$$

The sulfidity value commonly used is from 20 to 30%; the higher values usually denote better chemical recovery. The specific gravity of the liquor will be approximately 1.2, and after clarification only small quantities of grit should be present. The liquor must be considered an abrasive that produces a high rate of wear on pump rotating elements. White liquor has a tendency to crystallize on internal surfaces of pipes and pumps, but there are no special viscosity problems and a pump head loss allowance of about 10% above that of water should be adequate. Group B pumps are recommended.

COOKING LIQUOR (SULFITE PROCESS)　　This is essentially an acidic solution, mainly calcium or sodium bisulfite, with an excess of sulfur dioxide present as sulfurous acid. Modern sulfite mills employ a variety of cooking liquors. A common method of preparing the cooking liquor consists of burning sulfur to form SO_2 and passing this through a packed tower of limestone or a lime solution so that calcium bisulfite is formed, along with about 1% of combined SO_2, the remainder being free SO_2 in amounts of about 4 to 5%. Some modern operations use liquors prepared with other bases, such as sodium, magnesium, and ammonia; in most such cases the liquor is highly corrosive. Until the widespread use of stainless steel, corrosion was a major problem. Up to 25% more stainless steel is used in sulfite mills than in sulfate mills. Care should be used in material selection because 316 or 317 stainless steel is not always suitable. Group C pumps are recommended.

BLOW TANK DISCHARGE　　As the liquor introduced with the chips into the digester combines with the noncellulose and hemicellulose fractions of the wood, it changes from white liquor to black liquor before reaching the blow tank. In addition, the sudden release of pressure frees the cellulose fibers from the other matter, so that the blow tank contains both raw stock (pulp) and black liquor. A stock pump, therefore, is required for this duty because the stock concentration is quite high.

BLACK LIQUOR　　For convenience these pumps are divided into three groups.

Weak Black Liquor (Total Solids Up to 20%)　　During washing, hot water is used to dissolve away the surplus organic matter from the pulp, and the liquor produced is termed *black liquor*. This liquor is a mixture of the lignins and carbohydrates in the original wood plus the cooking chemicals: it is alkaline with a solids content of 14 to 16% in a sulfate mill. The temperature will be about 180 to 190°F (82 to 87°C), and the specific gravity about 1.08. Washing is usually carried out with a minimum of three countercurrent stages, and the solids content given above is representative of the liquor leaving the stages nearest the inlet, that is, where it is most concentrated. The quantity of recirculated liquor is quite high, and many mills have found group A pumps with stainless trim to be satisfactory and economical. With the low solids content, there are no special viscosity problems. This may be seen from Fig. 2.
　　With sulfite mills group B pumps should be used because of the acidity of the liquor.

Black Liquor with Total Solids of 20 to 50%　　This liquor is formed by the evaporation of water from weak black liquor. The concentration is accomplished in multiple-effect evaporators, which usually discharge liquor with about 50% total solids at close to 200°F (93°C). In some odor-free installations, the solids concentration is much higher. Because of the nature of the evaporation, special pumps are usually required.

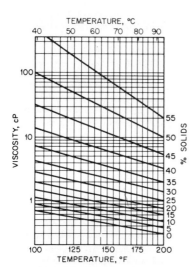

FIG. 2　Black liquor viscosity.

Liquor containing up to 50% solids is reasonably easy to pump, but allowance must be made for viscosity effects. In noting the values in Fig. 2, it should be remembered that the plant must often start up cold, so that cold liquor with a higher viscosity may have to be pumped. The liquor specific gravity rises during evaporation from about 1.1 to 1.25. Group B pumps are recommended.

Black Liquor with Total Solids of 50 to 65%　　This is often referred to as heavy black liquor because the specific gravity rises to 1.35. The liquor is formed by further evaporation, either in the multiple-effect units or by contact evaporators using hot flue gas.

　　From a pumping standpoint, this liquor is probably the most difficult of all liquids to pump satisfactorily in pulp and paper mills. Continuous operation requires careful attention to pump maintenance. Steaming out at regular intervals is particularly important.

No accurate figures are available for the viscosity of liquids having a solids concentration above 55% as there is wide variation in the liquors produced from different wood species and also in the liquors from the same wood of different ages. Hardwood species produce a more viscous liquor, especially eucalyptus, as well as more liquor per ton of pulp produced. Black liquor produced from straw pulping is even more viscous and, in addition, causes the deposition of silica on the walls of pumps and piping. An approximation of the viscosity of straw mill heavy black liquor may be determined from published figures, which give viscosities up to 8000 Redwood seconds. This is probably at least 50% higher than liquor from normal long-fibered softwood.

During recovery, liquor is sprayed into the furnace for evaporation to dryness and burning. Prior to this the make-up chemical (sodium sulfate or salt cake) is added and reduced to Na_2S in the reducing atmosphere of the furnace.

Little is known with certainty about heavy black liquor, but it does not seem to be very corrosive, and carbon steel is often used for pipework, although stainless steel pumps are fairly common. The pumps are subjected to severe duties—notably high heads, lumpy material, high temperature and pressure, and continuous service. Group B pumps are almost universally specified, often with casings of more wear-resistant material such as Worthite or alloy 20. In some cases, mills making straw pulp have not found suitable centrifugal pumps and have had to resort to gear pumps because of the very high viscosity of the liquor.

GREEN LIQUOR   Green liquor is a solution of sodium carbonate and sodium sulfide plus other elements and compounds. One of these other compounds is iron sulfide in a colloidal form, which produces a greenish color. The liquor is formed by dissolving smelt from the causticizing process. Severe erosion takes place in green-liquor pumps, primarily because of the violent action inside the dissolving tank but also because of the gritty matter always present. Green liquor also builds up on the walls of pumps and piping, causing high frictional losses. The specific gravity is usually about 1.2, and an allowance of about 20% should be made for viscosity. Group B pumps are recommended for this service.

LIME SLURRIES   In causticizing, various solutions and slurries are present which, apart from causing excessive wear in standard pumps, do not cause any problems. Thus any normal slurry pump should prove satisfactory. In sulfate mills the lime mud formed during green liquor causticizing presents the most serious problem, for approximately 1000 lb (500 kg) of mud may be formed for each ton (1000 kg) of pulp produced. Solids loads above 35% can occur, and frequent blockages are likely unless pumps are selected for minimizing downtime. Often special diaphragm pumps are used, but where a mild slurry duty is involved, group A pumps should prove satisfactory provided the wearing parts can be readily replaced; many of the parts will have a life of less than 12 months.

BLEACH PLANT LIQUOR   Most bleached pulp mills today use at least four stages of bleaching, and often six or more. Bleaching is used to remove residual lignins or to convert them to compounds that are stable as to color and heat. The stages used include chlorination, either by hypochlorite or chlorine dioxide (usually two stages), with an alkali extraction washing stage between. On occasions oxygen is also used. Bleach plant chemicals are usually prepared in the mill so that solutions such as chlorine water, sulfuric acid, sodium chlorate, sodium chloride, sodium hydroxide, calcium hypochlorite, and chlorine dioxide all have to be pumped.

It cannot be emphasized too strongly that materials of construction are of vital importance in the chemical preparation area of the bleach plant.

In addition to the standard chemicals, some of the common pulp mill bleach substances, together with some chemical preparation systems, are as follows.

CHLORINE   This is usually delivered to the mill in tank cars but is always vaporized to a gas before use.

CHLORINE WATER (HYPOCHLOROUS AND HYDROCHLORIC ACID)   Concentrations cover the range from pH 2 to 10 for bleaching pulp. In some cases the gas is mixed directly with pulp in special mixers. Group C lined pumps are essential.

SODIUM HYPOCHLORITE AND CALCIUM HYPOCHLORITE   This mixture is made in the mill by permitting chlorine to react with either sodium or calcium hydroxide concentrated caustic (70%) diluted to 5 to 6% before chlorination. Calcium hypochlorite is made from a 10% solution of slaked lime at temperatures up to 150°F (66°C), but not normally exceeding 70°F (21°C). These liquors are corrosive to steel, and therefore rubber or epoxy linings are necessary when handling solutions *to* the bleach plant; *after* bleaching the filtrate may still have residual hydrochloric acid.

CHLORINE DIOXIDE   This is the most common bleach solution used because it gives an excellent brightness to the pulp and, despite corrosion problems, is usually cheaper than other bleach solutions.

After generation of the gas, during which absolute cleanliness is vital, the gas is stripped in a packed tower as an aqueous solution and stored in plastic tanks made of special resins that resist chemical attack. In modern plants an increasing use is made of glass-reinforced plastic with selected resins for piping, valves, and pump linings. This is usually a cheaper alternative than the use of exotic metals, such as titanium, for pumps. Pumps must be group C, and stainless steel is not satisfactory. Solution strengths of up to 8 g/liter are used.

SODIUM PEROXIDE AND HYDROGEN PEROXIDE   These are used for bleaching groundwood pulp. Typical solutions contain sodium silicate (5%), sodium peroxide (2%), and sulfuric acid (1.5%). The latter is for controlling the pH of the liquor. Concentrations of bleach liquors are up to 15%. Temperatures are usually less than 90°F (32°C). Group C pumps are necessary.

WASH LIQUORS   In general, the filtrate from bleach washing stages will exhibit at least some of the properties of the stage immediately before washing, owing to slight excesses of chemical present. Filtrates are collected in corrosion-resistant pipes and vessels, usually made from glass-reinforced plastic, and the pumps used will be either group B or C, depending on the stage in question. The filtrate from the chlorine dioxide stages may be pumped with a 317 stainless steel case and trim pump because the filtrate is not as corrosive as the bleach solution.

Spent acid from chemical preparation plants is also highly corrosive, and usually stainless steel is not satisfactory for use with it.

Effluent from the bleach plant, on the other hand, is usually a mixture of several liquors, and experience has shown that 317 stainless steel is a suitable material for pumps that handle it.

CHLORINE DIOXIDE PREPARATION; SODIUM CHLORATE   Chlorine dioxide is used with the Mathieson process and is produced by permitting sodium chlorate to react with sulfuric acid in a vessel into which sulfur dioxide is introduced in controlled quantities. Sodium chlorate solutions are usually from 43 to 46%, at which strength the specific gravity is about 1.38. Stainless steel pumps may be used, but epoxy-resin-lined pumps are superior.

FOUL CONDENSATE   This arises from the evaporation of water from black liquor at the multiple-effect evaporators, as these units flash vapor from the liquor in one stage and use this to evaporate the liquid in the next stage. The vapor when condensed contains some carry-over from the black liquor, and thus the condensate is contaminated and corrosive. When a nickel cast iron casing and stainless trim are used, group A pumps should be satisfactory. Some liquors produce very corrosive vapors, and a stainless casing pump may prove necessary. Group B pumps are recommended.

Stock   Stock is the term applied to the suspension of cellulose fiber in water. It first appears either after grinding (in the case of mechanical pulp) or after the blow tank (in the case of chemical pulp). The quantities of stock required for a given pulp or paper production are shown in Fig. 3.

After the separation of chemicals or impurities by washing and screening, the stock is given a mechanical treatment known as either beating or refining, depending on the nature of the treatment. This enhances the sheet properties. Additives such as starch, clay fillers, alum, and size are introduced to impart special characteristics, depending on the end use of the product.

Over the range of stock in normal use, the specific gravity may be considered constant for all practical purposes, with a value equal to that of water at the appropriate temperature.

Cellulose fibers have a tendency to float in water, and constant agitation is required to ensure

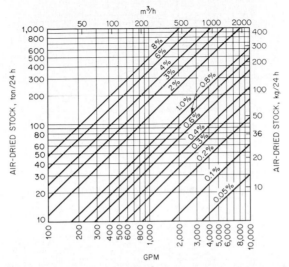

FIG. 3 Stock quantity conversion chart.

that stratification does not occur in storage. Agitation, however, can also introduce air, to the detriment of the stock.

The pH of stock varies over a wide range—from as low as 1.0 during some bleaching processes to 11.0 with others. In the paper machine room, the pH of the stock will usually range from 4 to 8. Thus from a corrosion viewpoint washed stock does not usually present special problems except when high-grade bleached products are produced. Stains will be caused by iron sulfides or oxides, and therefore stainless steel must be used—frequently 304 for washed stock, but 316 or 317 within the bleach plant before washing or where bleach liquor is likely to be present with the stock.

Unbleached paper mills generally do not experience corrosion with washed stock, except in the case of groundwood mills, where the pH is usually lower than in chemical pulp mills.

FIBER CHARACTERISTICS   Stock made from softwoods will have a predominance of fibers 2.8 to 3.5 mm long and 0.25 to 0.3 mm wide; fibers from hardwoods will be about 1.0 to 1.3 mm long and 0.1 mm wide. Straw fiber will be still shorter—0.75 mm on the average—but flax can have fibers up to 9.0 mm long. These figures are typical and are of interest because of their effect on pump performance.

Consistency   This is the amount of dry fiber content in the stock, expressed as a percentage by weight. Typical values will vary from about 0.1% for the feed to the headbox of a special paper machine to 14% for stock between some bleaching stages or in high-density towers. The critical stock consistency in the selection of pumps is between 2 and 3%. Up to the 2 to 3% level, pumps may be selected on the basis of their water performance. Above 3%, pump performance decreases rapidly with increasing consistency. The magnitude of the correction will depend not only on the properties of the stock but also on the type, size, and specific speed of the pump. Figure 4 illustrates the water performance of a typical stock pump and the effect of stock consistency on that performance.

Freeness   When stock is beaten, or refined, it acquires an affinity for water, and the longer the stock is beaten, the longer the water retention period. The retention of water by the stock increases the friction factor of the flow of stock. In pipe flow the increased friction factor is usually not significant because the pipe velocities are low. At higher velocities in a pump, however, a heavily beaten stock with a very low freeness value is very slippery, and the stock may be very difficult to pump.

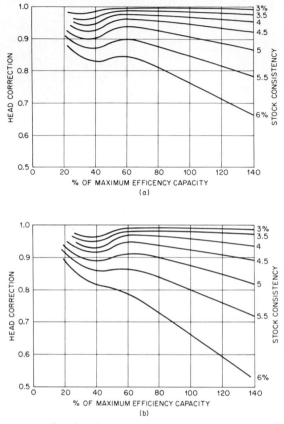

FIG. 4   Effect of stock consistency on water performance: (*a*) groundwood stock, (*b*) other stocks.

Freeness is often measured by an instrument called the Canadian standard freeness tester, and the range of values covers a scale from 0 to 900, with a higher freeness value indicating a less refined stock and thus a lesser affinity for water. This instrument measures the amount of water drained from a sample of stock under a regularly decreasing head. Its use is recommended by the Technical Association of the Pulp and Paper Industry (TAPPI), and it is commonly employed in North American mills. Another measurement of freeness, frequently used in Europe, is the Schopper Riegler system. In this system, increasing freeness values indicate increasing affinity for water (just the opposite of the CSF system).

FLOW CHARACTERISTICS   To illustrate the basic characteristics of stock, some typical curves are shown in Fig. 5.

STOCK PUMPS   In stock pumps, viscosity may be a problem but consistency is not a major problem until a value of about 6% is reached. The essential requirement is to get the stock to the pump impeller, and every effort should be made to keep the piping as large and straight as possible.

Above 6% consistency, special pumps are required, and they are usually of the screw type. Air entrainment in the stock will reduce pump output. Air entrainment occurs from agitation in the chests, from flow over weirs, and from flow through restricted openings. How air entrained in water and in stock affects pump performance is shown in Fig. 6.

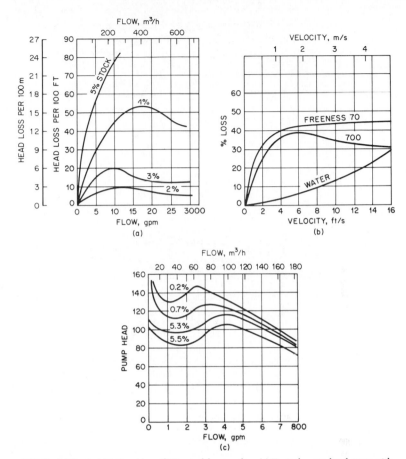

FIG. 5 (*a*) Typical frictional loss. (*b*) Typical freeness loss. (*c*) Typical pump head curve with stock consistencies.

Piping Arrangement　Piping should be as straight and short as possible. This is particularly important on the suction side of the pump to prevent dewatering of the stock. The diameter of the suction piping should be at least one pipe size larger than the diameter of the pump suction and should project into the stock chest. The inlet end of the suction pipe should be cut at an angle, and the bottom of the pipe should be at least 1½ pipe diameters from the bottom of the chest. With the long side of the pipe on top, the probabiliy of drawing air into the suction of the pump through vortices is reduced. Some manufacturers provide a lump breaker or screw feeder at the suction side of the pump for pumping stock above 4 to 5% consistency.

Size of Pumps　It is important to estimate the performance requirements of stock pumps as accurately as possible. Oversize pumps can cause an unbalanced radial thrust on the impeller, excessive wear of the sleeves and glands, and an increase in the risk of cavitation.

ECONOMICS AND PUMP SELECTION

The normal economic considerations of any continuous process apply equally well to pulp and paper mills, with a few points of difference. In pump installations the improved cost figures that

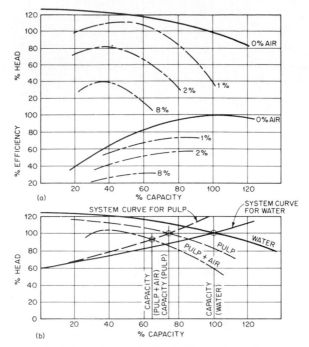

FIG. 6 (a) Reduction in pump head and efficiency when water contains entrained air. (b) Reduction in pump capacity when pulp contains entrained air.

are possible from larger units are limited to some extent by the manufacturer's standard size units. Because the industry is capital intensive, the overriding factor in any pump installation is reliability. To achieve this, it does not necessarily follow that it is better to have two pumps installed, with one as a standby. One properly designed and serviced unit may well be better than two unknown units; this is especially true where the pump is in a portion of the process that cannot be interrupted without serious losses, either in raw materials or in quality of the finished product.

A duplication of pumps means complications in extra valves, pipework, connections for steam and viscous liquids, electric motors, cables, and starters. The result is that, in modern mills with good machinery and materials of construction, there is a strong tendency away from the duplication of pumps because of increased costs and questionable reliability. It follows that the important thing is to select the right pump and the right duty point in a particular range. In the case of stock pumps, this is very important because the shape of the head-capacity curve can be changed by the consistency of the stock (Fig. 6).

There are some 200 pumps in a modern pulp and paper mill, but the cost of these pumps is probably less than 5% of the total equipment cost. It is unwise, therefore, to jeopardize mill reliability by compromising pump quality. Corrosion and erosion are major factors in pump life, and even with the best materials the life of some components in severe service may be 12 months or less. Moreover, the power used by pumps is usually less than one-third of the mill demand. If one remembers that the cost of total power absorbed in a mill is only around 4%, even a 50% reduction in pump power will still be less than 1% net.

Efficiency The best point at which to operate a pump is, of course, its maximum efficiency, but this is not always possible, particularly in the case of stock pumps. In any event, there will be a reduction in the water efficiency because of the viscosity effects in liquor and stock pumps. Pumps designed to produce maximum efficiency on water are not necessarily the optimum selection for a pulp or paper mill. For example, open impellers and excess clearances reduce the efficiency, yet these factors are much more important in stock pumps than efficiency. Another important consideration is speed. Stock pumps should be chosen to run at as low a speed as possible to achieve stable operation, and this speed may not produce an efficient pump. The shape of the pump performance curve is much more important than the best efficiency quoted by a manufacturer. A flat or unstable head curve may produce surging or instability in the pump output. Good pump selection, therefore, must emphasize reliability as the first consideration and efficiency and costs as secondary considerations.

Pump Speeds   Most of the pump duties in a mill can be accomplished by single-stage pumps and four-pole motor speeds. For liquids other than water, two-pole motor speeds should be avoided if possible. For special duties, including stock pumping, six- or eight-pole motors may be required unless some indirect or variable speed is used. While it is true that lower speeds mean larger pumps and more expensive electric motors, lower speeds are justified because of reduced maintenance and greater reliability. Some deviation from these speeds may be necessary for pumps generating heads in excess of 150 ft (46 m), but this can often be taken care of by a larger impeller rather than a higher shaft speed.

As motor power increases, the use of variable-speed pumps becomes more attractive, and stock pumps ahead of a larger paper machine may require a prime mover in excess of 500 hp (370 kW). In such cases variable speeds can usually be justified, and the use of thyristors, steam turbines, and hydraulic couplings should be considered.

For pumps of lesser power, experience in several modern mills demonstrates that with conservative design, V-belt drives are reliable and give excellent service.

Multistage Pumps   Except with boiler feedwater, the use of multistage pumps should be avoided. This is particularly true for stock and viscous liquors. The complication of the pump design makes such units unacceptable for these services.

Pipeline Systems   With black liquor, green liquor, and similar high-viscosity liquors, adequate provision must be made for steaming out and subsequent liquor drainage. The pumps must also be included in this system. Although the liquor pumps should be designed to pass some solid matter, motorized strainers should be used on the pumps for cyclone evaporators and recovery boiler-fuel pumps because both pumps discharge to spray nozzles. Dead pockets and other areas where liquor can collect should be avoided since solids from the liquor will build up in these areas and possibly break away to block pipelines or pump impellers. For protection at shutdowns, even for short periods, steaming out is essential.

Positive Displacement Pumps   Apart from high-density stock pumps, the principal use of the positive displacement pump is for consistency control of stocks above 5%. The normal measuring device used is quite satisfactory at low consistencies but is less reliable at the higher values. More satisfactory control may be achieved by using a screw pump, where the power is proportional to the pulp consistency at constant flow. Such pumps have been very reliable on consistency control.

Digester Circulating Pumps   Digester circulating pumps are used with indirectly heated batch digesters to circulate the liquor at the digester pressure and temperature. Maintenance problems are common on these pumps because heads can be as high as 150 lb/in^2 (1034 kPa) and temperatures as high as 350°F (177°C). In addition, the circulating liquor contains some raw pulp even though screens are fitted to the digester outlets. Pumps for this service, therefore, should be of very heavy construction and have open impellers and water-cooled stuffing box assemblies. In addition, there is often considerable pipework involved, for a digester may easily be 60 ft (18 m) high and pipe loads are often imposed on the pumps. Expansion joints and long-radius bends are used, but it is desirable to support the pumps on springs.

Pumps for Heavy Black Liquor above 60% Solids   A typical pump for this service is shown in Fig. 7. An open impeller in a 316 stainless steel casing is recommended. A heavy sleeved shaft of 316 stainless steel with ample clearance between the rotating parts is also required for satisfactory operation. These pumps may be required to handle black liquor up to 8000 Red-

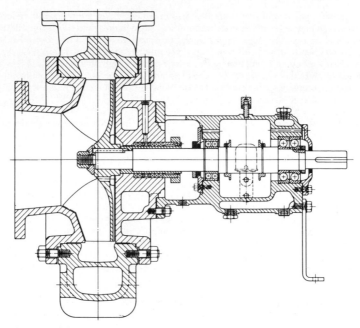

FIG. 7   Black liquor pump.

wood seconds viscosity and should be provided with water cooling. Steam jacketing is not always satisfactory because the liquor may tend to bake on the walls of the casing. Pump speeds should be below 1800 rpm, if possible.

Multiple-Effect Evaporator Pumps   Pumps in this service often operate in cavitation owing to problems in regulating the flow between evaporation stages. Level control valves in the suction line to the pump can alleviate this problem, but cavitation can still be expected in the pump. A self-priming pump may give longer life of the rotating elements than the condensate pump usually used on this service.

Diaphragm Pumps   Diaphragm pumps are used in pumping lime mud slurries of high concentration. They consist of a rubber or neoprene diaphgram with a pulsating air supply on one side, controlled by a timer, and the slurry on the other.

Stock Pumps for Intermediate Consistencies   Representative of pumps in this category are those provided with a lump breaker and/or screw feeder at the pump inlet. The casing is often also split at 45° to permit easy access in the event of a blockage. Stock at 6% can be easily handled provided proper care is taken with suction piping.

Vacuum Pumps   These are used to extract water from the sheet on the fourdrinier wire and at the suction presses by means of a vacuum up to a maximum of about 25 inHg (635 mmHg). Approximately 40,000 lb (18,100 kg) of water is extracted by this means for every ton (1000 kg) of paper produced, and this water is removed by entrainment with the air handled by the vacuum pumps. Frequently water separators are used to remove water; their use is a matter of economics, as a reduction of up to 10% in power may be achieved.

Vacuum pumps are of three basic types:

1. Water ring
2. Positive displacement
3. Centrifugal or axial-flow

Many engineers prefer the water-ring type, probably because of its simplicity. In general, however, this type uses more power, mainly because of the heating of the circulating water, which is then discharged to a drain.

Centrifugal and axial-flow machines must be provided with water separators, but they are more efficient over all, particularly when the heat of compression is used in the machine room ventilating system. The machines run at high speed and are usually driven by a steam turbine.

The paper machine system requires vacuums at different levels, from a few inches (millimeters) of mercury to the maximum possible. Often pumps are connected to a common header, and orifice plates are used to divide the flow to ensure some measure of standby capacity. This involves throttling, however, and may create flow problems unless quantities are carefully estimated. The axial-flow machine permits extraction at any point along the rotor within fairly close limits and requires an accurate estimation of the quantities and specific vacuum required. A standby for axial-flow machines cannot usually be justified.

Stock and Liquor Pump Standardization   Throughout the mill the duties of many pumps are similar, but different materials of construction may be used. If, at a slightly extra initial cost, the rotating elements of the pump can be standardized, this will reduce spare-parts inventories. This is also an advantage when purchasing pumps for a new mill. For example, if a standard arrangement consists of a complete rotating element, including bearings, only the impeller size and material need be different. For this reason the nonstock pumps are divided into two groups, A and B. Standardization is an additional reason for recommending that all pumps be stainless-fitted. Obviously, large pumps need individual evaluation. Standardization of stock pumps is less feasible, but up to about 3% stock, similar pumps can usually be specified.

Specification of Design Margins   In many cases an excess margin on head and capacity is specified. Where margins are excessive, mechanical troubles and cavitation often occur. Because pump manufacturers have different standards, it is not realistic to use a fixed formula for margins, if only because the accuracy of the estimate will vary with the liquor concerned. However, experience in pump selection indicates that the following guidelines should give reasonable results in most cases:

1. Carefully estimate pump duty, using the best available data and the actual piping configuration to be used.
2. Allow in the head loss estimated for control valves the increased head loss for the partially open position.
3. Using the above figures, add 10% to the flows at the normal load and 10% to the head at the maximum duty.
4. Compare the new flow figure with the maximum flow anticipated. If the new figure is greater, make no adjustment to the figure obtained in item 3. If the new figure is less, add 5% to the maximum flow anticipated.
5. To the figures derived from item 4, add a further 10% for flow and 5% for head to allow for the decrease in pump performance with time.

Notes   These margins should be applied before the motor is selected. When considering motor power, allow for the maximum impeller diameter.

Using the figures derived from item 5, study the manufacturer's pump curve and note the impeller diameter and range for the pump. If the duty point falls near the end of the curve and if the impeller is the largest that can be fitted in the casing, it is advisable to select the next size larger pump.

SECTION 9.9
FOOD AND BEVERAGE

JAMES L. COSTIGAN

INDUSTRY STANDARDS

Centrifugal, rotary, and reciprocating pumps are used throughout the food and beverage industry. Unlike other industrial applications, pumps for the food and beverage industries must meet rigid sanitation codes, known in the industry as the 3-A standards.° These standards have been established by the following organizations:

1. The International Association of Milk, Food, and Environment Sanitarians
2. The U.S. Public Health Service
3. The Dairy Industry Committee, composed of representatives from the following:
   American Butter Institute
   American Dry Milk Institute
   Dairy and Food Industries Supply Association
   Evaporated Milk Association
   International Association of Ice Cream Manufacturers
   Milk Industry Foundation
   National Cheese Institute
   National Creameries Association

The 3-A standards were established originally for the dairy industry and are rigidly enforced by the local sanitarian. The baking, bottling, and packing industries have also adopted these standards with modifications to suit their particular requirements.

The 3-A standards specify that the wetted parts of any pump must be furnished in a type-300 series stainless steel or in an equivalent material of the same corrosion-resistant properties. In the case of centrifugal pumps, this means that the impeller, casing, backplate, and seals must be of stainless steel. Cracks or crevices in any of the wetted parts are not permitted. The finish and minimum radius of curvature of any wetted surface are specified by the standards. Internal threads are not permitted except in designs where no alternative fastener is possible for the proper

---

° Available from the International Association of Milk, Food, and Environmental Sanitarians, Box 437, Shelbyville, Indiana 46176.

FIG. 1 A sanitary version of an external balanced seal. (Tri-Clover Division, Ladish)

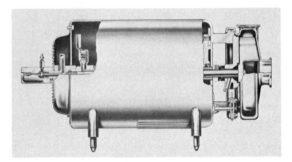

FIG. 2 An approved cleaned-in-place seal. (Tri-Clover Division, Ladish)

FIG. 3 An O-ring seal. (Waukesha Foundry)

functioning of the pump. The standards state that the pump must be designed in such a way that the wetted parts can be readily disassembled for inspection and cleaning. The purpose of these specifications is twofold: (1) to prevent the accumulation of food or beverage in isolated points and its ultimate putrefaction and (2) to allow for the cleaning of the internal components of the pump as quickly and efficiently as possible.

PUMP TYPES

Single-stage end-suction pumps are widely used throughout the food and beverage industry. They may be close-coupled or mounted on a baseplate and coupled to a separate motor. Pumps of this type were formerly used almost exclusively for transfer service, but in modern plants they are applied throughout the process cycle. The use of variable-speed motors and controls has simplified the problem of the variable-flow requirements of such process machines as separators, clarifiers, and filters. Variable-speed pumps are also used as booster pumps to the timing pumps of plate pasteurizers.

Many of the heavy, viscous products handled in the processing of food cannot be pumped effectively by centrifugal pumps. Cottage cheese and baking doughs are typical examples. Positive displacement pumps are required for this type of service, and a lobe-type rotary pump with timing gears is usually selected. Internal-gear, single-screw, and flexible-vane pumps are used to a lesser extent. Sanitary lobe-type pumps are limited to 150 lb/in^2 (1034 kPa) discharge pressure, flexible-vane and internal-gear pumps to 50 lb/in^2 (345 kPa), and single-screw pumps to 90 lb/in^2 (620 kPa).

The rotary lobe pump is also used where processing procedures require its use. An example of this is the timing pump used on a high-temperature, short-time pasteurizer. The flow rate is critical, and the system is carefully timed by the sanitarian. The drive is then sealed to prevent any unauthorized increase in the flow rate.

MECHANICAL SEALS

Mechanical seals are used almost exclusively throughout the food and beverage industry. An acceptable seal must be not only simple in construction but also readily accessible for the take-down cleaning. A sanitary version of an external balanced seal is shown in Fig. 1. An alternative design, shown in Fig. 2, can be cleaned in place, whereas that shown in Fig. 1 must be removed from the pump for cleaning. It is now permissible to clean pipelines and some equipment in place, provided strict sanitary controls are maintained. This procedure is known in the industry as CIP (cleaned in place), as opposed to COP (cleaned out of place).

The approved mechanical seal where CIP is permissible consists of a single carbon ring between the impeller and the backplate. It is loaded by a spring coupling driving the integral impeller-shaft combination. The seal is disengaged for cleaning by a manually or pneumatically operated cylinder. The cylinder presses against a thrust bearing on the shaft and opens the seal by pushing the shaft and impeller forward. Sanitizers are circulated through the pump following the cleaning, rinse, and detergent cycles. A separate CIP supply pump is used for circulation during cleaning and sanitizing. The preferred procedure is to coordinate the cleaning cycle of the pump with the cleaning cycle of the valves, fittings, and pipeline.

The sealing mechanism of the rotary lobe pump can be either an O-ring seal, as shown in Fig. 3, or a mechanical seal, as shown in Fig. 4.

A wide variety of food and dairy products are marketed in cans. Aseptic canning must be performed under sterile conditions. Leakage of nonsterile air through the connections or seals of the pumps used in canning can be a source of contamination. For this service the pumps must be of a special construction to prevent air leakage. Figures 5 and 6 show a rotary pump and a centrifugal pump seal designed for canning service. Steam or a sterile liquid is used as a sealing medium to prevent the entrance of air into the pumped fluid.

FIG. 4   A mechanical seal. (Tri-Clover Division, Ladish)

HOMOGENIZER

The homogenizer as used by the food industry incorporates a reciprocating pump—usually a triplex. On a high-temperature, short-time pasteurizer, the homogenizer can be used not only as a homogenizer but also as a timing pump. The homogenizer can also be used as a high-pressure pump for the jets in spray-drying operations by the simple expedient of removing the homogenizer valves.

PUMP DRIVES

Standard electric motors are used as pump drives in the food and beverage industry. The totally enclosed fan-cooled motor is preferred for cleanliness. To facilitate cleaning of the fan, the canopy

FIG. 5   A rotary pump. (Waukesha Foundry)

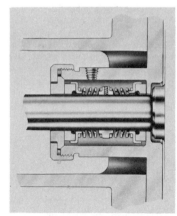

FIG. 6   A centrifugal pump seal. (Tri-Clover Division, Ladish)

FIG. 7 A typical pump and valve installation. (Tri-Clover Division, Ladish)

should be secured with wing nuts or wing bolts. It is also desirable to specify a motor with a fan that can be readily removed from the shaft. The motor and pump of a close-coupled unit must be installed at least 2 in (50 mm) above the floor. The baseplate of a direct-coupled unit must be grouted to the floor or supported on piers of minimum height. The baseplate must be of a solid construction without pockets and sealed to the motor to facilitate cleaning.

A typical pump and valve installation in a modern food plant is shown in Fig. 7.

SECTION 9.10
MINING

W. D. HAENTJENS

It might be asked why pumps used in mining services receive a separate classification when the types involved do not differ in principle or in general appearance from those used in fresh-water service. The answer is the need for utmost reliability and an ability to withstand both corrosive and abrasive waters. Not all mine waters are corrosive, but almost all mine waters from active mines are abrasive because of the suspended solids from mining operations. Much can be done to limit the amount of solids handled, as noted later under Sumps, but the solution generally requires a compromise based on economics. In other words, removal of most coarse solids is usually justified, but removal of fine solids is generally impractical. Thus a heavy-duty pump has evolved which warrants the classification of mine pump.

PUMPING CONDITIONS AND PUMP TYPES

The general pumping conditions in mining services can be determined from a consideration of the types of mines and the materials being mined. For example, the broadest category would be a division between open-pit and deep mines. Open-pit mines seldom exceed 600 ft (183 m) in depth, and so the pumping heads generally are not much greater than this unless very long discharge lines are involved. Unless the open-pit mine is located in an arid region, the greatest pumping load is produced not from groundwater but from rainfall. Unless a certain increase in water level can be tolerated at the bottom of the pit, the maximum rainfall rate and of the drainage area involved will determine the required pumping capacity. These flow rates are usually large, and the combination of large capacities with moderate heads generally requires a double-suction pump, either horizontal or vertical. If the capacity is large enough for good hydraulic design, a single-stage pump is preferred. Where the capacity is lower, two-stage units are common. Because of the fine solids generally present, operating speeds are usually limited to 1800 rpm. This puts a limit on the head per stage and thus minimizes wear.

The development of large overhung-shaft vertical pumps has been a boon to open-pit mines, for these units can easily be raft-mounted. The availability of large billets of plastic foam material has simplified small raft construction and made them virtually unsinkable. The weight of the water in the pump as well as in the discharge line to the point where an off-barge float contributes to the support of the hose must be considered. The number of personnel who might be aboard at any one time must be considered, as well as the possibility that all persons might be grouped together at one side. The forces involved in either overturning or righting a barge must be determined. With the float riding level, the center of gravity and the center of buoyancy are in the same line. If someone stands on one side of the barge, the barge tilts and more of it becomes submerged on that side. If there is sufficient freeboard, there is increased submergence and the barge is buoyed up by a greater force on that side because of the increased submergence. The center of gravity of the pump now acts so that its vector is displaced from the geometric center of the barge. As long as there is enough freeboard and buoyancy, the shift in the center of the buoyancy must be such as to have a righting moment. If there is insufficient freeboard, then the center of gravity vector may be outside the center of buoyancy and the barge will overturn. The buoying force is dependent on the volume of displaced liquid, and once the entire section is submerged there is no increase in buoying force.

Large barges have been designed, including some for the installation of four 700-hp (527-kW) or three 1500-hp (1119-kW) pumps. Overhung-shaft pumps are available in sizes to 1500 hp (119 kW) with either single- or double-suction design. Where the pumping conditions permit, a single-suction top inlet is preferred because the pump will not draw sand and mud from the bottom of the pit at minimum water level.

The selection of a single- or double-suction pump is based on the capacity-head combination at 1800 rpm or at the maximum operating speed. For a given set of hydraulic conditions of head and capacity, the operating speed of a double-suction pump can be increased by a factor of 1.41, that is, $\sqrt{2}$ times capacity in the specific speed formula:

$$N_s = \frac{N\sqrt{Q}}{H^{3/4}}$$

Thus, in cases where a 1200-rpm motor might be required by a single-suction design, an 1800-rpm motor, which is smaller, lighter, and cheaper, could be used for a double-suction design.

The wide variation in required capacity for pumps used in open-pit mines raises the problem of parallel pumping. The head at the maximum capacity is frequently much greater than that during normal pumping, and cavitation may occur during single-pump operation. There are a number of solutions. The best solution, although not necessarily the cheapest, is to use a separate discharge line for each pump. If the discharge line is short, this may be practical; if not, groups of two pumps can be combined in a single pipeline. The alternatives should be carefully examined by a system head analysis. An alternate solution with a single discharge line is a variable-speed drive for one or more pumps. It is seldom necessary to have all pumps with variable-speed drives.

Pumps used in underground mines must produce heads above 500 ft (152 m) since that is the maximum depth of an open-pit mine. Flow rates are generally lower than in an open-pit mine because a heavy rainfall does not produce an immediate heavy flow of water. Underground mining operations also attempt to divert as much surface water as possible away from the mine. When this is done, flow rates for individual pumps range from 1000 to 5000 gpm (227 to 1136 m³/h). Although an underground pump station might be designed for a maximum inflow of 20,000 gpm (4542 m³/h), it is unlikely that pumps larger than 5000 gpm (1136 m³/h) would be selected. There are a number of reasons for this limitation. These include the size of the pump and motor relative to the size of the mine shaft and haulage way, the maximum weight of the pump and motor that can be handled and serviced underground, the maximum water storage capacity in the sumps, the effect of continuous versus timed (off-peak power) pumping, and the consequences of a power failure. The effects of a power failure require a careful analysis to determine the normal sump level and emergency storage capacity. Electrical inrush limitations frequently determine the capacity of individual pumps. For surface-mounted transformers, long power lines may limit the motor size. Although reduced-voltage motor starters may be used, the cost can rarely be justified.

The maximum size of the motor and total pumping capacity may not be sufficient information to make a decision as to individual pump capacity. Obviously, the pumping head must be considered. If the mine is less than 1200 ft (366 m) deep, there may be only one pump station, although many mines operate at higher heads in a single lift. It is necessary to consider where the greatest water inflow occurs. If it should be at the 300-ft (91-m) level, it would be foolish to let all the water go to the bottom of the mine to be pumped out. The principal pump station should then be at the 300-ft (91-m) level, with a smaller station at the bottom of the mine pumping to the 300-ft (91-m) level or directly to the surface.

The depth of 1200 ft (366 m) is selected as a guide point since the main consideration is the suitability of 300-lb/in² (2000-kPa) fittings. If 600-lb/in² (4000-kPa) fittings must be used, the pumps and pipeline will be much more expensive. Again, this must be compared with the cost of additional pumping stations at 1000- to 1200-ft (305- to 365-m) intervals. Some studies may show that 2000-ft (610-m) pumping intervals are justified.

When the pumping head exceeds 2400 ft (730 m), it is difficult to obtain satisfactory horizontally split-case multistage pumps, and a decision must be made as to the desirability of pumping in increments or going to barrel pumps, which are available for very high pressure and were developed for high-pressure boiler-feed pump service. Since most barrel pumps for mine service operate at high heads per stage with close clearances, the mine water must contain a minimum amount of abrasive particles in suspension.

The basic decision in the design of the pump room is whether the sumps will be above or below the pumps. This will determine the type of priming system and the complexity of automatic control. The difference in available net positive suction head may also restrict the selection of pumps. A further decision is that of discharge line size. Line velocity should be kept below 10 ft/s (3 m/s) and generally in the neighborhood of 6 to 8 ft/s (1.8 to 2.4 m/s). This may not be possible if there is little room in the shaft. If stainless steel or lined pipe is used, higher velocities may be tolerated to reduce the cost of the piping. With higher velocities, however, the risk of damage from waterhammer must be evaluated.

The benefits of a sump below the pump room level and a positive prime of the pumps can be achieved with low-head—in the range of 20 to 50 ft (6 to 15 m)—vertical turbine pumps. What may appear as a complication is in effect a saving because pump priming and automatic control are simplified. Occasionally a combination of these arrangements is used for maximum safety.

An opposed-impeller pump with a horizontally split case is used in most mines for underground operation. A principal reason for this is the ease of servicing and the rapidity with which a pump can be repaired. Most mines keep a completely assembled spare rotating element on hand. This includes the stationary parts (such as casing rings, partition rings) and the bearings and pump half coupling. Removal of the top half of the pump casing permits replacement of the worn rotating element with a new or reconditioned unit. The worn unit can then be taken to the surface for repair.

Although pumps with a vertically split case are employed in some foreign mines, the time required for servicing is generally greater, and except for barrel designs, they have not found wide acceptance in the United States. Another reason for this lack of acceptance is that the impellers all face in the same direction and produce a large hydraulic axial thrust. This thrust can be balanced by various devices, such as balancing drums or pistons, but such devices are notoriously affected by abrasive solids in the water.

If the mine water cannot be economically clarified for pumping with conventional mine pumps, it is occasionally necessary to consider alternate arrangements. The use of slurry pumps in series at one location is a possible solution. Where there is no supply of clean water available for flushing the stuffing boxes, some mines have found it advantageous to use vertical overhung-shaft pumps in vertical open sumps at elevation intervals of 300 to 400 ft (91 to 122 m). These pumps do not have a stuffing box, and so packing or seal problems are eliminated. Most underground pumps use packed stuffing boxes, although mechanical seals are available. Because of the dirty water usually present, the mechanical seals must be flushed, and thus some of their advantage is lost. The principal objection to mechanical seals in this service is that a seal failure requires a complete pump disassembly, whereas a stuffing box failure can be corrected simply by repacking the stuffing box.

MATERIALS OF CONSTRUCTION

Most water pumps perform quite satisfactorily with bronze impellers, wearing rings, and shaft sleeves, but the dirty or corrosive waters in mine service require superior materials. Since mine waters range from neutral (pH of approximately 7) to severely acidic (as low as pH 1.5 in some coal mines) and to very basic (as occurs in limestone mines, etc.), there is no universally best material. The choice obviously is the lowest-priced material which gives satisfactory service life.

Many times the decision must be made not on the basis of the best available materials but on the basis of which material will best withstand the abrasive conditions during the intended life of the project. If there is a five-year anticipated life of the mine, there is little advantage in selecting materials which will last twice as long. Conversely, a mine with a projected life of 25 years would require the best commercially available materials. There are also the exotic alloys, but their use is seldom justified.

Assuming that there will always be a slight amount of solids present, the metallurgy given in Table 1 should be considered.

Care should be exercised when selecting alloys because many stainless steels do not have as great a strength as carbon steel. The same applies to bolting. Thus a pump which is rated for high-pressure service with a carbon steel or alloy steel casing and bolting may not be suitable when made of bronze or stainless steel.

Although ceramic-coated shaft sleeves are excellent in many applications, it should be remembered that ceramic coatings are porous and that the base metal must be able to withstand the environment. Also, some ceramics will not be suitable in strongly basic water. Others, however, are suitable, and so a general specification for ceramic coating should never be made. A plasma-applied ceramic is generally denser and more serviceable than one applied by a simple flame spray.

HYDRAULIC CONDITIONS

Low-Head Pumps   For low-head pumps, up to approximately 150 ft (46 m), the installation must be carefully checked to prevent cavitation during periods of low-head operation. This is particularly important during one-pump operation on a parallel system. For example, if two 5000-gpm (1136-m^3/h) pumps are discharging into a common discharge line against a static head of 80 ft (24 m) and a frictional head of 60 ft (18 m), the frictional head will be only 15 ft (4.6 m) when one pump operates alone at 5000 gpm (1136 m^3/h). The total head will now be only 95 ft (69 m), and the single pump will carry out to a much higher capacity. Unless sufficient NPSH is available for the single-pump runout point, the pump will cavitate. Although the two pumps should be selected for 5000 gpm (1136 m^3/h) at 140 ft (43 m) total head, the required NPSH should be determined not at 5000 gpm (1136 m^3/h) but at the capacity corresponding to the intersection of the pump and system curves.

High-Head Pumps   For high-head pumps, 1000 ft (305 m) head or more, the risk of cavitation for single-pump operation on a parallel system is less than for low-head pumps. For example, if the static head is 1000 ft (305 m) and the frictional head is 60 ft (18 m) when two 5000-gpm (1136-m^3/h) pumps are operating, the total head will decrease from 1060 ft (323 m) to only 1015 ft (309 m) when one pump operates at 5000 gpm (1136 m^3/h). This means that, for either one- or two-pump operation, the capacity of each pump will be approximately the same and the risk of runout cavitation is minimal.

Waterhammer and Pressure Pulsations   A waterhammer analysis should be made of both high- and low-pressure pumping systems. While the transient pressure pulsations are related to the rate of change of velocity rather than the magnitude of the steady-state condition, mine experience indicates that waterhammer problems can be anticipated when pipe velocities exceed 10 ft/s (3 m/s). In high-pressure pumping systems it is not unusual for transient pressure pulsations to be as high as 300 lb/in^2 (2068 kPa) above or below the steady-state pressure.

Transient pressure pulsations have been experienced in low-pressure pumping systems. The danger here is that the low-pressure portion of the cycle will fall below atmospheric pressure and the pipe will collapse.

While any pumping system for mine service should be analyzed in detail for transient pressure pulsations, experience has shown that adequate air bottles have proved to be one of the most effective and least expensive means of surge suppression. Slow-closing valves and flywheels have been used, but they must be sized correctly. This is especially true with high-speed pumps since these units possess little rotational inertia and will decelerate very rapidly on shutdown, with accompanying high-pressure surge.

TABLE 1 Materials of Construction for Mine Pumps

	Neutral waters		Acidic waters		Basic waters	
	Moderate heads	High heads	Moderate heads	High heads	Moderate heads	High heads
Casing	Cast iron	Ductile iron/cast steel	316 S.S./alloy 20	17-4 PH PH55A	Cast iron	Ductile iron/cast steel
Impeller	28% Cr	28% Cr	PH55A/17-4PH CD-4MCu	PH55A/17-4PH CD4-MCu	28% Cr	28% Cr
Wear rings	28% Cr	28% Cr	Same as impeller	Same as impeller	28% Cr	28% Cr
Shaft sleeve	28% Cr or 303 S.S. ceramic-coated	28% Cr or 303 S.S. ceramic-coated	316 or alloy 20 ceramic-coated PH55A, etc.	316 or alloy 20 ceramic-coated PH55A, etc.	28% Cr or 303 S.S. ceramic-coated	28% Cr or 303 S.S. ceramic coated
Shaft	Carbon steel	High-tensile alloy steel	316 S.S./alloy 20	17-4 PH 17-4 PH	Carbon steel	High-tensile alloy steel
	Alloy 20	21% Cr	29% Ni	2.5 Mo		
	CD4-MCu	26% Cr	5% Ni	2.0% Mo		
	28% Cr	28% Cr	—	—		
	304 S.S.	19% Cr	10% Ni	—		
	316 S.S.	19% Cr	10% Ni	2.5% Mo		
	17-4 PH	17% Cr	4% Ni	—		Hardenable
	PH55A	20% Cr	10% Ni	3.5% Mo		Hardenable

SUMPS

Permanent sumps are seldom used in open-pit mines because the sump area generally moves as the mining operation progresses. This means that the pump station must be portable, and the installation of the pumps on a barge provides the most convenient arrangement. Either horizontal or vertical pumps may be used, but vertical pumps eliminate the need for priming equipment. If the vertical pumps are of the overhung-shaft design, the stuffing box may be eliminated. This is important if the water is dirty. Although small cyclones can be used to clarify the water for pumps which have stuffing boxes, care must be taken to prevent leaves and other trash from blocking the gland water line. In freezing climates, special provision must be made to prevent the suction line, pump, and even the barge itself from freezing in place.

Underground mine sumps present special problems because they function not only as sumps but also as clarifiers. A well-designed sump is a good clarifier, but all too frequently inadequate provisions are made for cleaning the sump. If it is not cleaned at regular intervals, the loss of storage capacity may be critical in the event of a power failure. Furthermore, a sump partially filled with solids does not give the proper retention time for clarification, and the solids are directed into the pump. Although few sumps can economically be made large enough for complete clarification, it is important that a large portion of the solids be removed. This is particularly true for 3600-rpm pumps because the high-speed generally produces a high head per stage and the high differential pressure between stages causes severe wear if abrasive solids are present. Some mines use conventional thickeners and flocculating agents in an attempt to keep a high concentration of solids from reaching the pump.

Some general rules should be considered in designing sumps for underground pump rooms:

1. Attempt to get a complete analysis of the water (from another portion of the mine or from an adjacent mine if necessary).

2. Analyze the sample for corrosive properties to determine the proper materials of construction for the pump.

3. Analyze the sample for possible scale buildup in the pipeline and pumps. Check the velocity effect, if any, on the buildup rate.

4. Determine the percentage of suspended solids in the sample, its screen analysis, and the settling rate for various fractions. Determine the sump dimensions necessary for removal of all solids and then for progressively larger solids in order to select the most economical size.

5. Compare the sump size as determined in rule 4 with the size required for physical storage capacity for (a) continuous pumping, (b) off-peak power pumping, (c) programmed pumping, and (d) storage during estimated maximum length of power interruption.

6. Calculate practical sump dimensions, considering the geologic conditions.

7. Install grit traps ahead of the sump to remove large, heavy solids. Consider methods for cleaning the grit traps.

8. Install trash screens to prevent wooden wedges, etc., from entering the sump.

9. Review sump cleaning methods and program. The best-designed sump is of no value if it is not cleaned. Compare mechanical cleaning methods with cost of parallel sumps.

10. Review the suction requirements of the pumps to be used. Because of altitude, temperature, distance from low water level to pump centerline, and suction line loss, the available NPSH may be inadequate for even an 1800-rpm pump. If a decision as to pump size, type, and speed has been made and an NPSH problem does exist, then a decision must be made either to use low-speed booster pumps or to lower the pump room level to below the sump level. From a safety standpoint, the use of a booster pump is preferable, although it does add another piece of equipment.

11. Where the storage capacity is inadequate to meet possible power failures, consider either vertical pumps—possibly up to 100 ft (30 m)—for the shaft bottom pumping up to the main pump station level or sealed pump rooms that can operate over wide variations in the sump level from a 15-ft (4.6-m) suction lift to a positive head of several hundred feet.

12. Determine the final design based on a compromise between the mine engineer (who wants maximum output), the electrical engineer (who wants small starting load), the geologist (who

wants small sump dimensions), and the mechanical engineer (who wants the most reliable and easily maintained equipment).

AUTOMATIC PUMP CONTROL

Few pump stations can be operated economically with manual control. Automatic control can be a simple float switch or a pressure switch, or it can be sufficiently complex to provide reliable operation under the most critical or adverse conditions. Automatic control can provide greater reliability, and its cost can be depreciated over only a few years. Furthermore, the automatic recording of flow rate, flow totalizing, and periods of operation provides valuable data for analyzing the performance of the pumping installation as well as the possible cost savings in pumping during off-peak power periods.

Where the safety of a mine is dependent on the reliable operation of the dewatering pumps and controls, the following minimum requirements should be considered:

1. There should be a sump level alarm for high water, both local and remote (at the surface).

2. Sump level control should be dependable. For example, electrodes are generally unreliable in waters which leave a conducting film.

3. The control should be programmed where more than one pump is installed. However, the use of an alternator is not always desirable because all pumps are exposed to the same degree of wear. It is preferable to have one standby pump programmed through a sequence selection switch to operate at least once per week.

4. Pump priming should be positive. Hydraulic devices should be combined with electric controls so that complete dependence is not on the electric control. The presence of water in the pump should be detected to prevent the starting of a dry pump.

5. A delay circuit should be provided to ensure complete priming.

6. The control should provide for at least three starting attempts (unless an overload has occurred).

7. The priming time should be limited (if under a suction lift system).

8. The control should provide for a restart in the event of a false loss of prime on start-up (suction lift system).

9. Pump and motor bearings should have thermostats to stop pumps in the event of bearing failure.

10. Vibration monitoring may be important. This is particularly true for vertical pumps.

11. Pressure controls should indicate normal pressure and fail-safe in the event of a loss of pressure (broken column line, etc.)

12. Flow indication (check valve flow switch) is needed to signal a shaft failure.

13. Remote indication (generally at the mine office) should provide at least an indication of operation and signal pump failure or high water. More detailed information may be transmitted.

14. For long distances, investigate the use of carrier-current indication schemes, together with signal multiplexing, etc.

15. Provide a method to test the control and alarm system.

DRIVERS

Open-pit mines use electrically driven mining equipment, such as drag lines and shovels, and the availability of power has permitted the use of electrically driven pumps. There are still many gasoline-driven or diesel-driven units, but the convenience of electric power, particularly for automatically controlled units, has increased the trend to electric drive. The availability of reliable high-voltage cable has made portable high-voltage equipment safe and economical. Pumps in open-pit service are seldom provided with sophisticated control or drive mechanisms. The primary requirements are reliability, portability, and wear resistance.

In locations where rainfall may be heavy and there is danger of power failure, a combination of electric drive and engine drive is used. The engine can be direct-coupled to the pump through a motor with a double-extended shaft or with a clutch between the engine and the motor. Automatic control is simple and reliable.

Although some steam-driven pumps still exist in underground service, their number is rapidly decreasing. Electric motor drive is the simplest for automatic control. Variable-speed units, however, are seldom used in underground service because the ratio of static head to total dynamic head is quite high. Thus the frictional loss is not a large percentage of the total head loss and not much advantage is gained by variable speed. The solution is usually a multiple-pump installation. This must be designed with care because it is possible to raise the frictional head to a point where an additional pump produces little additional capacity. Multiple discharge lines are the answer and are frequently used for safety reasons. In normal service, all discharge lines are used in parallel, although conservative design allows each line to handle the required capacity.

As with all pumping installations, a complete set of system-head curves must be prepared to analyze the power requirements under all conditions.

Motor enclosures are important in underground service. Because of the high humidity, special insulation (epoxy encapsulated, etc.) should be specified. Dripproof enclosures are the minimum requirement, with weather-protected Type I a preferred construction. Heaters should also be provided. Screens should be installed to prevent the entrance of rats. Winding temperature detectors, bearing thermostats, and ground-fault detectors are recommended in mine service and should be incorporated in the pump-control and alarm circuits.

Although the starting torque of a centrifugal pump is low, in some cases the available torque may have to be checked. A normal-torque motor should be suitable for pumps in the range of 500 to 3000 (10 to 60) specific speed. High-specific-speed pumps, however, have the highest power at shutoff, and it will be necessary to examine the starting arrangements for such pumps.

Starting a pump against a long empty pipeline may present overload problems, and repeated starts may be necessary. In such cases the number of permissible starts per hour should be checked. Winding temperature detectors are important in such applications.

Although reduced-voltage starters may be required in some instances, most modern mines have electrical facilities designed for across-the-line starting. This is preferable because the starting equipment is cheaper. Before deciding on a reduced-voltage starter, the effect of the starting load on the transformer and line impedance should be checked. It may be that the voltage drop will eliminate the requirement for reduced-voltage starters. On the other hand, the effect on the primary side should be checked so that the voltage drop is not so large as to drop out other equipment.

Synchronous-motor drives are seldom used unless they are large (generally at least 1000 hp) (750 kW) and then only if they are in relatively continuous service. Under such conditions they can be operated under "leading current" conditions for power factor correction. Smaller installations frequently provide capacitors at the pump installation to provide the necessary correction for a particular installation.

Surge protection from lightning should not be overlooked. Some locations are particularly susceptible to lightning damage, especially to long surface lines. Lightning arresters should be provided at the surface, and surge arresters should be mounted at the motor location.

SECTION 9.11
MARINE

G. W. SOETE

Marine pumping services can be divided into two broad categories: (1) those related to the boiler-feed cycle of the power plant and (2) those auxiliary services that support the power plant, hotel load, loading and unloading of the ship, and, in the case of naval vessels, the armament. Centrifugal and rotary displacement pumps predominate in marine services. Reciprocating steam pumps, fixed-stroke and variable-stroke power pumps, and regenerative pumps are used to a lesser degree.

BOILER-FEED SYSTEM

Condensate   The condensate pump receives water from the condenser hot well, which is normally under vacuum. The piping from the hot well to the pump must be short and direct to minimize frictional losses. As the hot well is already at a very low point in the vessel, the condensate pump can be positioned only 6 to 24 in (152 to 610 mm) below the hot well. The condensate pump must operate under a relatively small positive submergence° (NPSH) and requires speeds of 1750 rpm or less. It must operate at conditions of varying capacity to suit the power plant load.

The chief means of control is to allow the pump to operate in the cavitation break, as shown by the curves in Fig. 1. At rated load, the pump operates at point R, at which time the hot well level provides a submergence as shown. This defined submergence provides the energy required to cause the condensate to flow to the suction area of the impeller. The pump can deliver only that specific capacity. The maximum submergence available must exceed the submergence required for operation at the rated capacity R. When less flow enters the hot well, such as at $0.9R$, the pump, operating at point R, will extract more water from the hot well than is entering. The hot-well level will be reduced, thus lowering the submergence at the pump suction. The pump capacity will be reduced to $0.9R$, coincident with the amount entering the hot well. At this lower flow rate, the system head requirement is reduced and the pump will develop the total head shown at point A. If the flow is reduced further, to point $0.7R$, the total head is reduced to point B on

---

°The term *submergence* is commonly used and refers to the static dimension between the liquid level and the suction nozzle of a vertical pump or to the impeller centerline of a horizontal pump. The preferred designation is net positive suction head (NPSH) at the impeller centerline, which is equal to the submergence less the frictional losses in the suction piping.

PUMP SERVICES

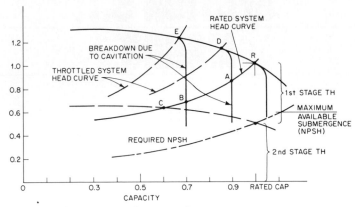

FIG. 1 Condensate pump performance and system curves.

the system head curve. When flow to the condenser increases, a rise in the hot-well level follows, and the higher level of submergence furnishes the additional energy needed to increase the flow to the pump suction. The pump responds by producing the corresponding increases in head and capacity.

Operation can be very stable, provided the flow to the condenser does not fluctuate violently and is not reduced below about 0.7R. While the pump is operating in the cavitation break, the impeller is partially filled with vapor. If the pump is a multi-staged type, it is possible to reduce the flow to the point where the first stage is completely filled with vapor, and only the remaining stage or stages develop head, as at point C. At lower flows, the second stage will become vapor-filled. Condensate service is severe under such conditions of fluctuating low values of submergence, and a close matching of the pump performance and system requirements is required.

At reduced flows it is usually necessary to open a bypass to permit the condensate to recirculate back to the hot well. This will reduce the noise and vibration that occur in the suction line and in the pump. The bypass may be controlled manually or automatically by the hot-well level. When the condensate is used as cooling water in the air-ejector condenser or other heat exchangers, the minimum flow is determined by those components rather than by the feedwater requirements, and the bypass may be thermostatically controlled.

It is possible to have the pump operate continuously on its head-capacity curve by throttling the discharge. This alters the system-head curve to such points as D and E in Fig. 1. A float control in the hot well maintains the water level between a preset maximum and minimum and throttles the pump discharge when the hot-well level falls or opens the discharge as the level rises.

Another arrangement combines the bypass method with the throttling control. It provides throttling of the discharge over the recommended range of operation from 70% to the rated capacity and also provides the additional flow at lower capacities to satisfy the cooling requirements of the heat exchangers. This combination control reduces the noise and vibration in the suction line and pump to a minimum.

The condensate is pumped from the condenser hot well to the direct-contact deaerating feedwater heater maintained at a pressure of 10 to 60 lb/in^2 (0.7 to 4 bar°). The elevation of the heater is determined by the submergence required by the boiler-feed pump when there is no booster pump. It is customary to pump the condensate discharge through an air-ejector condenser and gland vapor condenser. This serves the dual purpose of providing a fresh-water cooling source for these heat exchangers and of preheating the condensate before it enters the deaerating feedwater heater.

Condensate pumps may be of one, two, or three stages, depending on the precise amount of total head required by a particular installation. The total head is the sum of the difference in pressure between the hot well and the heater, the difference in their elevation, and the frictional resistance of the piping and the heat exchangers in the discharge line.

° 1 bar = 10^5 Pa. For a discussion of bar, see *SI Units—A Commentary* in the front matter.

Boiler Feed   The boiler-feed pump receives water from the deaerating feedwater heater. The suction pressure on the boiler-feed pump is equal to the sum of the pressure in the heater—usually 10 to 60 lb/in² (0.7 to 4 bar)—and the submergence of the heater water level above the feed-pump suction, less the frictional losses in the suction piping. The exact suction requirement will depend on the NPSH required by a particular pump. Boiler-feed pumps customarily run at 3500 rpm when motor-driven and at higher speeds, up to 12,000 rpm, when driven by steam turbines, by the main propulsion turbine, or by the turbogenerator turbine. The relatively high speed of the pump and the higher pumping temperature require a higher NPSH than with the condensate pump. In addition, the feed pump cannot be allowed to operate in the cavitation break, but is required to pump at a constant or variable discharge pressure under widely fluctuating capacities.

Figure 2 shows a typical system head-capacity curve for a marine boiler-feed pump. A pump running at constant speed is shown at *a*. At all capacities below the rated capacity, there is considerable power loss because of the excess pressure produced by the pump relative to the system head requirement. For this reason constant-speed pumps are selected only for low-capacity services, such as auxiliary feed and waste heat boilers, or where the simpler motor drive offers a degree of reliability beyond that of the turbine drive. A saving in power is realized by running the feed pump at varying speed and constant pressure, as at *b*. The variation in speed is obtained by using a constant-pressure governor which controls the turbine. The most economical arrangement permits the pump to operate at a varying speed while producing a discharge pressure coincident with the system-head curve, as at *c*. The driving turbine is controlled by a differential pressure governor.

Merchant vessels operating for long periods at constant load provide the special cases where the feed pump may be driven directly by the turbogenerator or the main propulsion turbine. A separate auxiliary feed pump must be installed for lower capacity or emergency conditions. A feed pump connected to the main turbine must have a special disengaging coupling to permit operation from an auxiliary turbine at some minimum speed when the main unit is at reduced speeds. System pressures up to 1500 lb/in² (100 bar) can be produced by one- and two-stage high-speed pumps. More stages are required for lower pump speeds and/or higher pressures.

A boiler-feed pump requires a recirculation line to limit the temperature rise of the water at low flows. A common practice is to limit the temperature rise to 15F° (8C°), but limits of 20 to 25 F° (10 to 14 C°) may be encountered. The amount of recirculation flow ranges from 5 to 15% of the rated capacity. It may be continuous or may be used only to satisfy the minimum flow condition, and this fact must be considered when sizing the pump. The method of pressure governing must be considered because a pump running at constant speed will require more recircu-

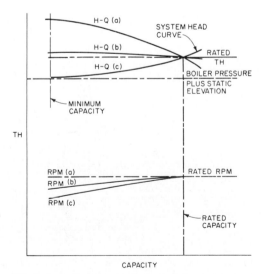

FIG. 2   Boiler-feed pump performance and system curve.

lation than a pump running at reduced speed on the system curve. The recirculation system requires a breakdown orifice, a special valve, or friction tubing to dissipate the pressure of the recirculation flow that is returned to the deaerating feedwater heater. For automatic operation, the recirculation valve is controlled by the signal from a flow transmitter connected to a flow orifice in the pump discharge line, or by a mechanical connection to a special discharge check valve.

Feedwater Booster   When the design requirements of the system and vessel do not permit the feedwater heater to be installed at an elevation that will provide sufficient NPSH for the feed pump, a booster pump must be installed. The booster acts essentially as the first stage of a feed pump, but at a lower speed—usually 1750 rpm. It may have one or two stages, and its design is similar to that of a condensate pump. Unlike the condensate pump, however, which may operate under cavitating conditions, the booster pump forms an integral part of the boiler-feed discharge system and must furnish a stable suction pressure to the feed pump. A recirculation line must be provided to bypass a minimum flow when the boiler-feed pump is not in use.

Pump Construction   A typical two-stage condensate pump is shown in Fig. 3. A vertical arrangement of the rotating element is preferred to provide favorable suction conditions at the first-stage impeller, located at the bottom, and to permit access to the bearings, coupling, and vertical driver. Wearing rings are fitted to the casing and impellers. The internal sleeve bearing is lubricated by the pumped condensate, and the upper external combined line and thrust bearing is of the grease-lubricated ball type. Water discharged from the first stage passes to the suction of the second stage through an external crossover. A vent connection is required at the suction flange so that any vapors may pass freely back to the condenser.

Booster pumps are similar to condensate pumps in construction. A single-stage condensate or booster pump would look like the unit shown in Fig. 3 with the upper stage omitted. Water-lubricated bearings are seldom used in booster pumps because of the higher pumping temperatures.

A typical boiler-feed pump is shown in Fig. 4. The pump is driven by a separate steam turbine. A governor is required to provide either constant-pressure or differential-pressure regulation. While one- and two-stage pumps may be direct-connected to the turbine, multistage pumps up to four stages are driven by a separate steam turbine through a flexible coupling.

Positive displacement pumps are used for temporary in-port or emergency boiler-feed service.

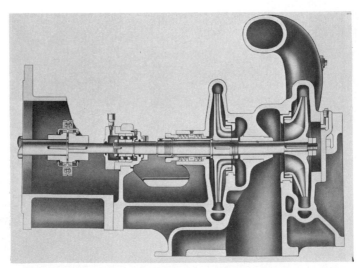

FIG. 3 Two-stage condensate pump. (Warren Pumps, subsidiary of Houdaille Industries)

FIG. 4   Boiler-feed pump driven by a steam turbine. (Worthington Pump)

Vertical steam reciprocating pumps are simple to operate, can be operated at variable speeds, and are suitable for severe suction conditions because of their self-priming characteristics. Motor-driven fixed-stroke power pumps may be used for low-capacity boiler-feed service, but their application is limited by the fact that they are essentially of constant capacity.

FRESH-WATER SERVICES

Condenser Vacuum   Motor-driven vacuum pumps operating on the liquid ring principle are frequently used in place of conventional steam-powered ejectors for extracting the saturated mixture of air and water vapor from the main and auxiliary condensers. This type of exhauster is classified as a rotary pump. Water is used as the sealing medium and serves as the liquid compressant.

A water-sealed vacuum pump is shown in Fig. 5. As can be seen, a bladed rotor revolves within the pump body. When sealing water is introduced, the rotor carries the liquid around the eccentric casing and forms a liquid ring that revolves at nearly the same speed as the rotor. The rotating liquid almost fills and then partly empties each rotor chamber once each revolution, setting up a piston action. As liquid passes through the diverging casing sector, it draws in air through the inlet port near the hub. As the fluid passes through the converging sector of the pump body, the liquid moves inward and the air and vapor are compressed and forced out through the discharge port near the hub. A portion of the liquid flows out with the compressed air and vapor and is removed in a mechanical separator. The liquid is cooled by circulation through a heat exchanger and is returned to the vacuum pump as make-up. Air and other uncondensed gases are discharged to the atmosphere.

Capacities of dry air with the water vapor required to produce saturation with 7.5 F° (4.2 C°) subcooling of the mixture range from 25 standard ft³/min (0.7 standard m³/min) at 1 inHg (25.4 mmHg) absolute to 65 standard ft³/min (1.84 standard m³/min) at 3.5 inHg (88.9 mmHg) absolute. Speeds of 1750 rpm and lower are common, and operation is continuous. Pumps of this type are used for main condenser vacuum as well as for auxiliary and evaporator plane vacuum. Vacuum pumps are also applied as gland seal exhausters to draw the mixture of air, condensate, and steam from the turbine glands.

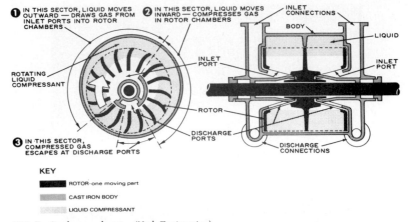

❶ IN THIS SECTOR, LIQUID MOVES OUTWARD — DRAWS GAS FROM INLET PORTS INTO ROTOR CHAMBERS

❷ IN THIS SECTOR, LIQUID MOVES INWARD — COMPRESSES GAS IN ROTOR CHAMBERS

INLET CONNECTIONS

BODY

LIQUID

INLET PORT

INLET PORT

ROTATING LIQUID COMPRESSANT

ROTOR

❸ IN THIS SECTOR, COMPRESSED GAS ESCAPES AT DISCHARGE PORTS

DISCHARGE PORTS

DISCHARGE CONNECTIONS

KEY

▮▮ ROTOR-one moving part

▨▨ CAST IRON BODY

░░ LIQUID COMPRESSANT

FIG. 5 Condenser exhauster. (Nash Engineering)

Drainage Fresh-water drainage service requires small-capacity centrifugal pumps that do not have the capability of being self-priming. Capacities to 350 gpm (80 m³/h) at total heads of 180 ft (55 m) and speeds of 3500 rpm are typical. The term *fresh water* implies that the collected water has not been contaminated and therefore may be reused in the condensate system with a minimum of treatment and deaeration. In practically all cases, the pump takes suction from a collecting tank and discharges to a receiver, such as the condenser hot well or fresh-water bottom. Operation is usually on-off and controlled by a float-level switch.

Pump construction may be of several motor-driven types, as shown in Fig. 6: (*a*) horizontal or vertical end-suction type, (*b*) horizontal or vertical in-line type, where the suction and discharge flanges provide the mounting, (*c*) horizontal or vertical tank-mounted type, where the suction flange serves as the support bracket and the pumping unit is mounted on the tank side, and (*d*) vertical end-suction type with an extended shaft, where the pumping unit is mounted on the tank top and the extended shaft allows the pump impeller to be at an appropriate submerged level near the bottom of the tank.

(*a*)

FIG. 6 Fresh-water pumps: (*a*) end-suction (Aurora Pump, unit of General Signal), (*b*) in-line (Fluid Handling Division, ITT), (*c*) pump mounted on tank side (Aurora Pump, unit of General Signal), (*d*) pump mounted on tank top (Taber Pump).

(b)

(c)

(d)

9.157

Distilling Plant Condensate   Pumps for this service are similar to the general type of fresh-water drain pumps except for several added features. The condensate pump must handle water at its vapor temperature and therefore requires a suitable vent at its suction near the impeller eye. In addition, the pump stuffing box must contain provision for positive sealing to ensure against loss of vacuum. Capacities to 350 gpm (80 m³/h) at total heads of 200 ft (61 m) are typical. Speeds up to 3500 rpm are used but are frequently lower because of the severe suction conditions.

Construction is similar to that shown in Fig. 3 (either one or two stages) and in Fig. 6 (a and c). Operation is usually continuous, with the pump operating in the cavitation break as described previously and as shown in Fig. 1.

Fresh-Water Supply   Pumps for this service are similar to the general type of fresh-water drain pump shown in Fig. 6 (a and b). Capacities to 350 gpm (80 m³/h) at total heads of 180 ft (55 m) at 3500 rpm are typical.

The operating system may use one pump running on an on-off service controlled by a pressure switch, with or without an air-charged accumulator tank connected to the discharge line, or it may use one pump on continuous service with one or more standby pumps controlled by pressure switches.

Circulating   Pumps for this service are similar to the general type of fresh-water drain pump shown in Fig. 6 (a and b). When relatively large capacities are required, double-suction types, either horizontal or vertical as shown in Fig. 7, are used.

Fresh-water circulating circuits are used for cooling purposes, namely, chilled water for air-conditioning systems, cooling of electronic components, and cooling of engine jackets, and for other auxiliaries where seawater must be excluded. Operation is usually continuous, with the system capacity controlled by a thermostat or by temperature-controlled throttling valves. The pump usually takes suction from a tank or standpipe of sufficient elevation to provide an adequate suction head. Conditions of service vary widely with capacities up to 1000 gpm (227 m³/h) at total heads up to 200 ft (61 m) and speeds up to 1750 rpm. At capacities up to 4000 gpm (908 m³/h), the total heads are lower, ranging down to 50 ft (15 m). For higher total heads, up to 400 ft (122 m), a speed of 3500 rpm is required, and capacities generally do not exceed 1000 gpm (227 m³/h). Pumping temperatures do not usually exceed 200°F (93°C).

SEAWATER SERVICES

Condenser Circulating   The main propulsion plant of a steam-powered vessel requires one or more large centrifugal pumps for condenser circulating. Capacities from 5000 gpm (1136 m³/h) at 90 ft (27 m) total head to 25,000 gpm (5678 m³/h) at 25 ft (8 m) total head are typical and can be produced by horizontal or vertical double-suction pumps. Speeds vary from 437 to 1150 rpm, depending upon pump capacity and head. A range from 14,000 gpm (3179 m³/h) total head to 30,000 gpm (6813 m³/h) at 15 ft (4.6 m) total head is common for vertical mixed-flow pumps. Speeds usually do not exceed 870 rpm. Where the combination of speed, capacity, and head results in a specific speed greater than 7500 rpm, the pump will be of the axial-flow or propeller type.

A typical circulating pump takes suction from a sea chest at the skin of the vessel and discharges to the inlet water box of the condenser. The circulated water leaving the condenser outlet is discharged overboard through another sea chest. A scoop may be fitted in a separate intake to the condenser. The scoop functions at higher ship speeds, during which time the pump may be secured or idling.

When more than one pump is installed, capacity may be varied by selecting the number of pumps operating, through speed regulation of turbine and dc motor-driven pumps and through speed selection of multispeed ac motors. Capacity may also be varied by throttling the discharge valve or by utilizing a bypass connecting the condenser outlet water box to the pump suction. Where an axial-flow pump is installed, it is also possible to consider the application of variable-pitch propeller blades for capacity regulation. A takeoff connection is frequently fitted at the pump discharge to divert a small amount of flow to the auxiliary cooling system under emergency conditions.

FIG. 7 Vertical double-suction pump. (DeLaval Turbine)

FIG. 8 Condenser circulating pump. (Warren Pumps, subsidiary of Houdaille Industries)

For typical construction features, refer to Fig. 7 for the double-suction pump and to Fig. 8 for the axial-flow pump. In the case of submersibles, the higher values of submergence require that special consideration be given to the design of the casing, seals, and thrust bearing.

Bilge and Ballast Centrifugal pumps for this service are usually of the double-suction type with a vertical shaft and with the motor or turbine mounted on top of the casing. Capacities vary up to 5000 gpm (1136 m³/h) at 90 ft (27 m) total head and at speeds up to 1150 rpm. Higher capacities at lower heads may also be required. A typical construction is shown in Fig. 7 for medium and larger sizes, and in Fig. 6a for very small sizes.

Bilge service requires a priming device, which may be mounted directly on the pump or may be a separate exhauster-type device. Vacuum pumps of the type depicted in Fig. 5 can be used for this purpose. A bilge connection may be fitted at the suction of the main condenser circulating pump and used for bilge evacuation under emergency conditions. Bilge pumps may be required to operate submerged; the driver must be mounted a suitable distance above the pump to prevent flooding or must be equipped with a protective submergence bell.

For rapid evacuation of small compartments, the bilge pump may be paired with one or more eductors. Normally the bilge pump discharges directly overboard, but for higher drainage rates it may discharge to an eductor either in the same compartment or in a remote compartment.

A unique type of bilge and ballast pump is shown in Fig. 9. This radial- or mixed flow centrifugal pump is driven by a water turbine. Capacities vary from 1000 gpm (227 m³/h) at 90 ft (27 m) total head to 2400 gpm (545 m³/h) at 50 ft (15 m) total head at variable turbine speeds. The turbine is driven by water at fire-main pressures.

The steam-driven reciprocating pump may be applied to bilge and ballast service as it is inherently self-priming and the speed can be easily regulated. Where steam is not available, motor-driven horizontal or vertical power pumps are used. They are usually of two- or three-cylinder design. Bilge and dirty ballast services can be handled by the same pump or group of pumps. For clean ballast service, a separate system and pumps are required.

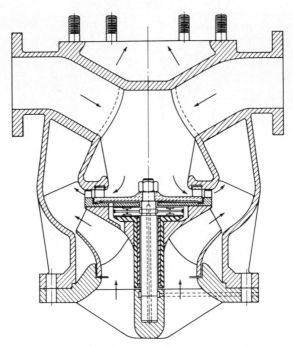

FIG. 9   Liquid turbine-driven pump. (Worthington Pump)

Fire Main   For this service centrifugal pumps are of the double-suction type (Fig. 7) and are mounted vertically or horizontally. The end-suction pump shown in Fig. 6a is also used. Capacities vary up to 1500 gpm (340 m³/h) at total heads from 150 to 400 ft (46 to 122 m) at speeds of 3500 rpm. For higher heads, a two-stage pump is used. Operation is continuous, with additional pumps placed in operation as the demand for higher volume increases. Pumps may be driven by a turbine, a diesel engine, or a gas turbine to ensure availability during a loss of electric power. The steam-driven reciprocating pump can also be applied for fire-main service.

Flushing   Pumps for this service are similar to those used for ballast and fire-main service. The capacity and total head seldom exceed 1000 gpm (227 m³/h) and 100 ft (30 m) at speeds of 3500 rpm. Operation is usually continuous for one pump, with additional pumps activated by pressure controls when the need arises. Flushing service may be provided by the fire pump, but operating at reduced pressures through a reducing valve.

Cargo Tank Cleaning   A water gun is used for cargo tank cleaning, and capacities vary up to 4000 gpm (908 m³/h) at heads of 100 to 400 ft (30 to 121 m), depending on the pressure required at the gun. The tank cleaning machine may be stationary, or it may be slowly rotated by air pressure or by water pressure available from the gun supply line. If the conditions of service permit, the clean ballast, fire-main, flushing, or auxiliary cooling seawater pumps may be used for tank cleaning. If not, then separate tank cleaning pumps must be installed. Pumps of the type shown in Fig. 6a would be used for the smaller capacities, and pumps of the type shown in Fig. 7 would be used for the larger capacities.

   Tank cleaning operations must meet all safety requirements in order to avoid explosions and must be coordinated with the gas-freeing operation.

Auxiliary Condenser Cooling   Pumps for this service are similar to fire and flushing pumps. Conditions of service vary greatly, depending on the service, with capacities up 4000 gpm

($908 \text{ m}^3/\text{h}$) and total heads up to 200 ft (61 m). Principal applications are for refrigeration condensers, turbogenerator condensers, and distilling plant condensers. These services may receive their principal supply from the main condenser circulating, fire, or flushing pump.

Operation is usually continuous at constant speed or at stepped speeds using a multispeed motor. Construction is similar to the end-suction type shown in Fig. 6a and the double-suction type in Fig. 7.

Distilling Plant Feed   End-suction pumps of the type shown in Fig. 6a with capacities up to 1000 gpm (227 m^3/h) and total heads of 200 ft (61 m) at 3500 rpm are widely used for this service. Regenerative pumps are used less frequently. Operation is continuous for both types.

Distilling Plant Brine   End-suction pumps of the type shown in Fig. 6a with capacities up to 1000 gpm (227 m^3/h) and total heads of 150 ft (46 m) at 3500 rpm are widely used for this service. Brine service is highly corrosive and erosive, and semiopen impellers are often used. The suction must be vented, and the stuffing box must have provision for sealing against a loss of vacuum. Operation is usually continuous, although in special cases, as with submersibles, the pump is multistaged and the operation is intermittent.

Sanitary   For this service a variety of pumps are used, depending upon the particular function. In any sanitary system the conservation of water and the prevention of the discharge of pollutants overboard are of paramount importance.

The collecting circuit that operates under a vacuum requires a vacuum pump. Vacuum pumps operating on the liquid ring principle (Fig. 5) and vane or piston positive displacement pumps are used. The vacuum is utilized to propel the liquid-solid plug to the holding tank. The containment system may involve holding only or may include a system for treatment. One or more pumps are used, depending on the specific system, for recirculating the treated water for flushing and for pumping overboard outside the contiguous limits or to a shoreside sewage connection. Pumps for this purpose are usually of the end-suction type (Fig. 6a).

Jet Propulsion   The combination of pump speed, capacity, and head required for jet propulsion corresponds to a specific speed that requires a pump of the mixed-flow design, as shown in Fig. 10, or of an axial-flow design. Wide market acceptance has been established for jet propulsion as the motive power for work boats, pleasure boats, and military and commercial transportation. The pumping circuit from suction to discharge is short, but it requires careful consideration to ensure proper suction conditions at variable speeds. The overall design emphasizes weight and space saving, reliability, and proper matching to the speed of the internal combustion engine or gas turbine.

Jet thrusters may be fitted as auxiliary devices to assist in maneuvering, positioning, station-

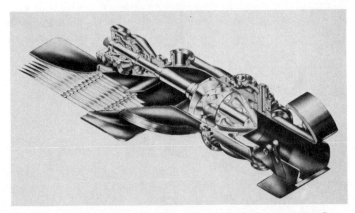

FIG. 10   Jet propulsion pump. (Rocketdyne Division, Rocketdyne International)

keeping, docking, and undocking. Installation may be (*a*) fixed, where the pump takes suction from the bottom of the hull and its discharge is directed to either side of the vessel, (*b*) fixed, with the pump mounted in an athwartship tunnel and discharging to either side of the vessel by reversible rotation or pitch, or (*c*) retractable, where the pump direction may be fixed or turned.

VISCOUS FLUID SERVICES

Cargo Double-suction pumps are used for this service to handle the medium-volume and high-volume rates required. Multistaged vertical mixed-flow pumps are used to a lesser extent to pump lower-volume rates against higher heads. Capacities extend up to 25,000 gpm (5678 m³/h) at 500 ft (152 m) total head for the single-stage double-suction pumps, and at lower capacities to 650 ft (198 m) total head for the vertical multistaged pumps.

Single-stage double-suction pumps are installed in a separate pump room with two or more pumps connected through manifolds to the suction piping to the cargo tanks. The suction piping consists of double valving, cleanable suction strainers, vent connections, and stripping connections. The pumps are connected to the discharge piping leading to the vessel's deck through manifolds. From the vessel's deck, connections are made to the on-shore receiving facility. Double valving is often used in the discharge piping along with a relief valve, a nonreturn check valve, and an adjustable valve for throttling the pumps.

Figure 11 shows a typical set of system conditions, where point *A* is considered to be the rated condition for a particular port. The pump must be selected to produce the total head equal to the sum of the elevation and the frictional head of the system piping at point *A*. In addition, the NPSH available at the pump suction must equal or exceed the NPSH required by the pump at point *A*. When cargo oil is being unloaded, the level in the cargo tanks will fall until the NPSH available is inadequate at capacity *A*. At this point the speed of the pump can be reduced to produce the new condition of head and capacity at point *B*, where a lower NPSH is required. The speed may be further decreased to accommodate the continued lowering of the section level, as at point *C*.

Lower capacities may also be obtained by throttling the discharge of the pump. Starting at the rating point *A* where the NPSH becomes inadequate, the system head curve is artificially changed by throttling the discharge, as at point *D*. As the suction level continues to fall, further throttling will decrease the capacity at point *E*, or else the pump speed may be reduced to produce the conditions at point *F*. A reduction in capacity is required to prevent cavitation, noise, vibration, and possible damage to the pump and system components.

Point *D* represents a set of conditions different from that at *A* which may apply to a particular port. When the pump is operating at point *D* and the NPSH available is reduced to the point of cavitation, the speed of the pump must be reduced to points *F* and *G*. At constant speed the discharge of the pump must be throttled to points *E* and *H*. The degree of throttling and/or speed

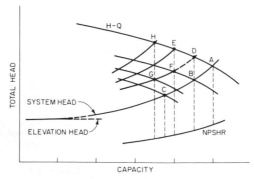

FIG. 11 Cargo pump performance and system curves.

control must be coordinated with the driver characteristics, the system requirements for the particular port, and the time available for unloading of the vessel. The variable-speed drive also has the advantage of providing adjustment in pump output for products of different viscosities and specific gravities.

Horizontal pumps are driven by a steam turbine and reduction gear located in the adjacent machinery space and connected by a drive shaft running in a bulkhead stuffing box. Vertical pumps are driven by vertical shafting extending to a machinery flat or deck above. The driver may be a vertical or horizontal turbine, motor, diesel engine, or gas turbine with a right-angle gear drive. Vertical multistaged pumps are mounted in the cargo tank and may be driven by a deck-mounted vertical or horizontal motor, an engine with a right-angle gear, or a hydraulic motor submerged in the cargo fluid.

The vertical multistage pump finds wide application in the pumping of crude and refined petroleum products, cryogenic fluids, and special chemical products. The canned motor pump with an isolated motor stator submerged in the fluid finds almost exclusive application for cryogenic fluids. This is an end-suction pump and may be mounted on the tank bottom or suspended in the discharge column. Capacities vary up to 10,000 gpm (2271 m^3/h) at 200 ft (61 m) total head. Both types employ a special first-stage impeller of the inducer type to meet the low values of NPSH encountered in these services. The single-stage canned pump may discharge to a second pump on deck for boosting the total head. Positive displacement pumps of the vane, screw, gear, and sliding-shoe types are used to a lesser extent for small capacities.

For all applications, some means must be provided to ensure complete unloading of the tanks. In older vessels a stripping system is used with vertical or horizontal reciprocating steam pumps or horizontal duplex motor-driven power pumps. Main cargo pumps can be made self-stripping by providing a liquid ring exhauster for removing the air and vapors during the last phase of unloading. The construction of modern large vessels includes a pipe tunnel near the keel; this eliminates the need for a separate stripping system. Horizontal double-suction and vertical multistage pumps equipped with a special priming valve and recirculating system can perform stripping operations. When the pump loses suction, vapors enter the impeller; the loss of prime opens the special priming valve and permits fluid to return to the suction recirculation tank of the horizontal pump or to the bell mouth suction of the vertical multistage pump. The pumping action resumes, and the cycle repeats itself until the stripping is completed.

All modern cargo vessels have an inert gas installation for freeing tanks and lines of explosive gases and toxic fumes. The system consists of a scrubber and gas cooler that use circulating water from an existing flushing or general service system or from a separate double-suction or end-suction pump.

Fuel Oil Service   Fuel pumps for marine steam generators are positive displacement rotary pumps. Two or more are installed. Operation is continuous at constant speed when the unit is motor-driven and continuous at variable speed when it is driven by a steam turbine controlled by a constant-pressure governor. Pump discharge pressures vary from 100 to 300 lb/in^2 (7 to 21 bar), although higher pressures are used for special cases.

The typical fuel oil service pump takes suction from either a low or a high connection at the service tank. A standby connection at a second tank is piped up for emergency use. A duplex strainer is fitted to the pump suction for protection. The discharge from the pump passes through a strainer, a heater, a meter, and control valves to the header at the boiler front. A recirculation line is provided to return fuel oil to the pump suction for control purposes, for heating oil during start-up, and to allow pump operation when all burners may be temporarily secured.

The rotary fuel oil pump for gas turbines is powered directly through auxiliary drive gears from the turbine or from a separate motor drive. The fuel oil pump for diesel engines takes suction from a service tank and discharges through a filter to the header supplying oil to the injection pumps of the engine. The injection pumps are of the plunger type, spring-loaded and cam-operated, and may be combined with the injector.

Fuel Oil Transfer   Positive displacement rotary pumps, horizontal or vertical steam-reciprocating pumps, and motor-driven power pumps are used for fuel oil transfer service. Capacities vary widely, depending on the precise function, and discharge pressures usually do not exceed 100 lb/in^2 (7 bar). Multistage mixed-flow pumps mounted in the tank may also be used for some services. Typical services include pumping to and from service tanks and to and from settling tanks, and assisting in maintaining vessel trim.

Lubricating Oil Service   Pumps are predominantly of the positive displacement gear type. Capacities vary widely, depending on the power rating of the main engine and reduction gears, or on the requirements of a particular auxiliary drive. Lubricating oil header pressure at bearings and gears is typically 10 to 12 lb/in^2 (0.7 to 0.8 bar) and requires a pump discharge pressure of approximately 50 lb/in^2 (3.4 bar) for a pressurized system where the pump discharges through a strainer and cooler to the lubricating oil headers. For a gravity system, the service tank must be located 25 to 35 ft (7.6 to 11 m) above the lubricating oil headers to provide the required 10 to 12 lb/in^2 (0.7 to 0.8 bar) of pressure. The pump discharges through a strainer and cooler to the overhead tank, where a constant level is maintained by an overflow. Oil temperatures are approximately 110 to 120°F (43 to 49°C).

The capacity of the service tanks varies from two times pump capacity for small auxiliaries to five times pump capacity for the main engine lubrication. The pump may be driven by an ac motor, by a steam turbine with a constant-pressure governor, by a dc motor, or by a direct drive from the main shaft or reduction gear. Power takeoff from the reduction gear is limited, as the capacity may be inadequate at low speeds. Separate standby and emergency pumps must also be installed.

Gas turbines and diesel engines use gear oil pumps driven directly from the crankshaft. Motor-driven gear or vane pumps must be provided for start-up and cool-down as well as for emergency use. All shipboard auxiliaries that have a pressurized lubricating oil system require one or more lubricating oil pumps of the gear or vane type. Stern-tube lubricating systems utilize a gravity tank to supply the static head to the bearings. The operation of the service pump to the tank may be continuous or intermittent.

Other Services   A wide variety of pump applications are related to the loading and unloading of specialty cargoes. Pumps are required for slurry cargoes to furnish high pressure at the jets that direct the solid-liquid mixture to collecting tanks. Slurry pumps unload the mixture from the tanks to the shore facility. Additional pumps are required to flush the slurry lines.

Vacuum pumps are used for pumping fish, for cement unloading, and for vacuum cleaning the holds of the ships and barges.

A completely independent hydraulic system fitted with one or more positive displacement rotary pumps may be used to power several hydraulic motors which in turn drive deep-well pumps for the unloading of such specialty cargoes as lubricating oil, petrochemicals, and solvents.

Variable displacement axial and radial piston pumps provide versatile control and instant reversibility for cargo-handling equipment and other hull machinery, including the following:

1. Steering gear
2. Pipe tensioners
3. Alignment stays
4. Hydraulic rams and jacks
5. Engagement cylinders
6. Interlocks
7. Selector valves
8. Flow control gates
9. Boom and shuttle positioners
10. Elevators
11. Watertight doors
12. Hatch covers
13. Stabilizers
14. Controllable-pitch propellers

FURTHER READING

American Bureau of Shipping: *Guide for Inert Gas Installations on Vessels Carrying Oil in Bulk*, ABS, New York, 1973.

Bourn, W. S.: *Shipboard Solid Waste Control*, ASME, New York, 1973.

Brandau, J. H. (ed.): *Symposium on Pumping Machinery for Marine Propulsion*, ASME, New York, 1968.

Brandau, J. H.: "Aspects of Performance Evaluation of Water-Jet Propulsion Systems and a Critical Review of the State of the Art," AIAA/SNAME paper 67–360, May 1967.

"Conversion of Centrifugal Pumps to Automatic Self-Priming Operation," Ship Repair and Maintenance International, Surrey, England, October–November 1973.

Fassell, W. M., and D. W. Bridges: "PURETEC System for Treatment of Shipboard Waste," *Marine Technol.*, July 1974.

Feck, A. W., and J. O. Sommerholder: "Cargo Pumping in Modern Tankers and Bulk Carriers," *Marine Technol.*, July 1967.

Garber, D. C.: "An Oil-Water Separator System," *Marine Technol.*, January 1974.

Harrington, R. L. (ed.): *Marine Engineering*, SNAME, New York, 1971.

Jones, R. M., C. W. Wright, and C. S. Smith: *A Study of Large Self-Unloading Vessels*, SNAME, New York, 1972.

Nielsen, R. A., and H. H. Kendall: "Stern Thruster Installation on the SS John Sherwin," *Marine Technol.*, January 1974.

Smith, I. W.: "Recent Trends in Hull Machinery," *Marine Technol.*, April 1974.

SECTION 9.12
HYDRAULIC PRESSES

A. B. ZEITLIN

TYPES OF PRESSES

Hydraulic presses, both vertical and horizontal, are used in many industrial technologies. Vertical press applications include forging presses with flat dies, used for hot work to break down ingots and shape them into rolls, pressure vessels (mandrel forgings), forged bars, rods, plates, etc.; forging presses with closed dies, used to process perheated billets into various shapes, such as aircraft bulkheads, engine supports, and main fuselage and wing beams; and upsetting presses, used for the production of items with elongated shafts—long hollow bushings, pipes, vessels, etc. Horizontal presses are used primarily for conventional hot extrusion. Cold hydrostatic extrusion is finding application in the production of various end products and bulk items, such as very thin wire in long strands.

ACCUMULATORS

Until about 1932 all power systems for hydraulic presses were operated with water. With the advent of reliable, fast, high-pressure oil pumps, there developed a trend to significantly less expensive oil pumps.

Installations requiring a relatively uniform and constant supply of hydraulic power are preferably designed with drives drawing the pressurized liquid directly from the pumps. In installations with high power (and liquid) peak demands of short duration, it is in most cases advantageous to arrange for a constant average flow of pressurized liquid from the pumps while providing equipment in which pressurized liquid can be stored in times of low demand and from which liquid can be drawn during demand peaks. These storage facilities are called accumulators and are illustrated in Fig. 1. A modern accumulator is buffered by a large pressurized gas (mostly air) cushion.

Until about 1960, the direct-contact surface between gas and liquid virtually precluded the use of oil in accumulator installations for larger presses. Gas diffuses into oil under pressure. When oil is discharged from the press at the end of the pressing cycle, the gas begins to bubble out of the liquid; the liquid foams, and because of its large volume, this foam is difficult to handle. This condition exists whether the pneumatic cushion is air or nitrogen. In addition, if air is used as the

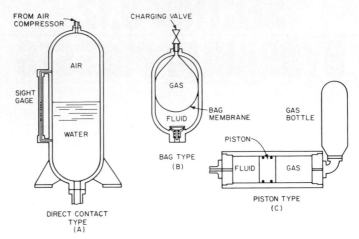

FIG. 1   Accumulators.

pneumatic cushion, the oxygen diffused into the oil oxidizes it, producing sludge and gum. (Spring- and weight-loaded accumulator installations have other serious drawbacks and have almost completely disappeared.)

A modification of the accumulator station uses a floating piston to separate oil from the pneumatic cushion (Fig. 1C). This design eliminates foaming and oxidizing, thus allowing the inclusion of accumulators into oil systems.

Power plants operated with water are used exclusively in special cases, for instance, where extreme precaution must be taken against fires caused by leaking oil. The increasing viscosity of oil precludes the use of oil at very high pressures.

CENTRIFUGAL VERSUS RECIPROCATING PUMPS

The competition between centrifugal and reciprocating pumps for hydraulic presses has been decisively won by reciprocating pumps. Under full load, centrifugal pumps have higher efficiency than reciprocating pumps (Fig. 2). However, in press installations, the power plants are idle a significant portion of the time. The losses during idling are only about 10% of full load for reciprocating pumps and well over 60 to 70% for centrifugal units. The total combined losses in installations with reciprocating pumps are less than half the losses in centrifugal pump plants, as shown in Figs. 3 and 4.

Reciprocating pumps for hydraulic presses which use water as the hydraulic medium are generally of the single-acting multiplunger type. Although there are large installations in operation using double-acting pumps, the faster vertical single-acting pump, which requires less floor space,

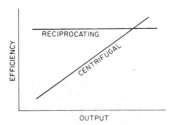

FIG. 2   Comparison of pump efficiency.

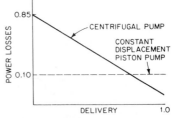

FIG. 3   Comparison of pump power losses.

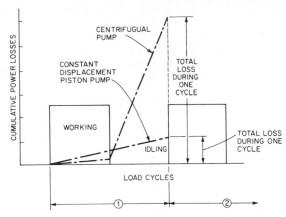

FIG. 4 Comparison of load cycling losses.

offers a considerable saving in capital expenditure and has proved itself dependable in service. In most cases this type of pump is used in conjunction with an accumulator, which allows the averaging of demand and thus a reduction in required pump capacity. There are, however, installations in which a vertical single-acting pump drives hydraulic cylinders directly.

Idling is controlled by remotely operated bypass valves or pump suction valve lifters which take command from the operator, accumulator level control, or some other sensor-monitoring press action.

Oil pumps most commonly used for the power strokes of presses are:

1. Vane rotary constant-delivery pumps, generally for pressures not in excess of 2500 lb/in^2 (17,200 kPa)

2. Piston rotary constant-delivery pumps

3. Piston rotary variable-delivery pumps

The constant-delivery pumps require a bypass system for idling. This is generally accomplished by an arrangement of directional control valves or by relief bypass valves (see above).

The variable-delivery pump has the advantage of providing efficient press speed control and idling by an adjustment of piston stroke. In cases of multiple pump application, constant- and variable-delivery pumps are often used jointly, their selection depending on the speed ranges required. It is possible to obtain these pumps designed for use with nonflammable liquid as a hydraulic medium.

Generally, both water and oil pumps in press applications are driven directly by electric motors, the motor speed matching the pump speed. Sometimes, on large water pumps, a geared speed reducer is installed between motor and pump.

OPERATING PRESSURES

Over the last 60 years, operating pressures in hydraulic presses have been slowly moving upward. The most popular level for water hydraulic installations is a rated pressure of 5000 lb/in^2 (34,500 kPa) and an actual operating pressure of 4500 lb/in^2 (31,000 kPa). Intensifiers are used to raise this pressure to between 6750 and 7500 lb/in^2 (46,500 and 51,700 kPa) in large presses, which corresponds to a rated power of from 7000 tons (62,300 N) upward. When used, the intensifiers are installed between the accumulators (and pump) on the low side and the press on the high side. A bypass is always provided. For very large presses, pressures of up to 15,000 lb/in^2 (10,300 kPa) have been suggested. In the Soviet Union, designers are using 15,000 lb/in^2 (10,300 kPa) regularly.

TABLE 1 Example of Power Plant Design for a Hydraulic Press

Standard procedure	Example
1. Water or oil can be used as the liquid in a hydraulic system of a vertical or horizontal press. The decision can be made in accordance with the general discussion in this section. The generation of full pressure during planishing is not required; because of rapid stroking, there would not be enough time to pressurize the liquid fully and expand the cylinder fully. The pull-back cylinders are kept continuously under pressure during planishing, thus reducing the time required for valving. The generation of about 10% of nominally rated pressure is considered sufficient for planishing.	Requirements: Press rating T = 6000 tons (53 MN) Pull-back rating T_R = 600 tons (5.3 MN) Total stroke S_t = 48 in (122 cm) Maximum pressing stroke S_p = 12 in (30 cm) Fast advance speed v_a = 360 in/min = 6 in/s (15 cm/s) minimum Fast return speed v_r = 360 in/min = 6 in/s (15 cm/s) minimum Pressing speed V_p required: At full 6000-ton (53-MN) rating: 120 in/min = 2 in/s (5 cm/s) maximum At $\frac{2}{3}$ = 4000-ton (36-MN) rating: 180 in/min = 3 in/s (8 cm/s) maximum At $\frac{1}{3}$ = 2000-ton (18-MN) rating: 360 in/min = 6 in/s (15 cm/s) maximum Rated operating cycles: Cogging: 15 c/min, each cycle consisting of Fast advance—4 in (10 cm) Pressing—4 in (10 cm) Return—8 in (20 cm) Planishing: 80 c/min, each cycle consisting of Advance—¼ in (0.64 cm) Pressing—¼ in (0.64 cm) Return—½ in (1.3 cm)
2. Select operating pressure p in pounds per square inch (megapascals) of the hydraulic system.	1. Selected: oil hydraulic system 2. Selected: p = 5000 lb/in² (34 MPa)
3. Determine the required effective pressing area A in square inches (square centimeters) of the main plunger of the press: $$A = \frac{T \times 2000}{p}$$	3. in USCS units $$A = \frac{6000 \times 2000}{5000} = 2400 \text{ in}^2$$

where T is the rated power in tons (newtons).

in SI units
$$\frac{53}{34} = 1.55 \text{ m}^2 = 15,500 \text{ cm}^2$$

4. Select the number of cylinders N_c comprising the main system of the press and their individual ratings $T_1, T_2, T_3, \ldots$.

4. Select 3 cylinders.
   Center ctyinder T_2 = 4000 tons (36 MN)
   Two side cylinders $T_1 = T_3$ = 1000 tons (8.9 MN)
   Operating all three cylinders: T = 6000 tons (53 MN)
   Operating center cylinder alone: 4000 tons (36 MN)
   Operating side cylinders alone: 2000 tons (18 MN)

5. Subdivide A in accordance with selected ratings and determine the individual diameters (see footnote):

$$D = \sqrt{\frac{A}{0.785}}$$

5. $A_1 = A_3 = \frac{1}{6} \times 2400 = 400 \text{ in}^2$ ($\frac{1}{6} \times 15,500 = 2580 \text{ cm}^2$)
   $A_2 = \frac{2}{3} \times 2400 = 1600 \text{ in}^2$ ($\frac{2}{3} \times 15,500 = 10,300 \text{ cm}^2$)

$$D_1 = D_3 = \sqrt{\frac{400}{0.785}} = 23 \text{ in}^a \text{ (58.4 cm)}$$

Adjusted area $A_1' = A_3'$ = 415 in^2 (2680 cm^2)

$$D_2 = \sqrt{\frac{1600}{0.785}} = 45 \text{ in}^a \text{ (114 cm)}$$

If any piston rods detract from the effective plunger area, their area should be added to A before determining D.

Adjusted area A_2' = 1590 in^2 (10,200 cm^2)
Total adjusted area A' = (2 × 415) + 1590 = 2420 in^2 [(2 × 2680) + 10,200 = 15,600 cm^2]

6. Using the total stroke S_t and the rated pressing stroke S_p, determine the geometric volumes V_t and V_p corresponding to these strokes.

6. $V_t = S_t \times A' = 48 \times 2420 = 116,160 \text{ in}^3$ (122 × 15,600 = 1,903,000 cm^3)
   For cogging $V_{pc} = 4 \times 2420 = 9680 \text{ in}^3$ per cogging stroke (10 × 15,600 = 156,000 cm^3)
   For planishing $V_{pp} = \frac{1}{4} \times 2420 = 605 \text{ in}^3$ per planishing stroke (0.635 × 15,600 = 9900 cm^3)

7. Using a general compressibility curve, determine the compressibility c of the liquid between atmospheric pressure and rated operating pressure.

7. For 5000 lb/in^2 (34 MPa), the oil compressibility may be assumed at c_0 = 1.5%.

---

^{a}It is customary to select diameters of large plungers in full-inch or ½-inch sizes; this requires adjustment of the area.

TABLE 1 Example of Power Plant Design for a Hydraulic Press (*cont.*)

Standard procedure	Example
8. Assume a reasonable approximate level of hoop stresses s_h along the inner surface of the cylinder barrel. The cylinder diameter expansion will be $$d_e = \frac{s_h}{E} \times D$$ where E = modulus of elasticity, assumed to be 30×10^6 lb/in^2 (207 GPa)	8. Assume $s_h = 30{,}000$ b/in^2 (207 MPa): Main cylinder $dem = \dfrac{30{,}000}{30 \times 10^6} \times 45 = 0.045$ in $\left(\dfrac{207}{207 \times 1000} \times 114 = 0.114 \text{ cm} \right)$ Side cylinders: $des = \dfrac{30{,}000}{30 \times 10^6} \times 23 = 0.023$ in $\left(\dfrac{207}{207 \times 1000} \times 58.4 = 0.058 \text{ cm} \right)$ where dem is the main cylinder diameter expansion and des is the side cylinder expansion.
9. On the basis of c_0 and s_h, determine the volume of liquid required to compensate for compression of liquid V_{cl} as well as for expansion of cylinder V_{ec}. During planishing, the pressures rise to only about 10% of rated; therefore the compressibility of the liquid and expansion of the cylinder may be neglected.	9. Compression of the liquid takes place along the entire length of the cylinder barrel. Assuming the length of the barrel to be approximately $S + \frac{1}{2}S$, or 48 $+ 25 = 72$ in $(122 + 61 = 183$ cm): $V_{cl} = 72 \times 2420 \times 0.015 = 2614$ in^3 $(183 \times 15{,}600 \times 0.015 = 42{,}800$ cm$^3)$ Omitting quantities small in the second order, $$V_{ec} = 72 \frac{\pi}{4} (2D \times \frac{30{,}000}{30 \times 10^6} \times D)$$ $= 72 \times 1.57 \times 0.001 \times D^2 = 0.113 \times D^2$ $(183 \times 1.57 \times 0.001 \times D^2$ $= 0.287 \times D^2)$ For the center cylinder: $V_{ec2} = 0.113 \times 45^2 = 229$ in^3 $\quad(0.287 \times 114^2 = 3730$ cm$^3)$

For the side cylinders:

$$V_{ec1} = V_{ec3} = 0.113 \times 23^2 = 60 \text{ in}^3 \qquad (0.287 \times 58.4^2 = 980 \text{ cm}^3)$$
$$V_{ec} = 229 + (2 \times 60) = 349 \text{ in}^3 \qquad [3730 + (2 \times 980) = 5700 \text{ cm}^3]$$

Total additional volume of liquid required.

$$2614 + 349 = 2963 \text{ in}^3 \text{ per stroke} \qquad (42,800 + 5700 = 48,500 \text{ cm}^3)$$

10. Determine the total amount of liquid V required per cogging stroke.

$$V = 9680 + 2963 = 12,643 \text{ in}^3 \qquad (156,000 + 48,500 = 204,500 \text{ cm}^3)$$

This value shows what additional burden can be imposed on the power plant by unnecessarily generous dimensioning of the total stroke.

11. Check whether specified speeds are compatible with the required number of cogging strokes. If not, increase the specified speeds (or reduce the number of strokes).

Fast advance time: % = 0.67 s
Pressing time: % = 2 s
Fast return time: % = 1.33 s
Valving time (3 switches) = 0.45 s
Total cycle time = 4.45 s

This time is too long to allow 15 cogging strokes per minute. Increase fast advance and fast return to 480 in/min = 8 in/s (20 cm/s) and pressing to 180 in/min = 3 in/s (8 cm/s). Then

Fast advance time: % = 0.5 s
Pressing time: % = 1.33 s
Fast return time: % = 1 s
Valving time = 0.45 s

12. $12,643 \times 0.2 = 2528 \text{ in}^3$ $\qquad (204,500 \times 0.2 = 40,900 \text{ cm}^3)$

12. Determine the liquid requirements for the pull-back stroke (Figs. 6 and 7). In general, the pull-back cylinders have an area equal to 10% of the area of the main cylinders and twice as long a stroke as the pressing stroke; in general, the required volume is therefore, with sufficient accuracy:

$$V \times 0.1 \times 2 = 0.2 \times V$$

TABLE 1 Example of Power Plant Design for a Hydraulic Press (*cont.*)

Standard procedure	Example

13. Determine the pumping requirements for the pump station with or without an accumulator. Without an accumulator:

$$\text{gpm} = \frac{V \times 60}{231 \times t_p} \qquad \left(m^3/h = \frac{V \times 60}{1000 \times t_p} \right)$$

where V is the volume in gallons (cubic meters) required for one pressing stroke and t_p is the pressing time in seconds.

With an accumulator:

$$\text{gpm (acc)} = \frac{V_{tc} \times 60}{231 \times t_c} \qquad \left(m^3/h = \frac{V_{tc} \times 3600}{10^6 \times t_c} \right)$$

where V_{tc} is the total volume in gallons (cubic meters) of pressurized liquid required during one cycle and t_c is the cycle time in seconds.

14. Determine the size of the accumulator required to store the accumulated volume V_s of liquid:

$$V_s = \text{gpm (acc)} \times \frac{t_c - t_p}{60} - V_R \qquad \left(m^3/h \times \frac{t_c - t_p}{3600} - V_R \right)$$

where V_R is the volume required for the return stroke in gallons (cubic meters).

Air volume required, based on 10% pressure fluctuation for isothermic condition, is $10V_s$. For adiabatic or, more often, polytropic conditions and 10% pressure fluctuation, the air volume is

$$V_{air} = \frac{V_s}{1.11^{1/n} - 1}$$

Although $n = 1.4$ is the exponent for adiabatic compression, $n = 1.3$ is considered satisfactory, even for severe and demanding conditions, which require full utilization of press stroke and a fast cycle. Based on $n = 1.3$,

$$V_{air} = 12V_s$$

---

13. Without an accumulator:

$$\frac{12,643 \times 60}{231 \times 1.33} = 2469 \text{ gpm} \qquad \left(\frac{204,500 \times 3600}{10^6 \times 1.33} = 553 \text{ m}^3/\text{h} \right)$$

With an accumulator:

$$\frac{(12.643 + 2,528) \times 60}{231 \times 4}$$

$$= 985 \text{ gpm (acc)} \qquad \left(\frac{(204,500 + 40,900) \times 3600}{10^6 \times 4} = 221 \text{ m}^3/\text{h (acc)} \right)$$

It is evident that the use of an accumulator will result in significant savings and in a substantial reduction of the peak load.

14. $V_s = 985 \times \dfrac{4 - 1.33}{60} - \dfrac{2528}{231} = 32.9 \text{ gal} = \dfrac{32.9}{7.48} = 4.4 \text{ ft}^3 \qquad \left(224 \times \dfrac{4 - 1.33}{3600} - \dfrac{40,900}{10^6} = 0.125 \text{ m}^3 \right)$

V_{air} (isothermic) $= 10 \times 32.9 = 329 \text{ gal} = 10 \times 4.4 = 44 \text{ ft}^3 \qquad (10 \times 0.125 = 1.25 \text{ m}^3)$

V_{air} (polytropic) $= 12 \times 32.9 = 395 \text{ gal} = \dfrac{395}{7.48} = 53 \text{ ft}^3 \qquad (12 \times 0.125 = 1.5 \text{ m}^3)$

Total accumulator volume $= 44 + 4.4 = 48.4 \text{ ft}^3 \qquad (1.25 + 0.125 = 1.38 \text{ m}^3)$

$= 53 + 4.4 = 57.4 \text{ ft}^3 \qquad (1.5 + 0.125 = 1.63 \text{ m}^3)$

15. Check planishing conditions.

15. Timing:

Fast advance: $\dfrac{\frac{1}{4}}{8} = 0.032$ s

Pressing: $\dfrac{\frac{1}{4}}{3} = 0.084$ s

Fast return: $\dfrac{\frac{1}{2}}{8} = 0.063$ s

Valving (2 switches) = 0.3 s

Total = 0.479 s

There will be more than enough time for 80 planishing strokes per minute. Only two valve switches are required for planishing because, during planishing, the pull-back cylinders are permanently connected to the pressure source.

During planishing, the pressure reaches only about 10% of rated; therefore compressibility and expansion require only 10% of the previously calculated amount. The pull-back system is constantly pressurized and does not require any liquid for compression of liquid and expansion of cylinder. The total requirements are thus

For the pressing stroke: $605 + 296 = 901$ in^3

$(9900 + 4850 = 14{,}750$ cm$^3)$

For the pull-back stroke: $605 \times 0.2 = 121$ in^3

$(9900 \times 0.2 = 1980$ cm$^3)$

Total $= 1022$ in^3 $(= 16{,}730$ cm$^3)$

gpm required $= \dfrac{1022}{231} \times 80 = 354$

m^3/h required $= (16{,}730 \times 10^{-6} \times 80$

$\times 60 = 80.3$

The selected accumulator and pump power plant will be more than sufficient for planishing. As a matter of fact, 100 or even 110 planishing strokes per minute will be feasible.

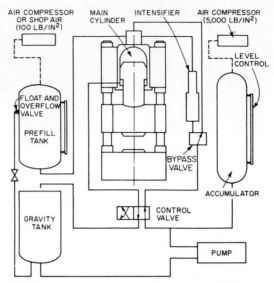

FIG. 5   Vertical forging press with double-acting cylinder.

In oil hydraulic installations, 3000 lb/in² (20,700 kPa) is the most popular pressure because of the availability of 3000-lb/in² (20,700-kPa) equipment from many reputable firms (pumps, valves, fittings, etc.) However, there are now on the market excellent units for pressures up to 6600 lb/in² (45,500 kPa), and the tendency in new installations is to provide 5000 lb/in² (34,500 kPa). Today, oil pumps for pressures as high as 9000 to 15,000 lb/in² (62,000 to 10,300 kPa) are being offered.

In the field of what is called ultrahigh-pressure equipment, pumps with direct action up to 225,000 lb/in² (1.55 MPa) are available today.

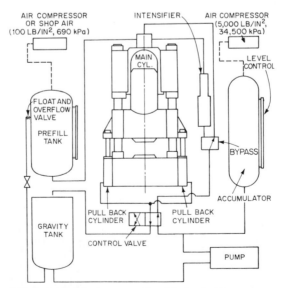

FIG. 6   Vertical forging press with pull-back cylinders.

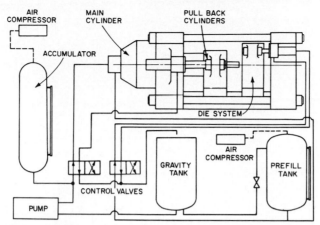

FIG. 7   Horizontal extrusion press.

DESIGN PROCEDURE

Considerations and calculations required to dimension the hydraulic power plant for a hydraulic press are given in Table 1. A standardized procedure for determining the basic design features of the power plant for a hydraulic press is offered in the left column. The right column gives an example of how the standard procedure should be used. The press selected for the example is an open die press for forging ingots into billets or bars through gradual reduction of cross section on narrow, flat dies which, at each stroke, penetrate several inches into the hot metal (cogging). The ingot is advanced and rotated, and, when it is close to the desired size, the surface is made smooth by the application of shallow-penetration—⅛- to ½-in (0.32- to 1.27-cm)—fast strokes (planishing) of the press (Fig. 5).

SECTION 9.13
REFRIGERATION, HEATING, AND AIR CONDITIONING

MELVIN A. RAMSEY

HEATING

Heat is usually generated at a central point and transferred to one or more points of use. The transfer may be by means of a liquid (usually water), which has its temperature increased at the source and gives up its heat at the point of use by reduction of its temperature. It may also be transferred by means of a vapor (usually steam), which changes from a liquid to a vapor at the source and gives up its heat at the point of use by condensation. Pumps may be required in both of these methods.

Hot Water Circulating A centrifugal pump best meets the requirements of this service. Water is usually used in a closed circuit so that there is no static head. The only resistance to flow is that from friction in the piping and fittings, the heater, the heating coils or radiators, and the control valves. In selecting the pump, the total flow resistance at the required flow rate should be calculated as accurately as possible, with some thought as to how much variation there might be as a result of inaccuracy of calculations or changes in the circuit because of installation conditions. It is not good practice to select a pump for a head or capacity considerably higher than that required, as this is likely to result in a higher noise level as well as increased power.

When hot water is used for radiation in a single circuit, through several radiators, the water temperature variation is usually only about 20 F° (11 C°) at the time of maximum requirements, so that there is not too great a difference in heat output between the first and last radiator in the circuit. With the flow rate based on water at 180° to 200°F (82 to 93°C) to heat air to about 75°F (24°C), a 10% reduction in the flow would have little effect, as the actual difference would increase to only 22 F° (12°C), and the reduction in the heat output of the radiator with 178°F (81°C) water would be only about 2%. Reference to Sec. 8.1, on the selection of pumps and the prediction of performance from the head-capacity pump curves and system head-flow rate curves, will show that a rather large undercalculation of circuit head loss would be necessary to produce a flow rate 10% less than desired.

Greater temperature differences are frequently used for other radiation circuits, and a reduced flow rate may have a greater temperature differential than in the single circuit. Whatever the condition, the pump should be selected only after full consideration of all the factors, and not by use of so-called safety factors, which are likely instead to be "trouble factors."

Air in the Circuit   Initially, the entire circuit will be full of air which must be displaced by the water. Arrangements should be provided to vent most of the air before the pump is operated. Even if all the air is eliminated at the start, more will be separated from the water when it is heated. Any water added later to replace that lost to evaporation will result in additional trapped air when the water is heated. Means must be provided for continuous air separation, but this cannot be accomplished by vents at high points in the piping because the flow is usually turbulent and the air is not separated at the top of the pipe.

A separator installed before the pump intake will remove the air circulating in the system. In a heating system, an air separation device is often provided at the point where the water leaves the boiler or other heating source. If the pump intake is immediately after this point, this is the point of lowest pressure and highest temperature in the system, and therefore it is the point where separation of air from the water can be most effectively achieved.[1]

If there are places in the system where the flow is not turbulent, air may accumulate and remain at these points and interfere with heat transfer. Automatic air vents should seldom or never be used. If they are used, it is important that they be located only where the pressure of the water is always above that of the surrounding air, whether the pump is operating or idle. Otherwise the air vent becomes an air intake.

Several important factors influence the choice of a pump for a hot water system with a number of separate heating coils, each having a separate control. Many systems in the past used three-way valves to change the flow from the coil to the bypass. When two-way valves are used, low-flow operation may occur for a large portion of the operating time. For this type of operation, there-fore, the pump selected should have a flat performance curve so that the head rise is limited at reduced flows. A very high head rise can cause problems when many of the valves are closed. Excessive flow rates through the coils and greater pressure differences across the control valves are some of the problems that can be avoided with a flat pump curve. A centrifugal pump should not operate very long with zero flow, for it would overheat. This condition is controlled by using one or two 3-way valves, a relief bypass, or a continuous small bleed between the supply and the return line. Whichever means is used to control minimum flow, the circuit must be able to dispose of the heat corresponding to the pump power at that operating condition, without reaching a temperature detrimental to the pump.

Types of Pumps   Many pumps for hot water circulation are for flow rates and heads in the range of in-line centrifugal pumps that are supported by the pipeline in which they are installed. Such pumps are available up to at least 5 hp (3.73 kW) and operate with good efficiency. More important than the type of pump are the performance and efficiency.

For greater flow rates and heads (and even for the smaller ones), the standard end-suction pump can be used. In the intermediate range, the use of an in-line or end-suction pump is a question not of one being better than the other but whether one or the other is better suited to the overall design and arrangement. Practically all the in-line or end-suction pumps for this ser-vice use seals instead of packing.

If the hot water system is of the medium- or high-temperature type, above 250°F (121°C), the pump must be carefully selected for the pressure and temperature at which it will operate.

Types of Water Circuit   There are several types of water circuits. Those shown in Figs. 1 and 2 are suitable for smaller systems and can be used for larger systems by having several of these circuits in parallel. The one shown in Fig. 3 is suitable for small or very large circuits, but the reverse return would add considerably to the cost if the circuit extended in one direction instead of in a practically closed loop as shown. For the extended circuit, a simple two-pipe circuit, with proper design for balancing, would be used.

There are a number of reasons for using other circuits, particularly primary-secondary pump-ing where the system is more extended or complicated, such as continuous circulation branches with controlled temperature. When a coil heats air, part or all of which may be below freezing, the velocity of the water in the tubes and its temperature at any point in the coil must be such that the temperature of the inside surface of the tube is not below freezing. The circuit shown in Fig. 4 makes this possible.

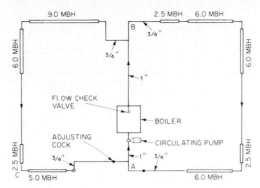

FIG. 1   A series loop system.

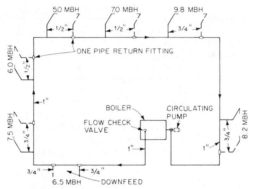

FIG. 2   A one-pipe system.

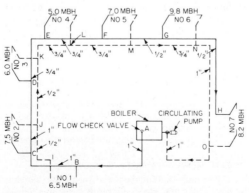

FIG. 3   A two-pipe reverse-return system.

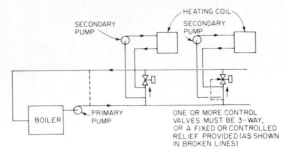

FIG. 4 A circuit with primary-secondary pumping to provide variable temperature at constant flow rate for two or more coils.

Primary-secondary pumping permits flow rates and temperatures in branch circuits to be different from those in the main circuit without the flow and pressure differences in the mains or branches having a significant effect on each other. There are many possible primary-secondary circuits to meet different requirements.

Steam Heating Systems No pumping is required with the smallest and simplest steam systems if there is sufficient level difference between the boiler and condensers (radiators, heating coils, etc.) to provide the required flow. When insufficient head exists between the level of the condensate in the condenser and the boiler to produce the required flow to the boiler, then a pump must be introduced to provide the required head. Since the condensate in the hot well will be at or near its saturation temperature and pressure, the only NPSH available to the pump will be the submergence less the losses in the piping between the hot well and the pump. A pump must be selected that will operate on these low values of NPSH without destructive cavitation.

In many cases, particularly for very large systems, vacuum pumps are used to remove both the condensate and air from the condensers. This permits smaller piping for the return of condensate and air, more positive removal of condensate from condensers, and, when high vacuums—above 20 in (0.5 m)—are possible, some control of the temperature at which the steam condenses. The use of vacuum return, particularly with higher vacuums, helps reduce the possibility of frozen heating coils exposed to outside air or to stratified outside and recirculated air. Vacuum return pumps are available for handling air and water. Vacuum, condensate, and boiler-feed pumps with condensate tanks are all available in package form.

Most condensate pumps are centrifugal. Vacuum pumps may be rotary, including a rotary type with a water seal and displacement arrangement.

Fuel Oil When oil burners are fairly far from the oil storage tank or when there are a number of burners at different locations in a building, a fuel oil circulating system is required. The flow rate is relatively low—1 gpm (3.8 liters/min) would provide more than 8,000,000 Btu/h (560,000 Gcal/s or 2,343,000 W)—and a small gear pump is usually used.

AIR CONDITIONING

Many air conditioning systems produce chilled water at a central location and distribute it to air cooling coils in various locations throughout the building or group of buildings. Centrifugal pumps are particularly well suited for this service.

The type of circuit and the number of pumps used require an evaluation of several factors:

1. The cooling requirements usually vary over a wide range.
2. Flow rate through a chiller must be kept above the low point where freezing would be possible and below the point where tube damage would result. Some methods of chiller capacity control require a constant flow rate through the chiller.

3. The temperature of the surface cooling the air must be low enough to control the relative humidity.[3] This limits the use of parallel circuits through chillers when one circuit may not be in operation and permits unchilled water to mix with water in the operating chiller. Under these conditions, it is difficult or impossible to attain a sufficiently low mixture temperature and the control of flow rate and water temperature in the air cooling coils is limited.[4]

4. Below-freezing air may sometimes pass over all or part of a coil. This condition would require a flow rate and water temperature adequate to keep the temperature of all water-side surfaces of tubes above freezing. Many water circuits are available to achieve the desired results. For control of the relative humidity, the air flow circuit must also be considered. Figure 5 shows a circuit for cooling coils with a variable air flow rate at constant air-leaving temperature and with two chillers in series. In addition, the two chillers are shown in parallel with a third chiller. This arrangement permits continuous flow through the coils to reduce the possibility of freezing when the average temperature of the air entering the coil is above freezing but the usual stratification results in a below-freezing temperature for some of the air entering the coil. The word *reduce* is used because full prevention requires appropriate air flow patterns, water velocities, and temperatures to assure that the water side of the surface will not be below freezing at any point in the coil. One of the coils is also arranged to add heat whenever the temperature of the air leaving the coil must be above that of the average air-entering temperature. Some circuits attempt to obtain the desired resutls from the circuit in Fig. 5 with fewer pumps. However, the use of fewer pumps, although it would reduce the cost slightly, would also require three-way instead of two-way valves, would make control somewhat more complicated, and would almost certainly result in greater power consumption. The circuit shown permits pump heads to match the requirements exactly. It also permits stopping an individual pump when flow is not required in one of the circuits; two-way valves 1, 2, and 3 will reduce pump circulation and the power of pump 3 at partial cooling load.

Air Separation and Removal   The methods for handling air with chilled water are about the same as those for hot water except that there is not usually a rise in temperature above that

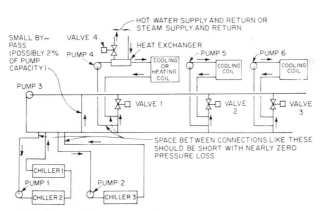

FIG. 5   Pump 3 does not operate unless pump 1 or pump 2 operates. Pump 1 operates only if chiller 1 or chiller 2 is required and operating. Pump 2 operates only if chiller 3 is required and operating. Pump 4 operates when air circulates over the coil to which the pump is connected, if cooling or heating is required, or if any air enters this coil below about 35°F (12°C). Pumps 5 and 6 operate in the same manner for the coils to which they are connected. Operation of pumps 4, 5, and 6 helps to equalize the temperature of air streams which enter the coils at different temperatures, and thus it may be desirable to operate these pumps continuously when air circulates over the coils. Valves 1 and 4 are interlocked, and so one must be closed before the other can open. Also, valve 1 should be prevented from opening if the temperature of the water in the pump 4 circuit is above about 90°F (32°C).

of the make-up water to produce additional separation of air. An expansion tank is required, but the reduced temperature difference requires a much smaller tank than with hot water.

Condenser Water Circulation Condenser water may be recirculated and cooled by passing through a cooling tower, or it may be pumped from a source such as a lake, ocean, or well.

Cooling Tower Water Centrifugal pumps are used for circulating cooling tower water. The circuit, which is open at the tower where the water falls or is sprayed through the air, transfers heat to the air before the water falls to the pan at the base of the tower. A pump then circulates the water through the condenser, as shown in Fig. 6. In this case the pump must operate against a head equal to the resistance of the condenser and piping plus the static head required to the tower from the water level in the pan.

Figure 7 shows a somewhat similar circuit except that here the condenser level is above the pan water level. The size of the pan of a standard cooling tower is sufficient to hold the water in the tower distribution system so that the pan will not overflow and waste water each time the pump is shut down. This capacity also assures that the pan will have enough water to provide the amount required above the pan level immediately after start-up, without waiting for the make-up which would be needed if there was any overflow when the pump stopped. When the condenser or much of the piping is above the pan overflow, the amount draining when the pump is stopped will exceed the pan capacity unless means are provided to keep the condenser and lines from draining. In Fig. 7 it will be noted that the line from the condenser drops below the pan level before rising at the tower. This keeps the condenser from draining by making it impossible for air to enter the system. This is effective for levels of a few feet but useless if the level difference approaches the barometric value. Such large level differences should be avoided, if possible, since they require special arrangements and controls.

When a cooling tower is to be used at low outside temperatures, it is necessary to avoid the circulation of any water outside unless the water temperature is well above freezing. The arrangement shown in Fig. 8 provides this protection. The inside tank must now provide the volume previously supplied by the pan, in addition to the volume of the piping from the tower to the level of the inside tank. Condensers or piping above the new overflow level must be treated as already described and illustrated in Fig. 7, or additional tank volume must be provided.

The only portion of the inside tank that will be available for the water that drains down after the pump stops is that above the operating level. This level is fixed by the height of liquid required to avoid cavitation at the inlet to the pump. The suction piping to the pump must remove only water from the tank without air entrainment.[5] The size of this pipe at the tank outlet should be determined not by pressure loss but by the velocity that can be attained from the available head. Exact data on this are not available, but the required velocity at the *vena contracta* (about 0.6 of the pipe cross-sectional area) can be calculated from $V = (2gh)^{1/2}$ where h is the height of the operating level above the *vena contracta*. The outlet from the tank should be at least as large as that from the cooling tower.

Well, Lake, or Seawater Centrifugal pumps are used for all of these services. The level from which the water is pumped is a critical factor. The level of the water in a well will be considerably lower during pumping than when the pump does not operate. When pumping is from a lake or from the ocean, the drawdown is usually not significant. When pumping is from a pit where the water flows by gravity, there will be a drawdown that will depend on the rate of pumping. With a seawater supply, there will be tidal variations. A lake supply may have seasonal level differences.

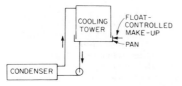

FIG. 6   Cooling tower with condenser below pan water level.

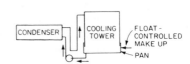

FIG. 7   Cooling tower with condenser above pan water level.

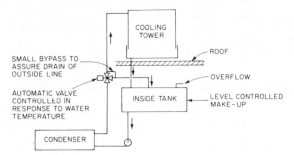

FIG. 8 Cooling tower with inside tank to permit operation when outside wet-bulb temperatures are below freezing.

All these factors must be taken into consideration in selecting the level for mounting the pump to assure that it will be filled with water during start-up. Check or foot valves may be used for this purpose. Also, the head of the water entering the pump at the time of highest flow rates must not be so low that the required NPSH is not available.

To assure proper pump operating conditions, the pump is frequently mounted below the lowest level expected during zero flow conditions, as well as below the lowest level expected at the greatest flow rate. These conditions may require a vertical turbine pump. The motor should be above the highest water level with a vertical shaft between the motor and the pump bowls, or the motor can be of the submerged type and connected directly to the pump bowls.

REFRIGERATION

For refrigeration systems with temperatures near or below freezing, pumps are often required for brine or refrigerant circulation. The transfer of lubrication oil also frequently requires pumps.

Brine Circulation   The word *brine*, as used in refrigeration, applies to any liquid which does not freeze at the temperatures at which it will be used and which transfers heat solely by a change in its temperature without a change in its physical state. As far as pumping is concerned, brine systems are very similar to systems for circulating chilled water or any liquid in a closed circuit. A centrifugal pump is the preferred choice for this service, but it must be constructed of materials suitable for the temperatures encountered. For some brines, the pump materials must be compatible with other metals in the system to avoid damage from galvanic corrosion.

Tightness is usually more important in a brine circulating system than in a chilled water system. This is true not only because of the higher cost of the brine but also because of problems caused by the entrance of minute amounts of moisture into the brine at very low temperatures.

Refrigerant Circulation   For a number of reasons—including pressure and level considerations as well as improvement of heat transfer— the refrigerant liquid is often circulated with a pump. The centrifugal pump is usually preferred for this purpose.

The liquid being pumped as a refrigerant may be the same one which is pumped as a brine. Whereas the material is all in liquid form throughout the brine circuit, some portion of it is in vapor form during its circulation as a refrigerant. In a refrigerant circulating system, most of the heat transfer is by evaporation, condensation, or both.

As there are changes from liquid to vapor, the liquid to be pumped must be saturated in some portion or portions of the circuit. Sufficient NPSH for the pump must be provided by the level of saturated liquid maintained in the tank where the liquid is collected. The level difference required for the NSPH must provide adequate margin to compensate for any temperature rise between the tank and the pump. This is an important consideration because the liquid temperature will usually be considerably lower than that of the air surrounding the pump intake pipe.[5]

When the pump is not operating, it may be warm and may contain much refrigerant in vapor form. It is usually necessary to provide a valved bypass from the pump discharge back to the tank so that gravity circulation can cool the pump and establish the prime.

Pumps for this service may require a double seal, with the space between the seals containing circulated refrigerant oil at an appropriate pressure. This will reduce the possibility of the loss of relatively expensive refrigerant and eliminate the entrance of any air or water vapor at pressures below atmospheric. A hermetic motor may also be used for this service and thus avoid the use of seals.

Lubricating Oil Transfer   Because the flow rates for lubricating oil transfer are rather low, the gear pump is usually preferred. The NPSH requirement is also critical here because, although the oil itself is well below the saturation temperature at the existing pressure, it contains liquid refrigerant in equilibrium with the refrigerant gas. Any temperature rise or pressure reduction will result in the separation of refrigerant vapor. It is important, therefore, to design the path for oil flow from the level in the tank where it is saturated with the same safeguards necessary for refrigerant pumping.

To reduce the oil pumping problem, the oil can be heated to a temperature above that of the ambient air and vented to a low pressure in the refrigerant circuit. This eliminates temperature rise in the pump as well as in the suction, with the corresponding reduction of available NPSH.

Usually, the oil flow is intermittent, and the best results are obtained by continuous pump operation discharging to a three-way solenoid valve. The discharge would be bypassed back to the tank whenever transfer from the tank is not required. This assures even temperature conditions and a pump free of vapor.

REFERENCES _____

1. *Tested Solutions to Design Problems in Air Conditioning and Refrigeration*, Industrial Press, New York, Sec. 3.

2. *Ibid.*, Sec. 10.

3. *Ibid.*, Sec. 1, pp. 19–38; Sec. 4, pp. 63–65.

4. *Ibid.*, Sec. 9, pp. 119–125.

5. *Ibid.*, Sec. 2, pp. 44–47.

SECTION 9.14
PUMPED STORAGE

GEORGE R. RICH

SIZE OF INSTALLATION

In the typical steam-based utility, it is the function of pumped storage to (1) furnish peaking capacity on the weekly load curve (Fig. 1) and (2) generate full pumped storage in an emergency for from 10 to 15 h, as required by the particular system. On the basis of comparative cost estimates, that size of installation is selected which is the most economical.

The principal features of a typical pumped storage project are shown schematically in Fig. 2. The overall efficiency is about 2:3, that is, 3 kW of pumping power will yield 2 kW of peak generation. The economy of the process stems from the fact that dump energy for pumping is worth about 3 mills/kW·h whereas peak energy is worth about 7 mills/kW·h. There is also an operating advantage. Because of the ease and rapidity with which it may be placed on line and becuase of its low maintenance charges, pumped storage is ideally suited to peaking operation. On the other hand, for maximum economy, modern high-pressure, high-temperature thermal plants should operate continuously near full load on the base portion of the load curve.

Table 1 shows a typical calculation to determine the reservoir capacity needed for sustaining the weekly load curve. The reservoir capacity to carry full load for a 10- to 15-h emergency is obtained simply by equating the electrical energy in kilowatt-hours in the load to the potential hydraulic energy stored in the upper reservoir.

SELECTION OF UNITS

At this stage of the basic engineering, it is necessary to make (in collaboration with the equipment manufacturers) at least a tentative selection of capacity, diameter, speed, and submergence for the turbomachine. This will be required for refinement of the calculations illustrated in Table 1 and also for the calculation of hydraulic transients which follows. In the head range most attractive for overall project economy, 500 to 1500 ft (152 to 457m), manufacturers are prepared to offer a single turbomachine that is capable of operating as a pump and whose direction of rotation is opposite that of a turbine. Electrical manufacturers offer a similar machine capable of operating as a synchronous motor for pumping and, in the opposite direction of rotation, as a generator. These machines are designated *pump turbines* and *generator motors*.

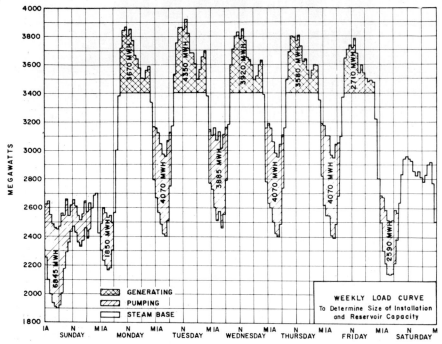

FIG. 1   Weekly load curve.

For best economy, the speed of the unit should be as high as is practicable without involving an objectionable degree of cavitation of the impeller under the assumed submergence below minimum tailwater. This speed is established (1) by model tests for cavitation at the hydraulic laboratories of the manufacturers and (2) by evaluating experience with similar prototype installations.

Figure 3 is an experience chart showing specific speed $N_s = \text{rpm}(\text{hp}^{1/2}/H^{5/4}$ versus head for the machine acting as a turbine, where H is total head, and Fig. 4 shows the specific speed $N_s = \text{rpm}Q^{1/2}/H^{3/4}$ for the machine acting as a pump. These charts presuppose moderate values of submergence, since unusually deep settings are uneconomical from the structural standpoint. Three curves are fitted to the installations shown, the equation of the curves being $N_s = K/H^{1/2}$. The depth of submergence may be verified by checking against the value of the cavitation constant $\sigma = (H_a - H_{vp} - H_s)/H$ given by the manufacturer's cavitation model test curves. Here H = total head, H_{vp} = vapor pressure, and H_a = atmospheric head. A typical curve of the family is shown in Fig. 5.

When the unit has been selected, manufacturers will furnish (in advance of bid invitations) prototype performance curves similar to Figs. 6, 7, 8a, and 8b. Figures 8a and 8b are designated four-quadrant synoptic charts and are required for the calculation of hydraulic transients. Figure 8b is for a 5.59-in (142-mm) gate opening. This is the largest gate opening at which the unit will be operating in the pumping cycle. Figure 8a, for an 8.94-in (227-mm) gate opening, is for operation on the turbine cycle only.

HYDRAULIC TRANSIENTS

In preparing purchase specifications for the generator motor, it is necessary to establish the maximum transient speed and the moment of inertia of the rotor, WR^2 (GD^2). Similarly, for the pump turbine and penstock, it is necessary to determine the maximum waterhammer. This primary

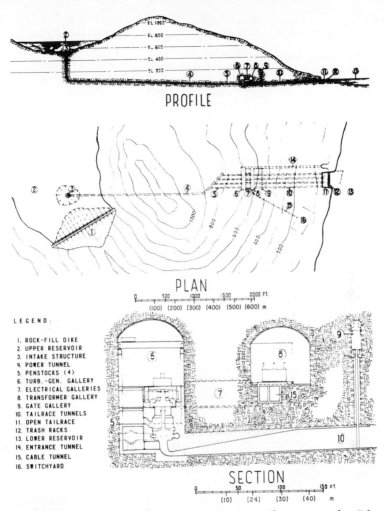

LEGEND:

1. ROCK-FILL DIKE
2. UPPER RESERVOIR
3. INTAKE STRUCTURE
4. POWER TUNNEL
5. PENSTOCKS (4)
6. TURB.-GEN. GALLERY
7. ELECTRICAL GALLERIES
8. TRANSFORMER GALLERY
9. GATE GALLERY
10. TAILRACE TUNNELS
11. OPEN TAILRACE
12. TRASH RACKS
13. LOWER RESERVOIR
14. ENTRANCE TUNNEL
15. CABLE TUNNEL
16. SWITCHYARD

FIG. 2 Schematic arrangement for a pumped storage project. Elevations are in feet. (1 ft = 0.3048 m)

calculation, as summarized in Fig. 9 and Table 2, is made by the trial-and-error method of arithmetic integration, using various trial values of WR^2 for the condition of full-load rejection on all units during the generating mode with the turbine gates assumed "stuck" in the full-gate position.

Upon instantaneous loss of load, the unit builds up overspeed. The increase in speed above normal causes a reduction in turbine discharge, which causes waterhammer, which in turn further increases the power delivered to the rotor. This pyramiding continues until the arrival of negative reflected water hammer from the upper reservoir. The head then decreases. The unit is then so much over speed that it begins to act as a brake, as shown by the four-quadrant synoptic chart (Fig. 8). As shown by Fig. 9 and Table 2, the process gradually damps down to the steady-state runaway speed and head. Many additional and more refined calculations are made later in the course of the design to establish the optimum governor time and rate of turbine gate closure, as given in detail in the works listed at the end of this section.

TABLE 1 Determination of Reservoir Capacity

Pertinent data
  Generating = rated net head = 900 ft (274 m)
    Generating capacity at rated head for 3 units = 525 MW
  Pumping = rated net head = 930 ft (283 m)
    Pumping power at rated head for 3 units = 555 MW
    Pumping-to-generating ratio = 3 to 2

Day		(1) Generating, MW	(2) Pumping, MW·h	(3) Equivalent generation (col. 2 × %), MW·h	(4) Net daily change in reservoir, MW·h	(5) Cumulative change in reservoir, MW·h	Remarks
Monday	PM	3670			−3670	−3670	
Tuesday	AM		4070	2710			
	PM	4350			−1640	−5310	
Wednesday	AM		3885	2585			
	PM	3920			−1335	−6645	
Thursday	AM		4070	2710			
	PM	3580			−870	−7515	Reservoir empty
Friday	AM		4070	2,710			
	PM	2710			0	−7ʌ	
Saturday			2590	1725	+1725	−5790	
Sunday			6845	4560	+4560	−1230	
Monday	AM		1850	1230	+1230	0	Reservoir full

Required reservoir capacity to sustain weekly load curve:

$$\frac{7515 \times 1000 \times 550 \times 3600}{62.4 \times 43{,}560 \times 0.746 \times 900 \times 0.85} = 9600 \text{ acre}\cdot\text{ft}$$

Hours of capacity at full load and 900 ft (274 m) head:

$$\frac{9600 \times 62.4 \times 43{,}560 \times 0.746 \times 900 \times 0.85}{525 \times 1000 \times 550 \times 3600} = \pm 14.3 \text{ h}$$

STARTING THE UNIT

The procedure for starting the unit is an essential feature of the design. There are three cases to be considered: (1) the pumping mode, (2) the conventional generating mode, and (3) rotating spinning reserve in the generating mode.

Pumping Mode If we attempted to start the unit as a pump, from rest and with the scroll case and draft tube filled, the power and inrush current would be excessive. Accordingly, the standard type of compressed air system is provided to depress the tailwater elevation to below the bottom of the impeller, with the wicket gates closed. The load to be overcome at starting then consists of the load of the rotating masses, which must be accelerated to synchronous speed, and the load due to windage. Owing to inevitable leakage past the wicket gates, this windage is substantially greater than that due to dry air. The main penstock valves must also be closed during starting, or else the leakage and "wet" windage would be still further increased at the much higher head.

For the larger units generally employed, a separate starting motor, of the induction type with wound rotor, is mounted directly above the main generator motor. It has not been found feasible, in the motor space available, to design an amortisseur winding capable of sustaining the heat from the inrush current resulting from across-the-line starting of the main generator motor, even at reduced voltage. For the smaller units, however, this may be accomplished. In rare instances,

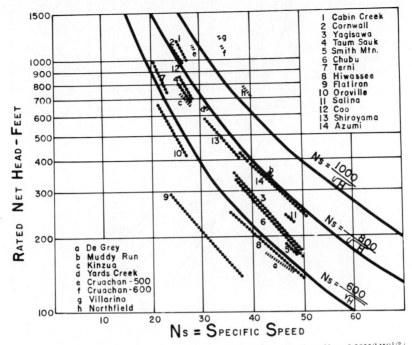

FIG. 3 Rated net head versus specific speed for a turbine. (1 ft = 0.3048 m; $N_s = 0.2622(\text{kW})^{1/2}/H^{5/4}$)

where a main unit is always available, this spare may be electrically coupled to the pumping unit and the two started from rest in back-to-back synchronism. The separate starting motor is usually sized to bring the main unit up to synchronous speed in about 10 min, as shown by Fig. 10.

When the unit has attained full speed, it is synchronized to the line, the compressed air is cut off, the tailwater rises to fill the draft tube, the wicket gates and main penstock valve are opened gradually to prevent shock, and pumping to the upper reservoir begins.

The maximum transient load on the generator motor thrust bearing occurs just as pumping begins. Prior to the advent of pumped storage, thrust bearings were designed to carry the weight of the rotating parts plus the hydraulic thrust at the steady-state condition. Now a greatly increased thrust of short duration must also be accommodated. Because of the short duration of this transient excess load, it may usually be carried safely by the bearing as designed for the steady-state requrement, depending on the detailed design of the bearing. It is now standard practice to provide a high-pressure oil pumping system to ensure that there will be a firm of oil between the bearing surfaces before the unit starts rotating.

Conventional Generation For generation in the conventional manner, the unit may be started from rest under its own power without assistance from the starting motor.

Rotating Spinning Reserve In considering the requirements for starting the unit for rotating spinning reserve, it will be assumed that the utility is a participant in a grid system of interconnection by means of extra-high-voltage (EHV) transmission lines of high capability. This means that, immediately upon loss of generation by the local utility, the EHV connection will carry the necessary load for the short time required (about 30 s) for the local pumped storage units to absorb full load. In readiness for just such an emergency, these local pump turbines will be motoring on the line in the generating direction, with the wicket gates and main penstock valves closed and with the tailwater depressed. The loads and procedure for bringing the units up

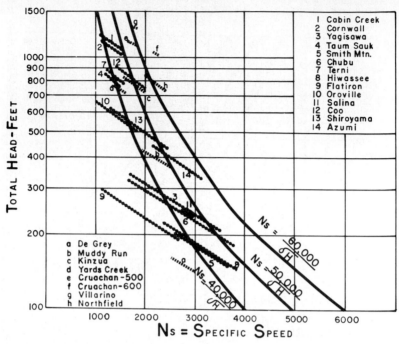

FIG. 4 Total head versus specific speed for a pump. (1 ft = 0.3048 m; $N_s = 51.65(\text{rpm})\sqrt{\text{m}^3/\text{s}}/H^{3/4}$)

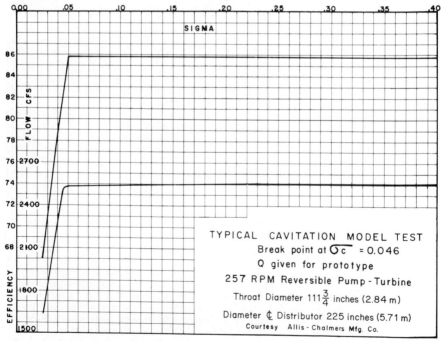

FIG. 5 Typical cavitation model test. (1 ft³/s = 0.0283 m³/s)

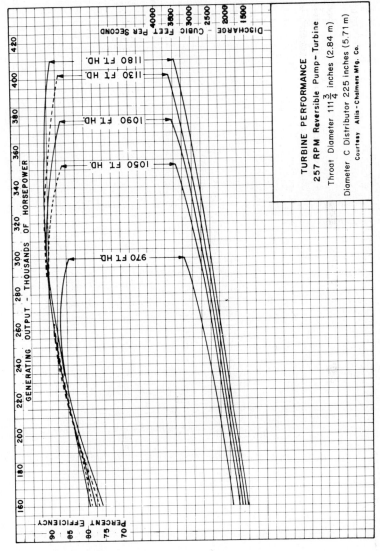

FIG. 6   Turbine performance. (1 ft³/s = 0.0283 m³/s; 1 ft = 0.3048 m; 1 hp = 0.746 kW)

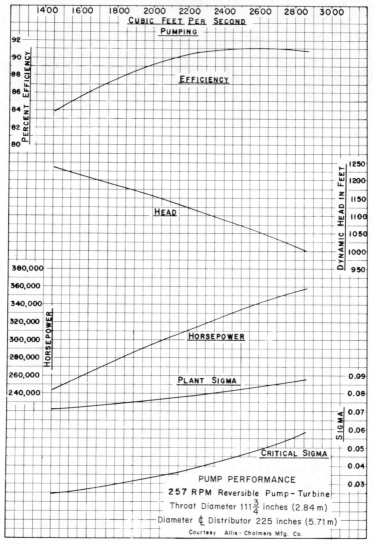

FIG. 7   Pump performance. (1 hp = 0.746 kW; 1 ft = 0.3048 m; 1 ft³/s = 0.0283 m³/s)

to speed for synchronizing to the line for rotating spinning reserve are identical with those given for operation in the pumping mode except that rotation is in the generating direction.

GOVERNOR TIME AND SURGE TANKS

In pumped storage plants, the time specified for the governor servomotors to move the wicket gates through a complete stroke is generally not less than 30 s. The reasons for this are (1) about 30 s is the minimum practicable time for penstock valve operation if the valve operating

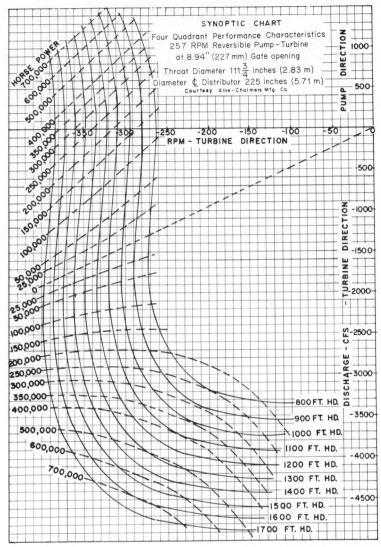

FIG. 8a Synoptic chart for an 8.94-in (227-mm) opening. (1 hp = 0.746 kW; 1 ft = 0.3048 m; 1 ft³/s = 0.0283 m³/s)

machinery is not to be made unduly complicated by the incorporation of dashpots and accessories and (2) one of the primary purposes of massive EHV connection is to carry loads from emergency outages for 30 to 60 s or more until the local pumped storage units take over. In the lengths of waterways that are permissible economically, penstock and tunnel velocities may usually be accelerated to full load in a much shorter time without excessive positive or negative waterhammer, so that surge tanks are not necessary. However, in the exceptional case of a long tailrace tunnel flowing as a closed conduit, under the relatively low head of tailwater, even a 30-s closure could be sufficient to produce negative waterhammer great enough to cause separation of the water column and damage to the unit. For such cases a surge tank[1] at the downstream face of the power

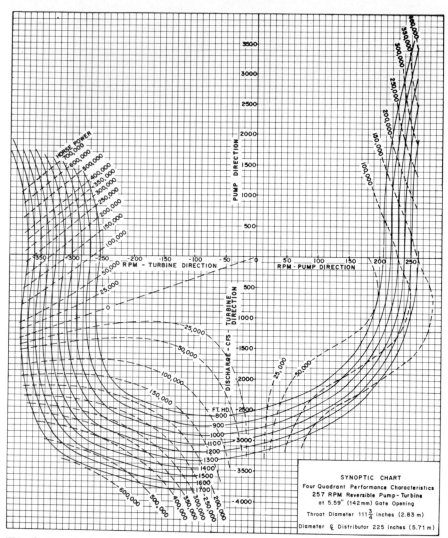

FIG. 8b Synoptic chart for a 5.59-in (142-mm) opening. (1 hp = 0.746 kW; 1 ft = 0.3048 m; 1 ft³/s = 0.0283 m³/s)

station is required. In a dual-purpose project for municipal water supply and by-product power, the length of the tunnel is dictated by water supply economics and may be as great as 40 to 50 miles (64 to 80.5 km). In such cases a surge tank on the upstream side of the power station may be indicated.

ECONOMIC DESIGN

A proposed pumped storage project for peaking service must be able to show a liberal margin of economic superiority over competing thermal types. This requires that the undertaking be

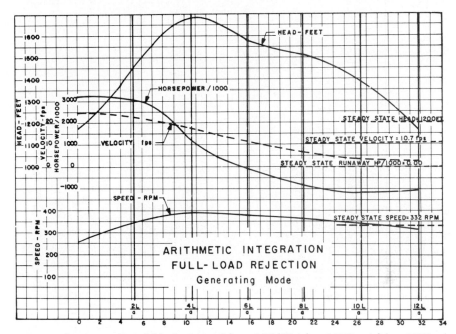

FIG. 9 Arithmetic integration, full load rejection. (1 hp = 0.746 kW; 1 ft = 0.3048 m; 1 ft/s = 0.3048 m/s; 1 ft³/s = 0.0283 m³/s)

designed in accordance with the so-called lean project concept, in which all elements are closely tailored to the specific purpose of pumped storage and many features considered standard in the conventional hydroelectric plant are found superfluous.

Selection of the proper site is of paramount importance. The rock should be strong and tight, so that no more than local grouting is needed to inhibit leakage and no drainage system for the concrete lining of the tunnels is required. The upper storage reservoir should preferably be located in a natural basin in elevated highlands so as to provide the requisite storage capacity by use of low rimdikes. It is a great advantage if the lower reservoir can be located on tidewater. The head, probably the most important natural feature, should be high, in the range from 500 to 150 ft. (152 to 457 m) High head means smaller water quantities and consequently smaller sizes for a given power output. High head also permits higher turbine speeds, lower torque, and a smaller generator. The intake need be only a bell mouth at the end of the supply tunnel; consequently, headgates, cranes, and accessories can be eliminated. The individual units should be of large capacity to ensure minimum equipment costs.

Table 3 shows a typical form for the economic comparison of pumped storage and competitive thermal peaking, and Table 4 is a typical form for the project cost estimate with its overheads.

Table 5 shows the range of machine requirements for a deep upper reservoir in which the head variation due to daily drawdown is relatively large. Note that the maximum electric motor load to develop the full hydraulic capability of the pump, occurring at the minimum head of 970 ft (296 m), is 297 MVA. However, this loading is of comparatively short duration and may be sustained at 80 C° temperature rise, which corresponds to 115% of normal rating. The normal rating at 60 C° temperature rise would then be 297/1.15, or 258 MVA, which is shown by the tabulation to be adequate to carry pumping and generating loadings of more protracted duration. This utilization of overload rating for Class B insulation affords substantial economies in electric machine cost.

TABLE 2 Arithmetic Integration at Full-Load Rejection (Generating Mode; Wicket Gates at 92% Opening)

Interval $\left(\dfrac{2L}{a}\right)$	Time, s	Trial V (ΔV), ft/s	ΔH, ft	ΣΔH, ft	Hf, ft	$H_o + H_f + \Sigma\Delta H$ = total H, ft	Trial speed, rpm	Q (8 units), ft³/s	Check V, ft/s	Hp, 8 units (000)	Average hp, 8 units (000)	N², rpm²	$N_2^2 - N_1^2$ rpm²	Check T, s
0	0	24.5 −1.9	—	—	−36.8	1173.2	257	28,100 (3,510)ᵃ	24.7	3200	—	66,000		
1	5.3	22.6 −5.3	274	274	−28.8	1455.2	342	25,800 (3220)	22.6	3040	3120	117,000	51,000	5.30
2	10.6	17.3 −6.0	766	492	−16.9	1685.1	390	19,700 (2460)	17.3	1180	2110	152,000	35,000	5.30
3	15.9	11.3 −4.7	867	375	−7.2	1577.8	378	12,900 (1610)	11.3	−104	538	142,900	9100	5.30
4	21.2	6.6 −3.25	680	305	−2.5	1512.5	368	7520 (940)	6.6	−784	−444	135,500	−7400	5.30
5	26.5	3.35 −0.85	470	165	−0.6	1374.4	345	3819 (477)	3.35	−1200	−992	119,000	−16,500	5.30
6	31.8	2.50	123	−42	−0.4	1167.0	316	2850 (256)	2.5	−1080	−1140	100,000	−19,000	5.30

$L = 12{,}322$ ft, $A = 1140$ ft², $a = 4650$ ft/s, $H_f = 0.0565V^2$, $WR^2 = 1023 \times 10^6$ (8 units), $2L/a = 5.3$s, $H_o = 1210$ ft, $\Delta H = a\,\Delta V/g = (4650 \times \Delta V)/32.2 = 144.5\,\Delta V$

$$\text{Check } \Delta T = \frac{4\pi WR^2(N_2^2 - N_1^2)}{2g \times \text{av. hp} \times 550 \times 3600} = \frac{4\pi^2 \times 1023 \times 10^6(N_2^2 - N_1^2)}{64.4 \text{ av. hp} \times 550 \times 3600} = \frac{320(N_2^2 - N_1^2)}{\text{av. hp}}$$

SI conversions: m/s = 0.3048 × ft/s; m = 0.3048 × ft; m³/s = 0.0283 × ft³/s kW = 0.746 × hp.

ᵃValues in parentheses are per unit-values.

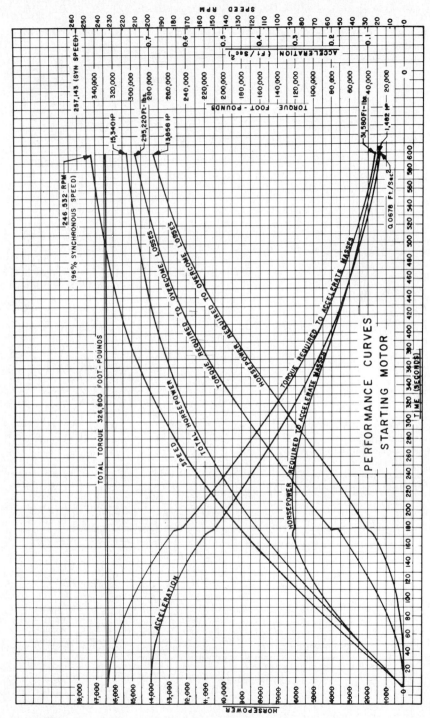

FIG. 10 Performance curves, starting motor. (1 hp = 0.746 kW; 1 ft·lb = 0.138 kg·m)

TABLE 3 Economic Evaluation of Hypothetical Pumped Storage Plant versus Conventional Steam Reheat Plant

	Reheat		Pumped storage	
Installed capacity, MW	3,600		3,600	
Number of units	4		16	
Plant investment (000 omitted)				
Generation	$450,000		$260,000	
Transmission	—		47,100	
Total	$450,000		$307,100	
Total per kilowatt	$ 125		$ 85	
Annual generation (10^6 kW·h)				
Peak load	1,300		1,300	
Base load	20,000			
Total	21,300		1,300	
Annual capacity factor, %	67.5		4.1	
Annual fixed charge rates, %				
Generating plant	13.35		10.90	
Transmission plant	—		11.35	
Annual costs	Total (000)	Per kW·yr	Total (000)	Per kW·yr
Fixed charges				
Generating plant	$ 60,076	$16.69	$28,340	$ 7.87
Transmission plant	—	—	5,346	1.49
Total	$ 60,076	$16.69	$33,686	$ 9.36
Fuel	59,044	16.40	5,018[a]	1.39
Operation and maintenance	9,900	2.75	3,492	.97
Total	$129,020	$35.84	$42,196	$11.72
Credit to reheat unit:				
Replacement for base				
Load generation	68,480	19.02		
Net cost of 3600-MW capacity and				
peak-load generation	$ 60,540	$16.82	$42,196	$11.72
Differential annual cost in favor of				
hypothetical plant			$ 9,172	$ 5.10

[a]Fuel cost for pumping energy.

TABLE 4 Hypothetical Pumped Storage Plant Cost Estimate—Summary

Account no.	Item	Cost
330	Land and land rights	$ 4,311,000
331	Structures and improvements	
−1	Powerhouse substructure	20,791,000
−2	Service bay substructure	2,023,000
−3	Powerhouse superstructure	553,200
−4	Cofferdam	600,000
332	Reservoirs, dams, and waterways	
−2	Reservoir clearing	123,000
−4	Dikes and embankments	25,162,000
−8	Intakes	1,203,000
−12	Power tunnels and waterways	76,526,500
	Tailrace	4,940,000

TABLE 4 Hypothetical Pumped Storage Plant Cost Estimate—Summary (*Cont.*)

Account no.	Item	Cost
333	Waterwheels, turbines, and generators	
−2	Fire protection system	86,000
−5	Motor generators	31,776,800
−8	Spherical valves	7,680,000
−9	Turbines and accessories	23,915,520
−11	Piezometer system	24,000
−13	Spiral case and draft tube unwatering systems	136,000
	Air depression system	208,000
	Draft tube racks, piers, and guides	2,071,600
334	Accessory electrical equipment	6,496,000
335	Miscellaneous power plant equipment	2,046,000
335	Roads, railroads, and bridges	2,393,600
352	Transmission plant—structures and improvements	2,226,500
353	Station equipment	12,119,400
		$222,412,120
	Contingencies, overhead, and engineering	32,587,880
	Total estimated cost of project	$260,000,000

TABLE 5 Operating Range of Hypothetical Project

	Reservoir elevation, ft			
	1000	1060	1120	1160
Tailwater elevation, ft	0	0	0	0
Gross head, ft	1000	1060	1120	1160
Generating cycle				
Frictional head loss, ft	30	30	30	30
Net head, ft	970	1030	1090	1130
Best gate turbine output, hp (000)	—	310		
Full-gate turbine output, hp (000)	310	—	380	400
Turbine output blocked at, hp (000)	—	—	350	350
Turbine efficiency, %	85	88.3	87.5	88.5
Turbine discharge, ft^3/s	3300	3000	3200	3100
Generator output at 98% efficiency, MW	227	227	256	256
Generator MVA (0.90 PF)	252	252	284	284
Pumping cycle				
Frictional head loss, ft	20	20	20	20
Net head, ft	1020	1080	1140	1180
Maximum possible discharge, ft^3/s	2700	2350	2000	1730
Pump power, hp (000)	350	324	298	277
Pump efficiency, %	89.5	89	87	84
Motor power, at 98% efficiency, MW	262	242	223	206
Motor MVA (0.90 PF)	297	275	252	234

SI conversions: m = 0.3048 × ft; kW = 0.746 × hp; m^3/s = 0.0283 × ft^3/s.

REFERENCE

1. Rich, G. R.: *Hydraulic Transients*, Dover, New York, 1963.

FURTHER READING

American Society of Mechanical Engineers: *Waterhammer in Pumped Storage Projects* ASME, New York 1965.

Chapin, W. S.: "The Niagara Power Project," *Civ. Eng.*, April 1961, p. 36.

Davis, C. V., and K. E. Sorensen: *Handbook of Applied Hydraulics*, McGraw-Hill, New York, 1969.

McCormack, W. J.: "Taum Sauk Pumped-Storage Project as a Peaking Plant," *Water Power*, June 1962.

Parmakian, J. *Waterhammer Analysis*, Dover, New York, 1963.

Rich, G. R., and W. B. Fisk: "The Lean Project Concept in the Economic Design of Pumped-Storage Hydroelectric Plants," paper presented at World Power Conference, 60, September 1964.

Rudulph, E. A.: "Taum Sauk Pumped Storage Power Project," *Civ. Eng.*, January 1963.

SECTION 9.15
NUCLEAR

9.15.1
NUCLEAR ELECTRIC GENERATION

W. M. WEPFER

W. M. WEPFER

APPLICATION

If there is a single item which distinguishes pumps for nuclear service from conventional pumps, that item is attention to safety. Cleaning up spills of radioactive fluid may be costly and require protection for personnel. Moreover, the functioning of some pumps is vital to certain emergency conditions. Most of the rules and regulatory standards for nuclear pumps are dedicated to the promotion of public safety. Pump builders must be fully knowledgeable of these procedures before they can offer their pumps in the nuclear market.

All nuclear power plants built for public utility service involve the generation of steam. It is usual for the plant to contain one major system, or loop, and a number of supporting systems. Since virtually every system requires some form of pumping, it is conventional to designate a pump by the name of the system with which it is identified. The principal methods of steam production are discussed below. For each, the main and auxiliary systems are listed and this is followed by a brief description of the pumping needs.

PWR Plants Pressurized water reactor (PWR) plants employ two separate main systems to generate steam. In the primary system, the water is circulated by *reactor coolant pumps* through the nuclear reactor and large steam generators. Overpressure is provided to prevent vapor formation. The secondary side of the steam generator provides nonradioactive steam to the turbogenerator. Typical primary water conditions are 2250 lb/in^2 gage (15.51 MPa) and 550°F (288°C). Table 1 lists the parameters of the more important pumps used in PWR plants.

Figure 1 is a simplified flow diagram of a PWR reactor coolant system. Only one primary loop is shown, but in practice plants use two, three, or four loops in parallel, each loop consisting of its own reactor coolant pump, steam generator, and optional stop valves. A single reactor and pressurizer supply all loops. The pressurizer provides the overpressure referred to above by maintaining a body of water at an elevated temperature such that its vapor pressure satisfies the primary loop pressure requirements.

In the PWR plant, high-pressure water circulates through the reactor core to remove the heat generated by the nuclear chain reaction. Figure 2 is an illustration of a typical PWR reactor vessel. The heated water exits from the reactor vessel and passes via the coolant loop piping to the primary side of the steam generators. Here it gives up its heat to the feedwater to produce steam for the turbogenerator. The primary-side cycle is completed when the reactor water is pumped back to the reactor by the reactor coolant pumps. The secondary-side cycle, isolated from the primary

TABLE 1 Typical Nuclear Pump Parameters in PWR Plants

Pump	Number per plant	Flow, gpm (m³/h)	Head, ft (m)	Design pressure, lb/in² (MPa)	Design temp., °F (°C)	Driver hp (kW)	Shaft	Length or height, including driver, in (mm)	Speed (nominal), rpm	Notes
Reactor coolant	4	100,000 (22,700)	290 (88)	2500 (17.2)	650 (343)	7000 (5220)	Vert.	305 (7750)	1200	
Component cooling water	3	4800 (1090)	250 (76)	200 (1.38)	200 (93)	450 (336)	Horiz.	110 (2790)	1800	
Residual heat removal	2	3800 (863)	350 (107)	600 (4.14)	400 (204)	500 (373)	Vert.	97 (2460)	1800	
Containment spray	2	2600 (590)	450 (137)	300 (2.07)	300 (148)	400 (298)	Horiz.	112 (2840)	1800	
Spent fuel pit cooling	3	4500 (1022)	150 (46)	150 (1.03)	200 (93)	250 (186)	Horiz.	87 (2210)	1800	
Charging (centrifugal)	2	120 (27)	5800 (1768)	2800 (19.3)	300 (148)	600 (448)	Horiz.	234 (5940)	4850	Gear drive
Charging (reciprocating)	1	98 (22)	5800 (1768)	2800 (19.3)	250 (121)	200 (149)	Horiz.	208 (5280)	205	Reciprocating
Safety injection	2	440 (100)	2680 (817)	1750 (12.07)	300 (148)	450 (336)	Horiz.	190 (4830)	3600	
Chilled water	2	400 (91)	150 (46)	150 (1.03)	200 (93)	40 (30)	Horiz.	52 (1320)	3600	

Spent resin sluicing	1	150 (34)	250 (76)	240 (1.66)	250 (121)	30 (22)	Horiz.	46 (1170)	3600
Reactor coolant drain tank	2	150 (34)	250 (76)	240 (1.66)	250 (121)	30 (22)	Horiz.	46 (1170)	3600
Boric acid transfer	2	100 (22.7)	200 (61)	150 (1.03)	200 (93)	15 (11)	Horiz.	45 (1140)	3600
Boron recycle evaporator feed	2	100 (22.7)	200 (61)	150 (1.03)	200 (93)	15 (11)	Horiz.	45 (1140)	3600
Boron injection recirculation	2	20 (4.5)	100 (30.5)	240 (1.66)	200 (93)	3 (2.2)	Horiz.	18 (460)	3600 Canned
Spent fuel pit skimmer	1	100 (22.7)	50 (15.2)	150 (1.03)	200 (93)	3 (2.2)	Horiz.	42 (1070)	1800
Refueling water purification	1	200 (45)	200 (61)	150 (1.03)	200 (93)	30 (22)	Horiz.	52 (1320)	1800
Waste processing system	5	100 (22.7)	200 (61)	150 (1.03)	200 (93)	15 (11)	Horiz.	45 (1140)	3600
Gas decay tank drain	1	10 (2.27)	90 (27)	150 (1.03)	180 (82)	3 (2.2)	Horiz.	15 (380)	3600
S. G. blowdown–spent resin sluice	1	110 (25)	165 (50)	150 (1.03)	100 (37)	15 (11)	Horiz.	39 (990)	3600

SOURCE: Westinghouse Electric.

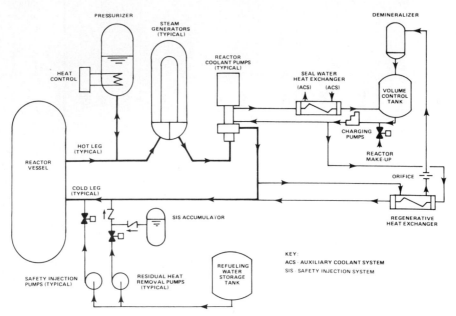

FIG. 1   Flow diagram for pressurized water reactor coolant system. (Westinghouse Electric)

loops, is completed when the *feedwater pumps* return water to the secondary side of the steam generator.

Since a reactor, once it has been critical, continues to generate heat even when shut down, *residual heat removal* (RHR) *pumps* are provided to circulate reactor water through coolers any time the reactor is inoperable and at low pressure, even during refueling. These pumps serve other functions also, as described below.

The chemical and volume control system (Fig. 3) performs a number of functions. Through *charging pumps*, which may be centrifugal or positive displacement or may include some of each type, the primary system can be filled and pressurized when cold. When the system is hot, the pumps are used to maintain the water level in the pressurizer and to replenish any fluid drawn from the primary loops by other systems. Additionally, the pumps supply clean water to the reactor coolant pump seals and are used to adjust the boric acid concentration in the reactor coolant water, which provides an auxiliary means of reactor power regulation. If positive displacement pumps are included in the chemical and volume control system, they are also used for hydrostatically testing the reactor primary coolant system. Where all the charging pumps are centrifugal, it is customary to provide a small positive displacement pump exclusively for this hydrostatic testing.

For cooling essential components and for supplying a variety of heat exchangers, a component cooling water system is provided. *Component cooling water pumps* circulate clean water at low system pressures for the purpose of cooling (1) primary water, which is continuously bled for purification, (2) main and auxiliary pump bearings and seals, (3) primary pump thermal barriers, (4) large motors, (5) the containment vessel, and (6) the spent fuel pit water.

When the primary pressure boundary is breached, elements of the emergency core cooling system (ECCS) are immediately activated. The primary function of the ECCS following a loss-of-coolant accident is to remove the stored and fission product decay heat from the reactor core. The safety injection system (Fig. 4) does most of this. Upon actuation of the safety injection signal, the charging pumps inject boric acid solution, which is stored in special tanks and continuously circulated by the *boron injection recirculation pumps*, into the reactor coolant system. At the same time, the *residual heat removal pumps* are started. These pumps take suction from a large refueling water storage tank and inject cold water into the reactor coolant circuit. To provide additional capacity, the *safety injection pumps* are started, taking suction from the cold water

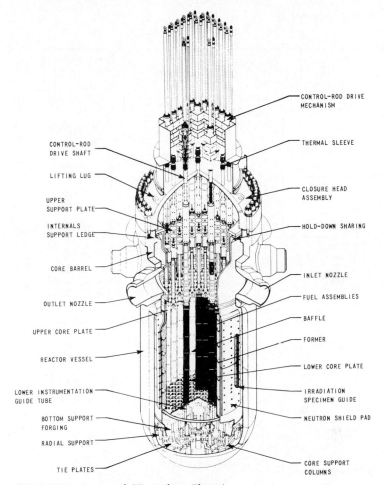

FIG. 2   PWR reactor vessel. (Westinghouse Electric)

in the refueling water storage tank and pumping this water into the reactor coolant system. If the large storage tank should run dry, these pumps will take suction from the containment sump.

Operated by a pressure signal, *containment spray pumps* condense any steam in the containment in order to lower the temperature and pressure in that environment. By taking suction from the containment sump, these pumps continue to circulate water through spray nozzles located near the top of the containment until the pressure has been reduced to an acceptable level.

All of the pumps and vital equipment associated with pipe rupture, that is, the emergency core cooling system, are supplied with diesel generator electrical power backup. If needed, the diesel generators will accept the various pump loads sequentially at intervals of a few seconds until all needed equipment is on line.

Other pumps serve other systems. *Spent fuel pit pumps* provide the necessary cooling of the fuel elements which have been removed from the reactor. Resin beds, which are a part of the water purification system, are flushed to a storage tank by *spent resin sluicing pumps*. An evaporator package, partly for removing boron from the primary water, is supplied by *recycle evaporator feed pumps*. *Chilled water pumps* supply the boron thermal regeneration system. Similarly, other pumps, some not listed in Table 1, support auxiliary systems.

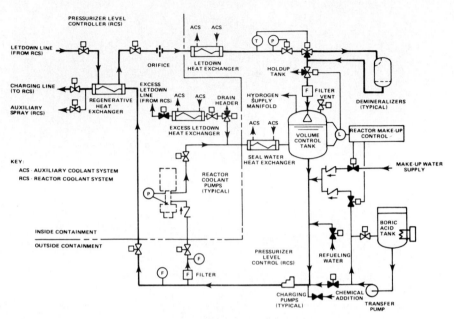

FIG. 3 Flow diagram for chemical and volume control system. (Westinghouse Electric)

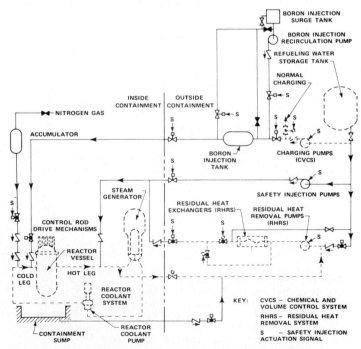

FIG. 4 Flow diagram for safety injection system. (Westinghouse Electric)

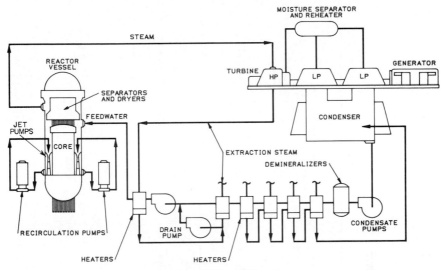

FIG. 5 Boiling water direct cycle reactor system. (General Electric)

BWR Plants In boiling water reactor (BWR) plants, active boiling takes place in the nuclear core and steam is piped to the turbogenerator (Fig. 5). Typical reactor water conditions are 1000 lb/in^2 gage (6.895 MPa) and 550°F (288°C). In the United States, the plant is usually arranged with a low-leakage containment vessel completely surrounding a dry well and a pressure-suppression pool (Fig. 6). The containment vessel is a cylindrical steel or concrete structure with an ellipsoidal dome and a flat bottom supported by a reinforced concrete mat. The containment forms a

FIG. 6 BWR containment and reactor vessel, showing one of two external recirculation pumps. (General Electric)

TABLE 2 Typical Nuclear Pump Parameters in BWR Plants

Pump	Number per plant	Flow, gpm (m³/h)[a]	Head, ft (m)	Design pressure, lb/in² (MPa)	Design temp., °F (°C)	Driver hp (kW)[a]	Shaft	Length or height, including driver, in. (mm)	Speed (nominal), rpm	Notes
Recirculation coolant	2	44000 (9993)	760 (231)	1675 (11.55)	575 (302)	8000 (5970)	Vert.	240 (6096)	1800	
RHR service water	4	7300 (1658)	115 (35)	150 (1.03)	150 (65)	300 (224)	Vert.	200 (5080)	900	
RHR	3	8520 (1935)	275 (84)	450 (3.10)	212 (100)	900 (670)	Vert.	350 (8890)	1800	
High-pressure core spray	1	1465 (333)	2600 (792)	1600 (11.03)	212 (100)	3000 (2240)	Vert.	500 (12700)	1800	
Low-pressure core spray	1	6000 (1363)	280 (85)	550 (3.79)	212 (100)	1750 (1310)	Vert.	400 (10160)	1800	
Closed cooling water	2	2040 (463)	110 (34)	150 (1.03)	212 (100)	60 (45)	Horiz.	100 (2540)	1800	
Reactor core isolation cooling	1	700 (159)	2600 (792)	1500 (10.34)	212 (100)	800 (597)	Horiz.	123 (3124)	4000	Turbine-driven variable-speed

Fuel pool cooling and cleanup	2	600 (136)	300 (91)	150 (1.03)	150 (65)	75 (56)	Horiz.	95 (2413)	1800	
Reactor water cleanup	2	150 (34)	500 (152)	1400 (9.65)	560 (293)	50 (37)	Horiz.	78 (1981)	3600	
Control rod drive hydr. system	2	95 (22)	3500 (1067)	1750 (12.07)	150 (65)	300 (224)	Horiz.	168 (4267)	1800	
Standby liquid control	2	40 (9)	2800 (853)	1400 (9.65)	150 (65)	40 (30)	Horiz.	60 (1524)	1800	Reciprocating pumps
Jet	20	10000 (2271)	80 (24)	N.A.	575 (302)	N.A.	Vert.	250 (6350)	N.A.	
Waste evaporator	2	9000 (2044)	50 (15)	50 (0.35)	274 (134)	150 (112)	Horiz.	120 (3048)	720	
Resin tank precoat	1	255 (58)	85 (26)	150 (1.03)	150 (65)	10 (7.5)	Horiz.	65 (1651)	1800	

[a]Vary depending on reactor power rating.
SOURCE: General Electric.

security barrier and prevents the escape of radioactive products to the atmosphere if an accident should occur.

Table 2 shows the principal pumps used in BWR plants together with significant characteristic data.

To assure a high reactor flow rate and to avoid local areas of core overheating, internal *jet pumps* have been used in all but the earliest U.S. BWR plants. These jet pumps are driven by large-volume, medium-head *recirculation pumps*. Variable flow rate is achieved either by flow control valves or by variable-speed motors driven by motor generator sets. The latest designs employ a flow control valve. The use of jet pumps decreases the size of the external loop piping and pumps and provides a core reflood capability in the event of pipe rupture. The main recirculation pumps are not required for emergency cooling. In normal operation the recirculation pumps operate in conjunction with the jet pumps, which are wholly contained in the reactor vessel. The purpose of the jet pumps is to increase the flow from the recirculation pumps at reduced head for reactor cooling. The jet pumps have no moving parts. In addition to their normal service, they also play a role in the natural circulation of the reactor water during emergency cooling.

Several subsystems operate in support of the recirculation system, and each contains one or more pumps. *Reactor water cleanup pumps* are used in a filter-demineralizer system to remove particulate and dissolved impurities from the reactor coolant. This system also removes excess water from the reactor. The control rods are operated hydraulically with water pressure provided by the *control rod drive pumps*. The pumps are located in an auxiliary building, and the fluid is piped to the control rod drive units, which are positioned directly under the reactor vessel. For those components which require constant cooling, such as the recirculation pump motors and other equipment located in the containment, auxiliary, fuel, or radwaste buildings, *closed cooling water pumps* furnish the necessary flow. These pumps are located in an auxiliary building to permit ready access for servicing if needed. The system is closed so that it can be isolated from an ultimate, usually raw water, heat sink, such as a river, lake, or ocean.

To cool the fuel stored under water in the fuel building and the water in the upper containment, separate *fuel pool cooling pumps* are provided in an independent system.

The emergency core cooling system is in reality an array of subsystems providing the necessary features, including redundancy, to protect the core in case of a significant malfunction. The high-pressure core spray system uses a vertical *high-pressure core spray pump*, motor-driven but backed by a diesel generator in event of loss of electric power. This pump, a single unit, provides the initial response when a small pipe breaks or an equivalent malfunction occurs. Should this system be inadequate to maintain reactor water level, the reactor vessel is automatically depressurized and the *low-pressure core spray pumps* supply additional capacity. As an added safeguard, the *RHR pumps* are used in a secondary-mode operation to inject cooling water directly into the reactor vessel. If steam should enter the containment region, the *RHR pumps* operate in another mode—as containment spray pumps—and are manually operated to condense the steam and thus reduce any potential pressure buildup in the containment. The *RHR pumps* function when needed to limit the temperature of the water in the suppression pool. The turbine-driven *reactor core isolation cooling pumps*, in a redundant and independent system, inject cool water into the reactor vessel. The *standby liquid control system pumps* inject boron solution into the reactor for alternative shutdown.

In addition to these systems, a number of support systems exist, most of which require some form of pumping. Without attempting to describe their systems, there are, for example, pumps for feedwater, condensate, chilled water, booster service, condenser service, demineralized water transfer, condensate transfer, dry-well drain, containment drain, concentrated borated water tank, water leg, precoat, radwaste, and sample station.

Fast Breeder (Liquid Metal) Plants   Many fast breeder plants currently in design or use make use of liquid sodium as the primary coolant. Often, a primary and a secondary system, both employing liquid sodium, are used, with a sodium-water steam generator in the secondary circuit to generate steam for the turbine. Circulating pumps in these systems operate at temperatures of approximately 1250°F (677°C) at relatively low system pressures. In addition, auxiliary liquid-metal pumps may be used to supply necessary subsystems.

MAIN COOLANT PUMPS

The term *main coolant pumps* as used here includes both the *recirculation pumps* in BWR plants and the *reactor coolant pumps* in the primary systems of PWR plants. As shown typically in Figs. 7, 8, 9, and 10, main coolant pumps use a vertical shaft with the impeller at the bottom and a drive motor coupled to the top end of the shaft.

Bearings   Main coolant pumps are usually single-bearing units, although one manufacturer provides a second guide-and-thrust oil-lubricated bearing just below the coupling. Within the pump, hydrodynamic water-lubricated bearings are conventionally used (one supplier using a hydrostatic type). The hydrodynamic bearing generally consists of a hardened sleeve journal shrunk on the shaft and a carbon-lined bearing with some type of self-aligning feature. Bearing diametral clearances are approximately 1.5 mils per inch of bearing diameter (0.0015 mm per millimeter). This rather large clearance is desirable because of thermal transient conditions. Where carbon bearings are used, a cooling mechanism must be provided because the carbon is not suited to long exposure in a hot environment. Many methods are available to accomplish this.

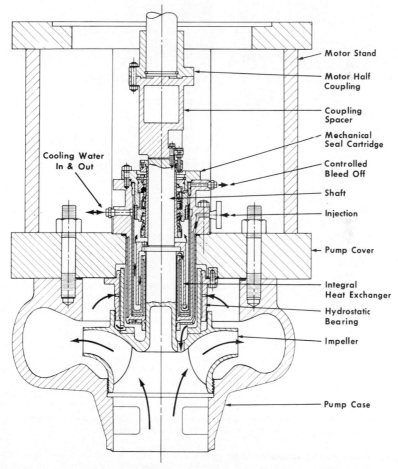

FIG. 7   Main coolant pump. (Byron Jackson Division, Borg-Warner)

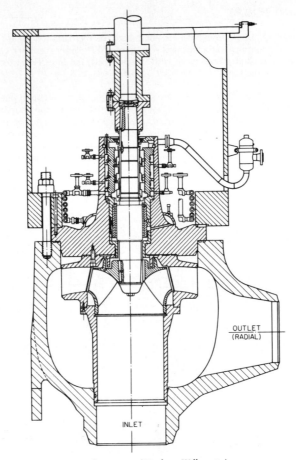

OUTLET
(RADIAL)

INLET

FIG. 8   Main coolant pump. (Bingham-Willamette)

With bearings adequately sized and operating in clean water, virtually trouble-free performance can be expected even with relatively frequent starts and stops.

The hydrostatic, or pressuirized, bearing is used where it is not convenient or desirable to provide bearing cooling, for the hydrostatic bearing can be designed to operate in reactor temperature water. Usually it will be larger than its hydrodynamic equivalent and somewhat more sensitive to starting because of the metal-to-metal rubbing which occurs until rotative speed has built up a small pressure to provide a water film clearance. At normal operating speed, however, a hydrostatic bearing will have a significantly larger lubricating film than a hydrodynamic bearing and can tolerate larger particulate matter without wear.

An exception to the above discussion occurs in both German and Swedish BWR designs, where the recirculation pumps are inverted and inserted directly into the bottom periphery of the reactor vessel and use wet winding motors. There are no jet pumps. Flow is varied by changing motor speed with solid-state power supplies, one for each recirculation pump.

Seals   Pump seals are invariably of the pressure-balanced type because of the high pressures involved. Several suppliers use pressure breakdown techniques to distribute the pressure either equally or in some desired proportion between two or more seals in series (Fig. 11). The technique is analogous to that of a potentiometer, where particular voltages may be obtained by selecting

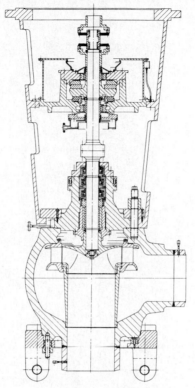

FIG. 9 Main coolant pump. (CE-KSB Pump)

the proper point along an electrical resistance. Also, as with a potentiometer, the flow through the primary resistance path must exceed the tapoff flow in order to maintain the system stability. For pressures in PWR systems, three series seals are frequently used (with each seal taking one-third of the overall pressure), and with lower-pressure BWRs two series seals are usually sufficient. A margin of safety is built into the systems such that pump operation can be continued even if one of the series seals fails.

In addition, a safety seal may also be installed as a backup. At least one manufacturer uses a high-pressure ceramic seal with a carbon seal backup followed by a vapor seal. Main coolant pump seals are usually supplied with reactor-grade water through a seal injection system. The advantage of using injection water is that it can be temperature-controlled and filtered. Most pumps, however, can be operated without it for at least reasonable time periods.

PRINCIPAL TYPES OF PUMPS

Most of the pumps in nuclear service are one- or two-stage centrifugal motor-driven pumps. Both vertical and horizontal types are used. Charging, safety injection, feedwater, and other high-head pumps are usually multistage motor-driven centrifugal units. Some requiring high power are turbine-driven. Double-suction designs are frequently used for RHR pumps, where service requires operation at low available NPSH. Reciprocating pumps find limited service in nuclear plants for make-up flow, seal injection flow, or chemical mixing service. Canned pumps are frequently used in subsystems where their zero-leakage capabilities can be exploited.

Figure 12 shows a multistage centrifugal pump used in charging and safety injection service.

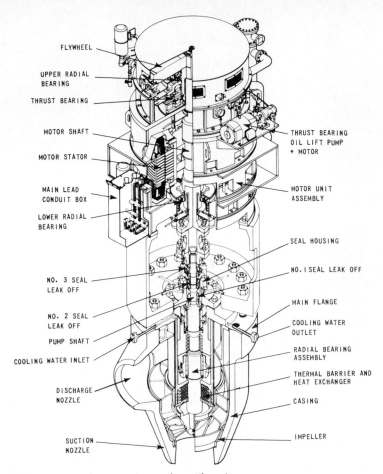

FIG. 10   Main coolant pump. (Westinghouse Electric)

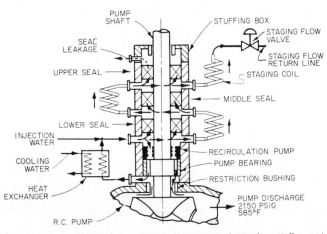

FIG. 11   Flow schematic for a reactor coolant pump seal. (Bingham-Willamette)

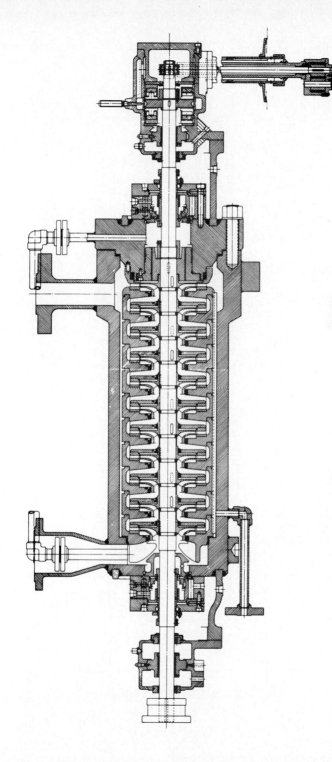

FIG. 12 Pump for PWR charging and safety injection service. (Pacific Pumps Division, Dresser Industries)

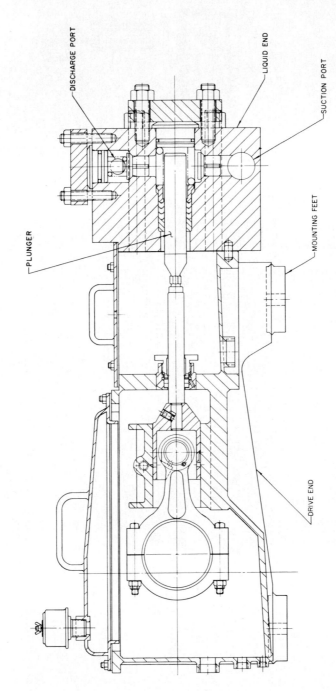

DISCHARGE PORT

LIQUID END

SUCTION PORT

PLUNGER

MOUNTING FEET

DRIVE END

FIG. 13   Reciprocating pump for nuclear service. (Gaulin)

9.218

Figure 13 illustrates a reciprocating pump often used for charging and hydrostatic test service. It is rated at 98 gpm (22 m³/h) at 5800 ft (1768 m) head. In Fig. 14, a vertical multistage unit is shown. This type of unit is common in RHR service. Figure 15 shows a single-stage pump of a type used in many service functions in a nuclear plant. Typically, its flow is 75 gpm (17 m³/h) with a developed head of 235 ft (71.6 m) at 3500 rpm.

MATERIALS OF CONSTRUCTION

Material Limitations   When pumps built to ASME standards are required, materials for the pressure-retaining parts must be selected from a list approved by the ASME. Section III (Nuclear Components) of the ASME Boiler and Pressure Vessel Code lists these materials, their allowable stresses, and the examination requirements which must be applied to ensure their suitability.

Typical Materials   Examples of acceptable materials for pressure-retaining boundary parts are shown in Table 3. They are suitable for all three ASME code classes; however, other acceptable materials may be restricted to use for a particular class. Hazardous and porous materials are generally avoided, as are materials such as cobalt, which, though normally harmless, may become hazardous from radioactive considerations (see below). Cobalt content is often limited in large stainless steel parts but is usually permitted in concentrated form in small areas, for example, where hard-facing is required.

Materials of construction should not be affected by the usual decontamination chemicals.

Nonpressure boundary parts may be made of conventional materials but will usually require the buyer's approval. Certain elastomers, such as ethylene propylene, which has

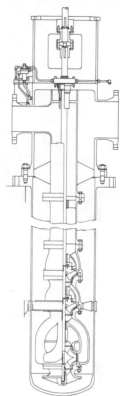

FIG. 14   Vertical multiple-stage pump. (Byron Jackson Division, Borg-Warner)

good radiation stability, are excellent for seal parts. Many fine grades of carbon-graphite are available for water-lubricated bearings and for mechanical seal facings.

SPECIAL REQUIREMENTS FOR NUCLEAR SERVICE PUMPS

It is in the area of special requirements that nuclear service pumps differ most widely from commercial products. These special requirements, described in greater detail below, far exceed the requirements of the general industrial field and illustrate the strong emphasis placed upon pressure integrity and pump operability.

Nuclear-grade pumps are designed, built, inspected, tested, and installed to rigid standards of the U.S. Nuclear Regulatory Commission, the American Society of Mechanical Engineers, the American National Standards Institute, and other regulatory agencies, such as state and local jurisdictional bodies. Established rules can be divided into two distinct categories: (1) those controlling integrity of the pressure-retaining boundary and (2) functional considerations.

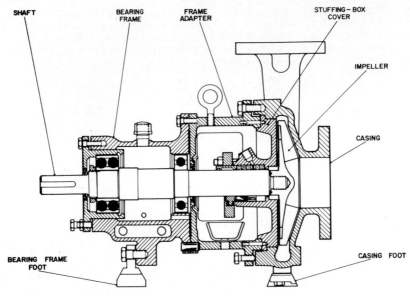

FIG. 15   Boric acid transfer pump. (Goulds Pumps)

The hydraulic design of nuclear pumps is the same as that of pumps in conventional service. For recirculation and reactor coolant pumps, a radial discharge is preferred by some users because it tends to simplify certain aspects of the plant design.

Vibration characteristics of pumps in nuclear service are especially important because of the relative inaccessibility of the equipment for checking and servicing and because safety requirements permit only limited outage of pumps; otherwise, the plant must come to standby condition.

An analytical or experimental determination of lateral and torsional natural frequencies is routine for most pump-driver combinations, and occasionally a transient analysis may be required.

Design Under ASME Code Rules   Under the rules established by the ASME Boiler and Pressure Vessel Code, Section III (Nuclear Components), the owner, such as a public utility, either directly or through an agent, prepares a design specification for a specific pump, which in code terms is a component.

TABLE 3   Materials for Pressure-Retaining Boundary Parts

Carbon steel	
Castings	SA-216, Gr WCA, WCB, WCC
Forgings	SA-105, Gr I, II
Plate	SA-515, Gr 55, 60, 65, 70
Bolting	SA-193, Gr B6, B7, B8, B16
Stainless steel	
Castings	SA-351, Gr CF8 (304), CF8M (316)
Forgings	SA-182, Gr 304, 316, 321, 347
Plate	SA-240, Gr 304, 316, 321, 347
Nonferrous	
A limited number of nonferrous materials are permitted.	

The manufacturer builds the pump to the design specification. Verification is provided through the combined efforts of the material supplier, manufacturer, insurance inspector, and state and local enforcement agencies.

The design specification requires compliance with the rules of the code with regard to the design of the pressure boundary and, in addition, includes supplementary requirements prescribed by the owner. Other standards are invoked as applicable to meet the safety and environmental requirements of the U.S. Nuclear Regulatory Commission. Functional needs may be included, and the code class to which the pump is to be built must be identified.

There are three ASME code classes for pumps. For the most critical service, a Class 1 pump is specified. Class 2 represents a pump serving a less critical system, and Class 3 is the lowest-class pump for nuclear service, except for a few pumps which are classified as nonnuclear class (NNC) but permitted in a nuclear power plant. It is the owner's responsibility to establish the pump class, with guidance provided by the Nuclear Regulatory Commission and the manufacturers.

For the pressure boundary evaluation, a Class 1 pump, by code rules, receives a detailed analysis by modern design techniques, supported, if necessary, by experimental stress analysis, and a stress report is prepared to document adherence to rules. Fatigue analyses of critical portions of the pump may be required, and behavior under all plant conditions, including accident, must be investigated. Class 2 pumps require less analysis and no stress report, but a data report certifying compliance with the code is required. For Class 3 pumps, wider latitude of design is permitted, but a certified data report is required. All three classes of pumps may be given the ASME N-stamp by a qualified pump builder upon completion of design, manufacture, inspection, and test.

Additionally, nuclear service pumps are usually examined in design for thermal steady-state and transient conditions, behavior under seismic disturbances, conditions of nozzle loadings imposed by system piping, means of support, and accessibility for service, in-service inspection, and replacement.

Rules for quality control during material procurement, manufacture, and test are also contained in the code. Nondestructive examination and document control are detailed, and a quality assurance program requiring a formal procedure manual must be prepared and implemented by the manufacturer. Periodic surveys by the ASME verify adherence to code rules. Local jurisdictional authorities provide day-to-day inspection services as required by the manufacturer.

Design of Noncoded Parts   Parts of the pump which are not classified as pressure boundary items or attachments to the pressure boundary are not covered by the ASME code. The owner's design specification may describe applicable requirements for these nonpressure parts, referencing other documents or excluding use of certain objectionable materials. If the pump is in critical service or is needed in emergency situations, certain ANSI standards may be invoked. Accompanying these rules will be extensive quality assurance and documentation to demonstrate compliance.

Considerations of Radioactivity   Radioactivity may become a serious consideration in the design of nuclear pumps because of the need for servicing the equipment. The water used in the primary system becomes contaminated with metallic elements through solubility, corrosion, and erosion. When circulated through the core region, the metallic elements become radioactive because of interaction with neutrons. These readioactive contaminants may, if soluble, remain in solution in the water or, if unsoluble, plate out on metal surfaces or become lodged in "crud traps," such as fit interfaces, screw threads, porous base metals, extremely rough surfaces, cracks, and certain types of weld configurations, such as socket welds. In one case, for instance, a pump impeller returned for overhaul defied attempts at decontamination until it was discovered that a repair had been made to a presumably integral wear ring by undercutting and shrinking on a new ring. The interface was barely perceptible, but once it had been found and the ring had been removed, the impeller was readily decontaminated.

Soluble contaminants are most easily removed by providing the pump with complete drainage features, that is, leaving no internal pockets which are not naturally drainable. Ease and speed of parts replacement are, therefore, also important items of design since they reduce the length of time service personnel are exposed to radiation.

The Nuclear Regulatory Commission has specified that no person shall be exposed to more than 100 mrem of radiation in seven consecutive days. A pump producing radioactivity to this extent could be brought into a conventional machine shop and handled in a conventional manner.

After prolonged service in a nuclear plant, however, a pump may emit radioactivity at a rate of, say, 2 to 50 rem/h. Hence provision for decontamination also becomes an important design characteristic of the nuclear service pump. It should also be mentioned that not all nuclear pumps operate in highly radioactive environments; for these pumps, radioactivity levels may be low, but the equipment will usually require some degree of decontamination before it can be freely handled.

SEISMIC DESIGN

The ASME code requires that a seismic analysis be performed on all Class 1 pumps. On Class 2 and Class 3 pumps, the code does not make such a requirement; however, it is not unusual for the owner to add this requirement to the design specification. See Subsec. 9.15.2 for more detailed information on this subject.

TESTING

Hydrostatic Testing   Hydrostatic testing in accordance with ASME code rules is conventional except that specific documentation is required.

Performance Testing   Performance of nuclear pumps is verified by procedures common to nonnuclear pumps. In addition, however, it may be necessary to demonstrate pump performance under some emergency condition, such as loss of cooling water or seal injection water. Also, a test under simulated total-loss-of-power conditions may be required, in which the reactor coolant pump must coast safely to a stop and withstand loss of both cooling and seal injection water for a finite time.

Periodic Testing   The ASME code, under its rules for in-service inspection (Section XI), requires periodic testing of certain pumps installed in a nuclear plant. The pumps affected by these rules are those associated with the safety systems and those which may be required to function during an emergency or a reactor shutdown. Examples of such pumps are those for core spray, residual heat removal, boron injection, and containment spray. If pumps cannot be tested in their usual circuit, they must be supplied with bypass loops which can be valved off.

The field testing procedure involves obtaining a baseline set of values for head, flow, speed (if variable), vibration, inlet pressure, and bearing temperatures. The flow quantity which sizes the bypass loop is not specified in the code, but in the practical sense, the flow must be adequate to prevent overheating of the pump and overloading of the bearings. Also, since it is an off-design point, consideration is given to the potential for rough operation due to impeller internal recirculation and possible low-flow cavitation damage. Code rules call for a 5-min running period every 3 months and longer times where it is necessary for bearing temperatures to stabilize. If the differences between periodic operating data and baseline data exceed permissible limits, the cause for the differences must be sought and corrected. Record-keeping is therefore a significant part of pump in-service testing.

9.15.2
NUCLEAR PUMP SEISMIC QUALIFICATIONS

D. NUTA

Nuclear electric generating structures and components are designed to resist the dynamic forces that could result from an earthquake at the site. This subsection covers the seismic qualifications of pumps classified according to function and according to importance in safe operation and public safety.

SEISMIC CLASSIFICATION OF PUMPS

Pumps which have to withstand a safe shutdown earthquake (SSE) and remain functional during and after such an event are designated seismic Catetory I. Pumps which are not required to remain functional during a SSE are designated nonseismic Category I and may be classified as follows:

1. Nonseismic Category I pumps located such that their failure or structural deformation might keep a seismic Category I piece of equipment or system from performing its intended function. The analysis methods and seismic input used on nonseismic Category I pumps to show that gross failure or excessive deformations will not occur are identical to those used for seismic Category I equipment.

2. Nonseismic Category I pumps located such that their failure or structural deformation cannot affect any seismic Category I system or component. Seismic-related requirements for these pumps, if any, are established using a less stringent seismic environment with the aim of minimizing possible replacement expenses following a credible event.

DEFINITIONS

Critical damping Minimum damping for which a vibrating system has no vibratory motion. For analysis purposes, damping is expressed as a percentage of critical damping. The amplification of input motion, or magnitude of response, is inversely proportional to the damping ratio of the

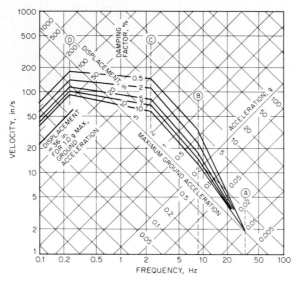

FIG. 1 Horizontal design response spectra scaled to 1g ground acceleration. (1 in = 2.54 cm; g = 32.17 ft/s² = 9.807 m/s²)

vibrating system. Figure 1 shows responses obtained for damping ratios which vary from 0.5 to 10% of critical damping.

Damping Dissipation of energy in a vibrating system. Damping is a function of many factors, including material characteristics, stress levels, and geometric configurations, and takes place as a result of the transformation of input energy in the form of seismic motion to heat, sound waves, or other energy forms. For analysis purposes, damping is assumed to be viscous, i.e., the damping force is proportional to velocity and acts in the opposite direction.

Degrees of freedom Coordinate(s) necessary to adequately describe the behavior of a structure or component. A node, or nodal point, may have up to six degrees of freedom, consisting of three translations and three rotations (such as Δ_x, Δ_y, Δ_z, θ_x, θ_y, θ_z, as shown in Fig. 2). This complex representation in terms of degrees of freedom is used when performing static analysis to obtain forces, stresses, or displacements.

Dynamic degrees of freedom A subset of degrees of freedom, selected to adequately describe the motion of a vibrating system. While a dynamic analysis may be conducted with the full complement of degrees of freedom that was used in the static analysis, a "condensation" is needed in order to reduce computational costs. Considering the lumped-mass mathematical model presented in Fig. 4 for the vertical pump shown, a nodal point, such as 15, will be allowed only two degrees of freedom when the seismic analysis is performed along the y axis, namely, Δ_y and θ_x (translation along y and rotation in the plane of the paper around the x axis).

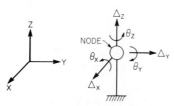

FIG. 2 Nodal degrees of freedom.

Finite element method Method of analysis which involves the idealization of a structure or component, such as a pump, as an assemblage of elements of finite size. This idealization is also called a mathematical model. One of the most commonly used mathematical models is the lumped-mass model, which consists of an assemblage of massless beam elements connected at nodal points and masses lumped such as to adequately represent the mass of the adjacent areas. Figures 3 and 4 contain a typical lumped-mass model of a vertical pump.

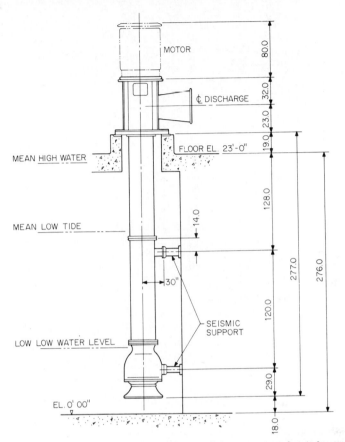

FIG. 3   Vertical pump geometry and layout. All dimensions in inches. (1 ft = 0.3048 m; 1 in = 2.54 cm)

Floor acceleration   Acceleration at a structural floor or any acceleration applicable at the equipment mounting location.

Ground acceleration   Acceleration of the ground associated with earthquake motion.

Modal analysis   A type of dynamic analysis in which the response of a vibrating system is obtained as a combination of responses of the system's mode shapes. (Specific requirements on the combination of modal responses is contained in U.S. Nuclear Regulatory Commission Guide 1.92.) Modal analysis may be used in conjunction with a response spectrum or time history analysis.

Natural frequency   Frequency at which a body vibrates due to its own physical characteristics and to the elastic restoring forces developed when it is displaced in a given direction and then released while restrained or supported at specified points.

Operating basis earthquake (OBE)   Earthquake which could reasonably be expected to affect a plant site during the operating life of the plant. The systems and components necessary for continued operation of a nuclear plant without undue risk to public health and safety are designed to remain functional when subjected to the vibratory ground motion produced by an operating basis earthquake.

Required response spectrum (RRS)   Response spectrum for which the equipment (pump assembly) has to be dynamically qualified.

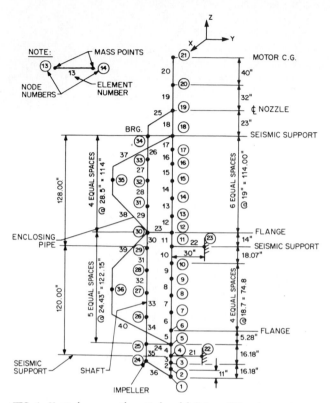

FIG. 4   Vertical pump mathematical model. (1 in = 2.54 cm)

Response spectrum   Plot of the maximum response (acceleration, velocity, displacement) of a series of damped single-degree-of-freedom (SDF) oscillators versus the natural frequencies (or, sometimes, periods) of the oscillators when subjected to the virbatory motion input (time history) at their supports. Figure 5 represents a plot of response spectra obtained for 0, 2, 5, and 10% of critical damping with an earthquake time history as input. The plot is on tripartite log paper, and only a limited number of SDF oscillators are shown (usually, more than 70 are needed to correctly define a spectrum).

In order to facilitate the understanding of how to use the Fig. 5 plots, the 0% of critical damping response spectrum is presented in Fig. 6. Assuming a SDF oscillator having a frequency of 2 cycles/s (Hz), the maximum displacement of the mass point when subjected to the earthquake record is 7 in (17.8 cm), the maximum velocity is 50 in/s (1.27 m/s), and the maximum acceleration is $2g$, where g = gravitational acceleration = 32.2 ft/s^2 (9.807 m/s^2). A line is drawn perpendicular to the frequency axis, and from the intersection of this line with the spectrum curve a line parallel to the acceleration lines will provide the acceleration $a \approx 2g$, a line parallel to the displacement lines will provide the displacement $d \approx 7$ in (17.8 cm), and a line perpendicular to the velocity axis will provide the velocity $v \approx 50$ in/s (1.27 m/s).

In general, response spectra are presented as plots of acceleration versus frequency. Given a time history of acceleration, such as the one presented in Fig. 7, the response spectrum is obtained as described above. The response acceleration at frequencies above 33 Hz (also called zero period acceleration) is equal to the peak acceleration of the time history record regardless of the damping ratio used (the value a is specified in Fig. 7).

Response spectrum analysis   A modal dynamic analysis which uses response spectra as vibratory input at the pump attachment points.

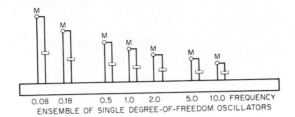

ENSEMBLE OF SINGLE DEGREE-OF-FREEDOM OSCILLATORS

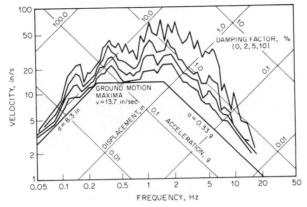

FIG. 5 Response spectra plots for various dampings. (1 in/s = 0.0254 m/s; 1 in = 2.54 cm; 1g = 32.17 ft/s² = 9.807 m/s²)

Safe shutdown earthquake (SSE)   Earthquake which produces the maximum vibratory ground motion for which certain structures, systems, and components are designed to remain functional. These structures, systems, and components are those necessary to assure (1) the integrity of the reactor coolant pressure boundary, (2) the capability to shut down the reactor and maintain it in a safe shutdown condition, or (3) the capability to prevent or mitigate the consequences of accidents which could result in potential offsite exposures to radioactive material. Various safety

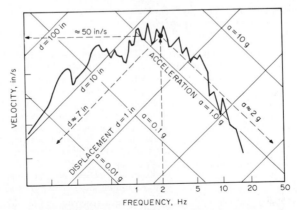

FIG. 6 Response displacement, velocity, and acceleration versus frequency. (1 in/s = 0.0254 m/s; 1 in = 2.54 cm; 1g = 32.17 ft/s² = 9.807 m/s²)

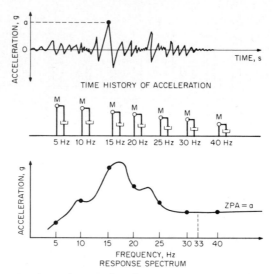

FIG. 7 Development of acceleration response spectrum. (1g = 32.17 ft/s^2 = 9.807 m/s^2)

classes are assigned to structures, systems, and components as a function of which of these three categories they belong to.

Single-degree-of-freedom oscillator (SDF)   Vibrating system consisting of a single node point, where the mass is concentrated and supported by a beam and spring element. If the single dynamic degree of freedom is along the axis of the beam, the frequency of the oscillator is

$$f = \frac{1}{2\pi} \sqrt{\frac{K}{M}}$$

where f = frequency, cycles/s or Hz
  $K = EA/L$ = spring stiffness, lb/in (N/mm)
  E = Young's modulus of elasticity of support beam, lb/in^2 (Pa)
  A = cross-sectional area of support beam, in^2 (mm^2)
  L = length of support beam in (mm)
  M = mass lb·s^2/in (N·s^2/mm)

With this idealization, the frequency of the SDF oscillator may be varied by changing the length of the beam element and maintaining E, A, and M constant (Fig. 8). A horizontal degree of freedom would be used in conjunction with beam bending and shearing.

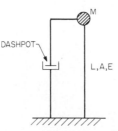

Damping effects are simulated by a dashpot, which can be assigned different damping ratios. While there is little effect on the frequency of the SDF oscillator when the dashpot is assigned damping ratios below 10 or 15% of critical damping, there is a significant effect when higher values are used. Thus, if the dashpot is assigned 100% of critical damping, the oscillator will behave as a rigid body, duplicating, without amplification, the support input motion.

Test response spectrum (TRS)   Response spectrum that is either theoretically constructed or derived using spectrum analysis based on the actual motion of the shake table (during dynamic testing).

FIG. 8 Single-degree-of-freedom oscillator.

Time history analysis   Analysis performed in the time domain using time history vibratory input. A direct integration or modal analysis may be performed.

Time history of acceleration   Variation of acceleration as a function of time during the postulated duration of the seismic event.

Zero period acceleration (ZPA)   Response acceleration at frequencies above 33 Hz.

Although these definitions will facilitate the understanding of the seismic considerations discussed herein, the reader is encouraged to review the works listed at the end of this subsection for a more complete list of terms and definitions used in the process of dynamic qualifications.

SEISMIC QUALIFICATION REQUIREMENTS

The seismic qualification of seismic Category I pumps is performed to demonstrate the ability of the equipment to perform its intended function during and after the time it is subjected to an SSE which follows a number of OBEs (present requirements are for five OBEs followed by one SSE). The simultaneous effects in two orthogonal horizontal directions and in the vertical direction have to be considered.

The seismic qualification could be achieved by

1. Predicting pump performance by analysis

2. Testing the pump under simulated seismic conditions.

3. Using a combination of tests and analyses

The information provided in the IEEE 344 standard on these three qualification methods is also applicable to pump assemblies.

The information presented hereafter either emphasizes some of the IEEE 344 requirements or is pertinent to pumps and pump assemblies in particular.

Qualification by Analysis   Qualification through analysis involves the development of a mathematical model of the pump assembly, dynamic analyses to establish equivalent static loads, and static analyses to generate forces, stresses, and displacements.

The mathematical model of the pump assembly and various supports and restraints is developed using sufficient degrees of freedom to adequately represent the behavior in two orthogonal horizontal directions and the vertical direction. The mathematical model is usually a lumped-mass model with the nodal and mass points located at points of discontinuity (sectional properties, connecting points, etc.) and augmented by additional nodal and mass points, the number and location of which will vary for various arrangements, geometries, or pump types.

In general, a mathematical model of a pump may be considered adequate when the addition of node points and dynamic degrees of freedom on any element or series of elements will not significantly change the dynamic responses (i.e., a change of less than 10%). A typical lumped-mass model of a pump assembly and motor is presented in Figs. 3, 4, and 9. (The reader is reminded that a node point and a dynamic degree of freedom are needed at any point where seismic response information is required.) The model should include, when applicable, the hydrodynamic mass, which represents contained water or the effect of submergence of the suction line. (In practice, the effect of immersion in liquids is rarely considered.) Although it is acceptable practice to incorporate the lumped-mass model of the pump into the dynamic analysis of the piping system and then perform a coupled analysis and establish nozzle loads (among others), the state of the art is to consider the pump assembly as a rigid support or anchor for the piping when analyzing the piping system. (This assumption is justified based on the frequency analysis of the pump assembly, which generally indicates frequencies in excess of 33 Hz.)

After developing the mathematical model of the pump assembly and prior to performing the dynamic analyses, a frequency analysis is performed. Since the three earthquake components (two horizontal, usually east-west and north-south, and one vertical) are usually analyzed separately,

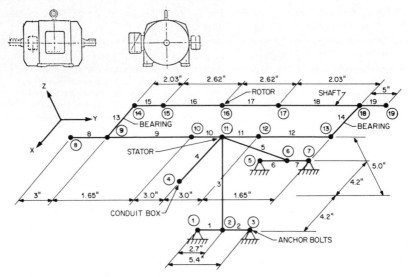

FIG. 9   Alternating-current motor outline and typical mathematical model. (1 in = 2.54 cm)

the frequency analysis will establish the frequency of the east-west, north-south, and vertical pump assembly models. If, for a given direction, the lowest frequency obtained is higher than the ZPA frequency (Fig. 7), then the pump is classified as rigid in that direction and seismic forces (equivalent static loads) are established by multiplying the various masses (concentrated at the mass and node points) by the ZPA of the spectra provided for the rigid direction. (If the pump has more than one support point and different spectra for a given direction where specified at the supports, the highest ZPA must be used).

If the lowest frequency established from the frequency analysis for a given direction is below the ZPA frequency, then the pump is classified as flexible in that direction and a dynamic analysis has to be performed in order to establish seismic responses and equivalent static loads. A response spectrum is usually determined, using the envelope of response spectra provided at the various supports for a given direction or time history analysis. These dynamic analyses are performed with computer programs in the public domain, such as ICES/Strudl, Stardyne, Ansys, and Nastran, which incorporate acceptable methods of combining modal responses when a modal analysis is performed, or with a multitude of other finite element programs which are either commercially available or developed specifically for the analyses. While certain sophistications may be employed, such as the use of a three-dimensional model and the simultaneous application of statistically independent time histories in three directions, it should be noted that a response spectrum analysis will establish the inertial effect only and that the effect of relative seismic displacement at the support points must be accounted for separately in order to establish the overall seismic effects. Further information regarding specific items related to dynamic analyses, such as damping ratios, combination of modal responses, and modeling, may be obtained from the works listed at the end of this subsection. In performing dynamic analyses, however, a thorough understanding, derived only from practice, is imperative.

Once dynamic analyses have been performed and equivalent static loads and other seismic information, such as displacements, established, static analyses are performed to determine seismic effects on the pump assembly. These are then combined with other loads in accordance with the prescribed load combinations. The finite element static analyses are performed with three-dimensional mathematical models. Seismic effects (forces, moments, displacements, stresses, reactions) in a given direction are obtained for each of the three components of the earthquake, and then the overall seismic effect for a given direction is obtained, usually by using the square root of the sum of the squares method.

Qualification by Testing   The equipment is tested by using the applicable seismic information (floor response spectrum or time history). The input test motion should have a TRS that closely envelopes the RRS, and the test should simulate the occurrence of five OBEs followed by one SSE.

Both the pump and the motor are tested under conditions that simulate as closely as possible actual operating conditions and operability. The operability of the pump, which is tested without the piping system attached and in the absence of any fluid, is established by the absence of undue large deflections during testing (which implies low stress levels) and by performance at the end of the test.

The input motion used must have a maximum acceleration equal to or greater than the ZPA and cannot include frequencies above the frequency associated with the ZPA. The input motion is applied to one vertical and one horizontal axis (resultant) simultaneously unless it can be proved that coupling effects are negligible. In general, multiaxis testing using independent random inputs is required.

Unless it has complete symmetry about its vertical axis, the equipment is rotated 90° and retested.

The mounting and support details used during testing should simulate the service mounting as closely as possible and should be designed such that the mounting will not cause dynamic coupling to the test assembly or alter the input motion.

Combination of Test and Analysis   If the motor assembly is tested separately and the pump is qualified by analysis, RRS at the motor assembly attachment points are developed from the pump assembly seismic analysis and used in the motor assembly testing. If both the pump assembly and the motor have to be tested but it is not feasible to test them together, dummy weights are used to replace the assembly part not included in the test and RRS at the location are established and used in testing the missing part.

DESIGN AND FUNCTIONAL REQUIREMENTS

Using the seismic responses obtained from the pump-motor assembly seismic analysis and the interface nozzle loads, derived from thermal, seismic, etc., analysis of the piping system, the stress levels for the applicable load combination and design conditions are verified to be within allowable limits. For a pump-motor assembly specified as in Section III of the ASME Boiler and Pressure Vessel Code, a loading condition including the OBE, the system operating transients, and other normal operating loads would be considered an upset condition and Level B stress limits would be used. A combination including the SSE, the design basis pipe break effects, and other normal operating conditions would be considered a faulted condition and Level D stress limits would be specified. The stress limits obtained from ASME Section III, Division 1, Subsection ND, are as shown in Table 1.

If no allowable stress limits are specified, no allowable stress increases should be used for load combinations including the OBE and for load combinations including the SSE, the allowable stresses may be increased by 50% to no more than $0.9F_y$, where F_y is the tension yield stress, or the critical stress for items other than tension.

In addition to the specific requirements for stress levels and/or functionality, the following specific items should be addressed as part of the seismic qualification of pump-motor assemblies:

- Secondary stress loading caused by differential movement of connected components should be determined.
- Displacements and deflections should be either measured during testing or calculated, if qualification is by analysis, in order to assure they are not excessive.
- Loads to be used in the design of the pump-motor assembly foundation and anchorage should be established.
- Seismic loads must be added to other normal shaft loads in order to verify stress levels, bearing loads, running clearances, and possible coupling misalignment.

TABLE 1 Stress Limits

Service limits	Stress limits (maximum normal stress)
Design and Level A	$\sigma_m \leq 1.0S$
	$(\sigma_m \text{ or } \sigma_L) + \sigma_b \leq 1.5S$
Level B	$\sigma_m \leq 1.1S$
	$(\sigma_m \text{ or } \sigma_L) + \sigma_b \leq 1.65S$
Level C	$\sigma_m \leq 1.5S$
	$(\sigma_m \text{ or } \sigma_L) + \sigma_b \leq 1.8S$
Level D	$\sigma_m \leq 2.0S$
	$(\sigma_m \text{ or } \sigma_L) + \sigma_b \leq 2.4S$

σ_m general membrane stress; equal to average stress across solid section under consideration; excludes discontinuities and concentrations and is produced only by pressure and other mechanical loads.

σ_L local membrane stress; same as σ_m except that σ_L includes effect of discontinuities.

σ_b bending stress; equal to linearly varying portion of stress across solid section under consideration; excludes discontinuities and concentrations and is produced only by mechanical loads.

S allowable stress, given in Tables I-7.0 and I-8.0 of the ASME code; corresponds to highest metal temperature of section under consideration during condition under consideration.

SOURCE: ASME Boiler and Pressure Vessel Code, Section III, Nuclear Power Plant Components, Division 1, Subsection ND, 1983, New York, N.Y.

- For electric motors, deflections or component dislocations should be closely monitored in order to assure that the rotor will not grind into the stator, that the fan blades will not damage the winding insulation, and that there is no loss of lubricant and no slipped winding or winding temperature detector.

DOCUMENTATION

Refer to IEEE 344 for a list of documents that might be required for dynamic qualification of the pump-motor assembly (by test or analysis) and for general guidance in preparing the pump assembly purchase specification.

FURTHER READING

U.S. Nuclear Regulatory Commission Guides   (available from NRC, Washington, D.C.)

Regulatory Guide 1.100, "Seismic Qualification of Electrical Equipment for Nuclear Power Plants."

Regulatory Guide 1.61, "Damping Values for Seismic Design of Nuclear Power Plants."

Regulatory Guide 1.70, "Standard Format and Content of Safety Analysis Reports of Nuclear Power Plants."

Regulatory Guide 1.89, "Qualification of Class 1E Equipment for Nuclear Power Plants."

Regulatory Guide 1.92, "Combining Modal Responses and Spatial Components in Seismic Response Analysis."

Industry Standards (available from The Institute of Electrical and Electronics Engineers, New York)

IEEE 323, "Qualifying Class 1E Equipment for Nuclear Power Generating Stations."

IEEE 344, "IEEE Recommended Practices for Seismic Qualification of Class 1E Equipment for Nuclear Power Generating Stations."

Other American Society of Civil Engineers: "Structural Analysis and Design of Nuclear Plant Facilities," Manuals and Reports on Engineering Practices, no. 58, New York, 1980.

SECTION 9.16
METERING

ROBERT H. GLANVILLE

METERING OR PROPORTIONING

Conventional reciprocating pumps can be adapted to function as metering or proportioning devices in the transfer of liquids. The principal adaptations are addition of a means of varying the pumping rate and predicting what that rate will be, which make the modified units suitable for use as final control elements in continuous-flow processes. Metering pumps are often employed where two or more liquids must be proportioned or where mixture ratios must be controlled. These effects are achieved by changing the displacement per stroke (by moving the crankpin by special linkages or by partial stroking) or by changing the stroking speed through the use of variable-speed transmissions or electric motors.

Three basic types of positive displacement reciprocating pumps, and several variations, are used for this service: packed plunger pumps, pumps with a mechanically actuated diaphragm, and pumps with a hydraulically actuated diaphragm.

Packed Plunger The packed plunger pump is the most commonly used type because of its relatively simple design and wide range of pressure capability. It is an adaptation of the conventional reciprocating transfer pump (Fig. 1). Its advantages are

1. Relatively low cost
2. Pressure capabiilty to 50,000 lb/in² (345 MPa) gage
3. Mechanical simplicity
4. Wide capacity range, from a few cubic centimeters per hour to 20 gpm (4.5 m³/h)
5. High accuracy, better than 1% over a 15:1 range
6. Least affected by changes in discharge pressure

Its disadvantages are

1. Packing leakage, making it unsuitable for corrosive or dangerous chemicals
2. Packing and plunger wear and the resulting need for gland adjustment
3. Inability to pump abrasive slurries or chemicals which crystallize

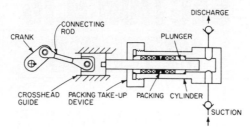

FIG. 1   Packed plunger pump.

Mechanically Actuated Diaphragm
The mechanically actuated diaphragm pump is commonly used for low-pressure service where freedom from leakage is important. This pump utilizes an unsupported diaphragm which is moved in the discharge direction by a cam and returned by a spring (Fig. 2). Its advantages are

1. Relatively low cost
2. Minimum maintenance at 6- to 12-month intervals
3. Zero chemical leakage
4. Ability to pump slurries and corrosive chemicals

Its disadvantages are

1. Discharge pressure limitation of 125 to 150 lb/in^2 (860 to 1030 kPa) gage
2. Accuracy in 5% range and as much as 10% zero shift with change from minimum to maximum discharge pressure
3. Capacity limit of 12 to 15 gph (0.045 to 0.057 m^3/h)

Hydraulically Actuated Diaphragm
The hydraulically actuated diaphragm pump is a hybrid design which provides the principal advantages of the other two types. A packed plunger is used to pulse hydraulic oil against the back side of the diaphragm. The reciprocating action thus imparted to the diaphragm causes it to pump in the normal manner without being subjected to high pressure differences. A flat diaphragm design is shown in Fig. 3 and a tubular diaphragm design in Fig. 4.

The advantages of the hydraulically actuated diaphragm pump are

1. Pressure capability to 5000 lb/in^2 (34.5 MPa) gage
2. Capacities to 20 gpm (4.5 m^3/h)
3. Minimum maintenance

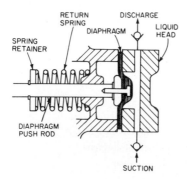

FIG. 2   Mechanically actuated diaphragm pump.

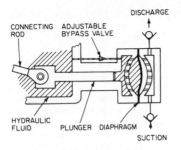

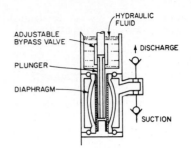

FIG. 3 Hydraulically actuated diaphragm pump with flat, circular diaphragm.

FIG. 4 Hydraulically actuated diaphragm pump with tubular diaphragm.

4. Zero chemical leakage

5. Ability to pump slurries and corrosive chemicals

6. Accuracy around 1% over 10:1 range

Its disadvantages are

1. Subject to predictable zero shift of 3 to 5% per 1000 lb/in² (6.9 MPa) gage

2. Higher cost

CAPACITY CONTROLS

There are five methods commonly used to adjust the capacity of metering pumps. The choice of which method to use is determined by the application for which the pump is intended.

Manually Adjustable While Stopped This control is generally found on packed plunger pumps of conventional design. Capacity changes are effected by moving the crankpin in or out of the crank arm while the pump is not in motion. This is the least expensive method and is used where frequent changes in pump displacement are not required.

Manually Adjustable While Running This feature is most frequently found on mechanically actuated diaphragm pumps, where it is accomplished by limiting the return stroke of the diaphragm with a micrometer screw. On packed plunger pumps, stroke adjustment while the pump is running is relatively complicated. Stroke length is set by some type of adjustable pivot, compound linkage, or tilting plate, which is manually positoned by turning a calibrated screw. On hydraulically actuated diaphragm pumps, relatively simple control is provided by manually adjustable valving which changes the amount of the intermediate liquid bypassed at each stroke.

Pneumatic Meeting pumps used in continuous processes must be controlled automatically. In pneumatic systems, the standard 3- to 15-lb/in² (21- to 103-kPa) gage air signal is utilized to actuate air cylinders of diaphragms directly connected to the stroke-adjusting mechanism.

Electric On electric control systems, stroke adjustment is through electric servos which actuate the machanical stroke-adjusting mechanism. These accept standard electronic control signals.

Variable Speed This method of adjusting capacity is achieved by driving a reciprocating pump with a variable-speed prime mover. Since it is necessary to reduce stroke rate to reduce delivery, discharge pulses are widely spaced when the pump is turning slowly. Surge chambers or holdup tanks are used when this factor is objectionable.

In control situations involving pH and chlorinization, two variables exist at once, e.g., flow rate and chemical demand. This is easily handled by a metering pump driven by a variable-speed prime mover. Flow rate can be adjusted by changing the speed of the pump, and chemical demand by changing its displacement.

SERVICE AND MAINTENANCE

Proportioning pumps utilizing manual capacity controls can be installed, serviced, and operated by plant personnel. Pneumatic capacity controls are more sophisticated, but once their construction and function are understood, maintenance and service should become routine. Electric capacity controls require a basic understanding of electric circuits for installation, operation, and routine service. Modular construction allows repair service by replacement of component groups, thus diminishing the task of trouble shooting.

All proportioning pumps utilize suction and discharge check valves. These require regular maintenance and service since they are used frequently, encounter corrosion, and must be kept in good working order to ensure accurate pump delivery. Service periods are greatly dependent on liquids pumped, pump operating speed, and daily running time. Usual service periods run from 30 days to 6 months or longer. Check valves are designed to facilitate service with a minimum of downtime, thus allowing replacement of wearing parts at minimum expense. Good design allows this service to be accomplished without breaking pipe connections to the pump.

Packed plunger pumps require periodic adjustments of the packing takeup device to compensate for packing and plunger wear. Packing has to be replaced periodically, and regular lubrication of bearings and wear points is required.

The lubricating oil in speed-reduction gearing, either separate integral units or built in, must be changed at 6-month intervals.

Diaphragm pumps usually require replacement of the diaphragm as part of routine service at 6-month intervals. On hydraulically actuated diaphragm pumps, the hydraulic fluid must also be changed.

Pneumatic and electric controls generally present no special maintanenace problems. The frequency of routine cleaning and lubrication is dependent on environmental conditions and should be consistent with general plant maintenance procedures.

INSTALLATION

Proper installation of metering pumps is very important if reliable pump operation is to be obtained.

NPSH must be kept as high as possible, and the manufacturer's recommendations as to pipe size and length, strainers, relief valves, and bypasses must be observed.

MATERIALS OF CONSTRUCTION

With the tremendous variety of liquid chemicals used in industry today, an all-inclusive guide to suitability of materials for pump construction is virtually impossible. For the great majority of common chemicals, pump manufacturers publish data on materials of construction. In general, packed plunger and hydraulically actuated diaphragm pumps are available as standard construction in mild steel, cast or ductile iron, stainless steel, and plastic. Mechanically actuated diaphragm pumps are usually available as standard construction in plastic and stainless steel. Almost any combination of materials can be furnished on special order.

Diaphragms are available as standard construction in Teflon, chemically resistant elastomers, and stainless steel from various manufacturers.

See also Sec. 3.3, Diaphragm Pumps, and Sec. 3.7, Displacement Pump Flow Control.

SECTION 9.17
SOLIDS PUMPING

9.17.1
HYDRAULIC TRANSPORT OF SOLIDS

R. A. HILL
P. E. SNOEK
R. L. GANDHI

Transportation of solids by slurry pipelines has a wide variety of industrial applications. These include dredging; conveyance of coal, ores, mineral concentrates, paper pulp, and concrete; disposal of tailings, fly ash, and other industrial waste products; and pipelining of municipal solid waste and sludges. Slurry pipelines are also used in in-plant applications involving agriculture, oil, food, chemicals, and mining. Ocean mining and the use of coal-oil and coal-water mixtures in power plants are other potential slurry applications receiving increased attention.

SLURRY CHARACTERISTICS

Depending upon the properties of the solid, properties of the liquid, pipe size, and flow velocity, different regimes, or modes, of flow can develop. Figure 1 is a qualitative representation of the various flow regimes. The major modes are homogeneous flow, heterogeneous flow, and flow with stationary bed. In addition, some researchers list flow with moving bed as a separate regime.[1-4] The solid-liquid interactive mechanisms are different in the different flow regimes. If there is more than one size of solid, smaller particles may be transported in homogeneous flow and larger particles in heterogeneous flow. This mode is referred to as compound flow.[5]

In the homogeneous flow regime, the solid particles are homogeneously distributed in the liquid. Homogeneous flow, or a close approximation to it, is encountered in slurries of high solids concentration and fine particle size. Such slurries often exhibit non-Newtonian rheology. The particle settling velocities are insignificant compared with the turbulent motion of the liquid. Typical examples of slurries which give rise to homogeneous flow at normal pipeline velocities—4 to 7.5 ft/s (1.2 to 2.3 m/s)—are sewage sludge, drilling muds, and concentrated suspensions of finely ground limestone and mineral concentrates.

Newitt et al.[1] proposed the following formula for estimating the velocity V_H in feet (meters) per second above which homogeneous flow occurs:

$$V_H = (1800 \, gDw)^{1/3} \tag{1}$$

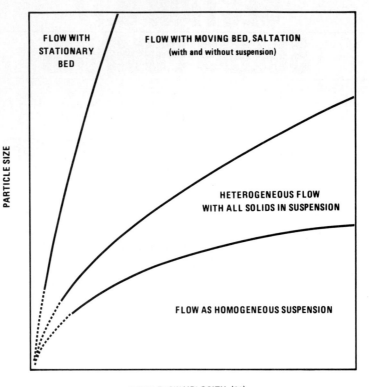

MEAN FLOW VELOCITY (V)

FIG. 1   Flow regimes for a given fluid, sediment, and pipe size. (The representation is qualitative.)

where g = acceleration due to gravity, 32.17 ft/s² (9.807 m/s²)
  D = pipe diameter, ft (m)
  w = settling velocity of particle in quiescent liquid, ft/s (m/s)

Wasp et al.[5] consider the flow to be homogeneous when the ratio of the concentration of solids C at 0.98 diameters above the pipe bottom to the concentration C_A at the axis of the pipe exceeds 0.8. The value of C/C_A is given as follows:

$$\log_{10} C/C_A = -1.8/BKU^* = -1.8/(BK \sqrt{\gamma_w/\rho_m}) \qquad (2)$$

where B = ratio of mass transfer coefficient to momentum transfer coefficient ($\simeq 1$)
  K = von Karman constant = 0.4
  U^* = friction velocity, ft/s (m/s)
   = $\sqrt{\tau_w/\rho_m}$
  τ_w = wall shear stress, lb/ft² (N/m²)
  ρ_m = density of slurry, slugs/ft³ (kg/m³)

In the heterogeneous flow regime, a solids concentration gradient exists along the vertical axis of a horizontal pipe. This type of flow occurs when the solid particles are coarse and of high density and the mean flow velocity is such that the particles cannot be uniformly suspended. Solids containing large particles, such as coarse coal and gravel, will travel in the form of a moving bed near the pipe bottom. Transport of coarse sand, coarse coal, and phosphate rock matrix are typical examples of heterogeneous flow.

Flow with a stationary bed occurs when the flow velocity is not sufficient to keep all the solids in motion. Some solids may be in heterogeneous flow above the stationary bed of solids. This type of flow should be avoided in commercial practice.

DESIGN PARAMETERS

The design of a slurry pipeline system requires the establishment of the following variables:

Solids concentration
Flow velocity
Pipe diameter
Frictional losses

Solids Concentration   All of these variables are governed by the size distribution and specific gravity of the solids. The concentration of solids will often be controlled by the upstream and/or downstream process conditions. If the solids concentration is fixed, the slurry characteristics can be easily determined. The solids concentration should be maintained as high as possible so as to reduce the amount of liquid required. In the case of slurries having a significant amount of solids smaller than 0.0017 in (44 μm) (325 Tyler mesh), a knowledge of the relationship between slurry rheology and concentration may be used to select a pumpable concentration. Figure 2

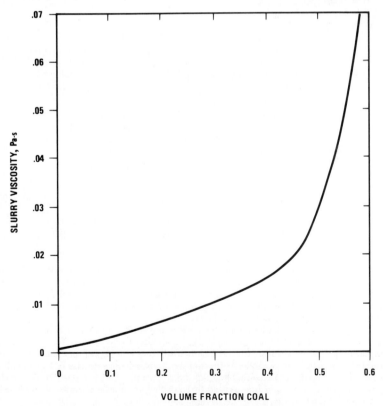

FIG. 2   Viscosity of coal slurry. (1 Pa·s = 2.089 × 10^{-2} lb·s/ft^2)

shows the variation in slurry viscosity with concentration for a typical coal slurry. It shows that beyond a concentration of 0.46 volume fraction solids, the slurry viscosity increases quite rapidly with an increase in solids concentration. It is advantageous to select a maximum slurry concentration below the point at which the rate of viscosity increase with slurry concentration becomes very high.

The slurry volume fraction solids, weight fraction solids, and slurry specific gravity are related as follows:

$$S_m = \phi(S_s - S_l) + S_l \tag{3}$$

$$S_m = \frac{1}{\dfrac{W}{S_s} + \dfrac{1 - W}{S_l}} \tag{4}$$

$$\phi = \frac{S_m}{S_s} W \tag{5}$$

$$\phi = \frac{W/S_s}{\dfrac{W}{S_s} + \dfrac{1 - W}{S_l}} \tag{6}$$

where S_m = slurry specific gravity
    ϕ = volume fraction solids
    S_s = solids specific gravity
    S_l = liquid specific gravity
    W = weight fraction solids

EXAMPLE 1   A 1.06-qt (1-liter) container is used to determine slurry concentration. The weight of the container full of slurry is 4.8 lb (2.20 kg), and the weight of the container full of water is 2.6 lb (1.20 kg). If the specific gravity S_s of the solids is 5.0, determine the volume and weight fraction solids.

Weight of 1.06 qt (1 liter) of water	= 2.2 lb (1.00 kg)
Weight of container full of water	= 2.6 lb (1.20 kg)
Weight of empty container	= 0.4 lb (0.20 kg)
Weight of container full of slurry	= 4.8 lb (2.20 kg)
Weight of 1.06 quart (1 liter) of slurry	= 4.4 lb (2.00 kg)

Specific gravity of slurry $S_m \qquad = \dfrac{4.4}{2.2} = 2.0$

Volume fraction solids $\phi \qquad = \dfrac{S_m - 1}{S_s - 1} = \dfrac{2 - 1}{5 - 1} = \dfrac{1}{4} = 0.25$

Weight fraction solids $W = \dfrac{S_s}{S_m}\phi = \dfrac{5}{2} \times 0.25 = 0.625$

Slurry Rheology   The presence of solids in a liquid increases the viscosity of the liquid. Theoretical formulas for determining this viscosity change have been derived for uniform-size particles at very low concentrations. In most practical applications the size of solids will not be uniform, and in most cases the solids concentration will be quite high. A rheometer may be used to determine the rheological properties of a slurry.

Rheology is the relationship between shear stress and shear strain under laminar flow conditions for a liquid. Rotational, or capillary tube, viscometers may be used for determining the rheological properties of a slurry. Figure 3 shows different rheological relationships commonly encountered in practice. At low solids concentration, most slurries give rise to Newtonian rheology. As shown in Fig. 3, for a Newtonian fluid, the relationship between shear stress τ and shear rate $\dot{\gamma}$ is expressed by a straight line passing through the origin. The viscosity of the Newtonian slurry is given by the slope $d\tau/d\dot{\gamma}$ of the straight line.

Slurries containing significant amounts of fines exhibit non-Newtonian rheological behavior at high concentrations. Most of these slurries give rise to shear-thinning behavior expressed by the

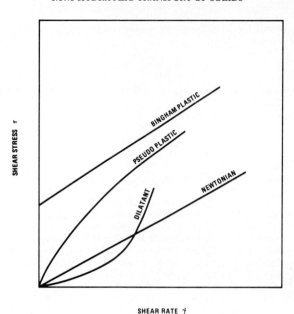

SHEAR RATE $\dot{\gamma}$

FIG. 3   Shear stress versus shear rate for several rheological models.

Bingham plastic or the pseudoplastic rheological model. Slurries made up of uniform-size particles may give rise to shear-thickening (dilatant) behavior at very high concentrations.

In many applications, a Bingham plastic rheological model is adequate. The plastic viscosity (coefficient of rigidity) and the yield stress of the slurry are determined from a plot of shear stress versus shear rate. The Bingham plastic model is expressed by the equation

$$\tau = \tau_y + \eta\dot{\gamma} \tag{7}$$

where τ = shear stress required to maintain a shear rate $\dot{\gamma}$, lb/ft^2 (N/m^2)
   τ_y = yield stress, lb/ft^2 (N/m^2)
   η = plastic viscosity, lb·s/ft^2 or slugs/ft·s (N·s/m^2 or Pa·s)
   $\dot{\gamma}$ = shear rate, s^{-1}

For a circular tube and a Newtonian liquid, the relationship between wall shear stress and shear rate is given by the Poiseuille formula:

$$\tau_w = \mu\,\frac{8V}{D} \tag{8}$$

where τ_w = wall shear stress, lb/ft^2 (N/m^2)
   μ = slurry viscosity, lb·s/ft^2 or slugs/ft·s (N·s/m^2 or Pa·s)
   V = average flow velocity, ft/s (m/s)
   D = tube diameter, ft (m)

For a Bingham plastic slurry, the relationship between wall shear stress and shear rate is given by the Buckingham equation:

$$\eta\,\frac{8V}{D} = \tau_w\left[1 - \frac{4\tau_y}{3\tau_w} + \frac{1}{3}\left(\frac{\tau_y}{\tau_w}\right)^4\right] \tag{9a}$$

which can be approximated by the equation

$$\tau_w = \frac{8V}{D}\left(\frac{\tau_y D}{6V} + \eta\right) \tag{9b}$$

Comparing Eq. 9b with the Newtonian rheological equation (Eq. 7), we can define the effective viscosity of a Bingham plastic fluid:

$$\mu_e = \frac{\tau_y D}{6V} + \eta \tag{10}$$

where μ_e is the effective viscosity. Since μ_e decreases as the shear rate, $8V/D$, increases, a Bingham plastic is called a shear-thinning liquid.

Equation 9b can be written as follows:

$$\tau_w = \frac{4\tau_y}{3} + \eta \frac{8V}{D} \tag{11}$$

Equation 11 shows that if τ_y is large, the wall shear stress will depend upon the velocity of flow to a much smaller extent than in the case of a Newtonian fluid.

Flow Velocity   The flow velocity should be such that solid particles do not form a bed on the bottom of a horizontal pipe. Since the particles in a slurry are kept in suspension by turbulence in the liquid, it is evident that the flow in a slurry pipeline should be turbulent. In certain in-plant applications with nonsettling slurries, laminar flow transport may be used.

The flow velocity should be higher than the laminar turbulent transition velocity. The transition velocity is defined by the critical Reynolds number below which laminar flow occurs. In the case of a Newtonian slurry, the Reynolds number should be maintained above 2000. Hanks[6] has developed a relationship between the critical Reynolds number N_{Re_c} and the Hedstrom number N_{He}, shown in Fig. 4 for Bingham plastic liquids. The Reynolds and Hedstrom numbers are given as follows:

$$N_{Re} = VD\rho/\eta \tag{12}$$

$$N_{He} = \frac{D^2 \rho \tau_y}{\eta^2} \tag{13}$$

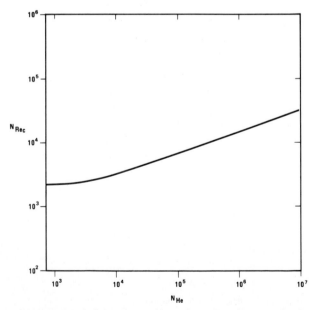

FIG. 4   Variation of critical Reynolds number with Hedstrom number for flow of Bingham plastic slurries in pipes.

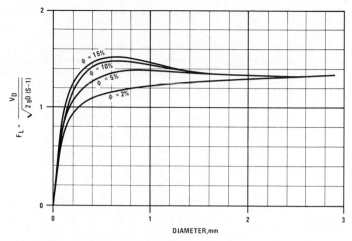

FIG. 5 Variation of F_L with slurry concentration and particle diameter (1 mm = 0.0394 in)

Even in the presence of turbulence, solids may settle out and form a bed. The velocity below which a bed of solids forms is called the deposition velocity. For uniform-size particles, the deposition velocity is estimated using Durand's formula:

$$V_D = F_L/\sqrt{2gD(S_s - 1)} \qquad (14)$$

where V_D = deposition velocity, ft/s (m/s)
$\quad F_L$ = a coefficient as shown in Fig. 5
$\quad g$ = acceleration due to gravity, 32.17 ft/s² (9.807 m/s²)
$\quad D$ = pipe diameter, ft (m)
$\quad S_s$ = solids specific gravity

Thomas[7] has given the following equations for estimating the deposition velocity:

$$\frac{w}{U_D^*} = 0.01\left(\frac{dU_D^*\rho_m}{\mu}\right)^{2.71} \qquad \text{if } d/\delta \le 1 \qquad (15)$$

$$\frac{w}{U_0^*} = 4.9\left(\frac{dU_0^*\rho_m}{\mu}\right)\left(\frac{\mu}{\rho_m DU_0^*}\right)^{0.6}(S_s - 1)^{0.23} \qquad \text{if } d/\delta > 1 \qquad (16)$$

$$\frac{U_D^* - U_0^*}{U_0^*} = 2.8\left(\frac{w}{U_*^*}\right)^{1/3}\tau^{1/2} \qquad (17)$$

where w = settling velocity of solid, ft/s (m/s)
$\quad U_D^*$ = frictional velocity at deposition, ft/s (m/s)
$\quad d$ = particle diameter, ft (m)
$\quad \rho_m$ = slurry density, slugs/ft³ (kg/m³)
$\quad \mu$ = slurry viscosity, lb·s/ft² or slugs/ft·s (N·s/m² or Pa·s)
$\quad \delta = \dfrac{5\mu}{\rho_m U_D^*}$ = thickness of laminar sublayer, ft (m)
$\quad U_0^*$ = frictional velocity at deposition for the limiting condition of infinite dilution, ft/s (m/s)
$\quad D$ = pipe diameter, ft (m)
$\quad S_s$ = solids specific gravity

A velocity in the range of 4 to 7 ft/s (1.2 to 2.3 m/s) is usually practical and economical. The frictional loss in a pipe increases approximately as the square of the flow velocity. It is therefore desirable to maintain the velocity as low as possible. Pipe abrasion may occur at velocities above about 10 ft/s (3 m/s).

EXAMPLE 2 Determine the frictional velocity at which deposition will occur in a 12-in-diameter (300-mm) pipeline carrying iron concentrate slurry at a concentration of 0.25 volume fraction solids. The mean particle diameter is 0.0016 in (40 μm), slurry viscosity is 2.088×10^{-4} slug/ft·s (0.01 Pa·s), solids settling velocity is 3.94×10^{-4} ft/s (0.00012 m/s), and solids specific gravity is 5.0.

Try Eq. 15:

$$\text{Slurry specific gravity} = 0.25(5 - 1) + 1 = 2$$
$$\text{Slurry density} = 3.88 \text{ slugs/ft}^3 \ (2000 \text{ kg/m}^3)$$
$$\text{Particle diameter} = 0.0016 \text{ in} = 0.00013 \text{ ft } (4 \times 10^{-5} \text{ m})$$

in USCS units
$$\frac{dU_D^*\rho_m}{\mu} = \frac{1.3 \times 10^{-4} U_D^* \times 3.88}{2.088 \times 10^{-4}} = 2.42 U_D^*$$

$$\frac{w}{U_D^*} = 0.01(2.42 U_D^*)^{2.71}$$

Since $w = 3.94 \times 10^{-4}$ ft/s,

$$U_D^* = 0.22 \text{ ft/s}$$
$$\delta = \frac{5 \times 6.72 \times 10^{-3}}{0.22 \times 124.7} = 1.22 \times 10^{-3} \text{ ft}$$
$$\frac{d}{\delta} = \frac{0.00013}{0.00122} < 1$$

in SI units
$$\frac{dU_D^*\rho_m}{\mu} = \frac{4 \times 10^{-5} \times U_D^* \times 2000}{0.01} = 8 U_D^*$$

$$\frac{w}{U_D^*} = 0.01(8 U_D^*)^{2.71}$$

Since $w = 1.2 \times 10^{-4}$ m/s,

$$U_D^* = 0.067 \text{ m/s}$$
$$\delta = \frac{5 \times 0.01}{0.067 \times 2000} = 3.73 \times 10^{-3} \text{ m}$$
$$\frac{d}{\delta} = \frac{4 \times 10^{-5}}{3.73 \times 10^{-3}} < 1$$

Therefore Eq. 15 is applicable, and

$$U_D^* = 0.22 \text{ ft/s } (6.6 \times 10^{-2} \text{ m/s})$$

EXAMPLE 3 Recompute Example 2 assuming that the mean particle diameter is 0.0118 in (300 μm), slurry viscosity is 4.177×10^{-5} slug/ft·s (0.002 Pa·s), and solids settling velocity is 0.13 ft/s (0.04 m/s).

Try Eq. 15:

$$(U_D^*)^{3.71} = 100w\left(\frac{\mu}{d\rho_m}\right)^{2.71}$$

in USCS units
$$(U_D^*)^{3.71} = 100 \times 0.13\left(\frac{4.177 \times 10^{-5}}{9.8 \times 10^{-4} \times 3.88}\right)^{2.71} = 6.38 \times 10^{-5}$$

$$U_D^* = 0.074 \text{ ft/s}$$
$$\delta = \frac{5\mu}{U_D^*\rho_m} = \frac{5 \times 4.177 \times 10^{-5}}{0.074 \times 3.88} = 7.27 \times 10^{-4} \text{ ft}$$
$$\frac{d}{\delta} = \frac{9.8 \times 10^{-4}}{7.27 \times 10^{-4}} > 1$$

in SI units $\quad (U_D^*)^{3.71} = 100 \times 0.04 \left(\dfrac{0.002}{3 \times 10^{-4} \times 2000} \right)^{2.71} = 7.75 \times 10^{-7}$

$$U_D^* = 0.0225 \text{ m/s}$$

$$\delta = \frac{5 \times 0.002}{0.0225 \times 2000} = 2.22 \times 10^{-4} \text{ m}$$

$$\frac{d}{\delta} = \frac{3 \times 10^{-4}}{2.22 \times 10^{-4}} > 1$$

Since d/δ is greater than unity, Eq. 15 is not applicable and it is necessary to try Eq. 16:

$$(U_0^*)^{1.4} = \frac{w \mu^{0.4} D^{0.6}}{4.9 d \, \rho_m^{0.4} (S_s - 1)^{0.23}}$$

in USCS units $\quad (U_0^*)^{1.4} = \dfrac{0.13 \times (4.177 \times 10^{-5})^{0.4} \times 1^{0.6}}{4.9 \times 9.8 \times 10^{-4} \times 3.88^{0.4} \times 4^{0.23}} = 0.2$

$$U_0^* = 0.32 \text{ ft/s}$$

Since

$$\frac{U_D^* - U_0^*}{U_0^*} = 2.8 \left(\frac{w}{U_0^*} \right)^{1/3} \phi^{1/2} = 1.0$$

$$\therefore \; U_D^* = 0.64 \text{ ft/s}$$

in SI Units $\quad (U_0^*)^{1.4} = \dfrac{0.04 \times 0.002^{0.4} \times 0.3^{0.6}}{4.9 \times 3 \times 10^{-4} \times 2000^{0.4} \times 4^{0.23}} = 0.038$

$$U_0^* = 0.097 \text{ m/s}$$

$$U_D^* = 0.194 \text{ m/s}$$

Pipe Diameter For a given solids through-put and slurry concentration, the slurry flow rate is fixed. A pipe diameter is selected such that the flow velocity is greater than both the laminar turbulent transition velocity and the deposition velocity. The following example illustrates the method of computation.

EXAMPLE 4 An iron concentrate slurry pipeline is required to transport 528 tons/h (480 metric tons/h) of iron concentrate at a concentration of 0.25 volume fraction solids. Determine the pipe diameter if the minimum operating velocity is 4.9 ft/s (1.5 m/s). The solids specific gravity is 5.0.

in USCS units $\quad$ Solids flow rate $= \dfrac{528 \times 2000}{3600 \times 5 \times 62.4} = 0.94 \text{ ft}^3/\text{s}$

Therefore

$$\text{Slurry flow rate} = 3.76 \text{ ft}^3/\text{s}$$

Since the flow velocity is 4.9 ft/s,

$$\text{Required pipe area} = 0.77 \text{ ft}^2$$
$$\text{Required pipe diameter} = 1 \text{ ft}$$

in SI units $\quad$ Solids flow rate $= \dfrac{480 \times 1000}{5 \times 1000} = 96 \text{ m}^3/\text{h}$

Therefore

$$\text{Slurry flow rate} = 384 \text{ m}^3/\text{h} = 0.11 \text{ m}^3/\text{s}$$

Since the flow velocity is 1.5 m/s,

$$\text{Required pipe area} = 0.071 \text{ m}^2$$
$$\text{Required pipe diameter} = 0.3 \text{ m}$$

ESTIMATE OF FRICTIONAL LOSSES _____

Homogeneous Flow   Depending upon the particle size and solids concentration, a slurry may exhibit Newtonian or non-Newtonian properties. Non-Newtonian flow is normally found with concentrated suspensions containing particles smaller than 0.002 in (50 μm). With Newtonian suspensions, the frictional loss is given as follows[1]:

$$\frac{i - i_w}{i_w \phi} = S_a - 1 \tag{18}$$

where i = frictional loss, feet (meters) of water per foot (meter) of pipe for slurry
   i_w = frictional loss, feet (meters) of water per foot (meter) of pipe for water at the same flow velocity
   S_s = solids specific gravity

Frictional losses can also be calculated using charts of friction factors and Reynolds numbers. The Reynolds number should be defined using the density and viscosity of the slurry. The rheological properties of non-Newtonian slurries are generally expressed by one of the following three models:

$$\text{Power law} \qquad \tau = K\dot{\gamma}^n \tag{19}$$

$$\text{Bingham plastic} \qquad \tau = \tau_y + \eta\dot{\gamma} \tag{20}$$

$$\text{Yield pseudoplastic} \qquad \tau = \tau_y + K\dot{\gamma}^n \tag{21}$$

where τ = shear stress required to maintain a shear rate $\dot{\gamma}$, lb/ft^2 (Pa)
   K = consistency index
   n = flow behavior index
   τ_y = yield stress, or shear stress require to initiate motion, lb/ft^2 (Pa)
   η = plastic viscosity, lb$\cdot$s/ft^2 or slugs/ft$\cdot$s (N$\cdot$s/m^2 or Pa$\cdot$s)

Dodge and Metzner[8] have defined a generalized Reynolds number such that the laminar pipe flow data are defined by the Newtonian relationship between friction factor and Reynolds number as follows:

$$f = 16/N_{Re}^* \tag{22}$$

$$N_{Re}^* = D^n V^{2-n} \rho_m / \zeta \tag{23}$$

$$\zeta = \frac{K(1 + 3n)^n 8^{n-1}}{4n} \tag{24}$$

where f = friction factor
   V = average flow velocity, ft/s (m/s)

For turbulent flow Dodge and Metzner have developed the following semitheoretical equation based on their data:

$$\frac{1}{f} = \frac{4}{n^{0.75}} \log\left(N_{Re}^* f^{1-n/2} - \frac{0.4}{n^{1.2}} \right) \tag{25}$$

This relationship is shown in graphical form in Fig. 6.
   The transition from laminar to turbulent flow occurs between generalized Reynolds number of 1900 and 2600. Hanks[6] has developed a method for estimating the transition Reynolds number.
   The Bingham plastic rheological model is more convenient than the pseudoplastic model because the friction factor for turbulent flow can be estimated using the Newtonian liquid flow curves, provided that the Reynolds number is defined as follows:

$$N_{Re} = VD\rho_m/\eta \tag{26}$$

Thomas[9] found that the curves of turbulent friction factor versus Reynolds number for the Bingham plastic suspensions he tested were lower than those given by the Newtonian relationship.

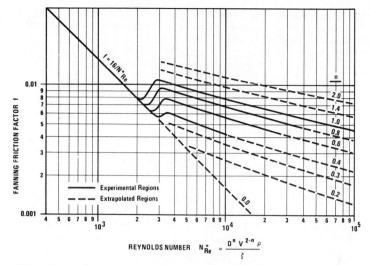

FIG. 6 Friction factor verus generalized Reynolds number diagram for pseudoplastic fluids.

He attributed this reduction in friction factor to the thickening of the laminar layer due to yield stress. Hanks and Dadia[10] have developed a friction factor versus Reynolds number diagram based on semitheoretical considerations (Fig. 7). The transition from laminar to turbulent flow can be estimated using the method proposed by Hanks.[6]

Hanks[11] has developed friction factor versus Reynolds number diagrams for yield psuedoplastic slurries. Cheng[12] has used this model successfully in predicting frictional losses for homogeneous suspensions.

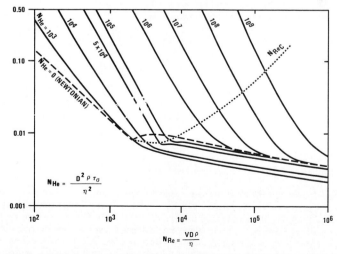

FIG. 7 Friction factor versus Reynolds number diagram for Bingham plastic fluids.

Heterogeneous Flow   The formula proposed by Durand[13] on the basis of extensive research carried out in France has been widely used for estimating frictional losses for heterogeneous flow of slurries. Durand's formula is

$$\frac{i - i_w}{\phi i_w} = \left[\frac{V^2 \sqrt{C_D}}{(S_s - 1)gD} \right]^{-1.5} \times 81 \tag{27}$$

where i = frictional loss for slurry, feet (meters) of water per foot (meter) of pipe
      i_w = frictional loss for water, feet (meters) of water per foot (meter) of pipe
      ϕ = volume fraction solids in slurry
      V = flow velocity, ft/s (m/s)
      C_D = drag coefficient of particles settling in fluid of infinite extent
      S_s = particle specific gravity
      g = acceleration due to gravity, 32.17 ft/s² (9.807 m/s²)
      D = pipe diameter, ft (m)

Zandi and Govatos[14] concluded from an examination of 2549 datum points that Durand's equation predicted observed head losses fairly well once the saltation data were separated from the heterogeneous flow data. Newitt et al.[1] have proposed the following relationship based on their experiments with a 1-in-diameter (25.4-mm) pipe:

$$\frac{i - i_w}{\phi i_w} = 1100 \times \frac{gD}{V^2} \times \frac{w}{V} \times (S_s - 1) \tag{28}$$

Turian and Yuan[15] have developed the following equation based on their work:

$$f - f_w = 0.5513 C^{0.8687} f_{fw}^{1.2} C_D^{-0.1677} \left[\frac{V^2}{gD(S_s - 1)} \right]^{-0.6938} \tag{29}$$

where $f = \dfrac{\Delta P}{2L\rho_m V^2}$ = friction factor for slurry

$\dfrac{\Delta P}{L}$ = frictional loss per unit length of pipe

f_w = friction factor for water

Saltation Flow   Long-distance slurry pipelines should not be designed to flow in this regime. However, coarse solids transport might produce saltation flow conditions.
     Newitt et al.[1] have developed the following formula based on their experiments:

$$\frac{i - i_w}{i_w \phi} = 66(S_s - 1)\frac{gD}{V^2} \tag{30}$$

Babcock[3] has proposed the following formula based on his tests:

$$\frac{i - i_w}{i_w \phi} = 60.6(S_s - 1)\frac{gD}{V^2} \tag{31}$$

Turian and Yuan[15] have proposed the following formula based on their tests:

$$f - f_w = 0.9857 \phi^{1.019} f_w^{1.046} C_D^{-0.4213} \left(\frac{V^2}{gD(S_s - 1)} \right)^{-1.354} \tag{32}$$

Of these, Newitt et al.'s formula has been used most.

Frictional Loss for Mixed Sizes   Most of the formulas for heterogeneous flow were developed for particles of uniform size. When a mixture of two or more size fractions is pumped, it is necessary to determine an average particle diameter for use in the various formulas. Use of a weighted average diameter, or drag coefficient, has been suggested by some researchers. A better approach is to divide the solid particles into two fractions: one that is carried in homogeneous flow and one that is carried in heterogeneous and saltation flow. The frictional losses for each fraction are computed separately using appropriate formula and then added together to obtain the total slurry frictional loss. Wasp et al.[5] and Gaessler[16] have successfully used this approach for correlating coal slurry data.

The method proposed by Wasp is an iterative procedure which works as follows:

1. Divide the total size fraction into a homogeneous part and a heterogeneous part.

2. Compute the frictional losses for the homogeneous part using the rheological properties of the slurry. Compute the frictional losses for the heterogeneous part using Durand's formula. The sum of the two parts gives an initial estimate of the slurry frictional losses.

3. Determine the C/C_A value of each size fraction based on the frictional losses estimated in step 2.

4. Based on the C/C_A value, determine the fraction of solids in the homogeneous phase and in the heterogeneous phase.

5. Recompute the frictional loss for the slurry as in step 2. If the new estimate closely agrees with the initial estimate, then use the new estimate as the final value of slurry frictional loss. If the new estimate significantly differs from the initial estimate, then repeat steps 3 through 5, each time using the newest estimate, until the difference between the initial and final values of estimated frictional losses is as small as desired.

Frictional Loss in Vertical Pipes In vertical pipe flow, there is no concentration gradient, and the slurry flow may be treated as homogeneous flow. For coarse particles, the frictional loss for the slurry has been found to be the same as that for water at the same velocity. For fine particles, the viscosity of the slurry should be considered in computing the frictional losses.

Frictional Losses in Inclined Pipes Worster and Denny[17] have proposed the following equation relating the frictional loss for inclined pipes with that for a horizontal pipe:

$$i_\theta = i_w + (i - i_w) \cos \theta \tag{33}$$

where i_θ = frictional loss in inclined pipe
  θ = angle of inclination of the pipe from horizontal

This equation suggests that the frictional losses in an inclined pipe are the same for both up and down flow and that they are smaller than the losses in a horizontal pipe. Experimental evidence presented by Kao and Hwang[18] shows that, in an inclined pipe with up flow, the frictional loss first increases and then decreases after the angle of incline reaches a certain magnitude. In the case of down flow, the frictional loss is less than that in a horizontal pipe.

EXAMPLE FRICTION CALCULATIONS _____

EXAMPLE 5 Determine the frictional loss and pump pressure requirement for transporting 528 tons/h (480 metric tons/h) of iron concentrate at a concentration of 0.25 volume fraction solids at a velocity of 4.9 ft/s (1.5 m/s). Assume that the slurry characteristics are the same as those given in Example 2. The pipeline is 12.4 mi (20 km) long. The elevation at the inlet is 656 ft (200 m), and that at the outlet is 492 ft (150 m). The pipe roughness is 0.002 in (0.05 mm), and the slurry viscosity is 2.089×10^{-4} slug/ft·s (0.01 Pa·s).

In Example 4 it was shown that a 12-in-diameter (0.3-m) pipe is required for the given through-put, slurry concentration, and flow velocity.

in USCS units $$N_{Re} = \frac{VD\rho}{\mu} = \frac{4.9 \times 1 \times 3.88}{2.08 \times 10^{-4}} = 90,000$$

$$\frac{\text{Pipe roughness}}{\text{Pipe diameter}} = \frac{0.002}{12} = 0.00017$$

$$f = 0.00479$$
$$U^* = V\sqrt{f/2} = 4.9\sqrt{0.00479/2} = 0.24 \text{ ft/s}$$

in SI units
$$N_{Re} = \frac{VD\rho}{\mu} = \frac{1.5 \times 0.3 \times 2000}{0.01} = 90{,}000$$

$$\frac{\text{Pipe roughness}}{\text{Pipe diameter}} = \frac{0.05}{300} = 0.00017$$

$$f = 0.00479$$
$$U^* = V\sqrt{f/2} = 1.5\sqrt{0.00479/2} = 0.073 \text{ m/s}$$

In Example 2, U_D^* was calculated to be 0.22 ft/s (0.067 m/s), which is less than the value of $U°$ corresponding to the operating velocity. Therefore, the operating velocity is greater than the deposition velocity. The frictional loss is

in USCS units
$$\frac{4fV^2}{2gD} = \frac{4 \times 0.00479 \times 4.9 \times 4.9}{2 \times 32.17 \times 1} = 0.0072 \text{ ft/ft}$$

Since the pipe length is 12.4 mi,

Total frictional loss = $0.0072 \times 12.4 \times 5280 = 471$ ft
Static head $\quad$ = outlet elevation − inlet elevation = $492 - 656 = -164$ ft
Pumping pressure = frictional loss + static head = $471 - 164 = 307$ ft

in SI units
$$\frac{4fV^2}{2gD} = \frac{4 \times 0.00479 \times 1.5 \times 1.5}{2 \times 9.807 \times 0.3} = 0.0072 \text{ m/m}$$

Since the pipe length is 20 km,

Total frictional loss = $0.0072 \times 20 \times 1000 = 144$ m
Static head $\quad$ = outlet elevation − inlet elevation = $150 - 200 = -50$ m
Pumping pressure = frictional loss + static head = $144 - 50 = 94$ m

The slurry specific gravity is 2.0, and therefore each foot (meter) of slurry head equals 0.866 lb/in² (19.6 kPa) pressure. Therefore

in USCS units $\qquad$ Pump pressure = 266 lb/in²
in SI units $\qquad$ Pump pressure = 1842 kPa

Four rubber-lined centrifugal slurry pumps operating in series may be used to develop the required pumping pressure. Assuming an efficiency of 65%, the required power for the pump drive can be estimated as follows:

in USCS units
$$\text{hp} = \frac{\text{flow in gpm} \times \text{pressure in lb/in}^2}{1714 \times \text{efficiency}}$$

$$\text{Flow in gpm} = \frac{\text{velocity in ft/s} \times (\text{diameter in in})^2}{0.4085} = 1727 \text{ gpm}$$

$$\text{hp} = \frac{1727 \times 266}{1714 \times 0.65} = 412$$

in SI units
$$\text{kW} = \frac{\text{flow in m}^3/\text{h} \times \text{pressure in kPa}}{3600 \times \text{efficiency}} = \frac{384 \times 1842}{3600 \times 0.65} = 302$$

EXAMPLE 6 Determine the pump discharge pressure required to pump a sand-water slurry for which the volume fraction solids is 0.2 and the solids specific gravity is 2.7. Solids properties are as follows:

Mean particle diam., in (mm)	Weight %	Settling velocity, ft/s (m/s)	Drag coefficient
0.0039 (1.0)	10	0.52 (0.16)	0.9
0.0031 (0.8)	30	0.43 (0.13)	1.1
0.0020 (0.5)	60	0.23 (0.07)	2.4

Slurry viscosity	$= 4.177 \times 10^{-5}$ slug/ft·s $(2 \times 10^{-3}$ Pa·s)
Pipe diameter	$= 0.66$ ft (0.2 m)
Design velocity	$= 13.1$ ft/s (4 m/s)
Pipe length (including fittings)	$= 984$ ft (300 m) horizontal, 65.6 ft (20 m) vertical
Inlet elevation	$= 984$ ft (300 m)
Outlet elevation	$= 1050$ ft (320 m)

Frictional Loss in Horizontal Pipe

The frictional loss in the horizontal pipe can be estimated using Eq. 27:

$$\frac{i - i_w}{i_w \phi} = 81 \left(\frac{V^2 \sqrt{C_D}}{(S_s - 1)gD} \right)^{-1.5} = 7.7 C_D^{-3/4}$$

Now compute the frictional loss i_w for water:

in USCS units N_{Re} (water) $= \dfrac{DV\rho_w}{\mu} = \dfrac{0.66 \times 13.1 \times 1.93}{2.08 \times 10^{-5}} = 8.0 \times 10^5$

With a pipe roughness of 0.002 in, the fanning friction factor f is 0.00308, and so

$$i_w = \frac{4fV^2}{2gD} = \frac{4 \times 0.00308 \times 13.1^2}{2 \times 32.2 \times 0.66} = 0.05 \text{ ft water per foot of pipe}$$

in SI units N_{Re} (water) $= \dfrac{DV\rho_w}{\mu} = \dfrac{0.2 \times 4 \times 1000}{10^{-3}} = 8.0 \times 10^5$

With a pipe roughness of 0.05 mm, the fanning friction factor f is 0.00308, and so

$$i_w = \frac{4fV^2}{2gD} = \frac{4 \times 0.00308 \times 4^2}{2 \times 9.8 \times 0.2} = 0.05 \text{ m water per meter of pipe}$$

Particle size, in (mm)	Wt %	ϕ	C_D	$C_D^{-3/4}$	$7.7\phi C_D^{-3/4}$	$i - i_w$
0.0039 (1)	10	0.025	0.9	1.08	0.208	0.010
0.0031 (0.8)	30	0.075	1.1	0.93	0.538	0.027
0.0020 (0.5)	60	0.150	2.4	0.52	0.599	0.030
						0.067

Frictional loss (horizontal) $= i_w + (i - i_w)$
$= 0.177$ ft (m) water per foot (meter) of pipe

Frictional Loss in Vertical Pipe

in USCS units Sp. gr. of slurry $= 0.2 \times 1.7 + 1 = 1.34$
$\rho_m = 1.34 \times 1.93 \times 2.59$ slugs/ft³

N_{Re} (slurry) $= \dfrac{DV\rho_m}{\mu} = \dfrac{0.66 \times 13.1 \times 2.59}{4.177 \times 10^{-5}} = 5.36 \times 10^5$

$f = 0.00397$

in SI units Sp. gr. of slurry $= 0.2 \times 1.7 + 1 = 1.34$
$\rho_m = 1.34 \times 1000 = 1340$ kg/m³

N_{Re} (slurry) $= \dfrac{DV\rho_m}{\mu} = 5.36 \times 10^5$

$f = 0.00397$

The frictional loss in the vertical pipe is then

$$\frac{4fV^2}{2gD} = 0.065 \text{ ft (m) slurry per foot (meter) of pipe}$$

$$= 0.087 \text{ ft (m) water per foot (meter) of pipe}$$

and the total frictional loss is

in USCS units

Total loss = $(0.117 \times 984) + (0.087 \times 65.6)$
= 120 ft water
= 90 ft slurry

in SI units

Total loss = $(0.117 \times 300) + (0.087 \times 20)$
= 36.8 m water
= 27.5 m slurry

The static head is

Outlet elevation − inlet elevation = 66 ft (20 m) slurry

The pump discharge head required is thus 156 ft (47.5m) slurry.

PIPING SYSTEM

When laying out a slurry piping system, the designer must consider:

- Piping flushing or draining on normal or emergency shutdown
- Replacement of high-wear areas near pump discharge, sharp bends, at and downstream of restrictions
- Rotation of straight horizontal sections (very coarse slurries)
- Access for unplugging
- Elimination of dead spaces at tees and tappings

These considerations are very important with heterogeneous slurries, and even with relatively fine slurries, general pipe abrasion becomes a consideration above about 10 ft/s (3 m/s) and is a major consideration at velocities above 13 ft/s (4 m/s). Rubber or polyurethane lining may be desirable for these situations. Extra-long-radius elbows are often used to minimize wear at changes of direction. Couplings that require grooved pipe ends are generally not used because they introduce a weak point when abrasion is a factor.

Valves, like pumps, must be designed for abrasive service. They preferably provide a full opening, do not depend on machined metal surfaces for closure, and do not have dead pockets that can fill with solids and restrict operation. Valves with restricted or circuitous openings create abrasion downstream. Many suitable low-pressure valves are available, often with rubber-to-rubber or metal-to-rubber sealing.

To date, high-pressure, large-diameter applications have used lubricated plug or ball valves, neither of which is ideally suited for slurry service. In some cases, plug valves are specified with stellite facing on the plug and internal body, to reduce wear. Most ball valves must be fitted with flush-and-drain connections so that solids can be removed from the body after operation.

CORROSION-EROSION

Loss of pipe metal due to corrosion and abrasion plays an important role in the design of a slurry pipeline system. A long-distance pipeline is usually designed to operate under homogeneous flow conditions at velocities below 10 ft/s (3 m/s) to prevent abrasion of the pipe wall. In the absence of abrasion, a mild steel pipeline without internal lining can be used. The mild steel pipe can be protected against corrosion by the use of corrosion inhibitors. Removal of dissolved oxygen and the use of alkaline solutions have been found to control corrosion.[19]

A short pipeline may be designed to operate in the heterogeneous flow regime, where abrasion of pipe wall occurs. Since a major part of the solids flow near the bottom of the pipe, the wear is expected to be maximum in this area. Rotation of the pipe to extend its life is normally used in this situation.

Various types of abrasion-resistant linings may be applied to mild steel pipe to extend its life in an abrasive environment. Rubber and polyurethane have been found to be suitable lining materials. Polyethylene and polybutylene pipes may also be used in corrosive and abrasive environment, provided the maximum operation pressures are below about 200 lb/in^2 (1380 kPa). Another possibility is to use polyethylene pipe as an insert into a steel line.

INSTRUMENTATION

The presence of solids in a slurry system complicates measurement of the system variables because many conventional primary measuring elements will be worn away or plugged by the solids. Wear may affect the measuring elements (such as orifices or turbine blades) or the system itself (e.g., as a result of turbulence created by the measuring element).

The need to avoid plugging of the sensing element or impulse lines is a major consideration in selecting measuring elements for slurry systems. This is a particularly troublesome problem at pressure tappings and can disable any element with small clearances or static zones.

Segregation of a slurry in the measuring element must also be considered. A radiation density meter or magnetic flowmeter will not read correctly in a horizontal pipe where solids are deposited. Fast-settling slurries will foil instruments requiring a side stream if flows are not maintained above critical velocity. Magnetic flowmeters have proved to be adequate, though expensive, primary flow rate indicators. Radiation density meters indicate a slurry concentration, again without flow restriction. However, experience shows that they require frequent calibration. Continuous-weighing devices, such as the Halliburton densometer, have been used, but they present the problem of taking a side stream and returning it to the main stream.

The difficulty in pressure measurement is in keeping the pressure taps from plugging. This is best overcome by using a diaphragm close-mounted to the pipe to separate the slurry from the pressure sensor. In some cases, a continuous backflush of pressure-sensing lines may be required.

Transmitters or gages should not be mounted on the piping around positive displacement pumps because they will often quickly fail from the vibration. Pressure devices can be supported separately with a capillary connecting element and disphragm.

ECONOMICS

Slurry pipelines can offer a number of advantages over competing modes of transport. These include minimum environmental impact, high degree of safety, low labor and energy requirements, and round-the-clock availability. However, the main reason slurry pipelines have been increasingly utilized by the mineral industry in the last two decades is that they offer a more economical means of transport.

Determining Factors For any mineral, the two main physical factors which determine the cost of slurry transport are pipeline distance and required through-put. Concerning the first factor, slurry pipelines become cheaper, measured on a cost per unit weight distance basis, as they become longer. This is because the costs of slurry preparation, water supply, and terminal handling become spread over a longer distance and therefore become relatively less important relative to the costs of pure pipelining.

Also, it should be noted that slurry pipelines are usually installed on a much more direct path between origin and terminal than is possible for alternate transport modes, such as railroads, trucks, and barges. Although there are slope limitations for slurry systems, these are considerably less restrictive than for railroad beds. For example, the Black Mesa coal pipeline was built with a maximum grade of 16% while unit trains seldom operate at slopes exceeding 3%. As a result, with a rail line there can be a "circuitry" increase of as much as 30% greater length.

With respect to through-put, slurry pipelines, like all pipelines in general, exhibit an economy

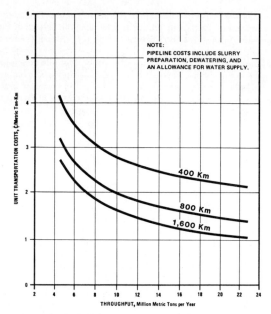

FIG. 8 Pipeline coal transportation cost (first quarter 1981, U.S. basis).

of scale, that is, the unit cost of transportation decreases with increasing volume. This is shown in Fig. 8, which gives some representative pipeline coal transportation costs in the United States. The effect of increasing through-put as well as length is quite evident. Except at very low through-puts, alternate modes of transport are not scale-sensitive; that is, increasing through-put does not result in decreasing unit costs since the capacity of a single unit train, truck, or barge tow is effectively fixed.

In addition to length and through-put, there are a number of other major physical factors which affect the cost of slurry transport. These include

- Route profile, particularly whether there is a net gain or loss in elevation between origin and terminal.
- Construction difficulty. Pipeline construction in urban areas or in areas requiring much rock blasting will obviously increase costs.
- Meteorological factors. Deep frost penetration usually translates into increased pipeline burial depths. Likewise, severe winter weather can also result in extensive station and slurry storage tank insulation needs.
- Size consist and specific gravity. Coarse solids and those with a high specific gravity tend to require higher transport velocities, which may result in increased pipeline wall abrasion. Linings, if required, can greatly increase system costs.
- Slurry corrosion. The material received from the mining facility may require extensive use of slurry inhibitors to limit pipe wall corrosion or avoid installation of a lining.
- Water availability.
- Power availability.
- Environmental restrictions.

Capital Charges   The total cost of slurry transportation is the sum of two components—capital charges and operating expenses. The capital cost includes the direct cost of all materials and installation as well as indirect costs for engineering, procurement, construction management, start-up, and owner's costs. The capital charge is a percentage of the total capital investment that must be paid each year to cover interest on borrowed funds, profit to the equity owners, depreciation charges, etc. These charges typically run from 15 to 25% of the total capital investment. Since large projects are usually financed mostly through debt, a major determinant will be the interest rate.

Operating expenses include the costs of power, labor, supplies, water, *ad valorem* taxes, insurance, legal expenses, royalties, etc. The proportion of expenses charged in each category will vary considerably, depending upon the specific system factors, but power, labor, and supplies will make up the large majority of costs. Typically, for a large, long-distance coal pipeline, they will account for more than 95% of the total annual operating expenses. Power is generally made up of electric power charges, as most pipeline systems constructed to date have used electric motors for driving the mainline pumps. However, diesel engines or even a gas turbine could find application in special situations. Labor costs include the operating and maintenance crews as well as the supporting accounting, legal, engineering, and management staff. Depending upon the degree of automation desired, intermediate pump stations can be operated unmanned, although maintenance help will be required. Supplies include replacement of expendable pump parts, lubricants, valve packing, corrosion inhibitors, etc. For most iron and copper concentrate systems, a high

TABLE 1 Historical Inflation Rates
GNP Deflator and Pipeline Construction Costs and Expenses

Year	GNPD index[a]	Pipeline indices			
		Construction[b]	Labor[c]	Power[d]	Supplies[e]
1961	69.3	182.0	3.27	102.8	91.4
1962	70.6	180.0	3.27	104.0	91.1
1963	71.6	175.0	3.40	103.2	91.3
1964	72.7	168.0	3.46	102.4	93.8
1965	74.3	176.0	3.54	102.4	96.4
1966	76.8	179.0	3.69	101.9	98.8
1067	79.0	181.0	3.85	102.1	100.0
1968	82.6	187.0	4.04	102.9	102.6
1969	86.7	191.0	4.26	104.4	108.5
1970	91.4	198.0	4.57	108.1	116.6
1971	96.0	211.0	4.87	114.7	118.7
1972	100.0	221.0	5.19	122.9	123.5
1973	105.8	235.0	5.50	132.4	132.8
1974	116.0	279.0	6.02	149.5	171.9
1975	127.2	343.0	6.92	181.3	185.6
1976	133.9	361.0	7.47	206.4	195.9
1977	139.8	374.0	8.14	225.8	209.0
1978	150.1	417.0	8.81	253.2	227.1
1979	162.8	458.0	9.52	264.5	259.3
1980	177.5	513.0	10.50	296.8	286.4
Rate of change	5.14	5.99	6.60	5.97	6.40
Rate of change / GNPD rate of change		1.17	1.28	1.16	1.25

Sources: [a]*Survey of Current Business and Economic Report of the President* (annual), Bureau of Economic Analysis, Department of Commerce, Washington, D.C.
[b]*Oil and Gas J.* (August) = I.C.C.
[c]*BLS: Employment and Earnings,* Transportation and Public Utilities Workers, SIC Code 46, Pipeline Transportation, Bureau of Labor Statistics, Washington, D.C.
[d]*WPI,* Industrial Power—West North Central, Code No. 0543-1411
[e]*Wholesale Price Index,* Metals and Metal Products, Code 10

slurry pH is maintained to control pipe wall corrosion. Lime or caustic is typically used. However, in all cases, an oxygen scavenger is also required for water-batching.

Cost Escalation Since a major pipeline system can take upwards of 4 years to complete after an initial feasibility study, and could be in operation for 15 to 50 years thereafter, one additional step must be taken to complete the economic evaluation. This step is to escalate both capital costs and operating expenses so that tariffs can be projected for the life of the pipeline systems. Although economic modeling is the preferred procedure, an escalation methodology that has been often adopted in the past assumes that escalation rates of specific pipeline costs are a function of the general inflation rate, as indicated by the implicit GNP deflator (GNPD). It also assumes that the relationships demonstrated in the past can be projected into the long-term future.[20]

Table 1 gives the historical record for pipeline construction costs, operating and maintenance components, and *ad valorem* taxes and insurance. The 20-year average rate of change shown for each item was calculated by a least squares trend analysis. Similar indices can be used to project cost escalation in processing plants (i.e., slurry preparation and terminal handling facilities). By utilizing these various indices for an assumed general inflation scenario, it is possible to project pipeline capital costs and operating expenses for any particular system.

In computing future tariffs, it must be remembered that there are two distinct periods in the life of a slurry pipeline system. One is the prestart period, when all costs will increase. The other is the period after the project is completed and the system is in operation, at which point the costs

TABLE 2 Worldwide Slurry Pipeline Systems

Material	System name or location	Through-put capacity, million tons/yr (million metric tons/yr)	Length, mi (km)	Start of operations
Coal	Consolidation Coal	1.3 (1.2)	108 (174)	1957[a]
	Russia	4.4 (4.0)	7 (11)	1966
	France (Merlebach)	1.7 (1.5)	6 (9)	1954
	Black Mesa	5.5 (5.0)	273 (439)	1970
Limestone	Columbia	1.0 (0.9)	6 (9)	1963
	Trinidad	0.6 (0.5)	6 (9)	1959
	Rugby Cement	1.7 (1.5)	57 (92)	1964
	Calaveras	1.5 (1.4)	17 (27)	1971
Magnetite	Savage River	2.5 (2.3)	53 (85)	1967
	Waipipi	1.1 (1.0)	4 (6) and 2(3)	1971
	Pena Colorada	1.8 (1.6)	30 (48)	1974
	Las Truchas	1.5 (1.4)	17 (27)	1975
	Sierra Grande	2.1 (1.9)	20 (32)	1977
	Kudremukh	8.2 (7.5)	42 (67)	1980
Hematite	Samarco	13.2 (12.0)	247 (398)	1978
Copper concentrate	Bougainville	1.1 (1.0)	17 (17)	1972
	West Irian	0.3 (0.3)	69 (111)	1973
	Pinto Valley	0.4 (0.4)	11 (18)	1974
Phosphate	Valep	2.2 (2.0)	70 (113)	1979
Gilsonite	Utah	0.4 (0.4)	72 (116)	1957[b]
Gold tailings	South Africa	1.1 (1.0)	21 (34)	N/A[a]
Nickel tailings	Australia (Western Mining)	0.1 (0.1)	4 (7)	1970
Copper tailings	Japan	0.6 (0.5)	44 (71)	1968
Kaolin	Georgia	0.7 (0.6)	16 (25)	N/A[a]

[a]Several systems in existence.

[b]Not currently operating.

related to the capital investment become fixed and only operating expenses (including *ad valorem* taxes and insurance) will escalate. For a long-distance slurry pipeline, initially about 60 to 80% of the pipeline tariff will represent capital-related charges and only a small portion of the tariff will be subject to escalation. Therefore slurry systems, as is true with most pipelines, provide a high degree of inflation protection.

EXISTING COMMERCIAL SYSTEMS

Slurry pipelines have been used for transporting dredgings and storm and sanitary sewage for a number of years. However, it was only in the 1950s that the technology was sufficiently developed to allow their use for long-distance transport in the mineral industry. During the decade from 1950 to 1960, a limestone pipeline was constructed in Trinidad, a gilsonite line in the United States, and coal pipelines in both France and the United States. As shown in Table 2, a large number of other applications have been added since this period, including systems for transporting iron concentrate, phosphate, copper concentrate, various tailings, and kaolin. Today there are more than 30 commercial long-distance pipelines in operation transporting solids for distances up to 273 mi (440 km) and having capacities up to 13×10^6 tons (12×10^6 metric tons) per year.

REFERENCES

1. Newitt, D. M., J. F. Richardson, M. Abbott, and R. B. Turtle: "Hydraulic Conveying of Solids in Pipes," *Trans. Instit. Chem. Eng.* 33:93 (1955).

2. Zandi, I.: *Advances in Solid-Liquid Flow in Pipes and Its Application*, Pergamon Press, New York, 1971.

3. Babcock, H. A.: "The State of the Art of Transporting Solids in Pipelines," paper presented at 48th national meeting of American Institute of Chemical Engineers, Denver, August 1962.

4. Vanoni, V. A.: *Sedimentation Engineering*, Manuals and Reports on Engineering Practice No. 54, ASCE, New York, 1975, pp. 245–278.

5. Wasp, E. J., J. P. Kenny, and R. L. Gandhi: *Solid-Liquid Flow: Slurry Pipeline Transportation*, Trans Tech, Zellerfeld, Germany, 1977.

6. Hanks, R. W.: "The Laminar-Turbulent Transition for Fluids with a Yield Stress," *AIChE.* 9(3):306 (1963).

7. Thomas, D. G.: "Transport Characteristics of Suspensions. Part VI: Minimum Transport Velocity for Large Particle Size Suspensions in Round Horizontal Pipes," *AIChE J.* 8(3):373 (1962).

8. Dodge, D. W., and A. B. Metzner: "Turbulent Flow of Non-Newtonian Systems," *AIChE J.* 5(2):189 (1959).

9. Thomas, D. G.: "Transport Characteristics of Suspensions. Part IV: Friction Loss of Concentrated Flocculated Suspensions in Turbulent Flow," *AIChE J.* 8(2):266 (1962).

10. Hanks, R. W., and B. H. Dadia: "Theoretical Analysis of the Turbulent Flow of Non-Newtonian Slurries in Pipes," *AIChE J.* 17(3):554 (1971).

11. Hanks, R. W.: "Low Reynolds Number Turbulent Pipeline Flow of Pseudohomogeneous Slurries," paper presented at Hydrotransport 5, BHRA Fluid Engineering, Cranfield, Bedford, England, May 1978.

12. Cheng, C. D.-H.: "A Design Procedure for Pipeline Flow of Non-Newtonian Dispersed Systems," paper presented at Hydrotransport 1, BHRA Fluid Engineering, Cranfield, Bedford, England, September 1970.

13. Durand, R.: "Basic Relationship of the Transportation of Solids in Pipes: Experimental Research," *Proc. Minnesota International Hydraulic Conference*, Minneapolis, Minnesota, September 1953, pp. 89–103.

14. Zandi, I., and G. Govatos: "Heterogeneous Flow of Solids in Pipelines," *J. Hydraulic Div. ASCE* **93**(HY3):145 (1967).

15. Turian, R. M., and T. Yuan: "Flow of Slurries in Pipelines," *AIChE J.* **23**(3):232 (1977).

16. Gaessler, H.: "Experimentelle and theoretische Untersuchungen über die Stromungsvorgange beim Transport von Fetstoffen in Flussigkeiten durch horizontale Rohrleitungen," doctoral dissertation, Technische Hochschule, Karlsruhe, West Germany, 1967.

17. Worster, R. L., and D. F. Denny: "Hydraulic Transport of Solid Materials in Pipes," *Proc. Instit. Mech. Eng.* **169**:563 (1955).

18. Kao, D. T. Y., and L. Y. Hwang: "Critical Slope for Slurry Pipeline Transporting Coal and Other Solid Particles, paper presented at Hydrotransport 6, BHRA Fluid Engineering, Cranfield, Bedford, England, September 1979.

19. Gandhi, R. L., B. L. Ricks, and T. C. Aude: "Control of Corrosion-Erosion in Slurry Pipelines," paper presented at 1st International Conference on Internal and External Protection of Pipes, BHRA Fluid Engineering, Cranfield, Bedford, England, September 1972.

20. Cucek, E. M., and P. E. Snoek: "Economics of Long-Distance Coal Slurry Pipelines: Methodology," paper presented at 22d annual meeting of American Association of Cost Engineers, San Francisco, July 1978.

FURTHER READING

Dauber, C. A., and N. F. Gill: "Dewatering Pipeline Coal Slurry," *J. Pipeline Div. ASCE,* October 1959, p. 1.

Fister, L. C., B. C. Finerty, and R. A. Hill: "Valep: The World's First Long-Distance Phosphate Concentrate Slurry Pipeline," paper presented at 4th International Technical Conference on Slurry Transportation, Las Vegas, March 1979.

Frey, D., J. Jonakin, and V. Caracristi: "Utilization of Pipeline Coal," *Combustion,* June 1962, p. 22.

Halvorsen, W. J.: "Slurry Pipeline Hydraulics Improved," *Oil and Gas J.,* March 22, 1976, p. 62.

Halvorsen, W. J.: "Operating Experience of the Ohio Coal Pipeline," *Coal, Today and Tomorrow,* June 1964, p. 18.

Hill, R. A., M. E. Jennings, and R. H. Derammelaere: "Samarco Iron Ore Slurry Pipeline," paper presented at 3d International Technical Conference on Slurry Transportation, Las Vegas, March 1978.

"Limestone Slurry Transported via 17.6-Mile Underground Pipeline," *Rock Products,* May 1972, p. 101.

McNamara, E. J.: "Operational Problems with a 67-Mile Copper Concentrate Slurry Pipeline," paper presented at 4th International Conference on Hydraulic Transport of Solids in Pipes, Banff, Alberta, May 1976.

Montfort, J. G.: "Black Mesa Coal Slurry Pipeline Is Economic and Technical Success," *Pipe Line Industry,* May 1972, p. 42.

Sweeney, W. T.: "Coarse Coal Transport," paper presented at 3d International Symposium on Freight Pipelines, Houston, February 1981.

9.17.2
CONSTRUCTION OF SOLIDS-HANDLING CENTRIFUGAL PUMPS

G. WILSON

ABRASION WEAR

A number of conflicting requirements have to be considered in selecting abrasion-resistant materials for pumps. The material may have to withstand not only abrasion but also high, moderate, or low impact, fatigue stresses, shock loads, and corrosion. The predominant factor that will cause wear must be understood and recognized so that the most suitable construction materials can be selected.

Abrasive wear in pumps is generalized into three types:

Gouging abrasion occurs when coarse particles impinge with such force that high-impact stresses are imposed, resulting in the tearing of sizable pieces from the wearing surfaces.

Grinding abrasion results from the crushing action on particles between two rubbing surfaces.

Erosion abrasion occurs when free-moving particles (sometimes parallel to the surface) at high or low velocities impinge on the wearing surface.

ABRASIVENESS OF SOLID-LIQUID MIXTURES

The rate of wear is related to the mixture being pumped and to the materials of construction. Wear increases with increasing particle size and increases rapidly when the particle hardness exceeds that of the metal surface being abraded. There is little to be gained from increasing the hardness of the metal unless it is to a level that exceeds that of the particles. The effective abrasion resistance of any metal will depend on its position on the Mohs or Knoop hardness scale. Figure 1 shows approximate hardness relationships for various common ore minerals and metals.

Sharp-edged particles cause greater wear than smooth, rounded particles. Wear increases with increasing particle concentration.

The particle's velocity, kinetic energy, and angle of impact are prime considerations in the selection of the material. Metals with high elastic limit are required to resist direct impact; metals with high hardness are used when there is relatively low impact, i.e., when the flow is almost parallel to the surface.

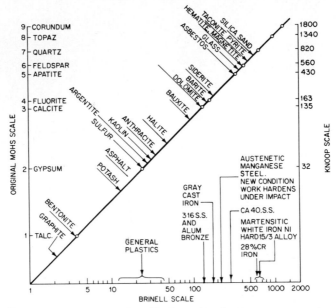

FIG. 1 Approximate comparison of hardness values of various common ore minerals and metals.

A simple expression, frequently quoted, is that the rate of wear is proportional to (particle velocity)^m, where m varies between 2.5 to 4.

It is essential to keep the kinetic energy of the particles in the mixture as low as possible in order to give good life; this will be achieved by reducing the total head requirements of the system or by staging the pumps in series.

CENTRIFUGAL PUMP MATERIALS

All the wetted parts in the pump are subjected to varying degrees of wear. In order of decreasing severity, these are impeller, suction side of casing adjacent to impeller, casing volute (particularly at the tongue and suction nozzle), and gland side of casing adjacent to impeller.

Austenetic manganese steel of 220 Brinell hardness (BHN) (12 to 14% manganese) is used for high-impact gouging abrasion. It has high toughness, work-hardens under impact, is free from high residual stresses in castings, and can be machined by orthodox methods and welded. This material is used for the construction of dredge and heavy-duty gravel pumps.

Martensitic white irons, such as Ni-Hard (4% nickel, 2% chrome) and 15/3 alloy (15% chrome, 3% molybdenum), are used for grinding and erosion abrasion where the particles are very coarse and the impact is moderately low; these irons are cast in the 550 to 650 BHN range. In special cases, 730 BHN can be obtained by heat treatment. The harder the material, the more brittle it will be. For alkaline mixtures, 28% chrome alloy iron of 550 BHN may be preferred because of its superior corrosion resistance. All of these hard irons are used in sand and gravel pumps as well as slurry pumps.

If natural rubber of 38 to 44 Shore A hardness is compatible with the mixture being pumped, it will far outlast any other metal or rubber elastomer within definite limits of particle size and velocity. Natural rubber has poor cut resistance and is confined to applications where the particles are 7 mesh or less. It is suitable for continuous operation up to 150°F (66°C) but has poor swell resistance and will not withstand attack from oils.

TABLE 1 Classification of Pumps According to Solid Particle Size

Tyler standard sieve series				
Aperture,				
in	mm	Mesh	Grade	General pump classification
3				
2				
1.5				
1.050	26.67			
0.883	22.43			
0.742	18.85		Scree shingle gravel	Austenitic manganese steel pumps / Dredge pump
0.624	15.85			
0.525	13.33			
0.441	11.20			
0.371	9.423			
0.321	7.925	2.5		Hard iron pumps
0.263	6.68	3		
0.221	5.613	3.5		Rubber-lined pumps, closed impeller; particles must be round / Sand and gravel pump
0.185	4.699	4		
0.156	3.962	5		
0.131	3.327	6		
0.110	2.794	7		
0.093	2.362	8		Rubber-lined pumps, closed impeller
0.078	1.981	9	Very coarse sand	Sand pump
0.065	1.651	10		
0.055	1.397	12		
0.046	1.168	14		
0.039	0.991	16	Coarse sand	
0.0328	0.833	20		
0.0276	0.701	24		
0.0232	0.589	28		Rubber-lined pumps, open impeller
0.0195	0.495	32	Medium sand	
0.0164	0.417	35		
0.0138	0.351	42		
0.0116	0.295	48		
0.0097	0.248	60		
0.0082	0.204	65	Fine sand	Slurry pump
0.0069	0.175	80		
0.0058	0.147	100		
0.0049	0.124	115		
0.0041	0.104	150		
0.0035	0.089	170		
0.0029	0.074	200	Silt	
0.0024	0.061	250		
0.0021	0.053	270		
0.0017	0.043	325		
0.0015	0.038	400		
	0.025	[a]500		
	0.020	[a]625		
	0.010	[a]1250	Pulverized	
	0.005	[a]2500		
	0.001	[a]12500		
			Mud clay	

[a]Theoretical value.

Although it is not as good as natural rubber, neoprene, when temperature is limited to 200°F (93°C), has given excellent abrasion resistance in oils without swelling and is the next most common elastomer chosen for pump linings.

Molding tolerances allowed in rubber parts are generally greater than those for other engineering materials because rubber shrinks after molding. The amount of shrinkage depends on the mix and may vary from batch to batch. There are various methods of molding the rubber, e.g., hand lay-up, compression, transfer, and injection. High-pressure compression molding has yielded good results. The majority of rubber-lined pumps are applied to pumping fines of less than 60 mesh; therefore side clearances of 0.06 in (1.5 mm) per side will minimize wear and compensate for shrinkage.

Rubber liners should be replaceable and clamped to the pump casing walls to reduce downtime and maintenance costs.

Polyurethane, which is more expensive than natural rubber, is becoming more popular in large pump construction because of the significantly lower cost of molds. This material gives excellent abrasion resistance when fines are pumped.

Silicon carbide ceramics, with hardness about 9.5 on the Mohs scale, in castable form are used in slurry pump construction and give excellent resistance to erosion abrasion where there is negligible impingement and impact. They possess excellent corrosion and temperature resistance but are extremely brittle and expensive to produce.

Rubbers, urethanes, and ceramics are used in sand and slurry pumps.

Table 1 shows a general classification for pump materials and design according to particle size.

CENTRIFUGAL PUMP DESIGNS

Elimination of wear is impossible; however, the life of the parts can be appreciably prolonged and the cost of maintenance reduced by a sound design approach. This means:

Construct the pump with good abrasion-resistant materials.

Provide generous wear allowances on all parts subjected to abrasion.

Adopt hydraulic designs which minimize wear.

Adopt mechanical designs which are suitable for the materials of construction and allow ready access to the wearing parts for renewal.

The casing of a centrifugal pump has a profound effect on performance. If designed incorrectly, it can waste a large part of the energy given to the flow by the impeller, resulting in low efficiency, high hydraulic forces, and excessive wear on both the casing and the impeller.

In a conventional spiral volute, the static pressure around the impeller is uniform only when there is a free vortex velocity distribution in the volute, i.e., when the spiral flow from the impeller coincides with that of the volute spiral. This occurs only at the best efficiency point (BEP). Departure from the free-vortex velocity distribution in the volute at partial capacity or overcapacities produces large circumferential pressure gradients at the impeller periphery, which are directly proportional to the specific gravity of the mixture being pumped. At partial capacity or overcapacities the flow will not tolerate the abrupt deflection at the tongue, and so severe energy losses and wear occur at the vicinity of the casing throat and on the impeller.

Concentric and semiconcentric volute designs produce more uniform static pressures at the impeller periphery and make the pump much less sensitive to off-BEP operation; consequently the hydraulic forces and wear are considerably reduced. The elimination of the tongue, in casings of these designs, prevents abrupt deflection of flow, minimizing turbulence, and the velocity in the casing volute and throat is reduced.

For these reasons concentric and semiconcentric casing designs are adopted for solids-handling pumps; however, the degree of concentricity must be reconciled with efficiency, hydraulic radial load, and wear for each specific speed. Up to $1200N_s$ specific speed ($23N_s$ in SI units of cubic meters per second, meters, and revolutions per minute), the castings can be fully concentric without too much sacrifice in efficiency. Above $1200N_s$ ($23N_s$), the casings can be semiconcentric; the angle of concentricity will progressively reduce as the specific speed increases. Semiconcentric designs can still be adopted in the $2000N_s$ ($39N_s$) range. See Subsecs. 2.2.1 and 2.3.1 for additional information on casings.

Support feet are not provided on the casings since they are frequently replaced and realignment would be necessary. The absence of feet on the hard metal casing makes the casing easier to cast and reduces machining. The ability to rotate the casing to give alternative choices of discharge nozzle positions has advantages when pumping in series.

Although they are more expensive to produce, closed-impeller designs are preferred to open impellers because the passages between the inlet and the outlet are usually relatively evenly worn. They are more robust and are not sensitive to falloff in performance when wear causes the clearance between the suction side shroud and the casing wall to increase.

The requirement for extra-thick impeller vanes causes severe restrictions at the impeller eye, and the number of vanes has to be reconciled with the impeller geometry, NPSH, and wear. Three or four vanes are usual, depending on the specific speed.

Impeller nuts are not recommended since they are subject to wear; it is good practice to screw the impeller onto the shaft in the direction opposite the rotation of the pump. Heavy-duty thread forms are used to withstand the torsional stresses, shock, and wear and tear of maintenance. Aluminum, nodular iron, or steel can be used for the rubber impeller skeleton; however, steel is the best choice since it will not fracture under impact and can be repaired by welding.

Sealing   Sealing of abrasive mixtures under pressure has long been a problem in centrifugal slurry pumps. Excessive gland leakage, costly maintenance, and even failure can occur if the seal is not effective. It is essential to keep abrasive particles in the mixture being pumped away from the seal, or at least to minimize the pressure to such an extent that excessive wear is prevented.

The rotational flow between a conventional impeller shroud (without sealing rings) and the casing wall is a forced vortex. In this case the angular velocity of the liquid in this vicinity will be approximately one-half the angular velocity of the impeller. Consequently, high pressures are obtained at the seal, combined with extreme axial hydraulic unbalance, both of which are directly proportional to the specific gravity of the mixture being pumped. Steps are taken to minimize these effects by the addition of hydrodynamic pumping vanes on the back shrouds of the impeller. Properly designed, vanes can effectively increase the angular velocity of the liquid to 95% of the impeller angular velocity. Pumping vanes on the impeller back shroud will therefore effectively reduce the pressure at the stuffing box and greatly reduce the axial hydraulic unbalance.

Unless precautions are taken to prevent, or at least minimize, the ingress of solids into the stuffing box area, rapid wear will result and the gland will require constant attention. There are many methods proposed for sealing the gland, all of which may work under certain limitations.

In slurry pumps the shaft is always protected at the stuffing box by a sleeve. A hook-type design is preferred since this has the effect of reducing the stress values on the shaft and compensates for differential expansion when high-temperature mixtures are being pumped. The choice of sleeve material must be based on operating experience and can range from centrifugally spun cast iron to vacuum-hardened stainless steel, stellite-faced and ceramic-faced.

Mechanical Seals   Face-type seals (even with tungsten carbide faces) that depend on an interface film for lubrication are not suitable for rough abrasive service. They require precision adjustment, they are sensitive to impact and shaft deflection, and they usually fail catastrophically. If failure occurs, the pump has to be stripped down in order to replace the seal. Split seals are complex and expensive.

Double seals with a clean barrier liquid have proved more successful, but again these are complex and expensive.

Packed Gland Seals   A combination of hydrodynamic sealing vanes to minimize the pressure at the shaft and an orthodox packed-box arrangement using split rings of graphited asbestos or Teflon-filled asbestos are recommended for abrasive slurries.

If dilution of the product being pumped can be accepted, it is strongly recommended that a clean liquid service be provided under pressure at the bottom of the stuffing box. The abrasives will be prevented from entering the box and causing wear to the packing and the sleeve.

Within certain limitations of suction pressure, it is possible to operate with a dry box. The sealing vanes will effectively prevent abrasives from reaching the box area, although a seal is still required to prevent leakage in the static condition. If a dry box is adopted and the suction pressure is high, fines will soon penetrate into the box and become embedded in the soft packing; eventually the packing will become saturated with the fines, and leakage due to wear on the sleeve will occur. The seal options should therefore be

1. A packed box with clean fluid for flushing connected to the lantern ring located at the bottom of the box. This causes dilution of the product being pumped (unsuitable for cement slurries).
2. A packed box with clean fluid for flushing connected to the lantern ring located two rings from the inboard end of the box. This causes less dilution, but greater wear will take place.
3. A packed box with a grease seal connected to the lantern ring located two rings from the inboard end of the box. This dry box arrangement causes no dilution of the product.

Wear Plates An examination of Figure 2 shows that impeller radial wearing rings have been eliminated from the design. This simplifies the whole of the liquid end construction and makes the pump amenable to external end clearance adjustment to compensate for wear. The worst wear occurs on the suction side, and all good slurry pump designs should incorporate an easily replaceable suction side wear plate or liner. A wear plate or liner is not always provided on the gland side adjacent to the impeller. However, this area is also subjected to wear, and despite the increased initial cost it is advisable to have one fitted.

Bearing Cartridge Assembly A calculated bearing life of more than 50,000 h is normal to meet the arduous conditions of service. This is appreciably higher than the specified life of conventional pump bearings. Referring to Fig. 2, it can be seen that the bearing cartridge assembly is one complete element which can be easily interchanged for another sealed element at short notice. Oil lubrication is preferred to grease because less attention is required. Grease lubrication should be provided as an option.

Slurry pumps are usually situated in a dust-laden atmosphere, and adequate precautions should be taken to prevent the ingress of liquid or dust into the housing. A labyrinth deflector is almost mandatory at the line bearing cover to expell gland leakage from the bearing.

PUMP DRIVES

Hard metal or rubber impellers are not usually reduced in diameter to meet a given condition of service. For this reason, the pumps are usually belt-driven. The required pump performance can be achieved by selecting the desired sheave ratio that will give the pump speed required.

The motors are often fitted directly above the pump; this saves floor space and minimizes the

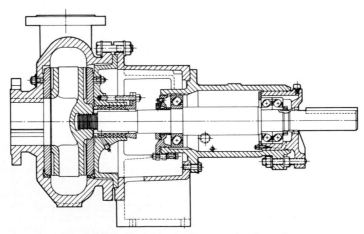

FIG. 2 Hard iron solids-handling pump for slurries, sand, and gravel.

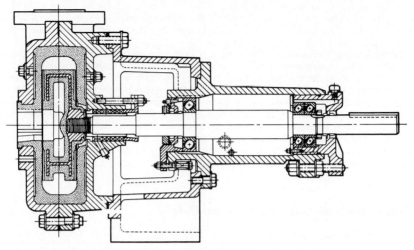

FIG. 3 Rubber-lined solids-handling pump for fines and sand.

risk of flooding. The largest motor frame that can be accommodated above the pump is determined by the bearing centers in the cartridge assembly and the shaft overhang at the drive end, as well as by the stability of the motor-supporting structure.

Figure 2 shows a typical hard iron slurry, sand, and gravel pump. Figure 3 shows a typical replaceable lined rubber pump for slurries and sand.

Pumps that handle large material and develop high heads are called *dredge pumps*. There are many designs on the market with large impeller diameters for operating at low rotational speeds to prolong the life of the parts. Pumps of this construction are very expensive.

The power required to drive the pump and the discharge pressure are directly proportional

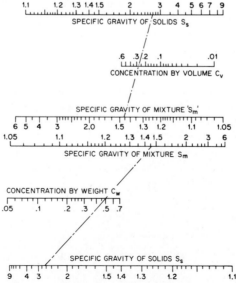

FIG. 4 Nomograph for the relationship between concentrations and specific gravities in aqueous slurries.

to the specific gravity of the mixture. Figure 4 shows a nomograph for the relationship of concentrations to specific gravities for aqueous slurries. Although solids are usually given as a percentage of the total mixture by weight, it is wise to cross-check its percentage by volume to ensure that a plug condition will not occur in the pump.

Particles will be conveyed in the pump as a heterogeneous suspension, i.e., in turbulence. Depending on the concentration and particle sizes, there can be an apparent viscosity effect as a result of the failure of the solid particles to follow the liquid path. Solids cannot possess or transmit pressure energy; they can do so only with kinetic energy, which is unrecoverable. Therefore, the total head and efficiency are bound to be lower. The predictions of these losses should be based on experience or calculated by the pump supplier. All performance curves for solids-handling pumps are for clear water only. A standard procedure often adopted is to rate the pump at 10% lower head than when pumping clear water and to select the motor for at least 30% above the calculated brake power requirement. See also Subsec. 9.17.1 on calculating system resistance of liquids containing solids.

OTHER TYPES OF SOLIDS-HANDLING PUMPS

Jet Pumps   See Chap. 4

Reciprocating Pumps   See Chap. 3 and Subsec. 9.17.3.

Diaphragm Pumps   See Sec. 3.3

FURTHER READING

Agostinelli, A., A. D. Nobles, and C. R. Mockridge: "An Experimental Investigation of Radial Thrust in Centrifugal Pumps," *Trans ASME, J. Eng. Power*, ser. A, **82**:120 (1960).

Aude, T. C., N. T. Cowper, T. L. Thomson, and E. J. Wasp: "Slurry Piping Systems: Trends, Design Methods, Guidelines," *Chem. Eng.*, June 28, 1971, pp. 74–90.

Bitter, J. G. A.: A Study of Erosion Phenomena, Part I, *Wear* **6**(3) (1964).

Lindley, P. B.: *Engineering Design with Natural Rubber*, Natural Rubber Producers Research Association.

Lipson, C.: *Wear Considerations in Design*, Prentice-Hall, Englewood Cliffs, N.J., 1967.

Stepanoff, A. J.: *Pumps and Blowers: Two-Phase Flow*, Wiley, New York, 1965.

Tsybaev, N.: "Wear-Resistant Rubber Linings for Pumps Delivering Abrasive Mixtures," *Tsvetnye Metally*, February, 1965.

Welte, A.: "Wear Phenomena in Centrifugal Dredging Pumps," *VDI Berichte* **75**:111 (1964).

Wilicenus, G. F.: *Fluid Mechanics of Turbomachinery*, Dover, New York, 1965.

Worster, R. C.: "The Flow in Pump Volutes and Its Effect on Centrifugal Pump Performance," *Ind. Mech. Eng.*

9.17.3
CONSTRUCTION OF SOLIDS-HANDLING DISPLACEMENT PUMPS

WILL SMITH

Positive displacement pumps are especially suited for many solids-pumping applications because constant high pressures and good efficiency are achieved over the full range of pump capacities. Low relative velocities between abrasive liquid-solids mixtures and the pump parts minimize erosion. With centrifugal pumps an increase in system resistance, such as a flow blockage caused by settled solids or the apparent viscosity increase characteristic of many slurries due to momentary capacity reduction, results in a self-defeating reduction in pump flow rate.

Positive displacement pumps are popular for a number of solids-handling applications. Each application presents peculiar demands for pump features.

Solids Transport Large-capacity positive displacement pumps at moderate pressures are used to pump coal and ores over relatively long distances. Solids for transport are usually slurried with water and are pumped at ambient temperatures. Double-acting horizontal piston pumps with large, "mud pump" valves and piston rings (Fig. 1) typify transport service.

Most pipelines have multiple pumping stations at intervals along the route, dictated by topography and by pipeline and pumping station first and operating cost balances. Centrifugal booster pumps are employed at the first station to deliver the prepared slurry to the displacement pumps at a sufficient pressure to satisfy their NPSH, including acceleration head, requirement. Subsequent stations are located at points where there is sufficient residual pipeline pressure to fulfill the displacement pump's suction requirements. The capacity of each station is adjusted by pump speed control so that the next station's inlet pressure is kept relatively constant.

Most stations use multiple positive displacement pumps in parallel, with at least one standby pump. Capacity modulation requires but one of the pumps to be operating with speed control although all are generally so fitted. Unlike centrifugal pumps, displacement pumps can provide full rated pressures at all speed-controlled capacities.

Process Pumping Some chemical and petroleum processes require pumping of solids to high pressures for process reactions. Typical of these are bauxite ore in hot caustic for alumina plants, ground coal in water or coal liquids for synthetic fuel production, and grain in water for alcohol production (Fig. 2).

Process streams involve a vast variety of liquids and solids. Concentrations may be as high as

FIG. 1   Horizontal double-acting duplex coal slurry pipeline pumps. (Black Mesa Pipeline, Oil Well Div. of U.S. Steel)

70% solids by weight. Temperatures can reach 800°F (427°C). Pressures from several hundred to many thousands of pounds per square inch are accommodated.

Other Specialty Services   A number of specialty services employ positive displacement pumps to handle solids. Some are relatively simple applications of standard catalog pumps, and others are single-purpose developments and thus have evolved with unique characteristics.

   Mud pumps are used to inject drilling mud, which transports cuttings out of wells and lubricates the drill bit during well drilling. Because of the limited duration of a drilling project, the life of expendable parts (packing, valves, rods) is compromised in favor of size and weight for portability and ease of overhaul. *Sludge disposal pumps* are common in sewerage treatment plants, and *tailings pumps* move solids out of mines.

PUMP CONSTRUCTION _____

Solids-handling displacement pumps differ from ordinary displacement pumps in the means used to alleviate the deleterious effects of the solids on the packing, the displacement element (whether

FIG. 2   Horizontal double-acting plunger steam pump for slurry of grain mash and water for alcohol production. (Worthington Pump)

plunger, piston, or diaphragm), and the suction and discharge check valves. To protect parts from the ravages of the solids, (1) particles are prevented from entering close clearances, (2) operation is at lower speeds to reduce the effects of erosion and abrasion, or (3) the susceptible parts are made of wear-resistant materials. Since sacrificial wear parts must be replaced more often than in clear liquid pumps, special design attention for ease of replacement of these parts is justified.

The designer's choice is between hard materials that resist wear and resilient materials that may accommodate the particles of solids. This choice is often limited by other factors, such as temperature and chemical compatibilities. Hard material used include

Tungsten carbide

Ceramics—chrome oxide and aluminum oxide

Hardenable stainless steel

Cobalt-nickel alloys

and some common soft materials are

Synthetic elastomers

Softer metals (sacrificial)

Unlike centrifugal pumps, positive displacement pumps are more prone to abrasive than erosive wear. This is due to the much lower relative velocities between the liquid and the pump parts. Centrifugal pumps must generate high relative velocities, typically greater than 100 ft/s (30 m/s), for the dynamic specific energy that is converted to discharge pressure. Positive displacement pump velocities are kept low, limited usually by the settling velocity of the slurry, which is on the order of less than 10 ft/s (3 m/s).

There are four types of positive displacement solids-handling pumps:

1. Reciprocating piston pumps
2. Reciprocating plunger pumps
3. Diaphragm pumps
4. Hydraulic displacement pumps

Valves Common to all types are suction and discharge check valves, the purpose of which is to allow the solids-laden liquid to flow into the pump from the suction line and into the discharge line from the pumping chamber while preventing backflow. The features of slurry valves include

1. Large areas for low flow velocity
2. Smooth, unobstructed passages to avoid trapping and build-up of solids
3. Special designs to enhance sealing against pressure in the closed position.
4. Special materials to minimize wear

Figure 3 illustrates some valves developed particularly for solids-handling pumps. Which valve is selected for a particular application depends on the abrasivity, viscosity, temperature, solids size, and uniformity of the slurry. It is important that the valve lift be sufficient to pass the largest particles of solids expected. Adequate sealing in the closed position is necessary to avoid excessive back leakage, which detracts from the volumetric efficiency of the pump.

Valve materials must meet temperature, corrosion, abrasion, and mechanical strength requirements. Materials for valves and seats range from cast iron and bronze through hardenable stainless steels to solid tungsten carbide. The cost of replacement or refurbishment over the expected life, including labor and downtime, should be balanced against first cost.

Packing Protection A distinguishing feature of solids-handling displacement pumps is the means used to avoid or minimize packing wear. Five different approaches are

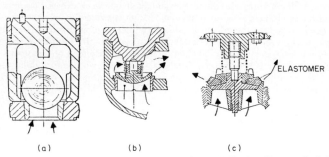

FIG. 3   Three types of slurry valves: (*a*) free-floating ball valve, (*b*) spring-loaded spherical (Rollo) valve, (*c*) spring-loaded elastomer-seal (mud) valve.

1. Use of conventional designs with special materials and operation at less demanding conditions, such as reduced speeds
2. Provision of a clean nonslurry environment for the packing rubbing surfaces
3. Separation of the slurry from the displacement element
4. Separation of the slurry from the pumping element
5. Total isolation of the slurry from the packing

CONVENTIONAL DESIGN WITH MODIFIED OPERATION   Piston pumps usually fall into this category. By using larger pumps at speeds that are lower than those normally found with clear fluids, velocity-sensitive wear rates are reduced. *Special piston rings* (Fig. 4) are used which tend to resist abrasion because they are made of zero clearance elastomeric materials and designed to scrape clean the cylinder wall. Such piston packing is a normal wear item and must be replaced at regular intervals.

CLEAN NONSLURRY ENVIRONMENT   This is accomplished by the injection of clear liquid (Fig. 5). Such injection may be either *continuous* or *synchronized* with the stroke of the plunger or piston. When effectively accomplished, flush injection presents essentially clear pumpage to the packing so that it attains normal clear liquid packing life. Flush liquid that enters the pumping chamber dilutes the pumpage and decreases the suction volumetric efficiency.

Clear, compatible flush liquid may be scarce or expensive. Therefore special stuffing box plus flush injection systems have been developed. Required flush rates are a function of pump type and size and system employed. Satisfactory operation has been attained with flush rates of about

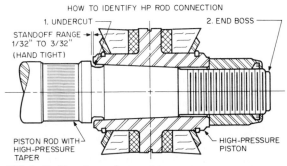

FIG. 4   Double-acting mud pump piston.

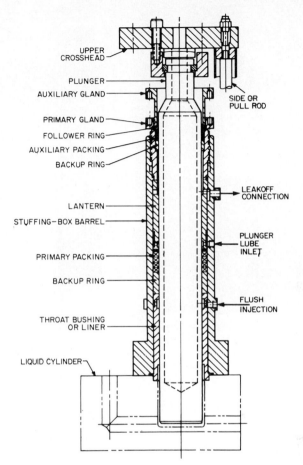

UPPER CROSSHEAD

PLUNGER

AUXILIARY GLAND

SIDE OR PULL ROD

PRIMARY GLAND

FOLLOWER RING

AUXILIARY PACKING

BACKUP RING

LEAKOFF CONNECTION

LANTERN

STUFFING-BOX BARREL

PLUNGER LUBE INLET

PRIMARY PACKING

BACKUP RING

FLUSH INJECTION

THROAT BUSHING OR LINER

LIQUID CYLINDER

FIG. 5   Typical flush injection.

2 to 5% of pumped flow and even less for very large pumps. If other than a positive displacement flush pump is used, flush flow will increase as packing and throat bushings wear.

Flushing prevents incursion of slurry particles by providing in clearances a liquid velocity counter to the unwanted slurry flow and by washing solids particles from the plunger or rod surfaces as they move toward the packing. Injection systems may be as simple as a constant-pressure liquid fed to a point between the packing and the throat bushing or as complex as a sophisticated synchronized injection system involving an additional positive displacement pump driven directly from the slurry pump drive train and independently furnishing a definite quantity of flush fluid to each plunger or piston during each pump stroke (Fig 6).

Combinations of positive, independent, synchronized pumped injection and auxiliary buffer liquids have been proposed for very special situations.

SEPARATION OF SLURRY FROM DISPLACEMENT ELEMENT   One means of reducing the concentration and temperature of slurry at the pump packing is the use of a liquid, or surge, leg (Fig. 7). The displacement element—plunger or piston—is separated from the suction and discharge valves by a length of pipe which is filled with clear liquid. This liquid leg communicates with the valve chambers such that, as the liquid ebbs and flows in response to the motion of the displacement element,

FIG. 6 Vertical single-acting triplex plunger pump with synchronized flush liquid injection. (Worthington Pump)

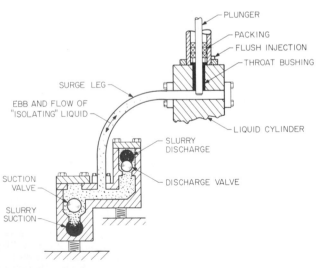

FIG. 7 Surge leg plunger pump.

it acts as a liquid piston, inducing and then expelling slurry through the suction and discharge valves. Make-up liquid is injected into the surge leg, preferably through a throat bushing isolating the displacement chamber from the packing so that there is theoretical flow away from the stuffing box. The surge leg thus offers the packing some degree of protection from the solids and from thermal effects.

Surge legs are, however, subject to the complete span of suction-to-discharge pressure change on each stroke. Therefore they must be designed for endurance fatigue loading. When high-temperature isolation is involved, thermal expansion and contraction must also be considered. While surge leg pumps have been demonstrated to increase piston or plunger packing life, they do not totally prevent solids migration from the pumping chamber. Therefore special construction materials, flush injection, and reduced speeds are still appropriate.

SEPARATION OF SLURRY FROM PUMPING ELEMENT   A special case of the liquid leg pump is the large, low-cycle chamber feeder (Fig. 8). While these units may or may not properly qualify as positive displacement pumps (because the motivating liquid may be delivered by a centrifugal pump), they do nevertheless have much in common with surge leg pumps.

The motivating liquid is pumped into a chamber and displaces a charge of slurry through suitable check valves. The chamber may have a moving separation barrier between the motivating and motivated liquids. When the slurry is expelled, controls cause valves to reverse the flow, filling the chamber again with slurry while voiding it of the clear liquid. By employing two or more chambers, a virtually continuous flow of slurry is obtained. Advantages of this scheme are the extra-low frequency of check valve operation, low slurry velocities, and high capacities. Centrifugal, positive displacement, or even compressed gas pumps can be used for motivation. Since the motivation pump is independent of the slurry system, it can operate at normal speed.

Two disadvantages of chamber feeders are their dependence on controls and switching valves in addition to the slurry valves and the fact that the systems are large.

TOTAL ISOLATION   Many types of diaphragm pumps which completely separate the slurry from the running gear of the pump have been designed (Fig. 9). Since the separating diaphragm is wetted on one side by the slurry, it is vulnerable to the mechanical, chemical, and thermal abuse

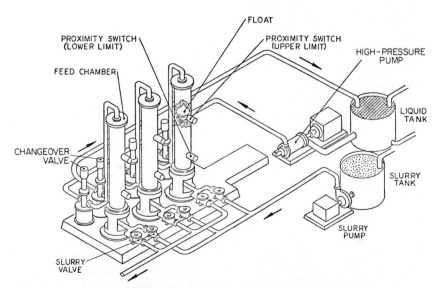

FIG. 8 Large, low-cycle chamber feeders separate the high-pressure pumping element from the slurry. (Hitachi)

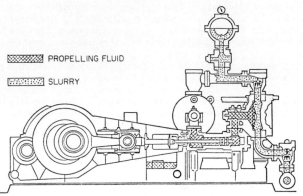

FIG. 9   Diaphragm pump isolates slurry from packing. (Geho)

that might otherwise apply to pistons, plungers, packing, etc. Of course, the diaphragm is also subjected to full reversal loads and must be designed for fatigue life. Both elastomeric and metallic diaphragms have been used.

The severity of diaphragm failure problems has been alleviated somewhat in some designs which employ double diaphragms with intervening barrier liquids. The thesis is that there is a much reduced probability of both diaphragms failing before detection and repair.

SECTION 9.18
OIL WELLS

JOHN R. BRENNAN

ARTIFICIAL LIFT OF OIL WELLS

Oil wells can range in depth from approximately 200 ft (60 m) to more than 20,000 ft (6000 m). They require pumps that are unique in their configuration. About 90% of pumping oil wells are equipped with either 2-in (50.8-mm) or $2\frac{7}{16}$ in (61.9-mm) ID tubing. Sucker rod pumps, hydraulic reciprocating pumps, or hydraulic jet pumps that must run inside this tubing cannot have their outside diameters exceed $1\frac{7}{8}$ in (47.6 mm) or $2\frac{5}{16}$ in (58.7 mm). These pumps are installed as close as possible to the bottom in order to achieve the maximum possible drawdown of fluid in the well. The output pressure required of the pump varies directly with the depth that the fluid must be lifted. For example, to compensate for the lift required and the fluid friction in the tubing, a 10,000-ft (3049-m) well might require a 4000-lb/in^2 (276-bar°) pump.

In a newly discovered field, wells will usually flow of their own accord and require no artificial lift. The volume from these wells is controlled by holding back pressure at the wellhead with a choke. The annulus between the production tubing and the oil well casing is closed off at the wellhead; thus all of the gas produced by the well flows up the production tubing with the oil. When the bottom hole pressure is no longer adequate to produce the well by natural flow, some form of artificial lift must be installed. This can be a sucker rod pump, a subsurface hydraulic reciprocating pump, a subsurface hydraulic jet pump, an electric submersible centrifugal pump, or a gas lift. The sucker rod pump is by far the most common form of artificial lift. More than 80% of the wells that must be artificially lifted are pumped with sucker rod pumps.

THE SUCKER ROD PUMPING SYSTEM

The sucker rod pumping system consists of a prime mover, a pumping unit, a polished rod with a stuffing box to seal off the pumped fluid pressure at the suface of the well, and a sucker rod string to transmit the reciprocating movement from the pumping unit to the sucker rod pump. The pumping unit is the mechanism that converts the rotary movement of the prime mover to the reciprocating movement needed to power the sucker rod pump. The sucker rod string consists

° 1 bar = 10^5 Pa. For a discussion of bar, see *SI Units—A Commentary* in the front matter.

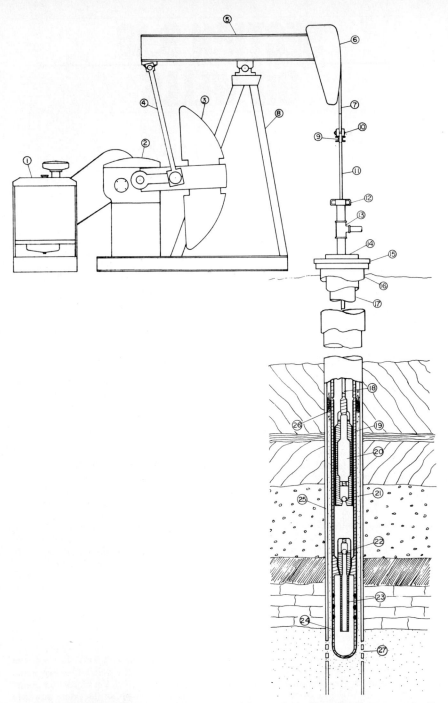

FIG. 1 Sucker rod pumping system: (1) prime mover, (2) gear reducer, (3) crank and counterweight, (4) pitman, (5) walking beam, (6) horsehead, (7) bridle, (8) Samson post, (9) carrier bar, (10) polished rod clamp, (11) polished rod, (12) stuffing box, (13) pumping tee, (14) tubing ring, (15) casing head, (16) casing surface string, (17) tubing string, (18) sucker rod, (19) pump barrel, (20) pump plunger, (21) traveling valve, (22) standing valve, (23) mosquito bill, (24) gas anchor, (25) casing oil string, (26) fluid level, (27) casing perforations. (National Supply)

of individual sucker rods of 25- or 30-ft (7.6- or 9.1-m) lengths which are connected with couplings when they are lowered into the well. This total length of rods connects the surface polished rod to the pump at the bottom and is called the sucker rod string. The complete sucker rod pumping system is illustrated in Fig. 1.

The operation of a sucker rod pump is illustrated in Fig. 2. The pump is submerged in fluid near the bottom of the oil well. As the sucker rod string and plunger make an upstroke, fluid from the well bore flows past the standing valve into the pump barrel. Also on this upstroke, the traveling valve is closed and the fluid above the plunger is pumped up the annulus between the sucker rod string and the tubing string. On the downstroke, the traveling valve is open and the standing valve is closed. The fluid in the pumping chamber between the traveling valve and standing valve is displaced into the annulus between the tubing and the sucker rod string above the plunger. The fluid in the annulus is lifted toward the surface only on the upstroke.

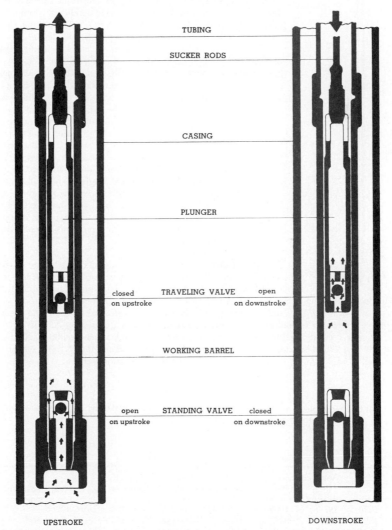

FIG. 2 Operation of a tubing pump.

The Sucker Rod Pump   The principal parts of the sucker rod pump are the barrel, plunger, traveling valve, and standing valve. These are seen in a cutaway section in Fig. 3. In shallow wells relatively free of sand, soft-packed plungers are frequently used. These have a number of fabric rings or cups which are expanded by the pumping pressure to a close fit with the barrel. In deep wells or in hard-to-pump shallow wells, the precision-fit metal barrel and plunger as shown in Fig.

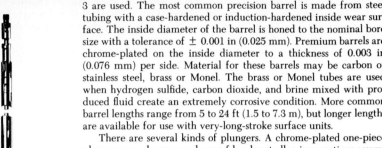

3 are used. The most common precision barrel is made from steel tubing with a case-hardened or induction-hardened inside wear surface. The inside diameter of the barrel is honed to the nominal bore size with a tolerance of ± 0.001 in (0.025 mm). Premium barrels are chrome-plated on the inside diameter to a thickness of 0.003 in (0.076 mm) per side. Material for these barrels may be carbon or stainless steel, brass or Monel. The brass or Monel tubes are used when hydrogen sulfide, carbon dioxide, and brine mixed with produced fluid create an extremely corrosive condition. More common barrel lengths range from 5 to 24 ft (1.5 to 7.3 m), but longer lengths are available for use with very-long-stroke surface units.

There are several kinds of plungers. A chrome-plated one-piece plunger or a plunger made up of hard cast alloy iron sections assembled over a steel plunger tube is usually used with the precision hardened steel barrel. A plunger that is hard-faced with a nickel-based spray metal material is usually used with the chrome-lined barrel, although the cast plunger can also be used there. Plungers range from 2 to 6 ft (0.6 to 1.8 m) in length, depending on the depth of the well. These plungers are all ground to a precision tolerance of + 0.0000, −0.0005 in (+0.000, −0.013 mm). The diametral fit between the inside of the barrel and the outside of the plunger ranges from 0.002 to 0.005 in (0.05 to 0.13 mm), depending on the quality of the well fluid and the diameter and length of the plunger.

The traveling valve and standing valve of the pump are simple ball-and-seat check valves. Type 440 hardened stainless steel materials are the most common, but in corrosive wells a cobalt-chromium-tungsten alloy is frequently used, and in very abrasive wells tungsten carbide seats and balls are used.

Sucker Rods   Sucker rods are manufactured from carbon or low-alloy steel. Table 1 lists the properties of the grades of rod used most frequently. Most sucker rods are manufactured in 25-ft (7.62-m) lengths, but a few areas use 30-ft (9.14-m) lengths. Both ends of the rods are upset° and externally threaded (Fig. 4). The upset ends

FIG. 3   Sucker rod pump.

also have a square for wrenching. Internally threaded couplings are used for connecting rods to make the sucker rod string. Maximum design stress for sucker rods is between 30,000 and 40,000 lb/in² (207 and 276 MPa), depending on the metallurgy of the rod and the corrosive environment of the well. Useful data for designing sucker rod strings are shown in Table 2.

°In oil field terminology, an upset is an enlarged portion formed on the end of the rod by forging. This allows for a joint that is stronger than the body of the rod.

TABLE 1   Mechanical Properties of Sucker Rods

API class	AISI steel	Yield point, 1000 lb/in² (MPa)	Tensile strength, 1000 lb/in² (MPa)	Brinell hardness	Heat treatment
C	1036	60–75 (414–517)	90–105 (621–724)	190–205	Normalized
M	4621	68–80 (469–552)	85–100 (586–689)	175–207	Normalized and tempered
D	4142	100–115 (689–793)	115–140 (793–965)	240–280	Normalized and tempered

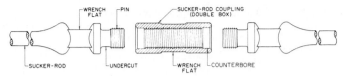

FIG. 4 A typical sucker rod joint. (API)

TABLE 2 Sucker Rod Data

Rod size, in (mm)	Pin size, in (mm)	Coupling OD, in (mm)	Cross-sectional area, in^2 (mm^2)	Dry weight, lb/ft (kg/ m)	Stretch factor C, USCS^a (SI)^b
⅝ (15.88)	¹⁵⁄₁₆ (23.81)	1⅝ (41.28)	0.307 (198)	1.15 (1.71)	1.322 (24.76)
¾ (19.05)	1¹⁄₁₆ (26.99)	1¹³⁄₁₆ (46.04)	0.442 (285)	1.63 (2.43)	0.919 (17.21)
⅞ (22.22)	1³⁄₁₆ (30.16)	2³⁄₁₆ (55.66)	0.601 (388)	2.18 (3.25)	0.676 (12.66)
1 (25.40)	1⅜ (34.92)	2⅜ (60.32)	0.785 (507)	2.94 (4.38)	0.517 (9.68)

^{a}in/1000 ft·1000 lb load.
^{b}mm/kn·kN load.

   Design of a single-size string for use in wells of shallow to moderate depth is a straightforward calculation and the tables and formulas included here are sufficient. In deeper wells tapered° strings of rods are used, frequently 1, ⅞, and ¾ in (25.4, 22.2, and 19.0 mm). Various methods are used for calculating the proper taper, but there are so many variables involved that computer programs are frequently used to provide the total system design.

Surface Pumping Units Rating standards of beam pumping units have been established by the American Petroleum Institute (API). There are fourteen gear reducer torque ratings established, and these can be combined with various structures and a variety of stroke lengths to supply a pumping unit matched to virtually any well condition. Table 3 shows these torque ratings, the maximum stroke length, and the maximum structure capacity of standard units.

Pumping Installation Calculations The pump bore (ID of the pump barrel) is selected based on the quantity of fluid to be produced. A trial calculation is made to select a pump bore and a pumping unit stroke length that will produce the required volume of fluid. A volume calculation at 100% efficiency is performed using the bore factor from Table 4 and the stroke length and design strokes-per-minute value shown in Table 3. This is multiplied by 0.7 to correct for anticipated pump efficiency under average well conditions.

WEIGHT OF FLUID ON PLUNGER   The trial calculation continues with the weight of the fluid on the plunger. Fresh water exerts a pressure of 0.433 lb/in^2 per vertical foot (9.79 kPa per vertical meter). The weight of the fluid on the plunger can be calculated with the formula

in USCS units $W_f = 0.433\, A_p D(\text{sp. gr.})$

in SI units $W_f = 0.00979\, A_p D(\text{sp. gr.})$

where W_f = weight (force) of fluid on plunger, lb (N)
      A_p = area of plunger (Table 4), in^2 (mm^2)
      D = fluid lift, distance from fluid level while pump is operating to ground surface, ft (m)
   sp. gr. = specific gravity of fluid pumped, ratio of density of fluid to density of water at 60°F (16°C)

   °When the sucker rod string is made up of several shorter strings, each of a different diameter, then the total string is called tapered since the smallest-diameter strings are at the bottom and the largest at the top. This is done to reduce the total weight and to maintain approximately the same stress levels top to bottom.

TABLE 3 Maximum Rating of Standard Pumping Units

Torque, in·lb (N·m)	Stroke length, in (m)	Structure capacity, lb (kN)	Design spm[a]
6,400 (8,677)	16 (0.406)	3,200 (14.23)	25
10,000 (13,560)	20 (0.508)	4,000 (17.79)	25
16,000 (21,690)	24 (0.610)	5,300 (23.58)	25
25,000 (33,800)	30 (0.762)	6,700 (29.80)	24
40,000 (54,230)	36 (0.914)	8,900 (39.59)	22
57,000 (77,280)	42 (1.07)	10,900 (48.49)	20
80,000 (108,400)	48 (1.22)	13,300 (59.16)	19
114,000 (154,600)	54 (1.37)	16,900 (75.17)	18
160,000 (216,900)	64 (1.63)	20,000 (88.96)	17
228,000 (309,000)	74 (1.88)	24,600 (109.4)	15
320,000 (433,900)	120 (3.05)	25,600 (113.9)	12
456,000 (618,300)	144 (3.66)	30,400 (135.2)	11
640,000 (867,700)	168 (4.27)	35,600 (158.4)	10
912,000 (1,237,000)	192 (4.88)	42,700 (189.9)	9

[a]Design spm (strokes per minute) is based on a Mills impulse load of 0.25 with a maximum spm of 25. This cycle rate may be exceeded; however, the values given here are considered good design practice.

WET WEIGHT OF RODS The next calculation is the weight of the rods. The dry weight (force) of the rods W_r in pounds is computed by multiplying the weight (force) per foot of the rods (Table 2) by the length of the rods. This weight (force) in newtons is kilograms per meter times the acceleration of gravity (9.807 m/s^2) times the length of the rod. Since the rods have buoyancy when immersed in a fluid, their wet weight (force) W_w can be calculated with the formula

$$W_w = W_r (1 - 0.128 \text{ sp. gr.})$$

DEAD WEIGHT OF RODS AND FLUID This value is needed in subsequent calculations. It is the weight (force) of fluid on the plunger plus the wet weight of the rods:

$$W_{df} = W_f + W_w$$

TABLE 4 Sucker Rod Pump Data

Pump bore, in (mm)	Plunger area, in^2 (mm^2)	Bore factor, USCS[a] (SI)[b]
1¹⁄₁₆ (26.99)	0.887 (572.1)	0.132 (0.826)
1¼ (31.75)	1.227 (791.7)	0.182 (1.139)
1½ (38.10)	1.767 (1140.)	0.262 (1.639)
1¾ (44.45)	2.405 (1552.)	0.357 (2.234)
2 (50.80)	3.142 (2027.)	0.466 (2.917)
2¼ (57.15)	3.976 (2565.)	0.590 (3.691)
2½ (63.50)	4.909 (3167.)	0.728 (4.555)
2¾ (69.85)	5.940 (3832.)	0.881 (5.512)
3¼ (82.55)	8.296 (5352.)	1.231 (7.702)
3¾ (95.75)	11.045 (7126.)	1.639 (10.255)

[a]Bore factor × stroke length in inches × spm = barrels/day (42-gal oil barrels).

[b]Bore factor × stroke length in meters × spm = cubic meters/day.

IMPULSE LOAD   The acceleration due to the reciprocating motion of the rods causes additional stresses in the sucker rod string. Twice during each stroke,° the rods are stopped and their motion reversed as they follow a sinusoidal acceleration pattern provided by the pumping unit. The resulting impulse load can be approximated using the Mills impulse load formula:

in USCS units
$$I = \frac{L \ (\text{spm})^2 \ W_r}{70,000}$$

in SI units
$$I = \frac{L \ (\text{spm})^2 W_r}{1780}$$

where I = impulse load, lb (N)
  L = length of polished rod stroke, in (m)
  spm = strokes per minute
  W_r = dry weight (force) of rods, lb (N)

PEAK POLISHED ROD LOAD   The peak polished rod load P is

$$P = W_{df} + I$$

This peak polished rod load is used in final sizing of the surface pumping unit. If a single size of rod is used, P divided by the cross-sectional area of the rod gives the maximum rod stress. In a tapered string of rods, P divided by the cross-sectional area of the top rod will also approximate the maximum rod stress unless the rod string design is such that the rod under maximum stress is not at the top but lower in the string.

MINIMUM POLISHED ROD LOAD   In addition to the peak polished rod load, a minimum polished rod load is also needed to determine the peak torque on the gearbox of the surface pumping unit. The minimum polished rod load P_m is

$$P_m = W_{df} - I$$

LOAD RANGE   Load range P_r is

$$P_r = P - P_m$$

PEAK TORQUE   The work transmitted from the prime mover through the gear reducer and crank applies a torque to the gears. Gear reducers are rated by the peak torque they can carry on a continuous basis. The peak torque T in inch-pounds (newton-meters) is proportional to the load range and may be calculated from the formula

$$T = 0.25 P_r L$$

The calculated peak torque must be lower than the torque rating, and the calculated peak polished rod load must be lower than the structure capacity rating of the unit selected (Table 3).

POLISHED ROD POWER   The polished rod power P_{pr} in horsepower (kilowatts) can be calculated from the formula

in USCS
$$P_{pr} = \frac{L \ (\text{spm}) \ P_r}{750,000}$$

in SI units
$$P_{pr} = \frac{L \ (\text{spm}) \ P_r}{113,600}$$

Since the polished rod power calculation does not take friction into account, the prime mover selected should have at least double the power of the calculated figure.

°In oil field terminology, a stroke is defined as the complete up and down cycle of the plunger. Distance traveled in one direction is stroke length.

TABLE 5 External Upset Tubing Data

Tubing size, in (mm)	Actual ID, in (mm)	Weight T&C,[a] lb/ft (kg/m)	Upset OD, in (mm)	Collar OD, in (mm)	Shrink factor K USCS (SI)
2⅜ (60.32)	1.995 (50.67)	4.63 (6.89)	2.594 (65.89)	3.063 (77.80)	0.313 (5.86)
2⅞ (73.02)	2.441 (62.00)	6.44 (9.58)	3.094 (78.59)	3.668 (93.17)	0.224 (4.20)
3½ (88.90)	2.992 (76.00)	9.27 (13.80)	3.750 (95.25)	4.500 (114.30)	0.151 (2.83)

[a]Threaded and coupled.

PLUNGER OVERTRAVEL   In accurately calculating pump displacement, it is necessary to calculate the actual plunger stroke, which is not the same as the polished rod stroke. The plunger gains stroke from overtravel. Long strings of rods are elastic and capable of stretching several feet. When the pumping unit stops the polished rod at the bottom of the surface stroke, the plunger continues to move downward because of inertia. This plunger overtravel OT in inches (meters) can be calculated from the formula

in USCS units
$$OT = \frac{(R_t \times \text{spm})^2 L}{50,000}$$

in SI units
$$OT = \frac{(R_t \times \text{spm})^2 L}{4645}$$

where R_t is the total length of rod string in 1000 feet (kilometers).

SUCKER ROD STRETCH, TUBING SHRINK   The sucker rod stretch starts when the plunger starts its upstroke and the fluid load is transferred from the tubing string to the sucker rod string. This occurs when the standing valve opens and the traveling valve closes. As the sucker rod string lengthens, the tubing string shortens. Both result in a stroke loss. The stroke loss due to rod stretch is shown in Table 2 as factor C, and the stroke loss due to tubing shrink is shown in Table 5 as factor K. Both are in inches per 1000 feet per 1000 pounds load (millimeters per kilometer per kilonewton).

NET PLUNGER STROKE   The formula below illustrates a method of computing the net plunger stroke for a tapered string of 1-, ⅞-, and ¾-in (25.4-, 22.2- and 19.0-mm) rods with the tubing suspended freely. If the tubing is anchored to the well casing near the pump, the KR_t factor will be zero.

in USCS units $L_p = L + OT - (CR_1 + CR_2 + CR_3 + KR_t)W_f \times 10^{-3}$

In SI units $L_p = L + OT - (CR_1 + CR_2 + CR_3 + KR_t)W_f \times 10^{-6}$

where L_p = net plunger stroke, in (m)
    R_1 = length of ¾-in (19.0-mm) rods, 1000 ft (km)
    R_2 = length of ⅞-in (22.2-mm) rods, 1000 ft (km)
    R_3 = length of 1-in (25.4-mm) rods, 1000 ft (km)
    R_t = length of tubing string, 1000 ft (km)

For a single-size rod string, zero values are assigned to the sizes not used.
    The pump displacement can now be calculated with the formula

in USCS units,   Barrels/day $= L_p \times$ bore factor $\times$ spm $\times$ 0.8

in SI units   Cubic meters/day $= L_p \times$ bore factor $\times$ spm $\times$ 0.8

The 0.8 factor assumes an 80% pump volumetric efficiency after correcting for sucker rod stretch and plunger overtravel. A lower efficiency should be assumed if the well is know to be gassy. If the pump displacement is more than required and the torque and structure capacity rating of the unit are not exceeded, the calculations should be repeated using the next smaller size unit.

FIG. 5   Winch pumping unit. (Bethlehem Steel)

COUNTERBALANCE   The pumping unit is supplied with a counterbalance in order to load the prime mover equally on the upstroke and the downstroke. The counterweights balance out the wet weight of the rods and half the weight of the fluid. All of the work in lifting is done on the upstroke. The energy stored in the counterweights during the downstroke supplies about half of the energy required for lifting the fluid. Since no fluid is lifted on the downstroke, the prime mover expends its energy lifting the unbalanced portion of the counterbalance load. Counterbalance effect CB is

$$CB = W_w + \tfrac{1}{2} W_f$$

The pumping unit pictured in Fig. 1 is known as a beam pumping unit. These are most common, but there are other types. One is a hydraulic surface unit, where a cylinder is set directly over the wellhead and a piston provides the reciprocating motion of the rods. These use commercial hydraulic pumps and control valves to supply the hydraulic power to the piston.

As shown in Table 3, standard sizes of beam pumping units have stroke lengths ranging from a little more than 1 ft (0.3 m) to a maximum of 16 ft (4.9 m). Recently a number of units have been marketed that permit stroke lengths up to 40 ft (12 m). The long stroke improves pump efficiency, prolongs sucker rod life, and reduces energy requiremens. The unit pictured in Fig. 5 uses wire line wound and unwound on a reversible drum to supply the reciprocating motion.

SUBSURFACE HYDRAULIC PUMPING SYSTEM

A complete hydraulic system consists of a fluid cleaning system, a power pump (usually a triplex), surface controls, a subsurface hydraulic pump, and tubing connecting the surface power pump to the subsurface pump. (See Sec. 3.1 for a complete description of power pumps.) Figure 6 is a schematic drawing of a complete multiple-well hydraulic system.

Because the pressurized power fluid supplies the energy needed to pump the well, the heavy sucker rod string and the heavy structure of the pumping unit required with the sucker rod pump are eliminated. Therefore, the hydraulic system can produce larger volumes of fluid from greater depths than the sucker rod pump.

The subsurface pump can be either a reciprocating piston pump or a jet pump. Both are usually made to be interchangeable in the same subsurface installation. Both pumps are made in what is called the free pump configuration, which means that the pump assembly can be pumped down into the well or pumped out of it by the same surface power fluid system that powers the pump during normal operation with the pump on the bottom. The free pump feature gives the hydraulic pump a distinct servicing advantage over the sucker rod pump, which requires a mast or derrick over the well and a pulling unit to pull the rods one or two at a time when the pump needs servicing.

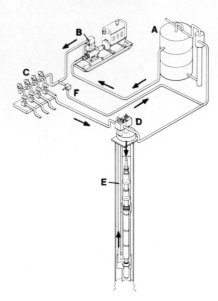

FIG. 6   Multiple-well hydraulic installation: (A) power fluid tank, (B) triplex pump, (C) manifold, (D) wellhead, (E) tubing, (F) bypass valve. (Kobe)

There are two basic types of hydraulic installations for the free pump: parallel free and casing free. Both result in a U-tube arrangement in which one leg of the U tube delivers the power fluid to the pump at the bottom and the other leg directs spent power fluid plus production back up to the surface. In the parallel free system, two parallel tubing strings° make up the U tube, and both are lowered into the well casing. In the casing free system, only one string is lowered into the well, and the annulus between this tubing and the well casing acts as the second leg of the U tube. Figure 7a illustrates installing a hydraulic pump in a parallel-free installation. A standing valve closes the bottom of the U tube, and the two strings are filled with fluid. The pump is inserted into the power fluid string, which is then securely capped, and the four-way valve is set to pump power fluid down the power fluid tubing. When the hydraulic pump reaches the bottom, a tapered bottom plug engages a seat on the standing valve and forms a pressure-tight seal. At the same time, an elastomer seal at the top of the pump enters a seal which now forces power fluid into the hydraulic pump (Fig. 7b) to start it pumping (as described later). When it is desired to retrieve the pump to replace worn parts, the position of the four-way valve is reversed and power fluid is directed down the production tubing (Fig. 7c.) The standing valve closes, the pump is unseated, and the circulation of fluid carries the pump to the surface, where it latches into the cap. The pressure is bled off from the U tube, and the wellhead cap with the worn pump is removed from the well.

The casing free installation (Fig. 8) employs a packer which anchors and seals the tubing to the casing just below the pump assembly. The annulus between the tubing and casing above the packer takes the place of the production string in the parallel free installation. In operation, the power fluid is pumped down the power fluid string as before, but the mixture of exhaust power fluid plus production is returned to the surface in the tubing casing annulus. The annulus below the standing valve is at pump suction pressure. The casing free hydraulic pump installation is more economical since it requires only one string of tubing. It also has the advantage that the return fluid passage up the annulus is very large and fluid frictional losses are very low.

° Individual joints of tubing coupled together to reach from the wellhead to the subsurface pump are called tubing strings.

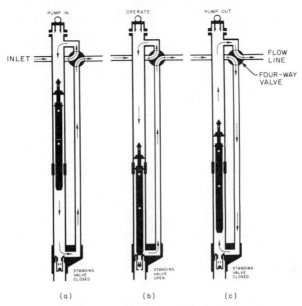

FIG. 7 Hydraulic parallel free pump installation.

Natural gas separates from the crude oil as it enters the well bore. In the parallel free installation, this gas rises to the surface in the annulus between the tubing strings and the casing, entering the flow line with the produced fluid at the wellhead. In the casing free installation, however, this annulus is used to return the liquid discharged from the pump, and all of the produced gas must be pumped with the produced liquid. The hydraulic pump is not very efficient in pumping gas. Therefore, in gassy wells, the pump efficiencies for the casing free system will be lower.

Both the reciprocating hydraulic pump and the jet pump are designed to operate on fluid from the well. Either crude oil or water may be used as the power fluid. Sand particles and other abrasives contained in the produced fluid are removed either by gravity separation in surface settling tanks or by centrifugal force in cyclone separators. The cleaned oil or water is then pressurized—typically to between 2500 and 4000 lb/in^2 (172 and 276 bar)—by a power pump at the surface, and then the high-pressure fluid is returned down the power fluid tubing to operate the subsurface hydraulic pump. The exhausted power fluid from the hydraulic pump mixes with the produced fluid from the well, and the mixture is returned to the surface in the parallel side string or the casing tubing annulus.

Figure 9 illustrates a typical central system. The power fluid is separated by settling in tanks. A triplex pump or a battery of triplex pumps manifolded together supplies the power fluid to a number of oil wells. Figure 10 shows a one-well unitized system, in which the combined stream of well production and power fluid is brought into a pressure vessel for a brief period for gas separation and solids settling. The selected fluid is then passed through a cyclone for cleaning before going to the surface pump. The abrasive-laden underflow from the cyclone and the surplus

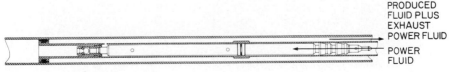

PRODUCED
FLUID PLUS
EXHAUST
POWER FLUID

POWER
FLUID

FIG. 8 Casing free installation.

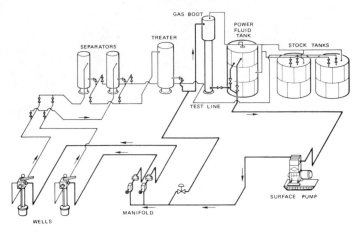

FIG. 9 Typical central system. (Kobe)

fluid from the pressure vessel then go into the flow line and are carried to the lease tanks. This fluid constitutes the well production.

Subsurface Reciprocating Hydraulic Pumps

A subsurface reciprocating hydraulic pump basically consists of an engine piston and cylinder with an engine reversing valve, and a pump barrel and plunger. These are assembled into one unit, and a polished rod connects the engine piston to the pump plunger so that the two reciprocate together.

Several designs of subsurface hydraulic reciprocating pumps are available. Figure 11a is a schematic of a double-acting pump. The engine valve directs high-pressure power fluid below the piston and opens the area above the piston to exhaust pressure on the upstroke. On the downstroke, it directs high-pressure power fluid above the piston and exhausts the power fluid below the piston to low pressure. The double-acting design exhausts an equal amount of power fluid from the engine and produced fluid from the pump on the upstroke and downstroke.

Figure 11b illustrates a balanced-design pump where the polished rod area is equal to half of the piston area. The underside of the piston is always connected to high-pressure power fluid. On

FIG. 10 One-well unitized system. (National Supply)

the downstroke, the engine valve directs high pressure on top of the piston. Since this area is larger than the bottom area of the piston, the unit makes a downstroke. On the upstroke, the engine valve exhausts the area above the piston to low pressure. The high pressure below the piston then causes the unit to make an upstroke. The balanced design exhausts all of the spent power fluid on the upstroke and all of the produced fluid displaced by the pump on the downstroke.

Figure 11c illustrates a single-acting pump where the polished rod area is small relative to the piston area. The underside of the piston is always connected to high-pressure power fluid. The engine valve directs high-pressure power fluid to the top of the piston on the downstroke and exhausts it to low pressure on the upstroke. The single-acting unit exhausts all of the exhaust power fluid and most of the produced fluid displaced by the pump on the upstroke. Only the displacement of the polished rod is exhausted on the downstroke.

In general, subsurface hydraulic reciprocating pumps are used in small- to medium-volume wells. When they are installed in 2⅜-in (60-mm) OD tubing, they are most commonly used to produce 25 to 500 barrels/day (4 to 80 m³/day); in 2⅞-in (73-mm) OD tubing, from 50 to 1000 barrels/day (8 to 158 m³/day); and in 3½-in (89-mm) OD tubing, from 100 to 1500 barrels/day (16 to 240 m³/day). Where conditions are right, these volumes can be exceeded, but in the higher volume ranges, the jet pump is usually a better application. Most subsurface hydraulic units have a maximum power fluid pressure rating of 4000 lb/in² (278 bar).

Reciprocating hydraulic pumps are available in several pressure ratios. For moderate-depth wells, the engine cylinder and the pump barrel can be of the same diameter. In deeper wells, a

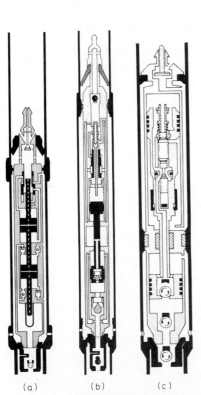

FIG. 11 Designs of subsurface hydraulic pumps: (a) double-acting, (b) balanced, (c) single-acting.

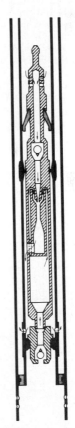

FIG. 12 Jet free pump.

pump plunger smaller than the engine cylinder is used; thus the operating pressure can be reduced proportionally, but a proportionally greater quantity of power fluid will be required. Hydraulic pumps are also offered with tandem engines and single pump for deep wells and with single engine and tandem pumps for shallow wells. The pump-to-engine-area ratio in these models varies from a low of 0.40 to a high of 2.00. There is no standardization of design among the various manufacturers, and the models of each are so diverse that no typical charts are offered here. Data can be obtained from the individual manufacturers.

As mentioned previously, the complete up and down cycle of the piston and plunger is called a stroke, and the distance traveled in one direction is called the stroke length. Stroke lengths offered vary from 1 to 5 ft (0.3 to 1.5 m). Stroke-per-minute ratings are from 200 on the shortest stroke to 50 on the longest stroke. The strokes are visible on the surface pressure gage since the pressure rises at each reversal when fluid flow is momentarily interrupted. Since the triplex power pump is positive displacement, the power fluid rate to the pump is usually controlled by bypassing the unneeded surplus. Some triplex pumps are equipped with a three-speed manual transmission that can be used to vary the speed of the surface pump. By selecting the proper rotational speed, the output volume of the surface pump can be adjusted closely to the requirements of the sub-surface pump. The triplex pump normally runs continuously, although occasionally in small-volume wells it will be run intermittently by a time clock.

Subsurface Jet Pumps

A jet pump has no moving parts, but it pumps by transferring momentum from a very-high-velocity—1000 ft/s (300m/s)—fluid jet to the pumped fluid (Fig. 12). At the nozzle exit, the jet creates a low-pressure area, thereby drawing in the fluid and carrying it to the throat of the pump, where the momentum transfer is completed. Then the combined stream enters the expanding tapered section of the diffuser, where the velocity head is converted back to a static head. The static head at this point is sufficient to lift the combined stream back to the surface. For a complete description of the general theory of jet pumps, see Chap. 4.

The jet pump has a broad range of application. It has been applied in wells as shallow as 800 ft (240 m) and as deep as 15,000 ft (4800 m). In general, a jet pump is capable of producing wells

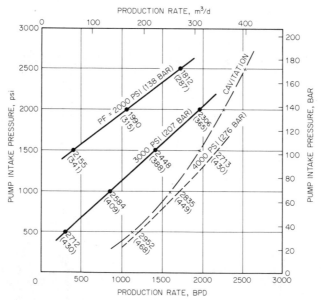

FIG. 13 Typical characteristic curves for a jet pump. Numbers adjacent to curves are power fluid in barrels (cubic meters) per day. Jet pump operation must be above and to the left of cavitation line.

in the following volume ranges: in 2⅜-in (60-mm) OD tubing, from 25 to 3000 barrels/day (4 to 475 m³/day); in 2⅞-in (73-mm) OD tubing, from 50 to 6000 barrels/day (8 to 950 m³/day); and in 3½-in (89-mm) OD tubing, from 100 to 12,000 barrels/day (16 to 1900 m³/day).

Jet pumps are less energy-efficient than reciprocating pumps. A well-designed jet pump may perform at approximately 40% hydraulic efficiency, as compared with over 90% efficiency for a reciprocating pump. However, jet pumps do have numerous advantages. With no moving parts, they have a high level of reliability, and sustained runs of several years are not uncommon. In addition to their high volume capability, jet pumps have a higher tolerance to abrasives in the produced fluid as well as in the power fluid. They are also more efficient in handling entrained gas in the pumped fluid.

Manufacturers offer a broad range of nozzle and throat sizes to cover the varied requirements of depths and volumes. Published data show the smallest nozzle requires 13.5 hydraulic hp (10 kW) and the largest requries 506 hydraulic hp (377 kW) at 4000 lb/in² (276 bar) power pump pressure. It requires a complex calculation to determine the optimum nozzle and throat sizes for any given set of well conditions. Manufacturers offer computer solutions, usually in graphical form. Variable factors, such as tubing friction, changing specific gravity due to gas, and changing well conditions, all contribute to the need of a graphical presentation. The chart in Fig. 13 shows a jet pump calculation for an 8000-ft (2440-m) well to predict the anticipated production volume versus pump intake pressure for power fluid pressures of 2000, 3000, and 4000 lb/in² (138, 207, and 276 bar). Note that at 4000 lb/in² (276 bar) power oil pressure, the pump would be cavitating, and this condition must be avoided. Cavitation occurs when the pressure at the entrance of the throat is less than the vapor pressure of the fluid pumped. The collapse of cavitation bubbles in the throat is so damaging that no known metal can resist destruction. Throat life in severe cavitation can be as short as 2 or 3 days. It may also be noted from the chart that the jet is very sensitive to pump intake pressure. Both the production rate and the overall hydraulic efficiency increase as pump intake (suction) pressure increases. A down-well pressure-recording instrument may be installed with the jet pump. It records pump intake pressure versus time for a 6-day period, at the end of which the jet pump with the instrument is pumped out and the pressure recording read and analyzed. This information is then used to determine if the nozzle and throat sizes are optimum and to select the power fluid pressure that will maximize production but still avoid cavitation.

Subsurface Progressing Cavity Pumps

A positive displacement screw-type pump (Sec. 3.4) can be used for handling the range of oil field fluids. Rotative speed can be varied to match well production with a smooth and steady delivery. No valves are required for pump operation. This type pump is well suited to handling gaseous formations.

As shown in Fig. 14, the rotor is a single rounded-cross-section external screw. The stator is a double internal helix molded of synthetic rubber. As the rotor turns, cavities form, and these cavities remain the same size as they progress from the bottom suction to the top discharge. The pump stator is suspended from a standard API tubing string and driven by standard API sucker rod. The pump is electric-motor-driven through belts and sheaves to obtain desired speeds.

Pumps of this type are available for oil well services (or for pumping fluids out of gas wells)

FIG. 14 Subsurface progressing cavity pump. (Fluids Handling Division, Robbins & Myers)

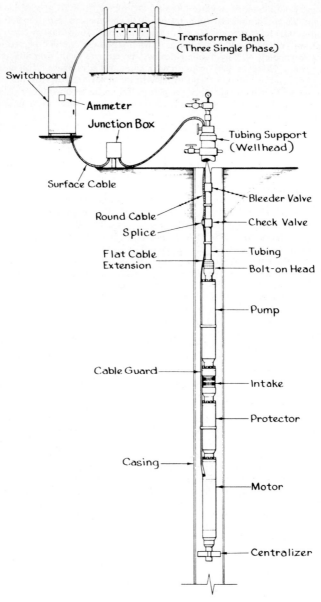

FIG. 15 Electric submersible centrifugal pump system. (PennWell Publishing)

to 3000 ft (914 m) or more. For pumping light or heavy crudes, pump sizes are available up to 100 barrels/day (16 m³/day), with speeds varying up to 550 rpm and power ratings to 5 hp (3.7 kW), depending on the well depth and flow required. A sucker rod size of ⅝ in (15.9 mm) is common.

ELECTRIC SUBMERSIBLE CENTRIFUGAL PUMPS

Electric submersible centrifugal pumps are adapted to a wide variety of pumping conditions. However, because of their high capacity, they are most frequently installed in wells where the volume of fluid to be produced exceeds the capacity of sucker rod or reciprocating hydraulic pumps. They are installed suspended from the discharge tubing and submerged in the well fluid. A three-conductor electric cable strapped to the discharge tubing transmits the power to the motor end of the pump.

The electric submersible pump system is illustrated schematically in Fig. 15. The electric motor and pump rotate at 3500 rpm for 60-Hz power and 2900 rpm for 50-Hz power. The electric motor is directly coupled to the pump with a seal section between the two. The motor is filled with an oil especially selected to provide high dielectric strength, lubrication for the bearings, and good thermal conductivity. The seal section isolates this fluid from the pump and allows for expansion and contraction of the fluid with temperature changes at the bottom of the well. The motor, which is the bottom element of the assembly, is usually installed in the well above the casing perforations where the fluid enters the well bore. This allows all of the fluid entering the pump suction to flow past the motor housing and provide the needed cooling.

The small diameter of the casing in which these pumps must be run severely limits design options. Length must be substituted for diameter to achieve the needed characteristics, and the motor design of the electric submersible pump barely resembles its surface counterpart. Single-motor assemblies range up to 32 ft (9.8 m) in length, and when power requirements exceed that of a single motor, two or more motors are assembled in tandem. The pumps must use many stages in order to gain the required head. The smallest-diameter pump—3.75-in (95.2-mm) OD—has 166 stages in an overall length of 12 ft (3.7 m). In deeper wells, two or more of these units can be assembled in tandem to gain the needed head. Manufacturers catalog pump combinations with as many as 400 stages. Table 6 gives typical data on the more common sizes. The numbers in the table are not limits, and manufacturers have available both smaller and larger sizes than those shown.

TABLE 6 Typical Characteristics Electric Submersible Centrifugal Pump

Well casing OD, in (mm)	Motor OD, in (mm)	Pump OD, in (mm)	Motor power, hp (kW)		Pump capacity, bbl/day (m³/day)		Pump head per stage, ft (m)	Efficiency, pump only, %
			Min	Max	Min	Max		
4.500	3.75	3.75	25	127	400	1,500	16	54
(114)	(95)	(95)	(18.7)	(94.7)	(64)	(238)	(5.2)	
5.500	4.56	4.56	120	240	400	2,800	23	61
(140)	(116)	(116)	(89.5)	(179)	(64)	(445)	(7.5)	
7.000	4.56	5.40	120	—	1,400	7,000	30	68
(178)	(116)	(137)	(89.5)	—	(223)	(1113)	(9.8)	
—	5.40	5.40	—	600	1,400	7,000	—	—
—	(137)	(137)	—	(488)	(223)	(1113)	—	
8.625	4.56	6.50	120	—	12,000	20,000	42	72
(219)	(116)	(165)	(89.5)	—	(1908)	(3180)	(13.8)	
—	5.40	6.50	—	600	12,000	20,000	—	—
—	(137)	(165)	—	(488)	(1908)	(3180)	—	
—	7.38	6.50	—	720	12,000	20,000	—	—
—	(188)	(165)	—	(537)	(1908)	(3180)	—	

OFFSHORE OIL WELL PUMPS

Most offshore oil well platforms are located in high-pressure fields where the wells will flow from natural pressure for several years. When the pressure declines, the wells are frequently produced with gas lift, where high-pressure gas is directed down the casing. This gas mixes with the well fluid at the bottom of the tubing, and the fluid-gas mixture lightens the fluid gradient in the tubing string to a point where the bottom hole pressure is sufficient to cause the well to flow. Where gas lift is not applicable, wells are pumped with subsurface submersible electric pumps, subsurface hydraulic reciprocating or jet pumps, or sucker rod pumping systems. A few wells are completed with the well head on the ocean floor. These must either flow or be gas-lifted, and the other systems are not applicable.

FURTHER READING

Brennan, J. R.: *Engineering Data and Production Calculations*, National Supply, Los Nietos, Calif., 1968.

Brown, K.: *The Technology of Artificial Lift Methods*, vol. 2b, Pennwell Publishing, Tulsa, Okla., 1980.

Primer of Oil and Gas Production, 3d ed., Johnson Printing, Dallas, 1976.

Shanahan, S. T.: *The Basics of Subsurface Oil Well Pumps*, BMW-Monarch, Gardena, Calif., 1975.

Wilson, P. M.: *Introduction to Hydraulic Pumping*, Kobe, Huntington Park, Calif., 1976.

SECTION 9.19
CRYOGENIC LIQUEFIED GAS SERVICE

L. R. SMITH

The regime of cryogenic technology has been generally taken to indicate temperatures colder than $-100°F$ $(-73°C)$. Fluids such as liquid oxygen, nitrogen, hydrogen, helium, argon, methane, and ethane, with normal boiling points below $-100°F$ $(-73°C)$, are called cryogenic fluids.

For the pump designer, the cryogenic regime requires consideration of the effect of low temperatures on the properties of construction materials and the effect of varying shrinkage rates on critical fits and clearances. The problem is further complicated by the fact that cryogenic fluids are stored at near atmospheric pressure and must be pumped at or near their normal boiling point, so that the only NPSH available is that due to the liquid level above the pump suction.

The history of commercial cryogenic pumping divides into two eras. The first, commencing in the early 1930s with the first liquid oxygen plant in the United States, was the period in which end-suction shaft seal pumps were developed and produced for pumping liquefied atmospheric gases, such as liquid oxygen, nitrogen, and argon. This industry grew until the late 1960s. Though continuing to grow, the explosive period appears to have ended.

The second era commenced in 1959 with the first transport of a commercial cargo of liquefied natural gas (LNG) from Lake Charles, La., across the Atlantic Ocean to England. This voyage, carrying an almost token quantity of 5000 m³ of LNG, inaugurated an era of international trade in liquefied hydrocarbon gases that has grown with astounding rapidity and seems still to be barely on the threshold of realizing its full potential.

ATMOSPHERIC GASES

Cryogenic pumps first came into being with the production of liquefied atmospheric gases in commercial quantities. The initial liquefaction of air and the subsequent separation of oxygen and nitrogen did not require pumps since the small volume of liquids produced could be readily transferred by pressure. Early pumps, where needed, were designed and built by the gas companies themselves and were primarily used to fill high-pressure bottles or cylinders. Flows were very low, and so positive displacement pumps were commonly used. These were notoriously inefficient, primarily because of the flashing that occurred during the intake stroke and led to a very low volumetric efficiency. However, the low volume of product being handled permitted this efficiency to be tolerated.

The World War II development of rocket engines that used liquid oxygen as an oxidizer to burn kerosene fuel and the postwar development of the large rocket booster required the development of centrifugal pumps to feed the propellants to the engine. These turbine-driven rocket engine pumps were adapted to electric motors for use in transferring the propellants from storage into the rocket's tanks because there were no high-capacity cryogenic pumps available from the industry. Although these early rocket pumps were also inefficient, they could be tolerated in the rocket because, since the fluid was burned immediately, temperature rise across the pump was not a problem.

Two other developments in the early postwar years increased the volume of liquefied gases in industrial applications. The first was the development of the basic oxygen furnace for making steel, and the second was the use of liquefied nitrogen in the fast-freezing of foods and as an inert atmosphere for heat-treating and chemical processes. These developments required the storage and transfer of large volumes of liquids, so that the boiloff loss due to pump inefficiency became a major economic factor in the profit and loss of the liquefaction company. The major companies instituted internal studies to reduce these losses, and soon the transfer pumps were brought under scrutiny.

Thus a two-pronged impetus developed to find pumps that would more efficiently and reliably transfer these valuable liquids; the first was the need to load military rockets with the coldest possible liquid with utmost reliability, and the second was from industry. Investigators studying the problem first turned to the producers of commercial pumps. However, since the pump volumes required were small and the technical problems formidable, they were unable to retain the interest of the major manufacturers and so turned to small producers of custom-designed pumps.

The earliest custom producer to enter the field was a builder of high-capacity in-flight refueling pumps for the military. One of these pumps was adapted to a commercial 60-Hz electric

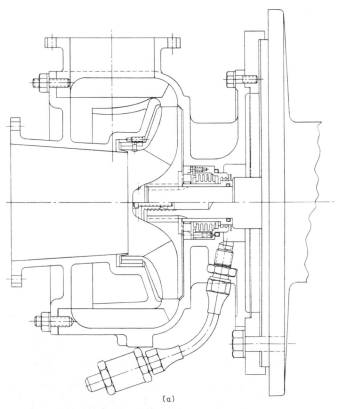

(a)

FIG. 1a In-flight refueling pump. (J. C. Carter Division, ITT)

FIG. 1b DC-motor-driven refueling pump with original flanges. (J. C. Carter Division, ITT)

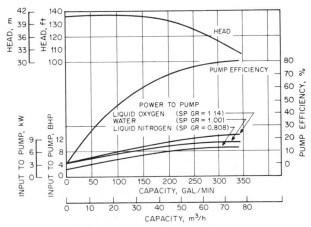

FIG. 1c Full commerical cryogenic configuration of dc-motor-driven refueling pump with ASA flanges. (J. C. Carter Division, ITT)

FIG. 2 Performance characteristics for a refueling pump at 3500 rpm. (J. C. Carter Division, ITT)

FIG. 3 Cryogenic pumps for loading liquid-propelled 1CBMs. (J. C. Carter Division, ITT)

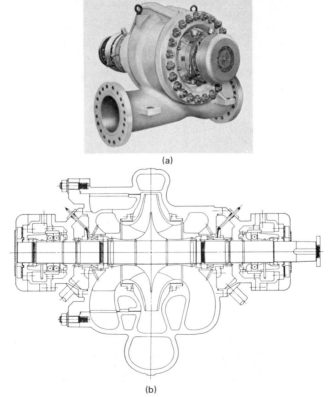

(a)

(b)

FIG. 4 (*a*) Double-suction cryogenic pump, (*b*) cross-sectional view. (Byron Jackson Division, Borg-Warner)

motor using a newly designed face-type shaft seal with the stationary face mounted on a stainless steel convoluted bellows. A cross section of this pump is shown in Fig. 1*a*.

The refueling pump was constructed of aluminum castings with bronze wear rings and stainless steel trim. These materials were good choices for cryogenic service as well, and so, aside from the seal, it was necessary only to change to standard pipe flanges on the inlet and discharge nozzles. Figure 1*b* shows this pump mounted on the original dc aircraft motor used on the KB29 airplane, transferred to a 60-Hz, 3-phase motor but still with the original flanges. Figure 1*c* shows the full commercial cryogenic configuration with ASA flanges and stainless steel thermal barrier/distance price for mounting the pump onto the motor. The pump, originally designed under the space and weight limitations of aircraft service, also fit the requirements of cryogenic pumping in that the low heat capacity and surface area resulted in short cool-down time and low heat leak. Since the pump was also very efficient (Fig. 2), the boiloff problem was greatly alleviated. The gas industry welcomed this pump, and it became one of the most popular transfer pumps for loading and unloading tank cars and trailers.

Following the lead set by this pump, several other small manufacturers began building similar pumps. As the industry grew, pumps of larger capacity and higher heads were built, covering the range from 4 gpm (0.9 m³/h), used in transferring liquid from rectifier columns to storage, to more than 5000 gpm (1125 m³/h), used for loading the early liquid-propelled ICBMs during the 1960s (Fig. 3). Only one of these latter systems was built. The pumps shown in Fig. 3 were about the largest end-suction close-coupled pumps ever built. The pump built to load the Saturn rocket at 10,000 gpm (2250 m³/h) was not close-coupled and was possibly the only double-suction cryogenic pump ever built (Fig. 4).

LIQUEFIED HYDROCARBON GASES

The cryogenic fluids encountered in this service are primarily LNG (a mixture that is normally more than 90% methane), ethylene, and ethane, along with the liquefied petroleum gases (LPG) propane and butane. A novel approach to pumping these fluids was introduced in 1959 with the application of the submerged electric-motor-driven pump. Since these fluids are excellent dielectrics, part of the pumped fluid stream can be directed through the motor to cool it and lubricate the bearings. There is no need to can or treat the windings with anything other than specially selected varnishes. Many advantages accrue to this design, such as

- No cool down requirement on pumps installed in tanks
- High inherent reliability due to protection from corrosion and humidity and elimination of shaft seal
- Low hazard due to 100% rich environment
- Minimal differential shrinkage problems
- Capability of directing rejected heat in accordance with designer's wishes

Also, the pump is close-coupled to the motor, eliminating the need for a long line shaft, line shaft bearing problems, and differential shrinkage problems. As a result, higher speeds can be used and pumps can be smaller and less expensive.

This construction has found many applications that have led to a variety of configurations, including the single and multistage units used in the liquefaction process plant to transfer to storage, load and unload cargo ships, and pump to high pressures for regasification. The pumps can be mounted in tubes inside the tanks so that they can be removed for service through the roof without emptying the tank, thus eliminating the need for an opening in the bottom of the tank. They are also mounted in suction barrels that can be maintained cold indefinitely on instant standby.

A typical cargo pump for use on board ships is shown in Fig. 5. This pump uses aluminum castings for all housings as well as for the pump impeller. The motor uses electrical iron laminations, a stainless steel shaft, and stainless steel ball bearings with special nonmetallic separators. The path of the top wear ring flow up through the motor and back into the tank is easily seen.

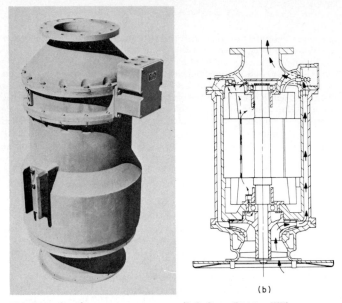

(b)

FIG. 5 On-board cryogenic cargo pumps. (J. C. Carter Division, ITT)

FIG. 6 Ship-loading cryogenic pump. (J. C. Carter Division, ITT)

This pump stands about 5 ft (1.5 m) high from the suction flange to the discharge flange. Flow is typically in the 5000-gpm (1125-m³/h) range. Motors with ratings up to 450 hp (336 kW) are available.

Other configurations are shown in Figs. 6 and 7. Figure 6 is a typical ship-loading pump provided with an 800-hp (597-kW) motor and capable of pumping more than 20,000 gpm (4500 m³/h). This type of pump is designed for mounting in a suction pot. Figure 7 shows a 1200-hp (894-kW) multistage pump being removed from the test tank. Pumps such as this are capable of more than 1000 lb/in² (69 bar°) discharge pressure pumping LNG and are in service with motors of 2500 hp (1860 kW).

FIG. 7 Multistage cryogenic pump on test stand. (J. C. Carter Division, ITT)

ENGINEERING PROBLEMS _____

The pumps described in the preceding pages, although widely diverse in configuration, still have a number of common problems. First of all there is the extreme, paralyzing cold. It is small wonder that the early pumps were considered successful when they did not freeze into immobility. The cold penetrated the motor, causing grease-lubricated bearings to seize and filling the inside of the motor with ice that locked the rotor fan blades and kept them from turning. The piping attached to the pumps shrank, distorting the pump casings into heavy rubbing contact with the impeller. Frost and ice entered the external side of the seal, immobilizing it so that leakage was uncontrollable. It was difficult to find materials for construction that did not become unusually brittle, and there were almost no data available on the total shrinkage of common materials at cryogenic temperatures.

The best source for data on the properties of materials at low temperatures is the *Cryogenic Materials Data Handbook*, available by requesting document AD609562 from the National Technical Information Service, U.S. Department of Commerce, Springfield, Va., 22161. More detailed data on the properties of particular materials are available in some instances in papers that have been presented at the Cryogenic Engineering Conference. These papers are published in *Advances in Cryogenic engineering*, K. D. Timmerhaus (ed.), published annually by Plenum Press, New York. Material suppliers can also provide information on specific problems, such as properties under high strain or dimensional stability under temperature cycling.

Other problems resulting from low-temperature operation are mainly mechanical and are surmounted by using thermal barriers, flexible sections of piping, and dry purge gas to prevent the ingress of moist air. Gasketing and lubrication problems are both currently being solved by materials which did not exist when the problems were originally encountered.

One common problem has to do with the characteristics of the fluids being pumped. Cryogenic fluids are stored in insulated tanks maintained at pressures only slightly above atmospheric, and the fluids become saturated at the storage pressure. Since it is not possible to obtain any pump elevation relative to the tank bottom, net positive suction head becomes a real problem. Various styles of inducers have been used to improve the suction performance of the pumps, but in nearly all cases some degradation of performance must be anticipated as the tank is drawn down to low levels and cavitation operation is inevitable. Fortunately, operation in cavitation is not as severe

° 1 bar = 10^5 Pa. For a discussion of bar, see *SI Units—A Commentary* in the front matter.

a problem in cryogenic fluids as it is in water, probably because of the much lower energy of the cryogenic fluid bubble, and so severe cavitation damage to pump parts is seldom encountered.

Testing of pumps for cryogenic service is difficult and requires a heavy investment in test loop equipment. It is also fraught with problems that are peculiar to testing and will not be present in actual operation. One such problem has to do with the recirculation operation in a test loop. The work put into a saturated fluid is usually released by allowing some of the fluid to boil away. Since it proves nearly impossible to extract all the vapor from the liquid, the test fluid specific gravity effectively decreases by an unknown amount. This makes it difficult to establish the exact performance that has been obtained since the data will be lower than true values based on the specific gravity of the unadulterated fluid. It has been found with careful testing that exact duplication of water head-capacity performance will be obtained on any cryogenic fluid, subject only to adjustment for the shrinkage in impeller diameter.

SECTION 9.20
WATER PRESSURE BOOSTER SYSTEMS

SATORU SHIKASHO

TYPES OF WATER PRESSURE BOOSTER SYSTEMS

Plumbing fixtures and equipment connected to a water source will function properly only when a minimum water supply pressure is consistently available. Whenever this minimum pressure cannot be maintained by the supply source, a means of boosting the water pressure should be considered.

The commonly used methods for boosting pressure are

1. Elevated gravity (tank) systems
2. Hydropneumatic tank systems
3. Variable-speed-drive centrifugal pump booster systems
4. Tankless constant-speed multiple centrifugal pump systems
5. Limited-storage, constant-speed multiple centrifugal pump systems

Table 1 indentifies the significant advantages and disadvantages of each system.

Although gravity and hydropneumatic tank systems have some distinct advantages, recent installations have favored constant-speed multiple-pump and variable-speed-drive systems. Since gravity and hydropneumatic tank systems are now seldom specified, they will not be discussed in detail here. Multiple-pump systems are well suited for high-rise buildings, apartment buildings, schools, and commercial installations. Variable-speed-drive systems are particularly suited for industrial applications, where control precision is required and maintenance capability for complex electronic apparatus is present.

It is important to remember that the primary function of a pressure booster pump for a building, whether variable- or constant-speed, is to maintain the desired system pressure over the entire design flow range. It is also important to recognize that the cold water service piping configuration is an open-loop system, not a closed-loop system as it is in heating and cooling systems. With an open-loop system, the static head above the pump is not canceled by the down leg of the piping, and therefore the pump must develop a head equal to or greater than the static head, even at zero flow (with suction pressure at design). Review Secs. 8.1 and 8.2 for further discussions on closed-loop systems.

TABLE 1 Comparison of Significant Factors for Pressure Booster Systems

System type	Advantages	Disadvantages
Elevated gravity tank	Simplicity, large storage capacity, low energy usage, reserve storage capacity for fire protection	Size and weight, water damage potential, freeze potential if roof-mounted, corrosion and contamination potential, limited pressure for floor immediately below tank, possible unsightly appearance, periodic cleaning and painting
Hydropneumatic tank	Location not critical, low energy usage, limited storage capacity, pump does not run when there is no demand, operation in optimum pump flow range	Requires compressed air source, corrosion and contamination potential, large pressure variation, relatively large, standby provision costly
Variable-speed drive: Fluid couplings	Simple controls, off-the-shelf motor, standby provision less costly than for above units	Slow response to sudden demand change, may require heat exchanger to cool drive, slip losses result in lower maximum speed and higher motor power, requires selective application, no water storage
Variable-speed drive: ac type	Low motor current inrush, precise pressure control, higher than 3600 rpm possible, few mechanical devices, large power capability at lower first cost	Complex electric circuitry, high initial cost for low-power units, may require special motors, motor low-speed limitation, rapidly changing technology, requires selective application
Tankless constant-speed multiple pump	Relatively low first cost, uses time-proven components, compact size, inherent partial standby capacity, extra standby capacity inexpensive, location not critical, good pressure regulation	Continuously running lead pump, difficult to accurately determine capacity split among pumps, no water storage, problems associated with low flow rates
Limited-storage constant-speed multiple pump	Shuts down during very low water demand, uses time-proven components, standby capacity inexpensive, location not critical, limited water storage, no air-to-water contamination with diaphragm tank	No significant disadvantages, tanks with high maximum working pressure may be difficult to obtain

Variable-Speed-Drive Pressure Booster Systems Initial variable-speed drives were fluid couplings, magnetic couplings, and liquid rheostat (wound rotor motor) drives. The coupling-type drives were driven by a constant-speed motor, with the coupling output shaft varying in speed. With the advent of solid-state electronics technology and circuit miniaturization, most variable-speed drives currently used for pressure booster applications are of the solid-state, ac, adjustable-speed type. These drives are discussed in Subsec. 6.2.2.

Variable-speed drives are usually specified for their low operating cost potential. To achieve the energy-savings goal, the system conditions should cause the drive speed to vary between 50 and 75% of full speed during most of the operating period.

The pump speed, head, and flow relationships are expressed by the affinity laws:

- Flow varies directly as speed:

$$\frac{G_2}{G_1} = \frac{R_2}{R_1} \tag{1}$$

- Head varies directly as the square of the speed:

$$\frac{H_2}{H_1} = \left(\frac{R_2}{R_1}\right)^2 \tag{2}$$

where G = flow rate, gpm (m^3/h)
$\quad R$ = pump speed, rpm
$\quad H$ = pump total head, ft (m)

The required pump total head at design conditions is

$$H = (H_d - H_s) + H_f \tag{3}$$

where H = required pump total head at design conditions, ft (m)
$\quad H_d$ = design system pressure at point of control, ft (m)
$\quad H_s$ = minimum design suction pressure, ft (m)
$\quad H_f$ = sum of all losses between H_d and H_s at design flow, ft (m)

At flow rates less than design conditions and suction pressures higher than the minimum design pressure, the pump head requirement is reduced by the change in pipe frictional losses H_f and the additional suction pressure H_s available. Also, because of the rising characteristic of the centrifugal pump performance curve as the flow decreases, an excess head is developed by the pump. All of these changes will alter the required pump total head and the pump speed from the design (maximum) values.

For a single operating pump unit, design maximum speed will be required at design minimum suction head and design maximum flow. Minimum speed will be required at maximum suction head and minimum flow. The procedure required to determine the speed range of the pump driver requires construction of the system-head curve, the pump head-capacity curve (see Sec. 8.1), and the affinity curve. Rearranging the affinity equations (Eqs. 1 and 2), the speed changes are calculated:

$$R_2 = R_1 \left(\frac{G_2}{G_1}\right) \tag{4}$$

$$R_2 = R_1 \left(\frac{H_2}{H_1}\right)^{1/2} \tag{5}$$

where the subscripts 1 and 2 represent the higher and lower speed values, respectively, for pump conditions at the same specific speed (see Subsec. 2.3.1) or along the same affinity line.

The following examples illustrate how changes in flow rate, pipe frictional losses, and suction head affect pump speed.

EXAMPLE 1   With no change in suction pressure.
First (design) conditions:

$$H_d = 193 \text{ ft } (58.8 \text{ m})$$
$$H_s = 50 \text{ ft } (15.2 \text{ m})$$
$$H_f = 7 \text{ ft } (2.1 \text{ m})$$
$$G = 190 \text{ gpm } (43.1 \text{ m}^3/\text{h})$$
$$R = 3500 \text{ rpm}$$
$$H = (193 - 50) + 7 = 150 \text{ ft } (45.7 \text{ m}) \text{ (Eq. 3)}$$

Second conditions:

$$H_d = 193 \text{ ft } (58.8 \text{ m})$$
$$H_s = 50 \text{ ft } (15.2 \text{ m})$$
$$H_f = 1.9 \text{ ft } (0.58 \text{ m})$$
$$G = 100 \text{ gpm } (22.7 \text{ m}^3/\text{h})$$
$$R = \text{to be calculated}$$
$$H = (193 - 50) + 1.9 = 144.9 \text{ ft } (44.18 \text{ m}) \text{ (Eq. 3)}$$

The 3500-rpm pump head-capacity curve, the system-head curves, and the affinity (square) curves are shown in Fig. 1. The operating points are lettered. Point A is for maximum flow at minimum suction head (design conditions). Point B is for lower flow at minimum suction head at a speed to be determined. The reduced piping losses between head points H_d and H_s are represented by the system-head curves.

To determine the approximate speed at point B, it is necessary to use trial and error since the flow and head ratios in Eqs. 4 and 5 are not known. The following procedure may be used to estimate the speed (Fig. 1):

1. Draw an affinity (square) curve passing through the zero-flow/zero-head point and the lower operating point (point B) and intersecting at head-capacity curve of known speed (point C).

2. Read the probable flow rate at point C, the point of intersection between the affinity curve and the pump curve of known speed.

3. Calculate the head relative to the flow rate read at point C, based on the square curve relationship:

$$H_1 = \frac{H_2}{\left(\dfrac{G_2}{G_1}\right)^2} \tag{6}$$

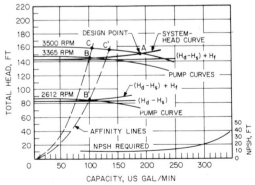

FIG. 1   System and pump curves to determine speed reduction for the 2-in (51-mm) pump in Examples 1 and 2. (ft × 0.3048 = m; gpm × 0.2271 = m³/h)

where subscripts 1 and 2 represent the values at the 3500-rpm and lower speeds (points C & B), respectively.

4. Compare the head calculated in step 3 with the head of the pump curve (point C) of known speed at the probable flow rate of step 2. If the values differ significantly, try another flow rate and repeat steps 3 and 4.

5. Using the accepted head from step 4, calculate the speed at the operating point (point B) using Eq. 4 or 5. The equation not used may then be used for verification.

For this example, the estimated flow rate at point A (step 2) is 104 gpm (23.6 m³/h) and the calculated head at the same point (steps 3 and 4) is

$$H_1 = \frac{144.9}{\left(\frac{100}{104}\right)^2} = 156.7 \text{ ft} \quad \text{or} \quad \frac{44.18}{\left(\frac{22.7}{23.6}\right)^2} = 47.76 \text{ m}$$

The speed at the operating point (point B) is

$$R_2 = 3500\left(\frac{100}{104}\right) \quad \text{or} \quad 3500\left(\frac{22.7}{23.6}\right) = 3365 \text{ rpm} \quad \text{(Eq. 4)}$$

Check:

$$R_2 = 3500\left(\frac{144.9}{156.7}\right)^{1/2} \quad \text{or} \quad 3500\left(\frac{44.17}{47.76}\right)^{1/2} = 3366 \text{ rpm} \quad \text{(Eq. 5)}$$

The speed change is

$$\frac{3500 - 3365}{3500} \times 100 = 3.9\%$$

EXAMPLE 2   With change in suction pressure.
First conditions: Same as Example 1.
Second conditions: Same as Example 1, except

$$H_s = 110 \text{ ft (33.5 m)}$$
$$H = (193 - 110) + 1.9 = 84.9 \text{ ft (25.88 m) (Eq. 3)}$$

Operating point B' and other curves related to this example are shown in Fig. 1. Again by trial and error, the flow rate and pump head at the intersection of the affinity curve and the 3500-rpm pump curve (point A') are 134 gpm (30.4 m³/h) and 152.4 ft (46.45 m). The speed at the operating point is:

$$R_2 = 3500\left(\frac{100}{134}\right) \quad \text{or} \quad 3500\left(\frac{22.7}{30.4}\right) = 2612 \text{ rpm} \quad \text{(Eq. 4)}$$

Check:

$$R_2 = 3500\left(\frac{84.9}{152.4}\right)^{1/2} \quad \text{or } 3500\left(\frac{25.88}{46.45}\right)^{1/2} = 2612 \text{ rpm} \quad \text{(Eq. 5)}$$

The speed change is

$$\frac{3500 - 2612}{3500} \times 100 = 25.4\%$$

These examples indicate that the speed of the pump is not significantly affected unless there is an appreciable change in suction pressure.

For optimum performance, it should be confirmed, after the speed has been determined, that the most often occurring flow demand range is ideally near the pump's maximum efficiency.

Generally, the design point at maximum speed should be selected to the right of the pump's best efficiency point. The operating flow range at different speeds should be within the hydraulically and mechanically stable ranges of the pump.

The pump shaft power at any specified speed is

in USCS units

$$\text{hp} = \frac{HG \text{ (sp. gr.)}}{3960E}$$

in SI units

$$\text{kW} = \frac{HG \text{ (sp. gr.)}}{367E}$$

(7)

where H = pump total head, ft (m)
G = flow rate, gpm (m³/h)
sp. gr. = specific gravity of fluid
E = pump efficiency, % (expressed in decimal equivalent)

H, G, and E are values from the pump performance curve for the specified speed and impeller size

The motor power is the same as the shaft power for pumps driven directly by the motor. For pumps driven through intermediate variable-speed couplings, the motor power must also include the drive slip and fixed losses.

When the required design flow exceeds the capacity of a single pump, several pumps, including possibly one constant-speed unit, can be arranged to operate in parallel. Two methods of staging the pumps are usually used:

1. The first pump is operated in the variable-speed mode until its maximum speed is reached; then the second pump, also a variable-speed unit, is energized. Now both pumps are operating in parallel at the same reduced speed. This sequence is repeated for the other pumps.

2. The first pump is operated in the variable-speed mode until its maximum speed is reached and is then locked in at this speed to operate as a constant-speed pump. The second pump is energized and operates in the variable-speed mode. The sequence is repeated for the other pumps. The speed variation of the second pump is relatively small since it must develop the same head as the first pump operating at maximum speed. Since the flow rate through the second pump is less, its speed is affected by the increase in total head available from the rise in the pump performance curve.

Which method of sequencing is selected depends on economics, equipment redundancy, and other considerations. For example, with the first method, both pumps must be furnished with variable-speed drive units. With the second method, only one variable-speed drive unit is required if it is an electronic type since the first pump is locked in by separate electric means.

The sole function of the pumps is to maintain constant pressure at the control point; therefore the controller selected must be pressure-actuated. The type (follower signal) of controller chosen depends on the type of speed change signal acceptable by the variable-speed control unit. For electronic motor speed controls, such as variable-voltage and variable-frequency units, typical follower signals are low-voltage dc, milliamp dc, 135-ohm potentiometer, and pneumatic.

Tankless Constant-Speed Multiple-Pump System The major components of this system (Fig. 2) are

1. One or more pumps, with two or three most common

2. A combination pressure-reducing and check valve (PRCV) for each pump (parallel-piped PRCVs are used with larger sizes; separate pressure-reducing valves and check valves may be used)

3. An automatic sequencing control panel

4. When factory assembled, a steel frame for the entire unit

(a)

(b)

(c)

FIG. 2 Constant-speed multiple-pump pressure booster systems: (*a*) horizontal end-suction volute pumps (ITT Bell and Gossett), (*b*) vertical end-suction volute pumps (ITT Bell and Gossett), (*c*) vertical turbine pumps with limited storage (SynchroFlo).

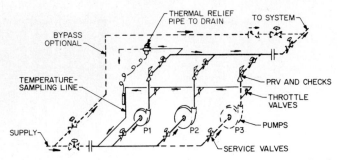

FIG. 3   Constant-speed multiple-pump pressure booster schematic. (ITT Bell and Gossett)

PIPING ARRANGEMENT AND FLOW PATH   A schematic of the piping arrangement and flow path in a constant-speed multiple-pump system is shown in Fig. 3. Supply water under fluctuating pressure enters the suction header and flows into the pump, where it is boosted to a higher pressure. This varying-high-pressure water enters the PRCV (Fig. 4), and the pressure is reduced to the constant pressure desired over the design flow range. Flow reversals through the idle and parallel pump circuits are prevented by the checking feature of the PRCV, which also dampens the pressure fluctuations caused by sudden flow changes.

When only minor changes in supply water pressure are anticipated, such as from a nonpressurized tank, and the pump head-capacity curve is relatively flat, silent check valves without pressure-reducing feature may be used. This will result in a slight increase in discharge pressure above the desired design constant pressure. When pumps of different sizes are used, care should be exercised to avoid having the higher-head pump force the lower-head pump to shut off or to operate at less than minimum design pump flow. A decrease in flow from a centrifugal pump must be accompanied by an increase in pump total head as required by the pump head-capacity curve. If check valves only are used, it is preferable to have all pumps and valves identical to avoid unbalanced flows.

Operating the pumps at shutoff (no flow) will cause the water temperature in the pump castings to rise (Subsec. 2.3.1). To keep the temperature in the pumps within a safe limit, the heated water is relieved through the thermal relief valve. This valve may be either a self-actuated (thermostatic) type or a solenoid valve actuated by a temperature controller.

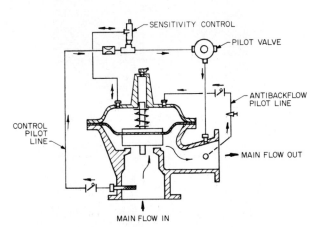

FIG. 4   Typical pressure-reducing and check valve to maintain constant system pressure and prevent backflow. (Cla-Val)

PUMP CONTROL PANEL   One of the advantages of a factory-assembled pressure booster package is the prewired, pretested, pretubed, mounted control panel which requires a minimum of field connections. In addition to providing for the proper sequencing of the pumps, the control panel should contain electrical interlocks for the operating and safety controls and circuit connections for remote control units.

Standard items and optional equipment vary considerably from one manufacturer to another. The components usually included with a standard panel are listed below.

Steel enclosure	Starters
Control transformer	Sequencing controllers
Control circuit protector	Pump failure interlocks
Selector switches	Minimum-run timers
Low-suction pressure control	Time delays
	Pilot lights

Optional features which may be available:

Power supply fused disconnects or circuit breakers	Low system pressure control
	Low water level control
Enclosure door interlock	Emergency power switchover
High water temperature control	Unit failure alarm
High system pressure control	Low-flow shutdown
Pump alternation	Misc. enclosure types
Program time switch	Power economizer circuit
Elapsed time meters	Additional pilot lights

Most factory-wired panels conform to one or more of the consumer safety agencies, such as Underwriters' Laboratories (UL), National Electrical Code (NFPA/NEC), and Canadian Standards Association (CSA), and are furnished with a label so indicating.

PUMP CONTROL SEQUENCE   A typical elementary wiring diagram provides the following sequences of events. With both pump selector switches in the auto mode and all safety and operating controls in run status, the lead pump starter is energized, starting pump 1. As the system water demand increases, a staging control switch, which senses motor current, flow, or system pressure, starts pump 2. Pump 2 continues to operate until the decreasing water demand causes the staging control switch to open, stopping pump 2. If the circuit is provided with a minimum-run timer, pump 2 will continue to run for the set time period regardless of the staging switch status. This timer prevents pump 2 from short-cycling during rapidly fluctuating demand periods.

For test purposes and emergency operation, both pumps may be operated by placing the selector switches in the *hand* mode. In this position, most of the safety and operating controls are bypassed.

Should pump starter 1 fail to operate because of an overload heater relay trip or starter malfunction, a failure interlock switch automatically starts pump 2.

Pump 1 will run continuously unless the circuit is provided with shutdown features, such as high-suction pressure control and/or low-flow shutdown control.

Many units are specified for manual or automatic alternation of the equally sized pumps. This feature is intended to equalize the wear among the pumps and associated components. Alternation of any pump designated for *standby duty* (emergency use) may not be prudent. The standby pump should be preserved until needed, similar to an emergency power generator.

Limited-Storage Constant-Speed Multiple-Pump System   The most notable shortcoming of the tankless multiple-pump system is the need for a continuously running pump, even at zero water usage. This drawback is remedied by the addition of a pressurized storage tank connected to the high-pressure side of the piping system with low-flow shutdown controls to stop the pumps.

The limited-storage system is not a scaled-down version of the hydropneumatic system. It differs in three important aspects:

1. The primary function of the tank is water storage.
2. Tank pressure is developed by the booster pump.
3. Air and water in the tank are separated by a flexible barrier (diaphragm, bladder) to eliminate direct interaction between the gas and water in the tank. The barrier also prevents the gas from escaping when the tank is emptied of water.

SEQUENCE OF OPERATION CONTROL CIRCUITRY   During periods of normal water usage, the operating sequence is that of the tankless unit. As the water flow approaches zero, a low-flow sensing device stops the pumps. The tank provides the water needs during the shutdown period until the water pressure diminishes to the minimum allowable value. At this time, a pressure switch starts the lead pump to restore the pressure and recharge the tank with water.

The low-flow device may be a flow switch, a pressure switch, or a temperature-actuated switch and must be capable of switching at flows below 5 gpm (1.14 m³/h).

The tank should be sized for sufficient drawdown volume (water available from tank during shutdown) to avoid excessive pump cycling.

RELATIONSHIPS BETWEEN TANK SIZE, DRAWDOWN VOLUME, AND PRESSURE   The approximate formula for sizing prepressurized diaphragm tanks at constant temperature is

$$V_t = \frac{V_e}{1 - \dfrac{P_f}{P_0}} \tag{8}$$

where V_t = tank volume, ft³ (m³)
V_e = change of gas volume in tank, ft³ (m³)
P_f = final gas pressure lb/in² (kPa) abs
P_0 = initial gas pressure, lb/in² (kPa) abs

The corresponding relationships of water volume and pressures in the tank are

- V_e is equal to the drawdown volume.
- P_f is the water pressure at the end of the drawdown cycle, the minimum allowable system pressure.
- P_0 is the water pressure at the beginning of the drawdown cycle or at the termination of the charging cycle.

TANK LOCATION   Since both P_f and Po are influenced by the static head above the tank and the available charging pressure, the location of the tank and point of connection to the sytem should be carefully selected. Connecting the tank at A in Fig. 5 provides the highest available tank charging pressure since this pressure is not affected by the reduction through the PRVs. However, there are several disadvantages to this point of connection:

1. The highest static pressure is applied to the tank since the pumps are usually located on the lowest floor of the building.
2. The tank may be subjected to high working pressure since any increase in suction pressure above the minimum design pressure is additive to the pump head.
3. The tank and all alternated lead pumps must be interconnected.
4. A silent check valve must be placed between the tank connection point and the pump discharge nozzle.

Connecting the tank in B in Fig. 5, downstream of the PRVs, will eliminate disadvantages 2, 3, and 4. The available charging pressure is less, and therefore a larger tank is required. However, tank location is not critical.

Locating the tank on the upper elevation of the building, such as at C, reduces the static head

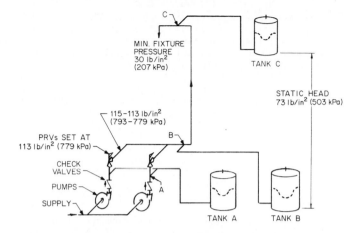

Tank Location Comparisons

Tank location	Drawdown volume		Tank size		Min. operating pressure		Max operating pressure	
	gal	(liters)	gal	(liters)	lb/in²	(kPa)	lb/in²	(kPa)
A	20	(76)	128	(484)	104	(717)	126[a]	(869)[a]
B	20	(76)	216	(818)	103	(710)	115	(793)
C	20	(76)	95	(360)	30	(207)	42	(290)

[a]Higher when suction pressure rises above design pressure.

FIG. 5   Relative effect of tank location on tank size and pressure for equal drawdown volumes. (ITT Bell and Gossett)

pressure on it; consequently, a smaller tank will suffice. The maximum design working pressure is also lower, and therefore it may be possible to use a tank with a lower pressure rating at less cost.

PRESSURE BOOSTER SYSTEM SIZING _____

Design Data   The proper sizing of a pressure booster system is subject to the accuracy of the design data available. Experience has indicated that the greatest single cause for unsatisfactory pressure booster performance is oversizing. Submitted data may be rough estimates or historical data not relevant to the location; therefore their source and accuracy should be scrutinized. Pertinent factors are the flow demand rate and hourly profile, pressure conditions, and types of building occupancy.

FLOW DEMAND RATE AND HOURLY PROFILE   Estimating water demand rate is one of the most difficult and controversial subjects. To determine the water demand, many designers refer to the classical Hunter curve (National Bureau of Standards, Report BMS79, by R. B. Hunter) or a modified version of it for lack of a more accurate method. Numerous studies have indicated that the Hunter method often results in very appreciable demand rate overstatement. It should be noted that the Hunter curve was intended for determining pipe sizes, not water demand.

The average hourly demand profile is helpful in proportioning the system demand rate among the pumps in the system. Optimum selection results in the lowest total energy consumption by the pumps while maintaining the demand requirements.

PRESSURE CONDITIONS   Four accurate pressure values are required: design system pressure, minimum suction pressure, maximum suction pressure, and minimum allowable system pressure. The difference between design system pressure and minimum suction pressure is used to determine the developed head of the pump at design flow. Inaccurate values will result in incorrect impeller sizing and incorrect motor power selection.

The design system pressure can be calculated with reasonable accuracy. However, when the water supply is from the municipal water main, obtaining accurate values of suction pressures for the proposed site of installation is very difficult because of the absence of recent pressure data. Often the minimum and maximum pressures are haphazardly estimated from average citywide values. A minimum suction pressure error as small as 5 lb/in² (35 kPa) could result in selection of the next larger motor.

The maximum suction pressure will determine the working pressures required of the pumps and piping and whether there is sufficient pressure to justify a bypass connection (Fig. 3) paralleling the pumps.

Knowing the minimum allowable system pressure is helpful in setting controls associated with this pressure.

TYPE OF OCCUPANCY   If the hourly demand profile is not available, the type of building occupancy could be used to determine the standby capacity requirement, whether low-flow shutdown is applicable, and the number of pumps for the system. Most pressure booster manufacturers provide data for determining demand, capacity splits, and pump sizing.

Pump Sizing Procedures   Information should be collected on system information and component performance. The system information required is:

$$Q_d = \text{total design capacity, gpm (m}^3\text{/h)}$$

$$P_d = \text{design pressure at system header outlet, lb/in}^2 \text{ (kPa)}$$

$$P_s = \text{minimum available pressure at suction header inlet, lb/in}^2 \text{ (kPa)}$$

$$Q_1, Q_2, Q_3 = \text{design capacity of individual pumps, gpm (m}^3\text{/h)}$$

Components performance data required are

• Pump performance characteristic curve
• Pressure-reducing valve (and check valve) flow chart° (Fig. 6)

Next the required pump total head at the maximum flow design condition should be calculated:

$$H = (H_d - H_s) + H_v + H_u \qquad (9)$$

where H = required total head, ft (m)
   H_d = design system pressure at PRCV outlet, ft (m)
   H_s = minimum design suction pressure, ft (m)
   H_v = PRCV pressure loss, ft (m)°
   H_u = sum of all unaccounted losses in the pressure booster package, ft (m)

For pressure booster applications, the PRCV is usually sized for a nominal pipe velocity between 8 and 18 ft/s (2.4 and 5.5 m/s). The pressure loss H_v is based on the valve being 80% open (Fig. 6). Operating the valve at less than full-open position is desirable for good pressure regulation. The term H_u represents the allowance for piping frictional losses between the suction header inlet and the system header outlet at design capacity, exclusive of the PRCV loss. Manufacturers of booster packages differ in determining this value. Some ignore it completely to compensate for usual oversizing as the result of inaccuracies in the design capacity and minimum suction pressure statements. Others assign a fixed value, about 5 ft (1.5 m), or a percentage, about 3%, of the difference of system and suction pressures ($H_d - H_s$). It is recommended that some value be used.

° Add loss through separate check valve if not integral with PRV.

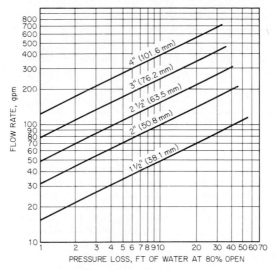

FIG. 6 Typical PRV or PRCV flow chart. (ft × 0.3048 = m; gpm × 0.2271 = m³/h; in × 25.4 = mm) (Cla-Val)

Then determine the best combination of pump and PRCV sizes. The selection will depend on whether the design criterion is least capital cost or lowest operating cost. Usually the smallest pump and PRCV combination will result in lowest first cost and the pump equipped with the smallest motor will yield the lowest operating cost. Various pump, PRCV, and motor combinations may be examined by plotting system curves (Fig. 7) for two or three PRCV sizes on one or more pump performance curves.

The pumps tentatively selected should exhibit the highest efficiency at the most frequent operating flow point on their head-capacity curves. When pumps are operated at suction pressures higher than design minimum, the PRCVs will throttle the excess suction head and maintain the same pump total head and constant design discharge pressure. When the flow demand is less than

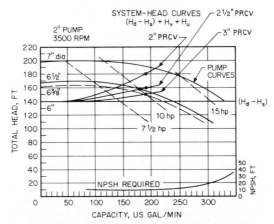

FIG. 7 System curves for PRCV sizing. (ft × 0.3048 = m; gpm × 0.2271 = m³/h; hp × 0.7457 = kW; in × 25.4 = mm) (ITT Bell and Gossett)

TABLE 2   Required Head for Pressure Reducing Valves

PRV size, in (mm)	Flow, gpm (m³/h)				
	0 (0)	50 (11)	100 (23)	150 (34)	190 (43)
2 (51)	140 (42.7)	143 (43.6)	151 (46.0)	166 (50.6)	181 (55.2)
2½ (64)	140 (42.7)	141 (43.0)	146 (44.5)	153 (46.6)	160 (48.8)
3 (76)	140 (42.7)	141 (43.0)	143 (43.6)	147 (44.8)	151 (46.0)

NOTE: All values are head in feet (meters).

the design maximum, characteristically, the head on the centrifugal pumps will increase to some higher value on the pump head-capacity curve. The PRCVs will reduce this excess head to maintain the constant design discharge pressure. Therefore, each pump operates along its own head-capacity curve, which is also the adjusted system-head curve.

When several pumps are to be used, especially pumps of different sizes, their head-capacity characteristics should be compatible. A pump having a shutoff total head less than the system total head at any operating point (with or without other pumps operating) will not pump. Also, pumps should be so selected and controls should be such that no one pump will operate at too low a flow, i.e., at a flow less than the minimum recommended for mechanical and hydraulical stability. If the minimum suction pressure is very low, the net positive suction head required by the pumps throughout their operating flow range must be checked (Subsec. 2.3.1).

PUMP AND PRCV SELECTION EXAMPLE

Total system design capacity	Q_d = 380 gpm (86.3 m³/h)
Design capacity	Q_1, Q_2 = 190 gpm (43.1 m³/h)
Design pressure at system header outlet	H_d = 210 ft (64.0 m)
Minimum pressure at suction header inlet	H_s = 70 ft at 100 ft NPSH (21.3 m at 30.5 m NPSH)
PRCV loss, Fig. 6	H_v = 36 ft (11 m) for trial 2-in (51-mm) valve
Allowance for internal frictional losses	H_u = 5 ft (1.5 m)

Calculate H:

in USCS units $H = (210 - 70) + 36 + 5 = 181$ ft° (Eq. 9)

in SI units $H = (64.0 - 21.3) + 11 + 1.5 = 55.2$ m

Using Table 2, calculate and tabulate H for all applicable PRCV sizes up to the design flow rate of 190 gpm (43.1 m³/h). H_v and H_u terms vary as the square of the flow rate.

Now locate H at 190 gpm (43.1 m³/h) on the applicable pump curve. Select the pump and PRCV combination that best satisfies the design criterion of least first cost or lowest energy input.

Plot H values on the selected pump curve to obtain the system curves for the three PRCV sizes (Fig. 7). the plots reveal that the 3-in (76-mm) valve, a 10-hp (7.5-kW) motor, and 6⅜-in diameter (162-mm) impeller are the proper selections.

Note that if the design capacity of the pump could be reduced to about 178 gpm (40.4 m³/h), a 2½-in (64-mm) PRCV with the impeller sized for 6½-in (165-mm) diameter may be used without additional power input. This choice reduces the cost and improves the PRCV performance at low flow rates.

° Due to the many approximations, decimal amounts are not used in the final results.

Using a 2-in (51-mm) PRCV would require a 15-hp (11-kW) motor operating at about 12.5 hp (9.3 kW) load. This choice is not energy effective.

The power required to drive the pump may be approximated by using the pump curve and Eq. 7. The available NPSH is in excess of the required pump NPSH (Fig. 7).

PUMP TYPES AND MATERIALS

Single-stage volute centrifugal pumps, followed by multistage vertical turbine diffuser pumps in a suction tank, are most commonly used for domestic water pressure booster systems. The volute pumps can be end-suction with a vertically split case, double-suction with an axially split case, or in-line. For high-pressure service, two-stage axially split case pumps or multistage vertical turbine pumps may be selected.

In selecting the type of pump best suited for the application, such inherent characteristics as low-flow recirculation, low-flow cavitation, high-flow cavitation, steepness of the pump curve, noise, and operating efficiencies should be evaluated. These factors are discussed in Subsec. 2.3.1.

The most often discussed factor in booster application is low-flow recirculation. Low pump flow conditions do occur; they cannot be completely designed out of the sytem. Most well-designed small pumps with positive (above atmospheric) suction pressure can operate safely in the low-flow region, with the only point of concern being water termperature rise at or near shutoff flow. Should radial thrust be a major consideration, double-volute pumps or diffuser (vertical turbine) pumps offer advantages.

Vertical pumps generally require less space. Pumps may be close-coupled to their drivers, or, if not, a coupling requiring careful alignment is required. Close-coupled pumps may require less maintenance, as there are no pump bearings. Shaft sealing may be either packing or mechanical seal.

All pumps and drivers in packaged units are mounted on a common steel frame, but individual bases for each pump and driver may be necessary for larger units.

The materials of construction for pumps, valves, and piping should be suitable for the water quality and conditions furnished to the system. Since water impurities vary from location to location, careful analysis is recommended. Federal and local agencies, such as the Food and Drug Administration, may also prescribe allowable materials.

FURTHER READING

American Society of Heating, Refrigerating, and Air-Conditioning Engineers: *Equipment Handbook*, ASHRAE, Atlanta, 1983.

ITT Bell and Gossett: "Domestic Water Service," Bulletin TEH-1175, Morton Grove, Ill., 1970.

ITT Fluid Handling Division: "Pressurized Expansion Tank Sizing/Installation," Bulletin TEH-981, Morton Grove, Ill., 1981.

Potter, P. J.: *Steam Power Plants*, Ronald Press, New York, 1949.

Steel, A.: *High Rise Plumbing Design*, Miramar, Los Angeles, 1975.

INTAKES
AND
SUCTION
PIPING

SECTION 10.1
INTAKES, SUCTION PIPING, AND STRAINERS

WILSON L. DORNAUS

The most critical part of a system involving pumps is the suction inlet, whether in the form of piping or open pit. A centrifugal pump that lacks proper pressure or flow patterns at its inlet will not respond properly or perform to its maximum capability. Uniformity of flow and flow control up to the point of impeller contact are most important. Part of this may be controlled by proper pump design, but the pit designer and suction piping designer have definite responsibilities to satisfy good pump operation. In open suction pit, wet-well design the flow must be as uniform as possible right up to contact with the pump suction bell or suction pipe, preferably without a change in direction or velocity.

Examples of dry-pit and wet-pit centrifugal pumps connected to open suction pits are shown in Fig. 1.

In dry-pit pumping, the suction pipe leading to the pump flange should not include elbows close to the pump in any plane, or other fittings which change flow direction and velocity and which may impart a spinning effect to the flow. Centrifugal pumps not designed for prerotation, either dry pit or wet pit, will suffer loss of efficiency and an increase in noise. Rotation with the impeller can result in a decrease in pump head. Rotation against the impeller can result in an increase in pump head and power, and possible driver overload. If the total system is to operate efficiently with minimum maintenance, close attention must be paid to the suction environment of the pumps.

COOLING WATER PUMP INTAKES

Purpose Either water-circulating systems must have a continuously renewable source, such as an ocean, lake, or river, or they must recirculate the same water, as from cooling ponds or cooling towers. Whichever type of pump is selected (wet pit or dry pit), the suction water will come from an open pit of some sort or from a pressure pipeline.

Types

ONCE-THROUGH: OCEAN, LAKE, OR RIVER SOURCE WET PIT This type calls for an intake structure, usually of concrete, to gather the water into a localized spot for pickup and to support the pumps. The optimum design will bring relatively clear water directly into the pump suction area at a low velocity.

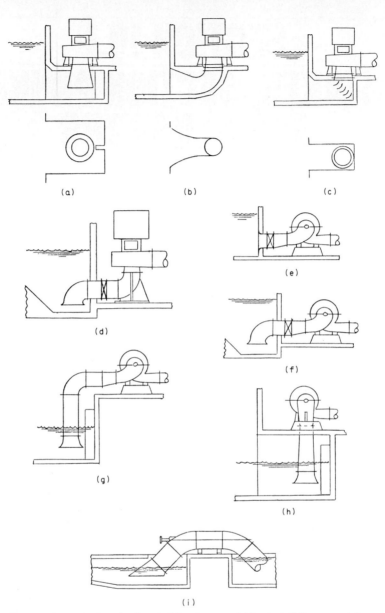

FIG. 1 Pump suction connections: (*a–d*) vertical dry-pit pumps. (*a*) Suction bell diameter and distance from bottom, side, and back walls designed to provide radial inflow, equal velocity distribution, minimum entrance loss. Proper spacing of bell and vertical baffle required to prevent underwater eddies and rotation at inlet. (*b*) To minimize excavation under pump, decreasing-type suction elbow can be formed in the foundation. (*c*) Width of suction inlet under pump minimized by use of turning vanes. (*d*) Smaller pumps provided with integral suction elbows and can be connected to wet well with commercial piping. Flared suction elbow is turned downward to obtain maximum submergence at inlet. Wet well sloped to suction inlet to prevent deposition of solids. (*e–i*) Horizontal dry-pit pumps. (*e*) Pump suction connected to larger pipe with eccentric reducer (double-suction pumps only). Wet-well connection should be bell mouth of a size to provide minimum required velocity. (*f*) Flared suction elbow turned downward for maximum submergence over inlet. Wet-well floor may be sloped toward inlet if solids are present. (*g*) Vertical suction pipe with bell inlet adequately submerged to prevent vortexing. Bell sized and spaced as in Figs. 1*a, h,* and *j.* (*h*) Vertical suction pipe connected to bottom-

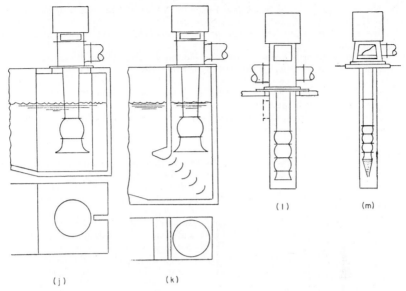

(l)   (m)

(j)   (k)

suction pump. Bell sized and shaped as in Figs. 1*a*, *g*, and *j*. (*i*) Design used for drainage over levees. Suction pipe designed to require minimum submergence and excavation. (*j–m*) Vertical wet-pit pumps. (*j*) Distance from bottom, side, and back walls designed to provide equal velocity distribution, with splitter to prevent vortexing. (*k*) Width of inlet minimized by use of turning vanes. (*l*) Canned suction pump with either above- or belowground suction connection. Can length to suit NPSH required. Vertical installation minimizes excavation area. (*m*) Deep-well pump takes suction from underground water source.

While it is recommended that the submergence of the pumps and the dimensions of the suction pit in the immediate vicinity of the pump suction inlet be as suggested by the pump manufacturer, preliminary intake drawings must usually be prepared for making studies and estimates and for writing specifications. During this preliminary stage of intake design for vertical wet-pit pumps (or dry-pit pumps having a vertical suction pipe belled at the entrance), the recommendations of the Hydraulic Institute should be followed. Figure 2 shows the basic layout, and Fig. 3 gives the average dimensions in relation to the flow required per pump. Recommendations are for pumps having capacities from 3000 to 300,000 gpm (700 to 70,000 m³/h) and are based on the capacity. If prolonged operation at a capacity greater than rated is expected, then intake pit dimensions should be based on this higher flow. Interpretation and limits of dimensions in Figs. 2 and 3 as stated in the *Hydraulic Institute Standards*, 13th ed., are

> The Dimension C is an average, based on an analysis of many pumps. Its final value should be specified by the pump manufacturer.
> Dimension B is a suggested maximum dimension which may be less depending on actual suction bell or bowl diameters in use by the pump manufacturer. The edge of the bell should be close to the back wall of the sump. When the position of the back wall is determined by the driving equipment or the discharge piping, Dimension B may become excessive and a "false" back wall should be installed.
> Dimension S is a minimum for the sump width for a single pump installation. This dimension can be increased, but if it is to be made smaller, the manufacturer should be consulted or a sump model test should be run to determine its adequacy.
> Dimension H is a minimum value based on the "normal low water level" at the pump suction bell, taking into consideration friction losses through the inlet screen and approach channel. This dimension can be considerably less momentarily or infrequently without excessive damage to the pump. It should be remembered, however, that this does not represent "submergence." Submergence is normally quoted as dimension H minus

C. This represents the physical height of water level above the bottom of the suction inlet. The actual submergence of the pump is something less than this, since the impeller eye is some distance above the bottom of the suction bell, possibly as much as 3 to 4 ft (0.9 to 1.2 m).

Dimensions Y and A are recommended minimum values. These dimensions can be as large as desired but should be limited to the restrictions indicated on the curve. If the design does not include a screen, dimension A should be considerably longer. The screen or gate widths should not be substantially less than S, and heights should not be less than H. If the main stream velocity is more than 2 ft/s (0.6 m/s), it may be necessary to construct straightening vanes in the approach channel, increase dimension A, conduct a sump model test of the installation, or work out some combination of these factors.

The 14th edition of the *Hydraulic Institute Standards* gives larger values for dimensions *C* and *Y* and smaller values for dimension *H* than does the 13th edition. This suggests more space between pump suction bell and pit floor, less suction bell submergence, and more distance from pump to screen. The author feels that caution should be used if revised values for *C* and *H* are followed. Rather, it is recommended that these dimensions be in conformance with criteria which follow in this section, including suction bell inlet velocity.

A preliminary intake design procedure that uses the Hydraulic Institute recommendations along with additional recommendations by the author is discussed here. Following selection of the intake pump manufacturer, final pit dimensions should be established based on the equipment supplier's recommendations. Should there be an appreciable variation in any of the dimensions from various sources, a model test of the intake is justified and recommended.

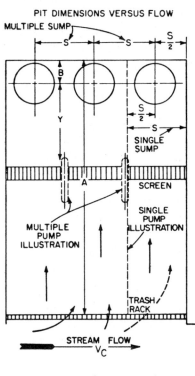

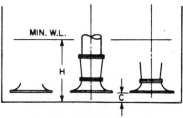

FIG. 2 Proportions for standard pump pit (see Fig. 3). (*Hydraulic Institute Standards*, 13th ed.)

1. Establish minimum water depth *H* required at maximum flow rate.

2. Calculate width of pump basin *S* required per pump. If the average velocity of approach is less than 1.25 ft/s (0.38 m/s) and the maximum flow rate and/or minimum water level occurs less than 10% of the total operating time, the velocity based on these limits can be increased to 1.25 ft/s (0.38 m/s), provided the velocity is constant and there is no flow directional change.

3. Calculate final width of pump basin if it can be made smaller than *S* per step 2. If suction bell diameter is known at this time, do not make width less than two bell diameters.

4. Recess pump location *Y* from the traveling screen. If the pump suction bell diameter is known at this time, calculate distance *Y* to be two diameters and use the larger of these two distances.

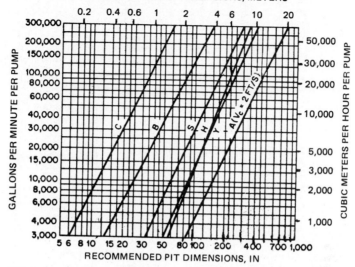

FIG. 3   Recommended pit dimensions for Fig. 2. (*Hydraulic Institute Standards*, 13th ed.)

5. Space bar racks (if used) dimension *A* from the pit back wall.

6. Project wing walls (walls between pumps if required structurally) into source area far enough to provide parallel uniform flow at pit velocity for a distance of one pit width prior to the trash racks.

It should be stressed that in all these steps the intent is to provide a cross-sectional inlet area such that straight-line flow at an average velocity of 1 ft/s (0.3 m/s) or less is admitted to the pump suction bell area.

The above design assumes the source area to be either a lake or a river with a maximum velocity of 2 ft/s (0.6 m/s). For higher river velocities, use correspondingly greater distances from source wing wall contact to trashrack or screen. (The choice of providing trashrack or screen, or both, is based on the type and amount of debris likely to be encountered at the inlet.)

An ocean inlet has additional requirements because of tidal action and variable-direction currents which may exist. For this reason it may be necessary to create a forebay inlet basin, independent of the pump pit and fed by a submerged inlet tube extending out into the ocean for some distance (Fig. 4). This tube may utilize normal pipe velocities, but the inlet must turn upward and the opening should be protected by a horizontal cap of such size and shape that the inlet water travels horizontally at a velocity high enough to scare fish away. The discharge into the forebay can be pipe velocity, since the turbulence will be dissipated by lower—2 ft/s (0.6 m/s)—outlet velocities through trash racks and traveling screens. From here on to the pump chamber, the velocity should be low and constant.

Any necessary change in inlet channel dimensions should be made gradually. Tapering walls should not diverge at more than a 14° included angle. If it is necessary to slope the floor, a maximum of 7° is recommended and, if possible, the floor should level off before reaching the pump area, as far back as possible. No sharp drops (waterfall effect) should be permitted.

If the inlet channel is a closed pipe with full-wetted perimeter, pipe velocities can be used up to within such a distance of the pump chamber that the tapered wall rule can be applied to the increase in pipe size as it discharges into the pump chamber at not over 1.5 ft/s (0.46 m/s) velocity into the pit.

Unless the inlet velocity into the pump chamber is kept below 1.5 ft/s (0.46 m/s), extensive

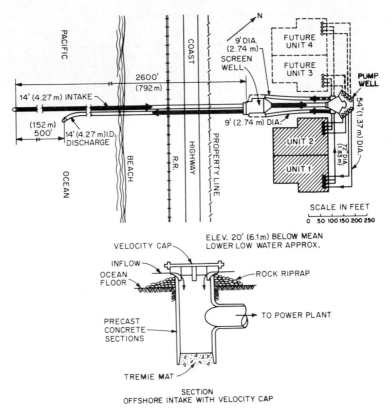

FIG. 4   Ocean intake with velocity cap to minimize fish pickup.

baffling and additional space will be necessary, which can be determined only by model pit testing.

RECIRCULATING SYSTEM—COOLING-TOWER   For cooling tower systems, the pump pit is normally adjacent to the tower basin. Since cooling tower basins are shallow, usually not more than 6 ft (1.8 m) water depth, and pumps will require more submergence than this, a sloping ramp or sharp dropoff will be required if pumps are to be placed close to the tower basin. This will save initial costs for excavation and concrete work, but the resulting bad flow conditions will cause endless pump problems.

Some small installations have been built with downslopes of 15° or more, but tower basin outlet velocity was 2 ft/s (0.6 m/s) or less and the downslope was widened as it approached the pump pit. For low flows, low velocity, and a fortunate arrangement of tower piers (lack of recirculation piping, etc.), these installations may be operating with some modicum of success. Others of similar design have proved disastrous on pump performance. Impellers and bearings suffer rapid deterioration under such conditions.

Some guidelines may be suggested. The best solution but obviously the most expensive is to build the pump pit far enough from the tower to obtain a flat channel bottom with average channel velocity of 1 ft/s (0.3 m/s) or less for a distance at least equal to A in Figs. 2 and 3, whether or not trashracks and/or screens are available.

The slope from the basin floor to the channel bottom should not be more than 7°. The basin exit width should be such as to make the exit velocity less than 2.5 ft/s (0.76 m/s). If several tower piers are in the exit path, reduce this velocity to 2 ft/s (0.6 m/s) or less and round off upstream sides of the piers and make the downstream sides ogival (tapered). Make the sidewalls of the downslope diverging so that the velocity at the bottom is 1 ft/s (0.3 m/s) or less.

Velocities should be based on the runout flow of one pump. All the above dimensions and velocities are limits, and if used all together with the maximum flow, good operation should follow provided the pump has been selected for reasonable velocities. It is obvious that a pump with high velocities (suction bell inlet, impeller eye, rotative speed) will require more submergence than a more conservative machine, and this will increase the cost by requiring a deeper pump pit farther from the tower basin.

If the best solution seems too expensive, or if topography will not allow the ideal arrangement, a model pit test is highly recommended. A model with a variable slope, variable sidewalls, and true representation of piers, pipes, and other obstructions should enable a satisfactory design to be achieved with a possible net reduction in overall cost. Actual pumps and valves should be used to create the flow in the pump pit, and extremes of flow and submergence should be included in the test program.

As an extreme design requirement, the basin flow might waterfall over a sharp drop into a subchannel, and the flow from here would have to be baffled before entering the pump approach channel. This design also requires model testing.

Sometimes the cooling towers are on a hill, so that a substantial drop in elevation provides pressure available nearer the flow area. If dry-pit pumps are used, this pressure can be utilized to reduce pump requirements. When this is done, the pump suction becomes a pipe, and the latter part of this section should be consulted.

RECIRCULATING SYSTEM—POND WET PIT   The intake structure in a cooling pond should be located as far as possible from the inlet pipe to the pond to generate the maximum cooling effect. If spray equipment is used, it should be so arranged in relation to the intake building that a minimum surface disturbance is encountered. Prevailing winds should be considered, and the building located on the lee side of the pond. If side and bottom areas might be easily disturbed to include silt in the flow, riprap should be applied to approach slopes as well as to bottom mats well beyond the inlet wing walls.

The handling of silt is usually not desirable in a pumping system. A high velocity through the pump will accelerate wear. At low velocities at other points in the system, silt will settle out and produce higher velocities and more wear as a result of area blockage. If space is available, a silt-settling basin can be constructed ahead of the inlet basin. Cross baffles should be provided to slow the inlet flow to a velocity of less than 1 ft/s (0.3 m/s). Most stream debris will settle out at this velocity, and the flow into the pump suction pit will be relatively clean, preventing the deposit of additional silt around the pump suction bell. If space is not available for silt beds in large-capacity installations, the main channel can be furnished with a weir across the flow path. The height of the weir should be selected to give an overflow depth above the weir of not more than one-third to one-fourth the water depth just preceding the weir. The velocity over the crest should not exceed the intake channel velocity. Weirs of this type are particularly effective when the intake channel is at right angles to the supply mainstream.

Another scheme to be considered when silt is a problem is a manifold of round suction pipe screens. This manifold is equivalent to laying out a number of well screens horizontally in sand beds and connecting them together as the spokes of a wheel into a hub, or any other manifold arrangement, similar to a shallow-water design (Fig. 14). If the area is sufficient to reduce the inlet velocity to below 0.5 ft/s (0.15 m/s), the screens will not plug up too quickly.

In areas where river levels vary considerably throughout the year, problems arise not only from silt and debris accumulation but from the structures required to prevent damage to motors and electric switch gear. If the required flow is moderate, a system known as the Ranney well (Fig. 5) can be constructed. A Ranney well is a concrete silo 13 ft (4 m) in diameter, which becomes the collecting basin and pump well. It may be situated in or near a river and can be partly below and partly above water level with settings up to 100 ft (30 m). Small perforated pipes radiate horizontally from the base of the well and tap into porous strata, bringing small flows of water into the main well. This water can then be pumped from the well with a deep-well wet-pit vertical pump.

Multiple Pumps   Some pump requirements are easily met with 100% capacity single pumps, but more reliability usually requires two 50 or three 33% pumps or, if the service is critical

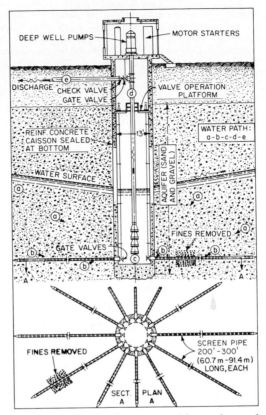

FIG. 5 Ranney well collects into a pool from underground strata. (Ranney Method Division, Pentron Industries)

enough, three 50 or four 33% pumps, etc. There are practical limits of size for various pump types, so that large flow demands will undoubtedly call for a multiple pump arrangement.

The problem arising from multiple pumps arranged in a common pit is from the probable nonuse of some of the pumps while others are operating. This causes variables in flow patterns which may lead to eddying and vortexing (Fig. 6). Installation of separating walls in the common pit may introduce additional problems, since the ends of the separating walls create eddy corners for the dead-end pocket at the unused pump. A back vent in the dividing wall will relieve this situation, provided it vents at the water surface (Fig. 7b left). If walls are extended past the screens and trashracks to a forebay, this problem will not occur, but the design has then become that of a single-pump basin.

The same velocity rules apply to multiple arrangements as to single-pump basins. Odd arrangements should be avoided even when they look invitingly symmetric—fan-shaped, round, radial, peripheral; all have directional problems that are not easily overcome.

A basic pit design consisting of a number of equally sized pumps in a common pit with flow entering parallel and straight in at 1 ft/s (0.3 m/s) or less would not need to be model-tested to assure reliability (Fig. 7a left). If separating walls are required for structural support and they are properly shaped and vented, no model will be required. The pumps should be located at the extreme rear of the pit so that the whole approach assumes the characteristics of a suction pipe. Individual pump manufacturers may vary the location of the pump relative to the pit bottom, velocity of inlet spacing, etc. It has been found that some of these variations require additional splitters or baffles below the pump, up the back wall behind the pump, or centered in the flow

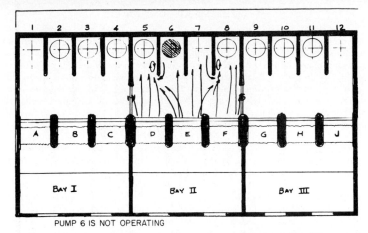

PUMP 6 IS NOT OPERATING

FIG. 6 Empty spaces on nonoperating pumps cause vortexing at wall ends because of flow reversal.

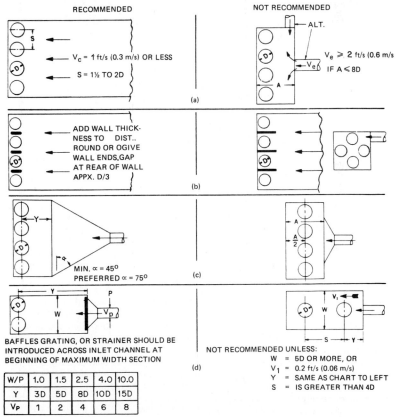

RECOMMENDED

NOT RECOMMENDED

$V_c = 1$ ft/s (0.3 m/s) OR LESS

S = 1½ TO 2D

$V_e \geqslant 2$ ft/s (0.6 m/s)

IF A ≤ 8D

(a)

ADD WALL THICK-NESS TO DIST.. ROUND OR OGIVE WALL ENDS,GAP AT REAR OF WALL APPX. D/3

(b)

MIN. α = 45°
PREFERRED α = 75°

(c)

BAFFLES GRATING, OR STRAINER SHOULD BE INTRODUCED ACROSS INLET CHANNEL AT BEGINNING OF MAXIMUM WIDTH SECTION

NOT RECOMMENDED UNLESS:
W   = 5D OR MORE, OR
V_1 = 0.2 ft/s (0.06 m/s)
Y   = SAME AS CHART TO LEFT
S   = IS GREATER THAN 4D

(d)

W/P	1.0	1.5	2.5	4.0	10.0
Y	3D	5D	8D	10D	15D
Vp	1	2	4	6	8

FIG. 7 Using basics of good pit design precludes the need for model testing. (*Hydraulic Institute Standards*, 13th ed.)

ahead of the pump. If so, a model test should be run and the additional pit cost weighed against other alternatives (changing pit shape, size, pump location, pump size and speed).

A dry-pit pump installation will have the pumps either in a dry well at or below wet-well water level (figs. 1e and 1f) or located directly above the wet well and using a suction lift (Figs. 1g and 1h), which calls for priming equipment. The additional cost of this equipment (vacuum pumps, etc.) may be partially offset by the additional space and valve requirement of the first option. In either case, the suction piping in the wet well should be treated in the same fashion as the pump suction bells in wet-pit installations as far as spacing, direction, and velocity of flow are concerned.

When the pump is installed at an elevation that may be below the suction pit water level, a valve must be installed at the suction of the pump. The temptation to reduce the inlet size in order to use a smaller valve should be avoided. Pipe size at a pump inlet may decrease into the pump down to the pump suction size, but it should not be reduced below that size and then have to increase again as it enters the pump. The pipe in the wet well should preferably have a bell end and project downward. The minimum water level above the top edge of the pipe or the lip of the bell should be at least 5 ft (1.5 m) for a recommended entrance velocity of 5 ft/s (1.5 m/s). The bell mouth should project downward to assure uniform inlet flow and to attain maximum submergence.

A wet-pit intake style that closely approximates a suction pipe arrangement and uses what is essentially a dry-pit pump has been expanded by the U.S. Department of the Interior, Bureau of Reclamation, to an elbow-type suction tube design. This incorporates a formed concrete suction inlet with a swing of 135° in the vertical plane and a gradual decrease in area to the suction eye of the pump (Fig. 8). The resulting design saves considerable excavation in the wet-well area, reduces losses into the pump, and allows a smaller, higher-speed pump to be used. With higher velocities, debris dropout is reduced so that silt buildup does not occur as readily and a smaller trashrack area remains effective longer. These inlets are independently self-sufficient and may be grouped into multiples as long as the forebay is designed to adhere to the basic design rules for wet-pit approach channels.

Considerations for either single- or multiple-pump pit design relate primarily to even flow and low velocity into the pump. At the leading edge of the pump impeller (suction) vanes, this velocity will be increased and the direction of flow violently changed. To make the transition from pit flow to pump flow is the work of the pump designer. Some pump designs include ribs in the suction bell, others do not. It is obvious that the transition from parallel flow at 1 ft/s (0.3 m/s) to right-angle rotating flow at 15 to 20 ft/s (4.6 to 6.1 m/s) requires a high degree of skill in matching the suction bell to impeller. If the pump is not designed to handle a wet-pit installation as described in previous paragraphs, a turning vane pit may be required.

Suction Pit Turning Vanes   Figure 1c illustrates how turning vanes are used to guide the flow of water into the inlet of a vertical volute dry-pit pump, and Fig. 1k shows the same for the suction bell of a vertical diffuser wet-pit pump. If these pumps were connected to an inlet sump without turning vanes, as illustrated in Figs. 1a and 1j, a wider and longer channel would be required to feed the suction bell from all directions. The pumps with turning vanes and narrower inlet channels may be desirable, as multiple pumps can be spaced closer together providing screen width requirements do not dictate spacing.

Comparing depth of pit bottom below pumps with and without turning vanes, the following is to be noted. The excavation beneath a vertical volute dry-pit pump with turning vanes can be slightly less than the excavation beneath the same pump with no vanes but with a suction bell. The velocity approach into the closed portion of the channel beneath the pump can be as high as approximately 3 ft/s (0.9 m/s) if turning vanes are used at the design flow but only 1.5 ft/s (0.46 m/s) if they are not. Although the suction bell design requires a wider channel than a vaned inlet (approximately two bell diameters), this is not wide enough and the channel must be made deeper to meet the lower velocity requirement for this type of inlet. When turning vanes are used with a vertical diffuser wet-pit pump, the pit must be excavated deeper than would be required if no vanes were used. This additional depth is required to form an elbow in the narrower channel and provide the necessary equal flow distribution to the impeller. Design inlet vane velocity is 3 ft/s (0.9 m/s).

The setting of the lip of the suction bell below design low-water level for volute dry-pit pumps

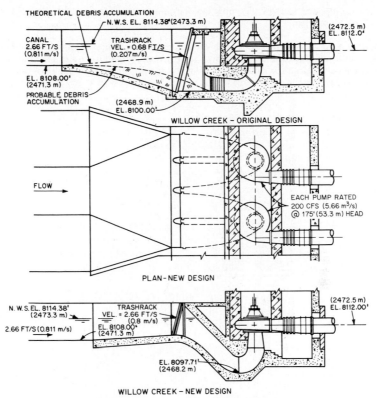

FIG. 8 Improved 135° design of Bureau of Reclamation elbow suction tube inlet. (U.S. Department of the Interior, Bureau of Reclamation)

must be the greater of the dimensions required to prevent vortexing (Hydraulic Institute dimensions), to provide adequate NPSH and to provide a level of water above the top of the impeller so that the unit is self-priming.

As an alternative to the use of turning vanes under a vertical volute dry-pit pump, the long-radius suction elbow inlet illustrated in Fig. 1b offers some advantage in reducing the width and depth of excavation under the pump. The inlet velocity to the long-radius elbow, which is usually formed in concrete, is preferably not greater than 3 ft/s (0.9 m/s) at the design flow.

A decision to use turning vanes should not be based on a guarantee that there will be an increase in pump efficiency. The design of the vanes—their number and spacing—is still an art, and it is difficult to prove pump performance in the field. Turning vanes can be effective in eliminating underwater vortices, sometimes a problem associated with suction bells without turning vanes It has been observed during model testing that, to prevent underwater vortexing, the suction bell inlet design illustrated in Fig. 1a must be placed closer to the back wall than normally recommended for open channel inlets similar to that in Fig. 1j. The flow of water into a closed channel from an open pit containing water of considerably greater depth creates an unequal flow pattern in which the maximum velocity is along the floor. Also, there is little, if any, flow to the back side of the suction bell down from the top of the channel. Unless the bell lip is close to the back wall, flow along the floor and from the front only will overshoot the inlet and roll over, back, and up into the bell, forming an underwater vortex. While turning vanes can prevent this, unless they are properly designed they may not prevent uneven flow distribution up into the pump impeller and may cause hydraulic and mechanical unbalance, which could result in noise, vibration, and accelerated wear of the pump bearings. For this reason, model testing of the turning vanes is recommended.

Screens and Trashracks   While economically it may not be practical to eliminate all refuse from a pumping system, it probably will be necessary to limit the size and amount of debris or sediment carried into the system. Depending on the probable source of debris, such as a river subject to flooding with considerable flotsam in the runoff and with a very loose bottom, or a lake

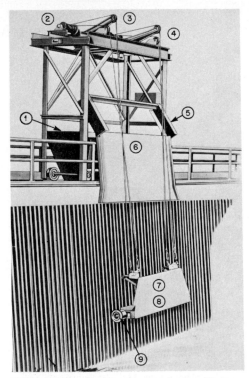

FIG. 9   Trashrack with raking mechanism: (1) Large enclosed trash hopper contains debris discharged by rake. Hinged door in end of hopper opens wide for debris removal. (2) Heavy-duty, single-drum hoist, push-button-controlled with two separate cables—one for carriage, one for rake teeth. (3) Walking beam actuated by hydraulic cylinder controls position of rake teeth. (4) Fixed sheave. Cable operating over stationary sheave raises and lowers rake carriage. (5) Discharge guide flanges assure positive positioning of rake over trash chute prior to dumping. (6) Dead plate, or apron, integral with superstructure guides rake to discharge point—prevents trash from falling off rake prematurely. Design permits operation over 3.5-ft (1.1-m) high hand rail. (7) Self-centering rack-guided carriage allows rake to ride over obstructions in water during lowering cycle. Debris of all types picked up on lifting cycle rather than forced to bottom of channel. (8) Rake mechanism assures positive removal of debris with maximum carrying capacity. Hydraulic relief valve provides automatic overload relief. Teeth automatically open if overload occurs, permitting load to drop off rake. No cable failures due to overload. (9) Wide, flanged rollers ride on at least two rack bars. (Envirex Inc., a Rexnord Company)

at constant level without disturbing inlet flows near the structure and with a solid bottom, the protection needed may include only a bar trashrack to avoid large floating objects or a raked trashrack plus rotating flushed fine-mesh screens. If sediment deposit is likely, a settling basin may be required.

The designer must note that, if this equipment is useful, it will pick up debris and gradually increase the velocity through the openings as the net area decreases with blockage. When this occurs at the trashrack, the water level differential will build up, causing a waterfall with increased velocity and turbulence on the pump side of the rack. In addition, the increase in velocity may pull more debris through the bars than can be tolerated. It is best to rake these racks (Fig. 9) frequently enough to keep the differential head across the rack below 6 in (0.15 m). The spacing of the bars should be such that objects which cannot be pumped will be excluded from passing through. This, in general, will call for the bar spacing to be in proportion to the size of the pump. A pump manufacturer can determine the maximum size sphere a pump will handle, and the bar spacing should be limited to 50% of that value. The size of the bar, the lateral distance between supports, and the pier spacing will influence the rate of debris accumulation and the allowable design differential head.

Rotating screens (Fig. 10) will remove trash of a much smaller size since the accumulation is continuously removed and the open area is kept fairly uniform. Finer screening than that required by the pump may be necessary in installations where the liquid pumped must

FIG. 10 Traveling water screen. (Envirex Inc., a Rexnord Company)

pass through small openings in equipment serviced, such as condenser tubes or spray nozzles (Fig. 11). Screens are usually installed in conjunction with trashracks so that large, heavy pieces will not have to be handled by the screens. Since velocity through the screen is limited to 2 ft/s (0.6 m/s) unless environmental considerations require lower velocities, the pit cross section may be determined by screen requirements. If flow is such that a maximum-width screen available would be too long (deep) for practical or economic reasons, two screens may be employed with a center pier. In this case the distance to the pump should be increased 50% over single-screen distance. Piers should be rounded (radiused) on the upstream side and ogived (tapered to a small radius) on the downstream side. Any corners at the side walls should be faired at small angles to the opening and wall to prevent pockets where eddies can form.

The trash collected by the screens must be disposed of. Traveling screens carry trash up into a hood above the operating floor, at which point a series of spray nozzles wash the trash into a trough leading into a disposal area (Fig. 12). The nozzles are supplied by pumps sized 200 to 300 gpm (45 to 68 m³/h) with pressures of 60 to 100 lb/in² (413 to 690 kPa). These pumps are normally deep-well turbine multistage units. They are located in the clear well, if possible, close to a wall. If this is not possible, they can be suspended in the circulating water pit, to one side and ahead of the main pumps. Care must be taken that they do not disturb the flow to the larger pumps. Their submergence requirements are usually less than those of the main pumps, and this allows use of the pump setting that gives the least interference with either pump flow.

It is possible to have these pumps also dewater the pit. This will require additional piping and valves and a more careful location of the pump since it will of necessity be close to the bottom of the pit. A small chamber off the main pit located far enough from the main pumps to avoid eddying will be required.

FIG. 11   Drum filter. (Green Bay Foundry and Machine Works)

The direction of the flow from the forebay through the screens and into the pump area should be continuous. Avoid right-angle screens, through which flow must change direction at least once and possibly twice. If screens must be at an angle to the flow into the pumps, increase the screen-to-pump distance by 100%. Environmental considerations may increase the possibility of problems in this area.

Environmental Considerations   Suction pit requirements will vary according to whether hydraulic or structural standpoints are being considered. Both of these may also be in conflict with environmental considerations.

A design to accommodate fish limitations was mentioned briefly in a previous paragraph. Fish react to a horizontal velocity but are not aware of a pull in a vertical direction. Thus, to keep

FIG. 12   Spray nozzle cleaning of baskets on traveling water screens. (Envirex Inc., a Rexnord Company)

them from entering the inlet, a horizontal flow must be established at a velocity low enough to permit fish to escape.

Intakes which take their flow directly from a river may have a high velocity that would trap fish. Even if the velocities are lowered to reasonable screen levels—2 ft/s (0.6 m/s)—fish may still be drawn into the screen area and carried up to trash disposal.

When the source of a water supply system is a body of water containing fish, steps must be taken to prevent undue disturbance and destruction of the fish. A site survey should determine

1. The intake location furthest from a natural feeding area and furthest from attractive, or "trap," areas

2. The number of species involved

3. The size range of each species—length and weight

4. The population of each species and whether they are anadromous or settled

Next, total flow, probable intake size, and the velocities at inlet, through screens, and at trashracks should be determined. Variations in flow throughout the year and temperature ranges in winter to summer should be available.

Sites for intakes should not be selected near feeding areas for large schools of fish (kelp beds, coral reefs, and similar attractive spots). Sheltered spots most suitable for intake flows may also be most attractive to fish.

The best source of information about local fish is marine biologists who have studied the local areas. They may not only have information on fish habits, feeding patterns, population, etc., but may also have test information about the fish swimming ability. If they do not already have this information, they can probably run a survey to develop the data.

The most difficult problem to overcome is related to small fish. Screen openings must be held to a minimum, and under velocity conditions small fish have much less swimming-sustaining ability than larger fish, both in velocity and in duration time. In a given stream flow (such as is generated by pumps with inlet water going through screens), a fish must have the ability to sustain a given speed against this flow for a certain length of time. When it weakens, it will fall into the current flow and will be impaled against the screen and destroyed. If the fish senses the velocity early enough and has an alternate route, it can use darting speed to escape or can follow another attraction (cross velocity flow into a separate chamber or a light attraction to the chamber) and be removed on an elevator or pumped out to a safe channel (Fig. 13). Migrating fish need a continuation channel to restore their interrupted journey.

In designing an intake, it is necessary to either (1) keep the velocity below 0.5 ft/s (0.15 m/s) through the screen to avoid drawing fish into the screen (note that for a tube inlet away from shore a horizontal *velocity cap* (Fig. 4) should be placed over the inlet to prevent fish from being subject to a vertical velocity and to maintain a horizontal velocity that will attract their attention and direct them away from the inlet) or (2) create a cross flow which will propel or attract the fish to one side of the inlet area, from where they can be directed into a bypass pool and lifted back to their own living area or sent around the plant to a downstream location or, if anadromous, to an upstream rendezvous, and (3) keep piers and screens flush across the inlet face to prevent attractive pockets in which fish can hide and be drawn into the screens when they weaken.

Inlet screen areas should be in small sections rather than one long face, so that a fish will not be trapped in the center and find it too far to swim to safety after it realizes its predicament. The maximum deterrent flow is about 10 ft/s (3 m/s), but this may be too much disturbance for the flow to the pumps. Smaller areas allow short-term limits for enticing fish away from the inlet, and survival will be much higher.

Fish congregating at an inlet or in a forebay pool can be crowded or herded to an outlet point by the use of vertical nets or horizontal screens. In a direct channel, horizontal moving screens can route fish past a sloping (relative to stream flow, say 35°) moving screen which directs fish to a narrow outlet at one side leading to an outlet channel away from the main inlet.

The Environmental Protection Agency has had a major impact on plant design since the passage of the Federal Water Control Act of 1972 (as amended). Enforcement of Section 316a regarding thermal effluent has softened somewhat as later studies indicate that the effects of heat distribution on marine life are variable. Section 316b, however, covers every aspect of best-known technology and is applicable to any facility using a water intake structure.

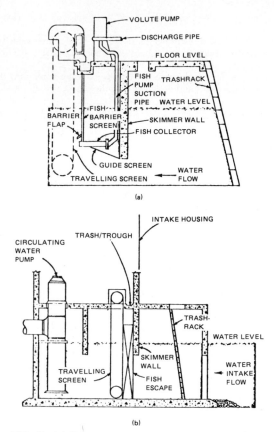

FIG. 13 (*a*) Fish pump prevents entrapment on traveling screen (Detroit-Edison). (*b*) Fish escape allows fish to bypass screen area.

Young fish must be kept from impinging on inlet screens. This can be accomplished by lowering the velocity through the inlet screens, by diverting or attracting the fish to other areas, by providing restraints at inlets, or by using fish buckets and elevators to remove the fish before they can enter the plant area—in short, by helping the fish avoid contact with the intake to the greatest extent possible. Obviously, trash and fish must be handled in separate areas, and screen wash pressures must be lowered to prevent harm to the fish.

Entrainment of marine organisms too small to be restrained by normal screens causes further problems in areas where such organisms normally develop. If the intake structure must be located in such an area, extensive information must be gathered at the site. Analyzing as much data as are available, in conjunction with plant flow and location requirements, may give the designer some idea of which equipment to select.

Small slot-width wedgewire screens (Fig. 14) are now being furnished in larger sizes than previously and, with their very low inlet velocity and backflushing capability, offer one good approach to reducing entrainment. Sand, chemical, and cloth filters may suit some situations, but backflushing and cleaning problems probably make them less attractive costwise.

Equipment may consist of any combination of

1. Stationary screens

   a. Bar racks in various attitudes (Fig. 9), with or without automatic rakes

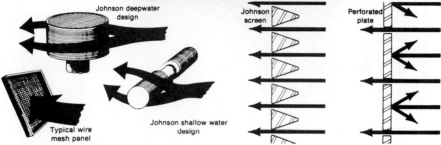

FIG. 14   V-shaped slotted screen provides velocity control to avoid attracting fish and good air or water back-flushing for debris cleanup. (Johnson Division, UOP)

   b. Air bubble, lighting, or electric barriers

   c. Underground, as in Ranney wells (Fig. 5)

2. Moving screens

   a. Vertical or horizontal (Fig. 10)

   b. Drum (Fig. 11)

3. Transport devices

   a. Elevators

   b. Pumps (Fig. 13)

   c. Baskets on vertical screens

   d. Attractive escape areas (Fig. 13)

   e. Velocity changes

4. Remote intakes

   a. Ocean outfall with velocity cap (Fig. 4)

   b. Ranney well (Fig. 5)

   If the source is a river, the angle of the intake structure relative to the direction of flow is important in modifying the impact of these design requirements. A case in point is shown in Fig. 15. The screen house at a low angle to the river flow (instead of the usual 90° inlet) allows the river current to provide a swim-by attitude for fish while a low-velocity screen approach is maintained. If the pump house is in line with the screen house, minimal disturbance will be felt by the

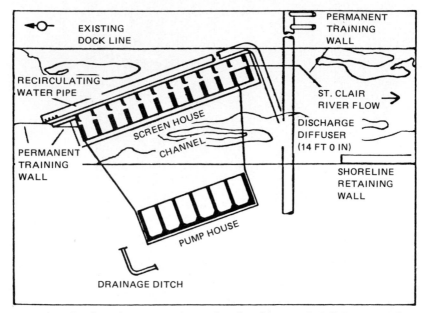

FIG. 15   Angling the intake structure to the river flow allows fish to swim by Belle River power plant. (Detroit-Edison)

pumps. Tendencies for eddies to form (and create vortices) can be minimized by placing wing walls upstream and downstream to control velocity.

Testing Model Pit Design   When the basic rules for good pump suction pit design are adhered to, no model test will be required to ensure proper operation of the pumps and pumping system. The substance of these rules is to keep a straight-in approach at a constant low velocity from the water source to the pump chamber. The Hydraulic Institute dimensions and charts satisfy these criteria for the average pump in general application.

Site layout problems may make the ideal solution impossible. Structural and environmental requirements may outweigh hydraulic requirements in some instances. When the ideal pit may not be possible or economically feasible, a model test should be considered. It should be noted that ideal dimensions are a composite covering not only a range of specific speeds but also a complex melding of pump design philosophies. Some variation from ideal dimensions should be expected from individual pump manufacturers.

Since pump manufacturers are not in a position to guarantee the pump pit design, differences of opinion between the structural and hydraulic pit design engineers and the pump design engineers may best be resolved by resorting to a model pit test (Fig. 16). Refer also to Sec. 10.2.

Vortexing   The real problems resulting from improper pit design occur largely on the water surface in the form of vortices, or cones, produced by localized eddies on the surface water. If this disturbance continues, the flow of water will carry the underwater part of the vortex down toward the pump suction bell and ultimately into the pump (Fig. 17). This introduces air into the impeller and will affect the mechanical radial balance of the impeller as a result of interruption of the normal solid-liquid flow pattern. This type of disturbance will produce hydraulic pulsations in the pump flow and mechanical overloading of bearings and impeller guides.

Underwater vortexing sometimes occurs in round pits or in pits where the pump suction bell is at some distance from the rear wall. Flow past the suction bell strikes the rear wall and rolls back toward the bell, forming an eddy which disturbs the normal flow into the pump. In a round pit, a cross baffle below the pump bell may reduce this effect. Where the pump is some distance

FIG. 16 Model pit test setup with fixed screens and pumps and valves for variable flow, to scale. (Byron Jackson Pump Division, Borg-Warner)

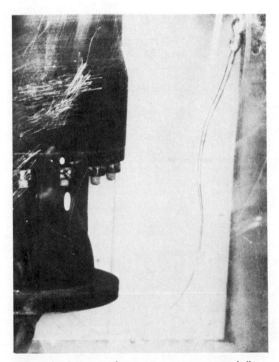

FIG. 17 Surface vortex drawing air into pump suction bell.

from the back wall, a wall can be installed near the pump, or a horizontal baffle at suction bell level behind the pump will also reduce the disturbance. In all cases, the distance between the suction bell and the bottom of the pit should not be more than one-half the suction bell diameter, and one-third the diameter is preferable.

The use of the suction bell diameter as a basis for spacing should be carefully evaluated. It will be seen that there is nothing magical in this relationship, especially when several pump manufacturers all use different bell diameters. The real criterion is the allowable velocity at the suction bell. It has been found that very-low-head pumps are much more sensitive to bell velocity over 6 ft/s (1.8 m/s). For example, the velocity head loss at the bell inlet with a high velocity may be such a large percentage of total head that efficiency could drop as much as 10%. A good design rule for safe operation can be related to pump head. For pumps having up to a 15-ft (4.6-m) head, the suction bell velocity should be held to 2½ ft/s (0.76 m/s); up to a 50-ft (15-m) head, 4 ft/s (1.2 m/s); and above a 50-ft (15-m) head, 5½ ft/s (1.7 m/s). These values should be used for any substantial amount of pumping, but for occasional short-term pumping they can be exceeded without destroying the pump.

Vortices may be broken up and effectually nullified by arrangements of baffles and vanes, or they may be prevented from occurring initially by a proper pit design. The only way to determine what baffling should be used, and its effectiveness, is by model testing. Methods for eliminating vortexing are shown in Fig. 18 and are further discussed in Sec. 10.2.

Vortices are usually generated when the flow direction of the liquid to be pumped changes or when there is high velocity past an obstruction, such as a gate inlet corner, screen pier, or dividing wall. In combination, these two causes invariably generate vortexes. For this reason the pump suction pit should be immediately preceded by a straight channel in which the velocity does not

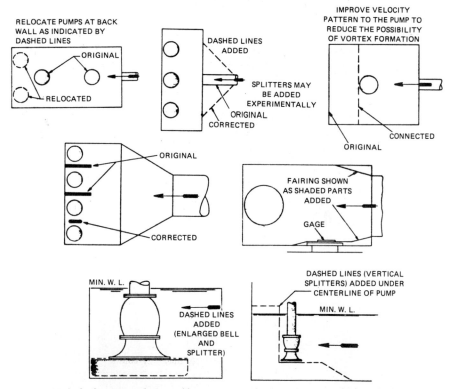

FIG. 18 Methods of correcting design problems in pits. (*Hydraulic Institute Standards*, 13th ed.)

exceed 1.25 ft/s (0.38 m/s). Satisfactorily operating pump pits with higher velocities are rare and should not be put into operation without the assurance of a model study.

An additional condition likely to generate vortexes is in a multiple pump pit with individual cells in which only a portion of the pumps will operate simultaneously. The dead space behind the nonoperating pump will have flowing water tending to reverse direction and form eddies. Eddies and vortices can be avoided by eliminating the walls or venting them at the rear and by positioning the pumps at the extreme rear of the pit, as well as by expensive modifications to the pit, such as splitter walls and baffles (Fig. 6).

Round pits tend to generate vortices, especially when the pump is centered in the pit. These vortices are usually centered around the pump column and are generated because of the eccentric inlet flow. Special cases of the round pit are tolerable either when a Ranney well (Fig. 5) is used or when there are booster pumps in the pipelines (Fig. 19). In the Ranney well, the ratio of pump size (and flow) to pit size (and capacity) is such that a very low velocity exists, as in a lake inlet. Water comes into the Ranney well all around the periphery. These conditions of direct flow and low velocity prevent vortexing. Booster pumps installed in a circular can (suction tank) must be centered in the can, and all inlet velocities to the can and flow in the can and into the pump must be uniform and high enough to provide fluid control This velocity will vary from 4 to 6.5 ft/s (0.76 to 1.98 m/s).

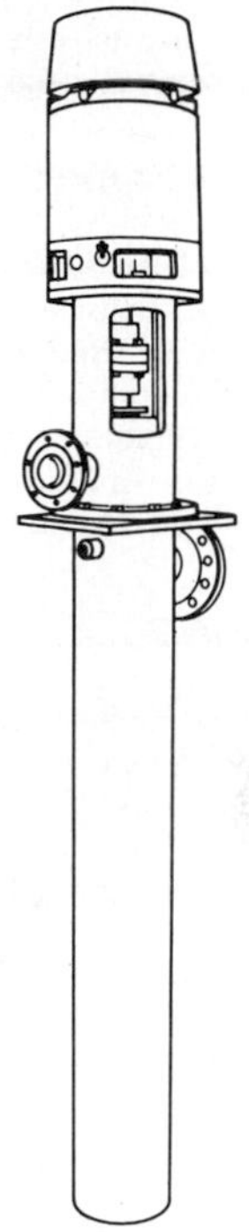

Vortices are not generated by a pump or pump impeller and so do not fall into clockwise

FIG. 19 Booster pump suspended in a steel well utilizes minute space for suction pit. (Byron Jackson Pump Division, Borg-Warner)

or counterclockwise rotation because of the pump rotation. Also, in a pump pit, vortices do not have a directional rotation induced by the rotation of the earth and therefore are not of opposite rotations above and below the equator, as is the case with "bathtub vortexing," which occurs in tanks being drained without pumps.

Submergence Centrifugal pumps in intake sumps must be submerged deeply enough to provide

1. A pressure sufficient to prevent cavitation in pump first-stage impellers, referred to as NPSH (it is assumed that the proper pump has been selected to perform satisfactorily with available NPSH)

2. Prevention of vortexing and associated pit flow problems detrimental to pump operation

A pump may have adequate submergence from a pressure standpoint and still be lacking in sufficient depth of cover above the suction inlet to prevent surface air from being drawn in. Any wet-pit pump must have its suction inlet submerged at all times, and for continuous pumping every pump will have a fixed minimum submergence requirement. Since this relates to velocity, there are two basic parameters: the suction inlet diameter and the depth of water above the inlet lip. As pump size (and flow) increase, the inlet velocity may stay constant as the bell diameter increases, but at the same time the impeller distance above the suction inlet becomes larger, so that a fixed submergence value would lead to increased surface velocity, peripheral drawdown, and an increase in air intake.

Basically, then, submergence must of necessity increase with pump size. For the final determination, some balance must be struck between submergence and pit width to satisfy an average flow velocity of 1 to 1¼ ft/s (0.3 to 0.4 m/s) and maintain reasonable economic balance between excavation costs, concrete costs, and screen costs without neglecting ecological requirements, and still fulfilling the primary need for circulating water in adequate quantities.

SMALL PUMPS TAKING SUCTION DIRECTLY FROM A SUMP

For pumps smaller than 3000 gpm (700 m³/h), the Hydraulic Institute suggests that recommendations for pit dimensions be obtained from pump manufacturers and that the inlet to the pump (or sump) be below minimum liquid level and away from the pump or pump inlet, so as not to impinge against the pump or pump inlet and cause rotation of the liquid. The usable volume of the pit in gallons (cubic meters) should equal or exceed two times the maximum capacity to be pumped in gallons per minute (cubic meters/hour) and sized to allow no more than three or four starts per hour per pump if pump service is to match outflow with pit inflow.

SUCTION PIPING

Single Pumps   Piping to the suction of a dry-pit centrifugal pump (Fig. 20) must be carefully worked out to provide a reasonably uniform velocity, straight-line flow, and adequate pressure and sealing against leakage, in or out. Air pockets just prior to entrance to the pump should be avoided, as well as down flow lines subject to sudden pressure changes. Air pockets can be prevented by proper elevations (Fig. 20a). Pressure surges can be controlled by gas bags, air tubes, etc., which may require a system pulsation study to determine possible need and solution.

In some cases, to provide the optimum flow pattern to avoid impeller disturbance, it may be necessary to have a straight run of pipe of as much as 8 small diameters immediately prior to the pump suction (e.g., following a short radius elbow or tee). Following a long radius elbow or a concentric reducer, a straight run of at least 3 large suction pipe diameters is recommended (Fig. 20c to f). The suction pipe should be at least as large as the pump suction. Valves should be in the larger section of piping. Care should be exercised in the use of eccentric reducers next to the pump suction nozzle. Although installing eccentric reducers with the flat side top will eliminate a potential air pocket, too large a change in diameters—more than 4 in (10 cm)—could result in a disturbed flow pattern to the impeller and cause vibration and rapid wear. Pipe venting and a concentric reducer are preferred rather than an eccentric reducer.

If it is necessary to provide a suction pipe at an angle to a double suction pump, the plane of the elbow or tee should not be in the same plane as the pump shaft if the fitting is closer than recommended to the pump suction (Fig. 20b). Rather, the fitting should be installed at right angles to the pump shaft and at the recommended distance upstream from the pump suction. An incorrect installation could result in an uneven flow to both sides of the double suction impeller. This could cause a reduction in capacity and efficiency, an increase in thrust on the bearing, noise, and possible impeller cavitation damage.

A dry-pit pump (Fig. 1g) may pull a suction lift and therefore will be located above the liquid source. All losses in piping and fittings will reduce the available suction pressure. Suction piping should be kept as simple and straightforward as possible. Any pipe flange joint or thread connection on the suction line should be gasketed or sealed to prevent air in-leakage, which would upset the vacuum and keep the pump from operating properly.

If expansion joints are required at a pump suction, an anchor should be interposed between the nozzle and the expansion joint to prevent additional forces from being transmitted to the pump case and upsetting rotating clearances. The same requirement applies to a sleeve coupling used to facilitate installation alignment.

Reciprocating pumps must have additional consideration because of the pulsating nature of their flow. Suction piping should be as short as possible and have as few turns as possible. Elbows should be long-radius. Pipe should be large enough to keep the velocity between 1 and 2 ft/s (0.3 and 0.6 m/s). Generally this will provide pipe one to two sizes larger than the pump nozzle.

High points that may collect vapor are to be avoided or, if necessary, properly vented. A pulsation dampener or suction bottle should be installed next to the pump inlet.

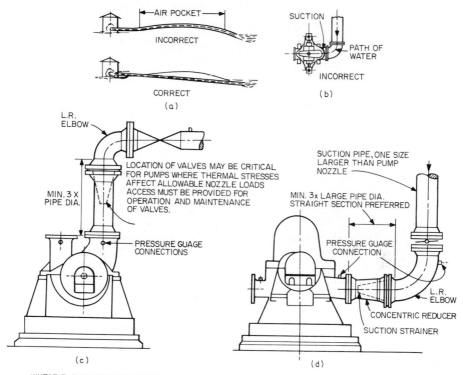

WHERE ELBOW REQUIRED, USE LONG-RADIUS ELBOW. SHORT-RADIUS ELBOW ACCEPTABLE IF 8 SMALL SUCTION PIPE DIAMETERS ARE USED BETWEEN ELBOW AND PUMP.

SUCTION PIPING FOR DOUBLE SUCTION PUMPS SHOULD NOT HAVE ELBOW NEAR PUMP BUT ELBOW IS PERMISSIBLE IF VERTICAL ONLY.

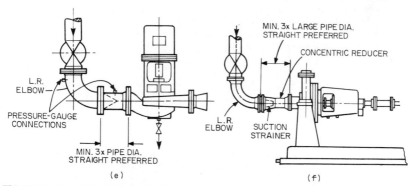

FIG. 20 Suction piping faults for dry-pit centrifugal pump. (*a*) Air pockets should be avoided. (*b*) Suction elbow should not be in a plane parallel to pump shaft. (*c*) Valve location may be critical to pump nozzle loading. (*d*),(*e*) Suction elbow should be one size larger than pump's, long radius and 3 large pipe diameters distant. (*d*),(*f*) Concentric reducer should be installed distant from pump.

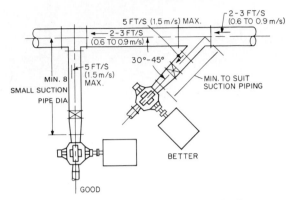

FIG. 21   Suction pipe header recommendations for dry-pit centrifugal pump.

Available NPSH should be sufficient to cover not only reciprocating pump requirements and frictional losses but also acceleration head (see Surge and Vibration below).

Manifold Systems   All comments relative to single pumps apply to manifold-pump systems, and additional points need to be noted.

In a suction manifold, the main-line flow should not be more than 3 ft/s (0.9 m/s). Branch outlets should be at 30 to 45° relative to main-line flow rather than 90°, and the velocity can increase to 5 ft/s (1.5 m/s) through a reducer (Fig. 21). With such velocities, branch outlets can be spaced to suit pump dimensions in order to avoid crowding. Also, if the angled manifold outlet is used, pumps can be set as close to the manifold as the elbow, valve, and tapered reducer will allow.

Manifold sections beyond each branch takeoff should be reduced to such a size that the velocity remains constant. One exception to this scheme is a tunnel suction (Fig. 22). The flow through the tunnel may operate independent of the pumps, which are suspended in boreholes drilled into the roof of the tunnel. Boreholes at least one-third tunnel diameter should be horizontally spaced at least 12 borehole diameters apart. Smaller ratios can be closer to a minimum of six diameters. Pump suction bells should be at least two borehole diameters above the tunnel roof. Velocities in the tunnel should be kept below 8 ft/s (2.4 m/s) for best pump performance.

NPSH   The net positive suction head so essential to correct and trouble-free pump operation is always reduced by losses in suction piping. An economic balance must be obtained between pump size and speed, required NPSH, pipe size, and suction velocity. If the suction source is a tank,

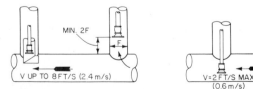

FIG. 22   Higher tunnel velocities require isolation of pump from direct flow to prevent distortion of close-clearance parts and shaft. (Hydraulic Institute Standards, 13th ed.)

such as a deaerator in a power plant, it is quite expensive to elevate the tank. Therefore the available NPSH is low, and so the pipe size from the tank to the pump should be large for low velocity and the length should be short for minimum losses. A cooling tower on a hill supplying water to a pump station below could have a much higher pipe velocity and longer length without forcing an extremely low NPSH requirement on the pump.

A dry-pit pump should be as close to the suction source as possible. When the NPSH required indicates a suction lift is possible, the most advantageous solution is to reduce the suction losses by increasing the pipe size, flaring the inlet bell, and keeping the pump suction eye of the first-stage impeller close to the minimum water level.

High-Pressure Inlets   Pumps in series build up pressure so that the second and following pumps will have a high-pressure suction pipe connection. This emphasizes the need for tight joints and flanges and careful welding. Expansion joints should not be used since the hydraulic force on the pump would be large, difficult to restrain, and perhaps impossible for the pump to handle without distortion. As NPSH will not be a problem, the pipe size can be kept small to minimize design problems and valve costs.

Effect on Pump Efficiency   Other than mechanical operation, the greatest effect of flow disturbance at the pump suction is on pump efficiency. The higher the pumping head, the less this effect becomes. For very-high-head pumps, a high velocity may be ignored completely unless there is an extremely large power evaluation factor. Greater attention should be given to the suction pipe design for pumps producing 100 ft (30 m) of head or less than for those with high head, as efficiency may be worth many dollars in power costs over the life of the plant and equipment.

Pump efficiency relates to the efficiency of the whole system, and so it may be well worthwhile to invest more money in a larger suction pipe.

Surge and Vibration   One of the possibilities arising from a power failure at a pump station is the reversal of a centrifugal pump if a valve fails to close, and its subsequent operation as a turbine. Under rated head, a pump will run from 20 to 60% above rated speed in the reverse direction. In its transition to that phase, the forward motion of the water is interrupted and gradually reversed. At the time of the power failure, the flow velocity in the suction pipe may not decelerate slowly enough to prevent a surge in the direction of the pump. The suction piping should be designed to withstand the resulting pressure rise since absolute integrity of valving and power supply will be too costly as a design parameter (see Chap. 8 for additional information on this subject).

Rotating machinery has a design vibration frequency which should be far enough removed from a system frequency to avoid sympathetic activity. Any combination of vibration frequencies near enough to each other to react will do so when a prime source, such as a pump, excites them. When a piping system has been designed to allow only very low stresses to be transmitted to a pump nozzle, it is quite vulnerable to vibration. It is usually good practice to analyze a suction system made up of pipe, valves, hangers, restraints, pump nozzle loads, pump speed and impeller configuration, foundation, and anchors to be sure the system is not "in tune." At the design stage it is relatively easy to change a valve or elbow location or to add a surge suppressor.

Reciprocating pumps are basically surge-producing machines. In particular, they require sufficient energy at suction to overcome pump required NPSH and pipe friction and also an energy called *acceleration head*. The pump energy must overcome the acceleration-deceleration pulsation flow in the suction end, which could lead to liquid flashing with pump noise and vibration. Surges large enough to rupture the pump cylinders may also be produced.

The total NPSH required by a reciprocating pump must include the NPSH required by the pump plus frictional loss from the suction pipe plus acceleration head. Acceleration head in feet (meters) is

$$H_a = \frac{LVNC}{gK}$$

TABLE 1 Wire Mesh Data

Screen mesh number, square weave	Equivalent in corduroy weave	Size (in)	Size (μm)	% open area
6		0.126	3327	57.2
8		0.090	2362	51.8
10		0.068	1651	46.2
12		0.060	1397	51.8
14		0.051	1168	51.0
16		0.045	991	53.0
20		0.033	833	43.6
24		0.0287	701	47.4
28		0.0227	589	40.4
30		0.0203	495	37.1
35		0.0176	417	37.9
40	12 × 64	0.0150	380	36.0
42		0.0138	351	33.6
50		0.0105	280	27.6
60	14 × 88	0.0077	246	21.3
65		0.0084	208	29.8
70		0.0068	190	22.7
80	24 × 110	0.0070	175	31.4
90	20 × 120	0.0056	159	25.4
100	20 × 150	0.0055	147	30.3
115		0.0051	124	30.0
120		0.0046	120	30.7
130		0.0043	115	31.1
140		0.0042	109	34.9
150		0.0041	104	37.4
160	20 × 200	0.0038	96	36.4
170	20 × 250	0.0035	88	35.1
180	20 × 300	0.0033	82	34.7
200	20 × 350	0.0029	74	33.6
120 × 330			70	
120 × 400			40	
120 × 600			30	
200 × 600			25	

1 μm = 10^{-6} m = 1 micron = 0.00004 in; 1 in = 25.4 mm.
SOURCE: Green Bay Foundry & Machine Works

where L = actual (not equivalent) length of suction pipe, ft (m)
$\quad\quad V$ = velocity of flow in suction line ft/s (m/s)
$\quad\quad N$ = pump speed, rpm
$\quad\quad C$ = pump constant, decreasing with number of cylinders from 0.2 to 0.04
$\quad\quad g$ = acceleration of gravity, 32.17 ft/s^2 (9.807 m/s^2)
$\quad\quad K$ = liquid constant: 2.5 for compressible fluids, 1.5 for water

The acceleration head can be reduced by installing a pulsation dampener in the suction line near the pump. See Sec. 3.1.

PIPE SCREENS AND STRAINERS

Except for special designs for handling slurries and other solids, most pumps are limited in their ability to handle dirt, trash, and solids of any size down to microparticles. Pump passageways (volutes, diffusers, impellers) handle generally spherical objects up to a "sticking size," but a severe problem may be caused by dust-size particles which lodge in sleeve bearings, bushings, and wearing rings and cause either rapid wear or rotation seizure. A wet-pit source may never be quite free from some degree of silt or sand. Once accepted by the pump, this liquid goes into a piping system, usually one with other pumps involved. The original pit pump may also be a dry-pit type. In either case, if the pumps of a system are to operate for long periods of time with minimum maintenance, wearing parts must be protected by fine screens or strainers.

In any screening operation, it is essential to remember that if the screen is necessary and does a good job, it will plug up and thus requires cleaning. In a wet pit, therefore, neither the suction bell of a wet-pit pump nor the suction bell going to a dry-pit pump should be basket-screened. Collection of debris will soon increase inlet velocity to unmanageable proportions, and the pump operation will suffer.

It follows that if the basic screening discussed previously does not provide liquid clear enough to be pumped, additional strainers must be put into the system. The need for such strainers, and the degree of particle reduction, are usually specified by the pump manufacturer. In some instances the system may require water as pure as it is possible to produce, as in boiler-feed service. For any situation, a variety of strainers are available.

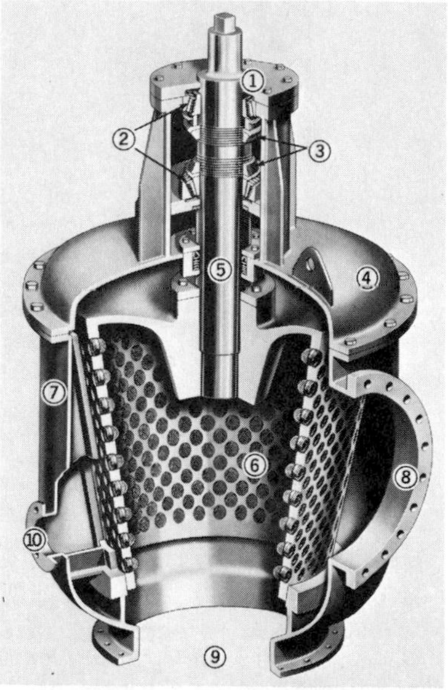

FIG. 23 Single flow strainer with provision for backwashing without interrupting service: (1) strainer drive, (2) strainer drum support bearings, (3) lock nuts for drum adjustment, (4) cover, (5) shaft, (6) drum—tapered for adjustment, (7) body, (8) inlet, (9) outlet, (10) backwash opening. (S.P. Kinney Engineers)

Drum rotating strainers (Fig. 11) can handle solids down to approximately 0.01 in (250 μm). Backflushing is used to renew the strainer area. The addition of woven wire cloth (Table 1) to such strainers may remove materials down to 0.1 in (25 μm). Woven wire drum strainers used in conjunction with sand filters will reduce the amount of backflushing required for the sand filters.

Where a finer degree of filtration is required, a line filter is used. Single flow filters require system shutdown to remove and clean baskets, but either a motorized automatic type of single flow filter (Fig. 23) or twin filters with alternate flow by valve arrangement (Fig. 24) permit uninterrupted system operation. When a particularly difficult situation exists, a battery of filters in series may be used.

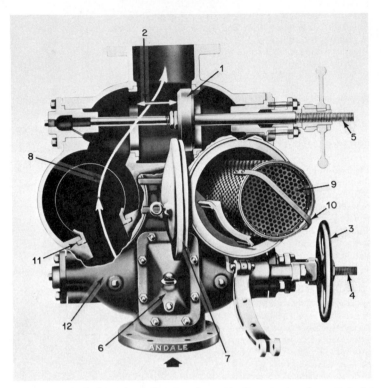

FIG. 24 Duplex basket strainer. Flow through one side allows cleaning of opposite basket while line is still in service. (1) positive flow diversion, (2) rapid switchover, (3) free-turning handwheels, (4) visual indication of valve position, (5) external acme threads, (6) internal access, (7) cover and clamp(s), (8) streamlined flow, (9) basket, (10) basket handle, (11) basket, (12) one-piece body construction. (Andale)

For areas where wear can be extremely critical, such as bearings and stuffing boxes, additional filtering effort is necessary, especially if the entrained silt is abrasive. A vortex filter using centrifugal flow action will remove particles down to a few micrometers. These filters have continuous sediment removal and should not require any maintenance (Subsec. 2.2.1, Fig. 78).

Some systems, such as boiler feedwater, are "closed" once they are in an operating cycle. Chemical and mechanical cleaning components are designed into the system to obtain a high degree of water purity. Before the system is ready to operate, however, the piping, tanks, valves, etc., will have a residue of weld spatter, metal chips, and other debris from the fabrication process. Temporary strainers are used in the suction lines of boiler-feed pumps and condensate pumps to gather up this debris. These strainers are in-line fabricated metal mesh or either a cone or box type. Feed pumps are most often equipped with cone strainers, pointing preferably upstream (Fig. 25). The length and diameter are set to conform to the suction pipe size and the mesh so that the velocity does not become excessive. These strainers, set between flanges in a pipe section, must be removed and cleaned whenever the differential pressure reaches 2 to 3 lb/in^2 (15 to 20 kPa). Condensate pump strainers are usually box strainers, a square section directly in front of the suction flange with a square box larger than the pipe diameter laced with eggcrate metal mesh (Fig. 26). This basket strainer can be lifted out of the box for cleaning as necessary. These strainers for collecting fabrication debris may be removed once the need disappears and the system will remain clean.

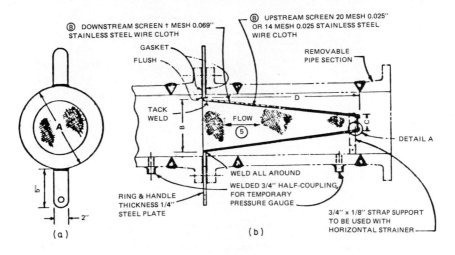

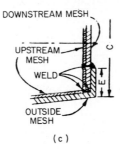

Flange or pipe size	A	B	C	D		E
				20 mesh 0.025-in wire, 0.025-in opening	14 mesh 0.025-in wire, 0.046-in opening	
3	5	2-¾	1-½	16	11	¼
4	6-³⁄₁₆	3-¼	1-½	16	11	¼
6	8-½	5	2	18	12	½
8	10-⅝	7-⅛	2	20	13	½
10	12-¾	9-⅛	3	23	14	½
12	15	11	3-½	24	15	½

FIG. 25 Cone strainer for start-up service in boiler-feed pump system. (*a*) cross section, (*b*) lengthwise section, (*c*) typical detail A (nose pointing downstream). All dimensions are in inches (1 in = 0.0254 m = 2.54 cm). Strainer ring material is carbon steel. When strainer has served its purpose, screen portion is removed, ID of ring is cut to suit ID of pipe, and ring is used as spacer. Cone strainers may be installed with nose pointing either upstream or downstream. The fine mesh screen should always be on the upstream surface of the strainer.

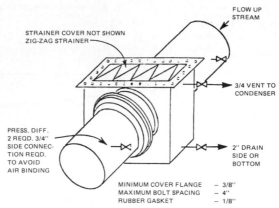

FIG. 26 Box strainer for temporary start-up service. Pressure differential connections are on top or side only. Upstream mesh is #16 stainless steel with 0.018-in (0.46-mm) wire diameter. Downstream mesh is #2 stainless steel with 0.063-in (1.6-mm) wire diameter. 1 in = 25.4 mm

FURTHER READING

Aten, R. E.: "Alternative Cooling Water Sources for the 'Water-Short' Midwestern States," paper presented at American Power Conference, Chicago, 1979.

Baker, O.: "Some Design Suggestions for Multiphase Flow in Pipelines," *Gas Age* **121**(12):34 (1958).

Bechtel Associates: "Survey and Performance Review of Fish Screening Systems," Job 8693, San Francisco, December 1970.

Dicmas, J. L.: "Effect of Intake Structure Modifications on the Hydraulic Performance of a Mixed Flow Pump," *Joint Symposium on Design and Operation of Fluid Machinery*, vol. 1, p. 403, June 1978.

Dicmas, J. L.: "Development of an Optimum Sump Design for Propeller and Mixed Flow Pumps," paper presented at ASME meeting, New York, 1967.

Dornaus, W. L.: "Stop Pump Problems before They Begin with Proper Pump Pit Design," *Power Eng.*, February 1960, p. 89.

Dornaus, W. L.: "Flow Characteristics of a Multiple-Cell Pump Basin, *Trans. ASME*, July 1958, p. 1129.

Fraser, W. H.: "Hydraulic Problems Encountered in Intake Structures of Vertical Wet Pit Pumps and Methods Leading to Their Solution," *Trans. ASME*, May 1973, p. 3.

Gegan, A. J., and J. T. Fong: "Upside-Down Pump Solves Design Puzzle," *Power*, February 1955, p. 84.

Hugley, D., et al.: Pulsation Primer: Seven Points, *Parts. Equip. Serv.*, May/June 1967 to May/June 1968.

Hydraulic Institute, *Standards for Centrifugal, Rotary and Reciprocating Pumps*, 13th and 14th eds., Cleveland, 1975 and 1983.

Langewis, C., Jr., and C. W. Gleeson: "Practical Hydraulics of Positive Displacement Pumps for High-Pressure Waterflood Installations," *J. Petrol. Technol.* **23**:173 (1971).

Mariner, L. T., and W. A. Hunsucker: "Ocean Cooling Water Systems for Two Thermal Plants," *ASCE Power Div. Proc.*, August 1959, pp. 65–85.

Messina, J. P.: "Periodic Noise in Circulating Water Pumps Traced to Underwater Vortices at Inlet," *Power*, September 1971, p. 70.

Mussalli, Y. G., P. Hofman, and E. P. Taft: "Influence of Fish Protection Considerations on the Design of Cooling Water Intakes," paper presented at Joint Symposium on Design and Operation of Fluid Machinery, Colorado State University, June 1978.

Paterson, I. S.: "Pump Intake Design Investigations," *MSJ Mech. Eng. Proc.* 31(1), London, January 1967.

"Proportions of Elbow-Type Suction Tubes for Large Pumps," *Bureau of Reclamation Report* HM-2, Denver, June 1964.

Scarola, J. A.: "American Power Plant Cooling Towers," *ASCE Power Div. Proc.*, New York, October 1959, pp. 119–132.

Scotton, L. N., and D. T. Anson II: "Protecting Aquatic Life at Plant Intakes," *Power*, January 1977, p. 74.

Stepanoff, A. J.: *Centrifugal and Axial Flow Pumps*, 2d ed., Wiley, New York, 1957.

Tsai, Y. J.: "Sediment Control for a Power Plant Intake Screenwell," paper presented at ASME meeting, New York, 1977.

Weltmer, W. W.: "Proper Suction Intakes Vital for Vertical Circulating Pumps," *Power Eng.*, June 1950, p. 74.

Whistler, A. M.: "New Look at Plunger Pump Suction Requirements," Publ. 69-PET-35, ASME, New York, 1969.

SECTION 10.2
INTAKE MODELING

JOHANNES LARSEN
MAHADEVAN PADMANABHAN

PROBLEMS ENCOUNTERED IN PUMP INTAKES

The various hydraulic problems associated with a pump pit include formation of surface and subsurface vortices, prerotation and swirl, and flow separation at or near the suction bell of either wet-pit or dry-pit centrifugal pumps (Figs. 1 and 2). Any of these problems can adversely affect pump performance by causing cavitation, vibrations, and/or loss of efficiency.[1] Usually there is more than a single reason for these problems, and the extent of the combined effects are seldom predictable by mathematical modeling. Formation of vortices, for example, even though dependent on suction pipe velocity and submergence, is strongly influenced by added circulation from vorticity sources, such as a nonuniform approach flow resulting from intake and approach channel geometries; rotational wakes shed from obstructions, such as columns or piers; and the velocity gradients resulting from boundary layers at the walls and floor.[2] The circulation contributed by these vorticity sources is unpredictable and strongly dependent on intake design and operating conditions, especially for large pumping units with multiple bays fed by a common approach channel. In these cases, physical modeling is the only way of predicting the behavior of the prototype with a reasonable degree of reliability.

Free Surface Vortices   This type of vortex is considered objectionable when it draws air bubbles or an air core into the pump inlet (Fig. 1a). It has been established that an air concentration in the suction pipe of from 3 to 5% can lower pump efficiency.[3] Also, air in the form of large bubbles or slugs can cause the impeller to vibrate. Strong surface vortices that do not draw air are also objectionable if they draw debris into the suction pipes and because their high rotational velocity can reduce the local pressure sufficiently to induce cavitation.

Subsurface Vortices   This type of vortex, also known as a submerged vortex (Figs. 2 and 3), usually originates from a floor or wall and is induced by vorticity produced in separation zones close to the pump entrance or, in wet-pit pumps, below the bell. The reduced pressure in the vortex core causes fluctuating load on the pump impeller along with associated vibration and noise,[4,5] increased possibility of cavitation, higher inlet losses, and decreased pump efficiency, especially when the core pressure is sufficiently low to release dissolved air or other gases from the fluid.

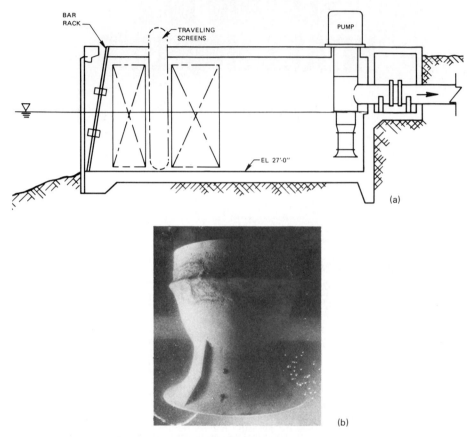

FIG. 1 (*a*) Typical vertical wet-pit pump intake. (*b*) free surface vortex can form at pump suction bell entrance due to combination of low submergence, high suction velocity, nonuniform approach flow, and/or other vorticity sources. (Alden Research Laboratory)

Prerotation and Swirl *Swirl* is a general term for any flow condition (due to vortexing or a pipe bend) where there is a tangential velocity component in addition to a usually predominating axial flow component. *Prerotation* is a specific term to denote a cross-sectional average swirl in the suction line of a pump or, in case of a vertical wet-pit pump, upstream of the impeller.

The prerotation angle θ is a measure of the strength of the tangential velocity component u_t relative to that of the axial velocity component u in the flow approaching the pump impeller; i.e., $\theta = \tan^{-1}(u_t/u)$. Adverse effects on the pump are decreased capacity and head when the rotation is in the direction of pump rotation and increased capacity and head when the rotation is opposite the pump rotation (antirotation). The increased capacity is associated with an increase in power requirement and may cause motor overheating.

Prerotation will influence pump performance since the flow approaching the impeller already has a rotational flow field which may oppose or add to the impeller rotation, depending on direction. The design of the pump vanes (i.e., shape and angle) assumes no prerotation, and the existence of prerotation implies flow separation along one side of the impeller vanes. The degree of prerotation that should be of concern is dependent on the type of pump and may not always be known. Prerotation could be quantified in a model by an average cross-sectional swirl angle, determined by detailed velocity measurements or by readings on a swirl meter. Since swirl decays along a pipe as a result of wall friction, internal fluid shear, and turbulence, the swirl meter in a model suction pipe should be located near the impeller.

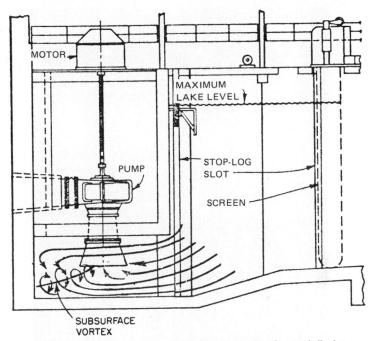

FIG. 2 Typical vertical dry-pit pump intake. Improper spacing of suction bell relative to floor and back wall caused subsurface vortexing. Model test revealed problem and correction. (Ref. 4)

Losses Leading to Insufficient NPSH   A poorly designed pump intake could result in large inlet losses. Losses caused by screens, poor entrance conditions, vortexing and swirl, and vortex suppression devices may add up to a value so great that the required NPSH of the pump is not satisfied. Increased inlet losses due to swirl have been reported in laboratory studies.[6] In a nuclear reactor residual-heat-removal sump model, inlet losses in a preliminary design wherein air core vortices and a high degree of swirl were present were 20% higher than in a revised design with no strong vortices and swirl, under similar pipe entrance geometry and flows. Since the degree of vorticity and swirl cannot be predicted and it is therefore not possible to calculate inlet losses reliably, they are usually obtained from model studies. With the experimentally derived values of the inlet losses, the NPSH available should be checked by recalculation.

DESIGN GUIDES

As a preliminary design guide (Sec. 10.1), published information may be used to establish pump sump dimensions and a minimum desired submergence.[7-10] Figures 1 and 2 of Sec. 10.1 show the basic layout of a pump sump and present typical dimensions in relation to the flow required per pump, as published by the Hydraulic Institute.[7] Some of these references may express sump dimensions in terms of bell mouth diameter, and Fig. 4 shows the typical relationship between design flow for a given wet-pit pump and the bell mouth diameter required to achieve reasonable and desirable velocities approaching the impeller. One should be aware that the dimensions given by design guides are considered applicable for the ideal condition of a simple, straight approach flow with a constant, low approach velocity to the pump sump.

The need for a physical model study still exists when site or operating conditions make the ideal condition impossible. For example, for a particular pump intake configuration, a hydraulic model study indicated that a strong submerged floor vortex existed with floor clearance of 0.5

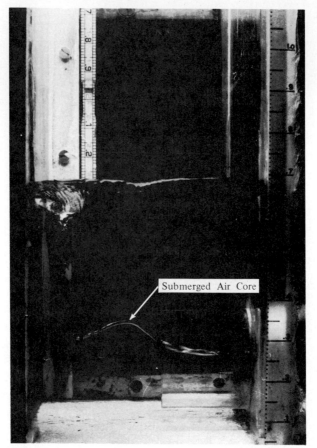

Submerged Air Core

FIG. 3 Submerged vortex at horizontal intake. (Alden Research Laboratory)

times the bell diameter (a usual design value) and that a reduced floor clearance of 0.3 times the bell diameter eliminated the vortex. Typically, for pump intakes of the following types, model studies should be considered essential:

1. Intakes with nonsymmetric approach flow, e.g., an offset in the approach channel
2. Intakes with multiple pump bays with a common approach channel and a variety of pump operating combinations
3. Intakes with pumps of capacities greater than 40,000 gpm (2.5 m³/s) per pump
4. Intakes with expanding approach channel
5. Intakes with possibilities of screen blockages and/or obstructions close to suction pipe entrance, e.g., reactor containment recirculation sumps, gate guides for dry-pit pumps

For items 1, 2, 4, and 5, a model study is recommended because of the unknown effects a non-uniform approach flow can have on vortexing and swirl. For item 3, considering the cost of large pump installations and the cost of backfit should problems occur, a model study is recommended to ensure proper pump operation.

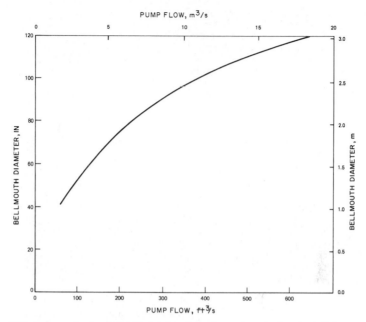

FIG. 4 Approximate relationship between design flow and required suction bell diameters for vertical pumps. (Alden Research Laboratory)

MODEL SIMILITUDE

The principle of dynamically similar fluid motion forms the basis for the design and operation of hydraulic models and the interpretation of experimental data. The basic concept of dynamic similarity is that two systems with geometrically similar boundaries have similar flow patterns at corresponding instants of time,[11,12] To achieve this, all individual forces acting on corresponding fluid elements must have the same ratios in the two geometrically similar systems. The condition required for complete similitude may be developed from Newton's second law of motion:

$$F_i = F_p + F_g + F_v + F_t \tag{1}$$

where F_i = inertial force, defined as mass m times acceleration a
  F_p = pressure force connected with or resulting from the motion
  F_g = gravitational force
  F_v = viscous force
  F_t = surface tension force

Additional forces, such as fluid compression, magnetic, or Coriolis forces, may be relevant under special circumstances, but generally these forces have little influence and are therefore not considered in the following development.

Equation 1 can be made dimensionless by dividing all the terms by F_i. Two systems of different size which are geometrically similar are dynamically similar if both satisfy the same dimensionless form of Eq. 1. We may write each of the forces of Eq. 1 as

$$F_p = \text{net pressure difference} \times \text{area} = \alpha_1 \Delta p L^2$$

$$F_g = \text{specific weight} \times \text{volume} = \alpha_2 \gamma L^3$$

$$F_v = \text{shear stress} \times \text{area} = (\alpha_3 \mu \Delta u / \Delta y)\,(\text{area}) = \alpha_3 \mu u L$$

$$F_t = \text{surface tension} \times \text{length} = \alpha_4 \sigma L$$

$$F_i = \text{density} \times \text{volume} \times \text{acceleration} = \alpha_5 \rho L^3 u^2 / L = \alpha_5 \rho u^2 L^2$$

where α_1, α_2, etc. = proportionality factors

$\qquad \Delta p$ = net pressure difference

$\qquad L$ = representative linear dimension

$\qquad \gamma$ = specific weight = ρg

$\qquad \mu$ = dynamic viscosity

$\qquad \Delta\mu/\Delta y$ = depthwise velocity gradient

$\qquad u$ = representative velocity

$\qquad \sigma$ = surface tension

$\qquad \rho$ = density

$\qquad g$ = acceleration due to gravity

Substituting the above terms in Eq. 1 and making it dimensionless by dividing by the inertial force, we obtain

$$\frac{\alpha_1}{\alpha_5} E^{-2} + \frac{\alpha_2}{\alpha_5} F^{-2} + \frac{\alpha_3}{\alpha_5} R^{-1} + \frac{\alpha_4}{\alpha_5} W^{-2} = 1 \qquad (2)$$

where

$$E = \frac{u}{\sqrt{\Delta p/\rho}} = \text{Euler number} \propto \frac{\text{inertial force}}{\text{pressure force}}$$

$$F = \frac{u}{\sqrt{gL}} = \text{Froude number} \propto \frac{\text{inertial force}}{\text{gravitational force}}$$

$$R = \frac{uL}{\mu/\rho} = \text{Reynolds number} \propto \frac{\text{inertial force}}{\text{viscous force}}$$

$$W = \frac{u}{\sqrt{\sigma/\rho L}} = \text{Weber number} \propto \frac{\text{inertial force}}{\text{surface tension force}}$$

Since the proportionality factors α_1, α_2, etc., are the same in model and prototype, complete dynamic similarity is achieved if the values of each dimensionless group E, F, R, and W are equal in model and prototype. In practice, this is difficult to achieve. For example, to have $F_{\text{model}} = F_{\text{prototype}}$ and $R_{\text{model}} = R_{\text{prototype}}$ requires either a 1:1 "model" or a fluid of very low kinematic viscosity in the reduced-scale model. Hence, the accepted approach is to select the predominant force and then design the model according to the appropriate dimensionless group. The influences of the other forces become secondary and are called scale effects.[11,12]

Froude Scaling   Pump intake models are generally designed and operated using Froude similarity since the flow is controlled by gravitational and inertial forces. The Froude number is therefore made equal in model and prototype:

$$F_r = F_m/F_p = 1 \qquad (3)$$

where m, p, and r denote model, prototype, and ratio between model and prototype.

In modeling a pump intake sump to study the formation of vortices, it is important to select a reasonably large geometric scale to achieve large Reynolds numbers. At large Reynolds number, energy loss coefficients usually behave asymptotically with Reynolds number. Hence, with $F_r = 1$ and a sufficiently high Reynolds number, the Euler number E will be equal in model and prototype. This implies that flow patterns and loss coefficients may be considered similar in model and prototype. From Eq. 3, the velocity, discharge, and time scales are

$$u_r = L_r^{0.5}$$

$$Q_r = L_r^2 u_r = L_r^{2.5}$$

$$t_r = L_r^{0.5}$$

For example, if a model-to-prototype scale of 1:10 is used, a prototype velocity of 1.0 ft/s (0.3 m/s) becomes a model velocity of $(1 \div 10)^{1/2} = 0.32$ ft/s (0.09 m/s). In the model, all physical dimensions of the pump bays and approach channel are scaled to the ratio 1:10. Submergence, being a linear dimension, is also scaled to 1:10. A flow of 100 gpm (0.006 m³/s) in the model corresponds to a flow of $100 \times 10^{2.5} = 31{,}645$ gpm (2 m³/s) in the prototype. Similitude parameters and laws are treated in detail in Refs. 11 and 12.

Unfortunately, an important phenomenon of interest, namely, free surface vortexing, is also influenced by viscous and perhaps surface tension forces, and the extent of these effects is not modeled properly in a Froude model, giving rise to scale effects. In a Froude model with the scaled flow conditions, the Reynolds number will be $L_r^{1.5}$ times the prototype Reynolds number at corresponding flow conditions, where L_r is the model-to-prototype scale ratio. Thus, for a $1:10$ scale model, a prototype Reynolds number of 10^6 would give Froude scale model Reynolds number of 3.16×10^4.

Liquid motion involving free surface vortex formation in pump intakes has been studied by several investigators.[6,13−17] Based on one of these studies, if the radial Reynolds number (defined as $Q/h\nu$, where Q is the pump discharge, h is the submergence of the suction bell entrance, and ν is the kinematic viscosity of the fluid) is greater than 3×10^4, viscous effects can be considered negligible.[16] Another study indicated that at pipe Reynolds numbers of $R = ud/\nu$ (d being the pipe diameter) above 3.2×10^4, viscous effects are negligible.[6] Above a Weber number of 11, surface tension effects may be negligible.[14]

DESIGN AND OPERATION OF MODELS

Model Scale   Scale effects are less as model size increases but construction and operation costs increase with model size, and so a compromise must be made. In general, the formation of vortices, both free surface and submerged, is highly responsive to approach flow patterns, and it is important to select a geometric scale that achieves Reynolds numbers large enough to keep the flow turbulent and to meet the fluid mechanic criteria for minimizing scale effects.[13] Also, one should consider other factors, such as access for instruments, accurate flow measurements, and ease of modification, in selecting a proper scale.

Information on preferred minimum values of Reynolds number and Weber number, discussed earlier, may be used in designing a model and deciding geometric scale. However, adhering to these limits does not, in itself, guarantee negligible scale effects in a Froude model since these limits are based on tests run under ideal laboratory conditions. In real situations there is usually more than one source of vorticity generation of unknown extent, and a generalization of scale effects for all cases would be inappropriate. To compensate for such unknown scale effects, a common practice is to test a model at higher-than-Froude scaled flows.

A special test procedure involving high temperatures and developed at the Alden Research Laboratory may be used to determine any scale effects and to project the model results to prototype ranges of Reynolds numbers.[2] The water temperature in the model is varied over a range, say 50 to 120°F (10 to 49°C), and flow velocities in the pipes are varied over a range of values, if possible, up to the prototype velocities. Vortexing and other flow patterns over a range of Reynolds numbers are obtained from these tests and can be used to evaluate any possible scale effects. A prediction of the prototype performance can be made based on these tests.

Extent of Model   It is very important to include a sufficient length of approach channel in the model since approach flow nonuniformities contribute greatly to vortex formation and swirl. The decision on what approach length should be included is usually based on the experience and engineering judgment of the model designer. The requirement is essentially the proper simulation of approach velocity profile to the intake and then on to the pump bays. In some cases, considering the cost and time involved in including a sufficient channel length in a pump intake model, a separate, smaller model can be built to determine the approach flow patterns to the intake and these flow patterns then simulated in the pump intake model.[18]

Figure 5 shows a $1:10$ geometric scale model of a three-bay pump intake.

Modeling of Screens and Gratings   In addition to providing protection from debris, screens suppress nonuniformities of approach flow. The aspects of flow through screens that are of concern in a model study are (1) energy loss in the fluid passing through the screen, (2) modification of the velocity profile, and (3) production of turbulence. As all these factors could affect vortex formation in a sump with approach flow directed through screens, a proper modeling of screen parameters is important.

The fluid passing through the screen loses energy at a rate proportional to the drop in pressure,

FIG. 5 A 1-to-10 scale model of a three-bay pump intake. (Alden Research Laboratory)

and this loss dictates the effectiveness of the screen in altering velocity profiles. The pressure drop across the screen is analogous to the drag induced by a row of cylinders in a flow field and can be expressed in terms of a pressure drop coefficient K (or, alternately, a drag coefficient), defined as[19]

$$K = \Delta p / 0.5 \rho u_a^2 = \Delta H / u_a^2 / 2g$$

where Δp = drop in pressure across screen
$\quad u_a$ = mean velocity of approach flow to screen
$\quad \Delta H$ = head loss across screen

The loss coefficient is a function of three variables: (1) screen pattern, (2) screen Reynolds number $R_s = u_a d_w / \nu$, where d_w is the wire diameter of the screen, and (3) solidity ratio S', the ratio of closed area to total area of screen (Sec. 8.1).

If S' and the wire mesh pattern are the same in the model and prototype screens, the corresponding values of K are a function of R_s only. This is analogous to the drag coefficient in a circular cylinder. At values of R_s greater than about 1000, K becomes practically independent of R_s.[19] However, for models with low approach flow velocity and fine wire screens, it is necessary to ascertain the influence of R_s on K for both model and prototype screens before selecting screens for the model which are to scale changes in velocity distribution.

Velocity modification equations relating the upstream and downstream velocity profiles usually indicate a linear relationship between the two, the shape and solidity ratio of the screen, and the value of K.[20] If wire shapes and solidity ratios are the same in model and prototype, it is possible to select a suitable wire diameter to keep the values of K approximately the same for the model and prototype screens in the corresponding Reynolds number ranges. This produces a head loss across the model screen which is scaled to the geometric scale of the model and which produces identical velocity modifications in model and prototype. Some model designers consider it a conservative approach to leave out the screens in the model altogether, under the assumption that this omission will only worsen the nonuniformity of the approach flow. This approach is not recommended unless it is not practically possible to select an appropriate model screen.

Modeling of Pumps   The exterior submerged surfaces of the pump bowl assembly and, for wet-pit pumps, the column including the bell mouth must be modeled to scale as well as the interior geometry from the bell mouth perimeter to the impeller eye. This is to ensure that flow patterns approaching the impeller are properly simulated.

Prerotation induced by the pump's rotating element is discussed in Secs. 2.3 and 10.1 and in Refs. 21 to 24. It has been shown that the rotating element does not affect upstream flow patterns when the pump is operated at design flow, and hence including the rotating element in a pump intake study is not necessary. When the pump is operated at less than rated discharge, a degree of swirl is induced upstream of the rotating element. This swirl increases rapidly at discharges below 45% of rated discharge and may affect vortex activity in the pump well and flow distribution to the pump.

Model Operation   Operating a model at the prototype suction pipe velocity is thought to be a conservative method to compensate for excessive viscous energy dissipation and the consequent less intense model vortices in a Froude model.[17] This method is often referred to as the Equal Velocity Rule. Operating a model at a higher than Froude scaled velocity should be considered a reasonable procedure for evaluating scale effects. However, operating a model based on the equal velocity rule may not be advisable unless the model is large enough, say at least a 1:4 scale model, since increasing the flow to many times the Froude scaled flow while keeping a scaled submergence could distort the approach flow patterns and turbulent intensities and thus cause unrealistic results. In general, a velocity increase to about 1.5 times the Froude scaled velocity can be considered reasonable. More appropriately, the information contained in Refs. 25 and 26 may be used to decide the velocity ratio for exaggeration, which often is considered a function of model scale. If the final recommended design does not show any coherent core vortices in a model, no large scale effects should be expected, and operating the model at higher than Froude scaled flows may be unnecessary in such cases.[27]

Model Cost   The cost of model studies varies considerably and is dependent on such factors as number of pump bays, number of operating conditions, and complexity of approach flow. In 1980 dollars, $25,000 to $70,000 can be expected to cover the range from simple one- or two-bay sumps to multibay installations with complex approach flow and several operating modes. For example, a 1:10 scale model of a three-bay sump with a total capacity of 240,000 gpm (15.1 m^3/ s) was tested for 39 operating modes (combinations of water level and pump and screen outages) and for unsteady flow distribution to the sump structure. The cost for model construction, testing with all modifications necessary to obtain satisfactory inflow conditions to the pumps, and preparation of a final report was $50,000.

MODEL OBSERVATION TECHNIQUES

Surface Vortices   Vortices in a model usually contain little energy, and therefore vortex circulation cannot readily be measured since any measuring device will affect the strength of the vortex. Therefore, vortex strength is best characterized by visual observation and classified by comparison with a vortex strength scale, such as the one shown in Fig. 6. Observations are carried out continuously for a representative period of time, usually one hour prototype, and the resulting data are analyzed in terms of the maximum observed vortex strength and in terms of persistence, i.e., the percentage of time a given vortex strength occurred.[28] Vortices of strength 5 and 6 are clearly objectionable in an intake design because of the effect air entrainment and localized pressure reductions has on pump performance. Usually in model studies, a vortex strength of 3 is considered the maximum allowable, as prototype strength may be slightly underestimated as a result of possible scale effects.

Subsurface Vortices   Vortices attached to fixed boundaries below the surface can be difficult to detect and may require a careful search with a dye tracer or an air-injection technique.

Submerged vortices are most commonly found below the center of wet-pit pumps and around splitter plates introduced to even out flow into the bell mouth. Pressure measurements made with electronic pressure transducers located in the floor under the pump may provide information on rapidly forming and disappearing vortices that cannot be detected with flow visualization. Subsurface vortices should, as far as practical, be eliminated from any design because their presence imposes fluctuating loads on the impeller and the resulting low-pressure areas may cause local cavitation damage to the pump.

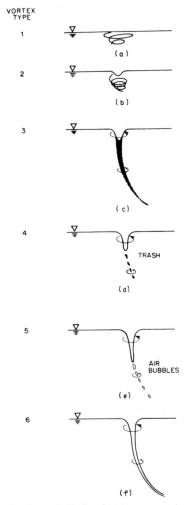

Prerotation The tangential velocity component may be measured by a swirl meter, or vortimeter, which is an impeller with zero pitch in the axial direction mounted axially in the pump column two to four column diameters downstream from the suction bell. A typical vortimeter is shown in Fig. 7. Vortimeter rotation is measured by either electronic or manual counting of rotation over a given time. Prerotation can also be determined from velocity traverses obtained with a two-dimensional pitot tube.

Intake Loss Coefficient Water manometers connected to piezometric taps along the pipe are used to measure the hydraulic gradient along the suction pipe. An intake loss coefficient is then determined by extrapolating the measured hydraulic gradient to the suction pipe inlet and computing the head loss $h_L = \Delta h - u^2/2g$, where Δh is the head loss between the pump pit water surface and the extrapolated pressure gradeline at the pipe inlet. With this procedure, the computed head losses will not include any pipe frictional losses and consequently any model pipe Reynolds number influence is avoided. At the same time, pressure measurements are done far downstream from the region of flow separation near the entrance and hence more accurate measurements of total pressure loss, including entrance losses, are obtained. The loss coefficient C_L is usually defined as $C_L = h_L/u^2/2g$, where $u^2/2g$ represents the velocity head in the pipe. The computed model loss coefficient is valid for the prototype, and hence the prototype inlet losses are C_L times the prototype pipe velocity.

Flow Distribution Radial flow under the perimeter of the pump bell is required to avoid vibrations and loss in efficiency caused by uneven impeller loading. Flow direction is usually indicated by six to eight yarn streamers equally spaced and mounted on wire supports 0.2 bell mouth diameter below the bell mouth perimeter. Figure 8 shows a typical installation. For an acceptable design, these streamers should not deviate more than 10 to 15° from

FIG. 6 Vortex strength scale: (*a*) Type 1, incoherent surface swirl, (*b*) type 2, surface dimple with coherent surface swirl, (*c*) type 3, dye core to intake with coherent swirl throughout water column, (*d*) type 4, vortex pulling floating trash, not air, (*e*) type 5, vortex pulling air bubbles to intake, (*f*) type 6, full air core to intake. (Alden Research Laboratory)

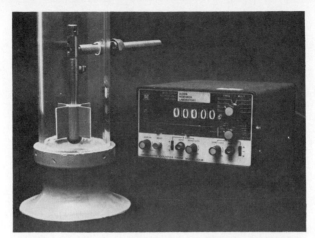

FIG. 7 Vortimeter for measuring swirl in a suction pipe. (Alden Research Laboratory)

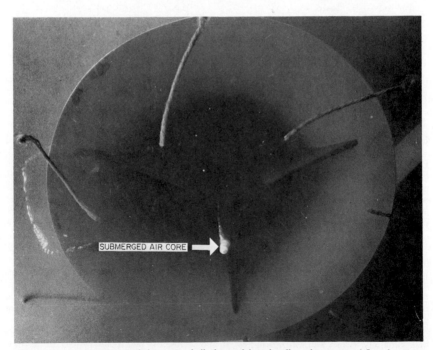

SUBMERGED AIR CORE

FIG. 8 Yarn streamers around the suction bell of a model intake allow observation of flow directions into the bell. (Alden Research Laboratory)

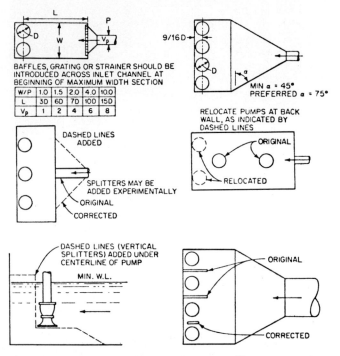

FIG. 9 Common method of eliminating vortexing problems in pump pits. (Ref. 7)

radial. Velocity distribution approaching the pump is usually measured with miniature propeller meters.

CORRECTIVE MEASURES

Problems in pump sumps, such as severe vortexing, intense swirl, or uneven flow distribution at the bell, can be reduced or eliminated by deriving proper corrective methods using models. Since no single remedial method to correct these problems exists, it is important to try several suitable methods in the model to derive an effective and practical one. Even though the pump bell velocity and submergence are very important parameters contributing to problems, it is seldom possible to change these values. Hence, the corrective measures are usually incorporated in the geometry of the pump pit and include the introduction of suitable appurtenant devices.

Severe free surface vortices may be broken up and effectively suppressed by arranging baffles and vanes to correct the rotational flow field due to the approach flow distribution. Relocation of pumps, introduction of horizontal grids below the water surface, changing of wall and floor clearances, improvements in approach channel configurations, and changes in lengths and spacing of piers are some of the common techniques for reducing vortex activity. Surface vortices can also be controlled by installing a curtain wall immediately upstream of the pump. Figure 9 indicates various methods of eliminating poor flow conditions as given in Ref. 7.

Submerged vortices are usually eliminated by instilling splitter vanes or floor cones under the bell. Variations in floor and wall clearances can also be effective and should be tried in the model.

The choice of corrective measures should always be made with the pump use in mind. For example, in testing pump intakes that will be used for sewerage, corrections that will allow trash build up, such as vanes or grids, should be avoided. It is also important that a continuous dialogue exist between the intake designer and the test laboratory so that the most practical and inexpensive solution for a specific installation can be obtained with a minimum of effort.

REFERENCES

1. Tullis, J. P.: "Modeling in Design of Pumping Pits," *J. Hydraul. Div. ASCE*, September 1979, p. 1053.

2. Durgin, W. W., and G. E. Hecker: "The Modeling of Vortices at Intake Structures," in *Proceedings of the ASCE, IAHR, and ASME* Joint Symposium on Design and Operation of Fluid Machinery, Colorado State University, Fort Collins, Colo., June 1978, vol. 1, p. 381.

3. Murakami, M., et al.: "Flow of Entrained Air in Centrifugal Pumps," in *Proceedings of the 13th Congress IAHR*, Kyoto, Japan, 1969, vol. 2, p. 71.

4. Messina, J. P.: "Periodic Noise in Circulating Water Pumps Traced to Underwater Vortices at Inlet," *Power*, September 1971, p. 70.

5. Denny, D. F.: "Experimental Study of Air-Entraining Vortices in Pump Sumps," in *Proceedings of the Institution of Mechanical Engineers*, London, 1956, p. 1.

6. Daggett, L. L., and G. H. Keulegan: "Similitude Conditions in Free Surface Vortex Formations," *J. Hydraul. Div. ASCE*, November 1974, p. 1565.

7. Hydraulic Institute: *Standards for Centrifugal, Rotary and Reciprocating Pumps*, 13th ed., Cleveland, 1975.

8. Chang, E., and M. J. Prosser: "Intake Design to Prevent Vortex Formation," in *Proceedings of the ASCE, IAHR, and ASME* Joint Symposium on Design and Operation of Fluid Machinery, Colorado State University, Fort Collins, Colo., June 1978, vol. 1, p. 393.

9. Sweeny, C. E., R. A. Elder, and D. Hay: "Pump Sump Design Experience: Summary," *J. Hydraul. Div. ASCE* March 1982, p. 361.

10. Reddy, Y. R., and J. A. Pickford: "Vortices at Intakes in Conventional Sumps," *Water Power* **24**(3):108 (1972).

11. Rouse, H.: *Handbook of Hydraulics*, Wiley, New York, 1950.

12. Daily, J. W., and D.R.F. Harleman: *Fluid Dynamics*, Addison-Wesley, Reading, Mass., 1966.

13. Anwar, H. O.: "Prevention of Vortices at Intakes," *Water Power*, October 1968, p. 393.

14. Jain, A. K., et al.: "Vortex Formation at Vertical Pipe Intakes," *J. Hydraul. Div. ASCE*, October 1978, p. 1429.

15. Hattersley, R. T.: "Hydraulic Design of Pump Intakes," *J. Hydraul. Div. ASCE*, March 1965, p. 223.

16. Anwar, H. O., et al.: "Similarity of Free Vortex at Horizontal Intake," *J. Hydraul. Res.* **2**:95 (1978).

17. Denny, D. F., and G.A.J. Young: "The Prevention of Vortices and Swirl at Intakes," Paper No. C1 in *Proceedings of IAHR 7th Congress*, Lisbon, 1957.

18. Larsen, J., and P. M. Bozoian: "Model Studies of Intake Pumphouse for the Perry Nuclear Power Plant Cooling Towers," report 153-77/M294EF, Alden Research Laboratory, Holden, Mass., 1977.

19. Padmanabhan, M., and S. Vigander: "Pressure Drop Due to Flow through Fine Mesh Screens," *J. Hydraul. Div. ASCE*, August 1978, p. 1191.

20. Baines, W. D., and E. G. Peterson: "An Investigation of Flow through Screens," *Trans. ASME* **73**:467 (1951).

21. Toyokura, T.: "Studies on the Characteristics of Axial Flow Pumps, Part 1," *Bull. Jap. Soc. Mech. Eng.* **4**(14):287 (1961).

22. Minami, S., et al.: "Experimental Study on Cavitation in Centrifugal Pump Impellers," *Bull Jap. Soc. Mech. Eng.* **3**(9):19 (1960).

23. Kurian, T., and H. C. Radhakrishna: "A Study of the Phenomena of Pre-Rotation at the Suction Side of Centrifugal Pumps," *Irrigation and Power*, July 1974, p. 381.

24. "Prerotation in Pumps Induced by the Impeller," internal technical memo, Alden Research Laboratory, Holden, Mass., July 1980.

25. Zajdlik, M.: "New Checking Mode of Model Parameters for Vortex Formation in Pump Tanks," in *Proceedings of the IAHR 17th Congress*, Baden-Baden, Germany, 1977, p. 379.

26. Chang, E.: "Scaling Laws for Air-Entraining Vortices," Report no. RR1519, British Hydromechanics Research Association, Cranford, Bedford, England, 1979.

27. Hecker, G. E.: "Model-Prototype Comparison of Free-Surface Vortices," *J. Hydraul. Div. ASCE*, October 1981, p. 1243.

28. Larsen, J., and B. J. Pennino: "Hydraulic Model Study of Jefferson Parish Drainage Pump Station No. 3," Report 68-81/M429F, Alden Research Laboratory, Holden, Mass., 1981.

FURTHER READING

Padmanabhan, M.: "Air Ingestion Due to Free-Surface Vortices," *J. Hydraulic Engineering, ASCE*, Dec. 1984, p. 1855.

Padmanabhan, M., and G. E. Hecker: "Scale Effects in Pump Sump Models," *J. Hydraulic Engineering, ASCE*, Nov. 1984, p. 1540.

SELECTING
AND
PURCHASING
PUMPS

—

S. I. Heisler

The sequence involved in obtaining a pumping system, following the initial decision that pumping equipment is required for a system and culminating with the purchasing of the equipment, can be divided into the following general steps:

Engineering of system requirements

Selection of pump and driver

Specification of pump

Bidding and negotiation

Evaluation of bids

Purchasing of selected pump

In the process of specifying pumping equipment, the engineer is required to determine system requirements, select the pump type, write the pump specification, and develop all information and data necessary to define the required equipment for the supplier.

Having completed this phase of work, the engineer is then ready to take the steps necessary to purchase the equipment. These steps include issuance of the specifications for bids or negotiations, evaluation of pump bids, analysis of purchasing conditions, selection of supplier, and release of all data necessary for purchase order issuance.

ENGINEERING OF SYSTEM REQUIREMENTS

The first decision the engineer must make is to determine the requirements and conditions under which the equipment will operate.

Fluid Type   One of the initial steps in the defining of the pumping equipment is the development of physical and chemical data on the fluid handled, such as viscosity, density, corrosiveness, lubricating properties, chemical stability, volatility, and amount of suspended particles. Depending upon the process and the system, some or all of these properties may have an important influence on pump and system design; for example, the degree of corrosiveness of the fluid will influence the engineer's choice of materials of construction; if the fluid contains solids in suspension, suitable types of pump seal designs and abrasion-resistant pump construction materials may have to be considered.

Fluid physical and chemical data of interest to the engineer should cover the entire expected operating range of the pumping equipment. The influence of such parameters as temperature, pressure, and time upon the fluid properties should also be considered.

System-Head Curves   The engineer should have a clear concept of the system in which the pumping equipment is to operate. A preliminary design of the system should be made and should include an equipment layout and a piping and instrumentation diagram (or other suitable diagram) showing the various flow paths, their preliminary size and length, elevation of system components, and all valves, equipment, piping specialties, etc., which establish the system-head losses.

The engineer should then determine the flow paths, flow rates, pressures, and temperatures for various system operating conditions and calculate line or conduit sizes and estimate pipe or conduit runs.

With this information, the engineer can develop system-head curves, which show the graphic relationship between flow and hydraulic losses in a pipe system. In calculating the hydraulic losses, the engineer must include adequate allowances for corrosion and scale deposits, etc., in the system over the life of the plant.

Since hydraulic losses are a function of flow rate, piping sizes, and layout, each flow path in a system will have its own characteristic curve. Care must be taken when specifying pump characteristics to take into account the characteristic curve of each possible flow path served by the equipment. In specifying pump equipment, it is convenient to add the effect of static pressure and elevation differences to the system-head curve to form a combined system-head curve. The resultant curve shows the total head required of the pumping equipment to overcome system resistance. The pump head must be at or above the combined system curve at all expected operating points and for all flow paths the pumping equipment is expected to serve. Refer to Secs. 8.1 and 8.2 for instructions for constructing system-head curves.

Alternate Modes of Operation The various modes of operation of the system are important considerations when specifying pumping equipment. Is operation of the pumping equipment to be continuous or intermittent? Is the flow or head to be fluctuating or constant? Will there be a great difference in flow and head requirements for different flow paths? These and other questions arising from different modes of operation greatly influence such decisions as to the number of pumps, their capacities, and whether booster pumps are needed in some flow paths.

The engineer should also consider the continuity of service expected of the pumping system. This factor will influence the decision on number, type, and capacity of installed spares and the quality expected of the equipment. Frequently, reliability considerations will dictate the use of multiple pumps, such as two full-size pumps and three half-size pumps or, where continuity is more important than full capacity, two half-size pumps. Where two half-size, three third-size, etc., pumps are used, loss of one pump will cause the others to run out on their system-head curves. This runout case should be evaluated when engineering the system and specifying the pump characteristics. This loss of a pump can occur not only by pump malfunction but by motor failure, external damage, loss of power supply, loss of control power, etc. The likelihood of these causes should be evaluated as part of the pump selection process.

Margins Pumps are frequently specified with margins over and above the normal rating. Over any long period of time, it is unlikely that a system can operate at steady state undisturbed by transient conditions, such as may be caused by changes in modes of operation, malfunction in system components, or electric system disturbances. It is necessary for the engineer specifying the pumping equipment to examine the probability and duration of such transients and to specify adequate margins to allow the equipment to undergo such transients without damaging effects. This involves also an evaluation of the combined effects of equipment cost, degree of criticality of the system, inconvenience due to unavailability of the equipment, and other economic and technological factors.

Some transients often considered in design are pressure and temperature fluctuations, electric voltage and frequency dips, and loss of cooling water. If the maintenance of continuous flow is important, then adequate margins must be allowed in the pump rating. For example, margins are added to the pump head and capacity to allow the pumping equipment to maintain rated flow in case of small electric frequency dips. In addition, certain design features may be included to allow the pumps to operate without damage through such transients as suction pressure dips, which can cause cavitation.

Pumps should not be purchased for capacities greatly (typically 15 to 20%) in excess of requirements. An oversized pump could operate at capacities less than those recommended by the manufacture, which could present mechanical and hydraulic problems.

Wear Wear is an ever-present factor in equipment and system design. No material which is handling fluids or used in contacting moving surfaces is free from wear. Thus, operating characteristics of both the pumping equipment and the system can be expected to change due to wear as time goes by. The engineer should assess the extent of such wear over the life of the plant and provide adequate margins in the system parameters so that the pumps can provide the expected flow even at the end of equipment life. Where abrasive or suspended materials are handled, pumps with replaceable liners are frequently specified. These liners are usually made of either resilient material, such as rubber compounds, or extremely hard alloys of cast iron. In addition, plastic linings (including impellers) are also frequently chosen for these types of services.

In some applications, especially in power plants, the expected pump life is specified as the same as plant life. However, the design life of a pump is a decision based on an evaluation of economic factors. The wear margin to be added is a function of such factors as mode of operation (continuous or intermittent) and fluid properties (abrasiveness, corrosiveness).

Future System Changes A final factor to be considered in the engineering system requirements for pumping equipment is the possibility of providing for future system changes. If the system changes can be predicted with any degree of certainty, then the system can be designed to enable the changes to be effected with minimum disturbance to operation. Thus it is important to review the possibilities and effects of such future system changes as well as provide pumping equipment to satisfy the immediate system requirements. The engineer should attempt to present future requirements based on projection of available data and then evaluate the possibility and desirability of designing the equipment to allow for the changes (such as providing extra flow or head margins or specifying a pump with impeller less than the maximum for a given casing size) versus the desirability of modifying the whole system, including the pumps, when the changes are made in the future. In any event, it must always be kept in mind that the equipment must operate satisfactorily in the *present* system, and this should be a factor in whatever evaluation is being made.

SELECTION OF PUMP AND DRIVER

The selection of the pump class and type for a particular application is influenced by such factors as system requirements, system layout, fluid characteristics, intended life, energy cost, code requirements, and materials of construction.

Basically, a pump is expected to fulfill the following functions: (1) pump a given capacity in a given length of time and (2) overcome the resistance in the form of head or pressure imposed by the system while providing the required capacity.

The behavior of the system has a very important bearing on the choice of pump. What are the required heads and capacities at different loads? Does the required head increase or decrease with changes in capacity? Does the required head remain substantially constant? These are some of the questions the engineer must answer.

Pump Characteristics Constant-speed reciprocating pumps are suitable for applications where the required capacity is expected to be constant over a wide range of system head variations. This type of pump is available in a wide range of design pressures, from low to the highest produced; however, the capacity is relatively small for the size of the equipment required.

That the output from a reciprocating pump will be pulsating is a factor to be considered. Where this is objectionable, rotary pumps may be required. However, the application of rotary pumps is limited to low to medium pressure ranges.

Centrifugal pumps are often used in variable-head, variable-capacity applications. Straight centrifugal pumps are generally used in low to medium capacity and medium to high pressure applications, while low-head, high-flow conditions suggest that an axial-flow pump may be more suitable. Mixed-flow impellers are used in intermediate situations.

It should be noted that reciprocating and rotary pumps are self-priming, but centrifugal pumps, unless specifically designed as such, are not. This may be an important consideration in certain applications.

In some cases the system layout can influence the decision on the choice of pump type. In general, centrifugal pumps will require less floor space than reciprocating pumps, and vertical pumps less floor space than horizontal pumps. However, more head room may be required for handling the vertical pumps during maintenance and installation.

Where the available NPSH is limited, such as when a saturated liquid is being handled, and the application calls for a centrifugal pump, the engineer may have to investigate the use of a vertical canned-suction centrifugal pump to gain adequate NPSH. In other cases the design may call for the installation of a pump immersed in the liquid handled, and here a vertical turbine pump may be advantageous.

Code Requirements The construction, ratings, and testing of most pumps normally used in industry are governed by codes such as the API or the Standards of the Hydraulic Institute. However, other codes of regulatory bodies may impose additional requirements which can affect both pump rating and construction. For example, the ASME Boiler and Pressure Vessel Code requires feed pumps to be capable of feeding the boiler when the highest set safety valves are discharging. In critical components for nuclear power plants, the Nuclear Regulatory Commission sometimes has imposed much more stringent quality assurance requirements than specified in normal industrial codes. Such additional requirements may affect the engineer's choice (see Sec. 9.15).

Fluid Characteristics Fluid characteristics, such as viscosity, density, volatility, chemical stability, and solids content, are also important factors for consideration. Sometimes exceptionally severe service may rule out some classes of pumps at once. For example, the handling of fluids having high solids content will exclude the use of reciprocating piston pumps or pumps with close clearances. Rotary pumps are suitable for use with viscous fluids, such as oil or grease, whereas centrifugal pumps can be used for both clean, clear fluids and fluids with high solids content. On the other hand, if it is undesirable for the process liquid to come into contact with the moving parts, diaphragm pumps may have to be used.

Pump Materials Materials are affected by both the pumped fluid and the environment. Resistance to corrosion and wear are two of the more important material properties in this regard, and the engineer should evaluate materials to determine which are most suitable and economical for the purpose intended. Often this becomes an evaluation of the desirability of specifying expensive long-life materials versus specifying cheaper materials which must be frequently replaced.

Operating factors, such as type of service (continuous or intermittent, critical or noncritical), running speed preferences (high or low), and intended life, will also influence the engineer's decision. For example, equipment used in continuous and/or critical service will generally demand heavier duty design and construction than equipment for intermittent and/or noncritical service. High-speed operation, if allowed, will permit the use of smaller, usually less expensive equipment. The life of the equipment cannot be predicted with certainty. For a given life expectancy, the engineer must evaluate the effects of materials of construction, design, severity of service, etc., before making a choice.

Driver Type The choice of driver type for the pumping equipment is as important as choosing the pump, for frequently the driver can cost more than the pump. Depending on the available energy sources, pumps may be driven by electric motors, steam turbiness, steam engines, gas turbines, or internal combustion engines. Also, pumps may be driven at constant speed or at variable speed. Variable speed can enable centrifugal pumps to operate along the system-characteristic curve and thus save on power for part-load operations.

Electric motor drives are usually used in constant-speed service unless a hydraulic coupling or other speed-varying device is introduced into the system. Internal combustion engine drives are usually chosen because of location (no electric power available), portability, or redundancy (loss of power backup) requirements. They can operate as either constant-speed or variable-speed drivers. Steam turbines, eddy-current couplings, adjustable-speed motors, fluid couplings, and gears and belts are frequently used where variable-speed operation is required.

In large, complex installations where the equipment is to be operated continuously, the decision as to type of driver and variability of pump speed should be based on a comparison of the total operating and capital costs for the pump system over the intended plant life for the several alternatives. Variable-speed operation would usually result in lower operating costs; however, the total first cost of the driving equipment to accomplish this would frequently be higher than for constant-speed equipment. The first cost should include cost of equipment, building space, etc., and the operating cost should include such factors as energy costs, maintenance, and replacement costs. This comparison usually results in the choosing of a pumping unit that provides the lowest cost per gallon pumped over the useful life of the plant.

It should be recognized that these are general guidelines and that there may be overriding nontechnical and noneconomic factors, such as prior satisfactory experience or excellent technical or service assistance, which may dictate the final choice of pumping equipment.

PUMP SPECIFICATIONS

Specification Types When selecting a pump, the first decision to be made is whether the procurement will be based upon a formal specification or whether some abbreviated form of requirements will be suitable. For relatively simple or inexpensive pumps, or for replacement pumps where duplication is desired, a specification is frequently not used. For inexpensive pumps the time and cost required to write a specification and obtain and analyze competitive bids frequently exceed the potential cost savings. In this case, and where the pump supplier is already established (replacement/duplication), a direct quotation is frequently requested from the supplier. It is important, when requesting this quotation, to have the principal requirements well defined and known to the supplier so that they can be properly included in the technical and priced offering. Thus, while a formal specification may not be appropriate, the purchaser should have the requirements well established. A pump data sheet (Figs. 1 and 2) is helpful in establishing these and can be used in lieu of a complete specification. Frequently the supplier can assist in developing and clarifying these requirements, although the engineer should recognize that, naturally, suppliers tend to favor their own equipment.

Where a formal specification is indicated, the type to be written is of fundamental importance. In most cases the specification will be of the performance type rather than the construction type. The performance specification basically establishes the performance which the pump must achieve and does not attempt to dictate pump design or construction methods, although certain details of construction are frequently established, particularly where choices may exist. For example where either leakoff or mechanical shaft seals may be offered, the performance specification usually states a preference. The performance specification, however, basically establishes "what," not "how."

The construction specification establishes in some detail the type of design, construction, and methods to be employed in designing the pump and certain other features which, if the performance specification is utilized, are left to the manufacturer's discretion. From the standpoint of legal responsibility, if a construction specification is used, manufacturers may respond and advise that, since the purchaser has established certain design features of the pump, the manufacturer cannot be responsible for the performance. It is therefore important that care be taken when writing a construction specification not to relieve the manufacturer of responsibility for applicability, suitability, and performance and that care also be exercised by the purchaser to avoid any unnecessary assumption of responsibility for the proper performance of the pump.

In short, unless there are unusual circumstances, it is far more appropriate to specify the pump on the basis of performance required rather than construction, unless the purchaser has a high degree of assurance that the requirements called out in the specification can be met and that the pump supplier will not be relieved of responsibility.

Codes and Standards   When specifying a pump, the codes and standards which apply are of major importance. Standards which relate to quality of materials should be referenced. ANSI, ASTM, MIL, or other industrial standards which establish such factors as metallurgy, dimensions, tolerances, and flange facing and drilling should be referenced where appropriate. Similarly, if a pump is to meet certain critical service requirements, there are in many cases industrial codes which apply to design, construction, and application. An example of this is the ASME Boiler and Pressure Vessel Code, Section III. For nuclear service pumps, which are pressure-retaining parts, the code establishes very specific requirements for control of construction materials and specifies certain of the pump design features and quality of manufacturing control. These codes in some cases are extremely detailed in specifying pump construction and are a rather well-defined specification in themselves. It is, of course, essential to establish the dimensional standards which apply, such as SI or English, the codes which may apply to the construction and fittings of the pump, and of course the industrial codes which apply to the application of the pump for the service intended.

In all cases, however, the engineer must review each reference to ensure that it does not introduce conflict. Some codes and standards include alternate choices of material or inspection methods, requiring selection by the engineer. Others may include cross references to additional codes which the engineer may wish to exclude.

Alternates   It is extremely difficult for a specification to cover all possible pumps offered by various manufacturers. Coupled with that is the problem faced by a potential user in remaining up to date with the changing state of the art and the development work being performed by manufacturers.

It is a good practice to allow manufacturers to offer alternatives. This gives them an opportunity to present their best offer and also gives the buyer the advantage of obtaining potentially

LINE NO.						
	Service_____ Req'n. No._____ Item No._____					
	Plant_____ Spec. No._____ No. Req'd._____					
	Location_____ P. O. No._____ Cost Code_____					
	Draw'g. Ref._____ (Layout)_____ (Process)_____ Mfr._____					
1	Liquid Pumped					
2	Viscosity (SSU) / Vapor Pressure (PSIA)					
3	Temperature (°F): Max. / Min. // Specific Gravity @ °F.					
4	Flow: Rating / Min. / Max. (GPM) (Lbs./Hr.)					
5	Suction Pressure @ (Flange) (Water Level) (PSIA)					
6	Discharge Pressure @ Flange (PSIA)					
7	Diff. Press.: Rating / Shutoff (Feet) (PSI)					
8	NPSH or Submergence: Available / Req'd @................ Elev. (Ft.)					
9	Type of Pump / Model / No. Stgs.					
10	RPM / Rotation (View from Motor Facing Pump)					
11	Efficiency / BHP at Rating / BHP Max. @...................GPM					
12	Impeller Diameter: Bid / Max. / Min.					
13	Imoeller: Eye Area / Periph. Vel.					
14	Working Press. Max./Hydrotest PSIG					
15	Clearance: Wear Ring / Bearing / Impeller					
16	Hydraulic Thrust: Rating / Max. / Up					
17	WK² / Speed-Torque: Rating / Speed-Torque Max. (Lb.Ft.²)/(Lb.Ft.)					
18	Suction: Size / Rating / Facing / Position					
19	Discharge: Size / Rating / Facing / Position					
20	Base Plate / Sole Plate					
21	Coupling: Type / Mfr. / Furn. By					
22	Suction: Strainer Mtl. / Splitter Mtl.					
23	Bearing Lube: Type / GPM / Pressure / Max. Micron					
24	Shaft Seal: Type / Sealing Conn. / Cooling (GPM)					
25	Coupling Guard Type (Horiz. only)					
26	Material: Case or Bowl					
27	(& Size) Barrel					
28	Shaft: Case or Bowl / (Dia.)					
29	Shaft Sleeve: Brg. / Stuff. Box					
30	Wear Ring: Case / Impeller					
31	Impellers / Liners					
32	VERT. ONLY { Disch. Head / Column (Dia. x Thickness)					
33	Shaft Encl. Tube / (Dia. x Thickness)					
34	Lineshaft / (Dia.) / (Brg. Spacing)					
35	Sleeve Bearing: Bottom / Bowl / Lineshaft					
36	Driver: Type (Motor—Turbine) (Solid—Hollow Sh.) / RPM / HP					
37	Furn. By / Weight / Dwg. Ref. / Mfr.					
38	Bearing Description / Thrust Rating					
39	Lubrication: Thrust / Radial / Cooling					
40	Drawing No.: Outline / Sectional / Performance Curve					
41	Net Weight: Pump / Removable / Rotating Elem.					
42	Inspection: Std. / Nuc. Class I, II or III / ASME III or VIII					
43	Testing: Ultrasonic / Eddy Cur. / Mag. Part. / Liq. Pen. / Radio.					
44	Hydrostatic / Witness // NPSH / Witness					
45	Performance / Witness / Field					
46	Quality Assurance: Mfr. Std. / Documented					
47	Seismic Design Req's: Class I / Class II					

FILL IN ALL BLANKS: IF NOT APPLICABLE MARK "NA"

NO.	DATE	REVISIONS		NOTES
△				
△				
△				

ORIGIN		**CENTRIFUGAL PUMP DATA SHEET**	JOB NO.
			DATA SHEET NUMBER	REV.

FIG. 1 Centrifugal pump data sheet.

attractive alternate offerings. However, the choice of whether or not to accept the alternates is fully up to the purchaser, who may choose to reject any and all bids, including alternates.

Bidding Documents The bidding documents for pumps normally consist of two major parts: technical specification and commercial terms. The technical specification establishes the performance requirements, materials of construction, and major technical features. The commercial terms include the contract language and cover such items as the location of the work, requirements for guarantees/warranties, shipping method, time of delivery, method of payment, normal inspection, and expediting requirements.

Service	Req'n. No.		Item No.	
Plant	Spec. No.		No. Req'd.	
Location	P.O. No.		Cost Code	
Draw'g. Ref.	(Layout)	(Process)	Mfr.	

Line	Description		
1	Liquid Pumped / Newtonian or Non-N.		
2	Viscosity (SSU) (CS): Norm. Temp. / Min. Temp. / Max. Temp.		
3	Vapor Press. (PSIA): Norm. Temp. / Min. Temp. / Max. Temp.		
4	Temperature (°F): Norm. / Min. / Max.		
5	Specific Gravity / Pour Point / Flash Point @.................°F		
6	Solids: Type / % by vol. / Abrasive / Shape / Distr. / Density		
7	Gases: Entr. / Dis'd. / Solubility (% by Vol. at 30″ Hg. abs.)		
8	Flow: Rating / Min. / Max. (GPM)		
9	Pressure: Outlet / Inlet / Diff. Max ΔP (PSIG)		
10	NPSH: Available / Req'd. @ ₵ Suction (PSIG)		
11	Type of Pump		
12	RPM / Rotation Facing Coupling		
13	Efficiency / BHP Rating / BHP Max		
14	Relief Valve: Internal - Setting / External - Setting (PSIG)		
15	Suction Conn.: Size / Rating / Facing / Position		
16	Discharge Conn.: Size / Rating / Facing / Position		
17	Mfr. Brg. No.: Radial / Thrust		
18	Base Plate: Pump (& Gear) Only / Comb. W. Driver (& Gear)		
19	Coupling: Mfr. / Model Size / Guard / Dr. Half Mtd.		
20	Seal: Packing Size, No., Type / Mech. Seal Mfr.. No.		
21	Lubrication: Main, Type / Secondary, Type / Max. Micron		
22	Cooling: Water GPM., Pr. / Quench / Steam Jacket		
23	Material: Case		
24	Rotors		
25	Shaft / Shaft Sleeve		
26	Throat Bushing		
27	Lantern Ring / Gland		
28	Relief Valve		
29	Base Plate		
30	Cooling Piping / Steam Piping		
31	Seal Piping / Lubrication Piping		
32	Driver: Type (Motor - Turbine — Gear) / RPM / HP		
33	Furn. By / Weight / Dwg. / Mfr.		
34	Bearing Description: Radial, No. / Thrust, No.		
35	Lubrication: Radial / Thrust / Cooling		
36	High Speed Cplg.: Mfr. / Model / Furn. By		
37	Drawings: Outline / Sectional / Perf. Curve		
38	Net Weight		
39	Inspection: Std. / QA		
40	Testing: Ultrasonic / Eddy Cur. / Mag. Part. / Liq. Pen. / Radiogr.		
41	Hydrostatic / Witness // NPSH / Witness		
42	Performance / Witness / Disassembly		
43	Quality Assurance: Mfr. Std. / Documented		
44			
45			
46			
47			

FILL IN ALL BLANKS: IF NOT APPLICABLE MARK "NA"

NO.	DATE	REVISIONS	NOTES
ORIGIN		ROTARY PUMP DATA SHEET	JOB NO.
			DATA SHEET NUMBER REV.

FIG. 2   Rotary pump data sheet.

Frequently the commercial terms and conditions are relegated to second place, especially when standard inexpensive pumps are being bought, but in many cases the commercial terms can assume more significance than many of the performance requirements.

Technical Specification   The technical specification should consist of a series of carefully defined and distinct sections. The more complete and specific the specification, the more competitive will be the bid prices. A typical specification might contain the following:

1. *Scope of Work.* Pump, baseplate, driver (if included), interconnecting piping, lubricating oil pump and piping, spare parts, instrumentation (pump-mounted), erection supervision

2. **Work Not Included.** Foundations, installation labor, anchor bolts, external piping, external wiring, motor starter

3. **Rating and Service Conditions.** Fluid pumped, chemical composition, temperature, flow, head, speed range preference, load conditions, overpressure, runout, off-standard operating requirements, transients

4. **Design and Construction.** (Care should be taken to provide latitude in this section, as this borders on dictating construction requirements.) Codes, standards, materials, type of casing, stage arrangement, balancing, nozzle orientation, special requirements for nozzle forces and moments (if known), weld-end standards, supports, vents and drains, bearing type, shaft seals, baseplates, interconnecting piping, resistance temperature detectors, instruments, insulation, appearance jacket

5. **Lubricating Oil System (if applicable).** System type, components, piping, mode of operation, interlocks, instrumentation

6. **Driver.** Motor voltage standards, power supply and regulation, local panel requirements, wiring standards, terminal boxes, electric devices; for internal combustion drivers, fuel type preferred (or required), number of cylinders, cooling system, speed governing, self-starting or manual, couplings or clutches, exhaust muffler

7. **Cleaning.** Cleaning, painting, preparation for shipment, allowable primers and finish coats, flange and nozzle protection, integral piping protection, storage requirements

8. **Performance Testing.** Satisfactory for the service, smooth-running, free of cavitation and vibration, shop tests (Hydraulic Institute standards) for pump and spare rotating elements, overspeed tests, hydrostatic tests, test curves, field testing

9. **Drawing and Data.** Drawings and data to be furnished, outline, speed versus torque curves, WK^2 data, instruction manuals, completed data sheets, recommended spare parts

10. **Tools.** One set of any special tools, including wheeled carriage for rotor if needed for servicing and maintenance

11. **Evaluation Basis.** Power, efficiency, proven design

Supplementing these may be technical specifications relating to other requirements of the order, such as specifications for the electric motor, steam turbine, or other type of driver; a specification on marking for shipment; a specification on painting; and requirements for any supplementary quality control testing (particularly important for pumps for nuclear services where quality control and quality assurance are critical requirements of the Nuclear Regulatory Commission).

In addition, it is important that any unusual requirements be listed in the technical specification so that the manufacturers are aware of them. Examples of these are special requirements for repair of defects in pump castings, a sketch of the intake arrangement for wet-pit applications, and special requirements regarding unique testing, for example, metallurgical testings which may be required during manufacture apart from performance testing.

It is helpful to the pump supplier to provide system-head curves, sketches of the piping system (dimensioned, if this is significant), listings of piping and accessories required, etc.

Pump data sheets are extremely useful in providing a summary of information to the bidder and also in allowing the ready comparison of bids by various manufacturers. Typical pump data sheets are shown in Figs. 1 and 2. As can be noted by inspecting these sheets, some of the items are filled in by the purchaser and the balance by the bidder to provide a complete summary of the characteristics of the pump, the materials to be furnished, accessories, weight, etc. The data sheets should be included with the technical specification.

Commercial Terms The commercial terms included with the bidding documents should cover the following information:

- General: name of buyer, place to which proposals are to be sent, information on ownership of documents, time allowed to bid, governing laws and regulations.

- Location of plant site. This establishes the geographic area in which the equipment is to perform and in a broad way the scope of the work. It should also state maximum temperatures, humidity, storage provisions (indoor or outdoor), and altitude (so that the motor drivers can be selected for the proper cooling).

- Definitions establishing buyer, agent, engineer, seller, whether "or equal" is intended to be exclusive or not.
- Proposal, which establishes the format of the proposal, number of copies, owner's right to accept or reject any bids, status of alternates.
- Schedule, including requirements for all drawings and design data submittals, manufacturing schedule, and equipment delivery.
- Acceptable terms of payment, retentions, etc.
- Transportation to and from point of use (or installation) is frequently a consideration since with very large equipment it may not be possible to ship by truck and it may be necessary to either barge- or rail-ship the material. In addition, it is important to establish the method of shipment which may be used so that the bidder can include the proper allowance for freight and to establish responsibility for the risk of loss. Thus, a bid could either include a freight allowance, that is, be FOB manufacturer's plant WFA (with freight allowed) to point of use or be FOB point of use, in which case freight is included. In either case, the risk of loss remaining with the seller and that assumed by the purchaser should be clearly stated.

SPECIAL CONSIDERATIONS

Performance Testing   An important part of any specification is the requirement for testing. Normally, small commercial pumps which are routinely produced by a manufacturer, up to about 6 in (150 mm), are tested on a sample-selection, quality control basis and from that, standardized curves of pump performance are available. Thus for pumps of this size it is not necessary to require certified tests unless the pumps are to be used in critical service, such as fire protection or boiler feed.

For larger pumps or pumps with more critical service requirements, however, a certified performance test should be required. This test requires that the manufacturer test the pump at several points on its performance curve to establish its exact head curve. Since it is necessary to assure that the pump driver is of the proper size, power curve must be furnished with the head curve. Occasionally, pumps for special services or extremely large pumps cannot be tested in the manufacturer's shop. Examples of this are very large low-head pumps for circulating water service, low-lift irrigation pumps, and pumps for liquid-metal service. A usual method for testing these pumps is to extrapolate the performance tests of a geometrically similar model (usually smaller) to the pump size under consideration and use that information for the development and engineering of the pump. The actual performance testing in this case takes place following installation of the pump. It is important that the purchaser and the supplier agree upon a proper (field) test method in some detail. This method should include the number of points at which the head curve will be determined, the applicable code, the specific method of traversing the pump discharge characteristics across the cross section of the discharge pipe, and the manner in which the head will be varied. Care should be taken in establishing this procedure to set forth the characteristics of the fluid and other variables which can affect the performance test. The specification should establish the performance testing requirements for the pump and whether or not it is necessary that the testing of the pump be witnessed. Witnessing and furnishing of certified test data (including the test work done) are frequently priced separately and if not specified can be a source of dispute between purchaser and supplier.

Pump Drivers   Pump drivers (motors, turbines, engines, etc.) can be purchased either with the pumps or separately. As a general rule, with pump drivers in the range of 50 hp (40 kW) and larger, consideration should be given to purchasing the driver separately since this can frequently result in cost savings. With small pumps, pumps using "monobloc" construction (where the pump is mounted on and supported by the motor), and pumps built to special codes, such as Underwriters' engine-powered fire pumps, there is not usually any cost advantage to buying the driver separately.

Where the driver is excluded from the pump scope of supply, the specification should require the pump supplier to determine the proper characteristics of the driver. This includes establishing the proper motor speed, sizing the driver for both accelerating and running loads, assuring end-float compatibility and/or thrust-bearing loadings (including direction), and selecting and fitting the couplings. If the driver is purchased separately and can be economically and conveniently shipped to the pump supplier's plant, the pump supplier should be required to mount the driver-half of the coupling as well as align and mount the driver (for common baseplate installations). For very large drivers or where it is costly or impractical to ship the driver to the pump supplier, it will be necessary to perform this work at the point of installation. To assure compatibility with the other drivers in the plant it is important to specify the driver enclosure type, insulation standards, and special features required, such as heaters and oversize junction boxes.

For steam turbine drivers, speed range, throttle pressure, steam quality, exhaust pressure and control method should also be specified.

When reciprocating pumps or drivers are specified, the bidders should be asked to furnish the magnitude and direction of the shaking forces, as well as weight and size, to permit preliminary evaluation of the foundation requirements.

NEMA (National Electrical Manufacturers Association) standards for certain type of motors provide a power overload allowance (service factor) of 15% based upon internal motor temperatures. It is good practice to conserve this margin for short-term overloads. Since it is based on the internal heating characteristics, short-term encroachment (starting surges, etc.) is frequently used to avoid increasing the motor to the next larger frame size. For large motors, 200 hp (150 kW) and over, where heat dissipation is more difficult, the motor nameplate usually carries starting frequency limitations. For very large motors, typically 1000 hp (750 kW) and larger, the purchaser should state starting frequency requirements.

Typical limitations for these very large motors will be three to five starts per hour but may vary due to the WK^2 of the load, the load torque, voltage applied, etc. The motor manufacturers should be consulted where there is any consideration for repetitive starting.

Synchronous motors are frequently used as drivers on large (generally low-speed) pumps. A rule of thumb often used is that synchronous motors should be considered for an application when the power is equal to or greater than the pump speed. Synchronous motors require more complex control equipment but can be tailored to the specific starting, pull-in (to synchronism), and pull-out (of synchronism) torques of the load. Since these motors operate at synchronous speed, pump-driver slip can be disregarded. In addition, these motors can be operated at unity power factor to avoid power factor changes in energy bills or, if desired, can be operated with leading power factors to improve the average power factor of the other equipment at the installation.

Intake   Vertical wet-pit pumps are sensitive to the geometry of their suction pit. Factors to consider include clearance beneath the bottom of the suction bell and the floor of the pit, spacing between pumps or between the pump and the pit walls (both side and rear), approach angle of the floor of the pit (including surging and surcharge), submergence, and lack of uniform approach flow.

The standards of the Hydraulic Institute include recommendations on the geometry of intakes. These as well as the recommendations of the pump manufacturer should be carefully reviewed. Suction piping, where complex or unusual, should be treated in a similar manner.

When specifying vertical wet-pit pumps, a layout of the installation should be furnished to the bidders for their information and comment. In many cases, if the geometry of the installation is not fixed, bidders can recommend small changes to improve pump performance. Where the geometry is fixed, it may be necessary to add antivortexing baffles, surge walls, or flow-directing vanes (or walls) to avoid pump operating problems.

For moderate or large installations where any design question exists, model testing should be considered. Several pump manufacturers offer this as a service, as do a number of universities and commercial testing laboratories. Responsibility for proper pump performance will rarely be assumed by the bidder when the intake pit is of nonoptimum size or shape. The use of model testing is usually resorted to in these cases also. (For a more complete discussion, see Chap. 10.)

Special Control Requirements   This is an area where care must be exercised to assure that pumps are not damaged by misapplication. Pumps that do not require more power at shutoff, that is, radial and mixed-flow pumps, can be started with a closed discharge valve and will not overload their drivers.

Propeller pumps require very high power and produce very high pressures (relative to design) at shutoff, and they are started with valves that are timed to be sufficiently open when the driver comes up to full speed. In systems with a large mass of water to accelerate, opening the pump discharge valve is of no value since the inertia of the water creates a shutoff condition or close to it. For these applications, a bypass or means of slowing the acceleration of the driver must be considered (see Sec. 8.1).

In addition to the question of starting control, consideration must be given to pump shutdown and closing of discharge valves. For systems with large-diameter piping, typically 24 in (600 mm) and larger, motor-operated butterfly valves are often used in the pump discharge in lieu of check valves. These butterfly valves act as isolation valves and by proper control can ensure that reverse flow does not occur when not wanted. A careful analysis must be made of the pump starting requirement, including the motor accelerating capability and the butterfly valve characteristics, that is, flow versus opening and opening angle versus port area.

Drawing and Data Requirements Form   This form indicates the type of drawings and data required to define requirements both for preliminary design purposes and for final information, that is, the as-built dimensions and the certifications which demonstrate that the pump meets the specified requirements. The schedule for information submittals is required to provide information on structural, electrical, or piping design necessary to incorporate the pump into the plant installation. A typical form is shown in Fig. 3.

Quality Assurance and Quality Control   Although the two are often used interchangeably, the term *quality assurance* is generally used to describe the overall programs and principles used to control quality and the term *quality control* is used to describe the methods, procedures, and practices which control activities.

Quality assurance principles include

1. Sufficient independence and authority for the QA/QC organization
2. Establishment of procedures to control activities affecting quality
3. Provision of documented evidence of the satisfactory performance of these activities
4. Loop-closing systems to identify, correct, and report on corrective actions taken on nonconforming items.

Quality control principles include

1. Control of purchased items
2. Control of manufacturing processes
3. Gage and tool control
4. Test control
5. Control of storage and shipping activities

With the increasingly costly economic consequences of equipment failures, outages, and so forth, reliability requirements are becoming more stringent and formal QA/QC programs are being applied by many companies and purchasers. One example of this is the 10 Code of Federal Regulations, Part 50, Appendix B, which establishes QA criteria for nuclear facilities. Its principles are generally applicable to all quality programs, although some items may not apply where activities are limited. (For example, if no welding, heat treating, nondestructive testing, or other special processes are involved, criteria IX control of special processes need not be applied.) Certain foreign governments have established QA programs for some industries.

An area of growing interest and applicability is the use of statistical methods to control quality. This is of particular use for repetitive manufacturing and can economically identify quality trends and permit corrective action before production results become rejectable.

A whole body of literature has developed on the subject, and the reader should obtain specialized assistance if working in this area.

This schedule of drawing and data requirements is to be fulfilled before rendering final invoices. See below for drawings required and dates due. Failure of Seller to comply with drawing and data requirements may result in order cancellation in the case of initial drawings, or final payment being withheld in the case of final drawings. Drawings are to be forwarded to:

Attention: ....

IN ADDITION, FORWARD WITH SHIPMENT, ONE SET OF ANY DRAWINGS NECESSARY FOR FIELD INSTALLATION. FORWARD COPY OF LETTER OF TRANSMITTAL TO:

	TYPE OF DRAWINGS AND OTHER REQUIREMENTS	APPROVAL BEFORE FAB (YES/NO)	KIND OF COPIES	NUMBER REQUIRED	
				INITIAL	FINAL
A	OUTLINE DIMENSIONS AND FOUNDATION REQUIREMENTS		TRANSPARENCY PRINTS		
B	CROSS SECTION WITH PARTS LISTS WITH PRICES		TRANSPARENCY PRINTS		
C	SHOP DETAIL DRAWINGS		TRANSPARENCY PRINTS		
D	CERTIFIED PERFORMANCE DATA AND TEST REPORTS		TRANSPARENCY PRINTS		
E	WIRING DIAGRAMS		TRANSPARENCY PRINTS		
F	CONTROL LOGIC DIAGRAMS		TRANSPARENCY PRINTS		
G	WELDING PROCEDURES		TRANSPARENCY PRINTS		
H	CODE CERTIFICATES, INSPECTION AND TEST REPORTS		ORIGINAL COPIES		
J	INSTRUCTIONS FOR ERECTION OR INSTALLATION, OPERATION AND MAINTENANCE		MANUALS OF EACH TYPE		
K	LIST OF RECOMMENDED SPARE PARTS FOR ONE YEAR'S OPERA- TION, WITH PRICES		LISTS		
L	COMPLETED DATA SHEETS		TRANSPARENCY		
M	MATERIAL CERTIFICATIONS				
N	MANUFACTURERS QUALITY CONTROL, INSPECTION AND TEST PROCEDURES AND REPORTS				

Seller's drawings will be reviewed and approved only as to arrangement and conformance to the specifications and related drawings, and approval shall not be construed to relieve or mitigate the Seller's responsibility for accuracy or adequacy and suitability of materials and/or equipment represented thereon.

Final drawings must be certified and must show adjacent to the title block, Buyer's equipment title and number. manufacturer's serial number and purchase order number. Initial transparencies must be made from faultless masters. Final transparencies shall be on wash-off Mylar. Additional drawing requirements will be specified in the Technical Specification.

Initial drawings required within..............days of receipt of firm order. Final drawings required within..............days of receipt of initial drawings, or within..............days of receipt of firm order if no initial drawings are requested. The finalized drawing transmittal requirement dates will be specified in the purchase order and will take precedence over the above.

DRAWINGS AND DATA REQUIREMENTS	JOB NO.	
	ATTACHMENT TO REQUISITION NUMBER	REV.

FIG. 3   Drawings and data requirements.

BIDDING AND NEGOTIATION

Public and Private Sector   Depending on whether the enterprise is in the public or private sector, there are certain requirements relating to the manner of bidding or negotiating for the award of major equipment contracts. Public bodies are required by law, in most instances, to take competitive bids from a broad range of bidders and to have public bid openings of their offerings. The bid openings are usually well documented and the bids recorded for public inspection. For work in the private sector, there are rarely such *legal* requirements, although in the case of certain regulated industries, such as public utilities, it is a normal procedural requirement that bids be taken. Similarly, many companies in the private sector require as a matter of policy that bids be taken for all substantial purchases.

Where bids are to be taken, a specification must be written in acocrdance with the guidelines previously presented. Where there is an option, it may be preferable to preselect a bidder and negotiate the price for the equipment. This is particularly true where the equipment is relatively inexpensive or desired as a replacement or duplicate for an existing piece. If negotiation is used, it is still important that the specification requirements be developed, that they be used as the basis for negotiations to ensure that major considerations and characteristics are not overlooked, and that a clear basis for understanding between the supplier and the purchaser be established. Thus the only real difference between a negotiation and competitive bidding lies in the formalization of the bid documents.

Bid List   The development of a bid list is extremely important for both the owner and the potential supplier. For the owner, the bid list provides a checklist of suppliers who are able to furnish pumps of the type and quality required on schedule and are in a position to offer the proper degree of support service necessary both during design of the installation and over the life of the equipment. This support service includes the ability to provide adequate engineering data on the pump to suit the schedules for the process engineering, piping, electrical, and foundation design, together with requirements and data for auxiliary systems of the equipment. In the case of very large equipment, installation supervision service may be part of the contract. In addition, the supplier may be called on to supervise the start-up of the installation and, over the life of the equipment, provide field service and spare parts.

Companies in the private sector can limit the bid list to selected bidders, on the basis of experience, investigation, or reputation. Prior satisfactory experience or reputation is frequently used as a guide to reducing the bid list to manageable size. Where complex or sophisticated equipment is involved, bidders should be prequalified to assure their ability to design and produce satisfactory equipment. This typically includes inspection of the manufacturer's shops and of the design, manufacturing, and quality assurance organization, as well as detailed reviews of operating installations previously supplied by the manufacturer.

It is convenient to try to limit the number of bids received to about five since this provides an adequate number for effective price competition without making the evaluation burdensome. To assure five bids, it is prudent to solicit inquiries from up to seven suppliers if there is expectation that some "no bids" will be received. This limitation on the number of bids presents a problem since typically there are many more qualified suppliers. The manner of selecting the bidders varies widely, but where qualified suppliers have indicated a genuine interest they should be allowed to bid even if the bid list is enlarged somewhat.

Bid Time   The preparation of bids for pumping equipment will take different amounts of time depending upon the complexity of the equipment and whether or not prime movers are included in the proposal requirement. As a guideline, Table 1 lists the typical times that vendors would like to have to prepare their proposal (from time of receipt of the bid request). This presumes that a request for proposal calls for a pump application that is relatively straightforward and does not require exhaustive analysis or research. A heavy work backlog in a company, or in the entire industry, may also extend the bid preparation time.

Bid extensions, if they are reasonable and require only modest changes in the bidding time,

TABLE 1   Bid Preparation Times

Application	Weeks required
Preengineered and conventional pumps, 6-in (150-mm) discharge and smaller	3
Pumps with 6- to 48-in (150- to 1200-mm) discharge; pumps with other than electric motor drivers	5
Large pumps	6 to 12
Special pumps	Consult bidders

should be granted to a qualified supplier. Occasionally pump suppliers have numerous bid requests falling due on the same date, and they may require an extension. If the bid due date is arbitrarily maintained, it is possible to lose a potentially favorable bid. If a bid extension is granted, all bidders should be notified so that they may have the advantage of the additional time to improve their proposals.

Since bid openings for public bodies are usually advertised long in advance, bid extensions are generally not granted in these cases.

Bidding of Alternates   Unless there are compelling reasons not to, the bidding of alternates should be allowed for the reasons cited earlier. The purchaser should maintain the right to reject any and all bids, including alternates.

Legal Restrictions   When working in the public sector particularly, there may be restrictions on the use of materials or equipment fabricated outside the United States. Frequently, where non-United States materials are allowed, penalties are assessed against the bidder for the use of these materials. In addition, in certain unique situations a bidder will be given an evaluation credit. One example of this has been the credit given to bidders on public works projects whose plants are located in depressed labor areas.

In certain large regulated utilities, it is sometimes preferred that awards for major equipment contracts be placed with companies in the service territory of the utility. There is usually no cost penalty in placing the order within the service territory; however, care must be taken to assure proper supplier cooperation if the plant design work is being performed elsewhere. This is particularly true for the timely furnishing of design information, coordination of motor or driver selection, and other interface areas.

EVALUATION OF BIDS

Perhaps the most important consideration after having specified the pumps is the manner of evaluating the bids. Evaluation should consider first cost, pump performance, guarantees, economic advantage of alternates, delivery, maintenance, installation service, etc. The sum of these several factors provides the purchaser with a broad base for the purchase decision and will assure more satisfactory performance over the life of the equipment.

It is convenient to separate the evaluation of bids into two categories: those that relate to price and those that relate to technical features. In this way the technical features which are important to performance but difficult to quantify are treated separately and can be compared with the cost differences from the price evaluation.

Cost   First cost is a major consideration in the evaluation of bids. In many cases where offerings may be almost identical because of a commonality among equipment supplied by various bidders, the only real difference may be cost. Where cost alone is considered, or where cost is of paramount concern, it is prudent to consider the difference in price between the first, second, and third bidders. In many cases it will be found that the difference in bid price may be tenths of percents and thus the absolute values of the differences when viewed in this perspective become less significant. Where this is the case, more detailed attention should be given to the technical factors.

More usually, however, and this is particularly true with large, complex equipment where relatively small details of performance can, over the life of the equipment, yield fairly high capitalized values, the technical factors require detailed consideration.

Efficiency   Evaluations should include an estimate of the effect of the different efficiencies quoted. This evaluation is normally made at the warranted point (usually full load), while efficiencies at other points are treated qualitatively unless extended periods of part-load operations are contemplated. The efficiency should be rated from *base*, the highest efficiency quoted, with an increasing evaluation penalty against pumps having decreasing efficiency. The quantification of the effect of the efficiency can be done in several ways:

1. A direct charge in dollars per year for one or more years for the difference in pump power required. For example, a 70-hp (52-kW) difference, measured at the motor terminals and thus

including the combined pump and driver efficiency effect, would have an evaluation value of $6580 based upon the following assumed factors:

> Electric motor driver
> 15 mills/kW·h power cost
> No demand charge
> Full-load operation 50 weeks/year, 7 days/week, 8 h/day
> 3-year amortization period

$$70 \times 0.746 \times {}^{15}\!/_{1000} \times 50 \times 7 \times 8 \times 3 = \$6580$$

2. A charge for the differential energy based upon the purchased cost of energy evaluated over the life of the pump installation using the present-worth technique. For this type of analysis, the required rate of interest on the investment is predetermined as well as the estimated economic life of the facility. From these values a present-worth factor can be determined from the standard tables of present worth given in any of the principal texts on engineering economics. For example, A 70-hp (52-kW) difference would have an evaluation (present-worth) value of $23,200 based upon the following factors:

> Electric motor driver
> 15 mills/kW·h power cost
> No demand charge
> Full-load operation 50 weeks/year, 7 days/week, 8 h/day
> 20-year economic life
> 7% interest rate

The present worth of 7% for 20 years is 10.594. The annual difference is multiplied by this to determine the present-worth value:

$$70 \times 0.746 \times {}^{15}\!/_{1000} \times 50 \times 7 \times 8 \times 10.6 = \$23,200$$

3. A charge for the cost of providing the capacity. For those purchasers who generate their own power, the calculation is similar to that above excepting that the efficiency cost is made up of two elements: the cost of the energy and the cost to provide the equipment to generate the power required. This latter cost is usually called the *capacity charge*. For example, with an energy cost of 9 mills/kW·h, a $200/kW capacity charge, and the previous values, the 70-hp (52-kW) difference has an evaluation value, on a present-worth basis, of $24,400:

$$70 \times 0.746 \times \left(\frac{9 \times 50 \times 7 \times 8 \times 10.6}{1000} + 200 \right) = \$24,400$$

For some very large companies, the method of calculating these values, the factors to be included, etc., are standardized and frequently more detailed than the foregoing examples.

Economic Life   The economic life of the equipment is a consideration in evaluation of any of the proposals, but it is an extremely difficult item to measure. To some extent, the weight of the equipment (discussed later in the chapter) is an indicator of its ruggedness and durability, all other things being equal. Another measure of the economic life is speed. Thus, if one pump operates at 1800 rpm while an alternate pump operates at 3600 rpm, assuming other construction details are roughly equivalent, it is likely that the 1800-rpm pump, because of its lower speed, will have a longer economic life and will be less likely to require premature replacement. Since economic life is extremely difficult to quantify, its evaluation is best left to the technical spread sheet, where it is dealt with qualitatively.

Alternates   For reasons cited earlier, the offering of alternates by the manufacturers should be encouraged. In the evaluation of these alternates, however, considerable care must be exercised since they may depart in many cases from the basic concept of the specification and thus will require careful review to assure compatibility. Since the purchaser will not have performed application engineering on the alternates during preparation of the specification, it will be necessary to perform this work as a part of the evaluation to ensure proper consideration of the effect of the alternate.

Shipping Cost   Shipping cost may be a factor in the evaluation, particularly if the pumps are quoted FOB manufacturer's plant with freight paid by the purchaser. The cost for shipping should be determined for each of the alternates, assuming they are quoted other than FOB point of installation or point of use, particularly if the pumps are large and require special handling. The shipping cost also becomes a factor in the assumption of responsibility for any damage to the pump prior to arrival at the point of use.

Delivery Time   An important element of the quotation is delivery time. The amount of time required to design and manufacture the pump, ready for delivery, after approval to manufacture, will vary among bidders but will be generally in accordance with the list shown in Table 2.

TABLE 2   Delivery Times

Application	Weeks required
Preengineered and conventional pumps 6-in (150-mm) discharge and smaller	0 (off the shelf) to 16
Pumps 6- to 48-in (150- to 1200-mm) discharge with other than electric motor drivers	12 to 48
Large pumps	26 to 78
Special pumps	Consult manufacturer

On large, complex projects, the scheduling of pump and prime mover delivery is often extremely critical, and it is necessary in many cases to stagger delivery of large pumps or perhaps components in multiple-pump installations and provide adequate time for certain fit-up and field assembly.

In general, deliveries that are earlier than scheduled are preferred; however, with very large pumping systems if materials are delivered too early (and invoices paid upon shipment), money is expended too early. The extra interest charges on that money should be considered. Early deliveries can also be troublesome if storage space at the point of use is not available. This can cause both storage and rehandling problems, with the storage problems causing extra costs and the rehandling ones not only causing extra cost but also affording a chance of damage to the equipment.

Spare Parts   It is important to evaluate the cost of spare parts. Since the manufacturer is usually in a better position to evaluate the spare parts requirements, the specification usually calls for the manufacturer to provide a price list of the spare parts which are recommended for one year's maintenance and operation. The cost of the spare parts should be reviewed, and if it appears that a bidder's spare parts requirements are either extremely high or extremely low, supplementary information or clarification should be obtained. Either a prebid or preaward conference with the bidders can be helpful to evaluate these requirements.

Maintenance Costs   Maintenance costs are extremely difficult to quantify. It is possible, however, using the estimated spare parts list to evaluate very roughly what the spare parts costs would be over one year and to estimate the number of hours required to install at least some of the principal spares. For instance, if the design quoted had two packed glands, it might be reasonable to assume that the two glands should be removed and the packing replaced once a year. If an alternate for a mechanical seal is offered, it would be reasonable to expect that the mechanical seal would have a life of perhaps three to five years in clear water service, and the cost, both parts and labor, evaluated on this basis.

Of more importance is the complexity of pump construction, arrangement of case, diffuser, bearings, etc., and evaluation of which pump would require more disassembly time and hence more labor for the same kind of maintenance.

Maintenance may also require the addition of lifting gear or special provisions in building structures, for example, hatches over the pumps. Thus, it may be necessary to install monorails or bridge cranes over the pumps and motors or perhaps pad eyes for chain falls or other lifting devices. In this case, if the pumps that are offered are significantly different, the fixed or permanently installed handling equipment required should be considered. Care should be taken, however, to also consider the possibility of using temporary equipment, for infrequent operations, rather than committing a large amount of capital to permanent installations which may be infrequently used.

Space and Auxiliary Requirements

For horizontal and vertical (other than wet-pit) pumps, space requirements may differ markedly among several pumps offered because of design or the type of prime mover offered. Thus, for example, if the specification allows bids of pumps at various speeds, a pump with a 3600-rpm motor would be much smaller than one with an 1800-rpm motor, and the space requirement for the 3600-rpm unit would be markedly smaller. Where pumps are offered having different physical features which affect the space required, a design study to compare the installed space requirements of the pumps should be performed. This study should be conducted prior to bidding and should include consideration of the equipment foundations, the volumetric requirements of the pumps which affect heights of building, excavation, etc. It should also include the requirements for weight-handling and servicing equipment discussed earlier, for access of personnel, and for platforms, handrails, and stairways needed to service the motor or driver. Space for motor control centers and control panels should also be considered.

A careful analysis of space becomes extremely important when alternates are offered. For instance, space studies prepared during the engineering analysis, which preceded the specification, may have been written around a horizontal centrifugal pump, and then a manufacturer may choose to offer, as an alternate, a vertical wet-pit pump for this same application. In this case it will be necessary to perform a careful engineering study of the installation of the vertical pump, including the cost associated with the additional foundations, underfloor piping, and special forming of underflow water conduits (if required). As part of the consideration of space requirements, it is necessary to develop an estimate of the *differential* cost for the bids and the alternates offered based upon an evaluation of the items listed in Table 3.

TABLE 3   Space Requirement Costs

Item	Typical cost, \$ (U.S., 1984)
Foundations, including embedments, anchor bolts, etc.	$160–450/yd^3$
Excavation	
Common	$3–7/yd^3$
Rock	$11–22/yd^3$
Building volume	$2–7/yd^3$
Additional or special motor control	Cost + 50–100%
Special heating and ventilating	Cost + 50–100%
Piping, valves, piping specialties	Cost + 100–150%
Handrails, stairways, ladders, other maintenance and access devices	1300–3400/ton
Weight-handling equipment	Cost + 50%
Instrumentation (cost is equipment cost)	Cost + 100–150%

Apart from the question of space and physical requirements, there is frequently a difference in the auxiliary requirements for the various pumps offered. Very large pumps will usually have separate, self-contained lubrication oil pump and reservoir systems. These may be driven by their own motors and include heaters, oil coolers, and other auxiliaries. For a proper evaluation it is important that these additional power or utility service requirements be clearly defined so that a proper comparison can be made. This is particularly significant where some manufacturers offer pumps with all this contained in the pump package and others offer it external to the pump.

Frequently, controls will be included with the pump package. Typically, the control voltage will be 120-V ac single-phase and motor voltage will be 480-V ac three-phase. For this case it is likely that *all* bidders would not offer a stepdown control transformer and its associated wiring even if the specification calls for it. This should be carefully reviewed and considered in the bid evaluation.

Similarly, with large pump packages there may be wiring and piping internal to the package which vary from one bidder to another. It is usually to the advantage of the purchaser to have fewer (and preferably single) points of connection of these services.

Weight of Equipment   While not a critical factor, weight can indicate the ruggedness and potential life of the equipment, as discussed earlier. In evaluating bids, it is important to consider the weight of the largest piece to be handled for both installation and maintenance. This is of great importance as equipment sizes increase. For example, in engineering a pump installation for the top floor of a high-rise building with pumps that weigh 9000 lb (4100 kg), the lifting ability of the tower cranes which the contractors expect to use must be considered; insufficient lifting ability could require rental of a special crane or disassembly and reassembly. For this case a pump weighing 5000 lb (2700 kg) offered by another manufacturer should be given an evaluation advantage. There are numerous instances of these kinds of items which are significant not only to the cost but also to the time required for installation.

Installation Service   Installation service should always be requested with a proposal when large or unusual pumps are purchased. The request should be specific about requiring the bidder to state the number of days of installation service time included, as well as the requirements for special tools. If the installation period is expected to be an extended one, arrangements for housing, transportation, and living expenses for the installation service representative should be clearly defined. Manufacturers will usually provide, with their bid installation, service of a specified and limited amount. Service over and above this, or extensions of service caused by delays not attributable to the supplier are usually quoted on a *per diem* basis. Frequently, these *per diem* rates are requested in case these services are needed. All these items should be established prior to placement of the purchase order, although they are usually evaluated as third- or fourth-order rank and rarely cause a change in evaluation results. The differences in the installation services offered by various bidders are usually not major, and if the installation service time offered by the lower bidders are reasonably equivalent, in general the purchaser need have no major concern.

Lowest Evaluated Bid   This reflects the effect on cost of all the items previously referred to. In addition, to evaluate the lowest evaluated bid there are other considerations which cannot be priced but which must be evaluated qualitatively. These include such items as extended guarantees and ability to accept larger impellers in the same case (margins for additional head), and prior relationship with a potential supplier. It is important that these be considered since in some cases the difference between the two or three lowest evaluated bids is small and these factors can assume increased significance. Evaluation of nonquantifiable factors is sensitive, however, as the concept of competitive bidding precludes buying other than the lowest evaluated price that meets the specification. In any case, these nonquantifiable factors should be reviewed and listed to assure that an item of significance is not overlooked.

Commercial Terms   The commercial terms offered by the suppliers can be more significant than the evaluated effect of technical differences. When evaluating the commercial terms, the following items should be considered.

Has the Period for Acceptance of the Proposal Expired?   With evaluations which have lengthy technical questions to resolve, some of the proposals may expire. Extensions of the acceptance period for the proposals should be obtained from the bidders. Bidders in general are willing to extend their proposals for a reasonable length of time under these circumstances.

What is the Price Basis Required and What Was Offered by the Bidder?   Purchasers prefer firm price bids while bidders, particularly on pumps requiring long delivery or development work, prefer to quote on a nonfirm basis, such as "price in effect at present date plus escalation" or "time and material."

Are the Payment Terms Offered by the Bidder Acceptable?   A typical preference by purchasers is for terms which pay 90% of the equipment price on delivery and retain 10% either for a year after shipment or until the satisfactory completion of acceptance testing. (For complex equipment, partial or progress payments may be allowed.) Frequently, bidders will wish smaller or shorter retentions to improve their cash flow position. In addition, purchasers will sometimes ask for deferred payment terms wherein the supplier receives no payment until completion of acceptance tests 1 year after installation.

Are the Shipment Terms Satisfactory?   If the equipment is bid FOB factory with full freight allowed (W/FFA) to point of use, will supplier prepare and handle any shipping claims for damage? Will purchaser handle traffic arrangements for shipment to point of use? Who is responsible for loss?

Does the Schedule of Shipments Match the Requirements of the Project?   Usually, payment terms are tied to shipments. Any delays caused by the supplier are paid for by the supplier; any delays caused by the purchaser are normally paid for by the purchaser.

Is Force Majeure Properly Defined in the Commercial Terms or Proposal?   There are frequently differences between purchasers and suppliers as to exactly what is included; these should be carefully reviewed.

Has the Bidder Agreed to Accept Other Commercial Terms Included with the Bidding Documents?   The other commercial terms may relate to rights of access to the fabrication shop for inspection or expediting and review of supplier's quality assurance. Where significant exceptions to the commercial terms are taken by the bidders, review by legal counsel should be obtained.

Guarantee/Warranty   The layperson, that is, the nonlawyer, must realize that the terms *guarantee* and *warranty* have lost some of the preciseness they once had. In the following comments, the word *guarantee* is used to mean *guarantee/warranty.*

An important consideration in evaluating bids is the question of the type of guarantee provided by the manufacturer. It is suggested that the purchaser, working with appropriate legal counsel, establish an acceptable clause on this subject to apply to the purchase of the equipment. It should be tailored to suit the purchaser's business philosophies and the law applicable to the business. This clause should then be included in the bidding documents, along with the other suitable commercial terms of purchase, with the request that bidders base their proposals on these terms.

Typically, guarantee clauses provide assurance of freedom from defects in material and workmanship for one year following installation and state that supplier will pay for any replacement parts needed during that year. For larger, more complex pumps, longer guarantees can usually be obtained, particularly if a pump is not put into service immediately after it is installed. Manufacturers are reluctant to provide extended guarantees, however, since these increase their financial risk by extending the period of their exposure for replacement. Therefore it is good practice to obtain the bidder's standard terms and to ask the bidder, as a bid alternate, for the price differential for guarantees longer than standard.

When equipment fails to perform in accordance with specified requirements, suppliers will typically offer new parts, field service as necessary, etc., to upgrade performance to meet specified requirements. In those cases where specified performance cannot be achieved, a remedy (usually financial) is stated. This may be a penalty in dollars per percent efficiency short of specified, dollars per additional unit of power required, etc.

Preaward Conference   For complex or costly orders, conferences with the evaluated low bidder are frequently held prior to placement of the order. The purpose of this meeting is to resolve open areas before award and thus ensure a clear understanding by both parties. In the event there are major differences which cannot be resolved, it is then possible to award to the second lowest bidder without invoking cancellation actions and thus avoiding embarrassment to both parties.

Sample Bid Evaluation   Figures 4a to 4d show a sample bid evaluation.

REV.	DESCRIPTION	ENG DR	CHK DEF	SUPV GK	MATL	APPROVALS	DATE XXX
0	ISSUED FOR CLIENT APPROVAL	ABC	DEF	GK			

NOTES: ENCIRCLED ITEMS ARE UNDESIRABLE. A CHECK ✓ INDICATES COMPLIANCE WITH SPECS

Description	1	2	3	4	5	6	SPEC
MANUFACTURER	MTM	PROSPER	FLJ	EUREKA			
NAME OF BIDDER	"	LUNOR ASSOC.	"				
PROPOSAL NO.	N-36473	LA 624	P4372-5F				
DATE OF PROPOSAL	11-14-71	11-9-71	11-16-71				
MFR'S. MODEL NO.	4XL39H	16-32NR	32-11CP9				
TYPE	7 STG.	9 STG.	6 STG.				
MOTOR MANUFACTURER	ELEC. PROD.						
ADDITIONS							
ONE SET OF ACCESSORIES	NOT REQ'D	INCL.	NOT REQ'D				REQ'D
ONE SET SPARE PARTS	INCL	+$1,550.	INCL.				
ONE SET SPECIAL TOOLS	NOT REQ'D	+180 ①	NOT REQ'D	NO			
ERECTION SUPERVISION	+$2,540	+1,130 ②	+2,540 ③				
FEATURES (PER UNIT)	(10 DAYS)	(5 DAYS) ④	(10 DAYS)	BID			ALT QUOTE
SAVINGS TO ELIMINATE UP THRUST	<3,440>	REQUIR'D	<3,620> ⑤				
SAVINGS WITH SHORTER BARREL	<2,330>	—	<2,840> ⑤				
ADD FOR HYD. TESTING	INCL	+200	INCL				
" INSTALL COUPLING	✓	+350 ③	+320				
" TEMP SUCTION STRAINER	✓	+500 ③	INCL				
" 11-13 CF, 1ST STG. IMP.	✓	DECLINED	INCL				
" 11-13 CF, SUCTION BELL	+3,140	TO BID	+2,650				
ENERGY PENALTY @ $107/HP (PER UNIT)	+12,650	BASE	+4,427				
TOTAL							
ADDITIONS & PENALTY (TWO UNITS)	+33,100	+8,090	+18,374				
PRICE FOB SHIPPING POINT (TWO)	$209,325	$239,760	$214,782				
ADDITIONS AS ITEMIZED ABOVE	33,100	8,090	18,374				
COMPARATIVE COST AT SOURCE	242,425	247,850	233,156				
TRANSPORTATION COST (ESTIMATED)	5,430	6,580	7,240				
COMPARATIVE COST AT DESTINATION	247,855	254,430	240,396				
TOTAL COST (2 UNITS)	247,855	254,430	240,396				
SHIPPING POINT	SERRA	UMPQRA	LAKEVIEW				
SHIPPING WEIGHT (TOTAL INCL. DRIVERS)	137,000 #	148,000 #	131,000 #				
DELIVERY (MO. AFTER PO)	12	14	12				12 FIRM
ESCALATION	NONE	NONE	NONE				

RECOMMEND: FLJ

REASON: LOWEST EVALUATED PRICE, MEETS SPEC., DELIVERY SATISFACTORY

① LETTER OF DEC. 3; ② LETTER OF DEC. 8; ③ LETTER OF NOV. 29; ④ DOUBLED FOR EVALUATION COMPARISON

SUMMARY OF BIDS
(PRICED)

CONDENSATE PUMPS

XYZ POWER COMPANY

JOB No XXXX		
P.O. NO.	**M-23**	REV.
	SHT. 1 OF 4	0

FIG. 4a   Condensate pump bid evaluation (sheet 1 of 4).

	MTM	PROSPER	FLJ	SPEC
PERFORMANCE				
RATED FLOW GPM	✓	✓	✓	6900 GPM
EFFICIENCY @ 6900 GPM	76.5%	83.0%	73.0%	
DESIGN FLOW GPM	✓	✓	✓	8400 GPM
DESIGN TEMP. °F	✓	✓	✓	112°F
SPEC. GRAVITY @ 112°F	0.991	✓	0.991	
DIFF. HEAD @ 8400 GPM	✓	✓	✓	1080' TDH
PRESS. @ MIN. FLOW PSIA	710	670 ②	610	675 PSIA
MIN. CONT. FLOW GPM	800	1700	800	
FLOW @ RUN OUT GPM	10,000	9,400	11,500	9400 GPM
NPSH REQ. @ 9400 GPM	15'	21.4'	15'	—
NPSH REQ @ DESIGN (8400GPM)	14.5'	17.6'	14.5'	—
BHP @ 6900 GPM HP	2700	2600 ②	2650	—
BHP @ 9400 GPM HP	2900	2880	3100	—
SPEED RPM	1180	1180	1180	900/1200
1ST CRITICAL SPEED RPM	NS	>9500	NS	—
Nº OF STAGES	7	9	6	—
WEIGHT ROT. PART LBS	NS	3700	3100	—
REVERSE FLOW RPM@100%HEAD	<125%	1250 RPM	<125%	—
DESIGN				
SUCTION: SIZE/FLANGE RATING	24"/150#	30"/150#	24"/150#	24" PREF.
DISCHARGE: SIZE/FLANGE RATING	18"/300#	18"/300#	18"/300#	16" PREF.
DIAM (O.D.) OF SHELL	42"	44"	42"	—
LENGTH OF SHELL	18'-6" ①	22'-5½"	20'-8"	—
WEIGHT OF PUMP (EACH) LBS	43,000	46,500	40,000	—
WEIGHT OF MOTOR LBS	26,000	23,000	26,000	—
SUCTION: DESIGN PRESS PSI	1.5×MAX.SUCT.	50	1.5×MAX.SUCT.	—
DISCHARGE: DESIGN PRESS. PSI	1.5×SHUT-OFF	650	1.5×SHUT-OFF	—
SUCTION: TEST PRESS. PSI	1.5×MAX.SUCT.	75	1.5×MAX.SUCT.	—
DISCHARGE: TEST PRESS. PSI	1.5×SHUT-OFF	975	1.5×SHUT-OFF	1.5×MAX.SHUT-OFF

① AS PER LETTER, MARCH 9, 1970; INCREASE IN AVAILABLE SUCTION HEAD BY 8 FT. ENABLES MTM TO SHORTEN LENGTH OF BARREL FROM 21'-6" TO 18'-6".

② ADJUSTED TO PROSPER LETTER MARCH 6, 1970

NOTES: ENCIRCLED ITEMS ARE UNDESIRABLE
A CHECK ✓ INDICATES COMPLIANCE WITH SPECS

Left margin: ISSUED FOR CLIENT APPROVAL — DESCRIPTION — REV. — DATE — APPROVALS — MATL — SUPV — CHK — DR — ENG

SUMMARY OF BIDS	JOB NO. XXXX	
CONDENSATE PUMPS	P.O. NO. **M-23**	REV. 0
XYZ POWER COMPANY	SHT. 2 OF 4	

FIG. 4b Condensate pump bid evaluation (sheet 2 of 4).

	MTM	PROSPER	FLJ	SPEC
CASE: MAT. OF SUCTION BELL	11-13Cr ST	C.I. (1)	11-13Cr. ST	11-13 Cr ST
MAT. OF WEAR RINGS	BRZ	C.I.	BRZ	—
MATERIAL OF CASE	A-48C.I.	A-48C.I.	A-48C.I.	—
STD. WALL THICKNESS / MIN.	1"/7/8"	1"/13/16"	1"/7/8"	—
VANES & WATERWAYS	STD ✓	AS CAST	STD ✓	—
WEAR RINGS SUCT / DISCH.	—/✓	✓/—	—/✓	—
CLEARANCE	0.02-0.024"	(1) NS	0.02-0.024"	—
IMPELLER: MAT. 1st STG. / OTHERS	11-13Cr ST/BRZ	11-13Cr ST/BRZ	11-13Cr ST/BRZ	11-13Cr ST/BRZ
TYPE 1st STG. / OTHERS	ENCL./SEMI-E	ENCL.	ENCL/SEMI-E	—
DIAM. 1st STG. DESIGN/MAX.	19 1/2"/20 5/8"	20 5/8"/20 5/8"	19 3/8"/20 11/16"	—
OTHER STG. DESIGN/MAX./MIN.	19 1/8"/20 7/16"	20"/20 5/8"/18 5/8"	19 3/8"/20 11/16"	
BALANCE-HYDR. / STAT. / DYN.	✓/—/✓	—/✓/✓	✓/—/✓	REQ'D.
WEAR RINGS SUCT. / DISCH.	—/✓	✓/—	—/✓	—
WR^2 LB FT^2	1250	650	1250	—
HYDR. THRUST MAX. DOWN	40,000	74700 (2)	40,000	—
HYDR. THRUST MAX. UP	NONE	22400 (2)	NONE	—
VELOCITY-EYE DESIGN/MAX.	15 1/2/23	17/19FPS	16/21	—
				—
BEARING: MATERIAL	BRZ	BRZ	BRZ	BRZ
BOTTOM LUBE / L/D	SELF/2.4	SELF/1.5 (3)	SELF/2.4	SELF
BOTTOM GROOVED / TYPE	✓/SLEEVE	✓/JOURN.	✓/SLEEVE	—
SERIES LUBE / L/D	SELF/2	SELF/1.0 (3)	SELF/2	SELF
SERIES GROOVED / TYPE	✓/SLEEVE	✓/JOURN.	✓/SLEEVE	—
TOP LUBE / L/D	SELF/2	SELF/1.0 (3)	SELF/2	SELF
TOP GROOVED / TYPE	✓/SLEEVE	✓/JOURN.	✓/SLEEVE	—
THROTTLE LUBE/L/D/GROOV.	SELF/2/✓	SELF/1.0 (3)/✓	SELF/2/✓	SELF
SHAFT: MAT. PUMP ELEMENT	SS410	SS416	SS410	SS400 SERIES
MAT. COLUMN	SS410	C. ST.	SS410	—
DIAM. / LONG BEARING SPAN	5"/48"	5 1/2"/45"	5"/48"	—
COMBINED STRESS MAX.	10,000	14,500PSI TORS. ONLY	10,000	—
SURFACE FINISH & BEAR'G.	32	32 RMS	32	GRD
SURFACE @ STUFF. BOX	32	125 RMS	32	GRD
SLEEVE @ STUFF. BOX / THRU GLAND	✓/✓	✓/✓	✓/✓	REQ'D/REQ'D.
PRESS @ STUFF. BOX	15PSI > SUCT.	15-20PSI > SUCT.	15PSI > SUCT.	—
SLEEVE MAT.	SS-420	ALLOY-ST.	SS-420	—

(1) AS PER PROSPER LETTER MARCH 4, 1970. (SPEC. REQUESTED 11-13Cr ST SUCTION BELL & 1st ST'G. IMPELLER MATERIAL)
(2) HIGH DOWN THRUST.
(3) SPEC. RECOMMENDS A MIN. RATIO OF 2. PROSPER DECLINES TO UPGRADE SUBJECT RATIO.

NOTES: ENCIRCLED ITEMS ARE UNDESIRABLE
A CHECK ✓ INDICATES COMPLIANCE WITH SPECS

REV.	DESCRIPTION	ENG	DR	CHK	SUPV	MATL	APPROVALS	DATE
0	ISSUED FOR CLIENT APPROVAL	ABC		DEF	JK		JM. WDP. R.	XXX

SUMMARY OF BIDS	JOB No XXXX	
CONDENSATE PUMPS	P.O. NO. M-23	REV.
XYZ POWER COMPANY	SHT. 3 OF 4	0

FIG. 4c Condensate pump bid evaluation (sheet 3 of 4).

	MTM	PROSPER	FLJ	SPEC
COUPLINGS: INTERNAL SPAN/TYPE	-/KEYED	45°/THREAD ①	-/KEYED	NOT SPEC'D
DRIVE: SOLID/FLANGED	✓/✓	✓/✓	✓/✓	-
ADJUST	✓	✓	✓	-
MAKE/SIZE	MTM/#5	PROSPER/15½	FLJ/#5	-
SUCT. BARREL: WALL/WANTED	3/8"/✓	¼"/✓	3/8"/✓	-
INT. VELOCITY DESIGN/MAX.	-/6	3½/4	-/6 ②	-
BASE PART FLANGE	-	57×57×2¼	-	-
GASKET MAT./TYPE	-/ROUND	-/FLAT	-/ROUND	-
MATERIAL	FAB. ST.	C. ST.	FAB. ST.	-
NOZZLE HEAD: NET AREA MOT. BARREL	-	52"	-	-
MAX. DIAM. BARREL	-	46"	-	-
BASE	-	52 × 1½	-	-
FASTENINGS: MAT. INTERNAL	ST	-	ST	-
EXTERNAL	ST	A-193-B7	ST	-
PERFORMANCE TEST IN SHOP	✓	NO	✓	REQ'D
FACSIMILE TEST/SAME PUMP TYPE	NOT REQ'D	✓/✓	NOT REQ'D	-
MECH. TEST IN SHOP, FULLY ASSEM.	✓	③	✓	MECH. ROT.
WELDING REQUIREMENTS TO ASME	✓	④	✓	-
COMPL. OF MATERIAL TO ASTM	✓	⑤	✓ ⑥	ASTM

① KEY DRIVEN COUPLINGS
ARE RECOMMENDED
② RECOMMENDED RANGE 4-6½ FPS
③ BOWL ASSEM. ONLY, ROTATED BY
HAND ONLY
④ WELDERS NOT CODE QUALIFIED
⑤ PHYS. & CHEM. PROPERTY ONLY
⑥ COMPLIES WITH ASTM PROPER-
TIES BUT DOES NOT SUPPLY
MILL TEST REPORTS.

NOTES: ENCIRCLED ITEMS ARE UNDESIRABLE
A CHECK ✓ INDICATES COMPLIANCE WITH SPECS

(left margin, bottom to top) ISSUED FOR CLIENT APPROVAL

APPROVALS		
MATL		
SUPV		
CHK		
DR		
ENG		
REV. DESCRIPTION		0

SUMMARY OF BIDS

CONDENSATE PUMPS

XYZ POWER COMPANY

JOB No. XXXX
P.O.NO. M-23
SHT. 4 OF 4
REV. 0

FIG. 4d Condensate pump bid evaluation (sheet 4 of 4).

FURTHER READING

American Society of Mechanical Engineers: *Boiler and Pressure Vessel Code*, Sec. III, ASME, New York, 1983.

"Flow of Fluids through Valves, Fittings and Pipe," Technical Paper 410, Crane, Chicago, 1980.

Grant, E. L., and W. G. Ireson: *Engineering Economy*, 4th ed., Ronald Press, New York, 1964.

Hydraulic Institue: *Standards for Centrifugal, Rotary and Reciprocating Pumps*, 14th ed., Cleveland, 1983.

Karassik, I. J.: *Engineers' Guide to Centrifugal Pumps*, McGraw-Hill, New York, 1964.

Martinez, S.: "Equipment Buying Decisions," *Chem. Eng.*, April 5, 1971, p. 146.

National Electrical Manufacturers Assocaition standards MG-1-12.47A, MG-1-14.30 to 1-14.42, and MG-1-20.43, NEMA, Washington, D.C.

Underwriters' standards, Underwriters' Laboratory, Northbrook, Ill.

Quality Assurance and Reliability

Burr, I. W.: *Elementary Statistical Quality Control*, Marcel Dekker, New York, 1979.

Henley, E. J., and H. Kumamoto: *Reliability Engineering and Risk Assessment*, Prentice Hall, Englewood Cliffs, N.J., 1981.

Juran, J. M., F. M. Gryna, Jr., and R. S. Bingham, Jr.: *Quality Control Handbook*, 3d ed., 1974, McGraw-Hill, New York, 1974.

Marguglio, B. W.: "Quality Systems in the Nuclear Industry," publication 616, American Society for Testing and Materials, Philadelphia, 1977.

10 Code of Federal Regulations, Part 50: "Quality Assurance Criteria for Nuclear Power Plants and Fuel Reprocessing Plants," Office of the Federal Register, National Archives and Records Service, General Services Administration, Washington, D.C.,.

INSTALLATION, OPERATION, AND MAINTENANCE

Igor J. Karassik

The information contained in this chapter is general and should be supplemented by the specific instructions prepared by the manufacturer of the pump in question.

INSTALLATION

Instruction Books   Instruction books are intended to help keep the pumps in an efficient and reliable condition at all times. It is necessary, therefore, that instruction books be available to all personnel involved in this function.

Preparation for Shipment   After a pump has been assembled in the manufacturer's shop, all flanges and exposed machined metal surfaces are cleaned of foreign matter and treated with an anticorrosion compound, such as grease, Vaseline, or heavy oil. For protection during shipment and erection, all pipe flanges, pipe openings, and nozzles are protected by wooden flange covers or by screwed-in plugs.

Usually the driver is delivered to the pump manufacturer, where it is assembled and aligned with the pump on a common baseplate. The baseplate is drilled for driver mounting, but the final doweling is performed in the field after final alignment. When size and weight permit, the unit is shipped assembled with pump and driver on the baseplate. If drivers are shipped directly to be mounted in the field, the baseplate should be drilled at the job site.

Care of Equipment in the Field   If the pumping equipment is received before it can be used, it should be stored in a dry location. The protective flange covers and coatings should remain on the pumps. The bearings and couplings must be carefully protected against sand, grit, and other foreign matter.

More thorough precautions are required if a pump must be stored for an extended period of time. It should be carefully dried internally with hot air or by a vacuum-producing device to avoid rusting of internal parts. Once free of moisture, the pump internals should be coated with a protective liquid, such as light oil, kerosene, or antifreeze. Preferably, all accessible parts, such as bearings and couplings, should be dismantled, dried, and coated with Vaseline or acid-free heavy oil and then properly tagged and stored.

If rust preventive has been used on stored parts, it should be removed completely before final installation, and the bearings should be relubricated.

Pump Location   Working space must be checked to assure adequate accessibility for maintenance. Axially split casing horizontal pumps require sufficient headroom to lift the upper half of the casing free of the rotor. The inner assembly of radially split multistage centrifugal pumps is removed axially (Subsec. 2.2.1, Fig. 23). Space must be provided so that the assembly can be pulled out without canting it. For large pumps with heavy casings and rotors, a traveling crane or other facility for attaching a hoist should be provided over the pump location.

Pumps should be located as close as practicable to the source of liquid supply. Whenever possible, the pump centerline should be placed below the level of the liquid in the suction reservoir.

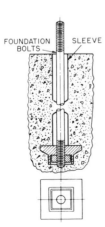

FOUNDATION   SLEEVE
BOLTS

Foundations   Foundations may consist of any structure heavy enough to afford permanent rigid support to the full area of the baseplate and to absorb any normal strains or shocks. Concrete foundations built up from solid ground are the most satisfactory. Although most pumping units are mounted on baseplates, very large equipment may be mounted directly on the foundations by using soleplates under the pump and driver feet. Misalignment is corrected with shims.

The space required by the pumping unit and the location of the foundation bolts are determined from the drawings supplied by the manufacturer. Each foundation bolt (Fig. 1) should be surrounded by a pipe sleeve three or four diameters larger than the bolt. After the concrete foundations are poured, the pipe is held solidly in place but the bolt may be moved to conform to the corresponding hole in the baseplate.

FIG. 1   Foundation bolt.

When a unit is mounted on steelwork or some other structure, it should be placed directly over or as near as possible to the main members, beams, or walls and should be supported so that the baseplate cannot be distorted or the alignment disturbed by any yielding or springing of the structure or of the baseplate.

Mounting of Vertical Wet-Pit Pumps
A curve ring or soleplate must be used as a bearing surface for the support flange of a vertical wet-pit pump. The mounting face must be machined because the curb ring or soleplate will be used in aligning the pump.

If the discharge pipe is located below the support flange of the pump (belowground discharge), the curb ring or soleplate must be large enough to pass the discharge elbow during assembly. A rectangular ring should be used (Fig. 2). If the discharge pipe is located above the support flange (aboveground discharge), a round curb ring or soleplate should be provided with clearance on its inner diameter to pass all sections of the pump below the support flange (Fig. 3). A typical method of arranging a grouted soleplate for vertical pumps is shown in Fig. 4.

If the discharge is belowground and an expansion joint is used, it is necessary to determine the moment that may be imposed on the structure. The pump casing should be attached securely to some rigid structural members with tie rods. If vertical wet-pit pumps are very long, some steadying device is required irrespective of the location of the discharge or of the type of pipe connection. Tie rods can be used to connect the unit to a wall, or a small clearance around a flange can be used to prevent excessive displacement of the pump in the horizontal plane.

Alignment
When a complete unit is assembled at the factory, the baseplate is placed on a flat, even surface; the pump and driver are mounted on the baseplate; and the coupling halves are accurately aligned, using shims under the pump and driver mounting surfaces where necessary. The pump is usually doweled to the baseplate at the factory, but the driver is left to be doweled after installation at the site.

The unit should be supported over the foundation by short strips of steel plate or shim stock close to the foundation bolts, allowing a space of ¾ to 2 in (2 to 5 cm) between the bottom of the baseplate and the top of the foundation for grouting. The shim stock should extend fully across

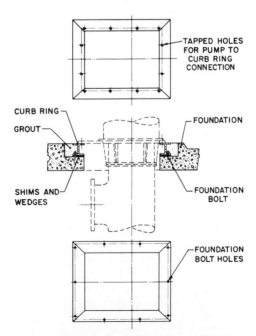

FIG. 2   Rectangular curb ring for belowground discharge vertical pump.

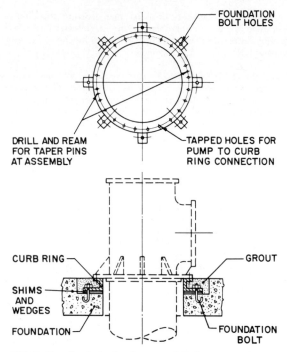

FIG. 3   Round curb ring for aboveground discharge vertical pump.

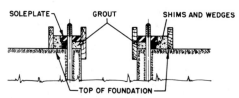

FIG. 4   Grouting form for vertical pump soleplate. (ANSI B-58.1 [AWWAE 101-61].)

the supporting edge of the baseplate. The coupling bolts should be removed before the unit is leveled and the coupling halves are aligned. Where possible, it is preferable to place the level on some exposed part of the pump shaft, sleeve, or planed surface of the pump casing. The steel supporting strips or shim stock under the baseplate should be adjusted until the pump shaft is level, the suction and discharge flanges are vertical or horizontal as required, and the pump is at the specified height and location. When the baseplate has been leveled, the nuts on the foundation bolts should be made handtight.

During this leveling operation, accurate alignment of the unbolted coupling halves must be maintained. A straightedge should be placed across the top and sides of the coupling, and at the same time the faces of the coupling halves should be checked with a tapered thickness gage or with feeler gages (Fig. 5) to see that they are parallel. For all alignment checks, including parallelism of coupling faces, both shafts should be pressed hard over to one side when taking readings.

When the peripheries of the coupling halves are true circles of equal diameter and the faces are flat and perpendicular to the shaft axes, exact alignment exists when the distance between the faces is the same at all points and when a straightedge lies squarely across the rims at any point. If the faces are not parallel, the thickness gage or feelers will show variation at different points.

If one coupling is higher than the other, the amount may be determined by the straightedge and feeler gages.

Sometimes coupling halves are not true circles or are not of identical diameter because of manufacturing tolerances. To check the trueness of either coupling half, rotate it while holding the other coupling half stationary and check the alignment at each quarter turn of the half being rotated. Then the half previously held stationary should be revolved and the alignment checked. A variation within manufacturing limits may be found in either of the half-couplings, and proper allowance for this must be made when aligning the unit.

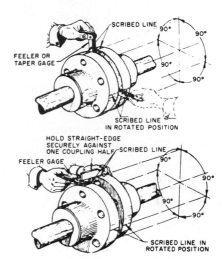

A more exact method for checking alignment that is recommended requires the use of a dial indicator. With the indicator bolted to the pump half of the coupling, both radial and axial alignment can be checked, as illustrated in Fig. 6. This method is called face-and-rim alignment. With the button resting on the periphery of the other coupling half, the dial should be set at zero and a mark chalked on the coupling half at the point where the button rests. For any check (top, bottom, or sides), both shafts should be rotated the same amount, that is, all readings on the dial should be made with the button on the chalk mark. The dial readings will indicate whether the driver must be raised, lowered, or moved to either side. After any movement, it is necessary to check that the coupling faces remain parallel to one another.

FIG. 5 Coupling alignment using feeler gages.

For example, if the dial reading at the starting point is set to zero and the diameterically opposite reading at the bottom or sides shows ± 0.020 in (± 0.508 mm), the driver must be raised or lowered by shimming or moved to one side or the other by half of this reading. The same procedure is used to align gear couplings, but the coupling covers must first be moved back out of the way and all measurements should be made on the coupling hubs.

When an extension coupling connects the pump to its driver, a dial indicator should be used to check the alignment (Fig. 7). The extension piece between the coupling halves should be removed to expose the coupling hubs. The coupling nut on the end of the shaft should be used to clamp a suitable extension arm or bracket long enough to extend across the space between the coupling hubs. The dial indicator is mounted on this arm, and alignment is checked for both concentricity of the hub diameters and parallelism of the hub faces. Changing the arm from one hub to the other provides an additional check. The dial extension bracket must be checked for sag, and readings must be corrected accordingly.

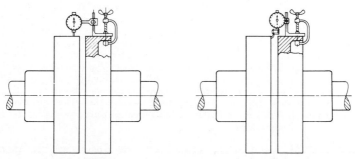

FIG. 6 Use of dial indicator for face-and-rim alignment of standard coupling.

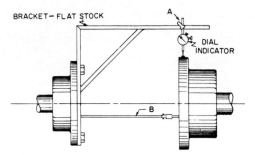

FIG. 7   Use of dial indicator for face-and-rim alignment of extension coupling.

The clearance between the faces of the coupling hubs and the ends of the shafts should be such that they cannot touch, rub, or exert a pull on either pump or driver. The amount of this clearance may vary with the size and type of coupling used. Sufficient clearance will allow unhampered endwise movement of the shaft of the driving element to the limit of its bearing clearance. On motor-driven units, the magnetic center of the motor will determine the running position of the motor half-coupling. This position should be checked by running the motor uncoupled. This will also permit checking the direction of rotation of the motor. If current is not available at the time of installation, move the motor shaft in both directions as far as the bearings will permit and adjust the shaft centrally between these limits. The unit should then be assembled with the correct gap between the coupling halves.

Large horizontal sleeve-bearing motors are not generally equipped with thrust bearings. The motor rotor is permitted to float, and as it will seek its magnetic center, an axial force of rather small magnitude can cause it to move off this center. Sometimes it will move enough to cause the shaft collar to contact and possibly damage the bearing. To avoid this, a limited-end-float coupling is used between the pump and the motor on all large units to restrict the motor rotor (Subsec. 6.3.1). The setting of axial clearances for such units should be given by the manufacturer in the instruction books and elevation drawings.

When the pump handles a liquid at other than ambient temperature or when it is driven by a steam turbine, the expansion of the pump or turbine at operating temperature will alter the vertical alignment. Alignment should be made at ambient temperature with suitable allowances for the changes in pump and driver centerlines after expansion. The final alignment must be made with the pump and driver at their normal temperatures and adjusted as required before the pump is placed into permanent service.

For large installations, particularly with steam-turbine-driven pumps, more sophisticated alignment methods are sometimes employed, using proximity probes and optical instruments. Such procedures permit checking the effect of temperature changes and machine strains caused by piping stresses while the unit is in operation. When such procedures are recommended, they are included with the manufacturer's instructions.

When the unit has been accurately leveled and aligned, the hold-down bolts should be gently and evenly tightened before grouting. The alignment must be rechecked after the suction and discharge piping has been bolted to the pump to test the effect of piping strains. This can be done by loosening the bolts and reading the movement of the pump, if any, with dial indicators.

The pump and driver alignment should be occasionally rechecked, for misalignment may develop from piping strains after a unit has been operating for some time. This is especially true when the pump handles hot liquids, as there may be a growth or change in the shape of the piping. Pipe flanges at the pump should be disconnected after a period of operation to check the effect of the expansion of the piping, and adjustments should be made to compensate for this.

For a further discussion of hot and cold alignment, face-and-rim versus reverse dial methods, measurement of dial bracket sag, and graphical alignment plotting procedure, refer to Subsec. 2.3.3.

Grouting   Ordinarily, the baseplate is grouted before the piping connections are made and before the alignment of the coupling halves is finally rechecked. The purpose of grouting is to prevent lateral shifting of the baseplate, to increase its mass to reduce vibration, and to fill in irregularities in the foundation.

The usual mixture for grouting a pump baseplate is composed of one part pure portland cement and two parts building sand, with sufficient water to cause the mixture to flow freely under the base (heavy cream consistency). To reduce settling, it is best to mix the grout and let it stand for a short period, then remix it thoroughly before use without adding any more water.

The top of the rough concrete foundation should be well saturated with water before grouting. A wooden form is built around the outside of the baseplate to retain the grout. Grout is added until the entire space under the baseplate is filled to the top of the underside (Fig. 8). The grout holes in the baseplate serve as vents to allow the air to escape. A stiff wire should be used through the grout holes to work the grout and release any air pockets.

The exposed surfaces of the grout should be covered with wet burlap to prevent cracking from too-rapid drying. When the grout is sufficiently set so that the forms can be removed, the exposed surfaces of the grout and foundations are finished smooth. When the grout is hard (72 h or longer), the hold-down bolts should be finally tightened and the coupling halves rechecked for alignment.

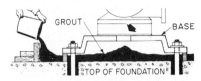

FIG. 8 Application of grouting. Grout is added until the entire space under the base is filled. Holes in the base (arrow) allow air to escape and permit working of the grout to release air pockets.

There is considerable controversy over whether the leveling strips or wedges should be removed after grouting. The best practice is to remove these in all cases for reciprocating machinery because pounding action or vibration will ultimately loosen the unit from the foundation. The space formerly occupied by shims or wedges must be regrouted. There is less danger in not removing the strips or wedges with rotating machinery, provided care is used in mixing the grout material and there is no shrinkage or drying. The strips or wedges can also be removed from a rotating unit; erectors can follow their own preference in this matter.

The pump and driver alignment must be rechecked thoroughly after the grout has hardened permanently, and at reasonable intervals thereafter.

Doweling of Pump and Driver When the pump handles hot liquids, doweling of both the pump and its driver should be delayed until the unit has been operated. A final recheck of alignment with the coupling bolts removed and with the pump and driver at operating temperature is advisable before doweling.

Large pumps handling hot liquids are usually doweled near the coupling end, allowing the pump to expand from that end out. Sometimes the other end is provided with a key and a keyway in the casing foot and the baseplate.

PIPING

Suction Piping The suction piping should be as direct and short as possible. If a long suction line is required, the pipe size should be increased to reduce frictional losses. (The exception to this recommendation is in the case of boiler-feed pumps, where difficulties may arise during transient conditions of load change if the suction piping volume is excessive. This is a special and complex subject, and the manufacturer should be consulted.) Where the pump must lift the liquid from a lower level, the suction piping should be laid out with a continual rise toward the pump, avoiding high spots in the line to prevent the formation of air pockets. Where a static suction head will exist, the pump suction piping should slope continuously downward to the pump.

Generally, the suction line is larger than the pump suction nozzle and eccentric reducers should be used. If the source of supply is located below the pump centerline, the reducer should be installed straight side up. If the source of supply is above the pump, the straight side of the reducer should be at the bottom (Fig. 9). Installing eccentric reducers with a change in diameters greater than 4 in (10 cm) could disturb the suction flow. If such a change is necessary, it is advisable to use properly vented concentric reducers.

Elbows and other fittings next to the pump suction should be carefully arranged, or the flow

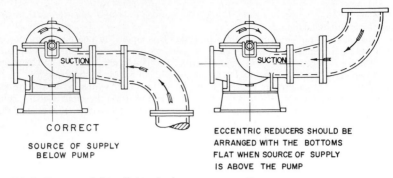

CORRECT

SOURCE OF SUPPLY
BELOW PUMP

ECCENTRIC REDUCERS SHOULD BE
ARRANGED WITH THE BOTTOMS
FLAT WHEN SOURCE OF SUPPLY
IS ABOVE THE PUMP

FIG. 9   Recommended installation of reducers at pump suction.

into the pump impeller will be disturbed. Long-radius elbows are preferred for suction lines because they create less friction and provide a more uniform flow distribution than standard elbows.

It is extremely important to avoid the formation of vortices at the suction of both wet-pit and dry-pit pump installations. For a discussion of this and other suction conditions recommendations, see Secs. 10.1 and 10.2.

Discharge Piping   Generally both a check valve and a gate valve are installed in the discharge line. The check valve is placed between the pump and the gate valve and protects the pump from reverse flow in the event of unexpected driver failure or from reverse flow from another operating pump. The gate valve is used when priming the pump or when shutting it down for inspection and repairs. Manually operated valves that are difficult to reach should be fitted with a sprocket rim wheel and chain. In some cases discharge gate valves are motorized and can be operated by remote control.

Piping Strains   Cast iron pumps are never provided with raised face flanges. If steel suction or discharge piping is used, the pipe flanges should be of the flat-face and not the raised-face type. Full-face gaskets must be used with cast iron pumps.

Piping should not impose excessive forces and moments on the pump to which it is connected, since these might spring the pump or pull it out of position. Piping flanges must be brought squarely together before the bolts are tightened. The suction and discharge piping and all associated valves, strainers, etc., should be supported and anchored near to but independent of the pump, so that no strain will be transmitted to the pump casing.

There are four factors to be considered in determining the effect of nozzle loads: material stress in pump nozzles resulting from forces and bending moments, distortion of internal moving parts affecting critical clearances, stresses in pump hold-down bolts, and distortion in pump supports and baseplates resulting in driver coupling misalignment. API Standard 610 (Centrifugal Pumps for General Refinery Services) provides guidelines for limiting the magnitude of nozzle loads and moments on pumps with suction nozzles 12 in (30 cm) and smaller and with casings constructed of steel or alloy steel.

With large pumps or when major temperature changes are expected, the pump manufacturer generally indicates to the user the maximum moments and forces that can be imposed on the pump by the piping. A typical diagram is illustrated in Fig. 10 for a radially split double-casing multistage pump with top suction and discharge.

Expansion Joints   Expansion joints are sometimes used in the discharge and suction piping to avoid transmitting any piping strains caused by misalignment or by expansion when hot liquids are handled. On occasion, expansion joints are formed by looping the pipe. More often, they are of the slip-joint or corrugated-diaphragm type. However, they transmit to the pump a force equal to the area of the expansion joint times the pressure in the pipe. These forces can be of very significant magnitude, and it is impractical to design the pump casings, baseplates, etc., to with-

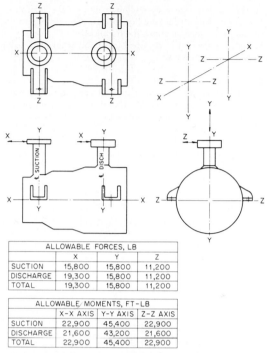

ALLOWABLE FORCES, LB			
	X	Y	Z
SUCTION	15,800	15,800	11,200
DISCHARGE	19,300	15,800	11,200
TOTAL	19,300	15,800	11,200

ALLOWABLE MOMENTS, FT-LB			
	X-X AXIS	Y-Y AXIS	Z-Z AXIS
SUCTION	22,900	45,400	22,900
DISCHARGE	21,600	43,200	21,600
TOTAL	22,900	45,400	22,900

FIG. 10 Diagram of typical permissible pipe stresses and moments for a radially split double-casing multistage pump with top suction and discharge. (1 N = 0.225 lb; 1 N·m = 0.737 ft·lb)

stand them. Consequently, when expansion joints are used, a suitable pipe anchor must be installed between them and the pump proper. Alternately, tie rods can be used to prevent the forces from being transmitted to the pump.

Suction Strainers   Except for certain special designs, pumps are not intended to handle liquid containing foreign matter. If the particles are sufficiently large, such foreign matter can clog the pump, reduce its capacity, or even render it altogether incapable of pumping. Small particles of foreign matter may cause damage by lodging between close running clearances. Therefore proper suction strainers may be required in the suction lines of pumps not specially designed to handle foreign matter.

In such an installation, the piping must first be thoroughly cleaned and flushed. The recommended practice is to flush all piping to waste before connecting it to the pump. Then a temporary strainer of appropriate size should be installed in the suction line as close to the pump as possible. This temporary strainer may have a finer mesh than the permanent strainer installed after the piping has been thoroughly cleaned of all possible mill scale or other foreign matter. The size of the mesh is generally recommended by the pump manufacturer. For further details on strainers, see Secs. 8.1 and 10.1.

Venting and Draining   Vent valves are generally installed at one or more high points of the pump casing waterways to provide a means of escape for air or vapor trapped in the casing. These valves are used during the priming of the pump or during operation if the pump should become air- or vapor-bound. In most cases these valves need not be piped up away from the pump because their use is infrequent, and the vented air or vapors can be allowed to escape into the surrounding atmosphere. On the other hand, vents from pumps handling flammable, toxic, or corrosive fluids must be connected in such a way that they endanger neither the operating personnel nor the installation. The suction vents of pumps taking liquids from closed vessels under vacuum must be piped to the source of the pump suction above the liquid level.

All drain and drip connections should be piped to a point where the leakage can be disposed of or collected for reuse if worth reclaiming.

Warm-Up Piping   When it is necessary for a pump to come up to operating temperature before it is started up or to keep it ready to start at rated temperature, provision should be made for a warm-up flow to pass through the pump. There are a number of arrangements used to accomplish this. If the pump operates under positive pressure on the suction, the pumped liquid can be permitted to drain out through the pump casing drain connection to some point at a pressure lower than the suction pressure (Fig. 11). Alternately, some liquid can be made to flow back from the discharge header through a jumper line around the check valve into the pump and out

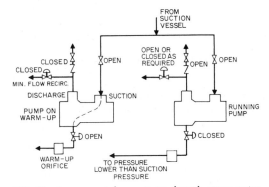

FIG. 11   Arrangement for warm-up through pump casing drain connection.

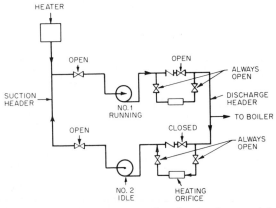

FIG. 12   Arrangement for warm-up through jumper line around discharge check valve.

into the suction header (Fig. 12). An orifice is provided in this jumper line to regulate the amount of warm-up flow. Care must be exercised in such an installation to maintain the suction valve open (unless the warm-up line valve is closed, as when the pump is to be dismantled) lest the entire pump, suction valve, and suction piping be subjected to full discharge pressure.

The manufacturer's recommendations should be sought in all cases as to the best means of providing an adequate warm-up procedure.

Relief Valves   Positive displacement pumps, such as rotary and reciprocating pumps, can develop discharge pressures much in excess of their maximum design pressures. To protect these pumps against excessive pressures when the discharge is throttled or shut off, a pressure relief valve must be used. Some pumps are provided with internal integral relief valves, but unless operation against a closed discharge is both infrequent and of very short duration, a relief valve with an external return connection must be used and the liquid from the relief valve must be piped back to the source of supply.

Surge Chambers   Generally, centrifugal pumps do not require surge chambers in their suction or discharge piping. Reciprocating pumps may have a suction and discharge piping layout that does not require compensation for variations in the flow velocity in the piping system. In many cases, however, reciprocating pump installations require surge chambers when the suction or discharge lines are of considerable length, when there is an appreciable static head on the discharge, when the liquid pumped is hot, or when it is desirable to smooth out variations in the discharge flow. The type, size, and arrangement of the surge chamber should be chosen on the basis of the manufacturer's recommendations. For more details, see Sec. 3.6.

Instrumentation   There are a number of instruments which are essential to maintaining a close check on the performance and condition of a pump. A compound pressure gage should be connected to the suction of the pump, and a pressure gage should be connected to its discharge at the pressure taps which may be provided in the suction and discharge flanges. The gages should be mounted in a convenient location so that they can be easily observed.

In addition, it is advisable to provide a flow-metering device. Depending upon the importance of the installation, indicating meters may be supplemented by recording attachments.

Whenever pumps incorporate various leakoff arrangements, such as a balancing device or pressure-reducing labyrinths, a check should be maintained on the quantity of these leakoffs by measuring orifices and differential gages installed in the leakoff lines.

Pumps operating in important or complex services or operating completely unattended by remote control may have additional instrumentation, such as speed indicators, vibration monitors, and bearing or casing temperature indicators. For more detail, see Subsecs. 2.3.2 and 2.3.3 and Sec. 3.6.

OPERATION

Pumps are generally selected for a given capacity and total head when operating at rated speed. These characteristics are referred to as "rated conditions of service" and, with few exceptions, represent those conditions at or near which the pump will operate the greatest part of the time. Positive displacement pumps cannot operate at any greater flows than rated except by increasing their speed; nor can they operate at lower flows except by reducing their operating speed or bypassing some of the flow back to the source of supply. (See Sec. 3.7.)

On the other hand, centrifugal pumps can operate over a wide range of capacities, from near zero flow to well beyond the rated capacity. Because a centrifugal pump will always operate at the intersection of its head-capacity and system-head curves, the pump operating capacity may be altered either by throttling the pump discharge (hence altering the system-head curve) or by varying the pump speed (changing the pump head-capacity curve). This makes the centrifugal pump very flexible in a wide range of services and applications which require the pump to operate at capacities and heads differing considerably from the rated conditions. There are, however, some limitations imposed upon such operation by hydraulic, mechanical, or thermodynamic considerations (Subsec. 2.3.1).

Operation of Centrifugal Pumps at Reduced Flows   There are certain minimum operating flows which must be imposed on centrifugal pumps for either hydraulic or mechanical reasons. Four limiting factors must be considered: radial thrust, temperature rise, internal recirculation, and shape of the brake horsepower curve.

Radial thrust is discussed in Subsecs. 2.2.1 and 2.3.1. For sustained operation, it is imperative to adhere to the minimum flow limits recommended by the pump manufacturer, which depend on the specific casing design of the pump.

The thermodynamic problem that arises when a centrifugal pump is operated at extremely reduced flows is caused by the heating up of the liquid handled. The difference between the brake horsepower consumed and the water horsepower developed represents the power losses in the pump, except for a small amount lost in the pump bearings. These power losses are converted to heat and transferred to the liquid passing through the pump.

If the pump were to operate against a completely closed valve, the power losses would be equal to the shutoff brake horsepower, and since there would be no flow through the pump, all this power would go into heating the small quantity of liquid contained in the pump casing. The pump casing would heat up, and a certain amount of heat would be dissipated by radiation and convection to the atmosphere. However, because the temperature rise in the liquid pumped could be quite rapid, it is generally safer to ignore the dissipation of heat through radiation and the absorption of heat by the casing. Calculations for determining the temperature rise in the pumped liquid are given in Subsec. 2.3.1. The maximum permissible temperature rise in a centrifugal pump varies over a wide range, depending on the type of service and installation. For hot-water pumps, as on boiler-feed service, it is generally advisable to limit the temperature rise to about 15 °F (8 °C). As a general rule, the minimum permissible flow to hold the temperature rise in boiler-feed pumps to this value is 30 gpm for each 100 bhp (9.13 m^3/h per 100 kW) at shutoff. When the pump handles cold water, the temperature rise may be permitted to reach 50 or even 100 °F (28 or 56 °C). The minimum capacity based on thermodynamic considerations is then established as that capacity at which the temperature rise is the maximum permitted. Means and controls used to provide the necessary minimum flows are described in Subsec. 2.3.4.

There are also hydraulic considerations which may affect the minimum flow at which a centrifugal pump can operate. In recent years correlation has been developed between operation at low flows and the appearance of hydraulic pulsations both in the suction and in the discharge of centrifugal impellers. It has been proved that these pulsations are caused by the development of an internal recirculation at the inlet and discharge of an impeller at certain flows below the best-efficiency capacity. This subject is treated in Subsecs. 2.3.1 and 2.3.2. The pump manufacturer's recommendations on minimum flows dictated by these considerations must be followed.

Priming   With very few exceptions, no centrifugal pump should ever be started until it is fully primed, that is, until it has been filled with the liquid pumped and all the air contained in the pump has been allowed to escape. The exceptions involve self-priming pumps and some special large-capacity, low-head, and low-speed installations where it is not practical to prime the pump prior to starting; the priming takes place almost simultaneously with the starting in these cases. For further details, see Sec. 2.4.

Reciprocating pumps of the piston or plunger type are in principle self-priming. However, if quick starting is required, priming connections should be piped to a supply above the pump.

Positive displacement pumps of the rotating type, such as rotary or screw pumps, have clearances that allow the liquid in the pump to drain back to the suction. When pumping low-viscosity liquids, the pump may completely dry out when it is idle. In such cases a foot valve may be used to help keep the pump primed. Alternately, a vacuum device may be used to prime the pump. When handling liquids of higher viscosity, foot valves are usually not required because liquid is retained in the clearances and acts as a seal when the pump is restarted. However, before the initial start of a rotating positive displacement pump, some of the liquid to be pumped should be introduced through the discharge side of the pump to wet the rotating element.

The various methods and arrangements used for priming pumps are described in Sec. 2.4.

Final Checks before Start-Up   A few last-minute checks are recommended before a pump is placed into service for its initial start. The bearing covers should be removed, and the bearings should be flushed with kerosene and thoroughly cleaned. They should then be filled with new lubricant in accordance with the manufacturer's recommendations.

With the coupling disconnected, the driver should be tested again for correct direction of rotation. Generally an arrow on the pump casing indicates the correct rotation.

It must be possible to rotate the rotor of a centrifugal pump by hand, and in the case of a pump handling hot liquids, the rotor must be free to rotate with the pump cold or hot. If the rotor is bound or even drags slightly, do not operate the pump until the cause of the trouble is determined and corrected.

Starting and Stopping Procedures The steps necessary to start a centrifugal pump depend upon its type and upon the service on which it is installed. For example, standby pumps are generally held ready for immediate starting. The suction and discharge gate valves are held open, and reverse flow through the pump is prevented by the check valve in the discharge line.

The methods followed in starting are greatly influenced by the shape of the power-capacity curve of the pump. High- and medium-head pumps (low and medium specific speeds) have power curves that rise from zero flow to the normal capacity condition. Such pumps should be started against a closed discharge valve to reduce the starting load on the driver. A check valve is equivalent to a closed valve for this purpose, as long as another pump is already on the line. The check valve will not lift until the pump being started comes up to a speed sufficient to generate a head high enough to lift the check valve from its seat. If a pump is started with a closed discharge valve, the recirculation bypass line must be open to prevent overheating.

Low-head pumps (high specific speed) of the mixed-flow and propeller type have power curves that rise sharply with a reduction in capacity; they should be started with the discharge valve wide open against a check valve, if required, to prevent backflow.

Assuming that the pump in question is motor-driven, that its shutoff power does not exceed the safe motor power, and that it is to be started against a closed gate valve, the starting procedure is as follows:

1. Prime the pump, opening the suction valve, closing the drains, etc., to prepare the pump for operation.
2. Open the valve in the cooling water supply to the bearings, where applicable.
3. Open the valve in the cooling water supply if the stuffing boxes are water-cooled.
4. Open the valve in the sealing liquid supply if the pump is so fitted.
5. Open the warm-up valve of a pump handling hot liquids if the pump is not normally kept at operating temperature. When the pump is warmed up, close the valve.
6. Open the valve in the recirculating line if the pump should not be operated against dead shutoff.
7. Start the motor.
8. Open the discharge valve slowly.
9. Observe the leakage from the stuffing boxes and adjust the sealing liquid valve for proper flow to ensure the lubrication of the packing. If the packing is new, do not tighten up on the gland immediately, but let the packing run in before reducing the leakage through the stuffing boxes.
10. Check the general mechanical operation of the pump and motor.
11. Close the valve in the recirculating line once there is sufficient flow through the pump to prevent overheating.

If the pump is to be started against a closed check valve with the discharge gate valve open, the steps are the same, except that the discharge gate valve is opened some time before the motor is started.

In certain cases the cooling water to the bearings and the sealing water to the seal cages are provided by the pump. This, of course, eliminates the need for the steps listed for the cooling and sealing supply.

Just as in starting a pump, the stopping procedure depends upon the type and service of the pump. Generally the steps followed to stop a pump which can operate against a closed gate valve are

1. Open the valve in the recirculating line.
2. Close the gate valve.
3. Stop the motor.
4. Open the warm-up valve if the pump is to be kept at operating temperature.
5. Close the valve in the cooling water supply to the bearings and to water-cooled stuffing boxes.

6. If the sealing liquid supply is not required while the pump is idle, close the valve in this supply line.

7. Close the suction valve, open the drain valves, etc., as required by the particular installation or if the pump is to be opened up for inspection.

If the pump is of a type which does not permit operation against a closed gate valve, steps 2 and 3 are reversed.

In general, the starting and stopping of steam-turbine-driven pumps require the same steps and sequence prescribed for a motor-driven pump. As a rule, steam turbines have various drains and seals which must be opened or closed before and after operation. Similarly, many turbines require warming up before starting. Finally some turbines require turning gear operation if they are kept on the line ready to start up. The operator should therefore follow the steps outlined by the turbine manufacturer in starting and stopping the turbine.

Most of the steps listed for starting and stopping centrifugal pumps are equally applicable to positive displacement pumps. There are, however, two notable exceptions:

1. Never operate a positive displacement pump against a closed discharge. If the gate valve on the discharge must be closed, always start the pump with the recirculation bypass valve open.

2. Always open the steam cylinder drain cocks of a steam reciprocating pump before starting, to allow condensate to escape and to prevent damage to the cylinder heads.

Auxiliary Services on Standby Pumps   Standby pumps are frequently started up from a remote location, and several methods of operation are available for the auxiliary services, such as the cooling water supply to the bearings or to water-cooled stuffing boxes:

1. A constant flow may be kept through the bearing jackets or oil coolers and through the stuffing box lantern rings, whether the pump is running or on standby service.

2. The service connections may be opened automatically whenever the pump is started up.

3. The service connections may be kept closed while the pump is idle, and the operator may be instructed to open them shortly after the pump has been put on the line automatically.

The choice among these methods must be dictated by the specific circumstances surrounding each case. There are, however, certain cases where sealing liquid supply to the pump stuffing boxes must be maintained whether the pump is running or not. This is the case when the pump handles a liquid which is corrosive to the packing or which may crystallize and deposit on the shaft sleeves. It is also the case when the sealing supply is used to prevent air infiltration into a pump when it is operating under a vacuum.

Restarting Motor-Driven Pumps after Power Failure   Assuming that power failure will not cause the pump to go into reverse rotation, that is, that a check valve will protect the pump against reverse flow, there is generally no reason why the pump should not be permitted to restart once current has been re-established. Whether the pump will start again automatically when power is restored will depend on the type of motor control used. (Subsection 2.3.1 and Sec. 8.1 give reasons why some pumps should not be started in reverse.)

Because pumps operating on a suction lift may lose their prime during the time that power is off, it is preferable to use starters with low load protection for such installations to prevent an automatic restart. This does not apply, of course, if the pumps are automatically primed or if some protection device is incorporated so that the pump cannot run unless it is primed.

MAINTENANCE

Because of the wide variation in pump types, sizes, designs, and materials of construction, these comments on maintenance are restricted to those types of pumps most commonly encountered. The manufacturer's instruction books must be carefully studied before any attempt is made to service a particular pump.

Daily Observation of Pump Operation   When operators are on constant duty, hourly and daily inspections should be made and any irregularities in the operation of a pump should be reported immediately. This applies particularly to changes in the sound of a running pump, abrupt changes in bearing temperatures, and stuffing box leakage. A check of the pressure gages and of the flowmeter, if installed, should be made hourly. If recording instruments are provided, a daily check should be made to determine whether the capacity, pressure, or power consumption indicates that further inspection is required.

Semiannual Inspection   The following should be done every six months. The free movement of stuffing box glands should be checked, gland bolts should be cleaned and oiled, and the packing should be inspected to determine whether it requires replacement. The pump and driver alignment should be checked and corrected if necessary. Oil-lubricated bearings should be drained and refilled with fresh oil. Grease-lubricated bearings should be checked to see that they contain the correct amount of grease and that it is still of suitable consistency.

Annual Inspection   A very thorough inspection should be made once a year. In addition to the semiannual procedure, bearings should be removed, cleaned, and examined for flaws. The bearings housings should be carefully cleaned. Antifriction bearings should be examined for scratches and wear. Immediately after cleaning and inspection, antifriction bearings should be coated with oil or grease.

The packing should be removed and the shaft sleeves—or shaft, if no sleeves are used—should be examined for wear.

When the coupling halves are disconnected for the alignment check, the vertical shaft movement of a pump with sleeve bearings should be checked at both ends with the packing removed. Any vertical movement exceeding 150% of the original clearance requires an investigation to determine the cause. The endplay allowed by the bearings should also be checked. If it exceeds that recommended by the manufacturer, the cause should be determined and corrected.

All auxiliary piping, such as drains, sealing water piping, and cooling water piping, should be checked and flushed. Auxiliary coolers should be flushed and cleaned.

The pump stuffing boxes should be repacked, and the pump and driver should be realigned and reconnected.

All instruments and flow-metering devices should be recalibrated, and the pump should be tested to determine whether proper performance is being obtained. If internal repairs are made, the pump should again be tested after completion of the repairs.

Complete Overhaul   It is difficult to make general rules about the frequency of complete pump overhauls, as it depends on the pump service, the pump construction and materials, the liquid handled, and the economic evaluation of overhaul costs versus the cost of power losses resulting from increased clearances or of unscheduled downtime. Some pumps on very severe service may need a complete overhaul monthly, while other applications require overhauls only every two to four years or even less frequently.

A pump should not be opened for inspection unless either factual or circumstantial evidence indicates that overhaul is necessary. Factual evidence implies that the pump performance has fallen off significantly or that noise or driver overload indicates trouble. Circumstantial evidence refers to past experience with the pump in question or with similar equipment on similar service.

In order to ensure rapid restoration to service in the event of an unexpected overhaul, an adequate store of spare parts should be maintained at all times.

The relative complexity of the repairs, the facilities available at the site, and many other factors enter into the decision whether the necessary repairs will be carried out at the installation or at the pump manufacturer's plant.

Spare and Repair Parts   The severity of the service in which a pump is used will determine, to a great extent, the minimum number of spare parts which should be carried in stock at the installation site. Unless prior experience is available, the pump manufacturer should be consulted on this subject. As an insurance against delays, spare parts should be purchased when the pump is. Depending upon the contemplated method of overhaul, certain replacement parts may have to be supplied either oversized or undersized instead of the size used in the original unit.

When ordering spare parts after a pump has been in service, the manufacturer should always

TABLE 1   Check Chart for Centrifugal Pump Troubles

Symptom	Possible cause of trouble (each number is defined in the list below)
Pump does not deliver water	1, 2, 3, 4, 6, 11, 14, 16, 17, 22, 23
Insufficient capacity delivered	2, 3, 4, 5, 6, 7, 8, 9, 10, 11, 14, 17, 20, 22, 23, 29, 30, 31
Insufficient pressure developed	5, 14, 16, 17, 20, 22, 29, 30, 31
Pump loses prime after starting	2, 3, 5, 6, 7, 8, 11, 12, 13
Pump requires excessive power	15, 16, 17, 18, 19, 20, 23, 24, 26, 27, 29, 33, 34, 37
Stuffing box leaks excessively	13, 24, 26, 32, 33, 34, 35, 36, 38, 39, 40
Packing has short life	12, 13, 24, 26, 28, 32, 33, 34, 35, 36, 37, 38, 39, 40
Pump vibrates or is noisy	2, 3, 4, 9, 10, 11, 21, 23, 24, 25, 26, 27, 28, 30, 35, 36, 41, 42, 43, 44, 45, 46, 47
Bearings have short life	24, 26, 27, 28, 35, 36, 41, 42, 43, 44, 45, 46, 47
Pump overheats and seizes	1, 4, 21, 22, 24, 27, 28, 35, 36, 41

Suction troubles

1. Pump not primed
2. Pump or suction pipe not completely filled with liquid
3. Suction lift too high
4. Insufficient margin between suction pressure and vapor pressure
5. Excessive amount of air or gas in liquid
6. Air pocket in suction line
7. Air leaks into suction line
8. Air leaks into pump through stuffing boxes
9. Foot valve too small
10. Foot valve partially clogged
11. Inlet of suction pipe insufficiently submerged
12. Water seal pipe plugged
13. Seal cage improperly located in stuffing box, preventing sealing fluid from entering space to form seal

System troubles

14. Speed too low
15. Speed too high
16. Direction of rotation wrong
17. Total head of system higher than design head of pump
18. Total head of system lower than pump design head
19. Specific gravity of liquid different from design
20. Viscosity of liquid different from design
21. Operation at very low capacity
22. Parallel operation of pumps unsuitable for such operation

Mechanical troubles

23. Foreign matter in impeller
24. Misalignment
25. Foundations not rigid
26. Shaft bent

Mechanical troubles (*cont.*)

27. Rotating part rubbing on stationary part
28. Bearings worn
29. Wearing rings worn
30. Impeller damaged
31. Casing gasket defective, permitting internal leakage
32. Shaft or shaft sleeves worn or scored at packing
33. Packing improperly installed
34. Type of packing incorrect for operating conditions
35. Shaft running off center because of worn bearings or misalignment
36. Rotor out of balance, causing vibration
37. Gland too tight, resulting in no flow of liquid to lubricate packing
38. Cooling liquid not being provided to water-cooled stuffing boxes
39. Excessive clearance at bottom of stuffing box between shaft and casing, causing packing to be forced into pump interior
40. Dirt or grit in sealing liquid, leading to scoring of shaft or shaft sleeve
41. Excessive thrust caused by mechanical failure inside pump or by failure of hydraulic balancing device, if any
42. Excessive grease or oil in antifriction bearing housing or lack of cooling, causing excessive bearing temperature
43. Lack of lubrication
44. Improper installation of antifriction bearings (damage during assembly, incorrect assembly of stacked bearings, use of unmatched bearings as a pair, etc.)
45. Dirt in bearings
46. Rusting of bearings from water in housing
47. Excessive cooling of water-cooled bearing, resulting in condensation of atmospheric moisture in bearing housing

be given the pump serial number and size (stamped on the nameplate). This information is essential in identifying the pump exactly and in furnishing repair parts of correct size and material.

Record of Inspections and Repairs   The working schedule of the semiannual and annual inspections should be entered on individual pump maintenance cards, which should contain a complete record of all the items requiring attention. These cards should also contain space for comments and observations on the conditions of the parts to be repaired or replaced, on the rate and appearance of wear, and on the repair methods followed. In many cases it is advisable to photograph badly worn parts before they are repaired.

In all cases complete records of the cost of maintenance and repairs should be kept for each pump, together with a record of its operating hours. A study of these records will generally reveal whether a change in materials or even a minor change in construction may not be the most economical course of action.

Diagnosis of Pump Troubles   Pump operating troubles may be either hydraulic or mechanical. In the first category, a pump may fail to deliver liquid, it may deliver an insufficient capacity or develop insufficient pressure, or it may lose its prime after starting. In the second category, it may consume excessive power, or symptoms of mechanical difficulties may develop at the stuffing boxes or at the bearings, or vibration, noise, or breakage of some pump parts may occur.

There is a definite interdependence between some difficulties of both categories. For example, increased wear at the running clearances must be classified as a mechanical trouble, but it will result in a reduction of the net pump capacity—a hydraulic symptom—without necessarily causing a mechanical breakdown or even excessive vibration. As a result, it is most useful to classify symptoms and causes separately and to list for each symptom a schedule of potential contributory causes. Such a diagnostic analysis is presented in Tables 1 to 4. Additionally, the following parts of this handbook should also be referred to for assistance in diagnosing pump hydraulic and mechanical problems and possible solutions: Subsecs. 2.2.2, 2.2.3, 2.3.2 and 2.3.3 and Secs. 3.6 and 8.4.

TABLE 2   Check Chart for Rotary Pump Troubles

Symptom	Possible cause of trouble (each number is defined in the list below)
Pump fails to discharge	1, 2, 3, 4, 5, 6, 8, 9, 16
Pump is noisy	6, 10, 11, 17, 18, 19
Pump wears rapidly	11, 12, 13, 20, 24
Pump not up to capacity	3, 5, 6, 7, 9, 16, 21, 22
Pump starts, then loses suction	1, 2, 6, 7, 10
Pump takes excessive power	14, 15, 17, 20, 23

Suction troubles

1. Pump not properly primed
2. Suction pipe not submerged
3. Strainer clogged
4. Foot valve leaking
5. Suction lift too high
6. Air leaking into suction
7. Suction pipe too small

System problems

8. Wrong direction of rotation
9. Low speed
10. Insufficient liquid supply
11. Excessive pressure
12. Grit or dirt in liquid

System problems (*cont.*)

13. Pump runs dry
14. Viscosity higher than specified
15. Obstruction in discharge line

Mechanical troubles

16. Pump worn
17. Drive shaft bent
18. Coupling out of balance or alignment
19. Relief valve chatter
20. Pipe strain on pump casing
21. Air leak at packing
22. Relief valve improperly seated
23. Packing too tight
24. Corrosion

TABLE 3 Check Chart for Reciprocating Pump Troubles

Symptom	Possible cause of trouble (each number is defined in the list below)
Liquid end noise	1, 2, 7, 8, 9, 10, 14, 15, 16
Power end noise	17, 18, 19, 20
Overheated power end	10, 19, 21, 22, 23, 24
Water in crankcase	25
Oil leak from crankcase	26, 27
Rapid packing or plunger wear	11, 12, 28, 29
Pitted valves or seats	3, 11, 30
Valves hanging up	31, 32
Leak at cylinder valve hole plugs	10, 13, 33, 34
Loss of prime	1, 4, 5, 6

Suction troubles

1. Insufficient suction pressure
2. Partial loss of prime
3. Cavitation
4. Lift too high
5. Leaking suction at foot valve
6. Acceleration head requirement too high

System problems

7. System shocks
8. Poorly supported piping, abrupt turns in piping, pipe too small, piping misaligned
9. Air in liquid
10. Overpressure or overspeed
11. Dirty liquid
12. Dirty environment
13. Water hammer

Mechanical troubles

14. Valves broken or badly worn
15. Packing worn
16. Obstruction under valve
17. Main bearings loose
18. Bearings worn
19. Oil level low
20. Plunger loose
21. Main bearings tight
22. Ventilation inadequate
23. Belts too tight
24. Driver misaligned
25. Condensation
26. Seals worn
27. Oil level too high
28. Pump not level and rigid
29. Packing loose
30. Corrosion
31. Valve binding
32. Valve spring broken
33. Cylinder plug loose
34. O-ring seal damaged

TABLE 4 Check Chart for Steam-Pump Troubles

Symptoms	Possible cause of trouble (each number is defined in the list below)
Pump does not develop rated pressure:	4, 5, 7, 8
Pump loses capacity after starting:	1, 2, 6
Pump vibrates:	9, 10, 11, 14
Pump has short strokes:	12, 13, 14
Pump operation is erratic:	1, 2, 3, 6

Suction troubles

1. Suction line leaks
2. Suction lift too high
3. Cavitation

System problems

4. Low steam pressure
5. High exhaust pressure
6. Entrained air or vapors in liquid

TABLE 4 Check Chart for Steam-Pump Troubles *(continued)*

Mechanical troubles	Mechanical troubles *(cont.)*
7. Worn piston rings in steam end	11. Piping not supported
8. Binding piston rings in liquid end	12. Excessive steam cushioning
9. Misalignment	13. Steam valves out of adjustment
10. Foundation not rigid	14. Liquid piston packing too tight

FURTHER READING

Hydraulic Institute: *Standards for Centrifugal, Rotary, and Reciprocating Pumps,* 14th ed., Cleveland, 1983.

Karassik, I. J., *Centrifugal Pump Clinic,* Marcel Dekker, New York, 1981.

Karassik, I. J., and R. Carter: *Centrifugal Pumps,* McGraw-Hill, New York, 1960.

PUMP TESTING

Wesley W. Beck

Ever since a device was first used to pump or lift water, there has been pump testing of one sort or another. Each improvement in pumping devices was accepted only after being tested, which was the proof of its worthiness. As pumping equipment has become more refined, so has the art of pump testing, both in the shop or laboratory and in the field. For very large pumps, model testing is being used to develop the optimum refinement in prototype design.

Every pump, regardless of size or classification, should be tested in some way before final acceptance by the purchaser. If not, the user does not have any way of knowing that all requirements have been fulfilled. What tests to run and what methods to use depend on the ultimate purpose of the tests, which normally have one of two objectives:

1. To check improvement in design or operation
2. To determine if contractual commitments have been met, thus making possible the comparison of specified, predicted, and actual performance

In most cases the manufacturer supplies a test report and certifies the characteristics of the pump being furnished. Even these can be given a cursory check by the customer from time to time to give a record of performance or an indication of the need for replacement or overhaul. If at all possible, the pump should be tested as installed, with repeat tests from time to time to check operation.

The main object of this chapter is to present a set of procedures and rules for conducting, computing, and reporting on tests of pumping units and for obtaining the head, capacity, power, efficiency, and suction requirements of a pump.

CLASSIFICATION OF TESTS

Pump tests should be classified as follows.

Shop tests are also called laboratory, manufacturer's, or factory acceptance tests. They are conducted in the pump manufacturer's plant under geometrically similar, ideal, and controlled conditions and are usually assumed to be the most accurate tests.

Field tests are made with the pumping unit installed in its exact environment and operating under existing field or ultimate conditions. The accuracy and reliability of field testing depend on the instrumentation used, installation, and advance planning during the design stages of the installation. By mutual agreement, field tests can be used as acceptance tests.

Index tests are a form of field testing usually made to serve as a standard of comparison for wear, changing conditions, or overhaul evaluation. Index tests should always be run by the same procedures, instruments, and personnel where possible, and a very accurate record and log of events should be kept to give as complete and comparable a history of the results as possible.

Model tests precede the design of the prototype and are usually quite accurate. They supplement or complement field tests of the prototype for which the model was made. The role of the model test must be clearly established as early in the design as possible, preferably in the specification or invitation to bid. Model tests may be used when very large units are involved, when the performances of several models must be compared, and when an advance indication of prototype design is required.

DEFINITIONS, SYMBOLS, AND UNITS

For a detailed discussion of letter symbols, definitions, description of terms, and table of letter symbols in general use, the user is referred to ASME Power Test Code[1] and to SI Units—A commentary on p. xxi of this handbook.

Standard Units Used in Pump Testing   The following definitions and quantities from the Hydraulic Institute standards[2] are used throughout the industry in pump testing.

VOLUME   The standard units of volume are the U.S. gallon and the cubic foot (cubic meter). The standard U.S. gallon contains 231.0 in^3 (0.00379 m^3), and 1 ft^3 = 7.4805 gal (0.028 m^3). Rate of flow is expressed in gallons per minute (cubic meters per hour), cubic feet per second, or million gallons per 24-h day. The specific weight (mass) w of pure water at 68°F (20°C), at sea level and 40° latitude, is 62.315 lb/ft^3 (0.998 kg/liter). For other temperatures or locations, proper specific weight corrections should be made. See Table 1 and the appropriate ASME power test codes.

HEAD   The unit for measuring head is the foot (meter). The relation between a pressure expressed in pounds per square inch (kilopascals) and one expressed in feet (meters) of head is

in USCS units
$$\text{Head, ft} = \text{lb/in}^2 \times \frac{144}{w}$$

in SI units
$$\text{Head, m} = \text{kPa} \times \frac{0.102}{w}$$

where w = specific weight of liquid being pumped under pumping conditions, lb/ft^3 (kg/l)

All pressure readings must be converted to feet (meters) of the liquid being pumped, referenced to a datum elevation, which is defined as follows. For a horizontal shaft unit, the datum elevation is the centerline of the pump shaft (Fig. 1). For vertical-shaft single-suction pumps, it is the entrance eye to the first-stage impeller (Fig. 2). For vertical-shaft double-suction pumps, it is the impeller-discharge horizontal centerline (Fig. 3).

VELOCITY HEAD   The velocity head h_v is computed from the average velocity V, obtained by dividing the flow by the pipe cross-sectional area. Velocity head is determined at the point of gage connection and is expressed by the formula

$$h_v = \frac{V^2}{2g}$$

where V = velocity in the pipe, ft/s (m/s)
g = acceleration due to gravity = 32.17 ft/s^2 (9.81 m/s^2) (Table 2)

FLOODED SUCTION   Flooded suction implies that the liquid must flow from an atmospherically vented source to the pump without the average or minimum pressure at the pump datum dropping below atmospheric pressure with the pump operating at specified capacity.

TOTAL SUCTION LIFT   Suction lift exists where the total suction head is below atmospheric pressure. Total suction lift h_s is the reading of a liquid manometer or pressure gage at the suction nozzle of the pump, converted to feet (meters) of liquid, and referred to datum minus the velocity head at the point of gage attachment.

TOTAL SUCTION HEAD   Suction head exists when the total suction head h_s is above atmospheric pressure. Total suction head is the reading of a gage at the suction of the pump converted to feet (meters) of liquid and referred to datum plus the velocity head at the point of gage attachment.

TOTAL DISCHARGE HEAD   Total discharge head h_d is the reading of a pressure gage at the discharge of the pump, converted to feet (meters) of liquid and referred to datum plus the velocity head at the point of gage attachment.

TOTAL HEAD   Total head H is the measure of the work increase per pound kilogram of liquid, imparted to the liquid by the pump, and is therefore the algebraic difference between the total discharge head and the total suction head. Total head, as determined on test where suction lift exists, is the sum of the total discharge head and total suction lift. Where positive suction head exists, the total head is the total discharge head minus the total suction head.

TABLE 1 Specific Weight of Water in Air

Latitude	Temperature, °F (°C)							
	32 (0)	40 (4.4)	50 (10)	60 (16)	70 (21)	80 (27)	90 (32)	100 (38)

At sea level

Latitude	32 (0)	40 (4.4)	50 (10)	60 (16)	70 (21)	80 (27)	90 (32)	100 (38)
0°	62.1741	62.1823	62.1654	62.1227	62.0578	61.9729	61.8701	61.7514
10	62.1840	62.1921	62.1753	62.1325	62.0677	61.9828	61.8800	61.7612
20	62.2125	62.2206	62.2038	62.1610	62.0961	62.0112	61.9083	61.7895
30	62.2562	62.2643	62.2475	62.2046	62.1397	62.0547	61.9518	61.8328
40	62.3098	62.3179	62.3011	62.2582	62.1932	62.1082	62.0051	61.8861
50	62.3670	62.3751	62.3582	62.3153	62.2503	62.1652	62.0620	61.9429
60	62.4208	62.4289	62.4120	62.3691	62.3040	62.2188	62.1156	61.9963
70	62.4647	62.4728	62.4559	62.4130	62.3478	62.2626	62.1593	62.0399

At 2000 ft

Latitude	32 (0)	40 (4.4)	50 (10)	60 (16)	70 (21)	80 (27)	90 (32)	100 (38)
0°	62.1665	62.1747	62.1578	62.1151	62.0502	61.9654	61.8626	61.7438
10	62.1764	62.1845	62.1677	62.1250	62.0601	61.9752	61.8724	61.7537
20	62.2049	62.2130	62.1962	62.1534	62.0885	62.0036	61.9008	61.7819
30	62.2486	62.2567	62.2399	62.1970	62.1321	62.0471	61.9442	61.8253
40	62.3022	62.3103	62.2935	62.2506	62.1856	62.1006	61.9976	61.8786
50	62.3594	62.3675	62.3506	62.3078	62.2427	62.1576	62.0545	61.9354
60	62.4132	62.4213	62.4044	62.3615	62.2964	62.2112	62.1080	61.9888
70	62.4571	62.4652	62.4484	62.4054	62.3402	62.2550	62.1517	62.0324

At 4000 ft

Latitude	32 (0)	40 (4.4)	50 (10)	60 (16)	70 (21)	80 (27)	90 (32)	100 (38)
0°	62.1588	62.1669	62.1501	62.1073	62.0424	61.9576	61.8549	61.7361
10	62.1686	62.1767	62.1599	62.1172	62.0523	61.9675	61.8647	61.7459
20	62.1971	62.2052	62.1884	62.1456	62.0807	61.9959	61.8930	61.7742
30	62.2408	62.2489	62.2321	62.1893	62.1243	62.0394	61.9365	61.8176
40	62.2944	62.3025	62.2857	62.2428	62.1779	62.0929	61.9899	61.8709
50	62.3516	62.3597	62.3429	62.3000	62.2349	62.1498	62.0467	61.9277
60	62.4054	62.4135	62.3967	62.3537	62.2886	62.2035	62.1003	61.9811
70	62.4493	62.4574	62.4406	62.3976	62.3325	62.2472	62.1440	62.0247

At 6000 ft

Latitude	32 (0)	40 (4.4)	50 (10)	60 (16)	70 (21)	80 (27)	90 (32)	100 (38)
0°	62.1508	62.1589	62.1421	62.0993	62.0345	61.9497	61.8469	61.7282
10	62.1607	62.1688	62.1520	62.1092	62.0444	61.9595	61.8568	61.7380
20	62.1891	62.1972	62.1804	62.1377	62.0728	61.9879	61.8851	61.7663
30	62.2328	62.2409	62.2241	62.1813	62.1164	62.0315	61.9286	61.8097
40	62.2864	62.2946	62.2777	62.2349	62.1699	62.0849	61.9819	61.8630
50	62.3436	62.3517	62.3349	62.2920	62.2270	62.1419	62.0388	61.9198
60	62.3974	62.4055	62.3887	62.3458	62.2807	62.1955	62.0924	61.9732
70	62.4413	62.4495	62.4326	62.3896	62.3245	62.2393	62.1361	61.0168

At 8000 ft

Latitude	32 (0)	40 (4.4)	50 (10)	60 (16)	70 (21)	80 (27)	90 (32)	100 (38)
0°	62.1426	62.1507	62.1339	62.0912	62.0264	61.9416	61.8388	61.7201
10	62.1525	62.1606	62.1438	62.1011	62.0362	61.9514	61.8487	61.7300
20	62.1810	62.1891	62.1723	62.1295	62.0646	61.9798	61.8770	61.7582
30	62.2246	62.2328	62.2159	62.1731	62.1082	62.0233	61.9205	61.8016
40	62.2783	62.2864	62.2696	62.2267	62.1618	62.0768	61.9738	61.8549
50	62.3354	62.3436	62.3267	62.2838	62.2188	62.1338	62.0307	61.9117
60	62.3893	62.3974	62.3805	62.3376	62.2725	62.1874	62.0843	61.9651
70	62.4332	62.4413	62.4244	62.3815	62.3164	62.2312	62.1280	62.0087

TABLE 1 Specific Weight of Water in Air (continued)

	Temperature, °F (°C)							
Latitude	32 (0)	40 (4.4)	50 (10)	60 (16)	70 (21)	80 (27)	90 (32)	100 (38)
	At 10,000 ft							
0°	62.1343	62.1424	62.1256	62.0828	62.0180	61.9333	61.8305	61.7119
10	62.1442	62.1523	62.1355	62.0927	62.0279	61.9431	61.8404	61.7217
20	62.1726	62.1807	62.1639	62.1212	62.0563	61.9715	61.8687	61.7500
30	62.2163	62.2244	62.2076	62.1648	62.0999	62.0150	61.9122	61.7934
40	62.2699	62.2781	62.2612	62.2184	62.1534	62.0685	61.9656	61.8466
50	62.3271	62.3352	62.3184	62.2755	62.2105	62.1255	62.0224	61.9034
60	62.3809	62.3890	62.3722	62.3293	62.2642	62.1791	62.0760	61.9568
70	62.4248	62.4330	62.4161	62.3732	62.3080	62.2229	62.1197	62.0005
	At 12,000 ft							
0°	62.1258	62.1339	62.1171	62.0743	62.0095	61.9248	61.8221	61.7035
10	62.1357	62.1438	62.1270	62.0842	62.0194	61.9347	61.8319	61.7133
20	62.1641	62.1722	62.1554	62.1127	62.0478	61.9630	61.8603	61.7416
30	62.2078	62.2159	62.1991	62.1563	62.0914	62.0066	61.9037	61.7849
40	62.2614	62.2695	62.2527	62.2099	62.1450	62.0600	62.0140	61.8382
50	62.3186	62.3267	62.3099	62.2670	62.2020	62.1170	62.0140	61.8950
60	62.3724	62.3805	62.3637	62.3208	62.2557	62.1706	62,0675	61.9484
70	62.4163	62.4245	62.4076	62.3647	62.2996	62.2144	62.1112	61.9920

NOTE: All values are in pounds per cubic foot; 1 lb/ft^3 = 16 kg/m^3; 1 ft = 0.3048 m.
Density of water from Source 1, p. 296.
Gravity formula from Source 4, p. 488:

$G_{\emptyset}$ = 980.616 (1 − 0.0026373 cos 2$\emptyset$ + 0.0000059 cos^2 2$\emptyset$)

Altitude correction = 0.0003086 × altitude in meters.
Standard gravity = 980.665 cm/s^2.
Density of air from Source 2:

Density = 0.001225 (1 − 0.0065H|288.16)$^{4.2561}$

where H is in meters.

SOURCES:
   1. *Smithsonian Physical Tables*, 9th rev. ed.
   2. National Advisory Committee for Aeronautics, TN 3182.
   3. American Society of Mechanical Engineers, PTC 2-1971.
   4. *Smithsonian Meteorological Tables*, 6th rev. ed.

NET POSITIVE SUCTION HEAD   The net positive suction head (NPSH) h_{sv} is the total suction head in feet (meters) of liquid absolute determined at the suction nozzle and referred to datum less the vapor pressure of the liquid in feet (meters) absolute.

DRIVER INPUT   The driver input ehp is the input to the driver expressed in horsepower (kilowatts). Usually this is electric input horsepower (kilowatts).

PUMP INPUT   Pump input bhp is the power delivered to the pump shaft and is designated as brake horsepower (brake kilowatts).

LIQUID OR WATER POWER   Water power whp (wkW) is the useful work delivered by the pump and is usually expressed by the formula

TABLE 2 Variation of Acceleration of Gravity with Latitude and Altitude

Latitude	Altitude above mean sea level, ft						
	0	2000	4000	6000	8000	10,000	12,000
0°	32.0878	32.0816	32.0754	32.0693	32.0631	32.0569	32.0508
10	32.0929	32.0867	32.0805	32.0744	32.0682	32.0620	32.0558
20	32.1076	32.1014	32.0952	32.0890	32.0829	32.0767	32.0705
30	32.1301	32.1239	32.1177	32.1115	32.1054	32.0992	32.0930
40	32.1577	32.1515	32.1454	32.1392	32.1330	32.1269	32.1207
50	32.1872	32.1810	32.1748	32.1687	32.1625	32.1563	32.1501
60	32.2149	32.2087	32.2026	32.1964	32.1902	32.1841	32.1779
70	32.2375	32.2314	32.2252	32.2190	32.2129	32.2067	32.2005

NOTE: All values are in feet per second per second; $1 \text{ ft/s}^2 = 0.3048 \text{ m/s}^2$; $1 \text{ ft} = 0.3048 \text{ m}$.
Gravity $= 980.616 (1 - 0.0026373 \cos 2\phi + 0.0000059 \cos^2 20)(1.0/30.48)$
Correction for altitude $= -0.003086 \text{ ft/s}^2/1{,}000 \text{ ft}$
The international standard value of gravity adopted by the International Commission on Weights and Measures is 980.665 cm/s^2 (32.17405 ft/s^2) at sea level and approximately latitude 45°.

SAMPLE COMPUTATION:

Given:
   Altitude $= 12{,}000$ ft (3657.60 m)
   Latitude $= 70°$
   Water temperature $= 40°F$ (4.4°C)
Altitude correction for gravity:
   $0.0003086 \times 3657.60 = 1.128735 \text{ cm/s}^2$
Gravity corrected for latitude and altitude:
   Gravity $= 980.616 (1 - 0.0026373 \cos 140° - 0.0000059 \cos^2 140°) - 1.128735 = 981.471784 \text{ cm/s}^2$
Density corrected for gravity:
   Density $= (0.9999983 \times 981.471784/980.6650 \times 1.000028) = 1.0007930 \text{ g/cm}^3$
Correcting for buoyancy of air:
   Density $= 1.0007930 - 0.0008491 = 0.9999438 \text{ g/cm}^3$
Density in USCS units:
   Density $= 0.9999438 \times 62.4279606 = 62.4244543 \text{ lb/ft}^3$

SOURCES:
   1. "Smithsonian Physical Tables," 9 Rev. Ed.
   2. American Society of Mechanical Engineers, PTC 2—1971

in USCS units
$$\text{whp} = \frac{(\text{sp. gr.})QH}{3960}$$

in SI units
$$\text{wkW} = 9.8QH(\text{sp. gr.})$$

where sp. gr. $=$ specific gravity of liquid, referred to water at 68°F (20°C)
   $Q =$ flow rate, gpm(m³/h)
   $H =$ total head, ft (m)

This formula is derived on p. 13.34.

EFFICIENCY   Pump efficiency E_p is the ratio of the power delivered by the pump to the power supplied to the pump shaft, that is, the ratio of the liquid power (also known as water power) to the brake power expressed in percent:

in USCS units
$$E_p = \frac{\text{whp}}{\text{bhp}} \times 100$$

in SI units
$$E_p = \frac{\text{wkW}}{\text{kW}} \times 100$$

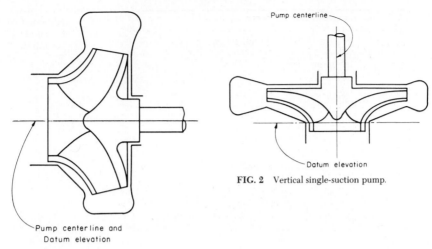

FIG. 2   Vertical single-suction pump.

FIG. 1   Horizontal pump.

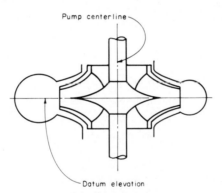

FIG. 3   Vertical double-suction pump.

Overall efficiency E_o is the ratio of the power delivered by the pump to the power supplied to the input side of the pump driver, that is, the ratio of the output power to the input power to the driver:

in USCS units

$$E_o = \frac{\text{whp}}{\text{ehp}} \times 100$$

in SI units

$$E_o = \frac{\text{wkW}}{\text{kW}} \times 100$$

Prime Mover Ratings   The prime movers for driving pumps are rated according to established standards. For example, for electric motor drivers, see the standards of the latest edition of the National Electrical Manufacturers Association.

ACCURACY AND TOLERANCES

Accuracy   The accuracy to which tests can be made depends on the instruments used, their proper installation, the skill of the test engineer, and the shop tests for the simulation of field

Quantity to be measured	Measuring device	Calibrated limit of accuracy plus or minus, %
Capacity	Venturi meter	¾
	Nozzle	1
	Pitot tube	1½
	Orifice	1¼
	Disk	2
	Piston	¼
	Volume or weight—tank	1
	Propeller meter	4
Head	Electric sounding line	¼
	Air line	½
	Liquid manometer, 3- to 5-in (75- to 127-mm) deflections	¾
	Liquid manometer, over 5-in (127-mm) deflections	½
	Bourdon gage—5 in (127-mm) min dial:	
	¼–½ full scale	1
	½–¾ full scale	¾
	Over ¾ scale	½
Power input	Watt-hour meter and stopwatch	1½
	Portable recording wattmeter	1½
	Test precision wattmeter:	
	¼–½ scale	¾
	½–¾ scale	½
	Over ¾ scale	¼
	Clamp on ammeter	4
Speed	Revolution counter and stopwatch	1¼
	Handheld tachometer	1¼
	Stroboscope	1½
	Automatic counter and stopwatch	½
Voltage	Test meter:	
	¼–½	1
	½–¾	¾
	¾–full	½
	Rectifier voltmeter	5

Source: ANSI B-58.1 (AWWAE 101-61).

FIG. 4   Limits of accuracy of pump test measuring devices in field use.

conditions. The test engineer must have sufficient knowledge of the characteristics and limitations of the test instruments to obtain maximum accuracy when using them, along with a thorough understanding of the pumps, prime movers, controls, and installation peculiarities to interpret the results. For shop testing, the acceptable deviations and fluctuations of the instrumented test readings are given in Table 2 of ASME Power Test Code.[1] These deviations are not to be misconstrued as tolerances, which must be spelled out in the specifications. The limits of accuracy of pump test measuring devices for use in field testing are shown in Fig. 4. Using these limits, the combined accuracy of the efficiency is the square root of the quantity [square of the head accuracy plus square of the capacity accuracy plus square of the power input accuracy]:

$$A_c = \sqrt{(\pm H^2) + (\pm Q^2) + (\pm \overline{\text{ehp}^2})} \quad \text{(percent)}$$

Pump speed and voltage are not required for efficiency computations, and so the values for these are not included in this formula.

Instrumentation   All instruments should be calibrated before the tests, and all calibration and correction data or curves should be prepared in advance. Where required, a certified calibration curve showing the calibration of the instrument, including any procedures for establishing a coefficient, should be furnished before testing begins. The specifications should be explicit in regard to a waiving of these calibration requirements. After testing, all instruments should be recalibrated. Any differences between before and after calibration values must be resolved either by retest or by acceptable variations being spelled out in the specifications.

Tolerances   The tolerances in pump performance permitted are usually given in the specifications. The user can and should make these requirements known before the order for pumping apparatus is placed. The test tolerances permitted by the Hydraulic Institute are quite commonly used; they state that no minus tolerance or margin shall be allowed on capacity, total head, or efficiency at the rated condition. Also, a plus tolerance of not more than 10% of rated capacity shall be allowed at the rated head and speed.

The tolerances are quite easy to meet; they protect the user from getting pumps that are too small to do the job and also from getting oversized pumps and drivers that would increase building, installation, and operating costs. These tolerances also give the manufacturer liberal leeway when impeller trim is required to meet the specified conditions.

TEST REQUISITES

Operating Conditions   The primary factors affecting the operation of a pump are the inlet (suction), outlet (discharge or total head), and speed. The secondary factors are physical and climatic variables, such as the temperature, viscosity, specific weight, and turbidity of the liquid being pumped and the elevation of the pumping system above sea level. In some installations it is impossible to measure discharge or even head accurately. In these instances good shop tests are essential. It follows then that, in order for the shop test to predict the field performance of a pump, the field operating, installation, and suction conditions should be simulated.

The inlet passages are critical, and the sump, where used, on the suction lift must be duplicated as closely as possible. During the shop tests no total suction head less than specified should be permitted, nor should the suction head exceed the specified amount in cases where cavitation or possibly operating "in the break" could occur.

For field installations above sea level, the difference in elevation between the shop test site and the field installation must be taken into account by reducing to the barometric pressure at the specified elevation. This is especially true if a suction lift or negative suction head is involved. Standard tables of barometric pressures are available for use in computing the data, and the tables to be used should be acceptable to all interested parties.

Cavitation Tests   Cavitation tests should be run if required by the specifications (provided such tests are needed and have not been previously conducted on similar pumps and certified by the manufacturer) or if needed to assume a successful pump installation.

The suction requirements that must be met by the pump are usually defined by the cavitation coefficient σ. Plant σ is defined as NPSHA (net positive suction head available) divided by total pump head per stage:

$$\sigma = \frac{\text{NPSHA}}{H}$$

Three typical arrangements for determining the cavitation characteristics of pumps are illustrated in Fig. 5. In Fig. 5A the suction is taken from a sump with a constant-level surface. The liquid is drawn first through a valve (throttle) and then through a section of pipe containing screens and straightening devices, such as vanes and baffles. This setup will dissipate the turbulence created by the suction valve and will also straighten the flow so that the pump suction flow will be relatively free from undue turbulence. In Fig. 5B the suction is taken from a relatively deep sump or well in which the water surface can be varied over a fairly large range to provide the designed variation in suction lift. In Figure 5C the suction is taken from a closed vessel in a closed loop in which the pressure level can be varied by a gas pressure over the liquid, by temperature of the liquid, or by a combination of these.

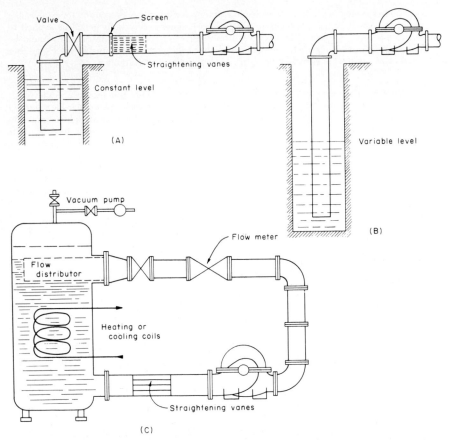

FIG. 5   Typical arrangements for determining cavitation characteristics. (Ref. 2)

By using one of these cavitation test arrangements, the critical value of σ, that is, the value at which cavitation will begin, can be found by one of the following two methods:

1. Constant speed and capacity vary the suction lift. Run the pump at constant speed and capacity with the suction lift varied to produce cavitation conditions. Plots of the head, efficiency, and power input against σ as shown in Fig. 6.

   When the values of σ are held high, the values of head, efficiency, and power should remain relatively constant. As σ is reduced, a point is reached when the curves break from the normal, indicating an unstable condition. This breakaway condition may and usually does impair the operation of the pump. The extent of impairment depends on the size, specific speed, and service of the pump and on the characteristics of the pumped fluid. A variation of this method is to plot results using capacities both greater and less than normal, as shown in Fig. 7.

2. Constant speed and suction lift vary the capacity. Run the pump at constant speed and suction lift and vary the capacity. For a given suction lift, the pumping head is plotted against capacity. A series of such tests will result in a family of curves, as shown in Fig. 8.

Where the plotted curve for any suction condition breaks away from the normal, cavitation has occurred. The value of σ may be calculated at the breakaway points by dividing NPSHA by total head H at the point under consideration.

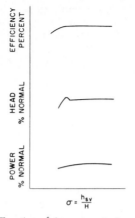

FIG. 6 Functions of sigma at constant capacity and speed; suction pressure varied.

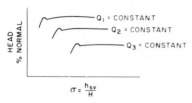

FIG. 7 Sigma capacities above and below normal; suction pressure varied.

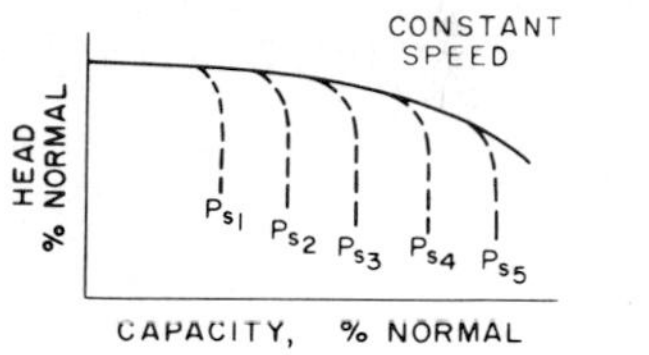

FIG. 8 Typical cavitation curves at constant speed and suction pressure.

TEST PROCEDURE

Agreements The specifications and contract should be very clear on any special points that must be covered by the pump testing. All interested parties shall be represented and given equal rights in regard to test date, setup, conditions, instrumentation, calibration of instruments, examination of pump and test setup, and accuracy of results and computations. Any controversial points or methods not provided for in the specifications should be resolved to the satisfaction of all interested parties before testing is begun.

In some special instances and by agreement between parties, an independent test expert may be engaged to take over full responsibility of the test stand and apparatus. This person will make all decisions after consultation with the interested parties and should be used only where an impasse is reached.

Normally, the manufacturer will establish the time and date for the pump tests. In some cases the specifications will cover such items as duration, limiting date, and notification time. Reasonable notice must be given to all official witnesses or representatives. A 30-day notice is preferred, and 1 week should be considered a bare minimum.

Any time limitation regarding correction of mechanical defects or equipment malfunctions that arise during testing must be resolved by mutual agreement.

Observers and Witnesses Representatives from each party to the contract shall have equal opportunity to attend the testing. Where more than one representative from one of the parties is present, then their function as observers, official witnesses, or representatives must be made clear before the start of testing. The number of representatives present from any one party shall not be a deciding factor when disagreements are being resolved. Any comments or constructive criticism from the observers and witnesses should be duly considered.

Inspection and Preliminary Operation   All interested parties shall make as complete an inspection as possible before, during, and after the test to determine compliance with specifications and correct connection of all instrumentation. The following items should be inspected before or during the test:

  Impeller and casing passages

  Pump and driver alignment

  Piezometer openings

  Electrical connections

  Lubricating devices and system

  Wearing ring and other clearances

  Stuffing box or mechanical seal adjustment and leakage

This is not a complete list of items and should be taken as only a guide for the interested parties.

Instruments installed on the pump to obtain the necessary test information shall not affect the pump operation or performance. If a question arises as to the effect an instrument has on the operation, it should be resolved by all parties. Where necessary, comparative preliminary tests can be conducted with the disputed equipment removed and then reinstalled. The dimensions at the piezometer connections on both the suction and discharge sides must be accurately determined to permit accurate determination of the velocity head correction.

On satisfactory completion of the preliminary inspection, the pump may be started. The pump and all instrumentation should then be checked for proper operation, scale readings, or evidence of malfunction. When all equipment and apparatus are functioning properly, a preliminary test run should be made. If possible, this run should be made at or near the rated condition. The correct procedures for observing and recording the data should be established during this run. Also, the time it takes to obtain steady test conditions is determined for use in the pump test runs. The acceptable deviations and fluctuations for test readings are given on pp. 13.7 and 13.8 under Accuracy.

Suggested Test Procedure

TIME AND DATE   Once tests have been decided upon, it is to the best interests of all parties to conduct them with the least delay. The contract normally will not give a time or date for the test but will specify a completion date for submission or approval of a final test report.

PERSONNEL   Pump testing, regardless of classification, should be carried out by personnel specially trained in the operation of the test equipment used. Representatives from each party to the contract shall be given equal opportunity to witness the test or tests and shall also have equal voice in commenting on the conduct of the tests or on compliance with specifications or code requirements where applicable.

SCHEDULE   A schedule should be agreed upon by all parties in advance of the test. The schedule should be as complete a program as possible and give some particulars on the range of test heads, discharge rates, and speed to be used. This schedule should be flexible and subject to change, especially after the preliminary runs have been made.

INSPECTION   The pump and test setup should be thoroughly inspected both before and after the tests. Special attention should be given to the hydraulic passages and pressure taps near the suction and discharge sections. Also, the discharge measuring device should be inspected.

CALIBRATION OF INSTRUMENTS   While the setup is being inspected, all measuring devices should be calibrated and adjusted as explained on pp. 13.8 and 13.9 under Instrumentation.

PRELIMINARY TESTS   After it has been determined that the test setup complies with the installation and specification requirements and that the instrumentation is properly installed, the pump is started. A sufficient number of preliminary test runs should be made to check the functioning of the test stand and all control and measuring devices. These preliminary tests also give the test personnel and representatives an opportunity to check and adjust the entire setup and serve as a basis for agreements on accuracy and compliance. Each test point is held until satisfactory stable conditions exist. The acceptable fluctuations in test readings are given on p. 13.8. It is suggested that preliminary computations be made, plotted, and analyzed prior to actual test runs. (*Note:* Most pump manufacturers will conduct a complete preliminary shop test to be sure of specifications and contractual compliance before inviting the purchaser's representatives to witness the official test.)

OFFICIAL TEST RUNS   The test points for the official test runs must be sufficient in number to establish the head-discharge curve over the specified range and to provide any other data needed to compute or plot the information required by the specifications. It is suggested that one test run be as near the rated condition as possible and that at least three runs be in the specified operating range of the pump.

LOGGING OF EVENTS   In most instances the official pump test is conducted by only two, three, or four test engineers, and no record other than test data is needed. In more complicated or important tests, it is sometimes very desirable to assign someone the task of recording and logging events as they occur. This is most important if reruns are required. Complete records, including any notes or comments on inspection and calibration, shall be kept of all data, readings, observations, and information relevant to the test. A suggested form for shop and field tests is shown in Fig. 28.

PRELIMINARY COMPUTATIONS   Sufficient preliminary computations should be made to determine whether all specification requirements have been met and whether reruns will be necessary.

RERUNS   When the preliminary computations indicate that reruns are necessary, they should be run immediately or as soon as possible after the official test runs and with the same personnel, instruments, and devices. Sometimes mechanical or electrical faults will necessitate a rerun. If after correction of these faults several reruns indicate a change, a complete retesting may be required. Any official representatives shall have the right to ask for a rerun or be shown to their satisfaction that a rerun is not required.

COMPUTATIONS, PLOTTING, REPORTS   These topics are discussed beginning on p. 13.34.

TEST MEASUREMENTS

Discharge   The choice of which method of discharge measurement to use should be made by agreement between all parties concerned. Some test codes and procedures in regular use permit or even recommend certain methods for model or shop testing but restrict their use in field or index testing. Some methods are more adaptable to the site conditions than others, and so the test engineers and interested parties should be completely familiar with the several methods applicable before settling on the one to be used.

The most commonly used methods of discharge measurement are those that use quantity

TABLE 3   Liquid Meters and Their Functions

Quantity meters	Rate-of-flow meters
Weighing meters	Differential pressure meters
Weighing tank	Venturi
Tilting trap	Nozzle
Volumetric meters	Orifice plate
Tank	Pitot tube
Reciprocating piston	Head area meters
Rotary piston	Weir
Nutating disk	Flume
	Current meters

meters and those that use rate-of-flow meters. Both meters are usually classified as liquid meters, and their functions are listed in Table 3.

QUANTITY METERS   The term *quantity* is here used to designate those meters in which the fluid passes through the primary element in successive and more or less completely isolated quantities, either weights or volumes, by alternately filling and emptying containers of known capacities. The secondary element of a quantity meter consists of a counter with suitably graduated dials for registering the total quantity that has passed through. Quantity meters are classified into two groups, weighing meters and volumetric meters.

Weighing Meters   There are two types of weighing meters: weighing tank and tilting trap. In the tilting trap meter, the equilibrium of a container is upset by a rise of the center of gravity as the container is filled. Weighing tank meters employ a container suspended from a counterbalanced scale beam. The weighing tank and the tilting trap are affected slightly by the temperature of the liquid but not enough to cause concern in normal testing.

Volumetric Meters   Volumetric meters measure volumes instead of weights. There are four types: tank, reciprocating piston, rotary piston, and nutating disk.

Tank meters are a very elementary form of meter of limited commercial importance. As the name implies, they consist of one or more tanks which are alternately filled and emptied. The height to which they are filled can be regulated manually or automatically. In some cases, the rising liquid operates a float which controls the inflow and outflow; in others, it starts a siphon. Occasionally, some tank meters have been erroneously classified as weighing meters.

Reciprocating piston meters use one or more members which have a reciprocating motion and operate in one or more fixed chambers. The quantity per cycle can be adjusted either by varying the magnitude of movement of one or more of the reciprocating members or by varying the relation between the primary and secondary elements.

Rotary (or oscillating) piston meters have one or more vanes which serve as pistons or movable partitions for separating the fluid segments. These vanes may be either flat or cylindrical and rotate within a cylindrical metering chamber. The axis of rotation of the vanes may or may not coincide with that of the chamber. The portion of the chamber in which the fluid is measured usually includes about 270°. In the remaining 90°, the vanes are returned to the starting position for closing off another segment of fluid. This may be accomplished by the use of an idle rotor or gear, a cam, or a radial partition. The vanes must make almost a wiping contact with the walls of the measuring chamber. The rotation of the vanes operates the counter.

Nutating disk meters have the disk mounted in a circular chamber with a conical roof and either a flat or conical floor. When in operation, the motion of the disk is such that the shaft on which it is mounted generates a cone with the apex down. However, the disk does not rotate about its own axis; this is prevented by a radial slot which fits about a radial partition extending in from the chamber sidewall nearly to the center. The peculiar motion of the disk is called *nutating*. The inlet and outlet openings are in the sidewall of the chamber on either side of the partition. These meters are usually adjusted by changing the relation between the primary and secondary elements.

RATE-OF-FLOW METERS   The term *rate of flow* is applied to all meters through which the fluid passes not in isolated quantities but in a continuous stream. The movement of this fluid stream through the primary element is directly or indirectly utilized to actuate the secondary element. The quantity of flow per unit time is derived from the interactions of the stream and the primary element, using physical laws supplemented by empirical relations.

In rate-of-flow meters, the functioning of the primary element depends upon some property of the fluid other than, or in addition to, volume or mass. This property may be kinetic energy (head meters), inertia (gate meters), specific heat (thermal meters), or the like. The secondary element senses a change in the property concerned and usually embodies some device which draws the necessary inferences automatically, so that the observer can read the rate of flow from a dial or chart. In some cases, the secondary element records pressures, such as static and differential, from which the rate of flow and time-quantity flow must be computed. In others, the secondary element not only indicates the rate of flow but also integrates it with respect to time and records the total quantity that has passed through the meter. In some cases, the indications of the secondary element are transmitted to a point some distance from the primary element.

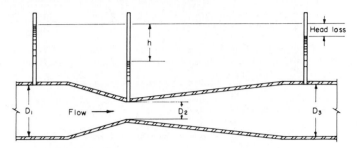

FIG. 9 Diagram of venturi meter.

Differential Pressure Meters With this group of meters the stream of fluid creates a pressure difference as it flows through the primary element. The magnitude of this pressure difference depends upon the speed and density of the fluid and features of the primary element.[3]

Flow in a pipeline, or closed pressure conduit, can be measured by a wide variety of methods, and the choice of method for a particular installation will depend upon prevailing conditions. The accuracy of flow measurements in pressure conduits made with properly selected, installed, and maintained measuring equipment, such as venturi meters, flow nozzles, orifice meters, and pitot tubes, can be very high.

The venturi meter (Fig. 9) is perhaps the most accurate flow measuring device that can be used in a water supply system. It contains no moving parts, requires very little maintenance, and causes very little head loss. Venturi meters operate on the principle that flow in a closed conduit system is faster through areas of small cross section (D_2 in Fig. 9) than through areas of large cross section (D_1). The total energy in the flow, consisting primarily of velocity head and pressure head, is essentially the same at D_1 and D_2. Thus the pressure must decrease in the constricted throat D_2, where the velocity is higher, and conversely must increase at D_1 upstream from the throat, where the velocity is lower. This reduction in pressure from the meter entrance to the meter throat is directly related to the rate of flow through the meter and is the measurement used to determine flow rate.

The coefficient of discharge for the venturi meter ranges from 0.935 for small-throat velocities and diameters to 0.988 for large-throat velocities and diameters. Equations for the venturi meter are

$$Q = \frac{CA_2\sqrt{2gh}}{\sqrt{1 - R^4}}$$

$$Q' = 3.118 \frac{CA_2'\sqrt{2gh}}{\sqrt{1 - R^4}}$$

where Q = rate of flow, ft³/s (m³/s)
  C = coefficient of discharge for meter
  A_2 = area of throat section, ft² (m²)
  g = acceleration of gravity, 32.17 ft/s² (9.81 m/s²)
  h = differential head of liquid between meter inlet and throat, ft (m)
  R = ratio of throat to inlet diameter (D_2/D_1)
  Q' = rate of flow, gpm
  A_2' = area of throat section, in²

Flow nozzles operate on the same basic principle as venturi meters. In effect, the flow nozzle is a venturi meter that has been simplified and shortened by omission of the long diffuser on the outlet side (Fig. 10). The streamlined entrance of the nozzle provides a straight cylindrical jet without contraction, so that the coefficient of discharge is almost the same as that for the venturi meter. In the flow nozzle the jet is allowed to expand of its own accord, and the high degree of turbulence created downstream from the nozzle causes a greater loss of head than occurs in the

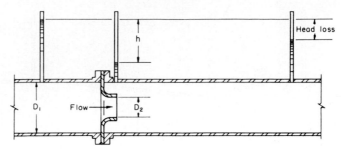

FIG. 10   Diagram of flow nozzle.

venturi meter, where the diffuser suppresses turbulence. The relationship of flow rate to head and flow nozzle dimensions is:

$$Q = \frac{CA_2\sqrt{2gh}}{\sqrt{1 - R^4}}$$

$$Q' = \frac{3.118CA_2'\sqrt{2gh}}{\sqrt{1 - R^4}}$$

where Q = rate of flow, ft³/s (m³/s)
   C = coefficient of discharge for nozzle
   A_2 = area of nozzle throat, ft² (m²)
   g = acceleration of gravity, 32.17 ft/s² (9.81 m/s²)
   h = head at or across the nozzle, ft (m)
   R = ratio of throat to inlet diameter (D_2/D_1)
   Q' = rate of flow, gpm
   A_2' = area of nozzle throat, in²

In an orifice meter, a thin plate orifice inserted across a pipeline is used for measuring flow in much the same manner as a flow nozzle (Fig. 11). The upstream pressure connection is often located about one pipe diameter upstream from the orifice plate. The pressure of the jet ranges from a minimum at the *vena contracta* (the smallest cross section of the jet) to a maximum at about four or five conduit diameters downstream from the orifice plate. The downstream pressure connection (the center connection in Fig. 11) is usually made at the *vena contracta* to obtain a large pressure differential across the orifice.

The pressure tap openings should be free from burrs and flush with the interior surfaces of the pipe. Equations for the orifice plate are

$$Q = \frac{CA_2\sqrt{2gh}}{\sqrt{1 - R^4}}$$

$$Q' = \frac{3.118CA_2'\sqrt{2gh}}{\sqrt{1 - R^4}}$$

where Q = rate of flow, ft³/s (m³/s)
   C = coefficient of discharge for orifice plate
   A_2 = area of orifice, ft² (m²)
   g = acceleration of gravity, 32.17 ft/s² (9.81 m/s²)
   h = head across the orifice plate, ft (m)
   R = ratio of throat to inlet diameter (D_2/D_1)
   Q' = rate of flow, gpm
   A_2' = area of orifice, in²

The principal disadvantage of orifice meters, compared with venturi meters and flow nozzles, is their greater loss of head. On the other hand, they are inexpensive and capable of producing accurate flow measurements.

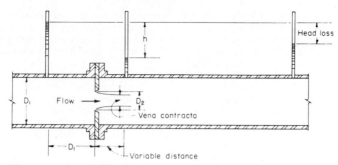

FIG. 11   Diagram of orifice meter.

It should be noted that the relationship of flow rate to head and dimensions of the metering section is identical for the venturi meter, flow nozzle, and orifice meter except that the coefficients of discharge vary.

Where it is impossible to employ one of the methods described above, the pitot tube is often used. A pitot tube in its simplest form consists of a tube with a right-angle bend which, when partly immersed with the bent part under water and pointed directly into the flow, indicates flow velocity by the distance water rises in the vertical stem. The pitot tube makes use of the difference between the static and total pressures at a single point.

The height of rise h of the water column above the water surface, expressed in feet (meters) and tenths of feet (millimeters), equals the velocity head $v^2/2g$. The velocity of flow v in feet (meters) per second may thus be determined from the relation $v = \sqrt{2gh}$.

In a more complete form known as the pitot static tube, the instrument consists of two separate, essentially parallel parts, one for indicating the sum of the pressure and velocity heads (total

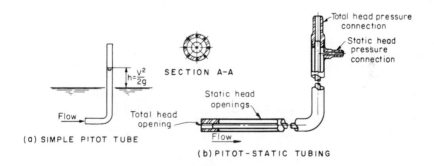

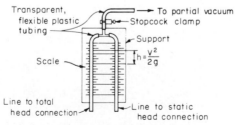

FIG. 12   Pitot tubes and manometer.

head) and the other for indicating only the pressure head. Manometers are commonly used to measure these heads, and the velocity head is obtained by subtracting the static head from the total head. A pressure transducer may be used instead of the manometer to measure the differential head. Oscillographic or digital recording of the electric signal from the transducer provides a continuous record of the changes in head.

The simple form of the pitot tube has little practical value for measuring discharges in open channels handling low-velocity flows because the distance the water in the manometer tube rises is difficult to measure. This limitation is overcome to a large extent by using a pressure transducer for the measurement and precise electronic equipment for the data readings.

The pitot static tube, on the other hand, works very well for this purpose if the tube is used with a differential manometer of the suction lift type (Fig. 12). In this manometer the two legs are joined at the top by a T that connects to a third line, in which a partial vacuum can be created. After the pitot tube has been bled to remove all air, water flows up through it into the manometer to the height desired for easy reading. Then the stopcock or clamp on the vacuum line is closed. The partial vacuum acts equally on the two legs and does not change the differential head. The velocity head h is then the difference between the total head reading and the static head reading. If desired, a pressure transducer can also be used for the head measurement.

Pitot tubes can be used to measure relatively high velocities in canals, and it is often possible to make satisfactory discharge measurements at drops, chutes, overfall crests, or other stations where the water flows rapidly and fairly large velocity heads occur. At low velocities, values of h become quite small. The pitot tube head error for a low velocity will lead to a much larger inaccuracy in discharge computation than the same error when the velocity is high. The velocity traverse with a pitot tube may be made in the same manner as with a current meter (discussed below).

Head Area Meters   The instruments used to measure flow in open conduits are normally classified as head area meters, the most common of which are weirs and flumes.

A weir is an overflow structure built across an open channel. Weirs are one of the oldest,

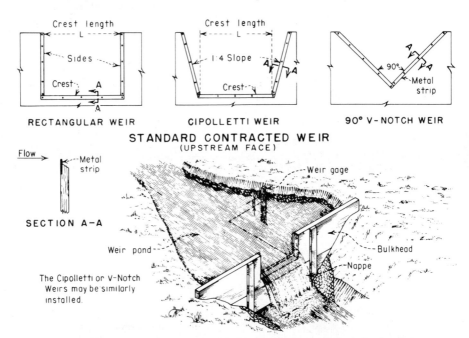

FIG. 13   Standard contracted weirs and temporary bulkhead with contracted rectangular weir discharging at free flow.

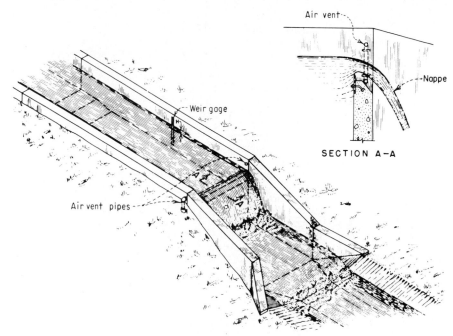

FIG. 14   Typical suppressed weir in a flume drop.

simplest, and most reliable structures for measuring the flow of water in canals and ditches. These structures can be easily inspected, and any improper operations can be quickly detected and corrected.

The discharge rates are determined by measuring the vertical distance from the crest of the overflow portion of the weir to the water surface in the pool upstream from the crest and referring to computation tables which apply to the size and shape of the weir. For standard tables to apply, the weir must have a regular shape and definite dimensions and must be placed in a bulkhead and pool of adequate size so that the system performs in a standard manner.

Weirs may be termed rectangular, trapezoidal, or triangular, depending upon the shape of the opening. In rectangular and trapezoidal weirs, the bottom edge of the opening is the crest and the side edges are called *sides* or *weir ends* (Figs. 13 and 14). The sheet of water leaving the weir crest is called the *nappe*. In certain submerged conditions, the under-nappe airspace must be ventilated to maintain near-atmospheric pressure.

The types of weirs most commonly used to measure water are

Sharp-crested and sharp-sided Cipolletti weirs

Sharp-sided 90° V-notch weirs

Sharp-crested contracted rectangular weirs

Sharp-crested suppressed rectangular weirs

For measuring water flow, the type of weir used has characteristics that make it suitable for a particular operating condition. In general, for best accuracy, a rectangular suppressed weir or a 90° V-notch weir should be used.

The discharge in second-feet (cubic meters per second) over the crest of a contracted rectangular weir, a suppressed rectangular weir, or a Cipolletti weir is determined by the head H in feet (meters) and the crest length L in feet (meters). The discharge of the standard 90° V-notch weir is determined directly by the head on the bottom of the V notch.

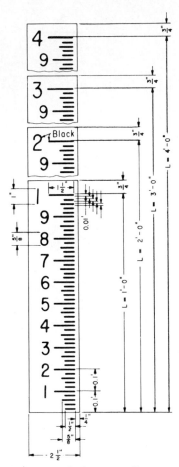

NOTES

Material of 18 gage
(U.S. Standard) metal
coated with substan-
tial thickness of porce-
lain enamel. Face of
gage is white. Numerals
and graduations are
black. Graduations are
sharp and accurate to
dimensions shown.
Length "L" represents
gate limits.
Gages may be made in
any length desired
using similar details.

FIG. 15   Standard weir, or staff, gage.

As the stream passes over the weir, the top surface curves downward. This curved surface, or drawdown, extends upstream a short distance from the weir notch. The head H must be measured at a point on the water surface in the weir pond beyond the effect of the drawdown. This distance should be at least four times the maximum head on the weir, and the same gage point should be used for lesser discharges. A staff gage (Fig. 15) having a graduated scale with the zero placed at the same elevation as the weir crest is usually provided for the head measurements.

Two widely used sets of formulas for computing discharge over standard contracted rectangular weirs are those of Smith[4] and Francis.[5] The formulas proposed by Smith require the use of coefficients of discharge that vary with the head of water on the weir and with the length of the weir. Consequently, the Smith formulas are somewhat inconvenient to use, although they are accurate for the ranges of coefficients usually given. For this type of weir operating under favorable conditions as prescribed in preceding paragraphs, the Francis formula when velocity of approach is neglected is

in USCS units $$Q = 3.33H^{3/2}(L - 0.2H)$$

in SI units $$Q = 1.837H^{3/2}(L - 0.2H)$$

(1)

and the formula when velocity of approach is included is

in USCS units $\qquad Q' = 3.33[(H + h)^{3/2} - h^{3/2}](L - 0.2H)$

in SI units $\qquad Q' = 1.837[((H + h)^{3/2} - h^{3/2})(L - 0.2H)]$ (2)

where Q = discharge neglecting velocity of approach, ft³/s (m³/s)
$\qquad Q'$ = discharge considering velocity of approach, ft³/s (m³/s)
$\qquad H$ = head on weir, ft (m)
$\qquad L$ = length of weir, ft (m)
$\qquad h$ = head due to velocity of approach $(v^2/2g)$, ft (m)

Note that the Francis formulas contain constant discharge coefficients which allow computation without the use of tables. A table of ⅔ powers, which provides values of $H^{3/2}$, $h^{3/2}$, and $(H + h)^{3/2}$ for convenience in computing discharge with the Francis formulas, may be found in hydraulic handbooks.

The principal formulas used for computing the discharge of the standard suppressed rectangular weir were also proposed by Smith and Francis. In the Smith formulas for suppressed weirs, as for contracted weirs, coefficients of discharge vary with weir head and length; therefore, these formulas are not convenient for use in computations without tables or coefficients.

The Francis formula for the standard suppressed rectangular weir neglecting velocity of approach is

in USCS units $\qquad Q = 3.33LH^{3/2}$

in SI units $\qquad Q = 1.857LH^{3/2}$ (3)

and that including velocity of approach is

in USCS units $\qquad Q' = 3.33L[(H + h)^{3/2} - h^{3/2}]$

in SI units $\qquad Q' = 1.837L[(H + h)^{3/2} - h^{3/2}]$ (4)

In these formulas the letters have the same significance as in the formulas for contracted rectangular weirs. The coefficient of discharge was obtained by Francis from the same general set of experiments as those used for the contracted rectangular weir. No tests have been made to determine the applicability of these formulas to weirs less than 4 ft (1.2 m) in length.

The Cipolletti weir is a contracted weir and must be installed as such to obtain reasonably correct and consistent discharge measurements. However, Cipolletti has compensated for the reduction in discharge due to end contractions by sloping the sides of the weir sufficiently to overcome the effect of contraction. The Cipolletti formula, in which the Francis coefficient is increased by about 1% and velocity of approach is neglected, is

in USCS units $\qquad Q = 3.367LH^{3/2}$

in SI units $\qquad Q = 1.858LH^{3/2}$ (5)

The formula including velocity of approach is

in USCS units $\qquad Q' = 3.367L(H + 1.5h)^{3/2}$

in SI units $\qquad Q' = 1.858L(H + 1.5h)^{3/2}$ (6)

where Q = discharge neglecting velocity of approach, ft³/s (m³/s)
$\qquad Q'$ = discharge considering velocity of approach, ft³/s (m³/s)
$\qquad H$ = head on weir, ft (m)
$\qquad L$ = length of weir, ft (m)
$\qquad h$ = head due to velocity of approach $(v^2/2g)$, ft (m)

The accuracy of measurements obtained with Cipolletti weirs and these formulas is inherently not as great as that obtained with suppressed rectangular or V-notch weirs.[6] It is, however, acceptable where no great precision is required.

There are several well-known formulas used to compute the discharge over 90° V-notch weirs. The most commonly used in the field of irrigation are the Cone formula and the Thomson formula. The Cone formula, considered by authorities to be more reliable for small weirs and for conditions generally encountered in measuring water for open channels, is

in USCS units $$Q = 2.49H^{2.48}$$ (7)

in SI units $$Q = 1.34H^{2.48}$$

where Q = discharge over weir, ft³/s (m³/s)
H = head on weir, ft (m)

Ordinarily V-notch weirs are not appreciably affected by velocity of approach. If the weir is installed with complete contraction, the velocity of approach will be low.

Flumes have a measuring section which is produced by contraction of the channel sidewalls or by raising of the bottom to form a hump, or by both. The Parshall flume[7] is the most common and best known measuring flume, especially in irrigation canals. It is a specially shaped open-channel flow section which may be installed in a canal, lateral, or ditch to measure the rate of flow of water. The flume has four significant advantages: (1) it can operate with relatively small head loss, (2) it is relatively insensitive to velocity of approach, (3) it has the capability of making good measurements with no submergence, moderate submergence, or even with considerable sub-mergence downstream, and (4) its velocity of flow is sufficiently high to virtually eliminate sediment deposition in the structure during operation.

Discharge through a Parshall flume can occur for two conditions of flow. The first, free flow, occurs when there is insufficient backwater depth to reduce the discharge rate. The second, sub-merged flow, occurs when the water surface downstream from the flume is far enough above the elevation of the flume crest to reduce the discharge. For free flow, only the flume head H_a at the upstream gage location is needed to determine the discharge from a standard table. The free-flow range includes some of the range which might ordinarily be considered submerged flow because Parshall flumes tolerate 50 to 80% submergence before the free-flow rate is measurably reduced. For submerged flows (when submergence is greater than 50 to 80%, depending upon flume size), both the upstream and downstream heads H_a and H_b are needed to determine the discharge (Fig. 16).

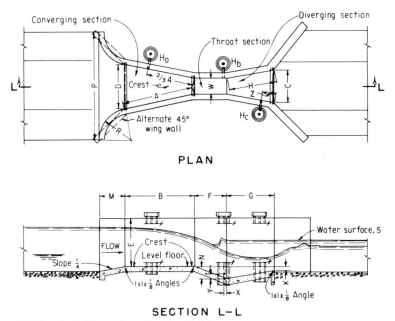

PLAN

SECTION L–L

FIG. 16a Parshall flume dimensions (sheet 1 of 2).

FIG. 16b Parshall flume dimensions (sheet 2 of 2.) (1 ft = 0.3048 m; ft³/s = 0.0283 m³/s)

1] Tolerance on throat width (w) ± $\frac{1}{64}$ inch; tolerance on other dimensions ± $\frac{1}{16}$ inch. Sidewalls of throat must be parallel and vertical.

2] From Colorado State University Technical Bulletin No. 61.

3] From U.S. Department of Agriculture Soil Conservation Circular No. 843.

4] From Colorado State University Bulletin No. 426-A.

W (FT IN)	Free-Flow Capacity Minimum Sec.-Ft.	Free-Flow Capacity Maximum Sec.-Ft.
0 1	.01	0.19
0 2	.02	.47
0 3	.03	1.13
0 6	.05	3.9
0 9	.09	8.9
1 0	.11	16.1
1 6	.15	24.6
2 0	.42	33.1
3 0	.61	50.4
4 0	1.3	67.9
5 0	1.6	85.6
6 0	2.6	103.5
7 0	3.0	121.4
8 0	3.5	139.5
10 0	6	200
12 0	8	350
15 0	8	600
20 0	10	1000
25 0	15	1200
30 0	15	1500
40 0	20	2000
50 0	25	3000

13.23

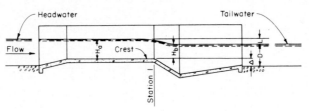

FIG. 17   Relationships of flow depth to flume crest elevation.

A distinct advantage of the Parshall flume is its ability to function as a flowmeter over a wide operating range with minimum head loss while requiring but a single head measurement for each discharge. The head loss is only about one-fourth of that needed to operate a weir having the same crest length. Another advantage is that the velocity of approach is automatically controlled if the correct size of flume is chosen and if the flume is used as it should be, that is, as an in-line structure.

Flumes are widely used because there is no easy way to alter the dimensions of flumes that have been constructed or change the device or channel to obtain an unfair proportion of water.

The main disadvantages of Parshall flumes are (1) they cannot be used in close-coupled combination structures consisting of turnout, control, and measuring device; (2) they are usually more expensive than weirs or submerged orifices; (3) they require a solid, watertight foundation; and (4) they require accurate workmanship for satisfactory construction and performance.

Parshall flume sizes are designated by the throat width W, and sizes range from 1 in (25.4 mm) for discharges as small as 0.01 ft^3/s (2.83 $\times$ 10^{-2} m^3/s) up to 50 ft (15 m) for discharges as large as 3000 ft^3/s (85 m^3/s).[8] Flumes may be built of wood, concrete, galvanized sheet metal, or other desired materials. Large flumes are usually constructed on the site, but smaller flumes may be purchased as prefabricated structures to be installed in one piece. Some flumes are available as lightweight shells, which are made rigid and immobile by placing concrete outside the walls and beneath the bottom. Larger flumes are used in rivers and large canals and streams; smaller ones are used for measuring farm deliveries or for row requirements in the farmer's field.

Flumes can operate in two modes: free flow and submerged flow. In free flow, the discharge depends solely upon the width of the throat W and the depth of water H_a at the gage point in the converging section (Figs. 16 and 17). Free-flow conditions in the flume are similar to those that occur at a weir or spillway crest in that water passing over the crest is not slowed by downstream conditions.

In submerged flow, other factors are operative. In most installations, when the discharge is increased above a critical value, the resistance to flow in the downstream channel becomes sufficient to reduce the velocity, increase the flow depth, and cause a backwater effect at the Parshall flume. It might be expected that the discharge would begin to be reduced as soon as the backwater level H_b exceeds the elevation of the flume crest; however, this is not the case. Calibration tests show that the discharge is not reduced until the submergence ratio H_b/H_a (expressed in percent) exceeds the following values:

50% for flumes 1, 2, and 3 in (25, 50, and 75 mm) wide

60% for flumes 6 and 9 in (152 and 229 mm) wide

70% for flumes 1 to 8 ft (0.3 to 2.4 m) wide

80% for flumes 8 to 50 ft (2.4 to 15.2 m) wide

The discharge equations for free flow over flumes are as follows. The equation which expresses the relationship between upstream head H_a and discharge Q for widths W from 1 to 8 ft (0.3 to 2.4 m) is

in USCS units

$$Q = 4WH_a^{1.522W^{0.026}}$$

in SI units

$$Q = 0.371W \left(\frac{Ha}{0.305} \right)^{1.57W^{0.026}}$$

If this equation is used to compute discharges through flumes ranging from 10 to 50 ft (3 to 15.7 m) wide, the computed discharges are always larger than actual discharges. Therefore a more accurate equation was developed for the large flumes:

in USCS units $$Q = (3.6875W + 2.5)H_a^{1.6}$$

in SI units $$Q = (2.293W + 0.474)H_a^{1.6}$$

The difference in computed discharges obtained by using the two equations for an 8-ft (2.4-m) flume is normally less than 1%; however, the difference becomes greater as the flume size increases. Because of the difficulties in regularly using these equations, discharge tables have been prepared for use with flumes 1 to 50 ft (0.3 to 15.2 m) wide.

Current Meters The essential features of a conventional current meter are a wheel which rotates when immersed in flowing water and a device for determining the number of revolutions of the wheel. For open-channel flow measurement, a type generally used is the Price meter with five or six conical cups. The relationship between the velocity of the water and the number of revolutions of the wheel per unit time is determined experimentally for each instrument for various velocities. Also, the operator must have considerable skill to obtain consistent satisfactory results.

A detailed explanation of the use of current meters is contained in Chap. 5 of Ref. 9.

OTHER METHODS The discharge measurement methods given above are the ones in common use; however, a number of other methods, some newer and more sophisticated, are well established. The use of these special methods is acceptable provided their limitations are recognized and all parties to the testing program are in agreement on their use. A few of these methods, not in any particular order, are:

Salt velocity

Salt dilution

Color velocity

Radioisotope

Acoustic flowmeters

Slope area

Deflection meters

Propeller meters

Float movement

Gates and sluices

Color dilution

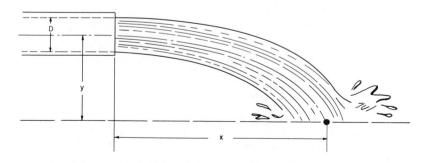

$$\text{CAPACITY, GPM} = \frac{2.45\,D^2\,x}{\sqrt{\dfrac{2y}{32.16}}}$$ $$\text{CAPACITY, m}^3/\text{h} = \frac{0.00283\,D^2\,x}{\sqrt{\dfrac{2y}{9.81}}}$$

Where as
D = Pipe Dia., Inches (mm)
x = Hor. Dist., Feet (m)
y = Vert. Dist., Feet (m)

FIG. 18 Approximating flow from a horizontal pipe.

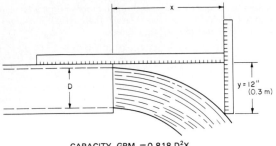

CAPACITY, GPM $= 0.818\ D^2 X$

CAPACITY, $m^3/h = 1.1336 \times 10^{-5} D^2 X$

D $=$ in (mm)

X $=$ in (mm)

APPROXIMATE CAPACITY, GPM,

FOR FULL FLOWING HORIZONTAL PIPES

STD. WT. STEEL PIPE, INSIDE DIA., IN.		DISTANCE X, IN., WHEN Y = 12"										
NOMINAL	ACTUAL	12	14	16	18	20	22	24	26	28	30	32
2	2.067	42	49	56	63	70	77	84	91	98	105	112
2½	2.469	60	70	80	90	100	110	120	130	140	150	160
3	3.068	93	108	123	139	154	169	185	200	216	231	246
4	4.026	159	186	212	239	266	292	318	345	372	398	425
5	5.047	250	292	334	376	417	459	501	543	585	627	668
6	6.065	362	422	482	542	602	662	722	782	842	902	962
8	7.981	627	732	837	942	1047	1150	1255	1360	1465	1570	1675
10	10.020	980	1145	1310	1475	1635	1800	1965	2130	2290	2455	2620
12	12.000	1415	1650	1890	2125	2360	2595	2830	3065	3300	3540	3775

FIG. 19 Simplified method for approximating flow from a horizonal pipe. (1 in = 25.4 mm; 1 gpm = 0.227 m³/h)

FIELD APPROXIMATING Often a field approximation of water flow from a pump discharge becomes necessary, especially if no other methods are practical or readily available. One of the accepted methods is by trajectory. The discharge from the pipe may be either vertical or horizontal, the principal difficulty being in measuring the coordinates of the flowing stream accurately. The pipes must be flowing full, and the accuracy of this method varies from 85 to 100%. Figure 18 illustrates the approximation from a horizontal pipe. This method can be further simplified by measuring to the top of the flowing stream and always measuring so that y equals 12 in (300 mm) and measuring the horizontal distance X in inches (millimeters), as illustrated in Fig. 19.

Figure 20 illustrates a method of measuring discharge from a vertical pipe.

Head Measurements Head is the quantity used to express the energy content of a liquid per unit weight of the liquid referred to any arbitrary datum. In terms of foot-pounds of energy per pound of liquid, all head quantities have the dimensions of feet of liquid. The unit for measuring head is the foot (meter). The relation between a pressure expressed in pounds per square inch (kilopascals) and one expressed in feet (meters) of head is

in USCS units
$$\text{Head, ft} = \text{lb/in}^2 \times \frac{2.31 \times 62.3}{W} = \text{lb/in}^2 \times \frac{2.31}{\text{sp. gr.}}$$

in SI units
$$\text{Head, m} = \frac{0.102 \times KPa}{W}$$

where $\quad$ = specific weight (mass), lb/ft³ (Kg/l)

sp. gr. = specific gravity of the liquid

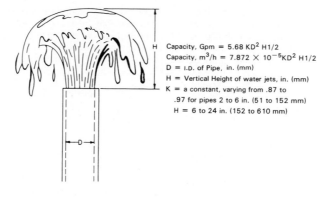

Capacity, Gpm = 5.68 KD^2 H1/2
Capacity, m^3/h = 7.872 × $10^{-5}KD^2$ H1/2
D = I.D. of Pipe, in. (mm)
H = Vertical Height of water jets, in. (mm)
K = a constant, varying from .87 to
.97 for pipes 2 to 6 in. (51 to 152 mm)
H = 6 to 24 in. (152 to 610 mm)

APPROXIMATE CAPACITY, G P M,

FOR FLOW FROM VERTICAL PIPES

NOMINAL	VERTICAL HEIGHT, H, OF WATER JET, IN.										
I.D. PIPE, IN.	3	3.5	4	4.5	5	5.5	6	7	8	10	12
2	38	41	44	47	50	53	56	61	65	74	82
3	81	89	96	103	109	114	120	132	141	160	177
4	137	151	163	174	185	195	205	222	240	269	299
6	318	349	378	405	430	455	480	520	560	635	700
8	567	623	684	730	776	821	868	945	1020	1150	1270
10	950	1055	1115	1200	1280	1350	1415	1530	1640	1840	2010

FIG. 20 Approximating flow from a vertical pipe. (1 in = 25.4 mm; 1 gpm = 0.227 m^3/h)

The following exerpt from the *Hydraulic Institute Standards* is used by the Bureau of Reclamation throughout its test program:

It is important that steady flow conditions exist at the point of instrument connection. For this reason, it is necessary that pressure or head measurement be taken on a section of pipe where the cross section is constant and straight. Five to ten diameters of straight pipe of unvarying cross section following any elbow or curved member, valve, or other obstruction, are necessary to insure steady flow conditions.

The following precautions should be taken in forming orifices for pressure-measuring instruments and for making connections. The orifice in the pipe should be flush with and normal to the wall of the water passage. The wall of the water passage should be smooth and of unvarying cross section. For a distance of at least 12 in (0.3 m) preceding the orifice, all tubercles and roughness should be removed with a file or emery cloth, if necessary. The orifice should be from ⅛ to ¼ in (3–6 mm) in diameter and two diameters long.

The edges of the orifice should be provided with a suitable radius tangential to the wall of the water passage and should be free from burrs or irregularities. Two pressure tap arrangements shown in Fig. 27 indicate taps in conformity with the above. Where more than one tap is required at a given measuring section, separate properly valved connections should be made. As an alternative, separate instruments should be provided.

Multiple orifices should not be connected to one instrument except on those metering devices, such as venturi meters, where proper calibrations have been made.

All leads from the orifice should be tight and as short and direct as possible. For dry-tube leads, suitable drain pots should be provided and a loop of sufficient height to keep the pumped liquid from entering the leads should be formed. For wet-tube leads, vent cocks for flushing should be provided at any high point or loop crest to assure that tubes do not become airbound. All instrument hose, piping, and fittings should be checked under pressure prior to test to assure that there are no leaks. Suitable damping devices may be used in the leads.

If these conditions cannot be satisfied at the point of measurement, it is recommended that four separate pressure taps be installed, equally spaced about the pipe, and that the pressure at that section be taken as the average of the four separate values. If the separate readings show a difference of static pressure that might affect the head beyond the contract tolerances, the installation should be corrected or an acceptable tolerance determined.

Figures 21 to 27 show suitable arrangements for various types of instruments and formulas for translating instrument readings into feet (meters) of liquid pumped, for expressing instrument head as elevation over a common datum, and for correcting these formulas for the velocity head existing in the suction and discharge pipes.

The datum is taken as the centerline of the pump for horizontal-shaft pumps and as the entrance eye of the impeller for vertical-shaft pumps.

The instruments, when practicable, are water columns or manometers for normal pressures and mercury manometers, bourdon gages, electric pressure transducers, or dead-weight gage testers for high pressure. When water columns are used, care should be taken to avoid errors due to the difference between the temperature of the water in the gage and that of the water in the pump.

DEFINITIONS AND SYMBOLS   The symbols used throughout this section for expressing and computing head are those used by the ASME Power Test Code[1] for pumps and the Hydraulic Institute standards[2]. The following symbols apply to Figs. 21 to 27 where temperature effects are negligible:

H = total head or dynamic head, ft (m) = the energy increase per pound (kilogram) of liquid imparted to the liquid by the pump and therefore the algebraic difference between total discharge head and total suction head ($H = h_d - h_s$); h_d and h_s are negative if the corresponding values at the datum elevation are below atmospheric pressure

h_{gd} = discharge gage reading, ft (m) H_2O

h_{gs} = suction gage reading, ft (m) H_2O

Both the discharge gage and the suction gage can be direct-reading water manometers, converted mercury manometers, or calibrated bourdon pressure gages.

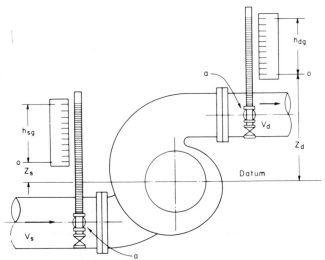

FIG. 21   Head measurement.

$$h_d = +h_{gd} + Z_d + \frac{V_d{}^2}{2g}$$

$$h_s = +h_{gs} + Z_s + \frac{V_s{}^2}{2g}$$

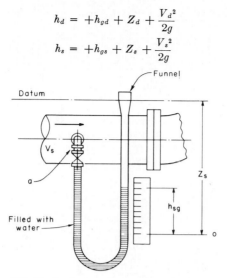

FIG. 22   Head measurement.

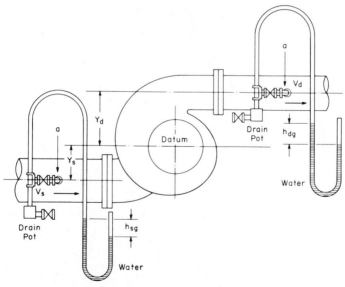

FIG. 23   Head measurement.

Z_d = elevation of discharge gage, zero above datum elevation, ft (m)
Z_s = elevation of suction gage, zero above datum elevation, ft (m)

The quantities Z_d and Z_s are negative if the gage zero is below the datum elevation.

Y_d = elevation of discharge gage connection to discharge pipe above datum elevation, ft (m)
Y_s = elevation of suction gage connection to suction pipe above datum elevation, ft (m)

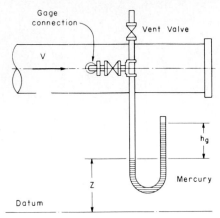

FIG. 24 Head measurement.

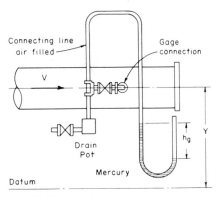

FIG. 25 Head measurement.

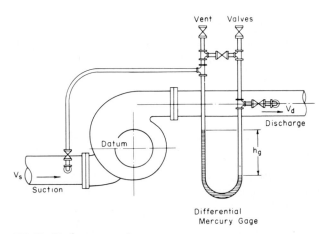

FIG. 26 Head measurement.

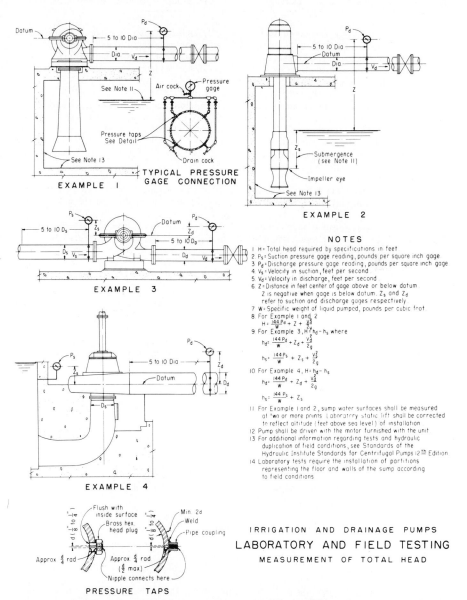

FIG. 27 Head measurement. In SI units, $h_d = [(0.102\,P_d/W) + Z_d + (V_d^2/2\,g)]$; $h_s = [(0.102\,P_s/W) + Z_s + (V_s^2/2\,g)]$. (ft $\times$ 0.3048 = m; psi $\times$ 6.894 = kPa; pcf $\times$ .016 = kg/liter)

The quantities Y_d and Y_s are negative if the gage connection to the pipe lies below the datum elevation.

V_d = average water velocity in discharge pipe at discharge gage connection, ft/s (m/s)
V_s = average water velocity in suction pipe at suction gage connection, ft/s (m/s)
h_d = total discharge head above atmospheric pressure at datum elevation, ft (m)
h_s = total suction head above atmospheric pressure at datum elevation, ft (m)
h_{vs} = velocity head in suction pipe $(V_s^2/2g)$, ft (m)
h_{vd} = velocity head in discharge pipe $(V_d^2/2g)$, ft (m)
NPSHA = net positive suction head available = total suction head in feet (meters) of liquid absolute, determined at suction nozzle and referred to datum less absolute vapor pressure of the liquid in feet (meters) of liquid pumped (NPSHA = $h_a - H_{vpa} + h_s$)
h_{sa} = total suction head absolute $(h_a + h_s)$
H_{vpa} = vapor pressure of liquid, ft (m) abs
h_a = atmosphere pressure, ft (m) abs

MEASURING HEAD WITH WATER GAGES   The following examples illustrate how to calculate the head in a centrifugal pump arrangement with gages either above or below atmospheric pressure.

EXAMPLE   The pressure at gage connection a in Fig. 21 is above atmospheric pressure, and the line between either the discharge or suction pipe and the corresponding gage is filled completely with water. The following equations apply:

$$h_d = +h_{gd} + Z_d + \frac{V_d^2}{2g}$$

$$h_s = +h_{gs} + Z_s + \frac{V_s^2}{2g}$$

Note 1:   The word *water* is used to represent the liquid being pumped. The provisions are applicable to the pumping of other liquids, provided the gages and connecting lines contain the liquid being pumped.

EXAMPLE   The pressure at gage connection a in Fig. 22 is below atmospheric pressure. The following equation applies:

$$h_s = h_{gs} - Z_s + \frac{V_s^2}{2g}$$

The negative sign of Z_s indicates that the gage zero is located below the datum.

EXAMPLE   The pressure at gage connection a in Fig. 23 is below atmospheric pressure, and the line between either the discharge or suction pipe and the corresponding gage is filled completely with air. The following equations apply:

$$h_d = -h_{gd} + Y_d + \frac{V_d^2}{2g}$$

$$h_s = -h_{gs} - Y_s + \frac{V_s^2}{2g}$$

Note 2:   If a connecting pipe is air-filled, it must be drained before a reading is made. Water cannot be used in the U tube if either h_{dg} or h_{ds} exceeds the height of the rising loop.

MEASURING HEAD WITH MERCURY GAGES   The following examples illustrate the use of mercury gages for measuring head in a centrifugal pump arrangement.

EXAMPLE   In Fig. 24, the gage pressure is above atmospheric pressure and the connecting line is completely filled with water. The following equation applies:

$$h = \frac{W_m}{W} h_g + Z + \frac{V^2}{2g}$$

where W_m = specific weight (mass) of mercury, lb/ft³ (kg/liter)
$\quad\quad W$ = specific weight (mass) of liquid pumped, lb/ft³ (kg/liter)
$\quad\quad h_g$ = suction or discharge gage reading, ft (m) Hg

The quantities h, Z, Y, and V without subscripts apply equally to suction and discharge head measurements.

EXAMPLE   The gage pressure in Fig. 25 is below atmospheric pressure, and the connecting line is completely filled with air, with a rising loop to prevent water from passing to the mercury column. The following equation applies

$$h = \frac{W_m}{W} h_g = Y + \frac{V^2}{2g}$$

MEASURING HEAD WITH DIFFERENTIAL MERCURY GAGES   Figure 26 indicates a centrifugal pump arrangement in which a differential mercury gage is used to measure head. When this type of gage is used and the connecting lines are completely filled with water, the correct equation is

$$H = \left(\frac{W_m}{W} - 1 \right) h_g + \frac{V_d^2}{2g} - \frac{V_s^2}{2g}$$

In addition to the differential gage, a separate suction gage can be used, as shown in Figs. 22 and 25. The equation in this case is

$$h_s = \frac{W_m}{W} h_{gs} - Z + \frac{V_s^2}{2g}$$

MEASURING HEAD WITH BOURDON GAGES   An example of a centrifugal pump arrangement that uses calibrated bourdon gages for head measurement is shown in Example 3 of Fig. 27, with the gage pressure above atmospheric pressure. The distances Z_s and Z_d are measured to the center of the gage and are negative if the center of the gage lies below the datum line.

MEASURING HEAD ON VERTICAL SUCTION PUMPS IN SUMPS AND CHANNELS   In vertical-shaft pumps drawing water from large open sumps and having inlet passages whose length does not exceed about three inlet opening diameters, such inlet pieces having been furnished as part of the pump, the total head should be the reading of the discharge connection in feet (meters) plus the vertical distance from the gage centerline to the free water level in the sump in feet (meters) (Example 2 of Fig. 27).

Power Measurement   The pump input power may be determined with a calibrated motor, a transmission dynamometer, or a torsion dynamometer. The *Hydraulic Institute Standards* are generally used as the basis for most power measurement procedures.

CALIBRATED MOTORS   When pump input power is to be determined with a calibrated motor, the power input should be measured at the terminals of the motor to exclude any line losses that may occur between the switchboard and the driver. Certified calibration curves of the motor must be obtained. The calibration should be conducted on the motor in question and not on a similar machine. Such calibrations must indicate the true input-output value of motor efficiency and not some conventional method of determining an arbitrary efficiency. Calibrated laboratory-type electric meters and transformers should be used to measure power input to all motors.

TRANSMISSION DYNAMOMETERS   The transmission, or torque-reaction, dynamometer consists of a cradled electric motor with its frame and field windings on one set of bearings and the rotating element on another set, so that the frame is free to rotate but is restrained by means of some weighting or measuring device.

When pump input power is to be determined with a transmission dynamometer, the unloaded and unlocked dynamometer must be properly balanced prior to the test at the same speed at which the test is to be run. The balance should be checked against standard weights. After the test the balance must be rechecked to assure that no change has taken place. In the event of an appreciable change, the test should be rerun. An accurate measurement of speed is essential and should not vary from the pump rated speed by more than 1%. Power input is calculated as shown below under "Computations."

TORSION DYNAMOMETERS   The torsion dynamometer consists of a length of shafting whose torsional strain when rotating at a given speed and delivering a given torque is measured by some standard method. When pump input power is to be determined with a torsion dynamometer, the unloaded dynamometer should be statically calibrated prior to the test. This is done by measuring the angular deflection for a given torque.

Immediately before and after the test, the torsion dynamometer must be calibrated dynamically at the rated speed. The best and simplest method to accomplish this is to use the actual job driver to supply power and use a suitable method of loading the driver over the entire range of the pump to obtain the necessary calibrations. The calibration of the torsion dynamometer after the pump tests should be within 0.5% of the original calibration. During the test runs the speed should not vary from the pump rated speed by more than 1%. The temperature of the torsion dynamometer during the test runs must be within 10 °F (6 °C) of the temperature when the dynamic calibrations were made. All torsion dynamometer calibrations should be witnessed and approved by all parties to the test. In the event of a variation greater than allowed, a rerun of the test must be made. Power input calculations are shown below.

Speed Measurement

The speed of the pump under test is determined by one of the following methods:

Revolution counter (manual or automatic)

Tachometer

Stroboscopic device

Electronic counter

In all cases the instruments used must be carefully calibrated before the test to demonstrate that they will produce the required speed readout to within the desired accuracy. Accepted accuracy is usually $\pm 0.1\%$. Should cyclic speed change result in power fluctuations, then at least five equally spaced, timed readings should be taken to give a satisfactory mean speed.

COMPUTATIONS

Pump Power

POWER OUTPUT   The water power, or useful work, done by the pump is found by the formula

in USCS units $\text{whp} = \dfrac{\text{lb of liquid pumped/min} \times \text{total head in ft of liquid}}{33{,}000}$

in SI units $\text{wkW} = \dfrac{\text{kg of liquid pumped/min} \times \text{total head in m of liquid}}{6131}$

If the liquid has a specific gravity of 1 and the specific weight of the liquid is 62.3 lb/ft^3 (1.0 kg/liter) at 68°F (20°C), the formula is

in USCS units $\text{whp} = \dfrac{\text{gpm} \times \text{head in ft}}{3960}$

in SI units $\text{wkW} = 9.8 \text{ m}^3/\text{h} \times \text{head in m}$

POWER INPUT   The brake power required to drive the pump is found by the formula

in USCS units

$$\text{bhp} = \frac{\text{gpm} \times \text{total head in ft}}{3960 \times \text{pump efficiency}}$$

in SI units

$$\text{bkW} = \frac{9.8 \text{ m}^3/\text{h} \times \text{total head in m}}{\text{pump efficiency}}$$

where pump efficiency is obtained from the formula

$$\text{Pump efficiency} = \frac{\text{output}}{\text{input}} = \frac{\text{whp}}{\text{bhp}} = \frac{\text{wkW}}{\text{bkW}}$$

The electric power input to the motor is

in USCS units

$$\text{ehp} = \frac{\text{bhp}}{\text{motor efficiency}} = \frac{\text{gpm} \times \text{head in ft}}{3960 \times \text{pump efficiency} \times \text{motor efficiency}}$$

in SI units

$$\text{kW} = \frac{9.8 \text{ m}^3/\text{h} \times \text{head in m}}{\text{pump efficiency} \times \text{motor efficiency}}$$

The kilowatt input to the motor is

$$\text{kW input} = \frac{\text{bhp} \times 0.746}{\text{motor efficiency}}$$

$$= \frac{\text{gpm} \times \text{head} \times 0.746}{3960 \times \text{pump efficiency} \times \text{motor efficiency}}$$

Pump Efficiency   The pump efficiency is

$$\text{Pump efficiency} = \frac{\text{output}}{\text{input}} = \frac{\text{whp}}{\text{bhp}} = \frac{\text{wkW}}{\text{bkW}}$$

For an electric-motor-driven pumping unit, the overall efficiency is

$$\text{Overall efficiency} = \text{pump efficiency} \times \text{motor efficiency}$$

In many specifications it is required that the actual job motor be used to drive the pump during shop or field testing. Using this test setup, the overall efficiency then becomes what is commonly called "wire-to-water" efficiency, which is expressed by the formula

$$\text{Overall efficiency} = \frac{\text{water power}}{\text{electric power input}} = \frac{\text{whp}}{\text{ehp}} = \frac{\text{wkW}}{\text{kW}}$$

Speed Adjustments   The best and standard practice in testing pump speed is to use the actual job motor to drive the pump under test. However, for purposes of plotting test results, it becomes necessary to correct the test values at test speed to rated pump speed. The rated pump speed should always be less than the test speed since even a small increase in speed beyond the test speed may result in going into an unstable zone of the pump. Also, it is recommended that the speed change from test speed to rated or specified speed not be greater than 3%.

To adjust pump capacity, head, power, and NPSH from the values recorded during test to another speed, the following formulas[2] should be used:

Capacity:

$$Q_2 = \frac{N_2}{N_1} \times Q_1$$

where Q_1 = capacity at test speed, gpm (m^3/h)
$\quad Q_2$ = capacity at rated speed, gpm (m^3/h)
$\quad N_1$ = test speed, rpm
$\quad N_2$ = rated speed, rpm

Head:

$$H_2 = \left(\frac{N_2}{N_1}\right)^2 \times H_1$$

where H_1 = head at test speed, ft (m)
  H_2 = head at rated speed, ft (m)

Power:

in USCS units $$hp_2 = \left(\frac{N_2}{N_1}\right)^3 \times hp_1$$

in SI units $$kW_2 = \left(\frac{N_2}{N_1}\right)^3 \times kW_1$$

where hp_1 = power at test speed, hp
  hp_2 = power at rated speed, hp
  kW_1 = power at test speed, kW
  kW_2 = power at rated speed, kW

$$\text{Net positive suction head: } NPSH_2 = \left(\frac{N_2}{N_1}\right)^2 \times NPSH_1$$

where $NPSH_1$ = net positive suction head at test speed, ft (m)
  $NPSH_2$ = net positive suction head at rated speed, ft (m)

Shop or field testing at either reduced or increased speed should be permitted only when absolutely no alternatives are available. It is recommended, if reduced or increased speed tests are used as official performance tests, that the specifications state the test head and speed and that the performance warranties be based on the specified head and speed conditions.

If reduced or increased speed tests are considered, then certain affinity laws must be observed to maintain hydraulic similarity between actual and test conditions. These affinity law relationships can be expressed by

$$\frac{Q_1}{Q} = \frac{N_1}{N} = \left(\frac{H_1}{H}\right)^{1/2}$$

where test = Q_1 = capacity and H_1 = head at N_1 = rpm
  actual = Q = capacity and H = head at N = rpm

RECORDS

Data There probably are as many test data forms as there are test laboratories. Each may or may not have an advantage for its particular application. A form for recording pump performance data is shown in Fig. 28.

The manufacturer's serial number, type, and size or other means of identification of each pump and driver involved in the test should be carefully recorded in order that mistakes in identity may be avoided. The dimensions and physical conditions not only of the machine tested but of all associated parts of the plant which have any important bearing on the outcome of a test, should be noted.

Normal practice suggests that one test run be either at or as near as possible to the rated condition and that at least three runs be within the specified range of heads.

Plotting Test Results In plotting curves from the test results, it should be kept in mind that any one point may be in error but that all the points should establish a trend.

Unless some external factor is introduced to cause an abrupt change, a smooth curve can be drawn for the points plotted, not necessarily through each and every one. Figure 29 is a graphic representation showing the determination of pump performance with the total head, power input, and efficiency in percent, all plotted on the same graph with the capacity as the abscissa.

Reports In some instances a preliminary report may be issued as part of a contract agreement. However, normally a final report is all that is required.

RECORD OF PUMP PERFORMANCE TEST

DATE OF TEST_____TEST NO._____CUSTOMER_____CUST. ORDER NO._____

MANUFACTURER'S ORDER NO._____PLANT_____UNIT NO._____

PROJECT_____

RATED CONDITIONS:

  CAPACITY, G.P.M. (m³/h) _____TOTAL HEAD, FEET (m)_____R.P.M._____

  OVERALL EFFICIENCY PERCENT_____RANGE OF HEAD_____

DRIVER:

  TYPE_____HORSEPOWER (kW)_____

  MANUFACTURER_____SERIAL NO._____TEST VOLTAGE_____

TEST EQUIPMENT:

  DISCHARGE MEASUREMENT METHOD_____CONVERSION FACTOR_____

  DISCHARGE GAGE_____CORRECTION_____SUCTION GAGE_____CORRECTION_____

  DIFFERENTIAL BETWEEN GAGES_____INSIDE DIAMETER SUCTION_____INSIDE DIAMETER DISCHARGE_____

PUMP DATA:

  TYPE OF PUMP_____SIZE_____NO. STAGES_____

  MANUFACTURER_____SERIAL NO._____SUCTION SIZE_____DISCHARGE SIZE_____

	RUN NO.	1	2	3	4	5	6	7	8	9	10
HEAD	PRESSURE, P.S.I. (kPa)										
	HEAD, FEET (m)										
	GAGE ₵ TO WATER LEVEL, FEET(m)										
	VELOCITY HEAD, FEET (m)										
	TOTAL HEAD, FEET (m)										
CAPACITY	READING										
	CONVERSION										
	FLOW, G.P.M. (m³/h)										
POWER DATA	MOTOR VOLTAGE										
	AMPERES										
	KILOWATTS										
	TOTAL HORSEPOWER INPUT										
	MOTOR EFFICIENCY, %										
	SPEED, R.P.M.										
	DYNAMOMETER										
	BRAKE HORSEPOWER (kW)										
	WATER HORSEPOWER (kW)										
	PUMP EFFICIENCY, %										
	OVERALL EFFICIENCY, %										

TESTED BY_____WITNESSED BY_____

TYPE OF TEST_____

(FIELD OR SHOP)

REMARKS:

_____

_____

_____

IRRIGATION AND DRAINAGE PUMPS
LABORATORY AND FIELD TESTING
PUMPING UNIT TEST DATA

FIG. 28 Record of pump performance test.

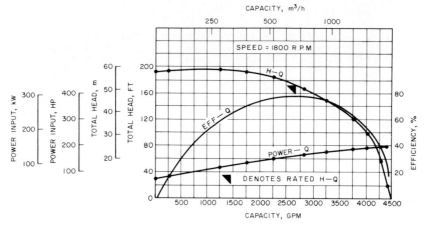

FIG. 29   Plotted pump performance.

On shop tests of relatively small pumps, the test log and plotted results constitute the entire report. Reports get progressively more involved as pumps become larger. Where model testing is used, the final report is a complete record of the agreements, inspections, personnel, calibration data, sample computations, tabulations, descriptions, discussions, etc.

MODEL TESTING

Models are tested for one or more of the following purposes:

1. To develop new ideas and new designs
2. To give the manufacturer a range of warranties for performance and efficiency
3. To give the buyer some assurance that requirements are being met
4. To replace or supplement field testing of a prototype
5. To compare performances of several models

Model testing in advance of final design and installation of a large unit not only provides advance assurance of satisfactory performance but allows for alterations in time for incorporation into the prototype.

Testing Procedures   Early editions of the *Hydraulic Institute Standards* state that, in the comparison of model and prototype heads, the minimum acceptable agreement is 80%. There are differences of opinion concerning this requirement. This percentage of prototype head has not been clearly substantiated either by theoretical considerations or by comparison of performance data from tests of models and their full-sized prototypes, and at present this ratio is in a transitional state.

The model should have complete geometric similarity to the prototype in all wetted parts between the intake and discharge sections of the pump. The model should be tested in the same horizontal- or vertical-shaft position as the prototype. The speed of the model should be such that, at the test head, the specific speed for each test run is the same as that of the installed unit or prototype. Unless otherwise specified, the suction head or suction lift should give the same (cavitation factor) value.

If model and prototype diameters are D_1 and D, respectively, then the model speed N_1 and capacity Q_1 under the test head H_1 must agree with the relations

$$\frac{N_1}{N} = \frac{D}{D_1} \sqrt{\frac{H_1}{H}} \quad \text{and} \quad \frac{Q_1}{Q} = \left(\frac{D_1}{D}\right)^2 \sqrt{\frac{H_1}{H}}$$

In testing a model of reduced size under the above conditions, complete hydraulic similarity will not be secured unless the relative roughness of the impeller and pump casing surfaces are the same. With the same surface texture in model and prototype, the model efficiency will be lower than that of the prototype, and greater relative clearances and shaft friction in the model will also reduce its efficiency.

The efficiency of a pump model can conveniently be stepped up to match the prototype efficiency by applying a formula of the same general form as the Moody formula used for hydraulic turbines:

$$\frac{1 - e_1}{1 - e} = \left(\frac{D}{D_1}\right)^n$$

The exponent n should be determined for a given laboratory and given type of pump on the basis of an adequate number of comparisons of the efficiencies of models and prototypes, with consistent surface finish in model and prototype. The *Hydraulic Institute Standards*[2] states that n has been found to vary from zero (when the surface roughness and clearances of the model and prototype are proportional to their size) to 0.26 (when the absolute roughness is the same in both).

When model tests are to serve as acceptance tests, it is generally recommended that the efficiency guarantees be stated in terms of model performance rather than in terms of calculated prototype performance. In the absence of such a provision, the efficiency stepup formula and the numerical value of its exponent should be clearly specified or agreed upon in advance of tests.

The *Hydraulic Institute Standards*[2] gives an example of model testing as follows.

EXAMPLE   A single-stage pump to deliver 200 ft^3/s (5.66 m^3/s) against a head of 400 ft (122 m) at 450 rpm and with a positive suction head, including velocity head, of 10 ft (3 m) has an impeller diameter of 6.8 ft (2.1m). The pump being too large for a shop or laboratory test, a model with an 18-in (0.46-m) impeller is to be tested at a reduced head at 320 ft (97.5 m). At what speed, capacity, and suction head should the test be run?

Applying the above relations:

in USCS units $N_1 = N \dfrac{D}{D_1} \sqrt{\dfrac{H_1}{H}} = 450 \left(\dfrac{6.8}{1.5}\right) \sqrt{\dfrac{320}{400}} = 1825$ rpm

$Q_1 = Q \left(\dfrac{D_1}{D}\right)^2 \sqrt{\dfrac{H_1}{H}} = 200 \left(\dfrac{1.5}{6.8}\right)^2 \sqrt{\dfrac{320}{400}} = 8.73$ ft^3/s $= 3920$ gpm

in SI units $N_1 = 450 \left(\dfrac{2.1}{0.46}\right)^2 \sqrt{\dfrac{97.5}{122}} = 1825$ rpm

$Q_1 = 5.66 \left(\dfrac{0.46}{2.1}\right)^2 \sqrt{\dfrac{97.5}{122}} = 0.247$ m^3/s

To check these results, the specific speed of the prototype is

in USCS units $N_s = N \dfrac{\sqrt{Q}}{H^{3/4}} = 450 \dfrac{\sqrt{200}}{400^{3/4}} = 71.2$ in the ft^3/s system

in SI units $N_s = \dfrac{450\sqrt{5.66}}{122^{3/4}} = 29$ in the m^3/s system and that of the model is

in USCS units $N_{s1} = 1825 \dfrac{\sqrt{8.73}}{320^{3/4}} = 71.2$ (or 1,510 in the gpm system)

in SI units $N_{s1} = 1825 \dfrac{\sqrt{0.247}}{97.5^{314}} = 29$

The cavitation factor σ for the field installation, assuming a water temperature of 80°F (27°C) as a maximum probable value and $H_b = 32.8$ ft (10 m) as in the first example, will be

in USCS units $\sigma = \dfrac{H_b - H_s}{H} = \dfrac{32.8 - 10}{400} = 0.057$

in SI units $\sigma = \dfrac{10 - 3}{122} = 0.057$

where $H_b = h_{sa} - H_{vpa}$ (absolute atmospheric pressure minus absolute vapor pressure)
       H_s = distance from datum to suction level

which should be the same in the test. With the water temperature approximately the same,

$$\sigma = \frac{H_b - H_{s1}}{H_1}$$

in USCS units $H_{s1} = H_b - \sigma H = 32.8 - (0.057)(320) = 14.6$ ft

in SI units $H_{s1} = 10 - (0.057)(97.5) = 4.4$ m

Hence the model should be tested with a positive suction head of 14.6 ft (4.4 m) to reproduce the field conditions.

Normally one of the requirements when using model tests as acceptance tests is to make sure true geometric similarity exists between model and installed prototype. True values of all required and specified dimension should be determined. The actual parts, areas, shape, clearances, and positions should be clearly understood by all parties. Also, the amount of permissible geometric deviation between prototype and model should be agreed to, in writing, before the test is begun.

OTHER OBSERVATIONS

Testing Noncentrifugal Pumps   The next largest class of pumps after centrifugal are displacement pumps. This classification includes reciprocating, rotary, screw, and other miscellaneous displacement pumps. Testing of these closely parallels the centrifugal procedures. Normally the capacities are smaller and the heads higher, but the objectives, methods, and measurements are all about the same. The *Hydraulic Institute Standards* thoroughly covers the testing of rotary and displacement pumps.

The testing of pumps not falling into the two broad classifications of centrifugal and displacement is usually very special, and each case must be treated separately. The test procedures are normally spelled out in the specifications; otherwise an agreement between all parties must be made before testing is started. The testing of eduction or jet pumps falls under this special category of testing.

Other Test Phenomena   When testing pumps, other phenomena of interest should also be checked and noted on the test record. The two phenomena normally reported on are vibration and noise. The acceptable limits of these plus instrumentation for measuring them are special and normally covered in the contract. If the pumps are to be installed in a special environment, this should also be taken into consideration during testing.

REFERENCES

1. American Society of Mechanical Engineers: "Power Test Code, Centrifugal Pumps," PTC 8.2, ASME, New York, 1965.

2. Hydraulic Institute: *Standards for Centrifugal, Rotary, and Reciprocating Pumps*, 14th ed., Cleveland, 1983.

3. American Society of Mechanical Engineers: *Fluid Meters: Their Theory and Application*, 6th ed., ASME, New York, 1971.

4. Smith, H., Jr.: *Hydraulics*, Wiley, New York, 1884.

5. Francis, J. B.: *Lowell Hydraulic Experiments*, Van Nostrand, New York, 1883.

6. Shen, J.: "A Preliminary Report on the Discharge Characteristics of Trapezoidal-Notch Thin-Plate Weirs," *U.S. Gel. Surv.*, 1959.

7. Parshall, R. L.: "Improving the Distribution of Water to Farmers by Use of the Parshall Measuring Flume," Bull. 488, Soil Conservation Service, U.S. Department of Agriculture, Washington, D.C., 1945.

8. Parshall, R. L.: "Measuring Water in Irrigation Channels with Parshall Flumes and Weirs," Bull. 843, Soil Conservation Service, U.S. Department of Agriculture, Washington, D.C., 1950.

9. *Water Measurement Manual*, 2d ed., Bureau of Reclamation, U.S. Department of Interior, Denver, Colorado, 1967.

FURTHER READING

American Water Works Association: "American Standards for Vertical Turbine Pumps," ANSI B-58.1 (AWWA E101-61) including appendix, AWWA.

International Conference on Advancement in Flow Measurement Techniques, BHRA Fluid Engineering, Cranfield, Bedford MK43 OAJ, England, 1981.

A·P·P·E·N·D·I·X

TECHNICAL DATA

Table 1a	Base Units of the International System of Units (SI) and Their Definitions	A.2
Table 1b	Supplementary SI Units and Their Definitions	A.2
Table 2	Derived Units of the SI System	A.3
Table 3	Prefixes of SI Multiple and Submultiple Units	A.3
Table 4	Velocity and Frictional Head Loss in Old Cast Iron Piping based on Hazen-Williams C = 100	A.4
Figure 1	Change in Hazen-Williams Coefficient C with Years of Service for Cast Iron Pipes Handling Soft, Clear, Unfiltered Water	A.6
Figure 2	Atmospheric Pressures for Altitudes up to 12,000 ft (3600 m)	A.7
Table 5	Velocity and Frictional Head Loss in New Schedule 40 (Standard Weight) Steel Piping Based on Darcy-Weisbach	A.8
Table 6	Viscosity of Common Liquids	A.10
Table 7	Viscosity Conversion Tables	A.22
Table 8	Properties of Water at Temperatures from 40 to 705.4°F (7.4 to 374.1°C)	A.26
Table 9	Corrosion Resistance Chart for Materials Commonly Used in Nonmetallic Pumps	A.27
Table 10	Conversions of USCS to SI Units	A.36

TABLE 1a   Base Units of the International System of Units (SI) and Their Definitions

Symbol	Unit	Definition
m	meter[a]	Unit of length equal to 1,650,763.73 wavelengths in vacuum of the radiation corresponding to the transition between the $2p^{10}$ and $5d^5$ levels of the krypton-86 atom.
kg	kilogram	Unit of mass equal to that of the international prototype of the kilogram.
s	second	Unit of time equal to the duration of 9,192,631,770 periods of the radiation corresponding to the transition between two hyperfine levels of the ground state of the cesium-133 atom.
A	ampere	Unit of electric current equal to that which, if maintained in two straight parallel conductors of infinite length and negligible cross section and placed 1 meter (m) apart in vacuum, produces between those conductors a force equal to 2×10^{-7} newtons per meter (N/m) of length.
K	kelvin	Unit of thermodynamic temperature equal to the fraction 1/273.16 of the thermodynamic temperature of the triple point of water.
cd	candela	Unit of luminous intensity equal to that (in the perpendicular direction) of a surface of $1/600,000$ m^2 of a blackbody at the temperature of freezing platinum under a pressure of 101,325 newtons per square meter (N/m^2).
mol	mole	Unit of substance equal to the moment of substance of a system which contains as many elementary entities[b] as there are atoms in 0.012 kg of carbon-12.

[a]Spelled metre in countries other than the United States.
[b]When the mole is used, the elementary entities must be specified and may be atoms, molecules, ions, electrons, other particles, or specified groups of each particles.

TABLE 1b   Supplementary SI Units and Their Definitions

Symbol	Unit	Definition
rad	radian	Unit of measure of a planar angle with its vertex at the center of a circle subtended by an arc equal in length to the radius.
sr	steradian	Unit of meausre of a solid angle with its vertex at the center of a sphere and enclosing an area of the spherical surface equal to that of a square with sides equal in length to the radius.

TABLE 2 Derived Units of the SI System

Quantity	Unit	Symbol	Formula
Acceleration	meters per second per second	m/s^2	—
Angular acceleration	radians per second per second	rad/s^2	—
Angular velocity	radians per second	rad/s	—
Area	square meters	m^2	—
Density	kilograms per cubic meter	kg/m^3	—
Energy	joules	J	$N \cdot m$
Force	newtons	N	$kg \cdot m/s^2$
Frequency	hertz	Hz	cycle/s
Power	watts	W	J/s
Pressure	pascals	Pa	N/m^2
Stress	newtons per square meter	N/m^2	—
Velocity	meters per second	m/s	—
Viscosity, dynamic	newton-seconds per square meter	$N \cdot s/m^2$	—
Viscosity, kinematic	square meters per second	m^2/s	—
Volume	cubic meters	m^3	—
Work	joules	J	$N \cdot m$

TABLE 3 Prefixes for SI Multiple and Submultiple Units

Prefix	SI symbol	Multiplication factor
exa	E	10^{18}
peta	P	10^{15}
tera	T	10^{12}
giga	G	10^{9}
mega	M	10^{6}
kilo	k	10^{3}
hecto	h	10^{2}
deka	da	10
deci	d	10^{-1}
centi	c	10^{-2}
milli	m	10^{-3}
micro	μ	10^{-6}
nano	n	10^{-9}
pico	p	10^{-12}
femto	f	10^{-15}
atto	a	10^{-18}

TABLE 4 Velocity and Frictional Head Loss in Old Cast Iron Piping based on Hazen-Williams $C = 100$

gpm[a]	3-in ID pipe[d]		4-in ID pipe		5-in ID pipe		6-in ID pipe		8-in ID pipe		10-in ID pipe		12-in ID pipe		14-in ID pipe		16-in ID pipe		18-in ID pipe		20-in ID pipe		gpm
	v[b]	f[c]	v	f	v	f	v	f	v	f	v	f	v	f	v	f	v	f	v	f	v	f	
30	1.36	0.534	0.77	0.131																			30
40	1.81	0.910	1.02	0.224																			40
50	2.27	1.38	1.28	0.338	0.82	0.114																	50
60	2.72	1.92	1.53	0.475	0.98	0.160																	60
70	3.18	2.56	1.79	0.631	1.14	0.213	0.79	0.088															70
80	3.63	3.28	2.04	0.808	1.31	0.273	0.91	0.112															80
90	4.08	4.08	2.30	1.01	1.47	0.339	1.02	0.139															90
100	4.54	4.96	2.55	1.22	1.63	0.412	1.14	0.170															100
125	5.68	7.50	3.19	1.85	2.04	0.623	1.42	0.256															125
150	6.81	10.5	3.83	2.59	2.47	0.874	1.70	0.360	0.96	0.89													150
175	7.95	14.0	4.47	3.44	2.86	1.16	1.99	0.478	1.12	0.118													175
200	9.08	17.9	5.10	4.41	3.27	1.49	2.27	0.613	1.28	0.151													200
225	10.2	22.3	5.74	5.48	3.68	1.85	2.55	0.762	1.44	0.188													225
250	11.3	27.1	6.38	6.67	4.08	2.25	2.84	0.926	1.60	0.228	1.02	0.077											250
275	12.5	32.3	7.02	7.96	4.50	2.68	3.12	1.11	1.76	0.272	1.12	0.092											275
300	13.6	37.9	7.65	9.34	4.90	3.13	3.41	1.30	1.91	0.320	1.23	0.108											300
350	15.9	50.4	8.93	12.4	5.72	4.20	3.97	1.73	2.23	0.425	1.43	0.144	0.99	0.059									350
400	18.2	64.6	10.2	15.9	6.54	5.38	4.54	2.21	2.55	0.545	1.63	0.184	1.13	0.076									400
450			11.5	19.8	7.36	6.68	5.10	2.75	2.87	0.678	1.84	0.228	1.28	0.094	0.94	0.044							450
500			12.8	24.1	8.18	8.12	5.68	3.34	3.19	0.823	2.04	0.278	1.42	0.114	1.04	0.054							500
550			14.0	28.7	8.99	9.69	6.24	3.99	3.51	0.982	2.24	0.331	1.56	0.136	1.15	0.064							550
600			15.3	33.7	9.81	11.4	6.81	4.68	3.82	1.15	2.45	0.389	1.70	0.160	1.25	0.076	0.96	0.039					600
650			16.6	39.1	10.6	13.2	7.38	5.43	4.15	1.34	2.65	0.452	1.84	0.186	1.36	0.088	1.04	0.046					650
700			17.9	44.9	11.4	15.1	7.94	6.23	4.47	1.53	2.86	0.518	1.99	0.214	1.46	0.100	1.12	0.052					700
750					12.3	17.2	8.51	7.08	4.78	1.74	3.06	0.589	2.13	0.242	1.56	0.114	1.20	0.060	0.95	0.034			750
800					13.1	19.4	9.08	7.98	5.10	1.97	3.26	0.666	2.27	0.273	1.67	0.129	1.28	0.067	1.01	0.038			800
900					14.7	24.1	10.2	9.92	5.74	2.44	3.67	0.825	2.55	0.339	1.88	0.160	1.44	0.084	1.13	0.047			900
1,000					16.3	29.3	11.4	12.1	6.38	2.97	4.08	1.00	2.83	0.412	2.08	0.195	1.60	0.102	1.26	0.057	1.02	0.034	1,000
1,100					18.0	35.0	12.5	14.4	7.02	3.55	4.50	1.20	3.12	0.492	2.29	0.232	1.76	0.121	1.39	0.068	1.12	0.041	1,100

Frictional head loss table for water flow in pipe (gpm vs. pipe size). The only printed pipe-size header on this page is "24-in ID pipe." For each pipe the left value of a pair is velocity[b] (ft/s) and the right value is frictional head loss[c] (ft per 100 ft). Columns are given in the order shown in the original (12-, 14-, 16-, 18-, 20-, 24-in, then 6-, 8-, 10-in).

gpm	V	hf	V	hf	V	hf	V	hf	V	hf	24-in ID pipe V	hf	V	hf	V	hf	V	hf	gpm
1,200	3.40	0.578	2.50	0.273	1.92	0.143	1.51	0.080	1.23	0.048			13.6	16.9	7.66	4.17	4.90	1.41	1,200
1,300	3.69	0.671	2.71	0.316	2.08	0.165	1.64	0.093	1.33	0.056			14.8	19.6	8.30	4.83	5.31	1.63	1,300
1,400	3.97	0.770	2.92	0.363	2.24	0.190	1.76	0.107	1.43	0.064			15.9	22.5	8.93	5.54	5.72	1.87	1,400
1,500	4.25	0.875	3.12	0.413	2.40	0.215	1.89	0.121	1.53	0.073	1.06	0.030	17.0	25.5	9.55	6.30	6.12	2.13	1,500
1,600	4.54	0.985	3.33	0.465	2.55	0.243	2.02	0.137	1.63	0.082	1.13	0.034	18.2	28.8	10.2	7.10	6.53	2.39	1,600
1,800	5.11	1.22	3.75	0.578	2.87	0.302	2.27	0.170	1.84	0.102	1.28	0.042			11.5	8.83	7.35	2.98	1,800
2,000	5.67	1.49	4.17	0.703	3.19	0.367	2.52	0.207	2.04	0.124	1.42	0.051			12.8	10.7	8.17	3.62	2,000
2,500	7.09	2.25	5.21	1.06	3.99	0.555	3.15	0.312	2.55	0.187	1.77	0.077			16.0	16.2	10.2	5.48	2,500
3,000	8.51	3.16	6.25	1.49	4.78	0.778	3.78	0.438	3.06	0.262	2.13	0.108			19.1	22.8	12.3	7.67	3,000
3,500	9.93	4.20	7.29	1.98	5.59	1.04	4.41	0.583	3.57	0.349	2.48	0.143					14.3	10.2	3,500
4,000	11.3	5.38	8.33	2.54	6.39	1.33	5.04	0.746	4.08	0.447	2.83	0.184					16.3	13.1	4,000
4,500	12.8	6.68	9.38	3.15	7.18	1.65	5.67	0.928	4.59	0.555	3.19	0.228					18.4	16.3	4,500
5,000	14.2	8.13	10.4	3.83	7.98	2.00	6.30	1.13	5.10	0.675	3.54	0.278							5,000
5,500	15.6	9.70	11.5	4.58	8.78	2.39	6.93	1.35	5.61	0.806	3.90	0.332							5,500
6,000	17.0	11.4	12.5	5.38	9.68	2.81	7.56	1.58	6.12	0.947	4.25	0.390							6,000
6,500	18.4	13.2	13.6	6.24	10.4	3.26	8.19	1.83	6.73	1.10	4.61	0.452							6,500
7,000	19.9	15.2	14.6	7.16	11.2	3.74	8.82	2.11	7.15	1.26	4.96	0.518							7,000
7,500			15.6	8.13	12.0	4.24	9.45	2.39	7.66	1.43	5.32	0.589							7,500
8,000			16.7	9.16	12.8	4.79	10.1	2.69	8.17	1.61	5.66	0.664							8,000
9,000			18.8	11.4	14.4	5.95	11.3	3.39	9.18	2.01	6.38	0.825							9,000
10,000					16.0	7.24	12.6	4.07	10.2	2.44	7.09	1.00							10,000
11,000					17.6	8.63	13.9	4.86	11.2	2.91	7.80	1.20							11,000
12,000					19.2	10.1	15.1	5.71	12.3	3.42	8.51	1.41							12,000
13,000							16.4	6.62	13.3	3.96	9.12	1.63							13,000
14,000							17.6	7.59	14.3	4.54	9.93	1.87							14,000
15,000							18.9	8.63	15.3	5.27	10.6	2.13							15,000
16,000									16.3	5.82	11.3	2.40							16,000
18,000									18.4	7.24	12.8	2.98							18,000
20,000											14.2	3.62							20,000
25,000											17.7	5.48							25,000

[a] 1 gpm = 6.31×10^{-5} m³/s = 0.227 m³/h = 0.0631 liters/s.

[b] Velocity, in feet per second. 1 ft/s = 0.305 m/s.

[c] Frictional head loss, in feet of water per 100 feet of pipe. 1 ft = 0.305 m. Friction values apply to cast iron pipes after 15 years of service handling average water. Based on Hazen-Williams formula with C = 100. See Fig. 1 for other values of C.

[d] 1 in = 25.4 mm.

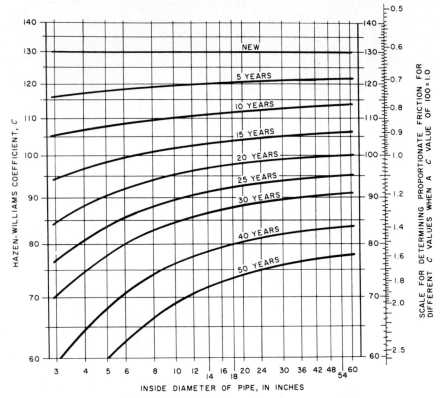

FIG. 1   Change in Hazen-Williams coefficient C with years of service, for cast iron pipes handling soft, clear, unfiltered water. To correct head loss from Table 4, which is based on $C = 100$, multiply by the conversion factor. For example, with a flow of 700 gpm through a 6-in pipe, the frictional head loss is 6.23 ft per 100 ft of pipe with $C = 100$. For $C = 130$, the conversion factor is 0.613; therefore, the frictional head loss is 6.23 × 0.613 = 3.82 ft per 100 ft of pipe. (See also Sec. 8.1.) (1 in = 25.4 mm)

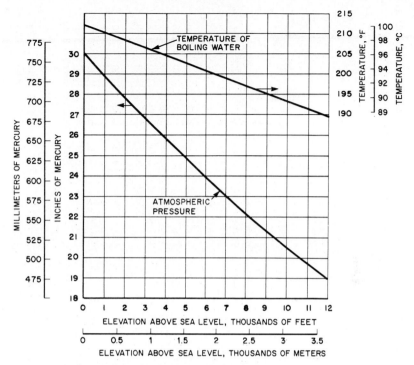

FIG. 2   Atmospheric pressures for altitudes up to 12,000 ft (3600 m).

TABLE 5 Velocity and Frictional Head Loss in New Schedule 40 (Standard Weight) Steel Piping Based on Darcy-Weisbach

gpm[a]	1-in pipe (1.049-in ID) v	f[d]	1¼-in pipe (1.380-in ID) v	f	1½-in pipe (1.610-in ID) v	f	2-in pipe (2.067-in ID) v	f	2½-in pipe (2.469-in ID) v	f	3-in pipe (3.068-in ID) v	f	3½-in pipe (3.548-in ID) v[b]	f[c]	4-in pipe (4.026-in ID) v	f	5-in pipe (5.047-in ID) v	f	6-in pipe (6.065-in ID) v	f	8-in pipe (7.981-in ID) v	f	gpm
1	0.37	0.11																					1
2	0.74	0.39	0.43	0.10																			2
3	1.11	0.82	0.64	0.21	0.47	0.10																	3
4	1.49	1.37	0.86	0.36	0.63	0.17																	4
5	1.86	2.08	1.07	0.54	0.79	0.26																	5
6	2.23	2.83	1.28	0.76	0.95	0.35	0.57	0.10															6
8	2.97	4.88	1.72	1.29	1.26	0.61	0.76	0.17															8
10	3.71	7.12	2.14	1.95	1.57	0.90	0.96	0.26	0.67	0.11													10
15	5.56	15.0	3.21	4.06	2.36	1.87	1.43	0.54	1.00	0.23	0.87	0.13											15
20	7.41	25.6	4.28	6.80	3.15	3.12	1.91	0.92	1.34	0.38													20
25			5.35	10.3	3.94	4.70	2.38	1.39	1.67	0.58	1.08	0.20	0.81	0.10									25
30			6.43	14.4	4.72	6.60	2.86	1.92	2.00	0.81	1.30	0.28	0.97	0.13									30
40					6.30	11.2	3.82	3.35	2.68	1.36	1.73	0.47	1.30	0.23	1.01	0.12							40
50					7.87	16.6	4.77	5.00	3.34	2.06	2.16	0.72	1.62	0.34	1.26	0.18							50
60							5.72	7.00	4.02	2.85	2.60	0.99	1.94	0.48	1.51	0.25							60
70							6.68	9.40	4.68	3.80	3.03	1.33	2.27	0.63	1.76	0.34	1.12	0.11					70
80							7.62	11.9	5.35	4.95	3.46	1.72	2.59	0.82	2.01	0.43	1.28	0.15					80
90							8.60	14.7	6.02	6.05	3.89	2.13	2.91	1.00	2.27	0.54	1.44	0.18					90
100							9.56	18.7	6.70	7.47	4.83	2.58	3.24	1.24	2.52	0.65	1.60	0.22	1.11	0.09			100
125									8.37	11.1	5.41	3.90	4.05	1.82	3.15	1.00	2.00	0.33	1.39	0.13			125
150									10.0	15.4	6.50	5.44	4.86	2.55	3.78	1.39	2.40	0.47	1.66	0.18			150
175									11.7	20.8	7.58	7.30	5.66	3.40	4.40	1.90	2.80	0.62	1.94	0.24			175
200											8.66	9.18	6.48	4.35	5.04	2.40	3.20	0.80	2.22	0.31			200
225											9.75	11.6	7.30	5.44	5.66	2.98	3.60	0.97	2.50	0.38	1.44	0.10	225
250											10.8	14.0	8.10	6.59	6.29	3.68	4.00	1.19	2.77	0.47	1.60	0.12	250

											10-in pipe (10.020-in ID)				
275	8.91	7.90	6.92	4.35	4.40	1.43	3.05	0.56	1.76	0.15			11.9	16.9	275
300	9.72	9.30	7.55	5.04	4.80	1.65	3.32	0.66	1.92	0.17			13.0	19.6	300
350	11.3	12.2	8.80	6.85	5.60	2.21	3.88	0.88	2.24	0.23					350
400	13.0	15.9	10.1	8.67	6.40	2.89	4.44	1.12	2.56	0.29	1.62	0.10			400
450	14.6	20.0	11.3	10.9	7.20	3.56	4.99	1.40	2.88	0.37	1.82	0.12			450
500			12.6	13.3	8.00	4.36	5.54	1.72	3.20	0.45	2.03	0.15			500
550			13.9	16.0	8.80	5.17	6.10	2.06	3.52	0.55	2.23	0.18			550
600			15.1	19.1	9.60	6.16	6.65	2.42	3.84	0.63	2.44	0.21			600
650					10.4	7.22	7.20	2.78	4.16	0.73	2.64	0.24			650
700					11.2	8.29	7.75	3.25	4.47	0.85	2.84	0.28			700
750					12.0	9.40	8.31	3.63	4.80	0.97	3.04	0.31			750
800					12.8	10.3	8.87	4.11	5.11	1.11	3.25	0.35			800
900					14.4	13.0	9.96	5.12	5.75	1.33	3.65	0.44			900
1,000					16.0	15.8	11.1	6.17	6.40	1.64	4.06	0.55			1,000
1,100					17.6	19.0	12.2	7.45	7.04	1.98	4.46	0.64			1,100
1,200							13.3	8.73	7.67	2.36	4.87	0.75			1,200
1,300							14.4	10.2	8.31	2.71	5.27	0.88			1,300
1,400							15.5	11.9	8.95	3.10	5.68	1.02			1,400
1,500							16.7	13.2	9.60	3.49	6.09	1.18			1,500
1,600							17.8	15.0	10.2	3.92	6.49	1.31			1,600
1,800							20.0	18.5	11.5	4.99	7.30	1.60			1,800
2,000									12.8	5.96	8.11	1.97			2,000
2,500									16.0	9.00	10.2	2.95			2,500
3,000									19.2	12.5	12.2	4.15			3,000
3,500									22.4	16.6	14.2	5.60			3,500
4,000											16.2	6.90			4,000
4,500											18.3	8.80			4,500
5,000											20.3	10.8			5,000
5,500											22.3	13.0			5,500
6,000											24.4	15.3			6,000

[a] 1 gpm = 6.31×10^{-5} m^3/s = 0.227 m^3/h = 0.0631 liters/s.

[b] Velocity in feet per second. 1 ft/s = 0.305 m/s.

[c] Frictional head loss, in feet of water per 100 feet of pipe. 1 ft = 0.305 m. Friction values apply to pipe carrying water.

[d] 1 in = 25.4 mm.

TABLE 6 Viscosity of Common Liquids
Typical SSU Values as Standardized by Reputable Authorities Are Shown in Boldface Type

Liquid	Sp. gr. at 60°F (15.6°C)^a	Viscosity		At °F	At °C
		SSU	Centistokes		
Freon	1.37 to 1.49 @ 70°F (21°C)		0.27–0.32	70	21
Glycerine (100%)	1.26 @ 68°F (20°C)	2950	648	68.6	20.3
		813	176	100	38
Glycol					
Propylene	1.038 @ 68°F (20°C)	240.6	52	70	21
Triethylene	1.125 @ 68°F (20°C)	185.7	40	70	21
Diethylene	1.12	149.7	32	70	21
Ethylene	1.125	88.4	17.8	70	21
Hydrochloric acid (31.5%)	1.05 @ 68°F (20°C)		1.9	68	20
Mercury	13.6		0.118	70	21
			0.11	100	38
Phenol (carbolic acid)	0.95 to 1.08	65	11.7	65	18
Silicate of soda	40 Baumé	365	79	100	38
	42 Baumé	637.6	138	100	38
Sulfuric acid (100%)	1.83	75.7	14.6	68	20
Fish and animal oils					
Bone oil	0.918	220	47.5	130	54
		65	11.6	212	100

	Density			Temperature	
Cod oil	0.928	**150** 95	32.1 19.4	100 130	38 54
Lard	0.96	**287** 160	62.1 34.3	100 130	38 54
Lard oil	0.912 to 0.925	**190 to 220** 112 to 128	41 to 47.5 23.4 to 27.1	100 130	30 54
Menhaden oil	0.933	**140** 90	29.8 18.2	100 130	38 54
Neatsfoot oil	0.917	**230** 130	49.7 27.5	100 130	38 54
Sperm oil	0.883	**110** 78	23.0 15.2	100 130	38 54
Whale oil	0.925	**163 to 184** 97 to 112	35 to 39.6 19.9 to 23.4	100 130	38 54

Mineral oils

Automobile crankcase oils (average midcontinent paraffin base):	Density			Temperature	
SAE 10	0.880 to 0.935[b]	**165 to 240** **90 to 120**	35.4 to 51.9 18.2 to 25.3	100 130	38 54
SAE 20	0.880 to 0.935[b]	**240 to 400** **120 to 185**	51.9 to 86.6 25.3 to 39.9	100 130	38 54
SAE 30	0.880 to 0.935[b]	**400 to 580**	86.6 to 125.5	100	38

TABLE 6 Viscosity of Common Liquids (*cont.*)

Liquid	Sp. gr. at 60°F (15.6°C)[a]	Viscosity SSU	Viscosity Centistokes	At °F	At °C
SAE 40	0.880 to 0.935[b]	**185 to 255**	39.9 to 55.1	130	54
		580 to 950	125.5 to 205.6	100	38
		255 to	55.1 to	130	54
		80	15.6	210	99
SAE 50	0.880 to 0.935[b]	950 to 1600	205.6 to 352	100	38
		80 to 105	15.6 to 21.6	210	99
SAE 60	0.880 to 0.935[b]	1600 to 2300	352 to 507	100	38
		105 to 125	21.6 to 26.2	210	99
SAE 70	0.880 to 0.935[b]	2300 to 3100	507 to 682	100	38
		125 to 150	26.2 to 31.8	210	99
SAE 10W	0.880 to 0.935[b]	**5000 to 10,000**	1100 to 2200	0	−18
SAE 20W	0.880 to 0.935[b]	**10,000 to 40,000**	2200 to 8800	0	−18
Automobile transmission lubricants					
SAE 80	0.880 to 0.935[b]	**100,000 max**	22,000 max	0	−18
SAE 90	0.880 to 0.935[b]	**800 to 1500**	173.2 to 324.7	100	38
		300 to 500	64.5 to 108.2	130	54
SAE 140	0.880 to 0.935[b]	950 to 2300	205.6 to 507	130	54
		120 to 200	25.1 to 42.9	210	99
SAE 250	0.880 to 0.935[b]	Over 2300	Over 507	130	54
		Over 200	Over 42.9	210	99

Crude oils

	Specific gravity				
Texas, Oklahoma	0.81 to 0.916	40 to 783 34.2 to 210	4.28 to 169.5 2.45 to 45.3	60 100	16 38
Wyoming, Montana	0.86 to 0.88	74 to 1215 46 to 320	14.1 to 263 6.16 to 69.3	60 100	16 38
California	0.78 to 0.92	40 to 4840 34 to 700	4.28 to 1063 2.4 to 151.5	60 100	16 38
Pennsylvania	0.8 to 0.85	46 to 216 38 to 86	6.16 to 46.7 3.64 to 17.2	60 100	16 38
Diesel engine lubricating oils (based on average midcontinent paraffin base)					
Federal specification no. 9110	0.880 to 0.935[b]	165 to 240 90 to 120	35.4 to 51.9 18.2 to 25.3	100 130	16 38
Federal specification no. 9170	0.880 to 0.935[b]	300 to 410 140 to 180	64.5 to 88.8 29.8 to 38.8	100 130	16 38
Federal specification no. 9250	0.880 to 0.935[b]	470 to 590 200 to 255	101.8 to 127.8 43.2 to 55.1	100 130	16 38
Federal specification no. 9370	0.880 to 0.935[b]	800 to 1,100 320 to 430	173.2 to 238.1 69.3 to 93.1	100 130	16 38
Federal specification no. 9500	0.880 to 0.935[b]	490 to 600 92 to 105	106.1 to 129.9 18.54 to 21.6	130 210	38 99

TABLE 6 Viscosity of Common Liquids (*cont.*)

Liquid	Sp. gr. at 60°F (15.6°C)[a]	Viscosity		At °F	At °C
		SSU	Centistokes		
Diesel fuel oils					
No. 2 D	0.82 to 0.95[b]	32.6 to 45.5 / 39	2 to 6 / 1 to 3.97	100 / 130	38 / 54
No. 3 D	0.82 to 0.95[b]	45.5 to 65 / 39 to 48	6 to 11.75 / 3.97 to 6.78	100 / 130	38 / 54
No. 4 D	0.82 to 0.95[b]	140 max / 70 max	29.8 max / 13.1 max	100 / 130	38 / 54
No. 5 D	0.82 to 0.95[b]	400 max / 165 max	86.6 max / 35.2 max	122 / 160	50 / 71
Fuel oils					
No. 1	0.82 to 0.95[b]	34 to 40 / 32 to 35	2.39 to 4.28 / 2.69	70 / 100	21 / 38
No. 2	0.82 to 0.95[b]	36 to 50 / 33 to 40	3.0 to 7.4 / 2.11 to 4.28	70 / 100	21 / 38
No. 3	0.82 to 0.95[b]	35 to 45 / 32.8 to 39	2.69 to 0.584 / 2.06 to 3.97	100 / 130	38 / 54
No. 5A	0.82 to 0.95[b]	50 to 125 / 42 to 72	7.4 to 26.4 / 4.91 to 13.73	100 / 130	38 / 54
No. 5B	0.82 to 0.95[b]	125 to 400 / 72 to 310	26.4 to 86.6 / 13.63 to 67.1	100 / 122 / 130	38 / 50 / 54

	Specific gravity			°F	°C
No. 6	0.82 to 0.95[b]	450 to **3000** / 175 to 780	97.4 to 660 / 37.5 to 172	122 / 160	50 / 71
Fuel oil, navy specification	0.989 max[b]	110 to **225** / 63 to 115	23 to 48.6 / 11.08 to 23.9	122 / 160	50 / 71
Fuel oil, navy II	1.0 max	**1500** max / 480 max	324.7 max / 104 max	122 / 160	50 / 71
Gasoline	0.68 to 0.74		0.46 to 0.88 / 0.40 to 0.71	60 / 100	16 / 38
Gasoline (natural)	76.5 degrees API		0.41	68	20
Gas oil	28 degrees API	73 / 50	13.9 / 7.4	70 / 100	21 / 38
Insulating oil Transformer, switches, and circuit breakers		115 max / **65** max	24.1 max / 11.75 max	70 / 100	21 / 38
Kerosene	0.78 to 0.82	35 / 32.6	2.69 / 2	68 / 100	20 / 38
Machine lubricating oil (average Pennsylvania paraffin base)					
Federal specification no. 8	0.880 to 0.935[b]	112 to 160 / 70 to **90**	23.4 to 34.3 / 13.1 to 18.2	100 / 130	38 / 54
Federal specification no. 10	0.880 to 0.935[b]	160 to 235 / **90** to **120**	34.3 to 50.8 / 18.2 to 25.3	100 / 130	38 / 54
Federal specification no. 20	0.880 to 0.935[b]	235 to 385 / **120** to **185**	50.8 to 83.4 / 25.3 to 39.9	100 / 130	38 / 54
Federal specification no. 30	0.880 to 0.935[b]	385 to 550 / 185 to 255	83.4 to 119 / 39.9 to 55.1	100 / 130	38 / 54

TABLE 6 Viscosity of Common Liquids (*cont.*)

Liquid	Sp. gr. at 60°F (15.6°C)^a	Viscosity		At °F	At °C
		SSU	Centistokes		
Mineral lard cutting oil					
Federal specification grade 1		**140 to 190** / 86 to 110	29.8 to 41 / 17.22 to 23	100 / 130	38 / 54
Federal specification grade 2		**190 to 220** / 110 to 125	41 to 47.5 / 23 to 26.4	100 / 130	38 / 54
Petrolatum	0.825	100 / 77	20.6 / 14.8	130 / 160	54 / 71
Turbine lubricating oil					
Federal specification (penn base)	0.91 average	**400 to 440** / **185 to 205**	86.6 to 95.2 / 39.9 to 44.3	100 / 130	38 / 54
Vegetable oils					
Castor oil	0.96 @ 68°F (20°C)	1200 to 1500 / **450 to 600**	259.8 to 324.7 / 97.4 to 129.9	100 / 130	38 / 54
China wood oil	0.943	**1425** / 580	308.5 / 125.5	69 / 100	21 / 38
Cocoanut oil	0.925	**140 to 148** / 76 to 80	29.8 to 31.6 / 14.69 to 15.7	100 / 130	38 / 54
Corn oil	0.924	135 / **54**	28.7 / 8.59	130 / 212	54 / 100

Name	Specific gravity				
Cotton seed oil	0.98 to 0.925	**176**	37.9	100	38
		100	20.6	130	54
Linseed oil, raw	0.925 to 0.939	**143**	30.5	100	38
		93	18.94	130	54
Olive oil	0.912 to 0.918	**200**	43.2	100	38
		115	24.1	130	54
Palm oil	0.924	**221**	47.8	100	38
		125	26.4	130	54
Peanut oil	0.920	**195**	42	100	38
		112	23.4	130	54
Rape seed oil	0.919	**250**	54.1	100	38
		145	31	130	54
Rosin oil	0.980	**1500**	324.7	100	38
		600	129.9	130	54
Rosin (wood)	1.09 avg.	500 to **20,000**	108.2 to 4,400	200	93
		1000 to 50,000	216.4 to 11,000	190	88
Sesame oil	0.923	**184**	39.6	100	38
		110	23	130	54
Soybean oil	0.927 to 0.98	**165**	35.4	100	38
		96	19.64	130	54
Turpentine	0.86 to 0.87	33	2.11	60	16
		32.6	2.0	100	38

TABLE 6 Viscosity of Common Liquids (*cont.*)

Liquid	Sp. gr. at 60°F (15.6°C)[a]	Viscosity		At °F	At °C
		SSU	Centistokes		
Sugar, syrups, molasses, etc.					
Corn syrups	1.4 to 1.47	5000 to 500,000 1500 to 60,000	1100 to 110,000 324.7 to 13,200	100 130	38 54
Glucose	1.35 to 1.44	35,000 to 100,000 4000 to 11,000	7700 to 22,000 880 to 2,420	100 150	38 66
Honey (raw)		340	73.6	100	38
Molasses "A" (first)	1.40 to 1.46	1300 to 23,000 700 to 8,000	281.1 to 5070 151.5 to 1760	100 130	38 54
Molasses "B" (second)	1.43 to 1.48	6400 to 60,000 3000 to 15,000	1,410 to 13,200 660 to 3300	100 130	38 54
Molasses "C" (blackstrap or final)	1.46 to 1.49	17,000 to 250,000 6000 to 75,000	2630 to 5500 1320 to 16,500	100 130	38 54
Sucrose solutions (sugar syrups) 60 Brix	1.29	230 92	49.7 18.7	70 100	21 38
62 Brix	1.30	310 111	67.1 23.2	70 100	21 38

62 Brix	1.30	310 111	67.1 23.2	70 100	21 38
64 Brix	1.31	440 148	95.2 31.6	70 100	21 38
66 Brix	1.326	650 195	140.7 42.0	70 100	21 38
68 Brix	1.338	1000 275	216.4 59.5	70 100	21 38
70 Brix	1.35	1650 400	364 86.6	70 100	21 38
72 Brix	1.36	2700 640	595 138.6	70 100	21 38
74 Brix	1.376	5500 1100	1210 238	70 100	21 38
76 Brix	1.39	10,000 2000	2200 440	70 100	21 38

Tars

Tar, coke oven	1.12+	3000 to 8000 650 to 1400	600 to 1760 140.7 to 308	71 100	21 38
Tar, gas house	1.16 to 1.30	15,000 to 300,000 2000 to 20,000	3300 to 66,000 440 to 4400	70 100	21 38

TABLE 6 Viscosity of Common Liquids (*cont.*)

Liquid	Sp. gr. at 60°F (15.6°C)[a]	Viscosity			
		SSU	Centistokes	At °F	At °C
Road tar					
Grade RT-2	1.07+	**200 to 300** 55 to 60	43.2 to 64.9 8.77 to 10.22	122 212	50 100
Grade RT-4	1.08+	**400 to 700** 65 to 75	86.6 to 154 11.63 to 14.28	122 212	50 100
Grade RT-6	1.09+	**1000 to 2000** 85 to 125	216.4 to 440 16.83 to 26.2	122 212	50 100
Grade RT-8	1.13+	**3000 to 8000** 150 to 225	660 to 1760 31.8 to 48.3	122 212	50 100
Grade RT-10	1.14+	**20,000 to 60,000** 250 to 400	4400 to 13,200 53.7 to 86.6	122 212	50 100
Grade RT-12	1.15+	**114,000 to 456,000** **500** to **800**	25,000 to 75,000 108.2 to 173.2	122 212	50 100
Pine tar	1.06	2500 500	559 108.2	100 132	38 56

Miscellaneous

	Specific gravity			Temp, °F	Temp, °C
Corn starch solutions					
22 Baumé	1.18	150 130	32.1 27.5	70 100	21 38
24 Baumé	1.20	600 440	129.8 95.2	70 100	21 38
25 Baumé	1.21	1400 800	303 173.2	70 100	21 38
Ink, printers	1.00 to 1.38	2500 to 10,000 1100 to 3000	550 to 2,200 238.1 to 660	100 130	38 54
Tallow	0.918 avg.	56	9.07	212	100
Milk	1.02 to 1.05		1.13	68	20
Varnish, spar	0.9	1425 650	313 143	68 100	20 38
Water, fresh	1.0		1.13 0.55	60 130	20 38

[a]Unless otherwise noted.
[b]Depends on origin or percent and type of solvent.

SOURCE: *Pipe Friction Manual*, 3d ed., Hydraulic Institute, Cleveland, 1961.

A.21

TABLE 7 Viscosity Conversion Table

This table gives an approximate comparison of various ratings so that if a viscosity is given in terms other than Saybolt seconds universal, it can be translated quickly by following horizontally to the SSU column.

Saybolt universal seconds SSU	Kinematic viscosity, cSt[a]	Approx. seconds, Mac Michael	Approx. Gardner Holt bubble	Seconds, Zahn cup 1	Seconds, Zahn cup 2	Seconds, Zahn cup 3	Seconds, Zahn cup 4	Seconds, Zahn cup 5	Seconds, Demmler cup 1	Seconds, Demmler cup 10	Approx. seconds, Stormer 100-g load	Approx. Seconds, Pratt and Lambert "F"
31	1.00	—	—	—	—	—	—	—	—	—	—	—
35	2.56	—	—	—	—	—	—	—	—	—	—	—
40	4.30	—	—	—	—	—	—	—	1.3	—	—	—
50	7.40	—	—	—	—	—	—	—	2.3	—	2.6	—
60	10.3	—	—	—	—	—	—	—	3.2	—	3.6	—
70	13.1	—	—	—	—	—	—	—	4.1	—	4.6	—
80	15.7	—	—	—	—	—	—	—	4.9	—	5.5	—
90	18.2	—	—	—	18	—	—	—	5.7	—	6.4	—
100	20.6	125	—	38	20	—	—	—	6.5	—	7.3	—
150	32.1	145	—	47	23	—	—	—	10.0	1.0	11.3	—
200	43.2	165	A	54	26	—	—	—	13.5	1.4	15.2	—
250	54.0	198	A	62	29	—	—	—	16.9	1.7	19	—
300	65.0	225	B	73	37	—	—	—	20.4	2.0	23	—
400	87.0	270	C	90	46	—	—	—	27.4	2.7	31	7
500	110.0	320	D	—	55	—	—	—	34.5	3.5	39	8
600	132	370	F	—	63	22.5	—	—	41	4.1	46	9
700	154	420	G	—	72	24.5	—	—	48	4.8	54	9.5
800	176	470	H	—	80	27	18	—	55	5.5	62	10.8
900	198	515	I	—	88	29	20	—	62	6.2	70	11.9
1000	220	570	M	—	—	40	28	138	69	6.9	77	12.4
1500	330	805	—	—	—	—	—	18	103	10.3	116	16.8
2000	440	1070	Q	—	—	51	34	24	137	13.7	154	22

A.22

2500	550	1325	T	—	63	41	29	172	17.2	193	27.6
3000	660	1690	U	—	75	48	33	206	20.6	232	33.7
4000	880	2110	V	—	—	63	43	275	27.5	308	45
5000	1100	2635	W	—	—	77	50	344	34.4	385	55.8
6000	1320	3145	X	—	—	—	65	413	41.3	462	65.5
7000	1540	3670	—	—	—	—	75	481	48	540	77
8000	1760	4170	Y	—	—	—	86	550	55	618	89
9000	1980	4700	—	—	—	—	96	620	62	695	102
10,000	2200	5220	Z	—	—	—	—	690	69	770	113
15,000	3300	7720	Z2	—	—	—	—	1030	103	1160	172
20,000	4400	10500	Z3	—	—	—	—	1370	137	1540	234

[a] Kinematic viscosity (in centistokes) = $\dfrac{\text{absolute viscosity (in centipoises)}}{\text{density}}$

When the SI units centistokes and centipoises are used, the density is numerically equal to the specific gravity. Therefore, the following will be sufficiently accurate for most calculations:

Kinematic viscosity (in centistokes) = $\dfrac{\text{absolute viscosity (in centipoises)}}{\text{specific gravity}}$

When USCS units are used, the density must be used rather than the specific gravity.

NOTE: For values of 70 cSt and above, use the conversion

$$\text{SSU} = \text{centistokes} \times 4.635$$

Above the range of this table and within the range of the viscosimeter, multiply the particular value by the following approximate factors to convert to SSU:

Viscosimeter	Factor	
Mac Michael	1.92	(approx.)
Demmler 1	14.6	
Demmler 10	146	
Stormer	13	(approx.)

SOURCE: *Pipe Friction Manual*, 3d ed., Hydraulic Institute, Cleveland, 1961.

A.23

TABLE 7 Viscosity Conversion Table (*cont.*)

This table gives an approximate comparison of various ratings so that if a viscosity is given in terms other than Saybolt seconds universal, it can be translated quickly by following horizontally to the SSU column.

Saybolt seconds universal SSU	Kinematic viscosity, cSt[a]	Saybolt seconds Furol SSF	Seconds, Redwood 1 (standard)	Seconds, Redwood 2 (admiralty)	Degrees, Engler	Degrees, Barbey	Seconds, Parlin cup 7	Seconds, Parlin cup 10	Seconds, Parlin cup 15	Seconds, Parlin 20	Seconds, Ford cup 3	Seconds, Ford cup 4
31	1.00	—	29	—	1.00	6200	—	—	—	—	—	—
35	2.56	—	32.1	—	1.16	2420	—	—	—	—	—	—
40	4.30	—	36.2	5.10	1.31	1440	—	—	—	—	—	—
50	7.40	—	44.3	5.83	1.58	838	—	—	—	—	—	—
60	10.3	—	52.3	6.77	1.88	618	—	—	—	—	—	—
70	13.1	12.95	60.9	7.60	2.17	483	—	—	—	—	—	—
80	15.7	13.70	69.2	8.44	2.45	404	—	—	—	—	—	—
90	18.2	14.44	77.6	9.30	2.73	348	—	—	—	—	—	—
100	20.6	15.24	85.6	10.12	3.02	307	—	—	—	—	—	—
150	32.1	19.30	128	14.48	4.48	195	—	—	—	—	—	—
200	43.2	23.5	170	18.90	5.92	144	40	—	—	—	—	—
250	54.0	28.0	212	23.45	7.35	114	46	—	—	—	—	—
300	65.0	32.5	254	28.0	8.79	95	52.5	15	6.0	3.0	30	20
400	87.60	41.9	338	37.1	11.70	70.8	66	21	7.2	3.2	42	28
500	110.0	51.6	423	46.2	14.60	56.4	79	25	7.8	3.4	50	34
600	132	61.4	508	55.4	17.50	47.0	92	30	8.5	3.6	58	40
700	154	71.1	592	64.6	20.45	40.3	106	35	9.0	3.9	67	45
800	176	81.0	677	73.8	23.35	35.2	120	39	9.8	4.1	74	50
900	198	91.0	762	83.0	26.30	31.3	135	41	10.7	4.3	82	57
1000	220	100.7	896	92.1	29.20	28.2	149	43	11.5	4.5	90	62
1500	330	150	1270	138.2	43.80	18.7	—	65	15.2	6.3	132	90
2000	440	200	1690	184.2	58.40	14.1	—	86	19.5	7.5	172	118
2500	550	250	2120	230	73.0	11.3	—	108	24	9	218	147

3000	660	300	2540	276	87.60	9.4	—	129	28.5	11	258	172
4000	880	400	3380	368	117.0	7.05	—	172	37	14	337	230
5000	1100	500	4230	461	146	5.64	—	215	47	18	425	290
6000	1320	600	5080	553	175	4.70	—	258	57	22	520	350
7000	1540	700	5920	645	204.5	4.03	—	300	67	25	600	410
8000	1760	800	6770	737	233.5	3.52	—	344	76	29	680	465
9000	1980	900	7620	829	263	3.13	—	387	86	32	780	520
10,000	2200	1000	8460	921	292	2.82	—	430	96	35	850	575
15,000	3300	1500	13,700	—	438	2.50	—	650	147	53	1280	860
20,000	4400	2000	18,400	—	584	1.40	—	860	203	70	1715	1150

aKinematic viscosity (in centistokes) = $\dfrac{\text{absolute viscosity (in centipoises)}}{\text{density}}$

When the SI units centistokes and centipoises are used, the density is numerically equal to the specific gravity. Therefore, the following will be sufficiently accurate for most calculations:

Kinematic viscosity (in centistokes) = $\dfrac{\text{absolute viscosity (in centipoises)}}{\text{specific gravity}}$

When USCS units are used, the density must be used rather than the specific gravity.

NOTE: For values of 70 cSt and above, use the conversion

$$\text{SSU} = \text{centistokes} \times 4.635$$

Above the range of this table and within the range of the viscosimeter, multiply the particular value by the following approximate factors to convert to SSU:

Viscosimeter	Factor
Saybolt Furol	10.
Redwood Standard	1.095
Redwood Admiralty	10.87
Engler, degrees	34.5

Viscosimeter	Factor
Parlin cup #15	98.2
Parlin cup #20	187.0
Ford cup #4	17.4

SOURCE: *Pipe Friction Manual*, 3d ed., Hydraulic Institute, Cleveland, 1961.

TABLE 8   Properties of Water at Temperatures from 40 to 705.4°F (7.4 to 374.1°C)

°F	°C	Specific volume[a], ft³/lb	Specific gravity 39.2°F reference	60°F reference	70°F reference	Wt[b], lb/ft³	Vapor pressure[c], lb/in²abs
40	4.4	0.01602	1.000	1.001	1.002	62.42	0.1217
50	10.0	0.01603	0.999	1.001	1.002	62.38	0.1781
60	15.6	0.01604	0.999	1.000	1.001	62.34	0.2563
70	21.1	0.01606	0.998	0.999	1.000	62.27	0.3631
80	26.7	0.01608	0.996	0.998	0.999	62.19	0.5069
90	32.2	0.01610	0.995	0.996	0.997	62.11	0.6982
100	37.8	0.01613	0.993	0.994	0.995	62.00	0.9492
120	48.9	0.01620	0.989	0.990	0.991	61.73	1.692
140	60.0	0.01629	0.983	0.985	0.986	61.39	2.889
160	71.1	0.01639	0.977	0.979	0.979	61.01	4.741
180	82.2	0.01651	0.970	0.972	0.973	60.57	7.510
200	93.3	0.01663	0.963	0.964	0.966	60.13	11.526
212	100.0	0.01672	0.958	0.959	0.960	59.81	14.696
220	104.4	0.01677	0.955	0.956	0.957	59.63	17.186
240	115.6	0.01692	0.947	0.948	0.949	59.10	24.97
260	126.7	0.01709	0.938	0.939	0.940	58.51	35.43
280	137.8	0.01726	0.928	0.929	0.930	58.00	49.20
300	148.9	0.01745	0.918	0.919	0.920	57.31	67.01
320	160.0	0.01765	0.908	0.909	0.910	56.66	89.66
340	171.1	0.01787	0.896	0.898	0.899	55.96	118.01
360	182.2	0.01811	0.885	0.886	0.887	55.22	153.04
380	193.3	0.01836	0.873	0.874	0.875	54.47	195.77
400	204.4	0.01864	0.859	0.860	0.862	53.65	247.31
420	215.6	0.01894	0.846	0.847	0.848	52.80	308.83
440	226.7	0.01926	0.832	0.833	0.834	51.92	381.59
460	237.8	0.0196	0.817	0.818	0.819	51.02	466.9
480	248.9	0.0200	0.801	0.802	0.803	50.00	566.1
500	260.0	0.0204	0.785	0.786	0.787	49.02	680.8
520	271.1	0.0209	0.765	0.766	0.767	47.85	812.4
540	282.2	0.0215	0.746	0.747	0.748	46.51	962.5
560	293.3	0.0221	0.726	0.727	0.728	45.3	1133.1
580	304.4	0.0228	0.703	0.704	0.704	43.9	1325.8
600	315.6	0.0236	0.678	0.679	0.680	42.3	1542.9
620	326.7	0.0247	0.649	0.650	0.650	40.5	1786.6
640	337.8	0.0260	0.617	0.618	0.618	38.5	2059.7
660	348.9	0.0278	0.577	0.577	0.578	36.0	2365.4
680	360.0	0.0305	0.525	0.526	0.527	32.8	2708.1
700	371.1	0.0369	0.434	0.435	0.435	27.1	3093.7
705.4	374.1	0.0503	0.319	0.319	0.320	19.9	3206.2

[a]1 m³/kg = 16.02 ft³/lb.
[b]1 kg/m³ = 0.06243 lb/ft³.
[c]1 kPa = 0.145 lb/in²; 1 bar = 14.5 lb/in².

SOURCE: Keenan & Keyes' steam table, *Pipe Friction Manual*, 3d ed., Hydraulic Institute, Cleveland, 1961.

TABLE 9 Corrosion Resistance Chart for Materials Commonly Used in Nonmetallic Pumps (See Sec. 5.2 for Explanation)

The four key polymers listed are:
411 V.E. (Dow Chemical vinyl ester FRP)
Epoxy (solvent-resistant, chemical grade)
382 Atlac® (ICI polyester resin)
Kynar® (Pennwalt Corporation's vinylidene fluoride resin)

Polymers are rated as follows:
S = satisfactory
NR = not resistant
Temp., °C = satisfactory up to that temperature

The four corrosion-resistant metals are:
316SS (CF-8M) (DIN 17440)
C-20 (UNS 8020)
Hastelloy C (57 Ni × 16 Cr × 16 Mo)
No. 4 titanium, chemical grade

Metal corrosion ratings are as follows:
A = less than 0.002 in/yr (0.051 mm/yr)
B = less than 0.020 in/yr (0.51 mm/yr)
C = between 0.020 and 0.050 in/yr (0.51 and 0.13 mm/yr)
D = not recommended

Chemical environment	% conc.	Temp. °F/°C	Polymers				Metals			
			411 V.E.	Epoxy	382 Atlac®	Kynar®	316SS	C-20	Hast C	Titan.
Acetic acid	0–10	150/66	S	S	S	S	—	—	—	—
Acetic acid	10–50	100/38	S	S	S	S	—	—	—	A
Acetic acid, glacial	75–100	—	NR	NR	NR	NR	B	A	A	A
Acetone	0–5	250/121	121°C	66°C	82°C	NR	A	A	B	A
Acetone	5–100		NR	NR	NR	NR	A	A	B	A
Acrylic acid	10–25		S	NR	S	NR	A	A	—	—
Adipic acid, soluble		100/38	82°C	S	—	S	B	—	A	A
Alcohol, ethyl	10	210/99	S	S	60°C	S	B	A	A	A
Alcohol, isopropyl	10	150/66	S	S	43°C	S	A	—	—	—
Alcohol, methyl	10	150/66	S	S	43°C	S	A	A	A	—
Allyl chloride	All	80/27	S	38°C	NR	S	D	A	A	A
Alum	All	210/99	S	S	S	S	—	A	A	—
Aluminum chloride	All	210/99	S	S	S	S	D	D	A	B
Aluminum fluoride	All	150/66	27°C	S	43°C	S	C	C	B	A
Aluminum hydroxide	100	150/66	82°C	S	S	S	B	A	B	A

TABLE 9 Corrosion Resistance Chart for Materials Commonly Used in Nonmetallic Pumps (See Sec. 5.2 for Explanation) *(cont.)*

Chemical environment	% conc.	Temp. °F/°C	Polymers				Metals			
			411 V.E.	Epoxy	382 Atlac®	Kynar®	316SS	C-20	Hast C	Titan.
Aluminum nitrate	10	150/66	S	99°C	S	S	A	—	—	A
	10–100	210/99	82°C	S	—	S	A	—	—	A
Aluminum potassium sulfate	All°	210/99	S	S	S	S	B	B	B	A
Ammonia (wet)	All	100/38	S	S	—	S	A	B	A	B
Ammonium bicarbonate cooking liquor	10–50	160/71	S	—	S	S	B	B	—	—
Ammonium bisulfite black liquor	—	150/66	S	—	S	S	—	—	—	—
Ammonium bisulfite	—	180/82	S	—	—	S	—	—	—	—
Ammonium carbonate	—	100/38	65°	93°	S	S	B	B	B	A
Ammonium chloride	All	210/99	S	S	S	S	C	B	A	A
Ammonium hydroxide	0–20	150/66	S	S	60°	S	B	A	A	A
Ammonium hydroxide	20–30	100/38	S	S	S	S	B	A	B	A
Ammonium nitrate	All	210/99	S	S	S	S	A	A	B	—
Ammonium phosphate	65	150/66	99°C	S	S	S	—	—	—	A
Ammonium sulfate	All	210/99	S	S	S	S	B	B	B	A
Amyl acetate	All	75/24	NR	S	S	S	A	A	A	A
Aniline	100	75/24	NR	S	NR	NR	B	B	B	C
Aniline sulfate	All	210/99	S	—	S	S	—	—	—	A
Arsenic acid	19 Be	180/82	S	S	S	S	B	B	B	B
Barium chloride	All	210/99	S	S	S	S	B	B	B	A
Barium hydroxide	10	210/99	65°C	S	Sat'd	S	A	A	B	B
Barium sulfate	All	210/99	S	S	Sat'd	S	B	B	A	B
Beer	—	120/49	S	99°C	43°C	S	A	A	A	B
Beet sugar liquor	—	180/82	S	—	S	S	A	A	—	A
Benzene	All	100/38	NR	S	NR	S	B	B	B	A
Benzene	5% in Kerosene	200/93	NR	S	—	S	B	B	B	A
Benzenesulfonic acid	50	210/99	65°C	—	S	21°C	B	B	B	B

Chemical	Conc.					Sol.				
Black liquor (pulp)	All	180/82	S	—	S	S	A	—	—	—
Bleach liquor (pulp)	100	180/82	S	—	—	S	—	—	—	A
Boiler blowdown	—	210/99	S	—	—	S	A	—	B	B
Borax	100	210/99	S	S	Sat'd	S	A	B	B	A
Boric acid	All	200/93	S	S	Sat'd	S	B	B	A	A
Brine	All	210/99	S	—	Sat'd	S	—	—	B	A
Bromine liquid	100	—	NR	NR	NR	S	—	—	—	A
Bromine wet gas	All	100/38	S	—	5%	S	—	B	B	A
Calcium bisulfite	All	180/82	S	99°C	S	S	A	B	B	B
Calcium carbonate	All	210/99	82°C	S	S	S	B	B	B	B
Calcium chlorate	All	210/99	S	S	S	S	B	B	B	B
Calcium chloride	All	200/93	S	S	S	S	B	A	A	A
Calcium hydroxide	0–50	200/93	S	S	NR	S	A	B	B	B
Calcium hydroxide	100	200/93	S	—	—	S	—	—	B	—
Calcium hypochlorite	0–20	160/71	S	NR	Max. Stable	S	—	B	B	A
Calcium nitrate	All	210/99	S	S	S	S	B	B	B	B
Calcium sulfate	All	210/99	S	S	S	S	B	B	A	A
Cane sugar water	All	100/38	82°C	S	S	S	A	A	B	B
Carbon tetrachloride	All	100/38	66°C	S	Sat'd 43°C	S	A	A	B	A
Carbonic acid	—	150/66	—	S	—	S	A	A	—	B
Cascade detergent	All	180/82	S	—	—	S	A	—	—	—
Chlorinated wax	All	180/82	S	—	—	S	—	A	—	A
Chlorine dioxide	15–100	150/66	S	—	—	S	C	B	B	B
Chlorine water	Sat'd	180/82	S	23°C	S	S	B	B	C	B
Chloracetic acid	0–50	100/38	S	—	S	S	—	—	C	B
Chlorobenzene	100	100/38	NR	S	NR	S	B	B	B	B
Chloroform	100	100/38	NR	S	NR	S	B	B	B	A
Chlorosulfonic acid	100	NR	NR	NR	NR	NR	C	D	D	A
Chromic acid	0–20	150/66	S	NR	43°C	S	D	D	D	B
Chromic acid	30–100	NR	NR	NR	NR	S	D	D	D	B

TABLE 9 Corrosion Resistance Chart for Materials Commonly Used in Nonmetallic Pumps (See Sec. 5.2 for Explanation) (*cont.*)

Chemical environment	% conc.	Temp. °F/°C	Polymers				Metals			
			411 V.E.	Epoxy	382 Atlac®	Kynar®	316SS	C-20	Hast C	Titan.
Citric acid	All	210/99	S	S	S	S	D	B	A	A
Copper chloride	All	210/99	S	S	S	S	—	D	A	B
Copper cyanide	All	210/99	S	—	S	S	B	B	B	A
Copper sulfate	All	210/99	S	S	S	S	B	B	A	A
Cornstarch	Slurry	210/99	S	—	S	S	A	—	—	—
Crude oil sour/sweet	100	210/99	S	S	S	S	A	A	—	A
Diacetone alcohol	100	150/66	S	S	—	21°C	A	—	—	—
Dichlorobenzene	100	100/38	NR	S	S	S	B	B	A	—
Dichloroethylene	100	75/24	NR	S	NR	S	B	B	B	A
Detergents, organic pH 12	100	150/66	S	S	—	S	A	—	B	A
Detergents, sulfonated	All	210/99	S	S	—	S	A	—	B	—
Diethylene glycol	100	180/82	S	—	—	NR	A	—	B	A
Dipropylene glycol	—	150/66	82°C	—	S	21°C	A	—	B	—
Dodecyl alcohol (lauryl)	100	150/66	S	—	S	S	A	—	—	—
Electrosol	5	150/66	49°C	—	S	S	A	—	—	—
Esters, fatty acid	100	182/82	S	S	S	NR	A	—	—	—
Ethyl acetate	—	150/66	NR	S	NR	NR	A	A	A	A
Ethyl chloride	100	75/24	NR	S	—	S	A	B	B	C
Ethylene glycol (dihydroxy)	All	200/93	99°C	99°C	S	S	A	A	B	—
Fatty acids	All	210/99	S	S	S	S	A	A	A	A
Ferric chloride	All	210/99	S	S	S	S	D	D	B	A
Ferrous sulfate	All	210/99	S	S	S	S	B	B	B	A
Ferrous chloride	All	210/99	S	S	S	S	C	D	B	A
Ferrous chloride and HCl	20:5	210/99	S	—	—	S	D	C	B	A
Ferrous nitrate	All	210/99	S	S	S	S	—	—	—	—

Chemical	Conc. %	Temp °F/°C								
Fertilizer 8-8-8	—	120/49	—	—	43°C	S	A	—	—	—
Fertilizer, urea	—	120/49	—	—	S	S	C	—	B	A
Fluosilicic acid	10–32	100/38	S	93°C	66°C	49°C	B	B	B	D
Formaldehyde	44	150/66	S	93°C	43°C	S	C	A	A	B
Formic acid	10	180/82	S	38°C	66°C	S	C	C	C	D
Formic acid	All	100/38	S	S	50% 43°C	S	C	A	C	D
Gluconic acid	50	180/82	S	—	43°C	S	A	B	A	A
Glycerine	100	220/104	99°C	99°C	S	S	A	A	A	A
Glycol, ethylene	All	200/93	99°C	99°C	S	S	B	B	A	A
Glycol, propylene	—	200/93	S	99°C	S	S	—	—	—	—
Gold plating solution	40	180/82	S	S	—	S	A	A	A	—
Hexylene glycol, alcohol	—	15/–66	S	S	—	S	D	D	C	A
Hydrobromic acid	18–50	180/82	66°C	66°C	NR	S	D	D	A	D
Hydrochloric acid	0–37	200/93	S	S	43°C	S	D	D	A	D
Hydrochloric acid and free Cl₂	All	180/82	S	—	—	S	D	D	D	D
Hydrocyanic acid	All	210/99	S	24°C	S	S	B	A	A	B
Hydrogen peroxide	10–30	150/66	S	S	38°C	S	B	B	B	B
Hydrogen sulfide in H₂O	All	180/82	S	—	Max Temp Stable	S	—	—	—	—
Hydrosulfite bleach	—	180/82	S	—	MST	S	—	—	—	A
Hypochlorous acid	10	180/82	S	93°C	MST	S	—	—	—	A
Hypochlorous acid	20	150/66	S	—	MST	S	—	—	—	—
Iron plating solution	—	180/82	S	S	—	S	A	A	A	A
Iron and steel cleaning bath, HCl and H₂SO₄	—	120/49	82°C	—	—	S	—	—	—	—
Lactic acid	All	210/99	S	—	S	21°C	B	B	B	B
Lead acetate	All	210/99	S	S	Sat'd 77°C	S	B	B	B	B
Lithium bromide	Sat'd	210/99	S	82°C	S	S	—	—	—	—
Lithium hypochlorite	All	150/66	S	—	—	S	A	—	—	A
Magnesium carbonate	All	180/82	S	79°C	66°C	S	A	A	A	A
Magnesium chloride	All	210/99	S	S	S	S	B	A	B	A
Magnesium hydroxide	100	210/99	S	S	S	S	B	B	B	A

TABLE 9 Corrosion Resistance Chart for Materials Commonly Used in Nonmetallic Pumps (See Sec. 5.2 for Explanation) (cont.)

Chemical environment	% conc.	Temp. °F/°C	Polymers				Metals			
			411 V.E.	Epoxy	382 Atlac®	Kynar®	316SS	C-20	Hast C	Titan.
Magnesium sulfate	—	200/93	S	S	S	S	B	A	A	A
Maleic acid	All	150/66	99°C	S	S	S	B	B	B	A
Mercuric chloride	All	210/99	S	—	S	S	—	—	—	—
Mercurous chloride	All	210/99	S	—	Sat'd S	S	—	—	—	—
Methyl isobutyl ketone	—	150/66	NR	S	NR	NR	A	A	A	A
Mineral oils	100	210/99	S	S	S	S	A	A	A	A
Naphtha	100	180/82	S	93°C	66°C	S	A	A	B	B
Naphthalene	100	150/66	99°C	S	—	S	A	A	A	A
Nickel chloride	All	210/99	S	S	S	S	D	B	B	B
Nickel nitrate	All	200/93	S	S	S	S	B	B	B	A
Nickel plating solution	—	180/82	S	—	S	S	A	—	—	—
Nickel sulfate	All	210/99	S	NR	S	S	B	B	B	B
Nitric acid	5	150/66	S	NR	S	S	A	A	A	A
Nitric acid	20	120/49	S	—	—	S	A	A	A	A
Oleic acid	All	210/99	S	93°C	S	S	B	A	B	D
Oxalic acid	All	210/99	S	S	S	49°C	D	C	B	D
Palmitric acid	100	210/99	S	—	S	S	A	A	—	—
Perchlorethylene	100	100/38	S	—	S	S	B	B	B	A
Perchloric acid	10	150/66	S	24°C	—	S	D	B	—	D
Peroxide bleach	—	210/99	—	—	—	—	—	—	—	—
Phenol	1	150/66	NR	S	43°C	S	—	A	A	—
Phosphoric acid	85–100	210/99	S	NR	80%	S	B	B	A	C
Phosphoric acid:hydrochloric acid sat'd w/Cl₂	15:9	210/99	S	—	—	S	—	—	A	—
Phosphorous acid	70	100/38	S	—	—	S	A	A	A	—
Polyphosphoric acid	—	210/99	S	—	115% S	S	—	—	—	—

Chemical	Conc.	Temp (°F/°C)								
Pickling acids (Sulfuric and hydrochloric)	All	200/93	S	S	—	S	—	—	B	—
Potassium aluminum sulfate	10–50	210/99	S	—	S	S	B	A	B	A
Potassium bicarbonate	25–50	150/66	S	99°C	10% S	S	B	B	B	A
Potassium carbonate	—	150/66	S	99°C	10% S	S	B	A	B	A
Potassium chloride	All	210/99	S	S	S	S	A	B	B	A
Potassium dichromate	All	210/99	S	S	S	S	A	B	B	A
Potassium gold cyanide	12	100/38	S	—	—	S	A	—	—	—
Potassium hydroxide	10–25	150/66	S	99°C	10% S	S	B	B	B	C
Potassium hydroxide	45	180/82	S	99°C	—	S	B	B	B	C
Potassium hydroxide: potassium cyanide, copper cyanide	2:3:8	180/82	S	—	—	S	—	—	—	—
Potassium nitrate	All	210/99	S	S	Sat'd S	S	B	B	B	A
Potassium permanganate	All	210/99	66°C	—	Sat'd S	S	B	B	B	A
Potassium persulfate	All	210/99	S	—	Sat'd S	S	A	A	—	—
Potassium pyrophosphate	60	130/54	S	—	—	S	A	A	—	—
Potassium sulfate	All	150/66	99°C	S	Sat'd S	S	A	A	B	—
Pulp and paper mill effluent	—	180/82	S	—	—	S	A	A	—	—
Seawater	100	210/99	82°C	S	S	S	B	B	A	A
Silver nitrate	—	210/99	S	S	S	S	B	B	A	A
Silver plating solution	10-sat'd	180/82	S	—	S	S	A	A	A	A
Sodium bicarbonate	15–20	180/82	S	99°C	S	S	A	A	B	A
Sodium bicarbonate:sodium carbonate	All	180/82	S	—	—	S	A	A	B	A
Sodium bisulfate	Sat'd	180/82	S	—	Sat'd S	S	D	A	C	A
Sodium bisulfite	10–35	210/99	S	S	S	S	B	B	B	A
Sodium carbonate	50–100	210/99	S	99°C	—	S	B	A	A	A
Sodium chlorate	—	180/82	S	—	71°C	S	B	B	B	A
Sodium chlorate:sodium chloride	—	210/99	S	S	50% S	S	B	B	B	A
Sodium chloride	Sat'd	210/99	S	—	—	S	D	B	A	A
Sodium chloride and Cl$_2$	Sat'd	180/82	S	—	S	S	D	B	A	A
Sodium chlorite	10–50	100/38	S	S	—	S	—	—	—	—
Sodium cyanide	All	210/99	S	—	5% S	S	D	A	A	A

TABLE 9 Corrosion Resistance Chart for Materials Commonly Used in Nonmetallic Pumps (See Sec. 5.2 for Explanation) (*cont.*)

Chemical environment	% conc.	Temp. °F/°C	Polymers				Metals			
			411 V.E.	Epoxy	382 Atlac®	Kynar®	316SS	C-20	Hast C	Titan.
Sodium fluorosilicate	All	120/49	S	—	S	S	B	—	—	—
Sodium hexametaphosphate	10	100/38	S	—	—	S	A	—	—	—
Sodium hydrosulfide	All	210/99	S	—	NR	S	A	A	—	—
Sodium hydroxide	0–10	180/82	S	S	66°C	S	B	B	B	D
Sodium hydroxide	25–50	210/99	S	S	50% S	S	B	B	B	D
Sodium hypochlorite	5–18	180/82	S	NR	MST	S	C	D	B	C
Sodium Hypochlorite::NaOH scrubbing Cl₂, ClO₂	5	120/49	S	NR	—	S	—	—	—	—
Sodium monophosphate	All	210/99	S	S	—	S	—	—	—	—
Sodium nitrate	All	210/99	S	S	S	S	A	A	B	A
Sodium nitrate	All	210/99	S	S	S	S	C	—	—	A
Sodium silicate	All	210/99	S	66°C	S	S	A	B	B	A
Sodium sulfate	All	210/99	S	S	Sat'd S	S	B	B	B	A
Sodium sulfite	All	210/99	S	S	S	S	B	B	B	A
Sodium tetraborate	Sat'd	180/82	S	—	77°C	S	A	—	—	—
Sorbitol solutions	All	160/71	S	—	66°C	S	A	—	—	—
Stannic chloride	All	210/99	S	S	66°C	S	D	A	B	B
Stannous chloride	All	210/99	S	—	S	S	D	A	B	B
Stearic acid	All	210/99	S	66°C	S	S	A	A	B	A
Sugar beet liquor	—	180/82	S	—	—	S	A	—	—	—
Sugar cane liquor	—	180/82	S	—	—	S	A	—	—	—
Sulfamic acid	10	210/99	S	S	66°C	NR	A	—	B	—
Sulfanilic acid	All	210/99	S	—	Sat'd 82°C	NR	—	—	B	—
Sulfated detergents	50–100	210/99	S	NR	S	S	B	A	B	—
Sulfite/sulfate liquors		200/93	S	NR	S	S	B	B	B	—
Sulfonated detergents	100	160/71	S	—	S	S	A	A	—	—
Sulfuric acid	25	210/99	S	S	S	S	D	B	—	D
Sulfuric acid	70	180/82	S	24°C	S	S	D	B	—	D

Material	Conc (%)	Temp, °F/°C								
Sulfuric acid	75	100/38	S	—	S	S	D	B	—	D
Sulfuric acid	75–100	210/99	NR	NR	—	49°C	D	B	—	D
Sulfuric acid:ferrous sulfate	Sat'd		S	—	—	S	—	B	—	—
Sulfurous acid	10	120/49	S	93°C	43°C	S	D	A	B	A
Superphosphoric acid	105	210/99	S	—	S	S	A	—	—	—
Tannic acid	All	210/99	S	93°C	S	S	A	B	B	A
Tartaric acid	All	210/99	S	S	S	S	A	B	B	A
Tin fluoborate plating solution	All	210/99	S	—	S	S	—	B	—	—
Tobias acid	All	210/99	S	—	S	NR	A	—	—	—
Toluene:toluol	100	150/66	27°C	S	NR	S	A	A	A	A
Toluenesulfonic acid	All	210/99	S	—	S	—	—	—	A	—
Trichloracetic acid	50	210/99	S	S	—	21°C	D	A	A	D
Trichloroethylene	—	150/66	NR	—	—	S	B	—	B	A
Tricresyl phosphate	—	160/71	S	—	—	NR	B	A	A	A
Trisodium phosphate	All	150/66	99°C	NR	50% S	S	B	B	A	—
Tydex 12 fluocculant	12	150/66	S	—	S	S	—	—	—	—
Ultrawet surfactants	All	150/66	S	—	—	S	A	—	—	—
Uran fertilizer	—	150/66	49°C	S	S	S	—	A	—	A
Urea	50	150/66	S	S	S	S	A	B	B	A
Urea:ammonium nitrate	—	120/49	S	S	—	S	—	—	—	—
Versene	—	120/49	S	—	43°C	S	A	A	A	B
Vinegar	100	150/66	99°C	—	S	S	A	—	—	A
Vinyl acetate	—	150/66	—	S	NR	S	B	—	—	—
Water, deionized	100	200/93	82°C	S	S	S	—	—	—	—
Water, demineralized	100	180/82	S	99°C	—	S	B	—	—	—
Water, distilled	100	200/93	82°C	S	S	S	—	A	A	—
Water, sea	100	210/99	82°C	S	S	S	B	A	A	A
Water liquor (pulp)	100	180/82	S	—	—	S	A	—	—	—
Zinc chloride	70	210/99	S	S	S	S	D	A	A	A
Zinc cyanide plating solution	—	180/82	S	—	—	S	—	—	—	A
Zinc electrolyte	—	150/66	S	—	—	S	—	B	—	A
Zinc sulfate	All	210/99	S	S	S	S	A	B	B	A

SOURCE: Robert D. Norton Co.

A.35

TABLE 10 Conversions of USCS to SI Units.
The first two digits of each numeral entry represent a power of 10. An asterisk follows each number which expresses an exact definition.

Acceleration

To convert from	to	multiply by
foot/second²	meter/second²	−01 3.048*
free fall, standard	meter/second²	+00 9.806 65*
gal (galileo)	meter/second²	−02 1.00*
inch/second²	meter/second²	−02 2.54*

Area

To convert from	to	multiply by
acre	meter²	+03 4.046 856 422 4*
circular mil	meter²	−10 5.067 074 8
foot²	meter²	−02 9.290 304*
inch²	meter²	−04 6.4516*
mile² (U.S. statute)	meter²	+06 2.589 988 110 336*
yard²	meter²	−01 8.361 273 6*

Density

To convert from	to	multiply by
gram/centimeter³	kilogram/meter³	+03 1.00*
lbm/inch³	kilogram/meter³	+04 2.767 990 5
lbm/foot³	kilogram/meter³	+01 1.601 846 3
slug/foot³	kilogram/meter³	+02 5.153 79

Energy

To convert from	to	multiply by
British thermal unit (ISO/TC 12)	joule	+03 1.055 06
British thermal unit (International Steam Table)	joule	+03 1.055 04
British thermal unit (mean)	joule	+03 1.055 87
British thermal unit (thermochemical)	joule	+03 1.054 350 264 488
British thermal unit (39°F)	joule	+03 1.059 67
British thermal unit (60°F)	joule	+03 1.054 68
calorie (International Steam Table)	joule	+00 4.1868
calorie (mean)	joule	+00 4.190 02
calorie (thermochemical)	joule	+00 4.184*
calorie (15°C)	joule	+00 4.185 80

Length

To convert from	to	multiply by
fathom	meter	+00 1.8288*
foot	meter	−01 3.048*
foot (U.S. survey)	meter	+00 1200/3937*
foot (U.S. survey)	meter	−01 3.048 006 096
furlong	meter	+02 2.011 68*
inch	meter	−02 2.54*
league (U.K. nautical)	meter	+03 5.559 552*
league (international nautical)	meter	+03 5.556*
league (statute)	meter	+03 4.828 032*
light-year	meter	+15 9.460 55
meter	wavelengths Kr 86	+06 1.650 763 73*
micron	meter	−06 1.00*
mil	meter	−05 2.54*
mile (U.S. statute)	meter	+03 1.609 344*
mile (U.K. nautical)	meter	+03 1.853 184*
mile (international nautical)	meter	+03 1.852*
mile (U.S. nautical)	meter	+03 1.852*
nautical mile (U.K.)	meter	+03 1.853 184*
nautical mile (international)	meter	+03 1.852*
nautical mile (U.S.)	meter	+03 1.852*
rod	meter	+00 5.0292*
statute mile (U.S.)	meter	+03 1.609 344*
yard	meter	−01 9.144*

Mass

To convert from	to	multiply by
carat (metric)	kilogram	−04 2.00*
grain	kilogram	−05 6.479 891*
gram	kilogram	−03 1.00*
kgf second² meter (mass)	kilogram	+00 9.806 65*
kilogram mass	kilogram	+00 1.00*
lbm (pound mass, avoirdupois)	kilogram	−01 4.535 923 7*
ounce mass (avoirdupois)	kilogram	−02 2.834 952 312 5*
ounce mass (troy or apothecary)	kilogram	−02 3.110 347 68*
pennyweight	kilogram	−03 1.555 173 84*
pound mass, lbm (avoirdupois)	kilogram	−01 4.535 923 7*
pound mass (troy or apothecary)	kilogram	−01 3.732 417 216*
slug	kilogram	+01 1.459 390 29
ton (assay)	kilogram	−02 2.916 666 6

Power

To convert from	to	multiply by
ton (long)	kilogram	+03 1.016 046 908 8*
ton (metric)	kilogram	+03 1.00*
ton (short, 2000 pound)	kilogram	+02 9.071 847 4*
Btu (thermochemical)/second	watt	+03 1.054 350 264 488
Btu (thermochemical)/minute	watt	+01 1.757 250 4
calorie (thermochemical)/second	watt	+00 4.184*
calorie (thermochemical)/minute	watt	−02 6.973 333 3
foot lbf/hour	watt	−04 3.766 161 0
foot lbf/minute	watt	−02 2.259 696 6
foot lbf/second	watt	+00 1.355 817 9
horsepower (550 foot lbf/second)	watt	+02 7.456 998 7
horsepower (boiler)	watt	+03 9.809 50
horsepower (electric)	watt	+02 7.46*
horsepower (metric)	watt	+02 7.354 99
horsepower (U.K.)	watt	+02 7.457
horsepower (water)	watt	+02 7.460 43
kilocalorie (thermochemical)/ minute	watt	+01 6.973 333 3
kilocalorie (thermochemical)/ second	watt	+03 4.184*
watt (international of 1948)	watt	+00 1.000 165

Pressure

To convert from	to	multiply by
atmosphere	newton/meter²	+05 1.013 25*
bar	newton/meter²	+05 1.00*
centimeter of mercury (0°C)	newton/meter²	+03 1.333 22
centimeter of water (4°C)	newton/meter²	+01 9.806 38
dyne/centimeter²	newton/meter²	−01 1.00*
foot of water (39.2°F)	newton/meter²	+03 2.988 98
inch of mercury (32°F)	newton/meter²	+03 3.386 389
inch of mercury (60°F)	newton/meter²	+03 3.376 85
inch of water (39.2°F)	newton/meter²	+02 2.490 82
inch of water (60°F)	newton/meter²	+02 2.4884
kgf/centimeter²	newton/meter²	+04 9.806 65*
kgf/meter²	newton/meter²	+00 9.806 65*
lbf/foot²	newton/meter²	+01 4.788 025 8
lbf/inch² (psi)	newton/meter²	+03 6.894 757 2
millibar	newton/meter²	+02 1.00*
millimeter of mercury (0°C)	newton/meter²	+02 1.333 224
pascal	newton/meter²	+00 1.00*
psi (lbf/inch²)	newton/meter²	+03 6.894 757 2

To convert from	to	multiply by
calorie (20°C)	joule	+00 4.181 90
calorie (kilogram, International Steam Table)	joule	+03 4.1868
calorie (kilogram, mean)	joule	+03 4.190 02
calorie (kilogram, thermochemical)	joule	+03 4.184*
electronvolt	joule	−19 1.602 10
erg	joule	−07 1.00*
foot pound force (ft·lbf)	joule	+00 1.355 817 9
foot poundal	joule	−02 4.214 011 0
joule (international of 1948)	joule	+00 1.000 165
kilocalorie (International Steam Table)	joule	+03 4.1868
kilocalorie (mean)	joule	+03 4.190 02
kilocalorie (thermochemical)	joule	+03 4.184*
kilowatthour	joule	+06 3.60*
kilowatthour (international of 1948)	joule	+06 3.600 59
ton (nuclear equivalent of TNT)	joule	+09 4.20
watthour	joule	+03 3.60*

Energy/Area Time

To convert from	to	multiply by
Btu (thermochemical)/foot² second	watt/meter²	+04 1.134 893 1
Btu (thermochemical)/foot² minute	watt/meter²	+02 1.891 488 5
Btu (thermochemical)/foot² hour	watt/meter²	+00 3.152 480 8
Btu (thermochemical)/inch² second	watt/meter²	+06 1.634 246 2
calorie (thermochemical)/cm² minute	watt/meter²	+02 6.973 333 3
erg/centimeter² second	watt/meter²	−03 1.00*
watt/centimeter²	watt/meter²	+04 1.00*

Force

To convert from	to	multiply by
dyne	newton	−05 1.00*
kilogram force (kgf)	newton	+00 9.806 65*
kilopond force	newton	+00 9.806 65*
kip	newton	+03 4.448 221 615 260 5*
lbf (pound force, avoirdupois)	newton	+00 4.448 221 615 260 5*
ounce force (avoirdupois)	newton	−01 2.780 138 5
pound force lbf (avoirdupois)	newton	+00 4.448 221 615 260 5*
poundal	newton	−01 1.382 549 543 76*

Speed

To convert from	to	multiply by
foot/hour	meter/second	−05 8.466 666 6
foot/minute	meter/second	−03 5.08°
foot/second	meter/second	−01 3.048°
inch/second	meter/second	−02 2.54°
kilometer/hour	meter/second	−01 2.777 777 8
knot (international)	meter/second	−01 5.144 444 444
mile/hour (U.S. statute)	meter/second	±01 4.4704°
mile/minute (U.S. statute)	meter/second	+01 2.682 24°
mile/second (U.S. statute)	meter/second	+03 1.609 344°

Temperature

To convert from	to	proceed as follows
Celsius	Kelvin	$t_K = t_C + 273.15$
Fahrenheit	Kelvin	$t_K = (5/9)(t_F + 459.67)$
Fahrenheit	Celsius	$t_C = (5/9)(t_F − 32)$
Rankine	Kelvin	$t_K = (5/9)t_R$

Viscosity

To convert from	to	multiply by
centistoke	meter²/second	−06 1.00°
stoke	meter²/second	−04 1.00°
foot² second	meter²/second	−02 9.290 304°
centipoise	newton second/meter²	−03 1.00°
lbm/foot second	newton second/meter²	+00 1.488 163 9
lbf second/foot²	newton second/meter²	+01 4.788 025 8
poise	newton second/meter²	−01 1.00°
poundal second/foot²	newton second/meter²	+00 1.488 163 9
slug/foot second	newton second/meter²	+01 4.788 025 8
rhe	meter²/newton second	+01 1.00°

Volume

To convert from	to	multiply by
acre foot	meter³	+03 1.233 481 9
barrel (petroleum, 42 gallons)	meter³	−01 1.589 873
bushel (U.S.)	meter³	−02 3.523 907 016 688°
cord	meter³	+00 3.624 556 3
cup	meter³	−04 2.365 882 365°
dram (U.S. fluid)	meter³	−06 3.696 691 195 312 5°
fluid ounce (U.S.)	meter³	−05 2.957 352 956 25°
foot³	meter³	−02 2.831 684 659 2°
gallon (U.K. liquid)	meter³	−03 4.546 087
gallon (U.S. dry)	meter³	−03 4.404 883 770 86°
gallon (U.S. liquid)	meter³	−03 3.785 411 784°
gill (U.S.)	meter³	−04 1.182 941 2
inch³	meter³	−05 1.638 706 4°
liter	meter³	−03 1.00°
ounce (U.S. fluid)	meter³	−05 2.957 352 956 25°
peck (U.S.)	meter³	−03 8.809 767 541 72°
pint (U.S. dry)	meter³	−04 5.506 104 713 575°
pint (U.S. liquid)	meter³	−04 4.731 764 73°
quart (U.S. dry)	meter³	−03 1.101 220 942 715°
quart (U.S. liquid)	meter³	−04 9.463 529 5
tablespoon	meter³	−05 1.478 676 478 125°
teaspoon	meter³	−06 4.928 921 593 75°
ton (register)	meter³	+00 2.831 684 659 2°
yard³	meter³	−01 7.645 548 579 84°

Volume/Unit Time

To convert from	to	multiply by
gallon/minute	meter³/second	−05 6.309 020
gallon/hour	meter³/hour	−01 2.271 247
gallon/minute	liter/second	−02 6.309 020

INDEX

INDEX

Note on the Use of Italics: To allow space in this second edition for the most up-to-date information, a small number of topics discussed in the first edition have been covered in less detail here. The first edition is now out of print, but it can still be obtained on loan from libraries. Moreover, many owners of this Handbook will also have a copy of the first edition on their shelves. For the convenience of readers who have access to a first edition, the page numbers of topics covered in more detail there are given below in italic type.

Abnormal operation, centrifugal pumps, 2.247 to 2.252
Abrasion wear:
  on centrifugal pumps, 9.261, 9.262
  materials for, 5.4, 5.6 to 5.7, 9.262 to 9.264
  on rotary pumps, 3.105
  on screw pumps, 3.74
Absolute velocity, 2.9, 2.196
Acceleration head:
  centrifugal pumps, 8.21 to 8.24
  reciprocating pumps, 3.9 to 3.11, 10.27, 10.29
Acceleration time, centrifugal pumps, 6.15 to 6.17, 8.21 to 8.24
Accelerometers, 2.288 to 2.294
Accumulators, 3.113
Acoustic filters, 3.112 to 3.113
Acoustic flowmeters, 13.25

Adjustable-speed belt drives (*see* Belt drives, adjustable-speed)
Adjustable-speed electric drives (*see* Electric motor speed controls)
Affinity laws, centrifugal pumps, 2.200 to 2.201
  (*See also* Similarity relations)
Air, entrained (*see* Entrained air)
Air admission, 2.239
Air chambers, 2.327, 8.94
Air conditioning, 9.182 to 9.185
  air separation and removal, 9.183 to 9.184
  circuit, 9.183
  condenser water circulation, 9.184
  cooling tower water, 9.184
  well, lake, or seawater, 9.184 to 9.185
Air-lift eductors, 4.22, 4.26
Air-lift pumps, 9.31
  table, 9.27

3

Air-separating tanks, 2.327
Air separations:
  air conditioning, 9.183 to 9.184
  heating systems, 9.180
  priming, 2.327
Air siphons, 4.22, 4.23
Alignment of pumps and drivers, 2.281
    to 2.287, 6.59, 6.60, 12.3
American National Standards Institute
    (ANSI), 9.95 to 9.99, 9.219, 9.221
American Petroleum Institute (API)
    Standard 610, 9.101 to 9.102
American Society of Mechanical Engi-
    neers (*see* ASME *entries*)
Archimedean screw, 1.2, 9.29
Ash pumps, 9.34
ASME Boiler and Pressure Vessel Code,
    Section III (Nuclear Compo-
    nents), 9.219 to 9.222
ASME N-stamp, 9.221
ASME Power Test Code, 13.8
Atmospheric pressure, table, A.7
Automatic bypass control system, 2.316
Axial-flow impeller (propeller), 1.3, 2.33,
    2.48, 2.49, 2.62
Axial thrust:
  centrifugal pumps, 2.59 to 2.66
    balancing, 2.63 to 2.66
    mixed- and axial-flow impellers, 2.62
    multistage pumps, 2.62 to 2.66
    semiopen impellers, 2.61 to 2.62
    single-stage pumps, 2.59 to 2.61
    vertical, 2.102, 2.104 to 2.106
  screw pumps, 3.62

Backflow preventers, 2.76
Backward-curved vanes, 2.197 to 2.198
Balancing devices, hydraulic:
  disks, 2.65 to 2.66
  drums, 2.63 to 2.66
Ballast pumps, 9.159
Barrel pumps, 2.45
Baseplates (*see* Bedplates and pump sup-
    ports)
Bearings:
  antifriction, 2.81, 2.83 to 2.84
    lubrication of, 2.85 to 2.86, 2.302,
      2.303
  babbitted, 2.82
  ball, 2.83 to 2.86
    frequencies generated by, 2.275 to
      2.280

Bearings (*Cont.*):
  journal, 2.166 to 2.192
  Kingsbury, 6.13, *2-101* (*see* index head-
    note)
  power pumps, 3.22 to 3.23
  roller, 2.83
  sleeve, 2.82, 6.13
  stiffness, 2.275 to 2.280
  vertical dry-pit pumps, 2.91
Bedplates and pump supports, 2.88 to
    2.90
  (*See also* Foundations)
Belt drives, adjustable-speed, 6.145 to
    6.152
  controls for, 6.150 to 6.151
  mounting arrangements of, 6.147
  operating principles of, 6.148
  power range of, 6.147
  rating basis of, 6.149 to 6.150
  service factoring of, 6.150
  speed range of, 6.145, 6.148
Bernoulli's theorem, 2.5, 2.7, 4.2, 4.3,
    8.5, 8.8 to 8.11
Bidding, 11.13 to 11.15
  evaluation of, 11.15 to 11.20
Bilge pumps, 9.159
Black liquor, 9.126, 9.133
Bladeless impellers, 9.28
Boiler circulating pumps, 9.78 to 9.80,
    9.82
Boiler-feed pumps:
  marine pumping system, 9.153 to 9.154
  steam power plants, 9.64 to 9.68
Boiler injectors, 4.26, 4.27
Boiling water reactor (BWR) plant, 9.209
    to 9.212
Booster pumps:
  marine feedwater, 9.154
  steam power plants, 9.68 to 9.70
  water, 9.20 to 9.21, 9.303 to 9.317
    limited-storage constant-speed multi-
      ple-pump system, 9.311 to 9.313
    tankless constant-speed multiple-
      pump system, 9.308 to 9.311,
      9.313 to 9.317
    variable-speed-drive pressure boost-
      er system, 9.305 to 9.309
Borehole pumps (*see* Vertical turbine
    pumps)
Boric acid transfer pumps, 9.205
Boron injection recirculation pumps,
    9.205
Bottom-suction pumps, 2.34, 2.40, 2.41

Bowl, centrifugal pumps, **2.36**
  assembly of, **2.95** to **2.98**
Brake horsepower:
  of centrifugal pumps, **2.7**, **2.8**, **2.195**,
    2.196
  of power pumps, **3.3** to **3.4**
  of rotary pumps, **3.94**
  of screw pumps, **3.73**
Branch-line pumping systems, **8.77** to
    8.84
  closed-loop system-head curves, **8.77**
    to **8.79**
  flow control in, **8.82** to **8.83**
  open-ended system head curves, **8.79**
    to **8.80**
  pump total head in, **8.84**
BWR (boiling water reactor) plant, **9.209**
    to **9.212**
Bypass (*see* Pump bypass)

Calibration, test instrument, **13.8**
Can (tank), vertical turbine pumps, **2.96**,
    10.5, **10.23**
Canned (sealless) motor pumps, **2.110** to
    2.112, **6.22** to **6.25**
Cantilever pumps, **2.101**
Capacity:
  of centrifugal pumps, **2.194**
  of drainage pumps, **9.43** to **9.44**
  of fluid current couplings, **6.118** to
    6.119, **6.123**
  of irrigation pumps, **9.43** to **9.44**
  of power pumps, **3.4**
  of rotary pumps, **3.93**
  of screw pumps, **3.60**, **3.71**
  of steam pumps, **3.42**
  testing of, **13.13**
Capacity regulation, centrifugal pumps,
    2.237 to **2.239**
  adjustable vanes, **2.238** to **2.239**
  air admission, **2.239**
  bypass regulation, **2.238**
  discharge throttling, **2.237**
  speed regulation, **2.238**
  suction throttling, **2.237** to **2.238**
  (*See also* Controls, pump)
Cargo pumps, **9.162** to **9.163**
Casing handholes, **2.41**
Casing suction head (cover), **2.41**
Casings, centrifugal pumps:
  axially split (horizontally), **2.34**, **2.37**,
    2.39, **2.42** to **2.45**

Casings, centrifugal pumps (*Cont.*):
  diffuser, **2.33** to **2.36**, **2.95** to **2.99**
    (*See also* Diffuser casing)
  double-volute (twin), **2.36**, **2.37**, **2.101**
  radially split (vertically), **2.34**, **2.39**,
    2.45
  volute, **2.33**, **2.35** to **2.37**
    (*See also* Volute casing)
Cavitation, **2.266** to **2.267**
  centrifugal pumps, **2.213** to **2.229**
  materials resistant to, **5.6**, **2.227**
  model tests of, **9.192**, **13.40**
  noise from, **8.104** to **8.105**
  pump turbine, **9.188**
  pumped storage, **9.188**
  screw pumps, **3.67**
  test of, **2.220** to **2.223**
  Thoma parameter for, **2.221**
    (*See also* Net positive suction head)
Cavitation erosion, **5.6**
Centerline support of casings, **2.90**
Centrifugal pumps, **2.1** to **2.330**
  abrasion wear on, **9.261** to **9.262**
  acceleration head, **8.21** to **8.24**
  acceleration time, **6.15** to **6.17**, **8.21** to
    8.24
  affinity laws, **2.200** to **2.201**
  axial thrust (*see* Axial thrust, centrifu-
    gal pumps)
  bearings for (*see* Bearings)
  bowl, **2.36**, **2.95** to **2.98**
  brake horsepower, **2.7**, **2.8**, **2.195**,
    2.196
  bypass relief, **2.305** to **2.318**, **8.80** to
    8.82
  capacity of, **2.194**
  capacity regulation (*see* Capacity regu-
    lation, centrifugal pumps)
  casings for (*see* Casings, centrifugal
    pumps)
  cavitation, **2.213** to **2.229**
  characteristic diagrams, complete,
    2.247 to **2.250**, **8.15** to **8.24**
  characteristics of, **2.196** to **2.213**, **2.247**
    to **2.252**
  classification of, **1.2** to **1.4**, **2.33** to **2.35**
  construction of, **2.33** to **2.192**
  design of, **2.3** to **2.32**
  efficiency of (*see* Efficiency, of centrif-
    ugal pumps)
  impellers for (*see* Impellers)
  major components of, **2.33** to **2.113**

Centrifugal pumps (*Cont.*):
  materials for, **9.93** to **9.95**
    (*See also* Materials of construction)
  minimum flow (*see* Minimum flow,
    centrifugal pumps)
  noise from (*see* Noise, pump)
  nomenclature, **2.33** to **2.35**
  packing, **2.114** to **2.126**
  performance: hydraulic, **2.194** to **2.273**
    mechanical, **2.274** to **2.304**
  petroleum industry (*see* Petroleum in-
    dustry, centrifugal pumps)
  priming (*see* Priming, centrifugal
    pumps)
  recirculation, **2.241** to **2.243**, **2.267**,
    2.305 to **2.318**
  selection of, **2.229** to **2.230**
  shafts for (*see* Shafts, centrifugal
    pumps)
  solids-handling, **9.261** to **9.268**
  starting (*see* Starting centrifugal
    pumps)
  stuffing boxes (*see* Stuffing boxes, cen-
    trifugal pumps)
  temperature rise, **2.243** to **2.244**, **12.11**,
    12.12
  theory of, **2.3** to **2.32**
  torque (*see* Torque, centrifugal pumps)
  troubleshooting, **2.294** to **2.301**, **12.16**
  vertical (*see* Vertical centrifugal
    pumps)
  vibration, **2.253**, **2.274** to **2.304**
  viscosity, effects of, **9.114** to **9.119**
Channels, open, flow in, **8.39**, **8.49** to
  8.52
Charging pumps, **9.204**, **9.206**
Chemical industry pumps, **9.83** to **9.99**
  corrosion of, **9.83** to **9.89**
  design of, **9.90** to **9.93**
  materials of construction for, **9.85** to
    9.89, **9.93** to **9.95**
  standards for, **9.95** to **9.98**
Chilled water pumps:
  air conditioning, **9.182**
  nuclear, **9.204**, **9.207**
Circulating pumps:
  boiler, **9.78** to **9.80**, **9.82**
  condenser (*see* Condenser circulating
    pumps)
  hot water, **9.179** to **9.180**
Classification of pumps, **1.1** to **1.5**
  displacement (reciprocating; rotary)
    (*see* Displacement pumps)

Classification of pumps (*Cont.*):
  dynamic (centrifugal), **1.2** to **1.5**, **2.33**
    to **2.35**
Clutches, **6-152** to **6-161** (*see* index head-
    note)
Coatings, **5.16** to **5.19**, **5.21**
Codes, **11.4** to **11.6**
Component cooling water pumps, **9.204**,
  9. 206
Compressibility factor, **3.5**
Condensate injection sealing, **2.159** to
  2.165
Condensate pumps, **9.70** to **9.75**
  marine distilling plant, **9.158**
  marine pumping system, **9.151** to **9.152**
  steam power plant, **9.70** to **9.75**
Condenser circulating pumps:
  air conditioning, **9.184**
  marine pumping system, **9.158**
  steam power plant, **9.76** to **9.78**
Conduits, noncircular, flow in, **8.39**, **8.48**
  to **8.51**
Connecting rod, power pumps, **3.20**
Construction of pumps (*see* Materials of
  construction)
Construction services, **10-137** to **10-142**
  (*see* index headnote)
Contact secondary controller, **6.99** to
  6.100
Containment spray pumps, **9.207**
Control rod drive pumps, **2.210**, **2.212**
Controls, pump, **2.237** to **2.239**, **7.1** to
  7.14
  alternators for, **7.8**
  boiler feedwater control, **7.11** to **7.13**
  building-water pressure control, **7.13**
    to **7.14**
  bypass for, **2.238**, **7.9**, **8.80** to **8.83**
  closed-loop control, **7.3**
  liquid-level sensors, **7.4** to **7.6**
  mining services, **9.149**
  on-off control, **7.3**
  open-loop control, **7.3**
  pressure sensors, **7.6** to **7.8**
  proportional control, **7.3**
  submergence control, **7.9**
  telemetry systems for, **7.8** to **7.9**
  throttling, **2.237** to **2.238**, **7.9**, **7.11**
  transducers and transmitters, **7.8** to **7.9**
  types of, **7.2**
  valve-throttling control, **7.10** to **7.11**
  (*See also* Capacity regulation, centrifu-
    gal pumps)

Conversion tables:
  USCS to SI units, **A**.36 to **A**.38
  viscosity, **A**.22 to **A**.25
Cooling tower water, air conditioning, 9.184
Core spray pumps, 9.210, 9.212
Corrosion:
  abrasive wear, **5**.6 to **5**.7
  cavitation erosion, **5**.6
  concentration cell (crevice), **9**.88
  erosion, **5**.4 to **5**.5, **9**.89, 9.254 to 9.255
  fatigue, **5**.5
  galvanic, **5**.5, **5**.13, **9**.89
  graphitization, **5**.7 to **5**.8
  intergranular, **5**.5 to **5**.6, **9**.88 to **9**.89
  leaching, **9**.89
  nonmetallic materials, **A**.27 to **A**.35
  pitting, **9**.88
  stress (cracking), **9**.88
  velocity, **5**.4
Cost, *9-60* to *9-70* (*see* index headnote)
Couplings:
  motor: direct, **6**.12 to **6**.13
    nonreverse ratchet, **6**.13
    self-release, **6**.13
  pump, **6**.153 to **6**.165
    continuously lubricated, **6**.155
    eddy-current (*see* Eddy-current couplings)
    flexible, **6**.155 to **6**.160
    fluid (*see* Fluid couplings)
    limited end float, **6**.159
    rigid, **6**.153 to **6**.155
    spacers, **6**.161
    universal joint, **6**.163, **6**.164
    vibration due to, **2**.275
Crankshaft, power pumps, **3**.19 to **3**.20
Critical speed, **2**.67
  shaft lateral, **2**.67 to **2**.69, **6**.164, **6**.165
  shaft torsional, **6**.164 to **6**.165
Crosshead, power pumps, **3**.21
Crossways, power pumps, **3**.21
Cryogenic liquid pumps, 9.295 to 9.302
  construction, 9.296, 9.298, 9.300
  performance, 9.297
  materials, 9.301
  testing, 9.302
Current meters, hydraulic, **13**.13, **13**.25
Curve shapes, centrifugal pumps, **2**.201, 2.203
Cutwater, **8**.106 to **8**.107
Cyclone separator, **2**.76

Darcy-Weisbach formula, **8**.32 to **8**.33
  table, **A**.8 to **A**.9
Datum plane (elevation), **2**.195, **8**.4 to **8**.8
Deep-well pumps (*see* Vertical turbine pumps)
Diaphragm, interstage, **2**.42 to **2**.45
Diaphragm pumps, **3**.49 to **3**.56
  hydraulically actuated, 9.236 to 9.237
  materials of construction, **3**.54 to **3**.55
  mechanically actuated, 9.236
  slurries, 9.134
Differential pressure, pump, **8**.8 to **8**.11
  (*See also* Head)
Differential head, **8**.8 to **8**.11
Differential pressure meters, **13**.13, **13**.15 to **13**.18
Diffuser casing, **2**.19
  centrifugal pumps, **2**.33 to **2**.36, **2**.95 to 2.99
  materials of construction, **5**.11
Diffuser pumps, **2**.33 to **2**.36, **2**.95 to **2**.99
Discharge recirculation, **2**.267
Disk friction, **2**.7, **2**.9, **2**.21
Displacement, **3**.4
Displacement pumps:
  classification of, **1**.2 to **1**.5
  diagnostics, **3**.114 to **3**.117
  flow control, **3**.119 to **3**.127
  instrumentation, **3**.114 to **3**.117
  operational problems, **3**.107, **3**.108
  pulsations, **3**.108, **3**.111 to **3**.112
  vibrations, **3**.114
  (*See also specific type*)
Displacement sensors, **2**.288 to **2**.294
Double-acting steam pumps, **3**.27
Double-casing pumps, **2**.45
Double-end screw pumps, **3**.62
Double-suction impellers:
  dry-pit pumps, **2**.33, **2**.48 to **2**.50
  wet-pit pumps, **2**.101
Double-volute (twin-volute) casing pumps, **2**.36, **2**.37, **2**.101
Doweling of pumps and drivers, **12**.7
Drainage pumps, 9.41 to 9.50
  capacity of, 9.43 to 9.44
  drivers for, 9.48, 9.49
  head of, 9.44 to 9.48
  marine pumping system, 9.156
  number required, 9.43
  setting of, 9.48
  station design for, 9.49, 9.50
  types of, 9.42 to 9.43

Dry-pit centrifugal pumps, **2.35**, **2.90** to **2.95**
  suction piping for, **10.3**, **10.4**, **10.24** to **10.29**
Dual-volute casing (*see* Double-volute casing pumps)
Duplex steam pumps, **3.27** to **3.28**
Dynamic pumps:
  classification of, **1.2** to **1.5**, **2.33** to **2.35**
  special variants of, *2-199* to *2-206* (*see* index headnote)
Dynamometers:
  torsion, **13.34**
  transmission, **13.33** to **13.34**

Eccentric straps, power pumps, **3.20**
Economics of pumping systems, *9-60* to *9-70* (*see* index headnote)
  paper mill service, **9.131** to **9.135**
  pumped storage project, **9.196** to **9.197**
  slurry pipelines, **9.255** to **9.259**
Eddy-current couplings, **6.81** to **6.91**
  applications of, **6.87** to **6.89**
  construction of, **6.85** to **6.86**
  controls for, **6.87**
  description of, **6.81**
  efficiency of, **6.83**
  fundamentals of, **6.81** to **6.83**
  load characteristics of, **6.83** to **6.85**
  ratings and sizes of, **6.89** to **6.91**
  slip loss in, **6.82** to **6.83**
  thrust capabilities of, **6.91**
Eddy-current displacement sensor, **2.288** to **2.291**
Eductors, **4.2** to **4.20**
  air-lift, **4.22**, **4.26**
  applications of, **4.8** to **4.20**
  deep-well, **4.16** to **4.18**
  definitions of, **4.2**
  design of, **4.2** to **4.8**
  general purpose of, **4.9** to **4.11**
  mixing, **4.11** to **4.13**
  multinozzle, **4.16**
  performance of, **4.8**
  priming (water-jet exhausters), **4.19** to **4.20**
  sand and mud, **4.13** to **4.15**
  solids-handling, **4.15**
  spindle proportioning, **4.13**
  theory of, **4.2** to **4.8**

Efficiency:
  calculation of, **13.35**
  of centrifugal pumps, **2.196**, **2.203**, **2.253** to **2.255**
  definition of, **13.6**
  of eddy-current couplings, **6.83**
  of electric motors, **6.17**
  of fluid couplings, **6.122** to **6.123**
  hydraulic, **2.7**, **2.20**
  hydraulic power, **6.177** to **6.178**
  impellers, **2.8** to **2.9**
  maximum, **2.22** to **2.23**
  mechanical: of power pumps, **3.6**
    of steam pumps, **3.45**
  model tests, **2.253** to **2.255**
  overall, **2.8**, **2.196**, **13.7**
  of paper stock pumps, **9.132** to **9.133**
  of rotary pumps, **3.95** to **3.96**
  scale correction, **13.39**
  specific speed relationship, **2.12**
  volumetric, **2.8**, **2.20**, **3.5**, **3.96**
  wire-to-liquid, **2.196**
  wire-to-water, **2.196**
Ejectors, **4.2**
  pneumatic, **9.25**, **9.27**, **9.28**
  priming, **2.322**
  (*See also* Jet pumps)
Elastomers, **5.16** to **5.17**
  rubberlike, **5.18**
Electric drives, single-unit adjustable-speed (*see* Electric motor speed controls)
Electric input horsepower (kilowatts), **13.5**, **13.35**
Electric motor speed controls, **6.92** to **6.109**
  ac adjustable-voltage drives, **6.92** to **6.94**
  adjustable-frequency drives, **6.100** to **6.102**
  comparison of, **6.105** to **6.109**
  contact secondary controls, **6.99** to **6.100**
  dc motors with SCR power supplies, **6.103** to **6.105**
    drive costs, **6.105**
  efficiency of, **6.105**
  heat loss, **6.108** to **6.109**
  power factor, **6.105**
  wound-rotor induction motors: contact secondary controls, **6.99** to **6.100**
    liquid rheostat controls, **6.94** to **6.96**

Electric motor speed controls (*Cont.*):
  Tirastat II secondary controls, **6.**96
    to **6.**99
Electric motor starters, **6.**18 to **6.**22
Electric motors:
  ac types of, **6.**3 to **6.**6
    squirrel-cage induction motor, **6.**3 to
      6.4, **6.**92 to **6.**94, **6.**100 to **6.**103
    synchronous motor, **6.**4, **6.**6
    wound-rotor induction motor, **6.**4,
      6.94 to **6.**100
  bearings, **6.**11
  controls for, **6.**18 to **6.**22
  coupling methods, **6.**12 to **6.**13
  dc types of, **6.**6 to **6.**9
  efficiency of, **6.**17
  enclosures, **6.**9 to **6.**11
  insulation, **6.**11 to **6.**12
  lubrication, **6.**11
  NEMA, **6.**4, **6.**11, **6.**17, **6.**19 to **6.**21
  performance of, **6.**14 to **6.**18
  power factor, **6.**17 to **6.**18
  sealless centrifugal pumps, **6.**22 to **6.**25
  service factor, **6.**17
  speed-torque characteristics of, **6.**4,
    6.93 to **6.**96
  total mass moment of inertia (MK^2),
    6.16
  total weight (force) moment of inertia
    (WK^2), **6.**15, **6.**16
Electromagnetic pumps, *2-205* (*see* index
    headnote)
Elevation head, **2.**5, **2.**195, **8.**6
End-suction pumps, **2.**34, **2.**39
Endurance limit, **5.**5, **5.**9
Energy in incompressible liquid, **2.**194
    to **2.**196, **8.**4 to **8.**8
  elevation head, **2.**5, **2.**195, **8.**6
  kinetic, **2.**5, **2.**194 to **2.**195
  potential head, **2.**195
  pressure head, **2.**5, **2.**194, **8.**5 to **8.**6
  total head, **2.**194, **2.**195, **8.**6 to **8.**8
  velocity head, **2.**5, **2.**194 to **2.**195, **8.**5,
    8.6
  (*See also* Head)
Energy gradient, **8.**11, **8.**50
Engines, internal combustion, **6.**44 to
    6.60
  air-intake systems for, **6.**55 to **6.**56
  alignment of, **6.**59 to **6.**60
  brake mean effective pressure, **6.**46
  cooling systems for, **6.**49 to **6.**55

Engines, internal combustion (*Cont.*):
  design variations of, **6.**44
  displacement of, **6.**46
  duty cycle, **6.**45
  exhaust systems for, **6.**56
  foundations for, **6.**59
  fuel systems for, **6.**47 to **6.**49
    diesel, **6.**48 to **6.**49
    gas, **6.**48
    gasoline, **6.**47
  ignition systems for, **6.**58
  piston speed of, **6.**46 to **6.**47
  power ratings, **6.**44 to **6.**47
  rotation of, **6.**47
  starting systems for, **6.**56 to **6.**58
  torque of, **6.**46
  torsional analysis of, **6.**60
  vibration isolation of, **6.**60
Entrained air:
  centrifugal pumps, **2.**226, **2.**228 to
    2.229, **10.**35, **10.**43
  screw pumps, **3.**69 to **3.**70
Entrained gas:
  rotary pumps, **3.**105
  screw pumps, **3.**75 to **3.**78
Equivalent diameter, **8.**52 to **8.**54
Equivalent length for valves and fittings,
    8.52 to **8.**53
Erosion:
  abrasive wear, **5.**6 to **5.**7, **9.**262 to **9.**264
  cavitation, **5.**6
  impeller vane tip, **5.**6
  (*See also* Corrosion)
Euler's equation, **2.**6 to **2.**7, **2.**11, **2.**196
Expansion joints, **12.**8 to **12.**9

Face and rim (*see* Alignment of pumps
    and drivers)
Fast breeder (liquid metal) plants, **9.**212
Field approximating, **13.**26
Field tests, **13.**2
Fire pumps, **9.**51 to **9.**59
  controllers, **9.**58
  drivers, **9.**54
  materials, **9.**54
  selection, **9.**52
  shutoff pressure, **9.**53
Firm capacity, definition of, **9.**28
Flexible member pumps, **3.**85 to **3.**86
Flexible tube pumps, **3.**85, **3.**87, **5.**23 to
    5.24

Float switch, **7.**4 to **7.**6
Floating ring seals, **2.**160 to **2.**161
Flow coefficient for valves (C_v;K_v), **7.**21,
    7.22, **8.**54 to **8.**56
Flow work, **2.**194, **8.**5, **8.**6
Flowmeters:
  acoustic, **13.**25
  flume, **13.**13, **13.**22
  horizontal discharge pipe, **13.**26
  losses, **8.**64
  nozzle, **8.**64, **13.**13; **13.**15
  nutating disk, **13.**13, **13.**14
  orifice, **8.**64, **13.**13, **13.**16
  piston, **13.**14
  pitot tube, **13.**13, **13.**17
  radioisotope, **13.**25
  salt velocity, **13.**25
  venturi, **8.**64, **13.**13, **13.**15
  vertical discharge pipe, **13.**26, **13.**27
  weir, **13.**13, **13.**18
Fluid couplings, **6.**110 to **6.**124
  capacity of, **6.**118 to **6.**119, **6.**123
  controllers for, **6.**123
  dimensions of, **6.**123 to **6.**124
  efficiency of, **6.**122 to **6.**123
  hydrodynamic drives, **6.**115 to **6.**116,
    6.122
  hydrokinetic drives, **6.**110 to **6.**115,
    6.120, **6.**122
  hydrostatic drives, **6.**117 to **6.**118
  hydroviscous drives, **6.**116 to **6.**117,
    6.122
  regulation of, **6.**119
  selection and pricing of, **6.**124
  turndown of, **6.**119 to **6.**120
Fluid drives (*See* Fluid couplings)
Flywheel effect, **8.**88
Food and beverage pumping, **9.**137 to
    9.141
  homogenizer, **9.**140
  industry standards for, **9.**137, **9.**139
  mechanical seals, **9.**139
  pump drives, **9.**140 to **9.**141
  pump types, **9.**139
Foot valves, **2.**319 to **2.**320
Forebay (*see* Intake, forebay)
Forward-curved vanes, **2.**200
Foundations:
  gears, **6.**135, **6.**140
  installation of, **12.**2 to **12.**3
  screw pumps, **3.**74
  vibrations due to, **2.**276 to **2.**280

Foundations (*Cont.*):
  (*See also* Bedplates and pump sup-
    ports)
Friction, pipe, **8.**31 to **8.**49, **A.**4 to **A.**6,
    A.8 to **A.**9
  (*See also* Head loss, pressure pipes)
Friction factors:
  Darcy-Weisbach formula, **8.**32 to **8.**33,
    A.8 to **A.**9
  Hazen-Williams formula, **8.**33, **8.**37,
    A.6
  laminar flow, **8.**31, **8.**32
  Manning formula, **8.**39
    table, **8.**49
  turbulent flow, **8.**31, **8.**32
Froude number, **10.**40
  (*See also* Intake modeling)
Froude scaling, **10.**40 to **10.**41
Fuel oil pumps:
  heating systems, **9.**182
  marine pumping systems, **9.**163

Galvanic corrosion, **5.**5, **5.**13, **9.**89
Gas, entrained (*see* Entrained gas)
Gas turbines (*see* Turbines, gas)
Gear pumps, **3.**81 to **3.**82
Gears, **6.**125 to **6.**144
  cross-axis, **6.**129 to **6.**130
    hypoid, **6.**125, **6.**129
    spiral-bevel, **6.**125, **6.**129
    straight-bevel, **6.**125, **6.**129
    worm, **6.**125, **6.**129 to **6.**130
    zerol, **6.**129
  epicyclic gear units, **6.**135
  foundation for, **6.**135, **6.**140
  installation of, **6.**135, **6.**140
  lubrication of, **6.**140, **6.**144
  materials and heat treatment for, **6.**130
    to **6.**131
  noise in, **6.**132
  packages, types of, **6.**133 to **6.**134
  parallel-shaft, **6.**125 to **6.**129
    continuous-tooth herringbone, **6.**129
    double helical, **6.**128
    helical versus spur, **6.**126 to **6.**127
  performance range of, **6.**133 to **6.**134
  troubleshooting, **6.**144
    table, **6.**141 to **6.**143
  use of, with pump drives, **6.**125
Gland plates:
  mechanical seals, **2.**147
  stuffing box, **2.**117

Graphitization, **5.**7 to **5.**8

Green liquor, **9.**127

Grout:
  foundation for, **12.**6 to **12.**7
  pumping, **3.**37

Hand-operated pumps, **3.**51

Hardness of material and erosion resistance, **5.**4

Hazen-Williams coefficient, change in,
  with age, **A.**6

Hazen-Williams formula:
  friction factor, **8.**33, **8.**37, **9.**11, **9.**12,
    A.6
  pipe friction, **A.**4 to **A.**5

Head:
  absolute, **8.**5
  acceleration (*see* Acceleration head)
  centrifugal pumps, **2.**11, **2.**194 to **2.**195
  definition of, **13.**26, **13.**28
  discharge, **2.**195, **9.**44 to **9.**48
  elevation, **2.**5, **2.**195, **8.**6
  gage, **8.**4 to **8.**8
  inertial, **8.**21 to **8.**24
  inlet, **2.**195
  measurement of, **13.**26 to **13.**33
  pool-to-pool, **9.**46
  pressure, **2.**5, **2.**194, **8.**5 to **8.**6
  pump rating, **8.**3
  pump total, **2.**194 to **2.**195, **8.**8 to **8.**11,
    9.44, **9.**48
    in branch-line systems, **8.**84
    calculation of, **8.**9 to **8.**11
    rating, **8.**8 to **8.**9
  shutoff, **2.**243 to **2.**247
  siphon, **8.**24 to **8.**31
  suction, **2.**195, **9.**44, **13.**3
    (*See also* Net positive suction head)
  suction lift, **2.**216, **9.**47 to **9.**48, **13.**3
  system-head curves, **8.**11 to **8.**15, **8.**77
    to **8.**84
  testing, **13.**28
  total, **2.**194 to **2.**195, **8.**6 to **8.**8, **9.**44 to
    9.48, **13.**3
  total dynamic, **2.**195
  transients in system, **8.**16, **8.**31
  variable system, **8.**14 to **8.**15
  varying temperature system, **8.**73 to
    8.75
  velocity (dynamic), **2.**5, **2.**194 to **2.**195,
    8.5, **8.**6
    testing, **13.**3

Head area meters, **13.**13, **13.**18 to **13.**25

Head-capacity curves, centrifugal pumps,
    2.196 to **2.**200
    (*See also* System-head curves)

Head coefficient, **2.**11

Head loss, **8.**31 to **8.**73
  bar racks, **8.**69 to **8.**73
  bends, **8.**63
  increasers, **8.**62 to **8.**63
  meters, **8.**64
  minor losses, **8.**52
  open channels, **8.**39, **8.**49 to **8.**52
  perforated plates, **8.**69 to **8.**73
  pressure pipes, **8.**31 to **8.**38
    tables, **A.**4 to **A.**6, **A.**8 to **A.**9
  pump suction elbows, **8.**64
  reducers, **8.**63
  screens, **8.**64 to **8.**68
  slurries, **9.**248 to **9.**254
  valves and standard fittings, **8.**52 to
    8.53
  viscous liquids, **8.**32, **8.**39

Heater drain pumps, **9.**75 to **9.**76

Heating, **9.**179 to **9.**182
  hot water circulating, **9.**179 to **9.**180
  pumps, types of, **9.**180
  steam heating systems, **9.**182
  water circuits, types of, **9.**180, **9.**182

Heterogeneous flow, **9.**250

Homogeneous flow, **2.**248 to **2.**249

Horsepower:
  brake (*see* Brake horsepower)
  liquid, **2.**195, **3.**73
  water (*see* Water horsepower)

Hose pumps, **5.**23 to **5.**24

Hydraulic balancing devices (*see* Balanc-
    ing devices, hydraulic)

Hydraulic couplings, (*see* Fluid cou-
    plings)

Hydraulic gradient, **8.**11, **8.**33, **8.**49, **8.**50

Hydraulic Institute Standards, **2.**216,
    10.5 to **10.**12, **10.**22

Hydraulic motors, **6.**172 to **6.**178

Hydraulic power:
  advantages of, **6.**173 to **6.**174
  efficiency of, **6.**177 to **6.**178
  for rotary pumps, **3.**86
  transmission systems, **6.**172 to **6.**178

Hydraulic presses (*see* Presses, hydraulic)

Hydraulic radius, **8.**33, **8.**48, **8.**49

Hydraulic turbines (*see* Turbines, hy-
    draulic)

Impeller design:
  discharge, **2.**14, **2.**25
  Francis vanes, **2.**17
  head coefficient, **2.**11, **2.**24
  inducer, **2.**15, **2.**48 to **2.**49
  inlet, **2.**15 to **2.**16, **2.**26 to **2.**27
  suction specific speed, **2.**15
  vane layout, **2.**16 to **2.**17
Impeller diameter reduction, **2.**205 to
    2.211
Impeller nomenclature, **2.**50, **2.**53
Impeller shaping:
  inlet vane tips, **2.**211 to **2.**213
  overfiling vane tips, **2.**211
  underfiling vane tips, **2.**211
Impellers, **2.**47 to **2.**53
  axial-flow (propeller), **1.**3, **2.**33, **2.**48,
    2.49, **2.**62
  backward-curved vane, **2.**197 to **2.**198
  Bernoulli's equation for, **2.**7 to **2.**8
  bladeless, **9.**28
  design of (*see* Impeller design)
  double-suction, **2.**33, **2.**48 to **2.**50, **2.**53
  efficiency of, **2.**8 to **2.**9
  enclosed (closed), **2.**34, **2.**48 to **2.**50
  Euler's equation for, **2.**6 to **2.**7
  forward-curved vane, **2.**200
  Francis-vane, **2.**48, **2.**49
  materials for, **5.**7
  mixed-flow, **1.**3, **2.**33, **2.**48, **2.**49, **2.**62,
    2.208, **2.**211, **5.**6, **9.**78
  nonclogging (sewage), **2.**49, **9.**28
  open, **2.**34, **2.**39 to **2.**40, **2.**49 to **2.**50
  radial-flow, **1.**3, **2.**33, **2.**48 to **2.**50, **2.**61
    to **2.**62
  radial-vane, **2.**198 to **2.**200
  reversed, **2.**252
  semiopen, **2.**49, **2.**50
  side plates, **2.**39 to **2.**40, **2.**50
  single-suction, **2.**33, **2.**48 to **2.**50
  straight-vane, **2.**48
Inclined rotor pumps, *2-201* (*see* index
    headnote)
Index tests, **13.**2
Inducers, **2.**15, **2.**48 to **2.**49, **2.**227 to
    2.228
Inductors, priming, **2.**320 to **2.**322
Inertia:
  liquid, **2.**235, **8.**21 to **8.**24
  moment of, **2.**232, **6.**15 to **6.**16, **8.**22
Injection-type shaft seals, centrifugal
    pump, **2.**159 to **2.**165
Injectors, **4.**2, **4.**26, **4.**27

Injectors (*Cont.*):
  boiler, **4.**26, **4.**27
  definition of, **4.**2
Inlet conditions, screw pumps, **3.**66
Inlets, pump (*see* Intakes)
In-line centrifugal pumps, **9.**104
Inspection for maintenance, **12.**15
Installation:
  alignment, **2.**281 to **2.**288, **6.**59 to **6.**60,
    6.140, **12.**3 to **12.**6
  care of equipment in the field, **12.**2
  doweling of pump and driver, **12.**7
  of engines, **6.**59 to **6.**60
  foundations, **12.**2 to **12.**3
  gears, **6.**135, **6.**140
  grouting, **12.**6 to **12.**7
  instruction books, **12.**2
  mounting of vertical pumps, **12.**3
  preparation of shipment, **12.**2
  pump location, **12.**2
  rotary pumps, **3.**103
  screw pumps, **3.**73
Instruction books, **12.**2
Instrumentation:
  centrifugal pumps, **2.**266 to **2.**304,
    12.11
  displacement pumps, **3.**114 to **3.**116
  testing, **13.**8
Intake modeling, **10.**35 to **10.**48
  cost of, **10.**43
  design and operation of, **10.**41 to **10.**43
  observation techniques in, **10.**43 to
    10.46
  reasons for, **10.**35
  similitudes for, **10.**39 to **10.**41
Intakes, **9.**3, **10.**3 to **10.**33
  back vents in, **10.**10
  environmental considerations of, **10.**16
    to **10.**20
  fish in, **10.**7, **10.**16 to **10.**20
  forebay, **10.**7, **10.**16 to **10.**20
  multiple pumps, wet-pit, **10.**9 to **10.**12
  once-through, open-pit, **10.**3
  Ranney well, **10.**9, **10.**23
  recirculating, open-pit, **10.**8, **10.**9
  screens and trashracks for, **10.**14 to
    10.16
  silt in, **10.**9, **10.**12
  turning vanes, **10.**12 to **10.**14
  velocity caps in, **10.**4
  vortices and eddies, **10.**20 to **10.**23
Intergranular corrosion, **5.**5 to **5.**6, **9.**88 to
    9.89

Internal combustion engines (*see* Engines, internal combustion)
Irrigation pumps, **9.41** to **9.50**
  capacity of, **9.43** to **9.44**
  head of, **9.44** to **9.48**
  number required, **9.43**
  prime movers, **9.48** to **9.49**
  setting of, **9.48**
  station design for, **9.49** to **9.50**
  types of, **9.42** to **9.43**
Isentropic bulk modulus of water, **3.109**, **3.110**
Isothermal secant bulk modulus of oils, **3.110**, **3.111**

Jet compressors, **4.2**
  (*See also* Jet pumps)
Jet propulsion, **9.161** to **9.162**
Jet pumps, **4.1** to **4.27**
  eductors (*see* Eductors)
  ejectors, **4.2**
  injectors, **4.2**, **4.26**, **4.27**
  jet compressor, **4.2**
  siphons, **4.2**, **4.20** to **4.22**
Jockey pumps, **9.52**, **9.54**, **9.58**, **9.59**
Kinetic energy (*see* Head, velocity)

Laminar flow, **8.31**, **8.32**
Laminated rotor pumps, **2-200** (*see* index headnote)
Lantern rings, **2.73** to **2.75**, **2.117**
Liquid end:
  power pumps, **3.12** to **3.17**
  steam pumps, **3.43**
Liquid metal (fast breeder) plants, **9.212**
Liquid rheostat controls, **6.94** to **6.96**
Lobe pumps, **3.83** to **3.84**
Lomakin effect, **2.69**
Losses:
  disk friction, **2.21**
  head (*see* Head loss)
  mechanical, **2.21**
LPG, **9.299**
Lubricating oil service, marine pumping systems, **9.164**
Lubrication:
  for gears, **6.140**, **6.144**
  for power pumps, **3.23**

Magnetic couplings (*see* Eddy-current couplings)

Magnetic drives, **6.166** to **6.171**
  configurations, **6.169**
  construction of, **6.166** to **6.167**
  cost, **6.171**
  eddy-current couplings, **6.167** to **6.169**
  magnet-to-magnet couplings, **6.169**
Maintenance:
  annual inspections, **12.15**
  check chart: centrifugal pump troubles, **2.294** to **2.301**, **12.16**
    reciprocating pump troubles, **12.18**
    rotary pump troubles, **12.17**
    steam pump troubles, **12.18** to **12.19**
  complete overhaul, **12.15**
  daily observation of pump operation, **12.15**
  gears, table, **6.144**
  metering pumps, **9.238**
  pump troubles, diagnosis of, **12.17**
  record of inspections and repairs, **12.17**
  semiannual inspections, **12.15**
  spare and repair pair, **12.15**, **12.17**
Manifolds, power pumps, **3.16**
Manning formula, **8.39**
  table, **8.49**
Marine pumping systems, **9.151** to **9.165**
  boiler-feed system, **9.151** to **9.155**
  fresh-water services, **9.155** to **9.158**
  seawater services, **9.158** to **9.162**
  viscous fluid services, **9.162**
Materials of construction:
  metallic pumps, **5.3** to **5.13**
    abrasive wear, **5.6** to **5.7**
    casings, **5.9**, **5.10**
    cavitation erosion, **2.227**, **2.228**, **5.6**
    chemical pumps, **9.85** to **9.88**
    corrosion fatigue, **5.5**
    diaphragm pumps, **3.54** to **3.55**
    diffusers, **5.11**
    erosion corrosion, **5.4** to **5.5**
    fatigue failures, **2.272**
    fire pumps, **9.54**
    graphitization, **5.7** to **5.8**
    impellers, **2.227**, **2.228**, **5.8** to **5.9**
    intergranular corrosion, **5.5** to **5.6**
    mining pumps, **9.145** to **9.146**
    nuclear pumps, **9.219** to **9.222**
    petroleum pumps, **9.104**, **9.105**, **9.109**, **9.112**
    power pumps, **3.13**, **3.15**
    rotary pumps, **3.87** to **3.88**
    screw pumps, **3.66**

Materials of construction,
  metallic (*Cont.*):
    shafts, **5.5**, **5.9**
    solids-handling pumps, **9.262**, **9.264**
    stuffing box sleeves, **5.4**
    wearing rings, **5.4**, **5.11** to **5.13**
  nonmetallic, **5.15** to **5.27**
    centrifugal pumps: dry-pit, **5.21** to
      5.23
      wet-pit, **5.19** to **5.21**
    corrosion resistance, **5.27**, **A.27** to
      A.35
    manufacturing process, **5.26** to **5.27**
    positive displacement, **5.23** to **5.24**
    properties of, **5.25**
    reasons for using, **5.15**
Mechanical performance, centrifugal
      pumps, **2.274** to **2.304**
Mechanical seals, **2.127** to **2.158**
Meridional velocity, **2.9**, **2.11**
Metering pumps, **9.235** to **9.238**
  controls for, **9.237** to **9.238**
  diaphragms, **3.51**
  hydraulically actuated diaphragm,
      9.236 to **9.237**
  maintenance of, **9.238**
  materials for, **9.238**
  mechanically actuated diaphragm,
      9.236
  packed plunger, **9.235**
  rotary pumps, **3.105**
Metric units (*see* SI units)
Minimum flow, centrifugal pumps,
      2.241, **2.243**, **2.305** to **2.318**,
      8.80, **8.82**
  control systems, **2.305** to **2.318**, **8.80**,
      8.82
Mining services, **9.143** to **9.150**
  drivers, **9.149** to **9.150**
  materials of construction, **9.145** to
      9.146
  pressure pulsations, **9.146**
  pump controls, **9.149**
  pump types, **9.143**, **9.146**
  sumps, **9.148** to **9.149**
  waterhammer, **9.146**
Minor losses, **8.52**
Mixed-flow impellers, **1.3**, **2.33**, **2.48**,
      2.62, **2.208**, **2.211**
  condenser circulating pumps, **9.78**
  materials, **5.6**
Model tests:
  cavitation, **9.192**, **13.40**

Model tests (*Cont.*):
  centrifugal pumps, efficiency of, **2.196**,
      2.203, **2.253** to **2.255**
  intakes (*see* Intake modeling)
  performance of, **13.2**
  procedures for, **13.38** to **14.40**
  pump noise, **8.116**
  purpose of, **13.38**
Modulated bypass system, **2.314**
Moody diagram, **8.33**, **8.34**
Moody formula for efficiency correction,
      13.39
Motive power, **4.2**
  (*See also* Jet pumps)
Multiple-screw pumps, **3.57**
Multistage centrifugal pumps, **2.34**
  casings for, **2.42**

National Electrical Manufacturers Asso-
      ciation (NEMA), **6.4**, **6.11**, **6.17**,
      6.19 to **6.21**
National Fire Protection Association
      (NFPA) publications, **9.51**
Net positive suction head (NPSH), **13.5**
  centrifugal pumps, **2.15**, **2.215** to **2.220**,
      9.64 to **9.68**, **9.73** to **9.74**, **10.23**, **10.26**
  condensate pumps, **9.151**
  materials, **2.227**, **5.3**, **5.6**
  power pumps, **3.9**
  reductions in, **2.224** to **2.227**
  screw pumps, **3.68**, **3.69**
  steam pumps, **3.46**
  testing, **2.220** to **2.223**
Newtonian (true) liquids, **8.32**
Noise, pump, **8.101** to **8.118**
  casing cutwater, **8.106** to **8.107**
  cavitation and flashing, **8.104** to **8.105**
  control of, **8.105** to **8.114**
  liquid sources, **8.102** to **8.105**
  measurement of, **8.114**
  mechanical sources, **8.102**
  model testing, **8.116**
  OSHA limits, **8.115**
  recirculation, **8.102**, **8.103**
Nomenclature, centrifugal pumps, **2.33**
      to **2.35**
Nonreverse ratchet, **6.13**
Nozzle locations, centrifugal pumps, **2.40**
      to **2.41**
Nozzle meters, head loss, **8.64**
Nozzles, design of, **13.15**
NPSH (*see* Net positive suction head)

Nuclear electric generation, 9.203 to
      9.222
   boiling water reactor (BWR) plants,
      9.209 to 9.212
   fast breeder plants, 9.212
   main coolant pumps, 9.213 to 9.215
   materials of construction, 9.219
   pressurized water reactor (PWR)
      plants, 9.203 to 9.208
   principal types of pumps, 9.215, 9.219
   radioactivity, 9.221 to 9.222
   seals, 9.214 to 9.215
   special design requirements, 9.219 to
      9.222
   testing, 9.222
Nuclear power plants, 9.203 to 9.222
Nuclear pump seismic qualifications,
      9.223 to 9.233
   definitions, 9.223 to 9.229
   design and functional requirements,
      9.231 to 9.232
   qualification by analysis, 9.229 to
      9.231
   qualification by testing, 9.231

Oil film journal bearing, centrifugal
      pumps, 2.166 to 2.192
Oil well pumps, 9.277 to 9.294
   offshore, 2.293 to 2.294
   submersible centrifugal pumps, 2.293
   subsurface hydraulic pumping system,
      2.285 to 2.293
   sucker rod pumping system, 9.277 to
      9.285
Oiler, constant-level, 2.86
Operation:
   parallel (*see* Parallel operation)
   reduced flows, 12.11
   rotary pumps, 3.103 to 3.105
   series (*see* Series operation)
   starting and stopping procedures,
      12.13
Orifice meters, 13.16
   head loss, 8.64
Orifices:
   for pressure-measuring instruments,
      13.27
   pressure reducing, 2.310 to 2.311
      (*See also* Pump bypass, centrifugal
      pumps)
   throttling, 8.73
OSHA noise levels, 8.115
Overall efficiency, definition of, 13.7

Packing:
   centrifugal pumps, 2.114 to 2.126
   power pumps, 3.14
   steam pumps, 3.25, 3.39, 3.40
Packless stuffing boxes (*see* Injection-
      type shaft seals, centrifugal
      pump)
Paper mill services, 9.121 to 9.135
   black liquor, 9.121, 9.126 to 9.127
   green liquor, 9.121, 9.127
   groundwood, 9.121, 9.123
   pump selection, 9.131 to 9.135
   sulfate process, 9.123 to 9.125
   sulfite process, 9.122, 9.126
   white liquor, 9.121, 9.125
Paper stock, 9.128 to 9.131
Paper stock pumps, 9.130 to 9.131
   efficiency of, 9.132 to 9.133
Parallel operation:
   centrifugal pumps, 2.239 to 2.240, 9.13
      to 9.14, 9.35 to 9.37
   displacement pumps, 3.126 to 3.127
Parshall flume, 13.22
Partially full pipes, flow in, 8.39, 8.49 to
      8.52
Performance life, 5.3
Peripheral velocity, 2.196
Petroleum industry, centrifugal pumps,
      9.101 to 9.119
   liquid petroleum gas (lpg), 9.113 to
      9.114
   materials for, 9.104 to 9.105
   pipeline, 9.105, 9.112
   reactor feed pumps, 9.113, 9.114
   refinery, 9.101 to 9.105
   viscosity correction, 9.114 to 9.119
   waterflood pumps, 9.112 to 9.113
Pipe roughness, 8.33
Piping:
   discharge, 2.272, 12.8
   expansion joints, 12.8 to 12.9
   mining, 9.145
   moments and forces, 12.8
   natural frequencies, 3.115
   optimum size, 9-67 (*see* index head-
      note)
   paper stock pumps, 9.131, 9.133
   strains, 12.8
   suction (*see* Suction piping)
   suction strainers, 12.9
   venting and draining, 12.9 to 12.10
   vibrations, 2.272, 3.114
   warm-up, 12.10

Piping (*Cont.*):
  waterhammer, **8.**87
Piston pumps, **3.**35
  circumferential, **3.**82 to **3.**83
  marine pumping systems, **9.**163
  rotary, **3.**84 to **3.**85
  variable-displacement, **6.**177
Piston-type liquid ends, **3.**35
Pistons:
  power pumps, **3.**8, **3.**14
  steam pumps, **3.**25
    speed, **3.**42
Pitot tube pumps, *2-203* (*see* index head-
    note)
Pitot tubes, **13.**13, **13.**17 to **13.**18
Plunger pumps, **3.**34, **3.**40, **9.**33
Plungers:
  power pumps, **3.**6, **3.**8, **3.**13, **9.**32 to
    9.33
  steam pumps, **3.**25
Pneumatic ejectors, **9.**25, **9.**27, **9.**28
Portable pumps, **6.**172, **9.**42 to **9.**43
Positive displacement pumps (*see* Dis-
    placement pumps)
Potential energy (*see* Head)
Power, centrifugal pumps, **2.**195 to **2.**196
  (*See also* Brake horsepower)
Power end, power pumps, **3.**17, **3.**19 to
    3.22
Power failure, **8.**89
Power measurement:
  calibrated motors, **13.**33
  test calculation, **13.**34
  torsion dynamometer, **13.**34
  transmission dynamometer, **13.**13 to
    13.34
Power pumps, **3.**3 to **3.**24
  acceleration head, **3.**9 to **3.**11
  applications of, **3.**23
  bearings, **3.**22 to **3.**23
  brake horsepower of, **3.**3 to **3.**4
  bypass relief, **3.**24, **3.**122
  capacity of, **3.**4
  charging, **9.**206
  connecting rods, **3.**20
  construction of, **3.**12
  crankshaft, **3.**19 to **3.**20
  crosshead, **3.**21
  crossways, **3.**21
  cylinder, **3.**12 to **3.**13
  cylinder liner, **3.**15
  displacement of, **3.**4
  duty service, **3.**24
  eccentric straps, **3.**20

Power pumps (*Cont.*):
  frame, **3.**17, **3.**19
  liquid end, **3.**12 to **3.**17
  liquid separation, **3.**11 to **3.**12
  lubrication, **3.**23
  manifolds, **3.**16 to **3.**17
  marine pumping systems, **9.**154
  mechanical efficiency of, **3.**6
  net positive suction head available
    (NPSHA), **3.**9
  net positive suction head required
    (NPSHR), **3.**9
  packing, **3.**14
  pistons, **3.**14
    number of, **3.**8
  plunger covers, **3.**17
  plungers load, **3.**6
  plungers, **3.**13, **9.**32 to **9.**33
    number of, **3.**8
  power end, **3.**17, **3.**19 to **3.**22
  pressure, **3.**4 to **3.**5
  pulsation dampeners, **3.**24
  pulsations, **3.**8 to **3.**9, **3.**24, **3.**107 to
    3.117
  relief value, **3.**24
  slip, **3.**5
  speed, **3.**6
  stuffing box, **3.**14
  stuffing box loss, **3.**6
  testing, **13.**40
  unbalanced reciprocating parts force, **3.**7
  unbalanced rotating parts force, **3.**7 to
    3.8
  unloaders, **3.**24
  valve covers, **3.**17
  valve loss, **3.**6
  valves, **3.**15 to **3.**16
  volumetric efficiency of, **3.**5 to **3.**6
  wrist pin, **3.**20 to **3.**21
Prerotation, centrifugal pumps, **2.**40,
    2.241
Present worth, **11.**16
Presses, hydraulic, **9.**167 to **9.**177
  accumulators, **9.**167 to **9.**168
  design procedure for, **9.**177
  operating pressure of, **9.**169 to **9.**170,
    9.176
  pumps for, **9.**168 to **9.**169
  types of, **9.**167
Pressure:
  power pumps, **3.**4 to **3.**5
  rotary pumps, **3.**93 to **3.**94
  screw pumps, **3.**60, **3.**73
Pressure head, **2.**5, **2.**194, **8.**5 to **8.**6

Pressure pulsations (*see* Pulsations)
Pressurized water reactor (PWR) plant,
    9.203 to **9.**208
Priming:
    automatic systems, **2.**324 to **2.**325
    central systems, **2.**323 to **2.**325
    centrifugal pumps, **2.**230, **2.**319 to
        2.330, **12.**12
    power pumps, **12.**12
    rotary pumps, **12.**12
    self-contained units, **2.**325
    sewage pumps, **2.**326 to **2.**327
    siphons (*see* Siphon systems)
    time required for, **2.**328 to **2.**329
Priming chambers, **2.**320
Priming inductors, **2.**320 to **2.**322
Priming valves, **2.**329
Progressing cavity pumps, **9.**33
Propeller pumps, **2.**98 to **2.**99, **9.**42, **9.**48
    condenser circulating, **8.**15, **8.**21, **9.**76
        to **9.**78
    materials for, **5.**5
Proportioning (*see* Metering pumps)
Pullout pump, **2.**99
Pulp mill (*see* Paper mill services)
Pulsations:
    centrifugal pumps, **2.**269, **8.**101 to
        8.118
    impeller recirculation, **2.**241 to **2.**243,
        2.269, **2.**272
    mining pumps, **9.**146
    power pumps, **3.**8 to **3.**9, **3.**24, **3.**107 to
        3.117
    screw pumps, **3.**77
    steam pumps, **3.**47
Pump bypass:
    centrifugal pumps, **2.**305 to **2.**318, **8.**80
        to **8.**81
    condensate pumps, **9.**152
    power pumps, **3.**24, **3.**122
    (*See also* Minimum flow, centrifugal
        pumps)
Pump control (*see* Control, pump)
Pump storage, **9.**187 to **9.**202
    economic design, **9.**196 to **9.**197
    governor time, **9.**194 to **9.**195
    hydraulic transients, **9.**188 to **9.**189
    selection of units, **9.**187 to **9.**188
    size of installation, **9.**187
    spinning reserve, **9.**191, **9.**194
Pumping system, **8.**4
    branch-line (*see* Branch-line pumping
        systems)
    economics of (*see* Economics of pump-
        ing systems)

Pump-out vanes, **2.**60, **2.**99
Purchasing pumps, **11.**1 to **11.**25
    bidding, **11.**13 to **11.**15
    codes, **11.**4 to **11.**6
    commercial terms, **11.**19 to **11.**20
    driver selection, **11.**5
    driver specifications, **11.**10 to **11.**11
    evaluation of bids, **11.**15 to **11.**20
        cost, **11.**15
        efficiency, **11.**15 to **11.**16
    materials, **11.**5
    selection of, **11.**4 to **11.**5
    specifications, **11.**5 to **11.**10
    system requirements, **11.**2 to **11.**4
        alternate modes of operation, **11.**3
        fluid type, **11.**2
        future system changes, **11.**3 to **11.**4
        head curves, **11.**2 to **11.**3
        margins, **11.**3
        wear, **11.**3
    testing, **11.**10
PWR (pressurized water reactor) plant,
    9.203 to **9.**208

Radial-flow impellers, **1.**3, **2.**33, **2.**48 to
    2.50
    axial thrust of, **2.**61 to **2.**62
Radial thrust (reaction), **2.**19, **2.**36 to
    2.37, **2.**42, **2.**244 to **2.**247
Radial vanes, **2.**198 to **2.**200
Radioactivity, **9.**221 to **9.**222
Rate-of-flow meters, **13.**13, **13.**14
Reactor coolant drain tank pumps, **9.**205
Reactor coolant pumps, **9.**203 to **9.**204,
    9.213 to **9.**215
Reactor core isolation system pumps,
    9.210, **9.**212
Reactor water cleanup pumps, **9.**211,
    9.212
Recirculation, centrifugal pumps, **2.**241
    to **2.**243, **2.**267, **2.**305 to **2.**318
Recirculation coolant pumps, **9.**210,
    9.212
Recycle evaporator feed pumps, **9.**205,
    9.207
Reduced flow troubles, centrifugal
    pumps, **2.**305
Refrigeration:
    brine circulation, **9.**185
    lubrication oil transfer, **9.**186
    refrigerant circulation, **9.**185 to **9.**186
Regenerative turbine pumps, *2-202* (*see*
    index headnote)
Relative velocity, **2.**9
Relief valves (*see* Valves, relief)

Residual heat removal (RHR) pumps, 9.204, 9.206, 9.210, 9.212

Resistance coefficients for valves and fittings (*see* Piping; valves)

Reverse dialing (*see* Alignment of pumps and drivers)

Reverse flow, centrifugal pumps, 2.255 to 2.259

Reverse-speed–torque characteristics, centrifugal pumps, 2.247 to 2.250, 8.20 to 8.21

Reversed impeller, 2.252

Reversible centrifugal pumps, *2-199* (*see* index headnote)

Revolution counter, 13.34

Reynolds number, 2.223, 2.254, 8.32, 8.33

Rotary pumps, 3.79 to 3.105
  abrasion wear on, 3.105
  applications of, 3.103 to 3.105
  brake horsepower of, 3.94
  capacity of, 3.93
  circumferential piston pumps, 3.82 to 3.83
  classification of, 1.4
  efficiency of, 3.95 to 3.96
  flexible member pumps, 3.85 to 3.86
  gear pumps, 3.81 to 3.82
  lobe pumps, 3.83 to 3.84
  materials of construction, 3.87 to 3.88
  multiple-rotor screw pumps, 3.82
  operating characteristics of, 3.88 to 3.103
  pressure, 3.93 to 3.94
  rigid rotor vane pumps, 3.84
  rotary piston pumps, 3.84 to 3.85
  screw and wheel pumps, 3.81

Rotating casing pumps, *2-203* (*see* index headnote)

Rotation, centrifugal pumps, 2.41
  incorrect, 2.253

Runaway speed, centrifugal pumps, 2.236 to 2.237, 2.250, 2.251, 8.20 to 8.21

Runout, centrifugal pumps, 9.13, 9.14

Safety injection pumps, 9.204, 9.206 to 9.207

Salt dilution test, 13.25

Salt velocity test, 13.25

Saltation flow, 9.250

Sanitary pumps, 3.86, 13.25
  (*See also* Food and beverage pumping)

Screens, 10.29 to 10.32
  head loss, 8.67 to 8.68

Screw pumps, 3.57 to 3.78
  abrasion wear on, 3.74
  advantages of, 3.59
  brake horsepower, 3.72
  capacity of, 3.60
  construction of, 3.61 to 3.66
  disadvantages of, 3.59
  double-end, 3.62
  entrained gas, effect of, 3.75 to 3.78
  installation and operation, 3.73
  multiple-screw pump, 3.57, 3.82
  net positive suction head, 3.68, 3.69
  noise, 3.74
  paper stock, 9.134
  performance of, 3.66 to 3.73
  pressure, 3.60, 3.73
  pulsations, 3.77
  screw and wheel, 3.81
  sewage service, 9.32, 9.33
  single-end, 3.62, 3.64
  single-screw pump, 3.57, 3.84
  starting, 3.74
  theory of, 3.59 to 3.61
  (*See also* Archimedean screw)

Scum pumps, 9.30 to 9.31

Seal cages, 2.73 to 2.75, 2.117, 3.80

Seal well, 8.24 to 8.31

Sealless (canned) pump motors, 2.110 to 2.112, 6.22 to 6.25

Seals:
  centrifugal pumps: injection shaft, 2.159 to 2.165
    mechanical, 2.80 to 2.81, 2.127 to 2.158
  nuclear pumps, 9.14, 9.15
  rotary pumps, 3.80
  screw pumps, 3.66

Selecting pumps, 1.3 to 1.5, 11.1 to 11.25
  (*See also* Purchasing pumps)

Self-priming pumps:
  centrifugal, 2.325 to 2.326
  marine, 9.159

Separation, liquid, 2.241 to 2.243

Series operation:
  centrifugal pumps, 2.41, 2.42, 2.240 to 2.241, 9.13 to 9.14
  displacement pumps, 3.125 to 3.127
  system-head curves, 8.15

Series units, centrifugal pumps, 2.41 to 2.42

Serrated throttle bushing, 2.159 to 2.160

Settling velocity, **9.**239
Sewage, **9.**23 to **9.**29
  collection of, **9.**23 to **9.**24
  nonclog pumps, **2.**91, **9.**24, **9.**28
  pump controls, **9.**37 to **9.**38
  pump drivers, **9.**37
  pump selection, **9.**34 to **9.**37
  pump station design, **9.**38 to **9.**39
  pumps used in treating, **9.**25 to **9.**37
  treatment of, **9.**24 to **9.**25
Shaft sleeves, **2.**70 to **2.**72
  materials for, **2.**72
  nuts for, **2.**71, **2.**72
Shafting, intermediate pumps, **6.**161 to
    6.165
  balancing of, **6.**165
  bearings for, **6.**163, **6.**164
  critical speed of, **6.**164 to **6.**165
  flexible drive shafts, **6.**163 to **6.**164
  floating shafts, **6.**161 to **6.**162
  rigid shafts, **6.**162 to **6.**163
  torsional stress in, **6.**164
Shafts, centrifugal pumps, **2.**66 to **2.**70
  deflection of, **2.**68 to **2.**69
  double-extended, **2.**69
  failures, **2.**269, **2.**271
  flexible, **2.**67 to **2.**68
  lateral critical speed of, **2.**67 to **2.**69,
      6.164, **6.**165
  rigid, **2.**67 to **2.**68
  sizing, **2.**69 to **2.**71
  stress in, **5.**5, **5.**9
  torsional critical speed of, **6.**164 to
      6.165
Shear-lift pumps, *2-200* (*see* index head-
    note)
Shutdown:
  centrifugal pumps, **12.**13
  screw pumps, **3.**74
Shutoff, centrifugal pumps, **2.**243 to
    2.247
SI (International System of Units) units,
    A.2, **A.**3
  conversion of, to U.S. units, table, **A.**36
Side-pot liquid ends, **3.**35
Side-suction centrifugal pumps, **2.**33,
    2.34, **2.**41
Sigma (cavitation), **2.**221, **13.**9, **13.**40
Similarity relations:
  geometric, **2.**200, **13.**38
  speed correction, **13.**35
Simplex steam pumps, **3.**27 to **3.**28
Single-acting steam pumps, **3.**27

Single-end screw pumps, **3.**62
Single-screw pumps, **3.**57
Single-stage centrifugal pumps, **2.**33,
    2.34
Single-suction impellers, **2.**33, **2.**48 to
    2.50
Single-volute casing pumps, **2.**36
Siphon head, **8.**24 to **8.**31
  systems, **8.**24 to **8.**31, **9.**46
  (*See also* Jet pumps)
Siphons, **8.**25
  air pumps, **4.**22
  jet pumps, **4.**2, **4.**20 to **4.**21
Sleeves, shaft, stuffing box, **2.**70 to **2.**72
Slip:
  centrifugal pumps, **2.**3, **2.**9, **2.**14
  power pumps, **3.**5
  rotary pumps, **3.**92 to **3.**93
  screw pumps, **3.**60, **3.**71
Slip couplings (*see* Eddy-current coup-
    lings)
Sludge pumps, **9.**30 to **9.**33
Slurries, pumping (*see* Solids, hydraulic
    transport of)
Slurry concentration vs. specific gravity,
    9.267
  economics, **9.**255 to **9.**259
  frictional losses, **9.**248 to **9.**254
Slurry pumps:
  diaphragm, **3.**55
  marine pumping system, **9.**164
  power pumps, **3.**17, **9.**269 to **9.**276
Soleplates, **2.**90
Solids, hydraulic transport of, **9.**239 to
    9.260
  economics, **9.**255 to **9.**259
  frictional loss, **9.**248 to **9.**254
  pipelines for, **9.**254
Solids-handling centrifugal pumps, **9.**261
    to **9.**268
Solids-handling displacement pumps,
    9.269 to **9.**276
Sound, velocity of, **3.**109 to **3.**111
Spare parts, **12.**15
Special-effect pumps, classification of,
    1.3
Specific speed centrifugal pumps, **1.**3,
    2.3, **2.**12, **2.**24, **2.**201, **2.**203 to
    2.204, **8.**15 to **8.**21
  effect of, on waterhammer, **8.**88
  mining services, **9.**144
  pump turbine, **9.**188
  suction, **2.**15, **2.**24, **2.**223 to **2.**224

Specific weight (gravity) of water, **13.4**
  table, **A.**26
Specifying pumps (*see* Purchasing
  pumps)
Speed:
  correction for tests, **13.**35
  measurement of, **13.**34
  paper stock pumps, **9.**133
  power pumps, **3.**6
  rotary pumps, **3.**93
  screw pumps, **3.**71
  steam pumps, **3.**42
Spent fuel pit pumps, **9.**204, **9.**207
Spent fuel pit skimmer pumps, **9.**205
Spent resin sluicing pumps, **9.**205, **9.**207
Stage pieces, **2.**44
Standby liquid control system pumps,
  9.211, **9.**212
Starting centrifugal pumps, **2.**230 to
  2.237, **6.**95, **8.**17 to **8.**21
  check valve, **2.**33 to **2.**36, **8.**19
  closed valve, **2.**33 to **2.**35, **8.**17 to **8.**19
  open valve, **2.**33 to **2.**36, **8.**19
  reverse rotation, **8.**20 to **8.**21
  (*See also* Priming; Siphons)
Starting pump turbines, **9.**190 to **9.**191,
  9.194
Starting screw pumps, **3.**74
Static head, **8.**6, **8.**14
  variable, **8.**14 to **8.**15
  (*See also* System-head curves)
Steam, saturated, table, **6.**37
Steam end of steam pumps, **3.**28 to **3.**34,
  3.43 to **3.**44
Steam power plant, **9.**61 to **9.**82
  boiler circulating pumps, **9.**78 to **9.**80,
    9.82
  boiler-feed pumps, **9.**64 to **9.**68
  booster pumps, **9.**68 to **9.**70
  condensate pumps, **9.**70 to **9.**75
  condenser circulating pumps, **9.**76 to
    9.78
  cycles, **9.**61 to **9.**62
  heater drain pumps, **9.**75 to **9.**76
  load reduction transients, **9.**64 to **9.**68
  pump services, **9.**62 to **9.**63
Steam pumps, **3.**25 to **3.**46
  capacity of, **3.**42
  direct-acting, **3.**27 to **3.**28
  duplex, **3.**27 to **3.**28
  flow characteristics of, **3.**47
  liquid end construction of, **3.**34 to **3.**40

Steam pumps (*Cont.*):
  marine pumping systems, **9.**155
  net positive suction head (NPSH), **3.**46
  packing, **3.**25, **3.**39, **3.**40
  performance of, **3.**40, **3.**42 to **3.**47
  pulsations, **3.**47
  simplex, **3.**27 to **3.**28
  steam consumption of, **3.**44
  steam end construction of, **3.**28 to **3.**34
  steam end materials of, **3.**34
  steam end operation of, **3.**27 to **3.**44
  steam valves, **3.**30 to **3.**34
  theory of, **3.**25, **3.**27
Steam turbines (*see* Turbines, steam)
Steel mills, *10-159* to *10-172* (*see* index
  headnote)
Stop pieces, **2.**42
Straight radial-vane high-speed pump,
  2.112 to **2.**113
Strainers, **10.**29 to **10.**32
Stroboscopic device, **13.**34
Strouhal equation, **8.**103
Stuffing box sealing, **2.**74 to **2.**77, **2.**117
  use of grease or oil for, **2.**76 to **2.**77
Stuffing boxes:
  centrifugal pumps, **2.**73 to **2.**80
    glands for, **2.**80, **2.**117
    lantern rings for, **2.**117
    packing for, **2.**114 to **2.**126
    pressure-reducing devices for, **2.**122
    seal cages for, **2.**73 to **2.**75, **2.**117
    sleeves, **2.**70 to **2.**72
    water-cooled, **2.**77 to **2.**78
  power pumps, **3.**6, **3.**14
  (*See also* Packing)
Submergence, wet-pit pumps, **10.**5 to
    10.12, **10.**23 to **10.**24, **10.**35 to
    10.48
  (*See also* Intakes)
Submergence control, **9.**73, **9.**75
Submersible pumps, **2.**108 to **2.**110
  motors for, **6.**22 to **6.**251, **9.**293
Sucker rod pumps, **9.**277 to **9.**285
Suction elbows, pump, **2.**90 to **2.**95, **8.**64
Suction head, **2.**195, **8.**8 to **8.**11, **9.**44,
    13.3
  (*See also* Net positive suction head)
Suction lift, **2.**116, **9.**47 to **9.**48
  testing, **13.**3
Suction piping, **10.**24 to **10.**29
  air pockets, **10.**24 to **10.**25
  manifolds, **10.**26

Suction Piping (*Cont.*):
  minimum straight length, **10.26**
  recommended velocities, **10.24** to
    10.26
  surge and vibration, **2.272**, **10.27** to
    10.29
  tunnels, **10.26** to **10.27**
Suction pits (*see* Intakes)
Suction recirculation, **2.267**
Suction specific speed, **2.15**, **2.24**, **2.223**
    to **2.224**
Suction strainer, **10.29** to **10.32**, **12.9**
Suction throttling, **2.237** to **2.238**
Sump (intake), **9.49**
  (*See also* Intakes)
Sump pumps, **2.101** to **2.102**
Surge suppressor, **8.93**
Surge tank, **8.87**, **8.94**, **12.11**
  pump turbine, **9.194** to **9.196**
System characteristics, **8.3** to **8.76**
System-head curves, **2.39** to **2.41**, **8.11** to
    8.14, **9.11** to **9.14**, **9.18**, **9.20**,
    9.35 to **9.37**
  branch-line, **8.77** to **8.84**
  fixed heads in, **8.12** to **8.14**
  negative, **8.13**
  pumps in series, **8.15**
  variable static heads in, **8.14**
  variable system resistance, **8.14** to **8.15**
  (*See also* Head)

Tachometer, **13.34**
Temperature of dry and saturated steam,
    table, **6.37**
Temperature rise, centrifugal pumps,
    2.243 to **2.244**, **12.11** to **12.12**
Testing, **13.2** to **13.41**
  accuracy and tolerances, **13.7**
  cavitation, **2.220**, **13.9**
  classification of pumps, **13.2**
  definitions, **13.2**
  measurements, **13.13**
  nuclear pumps, **9.222**
  procedure of, **13.11**
  records, **13.36**
Theory, centrifugal pumps:
  actual total head, **2.7**
  Bernoulli's equation, **2.5**
  Euler's equation, **2.6**
  hydraulic efficiency, **2.7**
    (*See also* Efficiency, centrifugal
      pump)

Theory, centrifugal pumps (*Cont.*):
  overall efficiency, **2.8**
  power input, **2.7**
  similitude, **2.12**
  slip factor, **2.9**
  specific speed, **2.12**
    (*See also* Specific speed, of centrifu-
      gal pumps)
  specific speed chart, **2.14**
  theoretical total head, **2.7**
  velocity triangles, **2.9**, **2.11**
  volumetric efficiency, **2.8**
  volute casing, **2.17**, **2.28**
Thixotropic liquids, **8.32**
Throttle control, condensate pumps,
    9.152
Thrust, axial (*see* Axial thrust)
Timed rotors, screw pumps, **3.61**, **3.64**
Tirastat II controller, **6.96** to **6.99**
Tolerances, testing, **13.7**
Tongue, casing, **2.35**, **2.213**
Top-suction centrifugal pumps, **2.34**,
    2.45
Torque, centrifugal pumps, **2.231** to
    2.236, **8.15** to **8.21**
  accelerating, **2.231** to **2.236**, **8.21** to
    8.24
  breakaway, **2.232**
Torque flow pumps, **9.27**, **9.31**
Total discharge head, testing, **13.3**
Total dynamic head, **2.195**
Total head, **2.194** to **2.195**, **8.6** to **8.8**,
    9.44 to **9.48**, **13.3**
  (*See also* Pump total head)
Total suction head, testing, **13.3**
Total-suction lift, testing, **13.3**
Transients:
  pumped storage, **9.188** to **9.189**
  in system heads, **8.15** to **8.31**, **8.88**
Troubleshooting:
  centrifugal pumps, **2.294** to **2.301**,
    12.16
  gears, **6.144**
    table, **6.141** to **6.143**
  reciprocating pumps, **12.18**
  rotary pumps, **12.17**
  steam pumps, **12.18**
True (Newtonian) liquids, **8.32**
Turbine pumps (*see* Diffuser pumps;
    Vertical turbine pumps)
Turbines:
  gas, **6.71** to **6.80**

Turbines, gas (*Cont.*):
  aircraft derivative, **6.**72
  applications of, **6.**76 to **6.**79
  classifications of, **6.**71 to **6.**72
  closed-cycle, **6.**71 to **6.**72
  combined cycle, **6.**71
  conventional, **6.**72
  environmental considerations; emis-
      sions, **6.**75 to **6.**76
    noise, **6.**73 to **6.**75
  fuels for, **6.**73
  open-cycle, **6.**71 to **6.**72
  ratings of, **6.**72 to **6.**73
  single-shaft, **6.**72
  split-shaft, **6.**72
  support systems for, **6.**76
  thermal efficiency of, **6.**71
  thermodynamic cycle for, **6.**71
  torsional analysis of, **6.**79
hydraulic, **6.**61 to **6.**70
  affinity laws for, **6.**69
  applications of, **6.**61 to **6.**62
  data for selection of, **6.**69 to **6.**70
  sigma for, **6.**68
  sizes and ratings of, **6.**66
  specific speed of, **6.**65
  speed characteristic for, **6.**69
  torque of, **6.**69
  types of, **6.**64 to **6.**67
    Francis, **6.**64 to **6.**67
    impulse or Pelton, **6.**64 to **6.**67
    propeller, **6.**64 to **6.**67
pumps used as, **2.**247 to **2.**252, **2.**255 to
    2.259, **9.**193
steam, **6.**26 to **6.**43
  construction of, **6.**29 to **6.**31
  cost evaluation of, **6.**43
  definitions of, **6.**26
  efficiency of, **6.**36 to **6.**37
  governor, controls for, **6.**31 to **6.**34
  reasons for use of, **6.**26 to **6.**27
  steam rate of, **6.**35, **6.**37
theory, **6.**35 to **6.**41
types of, **6.**27 to **6.**29
  Curtis, **6.**35 to **6.**37
  impulse, **6.**28
  multiple-stage, **6.**28
  single-stage, **6.**27 to **6.**28
Turbulent flow, **8.**31, **8.**32
Twin-volute casing pumps (*see* Double-
    volute casing pumps)

Unbalance:
  hydraulic, **2.**302
  impeller, **2.**301
Underwriters Laboratories, **9.**52
USCI to SI units, A.36
U.S. Nuclear Regulatory Commission,
      9.219 to **9.**221, **9.**232
Unloaders, power pumps, **3.**24
Untimed rotors, screw pumps, **3.**61, **3.**65
Upsurge, **2.**250, **2.**251

Vacuum pumps:
  dry, **2.**322
  ejectors, **2.**322
  heating systems, **9.**182
  marine pumping systems, **9.**155, **9.**164
  paper mills, **9.**134
  wet, **2.**322 to **2.**323
Valves:
  air-release, **8.**28, **9.**44
  control, **7.**15 to **7.**28
    actuators for, **7.**22 to **7.**23
    cage-type, **7.**16 to **7.**18
    critical flow factor (C_f) for, **7.**22
    flow coefficient (C_v; K_v) for, **7.**21,
        7.22, **8.**54 to **8.**56
    multiple orifices in series-type, **7.**19
    positioners for, **7.**23 to **7.**28
    rangeability of, **7.**20 to **7.**21
    recovery coefficient (K_m) for, **7.**21 to
        7.22
  power pumps, **3.**6, **3.**15 to **3.**16
  pressure reducing, **9.**310, **9.**314, **9.**317
  priming, **2.**239
  relief: power pumps, **3.**24, **12.**11
    rotary pumps, **3.**80, **3.**104
    steam pumps, **3.**25, **3.**30, **3.**39
Vane angle, **2.**9, **2.**11, **2.**14
Vane frequency, **8.**98, **8.**104, **8.**107
Vane pump:
  balanced, fixed displacement, **6.**174 to
      6.177
  rigid rotor, **3.**84
Vane-tip clearance, **2.**204 to **2.**205, **8.**107
Vane-tip erosion, **5.**6
Vapor pressure, **3.**67, A.26
Variable-pitch propeller pumps, **9.**42
Velocity cap (*see* Intakes)
Velocity head (*see* Head, velocity)
Velocity sensors, **2.**288, **2.**294

Velocity triangles, **2.**7, **2.**9, **2.**25, **2.**196
Vent valves, **2.**324
Venting and draining, **12.**9
Venturi meters:
  design of, **13.**15
  head loss, **8.**64
Vertical centrifugal pumps, **2.**90 to **2.**107
  dry-pit, **2.**90 to **2.**95
  wet-pit, **2.**95 to **2.**107, **9.**76 to **9.**78
    double-suction impellers, **2.**107
    (*See also* Propeller pumps; Vertical
      turbine pumps)
Vertical turbine pumps, **2.**95 to **2.**98, **9.**21
  to **9.**22, **9.**42
Vibration:
  centrifugal pumps, **2.**253, **2.**274 to
    2.304
  diagnostics, **2.**294 to **2.**301
  instruments, **2.**288 to **2.**294
  nuclear plants, **9.**220
  piping, **3.**107 to **3.**117
  reciprocating pumps, **3.**107 to **3.**117
Viscosity, **8.**31, **8.**32, **8.**39 to **8.**47
  effects of, on pump performance: cen-
    trifugal pumps, **9.**114 to **9.**119
    gear pumps, **9.**119
    power pumps, **3.**6, **9.**119
    rotary pumps, **3.**92, **3.**93, **9.**119
    screw pumps, **3.**66, **3.**67. **3.**70, **9.**119
  table **A.**10 to **A.**25
Volumetric efficiency, **2.**8, **2.**20, **3.**96
  power pumps, **3.**5
Volumetric meters, **13.**13, **13.**14
Volute casing:
  centrifugal pumps, **2.**33, **2.**35 to **2.**37
  design of, **2.**17 to **2.**19, **2.**28
  materials for, **5.**15 to **5.**27
  (*See also* Volute pumps)
Volute pumps, **2.**33, **2.**99 to **2.**102, **9.**42
Vortex pumps, **2.**100 to **2.**101, **2.**259 to
    2.262, **9.**27, **9.**31
Vortices, **10.**35, **10.**43, **10.**44
  (*See also* Intake modeling)

Warm-up piping, **12.**10
Water, properties of, **A.**26
Water horsepower (power):
  centrifugal pumps, **2.**8, **2.**195 to **2.**196
  definition of, **13.**5
  rotary pumps, **3.**95
  steam pumps, **3.**44
  test calculation, **13.**34

Water supply, **9.**3 to **9.**22
  booster pumps for, **9.**20 to **9.**21
  consumption of, **9.**5 to **9.**8
  pump stations for, **9.**8 to **9.**14
  sources of, **9.**3 to **9.**4
  treatment plant pumps for, **9.**14 to **9.**20
  uses of, **9.**4 to **9.**8
  well pumps for, **9.**21
Water seal units, **2.**76
Waterhammer, **8.**85 to **8.**100
  air chamber, **8.**94
  available solutions, **8.**88
  centrifugal pumps, **8.**85 to **8.**96
  discharge line profile, **8.**87
  mining services, **9.**146
  power failure, **2.**50, **2.**51, **8.**89
  power pumps, **3.**17
  surge tank, **8.**87, **8.**94, **8.**95
  water column separation, **8.**94
Wear plates, **9.**266
Wearing rings, **2.**53 to **2.**59
  back, **2.**60, **2.**62
  casing, **2.**53
  clearances in, **2.**58 to **2.**59, **2.**204 to
    2.205
  dam-type, **2.**56
  double, **2.**54
  flat, **2.**53, **2.**54
  impeller, **2.**53
  L-type, **2.**54 to **2.**56, **2.**204, **2.**205
  labyrinth, **2.**55
  location of, **2.**56 to **2.**57
  material of, **5.**4, **5.**11
  mounting of, **2.**57 to **2.**58
  nomenclature, **2.**53 to **2.**54
  nozzle-type, **2.**55
  renewable, **2.**53
  step-type, **2.**54
  water-flushed, **2.**56
Weighing meters, **13.**13, **13.**14
Weir meters, **13.**13, **13.**18
Well pumps (*see* Vertical turbine pumps)
Wet-pit centrifugal pumps, **2.**95 to **2.**107,
    9.76 to **9.**78
  head measurement for, **13.**33
  installation of, **12.**3
  intakes for, (*see* Intakes)
  submergence (*see* Submergence, wet-
    pit pumps)
Wet-well design, **9.**35
  (*See also* Intake modeling; Intakes)
WK^2 (WR^2) (*see* Inertia, moment of)
Wrist pin, power pumps, **3.**20

ABOUT THE EDITORS

Igor J. Karassik, Chief Consulting Engineer of the Worthington Division, Dresser Industries, received his B.S. and M.S. in Mechanical Engineering from Carnegie Institute of Technology. He is a Life Fellow of the American Society of Mechanical Engineers and recipient of the first ASME Henry R. Worthington Medal (1980). Since 1936 he has written over 500 articles on centrifugal pumps and allied subjects, which have appeared in over 1500 technical publications all over the world. In addition, he has authored or coauthored five books on pump technology.

The late *W. C. Krutzsch* was Director of Research and Development, Engineered Products, for the Worthington Pump Group, McGraw-Edison Company (now Dresser Industries). Holder of a B.S. in Mechanical Engineering from Massachusetts Institute of Technology, he was Chairman of the Northern New Jersey Section of ASME and a well-known author.

Warren H. Fraser, Chief Design Engineer, Engineered Products Operation, Worthington Division, Dresser Industries, holds a B.M.E. from Cornell and is a frequent contributor to the professional journals on centrifugal pumps and pump systems.

Joseph P. Messina, Principal Staff Engineer, Public Service Electric and Gas Company, Adjunct Instructor at the New Jersey Institute of Technology, and consultant, holds a B.S. in Mechanical Engineering and an M.S. in Civil Engineering from the New Jersey Institute of Technology. Like his coeditors, he is a well-known contributor to the technical journals on pump-related subjects.